Sorghum and the Millets: Their Composition and Nutritive Value

Frontispiece. Traditional dehulling by hand-pounding.

Sorghum and the Millets: Their Composition and Nutritive Value

JOSEPH H. HULSE
EVANGELINE M. LAING
ODETTE E. PEARSON

International Development Research Centre
Box 8500, Ottawa, Canada K1G 3H9

Including a discussion of the chemistry
and biosynthesis of polyphenols
by R. K. Gupta and E. Haslam

1980

ACADEMIC PRESS
A Subsidiary of Harcourt Brace Jovanovich, Publishers
London New York Toronto Sydney San Francisco

ACADEMIC PRESS INC. (LONDON) LTD.
24/28 Oval Road,
London NW1

United States Edition published by
ACADEMIC PRESS INC.
111 Fifth Avenue
New York, New York 10003

British Library Cataloguing in Publication Data

Hulse, Joseph Harry
Sorghum and the millets.
1. Millet 2. Sorghum
I. Title II. Laing, E M III. Pearson, O E
633'.171 SB191.M5 79-40871

ISBN 0-12-361350-7

Text set in 10/11 pt Monophoto Times New Roman

Printed in Great Britain by
Thomson Litho Ltd, East Kilbride, Scotland

Foreword

Sorghum and millets constitute a major source of energy and protein for millions of people in Asia and Africa. According to the statistics of the Food and Agriculture Organization of the United Nations, the average annual global production of sorghum during the triennium 1977–79 was 68·7 million tonnes. Over 52 million hectares were planted each year with numerous varieties and hybrids of sorghum. The annual millet production during the same period was 34·3 million tonnes from an area of 53·8 million hectares. Thus, sorghum and millets contribute annually over 100 million tonnes to the global food budget.

In the western hemisphere, these crops are mainly grown for feed grain production. Like maize, sorghum and millets also offer opportunities for industrial utilization. The food, fodder, feed and industrial uses of these crops make them important in the agrarian economy of the developing regions of Africa and Asia having low rainfall and limited irrigation resources.

It has been estimated that at least 500 million people in developing countries suffer from chronic undernourishment. Many of these people live in areas where sorghum and millets are important food crops. The improvement of the yield potential as well as nutritive qualities of these grains can hence make a valuable contribution towards minimizing malnutrition. The availability of considerable genetic variability in these crops makes planned breeding work possible. The genetic collection of sorghum at the International Crops Research Institute for the Semi-arid Tropics (ICRISAT) at Hyderabad in India now contains over 17 000 accessions. Similarly, the collection of millets contains over 14 000 genotypes.

In the past 20 years, the ceiling to yield in sorghum and pearl millet has been raised substantially through the commercial use of hybrids. Gene–cytoplasmic sterility-restorer systems have added a new dimension to yield improvement in these crops. The plant type concept involving dwarfing genes has been superimposed over heterosis. Considerable progress has also been made in incorporating resistance to major diseases and pests. The genetic upgrading of nutritive properties is, however, yet to receive integrated attention. This is why the present book is such a welcome one.

It is now well known that the first limiting amino acid in most cereals is lysine. The predominant prolamine fraction renders the endosperm proteins of sorghum and pearl millet to be of low quality. Amino acid antagonism in sorghum resulting from excess leucine depressing the utilization of isoleucine is known to cause pellagra, a nutritional disorder. In millet, excessive arginine may depress the utilization of lysine. Besides, the presence of antinutritional factors such as tannins in bran proteins further impair the quality of protein. In minor millets, data on nutritional quality are limited. While lysine may not be limiting in *Panicum mileaceum* and *Eleusine coracana* there is little information on digestibility or existence of anti-nutritional factors.

The discovery of Opaque-2 in maize in 1964, the Lys and 1 508 mutant genes in barley during 1970 and 1973 and the *hl* gene in sorghum during 1973 are parallel findings involving changes in the relative proportion of different solubility fractions of protein. These findings are of considerable interest in the nutritional upgradation of these cereals. There is also evidence that induced mutations may help to shift the amino acid profile of seed proteins in the desired direction. The occurrence of β-carotene in yellow endosperm sorghums and certain varieties of millets is also known. Genetic rectification of nutritional deficiencies in grains as well as the planned improvement of the nutritional value of staples can hence be undertaken now with some hope of success.

The nutritional criteria for cereal improvement are better understood today and powerful analytical methods are available. With the discovery of spontaneous mutants together with induced changes, the genetic variability available is great enough to bring about modifications in seed proteins. We are possibly on the threshold of substantial improvement of nutritional quality in sorghum and millets. The joint efforts of the breeder and nutritionist should help to accelerate the pace of progress in the development of high yielding cum high quality cum high stability varieties.

It is, thus, timely that the authors have written this volume on sorghum and millets. The volume furnishes a comprehensive coverage of all nutritional aspects of sorghum and millets, their processing and utilization. Since this book represents the first comprehensive and integrated treatment of all aspects of the chemical composition, biological quality as well as post-harvest technology of sorghum and millets, it will be of great value to all interested in the improvement of these crops. The authors hence deserve our sincere gratitude for undertaking this labour of love and accomplishing their aim so well.

February, 1980
New Delhi

M. S. Swaminathan
Secretary to the Government of India
Ministry of Agriculture and Irrigation

); the detailed information being presented in Chapters 2 through 6, ile Chapter 1 consists of a summary and critical review of the subsequent apters.

February 1980

Joseph H. Hulse
Evangeline M. Laing
Odette E. Pearson

Contents

List of Abbreviations

AACC	American Association of Cereal Chemists
ac	acre
Ala	alanine
AOAC	Association of Official Analytical Chemists (pre 1965, Association of Official Agricultural Chemists)
Arg	arginine
Asp	aspartic acid
bu	bushel
CE	catechin equivalent
CNRA	Centre national de recherche agronomique
Cys	cysteine, cystine
CIMMYT	Centro Internacional de Mejoramiento de Mais y Trigo
CNND	Committee on Nutrition for National Defense (Bethesda, Maryland, USA)
DWB	dry weight (matter) basis
FAO	Food and Agriculture Organization of the United Nations
FER	feed efficiency ratio
F/G	feed/grain quotient
GCA	general combining ability
Glu	glutamic acid
Gly	glycine
GE	gross energy
ha	hectare
His	histidine
hl	hectolitre
IARI	Indian Agricultural Research Institute
ICAR	Indian Council of Agricultural Research
ICARDA	International Centre for Agricultural Research in Dry Areas

ICC	International Association for Cereal Chemistry
ICRISAT	International Crops Research Institute for the Semi-Arid Tropics
IDRC	International Development Research Centre
Ile	isoleucine
INCAP	Instituto de Nutrición de Centro América y Panamá (Guatemala)
IS	International (formerly Indian) Sorghum
IU	International Unit
IUB	International Union of Biochemistry
IUPAC	International Union of Pure and Applied Chemistry
Leu	leucine
Lys	lysine
Leu/Ile	leucine/isoleucine quotient
Met	methionine
ME	metabolizable energy
NFE	nitrogen-free extract
ORANA	Organisme de recherches sur l'alimentation et la nutrition africaines
PER	Protein Efficiency Ratio
Phe	phenylalanine
Pro	proline
SCA	specific combining ability
SD	standard deviation
SEM	scanning electron microscope
Ser	serine
sp gr	specific gravity
t	tonne(s) (metric)
Thr	threonine
Trp	tryptophan
Tyr	tyrosine
UNDP	United Nations Development Programme
Val	valine
v/v	volume ratio
WHO	World Health Organization
w/v	weight/volume
w/w	weight ratio

Acknowledgements

Both in retrieving and reviewing the literature of relevance we have been greatly assisted by many sympathetic friends and colleagues. We are indeed deeply grateful to all who are named in the following acknowledgements and to many others who, through their advice and critical comment have helped to make this publication possible: Dr B. L. Amla, Dr J. D. Axtell, Dr E. C. Bate-Smith, Dr R. Bressani, Dr Glen Burton, Dr W. Clayton, Dr B. M. Craig, Dr R. H. Dobbs, Dr Hugh Doggett, Mrs Joyce Doughty, Dr Leland House, Dr R. Jambunathan, Mrs Elaine Johnson, Professor A. R. Mathieson, Dr L. Novellie, Dr O. L. Oke, Dr P. N. Okoh, Dr P. Pushpamma, Dr R. D. Reichert, Dr Ken Riley, Dr Lloyd W. Rooney, Dr W. M. Ross, Dr A. R. Saeed, Dr A. D. Shepherd, Dr S. G. Srikantia, Dr A. Sumner, Dr M. S. Swaminathan, Dr L. W. Swindale, Dr P. C. Williams, Dr C. G. Youngs.

Our most grateful thanks are due to Mrs Wanda Green, Mlle Denise Bérubé, Miss Julie Tranca and Mlle Claudette Boissonneault in Canada, and to Mrs Barbara Gotch, Mrs Margaret Honour, Mrs Janice Lee and Miss Joyce Wilkerson in the United Kingdom, for their unfailing patience and skill in preparing the typescript for publication.

Finally our warm thanks to Mrs K. Suggate to whom we are indebted for the preparation of the subject index.

To Jane and Hugh Doggett
in recognition of their devotion
to the people of the semi-arid tropics

CHAPTER 1

Summary and Commentary

Introduction

This first chapter consists of a brief summary and commentary upon the more detailed subsequent text. Chapter 1 is not intended as an index to the later chapters and does not follow the same sequence. It begins with a discussion of the known nutritional problems of the people of the semi-arid tropics and the special consideration that needs to be given to those whose diets consist largely of sorghum and the millets. There follows a brief commentary upon methods of analysis and biological assay of the constituents of the cereals under review, followed by a summary of what appear as the principal findings.

The content of Chapter 1 is summarized in the following sequence of constituents:

- protein and amino acids
- carbohydrates
- lipids
- vitamins
- minerals
- nutritional inhibitors

and concludes with recommendations for further research and investigation.

Subsequent chapters are more detailed and each deals with different influences upon the composition and characters of each grain. Chapter 2 describes the origins, geographical distribution, main areas of production and the plant and seed structure of the cereals under discussion, together with the proportions of the major and minor constituents of the grains as reported by different analysts. Results of biological evaluations with rodents, poultry, nonruminant and ruminant animals are also reported. Chapter 2 also contains all the data from proximate, amino acid, vitamin and mineral analyses, summarized in tabular form after Chapter 6. Each table includes: (1) a range of ranges—the highest and lowest results reported from all the literature reviewed, and (2) a range of means—the highest and lowest mean values reported among workers who analysed several samples. It is the authors' belief, recognizing the uncertainty of genetic origin of many of the samples analysed, that a range of means provides a more conservative indication of the range of diversity than a range of ranges (i.e. the highest and the lowest analytical result reported from all sources).

Chapter 3 describes the influence of genetic, environmental and agronomic factors on grain composition and constituents of nutritional importance. Chapter 4 deals with nutritional inhibitors and toxic factors and, we believe, presents new and original information on the nature and composition of the sorghum polyphenols that are of demonstrable nutritional importance.

Chapter 5 describes the effect of industrial, laboratory and domestic processing conditions upon the composition of the resultant products. The processes reviewed include decortication, milling, grinding, cooking, malting, brewing and conversion to animal feeds.

The final chapter, Chapter 6, discusses interactions between sorghum and the millets with other nutrient sources, with particular reference to the effect of various supplements upon the biological quality of typical poor Indian diets.

An appendix includes classified lists of the cereals discussed, giving both Latin botanical names together with the vernacular name by which each is designated in different parts of the world. We make no claim that the vernacular list is complete; only those names encountered in the literature have we tried to interpret.

In an earlier publication (Hulse and Laing, 1974) a plea was made for greater uniformity and standardization of methods of analysis and biological evaluation, and the conventions by which results are computed and expressed.

To enable readers to compare results among the sources reviewed, we have recalculated many of the original data to a standard system of units. Almost all proximate analyses are presented on a dry weight basis (DWB) and all amino acids as mg/g N and/or mg/100 g sample. Where the information was insufficient to permit recalculation, the data are given as originally presented.

Throughout the text the word "significant" means "statistically significant" as defined by those who did the research. Levels of significance have not been cited. Since the subject material comprehends many scientific disciplines it is possible that terms used will not be familiar to all who may read the text. Where in the judgment of the authors an expression appears widely esoteric, a definition has been offered. Some lack of uniformity is evident in the use of the terms "variety", "hybrid", "line" and "cultivar". "Variety" may be described as a relatively stable population in which individuals, though heterogeneous, bear some phenotypic similarity. "Hybrid" is the first generation of genetically uniform parents. A "line" is the name given to each of the entries in a plant breeding programme. A "cultivar" may be described as a genetically uniform type of plant which is cultivated and has found, or will find its way into the farmers' fields.

More has been published on sorghum, and to a lesser extent on pearl millet, than on common, foxtail, little, finger, Japanese barnyard, kodo, the other minor millets and teff. Also, more investigators seem to have been attracted to the study of protein and lysine, than to other components or characters. Polyphenols (tannins) in sorghum and their influence upon proteins have been extensively reviewed.

The Semi-arid Tropics (SAT)

Probably the most important character of sorghum and the millets reviewed is their ability to tolerate and survive under conditions of continuous or intermittent drought that results from low and uncertain rainfall. Their principal homes are to be found in the SAT of Africa and Asia.* In Africa, the SAT include countries surrounding the Sahara, in addition to large areas of Ethiopia, Kenya, Tanzania, Uganda, Zambia, Rhodesia, Mozambique, Somalia, Botswana, Lesotho and Swaziland. The Indian SAT include the States of Maharashtra, Andhra Pradesh, Kanataka, Tamil Nadu, parts of Rajasthan, Madhya Pradesh, Bihar and Orissa. Other countries of Asia are subject to semi-arid conditions during seasons of low rainfall and sorghum is gradually gaining acceptance among Asian farmers in dryland multiple cropping systems. Sorghums tolerant to low temperatures and high altitudes are gradually finding a place in Mexico, Brazil and other Latin American countries, in addition to their natural habitat in Ethiopia.

The SAT have been characterized as areas of "famine or potential famine"; "food priority countries with low incomes, inadequate diets and recurring food grain deficits"; and "countries with high infant mortality and morbidity". For many of the rural poor of the SAT, sorghum or one of the millets provides the major source of calories and protein. Of the 14 SAT countries in Africa, all but three were considered by the UN World Food Conference in 1974 to be deficient in dietary energy (Table 1.1). Recognizing the wide disparity in food availability among seasons, regions, communities, and among and within families, it is probable that the poorest and least privileged, particularly those who suffer chronic enteric and other diseases, are susceptible not only to calorie but to protein and other nutrient deficiences.

Research in the SAT

The severe droughts and resulting crop failures of the early 1970s drew world-wide attention to the plight of the people of the SAT. During the past 5 years, two international agricultural research centres (IARCs), the International Crops Research Institute for the Semi-Arid Tropics (ICRISAT) and the International Centre for Agricultural Research in Dry Areas (ICARDA) have been created with headquarters in Hyderabad, India and the Middle East respectively. A third IARC, the International Livestock Centre for Africa (ILCA), was also created by an international consortium

* (Authors' note: It is recognized that some country names have changed since references cited in this text were published. The authors have therefore used the name by which each country of origin was known at the time the referenced paper was published.)

of donors to improve systems of animal production throughout the region. In addition, several agencies have cooperated to create the Group for Assistance on the Storage of Grains in Africa (GASGA) which, as its name suggests, is dedicated to reducing post-harvest grain losses.

In addition, the established research programmes of the Indian Council of Agricultural Research (ICAR), agricultural research institutes in Ethiopia and Nigeria, and the national programmes of Senegal, Sudan, Uganda, Kenya and Tanzania, and other SAT countries have contributed much to our present knowledge of what influences sorghum and millet composition. One must also pay tribute to the valuable ongoing research at Purdue, Texas A&M and other universities in the USA.

Nutrition in the SAT

The long droughts of the 1970s so reduced harvests, depleted grain reserves and decimated livestock herds that by 1973 the Sahel required massive food aid from abroad to avert widespread starvation. These tragic circumstances brought before the whole world the depressed conditions that have long prevailed in the Sahel, where even at the best of times nutritional adequacy has barely kept pace with population increase. Until the results of many outstanding research programmes become manifest in farmers' fields and until governments of the region assign greater priority to improving the lot of the rural and farming sectors, continued dependence upon offshore food supplies seems inevitable.

In Haute Volta, Mali, Niger and Senegal and Tchad, cereals make up more than 70% of the calories consumed. A serious state of malnutrition has long existed in many parts of the region, especially among young children. Mortality rates from the common communicable diseases, especially measles, are higher and growth rates among children tend to be significantly lower than at comparable ages in more privileged communities. Among the adult population, on whose income generating ability the family's livelihood depends, protein and caloric restrictions are most often reflected in reduced work output, which in turn imposes further dietary restrictions on family members. The interrelation between diet and productivity is illustrated in Fig. 1.I. The following SAT countries are believed to be among those with an average *per capita* GNP of $150 (US) per annum: Sudan, Mali, Niger, Haute Volta, Burundi, Ethiopia, Somalia, Tchad, Lesotho and Swaziland. In most of the states of India where sorghum and the millets are important, extreme poverty is widespread.

Scrimshaw (1977) using Reutlinger and Selowsky's data, suggested that people in the LDCs on lowest incomes not only have a lower dietary intake but the quantity and quality of protein consumed declines in parallel with disposable income. The poorer socioeconomic groups tend to be most vulnerable to acute and chronic infections and therefore express a greater need for a higher intake of nutritionally efficient protein.

Available infant morbidity and mortality statistics provide an uncertain

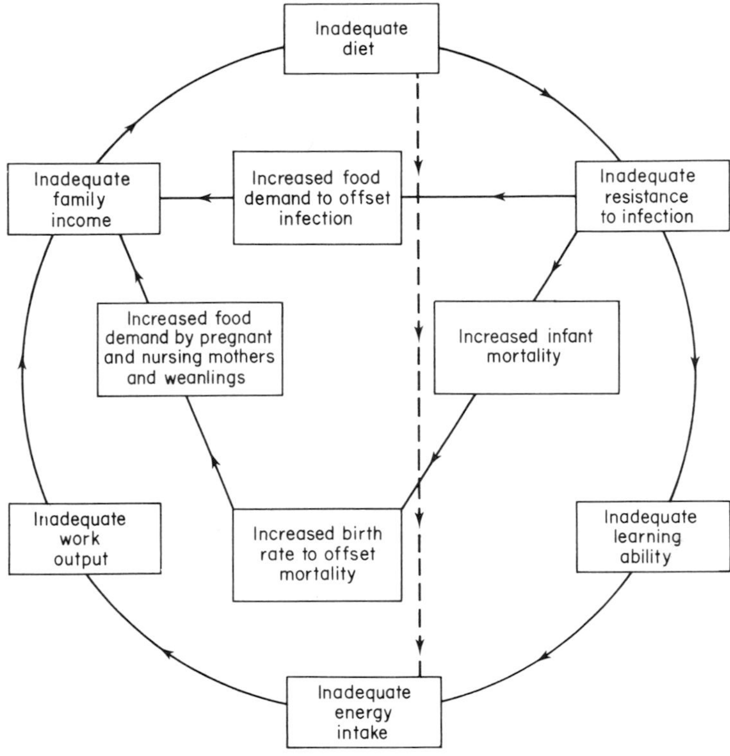

Fig. 1.I The vicious circle

record of disease incidence and case fatality ratios. Whereas infant mortality is frequently attributed to infectious diseases, including malaria, measles and gastroenteritis, it is suspected in many instances that malnutrition and more particularly protein malnutrition may well be the major precipitating factor. Malnutrition reduces human resistance to infection, particularly among young children. In some regions of Senegal, measles are said to be responsible for over 50% of deaths among children aged 1–4 (Stanbury and Childs, 1974). However, short of having children confined to a metabolic ward, it is difficult to identify specific protein malnutrition or kwashiorkor as distinct from protein calorie insufficiency or marasmus. (Kwashiorkor is generally recognized by extremely low serum albumin levels and marasmus by general wasting of the body tissue.)

Konczacki (1972) estimated that in the sub-Saharan region of Africa, 20–40% of children under the age of 6 years suffer from protein calorie malnutrition (PCM). In their publication on health, nutrition and population of the Sahel, Stanbury and Childs (1974) reported that based on

Stuart-Meredith weight/height mean, in Mali, Mauritania, Niger and Haute Volta, 26, 14, 8 and 10% respectively of children were below the acute malnutrition threshold. In India, protein calorie deficiency has been identified among many segments of the poorer population, particularly among children below 5 years of age (Pushpamma and Devi, personal communication, 1977).

The FAO/WHO Joint Expert Committee on Nutrition discussed the diagnosis and accurate recording of protein calorie malnutrition. The committee noted that in low income countries the prevalence of severe PCM appeared to range from 0–7·6% in children below 5 years of age, while the prevalence of moderate forms of PCM lay between 4 and 43% (WHO, 1972).

Throughout the Sahel food availability patterns change from season to season. From December to mid March, immediately following harvest, there is a period of comparative abundance through the dry season. From April to July, food supplies dwindle and from August through September, the winter months, serious grain shortages often appear. It is not unusual for sorghum and millet growers to sell their surplus grain to traders immediately after the harvest, then be obliged to rebuy at inflated prices before the next harvest.

A survey carried out in two Senegalese villages revealed that during the August/September period average calorie intakes were 25% below FAO/WHO recommended levels (Hellegouarch *et al.*, 1967).

Though accurate, reliable and comprehensive statistics are not easy to come by, the available evidence supports the view that the nutritional well being of the people of the SAT will continue to be a matter worthy of international concern for many years to come.

Recognizing the inferior nutritional quality of sorghum and several of the millets, and the heavy dependence upon these grains by the poor people of the SAT, one may feel justified in recommending to plant breeders that they give some attention to nutritional quality, in addition to those factors that influence yield and agronomic performance.

Several factors contribute to the difficulty of determining the degree and extent of PCM throughout the SAT. First, the estimates of production, particularly on small subsistence farms are far from complete. Second, it is probable that grain distribution and consumption vary widely throughout the SAT among seasons, regions, communities and families, and even within families. Third, the nutritional quality of sorghum and the millets, while generally of a low order, appears highly variable. Fourth, we know little about the quantities eaten and the resultant benefit of other plant and animal nutrients used to supplement the predominantly cereal diets. Fifth, more exact clinical studies are needed to determine the retention of essential nutrients by young children fed poor sorghum and millet diets, particularly those suffering or recovering from enteric or other infectious and debilitating diseases.

At present it seems that authorities are unable to agree upon what is the minimum safe protein intake required by different age groups and conditions of men or women living in the SAT (or for that matter anywhere else in the world). Since the mid 1950s several expert committees sponsored by FAO and WHO have considered human energy and protein requirements. In April 1971, the Joint FAO/WHO Expert Committee recommended (WHO, 1973a) levels of dietary protein intake roughly 20% lower than the 1963 Joint Committee. In 1963 the Committee recommended (WHO, 1965) a reference protein intake of 0·71 g/kg/day for adults which was revised in 1971 to 0·57 g/kg/day. Consequently, it became the conventional wisdom that the Protein Problem of the 1960s had disappeared and that protein malnutrition was no longer a matter for international concern. The revised 1971 FAO/WHO recommendations led some nutritionists to conclude that except where starchy roots and tubers predominate in the diet, protein levels would be adequate if caloric needs were met.

However, according to Scrimshaw (1976), there is increasing evidence that, when applied to the LDCs, the levels of intake recommended by the FAO/WHO Committee in 1971 may be too low.

Infants, pre-school children, pregnant and lactating women are the most vulnerable groups of nutritional concern. In response to the contention that protein calorie malnutrition can be prevented simply by increasing children's total intake of a traditional diet, Scrimshaw points out that many young children lack the capacity to ingest the volume of food necessary to satisfy their nutrient requirements. Furthermore, the dietary protein demand of young children in the tropics is aggravated first because of reduced protein absorption caused by intestinal parasites and chronic damage to the gastrointestinal tract from repeated infections, and second as a result of a higher than normal protein demand during the post-infective and recovery phases of various diseases. Body protein loss results from the stress response in which amino acids are mobilized from the protein in the lean body tissues for use by the liver in synthesizing glucose. Daily protein losses as high as 1·2 g/kg body weight are not uncommon during acute infection. Following serious infection, increased protein is needed to replace depleted body tissue.

A higher caloric and protein demand results from impaired intestinal absorption and from febrile infections. Moreover, whereas a deficit in calories may be compensated by reduced activity, the body has no comparable mechanism to offset protein deficiency. Children recovering from acute infectious diseases may require 40% more protein than is needed for normal health and growth.

Among the people of the SAT, cereals furnish a major source of dietary protein. The 1975 Joint FAO/WHO Committee report on energy and protein requirements (FAO/WHO, 1975) stated that safe levels of protein intake are based on the reference protein of milk or egg protein. Therefore, if other proteins are considered, a correction factor for protein quality and digestibility must be applied. The FAO/WHO committee recommended the following formula to arrive at desirable levels of intake of proteins other than milk or egg.

$$\text{Desirable intake} = R \times \frac{100}{\text{Amino Acid Score}} \times \frac{100}{\text{Protein digestibility}}$$

where R = recommended level of egg or milk protein.
The AA Score is defined later in this chapter.

From Table 1.2 the median Amino Acid Score for normal sorghum is roughly 40. Various studies report the relative digestibility of sorghum and millet protein to be below 50%. On this basis the desirable protein intake from an all cereal diet would be 5 × the reference protein.

$$R \times \frac{100}{40} \times \frac{100}{50}$$

$$= R \times 5{\cdot}0.$$

Thus the desirable protein intakes from an all sorghum or millet diet for different ages and conditions of men and women, in comparison with the FAO/WHO recommendations for egg or milk protein would be as tabulated in Table 1.3.

At 50% digestibility, assuming a protein content (DWB) of 10%, it would be necessary for a child of 5 years to consume 1 kg and a lactating woman 3·3 kg of sorghum or millet daily to satisfy their protein requirements.

It is freely admitted that few in the SAT will survive solely on sorghum or millets. Nevertheless, some of those who have discounted the need to improve the nutritional quality of sorghum and the millets appear to have overlooked their marked nutritional inferiority relative to accepted reference standards of protein quality in comparison with many other cereals.

Several vitamin and mineral deficiencies are believed prevalent in the SAT, though as with PCM, it is difficult to ascertain a precise estimate of their incidence. Where cereals are the staple foods, vitamin A, C, D and riboflavin deficiencies have been reported. In Ethiopia, rickets are especially evident among children from 6 months to 5 years of age. Niacin and riboflavin deficiencies have been reported in the Indian SAT. Calcium, iron and iodine levels are also low in the diets of SAT populations. Anaemia is frequently the compounded effect of inadequate diets and heavy infestation of blood and intestinal parasites. One clinical investigator reported a 12·5% incidence rate of anaemia among patients in a paediatrics ward in Ouagadougou, Haute Volta.

From evidence referred to later in this chapter, it is possible that zinc deficiency among people of the SAT may be of more serious consequence than is generally recognized.

In consideration of the foregoing, together with the knowledge that the SAT include some of the world's poorest and least privileged people, it is difficult to support the contention that all nutritional ills will disappear simply by providing more carbohydrate. The nutritional quality of the limited and uncertain food supply and diet of the world's least privileged peoples is invariably more critical and of greater consequence than is the case with inhabitants of wealthier countries.

Methods of Analysis

As reported earlier by Hulse and Laing (1974), the literature relevant to cereal analysis reflects a marked lack of methodological standardization. Consequently, the now defunct Protein-Calorie Advisory Group of the United Nations (PAG), in cooperation with several other international agencies, formulated PAG Guideline No. 16 on protein methods for cereal breeders as related to human nutritional requirements (Protein–Calorie Advisory Group, 1975). This guideline and the methods it recommends deserve the attention of all analysts in cereal breeding stations. Subsequently, a similar set of guidelines was published for legume breeders (Hulse *et al.* 1977). Purdue University (1977) has also published a useful handbook describing in detail various recommended chemical and biological methods of analysis for grain and forage sorghum. It would make for easier universal comparison if both the methods by which cereal grain composition is determined and the convention and units in which the results are reported were standardized. In the subsequent text almost all proximate analyses of major components such as protein, carbohydrate and lipid are expressed as g/100 g of sample (DWB).

Protein Methods

The Kjeldahl method is recommended as the standard reference for the determination of protein nitrogen and results are best expressed as g N/100 g sample (DWB). Williams (1974) has detailed the many potential sources of error in protein nitrogen determination and his paper is recommended to be read together with PAG Guideline No. 16.

Tkachuk (1977) described the method by which nitrogen to protein conversion factors are best calculated. After determining the amino acid composition of several samples of pearl millet, sorghum and teff, Tkachuk concluded that N × 5·7 is more accurate than N × 6·25 for all of the cereal grains under consideration and also for many tropical food grain legumes. Since most analysts reporting in the literature reviewed have used N × 6·25, the protein content of sorghum, the millets and most legumes of the SAT is probably overestimated by about 10%.

Given the many sources of error described by Williams (1974), it is needless to quote protein to more than a single decimal place. Also, when breeders select for "high-protein" genotypes among lines grown under identical conditions, any difference less than 1% in analysed protein content cannot be regarded as an indication of a significant genotypic difference.

The methods of determining grain protein by dye binding or near infrared reflectance spectroscopy (NIRS) together with possible sources of error are described in Hulse *et al.* (1977) and the PAG Guideline No. 16. Analysts who report results from the Udy dye binding method would provide a service to readers by quoting their results as estimated protein or N/100 g sample (DWB) rather than as an obscure mathematical ratio.

Plant breeders who wish to select high protein genotypes may find it more useful to express their results as mg N/seed rather than mg N/sample since the latter can be influenced by environment and agronomic factors. It is however recommended to researchers who select for high seed protein content that yield of harvested grain/ha, hectolitre and 100 kernel weights be quoted in addition to protein content since high protein (% DWB) may simply reflect a comparatively small endosperm.

Some analysts express their results as yield (kg) of protein/ha. This may be a useful statistic for pasture agronomists and animal nutritionists, but it has little meaning in human nutrition since human diets are not rationed according to a given area of land.

Amino Acids

The recommended methods of determining essential amino acids in cereal grains and the need for care and accuracy during hydrolysis are described in PAG Guideline No. 16, Hulse *et al.* (1977), and McLaughlan and Campbell (1974) (in Hulse and Laing, 1974). Throughout the text amino acid contents quoted have been expressed as mg/g N or mg/100 g sample (DWB), a convention recommended for general adoption.

The Amino Acid Score is a useful guide to the potential nutritive value of the cereal protein and is the analytically determined level of the first-limiting amino acid expressed as a percentage of the level of the same amino acid recommended in the FAO/WHO (WHO, 1973a) Provisional Reference Pattern (Table 1.2).

$$\text{Amino Acid Score} = \frac{\text{FLAAP (mg/g N)}}{\text{FLAAR (mg/g N)}} \times 100$$

where FLAAP is the amount of the first-limiting amino acid present and FLAAR is the amount of the first-limiting amino acid recommended in the FAO/WHO 1973 Provisional Reference Pattern (WHO, 1973a).

In the case of sorghum and the millets, lysine being first-limiting, the amino acid score is the level of lysine determined (mg/g N) expressed as a percentage of 340 mg/g N, the level of lysine recommended in the FAO/WHO reference pattern. From Table 1.4 it can be seen that mean lysine values in protein range from 71–212 mg/g N and therefore the amino acid (lysine) scores range from 21–62 with a median value of about 40.

Protein Fractionation Methods

Several references to protein fractionation appear in the literature reviewed.

Attempts to fractionate cereal protein began almost from the time protein was recognized as a unique class of substances. Early in the nineteenth century an Italian chemist Beccari washed out wheat starch and isolated gluten. About 80 years later, another Italian, Taddei, fractionated gluten into two components which he named: (a) "gliadin"—soluble in aqueous alcohol and (b) "glutenin" an alcohol-insoluble fraction.

Late in the nineteenth century an American, Osborne, fractionated wheat proteins by progressive treatment with a series of solvents. The Osborne fractionation method with various modifications is regarded as a classical standard among cereal chemists. Osborne's fractions were designated: (1) albumin—soluble in water; (2) globulin—soluble in 0·5N NaCl; (3) prolamine—soluble in 70% ethanol; (4) soluble glutelin—soluble in NaOH; and (5) insoluble glutelin—the residue. The essentials of the procedure are described by Osborne (1924).

Because the choice of solvents and conditions among analysts appear somewhat arbitrary and empirical, results of the various fractionation experiments reported should be regarded as comparative not absolute. Different workers used solvents differing in concentration, temperature and time of contact with the substrate. Therefore, though each may use the same fraction designations (albumin, globulin, prolamine, glutelin) results from different analysts are not comparable if each used a different solvent system. Factors which influence the proportion of the soluble fractions include: particle size or fineness of the flour extracted (possibly related to the degree of cell damage before fractionation); solvent temperature and concentration; moisture content; age and condition of storage before extraction (globulin fractions tend to be higher from dry grain but decline, while glutelin fractions increase, with time and storage temperature); pH (alcohol- and salt-soluble fractions increase and glutelins decrease with increasing pH); sequence of the solvents (the ethanol-soluble fraction is normally higher if extracted ahead of the NaCl-soluble fraction).

Other research workers proposed solvent systems markedly different to those of Osborne with consequently incomparable results. For example, protein solubility varies among different inorganic ions, the "salt-soluble fraction" increasing from Na to K to Mg, and from SO_4 to Cl to I. The salt-soluble fraction may increase by a factor of three between KCl and KI. The alcohol-soluble fraction varies among alcohols, increasing from ethanol to *iso*-propanol to *n*-propanol. In general, for any alcohol the soluble fraction increases with temperature, concentration and time of contact.

Acetic and other organic acids sometimes replace alkali following alcohol extraction, though seed proteins tend to be more greatly affected by acids than alkalis and therefore NaOH is generally preferred to acetic acid. Alkalis however tend to hydrolyse amide groups, to disrupt disulphide linkages and possibly some peptide bonds. An increase in the total soluble fraction can result from the addition of a reducing agent to the alcohol. Landry and Moureaux (1970) added 0·6 to 1% 2-mercaptoethanol to *iso*-propanol and increased the "kafirin" (prolamine) fraction probably by disrupting disulphide bonds. There is some difference whether this *iso*-propanol + 2 ME fraction is in fact a glutelin or a prolamine. Given the high proline, glutamic acid and amide nitrogen content, it probably should be regarded as a prolamine, though different from what would be extracted by Osborne's 70% ethanol.

It is probable that fractions and their constituents undergo varying degrees of chemical change during extraction. The proportion of total soluble fractions may be decreased through interaction between protein with phytin and/or polyphenols. The proportion of the residue following alcohol extraction can be increased in a borate buffer (pH 10) by successive additions of 0·6% mercaptoethanol and 0·5% sodium dodecyl sulphate (SDS) or sodium lauryl sulphate (SLS) which are known to release nucleic acids and lipids from protein.

Given the high degree of variability among methods based upon the Osborne technique and the complexity of the Landry and Moureaux method, it is evident that unless the conditions of sample preparation and all subsequent operations are fastidiously and precisely standardized, many sources of variability are possible and readers must exercise considerable caution in comparing the proportions of protein fractions reported by different research workers. Though the fractions may bear the same name (e.g. glutelin or prolamine) their nature and proportion will be influenced by the conditions of extraction.

Methods of Lipid Analysis

A similar degree of caution is needed in comparing reported levels of lipids (g/100 g sample DWB), particularly where the nature of the solvent and the conditions of extraction are not defined. Subsamples from the same sample of sorghum showed a lipid content of (1) 2·3% when extracted by petroleum ether (B.P. 40–60°C) and (2) 7·1% when extracted by tetrachlorethane.

Methods for Carbohydrate

Total carbohydrate, whether determined by difference or by direct analysis, and free sugars are reported as g/100 g total sample (DWB). Amylose and amylopectin are best reported as g/100 g total starch. In some cases analysts report amylose and amylopectin as g/100 g sample. Given the comparatively wide range of carbohydrate content reported by different analysts (the carbohydrate means in Table 1.4 range from 70–90% for sorghum and 62–89% for pearl millet) amylose as % sample may as much reflect variation in starch content as variability in the amylose/amylopectin ratio.

Methods for Mineral Elements

Results quoted for calcium, phosphorus, iron and other inorganic elements must also be viewed with some reservation. Though most results included are from one or other of the AOAC recommended methods, at least 14 different methods for calcium, 23 for phosphorus, and 17 for iron are reported throughout the text. Furthermore, particularly in the case of calcium and iron, it is difficult to decide what proportion of that determined by analysis was present as an intrinsic component of the original grain and how much resulted from contamination before and during cooking. Purdue University (1977) prescribes specific methods for Ca, Mg, P, Co and Mn in sorghum.

Methods for Vitamins

A wide variety of methods have been reported in the determination of vitamins. It is the authors' recommendation that for the B vitamins the Official Methods of the American Association of Cereal Chemists be adopted and for other vitamins, analysts use the Official Methods of the Association of Vitamin Chemists. It is recommended that both mineral elements and vitamins be reported as mg/100 g sample, (DWB).

Methods for Biological Evaluation

For a discussion of alternative methods of biological assay and evaluation, readers are referred to PAG Guideline No. 16, McLaughlan and Campbell (1974), and to Hulse *et al.* (1977). Chemical, physical and *in vitro* methods of analyses provide general and specific information on the composition of the cereal grains. They do not however indicate what proportion of the protein nitrogen, minerals, vitamins and other nutrients are in a form that is assimilable and digestible by the human or other animal to which they are fed. Methods of biological evaluation seek to assay the useful nutritional value of food in the form in which it is eaten. Therefore, such studies are best undertaken using the animal for which the food is intended. If the expense can be tolerated, direct biological evaluation can be carried out with farm animals confined to laboratory conditions and subjected to varying degrees of discomfort while digestibility and proportional retention of different nutrients are determined.

Similar studies using human subjects are difficult if not impossible among people as they go about their normal working lives. Nitrogen and other nutrient balance determinations (the difference between what is ingested and what is excreted) are best carried out on humans confined to a metabolic ward of a hospital or an equivalently controlled environment. There are few if any direct biochemical tests of nutritional adequacy that can be carried out among members of a free moving society. Though symptoms of severe kwashiorkor or marasmus are comparatively readily identifiable, subclinical levels of malnutrition are not easily recognized among poor rural communities in their normal habitat. Also, it is often difficult to isolate symptoms directly attributable to protein deficiency from those caused by inadequate calorie intake. Consequently, nutritionists now tend to use the term protein-calorie malnutrition (PCM) which embraces a broad spectrum of deficiency conditions.

A variety of anthropometric methods of indirect determination of nutritional adequacy have been reported including weight, height, and head circumference, but since the available standards of comparison are often derived from communities of different ethnic origin, these indirect indices of nutritional condition may need to be interpreted with some caution.

Consequently, it is the custom to use laboratory animals, particularly inbred strains of rats, to classify cereal grains according to their nutritional quality. Though some research workers have devoted their attention to mineral and vitamin balances and deficiencies, greatest attention has been given to protein nitrogen. The methods recommended by various expert panels are to be found in PAG Guideline No. 16, Hulse *et al.* (1977) and McLaughlan and Campbell (1974).

Nitrogen balance or retention is the difference between the total nitrogen intake and that which is excreted in the faeces and urine, and as stated above, accurate determination requires confinement of the test subjects. The most widely used method of protein evaluation is the Protein Efficiency Ratio (PER), the rate of body weight gain/g protein ingested from diets containing a fixed (usually between 8 and 10%) proportion of protein fed to young rats of like sex. It is customary to use casein as the standard of control, and the generally accepted method of PER determination requires that the diets compared be isonitrogenous.

A second method, which determines Net Protein Ratio (NPR) is similar to PER but includes a bottom control group of rats fed a protein free diet. The NPR though similar to PER, is calculated by subtracting the mean weight loss of the nonprotein group from the mean weight gain of those fed the test protein. Results of body weight gains from rats fed a single level of protein nitrogen can be deceptive in that proteins of comparatively low nutritional quality, such as wheat gluten, often give higher PER values when fed at low protein intakes (e.g. 4% of the diet) than when fed at higher levels (e.g. 10% of the diet). Consequently, when adequate quantities of the cereal are available and laboratory facilities permit, a third method of protein evaluation based on weight gain, known as the Relative Protein Value (RPV) or Slope Ratio Assay, in which protein is fed at three different levels, is recommended (PAG Guideline No. 16). The slope of the growth response line of the test protein is expressed as a ratio of the slope of a corresponding slope for a lactalbumin reference protein.

Net Protein Utilization (NPU) is similar to NPR except that body nitrogen rather than body weight is determined. Because of the carcass nitrogen analyses required, the NPU assay is less convenient than NPR, particularly if a large number of samples are to be compared. In general, NPR and NPU from the same samples correlate well, though there is some evidence that NPU may be more variable than NPR.

The relative merits of these methods are reviewed in the publications referred to above and a detailed commentary will not be offered in this publication. Nevertheless, it is evident from the literature reviewed that PER is most widely used and that some research workers may misconceive the significance of PER values. PER values do not bear a linear relation to one another; a protein that shows a PER of 2·0 is not necessarily twice as nutritious as a protein with a PER of 1·0.

A useful attribute of PER is that it serves to indicate the availability of the first-limiting amino acid. Under equivalent conditions, PER values tend to increase with additions of the limiting amino acid until it is no longer limiting. While in this context PER can provide a useful index of protein quality, the method tends to be overexploited. In several instances reviewed, the results of the PER were predictable. There is ample evidence to show that when lysine, whether synthetic or present in a natural high-lysine source, is added progressively to a pure cereal, the PER will continue to rise until lysine ceases to be limiting. It is unnecessary to reconfirm this by PER in every cereal supplementation experiment.

The PER is a comparative index, not an absolute measurement and ranking by PER cannot with confidence be extended beyond a single experiment at a given time. Within a particular experiment the PER index simply ranks different sources of protein according to their apparent influence upon the rate of weight gain of young rats. Ranking of proteins by PER is valid only when derived from isonitrogenous diets fed at the same time and place and under the same conditions.

A PER determined in one experiment at one time and place cannot with any accuracy or statistical meaning be ranked against calculated PERs determined from similar or different protein sources at another time, another place, or under different conditions. In Chapter 6 the results are tabulated from a number of experiments in which the effect of adding different protein sources to poor diets based upon sorghum or one of the millets are compared. In several of the experiments the PER serves as the index of comparison. The PER values ascribed to the various diets with their protein supplements are comparable only within each separate experiment, not between or among experiments.

Biological Evaluation with Farm Animals

From nutritional studies with farm animals, many different indices are reported including both *in vitro* and *in vivo* determinations of nutrient digestibility and retention. There is evident a marked lack of standardization among methods for the determination of overall feed and specific nutrient efficiency. By the time this text is published. a working group sponsored by the International Unions of Nutrition (IUNS), of Food Science and Technology (IUFoST), and the IDRC, will probably have recommended standard methods of analyses and evaluation for animal feeds.

One index which may cause some confusion to the unfamiliar is the Feed/Gain (F/G) ratio or quotient which is the weight of feed (DWB) required to produce unit weight gain. Consequently, the F/G ratio runs opposite to the PER in that in the former, the lower the value the higher the feed efficiency while in the latter, the higher the value the higher the comparative efficiency. The Feed Efficiency Ratio (FER) is similar to PER in that it records the ratio of animal body weight gain/unit of feed (DWB) intake. It will be noticed however that some investigators appear to use F/G and FER interchangeably rather than regarding one as the reciprocal of the other.

Several factors militate against a straightforward comparison from different farm animal feeding experiments of the cereals under review. First, it is not customary to feed animals diets composed entirely of a single cereal grain. Second, it is difficult to compose and compare animal diets that are both isocaloric and isonitrogenous and identical in all respects other than the cereals under test. Third, the nutritional quality of cereals is altered by processing and it is extremely difficult, using standard equipment, to establish and maintain precisely constant processing conditions. It is also true that formulators of animal feeds are, understandably, as much concerned with the economics of feed efficiency as with the ideal physiological well being of the farm animal. In addition, experiments with farm animals are expensive of money, time and space, and therefore in few if any of the experiments reviewed were all relevant factors comprehensively examined. Consequently, many of the experiments reported show evidence of compromise between what is desirable and what was possible and many reported results must therefore be regarded as indicative rather than definitive estimates of nutritional quality.

Summary and Commentary on Reported Results

The summary of results that immediately follows attempts only to highlight whatever seems of greatest significance, to suggest what appears obscure and uncertain yet deserves greater attention, and finally to recommend some priorities for future investigation. This summary does not follow the same sequence as the subsequent chapters: it is organized by grain constituents in the order given on p. 34, drawing together from each of the detailed chapters whatever seems of greatest relevance.

General Composition

Those who have studied grain structure either by dissection or microscopic examination, report that the embryo comprises roughly 10%, the pericarp (bran layers) about 8%, and the endosperm more than 80% of the mature sorghum grain. The relative proportions may vary with genetic background, environment and degree of maturity. The embryo is rich in protein, lipid, mineral and B vitamin content. Removal of the outer pericarp proportionally increases the protein and reduces the cellulose, lipid and mineral content of the residual grain.

The grain structures of the millets have been less thoroughly examined.

Some general structural similarities seem to exist between the grains of sorghum and pearl millet. An important feature of the minor millets is their high cellulosic outer seed coverings described variously in the text as hulls, husks and seed coats. In some instances for example Japanese barnyard millet, the mature grain is not exserted from and therefore remains enclosed by a hardened lemma and palea.

The proteins in sorghum and the millets can be divided into two main classes, (a) the structural proteins present in the embryo which are of relatively constant and immutable composition and (b) the storage proteins in the endosperm and surrounding aleurone layer which provide the food for the young plant and vary in composition among types according to genetic and agronomic history. As the seed matures and in general as the proportion of protein nitrogen in the endosperm increases, the composition of the endosperm storage proteins changes, glutamic acid and proline increase, lysine, other basic amino acids and methionine decrease in proportion to total nitrogen.

Moving inwards from the aleurone layer to the centre of the endosperm, nitrogen and lysine, each as a proportion of total mass, tend to decrease; but lysine as a proportion of total nitrogen tends to increase. The endosperm protein in sorghum, pearl millet, common millet and some of the minor millets exists both as a continuous matrix and in the form of protein bodies embedded in the matrix, these bodies being particularly high in proline and glutamic acid, and generally free from starch.

Sorghum endosperms may be classified according to several characters. The first takes account of the relative proportion of corneous and floury components, the former, as its name suggests, being hard and vitreous like horn, the latter soft and floury. The proportions vary significantly among types, but in general the corneous is in greater proportion in the outer layers; the floury predominates nearer the centre of the endosperm. The corneous contains higher proportions of protein, and proline and glutamic acid as percent protein. The floury is richer in lysine as % protein. In common with other cereals, the proteins of sorghum and millets are predominantly globular in molecular conformation.

Another classification takes note of the starch character and divides sorghums between waxy, in which the starch is higher in amylopectin, and nonwaxy, in which the starch contains 20% or more of amylose.

The range of proximate and specific nutrient compositions reported by different workers are presented in Chapter 2 in a series of summary tables. These tables present the total range for each constituent (i.e. the lowest and the highest values reported throughout the literature reviewed), together with what is probably a more meaningful statistic, the range of means, which is the range of each mean value reported from the analysis of a number of samples. From all the literature reviewed the protein values (N × 5·7 DWB) range from 5·0–19·3% whereas the mean values reported by different analysts range from 7·1–14·2%. A summary of the range of means of the main constituents in each of the cereals reviewed is given in Table 1.4.

The most notable results in Table 1.4 include the comparatively wide range of means in most components among all the grains extensively analysed. Worthy of special note are the wide ranges of calcium, phosphorus and iron in all the cereals, and of the B vitamins in common millet. Of particular nutritional interest are the high fibre contents among all the minor millets, attributable to the heavy and strongly adhesive outer seed coat layers, a character common to all of them. While it seems possible that a thick pericarp and/or an adherent hull may serve to protect the grain from unwarranted attack, plant breeders might usefully explore to what degree the proportion of high cellulosic material might be reduced by genetic selection.

Protein Composition

In all these cereals lysine is the first-limiting amino acid. An overall comparison of the amino acid scores suggests however that the protein composition of pearl millet, common millet, finger millet and kodo millet is nutritionally superior to normal sorghum.

Whereas the pericarp in sorghum and pearl millet is reportedly lower in average protein content than the whole grain, the pericarp of finger millet generally contains about twice the proportion of protein present in the whole grain. Nonetheless, the millet grains being exceptionally small, considerable skill is required to make a precise separation of the microscopically identifiable components. Within any of the cereals reviewed, smaller seed types tend to be higher in % protein, probably because of the relatively larger proportion of embryo and aleurone layers present, and in some instances because of poorly filled out endosperms.

Protein content and composition within the endosperm is influenced by both genetic background and the environment and conditions of growth. Grain protein content is directly controlled by the plant's capacity to take up and transfer nitrogen from the roots and leaves to the seed. Among ICRISAT lines, nitrogen uptake/plant varied from 0·22–1·14 g and Nitrogen Transfer Efficiency (NTE) from 58–87% at any given total nitrogen uptake, which suggests the possibility of identifying lines with a high protein potential from genotypes that combine high nitrogen uptake with high NTE.

Within populations of similar genetic background, protein content is generally inversely related to grain yield. The results reported suggest that many hybrids manifest their vigour more in high grain yield than in a higher than parental protein content. Several workers reported that the heritability of protein content is controlled by two groups of genes though heritability estimates among the research reviewed ranged from 9–85%. Heterosis might be simply defined as a significant difference in the mean of any character (e.g. protein content) among the offspring compared with the mean (or midparent value) of the same character among the parents. Some workers report a positive (i.e. a significant increase) and others a negative (i.e. a significant decrease) heterosis for protein content and composition. From statistical analysis of their results some workers attribute positive heterosis for protein to the influence of one or more dominant genes, others to additive gene action. Some reports suggest that significant protein increase between parent and offspring results from genetic interaction manifested in a General Combining Ability (GCA), by others it is attributed to Specific Combining Ability (SCA). It seems probable that either GCA or SCA may give rise to positive or negative heterosis dependent upon the genetic makeup of the parents that are brought together. To develop high protein cultivars, one of the most experienced of the world's sorghum breeders recommended a programme of recurrent selection and hybridization, using composites of two or more populations each of which has demonstrated significant, positive heterosis. A population breeding programme that includes the full range of the three taxonomic sorghum groups is more likely to be effective than the traditional method of crossing a comparatively narrow range of genotypes.

Several workers report significant correlations between protein content and various phenotypic characters. In most instances it seems these are more coincidental than real and therefore to identify high-protein genotypes, seed samples must be analysed for nitrogen content by one of the methods recommended in PAG Guideline No. 16.

Comparatively little has been published concerning the genetic control of protein content among the minor millets. Work in the Soviet Union suggests that protein content in common and foxtail millets may be increased by chemical mutagens. Though protein content and grain yield were negatively correlated among the mutants, repeated selection generated some strains higher in yield and protein content than the accepted standard.

Protein content in all the cereals under consideration is markedly influenced by the environmental conditions under which the crop is grown. Some authors report the genotypic influence to be greater than the environmental and others the reverse. There is evidence to suggest that protein content and composition may be influenced by genotype × environment interaction. Many results indicate that a particular genotype grown at different locations, or at the same location in different years or during different seasons, may produce grains significantly different in protein content and composition. Significant correlations were reported between several agronomic factors, including various fertilizer treatments, and grain protein content and composition. Many of the results reflect the prevailing environmental conditions rather than any general biochemical relation or interaction.

The effects of different forms of nitrogen fertilizer upon protein content vary considerably according to conditions. Endosperm protein content is almost certainly more influenced by the efficiency of nitrogen uptake and translocation to the seed than by the level and form of nitrogen applied to the soil. Nitrogen uptake and transfer efficiency is affected by the time of fertilizer application, rainfall patterns and soil moisture. Though results vary among sources, nitrogen fertilizer applied to the soil appears more often to produce a higher grain yield than a higher grain protein content. Results reported simply as yield of protein/ha can be misleading since higher protein yields are often more attributable to higher grain yields than to increased grain protein content.

A relatively consistent finding with sorghum and pearl millet is that nitrogen applied as a foliar spray results in a greater increase in grain protein content than nitrogen applied to the soil. Several workers, particularly those working on poor tropical soils, reported increases in both grain yield and grain protein content in pearl millet, finger millet and common millet with added nitrogen fertilizer. In all cases where amino acid analyses were provided and where added nitrogen fertilizer resulted in a higher protein content, the higher protein was proportionately higher in proline and glutamine and lower in lysine. In common with all other essential amino acids, lysine as mg/100 g of grain increased with increase in grain protein content.

Most of the protein and amino acid data published by plant breeders, agronomists and their colleagues derive from the analyses of whole grains. In Chapter 5, the influence of commercial and domestic processing on composition is detailed. Unfortunately, most of the information available relates to sorghum, very few processing studies on the millets having been published. Furthermore, in many of the laboratory processing studies reported, the history, origin, condition and quality of the grain processed is not described.

Extraction rate is the term used by millers to indicate the percentage by weight of the original grain which is retained after milling. Thus, a 90% extraction rate would indicate that 10% of the grain, composed largely of the outer bran layers, has been discarded and that 90% was retained.

The composition of milled fractions varies markedly at closely similar extraction rates when derived from different milling processes. It is generally true however that the lower the % extraction rate, the lower the content of protein, lipid, fibre, minerals and B vitamins. Several reports indicate greater protein losses from traditional milling processes, particularly those that involve soaking and wet-milling, than from mechanical processes such as disc or plate mills and abrasion processes. Controlled abrasion milling, particularly of corneous endosperm types appears to reduce protein content less severely than either traditional pounding or break and reduction roller-milling. Controlled abrasion that removes the lowest protein fraction from the outer pericarp of sorghum yields a 90% extraction product higher in average protein than the original grain. With continued abrasion the highest protein fraction comes with the second layer removed which represents 7–9% of the whole grain. The lowest protein is found in the final fraction from the centre of the sorghum endosperm. Though less well supported by experimental evidence, a similar pattern appears true with pearl millet.

The control of protein content by abrasion milling is more difficult in floury than corneous endosperm types because of the tendency of the former to chip or fracture. Sorghum grains that carry a very thin pericarp around a corneous endosperm tended to yield the highest extraction of high protein flour.

Sorghum milled by break and reduction systems produced a different sequence of fractions than abrasion milling, the more friable floury fraction from the centre of the endosperm being liberated first. In one report, break and reduction milling of sorghum produced a soft floury fraction containing 6·5% and a corneous fraction containing 20·5% protein. The extraction rate from sorghum by break and reduction milling is reported in most cases as below 60%.

Among the minor millets, particularly those which carry a heavy high-fibre seed coat, abrasive dehulling significantly increases protein and reduces lipid and fibre in the fraction retained. Removal of the outer pericarp of pearl millet and common millet by light abrasion increased the protein content of the retained high extraction fraction.

Protein Concentration

Fractions widely different in protein content can be isolated from the same grain by grinding the endosperm to an extremely fine particle size in, for example, an attrition mill, followed by air classification in which a mechanically induced centrifugal force imposed on the fine particles is opposed by a centripetal air drag. The relative magnitude of the opposing forces determines the particle size (effective mass) above and below which the fractions will separate. The critical cut size to gain maximum protein shift appears to be significantly lower for sorghum (*c.* 14 μm) than for wheat (*c.* 17 μm). The largest protein shift from a range of sorghum types was obtained from floury endosperms, the highest protein fractions containing between 4 and 5 times the percentage protein in the lowest fractions. Flour from corneous endosperms, in common with flour from hard wheats, is susceptible to less dramatic protein shifts, when air classified.

Protein Fractionation by Solvents

Different workers report widely variable results from the fractionation of sorghum and millet protein by sequential solvent extraction whether by modifications of the Osborne method or the Landry and Moureaux system. For reasons given earlier, it is difficult to determine to what extent differences in the size of fractions are attributable to intrinsic differences among grain types and to what extent differences in methodology are responsible. In general, the Osborne solvent sequence leaves a larger final insoluble residue from sorghum, finger millet, Japanese barnyard, kodo and the other minor millets, than from pearl millet or wheat. Some investigators reported as little as 30% of the total nitrogen in sorghum to have been extracted by the Osborne sequence compared with close to 80% from some pearl millet cultivars. Others reported as much as 84% of the total sorghum nitrogen to be solubilized by a modification of the Osborne system. In general, the literature reviewed strongly indicated a lower overall solubility among sorghums and finger millet than for example wheat, when subjected to the Osborne sequence.

Some workers reported differences in protein solubility patterns among different fractions of the grain, the endosperm protein showing lower overall solubility than the whole grain protein. Lower solubility in the endosperm protein has been reported for both sorghum and finger millet and tends to be more evident in endosperms comparatively high in protein content.

Earlier in this chapter, several possible causes of differences in protein insolubility were suggested, including reactions with polyphenols, phytin and lipids. It has also been reported that some portion of the prolamine in sorghum and possibly finger millet may react with ethanol to produce an insoluble gel which presumably would eventually appear in the insoluble fraction. Examinations of the prolamines of sorghum, including IR spectra suggest that the polypeptide backbone chain may be insulated by a surrounding environment of nonpolar side chains that inhibit dispersion in polar solvents and hydrogen bond formation.

As already outlined, the Landry and Moureaux system employs different sequences of solvents than the Osborne and therefore though the fraction names (albumin, globulin, prolamine, glutelin) are identical, the fractions are not comparable between the two systems. Since Landry and Moureaux use more potent or disruptive solvents, their method generally results in a smaller residual insoluble fraction than the Osborne system. Whether the higher total protein solubility from Landry and Moureaux is attributable to increased disruption of disulphide bonds, separation of protein from lipoprotein or from phytin protein complexes, increased dispersibility of the prolamines through the surfactants SDS or SLS, or some combination of these influences, is not apparent from the literature reviewed.

As described later, the comparatively low overall solubility of the protein present in high tannin sorghums is increased in the Landry and Moureaux system only after polyphenols have been removed by alkali treatment.

Several references to disc and gel electrophoresis of sorghum protein were encountered. One report indicated that disc electrophoretic patterns from kafir sorghums were significantly different from other sorghum races, but none appeared to throw any light upon nutritional differences among cultivars.

A study of nitrogen solubility showed a marked contrast between sorghum and pearl millet on the one hand and wheat on the other. Typically, wheat protein showed highest solubility at extreme pH with a marked drop towards the isoelectric point. Sorghum and pearl millet protein solubility was reported to be low at all pH levels with two low solubility points suggestive of two isoelectric points.

Amino Acid Content

As stated above, lysine is the first-limiting amino acid in all the cereals under review, except in one or two instances in which high-tannin sorghum was fed to chicks when methionine appeared first-limiting, possibly because of an *o*-methylation reaction between the methionine and the polyphenol. Methyl donors such as choline and methionine are believed to be essential for the detoxification of gallic acid to 4-*o*-methyl gallic acid and though no hydrolysable polymer of gallic acid or related polyphenols has been isolated from sorghum, it is possible that methyl donors may react *in vivo* to eliminate flavonoids and other polyphenols. Therefore, unless plentifully supplied, methionine might appear deficient in a sorghum diet high in polyphenols.

There are two significant correlations within grain types of similar genetic background: (a) a negative correlation between protein (% DWB) and lysine (mg/g N); and (b) a positive correlation between protein (% DWB) and lysine (mg/100 g of whole grain). Essentially similar correlations are evident between protein and lysine in the endosperm.

As can be seen from Table 1.4, lysine (mg/g N means range from 71–212 in sorghum, from 109–297 in pearl millet, from 89–266 in common millet, and from 160–262 in finger millet. The Amino Acid Scores range from 21–62; 32–87; 26–78; and from 47–77 respectively. Within the grain of each cereal lysine (mg/g N) is highest in the embryo and in the second fraction removed by abrasion milling. In the endosperm lysine (mg/g N) is significantly higher in the floury than in the corneous component. In general an opaque floury endosperm is indicative of protein high in lysine but since those classed as corneous endosperms may appear floury when immature, selection for floury high lysine types must be made from mature grains. In all fractions of sorghum and millet grains proline and glutamine (mg/g N) tend to be inversely correlated with lysine.

Among the Osborne fractions, lysine is highest in the albumin, globulin, and glutelin fractions. From Landry and Moureaux, lysine is highest in the first (NaCl) and last (borate buffer plus Na lauryl sulphate plus 2-mercaptoethanol) fractions.

High-lysine Genes

Normal sorghum, foxtail, little and Japanese barnyard millets appear to be among the lowest in lysine (mg/g N) of all the known cereal grains. In recent years, however, sorghums that contain endosperm proteins with higher than normal lysine contents have been found to occur naturally in Ethiopia, and in the USA have been induced by chemical mutation. In comparison with normal sorghums the naturally occurring high-lysine (*hl*) cultivars contain a larger embryo, higher lipid, higher lysine (as % N and % grain) a lower leucine:isoleucine ratio, higher total sugar, lower starch, and lower amylose as % starch.

The high-lysine genotypes also bear a marked indentation in the surface of the kernel indicative of a smaller than average endosperm. Grains of the chemical mutagens high in lysine though also opaque, are otherwise more nearly normal in appearance, in starch, amylose, sugar and lipid contents.

The high lysine character is believed to be controlled by a simple recessive allele which appears to suppress prolamine formation in the endosperm storage proteins. When fractionated by the Landry and Moureaux method, the high-lysine sorghum protein produced a larger fraction I (albumin plus globulin) and fraction V (glutelin residue).

Though in normal sorghums lysine (mg/g N) may be affected by environment, it is reported that both the naturally occurring and chemically induced high-lysine character are comparatively independent of environment. Biological evaluation of the natural high-lysine sorghum and the chemical mutagen showed significantly improved rates of growth and nitrogen retention among both rats and chicks.

At the time of writing (1978), research is in progress at several centres to try to combine one or both of the high-lysine prolamine depressant genes, with other desirable characters such as high grain yield, well developed endosperms, genotypic and phenotypic stability.

Some instability in the inheritance of the high lysine trait is evident, though one worker reported the development of advanced generation bulk populations that have stablized to produce a few lines moderately high in lysine from the natural (*hl*) genes with nondented though comparatively small seeds. Crosses between the chemically induced high-lysine mutant with lines derived from natural *hl* parents produced a higher than normal frequency of progeny high in lysine. To some degree the high-lysine character appeared more stable after several generations of selfing (i.e. self-pollinating). Since among the superior lines from these experiments, correlation between seed weight and lysine content appeared comparatively low, there seems to be a reasonable possibility of combining a higher than average lysine (as mg/g N) with other desirable grain characters.

Given the marked inferiority in nutritional quality of sorghum in comparison with most other cereals, a research effort to this end would seem amply justified.

Protein Digestibility

Among normal sorghums and the millets, protein digestibility is reported to vary widely. When fed to children in what are described as poor Indian diets, the digestibility of both sorghum and finger millet protein was of the order of 50%. A high excretion of nitrogen, calcium and phosphorus was reported from both grains. The nutritional quality of typically poor diets high in sorghum or finger millet, whether judged by weight gain or nitrogen retention was improved by the addition of lysine either in synthetic form or by partial replacement of sorghum or the millet with rice, various legume flours or animal protein. In a mixed sorghum/rice diet fed to young children, as the proportion of sorghum increased, protein digestibility and nitrogen retention declined. Similar results were obtained with finger millet and common millet.

Protein digestibility varied among milled fractions of the grain, and was reported to be inversely related to the proportion of coarse bran removed during milling. In one instance, the protein digestibility of a high extraction milled sorghum was estimated as 20% compared with 77% from a common wheat at an equivalent extraction rate.

Among the various soluble fractions, prolamine appears least digestible, consequently higher protein digestibility and nitrogen retention was observed in animals fed floury in comparison with others fed corneous endosperms. Wet processing of sorghum and millets, traditional among several communities in the SAT, tends to reduce not only total protein but also the biological value, possibly because of the preferential removal of the water and salt soluble protein fractions that are highest in lysine.

Among both monogastric and ruminant farm animals, total digestibility and protein digestibility vary significantly among sorghums. It is probable that many of the observed differences, though in part attributable to differences in nutrient content among the sorghums used, resulted from the widely different proportions and compositions of other ingredients in the diets compared. Several reports indicated that total and protein digestibilities in ruminants were higher from the waxy high amylopectin than normal sorghum types. Why this is so is not clearly apparent though there is some suggestion that point hydrolysis occurs in the waxy endosperm starch.

In general, most of the standard methods of processing, including grinding, partial cooking and pelleting, improved feed efficiency of sorghum and the millets when fed to either monogastric or ruminant animals. The adverse effect of tannin on protein digestibility was in some degree reduced when the grains were ensilaged.

Lipid Content

As stated earlier, the lipid extracted varies with the solvent system used. Invariably, lipid content is highest in the embryo and lowest in the endosperm. Consequently, total lipid is lowest in flours of lowest extraction rate. Roughly 80% of the lipid in sorghum, pearl, foxtail and common millets appear to be unsaturated with linoleic acid present in highest proportion. Some analyses suggest that linoleic is in higher proportion in the minor millet lipids than in sorghum but the tabulated data in Chapter 2 show wide variations.

Carbohydrate Content

Among all the cereals reviewed, total carbohydrate and starch are inversely correlated with protein content. Excluding sorghum and pearl millet, comparatively little is published on the nature and composition of the carbohydrates. The starch granules of pearl millet appear smaller but otherwise similar to maize and sorghum starch. The recovery of starch from sorghum by wet processing appears inversely related to the proportion of high protein peripheral endosperm cells present. Among most normal (nonwaxy) sorghums, amylose was reported to range from 20–30% and in certain fodder varieties from 12–13% of the starch. Among pearl millets, amylose represented between 18 and 25% and among common millets between 23 and 27% of the starch present.

It has been suggested that, as with rice, the cooking characteristics of sorghum and possibly the millets, may be influenced by the relative proportions of amylose and amylopectin present. At present there appears little evidence to support or refute this theory but it is a line of study that deserves to be pursued. Though it is not a primary consideration of this publication, the authors strongly recommend that more study be devoted to whatever factors may influence those properties that affect technological performance, utility and acceptability as judged by major users and consumers of sorghum and the millets. Grain characterized by superior nutritional and agronomic characteristics is of little value if it is unacceptable to users and consumers.

Free Sugars

From the limited evidence available, glucose appears to be the principal free sugar in sorghum and most of the millets. The total sugar content of high-lysine sorghums is roughly 2·5 times that in normal lines.

Cellulose and Pentosans

Compared to most other cereal grains cellulose and pentosan content is exceptionally high in Japanese barnyard, kodo, foxtail, common and little millets, and in Job's tears. Pentosans are in highest proportion in the pericarp and germ cell walls. Crude fibre content is significantly reduced by removal of the outer pericarp from sorghum and all of the millets. Faecal volume and weight is increased by greater ingestion of the outer pericarp of all these cereals. Limited evidence suggests that the pentosans in wheat bran are more digestible by humans than the pentosans of sorghum or pearl millet.

Mineral Elements

Nutritional studies among people of the SAT show evidence of deficiences in calcium, iron and zinc, all of which elements are rendered insoluble by phytic acid. In Table 1.4 calcium, phosphorus and iron contents are shown to vary widely among all the cereals under review. Calcium, phosphorus, including phytin phosphorus and iron contents may be significantly reduced by removal of the outer pericarp, though the highest concentration of phytate occurs in the germ.

Different workers suggest that calcium content may be genetically influenced in sorghum, pearl millet and finger millet. Given the general low level of retention (below 20% of intake) of calcium by humans fed diets high in sorghum or finger millet, there seems little purpose in breeding to select for high calcium genotypes (if indeed such exist) in either of these grains or the other millets. Several studies report that both phosphorus and iron are poorly retained in diets high in sorghum and the millets, particularly when fed to young children. Iron may also be rendered unavailable by tannins.

Vitamin Content

Research workers do not totally agree whether the B vitamins thiamine, niacin and riboflavin in sorghum and the millets are influenced genetically or by environment. Table 1.4 shows a wide range of results from different analyses. It is reported that some proportion of the niacin present in sorghum may not be biologically available. Chromatographic separation of methanol extracts of some sorghums have indicated that less than 60% of the niacin present is in a free and available form. Some workers found that niacin in sorghum could be converted to the free form by treatment with lime water and other alkalis. Other workers were unable to confirm this finding.

Proportions of the three B vitamins are reduced by removal of the pericarp layers and significant losses of all three are reported following traditional milling, and of thiamine and riboflavin during cooking. Riboflavin significantly increases during germination (malting) of sorghum and several of the millets. On a dry weight basis, riboflavin is generally higher in sorghum beer than in the original grain.

The mean values of niacin present in whole grain sorghum reported by different analysts ranged from 2·9–6·4 mg/100 g. In spite of the apparent adequacy of niacin and tryptophan from most of the analyses reported, pellagra-like symptoms have been reported from India among humans and animals fed high sorghum diets. The Indian workers noted in many sorghums a significantly higher leucine/isoleucine ratio than in other common cereals and advanced the hypothesis that the comparatively high level of leucine in sorghum was in some manner responsible for niacin deficiency. Symptoms typical of pellagra as induced by niacin deficiency, including black tongue, were reported in dogs and monkeys fed high sorghum diets. The Indian scientists reported that the niacin deficiency symptoms were aggravated by increasing leucine and reduced by adding isoleucine to the diet. Thus they postulated that the leucine/isoleucine ratio in the diet affects tryptophan-niacin metabolism. It is suggested that a leucine/isoleucine ratio higher than 3/1 should be regarded as potentially deleterious. Though similar or supporting results do not appear to have been reported elsewhere, and in spite of the fact that any relation between leucine or isoleucine and niacin is not readily evident from their chemical structures, the Indian research should be taken seriously. In Table 1.4 the range of means for the ratio of leucine/isoleucine is presented for all the cereals under study. It should be noted that the leucine/isoleucine ratio is generally lower among the progeny of high lysine genotypes than among normal sorghums.

Though absent from the harvested dry grains, ascorbic acid is reported to develop during germination of several of these cereals.

Tocopherol vitamin E) and vitamin K are both present in the embryo of all the cereals. β-Carotene was reported in greater proportion in yellow endosperm than white endosperm sorghums though in neither case was the concentration likely to be of nutritional importance.

Polyphenols

A recent seminar reviewed this subject in detail (Hulse, 1979). Apart from protein, more has been published on the subject of polyphenolic tannins than on any other component of sorghum. Phenolics and their oxidation products have long been known to react with proteins, at least three modes of reaction being possible: (a) hydrogen bonds between OH groups in the tannins and receptor groups (e.g. NH, SH and OH) in the protein; (b) ionic bonds between anionic groups in the tannins and cationic groups in the protein; and (c) covalent linkages between quinones and various reactive groups in the protein. According to Loomis (1969) as much as 33% of protein (DWB) may be bound to phenolics through hydrogen bonding by which mechanism plant enzymes may be inactivated or precipitated. Flavonoids have been shown to combine with α-amylase, and other simple phenols can bind to blood proteins. Condensed tannins, which occur in sorghum, appear to produce more stable complexes than hydrolysable tannins which do not. A familiar characteristic of polyphenolic tannins is their oral astringency, a sensation that probably results from their reaction with the glycoproteins of the saliva. It is believed that some polyphenols, if left free (unbound) and in direct contact for a sufficient time, can cause permanent damage to the mucosa.

All known vegetable tannins are polyphenols: all plant phenols are not tannins. Various polyphenols including apigenidin, luteolinidin and kaempferol have been reported present in sorghum by different analysts. Recently (1976), a flavonoid, glucosyl vitexin, was isolated from pearl millet.

In spite of the widespread interest in polyphenolic tannins in sorghum, it was not until 1976 that what is probably the most important condensed tannin was isolated and identified. The research that led up to its isolation and identification is described in the section of Chapter 4 written by Gupta and Haslam. The tannin isolated is classified chemically as a polymeric procyanidin, a substance typical of a family of related polyphenols often found in plants that display a woody habit of growth.

Polyphenols in the seed coats increase as the growing sorghum grain matures. Some reports state that most rapid development occurs between 35 and 42 days after anthesis. There is some evidence to suggest that bagging the panicles of high tannin genotypes reduces significantly the formation of polyphenols which suggests, but does not confirm, that some stage of the synthesis may be energized photosynthetically.

High-tannin content is usually associated with a deep brown or reddish brown seed colour. The literature reveals many inconsistencies between chemical or biochemical indications of high tannin and pigment intensity, whether determined by specific reflectance or absorbance of the extracted pigments. Such inconsistencies are not remarkable bearing in mind that the nature and composition of biologically active sorghum polyphenols was unknown and that many of the standards of analytical comparison were inappropriate. For example, many analysts used "tannin" or "tannic acid" (probably impure mixtures of hydrolysable tannins) as comparative standards for chemical and biological comparison with high tannin sorghums. The authors have uncovered no evidence that any such hydrolysable tannins exist in sorghum or in any of the millets.

Various substances that react with phenolics to produce colour complexes have been described, the most commonly used being vanillin which in the presence of hydrochloric acid reacts with *m*-dihydroxy phenols to produce a colour reaction. Using catechin as a standard the intensity is reported in Catechin Equivalents (CE). Both the polymeric procyanidin isolated by Gupta and Haslam and other *m*-dihydroxy phenols react with vanillin HCl, consequently the method is not specific for sorghum polyphenols.

Gupta and Haslam have recently reported (personal communication), (a) that the modified vanillin-HCl method is influenced by (1) the temperature of reaction, (2) the relative concentrations of vanillin and HCl; (b) that reaction kinetics are markedly different among reactive phenolic substrates of different biological origin and composition; (c) that the method measures not only tannins but also low molecular weight phenols with an unsubstituted resorcinol or phloroglucinol nucleus.

The rate and pattern of colour formation with vanillin HCl is reported to vary among *m*-hydroxyphenols. Consequently, it is essential that the standard of comparison be chemically identical with the phenol being assayed. Gupta and Haslam found that when catechin was used as the standard the V-HCl test produced a much higher estimate of polyphenol content than when the standard was the isolated sorghum procyanidin. It is not evident to what extent low molecular weight phenols present in tannins form pigments that contribute to CE values but are not biologically active.

Because earlier methods of analysis were imprecise, it is not absolutely certain which of the pericarp or the testa is the site of greatest tannin concentration. From recent evidence using the procyanidin polymer isolated from sorghum as the reference standard, among a limited range of genotypes those containing both a pigmented testa and a pigmented pericarp were highest in polyphenol tannin. Those with either a pigmented testa or a pigmented pericarp were very significantly lower in polyphenol content than genotypes with both a coloured testa and a coloured pericarp.

It is probable that the high-tannin character in sorghum, which is genetically controlled, provides the sorghum grain with a defensive mechanism against attack by birds, some insects and microorganisms. In the USA the BR (bird-resistant) prefix is commonly used to designate sorghums believed to be high in tannin content. There seems little doubt that the polyphenols in sorghum reduce both total and protein digestibility, and that they inhibit the activity of various enzyme systems including amylases and possibly lipases and proteases. When fed to laboratory and farm animals, sorghum tannins reduce overall biological value and feed efficiency, and increase the excretion of faecal nitrogen. There is evidence to suggest that in addition to general reactions with proteins, sorghum phenols may react specifically with methionine. Though *ortho*-methylation may provide a specific mechanism by which herbivores detoxify plant polyphenols, proteins in various forms can interact with ingested phenols to reduce their toxicity. Consequently, a higher than normal protein intake seems desirable in diets high in polyphenols.

Apart from their effect on total digestibility and nitrogen balance, comparatively little is reported on the effect of sorghum tannins in human nutrition. On the basis of circumstantial statistical evidence, one author suggested that tannins may be responsible for oesophageal cancer among high sorghum consumers in East Africa. It seems probable however that the reported incidence of oesophageal cancer may be attributable more to zinc deficiency than to polyphenol toxicity. As described elsewhere, zinc absorption and retention in humans is inhibited by phytic acid in the diet. Several workers have demonstrated a significant correlation between zinc deficiency and oesophageal cancer, particularly when accompanied by a moderate to high alcohol consumption.

While the adverse effects of sorghum polyphenols may be reduced by absorbants such as nylon or PVP, or by acid or alkaline hydrolysis, the most practically effective manner seems to be to remove the high tannin seed coats by alkali or abrasive decortication. Now that a principal tannin has been chemically identified, more research is required to determine to what extent the adverse nutritional effects of sorghum tannin can be reduced by abrasive removal of the seed coats and to what extent this would reduce the content of essential and desirable nutrients.

Several traditional methods among Africans are used to reduce the effects of polyphenols. In some communities the sorghum grains are soaked in sour milk or in water with tamarind seeds in which case the lactic or tartaric acids, respectively, probably bleach the seed coat pigments. Whether they reduce the tannins' biological effect is not evident. Other communities germinate the grains in wet wood ash under which conditions the alkali probably hydrolyses the tannins present. Whether other essential nutrients are destroyed under these alkaline conditions is not reported.

A process in Southern Africa claims that mono-acidic carbonyl compounds such as formaldehyde will reduce the polyphenol content of the milled products of sorghum. Though it may be applicable in large industrial processes, this invention does not seem technologically suitable or economic for traditional or other small scale processing of high-tannin sorghums.

The more reliable indicators of adverse physiological activity by tannins in sorghum include *in vitro* dry matter disappearance, and *in vivo* nitrogen digestibility and retention.

Phytate Phosphorus

Phytate phosphorus may be of greater nutritional importance to those who subsist on sorghum and the millets than is generally realized. Different analysts reported phytates to range between 45 and 95% of total phosphorus in sorghum. Phytic acid forms insoluble compounds with calcium, iron, zinc and several minor elements such as manganese and cobalt, and is probably responsible for the low retention and high excretion of ingested calcium and iron. It is not clear to what extent phytin phosphorus content is genetically controlled and therefore whether it can be reduced by selection measures. One difficulty for breeders who wish to select for low phytin content is the lack of a simple analytical method specific for phytic acid. Since phytin phosphorus occurs in highest concentration in the embryo and pericarp, the proportion in the diet can be reduced by controlled abrasion milling.

Other Toxins

There are reports of an unidentified fat-soluble toxin in kodo millet. Since it appeared most commonly when kodo millet was harvested during wet or humid conditions, the toxin was conceivably some form of mycotoxin.

Conclusion

Based upon the foregoing and the more detailed text that follows, it would be unwise to propose recommendations that apply to all sorghum and millet breeding and utilization research programmes. In economically developed countries, where the grains are used mainly in animal feeds, deficiencies in nutrient content and the presence of nutritional inhibitors can be counteracted by processing and/or supplementation.

In those countries where these cereals provide most of the dietary calories, overall nutritional quality must be considered important by everyone concerned. It is quite unsatisfactory to base policy judgements upon published statistics of production and availability, most of which are probably crude averages that give no indication of wide variations among years, seasons, regions, countries, communities, families and individuals.

It is clear that a universally ideal sorghum or millet cannot be defined since several desirable characters appear inversely related to others. Consequently, under most conditions, some compromise among desired attributes will be necessary. Wherever sorghum or millet is but one ingredient of a widely mixed and variable diet, one need have less concern for protein quality and mineral content than is desirable among people who rely heavily upon these cereals for essential nutrients.

Recognizing the inferior nutritional quality of normal sorghum protein, the inhibition of protein by polyphenols, together with nitrogen losses that occur during domestic and small industrial processing and, probably, in storage, it is recommended that research be continued to stabilize a higher than average lysine in combination with an average (*c.* 10% N × 5·7) protein content.

Research institutions in technologically advanced countries could helpfully examine (a) the role and range of phytin phosphorus among genotypes and (b) whether polymeric procyanidins, chemically similar to the one identified by Gupta and Haslam, are present in all high tannin sorghums and what hereditary and environmental factors influence their formation in the testa and/or the pericarp as the grains mature.

The growing interest among communities of the SAT in centralized milling of these cereals, together with the demonstrated opportunity for combining processed sorghum and millet with other cereals and legumes in composite flours, requires a more intensive study of the structure and other physical and biochemical properties of these grains. The influence of grain structure and composition on processing characteristics and acceptability, and of typical processing and cooking procedures on the fate of essential nutrients also deserve much greater attention.

It cannot be overemphasized that sorghum and the millets are of immensely greater nutritional importance in the diets of poor people in the SAT than is wheat in the diets of North Americans. In the SAT, sorghum and the millets cannot be replaced by wheat or other cereals except by importation. These cereals are unique in their tolerance of adverse conditions, conditions wholly inimical to the cultivation of other grains. Consequently, sorghum and the millets are deserving of a much greater research and development investment in the future than they have received in the past.

CHAPTER 2

History and Nature of Sorghum and the Millets; General Chemical Composition and Biological Quality

---

Sorghum and the millets are of much greater importance in human diets, particularly those of the poorest people of the semi-arid tropics, than the attention they have received would suggest. Though these cereals have been important staples in the semi-arid tropics for many centuries, there appears to be no reliable historical record of their origin or pattern of dispersion. Because they have been grown for so long in so many countries, mainly by small-holder cultivators, it is not surprising that the common and vernacular names by which they are known are many and various. In some records, no distinction is made between sorghum and the millets: production statistics quoted, even by international authorities, often group the cereals together.

Nomenclature

The origin of the name "sorghum" remains obscure. In medieval Latin it appears to have been known as "surgo", and may therefore have been derived from the Latin verb "surgere" meaning "to rise". On the other hand, it seems reasonably certain that "millet" was derived from the Latin "millesimum", meaning a thousandth part, the diminutive "mil" often being used to denote something which is extremely small. Millet is therefore a name applied generally to a number of cereals characterized by their small seeds.

Any one species may be described by several names according to the lingua franca or the place of cultivation. For example, *Panicum miliaceum* is variously known (in English) as proso millet (or simply proso), common millet, panicled millet, hog millet and broomcorn millet (not to be confused with the sorghum "broomcorn"); names in other languages may appear as literal translations, such as, in French, "mil commun" (equivalent to "common millet"), or in unrelated terms such as "mil indien".

Among Indian and African languages, vernaculars and dialects, the names for sorghum and the millets appear in many variations, some of which may reflect the inadequacy of English spelling to reproduce accurately or consistently the nuances of the original.

In Appendix I Part 1, is listed in alphabetical order those colloquial and vernacular names found in the original papers examined, together with what we, the authors, believe are the appropriate botanical names. In Part 2 of that appendix the same information appears, listed in alphabetical order by botanical name. Those readers who wish to pursue in more detail the origin, history and relevance of the vernacular names used in India and Africa are referred to the papers of Bono (1973), Gupta and Dutta (1967), Krishnaswamy (1962) and Porteres (1955, 1958abc, 1959abc). Lists of vernacular names with their botanical equivalents also appear in Adrian and Jacquot (1964), National Research Council (1961) (appendix), Rachie (1965, 1974) and Rachie and Peters (1977).

Below are the names chosen by the authors and used throughout the main text of this book, together with what the authors believe are the correct botanical names.

Term used in text	Botanical name
Sorghum	*Sorghum bicolor* (L) Moench
Pearl millet	*Pennisetum typhoides*
Finger Millet	*Eleusine coracana*
Foxtail millet	*Setaria italica*
Common millet	*Panicum miliaceum*
Little millet	*Panicum miliare*
Kodo millet	*Paspalum scrobiculatum*
Japanese barnyard millet	*Echinochloa frumentacea*
Browntop millet	*Bracharia ramosa*

It should be noted that different authors use different botanical terms for the same grain. For example, cultivated sorghum is described as *Sorghum bicolor* (L.) Moench in many English language papers and as *Sorghum vulgare* in much of the French language literature.

Teff (*Eragrostis tef*) is not strictly a millet, but is included because it is an extremely important cereal crop in Ethiopia, the only country in which it is known to be widely grown. Other small grains of minor economic consequence include *Digitaria exilis* and *Digitaria iburua*, both also referred to as fonio; Job's tears (*Coix lachryma-jobi*); and swamp grass (*Echinochloa stagnina*) and *Amaranthus paniculatus*. Each of these is referred to in the main text by the Latin name with the exception of *Coix lachryma-jobi*, for which "Job's tears" is used.

Very little information has been found on the nutritional value of these minor cereals. They are included as millets because they are characterized by small grains. Such information as has been traced on these small grains, albeit incomplete, is included in this review since they are used for human food in several developing countries.

Classification of Sorghum

In 1936, J. D. Snowden published his classification of cultivated sorghum and it has provided the basis for many later schemes. Doggett (1970) presented a table in which the cultivated sorghums are listed according to Snowden's classification and the geographical areas in Africa with which they are associated. Murty *et al.* (1967a) classified and catalogued a world collection of sorghum using a modification of Snowden's system. Jakushevsky (1969) also suggested a modified Snowden classification scheme.

Harlan and de Wet (1972) published a simplified classification based on Snowden's which has been checked against some 10 000 head samples, and correlated against the Snowden collection at the Royal Botanic Gardens at Kew, Surrey, England. Under their scheme, *Sorghum bicolor* (L.) Moench is partitioned into the following races:

Sorghum bicolor spp. *bicolor* cultivated races:

basic races	hybrid races (all combinations of basic races)
race (1) bicolor	race (6) guinea–bicolor
race (2) guinea	race (7) caudatum–bicolor
race (3) caudatum	race (8) kafir–bicolor
race (4) kafir	race (9) durra–bicolor
race (5) durra	race (10) guinea–caudatum
	race (11) guinea–kafir
	race (12) guinea–durra
	race (13) kafir–caudatum
	race (14) durra–caudatum
	race (15) kafir–durra

Sorghum bicolor spp. *arundinaceum* spontaneous races:

race (1)	arundinaceum
race (2)	aethiopicum
race (3)	virgatum
race (4)	verticilliflorum
race (5)	propinquum
race (6)	shattercane

According to Harlan (1972) the geographical distribution of the basic races is, in general terms:

bicolor	all over Africa
guinea	West Africa, Malawi, Tanzania, India
caudatum	Eastern Nigeria to Eastern Sudan, Uganda
kafir	East Africa (south of equator), Southern Africa
durra	Ethiopia, zones of Africa near the Sahara, India

It is outside the scope of this review to discuss critically alternative classification schemes. Nonetheless, the authors consider it necessary to include the above in order to assist the reader to a better understanding of the literature subsequently reviewed. Since the concern of the authors is with those who depend largely upon sorghum and the millets for their nutritional well-being, it is strongly recommended that greater attention be given to a possible chemo-taxonomic rather than a purely botanical taxonomic system of classification.

Harlan (1972) pointed out that the Nigerian kauras are durra–caudatums; the zera–zeras and hegaris are caudatums. What is known as feterita in the Sudan ranges from guinea–caudatum through caudatum to durra–caudatum. The feterita introduced into the USA was a durra–caudatum. Some of the descriptions applied to sorghum grains in the earlier literature may not be accurate and should be viewed cautiously.

Martin (1970) suggested that since many varieties now in use are derived from crosses between two or among more groups, the old classification groups cannot easily be universally applied and that in the USA the trend towards hybrid sorghums will probably eliminate the need for the older (variety) classification system. The world collection of genetic stocks could best be categorized, Martin (1970) suggested, on the basis of useful genetic characters such as:*

(1) plant height, colour, maturity and photosensitivity
(2) stalk juiciness, sweetness, strength, tillering capacity and branching tendency
(3) panicle size, density and recurving tendency
(4) seed colour, size, texture, shattering tendency and threshability and
(5) resistance to specific diseases, specific insects, birds, heat, cold and lodging.

Historical Development

Though the area they cover is comparable to or greater than most other internationally important cereals, sorghum and the millets have received scant attention when compared with the grains prominent in international trade such as wheat, rice and maize (IDRC, 1977). Among the comparatively sparse literature sources, at least three basic texts on sorghum deserve serious study: Doggett (1970), Wall and Ross (1970) and as editors, Rao and House (1972), all of which deal mainly with the botany, breeding, agronomy, and to a lesser extent, the use of sorghum as food and feed. Arnon (1972), Leonard and Martin (1963) and Kramer and Matz (1969) have written on sorghum in general texts on cereal crops. The nutritional aspects were considered by Adrian and Jacquot (1964) in a French publication now out of print. The Rockefeller Foundation (1967a, 1973), and Chaugale *et al.* (1955a) have published bibliographies.

* Authors' note: it is interesting to observe that Martin's proposal makes no reference to nutritional quality.)

A detailed account of the history and origin of sorghum is presented by Doggett (1970). Briefly, it would appear that sorghum had its origins in the region of Africa now known as Ethiopia where pearl millet, finger millet and teff have been cultivated over many centuries.

The term "sorghum" includes at least four groups of cultivated annual plants: (1) grain sorghum, (2) sweet sorghum or sorgo, grown as forage and for animal feed, and also for the juice in the stem, (3) sudangrass also grown mainly for pasture, hay and silage and (4) broomcorn, also known as broom-millet, grown, as the name suggests, for the making of brooms (Doggett, 1976).

There may be some overlapping among groups, and some types may be grown for both forage and grain. This publication comprehends almost exclusively the cultivated grain sorghums grown for human consumption. Readers interested in the other three groups are referred to Doggett (1970) and Wall and Ross (1970).

The first known record of cultivated sorghum appeared in a 2700-year-old carving found in an Assyrian palace (Anderson and Martin, 1949, cited by Arnon, 1972). Though there appears little doubt that the first cultivated sorghums appeared in Africa, it is uncertain though possible that they were developed in Ethiopia and the surrounding countries (Doggett, 1976). The ancestors of what we now know as sorghum may have been carried across North Africa to West Africa about 3000 B.C., at which time the Sahara was possibly better watered than it is today. It is probable that cultivated sorghum was carried to the region of the Upper Niger River many centuries ago and that browntop millet (*Brachiaria ramosa*) and the two fonio millets (*Digitaria exilis* and *Digitaria iburua*) were developed, at least in part, by the early Mande people (Doggett, 1976).

The sorghum race guinea was diversified in West Africa, and the race durra in the Ethiopia–Sudan area. Race durra appears to have been carried to the Near East and to India. In East Africa, many of the white grains displayed characteristics of the guinea race and this association may have been present in the original material which moved into East and West Africa. Agriculture was practised at an early date, certainly in the highland areas of Kenya, and probably also in similar regions of Uganda and Tanzania.

The Bantu people, who originated in the Cameroons, worked along the southern edge of the Congo forests and it is probably at the eastern end of the forest belt that they first encountered sorghum; the race kafir is associated with the Bantu people from Tanzania southwards.

The caudatum race is associated with the Nilotic and Nilo-Hamitic peoples of Uganda and Western Kenya and the race is also important in the Sudan, Tchad and Nigeria.

Race bicolor is widely scattered throughout Africa; it is characteristically low yielding, generally with poor grain quality. It may have been collected and distributed because of its sweet juicy stalk and its suitability for making brooms (Harlan and de Wet, 1972).

Each of the main races of sorghum appears to be associated with one or other of the wild species listed below:

wild race	cultivated race
arundinaceum	guinea
aethiopicum	bicolor, durra
verticilliflorum	kafir, caudatum

Independent origins for the cultivated sorghums from the wild races have been proposed by Snowden (1936) but Doggett (1976) considered it more probable that the early cultivated sorghum forms were transferred from place to place as agricultural practices diversified. Doggett (1976) is of the opinion that the wild races frequently appear as weeds and that much interaction is to be found among the wild and cultivated races.

Historical records suggest that sorghum reached India from East Africa soon after 2000 B.C. It may have been carried from East Africa to Arabia, and from Arabia, with the dhow traffic, it probably spread around the Persian Gulf and into the Near East. The spread along the coast of Southeast Asia and around to China may have taken place about the beginning of the Christian era, but it is also possible that sorghum arrived much earlier in China via the silk trade routes.

Grain sorghum appears to have arrived in America as "guinea corn" from West Africa with the slave trade. Brown and white durra were introduced from North Africa about 1874, milo about 1880, feterita in 1906, hegari in 1908. Kafirs were introduced from South Africa about 1876 (Doggett, 1976). Sorghum spread to Central and South America where its importance in semi-arid areas is increasing but mostly as a feed grain. Sorghum adapted to high altitudes and low night temperatures has been developed in Mexico (Doggett, 1976, CIMMYT, 1976).

The basic types of grain sorghum recognized in the USA are milo, kafir, durra, hegari, feterita, shallu and kaoliang. For many years milo and kafir were the groups most commonly grown and the term "milo" is often used synonymously, though not necessarily correctly, for grain sorghums. "Milo maize", "kafir corn" and even "maize" are sometimes inaccurately applied to grain sorghums (Ross and Webster, 1970). In the USA, the first improved varieties were derived from selections of milo and kafir and included Pink Kafir, Blackhull Kafir, and Dwarf Yellow Milo (Ross and Webster, 1970). All were tall. Cultivars that could be combine harvested became popular during the early 1930s and included Wheatland and the stiff-stalked types, Martin, Westland, Plainsman, Caprock and Midland.

Prior to the introduction of hybrids during the 1950s, these cultivars, together with Combine 7078, Combine Kafir 60, Dwarf Kafir 44 14, Early Hegari, Norghum, Redbine 60, Redbine 66, Redlan and Reliance were the most widely grown. Hybrid seed distribution began in 1957 and by the early 1960s, almost all the USA sorghum acreage was given to hybrids.

Hybrids from the USA agricultural experiment stations are classified and designated by a number based on maturity in relation to standard varieties.

These are shown below:

maturity class	series
earlier than Norghum	300
Norghum	400
Reliance	500
Martin	600
Plainsman	700
Dwarf Kafir	800

The 600 series appears the most common. Hybrid numbers may carry a State prefix such as OK for Oklahoma and NB for Nebraska, indicating that interest is limited to one State; or they may bear the prefix RS (regional sorghum) indicating suitability for wider distribution. Commercial seed companies may use either maturity numbers or a company designation (Ross and Webster, 1970). During World War II, waxy starch types high in amylopectin were developed in the USA for the production of industrial starch.

Hybridization offers a significantly higher yield potential. Under favourable conditions of soil fertility, the following yields are reportedly attainable in the USA: (1) 1000–2000 kg/ha with limited soil moisture, (2) 2500–4000 kg/ha with moderate soil moisture and (3) 8400–10 000 kg/ha with adequate soil moisture (Arnon, 1972).

High yielding hybrid sorghum varieties have been developed in India, two frequently mentioned being CSH 1 and CSH 5. NK 300, a commercial hybrid from the USA, is grown over a wide area but in general, the USA hybrids are grown mainly for animal feed.

The literature reviewed, together with personal communications to the authors suggest that some of the most impressive sorghum grain yields have been achieved in the USA. The authors' observations also indicate that many of the more productive US lines have been less successful in African and Asian countries for a variety of reasons including inadequate resistance to pests, parasites, and diseases, together with not clearly defined, but nevertheless unsuitable, functional properties. According to Doggett (1976) the best grain yields result from sorghum plants about 2 m tall. Dwarf-type sorghums, satisfactory for combine harvesting in the USA, are not widely accepted in Africa.

Geographical Distribution and Production

Sorghum and the millets are the most important food crops of the arid and semi-arid tropics. "Semi-arid" cannot be precisely defined, but in general, it describes those regions in which evapo-transpiration exceeds rainfall for more than half the year. The territories classified as semi-arid tropics include large areas of Africa, embracing most of the countries surrounding the Sahara, much of East Africa, a large area of central India and some regions of Southeast Asia and South America. Over most of the semi-arid tropics, rainfall distribution and frequency is variable and unpredictable as exemplified by the severe droughts that occurred in the Sahelian zone of Africa and much of India during the first half of this decade. It is believed that 90% of the rural people of the Sahelian zone rely upon sorghum and millets as their principal source of calories and together with certain food legumes, for most of their protein.

The sorghum and millets production zone almost encircles the earth beginning in China, through India, and extending to most of Africa, particularly close to the 15th parallel, the southern United States and Latin America. In Europe, sorghum does not extend much further than the 44th parallel north but millets are reportedly grown in Russia as far north as the 53rd latitude. While most sorghum and millets are probably grown at low altitudes, sorghums tolerant to high altitudes are found in Ethiopia and have recently been adapted in Mexico to altitudes above 2200 m (CIMMYT, 1977). Some cultivars of finger millet are also adapted to high altitude conditions in Asia, largely in the foothills of the Himalayas, and in Africa (Adrian and Jacquot, 1964; Purseglove, 1972).

It is probable that the total land area under sorghum and millets exceeds 70 million ha, greater than that devoted to maize. In developing countries, the area under sorghum and millets considerably exceeds that under maize and whereas the average yield of maize in developing countries is close to 1.25 tonnes/ha, the average yield of sorghum and millets is probably less than 0·5 t/ha. In the USA where sorghum is grown as a feed grain, average yields are seven times those achieved in most developing countries (IDRC, 1977).

Doggett (1970), based upon the theoretical data of Loomis and Williams (1963), estimated that a grain sorghum producing approximately equal amounts of dry matter as grain and forage, theoretically has the capacity to yield around 40 t of grain/ha. Doggett stated that Loomis and Williams found from some well authenticated crop yields that the mean growth rate of sorghum was exceeded only by Napier grass (*Pennisetum purpureum*). From West Africa reports have been received of yields in excess of 8 t/ha from breeders' sorghum lines that mature in less than 90 days. The opportunity therefore clearly exists to realize significantly higher yields of sorghum in the future than are common at present. As Doggett stated "we have in the sorghum plant an excellent carbohydrate factory; one that is better than most".

The production in 1976 (grain harvested, 10^3 t), area harvested (10^3 ha) and yield (kg/ha) of sorghum and millets as presented by the FAO Production Yearbook 30 (1976) are shown for various countries, excluding continental China, in Table 2.1.

In this table, sorghum covers various types embraced by *Sorghum vulgare*, including kafir corn, milo, durra, jowar and many others described colloquially (see Appendix I) and may also include Sudangrass and Columbus grass (*S. almun*). The data presented under millets included *Pennisetum glaucum* or *typhoides* (pearl millet), *Eleusine coracana* (finger millet), *Panicum miliaceum* (common millet), *Setaria italica* (foxtail millet) and *Echinochloa frumentacea* (barnyard millet), but as many of the countries reported sorghum and the millets together, where this occurred the combined data were presented in the FAO Production Yearbook under millets.

Arnould and Miche (1971) stated that sorghum and the millets provided roughly 16% of total world cereal consumption for humans. Among developing countries, these cereals provided at least 25% of the total cereal intake. Arnould and Miche citing as their source Document GR 70/7 of the thirteenth Session of the Cereal Study Group (1970), showed the percentage distribution of sorghum usage (mean of the years 1966-67/1968-69) as:

	Developed countries (%)	Developing countries (%)	Total all countries (%)
Human nutrition	1·1	82·5	53·3
Animal feed	95·8	7·7	39·4
Fermented beverages and industrial uses	1·6	2·3	2·0
Seed and losses	1·5	7·5	5·3

Clearly sorghum is important as food for people in developing countries and as feed for animals among more affluent communities.

A breakdown of the data by countries is shown in Table 2.2 with the consumption (*not* production) in 10^3 t.

Recent reliable information on worldwide and regional production of pearl millet, common millet and foxtail millet, is not easily found. Equally, data which one can quote with confidence are hard to come by on the production of sorghum in continental China and Manchuria but it has been estimated to equal 25% of total world production. Though the Maoist regime has discouraged production of the fermented beverage "sanshu" made from dark, glutinous millet grains (Arnould and Miche, 1971), kaoliang persists as an important cereal in those Chinese provinces in which wheat and rice are not dominant.

Africa

Bouchet (1963) reviewed the production of sorghum and pearl millet in Mali. At the time of his study he stated that the crops together represented about 58% of the cultivated areas with pearl millet occupying a marginally greater area than sorghum. Bouchet illustrated the difficulty of arriving at accurate estimates of production in a country where cultivation is widely dispersed and most of the crop consumed where grown. Varietal choice was greatly influenced by local preference and environment. The dwarf types of sorghum introduced from the USA yielded well (Early Hegari produced 1520 kg/ha without fertilizer) but their bitter taste and unfamiliar physical characters made them unacceptable to consumers. Only Shallu approached acceptability.

Idusogie (1971) reported on the crop yields and nutritive value of sorghum and pearl millet grown in Nigeria. The data on yields were obtained from FAO (1966), Oyenuga (1967) and the Annual Reports of the Rockefeller Nutrition Research Project, Ibadan (1963–1967). From these, nutrient values were calculated on the "as purchased" basis, applying the data of Platt (1962), Orr and Watt (1957), and FAO (1963). The values, /ha, of kilocalories, protein (N × 6·25), the amino acids lysine, methionine with cystine, threonine and tryptophan, the minerals calcium and iron and the vitamins thiamine, riboflavin and niacin, are shown in Table 2.3.

According to Idusogie, improved agricultural practices would improve yields, and increase the nutrients provided /ha. He computed a table showing the effect of such improved agricultural practices, but the methods of attaining them were not discussed. Idusogie reported the calculated utilizable protein quality (NDpCal %) (Platt and Miller, 1959; Platt *et al.*, 1961), as about 6·8 for sorghum and 7·4 for pearl millet. Idusogie considered the cereals to be good nutritive sources but believed that in parts of Nigeria, and elsewhere in Africa, traditional methods of food preparation might lead to considerable losses of some nutrients. This matter is discussed more fully in Chapter 5.

India

About 35% of the world area grown to sorghum and millet crops is in India, but that country produces only about 16% of the total grain production of these crops (Rachie, 1966). Almost all the grain produced, except that used for seed, is for human consumption and about one-third of the Indian population depends on one or more of the sorghums and millets as their principal dietary staple (Rachie, 1966). Production and yield for 1976–77 by State is given in Table 2.4 (Swaminathan, 1977, personal communication).

Ryan *et al.* (1975) citing National Institute of Nutrition (1974) stated that in Andhra Pradesh, Maharashtra and Karnataka the lowest income rural families consumed 455, 262 and 263 g of edible sorghum/consumption unit/day. (A "consumption unit" is the number of adult male equivalent consumers in the family calculated by giving women and children weights of 0·9 and 0·2–0·6 respectively).

Andrews *et al.* (1975) stated that during 1973–74 the area planted and production of millets in India were:

	Area (10^3 ha)	Production (10^3 t)
Pearl millet	13 600	7100
Finger millet	2400	2100
Other small millets	4500	1900

Andrews *et al.* considered that the proportion of pearl millet to other millets was probably similar in Africa, and pearl millet therefore represented about 70% of the millet acreage in the developing world, or about 20–50 million ha. Grain yields of pearl millet were about 500–700 kg/ha. High yields with maturity in less than 90 days made possible three generations/year.

Description

In the text which immediately follows sorghum is discussed in detail, and the millets are considered later in the chapter. Where, however, work related to the millets is compared with that related to sorghum, such work is considered under sorghum.

Sorghum

A simple diagram of the sorghum plant and its components is presented in Figure 2.I. Detailed descriptions are to be found in Doggett (1970) and Purseglove (1972).

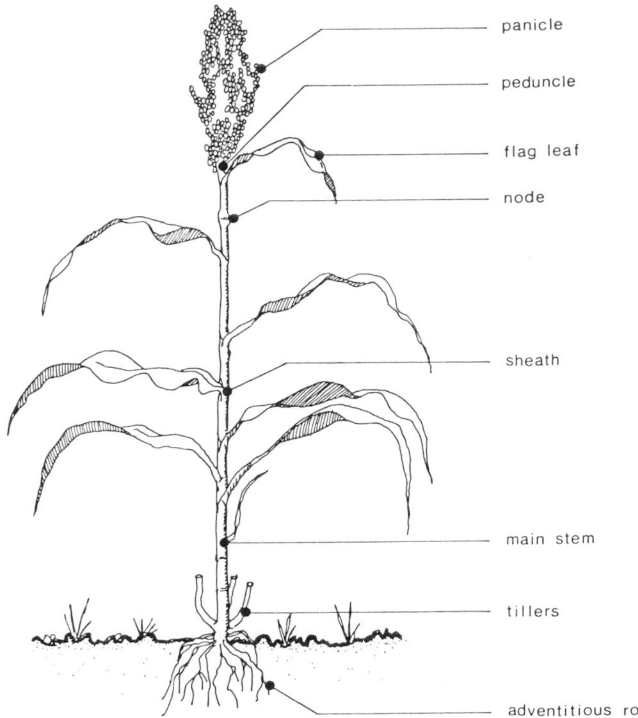

Fig. 2.I Diagram of the sorghum plant and its components. Illustration by Anthony B. Pearson.

Sorghum is a cereal of remarkable genetic diversity illustrated by the many forms in which it appears (see Fig. 2.II). More than 14 000 entries are to be found in the ICRISAT collection (ICRISAT, 1976). The inflorescence or head of sorghum consists of a panicle ranging from 75 to 500 mm in length and 40 to 200 mm in width. Each panicle contains between 800 and 3000 kernels enclosed in floral tracts (lemma and palea) which usually thresh free of the kernel during harvesting. The kernel, lemma and palea are located inside the glumes (Rooney, 1973). The glumes are black, brown, red or tan-coloured (Arnon, 1972). The seeds themselves are red, white, yellow or brown, the pigments being located in the pericarp and/or the testa. The 1000-seed weight ranges from 20 to 40 g (Arnon, 1972; Adrian and Jacquot, 1964).

The Indian Council of Agricultural Research (ICAR) (1970) stated the white pearly seeded types are preferred for human consumption in India. Possibly because of the process of natural selection and their greater resistance to predation by birds, brown-seeded varieties are more widespread in Africa.

The most attractive feature of sorghum and several of the millets is their capacity to survive and yield grain during continuous or intermittent drought stress. Sorghum can remain dormant during periods of stress and renew growth when conditions are more favourable. There is evidence (Simpson, personal communication, 1977) that the ability of sorghum to survive drought stress is hormonally controlled. Sorghum is more tolerant of flooding than maize but does not grow at its best under prolonged wet conditions. Grain sorghum grows successfully on many soil types but best on medium-textured, light-textured or sandy soils, and less satisfactorily on clay or heavy-textured soils. It tolerates medium–high pH conditions in the soil (Ross and Webster, 1970).

Grain sorghum feeds heavily on soil nutrients, nitrogen being a primary need with phosphorus and potassium somewhat less critical (Ross and Webster, 1970). A sorghum crop yielding 6000 kg/ha reportedly removes about 105 kg of nitrogen, and 15 kg each of potassium and phosphorus from the soil (Ross and Eastin, 1972). Sorghum is grown in a wide diversity of cropping patterns, including monocropping, intercropping with legumes, oilseeds and other cereals, and in many rotational sequences (IDRC, 1977). In monsoonal and other seasonally high rainfall areas, sorghum is gaining increasing recognition as a dry season crop.

Structure of the Grain

Though the stem, leaves and other components may be used in animal feeds, for structural and other non-food purposes, only the grain is of significant interest in human nutrition. The structure of the grain is illustrated in Figure 2.III and has been described by several workers including Rooney and Clark (1968), Rooney (1973), Sullins and Rooney (1974, 1975), York (1976), Hubbard *et al.* (1950) and Swanson (1928). The

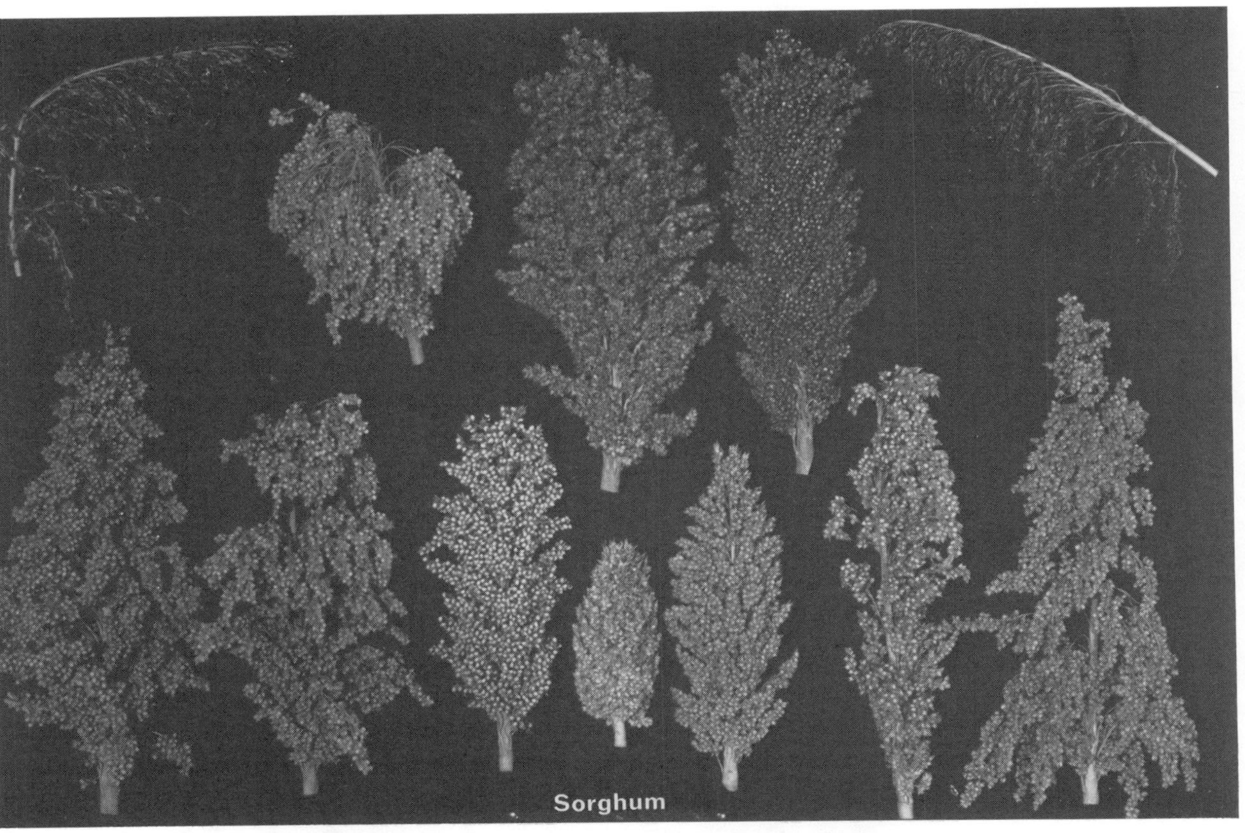

Fig. 2.II Panicles of sorghum illustrating the diversity of head types.

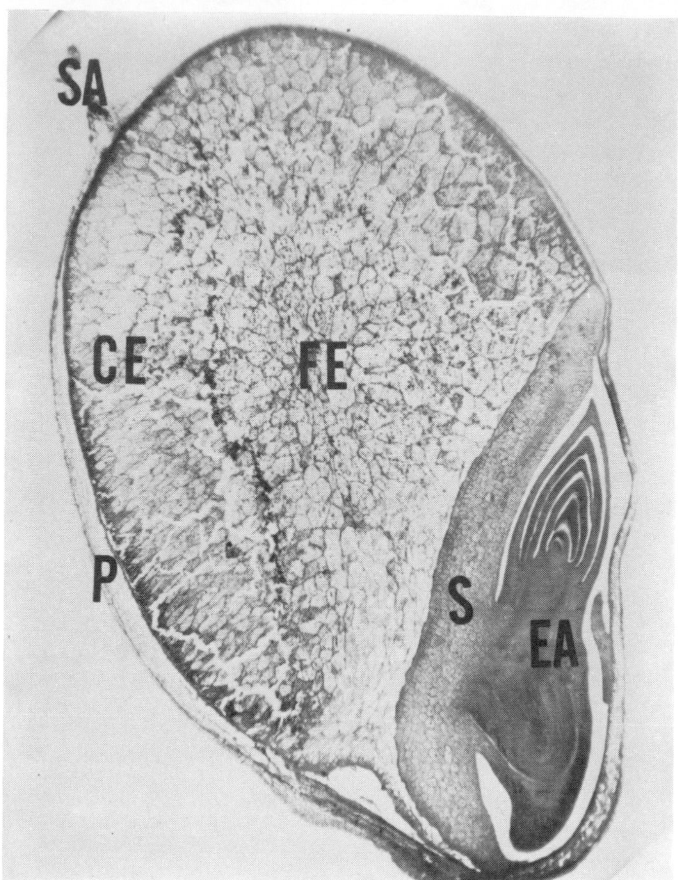

Fig. 2. III Cross section of the sorghum grain, P = pericarp, CE = corneous endosperm, FE = floury endosperm, SA = stylar area, S = scutellum and EA = embryonic axis. Courtesy of Dr Lloyd W. Rooney, Texas A & M University.

authors are grateful to Dr. John Axtell of Purdue University, Drs Rooney and Sullins and Miss Mary E. Glover of Texas A & M University for helpful personal communications on the subject of kernel structure.

Among authors, one finds inconsistency in nomenclature, particularly in the names applied to the layers which surround the endosperm. These layers are variously described as "seed coats", "bran layers" or "outer covering(s)".

There appears to be general agreement that the sorghum kernel, or caryopsis is roughly spherical in shape and composed of three main components: (1) the seed coats—the outer covering; (2) the embryo or germ and (3) the endosperm—the storage tissue that provides food for the young plant after germination.

(1) Seed Coats

The seed coats consist of the fused pericarp and testa.

(a) Pericarp

The extreme outer layer (see Fig 2.III) is the pericarp, the original ovary wall which is surrounded by a waxy cuticle and is composed of three or four different layers—the outermost being the epicarp (some name it the epidermis) which usually consists of two or three layers of elongated cells, that may or may not be pigmented; the hypoderm, not always differentiated from the epicarp; the mesocarp (the middle layer) and the endocarp.

The epicarp usually consists of two or three elongated rectangular cells which may be pigmented, the presence or absence of pigments being genetically controlled. If the kernel is enclosed in highly pigmented glumes, the pigments may migrate into the endosperm. The middle layer, the mesocarp, though in general the thickest layer of the pericarp, can vary significantly in thickness among genotypes. The mesocarp cells are generally thin-walled and under polarized light may be seen to contain polygonal starch cells.

The inner layer of the pericarp, the endocarp, consists of cross and tube cells, the former being long and narrow and set at right angles to the kernel's principal axis. The tube cells are roughly 200 μm long and oriented with their principal axis parallel to the principal axis of the kernel. It has been suggested that the endocarp cells are located at the point of breakage where the pericarp (bran) is removed during mechanical milling (Sullins and Rooney, 1975).

(b) Testa

The component of the seed subject to greatest controversy is the layer immediately below the endocarp and surrounding the endosperm. Henceforward in this text, it will be referred to as the "testa". It has, however, been variously called "subcoat", "hyaline layer", "nucellus", "undercoat", "integument", "seed coat" and "testa". Wall and Ross, (1970) stated that the testa is always present in the ovary as an inner-integument with a definite structure. As the kernel matures and the endosperm expands, the cellular configuration of the young testa may give way to a continuous layer or in some instances it may totally or partially disappear. The presence or absence of the testa appears to be controlled by the B_1 and B_2 genes and both genes must be dominant for the testa to remain. If either gene is recessive the rudimentary testa cells disintegrate and are reabsorbed. Consequently, some sorghum lines contain a complete testa, from some the testa seems absent, and some, the genetic control of which is not clearly understood, contain a partial testa. The partial testa occurs sporadically around the kernal and may or may not contain spots of pigment. Where a complete testa exists, it may vary in thickness among different sorghum lines and from one area of the kernel to another. It is often thickest at the crown of the kernel and thinnest over the embryo. The testa may or may not be pigmented, a feature which appears to be also genetically controlled.

Some workers however believe that colourless or transparent testas do not in fact exist in sorghums. In a personal communication, Axtell (1977) stated that at Purdue University he and his coworkers had never found a sorghum line composed of a coloured pericarp and a transparent testa. In all lines of their experience, testas were either pigmented or absent. Research in India (Ayyangar and Krishnaswami, 1941) quoted by York (1976) implied that transparent testas can in fact occur in combination with a pigmented pericarp. Axtell pointed out however that it may be difficult to bring about a clean separation of the pericarp layers from the testa and therefore to confirm positively whether or not a colourless testa exists below a coloured pericarp.

Schertz and Stephens (1966) stated that among the genotypes they examined, pericarp colour in sorghum was controlled by two independent sets of genes. Genes *RYI* controlled pericarp colour and had no effect on the presence or absence of a pigmented testa. In the studies described, colour ranged from white to various shades of red. Genes B_1B_2S controlled the presence or absence of the pigmented testa and also affected pericarp colour. When both *B* genes were dominant a pigmented testa was present; when either *B* gene was recessive the pigmented testa was absent and when both *B* genes were dominant in combination with a dominant *S*, the pericarp colour was brown regardless of the presence of the *RYI* genes.

As is described in more detail in Chapter 4, the pigmentation of both the pericarp and the testa is generally attributed to the presence of polyphenols (tannins). Some workers have expressed the view to the authors that the greatest concentration of polyphenols in sorghums exists in the testa layer. While this may be so, there appears to be no indisputable evidence to support the contention.

Glover (1977) stated that tannin content is influenced by pericarp colour, presence of a testa, extent of a testa and the environment in which development occurs.

The mode of influence of the agroenvironment on the nature and level of polyphenols in sorghum is not known to the authors but in a study by Tipton *et al.* (1974), six different hybrids grown during two separate years at several locations indicated that the year, location and hybrid grown affected the amount of tannin present in the kernel and in each hybrid there was a specific level of maturity of the caryopsis at which the highest tannin level in the kernel occurred.

Glover (1977) illustrated results of research into the structure of sorghum kernels with photomicrographs (see Fig. 2.IV–2.VII). Glover stated that where it occurred, the testa appeared as a pigmented strip just above the

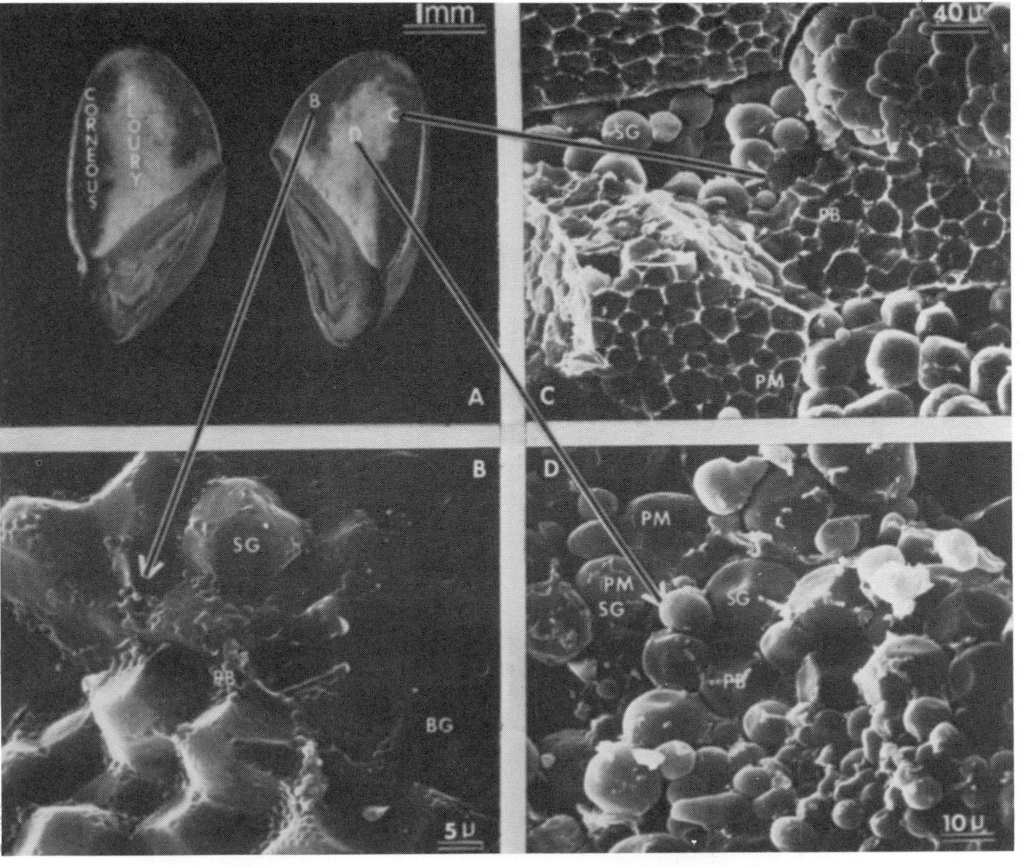

Fig. 2.IV General morphology of the sorghum kernel. Courtesy of Miss E. Glover, Texas A & M University.

aleurone layer. In many cases (Fig. 2.V,B.) the testa was two-toned, the darker colour being generally closest to the pericarp. This darker layer was reddish brown while the lower one was yellowish brown. The scanning electron micrograph (Fig. 2.V,D). shows the testa as a solid noncellular band. The presence or absence of a testa did not appear to affect the aleurone layer. In some cases (Fig. 2.V,A,B) the aleurone layer appeared as a single cell strip with little or no pigmentation. In another case (Fig, 2.V,C)

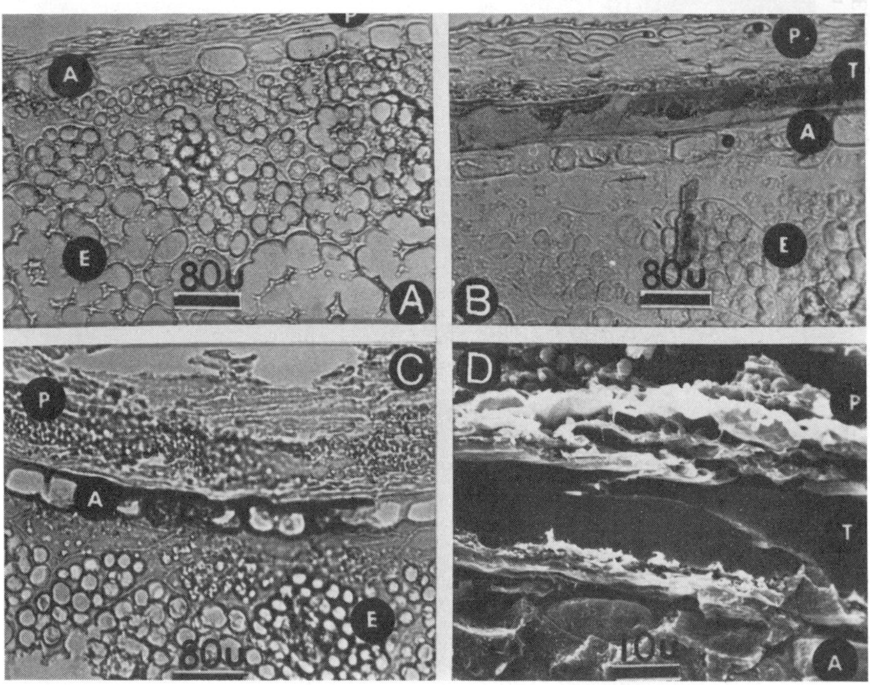

Fig. 2.V Cross sections of sorghum kernel. (A) White pericarp without testa: P, pericarp; A, aleurone; and E, endosperm. (B) White pericarp with testa: T, testa. (C) Red pericarp without testa. (D) Scanning electron photomicrograph of the testa (900 ×). Courtesy of Miss Mary E. Glover, Texas A & M University.

the aleurone layer was highly pigmented; this section originated from a kernel with a red pericarp and no testa. Glover stated that colour in the pericarp resulted mainly from pigments in the epicarp region.

Glover (1977) also described several kernels with partial testas and stated that each kernel with a partial testa was unique in that the testa did not occur in any set pattern. In one instance, the testa started as a thick layer, about half the size of the pericarp, but as the testa continued, the layer became first thinner, then less compacted and finally faded away. The testa varied in thickness over several areas of the kernel.

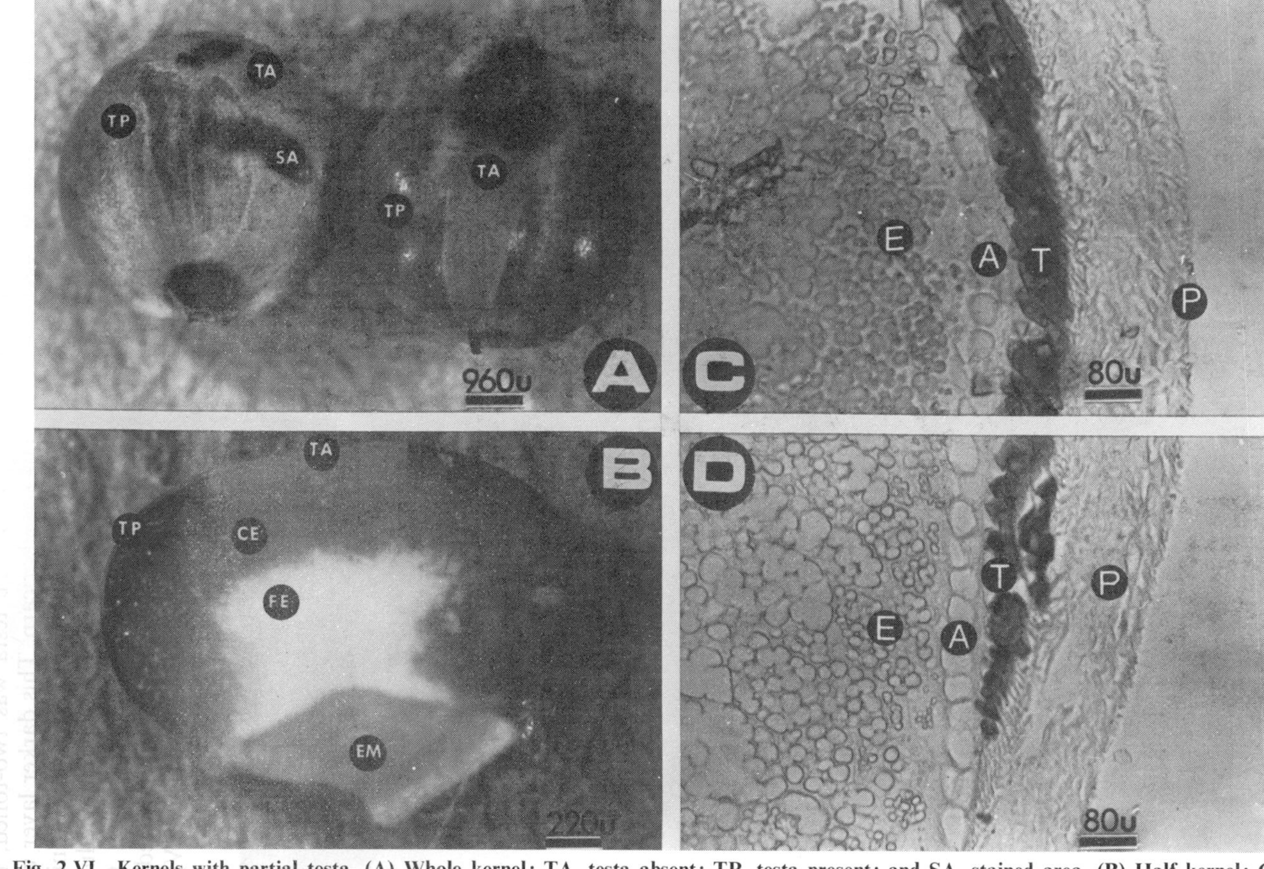

Fig. 2.VI Kernels with partial testa. (A) Whole kernel: TA, testa absent; TP, testa present; and SA, stained area. (B) Half kernel: CE, corneous endosperm; FE, floury endosperm; and EM, embryo. (C & D) Cross sections of area where testa ends: P, pericarp; T, testa; A, aleurone; and E. endosperm. Courtesy of Miss Mary E. Glover, Texas A& M University.

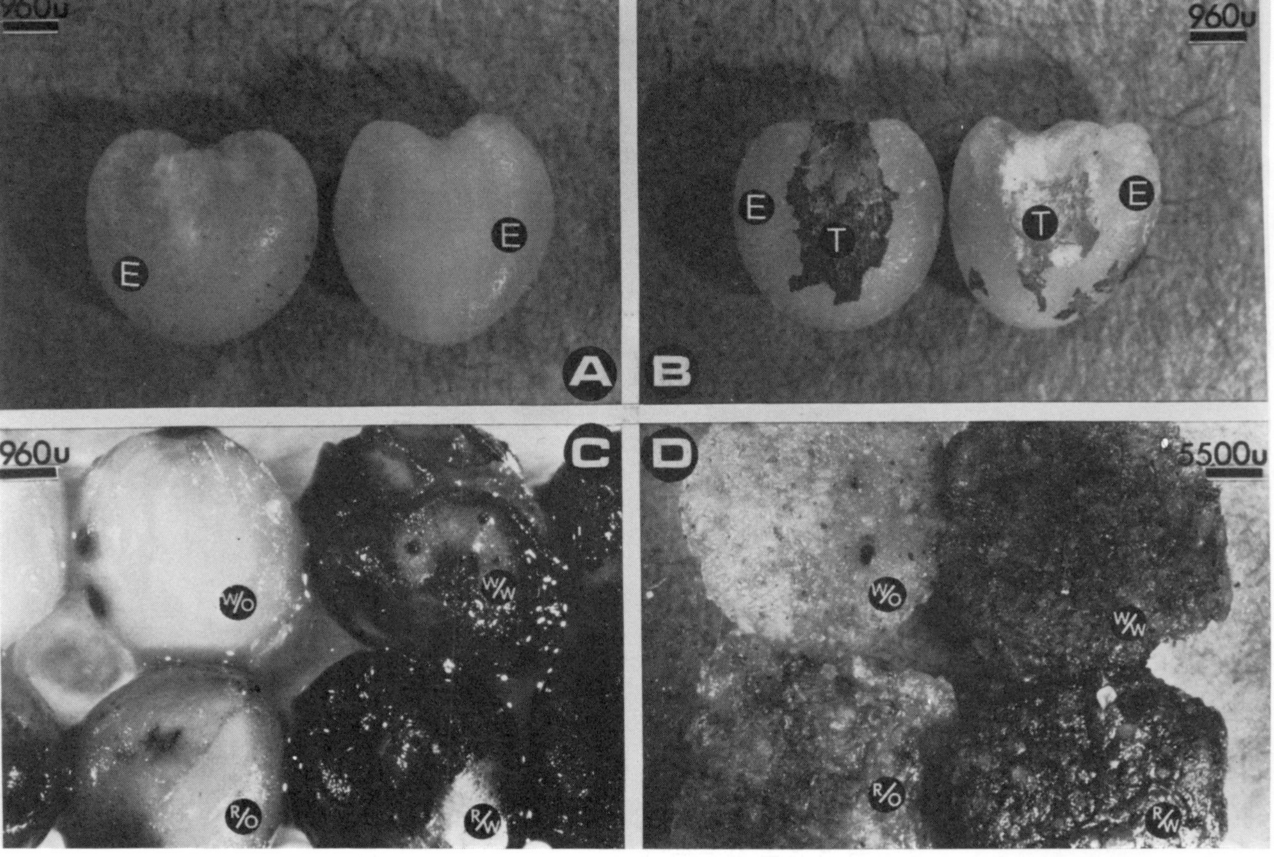

Fig. 2.VII (A) White pericarp, without testa after pearling: E, endosperm. (B) White pericarp, with testa after pearling: T, testa. (C) Whole kernels forty minutes after 5 N NaOH treatment: W/O, white pericarp, without testa; W/W, white pericarp, with testa; R/O, red pericarp without testa; and R/W, red pericarp with testa. (D) Ground grain 40 min after 5 N NaOH treatment. Courtesy of Miss Mary E. Glover, Texas A & M University.

(2) Embryo

The embryo consists of a large scutellum, an embryonic axis, a plumule and a primary root. The scutellum is the flattened portion that serves as an absorptive organ. There is a cementing layer between the scutellum and the endosperm with glands protruding from the scutellum into the endosperm. The embryo is relatively firmly embedded and difficult to remove by dry-milling (Rooney, 1973).

(3) Endosperm

The endosperm represents the largest proportion of the kernel and consists of an aleurone layer—a single continuous layer of cells around the extreme edge of the endosperm peripheral, corneous and floury zones.

The peripheral layer is made up of cells containing a high proportion of protein. The corneous layer, beneath the peripheral layer, contains less protein and a higher proportion of starch than the peripheral. The corneous is also termed the flinty, hard or horny endosperm, and is translucent in appearance. Inside the corneous layer is found the floury or soft endosperm layer, which is lowest in protein.

The texture of the endosperm varies even within a sample (Rooney, 1973) and is influenced by heredity and environment. Whether a sorghum is described as "floury" or "corneous" depends on the ratio of floury (soft) to corneous (hard) component within the kernel. Often the "floury" sorghums contain a brown pericarp and pigmented thick testa "seed coat" (Rooney, 1973). (See Fig 2.VIII.)

Bidwell (1918) dissected by hand a water-soaked sample of dwarf Blackhull Kafir (CI 340) sorghum into bran, germ and endosperm. The endosperm was ground in a coffee mill and classified into material that (a) passed a 20-mesh sieve, (b) passed a 40-mesh sieve (starchy endosperm) and (c) remained on the 40-mesh sieve, described as corneous endosperm. The starchy endosperm was described as "floury" in appearance, the corneous was said to look like sand. Further work was reported by Bidwell *et al.* (1922) on Dwarf Milo (CI 332) and Feterita (CI 182). The proximate composition (AOAC, no date, starch by a diastase method; no other detail given) is shown in Table 2.5, compared with an analysis by Hopkins *et al.* (1903) of fractions of the maize kernel. In general, the protein content of whole grain feterita was higher than the other grains and in each case the corneous endosperm contained more protein than the floury endosperm.

Hubbard *et al.* (1950) described the composition of component parts of the sorghum kernel, obtained by hand dissection after tempering and soaking in distilled water. Five varieties were chosen; Westland, Midland and Martin (yellow), Cody (waxy kafir) and Pink Kafir, all grown in 1945.

The whole grain and separated fractions: endosperm, germ and bran, were analysed for moisture, ash, N and lipids (AOAC, 1945), starch (polarimetrically), riboflavin (Snell and Strong, 1939), pantothenic acid and

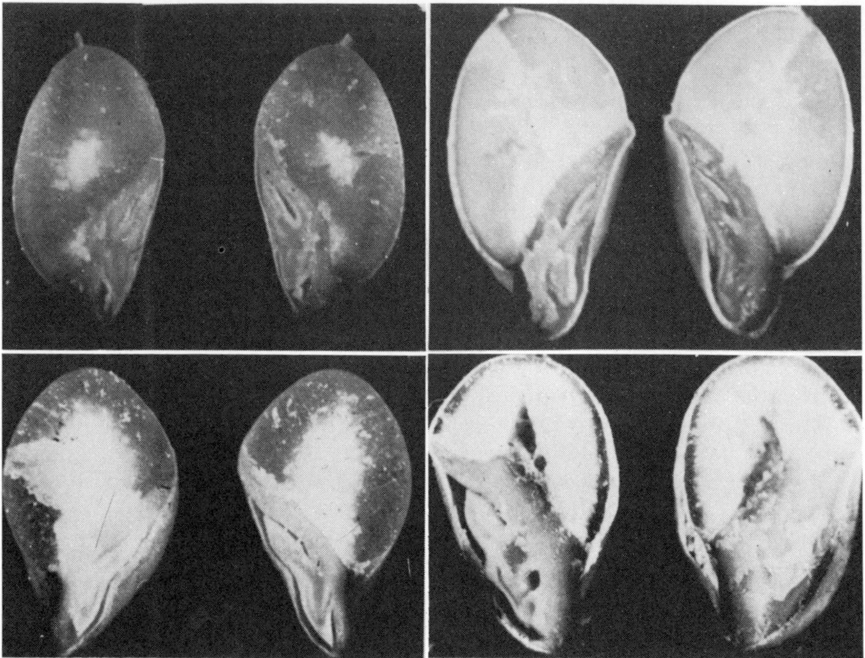

Fig. 2.VIII Longtitudinal sections of sorghum kernels illustrating endosperm texture and type (4·5 ×). Top left, corneous, SC 301; top right, waxy, TX 615; lower left, normal (intermediate), Kafir 60; and lower right, floury, NSA 740. Courtesy of Dr Lloyd W. Rooney, Texas A & M University.

biotin (Skeggs and Wright, 1944), and pyridoxine (Atkin *et al.*, 1943). Crude wax in the whole grain was determined by a described procedure using hot benzene as solvent. The means and ranges or analyses of composites are shown in Table 2.6

The proportion of endosperm ranged from 81·1 to 84·6%, the germ from 7·8 to 12·1% and the bran from 7·3 to 9·3%. These results agreed in general with the findings of Bidwell (1918) and Bidwell *et al.* (1922). Rooney (1973) stated that the results reported by Hubbard *et al.* (1950) might underestimate the proportion of pericarp and overestimate the endosperm content since the pericarp fraction appeared to have been separated at the mesocarp layer. Doggett (personal communication) stated that while the Hubbard *et al.* data may be typical of US grain sorghums, the range of proportional variation would be much greater among the international sorghum collection. Microscopic examination of the germ fraction by Hubbard *et al.* (1950) showed that perhaps as much as one-third of the starch resulted from incomplete separation of the endosperm. Microscopic examination also showed that under the tempering and soaking conditions applied, the bran separated within the starchy mesocarp and consisted of the cuticle, epidermis, hypoderm and most of the mesocarp. The inner fragments of the mesocarp, the nucellar layer (testa) and the aleurone remained with the endosperm fraction. The range of vitamin content (as determined microbiologically) was much wider among different sorghum types than the proximate analysis, but not as wide as reported by Tanner *et al.* (1947). The proportion of each fractional constituent is presented in Table 2.6.

Hahn (1969) commented that these data were useful in judging the purity of fractionation obtained by different milling procedures. Low ash and oil contents would indicate "clean" endosperm and an oil content of about 28% would be a reliable index of germ purity. (Authors' note: while 28% lipid in the germ agrees with Hubbard's findings, the range of lipid contents reported among the literature reviewed suggests Hahn's index of germ purity may not be universally applicable.)

Jambunathan and Mertz (1973a) separated the endosperm, germ and pericarp from four inbred lines of sorghum, two, IS 0062 and IS 3982, described as "low-tannin" and two, IS 6992 and IS 2283 as high-tannin. The grains were soaked in cold water for 2 h then dissected by hand scalpel (Hubbard *et al.*, 1950). Their results are given in Table 2.7.

High-lysine Sorghum

Sullins *et al.* (1975) used the SEM to study the endosperm structure of the high-lysine sorghum lines IS 11167, IS 11758 (Ethiopian lines), SC 1030 5 4 (reported by Rosenow, personal communication to Sullins *et al.*) BC_1 selection from IS 11790. A comparison of the protein (Kjeldahl N × 6·25), weight of 1000 kernels (g) and amino acid content in protein (ion-exchange chromatography) of whole and shrunken grains from the same, individual head (SC 1030 5 4) are shown in Table 2.8. The shrunken kernels were less than two-thirds the weight of normal kernels. In the shrunken kernels, protein was 2% higher, lysine higher, tryptophan slightly higher, proline and glutamic acid lower than in normal kernels.

The shrunken kernels contained a soft, floury endosperm, the centre of which in most kernels was hollow; voids, or air cells appeared between the starch granules and the protein matrix. The shrunken endosperm showed a marked reduction in the number of protein bodies (granules) compared with normal endosperms. Sullins *et al.* suggested that SEM might be used to select sorghums with plump kernels containg a high lysine content.

Official Standards

The only official standards for grain sorghum traced by the authors, are those laid down in (a) the USA and (b) the Republic of South Africa.

The Official Grain Standards for the United States (US Department of Agriculture revised 1970 cited by Rooney, 1973) are for feed grains only. They define grain sorghums as follows: "grain sorghums shall be any grain which, before the removal of dockage, consists of 50% or more of whole kernels of grain sorghum, which contains not more than 10% of other grains for which standards have been established".

The standards are for five grades and four classes. The grade is designated for a particular sample based on the factor which puts the sample in the lowest grade. It is arrived at by consideration of minimum test weight in pounds/bushel (e.g. 57 lb for grade 1, 51 lb for grade 4); moisture (13% for grade 1, 18% for grade 4); and total damaged kernels (2% for grade 1, 15% for grade 4). These standards encourage adulteration of the grain because much higher levels of broken kernels and foreign material are permitted in sorghum than for comparable grades of maize and wheat (Rooney, 1973).

The four classes are based on the visual appearance of the kernels: (1) yellow, (2) white, (3) brown and (4) mixed. The appearance of grain sorghum is influenced by combinations of pericarp colour, opaqueness or translucence of the pericarp (due to differences in mesocarp thickness), types of endosperm and presence or absence of a testa.

Class (1) yellow, includes kernels with some red in the pericarp and is the main type of grain produced in the USA. According to Rooney (1973), the kernels do not contain a testa, though remnants of the original testa may be present. Yellow grain sorghums yield flour or grits with an off-white or reddish-brown hue. Most yellow endosperm sorghums are in this class but the name does not relate to yellow endosperm sorghums.

Class (2) white grain sorghums, have kernels with a white or colourless pericarp and no testa. They are produced in the USA in only small quantities. Kernels with a coloured testa are classed as brown grain sorghums.

Class (3), brown grain sorghums, have a brown pericarp and are high in tannin content. Kernels with a pigmented testa are classed as brown regardless of pericarp colour. Brown sorghums are inferior for most purposes but their production in the USA is largely confined to the southeastern states.

Class (4), mixed, is self-explanatory.

There is at present (Rooney, 1973) no precise standard by which to identify yellow-endosperm grain sorghum, though it is known that the endosperm owes its appearance to the presence of carotenoid pigments. True yellow endosperm sorghums are surrounded by a thin, translucent pericarp which gives the grain an external yellow appearance. However, the yellow colour in the endosperm is controlled independently of pericarp colour, and a yellow endosperm may be present together with a white, red or brown pericarp. A coloured pericarp obscures the yellow pigment and hence changes the external appearance and the consequent market classification. Commercial sorghum hybrids with a red pericarp over a yellow endosperm are known as heteroyellow endosperm hybrids (Rooney, 1973).

In the Republic of South Africa, the regulations (South African Government, 1968) define "kaffircorn" as follows: "kaffircorn shall mean the seed of all grain sorghums excluding broom corn, hay sorghums and sweet sorghums". The regulations define "defective", "weather-stained", "unthreshed" and other such terms. There are four classes of kaffircorn:

(a) class KR, consisting of red kaffircorn of the varieties Barnyard Red, Short Red and others with a similar red colour—excluding Radar, which do not have a horny endosperm like, for example, Milo and Martin or a dark nucellar layer (testa) like, for example, Swazi Red, Ninety Day Red and Birdproof.

(b) Class KM, consisting of kaffircorn which varies in colour from light red to reddish brown but which does not have a dark nucellar layer, for example, Radar, Red Mixed, Red Milo, Ntuli Red, Martin and Framida, or kaffircorn which does not conform to the requirements for any of the grades set out in the table in sub-regulation (3) of regulation 2.*

(c) Class KW, consisting of white kaffircorn, which does not have a dark nucellar layer, for example ordinary white, White Milo and Dura.

(d) Class KF, consisting of all kaffircorn with a dark nucellar layer, for example White and Red Hegari, Swazi Red, Ninety Day Red and Birdproof, or kaffircorn which does not conform to the requirements for any of the grades set out in the table in sub-regulation (3) of regulation 2 for classes KR, KM and KW.

There are four grades, (1), (2), (3) and "sample" which lay down maximum permitted percentages of "defective", "unthreshed", "weather-stained" kaffircorn, and admixtures of kaffircorn of another class, or another colour, or of "foreign matter" as defined. The permitted grades are for class KW (1), classes KR and KM (1) or (2) and class KF (1), (2) or (3). "Sample" grade covers kaffircorn which does not conform to the requirements of grades (1), (2) and (3) and may be smut blackened, have an objectionable odour and contain *Datura* spp. seed. A sample of KR1 or KW1 (the highest grades), for example, may contain 5% defective kaffircorn, 4% unthreshed kaffircorn, 4% of another colour, 4% of another class, 1·5% foreign matter and 50% weather-stained kaffircorn.

The presence of "purplish anthocyanic blotches in or on the pericarp" shall not be deemed to effect [*sic*] the colour of kaffircorn which is otherwise white or red.

Essential Nutrients and Evaluation Methods

In this section an indication only will be given of the general importance in human diets of protein and other essential nutrients, and of the methods by which their content and availability may be measured. The contribution made by sorghum and the millets is discussed more fully later in the text, and, where individual papers are discussed, reference is made to the specific methods of evaluation, where known.

* (Authors' note: this last is the regulation defining the grades requirements.)

Protein and Amino Acids

The name "protein" was probably proposed by the German scientist Mülder, on the suggestion of the Swedish chemist Berzelius. The word is etymologically derived from the Greek πρωτειοσ meaning "primary" and was chosen because protein is the primary and fundamental material of most living organisms. "Protein" is the name given to a class of organic compounds made up of linked amino acids composed of nitrogen and other elements. Amino acids are the essential building materials of which all animal tissues and organs are composed and maintained.

Protein is essential to the growth and restoration of body tissues and is particularly important in the diets of "vulnerable groups" namely young children, pregnant and lactating women, and in rebuilding the human fabric after serious illness or injury. Protein is an essential component of a balanced diet which must also be adequate in energy sources. If a diet is deficient in carbohydrate and fat, the limited protein present may be used as a source of energy so leading to an aggravated state of protein malnutrition. The human requirements for protein and energy have been widely discussed, notably by the Joint FAO/WHO *Ad Hoc* Expert Committee (WHO, 1973a) and summarized in a Handbook of Human Nutritional Requirements (WHO, 1974).

Different proteins are composed of different combinations of amino acids. Some of these amino acids can be synthesized by living organisms from other nitrogenous material, the amino acids which can be synthesized varying among different higher animals. Those which cannot be synthesized *in vivo* are known as the "essential" or "indispensable" amino acids and must be provided by the food which is eaten. Of the 18 edible amino acids, eight are now accepted as "essential" for adults: isoleucine, leucine, lysine, methionine, phenylalanine, threonine, tryptophan and valine; in addition, infants are belived to require histidine (WHO, 1973a).

The first-limiting amino acid is the essential amino acid in which any given protein is most seriously deficient. As in most cereals, the first-limiting amino acid in sorghum and the millets is lysine. The relative requirement for lysine is greater when body tissue is being laid down, as during growth of infants, than for normal tissue maintenance in adults, the turnover of the acid being then relatively low. Consequently, the lysine content is of crucial importance in diets composed largely of cereals (Jansen, 1972). The overall nutritional quality, and in particular the lysine content and availability, is likely to be of especial importance in the diets of young African and Asian children in whose diets sorghum (and/or millets) form a high proportion.

Because the protein value of a food depends primarily on the content of its limiting amino acid, which will vary, protein evaluation for human consumption has proved difficult and controversial. As examples of this variability, in the literature examined, the mean values of lysine, expressed as mg/g total N varied from 71 to 212 in sorghum (Table 2.13), from 109 to 297 in pearl millet (Table 2.119) and from 160 to 262 in finger millet (Table 2.130).

These matters, including those of biological evaluation in relation to cereals were discussed by McLaughlan and Campbell (1974). The recommendations then made led to the setting up of a working group, jointly sponsored by the United Nations Protein Advisory Group and IDRC, which prepared the PAG Guideline No. 16 (Protein-Calorie Advisory Group 1975) on protein methods for cereal breeders as related to human nutrition. The guideline has also been published by the American Association of Cereal Chemists (Pomeranz, 1976).

Williams (1974) discussed the sources of error in protein testing by analytical methods especially in the Kjeldahl and Udy procedures (1977) and reviewed methods including infrared reflectance spectroscopy for testing grains and related products for protein content.

Pomeranz and Moore (1975) defined the reliability of an analytical method as depending on its (1) specificity, (2) accuracy, (3) precision and (4) sensitivity. They compared (a) the Kjeldahl (AACC, 1962), (b) the biuret (Greenaway and Johnson, 1974), (c) dye binding (Udy, 1971), (d) infrared reflectance (i) Dickey-John Model GAC-2 and (ii) Neotec Corporation Grain Quality Analyser Model GQA, both as instructed by the manufacturers and (e) alkaline distillation (Ronalds, 1974), using ten hard red winter wheat cultivars from 11 locations in the Great Plains area of the United States.

The results of the Duncan multiple range test showed no significant differences, at the 5% level, among the six methods for the varietal means; there were significant differences only for the samples from Clovis, New Mexico, irrigated, for the location means.

Using the reliability criteria suggested, the (a) Kjeldahl, (b) biuret, (c) dye binding, (d) IR reflectance, (e) alkaline distillation, (f) Dumas (Fiedler *et al.*, 1973; Hamlyn and Gasser, 1970; Mertz, 1974) and (g) neutron activation (Doty *et al.*, 1970) methods were graded satisfactory or unsatisfactory, subjectively, and the cost of the equipment (neutron activation in particular has a high initial cost), speed and usefulness in various situations, were not considered. The Kjeldahl method was judged the most satisfactory overall.

Pomeranz and Moore emphasized that none of the methods was truly accurate in that none measured the true protein content. For practical purposes, in developing new methods for protein determination, analysts are interested in those methods which show the best agreement with the Kjeldahl procedure rather than in those which give an accurate estimate of protein.

At a symposium on measuring protein quality for human nutrition held at the Sixty-first Annual Meeting of the AACC in 1976, four lengthy reviews were presented. Young *et al.* (1977) discussed the many problems involved when human subjects are used, and stressed the importance of multiple levels of test protein intake for estimation of protein quality. They considered that protein quality was more precisely estimated from an evaluation of the intersection of the N balance response curve with the line of N equilibrium as obtained with the test protein and compared with that of a reference protein such as egg or milk protein. Young *et al.* defined this measurement as the Relative Nitrogen Requirement (RNR) and reported the RNR of a soy protein isolate as 0·83 and of whole ground wheat protein as 0·68 when measured in young adult men.

Steinke (1977) reviewed the many factors, other than essential amino acid content, which can influence the results obtained with the rat PER method as conducted by the AOAC procedures, factors such as animal quality, strain of rat, acclimation period, instability of the diet, (e.g. the development of rancidity), balancing the diet nutritionally (e.g. with mineral mix), diet hydration, selection from the diet by the rats and problems arising from samples of mixed nutrient sources.

Bodwell (1977) reviewed the problems of applying protein nutritional data derived from rat experiments to humans. In the few reported studies in which rats and humans have been given the same preparations of protein (none relating to sorghum or the millets) agreement between the assays has been poor.

Hackler (1977) reviewed rat bioassay procedures and concluded that none was really appropriate for testing protein quality for human consumption but that for routine testing the most practical procedures were the relative nitrogen utilization (RNU), a relative protein efficiency ratio and PER.

Nitrogen to Protein Conversion Factors

In practically all the experiments reported in this review, protein content has been determined by the Kjeldahl method as specified by the AOAC, or some modification of this method. The N content as determined is a mixture of protein and nonprotein N and to arrive at the protein content of any cereal a conversion factor is employed based on the N the particular cereal protein contains. In virtually all the papers reviewed 6·25 was the factor employed, and this is in accord with WHO (1973a) which does not specify a particular conversion factor for sorghum and the millets but classes them as "other foods" with the factor 6·25. This factor is also used in FAO (1970).

Tkachuk in a personal communication (1975) gave N to protein factors derived from the amino acid composition of sorghum, pearl millet and teff as:

	Uncorrected factor	Corrected factor
Sorghum	5·9	5·8
Pearl millet	5·7	5·6
Teff	5·8	5·6

The corrected factor accounts for nonprotein N in the sample and is, scientifically, the more accurate value. In describing the calculation of such factors Tkachuk (1977) recommended that the factor of 5·7 might serve as the standard N/P factor for the reporting of protein content in all commodities of plant origin likely to be used for formulating food for human consumption, and for animal feeds.

In Tables 2.11 and 2.13 (sorghum), 2.118 and 2.119 (pearl millet), 2.129 and 2.130 (finger millet), 2.136 and 2.137 (foxtail millet), 2.143 and 2.144 (common millet), 2.145 and 2.146 (little millet), 2.147 and 2.148 (Japanese barnyard millet) and 2.149 and 2.150 (kodo millet), the protein content is shown as N × 6·25 but the means and ranges are also expressed recalculated by the authors, using the factor 5·7.

Energy

Energy, in nutritional terms, provides the ability to carry out work and to maintain life and the bodily functions. The body can derive its energy from carbohydrates, protein, fat or alcohol, though the last is not recommended as a primary source.

The energy value of a food has, until recently, been expressed in heat units and reported in kilocalories, correctly written by nutritionists as Calories with a capital "C" but more often as calories, with a small c. Under the metric system (in the form known as Système International d'Unités or SI) the unit of energy is the joule (J) and this is gradually being substituted for the calorie. It is probable in the future that the kilojoule (kJ) or the megajoule (MJ) will replace the kilocalorie in the nutritionist's vocabulary. In the present review, the kcal has been retained as it is the unit used in virtually all the papers reviewed. For the purpose of comparison, the conversion factors (WHO, 1973a) are given below:

1 kcal = 4·184 kJ	1 kJ = 0·239 kcal
1000 kcal = 4184 kJ	1000 kJ = 239 kcal
1000 kcal = 4·184 MJ	1 MJ = 239 kcal

Since the necessary energy content of human diets exceeds 1000 kJ, one may expect the megajoule (MJ) to become the unit in which physiological energy is expressed in the future (WHO, 1973a).

The energy available to the body is the gross energy of the diet minus the losses in urine, faeces and through respiration. The World Health Organization (WHO, 1973a Annex 2) recommended factors to be applied in calculating energy values of ingested nutrients. The millets were not specifically mentioned, but the WHO recommendations for sorghum and other cereals are shown in Table 2.9. Wide variations among different samples of the same cereal species are demonstrable, particularly among sorghums, as is described later in the text.

As an authoritative guide to nutrient requirements, readers are referred to a publication by the World Health Organization (WHO, 1974) which summarizes present knowledge, and proposes specific recommendations concerning nutrient intake.

Lipids

Lipid (from the Greek *λιποσ* meaning fat) is the generic name for all naturally occurring fats, oils, waxes and related substances found in living tissues.

Vegetable fats (solid at normal temperatures) and oils (liquid at normal temperatures) are compounds of glycerol with fatty acids and are usually triglycerides, in which three fatty acid units are combined with one unit of glycerol. Waxes are compounds of fatty acids and polyhydric alcohols other than glycerol. Sorghums and millets contain both oils and waxes.

Fatty acids may be saturated, in which case they are relatively stable and resistant to oxidation, or unsaturated in which case they are relatively susceptible to oxidation. The fatty acids in cereal seeds are mostly unsaturated and include oleic, linoleic and linolenic acids. The last two are essential fatty acids in that the body cannot synthesize them but must ingest them from the diet. Linoleic and linolenic are also known as polyunsaturated in that they contain two or more double bonds within the carbon chain. There is an extensive and widely publicized literature that deals with the possible relation between the fatty acid composition of the diet and cardiovascular and other degenerative diseases (Dayton 1975, Dayton *et al*., 1970, Keys, 1970).

The composition of dietary fats and oils, their role in human nutrition including recommended levels of intake were reviewed by an expert working group of FAO and WHO (FAO, 1977). The working group classified fats into (a) storage fats, mainly triglycerides accumulated in specific depots of living tissues; which act as energy reserves and (b) structural fats, mainly phospholipids and cholesterol, present in high concentration in the brain. While the composition of storage fats may vary with diet, the composition of the structural phosphoglycerides are generally highly consistent and specific to the tissue and species in which they occur.

Carbohydrates

For a comprehensive review of carbohydrate chemistry the reader is referred to Kerr (1950), and Radley (1968). The three major groups of carbohydrates are sugars, starches and cellulose and related materials. More than half the solid material in cereals consists of carbohydrates though not all carbohydrates are available to (i.e. digestible by) the human body.

Starch is composed of large numbers of units of the monosaccharide, glucose, linked to form straight chains (amylose) and branched chains (amylopectin). The relative proportions of amylose and amylopectin influence the physical properties and behaviour of cereal starches. Starch is stored in the seeds of cereals in the form of granules, the shape and size of which are characteristic of each plant species. The size and appearance of starch granules vary both within and among species.

The cell walls which enclose the starch are formed of cellulose which is indigestible by humans and other monogastric (single stomach) animals though cellulose can be digested by digastric ruminant animals such as cattle, sheep and goats. Starch granules may form suspensions but they do not dissolve in cold water and raw starch is not therefore easily digested. If heated in the presence of water, starch granules swell in varying degrees and eventually gelatinize to form a paste or gel.

Maltose is a disaccharide, a combination of two glucose molecules. It is formed from starch when grain germinates or when cereals are "malted". (See Chapter 5.)

Though WHO (1973a) recommended that carbohydrate content be measured directly and expressed as available monosaccharides, in most published analyses percentage carbohydrate is calculated and reported "by difference", that is by subtracting the analysed combined percentages of moisture, protein, lipids, fibre and mineral matter from 100.

Inorganic Elements

Certain inorganic substances are needed to provide material for growth and repair or to regulate the essential physiological and biochemical processes. Calcium, phosphorus, sulphur, potassium, sodium, chlorine and magnesium are present in the human body in relatively large amounts. Iron, zinc, copper, chromium, molybdenum, manganese and cobalt are present in smaller quantities and are generally known as "trace elements".

Much is yet to be discovered of the human body's requirement for inorganic elements: of the minimum levels needed, of the potential harmfulness of excessive intakes, of the mechanism of absorption, and of the relation and interaction between the inorganic elements and other dietary constituents. For example, calcium, iron and also zinc may be rendered less available to the body by phytic acid present in cereals, a phenomenon discussed more fully in Chapter 4.

For a fuller account of mineral elements in human diets, the reader is referred to the World Health Organization Technical Reports on calcium (WHO, 1962), iron (WHO, 1970) and trace elements (WHO, 1973b), and the Handbook on Nutritional Requirements (WHO, 1974); a review article on minerals and trace elements by Cuthbertson (1973); Health and Welfare Canada (1976); and the Manual of Nutrition (Ministry of Agriculture, Fisheries and Food (UK) 1976).

Vitamins

Early in the twentieth century it was discovered that humans, and other animals, needed small amounts of certain naturally occurring organic substances, most of which the body could not synthesize. Because the first of these substances to be recognized was thought to be an amine, the word "vitamine" was coined in 1912 by Casimir Funk from the Latin "vita" meaning life, and "amine" which describes a specific organic structure. Later the terminal "e" disappeared. As, in course of time, each of the vitamins was chemically identified and synthesized, the former alphabetical designation (vitamin A, B, C, etc.) gave way to more precise chemical names.

Human requirements of the vitamins, their availability from foods, absorption by the body, and role in human nutrition have been considered by Joint FAO/WHO Expert Groups and readers are referred to WHO (1967) covering vitamin A (retinol), thiamine, riboflavin and niacin; and WHO (1970) covering ascorbic acid, vitamin D, vitamin B_{12} and folate. A useful summary appears in the Handbook of Human Nutritional Requirements (WHO, 1974). A recent review by Truswell (1976) compared recommended nutrient intakes in several countries.

The vitamins can be divided between those soluble in water and those soluble in lipids and nonpolar solvents.

Water-soluble Vitamins

Vitamin B group

The members of this group, all soluble in water, are essential to human metabolism and must be obtained from the diet. The group includes a number of different substances, the most important being: thiamine (vitamin B_1); riboflavin (formerly known as vitamin B_2); and niacin (also known as nicotinic acid). The group also includes folate (folic acid); vitamin B_{12} (cyanocobalamin); pyridoxine (vitamin B_6); pantothenic acid; and biotin. Human requirements are summarized in WHO (1974). Vitamin B_{12} is found only in foods of animal origin, but cereals contain most other B-group vitamins.

Thiamine (*vitamin B_1*). The amount of thiamine required is related to the amount of carbohydrate in the diet, but the minimum daily requirement is about 1 mg. A deficiency may be manifested in retarded growth in children, irritability, a special type of peripheral neuritis, cardiac abnormalities and, in extreme cases, the disease beriberi. The richest natural sources are fresh yeast and cereal germs. Consequently, cereals contribute a high proportion of the thiamine in the diet in the developing countries (see Table 2.10). The vitamin is labile to moist heat and losses occur in cooking and baking. Thiamine is destroyed by alkali.

Niacin (*nicotinic acid*). Niacin is converted to its active form, niacinamide (nicotinamide), in the body. Niacinamide is also formed *in vivo* from the amino acid tryptophan, and the body may therefore obtain its supply from niacin or from tryptophan in the diet.

Niacin, like thiamine and riboflavin, is required for the body processes which convert food into energy. The niacin present in cereals is not always nutritionally available; in cereals it may exist in a bound form, e.g. niacytin, though it is generally in available form in most pulses.

Because it may be derived from tryptophan, the term "niacin equivalent" (or "nicotinic acid equivalent") has been introduced and by definition: one niacin equivalent is equal to either 1 mg niacin or 60 mg tryptophan (WHO, 1967). The bound form of niacin, nyacitin, represents a high proportion of, and consequently markedly affects the availability of, the niacin present in cereals. If the cereal is also deficient in tryptophan, this essential amino acid may not be converted to niacin. To what extent this phenomenon is of importance in human nutrition was not certain at the time of the World Health Organization report (WHO, 1967). Both niacin and tryptophan are stable to normal cooking and food processing, both in the dry state and in solution.

Niacin deficiency leads to retarded growth in children, dermatitis, diarrhoea, soreness of the tongue, mental confusion and eventually to the disease pellagra which culminates in dementia and death. The deficiency disease pellagra is discussed more fully in Chapter 4 since pellagra, or a condition closely resembling it, may occur when sorghum is the staple cereal. The daily adult requirement for niacin is between 10 and 20 mg.

Intakes of niacin among the population of certain countries are shown in Table 2.10. Intakes are lowest in India and the Far East where rice is the principal source; higher in Africa where pulses may provide the main intake; and higher still in Latin America and the Near East (WHO, 1967).

Riboflavin. Riboflavin, like thiamine, is involved in the biochemistry of energy conversion and in protein metabolism. The incidence of deficiency is widespread in South and Southeast Asia, Africa and Latin America. Growth is retarded in children, there may be thickening and cracking of the skin around the mouth, soreness of the tongue, and some misting of the eyes. The causes and prevention of riboflavin deficiency have apparently received relatively little attention since the sufferer is seldom incapacitated (WHO, 1967). The daily requirement is between 1 and 3 mg.

Although cereals are not particularly rich in riboflavin, in many developing countries they are an important source of the vitamin because of the quantity eaten (see Table 2.10). Riboflavin, which is located in the bran layers tends to be removed during milling. Riboflavin survives normal cooking but cooking in an alkaline medium accelerates destruction. The vitamin is sensitive to ultraviolet radiation and sun-drying may cause considerable destruction (WHO, 1967).

Other B-group vitamins. Folate (folic acid: folacin), cyanocobalamin (vitamin B_{12}), pyridoxine (vitamin B_6), pantothenic acid and biotin are reasonably widely distributed in natural foods but information on their occurrence, absorption and human requirements is scant compared to that on thiamine, niacin and riboflavin.

Ascorbic acid (vitamin C)

Ascorbic acid plays a complex role in the metabolic functions of man. A deficiency retards the growth of children and results in such manifestations of ill health as infected gums and the failure of wounds to heal. Scurvy results from a total deficiency. The minimum daily requirement is between 15 and 30 mg.

The richest sources are fresh fruit and leafy vegetables. The vitamin is destroyed by cooking and readily oxidized in neutral and alkaline solutions, particularly in the presence of traces of copper and other heavy metal catalysts.

Mature dry cereal grains contain no detectable ascorbic acid, but the vitamin is significantly presented in sprouted (germinated) cereal and legume seeds.

Fat-soluble vitamins

Retinol (vitamin A_1)

Retinol is found principally in fish liver oils, dairy products and green vegetables. Though it is believed to influence vision, its specific role in human metabolism and the level of human requirement is not fully understood (WHO, 1967). Among the wealthier nations deficiency symptoms are seldom seen but clinical retinol deficiency is known to occur in developing countries of Asia, the Near and Far East.

Earlier literature refers to "international units" of vitamin A. Since crystalline vitamin A_1 alcohol (retinol) became available as a reference standard, the recommended unit of measurement is μg of retinol. (One International Unit of vitamin A is equivalent to 0·3 μg retinol.) The daily requirement of adults is not exactly known but it is believed to be of the order of 1 mg.

The principal sources of retinol in the diets of the developing countries are fruits, green leafy and yellow vegetables. Few, if any, natural diets contain exclusively retinol, and the body derives most of its retinol by *in vivo* conversion of the carotene precursors in the diet. Of the carotenoids, β-carotene displays the highest biological activity. It is not as available as retinol, and in human diets 1 μg β-carotene is roughly equivalent in biological activity to 0·167 μg retinol or, conversely, 1 μg retinol is equivalent to 6 μg β-carotene. Some carotenoids display retinol activity, others do not. Mixed carotenoids may have retinol activity roughly one-half that of β-carotene (WHO, 1967, 1974).

Although cereals are not generally an important source of retinol, in Guatemala about 38%, in India about 12% and in Senegal about 30% of the retinol intake is estimated to be derived from cereals (WHO, 1967). The carotene content of yellow endosperm sorghum is discussed later.

Observations on experimental animals have indicated that protein deficiency impairs carotene utilization, and since in those areas where protein-calorie deficiency is observed, carotene forms the major source of retinol in diets, the World Health Organization (WHO, 1967) suggested that protein requirements and retinol requirements should be considered together. Both retinol and carotene are stable to heat but sensitive to oxidation.

Cholecalciferol (vitamin D_3)

Cholecalciferol plays an essential role in the absorption and laying down of calcium and phosphorus in bones. Extreme deficiency causes rickets. The vitamin is supplied mainly from fatty fish, eggs, liver and dairy products, and from the action of sunlight on human skin.

Vitamin E (tocopherol)

Vitamin E is essential to human nutritional well-being though its precise metabolic function is not clearly understood. Human minimum requirements are not known to the authors. The richest natural sources are the oils present in cereal germs.

Vitamin K

Vitamin K is essential for the production of prothrombin and therefore to the normal coagulation of blood. It is found in green vegetables and cereals.

Chemical Composition of Sorghum

In the text which follows, the data on the proximate composition of sorghum derived from numerous sources are presented in Table 2.11. This table includes data from papers containing little detail, together with those that state the types of sorghum used and the methods of analysis. The less detailed sources are shown at the foot of Table 2.11 as "representative values and ranges (not means)" and include: Adrian and Jacquot (1964); Bredon (1961) cited by Doggett (1970); Casey and Lorenz (1977); Chughtai and Waheed Khan (1960); Doggett (1970); Fabriani (1939); Ferrara and Paolis (1960); Fiorentini (1942); Food and Agriculture Organization (FAO, 1949); Jadhav *et al.* (1975); Kurien *et al.* (1960c); Munsell *et al.* (1949, 1950abcd); Oyenuga and Fetuga (1975); Platt (1962); Pushpamma (1968); Ranganathan *et al.* (1937); Rooney (1973); Rooney and Clark (1968); Rooney *et al.* (1970); Wall and Blessin (1969); and Watson (1971).

Practically all the authors reviewed employed the factor of 6·25 to convert N to protein; in the table this factor has been retained. At the foot of the table the range of the means and the range of the ranges are shown, and in the case of protein, a recalculation of these ranges using the factor 5·7 is also presented.

Representative analytical values quoted from Platt (1962) for whole grains of sorghum spp., pearl millet, finger millet, common millet, foxtail millet, kodo millet, *Digitaria exilis*, teff and Job's tears, compared with wheat, maize and rice, are presented in Table 2.12.

Africa

Shepherd *et al.* (1971–72) reported the proximate composition, protein (Kjeldahl, N × 6·25), crude fat (petroleum ether extract), crude fibre, and ash of 61 varieties of sorghum grown in East Africa in 1970 and 42 varieties grown in 1971 (see Table 2.11).

In 1971, the ranges were:

crude protein (%)	8.9 (IS 2584)–16·6 (IS 815)
crude fat (%)	2·5 (E 6432)–5·1 (E 6434)
crude fibre (%)	1.3 (IS 9816)–3·5 (E 6125)
ash (%)	1·1 (E 285, IS 9816)–2·7 (IS 815)

Yousif and Magboul (1972) determined the proximate analysis (AACC, 1962) of 15 samples of different varieties of dura (sorghum) grown in different parts of the Sudan. The samples were obtained from local markets and through public health authorities and the varieties were authenticated. In reporting, Yousif and Magboul expressed protein as Kjeldahl N × 5·7 as "most food analysts feel that a factor of 5·7 is more applicable [than 6·25] in the case of cereal protein". To facilitate comparison with other data, the Yousif and Magboul values have been converted to N × 6·25 and expressed DWB in Table 2.11.

Mayo, the most popular because of its white flour, contained the lowest and Fareek the highest protein content. Feterita, also high in protein was lowest in fat and highest in iron. The highest Ca content was found in Wazn'ashara. Fareek was also high in Ca and highest in P.

Wehmeyer (1969), at the National Nutritional Research Institute, Pretoria, determined the proximate composition of 11 varieties of sorghum grown at different locations in South Africa (see Table 2.11). Barnard Red was highest and TE 66 lowest in protein but highest in lipid, ash and carbohydrate.

India

Naik and Abhyanker (1955) analysed (AOAC, no date) nine improved strains of sorghum grown in Bombay State from the Dharwar varieties Bilichigam, Fulgar White, Fulgar Yellow, Nandyal, together with the Viramagam variety Chasatio and its local strain; a Surad variety and its local strain; Gauhala; Jowar No. 8 with a Broach local variety; plus a Mohol variety M35 1 and a Jalgaon variety, Aispuri.

The Dharwar varieties contained 6–9% protein, the local strain being slightly higher than the improved strain. Improved Nandyal was lowest in protein 6·3%. In 1947–48, improved Chasatio from Viramgam had 17·3% crude protein (compared with 9·5% in the local strain) but in succeeding generations it gradually declined to 13·5% in 1950–51.

Sur *et al.* (1955) reported the proximate composition (AOAC, 1950) of six numerically designated pure bred strains of sorghum obtained from the Department of Agriculture, Mysore. (See Table 2.11.)

Devadas *et al.* (1966) analysed four hybrid strains of sorghum grown at Coimbatore, MS × IS 84, MS × IS 3691, MS × IS 3687 and MS × IS 2930, and the most popular local variety, CO 18, for protein, ash and Ca (AOAC, 1960), P (Fiske and Subbarow, 1925) and Fe (Wong's method, no reference), and caloric value by bomb calorimeter (Parr Instrument Co, 1960). The only significant result was the higher protein content, 10·4%, of CO 18, against a mean of 8·5%.

Cooking quality was compared by boiling (uppuma and pongal) deep fat frying (fritters), pan broiling (dosai) and steaming (idli). A taste panel decided that CO 18 ranked highest, MS × IS 84 second, and MS × IS 3687 last.

Fourteen pure bred improved sorghum strains from the Millet Breeding Station of Mysore State and the Regional Station of IARI at Hyderabad, Yannigar 2, Annigeri 1, Nugati, GS 560 1 1, K Jowar, H 1, Bellary, Maldandi, CSH 1, CK 60B, CO 11, 3691 B, CSH 2, M 35 1 and Swarna, were analysed for protein (N × 6·25) ether extractives and ash (AOAC, 1950), total P (Fiske and Subbarow, 1925) and amylose (Sowbhagya and Bhattacharya, 1971) by Viraktamath *et al.* (1972).

The mean and ranges for protein, lipids and ash are shown in Table 2.11. The highest and lowest were:

	Highest	Lowest
Protein	CO 11 (Hyderabad)	Swarna (Hyderabad) Yannigar 2 (Raichor) Annigeri 1 (Raichor)
Lipid	Swarna (Hyderabad)	3691 B (Hyderabad)
Amylose	M 35 1 (Hyderabad)	Swarna (Hyderabad)
Ash	Swarna (Hyderabad)	GS 560 1 1 (Raichor)
Phosphorus	CSH 1 (Hyderabad)	3691 B (Hyderabad)

When pearled in a laboratory machine, K Jowar and CO 11 required two to three times longer to remove the husk completely. Eight varieties suffered low milling breakage and gave high total yield: CSH 2, M 35 1, Swarna, CSH 1, CK 60B, Maldandi, Yannigar 2 and Mugati. Grain hardness (Kiya hardness tester) was not correlated with milling breakage. Whiteness (photo-volt reflectance meter) increased with pearling, the degree depending on the bran pigmentation before milling. CO 11 became whitest after pearling.

All the varieties except K Jowar were suitable for making bhakri (dry pancake). Yannigar 2 and Swarna cooked into softer grains than the other varieties but they were not as soft as cooked rice after cooking for 30 min and required pressure cooking at 15 lb/in^2 to become acceptable. Loss of solids into cooking water varied from 9 (3691 B) to 26% (Yannigar 2).

Europe

Kandel (1954) reported the proximate composition of the varieties Black Amber, Coes Improved, Early Milo, Freemont, Hegari, Korai barna, Leoti ped, Midland, Norkan, Raxorange, Kaoliang and Fekete Cukor, grown in Hungary on the same experimental plot and under optimal conditions. (See Table 2.11.)

The highest and lowest were:

	Highest	Lowest
Protein	Korai barna	Hegari
Lipid	Fekete cukor Early Milo	Black Amber
Starch	Black Amber	Leoti ped
Glucose (after hydrolysis)	Early Milo	Leoti ped
Crude fibre	Korai barna	Hegari
Calcium	Kaoliang	Raxorange
Total Phosphorus	Kaoliang Fekete cukor	Black Amber

North America

The National Academy of Sciences (1971) in their Atlas of Nutritional Data on United States and Canadian Feeds, gave the chemical composition (methods not cited) of Atlas, Darso, Durra, Feterita, Hegari, Kafir, Kalo, Kaoliang, Milo, Norghum, Sagrain, Schrock and Shallu. (See Table 2.11.)

The highest and lowest contents were:

	Highest	Lowest
Protein	Shallu	Sagrain
Lipid	Kaoliang	Hegari
Carbohydrate	Hegari	Shallu
Fibre	Schrock	Kalo, Kaoliang
Ash	Durra	Darso

In 1944, Heller and Seiglinger reported the protein, fat, crude fibre, nitrogen-free extract, ash, Ca and P (AOAC, no date) of 29 named varieties of sorghum grown in 1941, 1942 and 1943 at Perkins and at Woodward in Oklahoma, USA, two locations 150 miles apart in the same latitude but differing in altitude and soil characteristics. The climatic conditions in 1941 and 1942 were similar and the results were averaged. The means and ranges for all varieties at both locations are shown in Table 2.11.

Variation in protein among varieties at one location tended to be lower than the variation within a variety between environments, e.g. Standard Kafir 9·9% in 1941–42 and 10·1% in 1943 at Woodward compared to 16·7% at Perkins in 1943.

The highest and lowest were:

	Highest	Lowest
Protein	Standard Kafir	Sumac Sorgo
Lipids	Sugar Drip Sorgo	Migari
Carbohydrate	Early Hegari	Standard Kafir
Fibre	Atlas	Atlas
	(depending on location)	
Ash	Red Kafir	Early Hegari

Barham *et al.* (1946) at Kansas Agricultural Experiment Station, determined proximate analyses (AOAC, no date) of 12 varieties of grain sorghum grown at Hayes (Feterita, Finney Milo, Cheyenne, Club, Hegari, Red Kafir, Wheatland, Early Sumac, Early Kalo, Atlas, Sorgo, Pink Kafir and Leoti Red Sorgo, one at Manhattan (Blackhull Kafir), and one at Stillwater (Schrock). (See Table 2.11.)

The highest and lowest were:

	Highest	Lowest
Protein	Feterita	Schrock
Lipids	Cheyenne	Hegari
Fibre	Leoti Red Sorgo	Cheyenne and Early Kalo
Ash	Leoti Red Sorgo	Schrock

South America

Capote *et al.* (1972) analysed 70 experimental grain sorghums, 22 commercial hybrids and 11 varieties, all grown in Venezuela. The mean values for proximate composition (AOAC, 1965), P (Fiske and Subbarow, 1925) and other inorganic elements (spectrophotometric) and tannin (Capote *et al.* 1971, no further reference given) appear below.

	Experimental hybrids	Commercial hybrids	Varieties
Moisture (%)	11·5	11·8	11·9
Crude Protein (%)	12·25	11·41	11·97
Fat (%)	2·35	2·38	2·39
Crude Fibre (%)	2·65	2·46	2·59
Starch (%)	63·10	66·02	55·16
Tannin (%)	0·38	0·49	0·49
Phosphorous (mg/100 g)	230	230	340
Potassium (mg/100 g)	360	380	540
Zinc (mg/100 g)	8	6	4

Differences between the means of hybrids and varieties were significant for starch, tannin, P, K and Zn.

Ferreira Oriá and de Lima (1975) reported the proximate analysis (AOAC, 1965) of 10 varieties of sorghum grown in Brazil during 1973. (See Table 2.11.) The highest and lowest were:

	Highest	Lowest
Protein	2201	R 1090
Lipids	2201	Nic 233, EA 201
Carbohydrates	R 1090	2201
Fibre	S 77	2201, EA 201
Ash	EA 201	RS NK 180

Protein and Amino Acid Composition

A large body of literature has been reviewed. In 1968 Coons published a bibliography of selected references on cereal grains in protein nutrition, covering the period 1910–1966; it included about 180 references to reports on sorghums and the millets, of which those relevant to this review are discussed later. Protein data relating to sorghum appear in the proximate analysis Table 2.11. In Table 2.13, the amino acid contents (means and ranges) are presented.

In many of the papers reviewed, the protein content has been presented, expressed as N × 6·25, and at the foot of Table 2.13 will be found the range of the means and the range of the ranges for the protein content and the various amino acids. These ranges are also presented recalculated using the factor N × 5·7. It will be observed that the ranges of protein in Table 2.11 and 2.13 differ, the upper value for protein being higher in Table 2.13 than in Table 2.11. In Table 2.13 those assays which were carried out microbiologically have been indicated since in general values obtained by microbiological assay are higher than those obtained by ion-exchange chromatography.

Some papers described the types of sorghum used, the methods of analysis and other relevant detail; other sources provided less detail. These latter include: Adrian and Jacquot (1964); Desikachar and De (1947); Deyoe *et al.* (1968); Food and Agriculture Organization (FAO, 1970); Nagpal and Bhatia (1971); Orok and Bowland (1974); Pion and Fauconneau (1969); Pushpamma (1968); Salunkhe *et al.* (1977); and Taira (1963a).*

Mixed Locations

Srinivasan *et al.* (1972) at Purdue University, Lafayette, Indiana, compared the amino acid composition of 522 lines of sorghum from the World

* (Authors' note: from the data collected, we have calculated the Amino Acid Scores based upon the Requirement Pattern of the World Health Organization (WHO, 1973a), and where possible the leucine/isoleucine quotient, the importance of which is discussed in Chapter 4.)

Collection divided into: (I) the 323 lines with 12% protein and above; and (II) the 199 lines with protein below 12%. They found no significant difference in average amino acid content between the two groups. The mean protein and amino acid composition of the 522 lines are presented in Table 2.13.

Overall correlations among amino acids, and between amino acids and protein, showed that lysine, cystine, threonine, arginine, histidine, glycine and serine were significantly, negatively correlated with protein. Lysine increased as histidine, arginine, aspartic acid, threonine, serine, glycine and valine increased and as glutamic acid, proline, cystine, alanine, leucine, tyrosine and phenylalanine decreased.

Among group I (higher protein), protein content increased as isoleucine and leucine concentration increased. A negative correlation was found between lysine and glutamic acid, alanine, tyrosine and phenylalanine; histidine and leucine were positively correlated.

Among group II (low protein), protein was positively correlated with threonine and valine.

Africa

Adda (1958) analysed (two-dimensional paper chromatography Thompson *et al.*, 1951, Thompson and Steward, 1951) the amino acid composition of 25 samples of sorghum of differing origins grown at Bambey in Senegal. (See Table 2.13.) *S. guineense* contained the highest lysine content, the mean of the four lines being 169 mg/g total N with a mean protein content of 11·4%. *S. cernum* and Dwarf Shallu were lowest in lysine.

Sixteen sorghum lines designated by numbers, and the varieties Fellah Rouge and Congossane examined at Bambey were reported by Bono and Vidal (1962) to have mean protein content (method not stated) (DWB) of 11·5%, ranging from 9·0 to 12·9%. Ten of the lines were guineense, four bicolor, three durra and one caffra. The highest protein contents, 12–13%, were in the guineense.

In general, the more vitreous the grain, (graded on a visual scale of 0–4) the higher the protein, but Bono and Vidal recommended that for selection purposes the two characters should be considered separately.

Haikerwal and Mathieson (1971b) at Ahmadu Bello University, Nigeria, separated by hand dissection the pericarp, endosperm and germ of kernels of the sorghum variety FF 5683 (Guinea). The percentage by weight of the whole kernel and the protein (Kjeldahl N × 5·83) of each fraction are shown below:

Fraction	% by weight of whole kernel	% protein in fraction
Germ	8·0	16·3
Endosperm	86·5	10·2
Pericarp	5·4	5·1
Entire kernel	100·0	10.4

The amino acid composition (auto-analyser following acid hydrolysis, excluding tryptophan) (mol % total) of the fractions is shown in Table 2.14.

From the amino acid analyses Haikerwal and Mathieson recommended a N to protein conversion factor of 5·83.

Haikerwal and Mathieson (1971b) reported a height and maturity hybridization trial in which four US varieties, Durra, Texas Blackhull Kafir, 100-day Milo and Combined Hegari were crossed with a local (Nigerian) variety, Shambul. The five parents and four crosses were grown in four blocks, each containing nine plots, assigned at random to the varieties.

Samples from three plants from each were analysed for total protein (Kjeldahl N × 5·83). Protein ranged from 8·8% (Milo × Shambul) to 11·6% (Durra) and variance among varieties was highly significant. There was a significant variation in protein content among different varieties grown under the same conditions. The protein content of the hybrids was intermediate between the parents, except Milo × Shambul in which the protein was below both parents. Durra × Shambul showed greatest increase in protein content over Shambul. The protein content of kernels from 148 plants in an F_3 population of Durra × Shambul ranged from 6·5 to 14·5%. A frequency distribution curve suggested that plants segregated into two major groups.

Kernels from Shambul, Durra and Durra × Shambul were analysed for amino acid composition (auto-analyser, excluding tryptophan). There were no significant differences among duplicate analyses or among plants. Almost all amino acids showed significant differences among replicates, and a number, especially aspartic acid, serine, glycine, alanine, half-cystine, valine, isoleucine and phenylalanine, among varieties.

There was significant variation in amino acid composition among different sorghum varieties and within varieties among experimental plots at one location.

Shepherd *et al.* (1971–2) analysed numbered samples of East African sorghums for amino acid composition (auto-analyser following acid hydrolysis) excluding tryptophan, cystine and methionine. Per cent N recoveries ranged from 80·0 to 110·8%. (See Table 2.13).

Khattab *et al.* (1972) examined eight sorghums commonly grown in the Sudan. The varieties Feterita (Gedarif), Feterita (Gezira), Mayo, Safra (White Nile), Zereizera, Gassabi and Abu 70 were grown at the Faculty of Agriculture Farm at Shambat in 1967, the variety Zinnari (East Kordofan) in East Kordofan was grown in sandy soil. Gassabi (12·5%) was highest and Feterita (Gedarif 9·4%) lowest in protein (micro-Kjeldahl N × 6·25).

The means and ranges of 13 amino acids (two-dimensional chromatography) are shown in Table 2.13.*

* (Authors' note: not all amino acids were determined in all varieties, the number of samples providing the mean and range therefore differed. Three of the values in mg/g total N as printed in the original, appear exceedingly high (threonine 765 in Safra, phenylalanine 796 in Zereizera and threonine 960 in Zinnari) and inconsistent with the values quoted in mg/100 g sample. No comment was made by Khattab *et al.* on these exceptionally high values which we, the present authors have omitted in calculating the mean and range in mg/g of total N.)

Mayo, highest in lysine (114 mg/g total N), was also highest in leucine and isoleucine, giving a leucine/isoleucine quotient of 3·3. Gassabi, leucine 441 and isoleucine 93 mg/g total N, giving a leucine/isoleucine quotient of 4·7, contained 64 mg/g total N of lysine.

Mehansho and Besrat (unpublished report, 1974) reported protein (N × 6·25, Pierce *et al.*, 1962) and nine essential amino acids (auto-analyser excluding cystine and tryptophan) of 44 sorghum grain varieties grown in Ethiopia. (See Table 2.13.) Protein ranged from 7·5 (Gato 994) to 13·1% (Martin). Gato 994, the variety that gave highest yield (92 quintal/ha) was lowest in protein, 7·5%, but the next highest yielder, Netch Shure, contained average protein of 10·6%. (Both were high in tannin: 2·65 and 2·90 CE/100 g dried sample respectively.)

In general, the high protein samples contained higher levels (mg/100 g sample) of all the essential amino acids determined. Lysine content (mg/g N) ranged from 160 (Abay 145) to 102 (Jij Wegere). Gato 1001, Gato 994 and WS 1763 were also high in lysine as % protein. The leucine/isoleucine quotients were, Jij Wegere 3·5; Gato 1001 3·1; Gato 994 3·2; and WS 1763 3·3.

Eastoe and Taylor (1974) reported the amino acid composition (amino acid analyser following acid hydrolysis, excluding tryptophan) of three samples of sorghum grown in the same climatic zone of Botswana, (A) light brown, protein 10·3%; (B) white, protein 8·7%; and (C) red, protein 10·4%, as shown in Table 2.13.

Although the protein in (B) was lower than in (A) and (C), the amino acid contents (% protein) were similar. The leucine/isoleucine quotient was 3·2.

India

Austin *et al.* (1972) examined 96 sorghum varieties and hybrids, grown under the (Indian) coordinated sorghum improvement programme, for protein (macro-Kjeldahl × 6·25) and lysine (Schaiberger and Ferrari, 1960 with modifications) % of grain and % of protein. They fell into four groups according to protein content: group I (protein below 10%) nine lines; group II (protein 10–12%) 24 lines; group III (protein above 12%) 30 lines; group IV (fodder types, small-seeded) 33 lines.

Among group I, lysine (mg/g N) varied from 77 to 211 with a mean of 138; in Group II from 74 to 207 with a mean of 112; in Group III from 45 to 139 with a mean of 92; and in Group IV from 64 to 184 with a mean of 96. The correlation between protein and lysine (% protein) was negative and significant but the relation was not demonstrably linear. IS 4532 in group II and IS 4592 in group III contained respectively 201 and 184 lysine suggesting the possibility of selection for high protein with high lysine in protein.

Among group I, AKP 1, N 1, and IS 5641 were low in lysine; in group II, CO 4, Early 56, IS 4532, IS 4533, IS 4989 and IS 5595, in group III, CSH 1,

IS 3691, CO 12, IS 4953, IS 4990 and IS 5609, in group IV, IS 4952 were higher in lysine.

Balasubramanian *et al.* (1952ab) and Balasubramanian and Ramachandran (1957) reported the amino acid composition (microbiological assay, Barton-Wright, 1946) of the sorghum varieties CO 1, CO 2, CO 4, CO 5 and CO 7, and of one bazaar sample. (See Table 2.13).

Sorghum CO 7 suffered a 19% loss and CO 5 a 32% loss in lysine of ground samples after storage for 10 months in well-stoppered glass jars with naphthalene as a preservative.

Chitre *et al.* (1956) reported the nitrogen (Kjeldahl) and the phenylalanine, leucine, isoleucine, lysine, valine and threonine (assayed microbiologically, Henderson and Snell, 1948), and chemically the methionine (Horn *et al.*, 1946), histidine (McPherson, 1942), arginine (Saskaguchi reaction, Block and Bolling, 1945) and tryptophan (Folin and Marenzie method, Block and Bolling, 1945) content of the pure bred strains of sorghum, N 1, CO 7 and H 1. (See Table 2.13)

The grain from different sorghum varieties and hybrids from germ plasm at Surat, India, was compared for grain size (1000-seed weight) protein content (AOAC, 1960) and colour (visual) by Chakravorty (1967).

The mean seed weights and protein contents (ranges in parentheses) were:

	Pearly-white grain		Coloured grain		Chalky-white grain	
No. of varieties or hybrids	14		23		5	
1000-seed weight (g)	32·2 (25·6–44·2)		29·6 (17·8–37·7)		25·6 (21·2–29·9)	
	MS × Taramo	Local Check BP 53	IS 474	IS 84 1	IS 1122 1	IS 2728
Protein (N × 6·25) (DWB) (%)	8·8 (7·5–11·7)		10·3 (8·1–14·2)		11·1 (7·5–14·1)	
	MS × Aispuri	MS × Jowar	IS 2942 2	IS 3906	IS 1122 1	IS 2728

There was no apparent relation between 1000-seed weight and protein content.

Chakravorty (1967) also reported the 1000-seed weight and protein content of large, medium and small seeds from single plant selections of four bazaar samples, IS 944, local check BP 53, and the hybrid MS × HC 39 142 Coddy line I, Coddy line II, Coddy line III, MS × Sel 124–3 (IS 2257) and MS × Nyethin IS 3541. In the same variety protein was higher in the large than in the small grains; among the hybrids the reverse was the case. In a personal communication, Doggett (1977) suggested the possibility of a different embryo/endosperm ratio between large and small grains. Characters of the hybrids will be dependent upon parentage; embryo and endosperm ("seed") size are independently inherited.

Pundarikakshudu and Seshadri (1969) reported that two hybrid strains of sorghum, MS 2219 × IS 3691, and MS 2219 × IS 3541 out-yielded CSH 1 and CO 18 in protein (AOAC, 1960) /ha. The results were:

Strain	Grain yield (kg/ha)	Protein production (kg/ha)
CSH 1	6000	473·4
MS 2219 × IS 3691	8000	756·8
MS 2219 × IS 3541	7000	638·4
CO 18	4500	458·5

Deosthale *et al.* (1970a) examined over 300 (undesignated) varieties of sorghum, from the collection at IARI for protein (micro-Kjeldahl N × 6·25/10% moisture) lysine and leucine (as % protein) (microbiological assay Steele *et al.*, 1949). The results appear in Table 2.13. Seventy varieties were found with lysine higher than normal and leucine lower than normal. A significant negative correlation was found between protein content and lysine (% protein), but no significant correlation was found between either protein and leucine or lysine and leucine. When the extreme values of leucine in six varieties were excluded, there was a significant correlation between protein and leucine.

Deosthale *et al.* (1970a) suggested that the ratio of leucine to lysine content would indicate the pellagragenic character of sorghum (see Chapter 4). Rice, wheat and pearl millet were known to be nonpellagragenic, and among these cereals the leucine/lysine quotient ranged from 2·0 to 3·8.

Among the sorghum varieties examined, the leucine/lysine quotient varied from 3·6 to 16·5 with an average of 7·0. Of the samples, 84% showed a quotient greater than 5·0 and only in 22 lines was it below 4.6, which Deosthale *et al.* (1970a) suggested as a safe upper limit. In three varieties, IS 4324, IS 516 and IS 4642, the quotients were near 3·6; IS 4324 was higher in protein than the other two.

Among the high protein lines, IS 5262, 8298, 5478 and 3950 were high in lysine and the leucine/lysine quotient was only slightly above 4·6.

Mali and Gupta (1974) compared the crude protein, ash, Ca, Fe, ether extract, crude fibre (AOAC, 1965), 100-seed weight and hardness (Tripathi *et al.*, 1971), total P (King, 1932), phytin P (McCance and Widdowson, 1935) and soluble carbohydrate (by difference) of 16 improved varieties of sorghum. Means and ranges of protein, ether extract and ash are shown in Table 2.11 and of the inorganic elements in Table 2.58.

IS 815, CSH 2 and Swarna were highest in 100-seed weight, 3·49, 3·47 and 3·14 g respectively. IS 2944 was highest, Swarna 1 lowest in crude protein. Mineral content varied; IS 2944 and IS 84 (I) were highest. Phytin P ranged from 33 to 52% of total P. IS 3924 and Swarna were highest, IS 3796 lowest in ether extract. IS 3691 was lowest and IS 815 highest in crude fibre.

The mean contents (% grain and % protein) of lysine, leucine and threonine (microbiologically) methionine and tryptophan (colorimetrically) with ranges, are shown in Table 2.13.

Variations among amino acids (mg/g total N) were significant for:

	Highest	Lowest
lysine	IS 3796	CSH 1
methionine	IS 3922 B	IS 2944
threonine	IS 3796	IS 1601 B
tryptophan	IS 3924	IS 1601 B

IS 3796 was comparatively high in lysine, threonine and tryptophan, though low in methionine.

Significant positive correlations were found between protein and lysine (% grain), protein and threonine (% grain), protein and tryptophan (% grain), protein and leucine (% grain), protein and methionine (% grain), protein and ash and protein and total P. There was a significant negative correlation between protein and lysine (% protein).

Nagarajan (1974) reported the protein and lysine content (analytical methods not stated) of the high yielding sorghum lines CSH 1, CSH 2, CSH 3 and Swarna. The mean and ranges are shown in Table 2.13.

N.G.P. Rao (personal communication, 1977) reported the protein (N × 6·25), lysine and leucine contents in the protein (microbiological assay) of 21 cultivars of sorghum grown at Hyderabad, India (see Table 2.15).

Jambunathan (1977) reported from ICRISAT the protein content (Technicon auto-analyser N × 6·25) of 6758 samples of sorghum from various locations grown during the rabi and kharif seasons and screened during 1976–77, as ranging from 6·0 to 21·1% with a mean of 12·1%. A high degree of correlation had been established between protein content as determined by the micro-Kjeldahl method and the values from the Technicon auto-analyser.

The content of lysine in the protein of the samples was estimated using dye binding capacity (DBC) values expressed as Udy Instrument Readings (UIR) and ranged from 69 to 219 mg/g total N with a mean of 116 mg/g total N.

A high degree of correlation had previously been obtained between lysine determined by amino acid analyser and DBC values. The frequency of the samples and lysine content are shown in Fig. 2.IX.

Europe

Zubaidov (1968c) reported 20 amino acids present in the extracted nonprotein N of eight different sorghum varieties. There were no significant differences in amino acid composition.

Central and South America

Bressani and Rios (1962) reported the proximate composition of 10

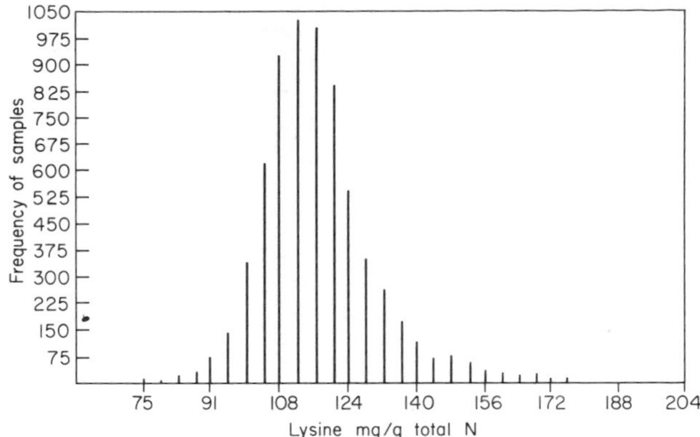

Fig. 2.IX Frequency distribution of lysine, as mg/g total N of 6758 samples of sorghum examined at ICRISAT, Hyderabad, India, 1976–77. Mean 116 mg/g total N, range 69–219, frequencies below 75 and above 176 were too few to report.

varieties and 13 hybrids of grain sorghum grown at one location in Guatemala in 1958 from seed imported from Texas. Two native varieties were also grown, 346908 V Criolla at the same location as the imported seeds, Saucite Criolla at another location. The means and ranges for protein (N × 6·25), ether extract, crude fibre and ash (AOAC, 1950) are shown in Table 2.11, Ca and P (Lowry and Lopez, 1946) in Table 2.58, and thiamine (modified thiochrome assay, Hennesey and Cerecedo, 1939), riboflavin (fluorometric, Hodson and Norris, 1939), and niacin (microbiological, Steele *et al.*, 1949) are shown in Table 2.59. The mean and ranges of 12 amino acids are shown in Table 2.13.

The two Guatemalan samples were higher in protein, thiamine and niacin and lower in ash, Ca, P and riboflavin than the imported varieties and hybrids. The variety Westland was highest in Ca and P.

Ferreira Oriá and de Lima (1975) reported the amino acid composition of 10 varieties of sorghum grown in Brazil. The means and ranges are presented in Table 2.13.

Martearena and Pérez (1977) examined in Venezuela 31 cultivars of sorghum as possible introductions, and reported the ranges found for Kjeldahl protein (no conversion factor stated), tryptophan (microbiological assay) and lysine (see Table 2.13). Two (undesignated) cultivars from India with protein content above 14%, one of which contained 213 mg/g total N lysine, appeared to justify commercial production.

United States of America

The content of 10 amino acids (mg/100 g grain, DWB) in the feed grain sorghums, Darso, Feterita, Hegari, Kafir and Shallu given in the Atlas of Nutritional Data on US and Canadian Feeds (National Academy of Sciences, 1971) is shown in Table 2.13. The methods of determination were not stated.

Deyoe and Shellenberger (1965) reported the crude protein content (AOAC, 1960) (N × 6·25%) of 15 sorghum hybrids grown at each of three locations in Kansas. The mean and range are presented in Table 2.13. Average protein contents at Hiawatha (11·1%) and Newton (11·4%) exceeded those at Manhattan (10·0%). The protein content among hybrids varied from a high average of 12·1% for 59 CH 5 to a low of 9·9% for KS 701. There were significant differences in protein among hybrids and among locations effects. These results were similar to those of Miller *et al.* (1964). (An anonymous author (Anon., 1966) in Nutrition Reviews commented that Deyoe and Shellenberger did not state whether their protein analysis was based on dry matter but he assumed that it was. However, Miller *et al.* (1964) with whom Deyoe is included as one of the authors, gave their protein on a 12·5% moisture basis.) The hybrids grown at Hiawatha and Manhattan were analysed for amino acid composition (excluding tryptophan) by auto-analyser following acid hydrolysis (Spackman *et al.*, 1958). Mean and ranges are shown in Table 2.13.

Amino acid (% sample) was significantly related to hybrid and location. Amino acid composition (% protein) was highly significantly related to hybrid but not to location.

Thirty-five genotypes, Martin B, Redlan B, IS 0062, 0075, 0187, 0418, 0616, 2288, 5437, 7125, 8165, 8361, the hybrid of Martin with each IS entry, the hybrid of Redlan with each IS entry, two commercial hybrids RS 610 and NK 300 and a commercial variety, Atlas, were grown in six locations in Indiana, Texas and Nebraska in 1970 and 1971 (Schaffert *et al.*, 1972ab). The agronomic traits, grain yield (adjusted to 14% moisture) days to median flower, plant height, protein (micro-Kjeldahl N × 6·25), lysine (% protein) (ion-exchange chromatography), oil (nuclear magnetic resonance), and 100-seed weight, were determined. The average values are shown in Table 2.16. The average values of locations and years for genotypes of the agronomic traits are shown in Table 2.17. The genotype—year—location interaction was significant for all variables except lysine (% protein). The genotype—location interaction was significant for all variables except grain yield, protein yield, lysine yield, lysine in sample and oil. The genotype—year interaction was significant for all variables except protein, lysine (% sample) and plant height.

Days to flower was correlated negatively with protein and positively with lysine (% protein). Protein was correlated negatively with lysine (% protein) and positively with lysine (% sample). Per cent oil and 100-seed weight were significantly positively correlated with protein but selection for increased oil content would tend to reduce grain yield.

Some 300 temperate (not photoperiod sensitive) lines of sorghum were selected from Puerto Rico and Lafayette, Indiana production of the World Collection in 1969 and 1970 (Oswalt and Srinivasan, 1973). The traits examined were: days to half flower, height of plant, 100-seed weight, pericarp colour, catechin equivalents (Burns, 1971), dye binding capacity, protein (micro-Kjeldahl N × 6·25, 10% H_2O), lysine (% protein) (ion-exchange chromatography), lysine (% sample), first replication grain yield (kg/ha of 10% moisture grain from centre row of 3 row plots of replication at Lafayette, Indiana, used for chemical analysis), average grain yield (kg/ha of 10% moisture grain from centre row of three row plots, the mean of three Indiana locations except where otherwise indicated). The group contained a high incidence of white pericarp genotypes, with complete, partial or no testa and yellow endosperm selections.

The average grain yield of the nine commercial hybrids included as checks was 7250 kg/ha, with a range of 6010–8090 kg/ha.

The average grain yield of 204 lines was 7050 kg/ha with a range of 4580–9860 kg/ha. The protein average of 8.9% was low compared to similar grain yield in previous years, but protein yields /ha were above average.

Pioneer 931 was high in lysine (399 mg/100 g sample), typical of immature grain, and among the lowest in yield (4970 kg/ha, first replication yield). The lysine in IS numbers 15391, 0071, 12102 and 16117 was comparatively high (299, 317, 300 and 331 mg/100 g sample respectively) for standard sorghum.

The relation between pericarp colour and catechin equivalents is discussed in Chapter 4.

Protein Solubility Fractionation

Osborne (1924) described his method, still widely used, by which seed proteins can be fractionated and categorized. Osborne's classification includes:

(1) albumins—soluble in water
(2) globulins—soluble in saline solution
(3) prolamines—soluble in alcohol
(4) glutelins—soluble in dilute alkali.

Only among the seeds of the Gramineae, which include sorghum, the millets, wheat and triticale, are prolamines and glutelins found in any quantity.

Nelson (1969a) stated that the prolamines of cereals can be divided into three groups on the basis of their amino acid composition:

(I) the gliadin of wheat, the hordein of barley and the secalin of rye;
(II) the zein of maize, the panicin of millet and the kafirin of sorghum;
(III) the avenin of oats.

The prolamine fraction of group II cereals is exceptionally low in lysine. Consequently, the increase in lysine as a result of breeding for prolamine suppression is likely to be greater in maize, sorghum and millet than among the cereals of groups I and III.

Osborne Fractions

Naik (1968) fractionated the proteins of ground, defatted samples of (1) VC 10 2, (2) M 35 1, (3) IS 84, (4) IS 3922 405, (5) K 5 and (6) IS 3691, by a modified Mendel Osborne method (Nagy *et al.*, 1941) and analysed the albumin, globulin, prolamine and glutelin fractions for nitrogen (Kjeldahl), lysine (lysine decarboxylase) and tryptophan (colorimetrically, Spies and Chambers, 1949). The results are presented in Table 2.18.

Prolamines and glutelins together accounted for between 63 and 70% and albumins and globulins for about 20% of total N, though the proportions of individual fractions varied significantly among the different lines. Prolamine, for example, ranged from 27·0% in M 35 1 to 43·1% in IS 3691, the latter also contained the lowest level, 0.045%, of albumin; of lysine, 91 mg/g total N; and of tryptophan 31 mg/g total N.

Tables 2.19 and 2.20 present the lysine and tryptophan content of each fraction expressed respectively as mg/g total N and as mg/100 g flour. The glutelins contained between 40 and 60%, the albumins and globulins together between 33 and 46%, and the prolamines less than 5% of the lysine in the grain. The albumin and globulin fractions together contained 37–56%, the glutelin 35–48% and the prolamines 6–10% of the tryptophan in the grain.

Ahuja *et al.* (1970) examined the amino acid composition (auto-analyser following acid hydrolysis; tryptophan, Spies and Chambers, 1949) of the sorghum varieties examined by Naik (1968). The most marked differences were between VC 10 2 (protein 12·4%, prolamine 28·5% of protein) and IS 3691 (protein 10·5%, prolamine 43·1% of protein). (The N to protein conversion factor was not stated). IS 3691 was higher in glutamic acid and proline, and lower in lysine and tryptophan than VC 10 2. The leucine/isoleucine quotient was 3·45 in IS 3691 compared to 2·5 in VC 10 2. These two varieties were fractionated (Naik, 1968) into albumin, globulin, prolamine and glutelin fractions, and the amino acid composition in the fractions determined. The results are presented in Table 2.21.

The prolamine from both varieties showed a similar amino acid composition, comparatively high in glutamic acid, proline and leucine and low in glycine, the basic amino acids, the sulphur amino acids and tryptophan. The leucine/isoleucine quotient exceeded four in the prolamine of both varieties. The amino acid compositions of the albumin and globulin fractions were more balanced, with higher concentrations of lysine and tryptophan and a leucine/isoleucine ratio as low as 1.5.

Pushpamma (1968) fractionated the protein of sorghum, pearl millet and finger millet samples obtained from the local market at Hyderabad. Two methods were used, (1) a percolation method of Maes (1962) as modified by Cagampang *et al.* (1966) and (2) by shaking, a described modification of the method of Baudet *et al.* (1966) as cited by Mosse (1966). Samples of rice, wheat and commercial (USA) sorghum were also extracted by the two methods. The results are presented in Table 2.22. Most striking are the differences between the two methods in the proportion of the total N extracted and the relative size of the glutelin fractions from the two sorghums analysed. By both methods of extraction a considerably higher proportion of total protein was extracted from pearl millet (81·4 and 72·4%) than from Indian sorghum (53·2 and 41·5%), from the US sorghum (54·8 and 38·2%)

and from finger millet (65·9 and 58·5%). Pearl millet contained greater amounts of albumins plus globulins (28·7 and 30·2%) than the other grains. The total albumins and globulins of the Indian and US sorghums were similar but the proportions were different. Prolamine was highest in wheat, followed in decreasing order, by pearl millet, maize, finger millet, sorghum and rice. By both methods, the prolamine fraction was higher in the Indian than the US sorghum. Pearl millet and Indian sorghum were lowest, rice followed by maize highest in glutelins.

Virupaksha and Sastry (1968) determined the crude protein (Kjeldahl N × 6·25) DWB of 44 genotypes from the World Sorghum Collection and of five local hybrids from the Agricultural Reasearch Station, Mysore State which included: M 35 1 Raichur, D 340 Siruguppa, M 35 1 Siruguppa, BS 81 3 Annigeri and CSH 1 Bijapur.

Among the World Collection samples, protein content ranged from 8.6% (1533 Conspituum) to 18·2% (361 Dochna) with a mean of 13·0%, and among the hybrids from 8·5% (M 35 1 Raichur) to 15·5% (CSH 1 Bijapur) with a mean of 12·1%. Lysine (mg/100 g whole seed—method of Selim 1965) ranged from 117 (1533 Conspituum) to 413 (160 Cernum) with a mean of 218; and among the hybrids from 186 (M 35 1 Raichur) to 300 (CSH 1 Bijapur) with a mean of 246. Lysine (mg/g total N) ranged among the World Collection samples from 81 (1094 tall white kafir) to 143 (219 Dur./Nig.) with a mean of 106, and among the hybrids from 114 (CSH 1 Bijapur) to 136 (M 35 1 Raichur) with a mean of 129. The solubility fractions of the endosperm protein (modified Osborne and Mendel, 1914) were determined for M 35 1 Siruguppa, BS 81 3 Annigeri, CSH 1 Bijapur, 160 Cernum and 361 Dochna. The results appear in Table 2.23. Extraction efficiency varied from 80·7 (361 Dochna) to 103·4% (BS 81 3 Annigeri). Prolamine and glutelin were the largest fractions and higher protein levels were associated mainly with an increase in the prolamine fraction.

The amino acid composition (excluding tryptophan) of the ground whole seed of 29 Cernum, 160 Cernum, 164 Subglab, 202 Subglab, 361 Dochna, 1533 Conspituum, BS 81 3 Annigeri, M 35 1 Siruguppa and CSH 1 Bijapur was determined by ion-exchange chromatography. The means and ranges for the genotypes and hybrids are shown in Table 2.24. In all samples, glutamic acid, proline, alanine and leucine contents were relatively high, with isoleucine/leucine quotient close to 4·0. The highest concentration of lysine in protein, 194 mg/g total N was reported in 160 Cernum. (Singh and Axtell (1973b) commented that a genetically controlled high-lysine character in 160 Cernum had not to their knowledge been confirmed by any other reports.)

The amino acid composition (ion-exchange chromatography) of the ground endosperm and of the globulin, prolamine and glutelin fractions of CSH 1 Bijapur are shown in Table 2.25. The amino acid composition of the globulin and prolamine fractions of 160 Cernum are also shown. The globulin fraction contained a more desirable distribution of essential amino acids than the ground endosperm and prolamine fraction, the glutamic acid, proline, alanine and leucine in the globulin fraction being about half what were found in the ground endosperm and prolamine fractions. Threonine, histidine, arginine, glycine, serine and aspartic acid were higher, tyrosine and alanine were lower, in the glutelin fraction than in the prolamine. Tryptophan was not determined. Glutamic acid, valine, leucine and phenylalanine were higher in the ground endosperm meal than in the individual protein fractions, suggesting a residual protein fraction "which cannot be extracted from the endosperm meal".

Doggett (personal communication, 1977) stated the possibility, in some cases, of a genotype × environment interaction. Some results indicated that during the Indian "rabi" season protein levels and prolamine content may be relatively low with lysine in protein comparatively high. At other seasons the same genotype may yield higher protein and prolamine, with consequent lower lysine.

Gupta and Gupta (1974) analysed the crude protein (Kjeldahl N × 6·25); lysine and leucine (microbiologically), methionine and tryptophan (colorimetrically) of composite samples of the pot-raised sorghums CSH 1 and Swarna taken at the early-milky (11·5 weeks after germination or 1 week after flowering), mid-milky (13·5 weeks after germination or 3 weeks after flowering) and mature stages.

Crude protein content was not substantially different between varieties (CSH 1, 14·0% and Swarna 13·7% at the mature stage), nor during the grain formation. Deyoe *et al.* (1970) also found little difference in the crude protein content of mature and immature grain. As the grain matured, lysine and tryptophan contents decreased, leucine increased and methionine increased from the early-milky to the mid-milky stage, and at maturity was intermediate between the early-milky and mid-milky levels.

During the immature stage, the major protein fractions (Nagy *et al.*, 1941) were albumin and globulin; at the mature stage prolamine and glutelin predominated. Details are shown below:

	Protein fraction as % of total fractionated protein.				Extraction efficiency
	Albumin	Globulin	Prolamine	Glutelin	(%)
CSH 1					
early-milky	57·93	12·87	12·87	16·30	78·1
mid-milky	33·17	20·73	23·04	23·04	91·0
mature	13·72	15·27	36·23	34·76	78·1
Swarna					
early-milky	56·31	12·72	13·57	17·38	81·5
mid-milky	30·18	19·14	25·96	24·70	79·5
mature	14·15	11·32	33·02	41·51	77·4

Gupta and Gupta (1974) suggested that, nutritionally, protein of the immature grain was superior to that of mature grain.

Skoch *et al.* (1970) studied the solubilities of five samples of the 1966 sorghum crop grown in test plots at Kansas State University Agricultural Experiment Station using the successive extraction technique of Osborne and Mendel (1914), Nagy *et al.* (1941), and Jimenez (1966). Asgrow's Rica, RS 625, and TE 66 (refered to as MO 6, MO 81 and MO 88 respectively) were grown in Riley County, and Asgrow's Rica and TE 66 (referred to as CO 6, CO 88) were grown in Colby County, under similar conditions of soil and fertility. Comparisons were made with Opaque-2 maize flour grown in 1967 in Northeast Kansas. The results appear in Table 2.26.

Only 10·6% of the maize protein remained insoluble, whereas the insoluble fractions of the sorghums ranged from 52·8 (MO 6) to 62·6% (CO 88). Total protein recovered (including that in the insoluble material) ranged from 84 to 98·6% in the sorghum samples, compared with 96·7% recovered from Opaque-2 maize. The average amino acid composition (auto-analyser following acid hydrolysis, tryptophan not determined) of the whole kernel and the solubility fractions is presented in Table 2.27. The values for sorghum were in agreement with those of Deyoe and Shellenberger (1965) and Waggle and Deyoe (1966).

The proline content of the albumin fraction increased and the leucine content decreased, with increased protein content. In the insoluble protein, lysine and glycine decreased and leucine and glutamic acid increased, with increased protein content. Lysine, arginine and glycine in the albumin and globulin fractions were nearly double those in the whole grain; the prolamine contained less lysine than the whole grain protein. Lysine, histidine, arginine, threonine, serine and glycine were several times higher in the glutelin than in the prolamine. Differences in the amino acid compositions of the insoluble protein followed the differences in the whole grain protein, regardless of % of protein solubilized.

The recovery of each amino acid was calculated by adding the weight (g) recovered from each fraction and comparing the total weight with the level of that amino acid in the whole grain (see Table 2.28). With the exception of lysine, which was significantly lower, recoveries were similar to those reported by Jimenez (1966).

Haikerwal and Mathieson (1971a) reported the extraction and fractionation of the protein of kernels of *Sorghum bicolor*, CV FF 5683, Guinea race, grown at the Institute for Agricultural Research, Zaria, Nigeria using various procedures. The method of Osborne (1924) applied both as separate, and stepwise, extractions indicated that sorghum protein consisted of albumins (aqueous extract) 8%; globulins (saline extract); 20% prolamines (alcohol extract) 30%; glutelins (soluble in 0·1% NaOH) 12%; and glutelins (bound or bonded, gelled by 0·2% NaOH) 30%.

Extraction with 8 M urea solubilized 60% of the protein, whereas 8 M urea plus 0·1 M sodium dodecyl sulphate extracted almost 100%. The stepwise method of Coates and Simmonds (1961) recommended for wheat flour extracted only 12·17% of sorghum protein. The stepwise procedure of Jones *et al.* (1959) for wheat gluten extracted only 0·08% of the total sorghum protein. The method of Mertz and Bressani (1957) developed for maize protein extracted only 13·7% of the total protein, 74·6% was not extracted and 11·7% was lost during dialysis.

Sephadex gel (G 100) filtration showed that the normal solvent extracts contained three major fractions: (1) low molecular weight, dialysable fraction; (2) high molecular weight proteins eluted around the void volume and (3) an intermediate fraction. The 8 M urea extracts yielded three main peaks (approximate molecular weights: 82 000; 46 000; and 2000 respectively) and six minor peaks. Disc electrophoresis of the 8 M urea extracts yielded three main bands and at least five minor ones, in agreement with the results from gel filtration. Disc electrophoresis of the albumin fraction showed no sharp band, probably because of too low a concentration to be detected; the globulin fraction showed one broad band; the prolamine fraction showed a single band near the origin; the glutelin fraction showed two unequal bands. These results agreed in general with those from gel filtration if it is assumed that the lowest molecular weight fractions do not show up in disc electrophoresis.

Ivancenko *et al.* (1972) reported that using a modification of Osborne's method, sorghum flour protein was fractionated into 42·9% prolamine, 23·2% glutelins, 2·95% albumins and 2·7% globulins. Seventeen amino acids were detected chromatographically in the flour and prolamine fraction. The fractions were separated electrophoretically with optimum separation at the highest voltages. Glutelin separation was the most difficult.

Schröder (1973) studied the extraction of N from Barnard Red sorghum, a variety used for making sorghum beer, (1) under conditions similar to those found in breweries and (2) under controlled laboratory conditions.

Little of the N present (only 0·65%) in sorghum of crude protein 12·7% (Daiber, 1970) was found in the beer, while the wastes contained a considerable amount, crude protein 28·9% (Novellie, 1957, unpublished results). Schröder (1973) suggested that the low N level in the beer might partly result from the low solubility of the N components. Of the N present, only 11·58% dissolved in water, 3·99% was nonprotein N and 7·59% water-soluble protein (albumins). The addition of salt increased the amount extracted to 16·03% (albumins plus globulins). The acid-soluble proteins constituted less than 3% of total N. Under controlled laboratory conditions, ethyl alcohol dissolved 18·7% and alkali (0·3 M KOH) 48·4% of the N. Because the nonprotein N extracted by alkali was so high, 20·3%, it was concluded that the alkali partially hydrolysed the protein. Urea extracted the same amount of protein as did water, and the level extracted was not increased by either dimethylsulphoxide or Tanif (1-fluoro-2-nitro-4-trimethylamino-benzene iodide). Sodium sulphite extracted only 17·5% of total N but in conjunction with sodium dodecyl sulphate (SDS), 80·1% was extracted: SDS with mercaptoethanol (ME) extracted 78·0% of total N. Repeated extractions by the various solvents produced little increase in solubles. When the SDS–ME extraction followed an aqueous extraction (water, salt, acid or alcoholic solution), the total yield for the two procedures did not approach that of a single SDS–ME extraction. When the SDS-ME extraction was performed first, the expected amount was solubilized but subsequent extractions with water, salt, acid, alcohol or alkali solubilized little additional N.

Schröder (1973) suggested that the solubilization could be affected by the presence of protein-binding components such as polyphenols. If the polyphenols and proteins were "compartmentalized" in the grain, then on the addition of water the two might react to form an insoluble complex. Schröder stated, however, that the concentration of polyphenols in the Barnard Red variety was comparatively low, about 0·3%, though the method of determination was not described.

Landry and Moureaux Fractions

The problems associated with the classical Osborne procedure when applied to the fractionation of the proteins of sorghum, led researchers at Purdue University, Lafayette, Indiana (Jambunathan and Mertz, 1973a) to employ the procedure of Landry and Moureaux (1970) using a sequence that had been applied to maize, to yield five fractions:

- I albumins and globulins (soluble in saline solution)
- II prolamines (soluble in 70% isopropanol)
- III glutelins (soluble in isopropanol + 2-mercaptoethanol)
- IV borate buffer + 2-mercaptoethanol
- V borate buffer + sodium lauryl sulphate + 2-mercaptoethanol

High-lysine Sorghum

Jambunathan and Mertz (1973b) and Jambunathan *et al.* (1975) reported the protein solubility fractions of normal sorghum (NL), defined as having rounded kernels with some corneous endosperm and low lysine content, and high-lysine (HL) lines with a floury and dented endosperm. The protein of ground samples of Redlan (NL), a cross Redlan × high-lysine IS 11758, IS 11758 (HL) and IS 11167 (HL) were fractionated (Landry and Moureaux, 1970; Jambunathan and Mertz, 1973a). The differences in protein distribution and lysine level (ion-exchange chromatography) are shown in Table 2.29.

The percentage of soluble N was higher in the fractions I (albumins, globulins and free amino acids) and V, and lower in fractions II (the prolamine kafirin) and III (alcohol-soluble glutelin) of HL sorghum compared to NL. The HL kernels were 0–36% higher in protein and 56–79% higher in lysine (mg/g total N) than NL kernels.

Fraction I was highest in lysine (Jambunathan and Mertz, unpublished data.) Fractions II and III together averaged 52·9% in NL and 33·9% in the HL sorghum, and were lowest in lysine. Fraction IV, N soluble in borate buffer and mercaptoethanol, was similar in NL and HL. Fraction V, N soluble in borate buffer, mercaptoethanol and sodium lauryl sulphate, was higher in HL than in NL.

The distribution patterns of the two HL lines, IS 11758 and IS 11167 were similar to those in the high-lysine F_2 kernels obtained from the Redlan × IS 11758 cross.

Studies on the ratio of normal to high-lysine kernels on segregating heads (Singh and Axtell, 1973a) indicated that the high-lysine character was simply inherited and could be transmitted by standard plant breeding procedures.

High-pigment Sorghum

Jambunathan *et al.* (1972) reported the solubility characteristics of two sorghums, a low-pigment IS 0062 and a high-pigment IS 2283. Percentage solubilities in five solvents similar to those used by Jambunathan and Mertz (1973a) are shown in Table 2.30 with the % of lysine soluble in the various fractions. More than 87% of the total N was solubilized.

Electrophoretic Patterns

Jones and Beckwith (1970) when studying the proximate composition and protein of three grain sorghums, OK 612, RS 626 and TE 77, and of their dry-mill fractions (for details see Chapter 5, Table 5.44), had reported that the prolamine of sorghum formed gels at low protein concentrations in a variety of solvents. Beckwith and Jones (1972) extracted the prolamine from grain sorghum flour following the methods of their earlier work (Jones and Beckwith, 1970) using mainly the hybrid TE 77. The prolamine fraction was dispersed in 95% ethanol and cooled to 9–10°C. About 55% of the prolamine N remained soluble. Using gel electrophoresis the fastest migrating component of the prolamine was isolated from a 0·5% w/v solution of decolorized prolamine in 60% v/v *t*-butanol–water containing 1·5 M GHCl to prevent gelling. The amino acid content of the TE 77 prolamine, the 95% ethanol fraction and the fast migrating component are shown in Table 2.31.

There was little difference in amino acid composition between the whole prolamine and the 95% ethanol fraction but the single component contained relatively large amounts of arginine, cystine and methionine but no detectable lysine or histidine.

Beckwith (1972) examined the electrophoretic patterns of glutelins from three sorghum hybrids, RS 626, OK 612 and TE 77. The insoluble glutelin proteins were isolated by a described method involving enzymatic degradation of starch (Paulis *et al.*, 1969). Beckwith reported the following findings: the glutelin isolates obtained after a single amylase treatment contained between 12 and 14% (Kjeldahl) N corresponding to between 75–85% protein (N × 6·25); the products were distinctly coloured suggesting the presence of protein pigment complexes; they probably also contained nonstarch carbohydrate impurities. The average amino acid composition of residues from the three sorghum hybrids are given in

Jones and Beckwith (1970). Table 2.32 presents the average amounts of selected amino acids found in glutelin and prolamine, accounting for more than 80% of the total proteins in sorghum flours. The glutelin contained appreciably more of the basic amino acids than the prolamine and as with other cereal proteins the glutamic acid and leucine contents of the glutelin were comparatively high (Paulis *et al.*, 1969; Wu and Dimler, 1963). Glutelin fractions from the three hybrids were not soluble in aqueous media, in 8 M urea solutions nor in dimethyl sulphoxide–water systems.

About 23% of the glutelin N could be dissolved in neutral, or weakly acidic 6 M guanidine hydrochloride (GHCl). Beckwith stated: "At present it is not known what contribution protein-pigment complexes make to the poor solubility of these fractions".

Reduction of disulphide bonds and alkylation of liberated sulphydryls in the glutenin increased N solubility in neutral 6 M GHCl to between 85 and 89%. The reduced proteins soluble in 6 M GHCl were also initially soluble in 8 M urea solutions but gelled on standing. Gel electrophoresis (Jones and Beckwith, 1970) whether of the native glutelin or the reduced glutelin, gave similar patterns for all three hybrids. All glutelin preparations contained material remaining at the gel origin after disulphide bond reduction.

Glutelin protein fractions of grain sorghum probably consisted of smaller protein units joined together through disulphide linkages. Sorghum might contain some substances which could become irreversibly bound to the protein. All fractions isolated by column chromatography (Sephadex G-150) showed a certain amount of discoloration when dissolved in urea solvents. Analyses of the amino acids of four column fractions (not presented in the paper) showed significant amounts of unknown acidic ninhydrin positive compounds. The nature of the complexes and means of preventing them are unknown.

Shechter and de Wet (1975) reported the comparative disc electrophoresis of acidic proteins, basic proteins and isozymes of esterase, malate dehydrogenase (MDH) and peroxidase performed with aqueous extracts of seeds from seven cultivars belonging to five races of *Sorghum bicolor* ssp. *bicolor*. The races, with the accession number of the cultivar, and the geographic origin are given below:

Race	Accession No.	Geographic origin of accession
bicolor	1616	35 miles east of Zinder, Niger
bicolor	1020	Kano, Nigeria
caudautum	1539	El-Obeid, Sudan
durra	769d	Sibol village, between Tiel and Gassane, Senegal
guinea	818c	17 miles south of Ilorin, Nigeria
kafir	2305	Maseru, Lesotho
kafir	1953	22 miles northwest of Tsumeh, South West Africa

Each cultivar yielded reproducible, characteristic patterns of distinct acidic and basic proteins. Cultivars belonging to the same race produced identical protein and isozyme patterns. Bicolor, caudatum, durra and guinea produced very similar acidic and basic protein patterns, and esterase, MDH and peroxidase isozyme patterns, though differences were observed among all races. All kafir patterns were significantly different from the patterns of other races. Shechter and de Wet, whose methods were fully described, suggested that comparative electrophoresis may provide a new source of taxonomic characters for investigating phenetic and phylogenetic relationships in sorghum. Later, the same authors confirmed and somewhat extended their earlier findings, commenting, however, that since the alcohol-soluble proteins of sorghum are relatively insoluble in water, these proteins also should be examined for interracial variability.

Zubaidov (1968a) reported that the electrophoretic patterns of the albumin (water-soluble protein) fractions from eight sorghum varieties were similar. At least five components were observed, two major components migrated to the positive pole and three minor components to the negative pole. However, the globulin fractions extracted by M NaCl showed differences among varieties (Zubaidov, 1968b) which were greater than differences between harvest years.

Klimenko and Zubaidov (1971) reported that glutelins extracted from sorghum seeds with 12% sodium salicylate separated into various fractions when chromatographed and during paper electrophoresis, and were not therefore homogeneous. Some of the fractions contained nucleic acids.

The tendency of sorghum prolamine solutions to gel (see, e.g. Jones and Beckwith, 1970, Haikerwal and Mathieson, 1971a) suggested to Wu *et al.* (1971) that a unique conformation might be present in sorghum prolamine. They therefore examined the optical rotatory dispersion (ORD) (Cary 60 recording spectropolarimeter), circular dichroism (CD) (CD accessory to Cary 60), and infrared spectra (Perkins-Elmer 621 grating infrared spectrophotometer) of the prolamines of four grain sorghum hybrids: OK 612, heteroyellow endosperm; RS 626, TE 77, described as "typical hybrids with considerable color", Funk G 766, white endosperm. The prolamines were extracted from defatted sorghum flour by 60% *t*-butanol after salt- and water-soluble proteins had been removed by the method of Jones and Beckwith (1970).

The extracts of OK 612 and Funk G 766 were slightly yellow and of RS 626 and TE 77, pink. Some of the coloured solutions were decolorized by activated carbon before use. The infrared absorption spectrum of prolamine in 60% *t*-butanol and D_2O showed the presence of α-helix and unordered structure, and the absence of β-structure. ORD data of the prolamines in 60% *t*-butanol gave α-helix content of 40–47% independent of hybrids and of colour of the prolamine solution. The α-helix content of the prolamines in 60% *t*-butanol + 1·5 M guanidine hydrochloride (G-HC1) was lowered to 34–40%, but was greatly reduced in 6M G-HC1. The CD and far ultraviolet ORD curves of decolorized prolamine showed α-helix content in agreement with that from ORD data. The high level of α-helix for sorghum prolamine supported the concept that hydrogen bonds between backbone polypeptide chains are protected from aqueous or other polar residues by a nonpolar environment resulting from the side chains.

Protein Bodies of Sorghum, Pearl Millet and Finger Millet

Adams *et al.* (1976) isolated the protein bodies of sorghum, pearl millet and finger millet by a described method of disrupting the starch–protein complex of mature seeds of sorghum, pearl millet and finger millet. The grain was laboratory milled, passed through a 1 mm sieve and defatted in acetone. The air dried acetone powder was suspended in 1% (w/v) sodium metabisulphite solution (1 part powder/5 parts solution) and stirred. The slurry was filtered and the filtrate centrifuged to separate starch and protein bodies. The protein bodies were dehydrated in acetone before chemical analysis, or resuspended in 0·1M citrate buffer at pH 2, before determination of enzyme activities or examination under an electron microscope. Under the light microscope very few starch granules could be seen within the protein bodies.

Total N and P (procedure of Thomas *et al.*, 1967), Ca, K and Mg (estimated by atomic absorption spectroscopy) in the whole grain and protein bodies (DWB) are shown in Table 2.33. The protein bodies did not appear to act as main repositories for the metals present in the grain. In all three grains the proportion of P was slightly higher in the protein bodies than in the whole grain.

Adams and Novellie (1975) cited by Adams *et al.* (1976) reported that sorghum protein bodies contained protease, phytase, α-glucosidase, β-glucosidase, β-galactosidase and several acid phosphatases. Almost all the protein present in the protein bodies isolated was insoluble in aqueous buffer. A higher enzymatic activity was found in the soluble fraction which suggested to Adams *et al.* either that the enzymes are more soluble than the nonenzyme protein, or the enzymes are present mainly as surface adherents of the protein bodies. The distribution of protease, phytase and protein in the soluble and insoluble fractions of sorghum, pearl millet and finger millet is shown in Table 2.34. Phytase activity was not detected in finger millet and was very low in sorghum and pearl millet.

Under the electron microscope the protein bodies of sorghum appeared mostly circular in section and varied in diameter from 0·5 to 3·5 μm. They displayed a discrete border possibly bounded by a membrane with an internal structure in which a darker-staining material was arranged in concentric rings. Pearl millet protein bodies ranged in cross-section diameter from about 1·5 to 4 μm with discrete edges possibly surrounded by a membrane. The internal structure showed numerous spots of darker-staining material arranged in concentric rings. They appeared to have invaginations and protruberances (not reported by Sullins and Rooney

(1977a)). The protein bodies of finger millet appeared as discrete organelles possibly surrounded by a membrane, varying in diameter from less than 1 μm to about 2 μm. In section they appeared to be relatively homogeneous but with a more darkly stained central area. In this respect, and in size, they appeared to resemble common millet as reported by Jones *et al.* (1970).

Adams *et al.* (1976) suggested that their method of separating protein bodies might be useful in the technological production of protein-rich materials from cereal grains.*

According to Adams *et al.* (1975) maltase (α-D-glucoside glucohydrolase) activity has been detected in a variety of ungerminated cereals including barley, buckwheat, rice, sorghum and maize. Sorghum maltase is insoluble in water, but active in the insoluble state. This insoluble maltase has been detected in two sub-cellular lytic bodies prepared from sorghum grain and designated protein bodies and spherosomes. These lytic bodies contain a very active maltase and are probably the sub-cellular location of sorghum maltase. Furthermore, a starch hydrolysing activity is also associated with these sub-cellular organelles and the combined carbohydrase activities are possibly of significance in the early stages of germination. The research reported indicated that protein bodies and spherosomes from sorghum contained carbohydrase activity against maltose, starch and *p*-nitrophenyl-α-D-glucoside. Maltase activities in sorghum and also in maize lytic bodies were very high; carbohydrase activities of lytic bodies from whole wheat, whole barley, sorghum aleurone, sorghum embryo and maize embryo were considerably lower. The pH response of sorghum lytic bodies was bimodal with an optimum in the range of 3·4–4·2 and a minimum or a shoulder near pH 3.8. Protein bodies from sorghum, maize, wheat and barley reduced the iodine-colouring capacity of soluble starch to give a purple colour typical of a β-limit dextrin. With spherosomes colour reduction was usually more rapid, eventually taking the breakdown of starch beyond the achroic point (the stage at which the solution is no longer coloured by iodine). The lytic bodies produced both maltose and glucose from starch except in the case of maize when only glucose was found. The data suggested that protein bodies contained a linked β-amylase-maltase system and that spherosomes contain a linked α-amylase-maltase system.

Seckinger and Wolf (1973) by scanning electron and transmission microscopy, examined the protein of the sorghum varieties RS 626, OK 612, TE 77, G 766, C 42Y, TX 09 and Cody waxy, and of eight experimental lines grown by the Agronomy Department at Purdue University.

Concentrations of protein within the endosperm varied with texture. Corneous cells of the subaleurone layer appeared to contain mostly protein with little or no starch while floury endosperm cells were composed mostly of loose starch with a small percentage of protein. Normal corneous

* (Authors' note: though the method may well be technologically sound, it is doubtful if it would be economic or of practical utility to the typical rural poor who depend upon sorghum and millets.)

endosperm embodied cells tightly packed with starch granules held together within a protein network. The protein network consisted of protein granules embedded in a protein matrix. The protein granules were found to be soluble in aqueous alcohol, therefore high in the prolamine fraction, the surrounding matrix protein contained more of the glutelin fraction.

Protein granules contained a nucleus with solubility properties similar to those of matrix protein. A laminar structure was suggested for the protein granules. Matrix protein showed no signs of an internal structure. Protein bodies of almost all grain sorghums examined were 2–3 μm in diameter and made up the major part of sorghum endosperm protein. One experimental line with higher than average lysine content contained smaller protein granules than the lower lysine lines which was consistent with the reported negative correlation between prolamine and lysine.

Lipids

In order to update the United States Department of Agriculture's tables of food composition, Weihrauch *et al.* (1976) reviewed the relevant literature since 1960 and reported representative data on total fat and fatty acids (g/100 g sample) in various food grains and cereal products. The terms "fat", "lipid" and "oil" were used interchangeably.

Weihrauch *et al.* stated that in cereal products, the fatty acid profile changed with the degree of extraction of total lipid. Consequently, their methods used solvent extractions that removed both polar and nonpolar lipids. Total lipids and fatty acids in sorghum, pearl millet, maize, rice, triticale and wheat are given in Table 2.35. Lipid concentration was highest in the germ, lowest in the endosperm and intermediate in the outer layers of the kernel. Cooking had little effect on the fatty acids and Weihrauch *et al.* suggested that the fatty acid content of processed foods could be estimated from analyses of the raw ingredients.

Weihrauch *et al.* reported that over 90% of the data required conversion from % fatty acid methyl esters to absolute units. The conversion factors used are shown in Table 2.36.

Price and Parsons (1975) reported the lipid content of cereal feed grains grown in the north central US, including: South Dakota 106 sorghum, Pioneer 3764 3X maize, North Dakota 203 triticale and Polk wheat. The proximate analysis (AOAC, 1970), total lipid composition (Price and Parsons, 1974, Parsons and Price, 1974), fatty acid composition of the total lipid extract (Price and Parsons, 1974), and the fatty acid content of the total lipid extract (AOAC, 1975; method 28.034) are presented in Table 2.37.

Linoleic was present in highest concentration. Thin layer chromatography demonstrated that phosphatidyl choline, phosphatidyl ethanolamine and lysophosphatidyl choline were the dominant phospholipids.

Adrian and Jacquot (1964) stated that sorghum oil was semi-liquid at ordinary temperatures and had a semi-drying character similar to maize.

The oils of foxtail millet and common millet were also similar to maize but higher in acid value and hydroxyl value. The fatty acid composition of the oils extracted from sorghum, foxtail millet and common millet did not appear to differ greatly. About 80% of the extractable lipids consisted of unsaturated fatty acids, principally oleic and linoleic; among the 15% saturated, palmitic and stearic were dominant. Linolenic acid was significantly present in the millets but in only minor amounts in sorghum.

Baldwin and Sniegowski (1951) described the composition of the lipids obtained from wet-milling an unspecified type of grain sorghum. The grain was steeped in dilute aqueous sulphur dioxide (about 0·2%) for about 48 h at 130°F, which removed most of the water-solubles, then roughly ground to liberate the germ which was removed by flotation. The washed germ was dried and the oil recovered by expelling (pressing) and/or solvent extraction. The data presented were from analysis of the crude expelled germ oil.

The germ-free portion of the grain was reground, the fibre separated by screening, and the fat extracted with hexane and carbon tetrachloride. The starch–gluten slurry (approximately 90% starch, 10% gluten) was separated on inclined tables, and the fat extracted from the starch with methanol, hexane and carbon tetrachloride, and from the gluten with hexane and carbon tetrachloride. The lipids obtained from the germ, starch, gluten and fibre were characterized (American Oil Chemists Society, no date) as shown in Table 2.38 in comparison with equivalent fractions derived from maize.

The fatty acids in the fibre lipids (see Table 2.39) were analysed spectroscopically (American Oil Chemists Society, 1949). The polyunsaturated constituents of the lipids in the germ, starch and gluten were determined spectrophotometrically (see Table 2.40).

The lipid composition differed among fractions but was similar between comparable fractions of sorghum and maize. Embryo lipids were the most unsaturated, contained least free fatty acids and least unsaponifiable matter. Starch lipids contained 71·5% (maize) and 91% (sorghum) free fatty acids with a high proportion of palmitic acid. Sorghum gluten and fibre lipids contained about 18%, and maize gluten and fibre lipids about 36%, unsaponifiable matter. The free fatty acids in sorghum gluten and fibre were each about 20%; 22% in maize gluten and 13% in maize fibre.

Stemler *et al.* (1976) reported the lipid components in the grain of 51 samples of five cultivated sorghum races: bicolor, caudatum, kafir, durra and guinea, originating in Africa, obtained from the World Sorghum Collection, grown out at Isabela, Puerto Rico. Seventeen wild sorghum samples from the collection of the Crop Evolution Laboratory, Urbana, Illinois, grown out at Urbana were also examined. All were analysed for oil and protein using a Dickey–John (Technicon) IR analyser; the composition of fatty acids was determined from methyl esters of the oil (Singh *et al.*, 1968) followed by gas chromatography. The results are shown in Table 2.41.

No significant differences in oil content were found among the cultivated and wild sorghums but the wild sorghums were significantly higher in protein than the cultivated sorghums. Since the weight/seed was three to four times greater for cultivated sorghums than for wild sorghums, Stemler *et al.* suggested that selection for increased grain size in cultivated cereals has resulted in increased content of carbohydrate rather than protein.

Kafir, caudatum and wild sorghums contained significantly higher levels of oleic acid than bicolor, durra and guinea, these latter three being significantly higher in linoleic acid than wild sorghums and kafir which in turn were higher than caudatum. Linoleic acid and oleic acid were negatively correlated. Stemler *et al.* found no significant correlation between linoleic acid content and (a) annual rainfall and (b) maximum temperatures during the growing season at the African locations where the wild and cultivated sorghums were originally collected.

Paul *et al.* (1972) reported the percentage composition of the free and bound neutral lipids of two improved strains of sorghum JS/73/53 and JS/2/53 from the Plant Breeding Department of the Punjab Agricultural University, Ludhiana, using the methods of Pruthi and Bhatia (1970). The results are presented in Table 2.42.

Bertoni *et al.* (1963) studied the physical and chemical characteristics, total tocopherol content and fatty acid composition of the germ oil from 11 varieties of sorghum cultivated in 1959–60, in the Argentine provinces of Buenos Aires, Cordoba, Santa Fe, La Pampa and Chaco. The germ was separated by steeping in water, disc-milling flotation and air drying. The yield of germ ranged between 2·9 and 4·2% of the grain, according to variety.

The ranges for yield of oil obtained by Soxhlet extraction using technical hexane and ethyl ether is shown in Table 2.43, together with the physico-chemical characteristics, the tocopherols expressed as α-tocopherol (Bertoni and Cattáneo, 1960), plus unsaponifiable matter and fatty acids determined spectrophotometrically (American Oil Chemists Society, 1960). The oil content of the germ ranged between 18·5 and 30·8% with most falling between 20 and 25%.

The fatty acid composition ranged in total saturated from 14·7 (Feterita Sel, Pergamino at Buenos Aires) to 23·2 (Martin Milo at Buenos Aires); in oleic from 13·8 (Martin Milo at Buenos Aires) to 37·3 (Feterita Sel, Pergamino at Buenos Aires); in linoleic from 46·2 (Feterita Sel, Pergamino at Buenos Aires) to 59·5 (Martin Milo at Buenos Aires) and in linolenic acid from 0·8 (Plainsman at Buenos Aires) to 3·0 (Martin at Cordoba).

Table 2.44 shows the range of proximate analyses of the residues remaining after hexane extraction. No reducing or invert sugars were found. The highest ash of 13·5% was from an unnamed sorghum variety grown in Cordoba, and the lowest 1·9% from Napalpi grown in Chaco. Most of the ash contents lay between 5·4 and 8·8%.

Kummerow (1946a) reported the composition of Western Blackhull sorghum from Fort Hays Experimental Station, Kansas, as containing 0·5% wax, extracted by Skellysolve B* and 2·5% oil, extracted by Skellysolve F.†

* Skellysolve B, essentially n-hexane, b.p. 60–68°.

† Skellysolve F, petroleum ether, b.p. 30–60°. Skelly Oil Co., Tulsa, Oklahoma, USA.

The composition of the wax, compared to carnauba wax is shown below:

	Carnauba wax	Sorghum wax
Non-saponifiable matter	54–55%	52%
Acid value	4–7	10·8
Melting points of:		
Wax	85°C	81°C
Fatty acids of wax	78–90°	78–82°
Hydrocarbon of wax	58–59°	74–79°
Alcohols of wax	79–87°	82–86°
Acetate derivative of alcohols	64–72°	65–69°

Sorghum grain oil was found to be similar to maize oil. The sorghum oil characteristics (AOAC, 1940) were:

acid value	3·14
saponification value	181
iodine value (Wijs)	119·0
thiocyanogen value	76·7
acetyl value	16·7
refractive index at 25°C	1·4718
unsaponifiable matter (%)	1·88

The composition of the fatty acids, from low temperature fractional distillation, is shown below compared to maize oil:

	Sorghum grain oil weight %	Maize grain oil weight %
Myristic	0·2	0·5
Palmitic	8·3	9·7
Stearic	5·8	3·6
Hexadecenoic	0·1	0·2
Oleic	36·2	30·4
Linoleic	49·4	55·6

Kummerow (1946b) subjected 10 grain sorghum varieties, all grown during the same season under identical conditions, to fractional solvent distillation. The wax was extracted with Skellysolve B, and the oil removed from the germ and endosperm with Skellysolve F.

The mean and ranges (in parentheses) for the wax and oil extracted from the different varieties, together with characters determined are shown in Table 2.45.

The melting point, neutralization equivalent, iodine and thiocyanogen values of the mixed fatty acids from the various oils were determined by methods of the AOAC (1940), the linoleic and oleic acid calculated from spectrophotometric analysis (Mitchell *et al.*, (1943) and from the iodine and thiocyanogen values (Riemenschneider *et al.*, 1941). The means and ranges are also shown in Table 2.45.

Cody and Feterita contained approximately the same amount of linoleic and oleic acids; the other varieties contained approximately 5% more linoleic acid than oleic. The two methods for determining linoleic and oleic acid agreed well and only the iodine and thiocyanogen values were presented.

The sorghum oils varied from light amber to green in colour. The induction period of the sorghum grain oils varied from 74 to 87 h compared with 70 h for the waxy maize oil and 53 h for a commercial maize oil, indicating that sorghum oils would be at least as stable as maize oil.*

The lipids extracted with various solvents from ground, waxfree, Western Blackhull Kafir, are shown in Table 2.46. Skellysolve F yielded the smallest (2·28%) and tetrachloroethane the largest (7·15%) percentage of lipids.

Rooney (1968) reported that the grain of 10 unnamed sorghum varieties, representing most (US) commercial grain sorghum hybrids, contained from 2·66 to 3·37% petroleum ether extractable lipids (free lipids). Extraction of the residue with water-saturated *n*-butanol yielded a further 0·13–0·28% lipids (bound lipids). The free lipids were fractionated by thin-layer chromatography and the relative proportions of hydrocarbons, triglycerides, free fatty acids, mono- and diglycerides, and polar compounds determined. The bound lipids were also fractionated by thin-layer chromatography and characterized against known compounds. The differences in lipid composition among varieties and among four parents with derived hybrids were comparatively small. Martin, Combine 7078 and Kafir 60, grown at the same location for three successive years, were similar in free and bound lipids though total lipids varied from year to year.

Wall and Blessin (1969) reported on the lipids present in (1) a commercial mixture of sorghum grain of medium hardness, (2) Combine 7078, a soft grain and (3) Martin, a hard grain. The nonpolar lipids in the three grains were separated by thin-layer chromatography. The major fraction consisted of triglycerides, with lesser amounts of hydrocarbons, sterol esters, fatty acids, diglycerides, sterols and phospholipids.

Sorghum Wax

Adrian and Jacquot (1964) reported that sorghum contained between 0·2 and 0·3% of wax extractable with petroleum ether.

Dalton and Mitchell (1959) extracted the wax of a Midland variety of sorghum with Skellysolve B and precipitated the wax with acetone. The purpose was to determine if sorghum wax could usefully replace carnauba wax. The crude wax was fractionated by adsorption on columns of tricalcium phosphate and silicic acid. X-ray diffraction studies indicated that each fraction was a mixture of homologues, rather than a single compound.

* (Authors' note: a long induction period indicates a low susceptibility to oxidative rancidity.)

The composition of the two waxes, which appeared significantly different, were as presented below:

	Sorghum wax (%)	Carnauba wax (%)
Paraffins	5	1
Esters	49	80
Free alcohols	46	12
Lactones	—	3
Resins	—	4

The results of Kummerow (1946a) are discussed on p. 96.

Carbohydrates

Francis and Smith (1916) reported the physical and chemical characteristics of the starches of the sorghums White Kafir, Red Kafir, Pink Kafir, White Milo, Yellow Milo, Brown Kaoliang, White Kaoliang, Feterita and Darso, in comparison with the starch of maize, sweet potato. Irish potato, arrowroot, Navy bean and wheat.

Feterita, the milos and kafir contained about 64% starch (Official Methods of Analysis of the US Bureau of Chemistry, Bulletin 107). The average size of the starch granules varied between 16–20 μm. The gelatinizing temperatures (thermoslide) varied from 64·7°C for Pink Kafir to 78°C for White Kaoliang but in most types was about 74°C, slightly higher than maize starch (71°C). Francis and Smith found no striking differences in appearance among the various sorghum starches.

MacMasters and Hilbert (1944) reported that the waxy character in sorghum was recessive and where growing conditions favoured cross-pollination, waxy cereals tended to revert to the nonwaxy type. MacMasters and Hilbert stated that two main waxy sorghum varieties were grown in the United States: Schrock (with a selected strain, Sagrain) and Leoti, the latter mainly for fodder. Samples of Sagrain and Leoti, with other waxy sorghum types, Kafir, Milo, Atlas and Club were compared with nonwaxy Blackhull Kafir, Milo and Pink Kafir grown over four years in various regions of the United States. The waxy types contained less starch, but more oil and protein than the nonwaxy.

In both Sagrain and Leoti, the pigment in the seed coat was water-soluble and strongly adsorbed by starch granules during starch separation in the laboratory. The colour of the starch from Sagrain was pale blue-grey, and from Leoti light tan. If most of the seed coat was removed by pearling before processing, the starch colour was much fainter. Sorghums in which the "nucellar layer" (testa) was resorbed before maturity were without pigment and yielded colourless starches but white sorghums, with a coloured testa produced starches with an "off-colour".

The waxy starches contained no amylose and only about half as much phosphorus as the nonwaxy except for Leoti which was high in ash and phosphorus, possibly owing to contamination from the closely adherent glumes.

The waxy starch pastes were translucent, flavourless, "long" and tacky with low rigidity and high viscosity compared to cassava starch. The nonwaxy pastes were opaque, with a characteristic flavour, and were "short" with high rigidity and low viscosity. Gelatinization temperature of the two types was similar. On standing, nonwaxy pastes or gels increased in opacity; waxy pastes retrograded slowly, if at all.

In 1946, Horan and Heider determined the protein (AOAC, 1940), crude fat (extraction with Skellysolve* F95) and starch (polarimetric, Earle and Milner, 1944) of 73 samples of nonwaxy sorghum and 28 samples of waxy grain and forage types, grown in Kansas, Nebraska, Oklahoma and Texas in 1944. One nonwaxy grain sorghum sample from Texas, Waxy Club × Bonita, SA 5590 4, had a high starch content of 81·8% and a low protein of 7·7% and one nonwaxy forage sample, Black Amber from Nebraska, had an exceptionally low starch content of 48.0%, but apart from these the range was relatively narrow, 61·3–78·5%. Apart from the above exceptions, the means and ranges for protein, fat and starch in the nonwaxy and waxy were practically identical. Using the method of Bates *et al.* (1943), the nonwaxy starches were found to contain 25–31% amylose.

Deatherage *et al.* (1955) reported the amylose content of a large number of starch bearing plants studied over 10 years at the Northern Utilization Research Branch, USDA. In the cereal seeds the pericarp and germ were separated from the endosperm, and the latter ground. The fat was extracted with 85% methanol before determination of amylose by potentiometric titration (Hilbert and MacMasters, 1946). The estimated 3% of fibre in the ground endosperm was neglected. The amylose content of each of five waxy varieties of US sorghum was 1.0%; the mean of 25 nonwaxy varieties was 24% ranging from 21 to 28%. From outside the US, only one waxy sorghum variety was found from Turkey, with amylose content of 2.0%. The remaining 179 samples, mainly from India and Turkey averaged 24% amylose with a range of 21–28%.

Waniska (1976) reported the amylose content (Williams *et al.*, 1970), range and mean, in samples of the flour of waxy, heterowaxy, and nonwaxy endosperm type sorghum as shown below:

	Endosperm type		
	Waxy	Heterowaxy	Nonwaxy
No. of samples	4	2	17
Amylose	2·2	12·3	16·4
(% DWB)	(1·6–2·7)	(11·2–13·4)	(13·5–18·6)

Williams (personal communication, 1978) reported the protein (Kjeldahl

* Skelly Oil Co., Tulsa, Oklahoma, USA.

N × 6·25) and amylose content in the sample of 100 sorghum lines of different types from ICRISAT Hyderabad. The results, mean and ranges, are shown in Table 2.47.

In the US, waxy endosperm sorghums high in amylopectin were specially bred for the production of waxy starch by wet-milling. There is now little commercial production of sorghum starch but several commercial sorghum hybrids contain waxy genes in their pedigree and there is evidence to suggest that some possess a superior feed value (Sullins and Rooney, 1974).

Davis and Harbers (1974) reported the endosperm structure of US grain sorghum of three endosperm types (1) CP 622 waxy, (2) C 42Y yellow and (3) ACCO 1023 bird-resistant, observed by scanning electron microscopy (SEM) (Hoseney *et al.*, 1974). Starch granules in the soft endosperm of (2) and (3) were smooth and spherical, those in (1) irregular in shape.

Harbers and Davis (1974a) reported that split kernels of three sorghum grain types taken from three waxy, four bird-resistant, and nine white, yellow, red or bronze (US) varieties were given to fasting rats from 1 to 36 hours before slaughter. The specific varieties of sorghum were not named. Kernels were collected from the stomach, jejunum, ileum, caecum and large intestine, and faecal samples were collected from rats in metabolism cages.

SEM revealed that little hydrolysis occurred in the stomach. Waxy sorghums showed point hydrolysis with no resistant surfaces and internal alternate layer erosion. Bird-resistant sorghums showed resistant surfaces with circular erosion at certain sites and combined alternate layer and continuous erosion internally. All other sorghums showed hydrolysis at selected surface sites with variable selected hydrolysis of starch layers. Cell walls surrounding starch granules impeded amylolysis but walls disrupted during grinding and offered no subsequent barrier to hydrolysis. Hydrolysis patterns in pigs were similar to those in rats.

Harbers and Davis (1974b) reported the observations by SEM of starch granule and endosperm cell wall digestion of split sorghum grain kernels, waxy CP 622, or yellow C42Y, recovered from the digestive tracts of rats and pigs. Little starch damage was found in granules recovered from the stomachs of rats. In samples recovered from the jejunum of the rat and the pig, most of the starch granules were removed from the surface of kernels and the cell walls were exposed. The radial erosion patterns of C42Y were similar to those obtained when the same variety was exposed to hydrolysis by rumen microorganisms and purified porcine α-amylase (Davis and Harbers, 1974). Point hydrolysis was observed in the waxy CP 622.

Near the end of the small intestine most of the starch granules had been removed from several layers of endosperm cells. The cell walls were not removed but seemed to shrink and flatten. The underlying starch seemed to have been hydrolysed even where there was no evidence of cell wall cleavage. It appeared therefore that digestive amylases had diffused through at least one cell wall to attack the underlying starch granules.

Kernels from the caecum showed evidence of little additional action. The pattern of starch granule hydrolysis from rat faecal samples did not much differ from hydrolysis in the large and small intestine. Hydrolysis of starch granules and changes in cell walls surrounding the starch granules were essentially similar in the small intestines of the rat and the pig but erosion as observed by SEM differed between waxy and yellow endosperm sorghums. Most of the pericarp remained unhydrolysed.

Harbers (1974) reported that split kernels of sorghum grain and forage sorghum starch granules hydrolysed in ruminally fistulated steers for 20–24 hours and then subjected to SEM showed three distinct patterns. Those with a testa layer showed extensive surface erosion with endocorrosion in certain random sites. All other types were attacked at susceptible sites except for waxy sorghum grain types the pattern in which was as described by Davis and Harbers (1974).

Under similar conditions, maize, "millet", and B-type barley, rye and wheat starches were hydrolysed at specific vulnerable surface sites by circular erosion patterns.

Split kernels of grain sorghum of three endosperm types (1) CP 622 waxy, (2) C42Y yellow and (3) ACCO 1023 bird-resistant were subjected to either rumen digestion for between 75 min and 8 h or hydrolysis by purified porcine α-amylase for 48 h and the results of the attack observed by SEM (Davis and Harbers, 1974).

After immersion in rumen fluid for as short a period as 75 min there was point-surface attack on the starch from (1), linear track hydrolysis on the starch from (2) and minor point hydrolysis on that from (3).

Split kernels suspended inside ruminally fistulated steers showed different modes of starch attack. Surface attack on (1) waxy starch was mainly point hydrolysis with alternate layers digested inside the starch granule. Starch in (2) was digested without surface attack. The bird-resistant type (3) showed a mixture of both types of hydrolysis on the surface with preferential layer digestion internally.

Starch from each variety, isolated by a simple wet-milling method (Goodwin, 1959) and purified by flotation and centrifugation was hydrolysed by porcine α-amylase at similar rates when the substrate was equal to or less than 1 mg starch/IU enzyme. At higher substrate: enzyme ratios, waxy sorghum starch (1) released more maltose equivalents than (2) and bird-resistant starch (3) was depressed, characteristic of substrate inhibition. This differentiation could be readily detected by SEM.

The mode of starch hydrolysis by rumen microorganisms and purified porcine α-amylase was similar for each hybrid.

Harbers (1975) examined by SEM structural changes in the starch granules of samples of mill run sorghums (1) steam flaked, (2) micronized and (3) popped, and the effect of amylolysis of that starch by purified porcine pancreatic α-amylase and rumen microflora obtained from fistulated steers. (For descriptions of these three processes see Chapter 5.)

Steam flaking either left starch granules intact or produced homogeneous, gelatinized clumps of several granules. Protein bodies remained intact but adhered to one another. Matrix protein stretched between starch granules. Rolling the steam flaked grain pressed ruptured starch granules together and crushed the starch into small nondescript pieces.

Popped starch granules expanded from the hilum to the edge and formed large, thin sheets that fused with adjacent granules. Non-expanded starch granules did not show the starch deterioration characteristics of granules produced by steam flaking. Protein bodies remained intact. Matrix protein seemed to disappear among the starch sheets.

Micronizing expanded most of the starch granules, forming sheets. Starch granules near the surface remained intact resembling those obtained from popping. Protein bodies appeared to remain intact; some seemed to have disappeared into the starch granules. Details of protein structure were virtually impossible to determine unless the granules were isolated, because starch readily expands under electron beams at high magnifications and produces distorted images.

Amylolysis, both from purified α-amylase and from rumen contents, occurred mainly along the edges of ruptured granules of steam flaked or micronized sorghum. In popped starch, amylolysis began on exposed edges of the starch sheets. After processing, sorghum grain starch became more vulnerable than intact granules to amylolysis.

Hoseney *et al.* (1974) examined by SEM the structure of seven sorghum varieties from the Sudan: (i) Dwarf White Milo, (ii) Feterita Gezera, (iii) Gassabi, (iv) Dabar, (v) Feterita Gedarif, (vi) Feterita Abu Diraira and (vii) Red Mugud. These he compared with four US-grown varieties and hybrids: (viii) a bulk red-seeded commercial sample, (ix) a white waxy sample (CP 622), (x) a brown-seeded, bird-resistant sample ACCO 1023 and (xi) a yellow-seeded sample (DeKalb C42Y). Only the bird-resistant ACCO 1023 and the three feterita samples contained a prominent testa layer.

The soft, opaque (floury) endosperm contained relatively large, inter-granular air spaces, round starch granules covered with a thin sheet of protein. Embedded in the protein sheet were relatively large spherical protein bodies. The hard, translucent (corneous) endosperm displayed a tightly packed structure with no air spaces. The starch granules were polygonal and covered with a thin protein matrix. Embedded in this protein matrix were protein bodies composed largely of kafirin, comparatively low in lysine. There appeared to exist a strong bond of adhesion between the starch granules and the surrounding protein in that, when fractured, breakage occurred more within the granules than at the starch-protein interface.

Mugud, red and large-seeded, contained a soft floury endosperm with protein bodies prominent in the outer layers near the aleurone cells. The starch was not polygonal and the protein bodies made practically no indentation in the starch. Gassabi displayed numerous protein bodies in a predominantly floury endosperm. Few protein bodies were found in the endosperm of a dwarf type from the Sudan (not otherwise named); there were numerous very small protein bodies in C42Y (USA hybrid), and numerous large bodies in the bulk USA sample.

The amino acid analysis (determined by auto-analyser after acid hydrolysis, tryptophan not included) for the US varieties and hybrids and the Sudan varieties are shown in Table 2.48. The Sudanese dwarf contained 188 mg lysine/g of total N and significantly less glutamic acid and leucine than the other samples. The amino acid values for the other samples were considered to be within normal variation.

Brethour and Duitsman (1965), Sherrod *et al.* (1969) and McCullough *et al.* (1972a) all reported that waxy sorghum gave higher feed efficiency than nonwaxy grain when fed to steers. To determine differences in endosperm texture which might be related to feeding value, Sullins and Rooney (1974) microscopically examined sorghums typified by a corneous endosperm (SC 301), a floury endosperm (NSA 740), and an intermediate endosperm—half corneous, half floury (Kafir 60). The starch of all three contained approximately 30% amylose and 70% amylopectin. A waxy-type endosperm—100% amylopectin (TX 615) was also examined. All lines were grown under comparable conditions at College Station, Texas in 1970. (See Fig 2.VIII.)

The 1000-kernel weight, test weight, hardness and density (Rooney and Sullins, 1970), moisture, protein, fat and ash (AACC, 1962), and starch determined by heat–gelatinization–glucoamylase hydrolysis (Norris, 1971) after extraction by the wet-milling procedure of Norris and Rooney (1970b) are shown in Table 2.49.

Microscopic examination of the waxy sorghum TX 615 showed it to contain the smallest proportion of peripheral endosperm. The waxy sections were also more rapidly solubilized than the nonwaxy by pronase and hog pancreas (α-amylase sources) and by buffered rumen fluid. NSA 704, Kafir 60 and SC 301 were nonwaxy and contained more peripheral endosperm than TX 615.

Sullins and Rooney (1975) reported upon microscopic and SEM studies of TX 3197 (nonwaxy endosperm kafir), its waxy endosperm counterpart TX 615, Texioca 63 and Texioca 54 (both waxy type kafirs), TX 378 (nonwaxy Redlan derivative) and its waxy counterpart SA 413. Except for the waxy genes, the grains of the waxy and nonwaxy counterparts were essentially similar. The 1000-kernel weight, test weight and hardness (Rooney and Sullins, 1970), protein, fat, and ash (AACC, 1962) are shown in Table 2.50.

Microscopic examination indicated that nonwaxy sorghum contained small starch granules embedded in a dense protein matrix located in the peripheral endosperm area. The waxy sorghums revealed a less dense peripheral endosperm containing larger starch granules with considerably less protein than the nonwaxy. One of the Texioca lines (it is not stated which) had a thicker, denser peripheral endosperm than the waxy TX 615, suggesting that differences occur among waxy lines.

Waxy, high amylopectin, starch granules were more susceptible to degradation by buffered hog pancreas α-amylase than nonwaxy starch granules and this, combined with the alteration in structure of waxy endosperm sorghum is believed by Sullins and Rooney (1974, 1975) to account for the improved feed efficiency of waxy grain over nonwaxy in feeding trials with ruminants.

Doggett (personal communication) stated that in general, yields of waxy types are lower than nonwaxy ("normal") types and that because waxy grains are softer they suffer greater loss in storage than do corneous, nonwaxy types.

Arora and Luthra (1972) at Hissar, India, examined the seeds of 19 strains of fodder sorghum, grown in the same field, for starch (Hassid and Neufeld, 1964 and Dubois *et al.*, 1956), total and reducing sugars (Dubois *et al.*, 1956 and Snell and Snell, 1953), amylose (Hassid and Neufeld, 1964, and Snell and Snell, 1953), amylopectin (difference between starch and amylose), tannin (Nierenstein, 1944, and Snell and Snell, 1953), mineral matter (AOAC, 1960) and protein (Kjeldahl N × 6·25).

Starch ranged from 57·4 to 70·6% and protein from 8·3 to 14·4%. IS 4906 was highest in starch and lowest in protein but IS 5459 with 14·4% protein also had a relatively high starch content of 64·0%. Amylose content was approximately 12–13% of the starch.

Total soluble sugars in the seed varied from 3·1 to 7·8%. The sugar content of JS 263 (regarded as a sweet variety) at 6·6% differed little from JS 20 (regarded as a nonsweet variety) at 5·9%. The latter contained 0·28% tannin, the former only 0·04%. Mineral matter in the seed varied from 1·1 to 4·1%.

Barham *et al.* (1946) examined 14 varieties of grain sorghum grown in Kansas and found, on the basis of starch viscosity (measured by rotating cylinder viscometer), that the sorghum grains fell into three groups: miscellaneous types, Feterita, Finney Milo, Cheyenne, Club Kafir and Hegari; Kafir types, Red Kafir, Wheatland, Early Sumac, Blackhull Kafir, Early Kalo, Atlas and Pink Kafir and waxy types, Leoti Red and Schrock.

The composition in starch, reducing sugars (as maltose) and nonreducing sugars (as sucrose) (AOAC methods, no date) at Kansas Agricultural Experiment Station, Manhattan of the 12 nonwaxy and the two waxy types, mean and ranges were:

	Nonwaxy	Waxy
Starch (%)	64·63	58·4
	(60·5–68·45)	(53·3–63·5)
Reducing sugars	0·32	0·33
(as maltose) (%)	(0·21–0·47)	(0·31–0·35)
Nonreducing sugars	1·2	0·69
(as sucrose) (%)	(1·02–1·34)	(0·39–0.98)

Kafir starches gave firm gels which were clearer and of better texture and gloss than those prepared from the miscellaneous starches. Pink Kafir was considered best for food use as judged by cooking tests with milk and fruit juices. Schrock starch most nearly resembled cassava starch in gel characteristics and in that respect was superior to the other waxy type, Leoti Red.

Miller and Burns (1970) compared the chemical, physical and organoleptic properties of 17 samples of grain sorghum. The samples, as designated by the supplier, with their endosperm characteristics, total starch (AOAC, 1965), amylose (Gilbert and Spragg, 1964), starch granule density (Beckman air comparison pycnometer) mean starch granule diameter (MacMasters, 1964) and organoleptic evaluation are given in Table 2.51.

Total starch ranged from 65 to 70% except for the one sugary endosperm (kafir type) sample which contained 58%. The waxy samples contained little amylose, the highest being 5·7% of total starch. The highest amylose content (34·9%) was found in a yellow endosperm hybrid. Density at 2·21 g/cm^3 was highest in the sugary endosperm sorghum and lowest (1·14) in a nonwaxy sample. The smallest starch granules, with a mean of 8·25 μm were in the pop sorghum, the largest 17·5 μm in a yellow endosperm F_1 hybrid. Correlation coefficients were positive and significant between % amylose and starch density, and between % total starch and % amylose; negative and significant between % total starch and starch density.*

Starch Gelatinization

Freeman *et al.* (1968) determined the gelatinization temperatures of starches from 85 samples of sorghum grain of diverse origin and type by a standard microscopic procedure (Schoch and Maywald, 1956; Watson, 1964) and found that, in general, waxy starches gelatinized at temperatures 2–3° higher than nonwaxy starches.

Martin and RS 610 (nonwaxy grain sorghum varieties), and Texioca 54 and the hybrid H 7919 (waxy varieties) were grown in each of four locations in Texas, three locations in the coastal bend of South Texas where rainfall was moderately high and irrigation not practised, and one location in the high plains of North Texas where irrigation was common and temperatures, especially at night, cooler. Starch synthesized by plants grown in North Texas gelatinized, on average, at temperatures 4° lower than from the same variety grown in South Texas.

The results indicated that starch gelatinization temperature may be influenced both by genotype and environment.

A six year breeding programme to transfer the property of low starch gelatinization (desirable for some applications) from the forage sorghums Atlas and Hoti to the grain types was unsuccessful.

Pentosans

Karim and Rooney (1972a) cited Edwards and Curtis (1943) as reporting a pentosan content ranging from 2 to 3% among several genetically similar

* (Authors' note: there are minor differences between the text and the table in the original paper. We have, in general, chosen from the text where a discrepancy appeared.)

varieties of sorghum. The former authors studied the pentosan content (Fraser *et al.*, 1956) of 31 unnamed varieties of grain sorghum grown under similar conditions in Texas and varying in endosperm characteristics: yellow, waxy, sugary, corneous, floury, normal and intermediate. Details are given in Table 2.52.

An analysis of variance indicated that the pentosan content of floury endosperm sorghum was not significantly different from corneous endosperm but was significantly different among the other groups. Pentosan contents varied significantly among varieties within each group.

The variety B 398 was used to determine the location of pentosan in the kernel (Johansen, 1940) after preparation for histochemical studies (Sass, 1945). The distribution is shown in Table 2.53. Pentosans were highest in the pericarp and germ; where they were concentrated in the cell walls. Pentosan content decreased from the periphery to the centre of the endosperm. Pentosan distribution in wheat was similar to sorghum (Elder *et al.*, 1953, cited by Karim and Rooney) but wheat endosperm (flour) contained 2·9 to 3·2% pentosan (Loska and Shellenberger, 1949, cited by Karim and Rooney, 1972a).

Nine varieties, SC 0283 6 and SA 216 (both corneous), SA 5875 6 1 (waxy) and the normal textures TX 403, 7078, B 398, B 3197, SA 394 and TX 09, differing in pentosan content and processing properties when laboratory milled (Rooney and Sullins, 1969) produced endosperm which did not differ significantly among the varieties in pentosan content (mean of 0·96%, range 0·82–1·20) confirming that differences in pentosan content in whole grain result from differences in the proportion of germ and pericarp present.

The variety RS 608 grown at eight locations in Texas in 1968 differed significantly in mean pentosan content according to location, though the range (3·05–3·74%) in value was relatively small. Four varieties B 3197, TX 09, NSA 740, and SC 283 were grown during 2 years at two locations in Texas. The pentosan content varied significantly by variety, location, and year. Variety × location, year × variety and location × year interactions were all significant.

Some 31 varieties were examined for kernel size and hardness (Rooney and Sullins, 1970), 1000-seed weight, density (pycnometer), test weight, protein and lipid (AOAC, 1970) and starch (Norris and Rooney, 1970a). The range for each group is shown in Table 2.54. In Table 2.55 the data for the dry-milling (Rooney and Sullins, 1969) and the wet-milling (Norris and Rooney, 1970b) properties of the 31 varieties are shown.

None of the properties measured appeared to be influenced by pentosan content when all 31 varieties were considered together. The only significant positive correlations were between pentosan content and total grit yield in the yellow endosperm group, and between pentosan content and kernel size index, hardness and total grit yield in the normal endosperm group. There was a significant negative correlation between pentosan content and starch content in the normal endosperm group. Pentosan content was not therefore regarded as a useful index for predicting milling characteristics.

Karim and Rooney (1972b) characterized the water-soluble (Lin and

Pomeranz, 1968) and alkali-soluble (Cartano and Juliano, 1970) pentosans extracted from the whole grain of the sorghum variety Martin and from the ground pericarp and endosperm obtained by dry-milling (Rooney and Sullins, 1969). Table 2.56 shows the properties and distribution of the pentosans.

The major pentosan fractions from the whole kernel and the endosperm were water-soluble; the pentosan in the pericarp was almost entirely alkali-soluble. The composition of pentosans from the sorghum grain were similar to those reported from rice (Cartano and Juliano, 1970) and wheat (Cole, 1967).

The hexoses, galactose and glucose, and the pentoses, xylose and arabinose, in the water-soluble and alkali-soluble pentosans were identified by paper chromatography and quantified (Dubois *et al.*, 1956). The results are given in Table 2.57. Ribose and mannose were not found. Glucose was the predominant sugar and pentosan preparations from the endosperm contained more glucose than those from whole grain and pericarp. This was reflected in the pentose/hexose ratio and was similar to results from barley and maize reported by Preece and Mackenzie (1952). Glucose was absent from rice bran pentosans (Cartano and Juliano, 1970) and maize pericarp pentosans (Wolf *et al.*, 1953) but the water-soluble and alkali-soluble pentosans in sorghum pericarp contained significant quantities of glucose and higher proportion of xylose and arabinose than those in the whole grain and endosperm.

In comparison with alkali-soluble pentosans, water-soluble pentosans contained more arabinose than xylose. Cartano and Juliano (1970) reported similar findings with milled rice.

Sugar

Watson (1967) cited unpublished data of Y. Hirata and S. A. Watson giving the average content of both waxy and nonwaxy sorghums as: 1.2% total sugars, 0·85% sucrose, 0·09% D-fructose and 0·11% raffinose. Sugary varieties contained 2·8% total sugar, the proportions of component sugars being similar to those in normal types.

Bhatia *et al.* (1972) reported the composition, 120 days after sowing, of the fodder varieties of sorghum, JS 263 (white grained, compact ear), JS 20 (purple or black grain, lax ear) in total water-soluble sugars (Yemm and Willis, 1954), free reducing sugars (Hulme and Narain, 1931), nonreducing sugars (by difference), starch (Clegg, 1956, as modified by Smith and Grotelneschen, 1966), free and bound glucose (Kline and Acree, 1930), free fructose (by calculation, free reducing sugars minus free glucose), bound fructose (by calculation, nonreducing sugars minus bound glucose).

The results (% DWB) are shown below:

	JS 263	JS 20
Free reducing sugars	0·85	2·48
Total water soluble sugars	2·21	3·75

Nonreducing sugars	1·36	1·27
Starch	75·20	73·75
Free glucose	0·60	1·80
Free fructose	0·25	0·68
Bound glucose	1·23	0·67
Bound fructose	0·13	0·60

Fibre

The physiological importance of fibre in human diets is of relatively recent interest. It has been proposed, specifically, that levels of fibre in the diet are important in the etiology of colonic cancer. Low dietary fibre intakes causing faecal stasis may prolong exposure of various compounds such as bile acids to bacterial flora with subsequent production of carcinogens. Mitchell and Eastwood (1976) however state that the correlation between dietary fibre and colonic cancer is not as yet clearly understood and requires more intensive investigation. The evidence suggests marked differences in physiological behaviour between vegetable fibres from different edible plant sources.

Clues to the etiology of colon carcinoma may be obtained from data on migrants from areas of low to high prevalence of colonic cancer. It has been reported that when nationals of a country with a low incidence migrate, they and their children assume the enhanced susceptibility to bowel diseases of nationals of their new country (Burkitt *et al*., 1974, Haenszel *et al*., 1973, as cited by Eastwood *et al*., 1976). Although higher colonic cancer rates in immigrants have been attributed to change of diet and greater consumption of more highly refined food products, faecal transit times of the migrants were not found to be significantly different. This observation suggests that the chemical composition of the faecal content of the bowel may in fact be more important than faecal volume.

Despite the observation that studies of the predisposing factors to colonic cancer identify diet as its primary cause and although nearly all major dietary components have been incriminated at some time or other, to assume a singular causal relationship is no doubt an oversimplification. There are ultimately many factors which can be linked to variations in risk levels for bowel cancer, not the least of which are age, sex, race and environment in addition to diet.

Inorganic elements

The mineral composition and chlorine content of sorghum reported by the following workers, are summarized together at the foot of Table 2·58:

Adrian and Jacquot (1964); Chughtai and Waheed Khan (1960); Doggett (1970) citing Miller (1958); Fabriani (1939); Fournier and Digaud (1948); Furr and Sherrod (1968); Gartner and Twist (1968); Munsell *et al.* (1950abc); Narayana Rao *et al.* (1961); Oke (1965ab); Pinta and Busson (1963); Platt (1962); Ranganathan *et al.* (1937); Twist *et al.* (1965); and Watson (1971).*

Apart from Ca, Fe and P (the last is discussed more fully in Chapter 4), there are few published data. For some minerals, only a single determination has been traced, that of Pinta and Busson (1963) who reported the presence at low levels of, in mg/100g: barium 0·08, beryllium < 0·05, bismuth < 0·05, gallium < 0·01, germanium < 0·01, lithium 0·07, rubidium 0·12, silver < 0·005, strontium 0·22 and vanadium < 0·01.

Naik and Abhyanker (1955) determined (AOAC, no date) the content of inorganic elements in nine improved strains of sorghum grown in Bombay State from the Dharwar varieties Bilichigam, Fulgar White, Fulgar Yellow, Nandyal, grown with local strains; the Viramgam variety Chasatio and its local strain; a Surad variety and its local strain, and Gauhala; and Jowar No. 8 with a Broach local variety. Also grown were a Mohol variety M 35 1 and a Jalgaon variety, Aispuri. The ranges and means are shown in Table 2.58.

The highest and lowest contents of Ca, P and Fe were found in the varities listed below:

	Highest		Lowest	
	Improved	Local	Improved	Local
Calcium	Chasatio	Broach	Surad Jowar	Fulgar Yellow
Phosphorus	Gauhala	Chasatio	Broach Jowar No. 8	Fulgar White
Iron	Broach Jowar No. 8	Broach	Fulgar Yellow	Fulgar Yellow

Sur *et al.* (1955) reported the Ca and P (AOAC, 1950) and Fe (Farrar, 1935) contents of six pure bred strains of sorghum obtained from the Department of Agriculture, Mysore, India (see Table 2.58). The widest ranges were shown by Ca and P.

Ferreira Oriá and de Lima (1975) reported the Ca (Ferro and Ham, 1957), P (Fiske and Subbarow, 1925), and Fe (Cramer *et al.*, 1950) of 10 varieties of sorghum grown in Brazil during 1973. The means and ranges are presented in Table 2.58. The highest and lowest were:

	Highest	Lowest
Calcium	R 1090	2201
Phosphorus	RS NK 180	Savanna 2
Iron	Rico	S 77

* (Authors' note: the full data in Table 2.58 are not entirely comparable because of variations in analytical methods among authors. The source of samples was not always stated.)

Barham *et al.* (1946) reported the mean composition (determined spectrographically) of the ash from 14 varieties of grain sorghum. The mean and ranges are shown in Table 2.58. Calcium was not determined; the highest levels of P were found in Early Sumac and the lowest in Leoti Red Sorgo.

Kramer and Matz (1969) reported the unpublished results of McNeill (1962) of the elemental analysis (semi-quantitative spectrography) of Redbine 66 (see Table 2.58).

Wehmeyer (1969), at the National Nutritional Research Institute, Pretoria, analysed the inorganic elements of 11 varieties of sorghum grown at various locations in South Africa (see Table 2.58). The highest and lowest in Ca, P and Fe were:

	Highest	Lowest
Calcium	Red Mixed	TE 66, NK 222
Phosphorus	Bird-proof	NK 300
Iron	NK 222	Barnard Red

Vitamins

In common with most other cereals, sorghum is comparatively rich in the B-group vitamins. Some vitamin A (carotene) is present but sorghum as normally eaten is not a source of either vitamin C or D. Ascorbic acid is however present in sprouted grain. Representative values, with ranges, are presented in Table 2.59 derived from the following sources: Blessin *et al.* (1958); Boas Fixsen and Roscoe (1937-38, 1940); Dakshinamurti (1955); De (1936a); Doggett (1970) citing Miller (1958); Fournier and Digaud (1948); Goldberg and Thorpe (1945); Lakshmiah and Ramasastri (1969); Munsell *et al.* (1950abc); Narayana Rao *et al.* (1961); Platt (1962); Ramasastri and Mohan (1969); Seljametov and Massino (1969) and Srinivasa Rao and Ramasastri (1969b).* Other references included in the summary table are discussed below.

Yousif and Magboul (1972) reported the thiamine, riboflavin and niacin content of Mayo, Feterita, Abu Sefera, Dahar and Houmecy, authenticated varieties of sorghum grown in the Sudan and analysed (unstated method) through the Nutrition Division of FAO. No significant varietal differences were shown.

Wehmeyer (1969), at the National Nutrition Research Institute, Pretoria, analysed the vitamin content of 11 varieties of sorghum grown at various locations in South Africa.

The extent to which variations may occur is illustrated by the cultivar TE 66 which demonstrated both the highest and the lowest thiamine contents.

Passmore and Sundararajan (1941) reported the thiamine content (thiochrome method) of six strains of sorghum, Perimanjal (yellow),

*(Authors' note: the methods of sampling, analyses and reporting appear to vary, and in many instances are not stated.)

Talaivirichan (white), Tella jonna (white), Patcha jonna (yellow), Vellai cholam (white) and Sen cholam (red) obtained from Coimbatore, India. The means and ranges are shown in Table 2.59, Patcha jonna was lowest and Talaivirichan highest.

Chitre *et al.* (1955) estimated thiamine, riboflavin contents (Chitre and Desai, 1955) and niacin (James *et al.*, 1947) of the sorghums N 1 and N 4 from Nandyal (rain-fed deep black cotton soil), CO 1, CO 7, CO 9 and CO 12 from Coimbatore (rain-fed or irrigated, red loamy) and H 1 from Bellary (as Nandyal). The means and ranges are shown in Table 2.59. CO 1 and CO 12 were highest in thiamine and niacin: N 1 highest in riboflavin. Chitre *et al.* (1955) concluded that both genetic and environmental factors may influence vitamin content.

Sur *et al.* (1955) reported the thiamine content (Swaminathan, 1942a) of six pure bred strains of sorghum from Mysore, India. The mean and ranges are shown in Table 2.59.

In 1941 and 1942, 29 varieties of grain sorghum were grown at Perkins and Woodward, Oklahoma, under different environmental conditions. The riboflavin, niacin and pantothenic acid (microbiological assay) contents were reported by Knox *et al.* (1944). Carotene content (method of Wall and Kelley, 1943) (one set of values only), was reported by Gross and Heller (1943).

Heller and Seiglinger (1944) stated that conditions in 1941 and 1942 were similar and therefore results were averaged. Means and ranges appear in Table 2.59.

There was little difference in riboflavin and pantothenic acid among groups but niacin varied widely. Darso averaged 2·1 mg/100 g at Perkins, 2·2 at Woodward: Migari averaged 8·4 at Perkins and 7·6 at Woodward. The Club kafirs and varieties with "crypto-brown" and "milo-yellow" pericarps were highest in niacin.

Sagrain, grown only at Perkins, was exceptionally high (0·5 mg/100 g) in carotene though in general, location appeared to have little effect.

Tanner *et al.* (1947) reported (microbiological assay) riboflavin, niacin, pantothenic acid, pyridoxine and biotin in 22 waxy and 20 nonwaxy sorghums grown in various regions of the USA. Means and ranges appear in Table 2.59.

Niacin content varied widely: the nonwaxy hybrid Wonder Club, thought to be a kafir × feterita cross, contained 6·52 mg/100 g. Three selections from crosses involving the waxy Cody (itself averaging 7·2) and Wonder Club resulted in progeny with higher niacin contents than their parents, 8·3, 8·3 and 9·2 mg/100 g.

Tanner *et al.* (1947) stated that among the B vitamins present, probably only niacin content could be increased by sorghum hybridisation.

Pant (1975) reported two high-lysine, high-protein Ethiopian sorghum lines, IS 11758 and IS 11167 that contained relatively large amounts of niacin. The table below compares the niacin levels of the two Ethiopian sorghum lines with those of common sorghums.

Sample	Sorghum line[a]	Total niacin (mg/100g) (ranges in parentheses)
High-lysine, high-protein sorghums	IS 11167 (1)	10·5 (mean of duplicate analyses)
	IS 11758 (17)	11·5 (9·27–13·45)
Normal sorghums	CSH (1)	2·9[b] (2·8–3.0)
	PJ 160 (1)	3·0[b] (2·85–3·2)
	G 4 (1)	4·9[b] (4·5–5·2)

[a] Numbers in parentheses indicate number of samples analysed.
[b] Mean of four replicate analyses.

Availability of Niacin

Ghosh *et al.* (1963) cited Kodicek (1940) and Krehl and Strong (1944) concerning the existence in cereal grains of an alkali-labile bound form of niacin. The bound form was relatively unavailable to microorganisms requiring niacin for their growth (Krehl and Strong 1944; Kodicek and Pepper 1948 and Das and Guha 1960); to rats (Chaudhuri and Kodicek, 1960); to pigs (Kodicek *et al.*, 1959; Luce *et al.*, 1967); and to man (Goldsmith *et al.*, 1956). The niacin could be released from the bound form by dilute alkali (Chaudhuri and Kodicek, 1960), but Das and Guha (1960) found that pepsin did not release the niacin, suggesting that bound niacin ingested in food would not be hydrolysed by the gastric juices.

Johnson (1945) and Krehl *et al.* (1946b) cited by Ghosh *et al.* (1963) recommended *Leuconostoc mesenteroides* ATCC No. 9135 as the best organism by which to assay niacin. The organism is apparently specific for niacin and cannot utilize either niacinamide or "bound" niacin.

Ghosh *et al.* (1963) described how ground seeds of pure bred sorghum and pearl millet from West Bengal were extracted for 45 min. with 0·1 N HCl over boiling water. The extract was cooled and centrifuged, the residue treated twice with 5–10 ml of 0·1 N HCl in the cold and centrifuged. The combined centrifugate was used for the microbiological assay (Johnson, 1945; Krehl *et al.*, 1946b) of niacin, "bound" niacin and niacinamide.

The niacin in pearl millet (and maize) was too little for accurate measurement, virtually all the niacin appearing to exist in the bound form. The niacinamide content of the sorghum was also too small for accurate measurement. A sample of chick pea contained all its niacin in the free form. The data are presented in Table 2.60.

Because tryptophan, in addition to being an essential amino acid, is a precursor of niacin (the "pellagra-preventive" vitamin), Adrian (1969) determined microbiologically (*L. arabinosus*) the content of both niacin and tryptophan in samples of sorghum from countries in the tropical zone of Africa. The results are shown below in mg/100 g of dry product.

	Niacin (total) (a)	Vitamin activity from tryptophan (b)	Potential vitamin activity (a)+(b)
Sorghum			
"Tchergeri"	4·40	1·82	6·22
"Bourgouri"	2·38	1·82	4·20
Mean	3·39	1·82	5·21

The World Health Organization (WHO, 1967) proposed 60 mg tryptophan as physiologically equivalent to 1 mg niacin.*

Adrian *et al.* (1970) described how niacin was assayed in a number of foods including four samples of sorghum (red sorghums RS 610 and AKS 614 from France, white sorghums from Niger and Morocco); two samples of pearl millet (from Niger and Cameroon), one sample of white little millet, and seven samples of maize. The samples were extracted by (A) cold water, (B) hot 0·1 N HCl, (C) 0·1 N HCl and boiling with 1 N NaOH and (D) autoclaving at 120° for 15 min. with 2 N HCl. *L. arabinosus* was the assay organism (Adrian, 1959).

Adrian *et al.* (1970) suggested fraction (A) contained "free" niacin. Extract (D) gave highest values from most samples and was regarded as representing total niacin. The results appear below.

	Minimum available (water extract (A)) (mg/100 g)	Total extract (D) (mg/100 g)
Sorghum	1·06	4·06
Pearl millet	0·89	2·46
Maize	1·09	2·42

Adrian *et al.* (1970) stated their opinion that a substantial proportion of niacin in cereals is "available". This was supported by Kodicek (1962), but disputed by Ghosh *et al.* (1963).†

Table 2.61 shows the amounts of niacin extractable by the various solvents, and the proportion of the whole these represent. AKS 614, a "bird-resistant" sorghum, contained the lowest total amount and the lowest proportion of "available" niacin extractable by water.

Carotene

The pigmented carotenoids in sorghum give yellow colour to eggs, the flesh of poultry and to the fat of beef (Blessin *et al.*, 1958), and provide a source of β-carotene which can be converted into retinol (vitamin A) in the human body.

* (Authors' note: this ratio might not apply when tryptophan is the first limiting amino acid.)

† (Authors' note: the reference to Kodicek (1962) quoted by Adrian *et al.* (1970) could not be traced.)

Blessin *et al.* (1958) extracted the total carotenes and xanthophylls (Cooley and Koehn, 1950) from yellow maize and several commercial grades of yellow milo, white kafir, and crosses of African yellow endosperm with US domestic feed sorghum from Nebraska. The carotenoids were separated by adsorption chromatography and characterized by spectrophotometry. The major carotenoids present in sorghum were identified as lutein, zeaxanthin and β-carotene, and the total carotene ranged from 0·02 in normal sorghum to 0·1 mg/100 g in yellow endosperm crosses. Carotenoids detected in yellow maize but not in sorghum included cryptoxanthin, hydroxy-α-carotene and α-carotene.

Blessin *et al.* (1962) reported that sorghum exposed to weathering after pollination contained 50% less carotenoids than seed heads protected by kraft paper bags. The presence of a coloured pericarp did not prevent loss of carotenoids from the endosperms of unbagged grains.

De (1936a) reported that the carotene content (spectrophotometric determination) of sorghum stored for two months in baskets or glass jars dropped from 0·16 to 0·01 mg/100 g.

Rooney (1973) commented that the carotenoids in yellow endosperm sorghum were not a significant source of dietary retinol.

Fehér (1971) reported that 14 varieties of sorghum grown in Hungary were in general low in β-carotene content (column chromatography) assuming that no less than 0·1 mg/100 g DWB is a desirable level. Six samples contained no β-carotene; six less than 0·1 mg/100 g. One variety (Szegedi 420) contained the high level of 0·628 mg/100 g.

Suryanarayana Rao *et al.* (1968) determined, by a described method based on hexane extraction, the β-carotene content of 57 yellow endosperm varieties of sorghum from Hyderabad, India. The means and ranges are presented in Table 2.59. The range was very wide from 0·097 mg/100 g for IS 7379 from Nigeria, to nil in eight of the samples, one of which, IS 2219 B, was also from Nigeria. The samples with β-carotene contents in the range 0·097–0·030 mg/100 g had protein contents within the 10–13% average range found among 350 other varieties of sorghum analysed at the Nutrition Research Laboratories, Hyderabad, India (unpublished results).

Biological Evaluation

Sorghum

Rats

Asia

Ranganathan (1935) compared whole wheat, pearl millet, finger millet, sorghum and polished rice in rat diets (two male and two female in each group). The cereals were finely ground, mixed to a dough and baked on a hot plate. Food intake was measured. The diets with urine and faeces excreted were periodically analysed (unstated methods) for Ca, Mg and P.

Weight gain was poor on all diets varying in descending order: pearl millet, sorghum, whole wheat, finger millet and polished rice. Calcium retention was good in rats fed whole wheat (73.5%), sorghum (70·4%) and polished rice (87·1%); relatively poor on pearl millet (49·5%) and finger millet (42·6%). Pearl millet, sorghum and finger millet diets produced negative Mg balance; whole wheat (50·3% retention) and polished rice (62.1%) diets produced positive Mg balance. Phosphorus balance was positive in all diets, the percentage retention being: whole wheat 59·3, pearl millet 54·1, sorghum 55·5, finger millet 42·4 and polished rice 66.4.

Swaminathan (1937a) reported the digestibility coefficients and biological value (Chick *et al.*, 1935ab) of samples of unstated origin of sorghum, pearl millet and finger millet fed to male rats weighing about 150 g, five to the group, protein in the diet about 5% (N × 6·25). The protein contents, digestibility coefficients and biological values, compared to rice and wheat, together with the available or net protein calculated from the formula:

$$\text{crude protein } \% \times \frac{\text{digestibility coefficient}}{100} \times \frac{\text{biological value}}{100}$$

are shown in Table 2.62.

On the basis of digestibility coefficients and biological values, the cereals ranged in the following descending order: rice, sorghum, pearl millet, finger millet and wheat. On the basis of available protein, the order was: sorghum, wheat, pearl millet, rice and finger millet.

Swaminathan (1937b) also examined similar cereals to those previously mentioned, but using the growth method, with young rats of 45–55 g body weight, three or four to the group, over 8 weeks, with cereal protein at about 5% in the diet. The "biological value" (PER) in these trials was calculated from the gain in body weight in relation to protein intake. The averages after 4 weeks and 8 weeks are given in Table 2.63.

Based on PER values after 8 weeks, the descending order of the cereals was: rice, wheat, pearl millet, sorghum and finger millet. Taken in conjunction with the earlier findings (Swaminathan, 1937a), the proteins of the cereals appeared more efficient for maintenance than for growth in rats.

Swaminathan (1938a) determined the total Kjeldahl N, protein N (using Stützer's reagent) and the nonprotein N (by difference) of various Indian foods including sorghum, pearl millet, finger millet and foxtail millet. The results are given in Table 2.64.

Swaminathan stated that in the determination of the digestibility coefficient and the biological value of the proteins, the total N content was taken into account. The loss of food N in digestion calculated from earlier experiments on rats (Swaminathan, 1937abc) is also shown in Table 2.64. Swaminathan commented that the method of estimating protein content by multiplying the total N content by the factor 6·25 was not fully justified but he did not suggest an alternative conversion factor.

Giri (1940) conducted mineral balance experiments in young albino rats, four or six to a group, to estimate the availability and retention of Ca and P from rice, finger millet, pearl millet and sorghum. The cereals, at 70% of the diet, constituted the main source of Ca and P. The diets, containing other essential nutrients, differed in Ca content but were essentially similar in P. The availability of Ca was 68% in finger millet, 89% in pearl millet and 84% in sorghum. The Ca in rice was too low to permit determination. The availability of P was 58% in finger millet, 74% in pearl millet, 67% in sorghum and 64% in rice. The availability of Ca and P in finger millet was determined at different levels of intake. Total Ca and P intake, the Ca/P ratio, and % Ca and P retention are shown in Table 2.65.

When the proportion of finger millet was reduced to about 40 and 20% of the experimental diet, Ca was insufficient for optimum retention during growth and the availability of Ca was about 87 and 84%, and of P 79 and 70%. Of the cereals examined, finger millet appeared the best source of Ca and P. Its Ca/P ratio was of the same order as milk, 1·3/1.

Kuppuswamy *et al.* (1958) cited by Narayana Rao *et al.* (1961), in a review of the nutritive value of certain cereals consumed in India listed the ranges of PERs and biological values of sorghum, pearl millet, finger millet, foxtail millet, common millet and little millet. These appear in Table 2.66.

Vasi and Desai (1976) studied the *in vitro* digestibility of market samples of several pulses and cereals, including sorghum, wheat and rice, grown in different parts of South Gujarat, India. The amino nitrogen digested was estimated by the Sorensen Formal Titration Method (King and Wooton, 1956, Mann and Saunder, 1960). The methods and results for the three cereals are shown in Table 2/67.

Vasi and Desai stated that although the pulses contained 2·5–3·5 times the protein present in the cereals, the percentage of amino acids obtained after 8 h digestion was higher from sorghum and rice than from the pulses.

Chitre and Vallury (1956ab) used pure bred samples of CO 1 sorghum as the sole source of the protein in 10% protein diets fed to weanling rats, six to the group, for 10 weeks, against 10% casein diets as control. The sorghum was fed in powdered raw form, or autoclaved at 15 lb/in^2 for 30 min. (Autoclaving was intended to destroy any proteolytic activity). Average weight changes, haemoglobin (Hb) levels (Sahli's haemoglobinometer) and plasma protein (Hawk *et al.*, 1947) are shown below:

Weight changes	Casein	Sorghum	
		Raw	Autoclaved
Weight changes (g)	+85·3	+15·1	−3·6
Hb (g/100 cm^3 blood)	14·87	9·17	8·89
Plasma protein (g/100 cm^3 blood)	5·85	3·90	2·95

Plasma protein deposition and haemoglobin were reduced by autoclaving.

The total (Kjeldahl) protein, fat (Soxhlet extraction) and glycogen (Good *et al.*, 1933, Hawk *et al.*, 1947) in the liver were also determined by Chitre and Vallury (1956b) using essentially the same diets as described by the same authors (1956a). The results in g/100 g fresh liver are shown below:

	Casein	Sorghum	
		Raw	Autoclaved
	(% in liver)	(% in liver)	(% in liver)
Protein	22·9	15·7	13·4
Fat	3·5	6·3	7·5
Glycogen	2·6	5·1	7·4

Chitre and Vallury (1956b) cited several workers, Tucker and Eckstein (1937), Channon *et al.* (1938), Best and Ridout (1940), du Vigneaud *et al.* (1941) and Keller *et al.* (1948), as having established that methionine and cystine can partly replace choline as lipotropic agents. In Chitre and Vallury's experiments, the methionine and cystine contents of sorghum-based diets were 278 and 167 mg/100 g. The levels of liver lipids were much higher than could be correlated with methionine content. Chitre and Vallury suggested other factors might be influencing the fat deposition or the amount of methionine available for growth, and lipotropic activity might not be correctly indicated by the chemical assay.

Liver protein followed the growth pattern. Higher levels of glycogen were formed from the sorghum diet than from the casein.

Autoclaving increased the deposition of fat and glycogen and decreased liver protein.

Misra *et al.* of the Vallabhbhai Patel Chest Institute, Dehli, reported in a series of papers (1974/75ab, 1974abcdefg, 1973ab, 1972/73) the effects of feeding defatted sorghum flour, at protein levels of (A) 5, (B) 10 and (C) 14·5% against (D) casein at 10% as control, to male rats weighing 90–100 g. over a period of six weeks. The diets were isocaloric, the protein levels being adjusted at the expense of potato starch. The sorghum was purchased from a local market and the ground flour was defatted by the following in sequence: (1) hot ethanol (twice), (2) acetone ether (thrice) and (3) ether (thrice) (Misra *et al.*, 1973b).

The protein content of the defatted flour (Kjeldahl method, conversion factor not stated) was 14·5% and carbohydrate 75% (anthrone method) (Misra *et al.*, 1972–73). The amino acid composition (method not stated), in mg/g total N, (assuming N to protein factor of 6.25): Ile 337, Leu 1006, Lys 169, Met 106, Cys 106, Phe 312, Tyr 175, Thr 225, Trp 69, Val 356, Arg 237 and His 119.

Compared with the controls, the rats fed the sorghum diets lost body weight but the weights of the brains were not significantly different. Brain lipids significantly increased in rats fed diet (B), and brain triglycerides on diets (A) and (B). Brain phospholipids did not change on diets (A) and (B) but were reduced on diet (C). Monophosphoinositides and polyphosphoinositides significantly increased on diet (A) and decreased on diets (B) and (C). Both phosphatidyl choline (PC) and phosphatidyl ethanolamine

(PE) significantly increased on diet (B) but PE showed significant reductions on diets (A) and (C) (Misra *et al.*, 1972–73).

Compared with the controls, there were significant increases in total lipids, glycerides, phospholipids, cholesterol and fatty acids in the livers of the rats fed sorghum. The cholesterol, fatty acid and glyceride increases were highest at intermediate levels of dietary protein. Liver phospholipids, PC and PE were inversely proportional to the amount of sorghum in the diet. The rats fed sorghum showed enhanced lipogenesis, cholesterol-genesis and incorporation of acetate–1–^{14}C into liver triglycerides, PC and PE compared with the controls (1974a).

The plasma lipid metabolism was studied using acetate–1–^{14}C, palmitate–1–^{14}C, glucose–U–^{14}C and $NaH_2{}^{32}PO_4$. Plasma total lipids and glycerides were significantly higher in all the rats fed sorghum. Plasma total and esterified cholesterol was lower on diet (A) and unaffected on diets (B) and (C). Plasma total phospholipids and lysophosphatidyl choline (LPC) and PC were reduced on diets (B) and (C). The incorporation of palmitate–1–^{14}C, acetate–1–^{14}C, and glucose–U–^{14}C into plasma triglycerides showed that the secretion of hepatic triglyceride (TG) into plasma was not impaired. The labelling pattern suggested that the availability of plasma cholesterol and phospholipids in the liver was not the limiting factor in the pathogenesis of fatty liver observed in rats on sorghum diets (Misra *et al.*, 1973a).

The liver DNA (deoxyribonucleic acid) and RNA (ribonucleic acid) of rats on diet (A) were significantly increased. Liver proteins of rats on diets (A) and (B) were significantly increased and plasma protein decreased. Incorporation of leucine–1–^{14}C into both liver and plasma proteins of sorghum fed rats was significantly higher than from casein. (The leucine content of liver was not determined but Misra *et al.* 1973b considered that since leucine is in excess in sorghum, the availability of leucine would not be limiting).

The incorporation of acetate–1–^{14}C, palmitate–1–^{14}C, and glucose–U–^{14}C, into adipose tissue esterified fatty acids was significantly increased in rats on diets (A) and (B) as compared to (D). Incorporation of $NaH_2{}^{32}PO_4$ into adipose tissue phospholipids of rats on diets (B) and (C) was significantly reduced compared to (D) (Misra *et al.*, 1974b). The incorporation of palmitate–1–^{14}C into liver TG, PC and PE of rats on diets (A), (B) and (C) was significantly greater than on diet (D). Incorporation of palmitate–1–^{14}C into plasma TG and into liver FC and EC of rats on diets (A), (B) and (C) was equal to, or slightly greater than, those on diet (A). The results suggested that the accumulation of TG in the livers of rats fed sorghum at different protein levels resulted from the greatly increased synthesis and unimpaired secretion of TG into plasma (Misra *et al.*, 1974c). Compared to rats fed the casein control diet, incorporation of glucose–U–^{14}C into the liver lipids of rats fed the sorghum diets was the same or higher; incorporation of glucose–U–^{14}C into liver TG, PC and PE was significantly higher (Misra *et al.*, 1974/75b).

Incorporation of $NaH_2{}^{32}PO_4$ into liver total phospholipids, PC and PE was significantly increased in rats on diets (A) and (C) and decreased in PC in rats on diet (B) as compared to diet (D). The effects of feeding sorghum on liver phospholipids are due to the quantity as well as the quality of the sorghum protein (Misra *et al.*, 1974d).

A significant increase in adrenal weight, total lipids, cholesterol, phospholipids and glycerides (mono- and triglycerides) was observed in the rats on diets (A), (B) and (C) compared to diet (D). Adrenal PE increased in all the rats fed sorghum but PC increased in rats fed diet (C) and decreased in rats fed diet (A).

Other phospholipid fractions (monophosphatidyl inositol, lysophosphatidyl ethanolamine, sphingomyeline, phosphatidic acid and polyglycerophosphatide) also showed significant alterations in the rats fed sorghum compared to the rats fed the control diet. Incorporation of acetate–1–^{14}C into adrenal lipids was lower and that of glucose–U–^{14}C, palmitate–1–^{14}C and $NaH_2{}^{32}PO_4$ was higher than the casein control (Misra *et al.*, 1974f).

A significant increase in mitochondrial triglycerides on diets (A) and (B), in microsomes on diets (A), (B) and (C), and in supernatant fractions on diets (A) and (C) compared to diet (D) was observed, together with a significant increase in total cholesterol in mitochondria and microsomes, and a significant decrease in supernatant fraction on diet (B). A significant increase in mitochondrial total phospholipids, PC and PE in rats fed diet (B) and a decrease in rats fed diet (A) was observed. In microsomes total phospholipids were increased in rats fed diet (B) and PC was increased in rats fed diet (C). Total phospholipids, PC and PE were significantly reduced in the supernatant fraction on diet (B).

The metabolism of the subcellular fractions of the lipids of the livers of the rats was studied by incorporation of acetate–1–^{14}C and $NaH_2{}^{32}PO_4$ (Misra *et al.*, 1974e).

Misra *et al.* (1974g) reported that, compared to young rats fed a casein diet at 10% protein level, young rats fed defatted sorghum flour at 10% protein level for a period of 6 weeks showed a significant increase in heart total lipids, free and esterified cholesterol, total phospholipids, PC, PE and phosphatidic acid plus polyglycerophosphatide fraction. Compared to the casein control, sorghum flour at 5 and 15% protein levels reduced free cholesterol, and increased triglycerides. Sorghum flour at the 5% protein level decreased total phospholipids, PC and phosphatidic acid plus polyglycerophosphatide. Compared to casein protein, increased incorporation of palmitate–1–^{14}C, glucose–U–^{14}C and $NaH_2{}^{32}PO_4$ was observed in the heart lipids of rats fed sorghum protein.

Misra *et al.* (1974–75a) reported that compared to rats fed the casein control diet at the 10% protein level, rats fed defatted sorghum flour at 5, 10 and 15% protein levels for a period of 6 weeks showed significantly increased total lipids, mono-, di-, and triglycerides, total phospholipids and PC and PE fractions in the aorta. Total, free and esterified cholesterol were significantly reduced in rats fed sorghum at 5 and 10% protein levels respectively, resulting in markedly reduced cholesterol : phospholipids ratios. Incorporation of glucose–U–^{14}C into total lipids in the aorta of rats fed sorghum at different protein levels was markedly reduced, and that of $NaH_2{}^{32}PO_4$ was not much different as compared to control. The incorporation of acetate–1–^{14}C and palmitate–1–^{14}C into aorta lipids of rats fed sorghum at 5 and 15% protein levels was higher than the control.

China

Li (1930) reported the biological value (method of Mitchell) of kaoliang (sorghum) as 56, of "millet" 57, of barley 64 and of rice 77, when fed to rats in a 10% protein diet.

Europe

Eggum (1970) determined the true digestibility (TD) (not defined, and methods not stated, but generally regarded as the proportion of food nitrogen that is retained, National Academy of Sciences, National Research Council, 1963a) of several feedstuffs, including sorghum and maize, and of the amino acids they contain. The results from baby pigs and rats presented in Table 2.68 indicate that since amino acids from the same protein source may be digested differently, it is difficult to estimate the availability of any particular amino acid from the digestibility of total nitrogen.

The TD of total N in pigs was 85·3 for sorghum, 90·2 for maize and 99·4 for casein, and in rats, 91·5 for sorghum, 87·6 for maize and 101·1 for casein.

Vermorel (1970) studied the energy and protein utilization by weanling rats of the sorghum hybrid INRA 450 (a hybrid of two American lines designated A 3052 and SD 102) obtained from the Station d'Amélioration des Plantes de Montpellier, France. The sorghum provided 85% of the diet, the protein value of which was adjusted by the addition of Norwegian herring flour, 40 g, and by synthetic amino acids, L-threonine 2·0 g, L-lysine 6·6 g, L-histidine 0·9 g, L-isoleucine 3·0 g, DL-methionine 0·8 g, all /kg of diet. The protein content (N × 6·25) of the diet was 15·4%, DWB. The energy and carbon balances were obtained by indirect calorimetry.

These studies were compared with similar experiments carried out using ordinary maize (Vermorel and Keller, 1967) and Opaque-2 maize (Vermorel, 1969).

Apparent digestibility from the INRA 450 diet was 89·8% and of N 87·7%. For the same metabolizable energy intake, the amount of energy retained was 6–8% lower and the amount of synthesized protein 6–10% lower than with the Opaque-2 maize previously studied. The net energy value of INRA 450 was 1–2% higher than that of Opaque-2 maize.

South America

Yañez *et al.* (1973) analysed five varieties of sorghum, dwarf Kaoliang, Ute, Supergene, Pawnee and Comanche, grown in Chile, for moisture, ash, N, ether extract and crude fibre (AOAC, 1965), Ca and Fe (Schmidt-Hebbel, 1966) and P (spectrophotometrically, Tausky and Shoor, 1953). Results are shown in Tables 2.11 and 2.58. Variations were small except for Ute which was low in ash (1·6%) and Ca (90 mg/100 g). Amino acid content means with ranges (auto-analyser following acid hydrolysis excluding trytophan) in grain and in protein are shown in Table 2.13. Variations were relatively small and the leucine/isoleucine quotient was approximately 3·0.

With rats, the biological value of a "pool" sample of sorghum was determined, the proximate analysis of which was compared to a sample of maize (Schmidt-Hebbel, 1969) and is shown below:

	"Pool" Sorghum	Maize
Protein (N × 6·25) (%)	10·2	11·8
Ether extract (%)	6·6	5·0
Nitrogen-free extract (%)	77·3	76·2
Crude fibre (%)	2·9	5·4
Ash (%)	3·0	1·6
Calcium (mg/100 g)	225	9
Phosphorus (mg/100 g)	510	372
Iron (mg/100 g)	6·1	3·7

Weanling albino rats of both sexes, 10 to the group, over 4 weeks, were fed a sorghum and a maize diet with a casein diet (10% protein) as a control (Chapman *et al.*, 1959).*

The intake, weight gain, PER and NPU_{op} (Platt and Miller, 1959) are shown below (Yañez *et al.*, 1973).

	Intake (g)	Weight gain (g)	PER	NPU_{op}
Sorghum	186	10·9	0·96	47
Maize	184	10·5	0·94	45
Casein	303	111·2	2·80	72[a]

[a] Not determined by Yañes *et al.* but taken from Tagle and Donoso (1965).

United States of America

Howe and Gilfillan (1970), at the Merck Institute for Therapeutic Research, New Jersey, studied the limiting factors to growth of rats fed all cereal diets

* (Authors' note: the level of protein in the sorghum and maize diets was not stated but was presumably also 10%.)

and compared market samples of rice, wheat, maize, sorghum and "millet". The diets and 35-day weight gains are shown in Table 2.69.

Only in the case of sorghum did the addition of lysine alone, diet (D) (iii), produce a significant improvement compared to diet (A) unsupplemented cereal. For the other cereals, vitamins and minerals were required.

Groups of rats were fed diets with purified casein providing the protein levels of 6, 8, 10 and 12%. When the 35-day weight gains were plotted against the equivalent weight gains from the cereal diets, supplemented with lysine, vitamins and minerals, diet (H), the cereals were approximately equally effective in promoting growth, except maize which was inferior. A second growth experiment was carried out with different samples (except for wheat) of the grains to which calcium carbonate at 1% of the diet was added.* The 35-day weight gains without and with calcium carbonate were:

	Weight gain (g)	
	without $CaCo_3$	with $CaCo_3$
Rice	7	8
Wheat	34	50
Maize	12	33
Sorghum	19	30
Millet	12	25

All weight gains, except from rice, were significantly greater than the control when calcium carbonate was added.

Howe and Gilfillan (1970) suggested their results indicated that vitamins and minerals, rather than lysine, were the limiting nutrients in human infants fed cereal diets, and that deficiency of Ca and other nonprotein nutrients may contribute to the poor utilization of the protein in cereal grains.

Two sorghum composites from Kansas State University, one (LP) "low" protein, 7·9% (N × 6·25) and one (HP) "high" protein, 11·8%, were compared in growth trials, five rats to the group, over 28 days (Waggle *et al.*, 1966). The amino acid content (auto-analyser following acid hydrolysis, excluding tryptophan) in the grain and in the protein is given in Table 2.70.

Six diets, all containing equal amounts of oil, vitamin and mineral premix, but otherwise varying, were fed. The composition of the diets and the average 4-week weight gains are shown in Table 2.71.

Except for leucine and phenylalanine, total essential amino acids were in smaller proportion in the HP than the LP sorghum protein, consequently the LP displayed a superior biological value. Rats on diets (A) and (D), containing equal amounts of sorghum but different protein levels, produced no significant difference in growth rates. Rats on diet (A) grew significantly faster than those on the isonitrogenous diet (B). Diet (C) (diet (B) with lysine, histidine and arginine added equivalent to diet (A)) produced increased growth but to a lesser extent than diet (A). The casein diets (E)

* (Authors' note: it is not clear from the original text what basal diet was used for the addition of calcium carbonate.)

and (F) produced significantly higher weight gains than the other four diets.

Heller and Green (1926) reported rat growth and reproduction trials in which 20 varieties of grain sorghum were compared with maize. The cereals represented about 93% of the rations which were supplemented with various additives including yeast, casein and cottonseed meal at levels of about 5%. It was concluded that the sorghums were comparable to maize and provided vitamins adequate for growth and reproduction but not for long term rearing.

Pond *et al.* (1958) fed Kafir 44 14 sorghum containing 9·1% crude protein, 0·2% lysine and 0·3% threonine (microbiological assay) in diets to weanling rats (eight to the group) in which (A) the sorghum contributed 93·85% of the diet (to provide 8·57% crude protein, 0·19% lysine and 0·28% threonine), against (B) a low-protein basal diet containing 11·0% casein, 81·85% sucrose, and 1% maize oil and (C) a control diet containing 21% casein, 71·85% sucrose and 1% maize oil. All diets contained salt and vitamin mixture.

The addition of 0·5% L-lysine and 0·2% DL-threonine to the basal diet (A) produced growth approximately equal to diet (B) but inferior to (C). The liver fat content of rats receiving ration (A) was significantly reduced by the added lysine and threonine. The addition of 0·2 or 0·3% DL-isoleucine, 0·05 or 0·10% DL-methionine, 0·1% DL-tryptophan or 0·02% DL-valine in any combinations had no effect on growth rate.

Sidransky (1960) reported earlier work in which pathologic lesions appeared in young rats force-fed for 3–7 days with purified diets devoid of single essential amino acids. The lesions resembled symptoms described as present in kwashiorkor victims.

Sidransky, at the National Cancer Institute, Bethesda, Maryland, therefore investigated the effect of 3-day forced feeding (method of Shay and Gruenstein, 1946) and 7-day *ad lib.* feeding of male and female rats, one month old, with diets based on maize, rice, wheat, cassava or sorghum grown in regions where kwashiorkor is familiar. The composition of the diets is shown below:

	% in diet	Protein in diet (%)
(A) Maize meal	74	6·1
(B) Rice flour	74	6·1
(C) White wheat flour	55·7	8·2
Sucrose	18·3	
(D) Cassava flour	74	1·1
(E) Sorghum, Kafir 44 14	74	9·0
All diets maize oil	17	
vitamin and mineral mix	9	

The diets were not isonitrogenous. A stock diet (F) was also included. The results, including body and organ weights of the animals are shown in Table 2.72 and the liver composition in Table 2.73. The force-fed rats developed periportal fatty liver, excess hepatic glycogen and atrophy of the spleen, pancreas and submaxillary gland within 3 days with lesions resembling those found with kwashiorkor. The changes were most marked with (A) maize, (B) rice, and (D) cassava diets and less marked with (C) wheat and (E) sorghum diets. The animals fed *ad lib.* over 7 days consumed less food and showed fewer pathologic changes; the difference in results was thus apparently related to the quantity of the diets consumed.

Three Combine sorghum grains: Shallu, pearly nonpigmented; Martin Maize red pigmented hard kernel; and Waxy Kafir, large opaque white grain with waxy carbohydrate, and wheat (of unspecified type) grown, harvested, ground and analysed at Texas Agricultural Experiment Station, Lubbock, were used in rat growth and reproduction trials (Lamb *et al.*, 1966). The basic diet contained, by weight, one-sixth whole dried milk, five-sixths ground grain, and 1% sodium chloride. A similar diet containing wheat, had been found by Sherman and Campbell (1924) to be adequate for reproduction and rearing of progeny through successive generations, though not adequate for optimum growth. The protein, fibre, ash, riboflavin, niacin, and thiamine (as analysed at Lubbock) contents are presented in Table 2.11 and Table 2.59.

The pairs of albino rats, 4-weeks old, designated P-1, fed wheat, Shallu and Kafir showed similar average weight gains which were significantly greater than those on Martin. The first litter F-1 rats on Shallu gained weight significantly faster than those fed wheat, Kafir or Martin; the latter were not significantly different. Weight gains of second litter F-1 rats fed Shallu or wheat were similar and significantly greater than Kafir or Martin. Martin produced significantly greater weight gains than Kafir. Weight gains of P-1, first litter F-1, and second litter F-1 rats on wheat did not differ significantly. Weight gains of P-1 and first litter F-1 rats on Shallu and Martin were similar and significantly greater than second litter F-1 rats. Average weight gain of P-1 rats on Kafir was significantly greater than first litter F-1 rats which was significantly greater than second litter F-1 rats. The gain in weight of F-1 young up to 3-weeks old was greater on the wheat or Shallu diet than on Martin or Kafir. On the wheat diet, F-2 weight gains were similar to F-1; on the sorghum diet F-2 had lower weight gains than F-1 rats. Only rats fed the Martin diet showed any outward signs of physical abnormality; P-1 females developed severe alopecia after their first litter as did their progeny.

Food consumption /g of total body weight and /g of weight gain of P-1 and F-1 rats are shown below:

			Food consumed	
Generation	Grain	No. of rats	g/g total Weight	g/g Gain
P-1	Wheat	21	0·71	4·54
	Shallu	21	0·59	4·22
	Martin	17	0·75	4·93
	Kafir	21	0·72	4·32

Generation	Grain	No. of rats	Food consumed g/g total Weight	g/g Gain
F-1	Wheat	83	0·92	4·22
	Shallu	45	1·06	4·93
	Martin	80	1·10	6·23
	Kafir	39	1·07	4·56

Although not shown in the table, Lamb *et al.* (1966) reported that first litter F-1 rats on the sorghum diets consumed less food per gram of grain and per gram of total weight than did second litter F-1 rats. The reproduction data are summarized in Table 2.74.

Two litters were produced by 95% of P-1 females fed wheat, by 90% of those fed Shallu and Martin, and by 73% of those fed Kafir. These differences were not significant. F-1 females fed wheat produced more litters from two breedings than those on Kafir, Shallu and Martin. Rats fed Kafir had significantly fewer litters than those on Shallu and Martin.

P-1 and F-1 females on wheat did not produce significantly different numbers of litters but F-1 females produced significantly fewer litters on Shallu, Martin and Kafir than P-1 females.

Complete litter loss immediately after birth was greatest on Shallu and least on Martin. Among the grain diets and between generations, average litter sizes were not significantly different nor were average size of surviving litters, so that differences in survival of progeny up to 3-weeks old were not apparently attributable to differences in litter size.

Significantly more progeny survived at 3 weeks from P-1 females on wheat than those on Kafir, Shallu or Martin. More progeny survived from P-1 females on Martin than those on Shallu, more from F-1 females on wheat than those on Shallu. Otherwise survival of F-1 progeny did not differ significantly among diets. Only F-1 rats fed Martin failed to rear their progeny to weaning age as successfully as P-1 rats.

From their results, Lamb *et al.* concluded that sorghum grains cannot, without nutritional supplements, be substituted for wheat. They also state that adequate nutritional comparisons cannot be made by short-term feeding experiments.

Nawar *et al.* (1970) evaluated the protein quality of 10 sorghum lines from the World Collection at Hyderabad but grown at Purdue University, in 21-day feeding trials with weanling male rats. The protein (N × 6·25) and amino acid contents (auto-analyser following acid hydrolysis) are shown in Table 2.13. Arginine was said to range from 149 to 312 mg/g total N. The mean leucine/isoleucine quotient was 3·0.

In the trials, weanling rats, six to the group, were fed diets in which the sorghums supplied 10% protein except for IS 1220 (9·4%) and IS 0948 (9·3%). The sorghum ranged from 77 to 93% of the diet, adjusted so that sorghum plus dextrin provided 93%, the remaining 7% was provided by an oil, vitamin and mineral mixture. Casein diets providing 10 and 20% protein were fed as controls.

Results, including food intake, weight gain and body composition (carcass N estimated from body water, method of Bender and Miller, 1953; liver lipid extracted with anhydrous ethyl ether) are shown in Table 2.75.

Coefficients of apparent digestibility varied widely, IS 2232 (77·3%) and IS 3552 (76·8%) being the nearest to 10% casein (87·1%) and 20% casein (88·1%). IS 3472 (48·9%) and IS 0948 (50·1%) were lowest, the remainder being intermediate. Differences in daily food consumption were highly significant. Rats ate more of the diets containing IS 2301 and 3552, and least of the poorly digestible IS 3472 and 0948. Differences in weight gain were also highly significant. Rats fed IS 2031 and 3552 gained 1·9 g/day (53% of casein); IS 0948 and 3472 produced average daily gains of only 0·5 and 0·3 g respectively.

Deposition of nitrogen in the liver followed the trend in weight gain; lipid in the liver varied inversely with lipid nitrogen.

Free niacin content (method Ghosh *et al.*, 1963) ranged from 0·5–0·6 mg/100 g and bound niacin from 2·2–2·4 mg/100 g of sorghum, higher than those reported by Ghosh *et al.*

The addition of 0·65% lysine alone or with methionine to IS 1220 produced a relatively small increase in rat weight gain but the addition of 0·65% lysine, 0·21% cystine and 0·16% threonine, quadrupled weight gain, tripled stored body nitrogen, doubled liver nitrogen and halved liver lipid.

A 13% protein diet containing 77% sorghum IS 2031 and 12% SMP gave weight gain and body nitrogen deposition equivalent to a 13% protein diet in which milk alone provided 37% of the diet. A 12% protein diet containing 63% IS 3552 and 12% SMP produced weight gains almost equivalent to a 20% casein diet.

Three of the sorghums discussed above, IS 0948, 0957 and 1220, with seven others, IS 0271, 0718, 3796, 4328, 7088, 8120 and 8382 were compared in slope-ratio assays (Hegsted and Worcester, 1967; Hegsted *et al.*, 1968). The protein content, relative nutritive value (RNV) and net available protein (NAP ≡ RNV × protein content) (Hegsted and Worcester, 1967) are shown in Table 2.76.

The confidence limits were wide and emphasize that even with 18 animals involved, as in these trials, the estimated value was subject to considerable error. IS 3796 (RNV 39%) appeared superior in RNV to IS 8120 (RNV 26%), and 0718 (RNV 21%). IS 8120 contained 12·8% protein, therefore its NAP (3·3%) was close to the other samples.

Twenty-seven varieties of sorghum grain grown under uniform conditions at Lubbock, Texas were analysed for protein (N × 6·25 DWB) (Experiment I) by Breuer and Dohm (1972). Mean and ranges are shown in Table 2.11. The highest was in waxy Double Dwarf Feterita, and the lowest in Early Hegari.

In Experiment II, Martin (B398), protein 14% was fed in three experimental diets, against a casein control diet to male weanling albino rats, six to the group, over 12 days. The diets are shown in Table 2.77, with the results of the rat feeding trials.

Although growth rate from diet (A) was only 12–13% that of (B), (C) and (D), the rats maintained a satisfactory appearance with a slight increase in feed during the experiment. Supplements of amino acids in diet (B) or casein diet (C) did not affect the apparent digestibility (Maynard and Loosli, 1962) of either the nonprotein organic matter or the protein in the sorghum when allowances were made for the digestibility of the amino acids and casein supplements. Breuer and Dohm therefore suggested that comparisons of nutritive value and digestibility could be made by feeding rats diets containing only sorghum grain supplemented with vitamins, minerals and essential fatty acids.

In Experiment III the basal diet (A) was compared with a similar diet containing maize (TX 60). The 10 varieties of sorghum and the maize were produced in 1965 under similar agronomic conditions. Protein (N × 6·25, 10% moisture basis) and amino acid composition (auto-analyser excluding tryptophan) mean and ranges, are shown in Table 2.13. The higher proteins reflect mainly the higher levels of leucine and nonessential amino acids (agreement with Vavich *et al.*, 1959 and Waggle *et al.*, 1966).

Since the grains provided all the protein and 90% of the dietary energy, the rat diets (fed over 28 days) were not isonitrogenous. Protein content (N × 6·25, 10% moisture basis) ranged, %, from 10·7 (Hegari) to 13·4 (Feterita); weight gains, g, from 0·47 (TX 414) to 1·27 (Martin); feed efficiency, g gain/g feed, from 0·04 (TX 414) to 0·10 (Martin); PER from 0·38 (TX 414) to 0·92 (Hegari); and true protein digestibility, %, from 74·4 (Martin) to 81·2 (Kafir).

Rats fed the maize diet gained fastest. Gains of rats fed SA 7078, Hegari, Caprock and Martin were significantly greater than those fed TX 414, Kafir, Feterita and Shallu. PERs showed a pattern similar to weight gains. Protein digestibility was lower among varieties that produced best growth and was negatively correlated with weight gain, feed intake and PER. Significant differences were found in the digestibility of nonprotein organic matter among the varieties ranging from 89·2 (Wheatland) to 92·2% (Redlan and TX 414); maize was intermediate at 91·0% and not significantly different from SA 7078, Caprock and Feterita.

In Experiment IV, sorghum varieties SA 7078, TX 414, Combine Kafir (B 3197), and Martin (B 398) and four hybrids produced from these varieties, RS 610 (7078 × Kafir), RS 626 (414 × Kafir) RS 608 (7078 × Martin) and RS 625 (414 × Martin), all produced in 1966 at the same location under similar conditions to the grains used in Experiment III were fed in rat trials over 24 days. The protein and amino acid composition, mean and ranges, are shown in Table 2.13.

The protein contents of the hybrids roughly equalled the means of the parents except in RS 610 which was below both parents. Differences in amino acid content were minor.

The results of feeding trials were similar to those in Experiment III. The hybrids closely resembled their parents in terms of rat weight gains.

Breuer and Dohm (1972) in conclusion stated that differences in nutritive value among the sorghum varieties appeared to be related more to the utilization of the protein than to the nonprotein organic matter presented. In agreement with Waggle *et al.* (1966), differences in rat weight gain were not correlated with either the protein or the lysine content of the grain. Greatest weight gains were associated with grains having the lowest protein digestibility. Breuer and Dohm suggested this may be due to amino acids released from the prolamine and/or glutelin fraction of protein having an imbalancing effect on the mixture of amino acids absorbed from the digestive tract. Nesheim and Carpenter (1967) cited by Breuer and Dohm had suggested that analysis for faecal N, as an estimate of true digestibility of dietary amino acids, might not be accurate since proteins might be digested in the large intestine resulting in the absorption of nitrogenous compounds other than amino acids. "Thus sufficiently large differences may arise in the true availability of the sorghum grain amino acids to account for the observed rat responses".

Breuer and Dohm suggested that there might be little advantage in producing sorghum grain containing high levels of digestible protein unless accompanied by an improvement in the amino acids balance. Alternatively supplementation of the protein appeared desirable.

Ilori and Conrad (1976), at Purdue, determined amino acid composition and by rat feeding compared the protein quality of the commercial sorghum RS 610 with six sorghum lines selected for either high protein or lysine content. The composition of the sorghum diets is shown in Table 2.78. The amino acid composition, mean and ranges, is presented in Table 2.13.

The first feeding trials used male weanling rats, four to each treatment. IS 0819 (A) and IS 2478 (D) produced significantly more rapid weight gains than the five other lines including RS 610 (see Table 2.78). (Fonseca (1970) cited by Ilori and Conrad reported that IS 0819 gave faster gains than RS 610 in chicks) IS 2190, (B), with a poor amino acid balance, was readily eaten; IS 2197 (C), with a good amino acid balance was poorly consumed; IS 0819 and 2478 each with a good amino acid balance were readily eaten and therefore produced greatest weight gains.

In a second metabolism experiment five rats were fed the diets shown in Table 2.78. Faecal and urinary N was determined (Kjeldahl AOAC, 1965). The digestibility and N retention are shown in Table 2.78.

Crude protein and dry matter digestibilities did not differ significantly among sorghums (Nawar *et al.*, 1970 using different sorghum lines reported differences in digestibility). N retention was significantly higher in rats fed IS 0819 and 2478.

Waxy Sorghum

El-Harith *et al.* (1976) reported from the United Kingdom that the gross caecal hypertrophy, accompanied by mortality, observed in rats fed diets containing 71% raw potato starch was not observed in rats fed maize, wheat, sorghum, rice or cassava starches. The sorghum starch obtained from the Sudan was described as dura.

Rooney and Sullins (1973) cited unpublished work by P. L. Lamar (1972) who fed waxy and nonwaxy sorghum grain to male weanling rats. Kafir 60 (nonwaxy) and TX 615 (waxy) were fed at equal levels to give respectively 9·2 and 10·3% protein in the diet. In a second experiment maize starch, nonwaxy sorghum starch and waxy sorghum starch were fed in isonitrogenous (10% protein) diets. In a third experiment, waxy and nonwaxy segregating heads from the variety SA 7536 were fed in diets providing 10% protein. No significant differences among diets in feed conversion or PER were observed though the waxy grain diets produced somewhat better feed efficiency than the nonwaxy.

Poultry

Africa

Fetuga (1977) conducted experiments at the University of Ibadan to determine the metabolizable energy (ME) and the metabolizable energy corrected for N retention (ME_n) for chicks, of maize, sorghum and "millet" (not otherwise characterized but probably pearl millet) collected from several locations in Nigeria. The proximate composition (AOAC, 1970), mean and ranges, are presented for sorghum in Table 2.11, and for millet, under pearl millet, in Table 2.118.

The reference diet contained, %, glucose monohydrate 48·0, basal mixture 51·5 and vitamin and mineral mix 0·5. The basal mixture contained, %, maize 33·7, fish meal 10·0, blood meal 15·5, groundnut meal 30·0, dicalcium phosphate 4·8, oyster shell 1·0 and rice bran 5·0. The test diets substituted (A) maize, (B) sorghum, (C) millet for 25% of the glucose monohydrate in the reference diet, DWB. Nitrogen content of the feed and faeces was determined by Kjeldahl (AOAC, 1970), combustible energy by bomb calorimetry, and ME calculated using the equation of Hill *et al.* (1960). The ME for glucose monohydrate was 15·41 MJ/kg (Fetuga and Oluyemi, 1976) and the value of 0·0344 MJ/g N (Hill *et al.*, 1960) was used to obtain MF_n The results are shown in Table 2.79.

Mean weight gain and efficiency of feed utilization were similar. Fetuga suggested that variations among the samples of the same cereals might be attributed to variations in proximate composition, particularly in crude fibre and lipid (ether extract) fractions.

In a further experiment the same birds were fed up to a period of 12-weeks old, but age was not found to have a significant effect on the ME values obtained although there was a tendency for the ME and ME_n values to increase with age.

Smith (1967) compared maize and sorghum grown in Rhodesia in pullet feeding trials starting with one day-old chicks and continuing for 375 days. The chick rations were approximately isonitrogenous (20%), the rearing rations (18%) and the laying rations (17%) protein. The grains provided 60% of the rations; (A) all maize, (B) 50 maize/50 sorghum and (C) all sorghum.

During the rearing stage (A), (B) and (C) were equal, on the laying rations, egg production and weight were similar from all diets though daily food consumption was greater on diets (B) and (C). The metabolizable energy intake was significantly lower in birds on diet (A).

Smith concluded that sorghum and maize were roughly equivalent in chick, grower and layer rations.

Wessels (1970a) in S. Africa, used sorghum cultivars grown under standardized conditions to determine the limiting amino acids for chicks.

In preliminary trials, taking N retention (Wessels and Bundock, 1968) as the criterion of available amino acids, lysine was declared the first limiting, methionine second and tryptophan probably third limiting amino acid in the cultivar K20 65.

The study extended to include a larger number of sorghum cultivars, using a diet composed of about 93% sorghum with vegetable oil, mineral and vitamins plus 0·3% L-lysine HCl. The N content of the diets varied according to sorghum protein content and with amino acid additions. Body moisture was determined as an index of N retention.

Of 23 cultivars compared, some of which contained tannin and others did not, seven cultivars, including DC 500F and Excell 505, did not respond to lysine supplementation. These two cultivars were therefore fed in a further trial; (a) unsupplemented and (b) supplemented with (i) L-lysine HCl 0·3%, (ii) L-arginine HCl 0·4%, (iii) DL-methionine 0·2%, (iv) L-tryptophan 0·1%, (v) DL-threonine 0·2% and (vi) glycine 0·4%. The effect on average adjusted chick body moisture per group is shown in Table 2.80.

In DC 500F methionine was first limiting. None of the amino acids appeared deficient in Excell 505 which gave better N retention than DC 500F even when the latter was supplemented with methionine.

Wessels (1970a) suggested that in view of reactions between tannins and methylating agents (Chang and Fuller, 1964), the methionine deficiency in DC 500F might result from its high "tannic acid" content.*

Wessels (1970b) continued his studies using a similar basic diet (1970a), a different breed of chick and different sorghum cultivars. The amino acids added included lysine, arginine, methionine, threonine, tryptophan and glycine. Lysine was first limiting in DC 36, NK 300, NK 222 and Sensako SK 52. First limiting in SK 2 appeared to be arginine and the second lysine. Methionine seemed to be second limiting in NK 300. Neither arginine, threonine, methionine nor tryptophan proved to be second limiting.

Wessels (1970b) cited du Preez and Wessels (1970), Vohra *et al.* (1966) and Chang and Fuller (1964) as being in agreement that "tannic acid"

* (Authors' note: it is possible that DC 500F was high in an unspecified tannin, it is most improbable that tannic acid was present (see Chapter 4)).

content in the range 0·24–0·46%, as found, was below the level likely to depress weight gain, feed intake or N retention in chickens (Vohra *et al.*, 1966). The available amino acids appeared to vary widely among different sorghum cultivars.

Asia

Singh and Barsaul (1977) compared maize, barley, sorghum and pearl millet mash in rations for White Leghorn (WL) and Rhode Island Red (RIR) chicks from day-old to day of slaughter (about 100 days). The cereals represented 40% of the diet, the remaining 60% being the same in all four diets. The time to slaughter, weight gain, feed consumed and F/G quotient are shown in Table 2/81.

The pearl millet rations produced a better performance than the other three cereals which showed practically the same results in each breed. Calculations for net profit made on the basis of market prices in Mathura, India during the experimental period showed a higher net profit from the use of pearl millet.

Australia

Connor *et al.* (1976) compared nine sorghum hybrids; Alpha, DeKalb E 57, Pacific 007, Pioneer 846, TX 610, TX 671, Pacific 303, TX 622 and Yates NK 207 grown at three regions in Queensland in 1971, and the first six hybrids grown at two of the regions in 1970, as a source of metabolizable energy (ME) for chicks. Maize hybrids were also grown. In feeding trials the test diets consisted of 60 parts of a basal diet (31% crude protein) and 40 parts of the grain under test. ME corrected for N equilibrium was calculated by the procedure of Sibbald and Slinger (1963). The means and standard deviations for ME (kcal/kg dry matter) for the 39 sorghum samples and 48 maize samples were respectively 3750 ± 239 and 3770 ± 154.

A 9×3 factorial comparison (of the nine hybrids grown in three regions) showed that the effects of hybrid and region, and the hybrid × region interaction, were significant. In the $6 \times 2 \times 2$ factorial comparison (six hybrids in two regions in two years) the effects of hybrid, region and year, and the hybrid × region and hybrid × year interactions were significant.

ME was also calculated in 20 samples from chemical composition following equation No. 5 of Sibbald *et al.* (1963) relating ME to the percentages of starch, sugar, crude protein and ether extract. The mean values found were (kcal/kg) 3530 for sorghum and 3470 for maize. Connor *et al.* considered that the correlations between ME and chemical composition were not sufficiently high to enable ME to be predicted with sufficient accuracy for practical application.

Europe

Seidler *et al.* (1964) in Poland, found in a balance trial with chicks using sorghum and maize of similar proximate composition, that protein, fibre and N-free extract were more digestible and N-balance lower, on sorghum based than maize based diets.

Turek *et al.* (1966) in Austria reported no unfavourable effect on weight, feed efficiency and mortality when sorghum replaced 10, 20 and 30% of the maize in broiler rations.

Herstad *et al.* (1966) in Norway, reported that chicks attained approximately equal weights on the following cereal-based diets:

		Proportion in diet (%)
(A)	Maize	58
(B)	Wheat	64
(C)	Maize	30
	Barley	30
(D)	Maize	32
	Oats	32
(E)	Maize	30
	Wheat	58
(F)	Maize	29
	Sorghum	29

The use of (G) 64% barley, (H) 67% oats or (I) 58% sorghum in the diets resulted in poorer growth. There was no detectable difference in general health or chick mortality among the different cereals.

Polidori *et al.* (1967) at the University of Catina, Sicily, fed grain sorghum as a replacement for maize in two series of diets to broilers. The first series of diets provided 3400 kcal metabolizable energy (ME)/kg, 24·5% protein from hatching to 5-weeks old and 22% protein from 5–8 weeks. In the second series, ME was about 2900 kcal and protein levels 21·5% and then 18%. The main cereals in all diets were (A) maize, (B) sorghum, (C) maize 2/sorghum 1 and (D) maize 1/sorghum 2.

Up to 5 weeks there were no significant differences but at 8 weeks, diets without sorghum gave greater weight and feed efficiency.

Bornstein and Bartov (1967) in Israel compared the nutritive value of sorghum with maize and fresh local grains with stored grains imported from the USA in broiler rations. Among five diets one grain constituted 100, 75, 50, 25 and 0% of the cereal, the other grain providing the balance. The diets were isonitrogenous, but not isocaloric, the cereal protein being supplemented with soybean meal, 3% fish meal and uniform supplementation with DL-methionine.

There was no consistent difference in growth rate and feed efficiency between sorghum and maize, or between local and imported grains. Skin pigmentation was increased with increasing levels of maize, especially with local maize with a xanthophyll content of 28 μg/g against 14·9 μg/g in imported maize.

Bornstein and Bartov (1967) commented that the literature comparing sorghum and maize in chick diets contained variable and contradictory results. The diets, birds and sorghums used in experiments in the 1930s and 1940s and early 1950s were not readily comparable to those in use in 1967. Records indicated that the average protein in sorghum from the USA declined from 11 to 9·5% between 1959 and 1965.

In 1971, Bornstein and Lipstein, with chicks from 5 to 8 and 22 to 29 days of age, compared the available sulphur amino acids in sorghum and maize supplemented with soybean meal. The crude protein contents of the main ingredients were:

	Maize	Sorghum	Soybean meal
Crude protein (%)			
Trial 1	8.5	9·7	47·0
Trial 2	9·1	9·5	44·8
Trial 3	8·2	9·0	45·0

The composition of the basal diets (A) maize and (B) sorghum, are shown in Table 2.82.

In each trial the maize or sorghum was supplemented with DL-methionine at levels of (1) 0·01%, (2) 0·06%, (3) 0·11%, (4) 0·31% to the sorghum and 0·26% to the maize diets. The latter were in excess of need.

The sulphur amino acids were reported as first- and only limiting amino acids for both diets (A) and (B). The estimated available sulphur amino acid content of sorghum protein was 2·62%, about 1·27% lower than maize.

Hassan (1974) in the United Kingdom, studied egg production from Thornber 606 laying hens fed grain sorghum with protein supplements including cottonseed meal, sesame meal and meat meal. Nitrogen retention and digestibility were determined for each diet, and of available lysine and available methionine on the meat meal used.

Though apparently adequate in protein and essential amino acids, the diets did not support satisfactory egg production. Supplementation with synthetic L-lysine plus DL-methionine, or with sodium glutamate, significantly improved egg production, live weight gain and feed conversion efficiency.

Bornstein and Lipstein (1972) reported from consecutive trials that Leghorn hens tended to produce smaller eggs from sorghum than maize diets. Supplementation of the sorghum diets with DL-methionine equalized egg weights. The calculated sulphur amino acid requirement for optimal egg size appeared to be about 560 mg/bird/day.

Whole grain maize contained more linoleic acid (2·3%) than sorghum (1·5%), a difference reflected in the fatty acid pattern of the yolk lipids.

Hornoiu *et al.* (1965) reported that the hybrid sorghums NK 210, NK 230 and X 3000 grown in Romania in 1961-62 could replace maize, wholly or partially, in the diets of pullets and broilers without affecting egg production or growth.

South America

Santoro *et al.* (1966) used feterita (*Sorghum caffrorum* var. *bicarinatum*), one of the most widely grown sorghums in Uruguay, to replace 14·4, 28·8, 43·2, 57·5 and 72·0% of the maize, barley and wheat in broiler rations. The ration containing 72·0% sorghum was withdrawn because of poor growth.

Except for the lowest level of sorghum, 14·4%, there was a progressive reduction in body weight with increase in feterita. Up to the 28·8% level the differences were not serious.

Mosquera *et al.* (1966) reported from Uruguay that substituting a mixture of cereals (yellow maize, barley and wheat) representing 72% of the ration for broiler chicks, by increasing percentages of the sorghum Martin Milo (14·4, 28·8, 43·2, 57·6% and complete replacement) did not affect growth, feed consumption or feed efficiency up to 12-weeks old. The rations supplemented with meat meal, liver meal and sunflower cake contained 20% protein.

United States of America

Quisenberry and Tanksley Jr (1970), in a long review of sorghum grain in chick and broiler diets, proposed starter and finisher diets containing approximately equal amounts of maize and sorghum supplemented by soybean meal, fish meal, alfalfa meal and maize gluten meal. They stated that poultry requirements for protein, lysine, threonine, tryptophan and methionine could not be met by sorghum without amino acid, mineral and vitamin supplementation.

Sanford *et al.* (1968) compared sorghum and maize in broiler diets of 17 and 21% protein, (A) all sorghum, (B) all maize and (C) 50 maize/50 sorghum. Essential amino acids were equalized by supplementation. At 17% protein diet (B) produced significantly better growth than (A). Sorghum supplemented with lysine and methionine improved feed utilization and growth and was significantly better than non-supplemented sorghum but not significantly different from maize. The 21% protein sorghum produced significantly better growth than the 17% protein; there was no significant difference between the 17 and 21% protein maize diets. Significantly better feed utilization was obtained with 21% protein than with 17% protein diets.

Sanford (1972) compared (A) yellow maize, (B) yellow endosperm sorghum and (C) ordinary sorghum in 24% protein starter diets, and 20% protein finisher diets fed to broiler chicks.

Pigmentation in the shanks and skin decreased in the order (A), (B), (C). Diet (A) produced the best rate of gain, followed by (B) then (C). In feed utilization efficiency, the order was (A), (C), (B).

Thayer *et al.* (1957) reported on the proximate composition (methods not stated) of 16 varieties of grain sorghum developed and grown in Oklahoma, where, under the semi-arid climate conditions of a large part of that state, sorghum is better adapted than maize. The range and means are given in Table 2.11. The varieties were fed in starter and grower rations to Barred Plymouth Rock and New Hampshire chicks as a total replacement for maize on an equal weight basis. Based on comparative body weight, feed consumption and mortality, most of the sorghums were considered superior to maize.

The sorghum varieties Redlan and DeKalb E 56A with two experimental strains of yellow endosperm sorghum Y 8 and Y 10 were compared with maize in two diets (A) high-nutrient and (B) low-nutrient (Ozment *et al.*, 1963). The cereals provided (A) about 30% and (B) about 60% of the diet and the crude protein in the diets was (A) 28% and (B) 23%.

Broilers fed (A) grew faster and consumed less total feed than those on (B). The birds fed the sorghum diets were equivalent in average body weight and consumed approximately the same amount of feed, as those fed maize.

There were differences among the sorghum varieties within the two diet groups. An average value for milo (see Table 2.11) had been used in formulating the diets. Subsequent analysis (methods not stated) showed differences in chemical composition (Table 2.83) among the samples used which could have accounted for differences in performance.

Waggle *et al.* (1967a) studied the relation between protein in sorghum grain, its nutritive value for the chick, and its amino acid composition.

Sorghum hybrid RS 610 containing (A) 8·3% protein (Kjeldahl N × 6·25) (low level) and (B) 10·5% (medium level); (C) a composite of protein content 12·1% (high level) produced equivalent weight gain and feed conversion when incorporated in isonitrogenous diets for White Rock chicks. When the diets contained equal quantities of grain and soybean meal of 43·7% protein, (C) produced significantly more gain than (A).

The amino acid composition (autoanalyser following acid hydrolysis) excluding tryptophan, of the three sorghum samples is shown in Table 2.13. Methionine and lysine were calculated to be the limiting amino acids in the diets.

Quisenberry *et al.* (1970) reviewed the work on sorghum in poultry diets conducted at Texas A & M University over the 12 years from 1946 to 1970 and concluded that sorghum grains provide an excellent source of energy for poultry. Differences in nutritive values among sorghum types were not entirely the result of differences in protein or amino acid patterns.

In chick diets in which the protein and energy content of the diets were equalized at approximately 20% and 1040 kcal/lb respectively by supplements of fish flour and refined cottonseed oil, DD 38 Milo produced significantly higher rates of growth than Early Hegari or 7078 Milo in faster growing male chicks: DD 38 Milo gave significantly higher growth than 7078 Milo in female chicks (Kemmerer and Heywang, 1965). DD 38 Milo (protein 9·8%) was supplemented with 1·0% fish flour and 1·0% cottonseed oil, Hegari (protein 7·8%) with 2·5% fish flour and 2·0% cottonseed oil; Martin Milo and 7078 Milo were not supplemented.

When maize and sorghum were compared at 63·5% of the diet, the protein content varying, Early Hegari (19·4% protein in diet), 7078 Milo (22.4%) and DD 38 Milo (21·9%) gave significantly lower growth rates than maize (20·1% protein in the diet) and Martin Milo (22·6% protein in the diet). Caprock Milo (of which no details were given) was found to be equal to Martin Milo and maize. Supplementation with 0·5% lysine HCl did not consistently improve the nutritive values.

No detectable amounts of "tannic acid", or cyanide, (unstated methods) were found.

Shoup *et al.* (1970a) determined the content of 17 amino acids and ammonia (ion-exchange chromatography) of Paymaster Kiowa grown at two locations, and TE 66, RS 6100, Asgrow Ranger A, Frontier 400C and DeKalb C44 b grown at one location in Kansas. The mean and ranges with protein contents (N × 6·25 AOAC, 1965) are shown in Table 2.13. Differences in amino acids among hybrids and locations were small.

In trials with Cobb's strain-cross White Rock chicks, each diet contributed 17·0% protein of which 5·87% protein was derived from the sorghum with the exception of one diet in which DeKalb C 44 b was used to provide 4·71% protein. The remainder of the protein was contributed by soybean meal, alfalfa meal and fish meal. Diets were maintained at uniform protein and amino acid levels by amino acid supplementation.

The results indicated no significant difference among hybrids on the weight gain of chicks. Reducing the crude protein content of the diet and rebalancing with pure amino acids did not affect performance significantly.

Deyoe *et al.* (1970) reported the test weight (lb/bu) and proximate composition (AOAC, 1965), and the amino acid content (ion-exchange chromatography) of the sorghum hybrids reported upon by Shoup *et al.* (1970a) but harvested when immature at the late-dough stage, 104 days from emergence. Samples of immature grain from a local elevator were also obtained, and samples of mature Frontier 400 C and Paymaster Kiowa were included. The mean composition and ranges are given in Table 2.11. Table 2.13 presents the amino acid composition. No difference was shown between mature and immature grain in crude protein; ash and fibre were higher in the immature. Ether extracts were lower in the lower test weight sorghums. Immature grain was higher in lysine, aspartic acid, glycine and methionine; lower in glutamic acid, proline and leucine than mature.

Strain-cross White Rock chicks were fed isonitrogenous (22% protein) diets in which mature or immature Frontier 400 C and Paymaster Kiowa represented about 60% of the diet. Soybean meal (45% protein), alfalfa meal (17% protein) and fish meal (60% protein) provided the remainder of the protein. Immature grain produced significantly poorer growth and feed conversion:

	Frontier 400 C		Paymaster Kiowa	
	Mature	Immature	Mature	Immature
Test weight, lb/bu	58·3	43·2	58·0	43·1
Weight gain 8 weeks (g)	1569	1508	1625	1497
F/G, 8 weeks	2·20	2·35	2·25	2·37

Deyoe *et al.* suggested the immature grain provided "less productive energy".

Stephenson *et al.* (1971) University of Arkansas determined the amino acid content, excluding tryptophan (Piez and Morris, 1960) (see Table 2.13), and the amino acid availability for the chick (Bragg *et al.*, 1969) of 24 grain sorghum hybrids grown on experimental plots under similar conditions and harvested at the same stage of maturity.

The amino acid content varied considerably among the hybrids especially in methionine and cystine, and in lysine, but to a lesser extent. The differences in availability among individual amino acids were greater than overall availability (see Table 2.84). The greatest extremes were observed in proline, 19·4% available in ARK 61002 but 93·3% available in RS 640. Availability of all amino acids was low in ARK 61002, GA 609, RS 617 and ARK 62003; overall availability was higher in Lindsey 755, RS 622, RS 640 and AK 30001R.

Quisenberry *et al.* (1970) compared (A) a red grain sorghum and (B) a yellow grain sorghum (DeKalb G 600) in laying hen diets, supplemented with soybean meal, to provide 14% protein and 973 kcal/lb productive energy. The control diets were (C) a similar diet containing maize, and (D) a maize diet supplemented with 5% fish meal to provide 16% protein. Diets (B) and (C) gave satisfactory body weight gain, egg production, egg size and feed efficiency. Diet (A) was unsatisfactory by all these criteria. Diet (D) gave poorer egg production, equal egg size and poorer feed eficiency than diet (C), which Quisenberry *et al.* suggested might have been due to poor quality fish meal.

Quisenberry *et al.* compared (E) a DeKalb yellow sorghum, (F) a DeKalb bronze sorghum and (G) a Northrup King yellow sorghum with (H) a maize-soya diet, all isonitrogenous (14% protein) and isocaloric (961 kcal/lb productive energy). All the grain sorghums were considered equal to the maize in egg production, egg size, feed efficiency and survival. There were no significant differences in feed consumption.

In addition, (I) a red grain sorghum 3758, (J) a white sorghum 765 W, (K) a yellow endosperm sorghum G 600, (L) a commercial red grain sorghum and (M) yellow maize were compared in layer diets (14% protein) grain–soy diets providing 938 kcal/lb productive energy. There was little difference among the diets; egg production was low and mortality was high, largely attributable to leucosis and Karel's disease.

Quisenberry and Tanksley Jr (1970) reviewed diets for laying hens and suggested rations for both heavy and light breeds, containing 50% or more of sorghum.

Sanford (1972) found no significant difference in % egg production regardless of whether laying hens were fed a 17% protein diet of (A) all maize, (B) all sorghum (not otherwise designated), (C) all wheat, (D) 50 maize/50 sorghum or (E) 50 wheat/50 sorghum. Hens on (C) required more feed/dozen eggs than on any of the other diets. Diet (A) produced the largest eggs and (C) the smallest with the other diets intermediate.

Sanford and Deyoe (1974) reported that 90% sprouted sorghum was not harmful when fed in 17% protein diets to laying hens and was equivalent to unsprouted grain for commercial egg production.

Guenthner and Carlson (1970) compared triticale, maize, wheat and sorghum in diets (the exact composition was not given) fed to laying hens. Two diets were formulated from each grain to contain 12·0 and 15·4% crude protein (N × 6·25). All the protein in the 12% wheat and triticale, essentially isocaloric, diets was provided by the cereals, the other diets being balanced with soybean meal. Methionine and lysine were added to provide a minimum of 0·52% methionine plus cystine and 0·5% lysine.

In comparison with the average egg production during the 3 weeks preceeding the 9 week test period, in the 15·4% protein series, egg production increased 4% with wheat and maize, 5% with sorghum and decreased 4% with triticale. In the 12% protein series, production decreased 13% with triticale, 10% with wheat, 7% with sorghum and increased 4% with maize. Hen performance was equivalent on both the maize diets; the 12% protein triticale, wheat and sorghum diets did not support egg production equivalent to maize.

Hens on both triticale diets and on the 12% wheat diet lost weight, indicating amino acid deficiences other than lysine and methionine plus cystine. The superior amino acid composition of the soybean meal that supplemented the maize and sorghum diets possibly influenced this result.

Various non-dietary environmental factors were found to affect liver fat accumulation in laying hens. Jensen *et al.* (1976) compared several cereals in isocaloric diets (containing soybean meal and other nutrients) fed over a 3-week period. The cereals included, %, in diet, (A) sorghum 68·2; (B) maize 70·2; (C) triticale 72·0; (D) wheat 67·3; (E) barley 60·0; (F) oats 55·5; and (G) rye 65·6.

Liver weights, % liver fat and total liver fat of hens fed diets (E), (F) and (G) were significantly lower than diets (A), (B) and (C). Diet (D) produced an intermediate result.

Peischel *et al.* (1976c) reported a 3 × 2 factorial design to evaluate diets for laying hens composed of (A) all maize, (B) all sorghum and (C) 50 sorghum/50 maize, each diet without, and with, 2% Brazilian fishmeal. During the 84 days of the trial the poorest egg production and feed conversions were from (B) without fishmeal and the best from (C) with fishmeal. Without fishmeal, the best egg production was from (C) 54·1%, with (B) lowest 48% and (A) intermediate 51·1%, the differences being significant.

Turkeys

United States of America

Quisenberry and Tanksley, Jr (1970) reviewed the literature on the use of sorghum in turkey diets and provided tables of pre-starter, starter and grower rations, containing high levels of sorghum, which have produced satisfactory growth in trials at Texas A & M University, USA.

Harris *et al.* (1966) and Waldroup *et al.* (1967) found that maize, wheat or sorghum grain could be used effectively in turkey poult diets when fed in balanced feeds. Pelleting significantly improved feed utilization with all grains.

Atkinson *et al.* (1974) found 18% protein superior to 15% protein rations for turkey reproduction. When used as the only grain source at both protein levels sorghum was superior to maize in egg production, and to a lesser extent in fertility and hatchability.

When day-old turkey poults were fed 28% protein diets in which sorghum, wheat or maize was the sole grain source, during 8 weeks, little or no difference between maize and sorghum diets was found in body weight, feed efficiency, or % mortality. Wheat was significantly superior to maize and sorghum in body weight gain and % mortality (Atkinson *et al.*, 1975).

Swine

Africa

Oyenuga and Fetuga (1975) in Nigeria compared (A) maize, (B) sorghum and (C) "millet" (possibly pearl millet) in feeding trials with pigs. The proximate analysis of the feeds (% DWB, AOAC, 1970) is shown in Table 2.85. In the first experiment, barrows were fed the cereal as the only source of nutrients. Faecal samples were analysed for proximate constituents (AOAC, 1970) and apparent digestibility of nutrients (National Academy of Sciences, National Research Council, 1963a). The results are presented also in Table 2.85.

In a second experiment 12 large white barrows were used in two consecutive metabolism trials to determine digestible energy (DE), metabolizable energy (ME) and ME corrected for nitrogen retention (ME_n): the ground cereals provided 98% of the diet and were supplemented with 2% vitamin and mineral premix. Feeds, faeces and urine were analysed for Kjeldahl N (AOAC, 1970) and combustible energy in a Gallenkamp oxygen ballistic bomb calorimeter. The ME_n values were calculated (National Academy of Sciences, National Research Council, 1966) assuming the energy equivalent of urinary nitrogen to be 6·77 kcal/g N retained (Diggs *et al.*, 1965). The results are shown in Table 2.85.

No significant differences wre found among the cereals in total digestible nutrients, which were high in GE, DE, ME and ME_n.

Asia

Takahashi *et al.* (1968) in Japan reported from digestibility studies with pigs, digestible crude protein, total digestible nutrients and digestible energy from maize and grain sorghum, as follows:

	Maize	Sorghum
Dry matter (%)	85·4	86·4
Digestible crude protein (%)	6·5	6·1
Total digestible nutrients (%)	80·0	81·2
Digestible energy (kcal/100 g)	356	358

Alcantara *et al.* (1970) fed Darso sorghum, grown in the Philippines, to female weanling pigs from an initial weight of about 10·52–90 kg live weight. Sorghum, maize and commercial feeds, each with a protein supplement were compared in three feeding regimes: (A) hand fed twice a day, (B) hand fed three times a day, all the pigs could eat in 30 min and (C) self-feeding with feed available at all times.

There were no significant differences in daily gain and feed efficiency among (A), (B) and (C) but (C) produced the highest daily gain and feed intake. The feed efficiency of ground maize and ground sorghum was significantly higher than either whole sorghum or the commercial feeds.

Han and Ha (1977) reported from Korea the results of feeding trials with growing-finishing swine over 12 weeks in which 25% of the starter diet consisted of, %, wheat bran 5·75, soybean meal 12·0, fishmeal 5·0, vitamins and minerals 2·25 and the remainder (A) maize 75, (B) maize 50/sorghum 25, (C) maize 25/sorghum 50 and (D) sorghum 75. The diets were approximately isonitrogenous ranging from 15·1 to 15·9%. A control commercial diet was also fed, protein 16·3%, the composition of which was not known to Han and Ha. The finishing diets were similar except for, %, wheat bran 12·25, soybean meal 7·0, fishmeal 4·0 and vitamins and minerals 1·75, giving a protein content ranging from 14·3 to 14·7%. The effects on weight gain and feed efficiency, digestibility, N retention, digestible energy and ME are presented in Table 2.86.

The substitution, partial or complete, of maize by sorghum did not affect growth rate but increased feed consumption and F/G. As the level of sorghum increased, digestibility of crude protein and fat, N retention and DE and ME, tended to decrease and digestibility of crude fibre improved; digestibilities of dry matter and NFE were not affected. Carcass quality improved, and cut yield percentage was approximately equal, as the level of sorghum increased. Han and Ha concluded that, although sorghum was somewhat inferior to maize in nutritive value, complete substitution of maize by sorghum was satisfactory if relative prices were comparable.

Australia

Beames *et al.* (1973) in Queensland, fed high-protein sorghum (13·3, 11·1 and 10·9%) and low-protein sorghum (7·4. 6·5 and 7·4%) to pigs from weaning to slaughter, in various diet formulations. The eight diets included soybean meal (protein 47·4, 48·1 and 48·5%) at four levels, 16·6, 13·8, 10·7 and 7·4%, providing dietary crude protein ranging from 18·6%–9·4%.

The level of the protein in the grain had no significant effect on growth rate or feed efficiency but as the soybean meal in the diet decreased growth rate and feed efficiency tended to decrease. Growth rate and feed efficiency were lower in pigs on the 10·7 and 7·4% soybean meal diets but only up to a live weight of 45 kg. In an experiment which commenced at 45 kg live weight, growth rate and feed efficiency were reduced only when the soybean meal was at the lowest level of 7·4%*.

Europe

Tschiderer (1966) reported that an Austrian-grown sorghum (crude protein 11·0%) fed as 60–67·5% replacement of maize in pig fattening rations resulted in reduced weight gain in varying degrees.

In pig fattening trials carried out in Belgium comparing yellow maize and sorghum imported from the USA, Vanschoubroek *et al.* (1964) found no significant differences between the cereals in live weight gain, feed utilization, and carcass quality. The maize and sorghum (crude protein 9·6 and 10·1% respectively) were present on an equal weight basis and represented 40% of the rations.

Angelov (1965) in Bulgaria reported that maize, providing 50% of pig-fattening rations, could be replaced by sorghum.

Castaing and Leuillet (1976) conducted trials in France in which sorghum replaced maize to the extent of (A) 25, (B) 50, (C) 75 and (D) 100% of 16–17% crude protein diets fed to growing–finishing pigs from a starting weight of 25 kg to a finishing weight of 103 kg. During the growing period sorghum showed slightly higher F/G than the all maize diet (E), but over the whole experimental period, no significant differences were observed. These results were more favourable to sorghum than those from previous trials (Castaing and Moal, 1973), when diets of maize and soybean meal, providing 18% crude protein, were fed to pigs with progressive substitution of the maize by "milo corn" of (A) O, (B) 40, (C) 70 and (D) 100%. A reduction in growth rate was observed as the proportion of sorghum increased. Simultaneously, the feed conversion ratio (kg feed/kg gain) increased: (A) 3·23, (B) 3·31, (C) 3·44 and (D) 3·51. Castaing and Leuillet (1976) recommended that sorghum be used moderately in the growing period but could be the sole cereal source in the fattening stage.

Lawrence (1968) conducted experiments in the United Kingdom in which high levels (85% in the starter diets fed up to 52 kg live weight and 90% in the finisher diets) of maize, flaked maize, sorghum, wheat and barley were fed to pigs from weaning to 90 kg.

* (Authors' note: in common with some other reported experiments, this appears more as a comparison of soy protein supplements and total protein in the diet than of the feed value of different sorghums).

The cereals compared on an equal weight basis were supplemented with equal amounts of fishmeal and soya meal. The oil, fibre and protein content of the diets therefore differed.

The general health of the pigs was good with the exception of those on the sorghum diet, three of which, at 27 kg live weight, became severely affected by general uncoordinated leg movement of the hind quarters, marked depression in appetite, and growth reduction. Because the symptoms resembled those of pantothenic acid deficiency, 25 mg each of pantothenic acid and niacin was added daily to the diet of all the pigs. Only after 2 months did the affected pigs return to normal. At slaughter, nerve degeneration in the spinal cords of all three pigs was apparent. The determined pantothenic acid content of the sorghum diet was 4·1 μg/g of 'free' calcium *d*-pantothenate compared to 6·7 in the control diet. All diets except the control were lower in pantothenic acid than recommended by the Agricultural Research Council (1966) and therefore pantothenic acid deficiency may not have been the main cause of the leg weakness in the pigs fed sorghum.

Up to 52 kg, growth rates did not differ significantly among diets but feed conversion for the maize and barley diets was superior to that for the wheat and sorghum. Overall, from start to slaughter, the maize diet was superior to the flaked maize, sorghum and control diet.

The highest percentage of lean in the carcass was found from the control and from barley.

Bokorov and Srećković (1964) in experiments conducted at Novi Sad, Yugoslavia found in rations for fattening pigs, maize gave the best weight gains followed by sorghum, with barley giving the poorest. The grains, the composition of which was not stated, were fed in rations that were essentially equal in protein.

Satisfactory weight gain, feed efficiency and carcass quality was obtained by combining maize and sorghum, barley and sorghum and barley, maize and sorghum.

Hornoiu *et al.* (1965, 1967) reported that a 50% replacement of maize with hybrid sorghums NK 210, NK 230 and X 3000 grown in Romania in 1961–62 produced 10–12% higher weight gains and 18% better feed consumption in pigs than maize alone.

Stafijcuk and Teljatnikov (1968) in Russia compared 212 samples of dehusked and 45 of whole grain sorghum in digestibility trials with sheep and pigs. The dehusked grain tended to be lower in ash and fibre and higher in N free extract that the whole grain. Digestibility/100 g sorghum was:

	Sheep	Pigs
Dehusked sorghum		
Feed units	123·7	125·4
Digestible crude protein	7·33	8·15

	Sheep	Pigs
Dehusked sorghum		
Whole grain sorghum		
Feed units	116·9	99·3
Digestible cruide protein	7·75	6·87

These values were said by Stafijcuk and Teljatnikov to be lower than those of maize for sheep and of barley for pigs.

South America

Gontijo *et al.* (1976) reported two experiments from Brazil to study (1) the effect of substituting sorghum, designated as the hybrid McNair 644, for maize in diets supplemented with soybean meal, minerals and vitamins and (2) the effect of supplementing 13% crude protein diets based on sorghum and soybean meal with 0·2 or 0·4% L-lysine, or 0·2% L-lysine + 0·1% DL-methionine, on the performance of growing swine. The crude protein content (AOAC, 1970) of the main ingredients was, %, maize 8·3 sorghum 7·8 and soybean meal 48·4.

In experiment 1, the proportions of cereal in the diets were, %, (A) maize 100, (B) maize 75/sorghum 25, (C) maize 50/sorghum 50, (D) maize 25/sorghum 75 and (E) sorghum 100. The diets for growing swine contained about 16% protein and for finishing swine about 13% protein. No significant differences were observed among diets in average daily weight gain or average daily feed consumption but there was a significant linear decrease in feed efficiency with increases in sorghum among pigs on the growing rations.

In experiment 2, reducing the crude protein content of the growing pig sorghum-based diet from 16 to 13% reduced the average daily gain from 0·71 to 0.57 kg. Supplementation with 0·2% L-lysine raised the average daily gain to 0·68 kg and was sufficient to counteract the 3% reduction in crude protein. The addition of 0·2% L-lysine + 0·1% DL-methionine or of 0·4% L-lysine gave no further advantage but raised the average daily gain to 0·68 and 0·70 kg respectively.

Gontijo *et al.* (1976) concluded that the feed value of sorghum for swine was 90–95% that of maize and that lysine was first-limiting.

United States of America

Quisenberry and Tanksley Jr (1970) reviewed the literature on the use of sorghum grain for feeding swine and commented that it provided an excellent feed when adequately supplemented, sorghum protein being deficient in lysine, methionine plus cystine, threonine and tryptophan. Sorghum seemed also deficient in Ca and P. Formulations incorporting levels of sorghum between 75–85% were suggested for gestation, lactation, grower, finisher and final finisher rations. For pig starter diets, lower levels of sorghum of about 38% were recommended.

Noland and Scott (1963) reported on the effect of substituting various grains obtained locally in Arkansas for maize in rations for growing–finishing swine. In the rations, soybean meal was the primary source of protein. On the basis of total replacement of the maize, sorghum gave improved weight gain but reduced feed efficiency.

Jensen *et al.* (1969) compared the feeding value of Opaque-2 maize (crude protein content 9·5%, DWB) with sorghum (crude protein content 8·5%) in diets in which the cereals provided about 97·0% of the diets, supplemented with 2·7% minerals and vitamins and lysine to provide roughly 0·49% lysine in each diet. In experiment 1, the pigs were fed ad lib. over 47 days, and in experiment 2, the animals were group-fed over 36 days or individually-fed over 49 days. The diets were fed in both meal and pellet form. The results are shown in Table 2.87. In both experiments sorghum produced lower gain/feed quotients than maize.

Luce *et al.* (1972) compared a hard red winter wheat (Triumph) and a yellow endosperm grain sorghum on equal weight and isonitrogenous basis, as meal or pelleted, with and without supplementation with L-lysine or soybean meal. The finishing swine rations contained different levels of 44% protein soybean meal. Pelleting improved gain by 7% and feed efficiency by 10% for both wheat and sorghum rations.

Klett (1973) in a shorter review agreed with Quisenberry and Tanksley Jr (1970) that the feeding value of sorghum grain was equal to that of maize and wheat for feeding swine.

Tanksley Jr (1973) reviewed research findings from 1963 and stated protein and foreign matter content of sorghum grain in Texas varied widely.

Cohen and Tanksley Jr (1973b) in trials with growing pigs fed a basal diet of heteroyellow sorghum grain plus 0·3% L-lysine, and test diets consisting of the basal diet with additions of (A) tryptophan, (B) threonine and methionine, (C) threonine and tryptophan, (D) methionine and tryptophan and (E) threonine, methionine and tryptophan. The results "strongly suggested" that for the growing pig threonine was second-limiting. Tryptophan appeared to be third-limiting followed by methionine.

Hodson *et al.* (1973) compared maize and sorghum in diets with soybean, at protein levels of 16, 14 and 12%, supplemented with L-lysine. There was no significant difference in weight gain between maize and sorghum diets but within each group, increase in protein and lysine supplementation increased weight gain. In feed efficiency, the maize–soy diets were significantly better than the sorghum–soy diets.

Tribble *et al.* (1972) fed maize, grain sorghum and wheat in pig rations balanced with soybean meal to provide 0·7% lysine initially and 0·5% after the pigs weighed 60 kg. The pigs fed maize gained significantly faster than those fed sorghum; maize and wheat were better than sorghum in F/G. Digestible protein and energy were significantly different and respectively: maize 80 and 83%, sorghum 73 and 80%, and wheat 84 and 86%.

Tonroy *et al.* (1973) determined the apparent P digestibility in growing swine fed sorghum in diets containing all the recommended (National

Research Council, 1968) nutrients, with the exception of P. The diets contained two levels of P (determined colorimetrically, Harris and Popat, 1954), 9·5% optimum or excess, and 0·3%. Low-P beef blood fibrin was the principal source of protein. Sorghum was compared with maize grits in diets equivalent in protein, Ca, P and Ca/P ratio. In the maize grits diets, dicalcium phosphate (USP grade) provided 86·0 and 91·8% of the total P in the 0·3 and 0·5% diets respectively. In the 0·3 and 0·5% sorghum diets dicalcium phosphate provided 31·6 and 59·1% of the total P, the sorghum providing 67·1 and 40·1% respectively. The apparent digestibility of the P in sorghum grain was calculated to be 4·5% for the 0·3% diet and 1·9% for the 0·5% diet, considered inadequate for the growing pig.

Eckert and Allee (1974) reported three growth and two N retention trials with pigs in which sorghum represented 96% of the rations. The results indicated that lysine was first-limiting, threonine second-limiting and that the third-limiting amino acid was not methionine but might be isoleucine.

Copelin *et al.* (1976a) used sorghum RS 671 (lysine 0·2%) in growing pig diets containing varying amounts of casein, maize starch and sorghum with all essential amino acids, except lysine, to provide 110% of the NRC requirements for the 10 kg pig.

The seven diets with lysine contents, %, were: (A) casein 0·553; (B) casein 0·465; (C) casein 0·377; (D) casein 0·509, sorghum 0·044; (E) casein 0·465, sorghum 0·088; (F) casein 0·421, sorghum 0·132 and (G) casein 0·377, sorghum 0·176. The available lysine was calculated to be 62·7% (Copelin *et al.*, 1978).

Similar experiments were reported by Copelin *et al.* (1976b) to estimate the availability of threonine in sorghum RS 671. Threonine present in seven isonitrogenous and isocaloric diets was, %, (A) casein 0·357; (B) casein 0·229; (C) casein 0·100; (D) casein 0·293, sorghum 0·064; (E) casein 0·229, sorghum 0·128; (F) casein 0·164, sorghum 0·193 and (G) casein 0·100, sorghum 0·257.

The available threonine in sorghum was calculated to be 88·9%.

Copelin *et al.* (1976c) estimated the availability of tryptophan in sorghum RS 671 from pig feeding trials reported above. Five isonitrogenous and isocaloric diets contained the following levels of tryptophan, %: (A) casein 0·105; (B) casein 0·087, sorghum 0·018; (C) casein 0·070, sorghum 0·035; (D) casein 0·052, sorghum 0·053 and (E) casein 0·035, sorghum 0·070.

Using statistical analyses similar to those used earlier, the availability of tryptophan in RS 671 sorghum was calculated to be 92·6%.

Copelin *et al.* (1974) compared (A) yellow maize, (B) nonyellow endosperm sorghum, (C) heteroyellow endosperm sorghum, (D) yellow endosperm sorghum and (E) low-protein (6·38%) yellow endosperm sorghum, all grown under similar agronomic conditions in feeding trials with growing–finishing swine. Soybean meal was added to each grain to give 16% protein in grower rations and 14% in finisher rations. Protein, lysine and ether extract (% grain, DWB) follow:

	Protein (%)	Lysine (mg/100 g grain)	Ether extract (%)
(A) Maize	8·9	240	3·79
(B) Nonyellow endosperm sorghum	8·9	190	3·13
(C) Heteroyellow endosperm sorghum	9·2	200	2·81
(D) Yellow endosperm sorghum	9·5	210	3·25
(E) Low protein yellow endosperm sorghum	6·4	170	2·64

There were no significant differences among diets in feed utilization or carcass characteristics indicating that low-protein sorghum is satisfactory in finisher diets if adequately supplemented.

Meadows *et al.* (1974) compared (A) yellow maize, (B) nonyellow endosperm sorghum, (C) heteroyellow endosperm sorghum and (D) yellow endosperm sorghum, all grown under similar dry land conditions in diets for growing-finishing swine. Soybean meal was added to provide 16% crude protein in the grower diet and a 14% in the finisher diet when mixed with the lowest protein grain, nonyellow sorghum. Crude protein, lysine, F/G and relative feed efficiencies (RFE) (assuming maize to be 100), are shown below:

	Crude protein (%)	Lysine (mg/100 g grain)	F/G	RFE
(A) Yellow maize	9·4	280	3·1	100
(B) Nonyellow sorghum	9·1	230	3·2	98·1
(C) Heteroyellow sorghum	9·7	250	3·3	95·0
(D) Yellow sorghum	9·4	250	3·3	94·6

No significant differences in average daily gain or feed intake were revealed but diet (A) gave better than the average F/G of the sorghum diets. There were no significant differences in carcass characteristics though loin area of pigs fed diets (A) and (D) were larger than those fed (B) and (C).

Orr *et al.* (1976) reported that pigs weaned at 4-weeks old and fed a 16% protein diet based on sorghum and soybean meal, supplemented with 0·1% or 0·2% L-lysine produced daily weight gain and feed efficiency similar to those fed a similar, unsupplemented, 18% protein diet.

Cohen and Tanksley Jr (1975) reported threonine as the second-limiting amino acid in sorghum for finishing swine; neither methionine nor tryptophan appeared to be third-limiting.

Cohen and Tanksley Jr (1976) in growing and finishing trials with pigs decided that threonine was second-limiting in sorghum; in growing pigs, tryptophan was more limiting than methionine; and in finishing pigs, the addition of neither methionine nor tryptophan to a threonine-supplemented diet produced a positive response, hence the third-limiting amino acid in sorghum for swine was not established.

In 1977 Cohen and Tanksley Jr reported experiments with growing pigs and with finishing pigs to establish the threonine requirement in the two sorghum–soybean meal diets. For the first, growing, diet the sorghum contained 8·5 % protein and the soybean meal 45·3 %; the proportions in the basal diet were: sorghum 79·6% and soybean meal 11·4%. In the finishing diet, the protein content of the sorghum was 9·0% and of the soybean meal 41·7%; the proportions in the basal diet were: sorghum 87·5% and soybean meal 2·5%. Supplemental threonine was added in six increments of 0·06% in the test diets. Regression analyses showed that the level of threonine required to optimize average daily gain and gain/feed in the 16% protein grower rations was 0·47%; in the 14% protein diets 0·39% was required to optimize average daily gain and 0·37% to optimize gain/feed.

Purser and Tanksley Jr (1976) in trials with growing pigs reported that tryptophan was third-limiting in sorghum, and that isoleucine and methionine appeared to be equally limiting.

Hillier *et al.* (1959) used certified seed of the varieties DeKalb F 62 A, Kafir 44 14, Darset, RS 610, Redlan and Amak R12, grown in Oklahoma, in feeding trials with gilts starting at 9–10-weeks old and continuing to 90 kg in weight. The analyses (methods not stated) and a summary of the results are presented in Table 2·88. Kafir 44 14 appeared to give highest feed efficiency and Darset the lowest. Differences among varieties in weight gain were not significant.

When the pigs were allowed to eat as much of each variety presented in rotation as they wished, the following was the order of preference, based on quantity eaten: RS 610, Kafir 44 14, Darset, Amak R12 and Redlan.

Cohen and Tanksley Jr (1973a) using growing swine compared the energy and protein digestibility of sorghums with different endosperm and starch characters.

The four varieties, grown in Texas were:

Variety	Endosperm type	Starch type	(N × 6·25) (%)
(A) NSA 740	Floury	Normal	15·0
(B) TX 3197	Intermediate	Normal	10·5
(C) SC 301–6	Corneous	Normal	13·0
(D) B TX 615	Intermediate	Waxy	11·5

Endosperm texture was determined by the visual rating system of Maxson *et al.* (1971). The grains on an equal weight basis provided 79% of the diets supplemented with 18·25% of 44% protein soybean meal. The proximate analyses (AOAC. 1970) and gross energy (bomb calorimeter) of the diets and feeding results are shown in Table 2.89.

Protein digestibility and N retention were not significantly different among the endosperms and starch types. Digestible energy of the intermediate endosperm (B) and (D) was higher than the floury. Digestible and metabolizable energy of normal and waxy starch were not significantly different.

Smith and Allee (1973) in feeding trials with pigs, compared three sorghum types (A) white endosperm, red seed coat, (B) heterozygous yellow endosperm, bronze seed coat and (C) homozygous yellow endosperm, yellow seed coat. The pigs showed a marked preference for (B), the heterozygous yellow endosperm.

In two feeding trials (1) for 28 days with pigs of starting weight 21·5 kg and (2) with pigs weighing from 57·0 to 100 kg, daily gain, feed/gain, carcass measurements, protein digestibility and N retention were not significantly different among endosperm types.

Ruminants

This review is primarily concerned with the use of sorghum and the millets in human nutrition, and since the digestive systems of the digastric animals, cattle and sheep, differ so greatly from those of the monogastric animals including man and rat, the authors have not reviewed the literature on ruminant nutrition in exhaustive detail. What follows may therefore be regarded as some fairly typical results.

Cattle

Miller *et al.* (1972) studied the digestibility of several sorghum varieties from the World Collection, and of selected lines, parents and hybrid combinations using the nylon-bag method (NBDMD) (Burton *et al.*, 1964) with fistulated steers and in the *in vitro* dry matter digestibility (IVDMD) method (Rooney *et al.*, 1970). Eight varieties grown in Puerto Rico in 1966, with kernel description and % of floury endosperm, with the 48 h NBDMD for the ground and the whole grain are shown in Table 2.90. The 22 varieties grown in Puerto Rico in 1967, with seed colour and percent floury endosperm, and with the 72 h NBDMD for the ground and whole grain are shown in Table 2.91.

All exotic (non-USA) floury endosperm sorghums were more digestible than the US varieties (B TX 406, RS 626, SA 5875 6 1, IS 6269). White grains with more than 70% floury endosperm produced the highest NBDMD values. Brown or purple grain, even with high floury endosperm, produced lower NBDMD values than white grains but higher than nonbrown grains of US varieties. The NBDMD values of whole and ground nonbrown hybrids were similar to the more digestible parent in each cross. Brown-seeded hybrids were generally significantly lower in digestibility than the lowest parent.

The modified Tilley and Terry IVDMD method (Rooney *et al.*, 1970) gave higher values than the one-stage method (Monson *et al.*, 1969) or the NBDMD method. NBDMD and IVDMD values for whole grain were highly correlated. Results from both IVDMD methods were significantly correlated for ground grain but IVDMD and NBDMD values for ground grain were not. Both IVDMD and NBDMD methods gave significant correlations for whole vs ground grain.

Klett (1973 reviewed sorghum in cattle feeding and concluded that adequately processed sorghum grain may be 95% as efficient as maize or wheat.

Australia. Morris (1970) reported the survival feeding of pregnant and lactating beef cows under conditions of severe drought in Australia. For the last 100 days of pregnancy 3 kg/head/day of grain sorghum was adequate for survival, though there was a mean total loss of 34 kg in body weight compared to 8 kg in cows fed 4 kg/head/day. Only 78% of the cows on the 3 kg rations survived to 70 days after parturition compared with 100% receiving 4 kg. Birth weight of calves was not different between the 3 and 4 kg groups.

Gartner *et al.* (1975) in Queensland, compared wheat, sorghum and "white French millet" (common millet) in feeding trials with steers. The diets consisted of 90% grain and 10% roughage and provided:

Diets	Protein
(A) Wheat and Rhodes grass	15·1%
(B) Sorghum and cottonseed hulls	14·0%
(C) Sorghum and lucerne	14·0%
(D) Millet and lucerne	15·6%

The proximate composition (DWB) was as follows:

	Wheat	Sorghum	Common millet
Crude protein (%)	16·2	8·2	15·7
Ether extract (%)	2·0	3·2	3·4
Crude fibre (%)	3·0	3·5	5·7
Nitrogen-free extract (%)	77·0	83·2	71·9
Ash (%)	1·8	1·9	3·3
Calcium (mg/100 g)	20	20	50
Phosphorus (mg/100 g)	400	350	280

Urea was added to (B) and (C) to raise the N × 6·25 to 14·0 %.

Steers on diet (C) showed mild to moderate bloat. A significantly lower growth rate among steers on diet (C) was reported. Feed conversions (kg dry matter/kg live weight gain) were respectively (A) 5·5, (B) 6·0, (C) 6·5 and (D) 5·6.

Europe. Hornoiu *et al.* (1965, 1967) reported that hybrid sorghum NK 210, NK 230 and X 3000 grown in Romania in 1961–62 could wholly replace maize in cattle and sheep feeds without altering milk and wool production, or weight gain.

United States of America. Sorghum grain in diets for beef and dairy cattle, and the rearing of calves, was reviewed by Hale (1970). Hale recommended that sorghum grains be processed by grinding, pelleting, rolling or flaking, otherwise much will be voided undigested in the faeces. (Processing methods for feeds are reviewed in Chapter 5). Sorghum may be used in high concentrate fattening rations (Hale, 1970, Smith *et al.*, 1968ab), and in feeding dairy cattle (Hale, 1970).

Rooney and Sullins (1974) reviewed work by Riggs (1970), Brethour and Duitsman (1965), McGinty (1969), Samford *et al.* (1970), and Nishimuta *et al.* (1969) on the differences in feedlot performance of ruminants fed sorghum of waxy, yellow or heteroyellow, and normal (nonyellow, nonwaxy) endosperm types and concluded that varietal differences in digestibility and feed efficiency existed but were difficult to detect. Bird-resistant high-tannin sorghums were less digestible and gave lower weight gains than non-bird-resistant. Waxy endosperms produced higher feed efficiencies in cattle than nonwaxy grains.

McCollough *et al.* (1972a) examined nine hybrid grain sorghums grown under similar conditions in Kansas. The varieties with yield, pericarp colour, endosperm description and proximate analyses are shown in Table 2.92.

The hybrids were fed separately, dry-rolled, in 90% concentrate rations (10% dehydrated alfalfa pellets) to 15 steers for 126 days. Feed efficiency (see Table 2.92) of the yellow endosperm hybrids, Funk's G 766W, ACCO R 109, Funk's G 522, Northrup King × 4087, was 10% better than the white endosperm hybrid RS 671 ised as control.

Four yellow endosperm hybrids, Funk's G 766W, ACCO 109, DeKalb E 57 and Northrup King 222, were grown in 1969 and 1970 under similar conditions in Manhattan, Kansas (McCollough and Schalles, 1972).

No significant differences in proximate analysis were found among hybrids or years. The grains were dry-rolled in all concentrate rations and fed to steers over 126 days during the winters of 1969–70 and 1970–71. Differences in weight gain, feed efficiency, or consumption among hybrids were not significant.

Eight hybrid sorghums and three hybrid maize, grown under similar conditions in Kansas, were dry-rolled and incorporated into all-concentrate rations for steers (McCollough, 1972). The colour, endosperm description, proximate composition, mineral matter, gross energy and feed efficiencies are given in Table 2.93.

There was no significant difference in average daily gain but differences among feed efficiencies of nine hybrids (excluding Funk's 3135 sorghum and Funk's 24554, high-lysine maize, fed under different conditions) were highly significant, E 57 giving the highest (7·68) and the bird-resistant ACCO 1023 the lowest (10·21). Yellow endosperm hybrids and regular maize (Halting × 9770) tended to higher feed efficiencies than white endosperm hybrids.

Using chromic oxide as a digestion indicator, the apparent coefficient of digestion of the hybrids was determined (McCollough and Brent, 1972). The protein digestibility of the bird-resistant ACCO 1C23 was significantly lower than the others. In general, digestibilities were higher for hybrid maize than hybrid sorghum, and lower for RS 671 (white endosperm) than for the yellow endosperm sorghums (McCollough *et al.*, 1972b).

Driedger and Riggs (1973) reported that 102 lines and hybrids of sorghum grains grown at three locations, Texas, Nebraska and Kansas, were significantly different in *in vitro* digestibility (related to steers) (Trei *et al.*, 1970) both among varieties and locations. A significant variety × location interaction was observed.

Sheep

Jordan *et al.* (1952) reported that the sorghum variety Norghum (dry matter 91·2%, crude protein 12·1%) compared favourably with maize in trials with lambs, providing daily weight gains of 0·22 lb/head compared with 0·27 lb/head from maize. Though a good deal of the sorghum was voided whole, digestibility compared favourably with that of maize.

Three sorghum grains, RS 610, RS 621 and Ute, each grown under three levels of N fertilizer, giving a crude protein range from 9·3 to 14·0, were compared in digestion trials with sheep. Apparent protein digestibility ranged from 62·1 to 78·6% and was directly related to crude protein in the grain.

In a second study, with eight parent lines of hybrid sorghums (varieties not stated) crude protein varied from 11·6 to 14·3%. Apparent protein digestibility ranged from 74·9 to 81·8% and was again directly related to crude protein.

In lamb fattening trials, Combine Kafir 60 produced 14·5% higher daily gains than Martin at similar daily feed intakes. Martin derivatives RS 608 and RS 625 produced slightly lower gains with lower feed efficiency than Kafir derivatives RS 610 and RS 626 (Riewe and Breuer, 1967).

Nishimuta *et al.* (1969) in sheep feeding trials compared Funk's 3758, a red grain, with waxy endosperm; Funk's G 766, a white grain, with yellow endosperm; and a regular red grain grown in the same area. The crude protein contents were 11·8% (commercial red grain), 11·7% (waxy endosperm Funk's 3758) and 11·3% (Funk's G 766) DWB. The grains were included, dry-rolled, at 72·4% of the ration.

Digestibility of organic matter and nonprotein organic matter, gross energy and N-free extract digestibility were significantly higher from the waxy and white grains than from the commercial red grain. Both crude and true protein digestibility, and N retention were higher from the waxy and white grain than the commercial grain.

In personal communications to Nishimuta *et al.*, W. H. Hale (1969) reported similar findings in digestibility trials with steers using the same grain varieties.

Sherrod and Albin (1973) in sheep digestibility studies compared waxy

(W), floury (F), intermediate low corneous (ILC), intermediate high corneous (IHC), regular (R) and bird-resistant (BR), represented by 20 individual (unnamed) sorghum types. The (R) grain was a commercial mixture. All grains were dry-rolled and fed in 70% grain plus 30% chopped alfalfa rations. Digestibility of the major energy components was highest in W and F types, followed by ILC and IHC, with BR lowest. Protein digestibility followed a parallel pattern except that F was lower than W.

Conclusion

This concludes the general review of the nature, composition, properties and nutritional value of sorghum. A similar review of the millets provides the remainder of this chapter. The influence of various specific genetic and environmental factors upon composition and nutritional value is recorded in Chapter 3.

History and Nature of the Millets

If the literature available is a fair indicator, the millets, particularly those other than pearl millet, have received scant attention when compared with that given to cultivation and utilization of sorghum. In 1975, Rachie (published by ICRISAT) reviewed the millets including relevant material from the French publication, now out of print, of Adrian and Jacquot (1964). Arnon (1972), Leonard and Martin (1963), and Matz (1969) have also written on the millets in general texts related to cereals. Bibliographies have been published by the Rockefeller Foundation (1967b) and Rachie (1974) on the millets and other small grains.

Description

The millets are small-seeded, annual cereal grasses, many of which are adapted to hot dry climates. The literature subsequently reviewed deals mainly with pearl millet (*Pennisetum typhoides*), finger millet (*Eleusine coracana*), foxtail millet (*Setaria italica*), common millet (*Panicum miliaceum*), little millet (*Panicum miliare*), Japanese barnyard millet (*Echinochloa frumentacea*), kodo millet (*Paspalum scrobiculatum*), fonio (*Digitaria exilis*), teff (*Eragrostis tef*), (*Amaranthus paniculatus*) and Job's tears (*Coix lachryma-jobi*). Typical heads of the first four are shown in Fig. 2.X.

Their ecology, structure, chemical composition and biological value are discussed separately under each species but, as already mentioned, where the work reviewed has included sorghum, together with one or more of the millets, it has in general been discussed under sorghum in the earlier part of this chapter.

Pearl Millet

Gopalkrishna *et al.* (1955a) published a bibliography of pearl millet

(*Pennisetum typhoides*). Ferraris (1973) reviewed the taxonomy, botany, breeding, physiology and agronomy of the crop. A classification was published by Murty *et al.* (1967b).

Description

Pearl millet is the most widely grown of all the millets. According to Purseglove (1972) pearl millet includes a number of cultivated races. Purseglove agreed with Bor (1960) that the cultivated races may be considered as a single collective species. Other workers have suggested that various races do not have a common origin but may have arisen from many wild species from different parts of Africa. Pearl millet almost certainly originated in western tropical Africa where the greatest number of both wild and cultivated forms occur. At least 2000 years ago the crop was carried to East and Central Africa and to India, where, because of its outstanding tolerance to drought, it became established in the drier environments.

Krishnaswamy (1962) stated that in India pearl millet is known by several vernacular names , the most common being bajra. In Africa, unlike India, many wild *Pennisetum* spp. are to be found. However, the African vernacular names for pearl millet are not used in India, though on the east coast of Africa, the name bajra was probably introduced by immigrants from the Indian sub-continent.

Pearl millet may be divided into two broad classes: (1) early millet, which matures within 60–95 days; and (2) late season millet which matures within 130–150 days. The many cultivars vary in several readily apparent characters, including height which ranges from 0·5–4 m, thickness and degree of branching of the stem, size and shape of the inflorescence, the length ranging from 15–140 cm, and size, shape and colour of the grain, which may be near white, pale yellow, brown, grey, slate blue or purple. (See Fig 2.X.) The seeds are exposed beyond the tips of the glumes and the lemma to give a beaded appearance (Arnon, 1972). The grain becomes free of lemma and palea at maturity so that the grains thresh free; the ovoid grains are about 3–4 mm long, much larger than those of other millets.

Pearl millet is one of the most drought tolerant of all cereals, more so than sorghum, the northern limit in West Africa for pearl millet being close to the 250 mm isohyet whereas the northern limit for sorghum is around the 375 mm isohyet. While surviving under conditions of lower total rainfall than sorghum, in general it is less tolerant to intermittent drought since, unlike sorghum, it does not seem to possess the facility of temporary dormancy during periods of drought. It grows well on light soils and under semi-arid conditions. Pearl millet is propagated by seed, and may be grown in pure stand or intercropped with legumes such as the cowpea (*Vigna unguiculata*).

Adrian and Jacquot (1964) reported the 1000-seed weights of *P. pycnostachyum* and *P. nigritarum* grown under identical conditions at CNRA, Bambey in Senegal as 8·15 and 7·75 g respectively.

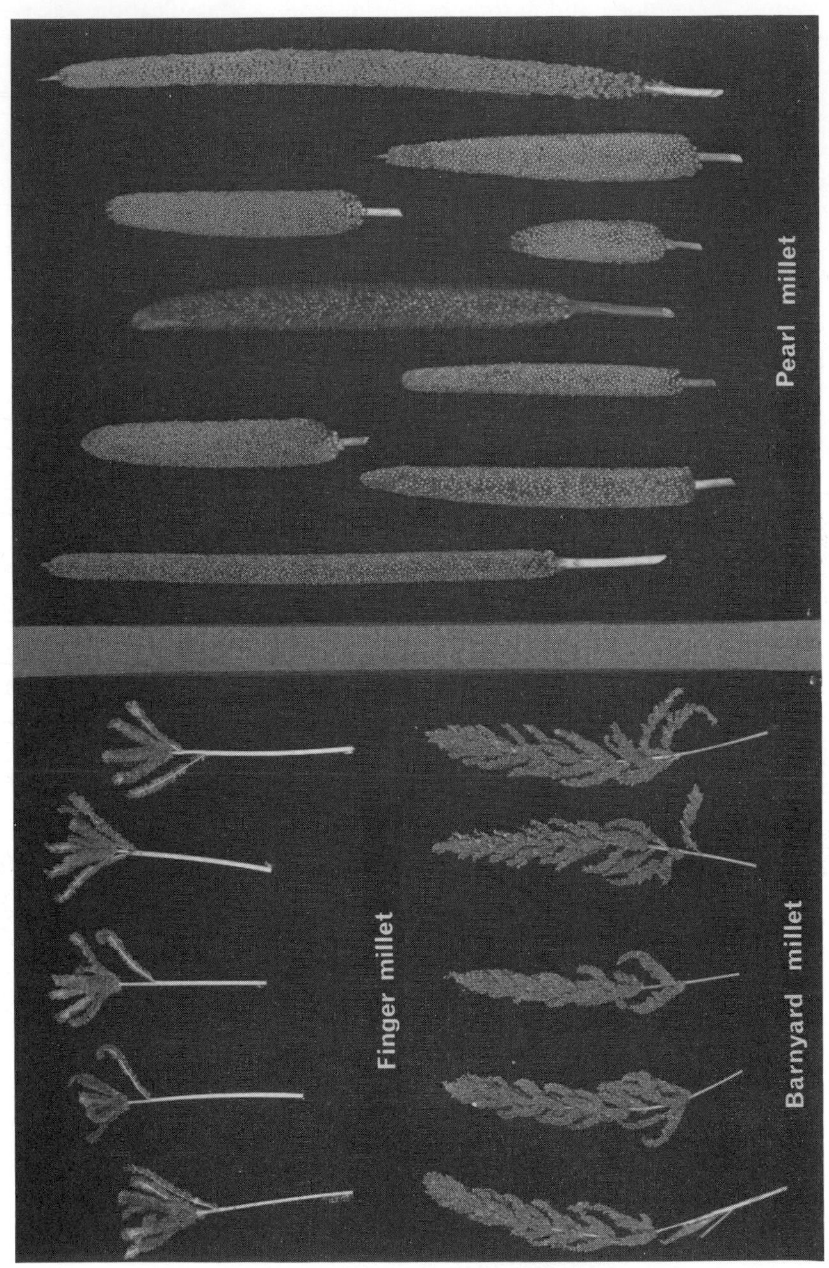
Pearl millet
Finger millet
Barnyard millet

Fig. 2.X Typical heads of millets. Proso or common millet is referred to as common millet in the text.

Adrian and Sayerse (1957) reported the average 1000-seed weight of pearl millet grown in Senegal as 6·5 g.

Jambunathan (personal communication, 1978) reported that among ICRISAT lines the 1000-seed weight ranged from 2·54 to 14·0 g with a mean of 7·96 g.

Structure

Badi *et al.* (1976) at Kansas State University examined the structure by SEM of F_3 and F_4 seeds of pearl millet derived from the cross Serere 17 × Tift 239. Two hybrid grain sorghums, a bulk red-seeded commercial sample and a yellow-seeded (C 42Y) sample were included for comparison. The millet kernels were cross-sectioned at the top of the germ with a razor blade which produced a fracture rather than a smooth cut. Samples of both grains were milled on a Quadrumat junior experimental mill to yield four fractions. The yield, moisture and protein (N × 6·25) are shown in Table 2.94. From both grains, grind-out yields (i.e. total flour fractions) were low and the flours grey in colour.

The amino acid composition of the whole grain samples (auto-analysed after acid hydrolysis) is shown in Table 2.119, and of the milled fractions in Table 2.95. The amino acid contents of the three millet flour fractions were similar.

The pearl millet kernels were about one-third the size of the sorghum kernels. The pericarp of pearl millet appeared to be similar to that of sorghum except that no starch granules were found in the millet mesocarp. The aleurone layer of the millet was a single cell thick.

The endosperm of both the sorghum and the millet contained (a) a hard, translucent and (b) a soft, opaque component. The hard component of the millet consisted of a tightly packed structure with no air spaces. The millet starch granules were polygonal and somewhat smaller (10 μm) than those of the sorghum. Protein bodies (1·5 μm) were embedded in the protein matrix that surrounded the starch and produced large indentations at the edge of the starch granules. The structure of the protein–starch bond in the hard endosperm was similar to that of maize (as reported by Robutti *et al.*, 1974) and sorghum (as reported by Hoseney *et al.*, 1974). The hila (0·5 μm) of pearl millet could be seen in many fractured starch granules and appeared to be smaller in diameter than those in sorghum. The soft endosperm of pearl millet contained relatively large air spaces and loosely packed, spherical starch granules.

The protein bodies in pearl millet were soluble in 70% ethanol at 60°C (prolamine fraction), and represented 30·7% of total protein. The protein bodies appeared to be unaffected by 100% *t*-butyl alcohol at room temperature but the protein matrix (about 27% of total protein) was dissolved. More vigorous extraction in a Waring blender rendered 44% of total protein soluble in 60% *t*-butyl alcohol and 42% of the total protein soluble in 70% ethanol at 60° %C. A second extraction with the same solvent (e.g. ethanol after ethanol), or switched solvents (e.g. ethanol after butanol) solubilized an additional 13% of total protein.

Badi *et al.* (1976) stated that the differences in protein solubility observed under the SEM were not apparent after vigorous (Waring blender) extraction of the millet protein. They emphasized that in view of the known diversity among pearl millets, their findings may not be representative of all pearl millet types.

Adams *et al.* (1976) isolated the protein bodies of pearl millet and examined their structure by electron microscopy. Their findings are discussed above under Sorghum.

Sullins and Rooney (1977a) at Texas A & M University, examined a sample of Tift 23 B normal pearl millet grain grown at Tifton, Georgia. Whole kernels were cut longitudinally with a razor blade. Some sections were treated with hog pancreas α-amylase (as described by Sullins and Rooney, 1975) to remove the starch. Other sections were treated with warm aqueous 70% ethanol for 30 min. The sections were examined under a Joel JSM–U3 SEM. Sections of isolated starch prepared by a laboratory wet-milling procedure (Norris and Rooney, 1970b) were examined using a Zeiss Universal microscope with bright field, differential interference contrast and phase contrast optics (see Fig. 2.XI and 2.XII).

In general, pearl millet kernels resembled sorghum, but were smaller in size with greater variability among kernels from a single spike; the relative proportion of germ to endosperm was higher in pearl millet than in sorghum.

The pericarp of both sorghum and pearl millet contained three layers, the epicarp, mesocarp and endocarp but pearl millet epicarp contained one or two cell layers and was thicker and more dense than that of sorghum. The mesocarp of pearl millet was free of starch granules; the endocarp contained cross cells and tube cells. The pericarp of pearl millet tended to break along the endocarp layer when scraped, during hand dissection or mechanical milling. Sullins and Rooney (1977a) suggested that the structure of the pericarp of pearl millet might partly account for its better resistance to weathering compared to sorghum. The pearl millet kernels examined did not appear to contain a testa although a thin membrane tissue was observed between the cross cells and aleurone cells. The aleurone layer was a single cell thick and extended around the entire kernel. Aleurone cells varied in size but averaged 17μm wide by 22 μm long.

The endosperm consisted of peripheral, corneous and floury areas, each composed mainly of starch granules embedded in a protein matrix containing protein bodies. The peripheral endosperm was about 2–5 cells thick depending on location in the kernel, the cells were small and uniform measuring about 21 μm wide and 40 μm long. The periphery contained the highest concentration of protein in the endosperm. The starch granules were smaller than those in the corneous area and increased in size towards the centre of the kernel. The corneous and floury endosperm cells were approximately 75 μm wide and 83 μm long. In the corneous area, the starch granules were angular in shape and tightly enmeshed by a network of protein which contained the spherical protein bodies. When the endosperm was fractured, the cells, including the cell walls, in the corneous area

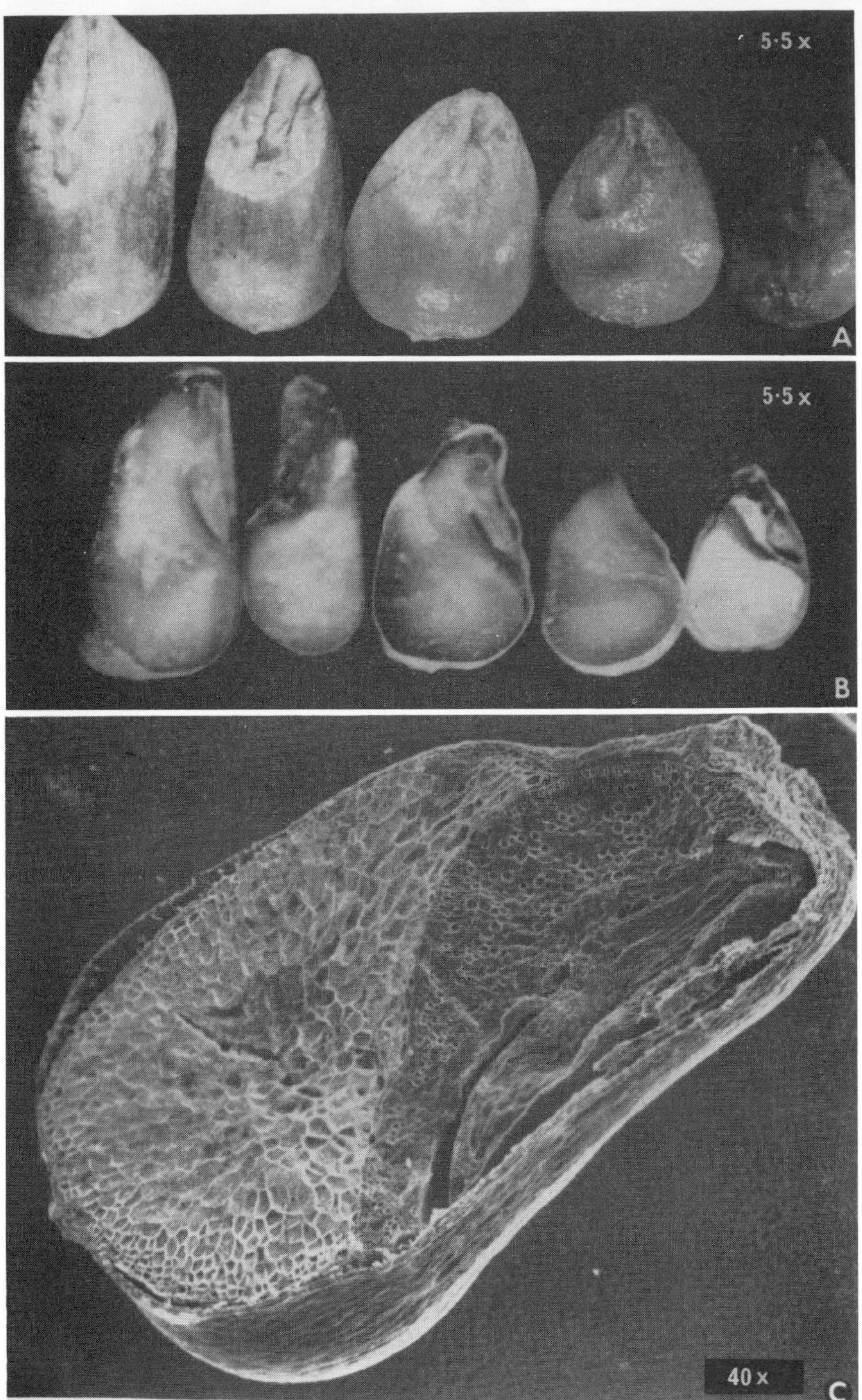

Fig. 2.XI (A) Whole kernels illustrating range in size of given spike of Tift 23B normal pearl millet; (B) long. section of pearl millet half kernels; (c) SEM of pearl millet half kernel treated with α-amylase to remove starch. Courtesy of Dr Lloyd W. Rooney, Texas A&M University.

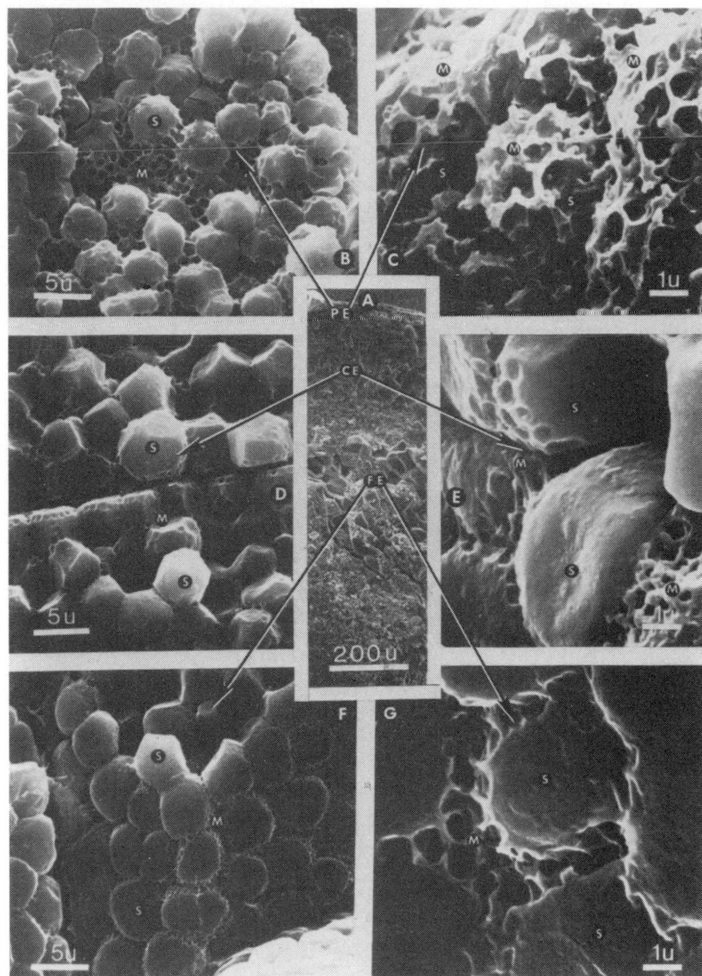

Fig. 2.XII SEM photographs of the peripheral, corneous and floury endosperm of pearl millet after treatment with hot aqueous 70% EtOH for 30 min. B and C = peripheral endosperm at 2400 and 8000 × respectively. D and E = corneous endosperm at 2400 and 8000 × respectively. F and G = floury endosperm at 2400 and 8000 × respectively.
S = starch; M = protein matrix. Courtesy of Dr Lloyd W. Rooney, Texas A & M University.

appeared to break in one plane. The lowest concentration of protein in the endosperm appeared to be in the corneous zone. (In sorghum, the lowest concentration of protein was found in the floury zone). Cells in the floury endosperm were less densely packed than in the corneous zone and the protein network appeared discontinuous. When fractured the cells broke along cell walls.

The protein bodies in the endosperm were extracted with warm aqueous 70% ethanol and were therefore considered to be mainly prolamines. As measured by the light microscope and SEM, the pearl millet protein ranged between 0·3 and 1·2 μm in diameter, and were spherical in shape. Sullins and Rooney (1977a) commented that their findings differed from those of Adams *et al.* (1976) who reported protuberances which could cause fusion or deformation of adjacent granules. The floury endosperm appeared to contain a greater number of protein bodies than the corneous endosperm. The protein bodies were largest in the peripheral endosperm and decreased slightly in size towards the centre. The peripheral and corneous endosperm of all the pearl millet samples examined contained extremely small (0·1 μm) spherical bodies that were not solubilized by warm aqueous alcohol. Similar bodies were not observed in sorghum. Sullins and Rooney suggested the structure of millet deserves further study.

Pearl millet starch granules ranged from 10 to 12 μm in diameter; sorghum from 20 to 30 μm. The granules in the corneous endosperm zone of pearl millet were polygonal, those in the floury endosperm more spherical. Indentations were found on the surface of granules in the corneous zone where protein bodies adhered. The starch of pearl millet was hydrolysed more slowly by hog pancreas α-amylase than that of sorghum. Sullins and Rooney (1976) suggested that this difference might also contribute to pearl millet's superior resistance to weathering compared to sorghum; it might also affect the availability of the energy in pearl millet.

Purseglove (1972) quoted Deshaprabhu (1966) as stating that the starch of pearl millet consisted of 68% amylopectin and 32% amylose.

Chemical Composition

The data on proximate composition, amino acid composition, inorganic elements and vitamins have been summarized and as far as possible are presented on a uniform basis in Tables 2.118, 2.119, 2.120 and 2.121 respectively. The data may not be uniformly comparable since different sources provided varying degrees of relevant information and background. Methods of sample preparation and analysis vary markedly among reporters and in some instances, were not clearly described. Data shown in the tables as "representative analytical values and ranges (not means) data from sources mentioned in the text" have been derived from the following sources:

Achutha Murthy *et al.* (1965); Adrian (1969); Adrian and Jacquot (1964); Adrian *et al.* (1970); Ambegaokar *et al.* (1964, 1965); Berulava (1950); Boas Fixsen and Roscoe (1937–38, 1940); Burton *et al.* (1972); Calder (1955, 1960); Chamberlain (1955); Chughtal and Waheed Khan (1960); Dakshinamurti (1955); De (1936a); Fabriani (1939); FAO (1949); Gayte-Sorbier *et al.* (1960); Goldberg and Thorp (1945); Hoover and Jayasuriya (1951); Johnson and Raymond (1964); Kadkol and Swaminathan (1955); Kadkol *et al.* (1954ab); Kalmykov *et al.* (1954); Lakshmiah and Ramasastri (1969);

Matz (1969); Narayana Rao *et al.* (1961); National Academy of Sciences (1971); Oke (1976, personal communication); Pinta and Busson (1963); Platt (1962); Portères (1955); Prugar and Vesela (1969); Qudrat-i-Khuda *et al.* (1960); Rahman *et al.* (1974); Ramasastri and Mohan (1969); Ranganathan *et al.* (1937); Srinivasa Rao and Ramasastri (1969b); and Watson (1971).

The tables also contain data from papers individually referenced which contain more specific information about the samples examined.

Various origins

During the 1965 and 1966 Kharif seasons, 163 pearl millet varieties were grown under similar conditions at the Punjab Agricultural University Farm, Ludhiana. The mean and ranges of protein (McKenzie and Wallace, 1954), P (Barton, 1948), Fe (1962 only) (Piper, 1966), fat (1965 only), and Ca (AOAC, 1960), DWB, were reported by Goswami *et al.* (1969ab, 1970ab) and are shown in Table 2.118 and 2.120.

Among the African entries (Goswami *et al.*, 1969a), protein ranged from 9·8 (IP 219) to 15·2% (IP 171); Ca from 20 (IP 171) to 47 mg/100 g (IP 199); crude fat from 3·5 (IP 231–4–116/1) to 8·0% (IP 233); and Fe from 2·1 (IP 227) to 8·3 mg/100 g (IP 139–48–46/1). Differences among varieties were significant for crude protein, ash, Ca, P, and Fe, and between years for ash, P and protein. It is suggested that protein, Ca, P, and Fe might all be improved by selection. A negative relation between protein and ash was the only statistically significant correlation coefficient reported.

Among 42 American entries (Goswami *et al.*, 1969b), crude protein varied from 10·5 (IP 849) to 14·4% (IP 890), P from 720 (IP 847) to 920 mg/100 g (Georgia 522), Ca from 28 (Georgia 44) to 53 mg/100 g (IP 938), ash from 1·7 (IP 836) to 2·5% (IP 938), crude fat from 4·0 (IP 834) to 7·6% (IP 862) and Fe from 3·3 (Georgia 483) to 9·4 mg/100 g (IP 831). Differences among varieties were significant for crude protein, P, Ca and ash, and between years for crude protein, P, Ca, and ash. Overall, IP 890, IP 919, Georgia 506 and Georgia 525 were regarded as of good nutritional quality. Among the 58 Indian inbred entries reported by Goswami *et al.* (1970a), the ranges were:
protein from 9·4 (S350–+–350–122–4–5–5) to 14·4% (S350–+–349–120–1–5–1), P from 700 (BIL 1) to 960 mg/100 g (D55 2), Ca from 25 (BIL 2A) to 46 mg/100 g (S350–25–4–B/A, ash from 1·6 (409/2) to 2·5% (103 B 5), crude fat from 2·8 (LL88C 1) to 7·8% (A1/3–14–430–4–1–1) and Fe from 3·5. (S350–+–350–122–4–5–5) to 11·7 mg/100 g (S350–361–122–2). The results were not very different from the African and American entries. Differences among varieties were significant for protein, P., Ca, ash and fat and between years for P. Overall, entries S350–+–349–120–1–5–1, S350–+–349–120–1–3–3, S350–+–349–119–3–1–1, ST29–2–2–1 and S350–19–5C were high in protein, P and fat, moderate in Ca, and Fe content, and were considered of good nutritional quality.

Among the 28 Indian varieties examined (Goswami *et al.*, 1970b) protein ranged from 10·3 (CO 4) to 13·2% (L 11), Fe from 3·1 (23A × BIL 3B) to 10·7 mg/100 g (CM 63), crude fat from 3·0 (CES 50) to 7·9% (S 530), ash from 1·6 (S 350) to 2·1% (RSK), Ca from 25 (A1/3) to 51 mg/100 g (AF 3) and P from 700 (Pusa Moti) to 910 mg/100 g (N 28 15 2).

Sawhney and Naik (1969) reported the proximate composition, Ca, Fe and P (AOAC, 1965), S (Basson, 1967), tryptophan (Spies and Chambers, 1949) and lysine (Naik 1968), of the pearl millet varieties S 530, A1/3, T 55, RSJ, Pusa Moti, $23D_2A$, 23A Tall, IP 48, HB 1, AK_1P, IP 47 and Tf23–AXJ88, obtained from the World Collection at the Indian Agricultural Research Institute. The means and ranges are shown in Tables 2.118, 2.119 and 2.120. Protein was highest in AK_1P (15·3%), IP 47 (17·0%) and Tf23–AXJ88 (20%). The embryo constituted 1·8 and 2·4% by weight of A1/3 and HB 1 respectively and contained roughly 10% of the total protein. Mean contents of lysine (166 mg/g total N) and tryptophan (68 mg/g total N) were higher than those reported by Aykroyd *et al.* (1966), which were 116 mg/g total N and 50 mg/g total N respectively. Crude fat, mean 5·5%, and Fe, 8·1 mg/100 g, were relatively high, Ca 54 mg/100 g, comparatively low. Protein, Ca, ash and crude fat varied significantly among varieties. Overall the varieties LM 38 39, RSK, N 207 and L 11 were considered best in nutritional quality.

Sharma and Goswami (1969) reported the composition (AOAC, 1960) of petroleum ether extracts of 25 pearl millet lines grown on sandy loam soil in the kharif season of 1968, including entries from Africa (3), USA (7), Indian inbreds (3), Indian varieties (6) and hybrids (6). The results are summarized in Table 2.96.

Acid value was positively correlated with saponification value. The high yielding hybrids were highest in lipid, lowest in acid value and intermediate in iodine value. Jellum and Powell (1971) extracted lipids from five inbred pearl millet lines using several solvent extraction systems including equal volumes of: (1) methanol/petroleum ether, (2) chloroform, methanol, petroleum ether, (3) chloroform, methanol and (4) petroleum ether followed by methyl ester synthesis and gas chromatography. As anticipated, there was greater variability among solvents than among millet lines.

Jellum and Powell (1971) used method (1) to study the fatty acid composition of 65 pearl millet lines grown at Tifton, Georgia, in 1968. The seed originated from Nigeria (31 lines), Sudan (1 line), Ethiopia (1 line), South Africa (5 lines), India (27 lines) and USA (1 line). The results are summarized in Table 2.97. Variations though significant were not as great as found in maize (Jellum, 1970). Stearic acid, ranged from 1·83% in a South African line to 7·95% in an Indian line. Oleic acid and linoleic acids were significantly negatively correlated. Analysis of different sizes of mature seeds of the variety Tiflate, showed linolenic acid increasing as seed size decreased. The fatty acid composition of four inbred lines during seed maturation showed most rapid and greatest changes during the first 16 days after anthesis. Changes in lipid composition were comparatively small from 24 days after anthesis.

Weihrauch *et al.* (1976) compared the fatty acid composition of pearl millet with sorghum (see Tables 2.35 and 2.36).

Africa

Shepherd *et al.* (1971–72) reported from East Africa the proximate composition (DWB), protein (Kjeldahl N × 5·83), crude fat (petroleum ether), crude fibre and ash of three lots of pearl millet from district trials. The means and ranges are shown in Table 2.118. Protein ranged between 11·5 and 13·8%, most clustered around 12%.

Samson and Adrian (1971) examined three races of pearl millet, Zongo, P3 Kolo and Tamangagi, from Eastern Niger. The grains varied in size within and among races. The proportions and 1000–seed weight appear in Table 2.98.

A large (14·5 mg) grain and a small (3·0 mg) grain of pearl millet were cut longitudinally, with a razor, after soaking in water. The surface area (cm^2 magnification × 26) and estimated proportions of germ and scutellum are shown in Table 2.99 compared with wheat.

The large grain was 90% greater in area than the small. Endosperm and scutellum were 125% greater and the germ 30% greater in the large than in the small grain. Proximate analysis (methods not specified), formic acid insoluble matter (cellulose plus lignin Guillemet and Jacquot, 1944), pentosans (Cerning, 1970); B group vitamins and amino acids (microbiological methods) (summarized in Table 2.118) were closely similar among the three races. Size of seed appeared to have more influence than race on composition. Table 2.100 shows the mean and ranges among various sieve fractions.

As seed size increased, protein increased from 7·9 to 9·2%, ash, cellulose and lignin plus pentosans decreased indicating a lower proportion of peripheral seed coats where the mineral and cellulosic material of cereals is concentrated (Schlesinger, 1942; Hinton, 1959). As grain size increased, thiamine increased, riboflavin and pantothenic acid decreased. Niacin varied slightly but not consistently. As in other grains (Hinton *et al.*, 1953), the B vitamins of pearl millet are concentrated in the germ, except thiamine which is highest in the scutellum. Excluding lysine, most amino acids (% protein) also increased with increase in grain size. Large grains contained 30% more tryptophan, 25% more leucine, 20% more isoleucine and valine and nearly 10% more methionine and threonine, but 30% less lysine than small grains.

Samson and Adrian (1971) stated that seed size is influenced by genetic and environmental conditions. Seed size is influenced genetically, but inadequate moisture may also result in seeds smaller than average (Doggett, personal communication, 1977).

Magboul (personal communication via Saeed, 1974) reported the mean proximate composition (AACC, 1962) of five market samples of pearl millet grown in the Sudan (see Tables 2.118 and 2.120). Differences in protein and lipid among samples were small but Ca ranged from 24·7 to 62·9 mg/100 g and P from 119 to 720 mg/100 g. Ca and P contents may be influenced by genetics and environment (Doggett, personal communication, 1977). The total "pellagra-preventive" vitamin activity (tryptophan and niacin) in two African samples of pearl millet was determined by Adrian (1969) (microbiological assay, *L. arabinosus*). The results are shown in mg/100 g below:

	Niacin (total) (a)	Vitamin activity from tryptophan (b)	Potential vitamin activity (a) + (b)
Pearl millet ("souna")	1·65	2·87	4·52
("mouri")	2·50	3·04	5·54
Mean	2·08	2·96	5·03

Asia

Rama Rao and Swaminathan (1953) reported the proximate composition including minerals (AOAC, 1950) and thiamine (Swaminathan, 1942a) of pearl millet varieties designated CO 1, CO 2, CO 3, CO 4, X 1 and X 2 (see Table 2.118, 2.120 and 2.121). Little variation was found: CO 3 had highest protein (14·3%), CO 1 highest Ca (49·7 mg/100 g) and X 2 highest P (391 mg/100 g) and thiamine (0·45 mg/100 g).

Murty (1969) reported the proximate composition (unstated methods) of the pearl millet hybrids HB 3, HB 4, HB 1 and a "local" variety grown under limited moisture. The data are shown in Table 2.101.

The Task Force on Integrated Rural Development (Government of India, 1973) declared HB 3 the most promising variety under moisture stress during the Kharif season. It yielded 2000 kg/ha with 80 kg N/ha compared to 950 kg/ha without N fertilizer.

Uprety and Austin (1972) analysed 14 high yielding pearl millet varieties grown on the Experimental Farm at the Indian Agricultural Research Institute during 1970–71 with fertilizer application of 40 kg N, 40 kg K and 20 kg P/ha. Mean and ranges of protein (micro-Kjeldahl N × 6·25), moisture, total sugar, crude fibre, ash, total P, phytin P (AOAC, 1960), fat, starch (by difference), Ca (Clifford and Winkler, 1954), Fe (Toth *et al.*, 1948) and yellow pigment, as carotene (AACC, 1962), are shown in Table 2.102 with protein in 1000 seeds plotted against 1000-seed weight.

23A × D 1328 and 23 D_2A × D 2031 with lower protein (12·6 and 13·7% respectively were higher in protein/1000 seeds because of higher 1000-seed weight, as were 23D_2A × D 1309, 23A × D 1317, HB 3, HB 1, 23D_2A × D 1381, HB 5 and HB 4. Crude fibre ranged from 1·3–3·0% with HB 3, HB 4, Hyb. D 356, Hyb, D 770 and 23D_2B containing less than 2·0%. Fat varied from 3·0 to 4·6%, HB 3 being highest. Hyb. D 770 was highest in Ca, 62 mg/100 g DWB and HB 1 lowest, 35 mg/100 g.

23D_2A × D 1309, 23A × D 1328 and 23D_2A × D 1381 (all 348 mg/100 g) were highest in P. 23D_2A × D 1309 also contained the highest phytin P,

221 mg/100 g, the lowest being HB 3 and Hyb. D 77, 35 mg/100 g. Fe varied from 2·1 (HB 4 and HB 5) to 5·2 mg/100 g (23A × D 1317). Carotene did not vary markedly among samples.

Kurien *et al.* (1961a) studied the distribution of protein, Ca and P between the husk and endosperm of pearl millet (the embryo was not mentioned). Thc whole grain was soaked in water for 24 h, ground to a fine paste and the husk separated by a 100-mesh sieve, the endosperm centrifuged and dried at 50–60°C. The results, DWB, are shown in Table 2.103.

Popli and Singh (1972) compared the protein (macro-Kjeldahl N × 6·25), moisture, fat, crude fibre, ash, Ca, P (AOAC, 1960) and NFE, by difference, of the high yielding pearl millet varieties HB 1, HB 4, T 55, HB 3, S 530, A1/3 and R559 obtained from the Haryana Agricultural University.

Means and ranges are shown in Tables 2.118 and 2.120. Variations were similar to those reported by Sawhney and Naik (1969), Goswami *et al.* (1969a, 1970ab), and Deosthale *et al.* (1971). Means and ranges of lysine, phenylalanine, valine, leucine, isoleucine, threonine (paper chromatography, Rao and Subramanian, 1970), tryptophan (Spies and Chambers, 1949) and methionine (Horn *et al.*, 1946), all mg/g N, appear in Table 2.119. The results were in agreement with those reported by Sawhney and Naik (1969) and Deosthale *et al.* (1971).

Agarwal and Sinha (1964) reported the composition of the light petroleum extract of "UP—isolated" variety of pearl millet. The oil was a clear orange colour and represented 5·1% of the grain. The composition is given in Table 2.104. The mixed fatty acids contained (by a urea adduct method) myristic 0·20, palmitic 10·80, stearic 0·28, oleic 53·84 and linoleic 34·88%. Linolenic was not found.

Pruthi and Bhatia (1970) examined two improved varieties of pearl millet, A1/3 and S 530, the seeds of which were obtained from the Plant Breeding Department of the Punjab Agricultural University. The free lipid content extracted with petroleum ether and chloroform methanol of S 530 was 5·23% and of A1/3, 5·15% and the bound lipid content (water-saturated *n*-butanol), 0·55% in S 530 and 0·47% in A1/3. In the nonpolar fraction (thin-layer chromatography) triglycerides formed the major component, 86·89% in A1/3 and 83·51% in S 530. Sterol esters, hydrocarbons, partial glycerides, free sterols, and free fatty acids together with unidentified components were also present. Among the polar lipids, lecithin was the major component (two-dimensional thin-layer chromatography); sterols containing glycolipids were also present in appreciable amounts.

Bhatia *et al.* (1972) at Punjab Agricultural University reported carbohydrate content of A1/3 (long-eared) and S 530 (bristle-eared) varieties of pearl millet 110 days after sowing. The analysis included total water-soluble sugars (Yemm and Willis, 1954), free reducing sugars (Hulme and Narain, 1931), nonreducing sugars (by difference), starch (Clegg, 1956, as modified by Smith and Grotelneschen, 1966), free and bound glucose (Kline and Acree, 1930), free fructose (by calculation, free reducing sugars minus free glucose), bound fructose (by calculation, nonreducing sugars minus bound glucose). The results are shown in Table 2.105.

Singh and Popli (1973) determined the amylose content of seven high yielding varieties of pearl millet, HB 1, HB 4, T 55, HB 3, S 530, A1/3 and R 559, at Haryana Agricultural University. Descending paper chromatography (Bealing and Bacon, 1953; Partridge, 1948; Bacon and Edelman, 1951), using *n*-butanol/acetic acid/water (4/1/5) identified fructose, glucose, sucrose, maltose and two higher oligosaccharides in extracts of HB 1, T 55, HB 3 and S 530. Fructose was not found in HB 4, A1/3 and R559. Reducing, nonreducing and total sugars, and diastatic activity (AOAC, 1960) also showed varietal differences. The means and ranges are presented in Table 2.106 with those for protein and total carbohydrates (by difference), starch (Hassid and Neufeld, 1964; Wolf, 1963), and amylose (Williams *et al.*, 1970).

Total and nonreducing sugars were highest in A1/3 and lowest in R 559. Reducing sugars were highest in S 530, HB 3 and A1/3. HB 1 and HB 4 were highest in protein (14·3 and 14·5% respectively) and S 530 and R 559 lowest (8·4 and 9·4%). Diastatic activity was highest in T 55, and lowest in HB 1. Diastatic activity and amylose content appeared to parallel one another, HB 1 being lowest and T 55 highest in both.

The relative digestibility of the varieties was studied by suspending a specific amount (100 g) of uncooked and cooked (boiled) flour and isolated starch from each of the varieties in 25 ml of 0·02 M glycerophosphate buffer (pH 6·9) and incubating with 25 mg of α-amylase (bacterial, BDH) at 37°C. The rate of hydrolysis was followed by estimating the reducing sugars found at 15 min intervals for 90 min. Amylolysis was highest in T 55 and lowest in HB 1, the rate increasing when the flour was cooked. Amylolysis of the isolated starch followed a varietal pattern similar to the flours. Varieties with higher amylose content were susceptible to faster attack by amylase, similar to results obtained with rice (Srinivasa and Rao, 1971) but in contrast to results with pulses (Ramasastri and Rao, 1969) and maize (Borchers, 1962) where varieties with high amylose content showed lower *in vitro* starch digestibility.

Chitre *et al.* (1955) estimated thiamine and riboflavin (own method) and niacin (James *et al.*, 1947) in pearl millet varieties CO 1 and CO 3 from Coimbatore (rain-fed or irrigated, red loamy soil).
The results (mg/100 g) are shown below:

	CO 1	CO 3
Thiamine	0·28	0·39
Niacin	3·32	4·43
Riboflavin	not reported	0·19

Deosthale *et al.* (1971) determined the Ca, P and Fe (AOAC, 1960) Zn, Cu, Mg, Mn and Mo (colorimetrically, Sandell, 1959) in high yielding pearl millet varieties HB 1, HB 2, HB 3, HB 4 and of an improved variety Pusa Moti. The means and ranges are shown in Table 2.120.

Phillippines

Solpico and Yambao (1966) compared a local pearl millet variety P 13163 and an unnamed variety with other millets (the results from finger millet, foxtail millet and common millet are discussed under later sections) at Los Baños, Laguna during wet and dry seasons between 1957 and 1962. The proximate analyses (unstated methods), average values, are shown in Table 2.118.

USA

Bade *et al.* (1976) characterized the lipid fractions (thin-layer chromatography) of F_3 and F_4 seed from a cross of Serere 17 × Tift 239 pearl millet at Kansas State University. Free lipids were extracted by petroleum ether and bound lipids with water saturated butanol from the petroleum ether extract. Sorghum and wheat were included for comparison. The major nonpolar components of the free lipids in all three grains were triglycerides. The bound nonpolar lipids contained the same components in different ratios as the free nonpolar lipids. The polar free lipid pattern from wheat differed significantly from pearl millet and sorghum, neither millet nor sorghum contained phosphatidyl, ethanolamine, digalactosyl diglycerides or phosphatidyl choline. The bound polar lipid patterns were similar in all three grains.

Protein and Amino Acid Composition

Various origins

Table 2.119 brings together analytical data from a number of sources including FAO (1970), Adrian and Jacquot (1964), and Pushpamma (1968).

Deosthale *et al.* (1971) determined protein (micro-Kjeldahl N × 6·25), and essential amino acids (microbiologically) in pearl millet including 14 Nigerian varieties, with HB 1, an Indian variety grown in Guntur, Andhra Pradesh, 22 Senegal varieties grown at Vizianagaram, Andhra Pradesh and 14 hybrids from Jamnagar, Gujarat State. Results are summarized in Table 2.119. The range of protein was significantly greater and the mean lower among the Nigerian than the Indian types. The Senegal types were significantly higher than the Indian in mean protein. Protein in the Nigerian was lower, and many of the Senegal varieties higher, than the extremes reported by Goswami *et al.* (1969a) for African entries. Among Nigerian, Senegal and Indian protein and lysine in protein appeared inversely related but only among the Nigerian was it significant.

Selected "high-protein" white and grey coloured derivatives of Baroda lines, with protein 20·0% or above, were assayed (Deosthale *et al.*, 1971) for lysine, threonine, methionine and tryptophan. The mean and ranges are shown in Table 2.107. Lysine and threonine contents were considerably lower than the values for common varieties reported by Aykroyd *et al.* (1966). Protein was inversely correlated with all four amino acids assayed but only between threonine and methionine was the correlation significant.

Africa

Six numbered samples of pearl millet, grown in East Africa were analysed for amino acid composition (excluding cystine, methionine and tryptophan) by auto-analyser following acid hydrolysis (Shepherd *et al.*, 1971–72). There were appreciable differences in % N recovery. Means and ranges are shown in Table 2.119. Two individual entries gave lysine value above 219 mg/g total N.

Adda (1958) analysed (two-dimensional paper chromatography) the amino acid composition of 16 lines of *Pennisetum* spp. grown at Bambey, Senegal. The species were described as *P. pycnostachyum* (varieties Saboya, Tiare Kaolak, Diohene, Dosso Migu, crosses with cinereum Louga and nigrita rum komandougou) and *P. nigritarum*. The results, together with protein (N × 6·25) are summarized in Table 2.119.

Tryptophan (assayed microbiologically) in pearl millet from Cameroon and Senegal was assayed by Adrian (1969). The mean and ranges are shown in Table 2.119.

Asia

Balasubramanian *et al.* (1952ab) and Balasubramanian and Ramachandran (1957) reported the amino acid composition (microbiological assay) of pearl millet varieties CO 1 and CO 3 (see Table 2.119).

Chitre *et al.* (1956) reported the Kjeldahl N and the amino acid composition (chemical or microbiological assay) of CO 1 pearl millet (see Table 2.119).

Swaminathan *et al.* (1969) reported wide variability in protein (N × 6·25) and lysine (enzymatic decarboxylation) of 12 strains of pearl millet obtained from the IARI (see Table 2.119). The local variety Bichpuri had the highest protein, and was recommended as breeding stock.

Deosthale *et al.* (1971) compared yield, protein content (macro-Kjeldahl N × 6·25) and lysine (% protein) (microbiological assay) in 15 varieties of pearl millet grown under limited moisture and 14 varieties grown under adequate moisture during kharif 1969 at New Delhi, India. The means and ranges are shown in Table 2.119.

Grain yields and protein of 23A × K 530–531, 23A × K 230D–1, 23A × K 561, LH 234, 23A × D 107, 23A × D 109 and $23D_2A$ × J 41 were considerably higher (37–85%) than those of HB 1, HB 2 and HB 3. No significant correlation was found between protein and grain yield but a significant inverse correlation was observed between protein and lysine in protein.

Deosthale *et al.* (1971) reported grain yield, protein content (N × 6·25) (%) and lysine (microbiological assay) in protein of 9 crosses of pearl millet double dwarf male sterile $23D_2A$ (high in protein), with D 170, D 110, D 151, D 356, D 448, D 7015, D 7016, D 7017, J 934–5 and in D 174 and HB 1. Lysine (% protein) varied over a narrow range and differed little from other pearl millets examined.

Austin *et al.* (1971) determined protein (N × 6·25 DWB) in 40 hybrids and varieties of pearl millet grown with 40 kg N fertilizer/ha (see Table 2.118). Hybrids with Tf 23A, L 101 or 18A as a parent contained highest protein.

Swaminathan *et al.* (1971) reported several new strains of pearl millet with lysine (auto-analyser) superior to the standard hybrids HB 1, HB 2, HB 3. Means and ranges are shown in Table 2.119.

Pokhriyal *et al.* (1977) reported the protein (N × 6·25) and amino acid composition (auto-analyser) of eight high yielding reconstituted pearl millet hybrids grown at the IARI, New Dehli during the crop season 1975 with fertilizer application N40/K40/P40 kg/ha. Of the hybrids (1) white grain hybrid, (2) old HB 3 (23A × J 104), (3) 5141 A × D 32, (4) 5141 A × J 104, (5) 5054 A × J 104, (6) Hyb. PHB 14 (L 111 A × PIB 228), (7) 5141 A × K 560 (1576) and (8) Hyb D 111, the highest levels of lysine, histidine, arginine, aspartic acid, threonine, alanine, methionine, valine, isoleucine, tyrosine and phenylalanine were found in the protein of hybrid (5) which had a leucine/isoleucine quotient of 1·9. The protein content was 12·3%, below the mean of the hybrids of 13·1%. The highest protein contents were 14·7%, hybrid (6), and 14·6%, hybrid (7). For means and ranges, see Table 2.119.

Jambunathan (1977) reported from ICRISAT that during 1976–77, the protein content (N × 6·25) in 4646 pearl millet samples, estimated by the Technicon auto-analyser, ranged from 5·8 to 20·9% with a mean of 10·6%. A high degree of correlation ($r = 0{\cdot}996$) had been established between protein content as determined by the micro-Kjeldahl method and the values from the Technicon auto-analyser. The content of lysine in the protein of 1392 of these samples, estimated by dye binding capacity and calculated from Udy Instrument Readings, ranged from 161 to 183 with a mean of 170 mg/g total N.

Nagarajan (1974) reported protein and lysine (methods not stated) of the high-yielding pearl millet varieties HB 1, HB 2, HB 3 and HB 4. For ranges and means see Table 2.119.

Taira (1962a) in Japan reported the amino acid content of pearl millet assayed microbiologically (see Table 2.119).

USA

Burton *et al.* (1972) reported the mean amino acid content (chromatographic analysis) of 23 inbred lines grown at Tifton, Georgia, with their ranges (see Table 2.119).

Protein Fractionation

Narayanamurti and Aiyar (1930) examined the alcohol soluble protein (designated typhoidin) from pearl millet (crude protein 9·3%) grown at Madras. The N (Kjeldahl) from the finely ground meal (40 mesh) following successive extractions was found to be:

Solvent	Nitrogen (%)
Water	13·04
Sodium chloride	8·69
Cold ethyl alcohol 70%	33·27
Hot ethyl alcohol 70%	6·95
Sodium hydroxide	9·06
Residue not accounted for	28·99

Several alternative methods of extracting alcohol-soluble prolamine were investigated, the procedure eventually selected produced a light grey powder mostly soluble in 75% alcohol.

Pushpamma (1968) fractionated the protein of samples of pearl millet from the local market at Hyderabad using two methods. The results compared with finger millet, Indian and US sorghum, wheat and rice, are presented in Table 2.22.

Sawhney and Naik (1969) reported the distribution of the protein among the protein fractions (modified Mendel-Osborne method of Naik, 1968) of the defatted flour from S 530, A1/3, T 55, RSJ, Pusa Moti, $23D_2A$, 23A Tall, IP 48, HB 1, AK_1P, IP 47 and TF23 AXJ88. The mean values and ranges are shown in Table 2.108.

The prolamines plus glutelins accounted for about 60%, albumins 15% and globulins 9 % of total protein but the relative proportions varied among lines. Albumin ranged from 26·5 % in AK_1P to 6·1 % in IP 47, and globulin from 14·7% in 23 A Tall (also high in albumin, 20·4%) to 3·5% in RSJ.

Three varieties differing in protein content, A1/3 (9·6%), RSJ (11·8%) and TF23 AXJ88 (20·1%) were analysed for amino acid composition (autoanalyser) in the grain and in the protein. The results are summarized in Table 2.119. The variety RSJ was high in prolamine (37%), glutamic acid, proline, alanine and leucine. A1/3, prolamine 20% of protein, was lower in leucine and higher in isoleucine than the other two varieties. The leucine/isoleucine quotients in the three varieties were: A1/3, 1·3, RSJ, 3·5 and TF23 AXJ88, 3·3. Sawhney and Naik stated that pellagra had not been reported among populations subsisting on pearl millet but pointed out that it was the prolamine fraction that was mainly responsible for high leucine/isoleucine quotients.

In TF23 AXJ88 with highest protein, lysine was low (117 mg/g total N). Only 80% of amino N could be extracted compared with 90% in the other two varieties and only 70% total protein was extracted by the various solvents. The amino acid composition of the protein fractions of HB 1 are shown in Table 2.109.

The albumins and globulins were higher in the basic amino acids, lysine, arginine and histidine and the sulphur amino acids, cysteine and methionine, than the prolamines and glutelins. The prolamines were high in glutamic acid, proline and leucine. The high level of tryptophan in the prolamine of HB 1 pearl millet contrasted with the relatively low levels found in the prolamines of maize and sorghum. The amino acid compositions of the glutelin and prolamine fractions were similar though the glutelin contained higher lysine and lower tryptophan. Cysteine and methionine together accounted for almost all the S in HB 1. The distribution of total S in the protein fractions of the 12 varieties was studied as an indication of cysteine plus methionine. The means and ranges are shown in Table 2.110.

The distribution varied among lines, but the albumins and globulins together contributed about 42% of the total S. Prolamine contributed on average, 27% of total S, but Pusa Moti contained 40% and $23D_2A$ 35%. The glutelin contributed on average 38% of total S though RSJ and Pusa Moti (both 18%) and HB 1 (25%) were relatively low.

Singh and Popli (1974) studied the protein fractions of the high yielding varieties of pearl millet HB 1, HB 4, T 55, HB 3, S 530, A1/3 and R 559 obtained from the Haryana Agricultural University, Hissar. Total protein (N × 6·25) (AOAC, 1965) was fractionated by a modified Osborne and Mendel method (Sastry and Virupaksha, 1967). Protein content (N × 6·25) ranged from 8·4 (S 530) to 14·5% (HB 4) with a mean of 11·7%. The fractions (% total protein) varied as shown below:

Albumin from 11·6 (R 559) to 15·2 (HB 1) mean 13·4
Globulin from 7·7 (R 559) to 14·7 (S 530), mean 10·1
Prolamine from 27·1 (S 530) to 35·2 (HB 4), mean 31·7
Glutelin from 18·6 (R 559) to 28·5 (HB 4), mean 24·1

Prolamine plus glutelin accounted for 50–60%, albumin and globulin respectively for about 13% and 10% of total protein. About 79% of total protein was extracted.

Polyacrylamide gel electrophoresis of the fractions (Davis, 1964) showed varietal differences in the number of bands observed. Singh and Popli (1974) stated that Coulson and Sim (1964, 1965) consider the electrophoretic differences to be unimportant, whereas Kelley and Koeing (1963) regard them as significant.

Biological Evaluation

Human beings

Kurien *et al.* (1961b) reported on the metabolism of N, Ca, and P in eight boys, aged 11–12 years, living in a boarding home in Mysore. The chemical composition (AOAC, 1950) and phytate P (McCance and Widdowson, 1935) is shown in Table 2.111. Pearl millet represented 62% of the "poor Indian" diet, fed over 15 days with collection of urine and faeces (Murthy *et al.*, 1954) during the last five days. All the subjects maintained a positive N balance, average daily intake being 8·67 g and retention 1·11 g. The excretion of N in the faeces was considered high (4·08 g) and the apparent digestibility of the protein was only 52·9%.

All subjects were in positive Ca balance (average daily intake 0·479 g; retention 0·092 g) and in positive P balance (average daily intake 1·346 g; retention 0·356 g). About 52% of the ingested P and 70% of the ingested Ca, were excreted in the faeces.

Rats

In 1930 Visco reported that young rats fed exclusively on pearl millet maintained healthy, normal development over 130 days. Rats, nursed by mothers fed exclusively on the grain, which after weaning were themselves fed solely with the grain did not grow as fast as others fed a mixed diet. The findings of Ranganathan (1935), Swaminathan (1937ab), Giri (1940), and Kuppuswamy *et al.* (1958) are discussed under sorghum.

The nutritive value of pearl millet from Coimbatore, Madras State, was examined by Rama Rao *et al.* (1953a). The proximate composition, minerals (AOAC, 1950) and thiamine (Swaminathan, 1942a) are shown in Table 2.112 with wheat. The diets were fed on an equal weight basis (92·8% of total diet) in trials with weanling male and female rats, six to the group, over eight weeks. (A) pearl millet and (B) wheat were the sole sources of protein, B-complex vitamins and minerals except for calcium lactate (1·2%), and shark-liver oil (1·0%) which provided the daily requirements of Ca, vitamins A and D. No significant differences were found in daily food intake and average weekly gain in weight between diets (A) and (B).

Typically poor Indian vegetarian diets based, (C) on pearl millet and (D) on wheat, were also compared. Diet (C), pearl millet was significantly superior to (D) in average daily intake and average weekly weight gain.

Chitre and Vallury (1956ab) compared pure bred samples of CO 3 pearl millet, CO 3 finger millet and CO 3 foxtail millet as sole sources of protein in diets fed to weanling rats, six to the group, for 10 weeks against 10% casein diets as control. All diets contained 10% protein except the finger millet which contained 7·5%. The grains were fed powdered raw, or autoclaved at 15 lb/in^2 for 30 min. Average weight changes, haemoglobin (Hb) levels (Sahli's haemoglobinometer) and plasma protein (Hawk *et al.*, 1947) are shown in Table 2.113.

Weight gains were higher after autoclaving in pearl and finger millet but lower in foxtail millet. Too few rats on the raw foxtail millet diets survived for results to be statistically significant. The essential amino acid content (chemical and microbiological assay) of the millet diets is shown in Table 2.114. Plasma protein haemoglobin was reduced in pearl millet and finger millet and increased in foxtail millet by autoclaving.

The total protein (Kjeldahl), fat (Soxhlet extraction) and glycogen (Good *et al.*, 1933; Hawk *et al.*, 1947) as g/100 g fresh liver were also determined by Chitre and Vallury (1956b) using essentially the same diets described above (1956a). The results are shown in Table 2.115. Chitre and Vallury (1956b) stated that several workers (Tucker and Eckstein, 1937; Channon *et al.*, 1938; Best and Ridout, 1940; du Vigneaud *et al.*, 1941; Keller *et al.*, 1948) had suggested that methionine and cystine could partly replace choline as lipotropic agents. The methionine and cystine contents of the millet diets in mg/100 g diet were:*

	Methionine	Cystine
Pearl millet	348	404
Finger millet	325	260
Foxtail millet	302	258

Liver lipids were higher from the millets than casein and were not correlated with methionine content. Chitre and Vallury (1956b) suggested that, as with sorghum, other factors may influence fat deposition or the amount of methionine available for growth. Also lipotropic activity may not be accurately indicated by their chemical assay method. Liver protein followed overall growth pattern. Higher liver glycogen resulted from the millet diets than from casein. Of all the diets examined (which included other cereals) finger millet produced the highest level of liver glycogen. Autoclaving increased the deposition of fat and glycogen and decreased liver protein from pearl millet and finger millet but increased liver protein and decreased (not significantly) fat and glycogen from foxtail millet.

Jansen *et al.* (1962) analysed a "combined" sample of teff, and a sample of pearl millet for essential amino acid content (column chromatography; tryptophan and cystine microbiologically). The results are shown in Table 2.119 for pearl millet and 2.153 for teff. In the trials 22-day old weanling male rats were fed over 4 weeks, 10 to the group on millet, 6 to the group on teff. The grains provided about 92% of the diet, 10% protein and 5% fat, DWB, against a control casein diet containing 10% protein and 5% fat. All diets included vitamin and salt mixtures. The diets, resulting weight gains and PERs are shown in Table 2.116. In all cases PER reflected lysine content.

Plasma protein (Goldstein and Scott, 1956) in rats fed diet (A) unsupplemented teff, was lower than those on (E) unsupplemented pearl millet. Lysine supplementation increased plasma protein but not equivalent to the casein controls. Plasma protein paralleled PER more closely than weight gain.

Goswami *et al.* (1969c) compared the high yielding pearl millet varieties (A) A 1/3 and (B) T 55; and three hybrids (C) 23A × BIL 3B, (D) 23A × BIL 1 and (E) 101A × BIL 3B grown under indentical conditions, by chemical analysis in rat feeding trials. Crude protein (Kjeldahl,

* (Authors' note: these values differ from those reported for similar diets in earlier experiments (Chitre and Vallury, 1965a).)

McKenzie and Wallace, 1954), crude fat, cellulose and ash (AOAC, 1960), lysine (Moore *et al.*, 1958, Moore and Stein, 1954ab) methionine (Horn *et al.*, 1946), and tryptophan (Smith and Agiza, 1951), are shown in Table 2.117.

The experimental diets consisted of groundnut oil 10 parts, salt mixture 4 parts, vitamin mixture 1 part, milled millet flour to give 6·4% crude protein in the diet and starch to make up to 100, fed against a control casein diet to male weanling rats, eight to the group, for 4 weeks. The protein intake, weight gain and PER are also shown in Table 2.117. Differences in weight gains, PER values and protein intake were significant. Diet (E) (101A × BIL 3B), highest in lysine, gave significantly higher PERs than other diets except (D) 23A × BIL 1; (C), lowest in lysine gave the lowest PER.

Food and Agriculture Organization (FAO, 1970) reported the mean PER of pearl millet as 1·73, ranging from 1·60 to 1·83.*

Kenney and Puar (1973) reported feeding pearl millet in 10% protein diets, with additions of oil, salts, fibre and vitamins to lactating rats and weanlings. The diets were: (A) pearl millet, (B) (A) + 0·58% lysine; (C) (B) + 0·18% methionine and (D) casein + 0·23% methionine.

Weanlings on all diets were equivalent in total brain lipid, cholesterol and brain weight. Body weights of weanlings were lower on (A), (B) and (C). At 7 weeks, body and brain weights of rats fed (B), (C) and (D) after weaning exceeded those fed diet (A) regardless of the maternal diets during lactation. Lysine supplementation (B) improved body and brain growth after weaning but methionine (C) had no further effect. Brain lipid and cholesterol paralleled brain weight.

Majumder *et al.* (1975) fed pearl millet diet (A) among several whole grain cereal flours, to weaned (22-day old), young (40-day old) and adult (90–100-day old) male rats to study growth and ascorbic acid metabolism. Wheat diet (B) and maize diet (C) were also studied against a fortified wheat diet (D) which included 15% casein and (E) a synthetic balanced diet which included 20% casein. The diets were fed at equivalent weights but were not isonitrogenous. After 8 weeks, the diets (A), (B) and (C) produced low weight gains, (B) highest with (A) and (C) very similar. Weight gain from diet (D) was significantly greater than from (A), (B) and (C). When all rats were fed diet (D), *ad lib.*, for 6 weeks a spurt in growth took place. The rats previously on diets (A), (B) and (C) did not attain the weight of those previously on diet (D). Final body weights of rats on diets (D) and (E) were highest throughout.

The growth and development of the liver, brain and kidney of rats on diets (A), (B) and (C) were similar and significantly lower than those on diet (D). After 8 weeks ascorbic acid (method of Chatterjee, 1970) in the livers and kidneys of weanling and young rats on diets (A), (B) and (C) was 40–60% lower, in the brain 25–35% lower and in blood plasma significantly lower than those on diet (D).

Guinea pigs were also used as experimental animals in the ascorbic acid

*(Authors' note: the FAO data are from different sources. See authors' comment on comparability of PER values in Chapter 1).

study because (stated Kenney and Puar, 1973) this species is unable to synthesize ascorbic acid. Diets (A), (B) and (C) could not be fed for more than 25 days because of weight loss and the death of 70–75% of the animals. A different lot of guinea pigs fed (A), (B) and (C) with additions of 5 mg of ascorbic acid/100 g body weight/day increased in weight but not to the same extent as those on diet (D) plus 1 mg of ascorbic acid/100 g body weight/day.

Majumder *et al.* (1975) stated that the findings confirmed an earlier report (Nandi *et al.*, 1974) that guinea pigs under stress need five times the normal ascorbic acid level to counteract the excess histamine released.

When casein was omitted from diet (D), the urinary, adrenal and plasma levels of ascorbic acid fell almost to those on diet (B), wheat alone, but the effect was not observed when casein was retained and vitamin and salt mixtures omitted. This was said to agree with the finding of Williams and Hughes (1972) that retention of ascorbic acid in the tissues of the guinea pig depends on the protein content of the diet.

Snehalatha and Reddy (1978) reported the biological availability to weanling rats of iron in certain cereals and pulses. The haemoglobin repletion technique was used with $FeSO_4$ as a reference standard. The bioavailability, %, (gained Hb–iron/iron intake) × 100 was in descending order: standard $FeSO_4$ diet 41, cowpea dahl 36, wheat 35, rice 22, pearl millet 20, pigeon pea dahl 18, finger millet and soybean equal 17.

Arnal-Peyrot and Adrian (1974) fractionated pearl millet by the method of Adrian *et al.* (1975) to obtain endosperm and a "protein fraction" which included the aleurone layer and the embryo. Pentosans, (method of Cerning and Guilbot, 1973), represented 2·2% of the endosperm and 11·9% of the protein fraction, as-is basis.

Pentosan digestibility was measured using 5 or 6 Wistar rats of a mean weight of 140 g. The diets were, (%) (A) endosperm 92·0, wheat gluten 1·0 and (B) "protein fraction" 45·0, wheat gluten 3·5 and starch 47·0. Both diets included (%) complete vitamin mix 1·0, complete mineral mix 2·0 and groundnut oil 5·0.* The effect on digestibility is shown below DWB.

	(A) Endosperm fraction	(B) "Protein fraction"
Intake total: g/rat/day	15·6	15·2
pentosans: mg/rat/day	371	891
Faeces, total: g/rat/day	0·88	2·16
pentosans: mg/rat/day	157	603
Digestibility[a]		
total	94·3	85·8
(Kjeldahl) N	88·6	74·0
pentosans	57.7	32·2

$$^{a}\text{Digestibility} = \frac{\text{intake} - \text{faecal output}}{\text{intake}} \times 100$$

* (Authors' note: it will be observed these total more than 100).

The lower digestibility of the pentosans in the protein fraction was due, in the view of Arnal-Peyrot and Adrian, to the higher content of fibre in that fraction, 7·5% of the whole grain compared to 1·15% of the whole grain, in the endosperm fraction.

Poultry

Mukherjee and Parthasarathy (1948a) reported that in feeding trials with Indian fowls, pearl millet and sorghum were superior in energy value to common millet and finger millet, with barley ranking last. The biological values of the five cereal proteins were reported by Mukherjee and Parthasarathy (1948b) as sorghum 56, pearl millet 85, finger millet 81, common millet 56 and barley 82.

Qureshi (1967) reported feeding trials in Pakistan with 449 chicks from hatching to 18-weeks old on starting and growing rations containing maize and pearl millet (of unstated origin) separately or in combinations. The rations with their percentage of maize and/or pearl millet, and the estimated total protein content are shown below:

Ration	(% of ration)	Total protein in ration (%)
(A) Maize	57·0	20·1
(B) Pearl millet	57·0	22·7
(C) Maize	28·5	
Pearl millet	28·5	21·7
(D) Maize	66·0	16·4
(E) Pearl millet	66·0	18·9
(F) Maize	33·0	
Pearl millet	33·0	17·6

No statistically significant differences in body weight gain and feed/gain quotient were found among rations. The mortality rate (average of three replicates) was 6·5% from (A) and (D); 10·9% from (B) and (E) and 9·3% from (C) and (F). The overall rate was 8·9%. Nutritional deficiences were not observed among chicks on rations (A), (C), (D) and (F) but 12 cases of weak legs (some severe) and stunted growth appeared from rations (B) and (E). Qureshi concluded that rations containing maize alone, or maize and pearl millet, are better than pearl millet alone for chicks to 18 weeks of age.

Sanford *et al.* (1973) reported that, at equal protein levels, pearl millet was equal to sorghum as judged by weight gain and feed efficiency in chicks.

Swine

Calder (1955) reported that up to 75% of the grain component of pearl millet (munga) grown in Rhodesia was satisfactory in pig feeds.

Sharda *et al.* (1972) reported that pigs fed on an all maize diet gained weight faster, with more efficient feed/gain quotients and higher protein efficiency ratios than those on a diet containing 75% maize and 25% pearl millet. At equivalent protein the feed conversion of pearl millet was 79% that of yellow maize. The addition of pearl millet did not adversely affect carcass composition or quality score.

Allee and Paulsen (1975) reported from (1) a growth trial and (2) a digestion trial with swine, that pearl millet containing 12·0% protein (N × 6·25) and 0·4% lysine (DWB) satisfactorily replaced all the maize in a maize–soybean meal diet, 16% protein and 0·77% lysine, without significant effect on feed intake, daily gain or feed/gain quotient. Digestibility of protein and energy in pearl millet were similar to maize in isonitrogenous diets containing soybean meal.

Finger Millet

Chaugale *et al.* (1955b) published a bibliography of finger millet (*Eleusine coracana*) and more recently Rachie and Peters (1977) issued from ICRISAT a review of the world literature covering the *Eleusines*.

Finger millet is an important staple food in East and Central Africa and in India. Purseglove (1972) quoting Mehra (1963) recognized two groups of cultivars: (1) the African highland types with grains enclosed within the florets and (2) Afro–Asiatic types with mature grains exposed out of the florets. The cultivars vary in many characters including height ranging from dwarf-types 40 cm to tall-types 1 m in height; colour of the vegetative organs varying from green to purple; type of inflorescence with straight and open spikes, incurved or close spikes and branched spikes resembling a cock's comb; length of spikes ranging from 3 to 13 cm; colour of grain which may vary from white, through orange-red, deep brown, purple to almost black. The Uganda region is regarded as the centre of its origin because of the long tradition there of religious and other ceremonies connected with its cultivation. It was probably taken to India more than 3000 years ago and though found in many other tropical countries, it has gained little importance outside Africa and India.

In Africa, it is usually grown at altitudes between 1000 and 2000 m in areas with an annual rainfall of between 900 and 1250 mm. It requires well distributed rainfall during the growing season with hot dry conditions for drying the grain at harvest. It does not readily tolerate either intermittent drought or water logging.

In India, finger millet, known as ragi, is most widely cultivated in hilly, lateritic soils in the 500–1000 mm rainfall belt of the tropics and subtropics (Rachie, 1975). In certain parts of India, two crops/year are grown, the first between May and August, the second between July and November. In Uganda, it sometimes follows cotton or sesame in shifting cultivation patterns or alternatively it may be intercropped with pigeon pea or sorghum. The time from sowing to maturity varies from 3·5 to 6 months depending upon genotype and environment. Purseglove stated that yields were variable and cited Sastri (1952) in quoting average yields in India between 600 and 800 kg/ha though yields in excess of 5000 kg/ha have been recorded. Purseglove also described how in Zambia, finger millet is grown in the "Citemene" system in which branches from roughly 4 ha of woodland are piled and burned on about 0·4 ha of land following which the finger millet seed is broadcast in the ash and reportedly produces reasonable yields (close to 2·5 t/ha) without any further cultivation. Though reliable production statistics are difficult to obtain, it is possible that finger millet cultivation in India covers close to 2·5 M ha, total production being roughly 1·5 M tons. According to Purseglove (1972) finger millet is the most important cereal crop in Uganda, occupying 0·4 M ha.

The grain is smaller than that of pearl millet (see Fig. 2.X). Adrian and Jacquot (1964) cited Sorre (1942) as reporting a 1000–seed weight of 2·5 g.

Structure

Adams *et al.* (1976) isolated the protein bodies of finger millet by a described method and examined their structure by electronmicroscopy. Their findings are discussed above under sorghum.

Chemical Composition

Africa

Shepherd *et al.* (1971–72) reported from East Africa the proximate composition (DWB), protein (Kjeldahl N × 5·83), crude fat (petroleum ether extractives); crude fibre and ash of 30 finger millets. The means and ranges are shown in Table 2.129. Half the samples contained about 8% protein. Crude fibre was higher than reported in similar trials of pearl millet, possibly because the seed coat constituted a larger proportion of the grain in the smaller seeded finger millet.

Asia

Solpico and Yambao (1966) reported that the foxtail millet varieties designated Siberian, White Wonder and glutinous, the common millet varieties, Early Fortune and Crown and finger millet strain A (seeds reddish-brown) and strain B (seeds brown) were compared in a series of trials at Los Baños, Laguna, Philippines in wet and dry seasons between 1957 and 1962. It was necessary to discard Early Fortune and Crown at an early stage owing to shattering and sterile grains, and another (undesignated) common millet variety was substituted. All the millets were introductions from abroad. The mean proximate analysis (unstated methods) is shown in Table 2.129. In overall performance (yield, disease resistance), the two finger millet strains performed best in both the wet and dry seasons.

Kadkol and Swaminathan (1954) reported the proximate analysis, Ca and

P (AOAC, 1950), Fe (Farrar, 1935), and thiamine (Swaminathan, 1942a) of the finger millet varieties RO 862, RO 870, RO 871, RO 883, K 1, RCO_2 (all brown), EC 4310 and Majjige ragi (both white). The mean and ranges are shown in Tables 2.129, 2.131 and 2.132. Variations in chemical composition among varieties were slight. EC 4310 had the highest protein (N × 6·25, 10·6%), thiamine and P. Majjige ragi was highest in Ca (430 mg/100 g). K1 was lowest in protein content (7·1%).

Deosthale *et al.* (1970b) examined 20 high yielding varieties of finger millet obtained from three millet research centres in Andhra Pradesh and one in Mysore State (Karnataka) India. The varieties ranged in yield from 600 (ES 11 and H 22) to 1800 kg/ha (Purna). The Task Force on Integrated Rural Development (Government of India, 1973) stated that Sharda was the most promising variety of finger millet under moisture stress in the Kharif season.

The varieties Aruna, Hamsa, Purna, Cauvery, Annapurna, VR 250 6, VZM 2, VR 18 a 1, VZM 1 and AKP 2, were analysed for protein, total ash, Ca, P and Fe (AOAC, 1965), Mg, Cu, Mn and Mo (colorimetrically, Sandell, 1959), thiamine (Srinivasa' Rao and Ramasastri, 1969a), riboflavin and niacin (microbiologically) and lysine (microbiologically).

The varieties ES 11, H 22, VR 7 were also analysed for protein, lysine and minerals and the varieties PR 248, PR 717, PR 722, PR 434, PR 202, CR 652 and CR 312, for protein and lysine. The means and ranges of the analyses and yields are presented in Tables 2.130, 2.131 and 2.132. Highest protein was found in Aruna, Hamsa, VR 250 6, VZM 2 and VR 18 a 1 which gave grain yields of 1200 kg/ha and above. The varieties were grown at different locations and under differing fertilizer treatments; most of the high-protein contents were grown under high fertilizer treatment. Lysine varied from 131 (Cauvery, VR 250 6, VZM 2) to 237 mg/g total N (CR 312); a negative correlation between lysine and protein was highly significant. The white Hamsa variety was highest in thiamine. Niacin ranged from 0·27 (VR 250 6) to 0·87 (Cauvery) and riboflavin from 0·02 (VZM 1) to 0·07 mg/100 g (Aruna, Hamsa, VR 250 6).

Fe ranged from 1·3 (Aruna) to 17·6 mg/100 g (VR 7); Hamsa was highest in Ca 661 and AKP 2, lowest, 253 mg/100 g. The P and Mg ranges were relatively narrow. Cu ranged from 0·32 (VR 7) to 1·0 (VR 18 a 1); Mn from 0·0011 (Aruna) to 0·021 (Hamsa) and Mo (in nine samples) from 0·011 (VZM 1) to 0·019 mg/100 g (Hamsa). No detectable amounts of Mo were found in four varieties grown at Hebbal, Karnataka.

Kamalanathan *et al.* (1971) analysed a sample of white finger millet CO 9 from Coimbatore, for protein (micro-Kjeldahl × 6·25), Ca, thiamine (thiochrome method) (AOAC, no date), Fe (colorimetrically Hawk *et al.*, 1965) and phytin P (Sundararajan, 1938). The results, on a sun-dried basis, are shown below, compared with a local brown variety.*

* (Authors' note: it is not clear from the text whether the local brown values (other than phytic acid) were also taken from Aykroyd *et al.* (1966) or determined).

	CO 9 (white)	Local (brown)
Yield kg/ha	4000	2600
Protein (%)	10·8	7·3
Thiamine (mg/100 g)	0·41	0·42
Ca (mg/100 g)	417	344
Fe (mg/100 g)	12	17
Phytic acid (mg/100 g)	208	209-246[a]

[a]Aykroyd *et al*, (1966); Sundararajan (1938)

Pore and Magar (1977) reported the proximate analysis, Ca (AOAC 1955), P (Fiske and Subbarow (1925), Fe (Elvehjem, 1930) and Cu (Chilton, 1953) of 36 finger millet cultivars obtained from the agricultural research centres of Maharashtra, Andhra Pradesh and Karnataka, and the Mo content (Johnson and Arkley, 1956) of the Maharashtra and Andhra Pradesh samples. The samples from Maharashtra were B 11, A 16, E 31, 'White Nagli', 'Red Compact Panicle' and 'Red Loose Panicle'; from Andhra Pradesh, AKP 2, AKP 7, PR 202, PR 248, VZM 1, VZM 2, VR 7, VR 18a 1, C 157, CR 652, CR 312 and CO 7; from Karnataka, Hamsa, Purna, HP numbers 7, 18, 35, 62, 69, 71, 81, 83, 86, 90, 91, 94, 102, 103, 105 and 114. The mean and ranges are presented in Tables 2.129 and 2.131. The samples from Karnataka showed higher protein contents, averaging 10·6%; those from Maharashtra averaged 7·1% and those from Andhra Pradesh 9·4%, though the highest protent content was shown by VZM 1 from Andhra Pradesh.

Indira and Naik (1971) examined the following species of millet obtained from the World Collection of IARI, New Dehli: finger millet, IE 903, IE 902, IE 901, IE 832 and IE 831: foxtail millet, ISE 711, ISE 263 and ISE 3; common millet IPM 1640 and IPM 1639; little millet, IPE 420, IPE 103, IPE 386 and IPE 108; kodo millet, IPS 141, IPS 26, IPS 19 and IPS 158; and Japanese barnyard millet, IEC 230, IEC 231 and IEC 1080. The grain samples were finely ground and analysed for protein, lipid, Fe, Ca, and P (AOAC, 1965), S (Basson, 1967), lysine (decarboxylase method, Naik, 1968) and tryptophan (colorimetrically, Spies and Chambers, 1949). The mean and ranges are shown for the millets in Tables 2.129, 2.130 and 2.131. Mineral contents were similar though finger millet was high in Ca. Only traces of Fe were found in the two varieties of common millet but IEC 231 and IEC 1800 varieties of Japanese barnyard millet contained very high levels of Fe (153 and 134 mg/100 g respectively) compared with 26 mg/100 g in IEC 230. Finger millet and little millet were lower in mean protein than the other species. Foxtail millet, little millet and kodo millet all contained over 4·0% of lipid.

Finger millet and common millet were relatively high, Japanese barnyard millet and little millet relatively low, in tryptophan. Common millet was relatively high in lysine.

Protein in defatted, finely-ground flour of the same species was fractionated (Naik, 1968; modification of the Mendel-Osborne method). Results are shown in Table 2.122.

There were wide variations among species in total extractable protein. Only 20% of the Japanese barnyard millet protein could be extracted compared to between 60 and 80% among the other species. Excluding Japanese barnyard millet, prolamines represented the major fraction. Albumins and globulins varied widely. Lysine and tryptophan among fractions (excluding Japanese barnyard millet) are shown in Table 2.123.

All the prolamine fractions were low in lysine and, except finger millet, in tryptophan. Prolamine contents of foxtail millet, common millet and little millet were higher than kodo millet and finger millet. Indira and Naik (1971) suggested that the higher lysine of common millet was attributable to high lysine in the glutelin and globulin fractions, not to depressed prolamine synthesis.

The amino acid compositions (auto-analyser) of finger millet IE 841 and of kodo millet IPS 394 are shown in Tables 2.130 and 2.150 respectively. The isoleucine of finger millet was high, the leucine/isoleucine quotient being about two, almost equivalent to rice and wheat; leucine/isoleucine in kodo millet was above three similar to sorghum and maize. Kodo millet also contained higher methionine and lysine than finger millet.

The *in vitro* protein digestibility in finger millet IE 841 and kodo millet IPS 394 was studied using a modified method of Fasold and Gundlach (1963): papain digestion and the time course of the liberation of amino acid N determined by ninhydrin. The proteins were extremely resistant to papain unless subjected to 15 min autoclaving or boiling as shown below:

	α-amino N as % of total N			
	Finger millet IE 841		Kodo millet IP 394	
	2h	8h	2h	8h
Cold	14·72	18·92	10·39	12·13
Boiling	24·02	44·13	19·02	20·39
Autoclaving	42·02	48·33	20·79	26·04

Indira and Naik suggested that to improve the biological quality of these millets, it is necessary to improve protein digestibility as well as amino acid balance.

Balakrishna Rao *et al.* (1973) determined the moisture, ash, and crude fibre (Kanwar and Chopra, 1959), protein (Kjeldahl N × 6·25), P, Fe and Ca from the hydrochloric acid extract of the ash; P (ammonium phosphomolybdate), Fe (colorimetrically) and Ca (volumetrically) of the improved varieties of finger millet Annapoorna, Poorna, Cauvery, Aruna, Hamsa, Hullubele, H 22, IE 8 1, IE 28, IE 312, IE 776, ES 11, EC 4840, ROH 2 and ROH 4, grown under identical conditions at Hebbal, Bangalore. The means and ranges are shown in Tables 2.129 and 2.131.

Poorna was lowest (4·7%) and IE 81 highest (9·9%) in protein. Hamsa (9·0%), Hullubele (9·3%), IE 776 (9·3%) were also relatively high in protein. Annapoorna (3·6%), Poorna (3·1%), Cauvery (3·2%) and ES 11 (3·5%) were low in crude fibre; Hamsa contained 4·2% and Hullubele was highest at 5·7%. Ash content varied from 2·0 in EC 4840–4·1% in H 22. Hullubele, IE 776, ES 11, Annapoorna and IE 28 all contained about 3·0% ash. P, Fe and Ca showed no consistent relation with the ash content. P ranged from 227 (Annapoorna) to 470 (IE 28), Fe from 2·5 (Cauvery) to 19·9 (ROE 2) and Ca from 203 (Annapoorna) to 690 mg/100 g (H 22). Aruna, IE 776 and ROH 4 were also relatively high in P, Annapoorna, IE 28, IE 776 and ES 11 in Fe, and Cauvery, Aruna and IE 776 in Ca. The colour of the grains varied from deep brown (Hullubele), to dull white (Hamsa). Overall, IE 776 was considered a nutritionally good variety.

Passmore and Sundararajan (1941) reported the mean thiamine content (thiochrome method) of 10 lines of finger millet from Coimbatore (see Table 2.132). The four named varieties, all brown in grain colour, were T/S Ragi, Gidda Aryan, Mutti ragi, and Poorasi Keraru: the other lines were not named. Both the highest of 0·65 and the lowest 0·27 mg/100 g were local crosses white in grain colour.

Chitre *et al.* (1955) reported the thiamine and riboflavin (own method), and the niacin (James *et al.*, 1947) of CO 2 finger millet from Coimbatore (rain-fed or irrigated, red loamy soil) as respectively 0·19, 2·5, and 0·10 mg/100 g.

Mahudeswaran and Ayyamperumal (1970) reported the thiamine, niacin and riboflavin contents (unstated methods of determination) of the finger millet brown varieties CO 1, CO 7, K 2, AKP 1, AKP 2, MS 8070, MS 8071, PLR 1 and EC 4437, and the white varieties CO 6, EC 1008, EC 1540, EC 4275, EC 4312, EC 4336, EC 4354, EC 4461, MS 171, MS 192 and MS 3894 (see Table 2.132). The mean 1000-seed weights and vitamin contents revealed little difference between the two groups.

Protein and Amino Acid Composition

Table 2.130 brings together analytical data from a number of sources including FAO (1970) a single sample determined chromatographically and three or four assayed microbiologically. The chromatographic values agree with those given by Gayte–Sorbier *et al.* (1960) who reported tryptophan (microbiological assay), as 75 mg/g total N.

Desikachar and De (1947) using a chemical method based on the determination of S, reported cystine 230 and methionine 280 mg/100 g total grain. Lal (1950) using a microbiological method obtained results on bazaar samples within the overall range for lysine, tryptophan and phenylalanine but lower in mg/100 g grain for histidine (26) and arginine (121). The other amino acids were not assayed. Taira (1963b) reported amino acid contents (microbiological assay) as shown in Table 2.130. Isoleucine, leucine and cystine are markedly lower than those of FAO (1970).

Six numbered samples of finger millet grown in East Africa, were analysed for amino acids (excluding cystine, methionine and tryptophan) by auto-analyser following acid hydrolysis (Shepherd *et al.*, 1971–72). There were appreciable differences in N recovery. Means and ranges are shown in Table 2.130. From India, Balasubramanian *et al.* (1952ab) and Balasubramanian and Ramachandran (1957) reported the amino acid composition (microbiological assay) of the finger millet varieties CO 1 and CO 12 and of a bazaar sample. Their data are presented in Table 2.130. A ground sample of CO 1 stored for 10 months in a sealed jar containing naphthalene as preservative, suffered a 19% loss in lysine content.

Mahudeswaran *et al.* (1966) compared seven white finger millet varieties, CO 6, EC 1008, EC 1540, EC 4275, EC 4336, EC 4354 and EC 4461 with three brown varieties CO 7, CO 1 and K 2 (see Table 2.129 for mean and ranges). The white types were superior in protein content (method of determination and N conversion factor not stated).

Kempanna and Kavallappa (1968) compared 19 cultivars of finger millet from the World Collection designated 116, 118, 128, 132, 136, 141, 142, 145, 149, 154, 160, 164, 171, 348, 419, Aruna, Purna, Annapurna and Majjige, and grown under similar conditions. Crude protein, crude fat, ash, P_2O_5 and CaO content (methods not stated) are shown in Tables 2.129 and 2.131.

Variations in protein and in crude fat were significant, other differences were not. Majjige (white) was highest in fat (1·7%), P_2O_5 (890 mg/100 g) and CaO (930 mg/100 g) and slightly below the mean in protein (10·0%). The brown grained 164 was highest in protein (12·7%) and above the mean in ash and minerals.

Mallanna and Rajashekara (1969) reported that from five varieties of finger millet drawn from the World Collection, IE 810 (later released as "Hamsa" for the Bangalore district) gave the highest yields equivalent to Purna. IE 810's proximate analysis, Ca, P and Fe contents (methods not stated), lysine and methionine contents (microbiological assay) (conducted at the Central Food Technological Research Institute, Mysore, India) are shown in Tables 2.129 and 2.131.

Nagarajan (1974) reported the protein and lysine content (analytical methods not stated) of the five high yielding varieties of finger millet, Hamsa, Purna, Cauvery, Aruna and Annapurna. Mean and ranges appear in Table 2.130.

Protein Fractionation

Narayana and Norris (1928) reported that the proteins of finger millet merited study in view of the cereal's importance in the diet of people living in Mysore State (now Karnataka). Finger millet contained from 6 to 11% crude protein (N × 6·25). A "considerable proportion" of the protein was in the form of an alcohol-soluble prolamine for which they suggested the name "eleusinin". Four described variations in extraction technique with 70% alcohol gave final products with much the same degree of purity but the percentage of protein was not stated. Damodaran (1931) reported that attempts to isolate the globulin and glutelin of finger millet in a reasonable state of purity were unsuccessful but the prolamine was extracted with boiling 73% alcohol.

Niyogi *et al.* (1934) using a variety of finger millet known as H 22 ragi and grown at the Hebbal Agricultural Farm, Bangalore, were unable to obtain a satisfactory yield of albumin and globulin (water and salt extracts respectively). The % proximate analysis (methods not stated) of the finger millet flour (ground to pass through a 40-mesh sieve) was reported as: % protein (N × 6·25) 8·4; ether extractives 1·7; crude fibre 3·4; ash 2·7; and carbohydrates (by difference) 83·8. The prolamine (eleusinin) extracted by essentially the same method as Narayana and Norris (1928), was in the form of a light grey powder.

Taira (1965) fractionated the protein of finger millet by differential solubility. The proportions of 18 amino acids mg/g total N (by microbiological assay), are shown in Table 2.124. The prolamine fraction contained more glutamic acid, leucine and proline, and less glycine, aspartic acid, lysine and arginine than the other protein fractions.

Pushpamma (1968) fractionated the protein of samples of finger millet from the local market at Hyderabad. Her findings compared with those for pearl millet, Indian and USA sorghum, and for wheat and rice, are given in Table 2.22.

Virupaksha *et al.* (1975) compared 12 varieties of finger millet from the germ plasm collection of the University of Agricultural Sciences, Hebbal, Bangalore:

ECW 854, ECW 955, HX 799, Hamsa, HPW 27 4, HPW 83 4 (white), EC 4840, Purna, HPB 7 6, HPB 20 5, HPB 1 8 and HPB 23 6 (brown). The designation HPB (brown) or HPW (white) identified crosses between the white seeded Hamsa and the brown seeded Purna. Samples were pounded by hand in a pestle and mortar and endosperm flour obtained by sieving through a 100-mesh screen. Whole seed flour and endosperm flour were defatted by water-saturated *n*-butanol. The defatted flour was air-dried and ground to a fine powder. The seed colour and protein (micro-Kjeldahl N × 6·25 DWB) of the whole seed and endosperm are shown in Table 2.125 in descending order of seed protein content.

Protein in the whole seed varied from 6·8 (HPB 23 6) to 11·0% (ECW 854); in the endosperm from 3·5 (HPB 23 6) to 6·3% (ECW 854), the white varieties having higher protein contents than the brown. Extraction of endosperm flour by the Osborne and Mendel (1914) procedure yielded only 25–30% of the total N. The Landry and Moureaux (1970) procedure applied to ECW 854, Hamsa, HPW 83 4, EC 4840, Purna and HPB 20 5 yielded 60–84% of total N depending on the variety, Hamsa endosperm flour being the lowest and HPW 83 4 whole seed the highest. For unexplained reasons, in each case the total extract, as % total N, was greater from the whole grain than from the endosperm flour. The solubility fractions of the six varieties are shown in Table 2.126.

Fraction I (albumins and globulins) constituted 8 to 15% of the total seed protein. Fraction II and III (prolamines) ranged from 35 to 50% of the total protein and 29 to 41% of the endosperm flour. Fraction V (stated as probably glutelin) represented 12 –28% of total protein in whole seed and 13–25% in the endosperm. The white grain varieties produced a higher prolamine and a lower glutelin extract than the brown grain varieties: the higher protein content of the white grain varieties appeared attributable to the higher prolamine fraction. Undefatted flour, or the protein fractions, of all 12 varieties were analysed for amino acid composition (mg/g total N) (auto-analyser following acid hydrolysis, excluding tryptophan). The mean and ranges are shown in Table 2.130.

There were considerable variations among the varieties: glutamic acid, proline, valine, isoleucine and leucine showing a wider range than the other amino acids, probably reflecting differences in prolamine among varieties. The leucine/isoleucine quotient was close to three.

A highly significant negative correlation was found between protein in the millet and lysine as per cent protein. The regression line indicated a decrease of 0·507 mg lysine for each 1% increase in protein. Methionine also decreased with increase in protein but less so than lysine, the regression line indicating a decrease of 0·306 mg methionine for each 1% increase in protein.

The amino acid composition of the endosperm protein of Hamsa, Purna, HPW 83 4 and HPB 20 5 showed a gross similarity to the amino acid composition of their respective whole seed protein; lysine, histidine, arginine and proline were slightly higher in the whole seed protein than in the endosperm protein.

Three protein fractions were extracted from Hamsa and Purna. Fraction A (albumin–globulin) from defatted endosperm flour extracted with 0·5 M NaC1 followed by distilled water, fraction B (prolamine) from the residual flour extracted with (a) 70% isopropyl alcohol plus 0·6% 2—mercaptoethanol followed by (b) borate buffer (pH 10) containing 0·5 M NaC1, 0·6% 2—mercaptoethanol and 0·5% sodium dodecyl sulphate to give fraction C (glutelin). Table 2.127 shows the amino acid composition of the whole seed protein, the endosperm protein and the protein fractions of Hamsa (white) and Purna (brown).

With the exception of cystine, the amino acid composition of the three protein fractions were similar for the two varieties. Virupaksha *et al.* (1975) suggested partial destruction of cystine during acid hydrolysis. Serine, threonine, and alanine in the three fractions were about the same as in the endosperm protein; aspartic acid and glycine in fractions A and C were similar to those in the endosperm. Glutamic acid, proline, valine, isoleucine and leucine were almost equal in the fraction B and in the unfractionated endosperm protein.

Glutamic acid was nearly 75% higher in fraction C (glutelin) than in fraction A (albumin–globulin). Histidine, proline, isoleucine, leucine and phenylalanine were also higher in fraction C than in fraction A. Fraction B (prolamine) contained higher levels of glutamic acid, proline, valine, isoleucine, leucine and phenylalanine than fractions A and C. Lysine, arginine and glycine were higher in fractions A and C than in B.

Biological Evaluation

Human beings

In 1955, Subramanyan *et al.* reported on the metabolism of N, Ca, and P in eight healthy adult males, from 24–32-years old, on a poor Indian diet based on finger millet (variety H 22), over a period of 12 days with the final 5 days used for the collection of urine and faeces (Murthy *et al.*, 1954). The composition (AOAC, 1950) of the finger millet which represented 69% of the diet, is shown in Tables 2.129 and 2.131.

All subjects maintained a positive N, Ca and P balance. Average daily N intake was 10·6 g and retention 1·1 g. The excretion of N in the faeces was high, 5·5 g; apparent protein digestibility was only 50%. Average daily intake of Ca was 3·4 g and retention 0·1 g. Average daily P intake was 2·3 g, retention 0·2 g. The average daily excretion of phytate P in the faeces (McCance and Widdowson, 1935) was only 0·3 g; about 85% of ingested phytate was hydrolysed.

Joseph *et al.* (1958ab) reported on the metabolism of N, Ca, and P in eight girls, between 9 and 10 years old, living in a boarding home in Mysore City, fed a poor Indian diet based on finger millet, the chemical composition of which (AOAC, 1950) is shown in Tables 2.129 and 2.131.

Finger millet represented about 64% of the total diet, fed during 15 days with collection of urine and faeces (Murthy *et al.*, 1954) over the last 5 days. All the subjects maintained positive N, Ca and P balance. The mean daily N intake was 4·51 g and retention 0·52 g. Mean daily excretion of N in the faeces was 2·11 g, amounting to about 47% of intake; apparent protein digestibility was only about 53%. The mean Ca intake was 1151 mg, retention 226 mg, and mean loss in the faeces 889 mg, about 77% of ingested Ca. Mean Ca retention corresponded to 20% of intake. Mean daily P intake was 887 mg, retention 135 mg, and mean daily loss in the faeces 634 mg, about 71% of ingested P. Mean P retention corresponded to 15% of intake.

Rats

Niyogi *et al.* (1934) studied finger millet flour in metabolic trials with rats, with 5% protein in the diet (N = 0·84%). The control diet contained 5·0% dried egg powder. Average digestibility and biological value (Mitchell, 1923–1924, Mitchell and Carman, 1926) of the protein in the two diets were 77·5% and 90·5% respectively.

Ramiah and Satyanarayana (1936) reported that, by rat growth and N balance methods, the nutritive value of white finger millet grain was superior to that of a brown type. The same authors (1937) reported the biological value of white finger millet protein as 91·5% superior to that of brown grain.

The findings of Ranganathan (1935), Swaminathan (1937ab), Giri (1940) and Kuppuswamy *et al.* (1958) are discussed under Sorghum.

Chitre and Vallury (1956ab) reported the results of rat feeding trials with CO 3 finger millet in comparison with CO 3 pearl millet and CO 3 foxtail millet. These are discussed under pearl millet.

Pore and Magar (1976a) studied B 11 variety of finger millet obtained from Kolaba, Maharashtra State, in feeding trials over 8 weeks with male weanling albino rats, 6 to the group. The proximate composition, and Ca (AOAC, 1955), P (Fiske and Subbarow, 1925) and Fe (Elvehjem, 1930) of the cleaned dehusked powdered millet is shown in Tables 2.129 and 2.131.

Three isonitrogenous diets contained: (A) 8·5% casein, 51% starch, 30% sucrose (B) 90·0% millet flour and (C) 90% defatted millet flour, all supplemented with 5% oil, 5% minerals and vitamins. PER, blood haemoglobin (acid haematin method, Helper, 1950, Sahli Hellige haemometer), serum protein (Lowry *et al.*, 1951), serum cholesterol (Sperry and Webb, 1950), phospholipids (Taurog *et al.*, 1944) and liver composition, lipids (Folch *et al.*, 1957) protein (Lowry *et al.*, 1951) and total cholesterol (Sperry and Webb, 1950) are shown in Table 2.128.

Diets (B) and (C) both produced higher liver lipid, lower haemoglobin and lower serum cholesterol than (A). Blood haemoglobin levels from diets (B) and (C) were similar to those reported by Chitre and Vallury (1956a) from raw and autoclaved finger millet diets. Both Kurien *et al.* (1958) and Guttikar *et al.* (1968) reported 14·1% haemoglobin from finger millet diets.

The PER of a finger millet diet with 6% protein was reported as 0·95 by Daniel *et al.* (1968a) and by Leela *et al.* (1965) as 1·12. Pore and Magar (1976b) reported that finger millet, when fed as the sole source of protein to male albino rats, against a control casein diet, led to a decrease in the serum lipid and total cholesterol levels and an increase in the total lipid level of the liver, heart, brain, kidney and spleen tissues. The protein contents of the diets, (%), were: (A) casein 6·75, (B) B 11 finger millet 6·75, (C) B 11, defatted, 6·75, (D) A 16, finger millet 5·67% and (E) A 16 defatted 5·67%. The increase in liver lipid was significant and in agreement with reports of accumulation of fat in the livers of rats fed a low protein diet (Aoyama *et al.*, 1973) and low quality protein diets (Aoyama *et al.*, 1969, Aoyama and Ashida, 1972).

Swine

Calder (1960) reported finger millet (rupoko) grown in Rhodesia more suitable for breeding stock than for young growing stock and recommended finger millet be limited to 50% of the grain in pig feeds.

Foxtail Millet

Foxtail millet (*Setaria italica*) is a short plant, the seeds are smaller than those of common millet and are enclosed in thin outer hulls which are removed during threshing, most of the inner hulls remaining attached to the seed (Arnon, 1972). See Fig 2.X. Adrian and Jacquot (1964) cited Sorre (1942) as reporting a 1000-seed weight of 2·0 g.

According to Purseglove (1972) foxtail millet was probably domesticated in Eastern Asia where it has been cultivated since ancient times. It is believed to have been one of the five sacred plants of ancient China since 2700 B.C. It is used as a food grain in Asia, Southeast Europe and North Africa, is the most important millet in Japan, and is widely cultivated in India. In Russia it is used for brewing beer; it is widely used for feeding crop grown both for hay and silage. Foxtail millet can be grown at low rainfall from sea-level to 2000 m, can tolerate a wide range of soils from light sands to heavy clays, and gives reasonable yields on comparatively infertile soils, though it does not cheerfully tolerate water logging or long periods of drought. Various types differ in time to maturity, (80–120 days); height of plant, usually between 1 and 1·5 m; length, breadth and compactness of the inflorescence (the length can vary from 7·5 to 25 cm), colour of grain, which varies from pearl yellow, through orange, red, brown to black.

It may be grown in pure stand, or mixed with other crops such as finger millet, or cotton. In Asia it is also grown as a catch crop if the paddy (rice) crop fails. A rainfed crop usually yields 800–900 kg/ha of grain (Purseglove, 1972). It is grown in many regions of China where yields exceeding 11000 kg/ha have been claimed (Rachie, 1975).

Katwe *et al.* (1955) published a bibliography of foxtail millet, and Malm and Rachie (1971) published a review of the world literature on the *Setaria*.

Structure

Lorenz (1977) examined by SEM the foxtail millet variety Golden German and a Chinese line PI 391638, and the common millet variety Abarr and an experimental line identified as Big Red, PI 346946, all grown at the Central Great Plains Field Station of the Colorado State University, Akron, California. Externally, the surface of the hull of both foxtail millets was covered with small peaks arranged symmetrically; the hulls of the common millets were quite smooth and even. Internally, the structures were essentially similar. The pericarp was thin. The aleurone layer was one cell thick; starch granules were angular and polygonal in the corneous endosperm and spherical in the floury endosperm. The starch granules ranged in size from 2 to 10 μm.

The starches were isolated from the millets and for comparison, from Chris wheat and Prolific rye by the method of Adkins and Greenwood (1966). The millet starches had higher water binding capacities and higher gelatinization temperatures than the wheat and rye starches. Swelling power values at 60°C were higher for wheat and rye starches than for the millet starches. At 60°C the millet starches were less soluble than the wheat or rye starches; at 90°C, the millet starches were more soluble than the rye starch and less soluble than the wheat starch except for the starch of the common millet Big Red which was more soluble. Amylograph viscosities at 92°C of both foxtail and common millet starches were higher than those of the wheat and rye starches.

Chemical Composition

Solpico and Yambao (1966) reported the proximate composition of three varieties of foxtail millet grown at Los Baños, Philippines between 1957 and 1962 in comparison with finger millet and common millet. (See Table 2.136). Ramanthan *et al.* (1975) reported the proximate analysis, (AOAC, 1960) of CO 3, CO 2, and Thenai 5307 varieties of foxtail millet. Means and ranges are shown in Tables 2.136 and 2.138 with other representative values. Chitre *et al.* (1955) estimated thiamine and riboflavin (own method) and niacin (James *et al.*, 1947) in foxtail millet varieties N 1 from Nandyal (rain-fed, deep black cotton soil), CO 1 and CO 3 from Coimbatore (rain-fed or irrigated, red loamy soil) and H 1 and H 2 from Bellary (similar to Nandyal). The mean and ranges were in mg/100 g; thiamine 0·43 0·32–0·53), niacin 3·7 (2·0–4·7) and riboflavin 0·12 (0·09–0·17). Other representative analytical values, ranges only, were: thiamine 0·32–0·6; niacin 0·4–4·7; riboflavin 0·05–0·17; total folic acid 0·015; and free folic acid 0·004.

Protein and Amino Acid Composition

Table 2.137 brings together data from various sources including FAO (1970) covering a single sample and presenting 12 amino acids, and Taira (1963a), both microbiological assay. Lee (1970) reported from Japan the limiting amino acids in foxtail millet as lysine and methionine. The amino acid scores (WHO, 1973a) from FAO (1970) and Taira (1963a) indicate lysine as first-limiting.

Chitre *et al.* (1956) reported (Kjeldahl) N and phenylalanine, leucine, isoleucine, lysine, valine and threonine (assayed microbiologically) and the methionine, histidine, arginine and tryptophan content (assayed chemically) of the pure strains of N 1 and H 2 foxtail millet. For means and ranges see Table 2.137.

Taira (1968) compared the amino acid composition (microbiological assay, Tamura *et al.*, 1952) of five nonglutinous types of foxtail millet: Kunamoto–Chima No. 1, Komemasari, Shinano No. 1, Daiohkoku and Rikuu No. 1 and five glutinous types: Showa–Mochi, Nekoashi, Takayama–Shiromachi, Tsugaru–Wase and Shiroawa, collected from the Nagano Agricultural Experiment Station, Japan. The samples were dehulled and ground to pass a 40-mesh sieve. Protein (macro-Kjeldahl N × 6·25), DWB, and amino acid composition, mean and ranges, are shown in Table 2.137. Amino acid composition did not differ between non-glutinous and glutinous types. The results were in accord with Mangay *et al.* (1957) though from Taira's lysine and threonine were higher and isoleucine lower.

Taira stated that the amino acid pattern of foxtail millet was generally similar to common millet, Japanese barnyard millet, pearl millet, sorghum and maize (Taira, 1962a, 1963b, 1966) and was characterized by higher alanine and leucine contents than other subfamilies of the Gramineae. Crude protein correlated positively with glutamic acid, proline, alanine, leucine and isoleucine and negatively with lysine, aspartic acid, arginine and cystine.

Indira and Naik (1971) reported the chemical composition of three varieties of foxtail millet, the distribution of protein and of lysine and tryptophan among solubility fractions (Osborne-Mendel) (see Table 2.123). The data are discussed in comparison with other millets under finger millet.

Protein Fractionation

Lo (1941) in Minneapolis determined the moisture, total protein (N × 6·25) and ash (AOAC, no date) of selected ground seeds of foxtail millet and common millet, (see Tables 2.136 and 2.143). The Osborne protein solubility fractions, with other analyses, are presented in Table 2.133.

The alcohol-soluble protein was higher and the alkali-soluble lower in the common millet than the foxtail millet. The water-soluble and salt-soluble protein and the insoluble N in the residue were similar. Glutelin (alkali-soluble) and prolamine (alcohol-soluble) were the principal proteins in the common and foxtail millets.

Kametaka (1952) reported that the glutinous foxtail millet Tochigi No. 1, contained, DWB, 11·4% crude protein (1·82% total N; 1·75% protein N), 2·9% crude fat and 1·4% ash. The % N extracted by successive solvents was (mean of two determinations), as % of total N: distilled water 9·5; 10% NaCl solution 8·4; 0·2% NaOH solution 17·65; 70% ethanol 57·15 and total 92·7. The prolamine (ethanol extract) contained, DWB, 16·65% N, 0·17% ash, 0·04% P; 1·4% S; 1·5% arginine; 1·4 histidine; and 0·3% lysine.

Taira (1962b) reported the amino acid content (microbiological assay) of the solubility fractions of Japanese barnyard millet (*Echinochloa crus-galli* var. *frumentacea*) and a foxtail millet. The values are presented in Table 2.134.

Lipids

Obara and Kihara (1973) reported the glucosylglyceride content of the glycolipid fraction obtained by treating the acetone-soluble lipid fraction of foxtail millet by Carter's method (no further reference available). Monoglucosylglyceride (MG) and diglucosylglyceride (DG) were isolated from the acetone-insoluble and the acetone-soluble fractions by column chromatography and thin layer chromatography. Both MG and DG were found to contain glucose. The acetone-insoluble fraction contained mainly linoleic

(35·4% in MG, 35·7% in DG) and palmitic acid (32·2% in MG, 38·9% in DG). MG and DG from the acetone-soluble fraction contained mainly linoleic (67·9% in MG, 34·9% in DG) and linolenic acid (16·3% in MG, 32·4% in DG).

The comments of Adrian and Jacquot (1964) are referred to under Sorghum.

Carbohydrates

Rahman *et al.* (1974) in Dacca reported starch, total sugar, reducing and nonreducing sugars (all AOAC, 1960) in market samples, with and without husk, as shown below:

	Foxtail millet	
	With husk (%)	Without husk (%)
Starch	62·5	73·1
Total sugar	2·9	1·4
Reducing sugar	0·8	1·2
Non-reducing sugar	2·1	0·2

Biological Evaluation

Rats

Foxtail millet, moisture 11·2% and crude protein 10·0% was fed at 5·3% protein, DWB, to adult rats (Swaminathan, 1937c). Using a N balance method biological value was 77 and the digestibility coefficient 91, compared to 90 and 89 respectively for SMP also fed at 5% protein.

Chitre and Ganapathy (1956) studied the nutritive value of foxtail millet by a balance sheet method (no reference) in weanling and adult rats and reported the biological value as 52. The essential amino acid composition (assayed microbiologically or colorimetrically) of the sample is shown in Table 2.137. Lysine was first-limiting.

Young rats fed unsupplemented foxtail millet (10% protein in diet) lost weight and ate less than those on a casein control diet. It was estimated (Borchers and Ackerson, 1947) that the millet contained $2{\cdot}98 \times 10^{-3}$ trypsin inhibitor units/ml of extract. No improvement in growth resulted from autoclaved millet. After supplementation with 0·4% lysine or 15% gelatin the growth rate was almost equal to the casein control. Supplementation with (A) 10, 15 and 20% groundnut cake and (B) 20% horse gram (*Dolichos biflorus*) also gave satisfactory growth. The horse gram was autoclaved to destroy trypsin inhibition.

Ganapathy *et al.* (1957) studied N retention, and amino acid content of the liver, kidney and muscles of young albino rats (40–45 g in weight) fed diets based on (A) casein, (B) foxtail millet and (C) millet plus 0·47% lysine HCl (equivalent to lysine in (A) at 10% protein). The composition of the diets was described by Chitre and Ganapathy (1956) who stated that the millet, at 83% of the diet, was the sole source of the 10% dietary protein. For N (Kjeldahl) balance data, see Table 2.135.

On diet (B) fed *ad lib.* N intake was 41·5 mg/rat/day, on diet (C), 66·3 mg and on diet (A), 63·1 mg/rat/day. Faecal N was appreciably higher from the millet than the casein diets possibly because of the 8·0% fibre content of the foxtail millet. Faecal N was (A) 15, (B) 37 and (C) 21% total N whether fed *ad lib.* or isocalorically. Diet (B) maintained positive N balance though growth was retarded. Lysine supplementation improved biological value, digestibility and increased N retention from 19% (diet B) to 31% (diet C).

Ganapathy and Chitre (1961) fed weanling rats, 12 to the group over 4 weeks, on diets with 10% protein provided by foxtail millet without and with 0·48% added lysine, against a casein control. Added lysine increased the mean haemoglobin values from 8·43 (unsupplemented) to 13·54 g/100 ml blood; protein increased from 2·98 to 4·78 g/100 ml plasma, compared to haemoglobin 15·74 and protein 5·5 g in rats fed casein. The higher plasma protein was mainly albumin.

Chitre and Vallury (1956ab) reported rat feeding trials with CO 3 foxtail millet, CO 3 pearl millet and CO 3 finger millet. The results are discussed above under pearl millet.

Results reported by Kuppuswamy *et al.* (1958) are discussed under sorghum.

Poultry

Kalmykov *et al.* (1954) reported digestibility of foxtail millet in poultry similar to that of common millet. The proximate composition of the millet lay within the ranges shown in Table 2.136.

Kumanov (1958) reported the total digestible nutrients of foxtail millet fed to poultry as 86·7% from ground, sieved grain, 75·1% from whole grain and 74·2% from ground grain.

Bishop and Taylor (1963) fed diets containing (A) about 80% canary seed (*Phalaris canariensis*), (B) small yellow foxtail millet and (C) large white common millet to Rhode Island Red × Light Sussex day-old chicks for 4 weeks. The diets were adequate in all nutrients or fed with single omissions of riboflavin, niacin, pyridoxine, pantothenic acid, biotin, folic acid and choline.

From the mean weights after 4 weeks, all three grains appeared deficient in riboflavin; canary seed was otherwise adequate. Foxtail millet appeared deficient in pyridoxine and biotin (four out of 10 chicks showed deficiency signs); common millet appeared deficient in niacin, biotin, folic acid and choline.

When the energy value of the diets was raised without altering the ratio of metabolizable energy to crude protein (by increasing casein and arachis oil at the expense of the seeds) and analysed for niacin, pyridoxine, biotin, folic acid and choline, canary seed appeared slightly deficient in biotin but supported adequate growth without added niacin probably because of its adequate tryptophan content. Foxtail millet appeared to be slightly deficient in folic acid, niacin, pyridoxine, biotin and choline. A 3/1/1 mixture of canary seed, foxtail millet and common millet was deficient only in biotin. Maize under similar conditions was deficient in niacin and choline, and slightly deficient in biotin.

Common Millet

Common millet (*Panicum miliaceum*) is a short plant with flattened seeds. A great many varieties differing in height of plant, compactness of panicles and colour of glumes are reportedly grown (Arnon, 1972).

Common millet, suited to a dry, continental climate, grows farther north than most other millets, and is cultivated mainly in Eastern Asia, including Mongolia, Manchuria, Japan, India, Eastern and Central Russia. It is also found in Arabia, Syria, Iran, Iraq and Afghanistan. Matz (1969) stated it to be the largest millet crop in the USA. According to a recent visitor from the Peoples Republic of China, it is one of the most popular cereals in Northern China, commanding a price equal to wheat. It is generally dehulled, then steamed or boiled in a manner similar to rice. To a lesser extent it is sweetened with sugar and eaten as a dessert or flaked. It is believed by some Chinese to be endowed with superior nutritive qualities. Purseglove (1972) stated that common millet is of ancient cultivation; it is the "milium" of the Romans and the true millet of history. It was cultivated by the early lake dwellers in Europe. It is believed to have been domesticated in Central and Eastern Asia and because of its ability to mature quickly was often grown by nomads.

Common millet displays possibly the lowest water requirement of any cereal. It matures in between 60 and 90 days and may be grown where the climate is too hot, the rainy season too short or the soil too poor for other cereals. Some cultivars are adapted to high altitudes and the northern limit of cultivation is a June isotherm of 17°C and a July isotherm of 20°C. It is a shallow rooted plant varying in height between 30 and 100 cm, its inflorescence consists of a slender panicle up to 45 cm long which may be open or compact. The caryopsis is generally white, oval and smooth. The millet is usually grown under rainfed conditions. It is pulled up by the roots as soon as the grain is ripe to avoid excessive shattering, and is threshed immediately. In India, yields are reportedly between 450 and 650 kg/ha when rainfed and up to 2 t/ha when irrigated. The crop appears to have received little attention from plant breeders.

Common millet contains a comparatively high percentage of indigestible fibre because the seeds are enclosed in the hulls, difficult to remove by conventional milling processes (Matz, 1969). Adrian and Jacquot (1964) cited Sorre (1942) as reporting an average 1000-seed weight of 5·0 g. (See Fig 2.X.). Its structure is discussed by Lorenz (1977) in comparison with foxtail millet (see p. 156).

Chemical Composition

Indira and Naik (1971) reported from India the chemical composition of two varieties of common millet, the distribution of protein, lysine and tryptophan among the solubility fractions (Osborne-Mendel) (See Table 2.123.) The data are discussed in comparison with other millets above under finger millet.

Belova and Denisenko (1965b) determined (spectrophotometrically) linoleic and linolenic acids in the unrefined and refined oil of the Russian common millet varieties Saratovskoe 853 and Podolyanskoe 24 273. The results are shown below:

	Linoleic acid (%)	Linolenic acid (%)
Saratovskoe 853		
Unrefined	52·30	5·32
Refined	53·16	3·60
Podolyanskoe 24 273		
Unrefined	51·30	4·30
Refined	51·65	1·24

A thiocyanatometric method indicated 3·2% linolenic acid in Podolyanskoe 24 273, otherwise results by the two methods were similar. Some isomerization of linolenic acid appeared to take place during refining.

Belova and Denisenko (1965c) precipitated a "steroid" from ether extracts of common millet varieties. Belova *et al.* (1970) reported that common millet seeds contained 1·8–3·9% lipids, according to variety, the highest proportion (24%) being in the embryo. The fatty acids of millet lipids were mainly unsaturated (78·4–82·1%), the saturated acids ranging from 17·9–21·6%. The major fatty acids (gas chromatography) were linoleic, oleic, linolenic, arachidonic, palmitic, stearic with $\leqslant 1\%$ of myristic, palmitoleic and saturated C_{15} and C_{20} acids.

The comments of Adrian and Jacquot (1964) on the lipids in common millet are discussed under sorghum.

Busareva *et al.* (1973) in Russia reported the gas chromatographic analysis of freshly harvested common millet. In both free and bound acids, linoleic acid predominated (69·6–72·4% in free, and 44·5–53·4% in bound). In bound lipids palmitic acid predominated (35·2–46·7%). Differences in lipid composition appeared among varieties.

Seit-Ablaeva *et al.* (1973) reported that the tocopherol content of common millet depended on variety, year of harvesting, area of cultivation and grain moisture content. During storage tocopherol slowly decreased. A comparison of GLC and thin-layer chromatography indicated the presence of α-, β-, γ-, δ- and η-tocopherols.

Nechaev *et al.* (1973) reported considerable variation in fatty acid composition (GLC) of the free lipids, glyceride fractions, and free fatty acids of millet grain during storage. Linoleic acid increased in free lipids but decreased significantly in the triglyceride fraction.

Salomatina and Olifson (1969) analysed six varieties of common millet and reported 11·3–13·6% saturated (mainly palmitic) acids, 19·9–23% oleic, 63–66·8% linoleic and up to 2·5% linolenic acid.

Salomatina (1969) reported that the tocopherol content, (determined colorimetrically), varied among five varieties of *Panicum* spp. from 56·7–92·0 mg/100 g. The dominant carotene was α-carotene.

Belova and Denisenko (1965a) examined the ether extract and the refined oil from the common millet varieties Saratovskoe 853 and Podolyanskoe 24 273, by column chromatography, methods of the (USSR) All-Union Scientific Research Institute of Oils (1962) and the All-Union Vitamin Scientific Research Institute (1954). No vitamin A was found. In the unrefined oils, only α-carotene was found, 10·5 mg/100 g in Saratovskoe 853 and 8·3 in Podolyanskoe 24 273. The refined oils contained amounts too low to measure. The unrefined oil from Saratovskoe 853 and Podolyanskoe 24 273 contained 87 and 96 mg/100 g tocopherol respectively. Refining reduced the amounts to 57 and 65 respectively.

Prugar and Vasela (1969) in Czechoslovakia reported thiamine (modified thiochrome method) in four samples of millet ranging from 0·58–0.68 mg/100 g grain, with a mean of 0·64. The variety Janaka Mana was 10% higher than Slovenské cervené.

Solpico and Yambao (1966) reported the proximate composition of an unnamed variety of common millet grown at Los Baños, Philippines between 1957 and 1962 in comparison with finger millet and foxtail millet. (See Table 2. 143.)

In the USA, Goodearl (1943) reported the composition (see Table 2.143) of Early Fortune (red) common millet to be as satisfactory as maize in the diets for laying pullets. Goodearl's values, Ca 264 mg/100 g and P 671 mg/100 g (DWB) are much higher than other values reported.

Deatherage *et al.* (1955) reported 27% amylose in the starch (Hilbert and MacMasters, 1946) of a sample of Early Fortune common millet.

Carbohydrates

Rahman *et al.* (1974) in Dacca reported the starch, total sugar, reducing sugar and nonreducing sugar content (AOAC, 1960) of market samples of common millet with husk as 52·1, 0·6, 0·4 and 0·2% respectively.

Inorganic elements

Few data have been found but Indira and Naik (1971) from India, reported two samples containing mg/100 g sample, Ca 15 in both samples, P 178 and

187, and Fe traces. Representative samples were also reported as ranging Ca 14-40, P 30-333 and Fe 5-9. Pinta and Busson (1963) reported K and Mg (flame spectrophotometry) in a single sample, mg/100 g sample, 430 and 160 respectively.

Vitamins

Representative values found ranged, mg/100 g sample, thiamine 0·04–0·78; niacin 0·3–2·33; riboflavin 0·07–0·38; single values, choline 4·4 and pantothenic acid 1·1.

Protein and Amino Acid Composition

Data from FAO (1970), Taira (1963a) and the National Academy of Sciences (1971) appear in Table 2.144.

In a collection of ecotypes of common millet, grown in the same location in Russia, the Mongol-Burjat group were highest (up to 17·6%) in protein (Jaros 1965). Some of the Eastern Asiatic types contained from 1·5 to 2·5 times the albumin and globulin present in other regionally adapted types. The starch of the Eastern types was higher in molecular weight and lower in amylose (2·0–6% of starch).

Kovalev *et al.* (1974) determined the amino acid composition (ion exchange chromatography) of the grits from five varieties of common millet widely used in Russia, Saratovskoe 853, Skorospeloe 66, Veselopodolyanskoe 367 and 38 and Khar'kovskoe 25. (See Table 2.144.)

Rakhimbaev (1967) reported the amino acid composition (one-dimensional descending paper chromatography, Golenkov, 1960, following acid hydrolysis, Wolfe and Fowden, 1957, alkaline hydrolysis for tryptophan Smirnova-Ikonnikova *et al.*, 1965) of Saratov 853, Ural'skoe Tonkoplenchatoe, Skorospeloe 66 and Dolinskoe 86 varieties of common millet. Of the millet grown in the Kazakh SSR area of Russia, 80% is Saratov 853. (Marusev and Il'in, 1965, described Saratov 853 and Skorospeloe 66 as having red grain, with 13–14% protein and about 5·0% oil; Saratov 853 had a 1000-seed weight of 7·1 g.) After separation of the glumes, the millet grain was milled to a fine flour, defatted with ether and ball-milled. About 95·5% of the N was extracted by a described method employing ethylene chlorohydrin. The amino acid composition is shown in Table 2.144.

A 20% difference in amino acid content among varieties was stated to be significant. Lysine and methionine in Saratov 853 and Dolinskoe 86 were almost double that of Ural'skoe Tonkoplenchatoe. Saratov 853 and Dolinskoe 86 were considered most desirable in amino acid composition, Skorospeloe 66 was next and Ural'skoe Tonkoplenchatoe poorest. The results were from a single harvest.

Janicki *et al.* (1973) reported the amino acid composition (auto-analyser following acid hydrolysis excluding tryptophan) of hulled samples of Polish common millet varieties, Strezeleckie Zielone and Strezeleckie Zolte. The mean and ranges are given in Table 2.144. The Kjeldhal N contents, DWB, were 2·11 and 1·99% respectively.

Janicki *et al.* (1973) stated their values for cystine + cysteine, and for methionine were comparable to those of Beza (1967) but considerably lower than those of Tkachuk and Irvine (1969) who reported cystine + cysteine as 112 mg/g N and methionine as 148 mg/g N. Tkachuk and Irvine (1969) also reported a high alanine content, 708 mg/g total N compared with Janicki *et al.* who found only 278 and 283 mg/g N in the two varieties. Both investigators reported high leucine, Janicki *et al.* 679 and 671 and Tkachuk and Irvine 820 mg/g total N. Both reported a Leu/Ile quotient over 3.

Protein Fractionation

Jones *et al.* (1970) identified protein bodies in white common millet by both light and electron microscopy. The protein bodies were globular measuring up to 2·5 μm in diameter. In the outer endosperm cells, some of the globular protein was embedded in an amorphous protein matrix but this was less evident towards the centre of the endosperm.

The millet was dehulled and hammer milled, the flour defatted with *n*-butyl alcohol (Jones and Dimler, 1962) and then extracted successively with (a) water, (b) 1% sodium chloride and (c) 60% *t*-butyl alcohol of 60% ethanol both at 60°C (Jones and Beckwith, 1970). Of the total N, 5% was extracted by water, a further 4% by NaCl and additional 52% by either of the alcohols. the proximate composition of the dehulled millet is shown in Table 2.143. The amino acid composition (auto-analyser after acid hydrolysis) of the dehulled grain and of the solubility fractions, is shown in Table 2.139. The prolamine contained less lysine, arginine and glycine, and more alanine, methionine and leucine than the albumins or globulins. The globulins were high in arginine.

Jones *et al.* (1970) assumed that all the ammonia arose from hydrolysis of the amide group of asparagine or glutamine and, therefore, that about 90% of the acidic amino acids in prolamine and the insoluble residues are in the amide form. Only about 40% of the water and salt soluble amino acids appeared as amides.

Lo (1941) in Minneapolis determined the moisture, total protein (N × 6·25) and ash (AOAC, no date) in flour of selected common millet. (See Table 2.143.) The Osborne solubility fractions and analysis appear in Table 2.133. Glutelin (alkali-soluble) and prolamine (alcohol-soluble) were the largest fractions.

Waldschmidt-Leitz and Metzner (1962) determined the amino acid composition (excluding tryptophan) of "panicin", the prolamine fraction of common millet. The results, compared with the zein of maize, are shown in Table 2.140.

Biological Evaluation

Human beings

Kies *et al.* (1975) compared (A) a whole grain commercial triticale flour, (B) TVP (a proprietary extruded defatted soybean product) and (C) a whole grain white-seeded common millet in N balance tests with 26 young men and women, average age 21 years. N intake during the initial 2-day depletion period was 0·8 g/ subject/day from the basal diet; 4·8 g during the 4-day adjustment period (4·0 g from SMP and 0·8 g from the basal diet); 4·8 g during the 4-day experimental period, (4·0 g from (A), (B) or (C) and 0·8 g from the basal diet). Energy intake was at the level necessary for each subject to maintain weight. Triticale and millet were fed as bread, TVP as casseroles (Kies and Fox, 1970, 1971). The amino acid compositions (auto-analyser) are shown in Table 2.141. Isonitrogenously, (B) TVP diet supplied nearly double the lysine in (A) triticale and nearly four times that of (C) millet. Millet however contained 50% more methionine than TVP or triticale. The N balance data (method of Linkswiler *et al.*, 1958) are shown in Table 2.142. The N balance difference between (A) and (B) was not significant. The relatively high faecal N excretion from (C) suggested low digestibility.

In a second study, also over 4 days, replacement of (D) 25% and (E) 50% of millet flour with triticale flour resulted in a progressive and significant improvement in N balance. Higher protein and lysine contents were provided by the triticale, and digestibility improved.

In a third study, nonspecific N, (urea) to give (F) 0·72 g N/day or (G) lysine to give 0·01 g N/day were added to diet (C). Total N intake then became (C) 4·80, (F) 5·52 and (G) 4·81 g N/day. The differences in N balance were statistically significant: (C)–0·64, (F)–0·21 and (G)–0·42. (F) urea supplementation, gave a significantly better N balance than (G), lysine alone, which suggested that the addition of a triticale to millet flour in the second study produced nutritional benefit from the increases in both total N and lysine.

In a fourth study, diet (B), soy TVP, was fed alone or with half the quantity, i.e. 26·49 g replaced by (H) triticale or (J) 106·84 g millet flour. (B) and (J) then provided an N intake of 4·8 and (H) 5·5 g/subject/day. (B) and (J) were not significantly different but diet (H) produced a significantly higher N balance than (B) or (J).

Rats

In 1937 Markuze reported the biological value (Osborne *et al.* 1919) of several Polish cereals including "millet" (possibly common millet) of 13·4% protein content (N × 6·25) DWB. The millet, fed as a meal, was the sole protein source in rat diets containing (A) 8·4% and (B) 12·8% protein. The mean PERs were (A) 0·95 and (B) 0·88. The biological value of the cereals evaluated was, in decreasing order: buckwheat > rolled oats > barley meal > rice = semolina = whole wheat > wheat flour > millet meal. The protein levels varied among the cereal diets.

The findings of Kuppuswamy *et al.* (1958) and of Howe and Gilfillan (1970) are discussed under Sorghum.

Tashiro and Maki (1977) prepared an alkali-extracted protein from common millet and compared it with the residue from the extraction proceedure, and the unextracted flour in various biological evaluations with rats, and *in vitro*. In growth trials with young rats, 5 to the group, the N source of the diets was: (A) millet flour (protein 11·2%), (B) alkali-extracted flour (protein 76·2%), (C) casein (protein 88·9%) and (D) gluten (protein 84·8%). The diets were fed to provide 10% protein, adjusted by varying the content of maize starch in the ration. Over 21 days, the rats on diet (A) failed to gain weight but the weight gain from (B) was higher than from (C). Diet (D) gave somewhat similar results to (A). The PER of (B) was 3·1 compared to (C), 2·8.

In biological value estimations with adult rats fed diets at 5% protein level, the BV values were (A) 60·8, (B) alkali-extracted (protein content 81·2%) 92·3, (C) 71·4, (D) 68·2 and (E), the residue from the extracted flour (protein content 7·8%) 70·5. Only in the case of (B), the alkali-extracted protein, were the differences significant.

In *in vitro* experiments, the alkali-extracted protein was significantly digested by pepsin but not by trypsin; with pepsin–pancreatin its digestibility was similar to that of gluten and casein.

Against the WHO (1973a) scoring pattern, the amino acid composition of the alkali-extracted protein, the millet flour and the residue scored 100, 26·7 and 9·1, respectively, with casein scoring 98·8 and gluten 28·5.

Chicks

Terpstra (1961) reported that young chicks did not grow well on the whole grain of common millet, analysed as: H_20 13·3, crude protein 11·1, fibre 7·3, fat 4·0, ash 3·8, NFE 60·5 (starch 57·7), Ca 0·02 and P 0·36%.

Two month old chickens tolerated whole grain, the digestibility of each component in ground meal and whole grain being determined as:

	Ground meal (%)	Whole grain (%)
Protein	85·1	84·5
Fat	75·1	80·5
Fibre	7·4	0
NFE	94·1	89·3
Starch	98·0	94·8
Organic matter	84·4	79·4

Burton and Milne (1961) in Queensland compared common white millet

(9·4% protein) with wheat, maize and sorghum in chicken rations. They found that common millet used in all-mash chicken starter rations, either as a meal or as a whole grain, did not depress growth and could therefore replace the other grains. Protein varied among samples: 10·5% was recommended for use in chicken mash

Swine

Erickson *et al.* (1963) reported common millet could be the major component in rations for growing pigs if the diet contained 16% protein; at 13·5% protein the addition of 0·33% L-lysine HCI was required. Pigs fed a 13·5% protein ration without added lysine did not gain weight satisfactorily and required considerably more feed:

	Without lysine	With lysine
Average weight gain (lb)	0·67	1·07
Feed/gain	5·29	3·30

Little Millet

Little millet (*Panicum miliare*) (*Panicum sumatrense*) is grown throughout India up to altitudes of 2100 m but is of little importance elsewhere. It has received comparatively little attention from plant breeders. The dehusked grain may be cooked like rice or milled into flour. The soft straw is palatable to cattle and the green plant useful as a quick growing fodder. Little millet appears to thrive under conditions that will sustain no other edible plant and will mature, even during famine years, in between 2·5 and 5 months. Yields are generally less than 0·5 t/ha but under favourable conditions, may reach close to 1 t/ha. The plant varies in height between 30 and 90 cm, its oblong panicle varies in length between 14 and 40 cm, the caryopsis is glabrous, striated and usually brown (Purseglove, 1972).

Little millet (*Panicum miliare*) tends to be confused with common millet (*Panicum miliaceum*) (Arnon, 1972). It is generally shorter and has smaller panicles and seeds than common millet.

Chemical Composition

Indira and Naik (1971) analysed several varieties of little millet. The findings are in Table 2.145. The distribution of protein, lysine and tryptophan among the solubility fractions. (Osborne-Mendel) in comparison with other millets (see Table 2.123) are discussed under Finger Millet. Ramanathan *et al.* (1975) reported the proximate analyses (AOAC, 1955) DWB, of five varieties of little millet (samai). These data are summarized in Table 2.145.

Representative analytical values for the vitamin content have been reported as mg/100 g, thiamine 0·3; niacin ranging from 1·0 to 10·9 and riboflavin 0·1. Adrian and Jacquot (1964) reported traces of carotene (vitamin A).

Protein and Amino Acid Composition

Ramachandran and Phansalkar (1956) reported the proximate composition and amino acid content (microbiological assay) of a pure seed sample of little millet (samai). The data are presented in Table 2.146 with those of Indira and Naik (1971).

Biological Evaluation

The findings of Kuppuswamy *et al.* (1958) relating to the biological value of little millet are discussed above under Sorghum.

Japanese Barnyard Millet

Japanese barnyard millet (*Echinochloa frumentacea*) is a short plant frequently grown in Egypt as a reclamation crop on land too saline for rice (Arnon, 1972).

Arnon (1972) cited Mann (1950) as stating it was the fastest growing of the millets and, under favourable moisture and temperature conditions, the grain would ripen within 45 days of sowing. Rachie (1975) stated that *Echinochloa* millets grow well in different seasons and at high elevations but require 3–4 months to mature.

Matz (1969) described *Echinochloa decompositum* as the Australian millet, used as food by aboriginals.*

Purseglove (1972) also described Japanese barnyard millet as the fastest growing of all millets, able to produce a crop in about 6 weeks. Though not of major importance, it is grown in the Orient and India and as a forage crop in the USA where it has been reported to produce eight harvests/year. It varies between 50 and 100 cm in height, the inflorescence consists of a panicle frequently tinged purple, bearing up to 15 lateral branches. It is normally grown as a rainfed crop but may be cultivated in water-logged areas and survive submersion. It yields between 700 and 800 kg of grain and 1·0–1·5 t/ha of straw. (See Fig. 2.X.)

Chemical Composition

The proximate analyses and inorganic elements of Japanese barnyard millet from various representative samples and Indira and Naik (1971) are shown in Table 2.147.

* (Authors' note: no references to the use of this seed as food or to its nutritive value have come to our notice).

Representative values for vitamins, in mg/100 g, were: thiamine 0·35; niacin 1·84 and riboflavin 0·03.

Protein and Amino Acid Composition

Table 2.148 gives the amino acid composition of single samples of Japanese barnyard millet, determined chromatographically and microbiologically, quoted by FAO (1970). The table also includes two samples, assayed microbiologically, by Taira (1963a) which closely resemble those from FAO (1970), and three samples of Indira and Naik (1971). The amino acid scores (WHO, 1973a) also appear in Table 2.148.

Osborne protein solubility fractions

Taira (1962b) reported the amino acid content (microbiological assay) in the solubility fractions of Japanese barnyard millet and foxtail millet. (See Table 2.134.)

Biological Evaluation

Rats

Wade and Ariyama (1966) reported rat trials in which Japanese barnyard millet was compared with polished rice, buckwheat and casein. The test groups, fed with a high (30%) level of protein, all showed cholesterol levels lower than others fed 5% protein. Almost no difference among the types of cereal protein was found but all showed higher cholesterol levels than rats on a standard diet or the 30% casein control. Serum cholesterol was lower on a restricted than on the *ad lib.* diet.

Kodo Millet

Kodo millet (*Paspalum scrobiculatum*) is a minor grain crop throughout India, reaching its greatest importance in the Deccan (Purseglove, 1972). Majmudar and Khunte (1955) stated that the cultivation is more or less confined to Gujarat, Karnataka and parts of Madras State, where it is an important crop, and, dependent on panicle characters, is classified into the groups Haria, Choudharia, Kodra and Haria-Choudharia.

Purseglove (1972) stated that cultivated kodo millet is an annual tufted grass which grows to 90 cm high. The grain may vary in colour from light red to dark grey. Some forms have been reported as being toxic to men and animals. The grain is enclosed in hard, corneous, persistent husks which are difficult to remove. Kodo millet is hardy and dought resistant, often grown with little attention on poor gravelly soils. The crop matures in 4–6 months and the yield of grain varies from 250 to 1000 kg/ha.

Chemical Composition

Majmudar and Khunte (1955) reported that improved strains 494 1 (Haria) and 80 2 (Haria-Choudharia) were developed at Nadiad and later at Baroda. For proximate analysis (methods not stated) see Table 2.149.

Indira and Naik (1971) reported the chemical composition of some varieties of kodo millet (see Table 2.150) and the distribution of the protein, lysine and tryptophan in the solubility fractions (Osborne-Mendel) of one variety (see Table 2.123).

Ramanathan *et al.* (1975) reported the proximate composition, DWB, (AOAC, 1955), of kodo millet, Varagu CO 2 (see Table 2.149).

Representative values for vitamin content were, in mg/100 g grain; thiamine 0·15–0·5; niacin 0·4–1·0 and riboflavin 0·03–0·07.

Protein and Amino Acid Composition

Table 2.150 presents the amino acid content in one sample (chromatographic analysis) and two samples (microbiological assay) of kodo millet (FAO, 1970). The amino acid scores (WHO, 1973a) are also included and the findings of Indira and Naik (1971).

Biological Evaluation

Rats

Kadkol *et al.* (1954b) reported the nutritional value to male and female weanling rats of kodo millet grown at Mysore (Karnataka) from seeds supplied from the Millet Specialist to the Government of Mysore at Coimbatore. The kodo millet (A) was fed as a meal of 11·6% protein (N × 6·25), prepared by hand pounding and winnowing, in poor vegetarian diets, in comparison with (B) wheat of 13·1% protein. The average daily food intake was little different at 7·4 and 7·7 g respectively, but the average daily gain at 5·20 g from (A) was significantly inferior to the 9·01 g from (B). The feeding trials followed the method of Rama Rao *et al.* (1953a) in which grains were fed at equal weights not isonitrogenously.

Digitaria exilis and Digitaria iburua

Digitaria exilis is grown throughout the savannah zone of West Africa from Senegal to Cameroon. In some areas of Guinea and Nigeria, it is the staple crop. Though not a main staple throughout the Sudan zone, it is widely grown where rainfall exceeds 400 mm. The very small seed may be used to make porridge alone or mixed with other cereals. It is also used for brewing beer. Purseglove (1972) stated it is sometimes considered the oldest indigenous West African cereal and its cultivation is believed to date back to 5000 B.C. It can grow on poor, shallow, or even rocky soils and the period to maturity varies among cultivars from 90 to 130 days. It grows to about 45 cm in height and the inflorescence consists of a panicle up to 15 cm in length. The grain is extremely small (2000 seeds weight about 1 g) and usually yellow in colour. Normally it is grown in pure stand but may be intercropped with sorghum or pearl millet. Yields vary between 600 and 800 kg/ha but more than 1 t/ha has been recorded. *Digitaria exilis* is also known as hungry rice, fonio, fundi and acha.

Digitaria iburua is cultivated mainly in Nigeria (in the regions of Zaria, Kano and Katsena), in Togo and Dahomey (Portères, 1959c). According to Purseglove (1972), its seeds are white. Both *Digitaria* are grown principally on intermediate elevations where the rainfall and heavier soils are more congenial than the surrounding savannah (Rachie, 1975).

Chemical Composition

Portères (1965) presented a comprehensive review of *Digitaria exilis* Stapf. known as fonio in West Africa and stated that relatively large proportions of sand are found in millets harvested in the semi-arid tropics. The proximate analysis of the whole grain and of the caryopses (dehusked and winnowed), derived from a number of literature sources including Portères' findings are shown in Table 2.151. The ranges in mineral content reported from various sources were: Ca 10-40, P 170-180 and Fe 3·4–3·7 mg/100 g.

Pinta and Busson (1963) reported inorganic elements found in *Digitaria exilis*, mostly from single samples, flame spectrophotometry—K, Mg, Na; polarography—Zn; the remainder by arc spectography, mg/100 g sample, Ag < 0·005, A1 18·55, Ba 2·8, Be < 0·05, Bi < 0·05, Co 0·33, Cr 1·35, Cu 1·5, Ga < 0·01, Ge < 0·10, K 160, Li 0·03, Mg 40, Mn 3·0, Mo 0·45, Na 20, Ni 1·5, Pb 1·1, Ru 0·16, Sn 0·2, Sr 0·25, Ti 0·2, V 0·2, V 0·015 and Zn 3·0.

Representative values for thiamine, total niacin and riboflavin were 0·3, 3·0 and 0·1 mg/100 g respectively.

Browntop Millet

Protein and Amino Acid Composition

Browntop millet appears to have received little attention as a food crop. Baptist and Perera (1956) reported that a millet grown in Sri Lanka referred to as little millet and known locally as meneri (*Panicum*) species, was identified at the Department of Agriculture as *Brachiaria ramosa*, formerly known as *Pancium ramosum* Koenig. Baptist and Perera (1956) suggested *Pothu meneri* as a name for this millet.

The grain was found to be rich in N, 3·18% of the wholemeal, DWB, of which 1·54% represented nonprotein N. Amino acids (microbiologically), with the Amino Acid Score (WHO, 1973a) where obtainable, and the leucine/isoleucine quotient are shown below:

	mg/g total N	Amino acid score	Leu/Ile quotient
Ile	212	85	2·7
Leu	569	129	
Lys	106	31	
Met	106	—	
Phe	206	—	
Thr	187	75	
Trp	69	115	
Val	350	113	

Teff

Teff (*Eragrostis tef*) is one of the major cereal crops of Ethiopia, where it is believed to have originated (Mengesha *et al.*, 1965, citing Vavilov, 1951). Grains found in the pyramid of Dassur constructed in 3359 B.C. are believed to be either teff or a closely related ancestor *Eragrostis pilosa* (Portères, 1958a). Some 280 species of teff are known to exist, and the plant is found in India, Australia, Kenya and South Africa, but only in Ethiopia is it cultivated as a major food crop, where it may represent half the total grain production. Its time to maturity varies from 2 to 6 months depending upon variety (Anon, 1969b).

Two main types of teff are recognized in Ethiopia, one with white and the other with brown or reddish seeds. The higher priced white teff is generally preferred, though it makes higher demands on the soil and is unsuited to high altitudes. Above 2500 m, brown teff is usually grown. Teff is cultivated at altitudes between 1700 and 2800 m but about 2000 m is considered optimum.

Teff is an annual tufted grass varying in height between 40 and 80 cm. The inflorescence consists of a loose open panicle between 15 and 35 cm long. The grain varies between 1 and 1·5 mm in length, between 0·75 and 1 mm in width and between 2500 and 3000 seeds weigh 1 g. According to Purseglove (1972) yields vary between 300 kg and 3000 kg/ha but 1 t/ha would be considered satisfactory.

Mengesha *et al.* (1965) stated that white teff is preferred by most Ethiopians particularly in injera, the traditional flat fermented bread. They examined over 300 head selections from the major teff producing regions of Ethiopia and reported seed yields between 3·6 and 15·25 g/plant.

Chemical Composition

Platt (1962) stated the average proximate composition (DWB) of teff as protein 9·5, lipid 2·5, carbohydrate 82, fibre 2·5 and ash 2·2%.

Protein and Amino Acid Composition

Table 2.153 shows the amino acid content determined chromatographically and microbiologically, of teff from various sources (FAO, 1970) together with the Amino Acid Score (WHO, 1973a).

Jansen *et al.* (1962) determined the protein and essential amino acid content of six teff varieties (ion-exchange; tryptophan and cystine, microbiologically). The protein contents were, N × 6·25, DWB, (%); Kay Teff Wolliso 12·1; Teff Gondar 11·2; Teff Flour Wolliso 12·6; Teff Kolla Duba 11·9; White Teff Jimma 9·7 and Red Teff Jimma 10·9. All six varieties were closely similar in amino acid content, for ranges and amino acid score (WHO, 1973a) see Table 2.153. All were first limiting in lysine; Kay Teff Wolliso being lowest, Red Teff Jimma highest in lysine. Jansen *et al.* (1962) stated that apart from low lysine, the amino acid balance compared favourably with whole egg.

Inorganic Elements

The Interdepartmental Committee on Nutrition for National Defense, Washington DC, in its Ethiopia Nutrition Survey in 1959 reported that the Fe and Ca contents of teff were much higher than those of wheat, barley and sorghum (Mengasha, 1966). The Fe level was challenged by Almgård (1963). Mengesha (1966) reported mineral analysis (DR emission spectrography) of flour samples from 12 white and 12 purple lines of teff collected in Ethiopia and grown at Purdue, compared with spring wheat (Axminister CI 8195), winter wheat (CI 13701), winter hull-less barley (Purdue 5627-A-12-11) and grain sorghum (mixed strains). Table 2.154 shows the analysis of (1) a pure strain of purple teff (code No. 106), (2) 12 mixed purple strains, (3) a pure strain of white teff (code No. 71), (4) 12 mixed white strains, (5) means of spring wheat, (6) means of winter wheat, (7) means of winter barley and (8) means of sorghum all expressed in mg/100 g sample.

Teff was higher in Ca, Fe, Cu, Zn, Al (except white teff), Ba and Na than the other cereals (except Na in winter barley). Strain variability was indicated in the Al content of purple teff and the K content of white teff. High Fe and Ca contents were confirmed.

Sufian and Pittwell (1968) determined Fe in market samples of white and red teff. Almgård (1963) cited by Sufian and Pittwell (1968) stated that all but 0·052–0·059% of the Fe in teff occurred in soil contamination. Sufian and Pittwell (1968) therefore washed the samples repeatedly in 0·1 M HCl and water before determining Fe (thiocyanate method). Their results indicated a true Fe content in dirt free teff grain of about 3·3 mg/100 g. Before cleaning, the mean Fe content was close to 50 mg/100 g.

Sufian and Pittwell (1968) suggested as explanations of the higher values reported by Mengesha (1966) (1) Fe content in teff is variable, (2) Mengesha's sample was contaminated by windblown dust embedded in the grain wall and (3) the outer seedcoat, as distinct from the husk, is richer in Fe than the inner grain.

Vitamins

Platt (1962) published representative values for thiamine, niacin and riboflavin in teff as 0·5, 2·0 and 0·1 mg/100 g respectively.

Biological Evaluation

Kihlberg and Ericson (1963) determined the amino acid composition (chromatography after acid hydrolysis; tryptophan and cystine microbiologically) of two varieties of teff, designated A and B. The amino acids and N content, and amino acid score (WHO, 1973a) of whole grain A and B and a flour milled from millet A appear in Table 2.155.

In three series of experiments male weanling rats were fed *ad lib.* over 21 days, on diets containing 91% teff plus vitamins, salts and oil, without and with additions of lysine, threonine, isoleucine and tryptophan. Results including nitrogen efficiency ratios (NER) (weight gain/g N intake) and liver lipids are shown in Table 2.156. In Series I, diet (B) teff flour A + 0·3% lysine gave faster weight gain, higher NER by 80% and lower liver fat than diet (A) teff flour A alone. Diet (C), added threonine, gave NER below diet (A). Diet (D), with lysine and threonine, gave NER 115% higher than diet (A) with markedly lower liver lipid.

In Series II, diet (C) addition of lysine and threonine, gave growth rate and NER higher than diet (H), casein. A separate experiment, not detailed in the table, appeared to establish that differences between diets (B) and (D) in Series I were not due solely to higher lysine in diet (D). This suggested that with the addition of 0·3% lysine HCl, lysine was no longer first-limiting.

In Series III, diet (L), added lysine and isoleucine was inferior to diet (K), added lysine and threonine. The addition of isoleucine to diets already containing lysine and threonine in diet (M) gave no significant effect. The addition of tryptophan, diet (N), improved NER over diet (K) but not significantly.

The experiments showed that lysine was first-limiting, threonine second and tryptophan possibly third.

Rats fed whole teff variety A and variety B developed greasy, coarse and "tufty" fur after a few days. As liver lipid was also high, presence of a toxic factor was suspected but heating the whole teff at 110°C for 60 min brought no improvement, nor could the observed effects be related to differences in lipid or iron content between the whole teff and the flour. Only 23 mg Fe/100 g was found in the teff sample compared to 90–100 mg/100 g reported by Ethiopia Nutrition Survey (1959).

The biological value of teff was also discussed by Jansen *et al.* (1962) under Pearl Millet (see Table 2.116).

FAO (1970) reported a personal communication from Payne (1967) giving the determined NPU of teff as 64 when fed to rats at 10% protein in the diet.

Amaranthus paniculatus

Amaranthus paniculatus, also known, in India, as rajgira, is a small seed, classified by Sahasrabuddhe (1925) as an "inferior millet" which is eaten, puffed, in certain parts of India (Subramanian and Srinivasan, 1951).

Chemical Composition and Biological Quality

Subramanian and Srinivasan (1951) reported the average proximate composition of seed grown in various Indian States as:

	(%)	DWB (assuming mean moisture of 10%)
Moisture	9–11	
Crude protein	14·5–16·0	15·9–17·6
Crude fibre	2·0	2·2
Carbohydrates	66·8	73·5
Starch (acid hydrolysis)	57·8	63·6
80% alcohol-solubles	6·9	7·6
Ash	3·6	4·0
(Ca 6%, P 18% of Ash)		

When the raw seed was fed to rats at 10% protein in the diet the digestibility coefficient, and the biological value (balance sheet method, Chick *et al.*, 1935a) were determined, respectively, as 80·4% and 74%, and the PER (Osborne *et al.*, 1919) of the raw seed was 2·12, roughly equivalent to casein (2·27), and the puffed seed 1·90.

Rats fed raw and puffed *Amaranthus paniculatus* seeds in comparison with rice on an equal weight basis gained 13·3 g (raw) and 10·0 g (puffed) weekly and 3·3 g on rice. The rice diet contained 8·0% protein and 0·05% Ca. Reproduction and lactation in the animals fed millet seeds were considered normal.

Ramachandran and Phansalkar (1956), who referred to the seed as rajkeera, assayed the essential amino acids (microbiologically) of a bazaar sample (Coonoor, South India) containing 5·2% ash and 9·8% protein (N × 6·25). The results with the amino acid score (WHO, 1973a) are shown below:

	Amino acid score	(mg/g total N)
Ile	172	430
Leu	114	501
Lys	150	512
Met	70	154
Phe	78	298
Thr	109	273

Trp	90	54
Val	122	379
Arg	—	923
His	—	182

It should be noted that in reporting amino acid contents and patterns, WHO (1973a) combined methionine with cystine, and phenylalanine with tyrosine. Ramachandran and Phansalkar (1956) reported methionine as being the first-limiting amino acid in comparison with whole egg. Since, however, they did not include cystine with methionine and tyrosine with phenylalanine, it is conceivable that the methionine and phenylalanine values are lower than they would be using the WHO convention. On the other hand, the values reported for lysine and threonine appear extraordinarily high in comparison with other cereals. In the light of the comparatively high PER and biological values, the data reported may well be reliable. The biological value (method of Block and Mitchell, 1946) was reported by Ramachandran and Phansalkar (1956) as 77.

Kurien (1967a) separated the husk from *Amaranthus paniculatus* seeds by steeping, grinding and separating it over a 100-mesh sieve. The protein, Ca and P in the separated husk, the endosperm, and the dried solids in the supernatant are shown in Table 2.157 (methods not stated, no reference to the embryo). The husk formed 20% of the whole grain and contained about 34% of total protein, 32% of Ca and 20% of P. The high proportion of husk might limit the usefulness of *Amaranthus paniculatus* in the diets of young children.

Job's Tears (Coix Lachryma-jobi)

A thin-shelled type of Job's tears, known as adlay, is a minor grain used as food in the Philippines, Zaire and South America (Rachie, 1975). This type is known as mayuen in China (Purseglove, 1972), and as kirundi in Sri Lanka (von Schaaffhausen, 1952).*

Most known varieties grow 1–2 m tall in areas suited for hill rice, from sea level to over 1500 m in the tropics. The inflorescence, one of which develops at the end of each peduncle from a leaf sheath, ranges in colour from white through yellow, red and purple to brown; it is pale brown in a soft shelled edible form. The crop matures in 140–160 days and normal yields are between 2000 and 4000 kg/ha of husked grain (Purseglove, 1972).

Its use tends to be limited by its long time to maturity, its susceptibility to attack by birds, its limited shelf life when dehulled, and its uneven yield (von Schaaffhausen, 1952).

Purseglove (1972) reported the proximate analysis of the husked grain of Job's tears (Adlay) as H_2O 10·8%, protein 13·6%, fat 6·1%, carbohydrate

* (Authors' note: Job's tears is not generally regarded as a "millet". It is included in this review because it is a small grain of importance in some diets. Comparatively little appears to be published on its nutritional value.)

58·5%, fibre 8·4% and ash 2·6%. Because of its high cellulose (fibre) content the grain is less suitable as a human food than many other cereals.

The grain may be dehulled with rice dehulling equipment (von Schaaffhausen, 1952). A variety with a brown elongated seed was developed by von Schaaffhausen (1952) who claimed for it more rapid maturity, higher yield and greater resistance to bird attack because its many stalks were thin and pliable and failed to give support to feeding birds.*

The proximate composition of Job's tears has been reported by Jacquot *et al.* (1965), Ramasastri and Mohan (1969), and Lema de Rocha (1950) (cited by Matz, 1969). The analytical methods are not known, but the ranges reported were, %: protein 13·6–17·5, lipid 6–7·3, carbohydrate 58·5–63·0, fibre 8·4 (one result) and ash 1·9–2·6.

Jacquot *et al.* (1955) reported that the bran of Job's tears contained 18·0% ash and 35·5% cellulose which, in their opinion, made it unsuitable for nonruminants.

Scaut (1962) reported that the seed envelope, representing 45% of the grain, was without food value for swine and totally inhibited the feed value of the endosperm. He also reported the following percentage proximate composition of the endosperm: protein 16·2, digestible protein 14·1, Et_2OH extract, 0·6, NFE 81·4, and ash 1·0. Mineral content, mg/100 g, Ca 40, P 150, Mg 60, K 140, Si 130 and Na 6.

Deatherage *et al.* (1955) reported the amylose content in the starch (method of Hilbert and MacMasters, 1946) of two samples of Job's tears as 24% and 27% respectively.

Table 2.158 shows the amino acid content, in protein and grains of Job's tears (chromatographically and microbiologically), presented by FAO (1970), with the amino acid score (WHO, 1973a). Values for amino acid content reported by Taira (1962a) fall within the ranges quoted.

Ramasastri and Mohan (1969) reported that the grain consumed in the North East Frontier Regions of India was rich in protein but deficient in tryptophan and lysine. They quoted the following mineral, vitamin and amino acid contents (methods not stated): Ca 23, thiamine 0·31, niacin 6·7, and riboflavin 0·15 mg/100 g; lysine 131, methionine 137 and tryptophan 19 mg/g total N. Tryptophan was below the lowest value quoted by FAO (1970).

De Paula Santos (1950–51) compared the biological value (BV) for rats (Mitchell's method) of the protein in one finely ground and one coarsely ground flour sample of Job's tears. The mean values, ranges in parentheses, were:

	Finely ground flour	Coarsely ground flour
Mean BV	31·3 (27·6–34·3)	21·7 (12·7–27·5)
Mean coefficient of digestibility	96·1	98·1
Mean weight increase (g/g protein eaten)	2·1	not reported

* (Authors' note: no nutritional data relevant to this type have been found.)

The mean weight increase from whole wheat protein was 2·7.

Jacquot *et al.* (1955) reported that rats fed a diet consisting solely of Job's tears over 15 days, displayed slow growth and positive balance of N, Ca and P. Digestibility was equivalent to wheat but Ca and P were poorly utilized, possibly because of a high phytic acid content. Even when supplemented with lysine and methionine, Job's tears produced inferior weight gains in rats to wheat. The BV was reported as 20.

Conclusion

It is recognised by the authors that the contents of the chapter now completed lack uniformity in scope and detail among the cereals reviewed. This follows from the considerably greater attention given to sorghum, and to a lesser extent to pearl millet, than to the other millets. The authors have tried to include all that is relevant and available in the published literature. The next chapter summarizes the influence of genetic history and environmental condition upon the various properties and characters referred to in Chapter 2.

CHAPTER 3

Genetic, Environmental and Agronomic Factors

Introduction

This chapter consists of a review of the literature in which (a) genetic, (b) environmental and (c) agronomic factors are discussed in relation to the nutritional quality of sorghum and each of the millets under consideration. The information is presented for each cereal separately, beginning with sorghum. Though a serious attempt has been made to identify and isolate the specific influence and effect of each of the three factors, it must be remembered that every plant is the consequence of the interaction of all three factors and therefore a clear separation of individual cause and effect is virtually impossible. Consequently, the reader will encounter overlap and intermingling of the three influences in the text that follows. Under each cereal, the review presents the effect of the three factors on (i) proximate analysis; (ii) total protein nitrogen; (iii) amino acid composition; (iv) other determinable nutrients; and (v) biological quality.

Genetic factors

Plant breeding consists essentially of (a) the collection of plant populations within each of which the probability of desired character combinations occurring is high and (b) the identification and selection from within these populations of parent plants possessing the characters desired. Such genetically controlled characters include yield, resistance to pests and disease, environmental adaptability and biochemical composition.

The larger the number of genetically controlled characters one tries to combine in a single plant, the more complex the breeding programme becomes. Consequently, the individual entries that make up the parent population from which a new cultivar is to be derived must, of necessity, differ in comparatively few genetically controlled characters.

Even using the techniques of population breeding, in which composites of freely intercrossing individuals are planted and the best of their progeny selected to form the next composite, the simultaneous combination of several desired characters is difficult to achieve, since each major character needs to be selected from one or more composite populations.

The range of genetic diversity among the cultivated sorghums and pearl millets is known to be considerable and, though they have been less intensively examined, the same probably holds true for the minor millets. Until recently, when Ethiopian sorghum lines carrying prolamine depressant, or high-lysine genes, were discovered, comparatively little effort had been made to exploit the range of genetic diversity among sorghums and millets in order to enhance their nutritional quality.

In the literature the terms "varieties", "cultivars", "hybrids", "lines" and "sample" are used but it is not always clear in what sense the various authors have employed each term.

Doggett (personal communication, 1976) stated that the term "variety" should be used according to its precise botanical meaning. He prefers the term "cultivar", either as "indigenous", "local", "traditional" or "improved". "Line" would refer to derivatives from self-pollinating crops or derivatives extracted by inbreeding (equivalent to selfing) from cross-pollinating crops. "Hybrid" means a cross between two distinct lines unless the term is modified as "three-way", "double cross" or "varietal". The last term "varietal cross" is in general use and it refers to a cross between "varieties" or cross-pollinating crops (cultivars), that is, a cross between two populations both of which are heterogeneous.

In reviewing the literature on the nutritional value of sorghum and the millets, the present authors have, of necessity, retained the terms used in the papers cited.

The world centre for research on sorghum and pearl millet is the International Crops Research Institute for the Semi-Arid Tropics (ICRISAT), which has its headquarters in Hyderabad, India, supported by a network of related projects in Africa, Asia and Latin America. In its Departmental Report for 1976–77, ICRISAT stated that their sorghum germ plasm bank includes a total of 15 037 accessions.

Prasada Rao (personal communication, 1977) lists the following as ICRISAT's accessions in 1977:

Sorghum	
(A) Collection covered by IS numbers (rejuvenated from the World Collection assembled by the Rockefeller Foundation)	11 778
(B) Collection not covered by IS numbers (assembled by ICRISAT)	3259
Total:	15 037
Pennisetum	
(A) Collection covered by IP numbers (rejuvenated and identified from the World Collection assembled by the Rockefeller Foundation)	1075
(B) Collection not covered by IP numbers (assembled by ICRISAT)	3973
Total:	5048

Minor millets	
Eleusine coracana	405
Setaria italica	41
Panicum miliaceum	46
Panicum miliare	24
Paspalum scrobiculatum	47
Echinochloa spp.	26
Total:	589

Prasada Rao goes on to say:

> In sorghum, ICRISAT was able to rejuvenate 11 778 IS numbers from the World Collection previously assembled by the Rockefeller Foundation. There are still 5000 IS numbers missing and the only available source is NSSL, Fort Collins, Colorado, USA. We expect at least some will be present in their inventory. If not this will remain a permanent gap. Sorghum IS Collection and the 3259 new accessions assembled by ICRISAT (shortly we are going to assign IS numbers to them) is in proper shape. They were purified, described, classified and evaluated. Documentation is under way.
>
> With regard to pennisetum we are still in the process of collecting the material as three-quarters of the World Collection (IP numbers) assembled by Rockefeller Foundation lost its identity and vigour. We are making efforts to get fresh collections. As per the recommendation of the IBPGR* Advisory Committee on sorghum and millets germplasm, ICRISAT is taking responsibility in the collection, storage, maintenance, evaluation, documentation and distribution of seed and information, in respect of sorghum and pennisetum and just collection and maintenance for three other minor millets; *Setaria italica, Eleusine coracana* and *Panicum miliaceum.*

According to Rachie (1970b) a collection of over 16 000 entries of sorghum germplasm was held by the All-India Coordinated Sorghum Improvement Project at the Andhra Pradesh Agricultural University Campus. A large collection of pearl millet types was held by the Indian Council of Agricultural Research in cooperation with the Rockefeller Foundation.

Sorghum

Total Protein

The ICRISAT Departmental Annual Report for 1975–76 stated that a variability study of various characters during the growth stage found that there were substantial differences among genotypes in a number of characters including rate of leaf production, grain yield, seed number, seed size and grain filling rate. Forty entries from élite selections were studied to define the range in variability in nitrogen (N) uptake and distribution, and possible influences on yield. Nitrogen uptake/plant varied from 0·22 to 1·14 g, and Nitrogen Transfer Efficiency (NTE) from 57·8 to 86·6%, at a given total N

* International Board for Plant Genetic Resources

uptake. Genotypes with an efficiency of greater than 77% were identified. There was a weak negative correlation between grain yield and grain N concentration. Grain N was highly correlated with total biomass, consequently large plants tended to contain more N. NTE and harvest index (the weight of seed grain as a proportion of total dry matter) were strongly correlated and it was concluded that a high harvest index requires a high NTE. Grain yield was positively correlated with total plant N, with grain N content, and NTE. Bearing in mind the comparatively weak negative correlation between grain yield and grain N concentration, it appears possible that high yielding genotypes carrying a high N concentration in the seed can be obtained through selection of types with above average N uptake and high transfer of N from the plant to the grain.

Kambal and Webster (1966) reported on hybrid vigour in 190 grain sorghum hybrids produced by crossing 10 male sterile lines with 19 fertility restorers, and grown with their parents over 2 years at Lincoln, Nebraska. Most of the female parents were kafirs or kafir-milo derivatives; the male parents included lines which had been selected from hybrid combinations of kafirs, milos and feterita. Parental lines were not specifically identified. The hybrid grains were lower in average crude protein content (DWB) than their parents, but because of a higher yield of grain, the hybrids produced 16% more crude protein/ha than their parents.

Four male sterile (A lines), Redlan (A 378), Martin (A 398), Wheatland (A 399) and Combine Kafir 60 (A 3197) were crossed with eight fertility restorer (R lines) developed from introductions originating in Ethiopia. Combine 7078 was crossed with the same females as a check (Malm, 1968). The material was grown over 1965 and 1966 at Clovis, New Mexico. General Combining Ability (GCA) was defined by Sprague and Tatum (1942) and Rojas and Sprague (1952) and cited by Abifarin and Pickett (1970) as the average performance of a line in hybrid combination. Specific Combining Ability (SCA) refers to those cases in which certain combinations do better or worse than would be expected from the average performance of the parental lines. In this study (Malm, 1968) both GCA and SCA effects were of importance in the expression of grain yield in hybrids and agreed with the results of Beil and Atkins (1967), Kambal and Webster (1965) and Niehaus and Pickett (1966). GCA effects were 20 × greater than SCA effects. Protein levels differed considerably between the 2 years. A correlation between years for protein percentage for all 36 hybrids gave a value of +0·39 (significant at the 5% level). A between-year correlation of the average protein % of hybrids of male parents was +0·71 (also significant), indicating that the relative protein production was consistent during the 2 years.

All sets of exotic hybrids, except two, were significantly higher in protein content than the check samples. Certain hybrids produced almost 15% protein in 1965 in which year the check samples averaged less than 11%. The line designated Martin (A 398) made a higher contribution to protein content among the hybrids than did other female parents.

The hybrids of the large-seeded parents (R_2 and R_3) produced 50% more protein/ha than the check hybrids. All sets of hybrids from the introductions had an average protein production higher than the check average.

In 1965, kernel weight and protein % were positively correlated, and protein was negatively correlated with grain yield and starch content. The combined figures for the two years showed a negative correlation between yield and protein. Only starch content and test weight were significantly and positively correlated in both years.

Liang *et al.* (1969b) and other earlier authors have defined heritability as the fraction of total variance in a segregating population attributable to additive genetic effects. The concept of heritability is associated with the relative influence of heredity and environment on variations in a character.

Heritability was estimated by parent–offspring regression, parent–offspring correlation, and variance component methods for 12 characters including grain yield, kernel weight and protein (Kjeldahl) in 36 lines from a cross Redlan × Martin (population 1) and 40 lines from a cross of Combine Kafir 60 × KS 7 (population 2). The F_3 lines of both populations were grown at Manhattan and Ashland, Kansas in 1966 and F_4 lines at the same location in 1967. The three methods of estimating heritability in each population provided similar estimates.

Grain yield had medium heritability values. Kernel weight appeared highly heritable in population 2, less so in population 1. Protein content was highly heritable in population 1, less so in population 2. Grain yield appeared to be negatively correlated with protein content.

Liang *et al.* (1969a) studied genotypic and phenotypic correlations among 12 characters, including protein content, in segregating populations and in pure lines of grain sorghum. The testing materials were in three groups:

(1) Thirty-six F_3 lines from the cross Redlan × Martin and 40 F_3 lines from the cross Combine Kafir 60 × KS 7 evaluated at two locations in Kansas in 1966.
(2) Fifteen groups of F_2 plants from 15 single crosses grown in two locations in each of the 2 years 1966 and 1967.
(3) Six pure lines (Redlan, Plainsman, Martin, Combine 7078, KS 7 and Combine Kafir 60) grown as group (2).

Grain yield was positively and significantly correlated with head weight, kernel number, days to 50% bloom and leaf number, and negatively correlated with % germination and protein content.

Protein content was positively correlated only with % germination both genotypically and phenotypically but it was not significant in Group (1). Selection for protein *per se* is more reliable than selections for other characters as an indication of protein level.

In 1968 Liang *et al.* reported on the heritable variation for grain yield, anthesis time and protein content in a six variety diallel of sorghum using

Redlan, Plainsman, Martin, Combine 7078, KS 7 and Combine Kafir 60 grown in Kansas at Manhattan (clay loam soil) and Ashland (sandy loam) in 1965. The method of Hayman (1954) was used for analysis of quantitative traits.

Significant interactions were observed between GCA and locations for yield, anthesis time and protein content, and between SCA and locations for protein content only. The significant interaction between GCA and SCA, and locations suggested that the breeding plan should seek a separate variety for each soil type.

At least four groups of genes were involved in controlling grain yield, and heritability estimate was only 13%. Redlan was the highest yielder and carried the most dominant genes; Combine Kafir 60 was the lowest yielder and carried the least dominant genes. Two groups of genes were estimated for protein and heritability estimate was intermediate, 43%, though dominance variation was small. Combine 7078 carried the most dominant genes and Martin had the highest protein content. Combine Kafir 60 carried the least dominant genes and Redlan had the lowest protein content. Only the regression coefficient for protein content was significantly less than one.

Crook and Casady (1974) produced 40 grain sorghum hybrids by crossing each of four cytoplasmic MS lines (CK 60, KS 24, Redlan, Martin) with 10 diverse restorer lines (TX 414, Club × Day 16, Early Hegari, Cache Feterita, Plainsman, Caprock, DDE Shallu, Roky 10, Migari and Roky 4). The hybrids and their parental lines were grown in 1972 at four locations in Kansas. The characters studied included % protein, grain yield, days to 50% bloom, height, panicle exsertion, leaf area, panicles/plant, kernel weight and test weight. Location significantly affected all characters in the hybrid population except test weight. The genetic variance was significant for all characters measured. The SCA estimated by the male × female interaction was significant for protein content, days to 50% bloom, height, panicle exsertion, panicles/plant, kernel weight and test weight.

The heritability estimate suggests the degree to which a specific character (e.g. grain N content) is genetically controlled and the possibility of it being transmitted from parent to offspring. A high heritability estimate suggests a high probability of inheritance. Heritability estimates of sorghum grain protein have ranged from 41 to 78% depending on the population and method of calculation (Liang *et al.*, 1968, 1969a). In the trials described heritability estimates for all characters were high when calculated from variance components. When derived from mid-parent–offspring correlations, heritability estimates were high for protein content, height and panicles/plant, medium for yield and kernel weight; and low for days to 50% bloom, panicle exsertion, leaf area and test weight. Protein content was negatively correlated with yield, days to 50% bloom, height, leaf area, panicles/plant and test weight. Yields of hybrids were positively correlated with days to 50% bloom, height, leaf area, panicles/plant, kernel weight and test weight, and negatively correlated with protein percentage and panicle exsertion.

Combine Kafir 60, male sterile (MS) line 2219, MS line 172, and MS line 3675 were used as female parents and various unnamed Indian and non-Indian varieties of sorghum as male parents in trials carried out at Coimbatore during 1965–69 (Rao *et al*., 1972). Local varieties CO 18, CO 11 and CO 1 were also grown. Samples were analysed for protein, fat, crude fibre, carbohydrates and minerals (AOAC, 1965).

Expressed as % mean values, the non-Indian varieties contained more protein and minerals than the local varieties which in turn contained more protein than the hybrids of the MS lines. The hybrids of MS lines 172 and 3675 showed the highest values for minerals, fat and crude fibre, and the hybrids of Combine Kafir 60 and MS line 2219 showed the highest values for carbohydrates. The mean protein contents reported were:

	(%)
Local Indian varieties	10·4
Exotic varieties	11·3
+Combine Kafir 60+hybrids	9·8
+MS 2219+hybrids	9·1
+MS 172+hybrids	9·8
+MS 3675+hybrids	10·0

Rao *et al.* (1972) considered that the hybrids of MS line 3675 with local or non-Indian varieties, or composites developed from local and non-Indian varieties, had greater potential to improve the quality characters than Combine Kafir 60 and MS lines 2219 and 172.

From Bulgaria, Dechev (1971b) reported that F_1–F_4 generations of sorghum RS 610 studied in 1966–68 at the Vasil Kolarov Institute showed no appreciable differences in grain composition. Differences between years were greater than between generations.

Protein and Oil Content

The variation and interrelations of protein content (micro-Kjeldahl N × 6·25), oil content (nuclear magnetic resonance) and 100-seed weight of 14 parental varieties of sorghum, with the F_1 and F_2 seed of 48 hybrids, planted at Ames, Iowa, were reported by Reich and Atkins (1971). Table 3.1 gives the values for the male and female parents.

Analyses of variance did not indicate significant differences in oil content within either the male or female parent group. Differences in protein content (% DWB) among the female parents were significant. The female parents differed significantly in 100-seed weight but the male parents did not.

The parental and F_1 seed was produced in the nursery but F_2 seeds were grown in an adjacent field trial, and the analyses of variance for the parents and F_1 were therefore calculated separately from those of the F_2 seeds. Mean oil percentage for the F_1 seeds was 3·39 (range 2·92–3·79) compared to 3·14 (range 2·77–3·45) for the parents. The mean and range in oil content for F_2 seed were similar to those for F_1 seed. Average mid-parent heterosis of the F_1 seed was 8% and of the F_2 seed 10%, higher than the parental mean. Reciprocal crosses were not included but the data did not suggest a strong maternal influence on oil content. Redlan (female parent) was lower in oil content than any of the male parents but the F_1 seed of all crosses with Redlan exceeded the mid-parent value. This trend was also evident in other crosses; however F_1 crosses of Martin (female parent which was higher in oil content than any of the male parents) always exceeded mid-parent oil percentage.

There was a difference between F_1 and F_2 seeds in analyses of variance. No variation among the eight F_1 hybrids of any given female parent was significant: variation among F_2 hybrids for any particular female parent was always significant.

Protein percentage of the F_1 seeds differed significantly from the parental seed and showed an average mid-parent heterosis of 15%. The protein range for the F_1 seed was 7·9–18·0% but most seeds were in the range 11–14%. Variation among F_1 hybrids was significant within each female parent except for Combine Kafir 60 and Wheatland hybrids. Among F_2 hybrids, variation within Westland Combine Kafir 60 and Reliance hybrids was significant but not within each of the other female parents. Mean protein content of the F_2 seed at 12·8% was about 8% less than that of the F_1 seed at 14·9%, but 5% above the parental mean.

Mean 100-seed weights for the parental, F_1 and F_2 seeds were 2·72, 3·23 and 3·07 g respectively. Mid-parent heterosis averaged 19% for the F_1 and 1·3% for the F_2 seed. Variance analyses showed that only the hybrid involving the Martin female parent differed significantly in the F_1 seed, but variations among hybrids within each female parent were significant in the F_2 seed.

Only oil vs protein and protein vs seed weight in the F_2 seed were significantly positively correlated. The results indicate that the simultaneous improvement of oil and protein content would be feasible.

The F_2 seed of Reliance × TX 74; Martin × TX 74 and Redlan × TX 07, were relatively high in oil, protein and 100-seed weight.

The F_2 data were also analysed by the Design II procedure of Comstock and Robinson (1948). Heritability, calculated by determining the proportion of total variance due to additive gene action, was 65% for oil content, 32% for protein content and 52% for 100-seed weight. Since the data were derived from a single experiment, the genotype × environment interactions could not be isolated from genetic influence. Nevertheless, the results appeared to justify genetic selection for oil and protein content, and 100-seed weight.

Mukuru *et al.* (1973) reported from the Department of Agronomy, Purdue University the estimated genetic components, heritability and genetic advance of protein, lysine and oil content in crosses among six parental lines of sorghum considered homozygous and unrelated. Four crosses were made at Purdue and from each cross 45 F_2 plants were selected and advanced to the third generation in Puerto Rico. From each F_3 progeny row, two plants were selected and selfed. All the 360 F_3 selections from the four crosses were evaluated in the fourth generation in one replication and two locations (whereabouts not stated). The segregating progenies from each cross were designated a population.

The characters for which data were recorded as a single observation for a plot were: days to flower, plant height, panicles/plot, grain yield (kg/ha), kernel weight and volume, protein (DWB), lysine (% sample), lysine (% protein), oil (DWB) and catechin equivalent values. % heterosis over mid-parent values was negative in all populations for kernel weight and volume and variable from population to population for other characters. Each population exhibited a wide range of variation for most characters in the fourth generation. Analyses of variance suggested significant heterozygosity in the F_2 sub-populations in the fourth generation for kernel weight and volume, % protein and % oil. For these characters selection within and among F_2 families in early generations would, according to Mukuru *et al.* (1973) be most effective.

Estimates of dominant genetic variances in all populations for all the characters were always larger than those for additive genetic variances indicating that while additive gene action was present, dominant gene action had greater influence.

Estimates of heritability for each character were variable from population to population, and heritabilities among the F_2 and F_3 sub-populations in the fourth generation for each character in a population were also variable. Estimates of heritability and genetic progress, though modest, suggested that substantial improvement could be made by selection, particularly for kernel weight and volume in most populations, protein in population III, lysine % of protein in population III and % oil in populations III and IV.

Kernel weight and volume were highly positively associated in all populations but both kernel weight and volume were significantly and negatively associated with (a) days to flower, (b) plant height, (c) kernels/panicle and (d) lysine (% of protein) in most populations. The phenotypic correlations between kernel weight and % oil were negative and significant in population IV; % protein and % oil were positively and significantly correlated. In all populations protein content and lysine (% of protein) were negatively and significantly correlated.

Mukuru *et al.* (1973) concluded:

> In this study, no consistent and favourable correlations of chemical traits with any plant or seed characters were found in all populations. However, kernel weight or volume could be used as a good indicator in selecting % protein in population I and % oil in population III while % oil could be used as a good indicator in selecting % protein in population IV.

From Russia, Sepel and Sepel (1969) reported that the protein content of most of the hybrids of crosses between the sorghum line Steril'noe (sterile) 8 with 12 cultivars was lower than in the parents; in five hybrids protein content exceeded either the pollinating parent or both parents. In oil content, most hybrids were intermediate between the parents.

When the MS line Nizkorosloe 813 was crossed with the same pollinators, most of the hybrids had higher protein and oil contents than the corresponding hybrids from Steril'noe 8. Analyses of 22 hybrids grown in 1965 showed that hybrids from high-protein parents had protein contents below the parents; those from high-protein and medium-protein parents showed intermediate protein; those from two medium-protein parents showed increased protein; the remainder showed lower values.

Most of these hybrids gave increased grain yields, the best giving nearly double the yield of protein/ha of the standard cultivar Kubanskoe Krasnoe 1667.

Protein and Amino Acid Composition

Considerable interest has been shown in improving the amino acid composition of sorghum, especially to increase the proportion of lysine in protein. The recently discovered "high-lysine" lines will be discussed later as a separate issue. In this section, the genetic influence upon the amino acid composition of "normal" sorghum is considered.

In a study of combining ability and heterosis for yield, protein, lysine and certain plant characters carried out at Purdue University, Abifarin and Pickett (1970) selected 14 sorghum restorer lines for good panicle size, irrespective of height and other characters. The female parents were Combine Kafir 60, Martin, Redlan and Wheatland. The male parents (IS numbers) were: 2822, 5437, 855, 129, 115, 6974, 3441, 8295, 3568, 3977, 2936, 2942 and two identified as (1) an inbred selection from Combine Kafir 60 × H 7910 and (2) an inbred selection from 7763/55H × 1308. The variety RS 610 was grown as the check standard.

The 14 characters measured were: days to maturity; leaf number (actual count); plant height (cm); leaf area index (cm^2); panicle exsertion (cm); panicle length (cm); threshing percentage (% grain in head); panicle grain weight (g); 100-seed weight (g); protein (micro-Kjeldahl × 6·25 as % of seed); lysine (method not stated, as % of protein); grain yield (kg/ha); protein yield (kg/ha); and lysine yield (kg/ha). The ranges of the means for grain yield, protein in seed, protein yield, lysine in protein and lysine yield for parents and hybrids are shown in Table 3.2.

Significant differences for all characters were found for the 18 parents, the 56 F_1 hybrids and RS 610, and for the 56 hybrids, but among the parents considered separately there were no significant differences for threshing percentage, panicle grain weight and 100-seed weight. For all the characters studied the GCA computed from males was highly significant but the GCA from females was not significant for leaf number, panicle grain weight, grain yield, protein yield and lysine yield. Male × female effects (SCA) were significant only for days to 50% bloom and plant height. Analysis of variance of the combining ability data indicated that all the characters studied were influenced more by additive genes than those of dominance and epistasis. In all the characters, male parents which were of greater diversity contributed more to the performance of the hybrids than the female parents.

Estimates of narrow sense heritabilities indicated lysine yield was lowest (41%) and plant height highest (99%). The heritability of protein content was 85%, protein yield 68% and lysine as % of protein 62%. The heritability of grain yield was 77%, panicle grain weight 84% and 100-seed weight 86%.

Heterosis was calculated for each character as the increase of the overall F_1 mean over the overall weighted parental mean and was positive and significant for all characters except protein % for which it was negative and significant. Heterosis for protein yield was 36%, lysine % of protein 3% and lysine yield 48%. The highest value was for panicle grain weight, 58%, followed by plant height 50%. Grain yield heterosis was 44%, and 100-seed weight 6%.

The phenotypic correlation coefficients for the 56 F_1 hybrids showed that grain yield was positively and significantly correlated with all the characters except panicle exsertion, panicle length, % protein and lysine % of protein. There was a near significant correlation between grain yield and lysine % and a high correlation between grain yield and height. Per cent protein was positively and highly significantly correlated with panicle exsertion, the only character thus correlated with protein: 100-seed weight roughly paralleled protein but was not significantly correlated. Lysine (% protein) was negatively and significantly correlated with protein; positively and significantly with number of days to half bloom, leaf number, flag leaf area and panicle length. The phenotypic correlation coefficient between grain yield and lysine yield was high (0·778), suggesting that yield and lysine could be improved together, but negative phenotypic, genotypic and environmental correlations occurred between yield and protein.

Abifarin and Pickett suggested that breeders might therefore select for good panicle exsertion and obtain satisfactory protein content, and select for late maturing, many leaves/plant and good panicle length with small seeds, to obtain satisfactory lysine content. However, since panicle exsertion was negatively correlated with yield, and lysine content is positively correlated with yield, it would appear more promising to select for high yield with higher than average lysine content.

Collins and Pickett (1972a) made estimates of GCA and SCA for grain yield, grain protein and lysine as % protein by evaluating 48 F_1 hybrids obtained by crossing 12 diverse IS lines, from the World Collection onto male sterile lines of Martin, Combine Kafir 60, Redlan and Wheatland. All were grown on the Purdue University Agronomy Farm during the summer of 1967. Yield (kg/ha), protein (micro-Kjeldahl N × 6·25), and lysine (ion-exchange chromatography) of the means of F_1 hybrids with one parent in common and the mean of the common parent are shown in Table 3.3.

Almost all the hybrids produced grain yields superior to either parent. Only four hybrids, IS 1220 × CK 60, IS 1220 × Redlan, IS 8255 × CK 60 and IS 8255 × Redlan were superior to both parents for protein. Some lines exhibited dominance for high % protein, others showed no dominance or dominance for low protein.

There was no evidence for heterosis for lysine (% of protein) among the hybrids. Differences existed among male lines for GCA for all three characters but among female lines only for protein and yield. The interaction between male and female effects, used as a measure of SCA, was not significant indicating additive gene action. Levels of significance estimates of the GCA effects indicated IS 1220, IS 8168 as >0·1 for yield; Martin, IS 8005, IS 8120 and IS 8255 as >0·05 for protein; and IS 0508, IS 1220, IS 3935 and IS 8301 as > 0·05 for lysine.

Among the female parents, Redlan and Wheatland were superior for yield to Martin and CK 60; among the male parents IS 1220 and IS 8329 had larger general effect. Martin was the best among female parents and IS 8120 the best among the male parents for protein. Male parents IS 0508, IS 8168 and IS 0718 showed dominance for high protein.

There were no significant differences among the female parents for lysine (% of protein); among the male parents IS 6898, IS 0508 and IS 1220 produced largest general effects. IS 1220 was most effective in transmitting high yields, high protein and high lysine (% of protein) to its hybrids.

Among the 65 genotypes, grain yield and % protein were slightly but significantly negatively correlated, % protein and % lysine were significantly and negatively correlated: yield and % lysine were not significantly correlated. In the opinion of Collins and Pickett (1972a) it should be possible to increase both lysine and protein in a high yielding line if all three were selected simultaneously, using for example, IS 1220.*

Collins and Pickett (1972b) selected nine sorghum lines for diversity in protein (micro-Kjeldahl N × 6·25) and lysine % protein (ion-exchange chromatography). They were crossed in all possible combinations and grown at Purdue University Agronomy Farm in 1968. The lines were IS 2319, IS 0075, IS 0508, Redlan, CK 60, Wheatland, IS 8364, Martin and IS 1259. The variety RS 610 was adopted as a standard check. Analyses of variance indicated significant levels of GCA and SCA for yield, % protein and lysine % protein. The ratio of GCA/SCA mean squares indicated that the additive gene action had a greater influence upon protein content than upon yield and lysine (% protein). Valid estimates of the additive and non-additive components of variance could not be obtained because only nine lines were used.

Grain yield ranged from 3100 to 1590 kg/ha. In general, higher yielding parents produced higher yielding hybrids but hybrids with IS 2319, a moderate yielder, as one parent, gave highest yields. Heterosis was more apparent from crosses among the higher yielding parents. The large

* (Authors' note: the analytical method states "percentage lysine in the protein was determined...". The abstract refers to "protein and lysine levels in the grain". The basic text simply refers to "per cent lysine". Judging by the values quoted, lysine as per cent protein is probably what was determined.)

negative SCA effects for the parents IS 2319, IS 0075 and IS 0508 suggested a greater combining ability than might be expected from their individual yields. GCA effects indicated the possibility of selecting superior segregates from crosses among these parents.

Protein among hybrids varied from 9·2 to 18·2%. In general, hybrids of the higher protein parents contained higher than average protein but the higher protein content of IS 0075 was not transmitted to its hybrids, suggesting that recessive genes control protein. Low protein contents were found among most of the crosses involving the low-protein IS 0508. Heterosis was demonstrated by the high-protein lines IS 1259 × Martin. Hybrids highest in lysine (% protein) were derived from a high-lysine parent and the lowest, from a low-lysine parent. Hybrids with IS 0075 as one parent were higher in lysine than anticipated. Crosses involving the highest protein line, IS 1259, were uniformly low in lysine. Although the hybrid value was for the most part equal to the mid-parent value (the two parental values added and divided by two), there were a few instances of negative heterosis. None of the hybrids was higher in lysine than the higher parent. Most of the estimates of SCA effects for parents were positive. Collins and Pickett (1972b) suggested that it should be possible to select for higher lysine (% protein) among inbred lines.

Genotype correlations between protein and yield, and between protein and lysine (% protein) were highly significantly negative. Yield and lysine (% protein) were positively correlated among genotypes. Among environments, yield and protein content tended to be inversely related. In general, genotypic influence was greater than environment. Collins (1970) had reported that single seed analysis of parent, F_1 and F_2 plants produced no evidence to suggest that environmental influences could dominate genetic differences in the seed.

Collins and Pickett (1972b) suggested that among the lines examined, IS 2319 was the best in generating high-protein, high-lysine progenies. IS 0508 produced high yielding F_1 seeds low in protein but high in lysine. IS 1259 gave F_1 hybrids high in protein content but relatively low in both yield and lysine.

Tripathi *et al.* (1971) studied the protein (Kjeldahl N × 6·25) content of 123 cultivars of sorghum and lysine (microbiological assay) % of grain and % of protein. Of the samples, 46 were analysed for threonine and leucine (microbiological assay) in the grain and in the protein, hardness (kg pressure required to crack the grain), 100-seed weight and specific gravity (flotation in 50% sodium thiosulphate solution).

The mean and ranges are shown in Table 3.4. Crude protein ranged from 7·6 (IS 9070) to 22·2% (IS 186) with 64 samples falling between 8 and 12%. Both IS 186 and IS 188 contained over 20% protein. Lysine ranged from 72 (IS 8893) to 184 (IS 5391) mg/g total N; threonine (57 samples) ranged from 100 (IS 7575) to 278 (IS 3566) mg/g total N; and leucine (64 samples) from 426 (IS 7538) to 985 (IS 5413) mg/g total N. The other characters also showed wide variability.

Among 78 samples, significant negative correlations were found between protein (DWB) and lysine and threonine (both as % of protein). Significant positive correlations were found between protein content and leucine, both as % of grain and as % of protein, and with grain hardness.

Grain hardness was positively correlated with lysine and threonine as % in grain, and with leucine both as % of grain and % of protein. The 100-seed weight was negatively correlated with the threonine as % of grain. Other correlations were not significant. The grains which floated in the specific gravity test showed a lower mean for percentage protein than did those grains which did not float.

Eight varieties of cultivated sorghum, IS 3922, 8353, 3817, 3688, 3126, 675, 10202 and 968, of diverse geographic origin were studied by Govil and Murty (1973a) for protein, lysine, 100-seed weight, grain size and colour in a full diallel analysis. Their taxonomic groups were described in Govil and Murty (1973b), q.v. The F_2 progeny were grown at Coimbatore and Delhi, India in 1968–69. Observations were recorded on five randomly selected plants from each experimental row for crude protein content in flour (micro-Kjeldahl % DWB); lysine (microbiological assay, % in protein); 100-seed weight; grain size (scored from 1 small to 10 boldest); grain colour (scored (1) white, (2) light yellow, (3) yellow, (4) light brown, (5) dark brown, (6) light red, (7) red, (8) dark red, (9) purple, (10) black); panicle length (grain-bearing portion); number of primary branches in the panicle; length of primary branch; number of nodes in the panicle; distance between two successive nodes in the panicle; number of seeds/primary branch; days to 50% flowering; plant height to flag leaf; number of seeds at time of panicle emergence; circumference of the stem above the fourth leaf from the base; leaf drying (scored (1) darkest green lines to (10) completely dry at time of harvest); and grain yield/panicle.

Variation due to GCA and SCA (Griffing 1956) was significant for all characters except panicle length in F_1 and seeds/branch and stem girth in F_1 and F_2 hybrids. Reciprocal effects were found for protein, lysine, 100-seed weight and grain colour. The GCA × locations interaction was significant for all characters in F_1 and F_2 except panicle length and number of nodes in F_1.

The mean sum of squares due to SCA × locations was significant for all grain characters in F_1 but not for grain size and colour in F_2. Reciprocal effects × locations was highly significant for lysine and seeds/branch in F_1 and for 100-seed weight and yield in F_2. There seemed to be a more pronounced interaction in F_2. Differences due to GCA were predominant over those due to SCA for all characters.

Heritability estimates for protein and lysine were 9–14%, and for yield, 14%. Gene action was mostly additive for lysine, 100-seed weight and grain size. Heritability estimates were more consistent and better in F_2 than in F_1 for 100-seed weight, grain size and colour. Reciprocal effects were large for lysine, 100-seed weight, grain size and colour. There were large genotype × location interactions attributed to the photoperiodic effect on the growth of sorghum. The four varieties IS 3922, 3688, 3126 and 10202 showed desirable traits based on GCA and SCA effects, see below:*

	Parent			
	IS 3922	IS 3688	IS 3126	IS 10202
Grain yield/panicle (g)	36·7	23·0	23·0	40·5
Protein (N × 6·25) (%)	12·4	11·8	14·9	12·7
Lysine (% of protein)	2·1	2·1	1·7	1·9
100-seed weight (g)	3·7	3·2	2·5	4·2
Grain size (1 small–10 bold)	5	5	4	6

Each of the eight parents were crossed with four new MS lines, Combine Kafir 60A, 172A, 176A and 3675A in a line × tester analysis (Kempthorne, 1957). Hybrids derived from the female parents did not differ much from the mean values of the 15 characters. Judging by the SCA effects, Govil and Murty suggested that crosses of each of the MS lines with IS 10202, IS 3688, IS 3922 and IS 8353 might provide basic material to exploit SCA effects. Since all the available MS in sorghum are closely related in origin, there was a need for new sources of cytoplasmic male sterility.

Govil and Murty (1973b) chose 24 varieties, from the World Collection of sorghum at the IARI, New Delhi, representing different geographical areas and taxonomic groups to study genetic diversity in the parents for protein content, endosperm characteristics and panicle length, and the nature of gene action for these traits, in three selected crosses.

Table 3.5 lists the varieties under IS numbers with their group; flowering time (days), plant height (cm); 100-seed weight (g); endosperm texture scored as (1) starchy, (2) waxy, (3) sugary, (4) partly glutinous and (5) highly glutinous; endosperm colour scored as (1) white, (2) light yellow, (3) yellow, (4) brown, (5) red and (6) light purple; grain size scored from (1) for the smallest to (10) for the boldest; crude protein content (micro-Kjeldahl in flour, DWB); leucine content (method not stated, probably expressed as % of protein); and lysine (microbiological assay, % of protein).

Following Tocher's method described by Rao (1952) the 24 varieties were grouped into 11 clusters as under:

Cluster	Varieties code No.	Cluster	Varieties code No.
I	7, 16, 19, 20	VII	3, 15
II	13, 17, 21	VIII	6, 23
III	1, 9, 10	IX	2
IV	4, 14, 18	X	5
V	8, 24	XI	11
VI	12, 22		

* (Authors' note: a comparison of these data with those in Govil and Murty 1973b show that they differ. The difference may be due to different samples but IS 3126 is puzzling. This, according to Govil and Murty, is a broomcorn normally grown for making brooms and not for grain. The two sets of data for IS 3126 differ markedly.)

Variety 2, cluster IX was a broomcorn; variety 5, cluster X was a starchy endosperm caudatum type from Kenya differing from West African caudatums; variety 11, cluster XI was a complex dwarf derivative combining features of durra and bicolor.

The "genetic distance" (not defined in the paper but probably the degree of genetic relation as estimated by correlations of characters) was maximum (49·9) between clusters IX and X, and minimum (13·1) between clusters IV and V. Intracluster distance ranged from 0·0 to 17·30, the maximum being in cluster VIII. The clustering pattern did not follow the geographical distribution. The 11 varieties from the USA were distributed over 10 clusters; the East African varieties over five, and the Indian varieties over three clusters. All the varieties in cluster II were from the USA and all those in cluster VII from Africa. The cluster means for the different characters indicated considerable differences, except for leucine.

The two varieties in cluster VII (IS 8353 caffrorum durra and IS 8191 caffrorum bird-proof) contained high protein, large seed, pearly endosperm and medium height, and were relatively early. They were also high in lysine and relatively low in leucine. Of the characters studied, the greatest divergence was shown by grain size, followed by protein content, lysine in protein, endosperm texture and leucine.

Six varieties were chosen to study additive, dominance and interaction effects for sucrose (Hudson's inversion method, not otherwise referenced), protein and lysine. The crosses, IS 4400 × IS 8850; IS 3126 × IS 3817; and IS 3688 × IS 3126, were chosen on the basis of maximum differences between the parents for sucrose, protein and lysine, respectively. In all the three crosses, heterosis was negligible for all characters, possibly because of the limited number of crosses examined for the analysis of generation means. In the three crosses, dominance of higher sucrose content was indicated. For protein, none of the components of epistasis was significant because of high sampling variances. Dominance effects were largest for protein and lysine.

Twelve lines with protein contents (Kjeldahl N × 6·25) ranging from 12 to 17% and lysine (auto-analyser % of protein) from 1·43 to 2·13% DWB, from six geographical regions were selected by Rana and Murty (1975) from the sorghum collection at IARI, New Delhi for a diallel and line × tester mating analysis (Kempthorne, 1957). The parents are described in Table 3.6.

Numbers 1–7 were used in the diallel test. In the line × tester analysis, No. 3, IS 9837, corneous endosperm, and No. 5, IS 10526, chalky endosperm, were used as female parents, with Nos. 8–12 as the male parents. The significant differences among parents were reflected in the variation among hybrids for protein and lysine. As will be seen from Table 3.7, there was both positive and negative heterosis for protein and lysine depending on the cross combination. Crosses with high protein (>16%) were low in lysine as % of protein, while high-lysine crosses (>2·0%) were generally low in protein.

The GCA effects of parents showed IS 9985 (caudatum kaura) and IS 10670 (caudatum kafir) were better combiners for protein, and IS 8191 and IS 10525 good combiners for lysine as % of protein, based on the results of both the diallel and line × tester experiments.

The phenotypic and genetic correlations indicated a positive association between protein and lysine in the grain but no such association was found in the hybrids. The negative association of protein with grain yield in hybrids was not found in the parents.

Lysine and yield showed a very low association suggesting the possibility of breeding for both high lysine and high yield. It is possible to breed for lines high in protein combined with moderately high lysine by using Nigerian and Sudanese material with high GCA effects such as the caudatum kaura, roxburghii shallu, and durra groups included in the World Collection. Rana and Murty state their results contrast with those of Collins and Pickett (1972b) who found a negative association between protein and lysine in protein and whose material consisted mainly of kafir and bicolor types. Collins and Pickett (1972a) had earlier found lower negative association between protein and lysine in the protein and the absence of any significant association between yield and lysine.

Two high-lysine lines from Ethiopia with 15·7 and 17·2% protein and floury endosperm were notable exceptions to the generally strong negative association between protein and lysine (% of protein).

Doggett (1970) recommended a programme of recurrent selection and hybridization using composites of two or more populations showing considerable heterosis as a means of increasing protein content in sorghum. A population breeding programme, including the full range of the three taxonomic groups would likely be more effective in generating superior cultivars and hybrids than the traditional methods of crossing a comparatively narrow range of genotypes.

Nanda and Rao (1975a), at Delhi, grew six dwarf non-Indian sorghum lines, IS 511, IS 859, IS 2954, IS 2031 and IS 3797, and Swarna; two Indian tall grain sorghum lines, IS 4522 and R 168; a fodder sorghum A 1 14 8 with high protein content in the grain, and a dwarf selection R 1661 from an exotic (non-Indian) × Indian cross IS 3678 × Maldandi, and the 45 possible F_1 hybrids, without reciprocals, in a randomized block design with four replications. The material was grouped into talls and dwarfs in each block and the plot consisted of a single row of 3m length spaced 75 cm apart with a spacing of 15 cm between plants. The protein content (micro-Kjeldahl N), amino acid (% of protein) (auto-analyser at Purdue University, tryptophan at National Institute of Nutrition, Hyderabad), β-carotene (Klett's colorimeter using 240 μm filter), 100-seed weight and grain hardness mean and ranges for the parents and hybrids are shown in Table 3.8.

There was a significant variation among the parents in protein content, yield attributes and glycine. Among the hybrids, significant differences existed for protein, yield attributes, threonine, methionine, isoleucine,

leucine, phenylalanine, tyrosine and valine. The parent vs hybrid comparison was significant for yield attributes, lysine, leucine, phenylalanine, glutamic acid, proline and glycine. Additive gene action contributed more to all attributes except methionine, isoleucine, leucine, phenylalanine, tyrosine, glutamic acid, proline and alanine where SCA was significant or predominated over GCA. For protein, lysine and threonine, GCA was significant.

Significant negative correlations were observed between:

protein, with yield, seed size and grain hardness;
lysine, with isoleucine, leucine and cystine;
leucine with methionine;
carotene with protein, methionine and cystine;
seed size with threonine;
yield with cystine.

Significant positive correlations were observed between:

Yield with seed size and grain hardness;
grain hardness with seed size and lysine;
methionine with lysine;
isoleucine with threonine;
leucine with protein, threonine and isoleucine;
β-carotene with lysine, threonine, isoleucine and leucine.

Nanda and Rao (1975a) stated that their results on inheritance of protein and lysine were in general agreement with those of Collins and Pickett (1972a,b) and on β-carotene with those of Singhania *et al.* (1970), and indicated opportunities for breeding for both yield and improved nutritional quality.

Nanda and Rao (1975b) using essentially the same material as for the work described in (1975a) carried out analysis of variance (Hayman, 1954) for amino acids (mg/g sample) in a 9×9 diallel cross of *Sorghum bicolor*. Differences were significant for all amino acids except cystine. The magnitude of additive variance was smaller than non-additive variance for most amino acids.

The negative F-values found for most of the amino acids indicated that recessive alleles were predominant except for lysine, methionine, histidine, and glycine where F-values were positive.

Statistical analysis of their results indicated that dominant genes were positive for all amino acids except cystine, methionine, lysine and arginine where the reverse was the case.

Vasudeva Rao and Goud (1976) reported, from the College of Agriculture, Dharwar-580005, India, the inheritance of protein in a five-parent complete diallel set of five sorghum inbreds built up from crossing IS 2226 (exotic dwarf), 14-8-81-1-1 and 14-5-17-12 (induced dwarf mutants of local tall GM2 3 1), Shallu (exotic dwarf) and Karad local (local tall).

Protein was recorded as macro-Kjeldahl N $\times$ 6·25 and the diallel table analysed for gene action by the method of Jinks and Hayman (1953). Combining ability, variances and effects were obtained by method 1, model I (fixed effects) of Griffing (1956).

There were highly significant differences among the entries and GCA and SCA variances were also highly significant with SCA much higher (149·92) than GCA (7·21). In the material studied, additive and dominance components appeared to be of equal importance in the inheritance of protein. Heritability was low. Karad local was the best parent for a future programme to increase sorghum protein; the cross IS 2226 × Karad local had a high specific effect for protein content.

Nair *et al.* (1974) studied protease and nitrate reductase activity in relation to the sorghum hybrids CSH 2 and CSH 3 and reported (1) high nitrate reductase activity in the pre-anthesis phase and (2) a rise in protease in the post-anthesis phases. Nair *et al.* considered the first apparently provided reduced N and the second was responsible for the transfer of leaf N to the developing seeds.

Johari *et al.* (1976) reported from IARI, New Delhi, India that protein fractionation studies (modified Mendel-Osborne method described by Nagy *et al.*, 1941) of CSH 2 sorghum grown on the Institute farm using ^{15}N-labelled urea and ammonium indicated that as the kernel matured, there was a considerable decrease in the proportion of albumin and increase in prolamine, glutelin and residue proteins. The amino acids, basic, neutral and acidic, were separated by high voltage electrophoresis (Srivastava and Mehta, 1976) after acid hydrolysis. Incorporation of ^{15}N into basic amino acids (lysine, arginine and histidine) was lower when compared to that in neutral and acidic amino acids at all stages of grain development.

Johari *et al.* (1977) followed their earlier work by incorporating leucine–[^{14}C] and studying the changes in the protein fractions of CSH 2 sorghum. Most of the label from the injected leucine was found in the glutelin and residue fraction towards the later stages of maturity, with a corresponding decrease in the albumin, globulin and prolamine. During grain development, lysine, aspartic acid and glycine decreased; leucine, proline, alanine, tyrosine, phenylalanine and cystine increased; serine, methionine, valine and isoleucine increased marginally (amino acid analysis by auto-analyser after acid hydrolysis). As the grain matured, the tannin content (Folin Denis method as modified in Christensen, 1974) increased from 0·40% at 17 days to 0·51% at 31 days, after ear emergence.

Christensen (1977) surveyed a total of 1851 sorghum varieties from northern Cameroon for their acid orange 12 dye-binding capacity (indication of the lysine content) of the sample and for their protein contents (Technicon).

There were significant differences among varieties in residual dye binding capacities. The highly significant difference between high and low groups indicated that residuals were heritable. The significant variation among varieties within groups indicated that adjustment of variety means by the genetic regression also gives a heritable residual.

The adjustment of protein content by the phenotypic regression failed to eliminate selection for protein, a failure attributable to the relative size of laboratory error. The use of replicated plots allows elimination of laboratory and environmental errors in protein and provides sufficient information so that genetic covariation of lysine and protein can be quantified and removed.

Natural High-lysine Mutants

Because of the characteristically low lysine content of all cereals, widespread searches are in progress to discover cultivars and mutants with higher than normal lysine in protein contents. As reported below, considerable success has been achieved with maize. The pursuit of high-lysine sorghums is in progress in several research centres.

High-lysine Maize

During the early 1960s the amino acid analysis of a large number of maize varieties led to the discovery of two important mutants, Opaque-2 (Mertz *et al.*, 1964) and Floury-2 (Nelson *et al.*, 1965). The mutants were named "opaque" and "floury" because when placed over a strong light source they appeared opaque in comparison with the more translucent normal maize; also the physical condition of the endosperm was described as floury in contradistinction to the hard, vitreous or corneous condition of normal maize. Both mutants contained substantially larger than previously encountered proportions of lysine and tryptophan, the limiting amino acids for conventional or normal maize. Both Opaque-2 and Floury-2 genes have been transferred into "high-lysine" maize cultivars. The mutant genes appear to suppress partially the synthesis of the alcohol-soluble prolamine fraction and to stimulate the synthesis of the water-soluble albumin and salt-soluble globulin fractions. The compositional differences occur in the storage proteins of the endosperm; no difference is apparent in amino acid composition of the germ protein between conventional cultivars and high-lysine mutants.

Vasal (1975) cited by Singh (1976) reported the use of genetic modifiers to obtain normal (hard endosperm) type maize kernels with the Opaque-2 gene.

The International Maize and Wheat Improvement Centre (CIMMYT) (1977) reported that during 1976 CIMMYT had developed 29 gene pools and 17 advanced populations carrying the Opaque-2 gene from most of which hard (as distinct from floury) endosperm types were being selected. Twenty-three experimental varieties and seven élite varieties all carrying the Opaque-2 gene, grown at several locations around the world all gave yields equal to or better than normal check varieties.

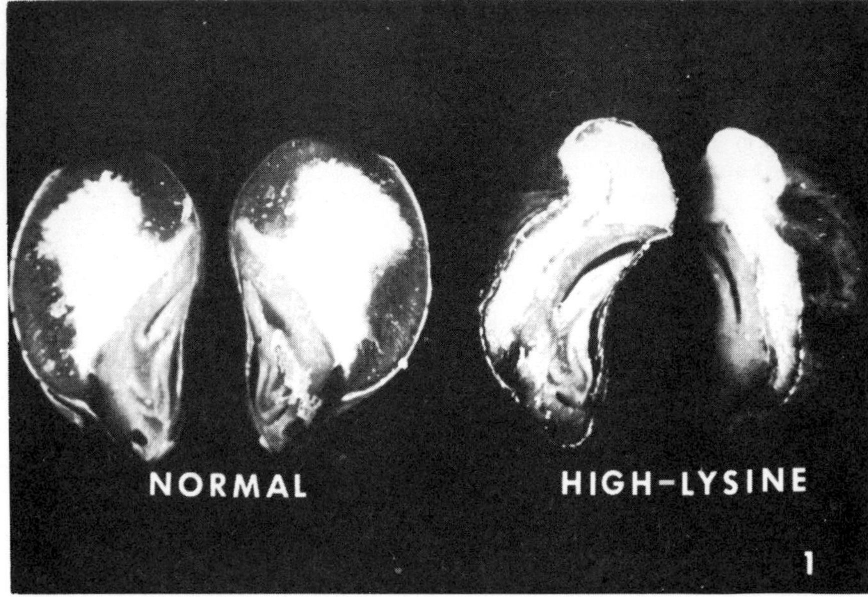

Fig. 3.I Longitudinal sections of normal and high-lysine sorghum kernels segregating on the same head, 6·6 ×. Note dented kernel of high-lysine type. By courtesy of Dr Lloyd W. Rooney, Texas A&M University.

High-lysine Sorghum

The purpose of the Purdue Project (Axtell, 1976) is to develop and release for use in developing countries lines of sorghum with genetically controlled higher protein content, improved amino acid balance and total digestibility. Thousands of lines from the World Sorghum Collection and various breeding programmes have been analysed for protein content and biological quality. The most promising are recombined to bring about further improvements in protein content and quality. Lines which do not produce seed at Lafayette, Indiana, are multiplied in Puerto Rico. During the past 4 years, the Purdue research team have identified two distinctly different genes which improve overall protein quality: (1) the naturally occurring high-lysine (*hl*) from Ethiopia; (2) mutant P 721, a chemically induced high-lysine mutant. It was Purdue's hypothesis that high-lysine content in sorghum might be associated with an opaque endosperm phenotype similar to that in Opaque-2 maize. Provided the pericarp and testa are not pigmented, samples can be screened over a light box for opacity.

Singh and Axtell (1973a,b) identified 62 floury endosperm lines from over 9000 lines of sorghum in the World Collection and, of these 62 lines, two of Ethiopian origin IS 11167 and IS 11758, were found in comparison with the remainder to be exceptionally high in lysine and relatively high in protein. The protein content (micro-Kjeldahl N × 6·25) (% DWB) reducing sugars, sucrose and water-soluble polysaccharides (Shannon, 1968) amylose and starch (Shuman and Plunkett, 1964), oil (nuclear magnetic resonance), catechin equivalents and seed characteristics of the two lines compared to a normal sorghum are shown in Table 3.9.

Singh and Axtell stated that floury grains contain a soft, chalky white endosperm. It is important to examine fully mature kernels since kernels of corneous sorghum types may appear floury when immature.

The starch concentration in the high-lysine lines was roughly equivalent to normal sorghum. In the high-lysine lines the relative proportions of reducing sugars and water-soluble polysaccharides was about equal to, but % sucrose content almost double, the normal values. It is interesting that the sucrose content of Opaque-2 maize is of the same order of difference (Barbosa, 1971, cited by Singh and Axtell, 1973a).

The endosperm of IS 11167 and IS 11758 was partially dented but the 100-seed weight of both lines was nearly equivalent to the average of 31 lines and hybrids of normal sorghum. The dented endosperm was also associated in F_1 panicles segregating for the *hl* gene, and may be due to a pleiotropic effect of the *hl* gene or it may be a consequence of linked modifier genes that alter the normal plump configuration of the sorghum grain. Singh and Axtell (1973a) stated: "the complete absence of vitreous starch in the *hl hl hl* endosperm tissue may influence the normal configuration of the grain".

Phenotypic classification of kernels was carried out on F_2 seeds from crosses between a Purdue random mating population, PP3R (ms_3) (normal plants) and each high-lysine line. Normal (low-lysine) kernels were plump with corneous endosperm, high-lysine kernels were partially dented with floury endosperms. All F_1 seeds obtained from the crosses of ms_3 with the two high lysine lines had corneous endosperms. Chi-square analysis of the F_2 segregation ratios indicated 3 corneous/1 floury for the progeny of both crosses and suggested that the high-lysine character is controlled by a simple recessive allele.

The amino acid composition (ion-exchange resin chromatography; methionine and cystine (method of Moore, 1963); tryptophan (method of Slump and Schreuder, 1969) were determined on the defatted endosperm tissue from floury and corneous F_2 seeds. The results are shown in Table 3.10.

The average lysine content (% of protein) in the endosperm of both crosses was 1·38% for the corneous and 2·53% for the floury kernels. The protein content in the embryo tissue of the normal kernels was 24·1% and in the floury kernels 24·4%. The lysine as % of embryo protein was 5·1 and 5·6% in the normal and floury endosperm respectively. The major effect of the *hl* allele appears to be on protein and lysine concentration in the endosperm tissue, in other words, it is the storage proteins which are most affected. (Singh and Axtell, 1973a).

The proportion of germ (separated by hand dissection) in the floury kernels was 16·0% compared to 10·2% in the corneous kernels (Singh and Axtell, 1973c). Lysine, arginine, aspartic acid, glycine, and tryptophan were consistently higher, and glutamic acid, proline, alanine and leucine consistently lower, in the floury endosperm. This is comparable to the shift in amino acid pattern reported for Opaque-2 maize by Nelson (1969b) and is consistent with the genetically controlled prolamine depressant factor (see Hulse and Laing, 1974 p. 43). The leucine/isoleucine quotient was lower in the floury kernels.

No information was presented on the yield potential of the two high-lysine lines. The average weight of 100 seeds of IS 11758 was 2·45 g and of IS 11167, 2·78 g compared to normal sorghum, 2·79 g (Singh and Axtell, 1973c). However, although the two Ethiopian grains were large in size, the endosperm was described as shrunken.

Singh and Axtell (1973c) determined the amino acid composition (% of protein) (auto-analyser except for methionine and cystine, method of Moore, 1963 and tryptophan method of Slump and Schreuder, 1969) of the defatted whole kernels of high lysine lines IS 11167 and IS 11758 grown at Lafayette, Indiana, and Puerto Rico and compared the results with the average of 522 sorghum lines (Srinivasan *et al*., 1972). The results are shown in Table 3.11. The high-lysine seeds were described as shrivelled.

In Table 3.12 the analysis of the defatted endosperm is presented to demonstrate the actual effect of the *hl* gene in the absence of any influence attributable to germ size. Lysine content of both the whole kernel and the endosperm of the high-lysine lines was substantially higher than the average lysine content of normal sorghum lines. There was also an increase in the thiamine and cystine (in the whole kernels from Lafayette), a decrease in leucine and a lower leucine/isoleucine ratio. The total sugar content of IS 11167 was nearly 2·5 × and of IS 11758 nearly double the level in normal sorghums. The subject of sugary endosperm sorghums is discussed more fully later in this chapter.

During the initial survey of the World Collection for high-lysine lines, many lines with kernel phenotype (partially-dented) similar to the high-lysine lines were identified. None had lysine values comparable to the high-lysine lines, the range being 1·68–2·18 lysine as % of protein. Lysine as % of sample varied from 0·24–0·32% compared to 0·56 and 0·59 in the two high-lysine lines. Singh and Axtell (1973c) suggested that this indicated that the increased lysine content of the high-lysine lines was not due to the dented or shrunken endosperm and was not a result of reduced endosperm size, in which there exists a higher than normal ratio of seed coats and embryo to endosperm.

Mertz (1974) fractionated the proteins of normal and high-lysine sorghum using the Landry and Moreaux (1970) method. The high-lysine values were the average of assays from five segregating heads using the high lysine (*hl hl hl*) kernels from F_2 heads derived from crosses between normal plants and high lysine sorghum line IS 11758. (See Table 3.13).

In the high-lysine sorghum there was an increase in Fractions I and V and a decrease in Fractions II and III. It was this change in the protein pattern that was primarily responsible for the increase in lysine content of the grain, not the development of any new proteins. There was a marked reduction in prolamine, and the endosperm was floury.

Doggett (1974a) stated that the two high-lysine Ethiopian lines IS 11167 and IS 11758, when grown at ICRISAT, Hyderabad produced shrivelled grain of low endosperm content.

Singh (1976) used six high-lysine sorghum lines, homozygous for the *hl* gene, of Ethiopian origin, to make 16 crosses (plastic bag technique of Schertz and Clarke, 1967) between IS 16199, IS 16210, IS 16227 and selected dwarf, early, hard endosperm, normal female lines at Poza Rica, Mexico during the winter of 1973–74. All F_1 hybrids were nearly as tall and late as their high-lysine parents. The resulting hybrids were grown at Tlaltzipan, Mexico the following season.

After threshing heads of each F_1 hybrid, four grain types were separated by inspection over a fluorescent glass screen:

(1) opaque-dented (no light passing through, floury endosperm)
(2) opaque-plump (no light passing through, floury endosperm)
(3) normal-corneous (completely translucent, corneous endosperm)
(4) modified (part opaque, part translucent; opaque fractions floury, translucent fractions corneous)

Selected F_2 seeds were planted at Poza Rica and Tlaltizapan during the 1974–75 season. Relatively short and early plants in each family were harvested and threshed individually. In subsequent generations only seed from plants with modified types was saved except for a few homozygous opaque-plump and dented types. Protein (micro-Kjeldahl; method of calculation not stated) and lysine % grain and in protein of some selected F_3 segregates are shown in Table 3.14.

The results from the early stages of this programme indicate higher lysine tends to be combined with lower protein.

The CIMMYT group (CIMMYT, 1976) reported that F_3 progenies of selected plants were planted at Tlaltizapan but because of unfavourable conditions, over 90% of the material was lost. The F_4 generation was grown at Poza Rica during 1976 and again less than 10% of the plants survived and produced ripe grain. Protein and lysine values for some selected samples are shown in Table 3.15. None of these appeared particularly promising.

Studies with the P 721* opaque mutant began during 1975 and 54 individual F_3 plant selections were grown from a single cross during 1976 at Tlaltizapan. Evaluation was also made of 27 F_2 families and three F_1 crosses using this mutant. CIMMYT (1976) stated that the stability in terms of grain structure and protein quality was yet to be determined.

In a later publication (CIMMYT, 1977) it is stated that during 1976 CIMMYT verified that the better F_3 crosses embodying Ethiopian (*hl*) lines as donors contained 3·0% lysine (% of protein). These lines were lower than normal in total protein but CIMMYT breeders believe they can achieve

* See "Induced high-lysine mutants," p. 236 *et seq.*

normal protein contents while retaining the higher lysine together with satisfactory yields. CIMMYT (1977) reported that their best cold-tolerant sorghum lines grown at several sites gave mean yields of 3·3 t/ha.

The high-lysine Ethiopian line of sorghum, IS 11758, grown at Rajendranagar and analysed at the National Institute of Nutrition Hyderabad, India (personal communication, 1974) was found to have a mean protein content (samples from 32 plants) of 13·6%. The mean lysine content (% protein) was 4·04 and ranged from 2·9 to 5·9. This confirmed the high-protein, high-lysine character of the strain under Indian conditions. The leucine content (% of protein) was also high, ranging from 10·1 to 15·9 with a mean value of 12·6.

Riley (1974, personal communication) reported that the Ethiopian variety Alemaya 70 contained 10·7% protein (N × 5·7, as-is basis) with lysine at 125 mg/g total N and 214 mg/100 g sample. A possible high-lysine variety, Watet Begunche contained 9·9% protein and lysine at 224 mg/g total N and 355 mg/100 g sample.

Ejeta (1977) reported from Purdue University that 1340 sorghum lines, mainly of Ethiopian origin, were screened by the dye binding capacity (DBC) for indication of high lysine. Of the 17 lines selected for further evaluation, IS 11932 and IS 12211 contained substantially higher protein and lysine (% of grain), and IS 12174, IS 12178, IS 12187, and IS 12195 showed a combination of high protein content and seed quality.

When 234 high-lysine (dented) and 204 normal (plump) sorghum varieties from the Wollo province in Ethiopia were compared, protein–lysine regression of both groups showed distinct differences between normal and high-lysine types. There was a high association between the phenotypic markers in the kernel (dented or plump) and prolamine (% of protein) in the germ plasm evaluated. The agronomic characteristics indicated wide variation among the high-lysine types suggesting to Ejeta that over the years, the *hl* gene may have stabilized in different genetic backgrounds through natural breeding.

Induced High-lysine Mutants

Genetic mutation may be induced by the irradiation of seeds with gamma or X-rays or by treatment with chemical mutagens. Mutations thus induced are essentially unpredictable and therefore mutation breeding calls for the treatment and subsequent examination of comparatively large quantities of seed for the presence and stability of the desired characters. Sree Ramulu (1975) in a review of mutation breeding of sorghum stated that relatively fewer reports had been found compared with wheat, barley and rice.

Naphade and Ghawghawe (1971) reported that seeds of NJ 156, M 22 5 16 and MCK 60 sorghums were irradiated with doses of gamma rays varying from 10 000 to 40 000 R. In both R_1 and R_2 generations, seeds of M 22 5 16 contained more protein than the other two varieties. In general, irradiation reduced the protein content of R_1 generation, but increased that of R_2. In both generations higher protein resulted at 30 000 R.

Mohan and Axtell (1974) and Axtell (1976) described the results of research at Purdue in which a high-lysine character was induced by treatment of sorghum with chemical mutagens. The parent lines used were 3-dwarf, photoperiod intensive lines with broad agronomic adaptability (Axtell, 1976). Since the pericarp was colorless and the endosperm translucent, the progeny from mutagen treatments could be screened for opaque endosperms over a light box. The selfed seed was treated with diethyl sulphate (DES) and the treated (M_1) seed was grown during 1972 at Lafayette. Each M_1 head bearing M_2 seeds was bagged prior to anthesis to ensure self-fertilization. M_2 seeds from each of the fertile heads were planted in single row plots during the winter of 1972–73 in Puerto Rico. M_2 heads were selfed and approximately 23 000 heads were harvested in the spring of 1973. Each head was threshed individually, cleaned and packaged in a coin envelope. A random sample of seeds was examined over a light box for opaque mutant kernels. Seed from each segregating head was separated into translucent corneous and opaque classes and given a putative mutant number (P number). For the most part, the ratio of opaque to corneous was 1/3, sometimes 1/9 and sometimes 1/15; in other cases no simple Mendelian ratios were observed. Heads homozygous for opaque seeds needed no separation. Whole kernel samples of opaque corneous and homozygous classes were defatted and analysed for protein content (micro-Kjeldahl N × 6·25) and lysine (ion-exchange chromatography) as % of protein and as % of sample DWB. The lysine and protein contents were compared between 470 putative opaque mutants and normal corneous control sib seeds. Among 206 putative opaques, lysine (% of sample) increased by 10–30%; 83 by 30–50% and 35 by greater than 50% compared to the normal sibs. The protein content among the opaque mutants ranged from 7·6 to 22·1% and lysine (% protein) from 1·37 to 4·93. Among 445 normal sibs average protein was 14·4%, lysine (% of protein) 1·9 and lysine (% of sample) 0·27.

Of the 35 (roughly 7·4% of total) opaque mutants in which the lysine concentration increased by more than 50% over the normal sibs, 28 were planted at Lafayette or Puerto Rico. From the results the high-lysine mutants were placed in two classes: class I phenotypically normal plants; and class II abnormal plants including those with chlorophyll deficiency, dwarfed or stunted plants. The class I high-lysine opaque mutants were studied further, particular attention being given to two mutants P 720 and P 721. The average lysine content (% of protein) of P 721 opaque was 2·8 at 17·5% protein compared with the normal P 721 sib average lysine 1·7%, at 15·4% protein. In P 720 opaque, lysine was 2·9%, protein 16·5% compared to the P 720 normal sib average lysine of 1·7% at 15·1% protein.

During 1974, P 721 was increased at Lafayette and 14 sib families were analysed. The average lysine was 3·9 (% of protein) and 0·432 (% of sample), protein content 13·9%. Analyses of eight families from P 721 normal sibs gave average lysine contents of 2·09 (% of protein), 0·27 (% of sample) at protein content of 12·1%, levels stated to be roughly the average for all lines in the World Sorghum Collection. From these results it would appear that the P 721 opaque has undergone a genetic mutation which alters the normal amino acid pattern in the sorghum endosperm.

Axtell (1976) described studies to determine the inheritance of the high-lysine P 721 opaque and its amino acid composition. A preliminary test for allelism involved making crosses between P 721 opaque and lines homozygous for the *hl* genes from Ethiopia. Analyses were made for protein content (micro-Kjeldahl AOAC, 1960, N × 6·25) and for amino acid content (auto-analyser following defatting and hydrolysis except for methionine, cystine and tryptophan analysed separately), methionine and cystine (method of Moore, 1963), tryptophan (method of Slump and Schreuder, 1969). Per cent oil in whole grain samples of P 721 was determined by NMR and catechin equivalents by the vanillin HCl (Burns, 1971) and modified vanillin HCl (Maxson and Rooney, 1972b).

Preliminary results suggested the high lysine character was simply inherited. A semi-dominant mode of inheritance was suggested for P 721 opaque similar to maize in which each successive dose of Floury-2 in the endosperm adds an additional increment of lysine. In a preliminary test for allelism with the high-lysine mutant IS 11758 and P 721 opaque, all F_1 seeds appeared to be corneous indicating the non-allelic nature of the two genes. There was no indication of an increase in lysine content which would be expected if these two mutant genes were allelic.

From the results obtained at Purdue, the P 721 opaque gene does not appear to affect the embryo proteins. Table 3.16 presents protein and lysine contents of the defatted embryos of both P 721 opaque and P 721 normal sib together with the proportions of embryo and endosperm in these two classes of kernels. It can be seen that the lysine content was similar in both classes of embryo though the protein content of the mutant was slightly higher. The embryo comprised 11·4% and 10·4% of the P 721 opaque and P 721 normal sib kernel respectively. The difference in lysine content was most pronounced in the endosperm. Lysine (% of protein) of the P 721 opaque 2·06% was 65% higher than that of the normal sib, 1·25%. The amino acid composition of defatted whole kernel and endosperm of P 721 opaque and normal is given in Table 3.17. The opaque is higher in lysine and argenine and lower in glutamic acid, alanine and leucine than the normal. Protein content was also higher in the opaque. The amino acid differences between P 721 opaque and normal were similar to those reported for Opaque-2 maize and the *hl* endosperm in sorghum. The potential improvements in nutritional quality were reflected in the increase in lysine and the reduced leucine to isoleucine ratio. The average test weight of P 721 opaque was 25% lower than the normal sib. Therefore the average protein content on a per seed basis was 15% lower in the opaque than in the normal sib. The oil content of the opaque was 22% higher than the normal sib.

A comparison of the chemically induced mutant P 721 opaque with the naturally occurring high-lysine line IS 11758 from Ethiopia indicated that, based on average results from two locations, P 721 was 9% lower in protein, lower in lysine by 6% (% of protein) and 17% (% of sample). However, the photoperiod insensitivity of P 721 should result in its being more widely adaptable than IS 11758 which suffers from the added disadvantage of having an orange-coloured seed.

Riley (personal communication, 1978, based on the findings reported in his PhD thesis) reported from ICRISAT that in the 1973 dry season, October–March, early generation segregating material was sown of crosses between the two Ethiopian high-lysine lines, IS 11758 and IS 11167 and selections from the Nebraska and Purdue populations, white seeded lines, high-protein lines, yellow endosperm lines, and high-oil and -sulphur lines. A year later, crosses between the two Ethiopian *hl* lines and three populations from Uganda were included.

Initial screening included (1) identification over a light box of floury endosperm types, (2) separation of plump from dented seeds and (3) removal of the pericarp prior to biochemical analysis. Subsequently, high U/P ratio segregates were selected from analysis of plump seed without removing the pericarp. (U/P ratio is a ratio of Udy (Udy, 1971) intensity reading divided by protein % (Technicon auto-analyser N × 6·25)). U/P ratios greater than 3·5 were classed by ICRISAT as high and equivalent to 2·3% lysine in protein. A good correlation between U/P ratio and lysine has been found by the ICRISAT Biochemistry Department (personal communication by R. Jambunathan to Riley).

Initially, high U/P ratio parents tended to produce high U/P progeny but subsequently, for three successive seasons, the frequency of high U/P segregates was very low. High-protein parental plants, with low U/P, gave a larger proportion of their progenies in the same class as their parents. High U/P parent plants generated a very low proportion of high U/P progeny. These results suggested that high U/P segregates are either more unstable in expressing the high-lysine character, or the recessive high-lysine gene is mutating to the dominant normal type in a large proportion of the plump seeded segregates.

In the 1975 dry season and the 1976 summer season 565 selections were made and planted at Hyderabad in the 1976 rainy season (July sowing); 85 of the entries sown had high U/P. Some 1875 phenotype selections were made, mainly from the high U/P entries, and subsequently planted in the 1977 summer season at Bhavanisagar. Analysis of remanent seed found 75 of these selections, mostly from high U/P parents, with high U/P ratios. A few high U/P parents resulted only in high U/P progeny. Of the 75 high U/P selections of Bhavanisagar, 50 of those with the best appearance from F_5 and F_8 generations were bulk harvested. Analyses of these bulked rows indicated that 40 maintained high U/P. All the lines from this Bhavanisagar sowing, including the checks, had light weight seeds and higher % protein which may have reduced their U/P ratios. Table 3.18 shows the protein, U/P ratio and 100-seed weight of the bulked rows with P 721 (the chemically-induced mutant) as check. The lines presented in Table 3.18 have lower U/P than P 721 but some seed weights, though relatively light, are higher than P 721. Seed size may be negatively associated with U/P ratio in these lines.

During the 1975 dry season P 721 was crossed with high U/P progenies from crosses carrying the Ethiopian *hl* genes. These *hl* derived segregants were also intercrossed among themselves. Phenotypic selection of the best appearing plants in the F_1, F_2 and F_3 generations was made mainly among those rows whose progenitors had high U/P ratios. Since each generation was grown during a different season, and on a different soil type and without consistent checks in all cases, it is difficult to separate environmental from generational or genetic effects.

The *hl* derived and P 721 F_1 lines and the *hl* derived intercross F_1 lines were grown during the 1976 rainy season at Hyderabad, on an alfisol with 200 kg N/ha in two applications. The P 721 crosses showed higher U/P, lower protein contents, and more floury endosperms than the *hl* derived intercrosses. The high U/P ratios of the P 721 crosses could have been due to the expression of the semi-dominant effect of the P 721 parent. The occurrence of high U/P ratio (3·5 and above) heads among the F_1 *hl* derived intercross group was only 1·0%. In each cross both parents had been selected for high-estimated lysine, and if the *hl* gene had been present in both parents it should have been expressed in the F_1.

The best appearing F_2 selections were grown at Bhavanisagar during the 1977 summer season on red sandy loam soil with 145 kg N applied in two doses. The F_2 gave a greater proportion of high U/P segregants than did the F_1. The P 721 crosses gave a higher percentage (66) of plants in the high U/P class compared to the *hl* × *hl* derived intercrosses (36%). A number of selections of the crosses of P 721 × *hl* derived lines gave U/P higher than the P 721 parent. Average seed weight was light but several selections with U/P ratios equal to P 721 had heavier seeds than the P 721 parent.

Comparison of the two groups of crosses in the F_3 during the 1977 rainy season at Hyderabad grown on a vertisol with 110 kg N/ha in one basal application confirmed that P 721 crosses tended to produce a greater proportion of segregants with high U/P, and also a good number of apparently high-lysine segregants after two generations of selfing. Seed weight was fairly constant from F_2 to F_3. Protein levels dropped which may have been an environmental effect since the protein content of the P 721 check dropped from 13·9% at Bhavanisagar to 10·8% at Hyderabad during the 1977 rainy season. The drop in U/P of P 721 from 4·4 to 3·5 might have been due to a selection of P 721 in which the high-lysine effect was lower, or the high-lysine effect of P 721 may be more variable in certain environments.

The means of protein, U/P ratio and seed weight, and correlations among these trials for three different generations of intercrossing and selection, all grown at Hyderabad during the 1977 rainy season on a vertisol with 110 kg N/ha are given in Table 3.19.

The selections found in each generation were classified into three groups according to their U/P ratios: (a) "very high"—4·0 or above, (b) "high"—3·5–4·0 and (c) "low and normal"—less than 3·5, as well as a total (d) of all 3 groups. These ratio groups occur across the progeny groups which are:

I. F_3 progenies of *hl* derived and P 721 crosses.
II. F_3 progenies of *hl* derived crosses, inter-crossed amongst themselves.
III. F_5 progenies from Purdue University of crosses between P 721 as the male and the female selected from populations which mostly did not contain the *hl* mutant.
IV. F_1 lines arising from two intercrosses, first from crossing between *hl* derived lines followed by intercrossing among selected F_2 lines. This is similar to full sib selection and intercrossing of the S_1 lines in a recurrent selection programme.

These data indicated that the F_3 lines from the *hl*-derived intercrosses (II) had the highest mean protein % but lowest U/P, while the second intercross F_1 lines (IV) had the lowest mean protein %, and a mean U/P equal to the other generations from P 721 crosses (I and III). In every progeny group the very high U/P groups (a) had higher protein % than the generation means, while the high U/P and low and normal U/P groups (b and c) had either lower mean protein % or mean protein % equal to the generation mean. This relation was also apparent in correlations between protein % and U/P. In all progeny groups of the P 721 crosses (I, III, IV) more positive correlations appeared in the higher U/P groups (a and b), while negative correlations were found in the low and normal U/P group (c). As protein % increased in the normal group (c) the lysine level decreased. This has been previously documented (Warsi and Wright, 1973). However, the correlations indicate that an increase in protein in the high U/P groups could be expected to result in higher lysine or at least no drop in lysine level as % of protein.

The mean U/P value in the *hl* derived intercross F_3 (II) is markedly lower than in the other groups. It appears that the *hl* gene may be lost in this group. The second intercross F_1 (IV), had a slightly lower mean U/P (3·37) than the later generation P 721 crosses (I and III), but apparently higher than the first P 721 intercross F_1 mean of 3·1. This suggested that some increase in mean U/P may have been achieved through the second intercross, but both groups of crosses must be grown at the same location to confirm this relation.

The analyses of 482 progeny showed no significant correlation between seed weight and protein %; the same was true for all progeny groups (I, II, III, IV). High and very high U/P selections in the second intercross F_1 (IV) and F_3 P 721 crosses (I) had increasingly negative correlations although in only one case was the correlation significant (−0·466). It would appear that there is either low or no linkage between protein % and seed weight in this material.

Correlations between seed weight and U/P ratio are negative and significant for all progeny groups except the *hl* derived intercross F_3 (II). However, these correlations are not large and at most account for only 14% of the variation. This should not prevent selection for high U/P segregates with good seed weight.

In summary, Riley reported that the appearance of instability in the inheritance of the high-lysine trait complicated its effective use but advanced generation bulks with nondented seed of crosses involving the Ethiopian high-lysine *hl* parents have stabilized to produce a few lines moderately high in estimated lysine but with rather small seeds. Crosses involving P 721 and lines from an *hl* parent appear to be promising. There appears to be greater opportunity to evaluate the expression of the high-lysine character in crosses after one or more generations of selfing. Correlations between seed weight, U/P ratio and protein content were generally low and linkage between these traits would not appear to be a major obstacle in making gains from selection.

Porter and Axtell (1977) reported inducing corneous high-lysine mutants of sorghum by treating seeds of P 721 with the chemical mutagen diethyl sulphate. Putative mutants were identified by screening M_2 seeds from individual self-pollinated M_1 heads over a light box. Putative mutants were grown in head rows in the M_3 and M_4 generations and self-pollinated heads harvested. Nonviable mutants or "mutants" classified as opaque in the M_3 generation were discarded. Corneous segregates at the M_2 and M_3 generations were compared to opaque samples from the same heads. Mutants at M_4 were compared to opaque P 721 and its normal-lysine, non-opaque sib. Approximately 70% of the mutants evaluated initially in M_2, M_3, or M_4 generations had lysine as % of protein greater than 2·5 (156 mg/g total N).

Fifty-six mutants were superior in protein quality (lysine greater than 2·5% or DBC greater than 40) in at least two generations. Thirteen of the 56 mutants which were homozygous and corneous at M_4 were selected. Lysine (% of protein) of the selected mutants averaged 2·65% at M_3 and 2·61% at M_4; lysine as % of sample averaged 0·36 (360 mg/100 g) in M_3, but because of reduced protein in M_4, lysine averaged 0·29 (290 mg/100 g) in that generation.

Mutants varied considerably in the degree of corneous character. Immature kernels were always less corneous than mature kernels. Most modified mutants with lysine approaching that of their opaque counterparts were intermediate in corneous character. Of the 13 homozygous corneous mutants selected, two were similar in appearance to the non-opaque, normal sib of P 721. Seven mutants were intermediate and four were only slightly modified in corneous character; Porter and Axtell suggested that mutants approaching the opaque in lysine content, but with modified, corneous endosperms should be pursued.

Rabson (1976) in reviewing the subject commented upon the apparent difficulty of combining high-lysine sorghum genotypes with appropriate genetic backgrounds in order to provide lines competitive in acceptability with normal cultivars.

Sullins and Rooney (1977b) reported that when high-lysine sorghum mutants SC 1030 5 4 and P 721 were compared to a normal sorghum (SC 0170 derivative), SC 1030 5 4 was lowest in test weight, 1000-kernel weight and density; P 721 was intermediate, and the normal sorghum highest. SC 1030 5 4 had a dented, floury endosperm and the protein contained over 219 mg/g total N compared to the normal grain's level of 128 mg/g total N. P 721 had a full, floury endosperm with lysine equal to 193 mg/g total N. Amino acid analysis of P 721 indicated an increase in methionine, threonine and tryptophan with a reduction in proline, glutamic acid and leucine. Microscopy revealed the dented mutant contained fewer and smaller protein bodies (alcohol-soluble prolamines) than the normal grain. P 721 was intermediate between the normal and the dented grain for size and number of protein bodies. Both SC 1030 5 4 and P 721 contained at least equivalent or higher protein than normal sorghum.

Sullins and Rooney recommend microscopic examination of protein body concentrations in high-lysine sorghum as a nondestructive means of screening kernels normal in appearance for high-lysine content.

Ejeta and Axtell (1977) stated that farmers in Wollo, Ethiopia, where the two naturally occurring high-lysine lines IS 11167 and IS 11758 originated, indicated that because of their sweet flavour these were generally consumed at the dough stage after roasting.

Eight strains of sorghum, the high-lysine IS 11758, IS 11167, YM 3, the normal BG 5, BG 6 and BG 10 (code numbers assigned by the Ethiopian Sorghum Project at Alemaya), all collected from Wollo, in November 1973, and P 721 opaque (0) with its normal sib parent P 721 normal (N), were planted on the Purdue Agronomy Farm during the summer of 1975. All plants were grown under long day until about knee-high after which a "growth-chamber"-like structure was made to cover the six photoperiod sensitive Ethiopian varieties. The selected photoperiod treatment was 8 h of light and 16 h of dark and continued for 34 days before termination at the boot stage when the plants were returned to long-day conditions until maturity.

Three random heads for each of the eight strains were sampled 21 days after flowering (DAF), 31 DAF selected to represent the optimum texture for roasting at the dough stage, and 61 DAF. Kernels were dissected into endosperm and germ fractions, and weighed separately. All samples were analysed for protein (micro-Kjeldahl $N \times 6{\cdot}25$) and lysine (ion-exchange chromatography following acid hydrolysis). Carbohydrate analysis was applied only to the endosperm fractions 31 DAF and 61 DAF and of two high-lysine and two normal varieties. Starch was determined by optical rotation, (Shuman and Plunkett, 1964), amylose by blue colour measurement on the same calcium chloride solution prepared for starch, total and reducing sugars (Hodge and Hofreiter, 1962), sucrose by difference between total and reducing sugars, water-soluble polysaccharide (WSP) concentration on a 10% ethyl alcohol extract on the residue from the total and reducing sugars extraction. Statistical analysis was conducted with four fixed factors: kinds (normal and high-lysine), varieties (8 in number) within kinds, maturity stages (3 in number) and three plants sampled for each maturity.

Highly significant differences for 50-kernel weight and 50-kernel endosperm weight were found among varieties and among maturities indicating that the high-lysine and the normal varieties behave differently for these characters at different stages of grain development. Means for 50-kernel weight and endosperm weight at 21 DAF were not significantly different, nor between high-lysine and normal varieties at 31 DAF except for YM 3 which had lower kernel weight in all three stages of grain development. As maturity proceeded, P 721 O was not significantly different from P 721 N in whole kernel weight and endosperm weight whereas the Ethiopian high-lysine varieties, because of their dented kernels showed significantly lower kernel and endosperm weights than their normal counterparts. Dry matter accumulation in the normal was significantly higher than in the high-lysine. This difference appeared to be due to some limitation on starch synthesis between 31 DAF and 61 DAF in the high-lysine Ethiopian varieties.

Mean squares from the analyses of variance for 50-kernel germ weight and germ % for varieties and maturities were highly significant. For germ %, kinds × maturity interactions were highly significant, and varieties × maturity interactions were significant. Duncan's multiple range test showed no difference in germ weight between the mean value of *hl* and normal varieties in all three stages of grain development. P 721 O showed no difference in germ %, calculated on the basis of whole kernel dry weight, but the Ethiopian *hl* varieties while not showing an absolute increase in germ weight produced differences in relative germ size (germ %) at 21, 31 and 61 DAF. Though not statistically significant, absolute germ size in the *hl* Ethiopian varieties was consistently greater than in the normal *hl* varieties. Overall, both germ weight and germ % for all genotypes increased as the season progressed.

At all three growth stages, P 721 O and P 721 N were not significantly different in H_2O content but the Ethiopian *hl* varieties were significantly higher in moisture than the normal varieties.

Analyses of variance on % protein over all three stages showed varieties and maturities to be highly significant, and kinds × maturity and varieties × maturity interactions to be significant. The mean comparison of high-lysine and normal genotype for % protein, lysine as mg/100 g sample and mg/g total N in the defatted endosperm and germ at 21 DAF, 31 DAF and 61 DAF are presented in Table 3.20.

The Ethiopian *hl* varieties at all stages were significantly higher in endosperm protein than the normal varieties; the highest endosperm protein, 15·3% (average of three *hl* lines) was obtained at 61 DAF. P 721 O was significantly lower in endosperm protein than P 721 N. Protein in the endosperm of the *hl* varieties from Ethiopia increased consistently during the maturity stages; in the normal varieties protein in the endosperm decreased between 21 DAF and 31 DAF and then increased to 61 DAF but at 61 DAF the absolute protein content of the *hl* varieties (average 15·3%) remained significantly higher than the normal varieties (average 11·8%).

P 721 N showed the same pattern of protein synthesis as the normal Ethiopian varieties.

The mean squares for varieties and maturities were significant for lysine both as % of protein and % of sample. Kinds × maturity interaction was significant for lysine as % of sample but not significant for lysine as % of protein. Variety × maturity interaction was significant for lysine both as % of protein and % of sample.

Lysine as % sample and % protein were not significantly different at 21 DAF between normal and *hl* varieties. P 721 O showed an increase over P 721 N for these characters in all three maturity stages. At 31 DAF, the milk-dough stage favoured by the farmers in Wollo, Ethiopia, lysine was maximum both as % sample and % protein in the *hl* lines. After 31 DAF both normal and *hl* lines decreased in lysine concentration but the *hl* lines remained significantly higher than the normal lines at 61 DAF.

The excellent performance in feeding trials of the *hl* lines IS 11167 and IS 11758 (Singh and Axtell, 1973a) could thus be attributed to the significantly higher % germ, higher absolute protein content and lysine as % protein. This combination makes the *hl* varieties from Ethiopia superior to normal varieties and unique relative to any endosperm mutant in sorghum, or maize.

The carbohydrate composition of the genotypes at 31 DAF is shown in Table 3.21.

P 721 O and P 721 N were not significantly different. The Ethiopian *hl* varieties contained significantly less starch than the normal varieties, amylose as % of starch was significantly lower, total sugar content was 5 × higher, reducing sugars were 3 × higher, sucrose was 10 × higher. These high levels could explain the special sweet flavour claimed by the Wollo farmers for their high-lysine sorghum. Water-soluble polysaccharides (WSP) were significantly higher in the *hl* varieties than in the normal at 31 DAF. If the same relation held true in mature kernels it would indicate a reduction in the synthesis of starch from WSP in *hl* varieties.

High-lysine Sorghum

Biological Evaluation

Rat trials

Singh and Axtell (1973a, d) compared two high-lysine sorghum lines, IS 11167 and IS 11758 with three normal lines, IS 2319, IS 2520 and IS 1269 in isonitrogenous diets for weanling rats at approximately 10% protein level against a protein diet at 13·3% protein fed over 28 days. Oswalt (1973b) at Purdue University had established IS 2319 as being of good nutritional quality. The 100-seed weights, of the high-lysine lines, compared to an average of 31 genotypes grown over 2 years at six locations (normal sorghum) were:

	(g)
IS 11167	2·78
IS 11758	2·45
Normal sorghum	2·75

The protein and lysine content of the grains and the rations, and the rat weight gain, feed efficiency (F/G), and Protein Efficiency Ratio (PER) are given in Table 3.22.

Weight gain of rats on the IS 11758 diet was nearly double that of the IS 2319 diet and 3 × higher than the normal average. The rats on IS 11167 gained 71% the weight of these on IS 11758 but this was still significantly superior to the normal sorghum and 36% better than IS 2319. PERs for both high-lysine lines were higher than for the average sorghum but lower than for casein.

In another trial high-lysine sorghum IS 11758 was compared over 28 days with normal sorghums, IS 2319 and IS 1484, and with Opaque-2 maize, normal maize and casein in diets which were not isonitrogenous and contained 94% whole grain, 2% vitamins and 4% minerals. The protein and lysine content of the grains and the diets, with rat weight gains and F/G are shown in Table 3.23.

The biological quality, as measured by rat weight gain, of IS 11758 was 4 × higher than the average of IS 2319 and IS 1484 and over 3 × higher than IS 2319. Is 11758 (18·4% protein) was equivalent to Opaque-2 maize (at 12·1% protein). The F/G value of IS 11758 indicated that less than half as much feed was required/unit gain as was needed on normal sorghum diets.

Feeding trials similar to those described above were conducted at a different laboratory with eight rats/group over 2 weeks using the high-lysine IS 11758, a sugary line IS 5376 and two normal sorghums IS 2319 and IS 2520 (Singh and Axtell, 1973d). The results at 2 weeks were comparable to the earlier findings at the same period.

High-lysine and Sugary Lines

Rats

Singh and Axtell (1973d) compared the five sugary lines IS 4526, 4668, 5376, 5614 and 5623 with one high-lysine line, IS 11758 and five normal lines, IS 2319, 5568, 2520, 0057, 8313, and 6901 all grown at the Purdue Agronomy Farm, during 1972, against a casein control, in diets for male weanling rats, six to the group over 21 days (except for IS 5623, fed for 12 days only) at a protein level of approximately 12·5%. The composition of the grain, in protein, lysine and oil, and of the seed in protein and lysine with the F/G and PER are shown in Table 3.24.

The weight gain at 3 weeks of the rats on the high-lysine line was 3 × higher than the mean value of the normal line, and very significantly higher than each normal line including IS 2319 known to be of good nutritional value. The gains from the sugary lines IS 4526, 5614 and 5623 were significantly better than the average of the normal but only IS 4526 was significantly better than IS 2319. Weight gains from the casein diet were all significantly higher than any of the sorghum lines. Feed consumption was significantly higher on the IS 11758 (high-lysine) and sugary lines 4526 and 5614 than on the normal sorghum and the sugary lines 4668 and 5376.

F/G* was nearly 3× higher for the normal sorghum lines than for the high-lysine line which was higher than the F/G for casein, but not significantly. Sugary lines IS 4526, 5614 and 5623 gave significantly lower F/G than the average of the normal sorghum lines. The PER of high-lysine IS 11758 was more than double that of the normal sorghums. Sugary IS 4526, 5614 and 5623 were significantly higher than the average normal sorghum; casein was significantly higher than all the sorghum lines.

In a second experiment feeds of the high-lysine IS 11167 and 11758, the sugary IS 4526, 4668, 5376, 5614 and 5623 and the normal IS 2319, 2520 and 1269 grown at Puerto Rico were compared with casein, all at approximately 10% protein level over 28 days. The composition of the grain, and of the feed, weight gain, feed consumed, F/G and PER are shown in Table 3.25.

Weight gain results were in line with those in the first experiment except that IS 5376 produced better gain in the second trial. Feed consumption, F/G and PER followed the same general pattern but high-lysine IS 11758 showed a 15% improvement over its PER in the first trial. Correlation coefficients for this isonitrogenous trial showed that lysine (as % of protein and % of sample) correlated significantly with rat weight gain, feed consumption and PER.

Axtell (1976) reported that the superior amino acid composition of the chemically induced P 721 opaque was reflected in biological evaluations. When isonitrogenous (10% protein) diets prepared from P 721 opaque were fed to weanling rats (21-day old male albino in groups of 20) the average rat weight gain was three times those fed diets containing the normal sib. The PER for P 721 was 1·8 compared to 0·9 for the normal sib. The F/G for the opaque was 5·7 compared to 10·6 for the normal sib. Consequently it might be anticipated that high-lysine sorghum would be superior to normal sorghum protein in the diets of both man and domestic animals.

The Purdue group (Axtell 1976) also compared *hl* sorghum IS 11758 with P 721 opaque in rat feeding experiments. When weanling rats were fed isonitrogenous (10% protein) diets the average gain in weight of rats fed P 721 opaque diets was 65% of those fed IS 11758 rations.

Chicks

In other biological evaluation studies day-old male White Mountain chicks were randomly sorted and fed four different isonitrogenous diets in which sorghum provided all of the protein. The four variables in the experiment were four different sorghums: (A) sorghum P 721 normal sib;

* (Authors' note: F/G is a confusing ratio; the higher its value, the poorer is the feed quality).

(B) a commercial sorghum RS 610; (C) IS 11758 the Ethiopian *hl* sorghum; and (D) P 721 opaque. The lysine content (g/100 g of protein) of the four diets were respectively: (A) 1·98; (B) 1·81; (C) 3·1; and (D) 2·74. F/G ratios were (A) 5·67, (B) 5·45, (C) 3·53 and (D) 3·84 indicating significantly superior weight gains of the chicks fed the two high-lysine sorghums IS 11758 and P 721 and the slight superiority of the naturally occurring IS 11758 over the chemical mutant P 721 opaque.

Featherston *et al.* (1975a) compared the high-lysine sorghum IS 11758, grown in Puerto Rico in the first half of 1974 with two "low-tannin" normal sorghum varieties, RS 671 and RS 610 in chick-feeding trials. The protein content (Kjeldahl, AOAC 1960) and lysine, methionine and cystine contents (auto-analyser following acid hydrolysis) of the grains are shown in Table 3.26.

In Experiment I, IS 11758 was compared with RS 671 in diets varying in protein content as shown in Table 3.27. The effect on weight gain and F/G is also shown.

In Experiment II, IS 11758 was compared with RS 610 in diets formulated to contain 15% protein. The diets and effect on weight gain and F/G are shown in Table 3.28.

In Experiment III, IS 11758 was compared with RS 610 supplemented with L-lysine HCl. The diets and effect on weight gain and F/G are shown in Table 3.29.

When all the protein in the diets was provided by sorghum, whether the sorghum was fed on an equal weight basis or an equal protein basis, the high-lysine sorghum produced weight gains approximately 3 × those from the normal sorghum with approximately 50% less feed required/unit gain. Supplementing the normal sorghum with lysine to attain the level in IS 11758 resulted in weight gains and feed efficiency comparable to those from IS 11758. Supplementation of both normal and high-lysine grain with 0·2% DL-methionine had little beneficial effect on growth. Chicks fed IS 11758 supplemented with safflower seed meal or soybean meal to provide 15% protein grew 2·5 and 1·9 × faster, respectively, and were significantly more efficient in feed utilization than chicks fed the normal sorghum similarly supplemented.

Featherston *et al.* (1975b) reported briefly that when a naturally occurring high-lysine sorghum gain IS 11758, and a chemically induced high-lysine mutant (P 721 opaque) were compared with commercial sorghum in diets for day-old chicks in which the sorghum provided all the protein, IS 11758 and P 721 O produced, respectively, 3 × and twice the weight gain obtained from normal sorghum with 40–50% less feed required/unit gain.

Nonprotein Nutrients

Waxy and Sugary Endosperm

Karper (1933) reported that waxy endosperm in sorghum stained red with iodine and was inherited as a simple Mendelian recessive to normal starchy endosperm.

In 1963, Karper and Quinby reported that sugary endosperm sorghum (the character called "dimpled" by Indian workers, Ayyangar *et al.*, 1936) wrinkled as they matured and varieties with corneous seeds and juicy stems wrinkled most. Sugary seeds were about 25% smaller than normal seeds and contained at least twice as much total sugars as normal seeds. The sugary (*su*) character was found to be a simple recessive to normal. No linkage was found between the *su* gene and waxy (*wx*) or between *su* and yellow y^2 seedlings.

Gorbet and Weibel (1972) studied the genetic relationships and inheritance of six endosperm types, normal (Wheatland, Bok 8); waxy (Dwarf Redlan); sugary (Sugary (R), Sugary (W), Sugary (F)); yellow (Boky 29, Roky 16); dent (W, white grain), dent (R, red grain); and defective (Defective (D), Defective (R)). These 12 lines were intercrossed in all possible combinations with reciprocals. Crosses were also made on MS lines of normal, waxy and yellow endosperm types to obtain additional F_1 plants for backcrossing.

The waxy, yellow and normal endosperm types all had plump kernels and normal appearance; sugary lines ranged from extremely shrivelled and translucent to plump with a depressed germ; Dent (W) had a dent in the upper back side of the kernel, Dent (R) frequently was not dented but was hollow in the middle of the endosperm; defective endosperm contained a collapsed caryopsis with an endosperm only one-quarter of the volume of normal. Defective endosperm appeared to be conditioned by recessive genes at a single locus. Dent, sugary and waxy endosperm were inherited as independent simple recessives. Several genes could be involved in the expression of yellow endosperm. No linkage was indicated.

Sugary endosperm in sorghum was first described by Patel and Patel (1928) in India cited by Singh and Axtell (1973b). The variety Vani had wrinkled seeds, pitted at the tips and somewhat flinty. In India it was eaten in the immature condition after parching.

Singh and Axtell (1973b) classified over 9000 lines in the World Collection of sorghum, based on the endosperm phenotype, as corneous, floury, waxy, sugary, yellow, or hollow types. Five sugary endosperm lines were identified, IS 4526, 4668, 5376, 5614 and 5623. Their phenotypic characteristics, origin, protein (micro-Kjeldahl N × 6·25 DWB) and lysine content (ion-exchange chromatography) are shown in Table 3.30.

The endosperm of sugary lines may be partly or fully glassy and translucent with a great range of variation in the dimpling of the kernel. The content of reducing sugars, sucrose and water-soluble polysaccharides (WSP) (Shannon, 1968), and of amylose and starch (Shuman and Plunkett, 1964) for the five sugary lines, two high-lysine lines, IS 11167 and IS 11758 and a normal sorghum IS 8313 are shown in Table 3.31.

Total carbohydrate content was calculated by adding the values for reducing sugars, sucrose, WSP and starch; total sugar by adding the values for reducing sugars and sucrose. The reducing sugar content of both high-lysine lines was equivalent to normal lines. Sugary lines IS 4608, 5376 and 5623 were extremely high in reducing sugars, ranging from 2·52 to 2·69% as compared to 0·34% in the normal IS 8313. Sugary lines IS 4526 and 5614 were low in reducing sugars but high in sucrose. IS 5623 was high in both sucrose and reducing sugars. Both high-lysine lines were high in sucrose content (3·08 and 2·61%) compared to 1·03% in normal sorghum; sucrose in the sugary line ranged from 0·68 to 2·1%.

The lowest total sugar content was in the sugary line IS 4526 (1·96%) and the highest in IS 5623 (4·39%). Total sugar in the high-lysine IS 11167 was nearly 2·5 × higher, and the high-lysine IS 11758 nearly twice as high as in normal sorghums.

The starch content was much reduced in all the sugary lines (ranging from 22·2% to 34·0%) except IS 4526 (58·5%) which was similar to normal sorghum (60·8%). The high-lysine lines were also comparable to normal sorghum in starch content. The amylose content of starch in the sugary lines was 22–62% higher than the normal line with IS 4526 the highest at 40·6% of starch. The high-lysine lines were equivalent to normal sorghum.

The WSP of the sugary lines IS 4668, 5376, 5614 and 5623 ranged from 29·4% to 39·6% compared to 1·1% in normal sorghum. The WSP of sugary line IS 4526 and the two high-lysine lines were very similar to normal sorghum.

The carbohydrate content of the sugary lines IS 4668, 5376, 5614 and 5623 differed totally from normal sorghum and there seems no doubt that they could be classified as sugar mutants. The line IS 4526 had 65% higher sucrose and 62% higher amylose content than normal but was equivalent to normal sorghum in reducing sugars, WSP and starch. It might therefore be classified either as sugary endosperm (similar to su_2 in maize) or as a high amylose sorghum.

Except for their higher sucrose the total carbohydrate content of the two high-lysine lines was comparable to normal sorghum content. This suggests that the high lysine content is not due to a pleiotropic effect of a block in carbohydrate synthesis.

Singh and Axtell (1973c) reported that the sugary lines IS 4526, 4668, 5376 and 5623 grown at Lafayette, Indiana and Puerto Rico, and IS 5614 grown only at Puerto Rico were all higher in lysine than the average of sorghum lines in the World Collection. (The dimpled endosperm leads to a bigger proportion of embryo in the seed and thus to a higher lysine content expressed both as % of seed and as % of protein). Methionine and cystine, measured at Lafayette only, were also higher than the average.

The F_2 segregation data for three sugary lines IS 4668, 5376 and 5614 × PP3Rms_3 and a back cross Redlan *ms* × IS 4526 supported the simple recessive mode of inheritance of the *su* gene reported by Karper and Quinby (1963) and Gorbet and Weibel (1972). The ratio of segregating kernels was 3 plump/1 mutant. The carbohydrate analysis of the endosperm of plump and mutant kernels is shown in Table 3.32.

Total sugar content in the mutant sugary kernels was 2–4 × that of the normal plump kernel. The average WSP of the sugary mutants was about 22% compared to a normal 0·89%. Starch content was 71% lower than normal endosperm tissue but there was no difference in amylose content. The carbohydrate analysis confirmed the classification of sugary lines in the World Collection.

β-Carotene

Worzella *et al.* (1965) studied the inheritance of β-carotene content (Booth, 1957; AOAC, 1960) in the grain of crosses between 10 sorghum lines that varied in β-carotene content. The parents and F_2 and F_3 lines were grown in the Beqa'a, Lebanon during 1963 and 1964. The parent lines varied in β-carotene content from 0·022 mg/100 g in Redlan and Wheatland (the only two named varieties) to 0·3 mg/100 g in line 66. The amount of β-carotene was influenced by environmental factors; line 8 in 1963 averaged 0·2 mg β-carotene and in 1964 0·17 mg/100 g. The mean β-carotene contents of the F_2s were intermediate to the mid-point of the parents in 1963 and below it in 1964. Positive correlation coefficients were obtained between endosperm colour and β-carotene content in the F_2. Relatively few genetic factors appeared to be involved in the inheritance of β-carotene content in varieties which contained high concentrations of β-carotene.

Singhania *et al.* (1970) reported the results of a full diallel set of crosses made between five inbred lines of sorghum which varied for total yellow pigments, sown in the summer of 1968 at IARI Regional Research Centre, Rajendranagar. Estimation of total yellow pigments (by extraction in petroleum ether plus acetone mixture (1/1) followed by photoelectric colorimeter reading using blue filter) was taken as a criterion for β-carotene content since "it is known that there is a high correlation between β-carotene content and total yellow pigments".

When the readings were plotted on a graph as suggested by Jinks (1954), the distances between the points suggested substantial genetic diversity. The regression line slope indicated that gene effects were of the additive type without much interaction. Singhania *et al.* indicated the feasibility of breeding and selecting for β-carotene content.

Niacin

Tanner *et al.* (1949) reported that the niacin content (microbiological assay) of the sorghum variety Westland ranged over 2 years from 4·3 to 4·9 mg/100 g, and of the variety Cody ranged over 5 years from 6·7 to 7·3 mg/100 g.

Seed from the F_1 generation of the Westland × Cody cross grown at Hays Kansas contained 4·6 mg niacin/100 g. Seed from 335 plants of the F_2 generation ranged from 3·8 to 10·4 mg/100 g. Analyses were made of 157 F_3 heads grown on 55 head rows. In one group only did the F_3 heads have an average niacin content below that of the F_2 heads and in one F_3 row a head containing 12·4 mg niacin/100 g was found; more than 5·0 mg/100 g higher than the Cody (initial high-niacin) parent.

Niacin content thus appeared to be an inherent varietal characteristic but was not associated with any observable plant character which might be used as a genetic tester.

Agronomic and Environmental Effects

Proximate Composition

In 1957 Adrian and Sayerse reported the proximate composition of samples of *Sorghum guineensis, S. durra* and *S. caffra* grown under standardized conditions, including about 12 t/ha manure (type unspecified) at Bambey, Senegal, and of Bassi sorghum of peasant cultivation obtained from the market at Dakar. The mean weight of 1000 seeds and specific gravity of the samples were:

	Mean weight 1000 seeds (g)	Specific gravity
Bassi	28·1	0·77
Bambey grown	29·0	0·74

About 60% of the weight could be attributed to the kernel and about 40% to the husk.

The moisture, ash, total reducing substances (after hydrolysis of polysaccharides), fat (carbon tetrachloride extraction), protein (Kjeldahl N × 5·83), minerals (AOAC, 1945), vitamins (microbiological assay) and amino acids in the protein (microbiological assay) of the four types of sorghum are shown in Table 3.33.

The Bambey grown *S. guineensis* and *S. durra* sorghums were appreciably higher in protein than the peasant cultivation. Phosphorus and sodium in all the Bambey grown samples, and calcium in *S. durra*, were higher than in Bassi. The Bambey samples contained more thiamine and niacin than the Bassi but there was little difference in riboflavin and pantothenic acid.

Genetic factors appeared to influence the quantity of protein in sorghum but there was remarkably little difference in the amino acid composition of the protein of the three types grown at Bambey suggesting to Adrian and Sayerse that genetic factors did not influence protein quality. In comparing peasant and experimental cultivation, threonine and phenylalanine were distinctly lower, and lysine somewhat lower, in Bassi than in Bambey grown sorghums.

Improvements in peasant cultivation could be made but it was uncertain whether the differences in composition found would have any physiological importance.

Effect of Location on Protein and Amino Acids

The agronomic traits of two high-lysine sorghums IS 11167 and IS 11758 grown at Lafayette, Indiana, and Puerto Rico were compared by Singh and Axtell (1973c) with the average of 31 genotypes grown over six locations and 2 years and are presented in Table 3.34. Environment influenced all the characters slightly, and plant height was substantially reduced in Puerto Rico.

The agronomic traits of five sugary lines from the World Sorghum Collection, IS 4526, 4668, 5376, 5614 and 5623 grown at Lafayette, Indiana, and Puerto Rico compared to an average of 31 normal sorghums grown over six locations and 2 years (Schaffert, 1972) are presented in Table 3.35 (data from Singh and Axtell, 1973c).

Lysine, as % of protein and as % of sample, and oil content, were higher in the sugary lines than in the normal, except for IS 4668 which was equivalent to the normal. The 100-seed weight was lower, at an average of 2·11 g, than the normal sorghum, at 2·75 g. Protein and lysine content per seed of the sugary lines were not higher than normal sorghum. IS 4526, 5614 and 5623 were non-restorer (B-lines) type while IS 4668 and 5376 were restorer (R-lines) type.

Ten brown-seeded sorghum hybrids, grown at four locations in Louisiana over the 2 years 1972–73 were analysed for crude protein (Haji-Hashim and Tipton, 1973). The hybrids with their mean protein content are shown in Table 3.36.

There were significant differences in protein content among the hybrids at each of the four locations. As an average of locations ACCO R-1093 was highest followed by RS 700 and Niagara Shoo Bird; the seven remaining varieties showed no significant differences except for KB Golden Grain BR which was significantly lowest in protein content. The locations also had a significant effect on % protein. The year effect was not significant but the year × location, location × hybrid and location × year × hybrid were all significant.

Per cent protein and protein yield are shown in Table 3.36. As an average of locations, ACCO 1093 again ranked first but RS 700 ranked below the other hybrids. The year × hybrid, year × location, location × hybrid and location × year × hybrid interactions were all significant.

The effect of different levels of N, P_2O_5, and K_2O fertilizer on the protein content and pounds of protein/acre of Funk BR 79 was studied at two locations over 1972–73. At one location, highest grain protein and protein yield resulted from 200 lb N, 60 lb P_2O_5 and 60 lb K_2O/acre (224 N, 67 P_2O_5 and 67 K_2O kg/ha). At the other location, the highest grain protein resulted from 150 lb N, 60 lb P_2O_5 and 60 lb K_2O/acre (168 N, 67 P_2O_5, 67 K_2O kg/ha); yield of protein was not significantly influenced by fertilizer application.

There was no consistent relation between % protein and grain yield but there was a positive association between % protein and protein yield and a very high positive association between grain yield and protein yield indicating that the most important factor influencing protein yield was the grain yield.

Rojas and Maranville (1975) conducted preliminary experiments to determine any relation which might exist between K and protein in sorghum seed from Nebraska Random Mating Population NP4.

The ranges and mean seed protein (Udy) as modified by Koening and Maranville (1970), and potassium (extraction method Isaac and Kerber, 1971, and atomic absorption spectrophotometer) of 80 observations are shown below:

	Range	Mean
Seed K (%)	0·261–0·664	0·447
Grain protein (%)	9·5–15·6	12·5

Correlation between K and protein, both as % grain, was highly significant and positive.

The effects of lodging on yield, protein and test weight of RS 626, medium maturity, and RS 671, medium late maturity, sorghums were studied by Larson and Maranville (1975, 1976) at the State Experimental Station, Mead, Nebraska. The experiment consisted of a control and (1975) five treatments: root lodging (1) 45° at heading, (2) early dough and (3) hard dough; stalk break of the peduncle at (4) early dough and (5) hard dough.

Later experiments (reported 1976) included a sixth treatment: stalk break at heading. The later experiments confirmed the earlier and showed that lodging decreased yield (14% moisture basis) test weight, and total protein production (Udy 1956, modified for sorghum by Maranville, 1970) compared to the control. Lodging tended to increase protein quantity in the grain, significantly in the earlier experiments but not in the later, probably as a result of the lower grain yield. Yields tended to be lower with earlier and increased severity of lodging, the lowest yield (31% lower than the control) being found with stalk break at heading.

Test weights were reduced by the lodging treatments and might be the primary reason for the reduction in grain yield.

There were no significant differences between the two hybrids.

The National Institute of Nutrition at Hyderabad (personal communication, 1974) reported that 34 sorghum lines grown in the kharif season showed lower protein contents than when grown in the summer season. Lysine, as % of protein, was higher in kharif crops than in summer crops in 18 out of 28 lines. Leucine content, however, showed no particular seasonal trend.

As the level of N fertilizer was increased, the protein content of the sorghum grain increased and the lysine content as % of protein decreased. No significant effect on protein content or quality resulted from the use of P and K fertilizers.

Brinsmead *et al.* (1970) reported that the late maturing sorghum strains,

TX 671, DeKalb E57, Pioneer 846 and NK 310, grown under irrigation at two sites in Australia, failed to show any marked yield advantage over the early maturing TX 610, TX 626 and NK 212. Grain N (method not stated) varied widely, ranging from 1·45 to 2·15% but the means for the various strains showed little difference, ranging from 1·77 to 1·91%. The early maturing strains were recommended in preference to the late maturing which required a larger supply of irrigation water and were more likely to suffer sorghum midge infestation.

Sorghum is grown in India in the rainy (kharif), winter (rabi) and summer seasons (Deosthale and Mohan, 1970). In the rainy season of 1967, grain samples of two hybrids, CSH 1 and CSH 2, two select lines IS 511 (S46) and IS 3797 (S370) and Swarna were obtained from 8 to 13 locations (not all the lines were grown in all the locations) where they had been grown under uniform recommended farm practices. Samples of traditional cultivars (local control) from 24 locations including the 13 referred to, were also obtained.

The mean protein content (Kjeldahl N × 6·25) with ranges in parentheses, for the five cultivars and the local controls are shown below:

	Protein (%)
Local controls	8·5 (6·0–13·7)
IS 511	9·1 (6·4–11·8)
IS 3797	9·1 (5·0–11·5)
CSH 1	8·6 (4·9–10·3)
CSH 2	8·6 (5·5–11·0)
Swarna	9·3 (6·5–11·4)

In the samples of the same variety from different locations, variations in protein content were significant e.g. Swarna 11·4% at Nanded, Maharashtra, and 6·5% at Karad, Maharashtra.

Average lysine and leucine content (microbiological assay) in the protein, means in parentheses were:

	Leucine (mg/g N)	Lysine (mg/g N)
Local controls	629 (513–746)	125 (72–175)
IS 511	606 (455–742)	106 (74–147)

IS 3797	631	116
	(569–674)	(73–160)
CSH 1	706	130
	(625–836)	(92–196)
CSH 2	734	146
	(645–827)	(105–266)
Swarna	668	116
	(554–841)	(76–166)

Varietal and locational differences in lysine (% protein) were negatively correlated with protein content. Varietal differences in leucine content were significant but not the effect of location on leucine. There was no significant correlation between protein content and leucine (%).

The traditional varieties, local controls, at the locations Nanded, Somanathpur, Siriguppa, Khargone, and Guntur contained relatively high levels of lysine, 144 mg/g N or above, and low levels of leucine, 578 mg/g N or lower.

Ramasastri and Mohan (1969) and Mohan and Deosthale (1969) reported on the protein, lysine and leucine as % protein (analytical methods not stated) of the sorghum varieties CSH 1, CSH 2, IS 511 (sel. 46), IS 3797 (sel. 370) and IS 3924 (sel. 413), grown with local varieties as a check at different (unspecified) locations in India. With the exception of IS 3924, the varieties appeared to be the same as those reported in Deosthale and Mohan (1970).

In 1964, Miller *et al.* reported that preliminary trials in 1958 indicated that the hybrid sorghums RS 590, RS 610, RS 630 and RS 650 grown at five locations in Kansas contained lower protein in the grain (AOAC, no date, $N \times 6{\cdot}25$ and corrected to 12·5% moisture) than the standard varieties Combine 7078, Plainsman, Redbine 60, Westland, Martin and Combine Kafir 60 grown at the same locations. Protein within varieties varied among locations.

In 1961 and 1962 more extensive trials were conducted. In 1961, 19 hybrids were grown at eight locations with differing soil conditions; protein ranged from 6·6 to 12·8%. Location differences were shown by a range in average protein from 8·2 to 11·3%.

In 1961, the hybrid Pioneer 851 with a mean of 10·5% protein differed from 8·2% grown on an irrigated test-plot, to 12·8% grown on a test-plot summer-fallowed in the previous season. From another test-plot also summer-fallowed in the previous season, Pioneer 851 yielded 10·7% protein. Hybrids high in protein content at one location tended to be relatively high in protein at other locations. Location effects and hybrid differences were both significant.

In 1962, 33 hybrids were grown at seven locations. Protein content in the grain ranged from 5·9% (Frontier 400 B) to 12·1% (TE 55) with an average range at all locations of 8·2–10·0%. As in 1961, the protein content within a hybrid differed from approximately 3·0 to 6·0 percentage points depending on location and variety. Both location differences and hybrid differences were highly significant.

The average protein percentages in 1961 and 1962 of nine hybrids grown at the same locations are shown below:

Hybrid	1961 (Protein %)	1962 (Protein %)
Genetic Giant R 103	10·2	9·1
KS 603	10·1	9·4
KS 602	10·0	9·5
KS 651	9·7	8·8
Genetic Giant R 106	9·6	9·2
RS 610	9·3	8·3
TE 77	9·0	8·4
RS 650	8·8	8·5
TE 66	8·8	8·2

The samples higher in protein in 1961 tended to be higher in 1962, though the rank order changed.

In 1961, the average protein content from samples grown on an irrigated test plot at the location Finney was, at 9·9%, lower than the average, 10·3% from a test plot area, also at Finney, summer-fallowed in the previous season. In 1962 the trend was reversed, presumably because the summer-fallowed test plot which had not been fertilized was more N deficient in 1962 than in 1961.

The effect of adding N fertilizer at 90 kg/ha increased the protein content of DeKalb F 62 A from 7·8 to 10·1%, and yield from 4·4 to 5·6 tonnes/ha; 134 kg N/ha increased protein content from 7·8 to 11·4% and yield from 4·4 to 5·8 tonnes/ha.

When the 1962 performance of 33 hybrids for yield and protein content was compared on irrigated plots, continuously cropped plots and summer-fallowed plots, there was a trend towards higher protein content as yield decreased. The means for all varieties under the different conditions were:

	Protein (%)	Yield (t/ha)
All locations	9·0	5·8
Irrigated	8·3	8·41
Continuously cropped	9·5	4·94
Summer-fallowed	9·5	4·27

The yield figures were based upon a bushel weight of 56 lb.

Busson *et al.* (1962) illustrated the relation between variety and environment in the production of protein in sorghum grain by comparing varieties grown at Parc du Pharo, Marseille, France and at Bambey, Senegal. The results follow:

Varieties	Protein content (N × 5·83% DWB)		
	Soil with fertilizer		Soil without fertilizer
	Pharo	Bambey	Bambey
	1960	1957	1959
54 28	16·9	14·4	12·17
54 3	16·1	13·95	—
50 55	16·1	14·2	—
50 5	15·5	13·9	—
AS 18H	14·0	11·8	10·95
51 63	13·8	12·7	—
Congossane	13·2	10·2	10·6
51 50	11·5	—	9·05

In the three series, the varieties ranked in essentially the same order, but those grown at Pharo were higher in N content than those at Bambey and, in general, those grown with added fertilizer were higher in protein content than those without.

The same authers compared different botanical types of sorghum (classified following Snowden), *Sorghum guineense*, *S. margaritiferum*, *S. conspicuum*, *S. roxburghii*, *S. gambicum*, *S. exsertum*, *S. notabile*, *S. caudatum*, *S. durra*, *S. cernuum* and *S. subglabrescens*. All originated in Africa, mostly from Senegal, South Africa and Tchad. The soil of Bambey was sandy, of the type known as Bior, but improved by the rotation groundnut–pearl millet–fallow and by annual additions of fertilizer. The rainfall at Bambey was about 650–700 mm.

The Kjeldahl N, and amino acid composition (auto-analyser), mean and ranges, are shown in Table 3.37.

It would appear that the composition of amino acids was influenced more by variety than environmental condition.

Worker and Ruckman (1968) studied variations in the protein content (N × 6·25%, DWB) of six varieties or selections, and 35 hybrids of sorghum grown at the Imperial Valley Field Station, El Centro, California, the South West desert area, for two years or more from 1961 through 1966.

Plants were two row seeded in April and July at 6·8 kg/ha on a 44 m bed. The crop received 8–10 irrigations. Soil type and cultural practices were the same for both plantings within a year and varied little between years. Nitrogen was applied as ammonium nitrate at 280–336 kg/ha prior to preparation of the seed bed. Air temperatures were recorded at 203 cm above soil level.

April plantings emerge in a cool soil and air temperature, and mature in warm soil and hot air temperature. July plantings are the reverse, being planted in warm soil and hot air temperature and maturing in cool soil and air temperature. The varieties and hybrids responded alike to each of the two environmental conditions. The April plantings averaged 10·1% protein (range 8·5–11·7%) and the July plantings 14·8% protein (range 10·7–21·5%).

Variation in protein content between years was slight, ranging from 9·4 to 11·0%. The difference among cultivars was significant.

The protein levels of all cultivars were higher, and varied more from year to year, in the July plantings than in the April plantings.

In the July plantings, averages ranged from a low of 11·7% in 1961 to a high of 18·4% in 1964. In 1961 protein content ranged from 10·8% for Pfister PAG 515 to 12·8% for DeKalb E 56 A; in 1964 the range was from 14·8% for IV 601817 to 21·5% for NK 222. The difference among cultivars was significant in each year.

When three cultivars, Meloland (variety) RS 610 and TE 66 (hybrids), were compared over 3 years, protein in the grain increased as yield decreased. There were more variations with the hybrids than in Meloland, an open-pollinated variety. Protein content correlated significantly with seed size and air temperature, and negatively with yield. Temperature appeared to influence protein content which is highest with cooler weather after anthesis.

Purdue University (1974) issued the first report on International Protein Yield and Quality Trials on grain sorghums. The mean protein content for all tropical and other locations was 11·0%. IS 2319 maintained the highest mean protein percentage (12·7%) and the lowest mean grain yield (2219 kg/ha) for all tropical locations.

Table 3.38 shows the grain yield ranking, the mean protein content with ranges, and the protein yield, mean and ranges of 24 entries from various locations. Protein yield reflected grain yield, and NK 300, BR 64 (US hybrids) and IS 4225 which ranked 1, 2 and 3 in grain yield ranked similarly in protein yield and had the greatest range in yield indicating their ability to perform well under satisfactory environmental conditions. The spread of the range in percent protein demonstrated the influence of environment, yield of plots and maturity on protein synthesis. IS 7822 with a spread of only 1·8 and 956036 with a spread of 2·1 showed the greatest stability in protein concentration and also ranked 10 and 9 respectively in protein yield.*

Protein and Amino Acids

Fertilizer Effects

Africa

Rai (1964) studied the effect of date of sowing, the varieties and the spacing of sorghum grown in the rainlands of the central clay plain of the Sudan on crude protein content and nitrogen accumulation.

* (Author's note: a Spearman rank correlation indicated that protein yield and grain yield were functionally and proportionally related at a 1% level of significance; protein % and protein yield were inversely related at a 5% level of significance; and protein % and grain yield were inversely related at a 5% level of significance.)

The soil of the area is fairly heavy montmorillonitic and calcareous clay, non-saline and alkaline. The average rainfall is about 700 mm, mostly from the end of June to early October. The rate of evaporation is low in the rainy season but increases rapidly from October onwards.

The experiments extended over 1959, 1960 and 1961. The crude protein (Kjeldahl N × 6·25) increased when the sowing date was delayed from the end of June to the end of August but yield decreased when sowing was delayed after the fourth week in July.

Neither crude protein content of grain nor plant population at harvest were significantly affected by different seed rates of 2·14 and 4·28 kg/ha.

The grain protein content and grain yield of the 11 varieties grown in 1960 are shown in Table 3.39.

The range was from 8·2 (Al Fadni)–13·0% (Gassabi). A negative relationship was observed between grain protein and grain yield.

Rai (1965a) also reported experiments with Um Benein variety of sorghum grown in the central rainlands of the Sudan under varying levels of N and P_2O_5 singly and in combination over 1959, 1960 and 1961. The sorghum was sown by the middle of July each year and continuously cropped. Phosphate had no effect on crude protein content and the application was significant on N accumulation only in 1961. The effect of N fertilizer application is shown in Table 3.40.

There was no apparent effect on crude protein in 1959 but an improvement over the control was seen in 1960 and 1961. Mean protein content in the grain generally decreased every year. Nitrogen application increased N accumulation in the grain in all three years.

The variety Wad Aker was grown with varying levels of N and P_2O_5 singly and in combination over the years 1959, 1960 and 1961 in a rotational system. The preceding crop in 1959 was sorghum, in 1960 2-year continuous sorghum, and in 1961 groundnut, and the dates of sowing were July 23, 6 and 7 respectively. Phosphate application had no significant effect on crude protein in the grain and on N accumulation only in 1960. The effect of N fertilizer application is shown in Table 3.41.

Nitrogen fertilizer increased considerably the crude protein of grain in 1960, when the higher levels were used, but had only a slight effect in 1961 and none in 1959. In all three years, N fertilizer increased N accumulation in the grain.

The results suggested that N fertilizer application could improve crude protein content but the precise conditions were not obvious.

Doggett (personal communication) points out that protein content is influenced by N taken up, not necessarily by N applied to the soil. The amount and distribution of rain can greatly affect N accumulation, as also can availability of P, and no doubt of other elements also.

The effect of the date of sowing, from July 7 to August 13, and of N and P_2O_5 fertilizer application on yield, and N and P accumulation in Wad Aker sorghum grown in the central rainlands of the Sudan was reported by Rai (1965b). The treatments and results are shown in Table 3.42.

Delay in the sowing date resulted in a marked decreased in yield and an increase in N and P content but the increase was not proportional to the decrease in yield. Phosphate did not affect N content of grain and N fertilizer substantially increased the N content of the grain of the early sown crops only.

The benefit of fertilizers in rain-fed sorghum can be obtained only when a favourable moisture supply is present.

Blondel (1970) described experiments carried out at Kotiary (Eastern Senegal) and Nioro du Rip (Sine-Saloum) in 1968 in which varying applications of N, 0, 50, 100, 150 and 200 kg/ha were made at sowing to sorghum, designated 51–69. The effect on grain yield and protein content is shown in Table 3.43.

The results showed that where the soil is poor in naturally occurring N, heavy application of N can increase yield without depressing N in the grain, so leading to tripling, and almost quadrupling, the yield of protein/ha.

Asia

Deosthale *et al.* (1972a) studied the effects of three Indian locations, Coimbatore, Karad and Parbhani on the protein (AOAC 1960), lysine, isoleucine, leucine, methionine and tryptophan contents (% protein, microbiological assay) of eight sorghum varieties CSH 1, CSH 2, Swarna, Ganeri 2, CO 9, Shenoli 4 2, G 3 and PJ8K. The results are presented in Table 3.44.

The protein contents (mean of three locations) varied from 8·1% (CSH 1) to 10·7% (G 3) with locational values of (mean of eight varieties) 10·5% at Coimbatore, 6·4% at Karad and 11·2% at Parbhani. The varietal effect and locational effect on protein content were both significant. The location × variety effect was also significant. The low protein samples at Karad contained higher concentrations of the five amino acids as % of protein. In the experiment as a whole, there were inverse correlations between protein and lysine, protein and methionine, and protein and tryptophan. Lysine and tryptophan were positively correlated but there was no significant correlation between lysine and leucine, protein and leucine, and protein and isoleucine.

The contents of Ca, P and Fe (AOAC, 1960), Mg, Cu, Mn, Zn and Mo (colorimetric, Sandell, 1959) in the same varieties were all, with the exception of copper, significantly affected by location (see Table 3.45). Calcium content in the Karad samples were significantly higher than in samples from Coimbatore and Parbhani. P, Mn and Mo contents were highest in the Coimbatore samples and lowest in those from Karad. Varietal differences in mineral contents were not significant. Fe was the only mineral significantly, at the 1% level, associated with protein.

In the second experiment, CSH 1 and Swarna were grown at Hyderabad under five levels of N fertilizer 0, 50, 100, 150 and 200 kg/ha, (N_0, N_{50}, N_{100}, N_{150}, N_{200}) added as ammonium sulphate just before sowing. The mean protein content of CSH 1 was 9·6% and of Swarna 12·0% with yields of 2344 and 1666 (units not stated). Protein content from applications up to 100 kg N/ha was not affected significantly nor were lysine, leucine, methionine and tryptophan as % protein.

Grain yield increased rapidly from applications up to N_{100} and thereafter more slowly from 750 (unit not stated, probably kg/ha) to 2680 at N_{200}. The protein contents from N_{150} and N_{200} were also significantly higher than from N_{100}. An increase in protein was associated with decreases in lysine and tryptophan both of which were significantly lower as % protein from N_{200} than from N_0 (see Table 3.46). The correlation between protein and lysine was negative and significant; lysine and tryptophan were positively correlated.

Swarna had a significantly higher content of Ca, and lower contents of P, Cu, Zn, Mn and Mo than CSH 1 but the mineral content was unaffected by N fertilizer except for Mn (mean of the two varieties) which was reduced significantly from 1·11–0·88 mg/100 g.

In the third experiment, CSH 1 and Swarna were grown under N levels of N_0, N_{60} and N_{120} and at three plant population levels, 90 000, 140 000 and 280 000 plants/ha. Protein content and leucine in the protein increased significantly with N fertilizer but lysine decreased. Plant population had no significant effect on protein, lysine or leucine contents. Swarna had a higher protein content (9·0%) than CSH 1 (8·0%) but the lysine and leucine contents of the two varieties were not significantly different. The N × variety × population interaction was not significant for protein, lysine or leucine. The most productive combination would be N_{120} and 140 000 plants/ha as this would give the highest output of grain and protein/unit area.

In a further experiment CSH 1 was grown under four levels of P_2O_5 (P_0, P_{40}, P_{80} and P_{120} kg/ha) and three levels of K_2O (0, 30 and 60 kg/ha) and no significant effect was found on the protein, lysine and leucine contents. K_2O had no significant effect on the mineral composition but P_2O_5 significantly increased Ca, Mg, P, Fe and Mn in the grain. Both P_2O_5 and K_2O additions progressively increased grain yield, K_2O to a greater extent than P_2O_5 (see Table 3.47).

Deosthale *et al.* (1972a) concluded that the genotype, location and the level of N fertilizer were all important factors influencing the quantity and quality of sorghum protein but the mineral composition was chiefly influenced by location and level of P_2O_5 fertilizer.

Warsi and Wright (1973) reported two experiments carried out in the rainy (kharif) seasons of 1968 and 1969 on the sandy loam soils of the Indian Agricultural Research Institute, New Delhi, on the effects of rates and methods of N application on the quality of the grain of CSH 1 sorghum. In 1968, two rates, 40 (N_{40}) and 80 (N_{80}) kg N/ha and three methods of application, were used with three additional treatments. In 1969, there were three rates, 60 (N_{60}), 120 (N_{120}) and 180 (N_{180}) kg N/ha and five methods of application, three as in 1968 and two additional. The sorghum was seeded at the rate of 12 kg/ha and received 11 kg P/ha as concentrated superphosphate (10% P) band placed. Treatments and results are shown in Table 3.48.

Protein (N × 6·25% corrected to 14% moisture) in the grain increased significantly in both seasons with an increase in the rate of N application ranging in 1968 from 9·3 to 11·8% and in 1969 from 9·1 to 11·4%, irrespective of the method of application. These findings were in conformity with those of Miller *et al.* (1964), Waggle *et al.* (1967b) and Reddy and Hussain (1968).

During both seasons foliar fertilization produced higher protein yield than other methods of application. Feeding N through foliage, M_3, increased the protein content in the grain by 3, 7, 2 and 6% over, respectively, the control, M_2, M_4 and M_5 applications through the soil in 1969.

The three split applications, M_4, increased protein by 22% over N_0, 6% over M_2 and 5% over M_5. N applied late in the season was stored in the grain to give a higher protein content but did not give higher grain yield.

Protein yield was increased by N application but foliar fertilization depressed protein yield with the higher rates of N, though N application through the soil tended to enhance it. Treatments N_{120} and N_{180}, half through soil and half through foliage, M_3, reduced protein yield by 19 and 13% respectively, compared to N_{60} applied in the same manner. The solubility fractions (modified Mendel-Osborne, Nagy *et al.*, 1941) for the six treatments are shown in Table 3.49. Albumin and globulin together accounting for 20% of total protein were little altered by N application. The prolamine increased with increasing levels of N, particularly at the N_{180} level. Foliar application, M_3, increased the prolamine fraction and the protein content when compared to N application all at the time of planting M_1. The glutelin fraction increased little beyond N_{60} and foliar application of N did not increase the glutelin compared to N_0.

The amino acid composition (chromatography, Moore and Stein, 1954a) as affected by N_{180} compared to N_0 is shown in Table 3.50.

N_{180} increased the protein content from 8·9 to 11·6%, and increased glutamic acid, proline, alanine, aspartic acid and cystine but decreased lysine, histidine, arginine and phenylalanine (mg/g N). The leucine/isoleucine ratio increased from 2·5 (N_0) to 3 (N_{180}).

Doggett (personal communication, 1977) states that the more N is taken up, the more prolamine is formed in the grain. The important times of N uptake are at floral initiation (influencing the potential number of seeds), and at grain filling (influencing N in the grain).

Singh and Bains (1973) reported an experiment conducted in the rainy (kharif) seasons of 1967 and 1968 at the Indian Agricultural Research Institute, New Delhi in which Swarna and CSH 1 sorghums were grown on well-drained sandy-loam soil of Yamuna alluvial origin under 0 (N_0), 60 (N_{60}), 120 (N_{120}) and 180 (N_{180}) kg N/ha and plant population rates of 272 000, 136 000 and 91 000/ha. Half the dose of N was broadcast before sowing and half top-dressed at the knee-high stage. P_2O_5 at 60 kg/ha and K at 42 kg/ha were added as basal dressing. In 1967, the total rainfall was 994 mm and no irrigation was required; in 1968 the rainfall was 530 mm and irrigation was given at the (a) panicle initiation and (b) grain-filling stages. The effect on grain yield and protein (method not stated) as an average of the 2 years, of variety, fertilizer treatment and plant population is shown in Table 3.51.

CSH 1 gave significantly higher (39% more) grain yield than Swarna with higher panicle weight, longer panicles and larger number of grains/panicle. Increasing applications of N up to N_{120} significantly increased grain yield which resulted from an increase in length and weight of panicle and in number of grains/panicle. Response equations showed that for yield improvement the economic optimum dose for CSH 1 would be N_{132} and for Swarna N_{140}.

Swarna contained 1% more protein in the grain than CSH 1. Application of N up to N_{180} improved the grain protein content over N_0.

As plant population decreased, yield decreased significantly but grain protein increased.

Ananda Rao and Reddy (1973) compared the effect of additions of N at 0 (N_0), 50 (N_{50}), 100 (N_{100}) and 150 (N_{150}) kg N/ha, as ammonium sulphate, applied half at sowing and half 35 days after sowing to Swarna, CSH 1 and to local types PJ7R and M35 1 grown on sandy-loam soil at Rajendranagar, Hyderabad in the summer of 1970. Phosphorus at 50 kg/ha as superphosphate and K at 30 kg/ha as potassium sulphate were applied as a basal dressing. Grain yield, 1000-seed weight and protein content (Kjeldahl $N \times 6{\cdot}25$) resulting from the various treatments are shown in Table 3.52. Grain yield increased significantly up to N_{100} in CSH 1 and Swarna but only up to N_{50} in PJ7R and M35 1. Variations in 1000-seed weight among the varieties were negligible. Crude protein increased with increasing levels of N from a mean of the varieties of 9·4% from N_0 to 11·7% from N_{150}. Albumin, globulin, prolamine and glutelin fractions were extracted by "the conventional shaking method described by Baudet *et al.* 1966". Only about 33% of the protein was extracted by the method and no significant differences in these fractions were noted among the varieties.

Roy and Wright (1973) reported experiments at the Indian Agricultural Research Institute, New Delhi in which CSH 1 hybrid sorghum was sown in July 1967 and 1968 to give a plant population of 120 000/ha. The soil was low in organic carbon, N and P and slightly alkaline. Fertilizer treatments were 0 (N_0), 60 (N_{60}) and 120 (N_{120}) kg N/ha and 0 (P_0) and 26 (P_{26}) kg P/ha. A uniform application of 42 kg K/ha was broadcast by machine prior to sowing. The P was drilled in a band approximately 5 cm deep and 5 cm to one side of the seed row. Half the N was broadcast at sowing and the remainder top-dressed when the crop was 60–75 cm high.

Treatment × year interaction was not significant and the data were therefore presented on a pooled basis. The grain yield and Kjeldahl N in the grain are shown in Table 3.53. N_{60} increased grain yield by 47% and grain

N by 14·5 %. Applications of N_{120} further increased yield by 8·8 % and grain N by 10·8 %. Nitrogen content in the grain was unaffected by the addition of P.

Sorghum CSH 1 was grown at Andra Pradesh Agricultural University, Hyderabad during the rabi (winter) season of 1966–67 under differing levels of N fertilizer and plant densities (Reddy and Hussain, 1968). A basal dose of farmyard manure at 12 t/ha was applied before final ploughing. Nitrogen (as ammonium sulphate) was added at the rate of 112 (N_{112}), 140 (N_{140}), 168 (N_{168}) and 196 (N_{196}) kg/ha, half at sowing 10 cm below the seed zone, and the other half 35 days after sowing 10 cm away from the plant row at a depth of 10 cm. P_2O_5 (as superphosphate) and K_2O (as potassium sulphate) were applied at half the N dosage with the first application. Twenty-one days after sowing, the crop was thinned to 1, 2 and 3 seedlings/hill, S_1, S_2 and S_3.

The effect of the treatments on grain yield (/hill and/hectare), 1000-seed weight and grain protein (Kjeldahl N) are shown in Table 3.54.

Maximum protein content in the grain was found at the highest level of N fertilizer (N_{196}) followed by N_{168}, N_{140} and N_{112} in that order. As plant density increased, protein content in the grain decreased.

The weight of 1000 seeds increased as N fertilizer level increased and decreased as plant density decreased. Grain yield/hill and/ha increased as N fertilizer level increased. The highest grain yields were obtained from the density of S_2. This density was not significantly superior to that of S_3 but both gave grain yields significantly higher than S_1.

No significant interaction was found between the levels of N fertilizer and plant density treatments for grain yield or protein content.

Gupta and Gupta (1972) examined the effect of varying levels of N fertilizer, 0, 75, 150 and 225 kg/ha (N_0, N_{75}, N_{150}, N_{225}) on the crude protein (Kjeldahl N × 6·25); true protein (total N minus nonprotein N × 6·25, nonprotein estimated by the method of Kulkarni and Schonie, 1956); lysine and leucine (microbiologically); methionine and tryptophan (colorimetrically); (see Table 3.55) and on the (Mendel-Osborne) protein fractions (Nagy *et al.* 1941) (see Table 3.56) of pot grown varieties, CSH 1 and Swarna. The optimum level of N fertilizer in improving crude protein and true protein was N_{150}; the varieties did not differ significantly from one another. Compared to the control, the concentration of amino acids in the grain increased with increasing N. Lysine in the protein increased at each level, leucine decreased until the highest level was reached when it rose above the control level, methionine and tryptophan were highest from N_{150}. The major protein fractions (see Table 3.56) were prolamines and glutelins; the increase in total protein from N application was associated mainly with an increase in prolamines. Gupta and Gupta (1972) stated that N application was not particularly effective in improving the quality of the protein.

In related experiments, reported by the same authors in 1975, the same varieties of sorghum were pot-grown (14 kg soil/pot) with a basal dose of 75 kg N, 60 kg P_2O_5 and 40 kg K_2O/ha, and urea applied as a foliar spray at the flowering stage, N_0 (water) N_2 (2% urea) and N_4 (4% urea). The grain yield, crude and true protein in the grain, and leucine, lysine, methionine and tryptophan in the grain and in the protein are shown in Table 3.57.

Treatment N_2 increased yield and protein content in relation to N_0; there was no further advantage from N_4. Leucine, lysine and methionine in the protein decreased as a result of urea spraying.

Akotar and Deshmukh (1974) reported experiments on the effect of the commercial insecticide BHC 10% on the uptake of N by PSH 2 [*sic*] sorghum in field trials on black cotton soil which had received before sowing a basal dose of 75 kg N, 50 kg P_2O_5, 25 kg K_2O and 55 cartloads of farmyard manure (the weight of FYM is not given, nor its composition)/ha. BHC was added at the levels of 0, 5, 10, 20 and 50 kg/ha and the effect on yield and (Kjeldahl) N in the grain is shown below:

Treatments BHC (kg/ha)	N in grain (kg/ha)	Yield (kg/ha)
0	16·73	2986
5	21·78	3385
10	25·94	3740
20	27·22	3949
40	22·83	3499
50	18·45	3014

Compared to the zero control, BHC at the rates of 10, 20 and 40 kg/ha significantly increased N in the grain but the increases from 5 kg/ha and 50 kg/ha were not significant. All additions of BHC except the highest significantly increased grain yield.

Palaniappan and Vijayakumar (1976) studied the effect of time of harvesting on the nutritional value of the sorghum cultivars CSH 5 and CSV 4 grown at the Tamil Nadu Agricultural University, Coimbatore in kharif, 1975. The fresh weight of 10 heads, weight and volume of 1000 fresh seeds adjusted to 14% H_2O, moisture, weight of grain from the 10 heads, protein content (N × 6·25) and total carbohydrate content (Somogyi, 1952) were measured 75 days after sowing and at 5-day intervals up to 105 days after sowing. There was little change in the values found after the ninetieth day and Palaniappan and Vijayamukar suggested that these particular cultivars reached their physiological maturity about 95 days after sowing and could then be harvested without loss in yield or nutritional quality.

Australia

The Allora District of the Darling Downs in Queensland, Australia was the site of trials reported by Littler (1967) from 1961 to 1965 with Alpha, an open-pollinated variety of sorghum, and TX 610 hybrid, grown on black earth soil in which the K and P content was adequate, with four rates of N fertilization:

N_0 (0 kg/ha)
N_{28} (28 kg/ha)
N_{56} (56 kg/ha)
N_{84} (84 kg/ha)

The fertilizer was applied in two forms; as urea at 63, 125 and 188 kg/ha, and as ammonium sulphate at 138, 276 and 413 kg/ha. Littler stated the results at equivalent N levels were closely similar.

In 1963 two trials were conducted in one of which sorghum followed sorghum and in the other sorghum followed "Panicum (*Setaria italica*)".*

In both trials fertilizer treatment gave a marked increase in both yield and grain protein (analytical method not stated) over no treatment, especially where sorghum followed sorghum. Economically, for yield, the best treatment was that of TX 610 and N_{188} of urea.

The improvement in protein content with fertilizer treatment is shown in Table 3.58.

Grain protein from the unfertilized plots of sorghum following "Panicum" was almost 3 × higher than from the unfertilized plots when sorghum followed sorghum. An increase in yield was also obtained.

In 1964, the sorghum was sown following a poor crop of "Panicum". The rainfall was more than sufficient and growth was impeded in the early stages by waterlogged soil. Ammonium sulphate which had given similar results to urea in the earlier years, in 1964 produced higher yields than urea but fertilizer treatment gave no significant increase in grain protein content. TX 610 outyielded Alpha but was inferior in grain protein.

In 1965 moisture was adequate to March but dry weather later affected the trial. Yield and % protein increased significantly from all fertilizer treatments and there was no significant difference between ammonium sulphate and urea.

MacKenzie *et al.* (1970) conducted sorghum trials in the Ord River Valley area of Australia where the soil type is heavy, calcareous and alkaline (pH7–9) Cununurra clay. The summers are hot and wet and the winters relatively warm and dry. TX 610 and Alpha grain sorghums were grown with sub-plots in which (1) the stubble was left and (2) the stubble was removed after harvest, and with sub-sub-plots treated with N at levels of 0, 56 and 112 kg/ha. Seeding rates were 9 kg/ha following side band application of superphosphate at the rate of 502 kg/ha. The crop was irrigated at 5–10-day intervals and the sorghums were grown for two cycles.

The grain yield of the TX 610 was 42 to 45% higher than the grain yield of Alpha in both cycles but grain yield was lower in cycle 2 than in cycle 1. Nitrogen application increased N content (% of grain) and yield.

Kondos (personal communication, 1978) studied varietal differences in protein content and essential amino acids, as affected by location and/or N application, in sorghum grain samples produced in southern Queensland,

* (Authors' note: if the attribution to *Setaria italica* is correct, then "Panicum" is presumably being used as a synonym for millet, a somewhat confusing use of nomenclature.)

Australia. Six sorghum varieties designated E 57, T 671, P 846, P 579, F 64 and C 44B were grown with four levels of N application, 0, 68, 136 and 273 kg/ha. The effect on protein (N × 6·25 air dry basis, McKenzie and Wallace 1954) is shown in Table 3.59.

The mean protein differences among varieties were relatively small and not significant but the range of individual values within varieties was significant. Two of the highest proteins occurred in sorghum grown with O fertilizer treatment. Statistical analysis of the data in Table 3.59 indicated that the protein content of sorghum grain in these experiments was not significantly influenced by variety or N application.

The mean values and ranges of essential amino acids (auto-analyser following acid hydrolysis, tryptophan colorimetrically, Miller, 1967) in 88 samples are presented in Table 3.60.

The maximum range in the concentration of essential amino acids was large and in several cases exceeded 50% of the mean value for each amino acid. Threonine showed the largest variation, 76·4% of the mean concentration of that amino acid. All three basic amino acids, lysine, histidine and arginine were negatively correlated with protein content whereas the other amino acids showed positive correlations. Only leucine and threonine were significantly affected as a result of N application and variety respectively.

Europe

Dechev (1971a) reported a field trial conducted at the B. Kolarov Higher Institute of Agriculture, Plovdiv, Bulgaria, under dry farming conditions during 1966–1968. The treatments were:

(1) no fertilizer (control)
(2) P_2O_5 at 60 kg/ha once at deep ploughing
(3) as (2) plus 60 kg N/ha at pre-sowing
(4) (2) plus 60 kg N/ha top dressing at first hoeing
(5) (2) plus 90 kg N/ha at pre-sowings
(6) (2) plus 90 kg N/ha top dressing at first hoeing
(7) (2) plus 60 kg N/ha at pre-sowing and 30 kg N/ha top dressing at first hoeing
(8) (2) plus 60 kg N/ha at pre-sowing and 60 kg N/ha top dressing at first hoeing.

Phosphorus alone did not increase sorghum grain yield and the lowest crude protein in the grain was obtained when P was applied alone. Grain yield was increased by 3–10% when N was applied as top dressing at first hoeing (treatments 4, 6, 7 and 8) compared to the pre-sowing application of N (treatments 3 and 5). N-P_2O_5 fertilizer application increased crude protein in the grain by 3–40%, ash content by 5–14% and carotene by 10–19% and reduced starch content by 1·6–6% compared to the unfertilized control.

In 1973 Dechev reported the effects of N fertilizer application to sorghum, without irrigation, during 1966–1968. The treatments were:

(1) no fertilizer (control)
(2) 60 kg N/ha, pre-sowing dressing
(3) 90 kg N/ha, pre-sowing dressing
(4) 60 kg N/ha, pre-sowing dressing plus 60 kg N/ha at first hoeing.

N fertilizer raised the crude protein content in the grain and reduced starch, the effect being more marked as the application level increased. Plants receiving no fertilizer accumulated 50–90% of the total amount of N, 54–70% of the P and 81–91% of the K; those receiving N fertilizer accumulated 65–95% of the N, 57–100% of the P and 93–99% of the potassium.

Central America

Aragón H. and Bressani (1965) examined the effect of different types of fertilizer on the protein quality of Hegari sorghum and a yellow and white variety of maize grown in Guatemala. The fertilizers were:

(1) nitrogen 19%; superphosphate (P_2O_5) 13%; potassium (K_2O) 7% (NPK)
(2) NPK + minor elements (a mixture of Ca, Mg, S, Fe and Mn with traces of B, Cl, I, Zn, Cu and Mo
(3) organic fertilizer (dried chicken manure)
(4) organic fertilizer + minor elements
(5) minor elements
(6) control

The effect on nitrogen content, ether extract, crude fibre and ash (AOAC, 1950) (10% moisture basis) are shown in Table 3.61.

N content increased significantly with treatments (1) and (3), ash with treatment (3). The amino acids, tryptophan, lysine, isoleucine and leucine (mg/g N) (microbiological assay) did not differ significantly among the treatments (see Table 3.61).

Feeding trials were carried out using male and female Wistar weanling rats, four of each sex to the group over 35 days. The basal diets consisted of the test sorghum, cleaned and ground, 1961 or 1962 crop, from the different treatments, adjusted with maize starch to bring all the diets to the same (unstated) protein level, and 90% of the whole diet. The remaining 10% of each diet consisted of 4% mineral mix, 5% cottonseed oil, 1% cod liver oil and vitamins solution. The weight gains and PER are shown in Table 3.61. The variations in weight gain were significant at the 1% level among the different diets. Diets based on sorghum from treatments (2), (4) and (5) containing the minor elements gave weight gains in both 1961 and 1962 superior to those based on treatments (1) and (3). The variations in PER among the different diets were significant. Taking both years together, the

PERs of the diets containing sorghum fertilized with the minor elements averaged 0·81 compared to an average of 0·71 for those, including the control, based on sorghum not fertilized with the minor elements. However, the PER of the control diet, treatment (6), was superior to those from treatments (1) NPK and (3) organic manure. In summary, N fertilization increased the amount of protein but decreased its biological quality for rats.

United States of America

Burleson *et al.* (1956) reported that Redbine 66 grain sorghum planted at 15·7 kg/ha in rows 50 cm apart in a Brennan fine sandy loam in the lower Rio Grande Valley of Texas with fertilizer applied 5–8 cm below and 25 cm to the side of the seed, gave significant increase in content and yield of protein in the grain.

The fertilizer was N (as ammonium nitrate, 33% N), phosphoric acid (as superphosphate, 45% P_2O_5) and potash (as muriate of potash 60% K_2O). The N and P were each applied at (1) 0, (2) 67 and (3) 134 kg/ha and the K at 0 and 67 kg/ha singly and in all possible combinations. The test area was irrigated five times.

Grain yield (at 15% moisture) increased from (1) 2461 to (2) 4837 kg/ha and to (3) 5578 kg/ha.

Protein (N × 6·25%) in the grain was increased from (1) 6·6 to (2) 7·9% and (3) 10·4%. The protein yields were: (1) 174, (2) 403 and (3) 638 kg/ha.

Waggle *et al.* (1967b) investigated the effect in Kansas of N fertilization levels of 0, 89·7 and 134·6 kg/ha applied as ammonium nitrate (35% N) at planting to hybrid sorghums RS 610 at Oberlin and Centralia, and to Dekalb F62A at Newton on amino acid composition (auto-analyser excluding tryptophan) in the grain and in the protein (N × 6·25%). The data are presented in Tables 3.62 and 3.63.

As the level of N fertilizer increased, the level of protein and of essential amino acids in the grain also increased significantly. However, amino acids in the protein did not increase proportionally with the protein. As N fertilizer level increased, there were significant increases (mg/g N) in glutamic acid, proline, alanine, isoleucine, leucine and phenylalanine and decreases in lysine, histidine, arginine, threonine and glycine; aspartic acid, serine, half-cystine, valine, methionine and tyrosine were not significantly altered by N fertilization. Only glutamic acid, alanine and leucine (mg/100 g grain) were significantly affected in the location experiments but, as mg/g total N, lysine, histidine, arginine, glutamic acid, glycine, alanine, half-cystine, methionine and leucine were all significantly affected.

Campbell and Pickett (1968) reported the effect of N fertilization on protein content and lysine (mg/g N) of 18 inbred lines from the World Sorghum Collection grown at Lafayette, Indiana, during the summer of 1966. In the first test, a uniform fertilizer treatment of approximately 28 kg K_2O, 14 kg N and 57 kg P_2O_5/ha was applied before planting. Three N rates, residual, 112 and 224 kg/ha were applied as ammonium nitrate in a band 5 cm deep and 5 cm from the row when the plants were 8 cm tall (296, 400 plants/ha). In the second test, the field in which plantings were grown had been brought to a very high fertility level over nine years. The amounts of P_2O_5 and K_2O available in the spring of 1965 were 488 kg/ha and 536 kg/ha respectively. In the fall of 1965, 672 kg K_2O/ha werc applied and before planting in 1966, 409 kg N/ha were added.

The varieties, with their protein content (micro-Kjeldahl N × 6·25), the weight of 100 seeds, yield/panicle, lysine (%) and protein × lysine are shown for both tests in Table 3.64 listed in descending order of protein content as found in Test 1.

In Test 1 (N fertilization) protein ranged from 11·7 to 16·3%, lysine from 102 to 157 mg/g N, and for protein × lysine from 1308 to 2567, grain weight/panicle from 6·1 to 75·1 g and 100 seed weight from 0·7 to 3·8 g. Several of the higher protein and lysine types were late maturing and had shrivelled seeds though some had mature seeds.

In Test 2 (high fertility conditions), the protein content ranged from 12·1 to 17·1%, lysine from 58 to 163 mg/g total N, protein × lysine from 702 to 2543, grain weight/panicle from 5·7 to 49·6 g and 100 seed weight from 0·7 to 3·5 g.

The highest grain weight per panicle of 75·1 g was found in RS 610 in Test 1; this cultivar was excluded from Test 2. The next highest grain weight/panicle in Test 1 was 69·9 g. There were significant differences among genotypes for 100-seed weight, grain weight/panicle, protein content, lysine in protein and protein × lysine in both the tests.

The combined analysis of variance for % protein, lysine in protein, protein × lysine and grain weight/panicle for the 18 strains at the two locations showed a significant interaction between location and genotypes for grain weight/panicle, lysine in protein and protein × lysine but no significant interaction for % protein. There was a significant location effect on each of the four characters. The mean values for thc 18 lines showed the % protein was 0·6% higher, lysine as % of protein 0·74% lower, protein × lysine 349 lower and grain weight/panicle 10·7 g lower when grown under high fertility conditions (Test 2). In the summer of 1966 severe drought conditions occurred from the time the plants were only a few cm tall until harvest and this might explain the relatively poor performance of the plants in Test 2. In Test 1 % protein and % lysine in protein were significantly and negatively correlated with grain weight/panicle.

Campbell and Pickett commented that protein and lysine in protein were negatively correlated but at a comparatively low level of significance. In Test 2, % protein was significantly and negatively correlated with the grain weight/panicle but the negative correlation of lysine in protein and grain weight/panicle was not significant. The relatively large genotypic differences for protein and lysine persisted under all fertility levels tested and were much more important than differences due to the difference in fertility treatments. No significant interactions of protein and lysine with fertility treatments were found.

Lutrick (1970) reported the effect of N fertilizer application on the yield and protein content of DeKalb BR 64 sorghum seeded at 11·2 kg/ha on April 16 1969 and April 3 1970 on a deep phase Red Bay fine sandy loam in Florida. The treatment and results are shown in Table 3.65. In 1969 rainfall was fairly well distributed but in 1970 rainfall in April and May was very low. Nevertheless, yields of grain from all treatments were comparable in both years. There were no significant differences in yield due to N application in 1969 and only between O and N application treatments in 1970. There were however marked increases in protein content (N × 6·25) at highest levels of N fertilizer; in 1970 protein increased by 40% from 8·26 to 11·40 with 448 kg N/ha.

Shoup (1970) reported that the sorghum hybrids Frontier 400C and Paymaster Kiowa grown at levels of 56, 112 and 224 kg N/ha on irrigated and non-irrigated land showed no significant differences in protein content and amino acid distributions.

Brawand and Hossner (1976) reported the effect of long-term farming practices on the protein (Kjeldahl N), P, K, Ca and Mg contents of the sorghum varieties NK 222 and NK 222A grown from 1968 to 1969 and 1970 to 1971, respectively, in Texas.

The soil at the experimental site was a Houston Black clay (fine montmorillonitic thermic family of Udic Pellustarts) with pH 7·8–8·1.

The treatments included continuous versus rotations of grain sorghum, wheat and cotton with and without N and P fertilizer and cattle manure. Most treatments had been applied annually for over 20 years. The lowest percentages of protein in the grain were associated with unfertilized plots and the highest levels from treatments which included 11·2 t of cattle manure plus N and P, but fertilizer effect on grain protein content appeared to be variable and inconsistent. The 1969–1971 data showed that grain protein content varied little while grain yields varied widely. The measurable response to fertilizer P was below expectation.

Day and Tucker (1977) described experiments at Tucson, Arizona, in which NK 280 sorghum was grown in (1) Comoro sandy loam soil and (2) Grabe silt loam soil with different irrigation and fertilizer treatments (a) well water with N 112, P 35 and K 70 kg/ha as control, (b) well water with N 224, P 73 and K 140 kg/ha, simulated municipal waste water and (c) treated municipal waste water from an activated sludge sewage plant, containing N 224, P 73 and K 140 kg/ha and with no added water or commercial fertilizer. Treatment (c) produced higher grain yields, 203 g compared to (b) 154 g and (a) 105 g, all per 0·28 m^2. The grain from all treatments contained similar proportions of total protein. The plant growth response to municipal waste water was similar in both types of soil. From soil type (1), the grain from treatments (b) and (c) contained less cystine, glycine and histidine than from treatment (a). All treatments in both soil types contained similar amounts of leucine, methionine, threonine and tyrosine.*

* (Authors' note: lysine was not mentioned).

Inorganic Elements

Kamalam (1964), in a field trial at Coimbatore, India, compared compost as the organic source of P and superphosphate as the inorganic source, both on an equal P basis, on the uptake of P by the sorghum varieties CO 1 and CO 18. The treatments and the effects on yield and grain P are shown in Table 3.66. In grain yield, CO 18 was significantly superior to CO 1. On an equal P basis, superphosphate was more effective than compost in raising grain P content. The maximum increase in grain yield for both CO 1 and CO 18 came from treatments (11) (16·8 kg/ha superphosphate and 33·6 kg/ha compost) and (13) 33·6 kg/ha superphosphate and 16·8 kg/ha compost) but the grain yield of CO 18 was double that of CO 1. The maximum uptake of P in the grain of both CO 1 and CO 18 came from treatment (13) but the uptake by CO 18 was nearly 3 × higher than that of CO 1. Both superphosphate and compost in some instances produced slight but irregular increases in potash, N and Ca content in the grain.

The National Institute of Nutrition at Hyderabad (personal communication, 1974) reported that 13 varieties of sorghum grown at two locations in the 1972 kharif season showed signficant locational differences in Mo and Zn content, and significant varietal differences in Fe content.

Vitamins

Deosthale *et al.* (1969) analysed the seeds of two yellow endosperm varieties of sorghum, IS 511 (sel. 46) and IS 3797 (sel. 370), grown at 33 locations representing nine states in India, for total β-carotene by a spectrophotometric procedure (Suryanarayana Rao *et al.*, 1968). A rapid screening test for total yellow pigment was also applied in which 2g of powdered sample was extracted overnight with 10 ml of petroleum ether/acetone mixture (1/1). The extract was separated and the optical density measured in a Klett-Summerson colorimeter with a No. 47 filter. The total yellow colour (Klett units) was plotted against the β-carotene content determined by the spectrophotometric procedure and showed a reasonable relationship.

The mean percentage moisture and protein (method not stated) and β-carotene content expressed as mg/100 g grain and as % of total yellow pigment for the two varieties are shown in Table 3.67. The β-carotene content of both varieties varied considerably at different locations. In IS 511 the content ranged from 0·01 mg/100 g at Parbhani to 0·07 mg/100 g at Hyderabad, the average for all locations being 0·04 mg/100 g. In IS 3797, the β-carotene ranged from 0·01 mg/100 g at Kalianpur to 0·06 mg/100 g at Bhavanipatna, the average for all locations being 0·03 mg/100 g. Moisture content and protein content also varied but there appeared to be no relation between β-carotene content and protein content. β-carotene content plotted as a percentage of total yellow pigment in the seed was fairly constant for each variety at several locations and Deosthale *et al.* suggested that the quick test for total yellow pigment could be used to screen varieties for β-carotene content.

Kapoor and Naik (1970) compared the β-carotene content, (column chromatography, Carotene Panel of Sub-Committee on Vitamin Estimation, 1955) and carotenoid pigments extracted by acetone/hexane (1/1), absorbancy measured at 430 μm in a Klett-Summerson colorimeter of three yellow endosperm sorghum varieties, IS 3922 (393), IS 511 (46) and IS 3924 (413) grown in pot-culture under differing levels of urea equivalent to O (N_0), 56 kg N/ha (N_{56}), 112 kg N/ha (N_{112}) and 168 kg N/ha (N_{168}). Each pot received a basal dose equivalent to 56 kg N/ha, 60 kg P_2O_5/ha, and 40 kg K_2O/ha. In addition to the soil treatments with urea, a spraying experiment was carried out in which one level of N only was used equivalent to 112 kg N/ha, half sprayed on the soil before sowing and half sprayed on the plants 70 days after sowing, after ear emergence. An equal quantity of water was sprayed on the control plants which received no urea. Soil application at the rate of 56 kg N/ha significantly increased the β-carotene content of all varieties over the control. In IS 3922 only the β-carotene content increased and not total carotenoids which also increased in the other two varieties. N_{112} did not effect a further increase and N_{168} reduced β-carotene and total carotenoids below N_0 in IS 3922 and IS 511. In IS 3924, the β-carotene content was very similar for N_{56}, N_{112} and N_{168}. The spray treatment, whether of urea or water, reduced β-carotene content below the control in all three varieties. Seeds were stored for 6 months at (a) refrigeration temperatures 0–4°C, (b) 20–25°C, (c) 28–32°C and (d) incubation temperature 37°C. β-carotene content was reduced during storage in all samples at all temperatures, but under refrigeration the reduction compared to the original control was about 30% in IS 3922, 24% in IS 511 and 40% in IS 3924 compared to losses of 60%, 40% and 90% respectively for storage at 28–32°C. Further reductions were suffered at temperatures of 37°C.

Kapoor and Naik (1970) also surveyed samples of sorghum lines IS 3922 (sel. 393) (four locations), IS 3922 (405) (three locations), IS 3797 (370) (eight locations), IS 511 (46) (six locations), IS 3924 (413) (three locations) and IS 84 (11), IS 944 (27), IS 504 (35), IS 815 (113), IS 2031 (133), IS 513 (53), IS 82 (1), IS 84 (22), IS 523 (67), IS 511 (51), IS 84 (19), IS 505 (38), IS 531 (77), IS 83 (8) and IS 3796 (361) (all one location only, Delhi). The β-carotene content ranged from 0·02 to 0·14 mg/100 g with a mean of 0·07 mg/100 g.

Ramasastri and Mohan (1969) reported that the β-carotene content (unstated method of determination) of the yellow endosperm sorghums, IS 511 and IS 3797 grown at more than 30 locations in India showed some variation, but it was not large within a region. In the case of both varieties the β-carotene content was related to the total yellow pigment in the endosperm (measured in Klett units). The data were presented graphically.

The National Institute of Nutrition at Hyderabad (personal communication, 1974) reported that none of the Indian yellow endosperm sorghum varieties investigated contained any significant level of β-carotene content. Of the 12 varieties found with β-carotene content ranging from

0·03 mg to 0·1 mg/100 g grain, none were of Indian origin; 10 were from Nigeria. Two of the lines IS 511 and IS 3797 with high β-carotene content were grown at 33 locations under uniform field trials during the kharif season of 1967. The mean values of β-carotene content were still high at about 0·04 mg and 0·03 mg/100 g respectively but the concentration varied considerably among locations. The maturity of the seeds and post-harvest weathering may have affected the carotene content.

Genetic Effects

Pearl Millet

Doggett (personal communication, 1977) stated that the pearl millet World Collection will probably have to be started again though the problem of how to maintain such a collection has not yet been satisfactorily resolved. Since pearl millet is a cross-pollinating crop it is possible to withdraw a very large number of different inbred lines from any given population. Doggett believes that an attempt will be made to maintain fairly large plots of populations collected from farmers' fields by intercrossing within the plots and also to maintain a set of inbred lines carrying valuable characters. The present pearl millet collection consists essentially of a number of useful inbred lines. Though collections are being made at a number of Indian institutions, and as stated above, at ICRISAT (see page 213), to the best of the authors' knowledge, no one has accepted the responsibility for establishing and maintaining a World Collection of the other millets, reviewed in this text.

Andrews and Majmudar (1975) stated that the range of genetic variability within the cultivated species of pearl millet is immense. Further variation exists in related species though successful crosses have been few. Protogyny, involving the hermaphrodite and staminate florets on the same inflorescence, allows selfing and the use of both artificial and natural cross-pollination techniques without the need for emasculation or genetic male sterility. The number of progeny which can be derived from a single plant or cross is large, one head will yield over 1000 seeds, and since tillering is common, several heads on one plant allow a single plant to be used simultaneously for different purposes.

Two main breeding methods are available: (1) traditional intercrossing, followed by inbreeding; and (2) recurrent selections in composites (populations), the latter method allows improvement in characters (such as grain yield) which are particularly dependent on the additive effects of minor genes (Andrews and Majmudar, 1975).

Protein and Amino Acids

Andrews *et al.* (1975) suggested that in breeding pearl millet it is difficult to select for very high levels of protein because of apparent negative correlation between yield and protein content. Selection for higher lysine is more feasible; the breeder can select for high grain yield, maintain moderate (*c*. 14%) protein, and at the same time seek to identify high-lysine lines.

Grain samples from pearl millet parents K1 4; PT 819/4, PT 852/2, PT 870 and PT 888 and their F_1 hybrids were drawn from two locations at the Agricultural College and Research Institute, Coimbatore, India in 1963 (Mahadevappa, 1967). The diallel cross data were analysed (Jinks and Hayman, 1953) and the genetic parameters estimated (Jinks, 1954). The mean protein percentages in the grain (Kjeldahl N × 6·25 DWB) from the 5 diallel cross for the two locations are shown in Table 3.68. Although, in general, the hybrids were either equal to or above the mid-parent value in protein content there was considerable environmental influence. In location 1, K1 4 × PT 819/4; PT 819/4 × PT 888 and PT 852/2 × PT 888 showed significant heterosis over mid-parent value but in location 2 the cross PT 852/2 × PT 888 gave a low protein value. The analysis of variance indicated considerable difference among parents and hybrids and the protein content showed no definite trend towards increase or decrease in the hybrids relative to the parents.

Estimates of components of variance revealed the absence of additivity as well as the dominance in both locations. From the graphical analysis epistatic gene action was conspicuous.

Phul and Athwal (1969) reported on the inheritance of grain size (1000-seed weight) and grain hardness (average weight in kg required to break the grains in a hand-operated hardness tester) and their relation to protein content (Kjeldahl N, McKenzie and Wallace, 1954) using seven generations of pearl millet, P_1, P_2, F_1, F_2, F_3, B_1, $(F_1 \times P_1)$, and $B_2(F_1 \times P_2)$, of a cross between two inbreds, SG 31 (small, soft grains) and L67B (large, hard grains). The cross was attempted during the normal season of 1965, the first generation was advanced at Coimbatore, India during December 1965 to February 1966; the F_2 generation with parents at F_1 was grown at Ludhiana in March 1966 as an off-season crop.

In grain size and grain hardness, the F_1 means exceeded the mean values of the superior parent showing dominance of large sized hard grain. Components of variance estimated by two methods (Mather, 1949: and Warner, 1952) showed similar trends, additive variance was not significant and there was a high degree of dominance. Among individual plants, the environment played an important role but was negligible where the mean value of rows were concerned. Heritability in the broad sense (total genotypic variance) was high (up to 80%) for both size and hardness. Heritability in the narrow sense was much lower. Phenotypically size and hardness were positively and significantly correlated, hardness and protein content significantly and negatively correlated. No other correlations were significant. The genotypic relationships were obscured by the environmental influence.

Phul and Gill (1970) reported on the diallel analysis of grain size, (1000-seed weight) and protein content (McKenzie and Wallace, 1954) of six inbred lines of pearl millet, SG 31, PAG 38, Ghana 73, Tf 18B, BIL 1 and BIL 3B. Only one way crosses were made. Parents with bold grains, Tf 18B and Ghana 73, were found to possess more dominant alleles; the gene action for grain size appears to be primarily additive. Phul and Gill stated their belief that the inheritance of protein content is complex. Tf 18B and SG 31 had high protein content and recessive alleles.

Phul *et al.* (1969) reporting from the Punjab Agricultural University crossed three Indian (SG 31, BIL 1, BIL 3B) two African (PAG 38, Ghana 73) and one American (Tf 18B) lines of pearl millet in all possible one-way combinations. The grain yield and protein content (McKenzie and Wallace, 1954) recorded on a plot basis for the parents are shown below:

	Grain yield	Protein content
SG 31	10·1	15·7
PAG 38	6·0	12·3
Ghana 73	13·6	14·2
TF 18B	19·9	17·5
BIL 1	19·4	11·5
BIL 3B	10·4	9·8

All the crosses exceeded the mid-parent values for grain yield by a considerable margin, ranging from 104% (SG 31 × Tf 18B) to 896% (PAF 38 × BIL 3B). In all crosses the protein content was less than the mean of the parents, the decrease ranging from 6·6% (SG 31 × BIL 3B)–42·8% (PAG 38 × BIL 3B) and Ghana 73 × Tf 18B.

Phul *et al.* (1969) commented that their results for protein are contrary to those obtained by Mahadevappa (1967) who also did not mention heterosis for grain yield; the differences in results might be due to differences in the genetic make up of the lines involved. The highest protein yield/plot (grain yield × protein content) of 1236 was shown by Tf 18B × BIL 3B and the lowest of 472 by SG 31 × Tf 18B.*

Nanda and Phul (1974) chose nine genetically diverse inbred lines of pearl millet PT 876, Ghana 73, C 630, BIL 3B, L4, 103 B, IP 832, 109 B, and 108 B for a diallel cross in 1968. The parents and 36 F_1s, without reciprocals, were grown at Punjab Agricultural University, Ludhiana during the off-season (February–June) and normal season (July–October) of 1969. Statistical analysis followed the methods of Hayman (1954), Jinks (1954) and Crumpacker and Allard (1962). Observations were recorded on plant height (cm), ear number/plant, ear length (cm), ear girth (cm), grain yield (g/plant for the normal season only) and grain protein % (method not stated). The analysis of variance for progeny means showed highly significant differences for all characters, except protein in the normal season. Parental variation was significant for all characters except ear number in

* (Authors' note: Phul *et al.* (1969) do not state the units in which grain yield and protein are expressed. It is probable that yield is in quintals (100 kg) but whether protein is % DWB, as-is, or corrected to some H_2O content is not stated.)

both seasons and grain yield in the normal season. The parents vs hybrids comparisons were highly significant for all characters indicating heterosis. The inheritance pattern for protein appeared to be mainly non-additive. The dominance variance was more pronounced compared to the additive though the latter was significant in both seasons. PT 876 and 109 B were the most dominant parents during the off-season; Ghana 73 and BIL 3B in the normal season. In the off-season, 108 B showed the highest concentration of recessive genes; 109 B in the normal season.

The results with regard to protein content were in agreement with those of Mahadevappa (1967) and Phul and Gill (1970).

Twenty pearl millet lines derived from dwarf D_1, D_2, D_3 and D_4 (Burton and Fortson 1966) and two lines from dwarf selections from IP 212 $F_2 \times$ IP 81 DF, and F 530 inbred for 1–9 generations were crossed to obtain a full diallel set of 231 F_1 hybrids without reciprocals (Harinarayana and Murty, 1970). IP 212 was of African origin; F 530 was Indian from the Punjab (Ahmad *et al.*, 1972). The F_1 hybrids and their parents were grown under nitrogen and irrigation and the seed analysed for crude protein (AOAC, 1965). Statistical analysis followed the methods of Griffing (1956), Kempthorne and Curnow (1961) and Fyfe and Gilbert (1963).

The range of variation in protein content was much higher in the hybrids, ranging from 11·1–20·6%, than in the parents, ranging from 14·0–18·4%, but the parents had much higher protein than many hybrids, and the overall mean of the parents and hybrids was 16·0%. No heterosis for protein content was found. The gene action was predominantly non-additive. The best general combiner was a derivitive of $(D_2 \times D_4)$ F_7, with a protein content of 16·5%. $(D_1 \times$ Improved Ghana) F_8, was a reasonably good general combiner and had the higher protein content, 18·4%. The cross between those two parents had a protein content of 20·0%.

Hybrids with more than 18% protein content were derivative D_2 and/or D_4 in which dwarfing was governed by one to two pairs of genes (Burton and Fortson, 1966). Yield could not be related to protein content indicating that recombinants could be obtained with desirable protein content, yield and plant height. The predominance of non-additive gene action for protein content indicated that the diversity of alleles rather than heterozygosity *per se* was important in bringing about adaptive changes.

Using essentially similar starting material, Ahmad *et al.* (1972) reported the results from growing a full diallel, without reciprocals in the F_1, F_2 and F_3 generations during the monsoon season of 1966, 1967 and 1968 at the IARI, New Delhi. A partial diallel cross ($s=11$) on the design of Kempthorne and Curnow (1961) was set up to obtain 121 ($n.s/2$) crosses in each generation (Ahmad *et al.* 1972). Significant differences were observed for crude protein content (AOAC, 1965) among parents and hybrids but differences between overall means of the parents and hybrids were negligible in all three generations. There was no apparent heterosis. The parental range of protein content was 12·5–18·4% (cf. Harinaryana and Murty, 1970 parental range 14·0–18·4%.) The range for the hybrids was 11·1–20·6% in

F_1 (as in Harinarayana and Murty, 1970) 11·5–17·7% in F_2 and 12·7–17·7% in F_3 generations.

Combining ability analyses for protein in grain showed that GCA and SCA were highly significant in each generation. While additive and non-additive components were both important, the non-additive gene action was predominant. This was supported by lower estimates of heritability in F_2 (7·77) and F_3 (8·11) compared to the F_1 (17·15). The five parents which were the best combiners in F_1 were also best in F_2 and F_3 and therefore appeared to be stable.

A derivative of ($D_2 \times D_4$) F_7, had the highest GCA with protein of 16·5% in F_1 (as found by Harinarayana and Murty, 1970), 15·5% in F_2 and 15·4% in F_3. A derivative of ($D_1 \times$ Improved Ghana) F_8, had the highest protein content, 18·4% in F_1 (as in Harinarayana and Murty, 1970), 15·0% in F_2 and F_3 and was also a good general combiner. The crosses between those parents had a protein content of 20·0% in F_1 (as in Harinarayana and Murty, 1970), the crosses between ($D_2 \times D_4$) F_7 and another (unspecified) line gave a protein content of 17·7% in F_2, and the cross between the same unspecified line and ($D_1 \times$ Improved Ghana) F_8 had a protein content of 16·7% in F_3. The predominance of non-additive gene action in each generation for protein indicated a diversity of alleles in the pearl millet populations study. The lack of heterosis for protein might be due to internal cancellation of the components of heterosis.

Twenty-two dwarf inbred parental lines and the 33 F_2 of a partial diallel cross ($s=3$) of pearl millet were grown at the IARI, New Delhi in 1967 (Ahmad and Murty, 1972). The finely powdered grain, hexane defatted, was analysed for protein (AOAC, 1965) tryptophan (colorimetric, Spies, 1950), methionine (colorimetric McCarthy and Paille, 1959) and total sulphur (Basson, 1967). Combining ability analysis (Kempthorne and Curnow, 1961) for protein, tryptophan, methionine and sulphur content showed that the variation due to GCA and SCA effects was highly significant. Both additive and non-additive components were important with non-additive gene action predominant for all the characters studied. Parents ($D_2 \times D_4$) F_7 and ($D \times$ Improved Ghana) F_8 had the highest GCA effect in F_2 and were consistent in F_3.

As reported in Ahmad *et al.* (1972), the best combiners for protein in F_2 and F_3 were: ($D_2 \times$ IP 81) F_4; ($D_1 \times$ Improved Ghana) F_8; ($D_4 \times$ IP 85) F_6; and ($D_2 \times$ IP 81) F_5.*

The best combiners for tryptophan (% flour and % protein) were ($D_4 \times$ S.530) F_7 and (D_2Oc) F_4; for methionine (as % of flour and of protein) were ($D_1 \times$ IP 81 Unk $\times$ IP 81) F_8 and ($D_4 \times$ IP 85) F_7; and for total sulphur (as % of flour) were ($D_2 \times$ IP 81) F_6 and (D_2Oc) F_4.

None of the parents was ideal for all the traits but (D_2Oc) F_4 and

* (Authors' note: Ahmad *et al.* (1972) used numbers for the parents with no line description. Ahmad and Murty (1972) gave the line description against a number. It is presumed that the numbers relate to the same line in both papers.)

($D_4 \times$ IP 85) F_6 appeared to be the best parents for the breeding programme described.

Protein and Inorganic Elements

The range, heterotic effect and group comparisons of grain yield, protein (Kjeldahl N), ash, Ca and Fe (AOAC, 1955) and P (King, 1932, modification of Fiske and Subbarow, 1925) of 25 Indian, 11 African and 5 American S_0 populations of pearl millet, and their top crosses, with Pusa Moti as a control and as a tester were studied by Lal *et al.* (1972) at the IARI, New Delhi. Pusa Moti (IP 81) was excellent for grain yield; none of its S_0 hybrids or other parents gave either higher significant yields or higher protein, P, Fe and ash. Eight populations were significantly superior to Pusa Moti for Ca. The mean values of African and American parents or hybrids and the control were significantly superior to the Indian group for grain yield with, in general, the African superior to the American. The Indian groups were higher than the other groups in protein, and to some extent, P, Fe and ash. Some of the American groups were higher in Ca.

Heterosis of the F_1 hybrids as measured by comparison with the mean of parents and over the superior parents, was positive and significant in a few combinations but negative heterosis was more pronounced. The best combinations were derived from non-Indian material. Six parental lines showed promise for grain yield, including material from Soroti, Uganda, the RSJ variety from Rajasthan and a dwarf line D_1 from America. Most of the parents were also promising for other nutritional characters. Four lines were promising for Ca; five for P; three for Fe, and five for ash.*

Significant and positive correlations were observed between the mean of the parents and the mean of the hybrids for ash content. Significant total correlation for the mean of all the populations was found between P and ash content.

Calcium

Adrian and Jacquot (1964) reported that genetic differences in the content of Ca existed in pearl millets grown under the same conditions at Bambey in Senegal. The range of Ca contents found was 13·0–49·0 mg/100 g of whole grain.

Common Millet

Konstantinov (1975) reported from Russia the results of crossing high-protein lines of foxtail millet with low-protein lines of common millet. The

* (Authors' note: it is not stated whether any of these lines were common in more than one nutritional character.)

protein content of the F_1 progeny was about mid parent value. In the F_2 there was greater variation in protein content which ranged from 14 to 20% depending on the type of panicle and colour of seed. The highest protein content in F_2 was found in the groups effusum and compactum. Creamy and yellow seeds were lower in protein than red seeds. In the F_3, protein content was generally intermediate in relation to the original strains. As a result of repeated individual selections for grain yield and protein content cultivars were obtained which showed improved protein yield/ha compared to the standard common millet cultivar Khar'kovskoe 25 as shown below:

	Average yield (kg/ha)	Protein content (%)	Protein yield (kg/ha)
Khar'kovskoe 25	3520	12·6	443
68 921	3680	13·5	497
71 258	3530	13·6	481

Konstantinov *et al.* (1975) reported from Russia that dry seeds of the common millet cultivar Mironovskoe 51 and of the stable line 70 6181 (Saratovskoe 853 × Veselopodolyanskoe 367 when treated with the chemical mutagens N-nitrosomethylurea (NMU) and N-nitrosoethylurea (NEU) produced M_2 mutants of (Kjeldahl) protein content as shown below:

	Mutagen (%)	Grain protein content (%)
Mironovskoe 51	—	14·3
	NEU 0·012	15·1
	0·025	16·0
	NMU 0·025	14·9
70 6181	—	16·1
	NEU 0·012	17·2
	0·025	16·9
	NMU 0·012	16·4
Khar'kovskoe 25 (standard)	—	13·1

The concentration of 0·025% NEU was significantly the most effective for inducing high-protein mutants in Mironovskoe 51. NEU was also most effective in raising the protein content of 70 6181 but the difference between the two concentrations was not significant.

In 1973, certain stable mutants were selected for grain yield, 1000-seed weight, protein content and milling characteristics. Compared to the standard, Khar'kovskoe 25, three mutants of Mironovskoe 51 were superior in yield, approximately equivalent in milling characteristics but lower in protein content, though superior to the untreated Mironovskoe 51 in that respect.

Some of the mutants in 70 6181 exceeded the untreated line in grain yield but were still inferior to the standard Khar'kovskoe 25. Some mutants were however superior to both the untreated line and the standard in protein content containing from 16·0 to 16·6% protein compared to the standard 12·9% and the untreated line 14·9%.

The correlation coefficient for protein content between the M_2 and M_3 mutants was significant indicating satisfactory inheritance of that character. The correlation coefficient between protein content and grain yield was −0·814, significant at the 1% level. Association between protein content and 1000-seed weight was low ($r = 0{\cdot}080$).

Konstantinov *et al.* (1977) have reported increased protein content in dry common millet seeds treated with ^{60}Co at doses of 10, 15 and 20 10^3 r and with 0·012 and 0·25% solutions of NMU and NEU.

Agronomic Effects

Pearl Millet

General Chemical Composition

Adrian and Sayerse (1957) reported the proximate composition of samples of *Pennisetum pycnostachyum*, *P. nigritarum* and *P. pycnostachyum* × *nigritarum* grown under standardized conditions which included the distribution of about 12 t/ha of manure (unspecified type) at Bambey, Senegal and of souna (early) and sanio (late) varieties of pearl millet of peasant cultivation obtained from the market at Dakar.

The mean weight of 1000 seeds and specific gravity of the samples were:

	Mean weight of 1000 seeds (g)	Specific gravity
Peasant cultivation	6·5	0·81
Bambey grown	8·2	0·82

About 60% of the seed weight could be attributed to the kernel and about 40% to the husk.

The moisture, ash (by incineration), total reducing substances (after hydrolysis of polysaccharides), fat (carbon tetrachloride extraction), protein (Kjeldahl N × 5·83), minerals (AOAC 1945), vitamins (microbiological assay) and amino acids in the protein (microbiological assay) from the five types of millet are shown in Table 3.69.

The Bambey grown millets were appreciably higher in protein than those of peasant cultivation. They were also higher in Ca, and, with the exception of *P. pycnostachyum* × *nigritarum*, in Na. There was little difference in thiamine, riboflavin and pantothenic acid, but the Bambey grown samples *P. pycnostachyum* and *P. nigritarum* were higher in niacin. Adrian and Sayerse commented that since the Bambey samples were grown under standardized conditions, differences among them could be attributed to genetic differences, but no comparison could be made between Bambey grown and peasant grown samples to establish whether differences were due to genetic or agronomic influences.

Threonine in the protein was notably higher in the Bambey sample and phenylalanine somewhat higher than in souna and sanio. The isoleucine content differed by 3–5% among the Bambey cultivated varieties and by 4–6% between souna and sanio.

Khanna (1966) compared the effect of urea at levels of 0, 11·2 and 22·4 kg N/ha applied as a foliar spray, 40, 47, 54 and 60 days after sowing; and 0, 22·4 and 44·8 kg N/ha as a soil application at the time of sowing on an unidentified variety of pearl millet grown on a loam soil, non-calcareous with a pH of 7·8 at Kalyanpur, near Kanpur, India. The 1000-seed weight, moisture, protein, fat, crude fibre and ash (AOAC, 1960), Ca and P (Piper 1950) and soluble carbohydrates by difference, are shown in Table 3.70. Urea application increased grain weight significantly except at the lower level of soil application. Protein content also increased significantly, the highest increase coming from 44·8 kg N/ha in the soil, but the effects on fat, fibre, ash and P_2O_5 were not significant. Ca content increased significantly at the 11·2 kg N/ha foliar application and the 22·4 kg N/ha soil application but otherwise remained the same as the control. Carbohydrate content decreased significantly compared to the controls with both levels of fertilizer and both methods of application.

Protein and Amino Acids

Africa

Busson *et al.* (1962) compared the improved Bambey, Senegal populations of *Pennisetum pychnostachyum* ('sanio') PC 19, PC 25, PC 14, PC 07, PC 10, PC 06, PC 03, PC 01, and local, non-improved types, obtained from Lambaye (15 km from Bambey), and Takhoum (near M'Bour and 100 km from Bambey). A single sample of *Pennisetum gambiense-nigritarum* ('souna') PC 28 cultivated at Bambey was included.

Rainfall was essentially the same at Bambey and Lambaye (650–700 mm) and higher at Takhoum (800–900 mm). The soil at both Lambaye and Takhoum was of the type known as Bior, very sandy with poor water retention and only slight capability of transferring mineral elements. The soil at Bambey was basically similar but having received each year 10 t/ha of farmyard manure (FYM) for the millet/sorghum cultivation, and being subject to regular rotation of groundnut/millet or sorghum/fallow it was considerably more fertile. The effect on yield in the same year in the three places may be seen from the following:

	(kg/ha)
Bambey	931
Takhoum	539
Lambaye	365

Busson *et al.* suggested that, having regard to the precision of the method, the ranges in amino acid content did not indicate great differences among the samples. Kjeldahl N ranged from 1·37 to 1·74% with a mean of 1·56 in sanio; the value for souna PC 28 at Bambey was 1·79. The mean lysine for sanio was 237 mg/g total N, ranging from 225 to 244; the value for PC 28 was 244 mg/g total N.

Blondel (1970) described experiments carried out at Kotiary, Eastern Senegal in 1967 in which a local variety of pearl millet was grown under varying application, at sowing, of 0, 50, 100, 150 and 200 kg N/ha. The effect on grain yield and protein content (N × 6·25%, DWB) is as presented in Table 3.71.

In 1968, the variety PC 28 was grown at Bambey, Senegal and in this trial, N was added at the level of (a) 50 kg/ha at sowing and (b) 50 kg/ha at sowing and 50 kg/ha at half-bloom. The effect on grain yield and protein content is shown in Table 3.72.

The results of both experiments indicated that on soil deficient in N, large applications of N fertilizer increased yield without depressing protein content in the grain, resulting in large increases in protein/ha. When 50 kg N/ha was applied at sowing and 50 kg N/ha at the end of half bloom, the protein/ha increased five-fold.

The distribution of N accumulated by PC 28 millet as affected by N fertilizer application was as shown below:

N accumulation	N application (kg/ha) 0	50 at sowing	50 at sowing 50 at half-bloom
Stalk (%)	25·6	24·5	18·0
Leaves (%)	21·5	23·5	20·7
Rachides and glumes (%)	18·3	14·7	10·4
Grain (%)	34·5	37·0	50·6
N accumulated (kg/ha)	26	51	88
Grain yield (kg/ha)	781	1333	2178

During 1973, a dry year in which the available rainfall at Bambey, Senegal was well distributed but 300 mm lower than the normal average of 650 mm, Ganry and Bideau (1974) studied the effect of N fertilizer on the pearl millet Souna III, a 90-day type and the most widely grown in the 500–700 mm rainfall zone. The levels of N fertilizer (as urea) were, in kg/ha:

N_0 (None), N_{30}, N_{60}, N_{90}, N_{120} and N_{150}. It was added, one-fifth at sowing, two-fifths after thinning and two-fifths at half-bloom. The applications were made (1) with and (2) without, compost (i.e. millet straw) at the rate of 15 t dry matter/ha ploughed into a depth of 20 cm at the end of the rainy season. It is not stated whether P was applied. Yield increased significantly up to N_{120}. With compost, it amounted to 3223 kg/ha, significantly higher than the 2937 kg/ha from N_{120} without compost. Even at the low level of N_{30} the improvement in yield from compost was significant (597 kg/ha). There was however, no evidence of a significant interaction between compost and N fertilizer. The effect of N_0, N_{60}, N_{90} and N_{150}, with and without compost, on the moisture, ash, protein, ($N \times 6{\cdot}25\%$), lipid, cellulose, Fe, Ca and P contents of the grain (the methods of determination are not stated) is shown in Table 3.73.

Only protein content showed a marked difference resulting from N fertilizer application, increasing in linear relation with the level of N fertilizer and with higher N from compost.

The effect on amino acid content, as % of pure protein, of the four treatments, with and without compost, is shown in Table 3.74.

N fertilization increased glutamic acid and leucine in the protein and decreased lysine, glycine, histidine, arginine and (very slightly) threonine. The overall decrease in total essential amino acids was less than 2·0% of total protein between N_0 and N_{150}.

Lysine, as % of grain, increased with N fertilization (see Table 3.74).

Growth trials were conducted with eight rats, four male and four female to the group, over one month, the diets consisting of the millet, supplemented with vitamins and minerals. The weight gains, protein consumption, and PER are shown in Table 3.75. The PER dropped from 1·77 (N_0 without compost) to 1·15 (N_{150} with compost) and 1·16 (N_{150} without compost). At N_{60} weight gain and N intake were higher and PER not greatly different from N_0.

Ganry and Bideau stated that the depressive effect of composting, without N fertilizer, on the growth of the rats at N_{90} and N_{150} compared to N_0 could not be accounted for by deficiency of lysine but was probably due to a reduction in other essential amino acids.

Asia

Shah and Mehta (1959b) compared the effect of no fertilizer and of (1) NH_4, Cl, (2) $(NH_4)_2SO_4$, (3) $(NH_4)_3SO_4NO_3$, (4) NH_4NO_3, (5) urea, (6) groundnut cake and (7) farmyard manure at levels to provide 22·4 and 44·8 kg N/ha on pearl millet grown on sandy loam (goradu) soil over a 2-year period and three crop seasons at the College Agronomy Farm Institute of Agriculture Anand Bombay. P was also applied as superphosphate to give 44·8 kg P_2O_5/ha. Rainfall in the first year was 1156 mm, 90% in July and August, and in the second year was only 548 mm, 75 % of which fell in July and August. The effect on grain yield and grain protein (Kjeldahl $N \times 6{\cdot}25$)

and protein yield/ha for the three season are shown in Table 3.76. In all three seasons significant increases in yield over the control without fertilizer were obtained from all N fertilizers, the maximum yield coming from (3) $(NH_4)_3SO_4NO_3$, (4) NH_4NO_3 and (5) urea, with lower yields from (1) NH_4Cl and (2) $(NH_4)_2SO_4$, and lowest yields of all from the organic fertilizers (6) and (7). P did not improve the yield. None of the fertilizers had any significant effect on the percentage crude protein content ($N \times 6{\cdot}25$) of the grain.

Sawhney and Naik (1969) reported the effect of four levels of N fertilizer (ammonium sulphate), in kg/ha, N_0, N_{44}, N_{88} and N_{176} on the pearl millet variety HB 1. The sample was obtained from the world collection maintained at IARI, Delhi 12. A basal dose of 20 kg P_2O_5 and 10 kg K_2O/ha was also applied. The protein (AOAC, 1965), S (Basson, 1967), lysine (Naik, 1968) and tryptophan (Spies and Chambers, 1949) are shown in Table 3.77.

N application increased the protein content over the control, the increase in protein from N_{88} and N_{176} being significant. It also increased S content. N_{44} significantly increased tryptophan in the grain and in the protein; N_{88} and N_{176} gave no additional increases. N_{44} and N_{88} increased lysine in the grain and reduced lysine in the protein; N_{176} reduced lysine in the grain and in the protein. The main effect of the treatments on the protein fractions (modified Mendel-Osborne method of Naik, 1968) was to increase the prolamine fraction by about 40% with a corresponding decrease in the proportions of albumin and globulin. The fertilizer treatment altered the distribution of S in the protein fractions. S increased in the albumin but since the albumin fraction itself decreased on N fertilization, the proportion of S in the albumin remained constant at about 20%. S decreased from 28 to 19% in the globulin fraction and decreased from 26 to 19% in the glutelin. In the prolamine fraction sulphur increased from 20 to 31% in the total S in the grain.

Tryptophan in the prolamine increased from 86 to 107 mg/100 g flour as a result of N fertilization and slightly decreased in the albumin, globulin and glutelin fractions from N_{44} and N_{88}. From N_{176} only, tryptophan increased in the defatted material to 30 mg/100 g from 21 mg/100 g flour in N_0. Lysine was not specifically discussed.

In pearl millet the prolamine fraction was relatively high in tryptophan compared to that fraction in maize and sorghum.

Kinra *et al.* (1970) compared the relative efficiency of foliar spray and soil applications of N as urea, to Kanpur, an open-pollinated variety of pearl millet grown over four years in the kharif season at IARI, Regional Station, Kanpur. The soil was loamy, poor in N and average in P and K. The crop was planted in rows, 45 cm apart and 15 cm between plants in the row. The urea was applied at 0 (N_0), 11·2 (N_{11}) and 22·4 (N_{22}) kg N/ha for the foliar spray and 22·4 (N_{22}) and 44·8 (N_{45}) kg N/ha for the soil application. There were four foliar sprayings, the first at the fourth leaf stage, thereafter weekly, with all completed before flower emergence.

The various fertilizer treatments had no significant effect on moisture, fat, crude fibre, total minerals or P contents in the grain (the methods of determination are not stated). Protein content however increased significantly as N fertilizer increased from both spray and soil application, from 9·2% (N_0) to a maximum of 11·0% for either N_{22} as foliar spray or N_{45} as soil application. The effect on grain yield and grain protein is shown in Table 3.78. Over the 4 years, the average increases over the control in grain yield due to foliar application were about 23% for N_{11} and 54% for N_{22} and 18% and 48% for N_{22} and N_{45} respectively from soil application.

Tomer (1970) reported that in the semi-arid area of Agra, pearl millet is the most important grain crop of the kharif season and is generally sown broadcast in a rough seed bed with little or no manure. At the Agricultural Research Station, Bichpuri, Agra, line and broadcast methods of sowing an "isolated variety" were compared with 55 kg N/ha broadcast (1) as $(NH_4)_2$ SO_4 at sowing, (2) as municipal compost at sowing, (3) 27·5 kg N/ha as $(NH_4)_2SO_4$ and 27·5 kg N/ha as compost, both at sowing and (4) 41·25 kg N/ha as compost at sowing and 13·75 kg N/ha as $(NH_4)_2SO_4$ as top dressing. Two seeding rates were also compared, 2·5 and 5·0 kg seed/ha. The crop followed chickpeas grown during the preceding rabi season and the soil was light loam, deficient in N but rich in P and K. The effect on test weight and on percent protein content in the grain (method not stated) are shown in Table 3.79.

Broadcast sowing appeared to result in a higher test weight and protein content than line sowing. Tomer proposed that this result could probably be attributed to the fact that in broadcast sowing the plants are better distributed over the total surface of the field, probably allowing for better uptake by the plants of the nutrients applied by the broadcast method.

Maximum protein content and test weight were obtained from (3), half N as $(NH_4)_2SO_4$ and half N as municipal compost both applied at sowing. Tomer suggested that under this treatment the N was more consistently available from the beginning. He also proposed that when all the N was applied as municipal compost at sowing, the N would not be available at an early stage.

No data on relative yields were quoted but the seed rate of 2·5 kg seed/ha gave significantly higher test weight than 5·0 kg seed/ha but the reduction in protein content was not significant as shown in Table 3.79.

Deosthale *et al.* (1972b) compared the effect of $(NH_4)_2SO_4$ at the rates of 0, 40, 80, 120, 160 and 200 kg N/ha applied half at the time of sowing and half as top-dressing after 30 days on the yield, protein content (Kjeldahl N × 6·25%) and amino acid composition (microbiological assay) of HB 1, HB 3, HB 4, KL 559 and D 356 varieties of pearl millet grown at Hyderabad in the rainy season of 1970. Superphosphate at the rate of 100 kg P_2O_5/ha and muriate of potash at the rate of 50 kg K_2O/ha were applied with the basal dose of N.

The values for grain yield and grain protein, together with lysine, leucine, isoleucine, threonine, methionine and tryptophan in the grain and in the protein are shown in Table 3.80.

With higher inputs of N fertilizer, the grain yield increased, maximum increase being at N_{120}. HB 1, HB 4 and D 356 produced higher grain yields with fertilizer applications than HB 3 and KL 559.

Protein content (% grain) increased significantly up to N_{120} and remained almost constant from N_{160} and N_{200}. A significant positive correlation ($r = 0{\cdot}855$) was found between protein content in the grain and level of N fertilizer.

At 0 kg N/ha the % grain protein contents of the varieties, DWB, were:

	(%)
HB 1	9·7
HB 3	11·1
HB 4	8·7
KL 559	8·5
D 356	9·4

The maximum protein content of N_{160} increased over the control by 59% in HB 4 and 61% in KL 559; at N_{200} protein increased by 46% in HB 1 and 79% in HB 3 and at N_{120} by 40% in D 356. The mean protein content of HB 3 was significantly higher than in the other varieties. Each amino acid in the grain was significantly increased with increase in N fertilizer.

Increase in N fertilizer paralleled an increase in leucine, isoleucine and tryptophan. Lysine, threonine and methionine decreased with N fertilizer but only the lysine decrease was significant. By Deosthale's calculations, varietal differences in lysine and tryptophan were significant at the 5% level and in leucine and methionine at 1% and 0·1% levels respectively. The highest amounts of lysine, isoleucine, threonine and methionine in the protein were found in D 356.

Deosthale proposed that since lysine, methionine and isoleucine all appeared to be correlated positively with threonine, identification of a variety of pearl millet with high threonine content might also ensure a relatively high level of the other three amino acids.

Kumar *et al.* (1973) compared the effect of N fertilizer (type not stated) and levels of 0, 20 and 40 kg N/ha, and three types of weed control, no weeding, hand weeding and spraying with 2,4-D, on the crude protein content of the pearl millet varieties RSK, Chadi, and a local Jodhpur variety grown at the Central Arid Zone Research Institute, Jodhpur, Rajasthan, during kharif 1963 and 1964. The NPK content of the soil was lower than optimum; in 1963 (a dry year) the rainfall was 165 mm and in 1964 620 mm. The effects on crude protein content are shown in Table 3.81.

Differences in protein content among the varieties were not significant but the protein percentages in 1963 were higher than in 1964. As N fertilizer levels increased, protein content increased; the increase over the control from N_{40} was significant in the 1964 season. Neither weed control treatments gave significant increases to protein content.

Inorganic Elements

Pearl millet, variety 207, was grown during three seasons, monsoon 1956 and 1957, and summer 1957, on sandy loam soil (goradu) in Bombay state, under various combinations of the fertilizers NH_4Cl, $(NH_4)_2SO_4$, $(NH_4)_3SO_4NO_3$, NH_4NO_3, urea, groundnut cake and farmyard manure; and analysed for Ca and Mg (US Salinity Laboratory, 1954), P (Winton and Winton, 1945), K (photometrically) and crude fat (ether extract) (Shah and Mehta, 1959a).

Ca and Mg in the seed appeared inversely related, with Mg present in higher amounts than Ca. In all three seasons the highest Mg content and the lowest Ca content occurred with highest NH_4Cl. The lowest Mg appeared with urea or NH_4NO_3 and the highest Ca with urea.

Shukla and Bhatia (1971) described research at the Haryana Agricultural University, Hissar on Fe and Mn uptake in two varieties of pearl millet, hybrid 1 (exotic) and T 55 (indigenous) grown on sandy-loam soil, at pH 7·8. All plots received a basal dose of 26 kg P/ha as superphosphate and 48 kg K/ha as muriate of potash together with N fertilizer (as NH_4NO_3) at levels of 0, 30, 60, 90, 120 and 150 kg N/ha. Fe and Mn concentrations (Johnson and Ulrich, 1959) in the grain are shown in Table 3.82.

Between varieties, Fe concentrations did not differ significantly; Fe and Mn increased significantly up to 90–120 kg N/ha above which both appeared to decline.

Desai and Zende (1972) compared hybrid pearl millet HB 1 and Pusa Moti, grown in a medium black soil (clay loam in texture) and calcareous soil (pH 7·4) at the Agricultural College Farm, Poona, with additions of 125 kg N, 50 kg P_2O_5, and 37·5 kg K_2O/ha, the source of the N differing among (1) $(NH_4)_2SO_4$, (2) NH_4Cl, (3) $(NH_4)_3SO_4NO_3$ and (4) urea. The effect on total N (Kjeldahl) (AOAC, 1950) and P, K, Ca, Mg, Fe and Mn (Piper 1960) are shown in Table 3.83. Desai and Zende stated that HB 1 contained significantly higher amounts of all the elements studied than Pusa Moti without fertilizer, and contents of all elements in both varieties increased significantly with all forms of N fertilizer. Urea and ammonium sulphate were the most effective.

Vitamins

Pai *et al.* (1958) reported that the thiamine content (thiochrome method) of pearl millet increased when fertilizers were applied: $(NH_4)_2SO_4$, NH_4Cl, superphosphate $(NH_4)_3SO_4NO_3$, NH_4NO_3 did not differ among themselves but organic manures, groundnut cake and farmyard manure, produced a greater increase than the inorganic. Farmyard manure resulted in the highest thiamine content in two seasons, probably because of its own small thiamine content. Thiamine content during the summer season was in general higher than the monsoon season. The maximum increase in thiamine content was 96% for the summer and 33% during the monsoon season.*

Curtis *et al.* (1966) analysed the carotene content (colorimetrically after separation from the xanthophylls) of six samples of 'dauro' a form of maiwa, a late-maturing pearl millet grown in Nigeria. The samples were collected from a farm in the foothills of the Jos Plateau and were chosen because the predominantly yellow endosperm somewhat resembled yellow maize and sorghum. The carotene content, with that of a sample of short Kaura sorghum and of hybrid V 10 maize, are shown below:

Sample	Carotene content (mg/100 g)
Millet 1	0·1
2	0·2
3	0·15
4	0·1
5	0·09
6	0·1
Hybrid V10 maize	0·4
Short kaura sorghum	0·06

The short kaura sample was old seed and not truly comparable.

Kapoor and Naik (1970) compared the β-carotene content (column chromatography, carotene panel of Sub-Committee on Vitamin Estimation, 1955) and carotenoid pigments extracted by acetone/hexane (1/1), absorbancy measured at 430 μm in a Klett-Summerson colorimeter of 38 varieties of pearl millet supplied by the Project Coordinator Millets, Indian Agricultural Research Institute, New Delhi. The varieties were IP 199, IP 2306, IP 2032, IP 2124 (four locations), IP 865, IP 2169, IP 971 (three locations), IP 2084, IP 2215, IP 2167, IP 2190, IP 856, IP 260, IP 2094, IP 76 (eight locations), IP 330, IP 912, IP 560, IP 550, IP 2280, IP 572, (six locations), IP 887, IP 969, 'local bijpuri' (three locations), IP 192, IP 322, IP 1136, IP 1816, IP 286, IP 505, IP 1951, IP 972, IP 196, IP 250, IP 310, IP 2120, IP 145 and IP 1993 (all at one location only, Delhi).

The mean β-carotene content was 0·03 mg/100 g with a range of 0·01 (IP 1951)–0·08 (IP 145), both at Delhi.

Finger Millet

Protein and Amino Acids

Venkataramana and Krishna Rao (1961) reported the effect of N, P and K fertilization on the variety AKP 6 of finger millet grown on a soil of sandy

* (Authors' note: absolute values have not been traced.)

texture and low fertility at the Agricultural College, Bapatla, Andhra Pradesh during 1954–1956. Fertilizer treatments with irrigation as required were:

(1) 0 (kg/ha)
(2) 50·4 N as ammonium sulphate
(3) 50·4 N (as ammonium sulphate) and 50·4 P_2O_5 (as superphosphate)
(4) 50·4 N (as ammonium sulphate) 50·4 P_2O_5 (as superphosphate) and 50·4 K_2O (as potassium sulphate)
(5) 50·4 N and 50·4 K_2O
(6) 50·4 K_2O and 50·4 P_2O_5
(7) 50·4 K_2O
(8) 50·4 P_2O_5
(9) 50·4 N (as farmyard manure)
(10) 50·4 N (as green manure)
(11) 50·4 N (as compost)

The fertilizers were used in two series; (a) with a basal dressing of farmyard manure (quantity not specified) and (b) without basal dressing. The treatments and their effects on N, P_2O_5 and K_2O (AOAC, 1950; Piper, 1947) in the grain are given in Table 3.84. The highest grain yield, 1649 kg/ha, was obtained from treatment (4), all three inorganic fertilizers with basal dressing. The relatively poor yields from treatments (9), (10) and (11) might be due to the N and other nutrients not being readily available. Treatments (2), (3), (4) and (5) all containing N as NH_4SO_4 and with basal dressings gave N contents in the grain of approximately 2·0% compared to 1·14% in the control. P_2O_5 and K_2O contents in the grain showed no consistent pattern.

In the locality of the experiment, the usual treatment given to finger millet was described as:

farmyard manure 5 cartloads/acre (12·5 cartloads/ha)
ammonium sulphate 50 lb/acre (56 kg/ha)
superphosphate 100 lb/ac (112 kg/ha)
potassium sulphate 40 lb/ac (45 kg/ha)

Sree Ramulu and Mariakulandai (1964) reported the effect of different levels of P fertilization applied as farmyard manure and superphosphate separately and in combination on the NPK and Ca content (AOAC, no date) of the grain of CO 1 and CO 7 finger millet grown at Coimbatore in 1959. Details of the fertilizer treatments and of their effects are given in Table 3.85. Without manure, the N content of CO 1 was 1·27% and of CO 7 1·38% DWB.

For the most part the N content of the grain increased with fertilizer. Combinations of low levels of farmyard manure with high levels of superphosphate (e.g. treatment (11) 11·2 kg N/ha as farmyard manure and 22·4 kg P/ha as superphosphate) produced the highest N in the grain, 1·36% N for CO 1 and 1·6% N for CO 7 (DWB).

Treatment (12) low farmyard manure and high superphosphate produced the same N content as the control without fertilizer, DWB.

The P content in the grain increased with P fertilizer, up to 22·4 kg P_2O_5/ha from farmyard manure and then decreased relative to the unfertilized control. In general P fertilization increased P content in the grain in both CO 1 and CO 7, the highest increases resulting from low levels of farmyard manure with high levels of superphosphate, (e.g. treatment 12 with 11·2 kg P/ha as farmyard manure and 33·6 kg P/ha). In all cases the Ca content of CO 1 appeared to fall with P fertilization.

Krishnamurthy (1968) observed H 22 finger millet grown in Bangalore during 1964 on medium red soil under rainfed conditions with five levels of N fertilizer applied all at planting as $(NH_4)_2SO_4$, N_0, N_{22}, N_{44}, N_{67} and N_{90} kg/ha. The protein (N × 6·25%) and Ca content (analytical methods not given) are shown below:

Fertilizer (kg/ha)	Protein (%)	Calcium (mg/100 g)
N_0	4·5	418
N_{22}	4·7	412
N_{44}	5·0	409
N_{67}	5·5	409
N_{90}	6·7	356

In 1965 Krishnamurthy compared the varieties H 22, Anna Purna, Caveri and Purna treated at planting with N fertilizer levels of N_0, N_{44}, N_{90} and N_{134} and as N_{90} plus P (as single superphosphate) 44 kg/ha. The resulting protein and Ca contents are shown in Table 3.86.

In 1974 Stabursvik and Heide reported the effect of N fertilizer (as calcium ammonium nitrate) applied at levels of 0, 50, 100 and 150 kg N/ha, with row spacings of 15, 30, 45 and 60 cm, on the protein content and amino acid composition of a local cultivar of finger millet grown on a highly weathered deep soil of the ferrallitic type (latosol) pH 5 and humus content 2% in the surface layer, at Makerere University Experimental Farm, Kabanyolo, Uganda during the second rains of 1968. In addition all experimental plots received a basal dressing of 150 kg P/ha (as single superphosphate) and 80 kg K/ha (as potassium sulphate). All fertilizers were applied as split applications (i) immediately after sowing and (ii) 35 days later. The resulting yields are shown in Table 3.87.

The main effect of row spacing and the interaction of spacings × N levels on grain yield were both highly significant. The effect of N was highest at the closest spacing (S15) and increased again at the widest spacing (S60). Optimal spacing for grain yield was 45 cm except for N_0 level where S30 was optimal. The increased yield at higher row spacings and N level arose mainly from increased head size and grain weight but these increases did not compensate for the reduction in head numbers in the 60 cm spacings. The effects on grain protein (Kjeldahl N × 6·25%, DWB) are shown in Table 3.87. Protein content increased with increasing N and row spacing, consistently and significantly. Rachie (personal communication, 1976) commented that improvement from N fertilization to 10·0, 12·0 and 13·0% crude protein at the highest level was surprisingly high for a millet which usually has a protein content of about 8·0%. The mean amino acid composition (auto-analyser following acid hydrolysis, tryptophan not determined) of two individual samples from different plots, each with a crude protein content of 6·0, 7·5, 9·0, 11·6, and 13·5% are shown in Table 3.88.

As protein content increased, lysine and the sulphur amino acids, methionine and cystine, expressed as a proportion of the total N decreased.

Stabursvik and Heide (1974) also conducted two pot-experiments in silica sand designed to study the effects of N and S fertilizers on protein content and amino acid composition. In Pot-experiment I, in addition to basal fertilizer applications, which included 5 p.p.m. S, N fertilizer was added as NH_4NO_3 at the rates of 10, 50 and 250 p.p.m. and S fertilizer as $Na_2SO_4 \cdot 10\,H_2O$ at 25 and 125 p.p.m. At the highest rate both N and S were applied in three split applications, at the start of the experiments, and 20 and 35 days after planting. This was a factorial experiment with the N rates and three S rates. In addition, a combination of 250 p.p.m. N and 12·5 p.p.m. S was included. The S/N ratios and the effect on crude protein and amino acid composition of the protein together with that of a local market sample of finger millet are shown in Table 3.89.

At 250 p.p.m. N, sulphur limited growth whereas at 10 and 50 p.p.m. N, all S levels seemed to encourage maximum N utilization. Grain size appeared not much affected but at 10 p.p.m. N level so few seeds developed that they had a more adequate N supply than those at 50 p.p.m. N.

Sulphur (determined turbidimetrically, Butters and Chenery, 1959) deficiency had a marked effect on methionine and cystine content in the protein. At 250 p.p.m. N, which gives a 'normal' seed protein content, methionine increased from 88 to 224 mg/g N and cystine from 66 to 138 mg/g N, when the S/N fertilizer ratio was raised from 1/50 to 1/10.

Analysis of a number of market samples of finger millet showed considerable variation in protein and methionine which suggested to the research workers that Uganda soils may vary markedly in S and N content.

In Pot-experiment II, the aim was to find the biologically optimal N level and optimum S/N ratio at the various N levels in terms of grain yield and quality. The basal fertilizer levels in the sand were doubled compared to Pot-experiment I, the extra nutrient being applied at heading. N at 250, 500 and 1000 p.p.m. combined with S to give S/N ratios of 1/20 and 1/10, 1/5 and 1/2 at all N concentrations were added in four split applications; at the start of the experiment, at 35 days, (half-way to heading), at 60 days (heading) and 3 weeks after heading. Crude protein content and amino acid composition of the protein are shown in Table 3.90. With low S/N ratios at 250 p.p.m. N, aspartic acid increased and methionine and cystine contents decreased. As N levels increased, the S/N ratio appeared to exert a declining influence. The results of Experiments I and II at the 250 p.p.m. N level were not absolutely comparable because of different fertilizer levels and the removal of seven additional plants at the 7-week stage in Experiment I which would have also removed some nutrient reserve.

Table 3.91 shows the effects of N and S fertilizers on N and S content in the seeds, and the proportion of S recovered as methionine and cystine S. Although seed S increased to some extent with increase in fertilizer S, there was a high degree of constancy of seed S within a wide range of N levels and S/N ratios. An S/N ratio of about 1/10 appeared to be almost optimal at normal fertilizer levels. No close relationship was found between total S and amino acid S in the seeds.

The effect of N and S fertilizer on sulphur amino acid yield is given in Table 3.92. At 250 p.p.m. N, there was a 10-fold increase in total methionine + cystine yield when the fertilizer S/N ratio increased from 1/50–1/10. A change from 1/20–1/10, values found under normal field conditions, doubled the sulphur amino acid yield. At 500 p.p.m. there was almost a 40% increase as the ratio increased from 1/20 to 1/10.

Stabursvik and Heide stated that under normal conditions, an S/N ratio of about 1/10 is required for quantitatively and qualitatively satisfactory seed protein yield in finger millet, which, because of its unusually high content of S-amino acids requires high levels of S. In the field experiment, methionine + cystine percentages increased more or less linearly up to the 8% protein level and then levelled off; lysine gradually decreased with increase in protein, the decrease starting at comparatively low protein levels. In the pot-experiments, in which seeds of protein contents lower and higher than those of the field experiments were produced, a similar relationship was found at the 1/2 and 1/10 S/N fertilizer ratios though with some rise in the methionine + cystine at the 1/2 ratio. In this series the S was in excess of optimum but in the 10–13% protein range the methionine increment was only half of that observed at the 'normal' protein levels. At the 1/20 ratio, the plants were deficient in S except at the 1000 p.p.m. N level when S was also sufficient for normal amino acid composition.

> The storage proteins of most cereals increase in prolamine content and decrease in lysine content with increase in N fertilizer levels (Michael, 1963; Sauberlich *et al*., 1953). In finger millet the storage proteins decrease in both lysine and S-amino acids with increase in N fertilizer.

Opinions differ as to whether soils in East Africa are or are not deficient in S (Stephens, 1970; Foster, 1972) but the increasing use of non S-containing N fertilizers may render S more limiting to crops such as finger millet which have high S requirements.

Inorganic Elements

Kumaraswamy and Venkataramanan (1974) reported pot-cultures on calcareous and non-calcareous red soils of Coimbatore to assess the response of the CO 10 variety of finger millet to P fertilization, using five treatments in kg/ha, (1) 0, (2) 13, (3) 26, (4) 39 and (5) 52 applying P as ^{32}P-labelled superphosphate.

In general, the performance of the millet was better in the non-calcareous than in the calcareous soil. P fertilization advanced the maturity of the crop by one week over the control, and the maximum grain yield was obtained at 39 kg P/ha.

Treatment P (kg/ha)	Yield (g/pot)	
	Calcareous red soil	Non-calcareous red soil
Control 0	4·42	6·72
39	9·61	12·76

The P content in the grain increased with increases in the level of P fertilizer as shown below, reaching its maximum on calcareous red soil at 39 kg P/ha:

Treatment P (kg/ha)	P in grain (mg/100 g)	
	Calcareous red soil	Non-calcareous red soil
Control 0	189	164
13	209	186
26	243	193
39	369	191
52	241	218

Kanaka Doss *et al.* (1975) conducted a pot-culture experiment to study the effect on yield and uptake of N, P_2O_5 and K_2O in CO 7 finger millet grown in four typical sandy soils (7 kg/pot) of Tamil Nadu without and with various fertilizer treatments. The basic analytical data of the soils are shown in Table 3.93. The dosages of fertilizers were at the level of 60 kg N/ha, 30 kg P_2O_5/ha and 30 kg K_2O/ha and the treatments were as shown in Table 3.94.

Soil samples from each pot were analysed at tillering, flowering and post harvest for available N (alkaline permanganate method), available P (Alsen's method), and available K (ammonium acetate extraction followed by flame photometry). Grain was analysed for total N, P and K. (P and K were estimated from the triple acid extract of the plant samples by colorimetry and flame photometry respectively). The grain yield, and uptake of N, P_2O_5 and K_2O by the grain are shown in Table 3.94. Available N decreased progressively and sharply up to the sixtieth day (flowering stage), less steeply thereafter. Available P also decreased progressively and sharply up to the sixtieth day but then remained constant. The major part of the loss of P was caused by leaching. Available K remained constant up to the sixtieth day and then showed a slight (unexplained) increase.

Thiruthuraipoondi and Cuddalore soils provided higher grain yields than the other two, treatment (4) urea with superphosphate, being the most successful. The uptake values of N, P_2O_5 and K_2O by the grain were closely correlated. Kanaka Doss *et al.* suggested that the high grain yield and high K uptake shown by the Thiruthuraipoondi and Cuddalore soils indicated that the increased yields could be related to the higher uptakes of P and K. They also considered that the high value of fine sand in Thiruthuraipoondi soil (88·4%) compared to 39·8% in Coimbatore soil contributed to the former's better grain yield and nutrient uptake.

Common Millet

Vatagin and Oksenko (1971) reported the effect on the proximate analysis of the common millet Gor'kovstoe 43, of N, P and K fertilizers and manure applied in spring before the first deep cultivation to plots of soddy-podzolic medium loamy soil; the preceding crop was potatoes. The mean values for three years are shown in Table 3.95. The highest protein content was obtained from manure, the other values showing little improvement over the control.

Conclusion

The foregoing illustrates the extent of the useful information reported on the manner in which chemical composition and nutritional quality may be influenced by genetic history and agronomic environment. There is a noticeable lack of uniformity both in the volume of data available among the cereals discussed and in the methodologies used by different authors. Little or nothing is reported on the minor millets relative to sorghum and, in less degree, pearl millet, which need and deserve very much more attention. Since such studies must of necessity be conducted by scientists at widely separate locations, it is essential, if their results are to be pooled and compared, that standard methods of analysis and evaluation be universally adopted.

The nutritional value of a cereal may be impaired by the presence of naturally occurring factors which interfere with its digestibility by human or other animals. In the following chapter the presence of such factors, in sorghum and the millets is considered and in particular the effect of the "tannins" in sorghum.

CHAPTER 4

Nutritional Inhibitors and Toxic Factors

Introduction

There appears to be a measure of belief, particularly among those who support the "natural food" or "health food" cults, that all "natural" foods are completely wholesome whereas many, if not all processed foods are to be viewed with suspicion. Many hundreds of naturally occurring plants contain substances that are harmful and many which have long been accepted as food crops contain what may variously be described as toxins, nutritional inhibitors, antinutrients or adverse factors. Cassava, which provides staple calories for about 300 million people, contains, a glycoside linamarin which is hydrolysed by the enzyme linamarinase, also present, to produce hydrocyanic (prussic) acid. Some types of potato (*Solanum tuberosum*) contain the cholinesterase inhibitor solanine, a glycoalkaloid which is highly toxic and can cause severe haemolytic damage in the gastrointestinal tract together with neurological disorders. Several food legumes contain both toxic substances and other compounds that inhibit the activity of essential enzymes such as trypsin. (For a more comprehensive review see Liener, 1969.)

Some of these naturally occurring substances are more or less toxic to everyone who eats enough of them. Favism (acute haemolytic anaemia) is induced in susceptible individuals by certain food legumes including *Vicia faba* (or *fava*) and seems to affect people of a particular ethnic background Allergens in food, of which the number seems infinite, may each affect comparatively few ultra-sensitive individuals.

The means by which to eliminate or depress the effect of toxic substances in food appears, in many cases, to have been discovered empirically by the people who had to subsist on them. More recently, plant breeders and food scientists have employed their respective talents and techniques in order to reduce toxic and antinutrient substances wherever they occur among important crops.

The distinction between a nutritional inhibitor and a toxic substance is indeed fine. The word "toxic" is defined as "poisoned or imbued with poison" and is derived from the Greek name for the poison used for smearing the tips of arrows (τοξικοσ, pertaining to the bow). A poisonous (or toxic) substance is one which when introduced or absorbed by a living organism, destroys life or injures health. Consequently, a nutritional inhibitor, if its inhibiting effect is sufficiently potent to cause serious malnutrition, might properly be described as "toxic".

While toxicity is difficult both to define and to quantify, the toxicity of any substance may be assessed and quoted in terms of its LD_{50}, the dose which causes mortality in (i.e. is lethal to) 50% of the animals tested. The LD_{50} is usually quoted in mg/kg of body weight, the lower the value the more toxic the substance. The LD_{50} of any substance may vary dependent upon the conditions of administration. For example, lower LD_{50} values generally result from intravenous and intramuscular injections than from oral intake; the age and state of health of the animal and the presence of other substances that may reduce or inhibit toxicity are also of importance.

The toxins and nutritional inhibitors reported in sorghums and the millets fall into two broad categories:

(a) intrinsic substances that occur naturally in the plant; these include polyphenols (often called tannins), phytates, cyanogenic glycosides, and possibly others not yet identified. In the absence of symptoms of acute or chronic toxicity, the presence of nutritional inhibitors may be detected by various symptoms including a lower rate of weight gain in the test animal than would be predicted by the quantity and chemical composition of the food ingested.

(b) substances produced by microbial or other infections or contaminations. Many of these, including ergot and mycotoxins, are as the name suggests, highly toxic.

In addition, imbalance among naturally occurring amino acids in sorghum protein may produce undesirable physiological effects. All of the known or suspected toxins and antinutrients reported to be found in sorghum and the millets are discussed in this chapter, together with the apparent adverse effect of certain naturally occurring amino acid combinations. Those that have received greatest attention, are the tannins and other polyphenols and in this context, the "bird-resistant" sorghums deserve special mention. The "bird-resistant" or "bird-proof" sorghums are characterized by grains that appear unpalatable to birds and thus tend to be grown where birds are a major predator.

Bird-resistance is discussed by Doggett (1970). In a later (1974) unpublished report, Doggett states that sorghum is not remarkably susceptible to bird damage if compared with rice, wheat or other small grains, except that it appears more readily attacked by the red-billed weaver bird, *Quelea quelea*, a predator comparable to the locust in the scope of its devastation. Doggett (1970) states that other birds, including sparrows, parrots and buntings, are known to prey upon sorghum in Africa and Asia. It is generally believed that bird-resistance among sorghum types is derived from a comparatively high proportion of polyphenols contained in the seed coat layers. It is suggested that because tannins react with glycoproteins of the saliva and produce an unpleasant sensation of astringency in the human mucosa, birds may be similarly repelled once having tasted a high tannin seed.

Polyphenols

It is not the purpose of this chapter to present a detailed account of the chemistry and physiological activities of all the known phenolic substances of plant origin. More detailed studies than this publication provides may be found in several other works (Haslam, 1966, Harborne, 1964, Ribereau-Gayon, 1972, Singleton and Kratzer, 1969, 1973). Nevertheless, a brief description of the chemistry, biochemistry and physiological activity of phenolic substances may be helpful to the general reader.

It will be well known to most readers that the functional group in phenols is a hydroxyl attached to one of the carbon atoms of a benzene ring. The formula of the simplest member, ordinary phenol, is C_6H_5OH represented by the structure (1) in Fig. 4.I. Three of its close relatives which are important to the subsequent text are resorcinol (2), catechol (3) and phloroglucinol (4).

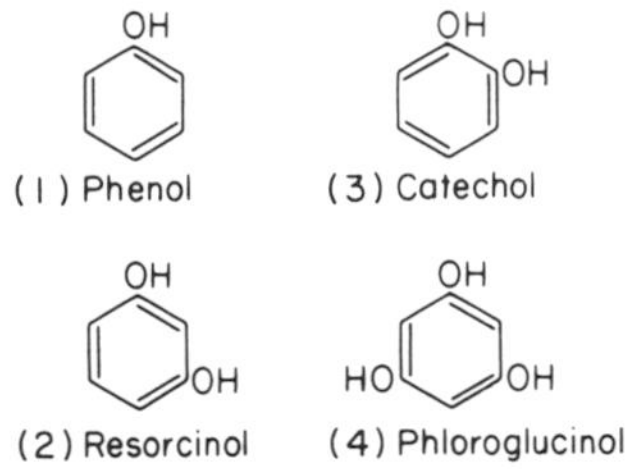

OH

(1) Phenol

OH OH

(3) Catechol

OH OH

(2) Resorcinol

HO OH OH

(4) Phloroglucinol

Fig. 4.I

Of particular importance to this text are the flavan-3-ols, the catechins. The term "catechin" refers specifically to the flavan-3-ol which has the 3′, 4′, 5, 7-pattern of hydroxylation. All these compounds contain two asymmetric carbon atoms (C_2 and C_3) thus giving four optical isomers. The four important configurations, shown in Fig. 4.II are: (+)-catechin (5); (−)-catechin (6); (−)-epicatechin (7); and (+)-epicatechin (8).

Physiological Activity of Phenolic Substances

Comparatively few known phenolic substances are produced in animals.

Those that are, including the steroidal oestrogens and the amino acid tyrosine together with certain phenolic amines involved in nerve action, probably have a specific physiological function. On the other hand, a great many polyphenols of plant origin are known and the reported minimum lethal does of a number of these have been tabulated by Singleton and Kratzer (1969). In general, it would appear that most herbivores have evolved the ability either to tolerate or to detoxify those plant polyphenols which they ingest and which appear nutritionally undesirable. Because, in general, phenols appear more acutely toxic to carnivores than to herbivores, it is tempting to suggest that animals can adapt to and detoxify undesirable plant polyphenols, and that human beings who over many generations have

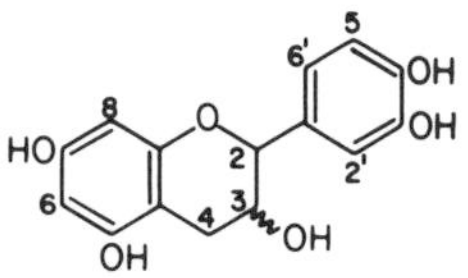

General structure of 3′, 4′, 5,7 - tetrahydroxyflavan - 3 - ol

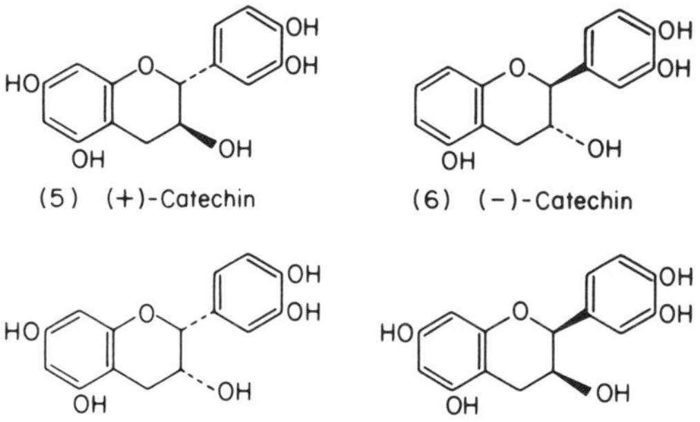

(5) (+)-Catechin

(6) (−)-Catechin

(7) (−)-Epicatechin

(8) (+)-Epicatechin

Fig. 4.II

subsisted on, for example, high-tannin sorghums, may have developed a higher than average tolerance and lower susceptibility to sorghum polyphenols. No such evidence has been found and one can do no more therefore than to speculate on the possibility. It is known however that certain microorganisms are able to modify and in some cases destroy polyphenols. It is also possible that the microflora of the alimentary tract may in some cases reduce the apparent toxicity of ingested polyphenols. Though the acute toxicity to rats of most known plant phenols is rated as slight (LD_{50} in a single oral does 500–5000 mg/kg) most phenols are more acutely toxic than ethanol, man's most common intoxicant.

It is possible that toxicity may be a protective function of some plant phenols. The bactericidal properties of simple phenol C_6H_5OH is a matter of common knowledge. The Rideal-Walker test which derives a coefficient of bactericidal efficiency for antiseptics uses simple phenol as a comparative standard. Hops (*Humulus lupulus*) are added to beer as much for the bactericidal effect of the polyphenolics humulone and lupulone they contain as for the flavour they impart. Both are effective against gram positive bacteria. Singleton and Kratzer (1969) stated that every phenolic substance possesses some antibacterial properties and that phenols, including those of plant origin, should be regarded as broad spectrum, though often weak, toxins to both microorganisms and animals. It is generally believed however that etherification and introduction of alkyl substituents into the benzene ring generally decrease the toxicity of phenolic substances to animals. This may explain why in some instances the apparent adverse effects of plant polyphenols are reduced by the addition of methyl donors such as methionine. By analogy one might infer that high levels of plant phenols might therefore tend to reduce the availability of dietary methionine.

The ability of many polyphenols to react with proteins to produce insoluble compounds has been known ever since the first piece of leather was produced. It is conceivable, and the observed evidence suggests, that in humans and other animals, body cell protein or ingested protein may be employed to bind up the phenols present in the diet thus rendering them insoluble and causing them to pass through the animal without being absorbed. Such a mechanism would be expected to reduce the protein nitrogen available to the animal. Many phenols are known to interfere with certain essential biochemical processes including the "uncoupling" of the oxidative phosphorylation essential to the synthesis of ATP concomitant with the oxidation of substrates derived from carbohydrate or fat. It is possible that this type of mechanism may in part explain the apparent impaired utilization of carbohydrate by animals fed high-tannin sorghums. It is also possible, alternatively or additionally, that the polyphenols present in sorghum react with and bind up the amylase essential to carbohydrate metabolism as they would with any other protein source.

Later in the chapter, particularly in the text presented by Dr Gupta and Dr E. Haslam of the Department of Chemistry, University of Sheffield, England, reference is made to comparative studies of different analytical methods prescribed for the detection and determination of phenols. Several of these rely upon the fact that distinctive reddish brown products are formed by the oxidation of phenols. A method which deserves special mention, since it is referred to more frequently than the rest, is the vanillin–hydrochloride method, the results of which are expressed in catechin equivalents (CE). The method relies upon the fact that vanillin (vanillaldehyde) in acid solution reacts with dihydroxy or trihydroxy phenols in which the hydroxyl groups are in the *meta* position relative to one another. Thus the vanillin hydrochloride reaction would take place with phenols containing a resorcinol but not with a catechol moiety. Thus, in the case of catechin, vanillaldehyde in acid solution would react with the A ring ((5) Fig. 4.II) which carries the resorcinol conformation but not with the B ring in which the hydroxyls are in the same positions as in catechol. The importance of this fact will become evident later. Since in the presence of hydrochloric acid some hydrolysis of a polyphenol polymer such as is described by Gupta and Haslam is likely to take place, but since it cannot be predicted to what extent hydrolysis will result in an increase in resorcinol type groupings, it is difficult to predict the number of vanillin–resorcinol condensation reactions that will take place and therefore one cannot regard the vanillin–hydrocholoride method either in its original or modified forms as a truly quantitative method for polyphenols in sorghum.

More work is clearly needed to determine more accurately the range and nature of the polyphenols present and what precisely happens, particularly to the polymers on acid hydrolysis.

Because of the complexity and esotericity of the subject, the authors invited Dr Edwin Haslam,* an authority on polyphenol chemistry, and his colleague, Dr Raj K. Gupta, to write an introduction to the subject and to describe their own research findings on the polyphenols of sorghum. This report follows:

Plant Polyphenols

Introduction

Most plant tissues contain a wide range of secondary products such as alkaloids, rare amino acids, terpenes, steroids and polyphenols, in addition to an array of intermediates of primary metabolism which are biosynthesized during the growth of the plant. The significance of these secondary plant products is a matter of wide speculation and it has been suggested that some may serve to protect the plant from pests, diseases and natural predators. In the case of phenols, it has been suggested that the evolutionary capacity to synthesize these substances was the plant's need (perhaps during senescence) to establish soil conditions more congenial for the growth of subsequent generations. However, with the important exception of the phenolic polymer lignin, which gives shape and form to the plant as it matures, the essential function of most plant polyphenols remains obscure.

The plant polyphenols are an extremely heterogeneous group of natural products. Many are biosynthesized via the shikimate pathway (a biochemical process in which shikimic acid is an obligate intermediary, Harborne, 1964) and occur in combination with sugars or analogous compounds either as glycosides or as phenolic esters. The majority are located in the cell vacuole in the plant's vegetative tissues. Knowledge of their distribution, particularly in quantitative terms, is still fragmentary but in some plants they make a substantial contribution to the dry weight of the tissue. Thus, for example, up to 40% of the dry matter of leaves of the tea plant *Camellia sinensis* is reported to be polyphenolic in character. A broad

* By Dr Raj K. Gupta and Dr Edwin Haslam, Department of Chemistry, University of Sheffield, Sheffield S3 78F, England.

classification of the major groups of plant polyphenols is shown below:

Plant polyphenols

Carbon skeleton	Class
C_6	simple phenols e.g. phenol C_6H_5OH
C_6–C_1	phenolic acids (hydrolysable tannins) e.g. gallic acid (9), ellagic acid (10) (Fig. 4.III)
C_6–C_2	hydroxyacetophenones, hydroxyphenyl acetic acids
C_6–C_3	hydroxycinnamic acids, coumarins, isocoumarins, chromones
$(C_6–C_3)_2$	lignans
$(C_6–C_3)_n$	lignin
C_6–C_3–C_6	flavanones, flavones, flavonols, anthocyanidins, flavanols, isoflavones, chalcones, aurones, e.g. flavan-3-ol (Fig. 4.II)
$(C_{15})_2$	procyanidins, biflavonyls
$(C_{15})_n$	polymeric procyanidins (condensed tannins)
C_6–C_1–C_6	benzophenones, xanthones
C_6–C_2–C_6	stilbenes
C_6, C_{10} and C_{14}	quinones

The range of compounds collectively designated as vegetable tannins is a distinctive group of plant polyphenols which have in common the characteristic property of precipitating proteins from aqueous media (Goldstein and Swain, 1963; 1965). This distinctive property has permitted their use for at least 2000 years as agents (tannins) for the conversion of raw animal hides to durable, impermeable leather. The characteristic astringent and bitter taste which results from the presence of vegetable tannins in a plant tissue may be ascribed similarly to the loss of lubrication in the mouth due to the cross-linking of glycoproteins in the mucous membranes by tannins (Bate-Smith, 1954a). Circumstantial evidence suggests that this may act as a defence mechanism for the plant against natural predators such as browsing animals.

Although it is generally agreed that polyphenols interact with proteins by multiple hydrogen bonding of the phenolic groups to sites on the protein molecule, there still remains considerable uncertainty and controversy about the exact nature of this association. Given our present state of knowledge, it may be difficult for plant scientists and nutritionists to describe exactly the polyphenols they imply in their use of the word “tannins”. Nevertheless, little progress will be made until these substances are precisely identified together with their physiological function within the host plant and upon the animal by which they are ingested.

Seguin probably introduced the name tannin in 1796 to describe the chemical constituents of various plant galls which were responsible for transforming fresh animal hides into leather. Subsequent work showed that these natural tanning agents possessed many of the characteristic properties of phenols, including the ability to form coloured complexes with iron salts, and to oxidize cold potassium permanganate.

Plant constituents were later identified from a large number of species which showed similar chemical properties but whose ability to precipitate proteins was never tested. Consequently, the definition of "tannins" was unfortunately enlarged in the general botanical and biochemical literature to cover a whole range of substances which clearly are polyphenols but not tannins. Thus, using these characteristic chemical properties of phenols as a guide, many authors have been led to classify plant phenols such as (+)-catechin (5) or chlorogenic acid as tannins. On the basis of their ability to precipitate proteins from aqueous solution, such compounds as these are patently not tannins and it is clear that much confusion has arisen from the simple error of equating tanning properties merely with polyphenolic character.

Although all classifications of tannins are necessarily arbitrary, the authors (Gupta and Haslam) believe that the following definition by Bate-Smith and Swain of natural vegetable tannins is perhaps the most useful one to follow at this point in time:

> water soluble phenolic compounds, having molecular weights between 500 and 3000 and, besides giving the usual phenolic reactions they have special properties such as the ability to precipitate alkaloids, gelatin, and other proteins.

This is a useful definition since it thus includes all molecules which are not tannins in the commercial sense of being economically important in the tanning of hides but excludes a large number of plant polyphenols and other substances which have been classified as tannins merely on the basis of certain chemical tests and colour reactions.

This discussion, although it highlights the difficulties inherent in trying to provide a coherent and workable definition of the word "tannin", nevertheless indicates the crux of the problem which must face any worker who sets out to determine the vegetable tannins present in any particular plant tissue. While accepting the Bate-Smith and Swain definition, one can state that while all vegetable tannins are polyphenols, not all plant polyphenols are vegetable tannins, and general reactions specific for particular orientations and arrangements of phenolic hydroxyl groups on an aromatic nucleus, or specific chemical reactions of phenols themselves are clearly inadequate to provide a universally comprehensive meaning to the term "tannin". Earlier failures to appreciate this point have therefore led to considerable confusion in the literature that relates to the role of tannins in human and other animal nutrition.

Vegetable Tannins

Vegetable tannins are structurally divisible into two major classes: (a) the hydrolysable and (b) the nonhydrolysable or condensed tannins (Haslam, 1966).

As their name suggests, the hydrolysable tannins are hydrolysed by acids

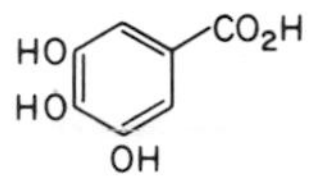

(9, gallic acid)

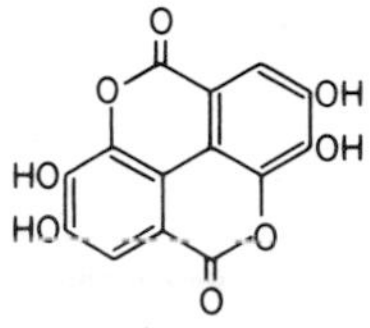

(10, ellagic acid)

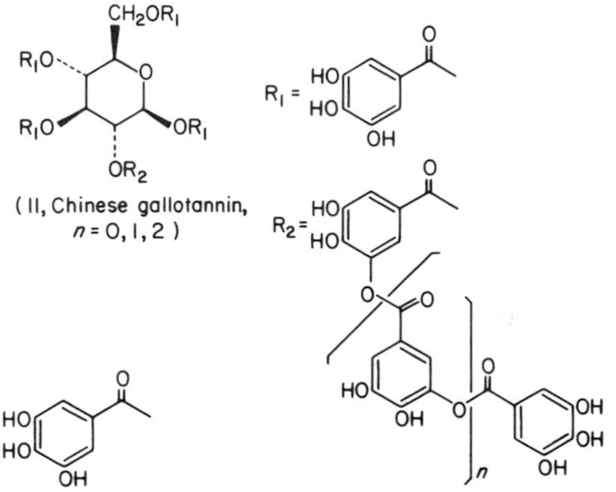

(11, Chinese gallotannin, $n = 0, 1, 2$)

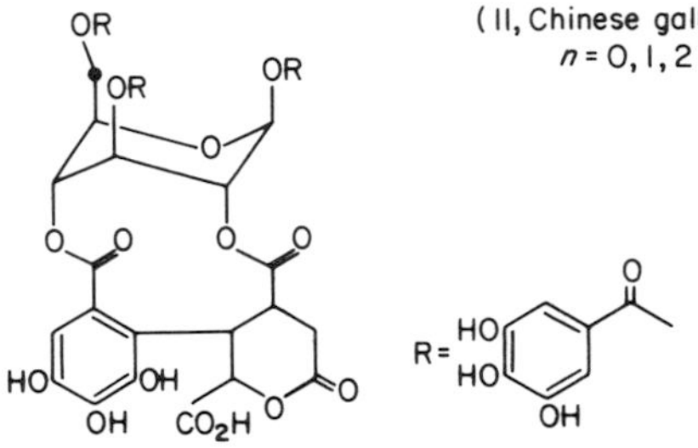

(12, chebulinic acid)

Fig. 4.III

or enzymes to smaller molecules such as gallic (9) and ellagic acids (10) plus glucose (Fig. 4.III). Typical hydrolysable tannins are the gallotannins from *Rhus semialata* (Chinese tree) and *R. typhina* (Sumach) (11), and the ellagitannin chebulinic acid (12) from the fruit of *Terminalia chebula* (Myrobalans). The gallotannin (11) is frequently referred to as Chinese gallotannin since it is predominantly derived from the leaf and twig galls on the Chinese tree. The plant normally metabolizes the polyphenol (11) but upon insect attack, this metabolite accumulates in substantial amounts in the resultant insect gall. In its unrefined form, it constitutes the familiar tannic acid of commerce. The gallotannins and ellagitannins are limited in their distribution in the plant kingdom and are closely related both structurally and biogenetically. Schmidt (1956) has put forward elegant biogenetic schemes which link the two groups.

The condensed or nonhydrolysable tannins are more complex and heterogeneous and little structural information concerning this class of plant polyphenol was available until comparatively recently when definitive structures for various condensed plant polyphenols were proposed by

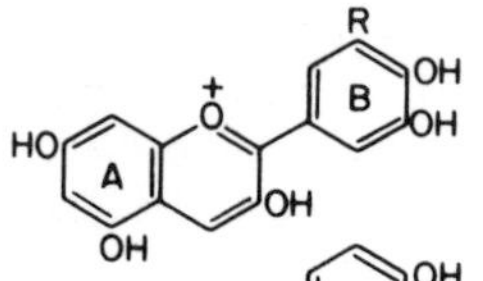

(13, R=H, cyanidin)
(14, R=OH, delphinidin)

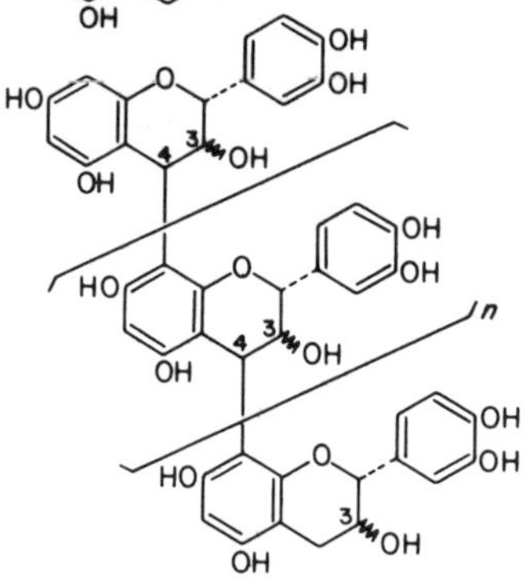

(15, procyanidins)
General structure: stereochemistry is variable at positions 3 and 4 in the flavan-3-ol structural units.

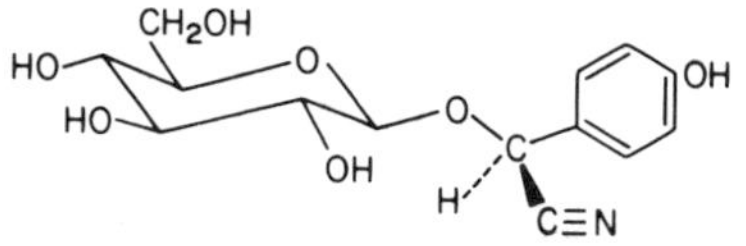

(16, dhurrin)

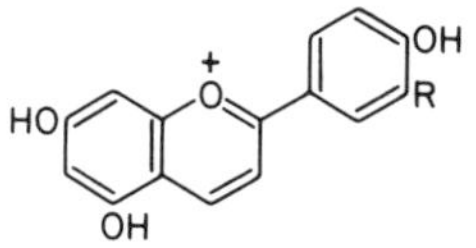

(17, R=OH, luteolinidin)
(18, R=H, apigenidin)

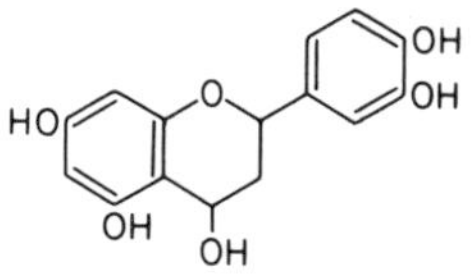

(19, luteoforol)

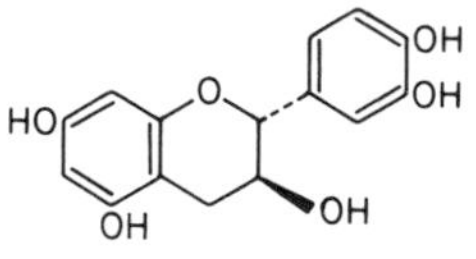

(5, (+)-catechin)

Fig. 4.IV

several schools (Drewes and Roux, 1964; Drewes *et al.*, 1967ab; Weinges *et al.*, 1968; Thompson *et al.*, 1972). The deep red colours developed from colourless materials present in many plant tissues upon treatment with mineral acid has long been recognized. Rosenheim (1920) named the class of substance responsible for this reaction "leucoanthocyanins". Subsequent examination of many plant "leucoanthocyanins" showed that they are mainly confined to plant tissues with a "woody" habit of growth and when treated with acids produce cyanidin (13) or delphinidin (14) (Bate-Smith and Lerner, 1954) (Fig. 4.IV). Bate-Smith and Lerner also suggested that these substances are probably most commonly responsible for a broad range of reactions in fruit and other plant tissues commonly attributed in the botanical literature to "tannins". Various workers during the last 15 years have examined the chemical properties of these polyphenols and, contrary to earlier suggestions, have shown that flavan-3-ol dimers (15, $n=0$) and higher oligomers (15, $n=1, 2\ldots$) are principally responsible for the properties outlined.

Willstatter and Tswett in Germany, Laborde in France and Robert Boyle about 1840 in England all noted that many plant tissues gave rise to a deep red colour—later shown to be that of the pigment cyanidin—when treated with acid. Rosenheim in 1920 began the first systematic investigation of these substances which he called "leucoanthocyanins" and in the 1930s Sir Robert and Lady Robinson in Oxford conducted the first surveys of plant materials in order to define their occurrence. In the 1950s further and much more extensive analytical work was developed by Bate-Smith (1954b) who noted that leucoanthocyanins were confined mainly to plants with a woody habit of growth and with Swain (Bate-Smith and Swain, 1953) he drew attention for the first time to the very close similarity in systematic distribution in plants of these compounds and those substances rather indefinitely defined in the literature of botany as tannins. Bate-Smith and Swain (1953) concluded that leucoanthocyanins were most commonly responsible for the broad range of reactions generally attritubed to tannins in plants, namely the precipitation of gelatin and alkaloids, astringent taste and the formation of amorphous phlobaphens with acid.

About this time, the first definitive work to determine structure began and since the molecules in question appeared to contain no carbohydrate the name was changed to "leucoanthocyanidin". Forsyth (Forsyth, 1952, Forsyth and Roberts, 1960) was the first investigator to show that the characteristic leucoanthocyanidin reactions of many plant tissues were probably due to flavan-3-ol oligomers. In the 1960s the nomenclature was further changed to "proanthocyanidins" following the proposals of Freudenberg and Weinges (1960). Proanthocyanidin is thus a generic term to denote all those substances isolable from plant tissues which give anthocyanidins when treated with acids. Procyanidins are particular members of this class of substance and give cyanidin (13) upon acid treatment; they are in addition the most common representatives of this class of polyphenol to be found in the plant kingdom.

Dimers such as (15) in which $n=0$ are barely able to precipitate proteins and on a molar basis they have a relative astringency only about 10% of Chinese gallotannin (11) (Fig. 4.III) (Haslam, 1974; Bate-Smith, 1973a). More complex procyanidins are known to be present in the seed coats of leguminous plants and in the vegetative tissues of various aquatic monocotyledons. These substances are normally difficult to extract from the plant by conventional procedures and hence are difficult to characterize. In some cases they appear to be associated with structural polymers within the tissue, but much further work on these materials is clearly essential.

Many procedures for the isolation of polyphenolic materials from plant materials have been proposed and reviewed (Haslam, 1966; Harborne,

1973; Ribereau-Gayon, 1972). As a general rule, it is desirable that whatever extraction procedure is adopted should utilize conditions sufficiently mild to isolate the spectrum of phenolic metabolites present in the plant tissue without significant destruction or transformation.

One should avoid the use of alkaline media since phenols, particularly those containing *ortho* or *para*—dihydroxy groups are highly susceptible to oxidation, particularly at high pH values. Since many phenols (e.g. the hydroxycinnamic acids *p*-coumaric, caffeic and ferulic, and sinapic acids and their derivatives) often appear as esters combined with sugars or related polyols, it is also advisable to avoid hot or boiling hydroxylic solvents and extremes of pH in the extraction process in order to minimize hydrolysis or acyl migration.

In the author's (E. Haslam) laboratory, the following general procedure has been employed successfully to extract polyphenols (molecular weight up to 1500) from fresh plant material such as fruit, root, leaf and stem tissue.

> Extract the plant material (4 times) with cold methanol in a blender, filter the plant debris free after each extraction and combine the methanol extracts. Remove some of the chlorophyll, fats and waxes from the methanol solution by extraction with light petroleum (2 times). Rotary evaporate the methanol solution at 30° to approximately one tenth of its original volume and dilute with water (4 to 5 times). Extract the aqueous solution with chloroform (2 times), rejecting the chloroform, and then with ethyl acetate (6 to 8 times). Dry the ethyl acetate solution (sodium sulphate) and finally evaporate at 25° to give the phenolic extract—usually as a brown gum or friable solid.

Thereafter, preliminary analysis of the extract should be undertaken to gain an insight into the range and type of phenols present. From this information it is then possible to select the separation procedure most suitable for the isolation of particular components. Analysis may be undertaken by two-dimensional paper chromatography (Haslam, 1966; Harborne, 1973; Ribereau-Gayon, 1972) using an aqueous solvent (6% acetic acid) and an organic solvent (butan-2-ol-acetic acid–water, 14/1/5, or *n*-butanol–acetic acid–water 4/1/5) to irrigate the paper (Whatman No. 2) in the two required directions. Phenols can be revealed on the paper by their characteristic fluorescence when examined under ultraviolet light (first in the absence and then the presence of ammonia vapour) and by numerous spray reagents (ferric chloride–potassium ferricyanide; vanillin–*p*-toluene sulphonic acid; diazotized *p*-nitroaniline; Gibbs reagent). No specific rules can be given for the selection of a particular isolation procedure but chromatography on polyamide, cellulose, Sephadex and counter-current distribution are the techniques which find most frequent application for the separation of simple phenols (Haslam, 1966; Harborne, 1973; Ribereau-Gayon, 1972).

Sorghum Polyphenols

The economic and agronomic importance of sorghum as a subsistence grain for people living in the semi-arid tropics has been emphasized elsewhere in this publication. Clearly its potential nutritional benefit is limited by the presence of tannins in some genotypes. These polyphenols give the grain an astringent taste and they are probably involved in the formation of the deep red to brown appearance which some grains assume as they ripen. Tannins may cause off-colour in products derived from sorghum grain. For example, during wet weather or during steeping the polyphenols can migrate from the seed coats into the endosperm and cause dark-coloured starch—either by complex formation with trace metals or by oxidation. Little definitive chemical research has been carried out on the phenolic compounds of sorghum and as is described later, the structure of whatever tannins are present has only recently been elucidated. Although it has not been reported in the seed coat, the phenolic cyanogenic glycoside dhurrin (16) (Fig. 4.IV) is reported (Eyjolfsson, 1970) as a constituent of the leaves of sorghum. Similarly, an unidentified acylated form of cyanidin-3-glucoside, luteolinidin (17) and apigenidin (18) (Fig. 4.IV) and glycosylated forms of these anthocyanidins have been reported as constituents of the first internode of sorghum (Stafford, 1965).

Bate-Smith (1969) and Bate-Smith and Rasper (1969) examined the tannins of a variety of sorghum grains and stated that these, when heated with acid, gave the very uncommon anthocyanidin luteolinidin (17) (Fig. 4.IV). Bate-Smith attributes this observation to the presence of a compound luteoforol (19) in the seed coat and further suggests that the red colour of the ripened seed coat and the red or black colour of the glumes result from the presence of complexes embodying the anthocyanidins (17 and/or 18) which arise from precursors such as luteoforol (19) in the green immature tissue. Bate-Smith, however, noted the presence of "leucocyanidin" in the grain of one variety of sorghum analysed, and an amylase inhibitor isolated from Leoti sorghum was later identified as a tannin of the same group (Strumeyer and Malin, 1969).

Strumeyer and Malin (1975) have reported a procedure for the isolation of condensed tannins from sorghum. The tannins were selectively adsorbed on Sephadex LH-20 in 95% ethanol and then removed in 50% aqueous acetone. Analysis showed the tannins from Leoti sorghum and GA 615 to consist of a series of polymeric polyphenols which gave cyanidin exclusively on acid hydrolysis.

More recent observations from the author's (E. Haslam) own laboratory (see later text) support these earlier reports and suggest that part, at least, of the astringency of high-tannin sorghum grain may be ascribed to the presence of procyanidins such as (15) based on the flavan-3-ol (+)-catechin (5).

Analysis of Polyphenols in Sorghum

Many scientists have reported the results of various methods by which the tannins in sorghum have been analysed. A critical evaluation of these methods has been made by Maxson and Rooney (1972a), who state that by "tannins", they mean polyphenols. At the same time, they succinctly highlight the current dilemma when they state: "We do not know what biochemical compounds are being measured, [because] each method measures different compounds".

Of the various procedures of tannin analysis examined by Maxson and Rooney (1972a), only one, which utilizes their efficiency as precipitants for proteins, is relatively specific for vegetable tannins as opposed to polyphenols in general. This method, gelatin precipitation, was found by Maxson and Rooney to be unreliable. Many of the remaining tests examined (including reactions with ferric ammonium salts, Folin-Denis reagent, arsenotungstic acid, vanillin-hydrochloric acid, and permanganate oxidation) are dependent either upon the ease of oxidation of phenols, or the ready formation of coloured complexes or condensation products. In this sense, it is unfortunate that these tests do not distinguish the polyphenols in a sorghum extract from those simple monomeric phenols such as (+)-catechin (5), which will respond to each of these reagents. Two other tests examined by Maxson and Rooney (1972a) were those of Bate-Smith and Rasper (1969) for the determination of luteoforol (19) (treatment in the cold with concentrated sulphuric acid in absolute methanol) and the methanol-hydrochloric test.

Bearing in mind the limitations outlined, of the seven tests examined, only three were judged by Maxson and Rooney to have potential use for the analysis of polyphenols and tannins in sorghum grain. In the sense that the polyphenol content may be related approximately to the tannin content, these methods appear to be the most reliable ones available at the present time. The tests selected were: (a) the Bate-Smith and Rasper procedure to detect luteoforol (19), the chemical basis of which is not clear; (b) the ferric ammonium sulphate method which, broadly speaking, detects all those polyphenols containing an *ortho*-dihydroxy or catechol grouping that yield a coloured complex such as (20) with an iron salt (Fig. 4.V); and (c) a modified vanillin-hydrochloric acid procedure formulated by Maxson and Rooney (1972a) in which those polyphenols with a *meta*-hydroxy or resorcinol group undergo condensation with vanillin to form a quinone–methide type of structure (21) (Fig. 4.V). Of these tests, that which showed least variation among samples analysed was the formation of a colour complex with ferric ammonium sulphate after extraction with urea. Maxson and Rooney (1972a) provide details of this test and the vanillin-hydrochloric acid method.

From the foregoing review the need is readily apparent for more precise and comprehensive research to determine the nature and range of the phenolic compounds present in sorghum grain, particularly those polyphenols that behave like vegetable tannins, and those that act as antinutrients and/or influence the palatability and acceptability of sorghum grain.

Research is needed to clarify the manner in which the seeming complex

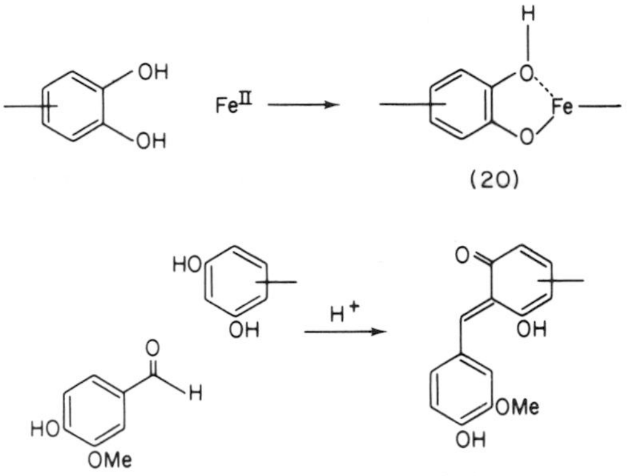

Fig. 4.V

range of phenolic metabolites change or are modified during the processes associated with ripening, giving rise to the characteristic colour and appearance of the mature seed coat(s) of many sorghums. Such research is seen as an absolute necessity before the physiological and biochemical consequences of the polyphenols presence in sorghum can be fully elucidated.

The following text describes in some detail the results of research carried out at the University of Sheffield by the authors (Gupta and Haslam). As described above, the structure of the condensed tannins has remained for 70 years or more a classical unsolved problem of organic chemistry and since a solution to the sorghum polyphenols problem is associated with a solution to the condensed tannin problem, it is convenient initially to present a review of work carried out in Sheffield over the past 10 years. Though they have been given a variety of names, the chemically preferred name for many of the condensed tannins found in the vegetative tissues of plants is "procyanidins", the name proposed by Freudenberg and Weinges in 1960, (see p. 307).

The skins of immature fruit are often rich in procyanidins which are probably the most significant and widely distributed of the complex polyphenols that are classified as condensed tannins, to be found in the plant kingdom. Plant procyanidins are oligomeric flavan-3-ols and they occur almost invariably in association with one or both of the flavan-3-ol monomers (+)-catechin (5) and (−)-epicatechin (7) (Fig. 4.II). Monomers and simple oligomers (procyanidins) occur free and unglycosylated. Biosynthetic studies (D. Jacques, C. T. Opie, L. J. Porter and E. Haslam, Sheffield University 1970–1976) show that the various distinctive patterns of procyanidins found in plants probably arise directly by reaction of one or both of the flavan-3-ols (5 or 7) utilizing their nucleophilic character at C-6 or C-8 with one or both of the carbocations (22 or 23) (Fig. 4.VI). Each of the four possible reactions can be reproduced exactly in the laboratory and

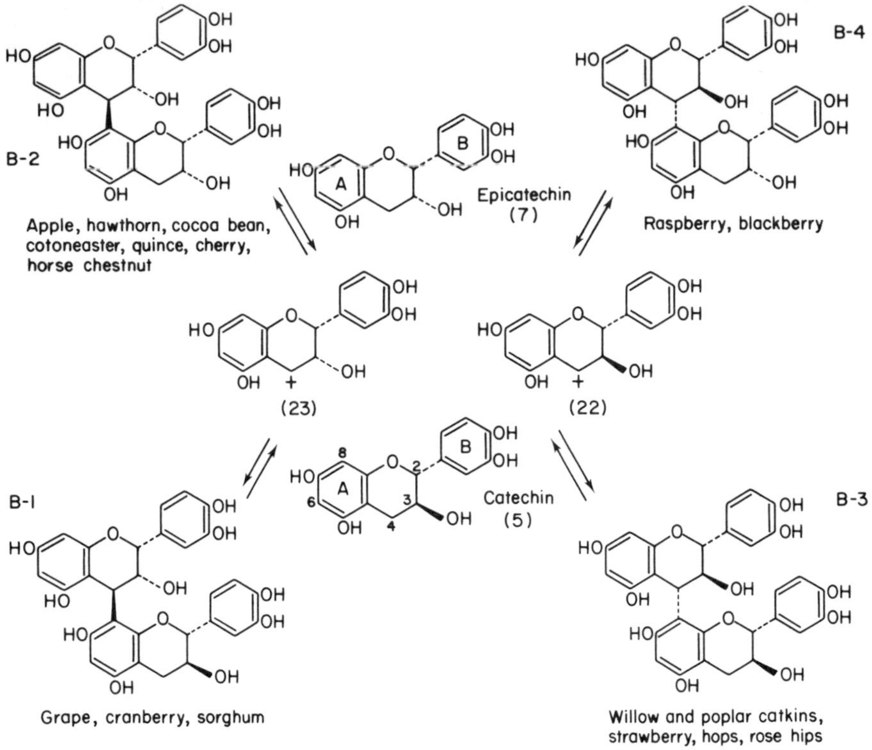

B-2 Apple, hawthorn, cocoa bean, cotoneaster, quince, cherry, horse chestnut

Epicatechin (7)

B-4 Raspberry, blackberry

(23) (22)

Catechin (5)

B-1 Grape, cranberry, sorghum

B-3 Willow and poplar catkins, strawberry, hops, rose hips

Fig. 4.VI Major procyanidin dimers: plant sources.

the pattern of products formed matches both qualitatively and quantitatively that found in in particular plants.

Work in Sheffield has concentrated upon the procyanidins found in the vegetative tissues of plants and for reasons of supply these have almost invariably turned out to be fruit-bearing plants. Initially a large scale survey of plants was carried out using two-dimensional paper chromoatography. In all cases, wherever procyanidins were found, so also were one or other or both of the flavan-3-ols, (+)-catechin (5) and (−)-epicatechin (7). The survey also revealed "procyanidin fingerprints" which could be classified into four principal categories. Plants were grouped according to their "procyanidin fingerprint" (Fig. 4. VI). Thus for example, the most commonly occurring "fingerprint" (as revealed by two-dimensional paper chromatography) found in dicotyledonous plants is probably that given by reaction of (−)-epicatechin (7) and the carbocation (23).

Two dimeric procyanidins are formed as well as trimeric, tetrameric and higher oligomeric forms. The dimer B-2 predominates (>80%) and may be readily identified. This pattern of procyanidins is found, for example, in *Malus* sp., *Crataequs* sp. and *Prunus* sp. Similar reactions occur between (5) and (22), (5) and (23) and (7) and (22) and the pattern of procyanidins formed is characteristic of different plant species as indicated in Fig. 4.VI. In each case trimers, tetramers and higher oligomers are formed but one dimer—respectively B-3, B-1 and B-4—predominates in the plant tissue.

The polymeric forms of procyanidins are presumably formed by reaction of the appropriate dimer with further carbocation in the manner previously described. They have the general structure as shown in (15), Fig. 4.VII,

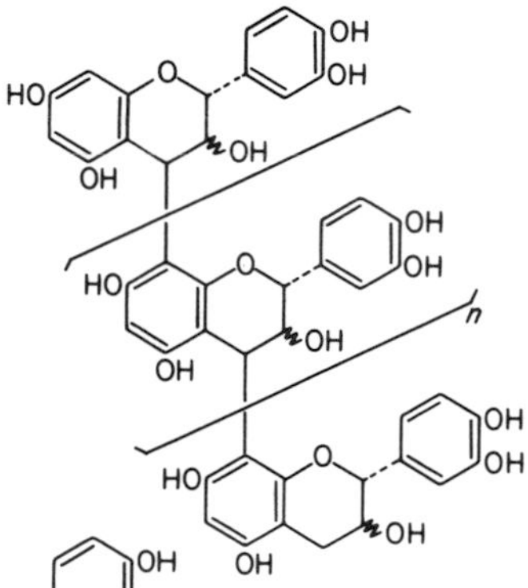

(15)

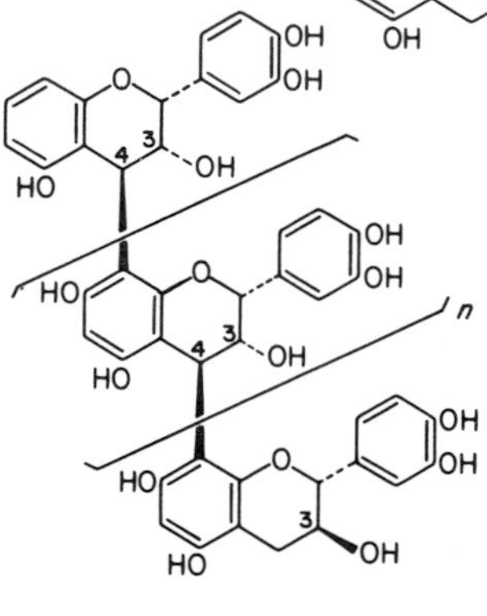

(24, n = 4–5)
Polymeric procyanidin from *Sorghum* - average structure

Fig. 4.VII

although, because of the apparent random nature of the biosynthetic reaction, a small proportion of the interflavan linkages will be C–4 to C–6 as opposed to the majority which will be (as depicted in Fig. 4.VII) C–4 to C–8. According to the accepted definition of vegetable tannins those molecular species where n is greater than three and less than 10 might be expected, assuming they are soluble, to act as vegetable tannins.

One highly characteristic fingerprint, and probably the most commonly encountered one, is that found in hawthorn (*Crataequs monogyna*). Co-occurring alongside (−)-epicatechin is one major procyanidin dimer (B-2, formally two (−)-epicatechin units, linked C–4 to C–8), a minor procyanidin dimer (B-5, formally two (−)-epicatechin units, linked C–4 to C–6), a major trimer and tetramer and various higher oligomers (Fig. 4.VI).

An alternative and again highly characteristic pattern is found in the sallow willow catkin (*Salix caprea*) where (+)-catechin, a major procyanidin dimer (B-3, formally two (+)-catechin units, linked C–4 to C–8), a minor procyanidin dimer (B-6, formally two (+)-catechin units, linked C–4 to C–6), a trimer, tetramer and higher oligomers constitute the fingerprint. The other two genetic situations with their characteristic fingerprints were found in raspberry and blackberry (*Rubus idaeus*, *R. fructicosus*)—(−)-epicatechin, procyanidin B-4 and associated procyanidins and in sorghum—(+)-catechin, procyanidin B-1 and associated oligomeric procyanidins. In passing it may be noted that these latter two categories are the most significant from a biosynthetic point of view. Thus it should be remarked that the two flavan-3-ol fragments which formally compose the major procyanidin dimer (B-4 or B-1) are of opposite absolute stereochemistry at C–3.

It was possible using this form of analysis to classify most procyanidin bearing plants according as to whether they belong to one or other of the four "pure" genetic forms or to any particular combination. For isolation purposes the most fruitful sources of the various procyanidins are indicated in Fig. 4.VI. Using a variety of isolation procedures (principally chromatography on Sephadex LH-20 and counter-current distribution) it has been possible to isolate all the dimeric and the trimeric procyanidins for chemical investigation. In some cases the yields are quite impressive. Thus 1 kg of fresh horse chestnut shells (*Aesculus hippocastanum* or *A.* x *carnea*) yields upwards of 1 g of procyanidin B-2.

Comparatively little is yet known concerning the higher oligomeric forms but the principles which govern the way in which they are constructed are thought to be the same as for the simple dimers and trimers. Hence it is assumed that their structures are essentially extensions of these simpler structures in which the molecular size is increased by the addition of further flavan units mainly through C–4 to C–8 linkages. From a physical point of view however this increase in molecular weight is accompanied by a decrease in solubility in aqueous media and special techniques may be necessary to isolate these larger molecules. In terms of their ability to precipitate proteins and hence their astringency to the palate the dimeric procyanidins have a relative astringency of some 10% compared to Chinese gallotannin (tannic acid). As the molecular size of the procyanidin increases so also does its astringency but as yet no reports have been encountered of studies that have identified the components of the procyanidin complex that contribute most to the astringent taste of plant materials.

Determination of the chemical structure of the procyanidins proved to be a challenging problem, particularly since the armamentarium of spectroscopic methods available to the organic chemist proved, in the early stages, to have limited value. Consequently, chemical procedures to deduce structure had to be evolved and although this often proves to be a more time consuming procedure, it has the invaluable bonus that one learns inevitably a great deal more about the properties and the chemistry of the substances under examination. A few aspects of this work are noted below.

Procyanidins are so named because they are chemically degradable to give the pigment cyanidin (13). This property arises from the lability towards acids of the interflavan bond (Fig. 4.VIII). In a procyanidin dimer acid catalysed fission of this bond gives the flavan-3-ol from the "lower half" and the flavan-4-yl carbocation from the "upper half". This extremely reactive intermediate unless it is trapped, decays by loss of a proton and a hydride ion to give the pigment cyanidin. This chemical reaction forms the

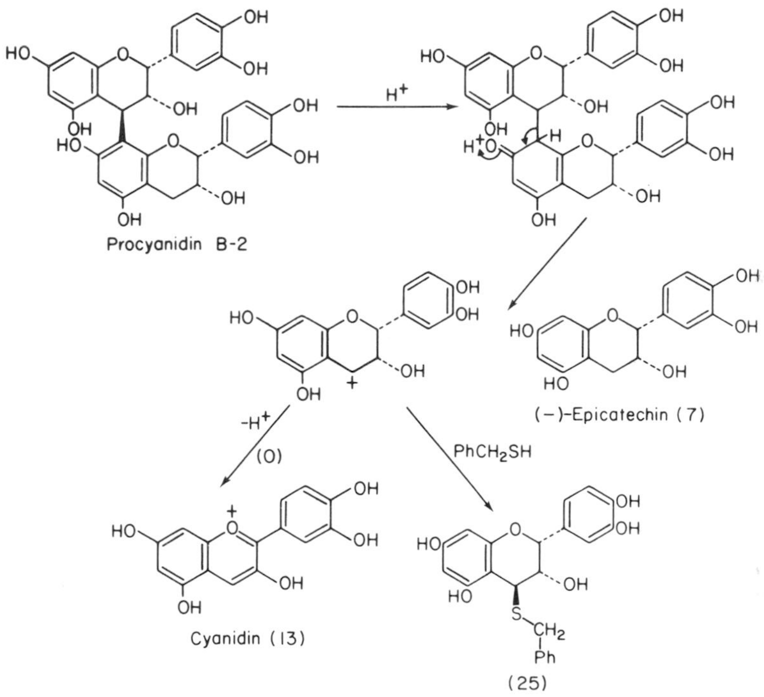

Fig. 4.VIII Procyanidin B-2: Acid catalysed degradation.

basis of the highly distinctive reaction by which the compounds are recognized and it will be noted is essentially the reverse of the biosynthetic reaction in which they are formed. The carbocation may however be trapped by a reactive nucleophile—such as a thiol, toluene-α-thiol, and this reaction forms the basis of the chemical degradation of procyanidins which has been utilized for the purpose of determining structure (Fig. 4.VIII). The flavan-4-yl-thioethers (25) may be isolated and characterized and hence the structure and composition of the procyanidin dimers determined.

Procyanidins are members of the flavonoid group of compounds and a great deal has already been adumbrated concerning the general pathways of flavonoid biosynthesis. Ring A of the flavonoid carbon skeleton (Fig. 4.VI) is derived from acetate (malonate) and ring B plus the three carbon atoms of the heterocyclic ring between A and B originate from cinnamate. Evidence for this theory comes predominatly from isotopic tracer studies and from some enzymic work in tissue cultures. The chalcone-dihydroflavone pair are the first recognizable intermediates on the pathway but steps from this stage to the various distinctive classes of flavonoid are somewhat uncertain. Fig. 4.IX shows one speculative pathway (route A),

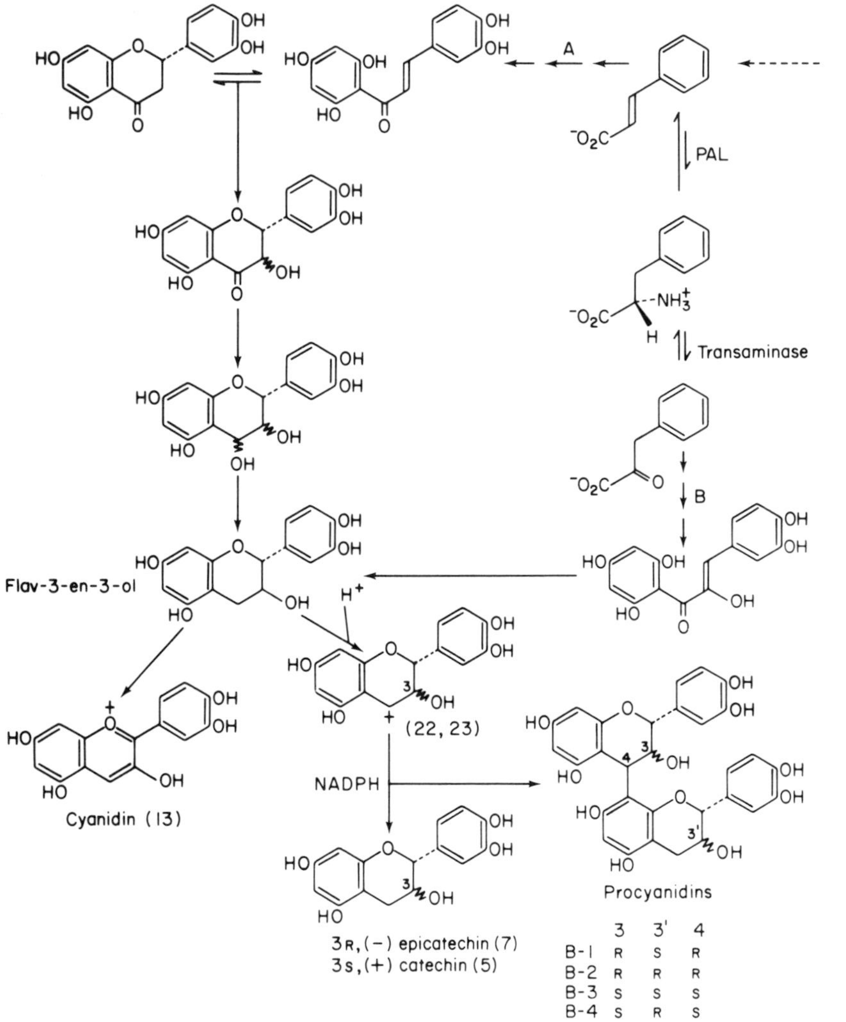

	3	3'	4
B-1	R	S	R
B-2	R	R	R
B-3	S	S	S
B-4	S	R	S

Fig. 4.IX Speculative pathways (route A and route B) to the various distinctive classes of flavonoid. (The nomenclature *R* and *S* shown for the procyanidins refers to their absolute stereochemistry at the positions indicated. A discussion is given by Cahn (1964).

based on the conventional intermediacy of the chalcone-dihydroflavone pair, to the flavan-3-ols and anthocyanidins which are the two principal groups of flavonoids which lack an oxygen atom at position 4 on the heterocyclic ring. The flav-3-en-3-ol is a key intermediate on this pathway; reduction would give the flavan-3-ol and oxidation the anthocyanidin. An alternative pathway, B, (Fig. 4.IX) to the flav-3-en-3-ol might, for example, be via the α-hydroxychalcone derived from the α-keto acid as the C_6–C_3 precursor.

The study of biosynthesis of the procyanidins has been conducted with a range of fruit-bearing plants—horse chestnut (*Aesculus hippocastanum*, *A.* x *carnea*), raspberry and blackberry (*Rubus* sp.) and male willow catkin (*Salix caprea*, *S. innorata*). The results, using a variety of labelled cinnamate precursors are summarized in Fig. 4.X and were broadly similar for all the plant species examined. A study of procyanidin metabolism in sorghum showed that when the seed coat of the grain was etiolated procyanidins were not metabolized but when the seed coat took on the green colour of unripe grain there also occurred a rapid phase of procyanidin synthesis, which was followed by a steady state situation in which procyanidin synthesis was small or zero. No such clear distinction was made with other plants but it was observed during the biosynthetic studies that the time of administration of the isotopically labelled precursors was critical. Tissue which was relatively mature showed little evidence of incorporation of the intermediates.

The results (Fig. 4.X) show that the C_6–C_3 carbon skeleton of the cinnamate precursor is incorporated intact into the flavan units, H_a was retained (although significantly only 80–90%), H_c was lost and H_b was retained (50% retention, one NIH shift occurring on the introduction of the *ortho*-dihydroxy orientation of phenolic groups). A further significant observation was that the two identical structural fragments which go

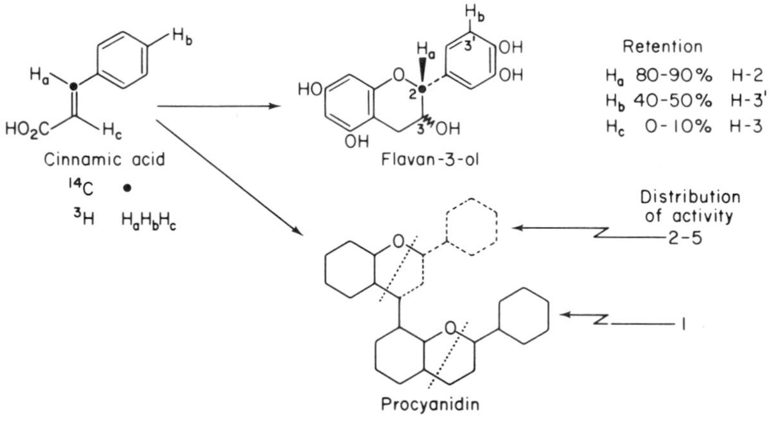

Fig. 4.X Procyanidin metabolism: experimental data.

towards the overall procyanidin dimer structure were labelled to different extents. This crucial feature of the evidence has been interpreted to show that the two flavan-3-ol type units of the procyanidin molecule are derived from *different* metabolic entities. Coupled with the strong circumstantial evidence which implies a very close connection to flavan-3-ol biosynthesis itself this evidence has been formulated to give a projected scheme of biosynthesis as indicated in Fig. 4.IX for the procyanidins.

If the reduction of the flav-3-en-3-ol to flavan-3-ol (5 or 7) is envisaged as a two-step process in which stereospecific proton addition to give the hybrid protonated species precedes stereospecific (*cis* or *trans*) delivery of hydride ion, or its equivalent from say nicotinamide adenine dinucleotide phosphate (NADPH), then the carbocations (22 or 23, 3S or 3R absolute stereochemistry) probably may derive from a situation in which the supply of biological reductant is rate-limiting. The carbocations would then be formed by leakage of the hybrid ion from the active site of the enzyme and, it is postulated, would then react with the final reduction product, the flavan-3-ol, still remaining in the vicinity of the enzyme to yield procyanidins. The trimers, tetramers and various higher oligomers and finally polymers which are also formed can then be visualized to result from reactions of the appropriate dimer and further carbocation in an exactly analogous manner.

Identification of Sorghum Polyphenols

Taking advantage of the experience acquired at Sheffield on the plant procyanidins, a research programme began during 1976 to isolate and determine the chemical nature of the "tannins" present in sorghum. Several samples of sorghum seed were initially analysed for polyphenols by paper chromatography of methanol extracts. The paper chromatographic pattern was relatively complex, but one of the "typical" procyanidin fingerprints was apparent in which the flavan-3-ol (+)-catechin (5) and the procyanidin dimer-procyanidin B-1 were identified on the paper chromatograms (Fig. 4.XI) along with various luteolinidin glycosides. Large scale extractions were made of a "high-tannin" sorghum obtained from Dr J. D. Axtell, Purdue University and identified as NK 300*. These extracts appeared to contain only polymeric procyanidins; the presence of procyanidin B-1 and (+)-catechin could not be detected. The reasons for this situation are still not entirely clear. Nevertheless, at the time of writing, (+)-catechin and procyanidin B-1 have not been isolated as distinct metabolites from any sorghum grain which has been made available to us for examination. Chemical work has concentrated therefore upon the polymeric procyanidin which was present in the "high-tannin" sorghums and which, from paper

*Other workers have reported the tannin content of NK 300 as: Wessels (1970ab) 0·37% (method of AOAC, 1955; Du Preez and Wessels, 1970); Rostagno *et al.*, (1973a) 0·66% (tannic acid equivalent, Folin-Denis method, Burns, 1963); Daiber (1975) 1·06% (total polyphenols, method Jerumanis, 1972).

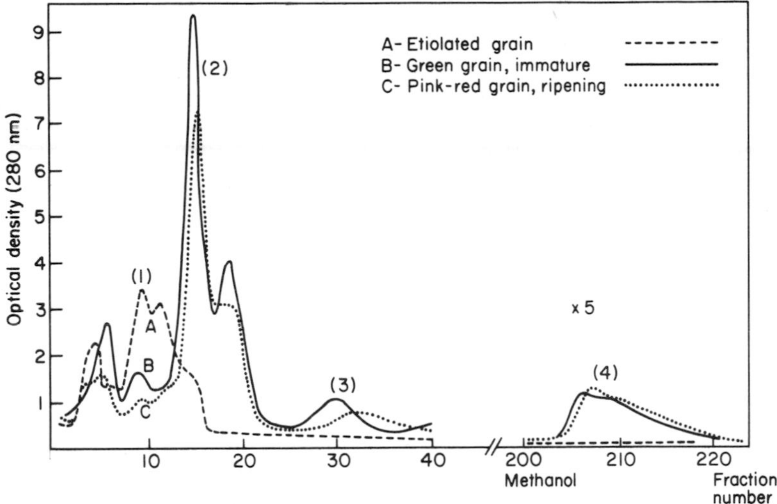

Fig. 4.XI Compounds identified in sorghum by paper chromatography included (1)—hydroxycinnamoyl esters; (2)—(+)-catechin; (3)—procyanidin B-1; (4)—oligomeric procyanidins.

chromatographic analysis, appears to be present in most—although not all—sorghum varieties we have examined. It appears to be present in brown-red seed-coated, but not white, sorghums.

On the basis of the previous work on procyanidins outlined above, however, the polymeric procyanidin in sorghum might be expected to be formed from (+)-catechin (5) and the carbocation (23) (Fig. 4.XII) in a multiple type of condensation via procyanidin B-1 (Fig. 4.VI) to give a polymeric structure of the general form (24). The subsequent chemical work carried out supports this proposal.

The polymeric procyanidin was isolated after chromatography of sorghum extracts on Sephadex LH-20 as a pale brown powder which analysed correctly for the polytetrahydroxyflavan-3-ol structure (24) (Fig. 4.XII). It contained, when pure, no nitrogen or sulphur. The polymer when treated with hydrochloric acid in ethanol at 60° gave cyanidin (13) and was degraded by two acid-catalysed procedures. Treatment with acid in the presence of phloroglucinol gave as the major product the adduct (26) [produced by capture of the carbocation (23) by the phloroglucinol] and (+)-catechin (5). A minor product was the phenol (27). The products of reaction were separated and identified by analysis, preparation of derivatives (acetate and methyl ether), and by ^{1}H NMR analysis. Treatment of the polymer with acid in the presence of toluene-α-thiol similarly gave as identifiable products (+)-catechin, and the two thioethers (25) and (28). These were characterized by ^{1}H NMR analysis and by the preparation of acetate and methyl ether derivatives. These products result from the random acid catalysed fission of the interflavan bonds in the polymer and capture by the thiol reagent of the carbocations (23) and (29). In later work

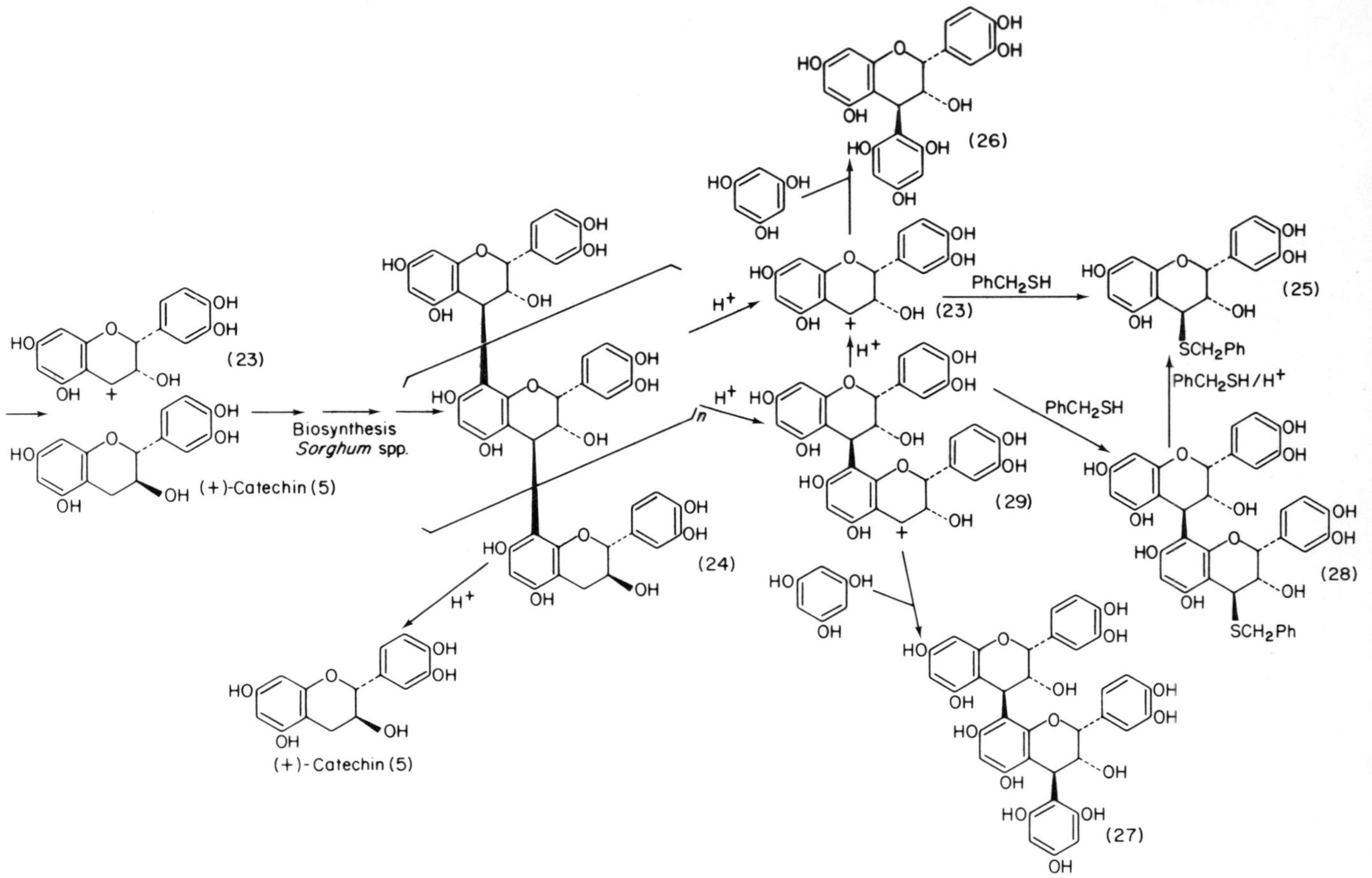

Fig. 4.XII Procyanidin polymer, sorghum.

the latter reaction was pursued to completion using excess toluene-α-thiol and long reaction times and in these instances the only recoverable products were (+)-catechin and the thioether (25). This reaction was subsequently used as a basis for the determination of the molecular weight of the procyanidin polymer. Thus the ratio of (+)-catechin (5, chain termination unit) and the thio ether (25, chain extension unit) obtained by typical degradation was 1/5 or 1/6 and this gives a chain length of 6 or 7 flavan-3-ol units in the polymer, (i.e. in formula (24), $n = 4$–5, the consequent molecular weight being 1700–2000). Two preliminary analyses using this method are reported below:

Sorghum IS 9115. Brown seed coat with red testa.	
(2R, 3S, 4S) -4-benzylthioflavan-3, 3′, 4′, 5, 7-pentaol	0·228 g
(+)-Catechin	0·051 g
Average molecular weight ~1700	
% Procyanidin polymer in seeds is ~0·4%	
Sorghum IS 3640. Brown seed coat.	
(2R, 3S, 4S) -4-benzylthioflavan-3, 3′, 4′, 5, 7-pentaol	0·569 g
(+)-Catechin	0·142 g
Average molecular weight ~1700	
% Procyanidin polymer in seeds is ~1·1%	

These results indicate the approximate order of molecular size of the procyanidin polymer in sorghum.

Analysis of "Tannin" Content of Sorghum Seed

Preliminary results and observations

The analyses carried out thus far on the "tannin" content of sorghum seed must be regarded as preliminary ones. However in terms of the convenience and ease of performance and in so far as a broad correlation exists in the results which have been obtained, three methods, 1, 2 and 3 have been selected for further investigations. These methods are all reasonably easy to perform although method 2 based on ultraviolet absorption is somewhat tedious and time-consuming.

Method 1: vanillin–hydrochloric acid. This is a variation on an old established procedure and is based on the measurement of nuclei of the phloroglucinol type by formation of a coloured chromophore with vanillin. The reaction is carried out with a methanol extract of sorghum and thus estimates the concentration of simple monomeric species, such as (+)-catechin, which are not "tannins" in addition to the higher oligomeric procyanidins which are thought to be "tannins". The reaction of vanillin catalysed by hydrochloric acid with (+)-catechin, procyanidins B-2 and B-3 and the polymeric procyanidin from sorghum NK 300 was investigated initially to determine its suitability as a quantitative measurement. One factor which was readily observed was that the reaction of vanillin with (+)-catechin and the various procyanidins is not the same in each case. The time course of the development of colour (λ—500 nm) is different and the time for maximum colour development is again, in each case, quite different. For this reason to estimate the "tannins" in a particular sample of sorghum the standard reference sample used in the estimation was a sample of the polymeric procyanidin previously isolated from sorghum NK 300. The procedure utilized in Method 1 is outlined below. The reagent was prepared by mixing just prior to use equal volumes of vanillin solution (4% in methanol) and hydrochloric acid (8% of 12 N in methanol). A standard calibration graph was prepared as follows.

Sorghum procyanidin polymer (NK 300, 25 mg) was dissolved in methanol (50 ml) and aliquots (1·0, 2·0, 3·0, 4·0 and 5·0 ml) were added to graduated flasks and the volumes made up, where necessary, to 5 ml with further methanol. From each flask solution (1·0 ml) was removed and added to the reagent solution (5 ml) at 30°. After 30 min the optical density of the solution was measured at 500 nm using the reagent in the blank cell. The measurements were all made in triplicate and a standard graph of optical density vs concentration constructed.

Determination of the tannin content was carried out as follows: sorghum seeds (5·0 g) were extracted with methanol (total volume 200 ml, 4 × 50 ml) in a high speed mixer. After each extraction the plant debris was filtered free and the methanol solutions combined. The methanol solution was reduced, by rotary evaporation at 30°, to a small volume and transferred to a graduated flask and the final volume of solution made up to 50 ml with methanol. This solution was then analysed by reaction of aliquots of differing concentration (1·0 ml) with the reagent (5·0 ml) in exactly the same way as described above. A second graph was thus constructed and from the slope of this graph and a comparison with the standard graph the concentration of tannin in the extract was determined.

Method 2: ultraviolet absorption method. In this procedure an extract of the polyphenols in the sorghum seed (5 g) was prepared as above in methanol (50 ml). An aliquot of this solution (2·0 ml) was taken and applied to a column of Sephadex LH-20 in ethanol (25 × 2·5 cm). The methanol solution was adsorbed onto the column and then the column was eluted with ethanol (500 ml)—by this procedure all the low molecular weight polyphenols were eluted from the column. The residue still retained on the column consists principally of the polymeric procyanidin and may be eluted with methanol (1000 ml). The methanol solution was concentrated and finally made up to 100 ml in a graduated flask. The optical density of this solution was measured at 280 nm and the amount of tannin determined from a standard concentration vs optical density graph prepared from authentic polymeric procyanidin (ex NK 300) earlier. It is important in this procedure that all solvents should be carefully redistilled.

Method 3: cyanidin coloration. This procedure makes use of the fundamental characteristic of procyanidins—namely the development of the pigment cyanidin with acid. Using this procedure all procyanidins (dimers → higher oligomers) react and are estimated. Whether dimers are "tannins" is questionable but from our observations their concentration is very low in all sorghum samples that we have examined.

Typically in this method a methanol extract (50 ml) of sorghum seeds (5 g) was prepared as outlined earlier. Aliquots (0·5 ml, 1·0, 1·5 and 2·0 ml) were taken, added to conical flasks (50 ml) and evaporated to dryness. Butan-1-ol (10 ml) containing hydrochloric acid (12N, 30% v/v) was added to each flask and the solutions heated at 115° for 2·5 hr. The solutions after cooling were transferred to graduated flasks and the volumes made up to 25 ml with butan-1-ol. The optical density of each solution was measured at 545 nm and a graph plotted of optical density vs concentration. Comparison with a standard calibration graph obtained analogously using the polymeric procyanidin from sorghum NK 300 then gave the percentage of tannin in the unknown.

Commentary

A summary of results is given in Tables 4.1, 4.2 and 4.3. These show the relative internal consistency which may be obtained by these methods and also demonstrate the differences in absolute values which the methods provide. In Table 4.2 a selection of results is given using the three analytical procedures described above and these are compared with values obtained by the toluene-α-thiol degradation described earlier (see p. 315). It will be noted that the latter procedure gives generally much lower absolute values and this has been ascribed to the difficulties of isolation inherent in this method on a small scale of operation. Finally in Table 4.3 some appreciation is given of the differences which result in the analysis if different standards are used in the analytical procedure. For the reasons outlined earlier authentic procyanidin polymer from sorghum is the only reliable standard which may be used for these analyses.

A major problem associated with the analysis of "tannins" in sorghum seed is what precisely constitutes the "tannins". Until much further fundamental work on the interaction of proteins and polyphenols has been carried out one can only base one's observations and conclusions upon the previous assumptions that those soluble polyphenols with molecular weights in the range 500–3000 constitute the "tannins" of plants. In the case of sorghum these are thought to consist of the various oligomeric procyanidins. Previous work indicates that in any plant which metabolizes pro(antho)cyanidins there is likely to exist in the tissue a range of molecules ranging from the simple "catechin" (flavan-3-ol) to complex oligomers based upon this basic structure. Biochemically it is not clear whether the metabolism of this type of polyphenol in a plant tissue constitutes a static or dynamic state. In some plants there is very little evidence to suggest that these substances are "turned over" metabolically and the present picture is rather that there is, during the growth of the tissue, a "burst" of proanthocyanidin synthesis which then either ceases or decreases to a vanishingly small level. This appears to be the case with sorghum. What then happens to the proanthocyanidins when the tissues mature and ripen is again completely unknown although some observations suggest that the balance of proanthocyanidins moves towards those with a higher molecular weight. Clearly therefore the analysis of "tannins" may be dependent upon the maturity of the sorghum seed and in this context it also appears probable that the ease of extraction of the polyphenols may be dependent on the physical state of the sorghum seed and hence on its maturity. All these factors contribute to the uncertainty of the analysis although it seems that the major uncertainty relates to what compound or compounds one should aim to analyse.

Conclusion

Sorghum contains a complex polymeric polyphenol of the procyanidin class in the seed coat. The condensed procyanidin isolated from a number of sorghum cultivars has a number average molecular weight of around 1700–2000, consisting on average of six or seven, 3, 3′, 4′, 5, 7-pentahydroxyflavan units linked by C–C bonds at the C_4 and (predominantly) C_8 positions in the flavan-3-ol molecule. All the available evidence and the known reactions of this class of molecule point to the fact that this polymeric procyanidin is the polyphenol responsible for what are the typical "tannin" reactions of sorghum grain affecting, for example, colour and its nutritional value. However it should be noted that it is still necessary and desirable to confirm that this assumption is correct.

Evidence from other areas of procyanidin biochemistry suggest that in the early stages of growth there occurs an initial synthesis of procyanidins and that as the tissue matures or ripens the balance of procyanidin metabolites changes from those with a low to those with a high molecular weight. This observation may well explain the general difficulty of finding the procyanidin dimer B-1 and associated low molecular weight procyanidins in all but a few samples of sorghum during this work. It also points to the need for an in depth study of the factors which influence and control polyphenol formation during plant growth, to what extent they are genetically determined, to what extent they change with the maturity of the seed and to what extent their concentration in the edible seed may be controlled by the plant breeder. All these variables must be studied to determine to what extent the tannin reactions of sorghum may be changed during growth and by the conditions of growth and breeding.

Sorghum and Millet Polyphenols

The following is a review of what was reported mainly before Gupta and Haslam began their research. Though some of the data may appear outdated given the recent findings, they are included for the sake of completeness.

Terminology and Analytical Methods

Throughout the text, the terms "tannin", "polyphenol" and "high- and low-pigment" are used as presented by the authors of the papers cited. It is also possible that the "tannins" reported varied quantitatively and qualitatively among the various sorghums referred to. Consequently, one must be cautious in making direct comparisons among the results of the different papers quoted.

In the light of Gupta and Haslam's scholarly presentation, no further attempt is made to review critically and comparatively the various methods reported for the determination of tannin content. Clearly most of them were borrowed from and relate to other sources of polyphenols and therefore are not precisely applicable to sorghums and millets. Nevertheless, for the sake of completeness, the painstaking efforts of many who have reported upon "tannins" in sorghum deserve to be included.

Herrett (1956) examined five varieties of grain sorghum: Martin, Darset, Redlan, 4414 and Wheatland in comparing four colorimetric methods and several ultraviolet methods. He concluded that only the Folin-Denis reagent and the arsenic tungstate reagent (prepared as described by himself) were suitable.

Maxson and Rooney (1972a), as quoted above by Gupta and Haslam, stated: "We do not know what biochemical compounds (tannins in sorghum) are being measured; each method measures different compounds". They evaluated 10 methods using four cultivars of sorghum all grown under similar conditions in 1968 at Lubbock, Texas: (A) GA 615, bird-resistant hybrid with brown testa and pericarp and normal endosperm; (B) B 398 (Martin), remnants of testa with reddish-brown pericarp, normal endosperm; (C) B 3197 (Kafir 60), no testa, white pericarp, normal endosperm; and (D) TX 2536 thin, white pericarp, no testa, yellow endosperm.

1. Ferric ammonium sulphate (FAS), as described in the paper, based on Mejbaum-Katzenellenbogen and Kudrewicz-Hubica (1966). (The use of two levels of urea (1a) 5 g and (1b) 50 g are reported).
2. Vanillin-hydrochloric acid (V-HCl), Burns (1963, 1971) as further modified and described in the paper (MV-HCl).
3. Ferric ammonium citrate, Burns (1963); reflux with water and reaction with ferric ammonium citrate.
4. A method for tannin in tea, AOAC (1965); boiling with water, titration with $KMnO_4$, precipitation with gelatin.
5. Modified Snell, Snell and Snell (1953); extraction with ethanol, isolation with lead acetate, reaction with arsenotungstic acid.

6. Methanolic-HCl, extraction with methanol: HCl (5/1) for 24h; absorbance determined at 465 nm.
7. Bate-Smith and Rasper (1969); extraction with 43% H_2SO_4 in methanol, absorbance determined at 465 nm.
8. Gelatin precipitation, several methods, based on precipitating gelatin from solution with tannin and measuring the protein remaining in solution.
9. Folin-Denis, Burns (1963); reflux 24h with urea, reaction with Folin-Denis reagent and saturated Na_2CO_3.
10. Folin-Denis, Schanderl (1970); reflux 5 h with water, reaction as Burns (1963).

The mean tannin content values and 95% confidence intervals for six of the methods are shown in Table 4.4. Data for the remaining methods were not reported.

In the light of the statistical analyses and from the problems encountered in the analyses, Maxson and Rooney (1972a) concluded that methods 3, 4, 5, 6, 8, 9 and 10 were unsuitable for the purpose of determining tannins in sorghum. Method 7 (Bate-Smith and Rasper, 1969) was rapid and reproducible but different varieties gave different absorption maxima.

Methods 1 (FAS) and 2 (MV-HCl) were chosen for further study with three sets of samples. Set I included grains of GA 615 grown at each of six locations during 5 years to determine if the methods could differentiate among tannin levels in a "high-tannin" sorghum grown under different conditions. Set II consisted of samples with and without pigmented testas and pericarps. Set III included samples varying widely in colour from white through lemon-yellow, red, brown and dark brown. Their results are summarized in Table 4.5.

It would appear that the two methods measured different substances in the grain: the "tannic acid equivalents" appeared considerably lower by the FAS than the MV-HCl "catechin equivalent" method. The MV-HCl method appeared faster than the FAS but produced significant day-to-day variations which, however, accounted for only 0·5% of the variability among set II. The FAS method, though more time consuming, displayed less variability.

Maxson and Rooney (1972a) stated that the Folin-Denis reagent was more responsive to certain nonphenolic substances than either the FAS or the MV-HCl method.

Maxson and Rooney (1972b) reported that both the MV-HCl and FAS methods detected differences in tannin content among cultivars and differences among replicates within a day. They suggested a significant cultivar × day interaction for the MV-HCl method but not for the FAS method.

Maxson *et al.* (1972) reported the tannic acid equivalents (FAS) and catechin equivalents (MV-HCl, Maxson and Rooney (1972b) in samples of GA 615, a brown-seeded, high-tannin sorghum hybrid, grown in 1968, 1969 and 1970 under dry-land conditions at each of six locations (exact whereabouts were not stated).

FAS values ranged from 0·78 to 1·10 mg/100 g and MV-HCl values from 2·71 to 4·32 mg/100 g. These values were significantly correlated. Both methods showed significant differences among locations: the MV-HCl method gave significant differences by years and suggested a location × year interaction. The specific environmental factors that appeared to affect tannin content by the methods compared were not elucidated.

Maxson *et al.* (1972) also examined four sorghum phenotypes, grouped I–IV, selected from near-isogenic lines to determine the effects of pericarp colour, plant colour, and pigmented testa on tannin content (among lines of similar genotype). The samples ranged in grain colour from white to red. Included for comparison were six phenotypes with pericarp colours of V brown, VI white and VII dark brown.

The phenotypes with their description and tannin content by the FAS and MV-HCl methods are shown in Table 4.6.

The FAS values for groups I–IV were not significantly different for red, yellow or white grains except where a testa was present; the MV-HCl values were significantly higher for red than for white or yellow grain. In grains in which a partial testa (i.e. one which did not surround the whole grain) was present, there was no apparent difference between red and white grains.

Both analytical methods detected significant differences between grains with and those without pigmented testas. MV-HCl values were significantly higher for plants with a red plant colour than for those with a tan plant colour.

The FAS method did not distinguish between red and white grain with and without a partial testa as found in group IV. The MV-HCl method showed a significant difference when the partial testa was present in red and white grain but no significant difference among red grain, red grain with a partial testa and white grain with a partial testa.

Maxson *et al.* (1972) also examined the effect on tannin content of the B_1 B_2 S genes which are believed to control the presence or absence in sorghum of the pigmented testa and pericarp (Schertz and Stephens, 1966). They describe 18 parental lines and hybrids together with genotypes known to possess these genes, grown at the College Station, Texas in 1970. Both the tannic acid equivalents (FAS) and catechin equivalents (MV-HCl) methods (Maxson and Rooney, 1972b) gave similar relative values for the 18 phenotypes with different B_1 B_2 S genotypes. The genotypes that produced a pigmented testa were highest in tannin content as defined by Maxson *et al.* (1972).

The FAS method indicated no significant difference among types without a pigmented testa, regardless of pericarp colour but the MV-HCl method produced higher values for types with a pigmented pericarp than for those with a white pericarp; the pigmented pericarp is controlled by genes *R Y I* (Schertz and Stephens, 1966).

Maxson *et al.* (1972) agreed with Burns (1963) in suggesting the two methods probably measure different phenolic compounds.

Bate-Smith (1973a) described a method for determining the astringency (ability to precipitate protein in the saliva) of extractable plant tannins. The method was based on the reaction of the tannins with the proteins of haemolysed blood and colorimetric determination of residual haemoglobin. Relative astringency is defined as the ratio of the concentration of the tannic acid to that of the tannin which causes the same degree of precipitation. The results of applying the methods to certain plants was described (Bate-Smith, 1973ab), but sorghum and millets were not examined.

Price and Butler (1977) proposed a method of determination for polyphenols in sorghum based upon the Prussian Blue reaction: the formation of a complex hexacyanoferrate $Fe_4(Fe(CN)_6)_2$ ion through the reduction of ferric to ferrous ions by polyphenols or other reducing substances. Aqueous extracts of the grain were mixed with $FeCl_3$ in HCl to which $K_3Fe\ (CN)_6$ was added. The intensity of colour, varying from deep blue, through blue, turquoise, green to pale green was said to reflect the level of tannins present. Spectrophotometric readings of optical density were made at 720 nm. Standard curves were developed using fresh solutions of commercial D-catechin. The Prussian Blue results were compared with catechin equivalents (CE) arrived at using the V-HCl method (Burns, 1971) but corrected by subtraction of a blank, this correction being necessary to compensate for the colour that develops from methanol, vanillin and HCl in the absence of any other phenolic substance. The comparative results are given in Table 4.7. Price and Butler stated that their Prussian Blue method makes no distinction between tannins and other polyphenols.*.

Price and Butler also presented standard curves for a variety of phenolic substances including gallic acid, *p*-dihydroquinone, quercetin and cyanidin. Since however none of these has been shown to be significantly present in sorghum or the millets, they are of little relevance to this review. Price and Butler made the useful observation that a wide degree of variation exists among different phenols in the degree to which they are oxidized in the Prussian Blue reaction; consequently the results obtained from mixed phenols may be difficult to interpret.†

Price and Butler (1977) emphasized that a blank reading must be subtracted from the CE value whether using the V-HCl (Burns 1971) or the modified V-HCl method (Maxson and Rooney, 1972b). They quote CE values uncorrected and corrected for a range of sorghums analysed by the V-HCl method. A few of their reported results follow:

* (Authors' note: it seems doubtful if the method will distinguish between phenols and any other reducing substances present).

†(Authors' note: it is not improbable that different sorghum samples may contain variable mixtures of phenolic substances; the Prussian Blue method should perhaps be applied with some caution).

Grain	CE (V-HCl)	CE corrected by subtracting blank
BR 64	3·42	2·45
IS 8193	2·68	2·11
IS 15526	1·42	1·01
IS 8687	0·70	0·02
IS 15991	0·60	0·39
IS 10486	0·57	0·00
IS 0339	0·36	0·00
IS 10562	0·20	0·00
IS 2042	0·15	0·00

Price and Butler (1977) stated the desirability of measuring only the polymeric tannins present in sorghum by their Prussian Blue method. They postulated that since condensed tannins are generally precipitated by NaCl, the difference between solubility in water and aqueous salt solutions might provide the basis for a Prussian Blue test specific for condensed tannins. After putting their hypothesis to the test, they reported a striking sensitivity of some sorghums to low salt concentrations, and that some varieties appeared to contain water soluble components "presumably tannin" [*sic*] which were not extracted in 0·2M NaCl. "All sorghum varieties tested contain other components, presumably anthocyanidins, whose extractability is unaffected by up to 1·0M sodium chloride."

Hagerman and Butler (1978) reported that the tannin content of crude extracts of several varieties of sorghum and partially purified tannins from the high-tannin sorghum BR 54, was determined by adding the sample to a standard solution of protein, isolating the insoluble tannin–protein complex, dissolving it in alkaline solution and measuring the absorbance at 510 nm after adding ferric chloride. Plots of absorbance as a function of the amount of tannin were linear for tannic acid and partially purified sorghum tannins for amounts of tannin ranging from 0·20 to 1·0 mg. Nontannin components of crude methanolic extracts of sorghum did not interfere with the assay. The results were qualitatively similar to results obtained with the vanillin assay (Burns, 1971). Hagerman and Butler stated the precipitation assay could be used to study the effects of pH and other parameters of tannin–protein interactions.

Phenolic Content in Sorghum

Barham *et al.* (1946) determined the tannin content (Snell and Snell, 1937 followed by Folin-Denis reagent) of 14 varieties grown on experimental plots in Kansas and Oklahoma. Atlas contained the lowest level (0·003%), and early Sumac (0·167%), Schrock (0·159%) and Leoti Red (0·158%) the highest.

Nasinec *et al.* (1966) determined the tannin content (colorimetrically, Horel, 1956) of seven varieties and line hybrids of sorghum, nine line hybrids of grain type *Sorghum vulgare* × *Sorghum sudanese* and five hybrids of *Sorghum vulgare* × *S. sudanensis* [*sic*] from a plant breeding programme in Czechoslovakia. The designations and tannin content are shown in Table 4.8.

Nasinec *et al.* stated that higher or lower tannin content was controlled genetically and, when crossed, the higher content in the grain was dominant. Within a variety, a negative correlation was found between size of seed (1000-seed weight) and tannin content; the bigger and heavier seeds having a lower content of tannin. The colour of the seeds did not always relate to the content of the tannin.

Nasinec *et al.* also separated the germ, pericarp and "perisperm" (this is probably meant to be "episperm" and refers to the testa) from the endosperm of samples of Early Hegari, Solary A, Solary ZH 6, Solary T and NK 135. The 1000-seed weight, percentage of germ with pericarp, and endosperm, with the tannin contents are shown in Table 4.9.

By the method used, the pericarp appeared to contain most of the tannin; the higher the total tannin content, the higher the proportion of tannins in the outer seed coat layers.

DuPreez and Wessels (1970) reported the "tannic acid" content (AOAC, 1955) of 14 varieties of sorghum grown in the Republic of South Africa. Details are shown in Table 4.10.*

At the time of publication, NK 222 and NK 300 were the cultivars most widely grown in South Africa. DC 500F was unusual in that methionine appeared as the first limiting amino acid (Wessels 1970b); at 0·45% it had the second highest tannin content of the samples examined. SSK 8 (0·94%) had the highest, and IC 500F (0·45%) the second highest tannin content.

Arora and Luthra (1972) found tannin contents (Nierenstein, 1944 and Snell and Snell, 1953) ranging from 0·23 to 3·73 mg/g in the seed of 19 grain sorghums grown for fodder at Hissar, India. (Details are shown in Table 4.11).

Pluenneke *et al.* (1973/4) determined the V-HCl tannin content of a light-coloured sorghum, DeKalb E 57 (0·5%), GA 700, a tan grain (3·8%), and a brown grain sorghum, DeKalb BR 64 (5·4%). They found no significant differences in total N which ranged (DWB) from 1·39 to 1·42% (nor in total protein which ranged from 9·0 to 10·7%). No significant difference among varieties was found in the content of 17 individual amino acids determined in acid hydrolysates. On this evidence, Pluenneke *et al.* suggested that protein quality was not related to tannin content. (No data were presented in this short report.)

Mehansho and Besrat (unpublished report, 1974) reported the tannin content (catechin equivalent (CE), Burns, 1971) of 44 varieties of Ethiopian

* (Authors' note: several authors refer to the "tannic acid" content of sorghum. We can find no published evidence that tannic acid, or any form of hydrolysable tannin has ever been isolated from sorghum).

sorghums grown at the Haile Sellassie I University, Ethiopia. (See Table 4.12).

Seven were high (above 1·0 CE) in tannin content, Leoti (5·20), Alemaya 108 (5·0), Abay 145 (4·0), Netch Shure (2·9), Gato 994 (2·65), WS 1763 (2·4) and WS 1641 (2·2). The varieties Gato 994 and Netch Shure were the highest yielding varieties reported by Gebrekidan (unpublished report, 1974).

In a later unpublished report, Mehansho and Besrat listed the protein (N × 6·25) and tannin content of 166 Ethiopian sorghum lines. The protein ranged from 7·1 to 15·7% with a mean value of 11·5%. Tannin contents ranged from 0·05 to 9·5 CE/100 g dry sample with a mean of 0·85. Twenty-five had tannin contents above 1·0 CE. Only 10 lines exceeded a CE of 3·5.

Jambunathan and Mertz (1973a) reported the protein (Kjeldahl N × 6·25) amino acid content by auto-analyser, tryptophan by ion-exchange chromatography, Slump and Schreuder, 1969), and tannin (Burns, 1971, CE) of six inbred lines of sorghum grown at the Purdue Agronomy Farm. The six lines examined were IS 0062, IS 3982, IS 2319 (low-pigment varieties, yellow or light yellow, or chalky white) IS 2283, IS 6992, IS 8165 (pigmented brown to red-brown and described as "bird-resistant"). Though the six lines varied in protein and tannin content, no major differences in amino acid content (as % protein) were noted either among the six lines or in comparison with the averages of 522 lines from the World Sorghum Collection maintained at Purdue University.

The endosperm, germ and pericarp (bran) fractions of IS 0062 and 3982 (low-pigment) and of IS 6992 and 2283 (high-pigment) were separated by soaking and manual dissection (method of Hubbard *et al.*, 1950). The proportion of pericarp (probably including the testa) was higher in the two high-pigmented sorghums; the low-pigmented samples had a higher endosperm content. The amino acid composition among the four lines was not different. The two low-pigmented lines (A) IS 0062 and (B) IS 3982, were compared with the two high-pigmented lines (C) IS 6992 and (D) IS 2283, on an equal weight basis in diets fed *ad lib.* to weanling rats over a period of 28 days.

In a further experiment, the high-pigmented IS 8165 (E) was compared with the low-pigmented IS 2319 (F), in isonitrogenous (8·8% protein) diets. The results are given in Table 4.13.

There was a marked difference in rate of weight gain among the different diets. The rats fed the high level tannin diet (D) lost weight. The weight gain differences may in part be attributable to the lower protein contents of diets (C) and (D). The influence of tannin content on weight gain was more apparent when the isonitrogenous diets (E) and (F) were compared. The average weight gain and the PERs were significantly higher among the rats fed low-tannin sorghums.

Jambunathan and Mertz (1973a) fractionated the defatted whole kernels, and hard-separated endosperms, of the same four lines used in diets (A), (B), (C) and (D) in rat feeding experiments following fractionation by the method of Landry and Moureaux (1970). The percentage of (micro-Kjeldahl) N in the five fractions obtained is shown in Table 4.14.

Fraction I represents the salt-soluble protein (albumins and globulins), fraction II the alcohol-soluble proteins (kafirin), fraction III the protein soluble in alcohol with a reducing agent, fraction IV (borate buffer + 2-mercaptoethanol) a class of proteins not previously reported for sorghum and fraction V, the glutelins.

The percentage of soluble N in the low-pigment sorghums was higher than in the high-pigment sorghums in fractions, I, II and III, and lower in fractions IV and V. The main difference in the distribution of protein between low-pigment and high-pigment sorghums occurred in fractions I and V. The distribution pattern was similar for the whole kernel and the endosperm but was less marked in the endosperm.

Fractions I, II and III were either colourless or pale yellow, with the whole kernels showing deeper colour than the endosperm. Fractions IV and V of the low-pigment IS 0062 and IS 3982 were respectively yellow and pale yellow; of the high-pigment IS 6992 and IS 2283, deep red and reddish-brown respectively.

Tannin determination on these extracts showed that fraction IV of IS 6992 and IS 2283 contained 5 ×, and fraction V twice as much tannin (CE) as the corresponding fraction of the low-pigment varieties.

Fractions I, II and III of the endosperm of the four sorghums were colourless, fractions IV and V of IS 0062 and IS 3982 were colourless and light yellow respectively, and of IS 6992 and IS 2283, red and yellow respectively.

Whole kernel samples of the high-pigment IS 6992 from a different lot from that used previously, and IS 2283 were alkali-treated (method of Blessin *et al.*, 1971) to remove the pericarp and testa containing the pigments. The dehulled samples were white and gave very low CE values. The fractionation pattern is shown in Table 4.15.

Dehlavi (1974) reported the relation among tannin content (V-HCl method, Burns 1971) and other characters in Combine Kafir 60 and Double Dwarf Early Shallu sorghum, both low in tannin content, and the F_1 and F_2 plants from their cross grown at the University of Georgia during 1972. There was a highly significant negative correlation between plant colour and head form; seed colour and tannin percentage; and seed colour and head form.* Nonsignificant correlations were found for plant colour and seed colour; plant colour and tannin percentage; and tannin content and germination percentage.

In the F_2 population variability for tannin content was much greater than in the F_1. The result of the F_2 phenotypic frequencies for testa colour were tested for goodness of fit to 9/7 genetic ratio by chi-square analysis

* (Authors' note: though Dehlavi's paper reports a significant negative correlation between "seed colour" and "tannin percentage", we believe this is possibly a printer's error since in all other cases encountered high tannin content is associated with dark-coloured seed coats).

and were shown to fit the 9/7 brown to white ratio. There was a highly significant positive correlation between seed colour and tannin content.

The V-HCl method was also used to determine tannin in GA 615 sorghum seeds collected 17, 22, 27, 32, 37 and 42 days after anthesis. Histological examination indicated that in the period following anthesis, mesocarp thickness decreased and testa thickness increased in relation to the other tissues; the epidermis and hypoderm did not change. Tannin content increased rapidly rising from 0·4% 17 days after anthesis to 4·6% at 42 days, the most rapid increase being found from 37 to 42 days after anthesis.

Guiragossian *et al.* (1978) grew four inbred lines of sorghum (1) P 721 N (normal, low-tannin); (2) P 721 O (opaque, high-lysine type derived from P 721 N by chemical mutagen treatment, Mohan, 1975); (3) IS 11167 (Ethiopian cultivar high in lysine, Singh and Axtell, 1973a); and (4) IS 4225 (high-tannin) at the Agricultural Experiment Station in Puerto Rico during the 1974–75 Winter–Spring season. The germ and endosperm were separated by hand dissection without removal of the pericarp. The protein (micro-Kjeldahl N × 6·25) and N distribution in the Landry and Moureaux fractions (method of Misra (P.S.) *et al.*, 1975) are presented in Table 4.16.

The mutant P 721 O in comparison with P 721 N shows a three-fold increase in fraction I and a decrease in fractions II and III. The distribution patterns of P 721 O and IS 11167 are very similar; the pattern of IS 4225, high-tannin differs from P 721 N.

The amino acid distribution in the kernels, endosperm and embryo are presented in Table 4.17. There was little difference between high-lysine IS 11167 and the high-lysine mutant P 721 O, and little or no difference among the embryo amino acids from P 721 N, IS 11167 and IS 4225.

The amino acid composition of the protein fractions is shown in Table 4.18. The amino acid levels in fractions I, IV and V of IS 11167 and IS 4225 were lower than in P 721 N and P 721 O because of lower total recovery of amino acids. Table 4.18 also shows the corrected values.

Individual lyophilized endosperm protein fractions subjected to SDS polyacrylamide gel electrophoreses revealed no significant differences among the proteins though an additional band associated with fraction V proteins was observed from IS 4225 (high-tannin). Guiragossian *et al.* considered this observation to be consistent with the premise that tannin–kafirin complexes appear in the same fraction with the glutelins.

Sorghum Pigmentation

Okano and Ohara (1935) reported the presence of a red colouring matter obtained from the methanol extract of hulls of kaoliang and suggested that the colouring matter could be phlobaphene. Another colouring matter, soluble in MeOH but not in acetone, appeared to be identical with durasantalin [*sic*]. The colouring matter of the hulls consisted of 20% apigenin (Okano *et al.*, 1934), 43% phlobaphene and 35% durasantalin. The colouring matter of the seed coat consisted of 15% durasantalin and 82% phlobaphene. Apigenin was not found in the seed coat.

Blessin *et al.* (1963) reported that certified seed of Martin, a yellow milo sorghum, when cut longitudinally, revealed five areas of pigmentation: (1) the dark brown seedtip; (2) the orange pericarp; (3) the light yellow embryo; (4) the yellow, yellow-orange or purple corneous endosperm; and (5) the white starchy endosperm. Treatment with 12N hydrochloric acid developed a magenta colour in the seedtip and pericarp, and in the corneous endosperm when purple. No colour change was observed in the embryo or starchy endosperm. Treatment with aqueous 0·1N sodium hydroxide developed a yellow colour in the pericarp and embryo and either a red-orange or purple colour in the seedtip. The magenta colour could be extracted when the whole grain was placed in aqueous 12N hydrochloric acid at room temperature, or with slight heat and shaking.

Martin, Midland and Westland Yellow Milo and Red Kafir were found to contain considerable amounts of "anthocyanogens". Little or none were found in White Kafir, waxy and yellow endosperm varieties nor in some pigmented sorghums, Pictoria Pink, Kershoma Purple and Pink Kafir. Blessin *et al.* (1963) therefore concluded that pericarp colour was not related to anthocyanogen content.*

Yasumatsu *et al.* (1965) subjected a methanol extract of the seed coats of commercial sorghum (not otherwise characterized) to column chromatography and obtained three chromogens (I, II, III) which on acid hydrolysis yielded the same two flavonoids for each of the three chromogens; one was a flavanone, probably eriodictyol, and the other an anthocyanidin, pelargonidin.

Misra and Seshadri (1967) stated that *Sorghum bicolor* var. *durra* seed material, with deep red or purple glumes, obtained from the Agricultural Research Institute, Coimbatore, contained in the ethanolic extract of its glumes, apigeninidin, luteolinidin, 7-O-methyl-luteolin and its glucoside, together with a polymeric pigment not identified. Caffeic acid was also isolated from the subsequent extract of the glumes.

Bate-Smith (1969) employed the method of Bate-Smith and Rasper (1969) to assay for luteoforol in the tissues of bronze and red seeds of a mixed sample of sorghum of African origin through a whole season's growth. Since the seeds analysed were hulled, the glumes of a different sample of sorghum, and also those from a genotype NK 300 A, bred in the United States and grown in Ghana, were also analysed.

* (Authors' note: the term "anthocyanogen", employed by Blessin *et al.* was suggested by Harris and Ricketts (1958) who stated that the term "leucoanthocyanin" has come to embrace different compounds which have in common only the property of yielding anthocyanidins on treatment with acids. They suggested therefore that the term "leucoanthocyanin" should be reserved for Δ^3-flavan-3-ols as described by Kuhn and Winterstein (1932), and that other compounds which yield anthocyanidins on heating with mineral acid should be termed "anthocyanogens". The term "procyanidin" now appears to be more generally accepted. See Gupta and Haslam above).

According to Bate-Smith the varying colour of the grain, and of the glumes of the mature grain appeared to be due to variations in the amount of colourless flavan-4-ol precursors of the pigments, claimed to be present in the green tissues of the plant before the ripening process begins. If such was the case, no biochemical process was identified by which the leucoanthocyanins were converted to the characteristically red pigments of the seed coat or the red and black pigments of the glumes.

Chakravorty (1969) examined a number of sorghum varieties at Sirsa (Haryana, India) for protein content (Kjeldahl N × 6·25) and colour of seed coat. Tannin content was also estimated by examining the colour produced by addition of dilute ferric chloride solution to a 1/1 alcohol and water extract of the grain flour. The depth of colour was arbitrarily graded 1–5, the figure 1 being allocated to a standard BP 53. The results are given in Table 4.19 arranged in order of protein content.

Four bazaar samples were also examined in comparison with bazaar sample BP 53 and these results are shown at the foot of the table.

Chakravorty stated that the coloured grains appeared to contain more protein than the white, but the estimated tannin content also tended to be higher in the higher protein grains.

McMillian *et al.* (1972) examined the bird-resistance of 142 sorghum lines from the collection at the Southern Grain Insects Research Laboratory, Tifton Georgia, and compared the resistance (on a damage rating scale of 0–10) with tannin (acidified vanillin method, Burns, 1971): seed colour (visual rating on a scale of 1–5: 1 white; 2 cloudy white; 3 tan; 4 reddish brown; and 5 brownish black); head exsertion; average number of days to 50% flower; and plant height.

The amount of bird damage was significantly negatively correlated with the amount of tannin in the seed, seed colour and plant height. No significant correlation was found between bird damage, sorghum head exsertion or days to 50% flower. A highly significant correlation was found between seed colour and tannin content. A large proportion of the bird damage occurred before the seeds became coloured suggesting that the developed tannin discouraged bird attack.

Sorghums were planted at two locations during 2 years, and though in 1969 the damage at one location was higher than at the other, the mean damage ratings for the two locations correlated significantly.

No satisfactory explanation was offered for the fact that some lines, though relatively low in "tannin" suffered only intermediate damage while other lines, relatively high in "tannin" were susceptible.

McMillian *et al.* (1972) suggested that the exceptions to the general correlation indicated that the nature and the quantity of tannin present may be influencing factors.

Oswalt and Srinivasan (1973) in a survey of some 300 photoperiod insensitive lines of sorghum grown during 1969–70 in Puerto Rico and Lafayette, Indiana from the World Collection, reported on the CE (Burns 1971) as related to seven pericarp colour classifications. Their findings are summarized in Table 4.20 listed under IS numbers, together with several Purdue base and commercial hybrids. None of the grains were graded grey or purple. No CE values greater than 1·0 were found in those classed as white sorghums. A straw-coloured sorghum, Purdue base No. 121169, displayed an exceptionally high CE of 6·97. Among the International Sorghums, IS 8583 at 3·99, IS 8544 at 4·77 and IS 8171 at 4·89 appeared unusually high in relation to the colour classification. The majority of the samples, 74 out of 87, fell below 1·0 and averaged 0·46. The yellow pericarp sorghums were in the main below 1·0 averaging 0·27, a lower value than the straw-coloured sorghums. The eight orange coloured sorghums were equally divided above and below 1·0. The single sample of brown pericarp displayed a CE of 4·40, and the four samples of red-brown pericarp averaged 4·74.

Pearl Millet Pigmentation

In a personal communication (1978) Reichert reported that the pH-sensitive pigments in pearl millet are important from an aesthetic but also perhaps from a nutritional point of view. Pearl millet seeds of many varieties are dark gray or yellow in colour. This pigmentation is markedly reduced by the traditional practice of soaking dehulled seeds in sour milk or in aqueous extract of tamarind (tartaric acid). This colour loss is simply a function of the pH and the effect can be duplicated by soaking seeds in dilute HCl or citric acid solutions. Dehulling is an important prerequisite to rapid absorption of acid. Without any dehulling, whitening of the seed requires up to 25 h because acid is only absorbed through areas of the seed around the embryo. Dehulled seeds whiten rapidly (5–10 min at 90% extraction) because acid is absorbed through all areas where the hull has been broken.

The major pH-sensitive pigments in pearl millet, which were found to be flavonoids, were isolated and identified mainly by the paper chromatographic techniques described by Mabry *et al.* (1970) Defatted millet flour was methanol extracted and the extract was subsequently developed (descending paper chromatography) in water/acetic acid (85/15) and *t*-butanol/acetic acid/H_2O (3/1/1). The methanol extraction removed the dark grey

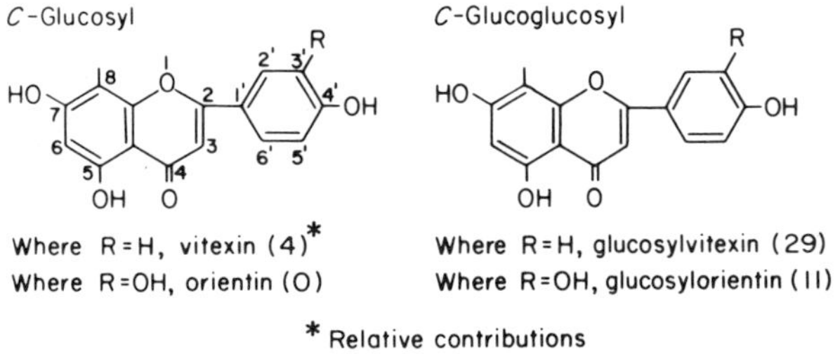

C-Glucosyl

Where R = H, vitexin (4)*
Where R = OH, orientin (O)

C-Glucoglucosyl

Where R = H, glucosylvitexin (29)
Where R = OH, glucosylorientin (11)

* Relative contributions

Fig. 4.XIII Structures and relative contributions of the major *c*-glycosyl flavonoids in millet.

pigmentation which is mainly situated in the peripheral region of the seed. The methanol extraction of millet also reduced the acid–base sensitivity of millet flour to a level comparable with hard red spring wheat flour.

The methanol extract was sensitive to pH; at high pH the extract turned yellow-green and in acid solution the extract was clear and colourless. The major compounds and their relative proportions responsible for this pH-sensitivity are illustrated in Fig. 4.XIII. These compounds are known as *c*-glycosylflavonoids by virtue of the presence of a sugar moiety (glucose) which is attached to the flavonoid nucleus via a carbon–carbon bond. This

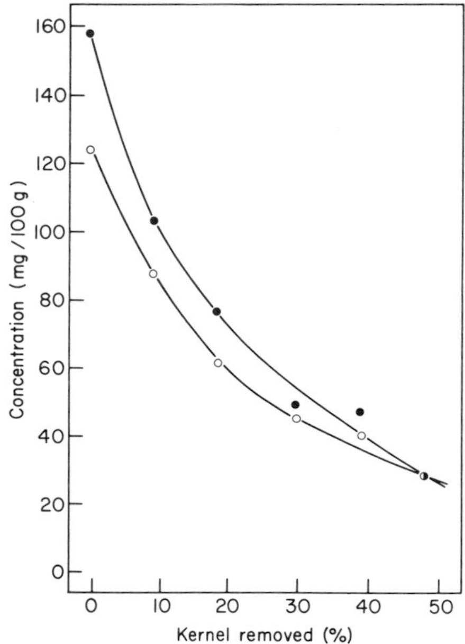

Fig. 4.XIV Concentration of phenolics in whole and dehulled millet flour. ○ = ***c*-glycosyl flavonoids.** ● = **alkali-labile ferulic acid in the methanol-extracted flour.**

bond is not broken by normal acid hydrolysis procedures or enzyme hydrolysis. Glucosylvitexin was the major *c*-glycosylflavonoid found by Reichert in millet. This compound has a glucose attached glucosidically to the glucose which is attached via the C–C bond to the flavonoid nucleus.

The methanol-extracted flour contained phenolic acids which were released only by alkaline hydrolysis (2N NaOH for 4 h at 25°C). The major alkali-labile phenolic acid was identified as ferulic acid by paper chromatography and TLC in comparison to the authentic compound.

Concentrations of total *c*-glycosylflavonoids and alkali-labile ferulic acid in whole and dehulled millet seeds is given in Fig. 4.XIV. These phenolics are present mainly in the peripheral endosperm regions and their concentrations decrease markedly on dehulling in a Strong-Scott barley pearler. By calculation the periphery of the seed comprising 10% of the kernel contains approximately 580 mg/100 g of alkali-labile ferulic acid and 390 mg/100 g of *c*-glycosylflavonoids.

Since these compounds are low molecular weight phenolics they do not bind or precipitate proteins as do tannins. Recent evidence has shown, however, that these types of phenolics may be nutritionally damaging. Horigome and Kandatsu (1968) exposed casein to various polyphenols undergoing enzymatic oxidation and established that the resulting brown products were nutritionally improverished. The first step in the mechanism whereby phenolics attach themselves to proteins is an oxidation of the phenol to a highly reactive quinone. For example, *o*-benzoquinone has been shown to react with the thioether group of methionine to yield 3,4-dihydroxyphenyl N-acetyl methionine sulphonium chloride (Vithayathil and Murthy, 1972) and is also thought to react with the ε-NH_2 group of lysine (Synge, 1975).

Polyphenol Development in Sorghum Seed

Prine *et al.* (1967) cited by Tipton *et al.* (1970) reported that as the sorghum grain matured, the tannin content in the seed was reduced. The latter authors were not able to confirm this finding. Three varieties, Lindsey BR 75 (bird-resistant); DeKalb E 57; and Asgrow Ranger B (non-bird-resistant), were analysed at: (1) the milk stage; (2) the dough stage; (3) 22% moisture; (4) 17% moisture; and (5) 16% moisture, for % tannic acid and total astringents (Louisiana State University, methods unstated). The comparative figures given in Table 4.21 show that while there was some reduction in tannic acid and total astringents in bird-resistant BR 75 between the milk and 16% moisture stage, the two other varieties showed no such reduction.

Nip and Burns (1969) at Texas A & M University characterized the pigments in three red varieties of sorghum grain: B 378, red with a shade of yellow, some black spots; TX 415, light red with a shade of yellow; and RS 671, red with a shade of yellow, some black spots. Whole kernels were cleaned and extracted with diethyl ether to remove the waxy surface coating. The pigments were extracted with acidified methanol (HCl-MeOH 1% v/v) and separated from the crude concentrate by descending paper chromatography from 1-butanol/2N HCl. These bands were rechromatographed, in descending manner, with different solvent systems suggested by Harborne (1967) and Stafford (1965) for the isolation of "anthocyanins."

In all three varieties orange pigmentation was found in the stylar area, the epicarp, the cross cell and tube cell layers of the pericarp. The mesocarp area was not coloured except that black or brown spots were visible in B 378 and RS 671. The results agreed with those of Blessin *et al.* (1963).

The chromatographic separation and purification process revealed three major pigments: two yellow (Y-1 and Y-2), and one orange (O-1) in each variety Other pigments were present in minor quantity and were not purified. Y-1 was probably a yellow anthocyanin and resembled the apigeninidin found by Stafford (1965); the R_F value suggested apigeninidin-5-glucoside. Y-2 was of the flavone and flavonol group; its R_F value suggested kaempferol-3-rutinosid-7-glucuronide. Authentic compounds were not available for direct comparison study. O-1 was of the anthocyanin type but was not further identified.

Nip and Burns (1971) examined six varieties and hybrids of sorghum with white seeds (ivory, with brown-coloured spots), ATX 607, Texioca 54, ATX 607 × Texioca 54, BTX 3197, RTX 2520 and BTX 3197 × RTX 2520.

The extraction and separation procedures described were essentially similar to those used in characterizing the red varieties (Nip and Burns, 1969) with some modification.

All six samples showed an orange colour in the stylar area, yellow in the epicarp and endocarp (the red varieties showed orange-red pigmentation). The mesocarp was colourless but the pericarp was orange in areas where brown or orange spots appeared on the kernel surface.

Four pigments, two yellow (Y-1 and Y-2), and two orange (O-1 and O-2), were separated by descending paper chromatography from each sample. Other yellow pigments were also observed in ATX 607, Texioca 54 and ATX 607 × Texioca 54, which did not separate in the solvent system used, and other orange pigments found in BTX 3197, RTX 2520 and BTX 3197 × RTX 2520 which also did not separate. The yellow pigments isolated from the red varieties (Nip and Burns, 1969) were not isolated in the white varieties.

The spectral properties, colour reactions, and R_F values indicated that two forms of unidentified apigeninidin (Y-1 and Y-2), and two forms of unidentified luteolinidin (O-1 and O-2) were present in each of the six samples.

Seed Germination and Mould Formation

Tsukunaga *et al.* (1932) described a method for the determination of small quantities of tannin in Manchurian sorghum and reported that kaoliang seeds contained 0·01–0·57% catechinic tannin. As determined, the tannin content of the seeds declined as the seed colour changed from brown, to red, to yellow and to white. The authors commented that the tannin content seemed to have some relation to germination and to resistance to disease.

Harris and Burns (1973) reported on the relationship between tannin content (V-HCl method, Burns, 1971) and pre-harvest seed moulding of 49 sorghum hybrids grown at two locations in Georgia during 1971. Seed tannin content was found to be strongly and negatively correlated with pre-harvest seed moulding.

Nelson and Cummins (1975) postulated that if high-tannin sorghum seed was more resistant to preharvest mould (as reported by Harris and Burns, 1973) and to preharvest germination (reported by Harris and Burns, 1970), comparatively low levels of propionic acid would be required to inhibit deterioration during storage of high-tannin grain.

Niagara Oro T sorghum of low-tannin content (Burns, 1971) 0·9 %, and DeKalb BR 64, high-tannin content 5·0%, DWB, were stored over 9 weeks at a moisture content of 20% with propionic acid at 0, 0·5, 1·0 and 1·5% (wet-weight basis) at fluctuating ambient temperature.

The high-tannin required less propionic acid (1·0%) than low-tannin sorghum (1·5%) for safe high-moisture storage. Low-tannin grain at high temperatures (29·5°C) spoiled faster than when stored at ambient temperatures. The results also indicated that higher levels of propionic acid were necessary for safe storage at higher temperatures.

Enzyme Inhibition in Sorghum

Kneen and Sandstedt (1946) cited by Miller and Kneen (1947) reported an amylase inhibitor in Leoti and Schrock varieties of sorghum; waxy and nonwaxy samples showing inhibitor activity to a similar degree. They found no inhibiting action in Kalo, Early Kalo, Pink Kafir, Atlas Sorgo, Waxy Kafir and Waxy Club.

Miller and Kneen (1947) determined inhibitor activity by a method based on the rate at which starch dextrinization was retarded. The results for samples of Schrock, Leoti Red, Leoti Red (grown at Hays, Kansas), Early Sumac (all showing inhibitor activity), and for Westland, Atlas, and Blackhull Kafir (no activity) are shown in Table 4.22. Experimentally milled fractions of the Hays, Kansas sample of Leoti Red appeared to contain a higher proportion of amylase inhibitor in the germ (1440 min dextrinization time) and bran (234·5 min) than in the flour (26 min). The amylase inhibitor in sorghum appeared to be extremely resistant to heat. A 1/10 distilled water extract of Leoti bran (pH 5·6) autoclaved at 15 lb/in^2 pressure for 1 h showed no reduction in inhibitor activity. The same process at pH 2·5 and 10·9 produced similar results.

Removal of the red-brown colour of the inhibitor solution by carbon black, eluted by 1·0 N NaOH appeared to destroy inhibitor activity. The pigment(s) associated with inhibitor activity was light-yellow in acid and red-brown in basic solution.

At pH below neutral the inhibitor was soluble in water, alcohol, acetone or concentrated ammonium sulphate solutions; on the basic side, it appeared to be precipitated by alcohol, acetone or calcium ions. It was insoluble in ether. A sample of Leoti grain germinated for three days at 30°C lost its inhibitory power.

Strumeyer and Malin (1969) followed up this finding of Miller and Kneen (1947) by fractionating Leoti Red sorghum using standard milling procedures and extracting the defatted germ fraction with 95% ethanol.

The inhibitor was found to have potent and indiscriminate behaviour towards a variety of enzymes including amylase. Starch did not prevent inactivation but bovine serum albumin appeared to interfere with the interaction of amylase and inhibitor.

A series of qualitative tests on the purified inhibitor suggested it to be a polyphenol, and its chromatographic and spectral properties were identical to those obtained with "authentic cyanidin". Strumeyer and Malin (1969) concluded that the material originally reported by Miller and Kneen (1947) as an amylase inhibitor, was a general protein denaturant, capable of inactivating a number of enzymes.

Filho (1974) at the Federal University of Ceara, Brazil, determined the trypsin inhibiting activity (caseinolytic assay using the Folin phenol reagent to detect trichloroacetic acid-soluble products) of an unnamed variety of unstated origin of *Sorghum bicolor* seeds ground to a fine powder.

One unit of trypsin inhibitor (TIU) was defined as the amount of inhibitor that reduced by 50% the activity of a preparation of trypsin that produced an absorbance of 0·500 at 750 nm, in the assay. Protein was determined by a microbiuret method (Goa, 1953). The trypsin inhibitors were readily extracted with distilled water. The aqueous extract was acidified with 4 N acetic acid to pH 4·0 at room temperature and the precipitated proteins separated by centrifugation. The clear supernatant was brought to pH 5·8 with 0·5 N sodium hydroxide. Dialysis of the acid supernatant indicated that about 60% of the TI activity was probably due to low (less than 6000 daltons) molecular weight compounds. The distribution of the protein and trypsin inhibiting activity is shown in Table 4.23.

The sorghum trypsin inhibitors, separated in a Sephadex G-100 column, showed no reduction in activity when solutions containing the activity peaks were heated at 100°C for 30 min.

Filho stated that further studies were required to establish the type of substance responsible for the TI activity in sorghum and suggested that at least in the case of low molecular weight fractions, the TI activity might be due to tannins present in sorghum and known to have, for example, a depressing effect on the growth of chicks (Rostagno *et al.*, 1973ab).

While studying the malting and brewing of sorghum beer Watson and Novellie (1974) reported sorghum grain α-glucosidase to be insoluble in water but limited extraction of the enzyme could be achieved from "normal" sorghum (i.e. non-bird-resistant sorghum, Barnard's Red) with NaCl under alkaline conditions and this extraction could be enhanced by the addition of papain. Maximum liberation was achieved using a combination of 8 M urea and 0·1 M sodium sulphite.

In contrast, the α-glucosidase from barley grain was not released by NaCl under alkaline conditions, was resistant to extraction at low concentrations of urea, and completely denatured at high concentrations.

Novellie (1959) had reported that the amylases in bird-resistant sorghum malts were not readily extracted by water. The amylases were however active in the insoluble state and could be extracted by a 2% solution of peptone.

Watson *et al.* (1974) seeking a diagnostic tool to identify bird-resistant cultivars of sorghum grain without recourse to malting and subsequent determination of diastatic power, found a good correlation between the degree of NaCl-insolubility of the grain α-glucosidase and the degree of water-insolubility of the malt amylases. Peptone, in the presence of NaCl, liberated NaCl-insoluble α-glucosidase from grain and similarly the water-insoluble amylases of malt were solubilized by peptone in water. However, maximum liberation of sorghum α-glucosidase was achieved only by using a combination of 8 M urea, 0·1 M sulphite, 5% peptone and 1% Triton.

The effect of peptone on the solubility in water of sorghum malt amylases and on the solubility in NaCl of the α-glucosidase of several cultivars is shown in Table 4.24. The results were the mean of samples of each grain and malt taken from seven different locations in South Africa. Environment was said not to play a significant part in determining the degree of enzyme solubility.

Since bird-resistant sorghums had, in preliminary experiments, been found to contain up to 10× higher tannin concentrations than non-bird-resistant sorghums, Watson *et al.* "speculated" that the insolubility of α-glucosidase and amylase might be due to formation of insoluble complexes between the tannins and the enzymes.

Watson (1975) reported that a bird-resistant cultivar (SSK 52) contained in the pericarp fraction an inhibitor of lactic acid (*Lactobacillus leichmannii*) fermentation. He suggested that this inhibitor was probably polyphenolic in character; the inhibition could be completely eliminated by the tannin-complexing agents PVP and Triton X 205.

The cultivar, when malted, also contained a substance or substances which inhibited alcoholic fermentation by *Saccharomyces cerevisiae* (brewers' yeast). This inhibitor occurred only in malted sorghum and could be partially depressed by prior conversion of malt suspensions before introduction of the yeast. It was not found in the whole grain, nor in the pericarp fraction. This inhibitory effect could not be offset by the addition of PVP or Triton. None of a range of phenolic compounds, tannic, gallic, caffeic and ellagic acids, or catechol inhibited alcoholic fermentation by yeast in 10% suspension of normal malt. Watson concluded that the yeast inhibitory factor was not a polyphenol.

Daiber (1975) described the four categories into which sorghums are classified in South Africa; KF, which includes all bird-resistant cultivars, is characterized by a distinct dark-coloured testa or nucellar layer, irrespective of pericarp colour; KR in which the grain is red and with a floury endosperm; KW which is white-coloured grain; and KM including all grains other than red or white without a dark testa. The mean solubility of the amylases from a range of samples in three of the classes over several seasons was:

Class	Amylase solubility (%)
KR	95·4
KM	89·7
KF (bird-resistant)	37·5

Daiber determined polyphenol content in sorghum using the urea-FAS procedure, the MV-HCl method (both Maxson and Rooney, 1972a) and the dimethylformamide (DMF)—ferric ammonium citrate (FAC) method of Jerumanis (1972). This last method, based on extraction of the grain with DMF, reaction with FAC and measurement of the absorbance at 525 nm was stated by Daiber to be the method accepted for the barley brewing industry by the European Brewing Convention (Bishop, 1972) as the most reliable and indicative analysis of those polyphenols causing protein precipitation in beer.* Daiber described a modification of the extraction procedure designed to make the method easier for routine analysis. The results of the three procedures using bird-resistant grain, non-bird-resistant grain and bird-resistant malt are shown in Table 4.25.

Daiber reported that tannic acid (Merck) inhibited enzymes in a manner similar to the polyphenols of bird-resistant sorghum but D-catechin (Sigma C1251) and cyanidin chloride (Fluka 28500) caused no enzyme inhibition. Under the prevailing experimental conditions the V-HCl method gave no reaction with tannic acid and Daiber stated his assumption that the substances responsible for the reactions recorded were "anthocyanogens". He further suggested that the DMF-FAC procedure distinguished between bird-resistant and non-bird-resistant grain but the FAS method did not.

Daiber tried to identify those cultivars that would affect (1) enzyme formation during matling; (2) enzyme activity in the germinating grain during souring and malting; and (3) the taste of sorghum beer. He therefore reacted finely ground grain directly with the enzymes essential for the mashing procedure. Results were affected by the ratio between sample size, enzyme concentration and concomitant substances, particularly proteins, which could all participate in complex formation.

In preliminary experiments, Daiber found the inhibitory material to be in the peripheral layers of bird-resistant grain with higher concentrations in abraded fractions comprising testa, nucellar layer and aleurone, than in fractions from the outer pericarp. The enzyme-inhibiting substance(s) was not evident in either the endosperm or the embryo. The reaction between the exogenous enzymes and the inhibitory substances was practically instantaneous, irreversible and specific for each cultivar. Addition of peptone, PVP or polyclar-AT before the addition of the enzymes, prevented the inhibitory effect. The inhibition of α-amylase, barley diastase, and sorghum malt extract by various sorghum cultivars is shown in Table 4.26.

*(Authors' note: the relevance of the method to the determination of sorghum polyphenols does not appear to have been demonstrated.)

Grain samples of 47 cultivars in the four classes produced in 1970/71 under identical conditions in the Transvaal were analysed for enzyme inhibition and polyphenol content. The results are shown in Table 4.27.

The non-bird-resistant classes KR, KM and KW were very similar in characteristics. The KF class differed significantly from the other three but the high standard deviation from the mean of the cultivars in the KF class suggested major differences within the class. A brewing trial indicated that within the KF cultivars, suitability as a brewing grain varied markedly and that satisfactory brewing properties could not be predicted by the colour of the testa. Using the method proposed by Jerumanis (1972), Daiber reported a correlation of $r = 0{\cdot}981$ between enzyme inhibition and total polyphenol content as determined by the DMF-FAC procedure which he considered represented the biologically active polyphenol fraction. Table 4.28 shows the cultivars in the KF class and the relation between enzyme inhibition and polyphenol content.

Daiber reported that the polyphenols of SSK 52, Brawis and DC 99 were more, and those of Fiva, FS 79 and DC 75 less inhibitory than expected from their analyses. Daiber suggested that enzyme inhibition was dependent on the "quality and nature" and not solely on the quantity of the polyphenols present; enzyme inhibition giving the best measure of the inhibitory polyphenols. For practical purposes, bird-resistant sorghum with more than 50% inhibition of diastatic activity or 1·1% total polyphenols should be excluded from malting.

Panels of Bantu and European tasters did not find any relation between perception of bitterness and total polyphenol content of beer. The pH of the beer samples ranged from 3·3 to 3·6. This high acidity was normal for sorghum beer. Brenner *et al.* (1956) (as cited by Daiber) found perception of bitterness to be related to pH. They reported that beer at pH 4·1 was perceived as pleasant and aromatic; the same beer adjusted to pH 4·9 was considered harsh, coarse and bitter.

Chibber *et al.* (1978) reported that crude tannin extracted from the hulls of BR 64 (milled on a Palyi mill) was purified by chromatography on a special dextran gel (LH-20 Pharmacia Fine Chemicals). At pH 2·0 (the pH of the gastric contents) the purified tannin had no harmful effect on purified pepsin but when the precursor of pepsin, pepsinogen, was placed at pH 2·0 in the presence of tannin only 40% as much pepsin activity was generated as when pepsinogen was allowed to activate alone at pH 2·0. The addition of tannin to the substrate casein did not affect its digestion by trypsin or chymotrypsin but if the enzymes were preincubated with the tannin before the addition of enzyme to casein, the activity of trypsin was reduced by over 90% and of chymotrypsin by about 30% of normal activity. Addition of tannin to deoxyribonuclease decreased its ability to break down deoxyribonucleic acid but the addition of tannin to deoxyribonucleic acid before its digestion by deoxyribonuclease nearly doubled the rate of deoxyribonucleic acid degradation. Chibber *et al.* suggested a positive explanation for the enhancing action of tannin on the deoxyribonucleic acid might be its ability to unwind the coiled helix in which the molecule exists, opening it up so that the enzyme could more readily approach the internal structure of the molecule.

"Tannins" and *in vitro* Digestibility

Cummins (1971) compared the IVDMD (Tilley and Terry, 1963) and the tannin content (V-HCl, Burns, 1963) of two sorghum hybrids. DeKalb BR 64 (bird-resistant) and DeKalb F 64 (non-bird-resistant), before and after ensiling. The varieties, genetically closely related, were grown at Experiment, Georgia under the same conditions, and harvested at the dough stage of maturity. The respective "tannin contents", before and after ensilage, of BR 64 and F 64 are shown below:

	Before ensilage		After ensilage	
	Tannin %	IVDMD	Tannin %	IVDMD
BR 64	10·5	50	4·0	68
F 64	4·2	65	4·0	70

Ensiling appeared to reduce the tannin content and to raise the IVDMD of the bird-resistant sorghum to the level of the non-bird-resistant type; no explanation was offered by Cummins for this phenomenon.

Dreyer and van Niekerk (1973) reported that protein digestibility, as measured by N balance tests in rats, was a more accurate index of the growth-suppressing properties of sorghum cultivars than tannin content as measured by the AOAC method (1965). Van Wyk *et al.* (1973), at the same institute in Pretoria, developed an *in vitro* method based on pepsin and pancreatin digestion.

The results (mean of three determinations) of applying the method to 49 cultivars are shown in Table 4.29 with the tannin content expressed as tannic acid.

While in general, "tannin" content appeared in inverse relation to *in vitro* protein digestibility, the relation was by no means constant or uniform. For example, where the tannin content appeared as 0·95%, "digestibility" ranged from 24 to 62%. Nevertheless, cultivars that demonstrated digestibilities less than 70% and tannic acid levels below 0·5% were stated by Dreyer and van Niekerk to be nutritionally superior.*

Oswalt (1973a) determined the yield, protein content (micro-Kjeldahl N × 6·25, AOAC, 1960), and IVDMD (modified Tilley and Terry, 1963) of the grain of DeKalb A 25, C 42a, BR 44, Pioneer 866, 846 and Taylor Evans Exp. 11105 and RS 610, grown under similar conditions on the Purdue Agronomy Farm, Lafayette, Indiana in 1969. The results, given in Table 4.30, showed significant difference in IVDMD among the varieties.

* (Authors' note: it seems possible that the discrepancies that exist between tannin content and digestibility are at least in part attributable to van Wyk *et al.* having used tannic acid as their standard for analysis).

Dark-coloured testas were found in those varieties showing lower IVDMD values.

Harris (1973) reported the yield,* seed colour, resistance to bird damage, grain tannin content (V-HCl method), IVDMD and protein content (methods not stated) of 19 varieties of grain sorghum grown over the 3 years 1969–71 at five locations in Georgia. The results are shown in Table 4.31 with the varieties listed in descending order of yield.

Harris *et al.* (1970) determined the tannin content (V-HCl, Burns, 1963) and the IVDMD (Tilley and Terry, 1963) of 43 varieties of sorghum, ranging in seed colour from yellow to brown, grown in two locations in Georgia in 1968. The varieties were not specified in detail, but at both locations the tannin content was related to the presence of brown colour and there was a highly significant and negative correlation between tannin content and IVDMD. At both locations, randomly selected panicles were bagged with plain kraft selfing bags. Bagging significantly reduced tannin percentages for hybrids with brown seed colour, and improved IVDMD.

As only 72–86% of the variability in IVDMD was accounted for by the correlation between tannin and IVDMD percentages, Harris *et al.* (1970) suggested that other factors were also influencing feed efficiency.

Harris (1973) observed that as RS 700, a sorghum with intermediate tannin, had a thick, opaque pericarp that effectively shaded the testa, breeding for thick opaque pericarps might be successful in producing sorghums with intermediate tannin and improved nutritional value. (See Table 4.31).

Armstrong *et al.* (1974b) compared the *in vitro* digestibility value (IVDV) (pepsin-pancreatin digestion, Akeson and Stahman, 1964) of the bird-resistant sorghums BR 64 and IS 8260 against (a) two non-bird-resistant varieties, RS 610 and RS 671 and (b) RS 671 and IS 8260 from which the tannins had been extracted (method of Armstrong *et al.*, 1974a). The IVDV values for RS 610 and RS 671 whether extracted or unextracted, were all close to 13%. The IVDV for BR 64 and IS 8260 were 4–5% but for the extracted IS 8260, increased to 17·8%.

These results showed similar trends to the findings noted in chick feeding studies (the lower biological values of bird-resistant sorghums; Rostagno *et al.*, 1973b) though the IVDV were low compared to *in vivo* results. Armstrong *et al.* (1974b) suggested that the test might be of value for comparative purposes and that an urgent need exists for a screening test by which to evaluate feed sorghums.

The protein (Kjeldahl N × 6·25) and tannin (CE, Burns, 1971) contents of 67 varieties of sorghum grown in the kharif season 1972 were reported by Arora and Luthra (1974) from Haryana Agricultural University, Hissar. Of these, 50 were of the yellow endosperm type, 6 white and 11 brown and dark brown.

In the yellow endosperm strains, protein contents varied from 9·2 to

*(Authors' note: unit/acre not stated, but assumed to be bushels).

15·7% with an average of 11·6% and tannin from 0·33 to 6·33 mg/g. In the white grains, protein varied from 8·3 to 14·3% with an average of 11·4% and tannin from 0·1 to 0·3 mg/g. In the brown grain protein varied from 9·4 to 15·7% with an average of 12·8%, and tannin from 30·0 to 62·0 mg/g.

The IVDMD (method of Tilley and Terry as modified by Barnes *et al.*, 1971) was applied to 17 varieties selected from the low-tannin (less than 0·3% CE) and the high-tannin (greater than 0·7% CE) varieties. The results are presented in Table 4.32.

IVDMD varied from 49·6% to 95·6% and there was a significant negative correlation between tannin content and IVDMD. The % digestibility declined to 6·36 with each unit increase of tannin content (CE/100 g). However, it was stated that total tannin content (as measured by CE) may not have been the only factor influencing IVDMD, since IS 2752 containing 5·9% tannin and 15·8% protein showed the lowest IVDMD percentage of 49·6%, while IS 7155 containing 6·2% tannin and 15·1% protein showed a higher IVDMD of 68·16%.

Arora and Luthra (1974) found a significantly larger amount of N in the residue from the IVDMD study of the high-tannin varieties than in the residue from the low-tannin varieties suggesting that the protein had become bound in the high-tannin varieties.

Schaffert *et al.* (1974a) determined the CE (V-HCl, Burns, 1963, 1971), the IVDMD (Tilley and Terry, 1963, 48 h fermentation) and the *in vitro* protein disappearance (IVPD, the difference between total N (AOAC, 1960) × 6·25 in the sample before and after *in vitro* fermentation) of 24 commercial, unnamed lines, plus 10 International Sorghum (IS) lines, Redlan B and 10 hybrids of Redlan A (a low-tannin line) and the IS lines.

IVDMD was also determined on sub-samples without and with additions of 10 and 20 mg urea/50 ml *in vitro* tube; as no significant difference was found between the two levels of urea, the results were averaged.

The results were not presented in detail, but the addition of urea reduced the regression coefficient of IVDMD on CE from −3·14 to −1·42. Urea more than doubled the % IVDMD of the entries with the highest CE. The regression coefficients for the IS lines and hybrids were calculated separately and gave similar results; a reduction from −4·01 to −1·62 on the addition of urea.

Schaffert *et al.* (1974a) suggested that these results indicated that tannin, as measured by CE, was not inhibiting digestion of the sorghum grain directly, but indirectly, by reducing the nitrogen available for bacterial growth and rapid substrate digestion. The increase in maximum IVDMD indicated that, even in low CE grain, there was sufficient N for maximum fermentation and IVDMD was consequently impaired.

In a second experiment, CE, IVDMD (48 h fermentation) and IVPD for IS 0062, 0148, 8165 and 0616, and their Redlan hybrid were determined. In all cases, high CE values were associated with low IVDMD and IVPD. IS 0062 had a colourless pericarp and CE value of 0·29; its F_1 hybrid with Redlan had a red pericarp inherited from Redlan and a CE of 0·30. The

IVDMD and IVPD of the hybrid without urea were significantly lower than those of IS 0062, suggesting that there might be other pigments or inhibitors in the pericarp not detected by the V-HCl method.

The addition of urea increased significantly both the IVDMD and the IVPD of the low-tannin genotypes and the IVDMD but not the IVPD of the high-tannin genotypes. These results suggested that the amount of digestible protein may be a major factor in limiting the utilization of high-tannin grains by mono- and digastric animals.

Anthony *et al.* (1975) reported that the IVDMD (method of Tilley and Terry, 1963) of 30 sorghum entries, distinguished as bird-resistant and non-bird-resistant but not otherwise characterized, showed the grain of the non-bird-resistant samples to be more digestible (73·7%) than the bird-resistant samples (57·1%).

Effect of "Tannins" on the Nutritive Value of Sorghum

Human beings

Apart from a very brief report by Belavady (1977) on studies carried out on human subjects using labelled iron no reported work has been found bearing directly on the effect of sorghum tannins in human nutrition. Using two types of sorghum with tannin contents, in mg/100 g sorghum, of (A) 25 and (B) 150, Belavady reported a marked reduction in Fe absorption from B compared to A. Morton (1970) observed that geographical zones where there was a high incidence of oesophageal cancer, such as Curacao and the Transkei area of South Africa, were also zones in which sorghum played an important part in the diet. She concluded that *if* tannins could be assumed to be the cause, and the dark-seeded sorghums the source, of oesophageal cancer, it might account for the pattern of oesophageal cancer zones in Africa. The evidence was far from conclusive, since other factors might well be influencing the incidence of oesophageal cancer. It does, however, illustrate the need to investigate more thoroughly the effects of the dark-seeded sorghums in human and animal nutrition.

Effect of added tannins to rat and chick diets

Several attempts have been reported to pin point the growth-depressing effects of tannin by adding various commercially available tannins to rat and chick diets. A few of these references in which tannins are added to diets other than those based on sorghum illustrate the problems which arise in even defining the "tannins". Even when the tannin is added to diets comparing bird-resistant or non-bird-resistant sorghums, the relevance of such additions to the nutritional influence of the naturally-occurring tannins in sorghum is doubtful.

Joslyn and Glick (1969) pointed out that the tannin used in reported experiments was often not characterized by the authors. In their own experiments they compared tannins from several sources, but not including sorghum, in the diet of weanling rats at levels of 2–10% in the diet. Mortality was greatest with galloyl glucose compounds and least with gallic acid, but was not related to growth depression.

Food intake was reduced and growth depressed in rats fed diets containing tannic acid (a commercial gallotannic acid powder, reagent grade) at 4, 5 and 8% levels in the diet (Glick and Joslyn, 1970a).

Not all the growth depression could be attributed to the reduced food intake. Supplementation with choline or methionine effected no improvement but supplementation with casein improved growth. The harmful effect of the tannic acid was most marked with the younger rats.

When a commercial crystalline gallic acid at a 5% level in the diet was compared with tannic acid at 5 and 8%, the gallic acid, but not the tannic acid at 5%, increased fat content in the rat livers.

The same authors (1970b) reported that hydrolysable gallotannin and condensed tannins from quebracho and grape seed increased significantly the excretion of N in the faeces of rats. Dietary catechin, gallic acid and ellagic acid had no such effect.

Experiments using ^{14}C-labelled casein and proteolytic enzyme assays of intestinal contents and pancreases, suggested that the excreted N was derived from endogenous sources rather than from dietary casein.

Mitjavila *et al.* (1971) utilized gallotannic acid (ether extract of Aleppo galls) in feeding trials with female weanling rats in the Institut de Physiologie, Toulouse. The diets were prepared in a semi-liquid form (Mitjavila and Derache, 1968) to overcome the reduced food intake caused by lack of palatability, and the tannic acid was added to provide 0, 32 and 64 g tannic acid/kg dry feed.

Feed consumption rose and growth was depressed as the level of tannic acid increased in the diet. At 64 g/kg the presence of tannic acid halved the efficiency of food utilization (i.e. doubled the F/G ratio).

Analysis of the carcass for lipid (petroleum ether extract), protein (N × 6·25) and water (difference between carcass weight and the sum of lipid weight and lipid-extracted dry weight) showed that the principal reason for the weight reduction was the marked reduction in the deposition of lipid.

Eggum and Christensen (1973) evaluated the effect of tannin (possibly tannic acid) when added at levels of 0, 0·5, 1·0 and 1·5% to diets for rats in which soybean meal was the protein source. True digestibility and Net Protein Utilization (measured as described by Eggum and Mercer, 1964) were decreased by the addition of tannin; biological value was unaffected by additions of 0·5 and 1·5% and increased slightly at the 1·0% level for reasons which were not understood.

The availability of all the individual amino acids, especially proline, glycine and glutamic acid, as estimated from the true digestibility of the 0 and 1·5% tannin groups by the faecal analysis method of Kuiken and Lyman (1948) was reduced in varying degrees by the addition of tannin. The availability of lysine was reduced from 93·3 to 81·1%. Ammonia which was 93·4% available with 0% tannin, was 64·9% available with 1·5% tannin, suggesting to Eggum and Christensen a reaction with the NH_2 groups.

Maxson and Shirley (1973) conducted experiments with weanling female rats to study the effect of tannic acid added to non-bird-resistant and bird-resistant sorghum (varieties not stated), together with the effect of sodium sulphate as a possible detoxifying agent for tannins, on weight gain, feed efficiency and reproductive efficiency.

Sorghum represented about 80% of the total diet which also included about 15% soybean meal of 49% protein content. The diets were (A) bird-resistant sorghum, (B) non-bird-resistant sorghum, (C) diet (A) plus 0·9% sodium sulphate, (D) diet (B) plus 0·9% sodium sulphate, (E) diet (A) with 3% tannic acid and (F) diet (B) with 3% tannic acid.

Added tannic acid in diets (E) and (F) depressed weight gain of weanling rats through the first 6 weeks of growth, but no effects were observed at the end of 10 weeks. The type of sorghum did not significantly affect growth but the diets lowest in tannins produced highest feed efficiencies. The addition of tannic acid decreased the percentage of the offspring weaned with both types of sorghum diets (E) and (F). The addition of sulphate had no statistically significant effect on weaning percentage or weaning weights of offsprings. Bird-resistant sorghum and non-bird-resistant sorghum diets gave equivalent rates of weight gain, but the non-bird-resistant sorghum resulted in higher feed efficiency.

Vohra *et al.* (1966) reported the growth depressing and toxic effects of tannic acid, chestnut oak bark extract, quercyl-quercitannic extract (from White Oak), quebracho extract, wattle bark extract, *d*-catechin and gallic acid in chick diets. High-tannin sorghum was not examined as such but one of the basal diets included sorghum of unstated composition.

Of the tannins examined, tannic acid was the most harmful, 0·5% resulting in growth-depression, 5% producing a 70% mortality after 7–11 days on the diet. Additions of methionine, choline, betaine or ornithine did not prevent growth-depression. No change was found in the liver lipids.

Rayudu *et al.* (1970a) from the University of California, reported the effect on average weight gain, F/G ratio and % mortality of chicks fed a basal diet which included 42% ground sorghum (of unstated composition), 20% ground maize, 20% soybean meal (50% protein) and 6·75% fishmeal (66% protein), with and without various levels of tannic acid, tannic acid (practical) (digallic acid), reagent grade gallic acid, pyrogallol and pyrocatechol.

At 0·1% level, pyrogallol and pyrocatechol depressed growth, and 100% mortality was observed at the 2% level. Tannic acid was less growth-depressing than pyrogallol and pyrocatechol, and digallic acid was less harmful than tannic acid at the 0·5% but not at the 1·0% level. Gallic acid produced the least growth depression.

Various mixtures of methionine, choline, arginine and ornithine partly alleviated the growth-depressing effect of tannic acid but were most effective against digallic acid. The fat content of chick livers increased from diets containing 1% gallic or tannic acid additions but not from 1% pyrogallol or 0·5% tannic acid.

Moreland *et al.* (1975) compared three diets for growing turkeys. (A) maize, (B) sorghum and (C) sorghum with tannic acid (at 14 g/kg diet); each diet was fed at two protein levels, 260 and 310 g/kg. All six diets were isocaloric and contained the same methionine/protein and choline/protein ratios. No significant differences were found in weight gain, food intake or feed conversion efficiency between the maize and sorghum (without tannic acid) diets at either protein level. When tannic acid was added, weight gain and food intake were significantly depressed at the lower dietary protein level. Food conversion ratio was not affected, suggesting that the poor growth resulted from low food consumption. Addition of tannic acid to sorghum at the higher protein level did not depress growth or reduce food intake significantly but food utilization was significantly less efficient which suggested an adverse effect not related to intake.

Peischel *et al.* (1976a) reported a study in which eight levels of tannic acid ranging from 0·0 to 1·67% of diet were added to chick diets. Body weight gain during the first 14 days was significantly lowered by increasing levels of tannic acid. There were significant decreases in feed consumption during the 15–28 days and in feed conversion during 0–28 days as the level of tannic acid increased. No significant differences were observed during the final 29–56 days of the trial.*

"Tannins" in Sorghum—Biological Tests

In biological studies of nutritional inhibitors it is clearly difficult, even in the laboratory, to study the effect of a single variable while keeping all others constant. When such investigations are conducted under commercial feeding conditions, a single variable experiment appears well-nigh impossible. Added to the fact that the nature and composition of the polyphenols present in sorghum were unknown to most of the experimenters, what follows can be regarded as mainly qualitatively indicative of the effects of tannins in animal diets.

Rat feeding trials

Most of the biological tests reported were carried out with rats. Maxson (E. D.) *et al.* (1973) compared various samples of sorghum, from plantings at College Station, Texas in 1970, including variations in pericarp colour, some with and some without pigmented testas. The designations of the samples with their B_1 B_2 S genotype (Schertz and Stephens, 1966), pericarp colour, presence or absence of testa, protein content (micro-Kjeldahl N × 6·25 DWB, AACC, 1962), lysine (by auto-analyser) as % of protein, tannin in rat diets (FAS and MV-HCl methods; Maxson and Rooney,

* (Authors' note: the composition of the diet was not stated).

1972b) and performance of male weanling rats (weight gain, feed intake, PER and IVDMD) are shown in Table 4.33.

The sorghum diets contained 9% protein from the sorghum and 1% protein from casein, (previous experiments had indicated that the inclusion of casein significantly reduced experimental error). The two control diets contained 10 and 5% cascin respectively.

Correlation coefficients between lysine content of the sorghum diets and weight gain, feed intake and PER were positive and highly significant; lysine content was not related to IVDMD. Negative correlation coefficients between IVDMD and FAS and MV-HCl values of the diets were significant and in agreement with previous IVDMD results (Harris *et al.*, 1970; McGinty, 1969). Maxson (E. D.) *et al.* (1973) attributed differences in IVDMD to the different FAS and MV-HCl (CE) values of the diets.

Low non-significant correlation coefficients were obtained when FAS and MV-HCl values were compared with weight gain, feed intake, PER and lysine content of the diet. (Oswalt *et al.*, 1972 found weight gain, expressed as % of check sample, negatively correlated with CE but the diets used were not isonitrogenous).

Maxson (E. D.) *et al.* (1973) also determined the α-amylase inhibition of sorghum lines, grown at College Station, Texas in 1970, using a modified method for α-amylase activity in barley malt (AACC, 1962) expressed as Sandstedt-Kneen-Blish (SKB) units.

All brown grains showed high levels (205 000–230 000 inhibition units) of α-amylase inhibition. White or red grain with a pigmented testa and grain without a pigmented testa did not inhibit α-amylase. Inhibition of α-amylase was associated with the same genes that produced brown pericarp, presence of a pigmented testa, and high tannin content. The inhibition was probably caused by the tannins produced as a result of a dominant gene at each B_1 B_2 S locus.

The effect of pericarp colour and presence or absence of a pigmented testa on weanling rat performance is shown in Table 4.34. White grain with a pigmented testa resulted in lower IVDMD (attributed to the higher tannin content), but higher feed consumption, greater weight gain and higher PER (attributed to higher content of lysine in protein) when compared to white grain without a pigmented testa.

Red grain with a pigmented testa gave the same IVDMD and similar tannin content, as red grain without a pigmented testa. Other differences in rat performance were attributed to differences in lysine content in protein. Grain with brown pericarp and a pigmented testa gave the lowest PER of all grain types and the lowest IVDMD when compared with grain that had a white or red pericarp and a pigmented testa. PER was not related to tannin content of the diets but some factor in brown grain, possibly related to enzyme inhibition or the formation of an indigestible protein–tannin complex, appeared to cause a reduction in PER. The reduction in IVDMD might be related to the high levels of α-amylase inhibition associated with the same gene combination that produced brown grain and high tannin content. No significant differences in IVDMD were produced by grain with white or red pericarp.

Samples from a planting at Lubbock, Texas in 1971 were selected to provide plants of red and tan plant colour and of white grain without a pigmented testa. Plant colour did not influence rat weight gain.

In 1966 Loomis and Battaile developed a technique in which insoluble polyvinylpyrrolidone (insoluble PVP) was used to remove phenols from what they described as H-bonded complexes with protein in plant tissues, and so obtained active soluble enzymes.

Jambunathan and Mertz (1973b) applied this technique before carrying out protein fractionation studies (the method was not stated but was probably Landry and Moureaux, 1970) with two high-tannin lines of sorghum, IS 2283 and IS 6992.

As previously reported (Jambunathan and Mertz, 1973a), high-tannin sorghums produced smaller fractions I, II and III and larger fractions IV and V when compared to low-tannin sorghums. With the addition of insoluble PVP the N distribution of IS 2283 and IS 6992 approached, but did not equal, the distribution patterns of low-tannin lines.

Three sorghum samples of different CE (shown in parentheses), (A) 05274 (0·3%); (B) 925120 (2·08%); and (C) RX 025440 (3·08%) alone; and mixed with PVP, (D), (E) and (F) to make 5% of the diet, were fed in isonitrogenous diets at 8·6% protein level to groups of six rats over 28 days. The addition of PVP, diets (E) and (F) had a beneficial effect on weight gain and PER of rats fed high-tannin diets, sorghums (B) and (C), but a harmful effect on those fed the low-tannin line. Jambunathan and Mertz (1973b) commented that the level of 5% PVP was chosen arbitrarily and may have been excessive for the low-tannin diet (A).

The same sorghum lines were also fed in isonitrogenous diets (9% protein) to groups of 10 weanling rats over 28 days, (a) without added lysine, and (b) with added L-lysine HCl to bring the total dietary lysine level to 1·0%. Details of the samples and results of the feeding tests are given in Table 4.35.

Supplementation with L-lysine HCl improved weight gains, increased feed consumption, and improved PER on all diets, the greatest improvement being shown by the low-tannin sorghum with PVP diet (D). On diet (C) rats lost weight but with lysine supplementation gained weight, the PER being about half that of the supplemented diet (F). As judged by general body composition and post-mortem liver condition, the diets supplemented with lysine were ranked in the following order: 1 (D), 2 (E), 3 (F), 4 (A), 5 (B) and (C). Earlier experiments at Purdue indicated that the growth rate of rats fed high-tannin sorghums (CE 4·6%) was *not* improved by the addition of L-lysine HCl.

Jambunathan and Mertz (1973b) pointed out the obscurity of the relationship between the CE value and the availability of sorghum protein to the rat. They suggested that some compounds in sorghum which react in the CE test may not be harmful to rats, in which case lysine supplementation of a high CE sorghum could improve rate of weight gain.

Jambunathan and Mertz (1973b) reported a cooperative feeding study, using the same sorghums, conducted at Purdue University with rats, at CIMMYT with meadow voles, at the University of Nebraska with mice, and at Washington State University with chicks, indicated that only the rat could differentiate between low- and high-tannin samples. The PER (rats) values of the sorghum compared to other cereals were as expected. With the vole the PER of sorghum was higher than that of Opaque-2 maize; with the mice the PER of sorghum was not significantly different from that of Opaque-2 maize. With the chick the PERs for the sorghums were not significantly different from one another and could not be differentiated from those given for wheat, triticale and Floury-2 maize.*

(A) IS 2319 and (B) 025042-1, described as low-tannin, and (C) IS 8165 described as high-tannin, were fed in 8·8% protein diets to groups of 10 male rats over 28 days. (Jambunathan and Mertz, 1973a, showed IS 2319 and IS 8165 as having CE values of 0·51 and 2·69% respectively.) The average PERs of the rats were: (A) 1·28, (B) 0·97 and (C) 0·71 and casein 3·34. Only (A) and (C) differed significantly.

A segregating F_2 population selected from a cross between a low-tannin sorghum, Purdue 954062 (0·15 CE), and a high-tannin line, IS 2942 (5·53 CE) were planted at Purdue Agronomy Farm in May 1972 (Cummings and Axtell, 1973a). The grain from 175 plants was classified for the presence or absence of testa by scraping the outer layers of several seeds from each sample and looking for a dark brown testa or subcoat. Each sample was analysed for tannin (CE) by the V-HCl method (Burns, 1963, 1971) slightly modified, and the MV-HCl method (Maxson and Rooney, 1972a). On the basis of the results, three groups of composite samples were prepared:

I coloured testa absent (no brown subcoat): CE values below 1·0 on both V-HCl and MV-HCl tests;

II coloured testa present; V-HCl values less than 1·0 CE, MV-HCl values greater than 2·0 CE; and

III coloured testa present; V-HCl and MV-HCl values both greater than 2·0 CE.

Thirty-nine samples which were not within the limits were excluded.

For the 60 samples without coloured testa, the mean V-HCl value was 0·428 CE, and the mean MV-HCl value was 0·705 CE. For the 115 samples with testa, the mean V-HCl value was 1·03 CE and the mean MV-HCl value was 3·36 CE. The V-HCl and MV-HCl tests were not significantly correlated when only the samples lacking coloured testa were compared but were highly significantly correlated among samples with a testa. For the

* (Authors' note: the unsuitability of the meadow vole (*Microtus pennsylvanicus*) as a test animal is well established (McDonald and Larter, 1972, unpublished data; Hulse and Laing, 1974).)

total 175 samples, the mean V-HCl value was 0·824 CE, the mean MV-HCl value was 2·451 CE and the correlation coefficient was highly significant.

The bulked samples were used in feeding trials with weanling rats, six to the group, with two replications of 14 days. Group I had a slightly higher concentration of white grains but otherwise the groups differed little in appearance. The composition of the sorghum samples in protein, lysine and CE is given in Table 4.36.

As each diet contained an equal weight of sorghum (94 g sorghum, 4 g mineral mix and 2 g vitamin mix) there were small variations in protein (11·3–11·7%) and lysine (1·74–1·84% protein). The composition of the diets and the effect on F/G and PER are given in Table 4.37.

Lysine was the first-limiting amino acid in all three diets. Average feed consumption was about the same for the three groups but rats on group III gained less and showed less efficient F/G and PER. Group I diet (no testa, V-HCl CE 0·24) and group II diet (testa present, V-HCl CE 0·52) performed equally well. Cummings and Axtell (1973a) concluded that the CE values as measured by the MV-HCl test were not reflected in the rat feeding trials while the V-HCl values were, and that the presence or absence of the testa was not a good indicator of sorghum nutritional quality. Group II represented 36% of the plants from a segregating F_2 population which had a testa and remained low in tannin (as measured by the V-HCl method). It was not known whether those hybrids retained the agronomic characteristics (bird-resistance, resistance to weathering) associated with a brown subcoat.

Cummings and Axtell (1973b) and Cummings (1973) carried out rat feeding trials with IS 8260, a dark brown, high-tannin (5·24 CE, V-HCl test, Burns, 1963, 1971) sorghum grown in the Purdue/AID Puerto Rico winter nursery in 1972. Whole grain, and grain dehulled by the sodium hydroxide method of Blessin (1971) to remove the pericarp and testa, provided the only source of protein in the isonitrogenous diets. The whole grain (CE 5·24) and the dehulled grain (CE 0·04) were mixed in varying proportions to give a gradient of tannin levels ranging from 0·11 to 4·83 CE.

All eight diets contained 1% vitamin mix and 4% mineral mix, and six were supplemented with 0·5% lysine as lysine-HCl. This addition brought the lysine in the sample to approximately 75% of the rat's requirements. Mertz (1969) cited by Cummings and Axtell (1973b) had estimated the lysine requirement of the rat to be 1% of sample. Weanling male rats, 10 to the group, feed and water *ad lib.*, were fed the lysine supplemented diets over 28 days. Because of a shortage of grain, the two unsupplemented diets were fed over only 14 days.

Alkali dehulling did not change lysine content (% protein) but appeared to increase protein by approximately 1·0%. Blessin (1971) and Oswalt (1973b) found no change in protein content with alkali dehulling. The substances measured by CE determinations were almost totally removed (or destroyed) by dehulling. The effect of dehulling the whole grain, without and with added lysine, on rat weight gain is shown in Table 4.38.

Dehulling (diets (A) and (B)) significantly increased weight gain, F/G and PER when compared to the whole grain; feed consumption remained the same when unsupplemented with lysine. Lysine supplementation of the whole grain (G) produced no significant improvement over (H) but gave marked improvement of the dehulled grain (diet (B)) in comparison with (A).

The addition of vitamin and mineral mix to the whole grain reduced the CE value from 5·24 to 4·38, a reduction greater than could be attributed to dilution alone and suggesting an interaction between one or more of the additives and one or more of the compounds responsive to the V-HCl test. Supplementation with lysine diet (G) led to a further reduction in CE to 3·66; possibly the result of a lysine–tannin interaction. The effect of the tannin gradient in the diets, which were near isonitrogenous but suboptimal in lysine, diets (B), (C), (D), (E), (F) and (G), on rat performance over 28 days is shown in Table 4.39.

The growth-depressing effects of tannin, even at the lower levels, diets (C) and (D) were marked. No threshold level was found and it was considered that any level of tannin in suboptimal protein diets was likely to reduce nutritional quality. PERs were significantly different among most of the diets, correlated more highly with other variables than F/G and were highly negatively correlated with CE levels.

Oswalt (1973a) compared the mean contents of nine essential amino acids of 36 low CE and 34 high CE sorghum selections, against the weanling rat requirement. The mean crude protein content of the low CE samples was 11·6% and of the high CE samples 10·3%. The minor differences in the proportions of amino acids between the two groups were consistent with the difference in crude protein.

Whether expressed as % of protein or as % of sample, lysine content was low in relation to rat requirement; arginine and leucine were present at double the required level. Lysine (% protein) was the only deficient amino acid. When expressed as % of sample, methionine + cystine, threonine, and possibly also isoleucine and phenylalanine, appeared deficient. Histidine and valine were adequate. Two selections, IS 2319 and IS 0129, each with a complete or partial testa but low CE values, and a high CE sample, IS 6992, were fed at near isonitrogenous levels to weanling rats over 14 days, (a) unsupplemented and (b) supplemented to bring the lysine content up to 125% of the requirement for normal rat growth. Supplementation improved the performance of rats on IS 2319 and IS 0129 diets but not of those on the high CE sample IS 6992 which, however, had given improved results from lysine supplementation, when fed dehulled.

The same low CE selections, supplemented with a combination of lysine, methionine, threonine and phenylalanine, all to levels equivalent to 125% of rat requirements, gave significantly increased rat gains, when compared with isonitrogenous rations, supplemented with lysine alone. A slight but insignificant improvement in weight gain resulted from supplementation of the high CE sample IS 6992 with all four amino acids compared to supplementation with lysine.

Examination of the data from 14-day feeding trials with weanling rats of 110 grain sorghum genotypes of varying CE concentrations and crude protein content showed a break in the spread of sample frequency, means for CE concentration and rat weight gain between the 66 samples below, and the 44 samples above, CE concentrations of 1·00 (Oswalt, 1973a). The rations were composed of 94% sorghum, 2% vitamin mix and 4% mineral mix and were not, therefore, isonitrogenous.

Rat weight gain, actual and as % of check sample, related to CE concentrations and IVDMD (Tilley and Terry, 1963) are shown in Table 4.40. IS 2319, crude protein 11·8% and CE value 0·51 was fed as a control.

Correlation of weight gain, as % of check, with lysine (% of protein) was negative in the low CE group. Amino acid analyses indicated that all genotypes contained less lysine than was required for normal weanling rat growth and a positive correlation of weight gain with lysine as % protein had been expected. As fed, in rations which were not isonitrogenous, correlation of weight gain with lysine (% of sample) was significantly positive in the low CE group. The high CE group did not show a significant correlation of weight gain with lysine (% of sample), indicative of the influence of high CE values on the nutritive quality.

Since the rat feeding trials were conducted with six rats/group, were not replicated, and were at different crude protein levels, Oswalt (1973a) commented that there was a high level of error associated with the data. Nevertheless, it was possible to identify extremes and select genotypes within the high and low CE groupings for more detailed isonitrogenous studies.

High CE values were associated with dark coloured testas and resulted in lower average rat weight gains. However, visual ratings, according to testa and pericarp characteristics, did not predict consistently CE concentrations and rat response. The correlation of rat weight gain as % check was positive and significant with IVDMD for the low CE samples and all samples together, and positive and significant at the 0·05% level for the high CE samples. There was a negative correlation between CE values and IVDMD for all samples and also among the high CE group, but not among the low CE group. IVDMD appeared to be a useful index of the combined effect of CE and crude protein content.

Table 4.41 presents data from sorghum samples grouped into (1) five near-isonitrogenous diets, and (2) sub-divided into high and low CE samples; it includes IVDMD values, lysine as % of sample and % of protein and rat weight gain.

Mean crude protein, and lysine (% protein) in the low and high CE groups were similar at any given level of protein content. Lysine (% of sample) increased as crude protein % increased, hence in the low CE samples, rat weight gain increased as crude protein and lysine (% of sample) increased.

The high CE groups were consistently lower than low CE groups in rat weight gains (except at 10·0–10·9% protein), rat weight gains as % of check, and IVDMD % for all levels of protein. Rat weight gains from the low CE samples increased as % protein and lysine content of the sample increased. In the high CE samples weight gain as % of check increased as % protein increased but at a lower level than from the low CE samples. IVDMD appeared as a useful index by which to identify the most suitable genotypes.

Oswalt (1973a) found no significant differences in weight gains, feed consumed or F/G when weanling rats were fed a low (0·48) CE sorghum, RS 610, without and with supplementation with methionine at the rate of 0·15 g/100 g of ration and/or choline at the rate of 0·20 g/100 g of ration.

When a high (8·10) CE sorghum, BR 64, was fed unsupplemented or supplemented with methionine or choline singly, rats showed significantly lower weight gains than those on RS 610 rations. Feed consumption was similar, indicating that palatability was not a factor. Supplementation of the BR 64 rations with methionine and choline together produced rat weight gains and an F/G similar to that from the RS 610 rations. The apparent benefit from combining methionine with choline was not understood.

Removal of the outer layers of the caryopsis with hot sodium hydroxide lowered the CE level of RS 610 from 0·48 to 0·10, and of BR 64 from 8·10 to 0·35, without significantly altering the crude protein content.

The amount of feed consumed by weanling rats fed the dehulled grain was not significantly different from the feed consumed by rats fed the whole grain but the rats on the whole BR 64 grain did not maintain starting weight. Rats on the dehulled BR 64 grain showed an F/G of 14·1 compared to ratios of 21·3 and 19·6 respectively for the whole and dehulled RS 610.

Supplementation with lysine of two high CE genotypes IS 6992 and IS 8260 did not significantly increase rat weight gains which did increase when dehulled samples were fed, with further increases with added lysine.

Featherston and Rogler (1975) extracted high-tannin BR 64 with alkali (Blessin *et al.*, 1971). The extracted grain was compared with unextracted grain at 93% of a rat diet that contained 2·0% maize oil, 4% vitamin and mineral premix, 0·75% L-lysine HCl and 0·25% glucose monohydrate and/or additives. Also added were 0·2% L-methionine or 0·255% methionine hydroxy analogue (MHA), equimolar to 0·2% methionine. Weight gain improved from 26·3 unextracted to 52·7 g extracted and feed efficiency from 9·56 to 5·20 but the addition of methionine or MHA had little effect. The results agreed with previous findings with chicks (Armstrong *et al.*, 1974a) except for the lack of response to methionine. (Armstrong *et al.*, 1973, 1974a).

Rats were fed the BR 64 sorghums, (A) extracted and (B) unextracted, and (C) RS 671, a commercial low-tannin sorghum, in a digestibility trial. The basal diets were similar to those used in the previous growth trial except that they were not supplemented with lysine. The feed consumed, weight of faeces, faecal N, dietary N and calories absorbed are shown in Table 4.42.

Feeding the high-tannin, unextracted BR 64 diet (B) did not result in decreased feed consumption. This agreed with the findings of Jambunathan and Mertz (1973a), and contrasted with the findings of Glick and Joslyn (1970a) who reported decreased feed consumption when tannic acid *per se* was added to a rat diet.

The weight of the faeces excreted by rats fed diet (B) was approximately double that of rats fed (A) and (C). Faecal N was higher from rats fed (B) than from (A) and (C) when expressed on a total N basis and as a % of faeces. Rats on diet (B) absorbed approximately 30% less N than those on (A) and (C).

Featherston and Rogler pointed out that Jambunathan and Mertz (1973a) had observed a different distribution of protein from high- and low-tannin sorghums when the proteins were fractionated following Landry and Moureaux (1970) and suggested that these findings indicated that the decreased N absorption from diet (B) was not due to the protein *per se* since the same sorghum, when extracted, gave protein of similar digestibility to that of the low-tannin RS 610.

Fewer calories were utilized from diet (B) than from (A) and (C) which were utilized about equally.

It appeared that protein was not the only component of sorghum grain where digestion and/or absorption was being adversely affected by the tannins. Differences in protein digestion accounted for only approximately one-third of the decrease in caloric utilization found from diet (B) compared to diet (A).

The two sorghums BR 64 (high-tannin) and RS 671 (low-tannin) were also compared, in chick diets, the diets being made isonitrogenous by glucose monohydrate. These diets were essentially similar to those used in the growth trial with rats.

Supplementation of diets with methionine or MHA resulted in a significantly greater weight gain and feed efficiency with the high CE BR 64 but an insignificant improvement with the low CE RS 671.

Neither isoleucine nor phenylalanine added on an equimolar basis had any significant influence on growth rate or feed efficiency. The addition of methionine had no significant effect on pancreas size. The reason for the apparent improvement by methionine supplementation of high-tannin sorghum in chick diets and its failure to improve rat diets was not explained.

Schaffert *et al.* (1974b) compared four varieties of sorghum described as IS 2549 (low-tannin, low-protein), IS 0062 (low-tannin, high-protein), IS 6992 (high-tannin, low-protein) and IS 1210 (high-tannin, high-protein) in male weanling rat feeding trials using soybean meal of 53·25% protein at 0, 5, 10 and 15% levels to replace maize starch in a basal diet of 79% sorghum, 4% mineral mix, 2% vitamin mix and 15% maize starch. The protein contents (Kjeldahl N × 6·25) of the sorghums, as % DWB were: IS 2549 8·08%; IS 0062 10·5%; IS 6992 7·5%; and IS 1210 8·5%.*

* (Authors' note: it is difficult to understand why IS 1210 was classified as a high-protein and IS 2549 as low-protein).

The low-tannin sorghums contained less than 1·0, and the high-tannin samples more than 4·0 CE.

Rat weight gains were significantly greater from the low-tannin, IS 2549 and IS 0062 sorghums, than from the high-tannin IS 6992 and IS 1210, at 0, 5 and 10% soybean supplementation levels, but not at 15%. Schaffert *et al.* (1974b) suggested this could indicate that the amount of digestible protein was not significantly different between the high- and low-tannin sorghums, or was above the rat requirement at the 15% level of soybean supplementation. Lysine was first-limiting even at the 15% supplementation level. Rat weight gains were significantly different among all levels of protein supplementation.

Schaffert *et al.* (1974b) stated that weanling rats fed rations with 0, 5 and 10% soybean meal could be used to detect differences between high- and low-tannin genotypes. In the unsupplemented rations, F/Gs were significantly lower for low-tannin sorghums than for the high-tannin, possibly due to reduced protein digestibility. There were no significant differences in F/G among any of the supplemented rations.

The significant effect of both grain tannin and grain protein appeared in only one of the high-tannin sorghums, IS 1210, which did not differ appreciably from the others in amino acid composition but contained a waxy, high amylopectin endosperm which may have influenced feed consumption and F/G without affecting weight gain. The other high-tannin sorghum, IS 6992, was intermediate in feed consumption indicating that the differences in F/G between the high- and low-tannin sorghums were due to factors other than feed intake.

Schaffert *et al.* (1974b) concluded that the increase in weight gain from additions of even small amounts of soybean meal to grain sorghum indicated that the effect of tannin was not toxic but related to the availability of protein in the ration.

Six sorghum cultivars, grown at an experimental farm in South Africa were compared in rat feeding trials (Dreyer and van Niekerk, 1973). The samples were selected on the basis of their tannic acid equivalents (AOAC, 1965) and were either "unextracted" (the whole grain ground into a relatively fine meal in a hammer mill) or "extracted", in which the dark-coloured outer layer of the grain was removed in a rice-pearling machine, followed by extraction with 60% v/v aqueous ethanol. The extracted outer layer portion after drying at room temperature, was recombined with the decorticated portion. Table 4.43 shows the Kjeldahl N and tannic acid equivalents of the meals.

The rat diets were the semi-synthetic type used routinely in protein quality studies at the National Food Research Institute, Pretoria. In the experimental diets the dextrin component was replaced weight for weight with (a) defatted whole egg to bring the protein content to 4% plus (b) sorghum meal to raise the protein level of the diet by 5·3% to a total of 9·3%. The experimental diets were thus equal in protein and total nutrient content but different in tannin and dextrin content. The control diets were similar except that none of the dextrin was replaced by sorghum meal. Egg protein was included to ensure a certain measure of growth and complete consumption of the food offered which, throughout the experimental period of 32 days, was at the level of 10% of rat body mass.

The experiments were designed to measure rat growth rate, the effect on rat liver lipids (petroleum ether extraction in a Soxhlet apparatus) and to estimate the digestibility of the sorghum protein (from N consumption and faecal N, Dreyer, 1962, 1973). The results for extracted and unextracted grain are shown in Table 4.43.

There were appreciable differences among cultivars in N content and in tannin content; extraction with 60% ethanol solution markedly reduced the tannin content. As the experimental diets were essentially equal in protein, Dreyer and van Niekerk (1973) considered that differences in the performance of the experimental animals were due to tannic acid content or to causes other than protein content. C 26 and C 45 with the highest levels of tannin content, gave improved rat growth rate and protein digestibility on extraction with ethanol, but still did not equal the performance of CI 34, the best of the samples with a tannin content, when extracted, of 0·31 (0·17 in the ration).

In general, ethanol extraction improved the digestibility of the protein but in the case of CI 23, with an unextracted tannin content of 0·87, extraction decreased protein digestibility.

Great variability was found in liver lipid contents and statistically significant differences were observed only in the groups fed the unextracted grains. Ethanol extraction reduced liver lipid in the case of C 45 but otherwise either increased the level or had no effect.

The most closely correlated coefficients were (a) growth rate and protein digestibility followed by (b) digestibility vs tannin content and (c) growth rate vs tannin content. Liver lipid content appeared not to be associated with any of the other variables. Dreyer and van Niekerk (1973) pointed out that the lack of perfect correlation between dietary tannin content and protein digestibility could be due to an effect on digestibility other than that caused by tannins, or due to the analytical method not measuring all the various tannins present in the grain. They considered protein digestibility a more accurate index of the growth-suppressing properties of sorghum than the tannin content. The data on liver lipid content could not be explained on the basis that there was detoxification of gallic acid through *o*-methylation.*

Dreyer (1974) compared the *in vitro* protein digestibility and the tannin content (data from van Wyk *et al.*, 1973) with the *in vivo* (rat) protein digestibility of 13 sorghum cultivars selected to represent the full range of available samples. The method for measuring the *in vivo* digestibility was essentially that employed by Dreyer and van Niekerk (1973) but the

* (Authors' note: this is a curious comment since no evidence is given to support the presence of gallic acid or any hydrolysable tannin containing it. Similarly, references to tannic acid content and tannic acid equivalent are misleading and almost certainly irrelevant.)

sorghum protein was fed at three levels, 0, 3·25 and 6·5%, to groups of five rats over an experimental period of 12 days. The results are shown in Table 4.44.

Dreyer (1974), from a rank correlation analysis of his data, concluded that *in vivo* digestibility of a sorghum cultivar might be estimated from either the grain tannin content (AOAC, 1965) or from the *in vitro* digestibility determination (van Wyk *et al.*, 1973). He also pointed out that *in vivo* digestibilities reported earler (Dreyer and van Niekerk, 1973) ranged from about 45 to 88% whereas in the present study, the range was 69–99%.

May and Nelson (1973) reported on the digestible and metabolizable energy content for rats of the sorghum varieties ARK 61002, AKS 614, RS 617, RS 608, TE 66 and ARK 62002. These varieties were selected on the basis of their average amino acid availability to chicks as determined by Stephenson *et al.* (1971). The experimental methods and the diet were as described in May and Nelson (1972). The treatments in each experiment were basal diet, and 50% basal diet plus 50% sorghum. Tannin was determined by a modification (not described) of the method of Lieberman *et al.* (1959) for determining chlorogenic acid, and N in feed, faeces and urine by AOAC (1965). Gross energy of feed and faeces was determined by a bomb calorimeter. Digestible energy (DE), in kcal/g was determined using the formula:

$$\text{DE} = (\text{g feed consumed} \times \text{GE/g feed}) - \frac{(\text{g faeces} \times \text{GE/g faeces})}{\text{g feed consumed}}$$

Gross energy of the urine was estimated from total urinary N using the equation (May and Nelson, 1972):

$$\text{ME} = \text{DE} - (\text{g urinary N} \times 7{\cdot}00)/6{\cdot}56/\text{g feed consumed}$$

Metabolizable energy (ME) was corrected for N balance using the same equation to calculate the energy value of the retained N. The DE and ME values for sorghum were determined using the equation of Sibbald *et al.* (1960). Correlations of tannin content with DE and ME corrected for N balance are shown in Table 4.45.

The DE values ranged from 3·49 kcal/g for ARK 61002 to 4·19 kcal/g for TE 66. The ME values, corrected for N retention, ranged from 3·30 kcal/g for ARK 61002 to 4·01 kcal/g for TE 66. Correlation between tannin content vs DE and ME was low. The DE and ME tended to increase as the availability of amino acids to chicks increased. May and Nelson (1973) did not know what if any relation existed between amino acid availability to chicks vs rats.

Martin-Tanguy *et al.* (1976) reported feeding trials with male rats, six or seven to the group, using sorghum samples harvested in 1967, 1970 and 1971 provided by the Station d'Amélioration des Plantes du Centre de Recherche de Montpellier and by the Coopérative Agricole Lauragaise de Castelnaudary dans l'Aude. The chemical composition of the sorghum diets is shown in Table 4.46.

Condensed tannin determinations were based on the transformation of the proanthocyanidins to coloured anthocyanidins in acid medium (Swain and Hillis, 1959). Cyanidins were colormetrically measured at 550 nm (Swain and Hillis, 1959). Tannin content was expressed in optical densities and defined in terms of the quantity (LA) (Troyer, 1964, Quesnel, 1968, Bate-Smith, 1973b). The proanthocyanidin polymerization index (Goldstein and Swain, 1963) was obtained following two chemical reactions: (a) the transformation of the leucoanthocyanidins to anthocyanidins (LA); and (b) the formation of a coloured complex with vanillin in an acid medium (V-HCl).

The effect of the diets on N balance and metabolizable energy is also shown in Table 4.46. The results indicated that energy and protein digestibility ranged from 79 to 90% and from 50 to 88% respectively. The CDU (coefficient of digestive utilization) values for energy and protein varied inversely and linearly with the level of procyanidins in the sorghum sample.

Chicks

McClymont and Duncan (1952) reported that sorghums described as Kalo, Caprock and Milo grown in Australia depressed food consumption and growth when substituted for wheat in chick diets. Supplementation with 13% liver meal, manganese or a mixture of B vitamins all failed to counter the adverse effects.

Thayer *et al.* (1957) reported that Oklahoma grain sorghums produced variable growth when fed as full replacement of maize but found no apparent relation between grain colour and growth rate.

Chang and Fuller (1964) at the University of Georgia, reported the tannin content, protein and fibre (AOAC, 1955) of Combine Kafir 60, Martin Milo, GA 609 (an F_1 hybrid of Combine Kafir 60 and double-dwarfed Early Shallu), Redbine 60, Combine Sagrain and NK 230 (this last, a discountinued variety). Metabolizable energy was determined by the method of Potter and Matterson (1960). Data are given in Table 4.47. In all the feeding trials with chicks, the grain sorghums constituted 50% of the diet and replaced an equal weight of maize which represented 55% of the basal ration which contained 36·5% of 50% protein soybean meal.

In the first trial, Combine Kafir 60 (0·1% tannin in diet), Martin Milo (2% tannin in diet) and GA 609 (0·8% tannin in diet) were compared with the maize basal diet (0% tannin in diet) and with the maize diet with additions of tannic acid (digallic acid) to give 0·1, 0·5, 1·0 and 2·0% tannin in the diet.

GA 609 (0·8% tannin in diet) gave poorest growth similar to 1·0% tannic acid in the diet. However, Combine Kafir 60 (0·1% in diet) gave a lower 7-week body weight gain than Martin Milo (0·2% in diet). Tannic acid reduced growth in proportion to the quantities added above 0·1% in the diet.

In a second trial, rations provided 23% protein and 1466 kcal of ME/lb.

As will be seen from Table 4.48, growth depression was directly related to sorghum tannin content.

Protein digestibility (Ekman *et al.*, 1949) was slightly depressed at highest tannin levels, but not sufficient to account for the observed growth retardation. Liver lipid and sorghum tannin contents were correlated.

Booth *et al.* (1959) compared commercial gallic acid and tannic acid, with other related compounds, in metabolic experiments with rats. They concluded that *O*-methylation, resulting in the formation of 4-*O*-methyl gallic acid, accounted for the major metabolite in the urine of rats ingesting gallic acid or tannic acid. Booth *et al.* (1961) reported that weanling rats fed a low-methionine, low-choline diet containing additions of 1% gallic acid, developed fatty livers: a sign of choline deficiency eliminated by added choline or methionine.

In Chang and Fuller's third trial, choline (200 mg/lb) and methionine as MHA were added to the basal maize–soybean diet with and without 1% tannic acid. The tannic acid significantly depressed growth, partially though significantly, alleviated by choline and MHA. In the absence of tannic acid, choline and MHA did not improve growth rate and only slightly improved feed conversion.

In a fourth trial, Combine Kafir 60 (0·2% tannin) and Combine Sagrain (2·0% tannin) were compared with and without additions of 400 mg choline and 900 mg MHA/lb of diet, at which levels Combine Sagrain was approximately equal in feeding value to Combine Kafir 60.

Unpublished data from Fuller, Chang and Potter, cited by Fuller *et al.* (1966) indicated 1% tannic acid in the diet was "inevitably toxic" at any level of added choline and methionine. Fuller *et al.* (1966) also determined (method not stated) the methionine content, as % of protein, of the varieties used by Chang and Fuller (1964). These are shown below:

Variety or hybrid	Tannin %	Methionine % of protein
Combine Kafir 60	0·2	1·50
Martin Milo	0·4	1·35
Redbine 60	0·4	1·39
GA 609	1·3	1·53
Combine Sagrain	2·0	1·60
NK 230	2·0	1·24

Combine Sagrain, highest in tannin, was also highest in methionine. Fuller *et al.* (1966) suggested that added methionine probably compensated for low levels of methionine in high-tannin sorghums.

Chang and Fuller concluded that tannins in sorghum differed chemically from the tannic acid (digallic acid) fed in the trials they described.

Fuller *et al.* (1966) stated that the grain sorghums grown in the south western US were predominantly of the milo type with red or yellow seed colour and compact heads. In the more humid south eastern States, grain sorghums with open heads and brown seedcoats predominated; open heads permit more rapid drying and the penetration of pesticide sprays. Fuller *et al.* (1966) examined 18 varieties of grain sorghum grown in different locations using the methods of Chang and Fuller (1964). Seed colour, tannin, protein, fat and fibre, and effect on weight gain and feed utilization of chicks are given in Table 4.49. The diets were equalized for protein and ME.

At 50% of the diet growth rate was depressed in tannin contents of 1·6% or more.

Fuller *et al.* (1966) also reported on tannin content in sorghum gluten meal and gluten feed from different commercial samples. In general, tannin was highest in the gluten feed that contained a higher proportion of seed coats.

Gerencsér *et al.* (1966) reported that when half the maize in chick rations was replaced by a Hungarian sorghum hybrid of 1·9% tannic acid content, the weight gain of (a) New Hampshire chicks was not affected but (b) Lohman hybrids, which generally grow more rapidly, was reduced. Total replacement of maize by sorghum in New Hampshires reduced weight gain by about 20% and feed conversion efficiency by 11%.

Connor *et al.* (1969) in Australia fed chicks with diets containing 70% sorghum. Tannin was determined by a modified Folin-Denis method (Pro, 1952). Diets were isonitrogenous by mixing grain of the same variety but of different N content.

In the first trial, the following diets were fed over 8 weeks:

(A) TX 630, tannin content 0·1%
(B) (A)+choline chloride (440 mg/kg)+DL-methionine (374 mg/kg)
(C) (B)+0·1% tannic acid
(D) (A)+1·0% tannic acid
(E) (B)+1·0% tannic acid
(F) Alpha (tannin content 0·2%)
(G) (F)+choline and methionine

The feed conversion ratios (FCR, not defined) did not differ on diets (A), (C) and (F); and were slightly improved, and similar, for (B) and (G). Diet (D) depressed weight gain, feed consumption and FCR. The effects were improved but not eliminated by the addition of choline and methionine, diet (E).

In the second trial, Sumac sorghum (tannin content 2·5%) was compared with TX 630, with and without choline and methionine. Sumac produced lower weight gains, feed consumption, and poorer feed conversion than TX 630; added choline and methionine did not improve FCR. In a third trial, TX 630 and Sumac, fed separately and mixed to give seriatim, tannin contents of 0·2%, 0·97, 1·73 and 2·5%, were compared with a 50/50 mixture of wheat and maize (tannin content 0%). Choline and methionine were also added at the levels previously used and at twice and three times those levels. FCR improved as tannin content declined and as choline and methionine increased, but even at the highest levels choline and methionine did not fully counteract the effect of tannin content.

Connor *et al.* (1969) commented that these results confirmed those of Chang and Fuller (1964) and of Vohra *et al.* (1966) but not the suggestion of Chang and Fuller (1964) that the tannin in sorghum induced a choline and/or methionine deficiency which could be overcome by providing high levels of those nutrients in the diet.

In four sorghums, RS 610, NK 300, BR 804 and BR 64, Rostagno *et al.* (1973b) determined amino acid availability for male chicks. Also included were RS 610 with tannic acid added to the tannic acid equivalent (Folin-Denis method, Burns, 1963) of NK 300, BR 804 and BR 64, together with a protein-free diet and the protein free diet plus tannic acid "equivalent to" BR 64. The protein and tannic acid contents of the diets are shown in Table 4.50.

Amino acid compositions of sorghums and excreta were determined by ion-exchange chromatography (Cromwell *et al.*, 1967). Dietary and excreta content were determined (Czarnocki *et al.*, 1961), and apparent and corrected amino acid digestibility calculated following Kleiber (1961). Tryptophan was not determined.

Diet I (1·4% tannic acid in the protein-free diet) produced a significant increase in the excretion of all 14 amino acids recorded. Glick and Joslyn (1970b) cited by Rostagno *et al.* (1973b) also found a three to four-fold increase in the level of proteolytic enzyme in the whole intestinal contents in rats fed tannic acid, and concluded that endogenous proteins accounted, at least in some part, for the high faecal N content from rats fed tannic acid. Rostagno *et al.* (1973b) considered that their results with chicks supported a similar conclusion. The apparent amino acid digestibilities were: diet (A) RS 610 73%; (B) NK 300 41·5%; (C) BR 804 25·6%; and (D) BR 64 22·2%.

These differences did not agree with Chang and Fuller (1964) who reported only a slight depression in protein digestibility between high-tannin and low-tannin sorghum.

The apparent digestibility of sorghum RS 610 with added tannic acid, diets (E), (F) and (G), were significantly higher than those for diets (B), (C) and (D) where the tannic acid equivalents* were naturally present in the sorghums. Rostagno *et al.* (1973b) could offer no explanation for this. Digestibility values of diets (A), (B), (C) and (D) corrected for endogenous N based on diet (H) followed the same trend. Rostagno *et al.* considered that apparent digestibility was a better guide than corrected digestibility since, whether exogenous or endogenous, the excreted amino acids were lost.

Rostagno *et al.* (1973a) conducted studies at Purdue University on sorghums of differing tannin contents (Folin-Denis method, Burns, 1963) in chick diets containing sub-optimal levels of protein (AOAC, 1960).

* (Authors' note: see footnote p. 361 on tannic acid equivalents.)

In the first trial the cereals used were:

	Protein (air-dried basis) (%)	Tannic acid equivalent (%)
RS 610	9·7	0·37
BR 64	9·0	1·57
NK 300	9·1	0·66
Maize	8·2	considered 0

The diets contained 70% sorghum plus 16·5% (50% protein) soybean meal. The diets were isonitrogenous at about 15% total protein.

Best weight gains and feed conversions were from diet (D) maize. Diets (A) RS 610 and (C) NK 300 gave poorer and similar weight gains but (C) a poorer feed conversion than (A). BR 64, diet (B), was lowest in weight gain and feed conversion. These findings agreed with Chang and Fuller (1964) and Connor *et al.* (1969).

In a second trial, the same sorghum varieties, but from a different year's crop, were compared with each other and maize on an equal weight basis, and with each other in isonitrogenous diets adjusted by glucose monohydrate. Tannic acid was added to the maize (I), RS 610 (G) and NK 300 (H) diets to tannic acid equivalent of BR 64 (B); all diets were fed on an equal weight basis. Protein, tannic acid equivalents, weight gain and feed gain ratio are shown in Table 4.51.

Weight gains from maize (D) were significantly higher than from sorghum diets (A), (B) and (C), but feed conversion of (D) and (A) (RS 610) were similar. Diets (A) and (C) were better than (B). NK 300 diets (C) and (H) were not as satisfactory as RS 610 diets (A) and (G); (C) was better than (B). Tannic acid reduced overall nutritional quality but no significant difference was apparent between maize diet (I) and RS 610 (G), and both were superior to (H) which was significantly poorer than BR 64 (B) with no added tannic acid but an equivalent tannic acid value.

The third trial was to discover whether older birds fed a bird-resistant sorghum for the first 3 weeks were less affected by BR 64 than birds reared on a maize–soybean meal diet. The results indicated that chicks which were adapted early in life to a higher tannin diet performed better on a high-tannin diet from 21 to 35 days when those reared for the first 21 days on a maize–soybean meal diet.

By the end of the 35 days, leg abnormalities were observed in the chicks and a subjective leg score was devised: 1—normal; 2—mild distortion; 3—severe distortion; and 4—inability to stand. The highest values included chicks which could not stand, and were noted for chicks fed BR 64 for the entire 35-day period, compared to a leg score of 1·17 for chicks fed maize from 0 to 3 weeks followed by BR 64 for the third–fifth weeks when only mild abnormalities were observed.

In the fourth trial, RS 610 (protein 10·5%) and BR 64 (protein 8·9%) were supplemented with essential amino acids (other than tryptophan) to raise the amino acids to the level supplied by the maize, and the maize (protein 9·6%) was supplemented with essential amino acids to raise the amino acid content to the level of RS 610. Excesses of amino acid in one grain above the other were not equalized. The cereals comprised 70% of the isonitrogenous diets.

Supplementation did not improve the maize diet but raised RS 610 to the level of the maize diet. Supplementation significantly improved BR 64 but the supplemented diet was poorer than the other diets. Significantly higher leg abnormalities were observed in the unsupplemented BR 64 diet (leg score 2·29) compared to the unsupplemented RS 610 diet (leg score 1·16). No leg abnormalities were observed in all other diets. No explanation was offered for the leg abnormalities but the comment was made that while the chick diets were supplemented with 1·5–2 × the chick's requirement for Mn, choline and niacin in addition to the amounts present in other ingredients, no additional biotin was added. Autoclaving sorghum RS 610 and BR 64 did not significantly improve weight gain or feed conversion.

Armstrong *et al.* (1973) compared high- and low-tannin sorghums without and with additions of DL-methionine, choline chloride, polyvinylpyrrolidone (PVP), lysine and carnitine. Rayudu *et al.* (1970b) cited by Armstrong *et al.* (1973), had reported that slurry forms of Na_2CO_3 and $Ca(OH)_2$, Tween 80 (a commercial surface-active agent) and PVP were effective in reducing the growth-depressing effect of tannic acid on chicks.

In the first trial, BR 64 protein (AOAC, 1960) 9·0%, tannic acid equivalent (Folin-Denis method, Burns, 1963) 1·57%, was compared with RS 610 protein 9·8%, tannic acid equivalent 0·33%, without and with commercial tannic acid to bring its tannic acid equivalent to that of BR 64. The diets provided sub-optimal protein and were isonitrogenous.

RS 610 produced significantly better weight gains and feed conversions than BR 64, or than RS 610 with added tannic acid. Supplementation of DL-methionine and choline chloride together significantly improved both grains, especially BR 64; however, chicks on the supplemented BR 64 diet did not perform better than those on the RS 610 unsupplemented diet. Leg abnormalities (leg score 2·0) were significant only in the BR 64 diet with methionine and choline.

In the second trial, sorghum BR 64 and IS 8260 (protein 7·4%, tannic acid equivalent 2·66%) were fed on an equal weight basis and supplemented, individually and in combination, with DL-methionine and choline on an isomethyl group basis. BR 64 and RS 610 were compared in diets on an isonitrogenous basis.

Methionine alone and with choline, produced significant improvement in weight gain and feed conversion in BR 64 and IS 8260 diets. The improvement to the RS 610 diet was not always significant. Choline alone had no effect which suggested that methionine was acting other than as a methyl donor. As in the first trial, methionine supplementation increased leg abnormalities but this was only significant with IS 8260 (leg score 2·0).

In the third trial, RS 671, protein 8·8% and tannic acid equivalent of 0·37%, and BR 64 were compared on an isonitrogenous basis in diets containing DL-methionine, PVP and/or commercial tannic acid at 1·0% in the diet. RS 610 was included as a positive control. RS 671 with 1·0% tannic acid in the diet produced results similar to those given by BR 64. Again, supplementation with methionine improved weight gain and feed conversion but the improvement in weight gain was not significant when RS 671 plus tannic acid plus methionine was compared with RS 671 plus tannic acid without methionine. Chicks fed RS 671 gave equal results to those fed RS 610. PVP at 1·0% in the diet almost completely alleviated the growth-depressing and feed conversion-depressing effects of BR 64 and of RS 671 plus 1·0% tannic acid.

In the fourth trial, additions of lysine and carnitine, alone and in combination, failed to alleviate the leg abnormalities observed in chicks on the IS 8260 plus methionine diet.

Using the alkali dehulling procedure of Blessin *et al.* (1971), Armstrong *et al.* (1974a) compared the feeding value for chicks of whole and extracted high-tannin sorghums. The protein (AOAC, 1960) and tannic acid equivalents (Folin-Denis method, Burns, 1963) of the various grains used are shown in Table 4.52. The alkali treatment loosened a large amount of pericarp which floated to the surface and was removed, producing a lower fibre content. Tannin content could be significantly reduced without total removal of the pericarp.

In the first trial, RS 671 (1) and IS 8260 (1) before and after alkali extraction were compared in isonitrogenous diets at sub-optimal protein levels. Untreated IS 8260 was poorer than untreated RS 671. Extraction significantly improved the F/G ratio of RS 671 and brought the F/G ratio of IS 8260 to the same level as the extracted RS 671.

In the second trial, using isonitrogenous but sub-optimal (15·3%) protein diets, intact RS 610 was compared with intact and extracted IS 8260 (1), with and without DL-methionine at 0·15% of the diets.

Supplementation with DL-methionine significantly improved weight gain and feed conversion of chicks fed intact IS 8260 (1), but not those fed extracted grain. These results agreed with those of Armstrong *et al.* (1973).

Methionine supplementation of the intact IS 8260 (1) increased leg abnormalities but there was only a slight increase when methionine was added to extracted IS 8260 (1).

In the third trial, BR 64 and IS 6992 were studied in addition to two different samples IS 8260 (2) and RS 671 (2) (see Table 4.52). The grains were compared on an isonitrogenous basis at sub-optimal levels of protein. The four extracted grains were compared with intact grains on an equal weight basis.

The low-tannin RS 671 (2) gave significantly better feed conversion than BR 64 (which gave the poorest result of all the grains), IS 8260 and IS 6992. In the extraction process, the protein content of RS 671 and BR 64 changed little, but IS 8260 and IS 6992 lost 1–2%, a loss believed to be partly due to the immaturity of the grains. This loss reduced the protein content of the diets fed on an equal weight basis and perhaps accounted for the finding that extracted IS 6992 resulted in poorer weight gains and similar feed conversion to intact IS 6992. IS 6992 gave the best results of any of the bird-resistant grains.

Fernandez *et al.* (1974) compared the nutritional value of a number of different cereal grains as protein sources for chicks using low (14%) protein diets. In the first series of tests, the basal protein premix contributed 8% and the grain under test contributed 6% of the protein. In the second series, 0·3% L-lysine was added, and in the third series, the diets were further supplemented with 50 p.p.m. of procaine penicillin.

The grain included three sorghums, IS 8165 described as high-tannin protein content (Kjeldahl N, factor not stated) 9·8%, IS 2319 (protein 11·7%) and S 0250421 (protein 11.6%), described as low-tannin.

No differences in average body weight and protein efficiency ratios were found among the sorghums, though the results obtained were much lower than those from the soybean meal and Opaque-2 maize diets. Fernandez *et al.* (1974) suggested that even after the correction of lysine deficiency, a strong nutritional deficiency or imbalance remained in the sorghum diets.

Nelson *et al.* (1975) studied the effect of tannin content (Burns, 1963) and dry matter digestion on energy utilization and average amino acid availability of 12 hybrid sorghums fed in chick diets. The colour of the seedcoat and endosperm, starch texture and tannin equivalents of the hybrids are listed in Table 4.53. Tannin equivalents were highest in hybrids with brown seedcoats. Dry matter digestion, amino acid availability (of 17 amino acids) (Bragg *et al.*, 1969), the gross energy (GE bomb calorimeter) and ME (Hill *et al.*, 1960) are also shown in Table 4.53.

Dry matter digestion, ME as % of GE, decreased significantly as tannin equivalents increased but there was no significant correlation between GE and tannin content. The correlations of dry matter digestion to amino acid availability (0·94) ME/g (0·98), and % GE utilized (0·99) were all highly significant. Amino acid availability and the % GE utilized were more closely related to the amount of dry matter digested than to tannin equivalents. Tannin content was more closely correlated with amino acid availability than with energy utilization.

The % GE utilized, the amino acid availability and dry matter digestion were similar for the four hybrids with the lowest tannin content, DeKalb E 57, RS 608, TE 66 and Niagara ORO 7, but GA 615 and Niagara Shoo Bird, both brown seeded and with tannin contents of 0·59% and 0·54% respectively, were comparable to the low-tannin hybrids in amino acid availability and utilization of energy. This finding agreed with that of Thayer *et al.* (1957) that seed coat colour was not a reliable indicator of nutritional quality. Colour of endosperm and starch texture did not appear to influence the nutritional quality of sorghum in chick diets.

Using similar materials to those of Nelson *et al.* (1975) in his chick bioassay tests, Ford (1977a) reported the content of total and available methionine, and the relative nutritional value (RNV) of the protein of 10 lines of sorghum of high- and low-tannin content (MV-HCl, Maxson and Rooney, 1972a) measured microbiologically with *Streptococcus zymogenes* as described by Ford (1962) except that the test samples were milled to pass an 80-mesh sieve and predigested with pronase instead of papain. "Reactive lysine" was determined by differential dye-binding (Hurrell and Carpenter, 1975). Available methionine values ranged from 32 to 96 mg/g total N (33–100%) and were highly correlated with tannin content as were the RNV values. Total methionine values varied little between samples, from 102 to 113 mg/g total N and were not related to tannin content. The same was true for "reactive lysine".

The effect of polyethylene glycol (PEG 4000) on the available methionine and RNV of one of the sorghum lines with a tannin content of 3·2% was reported by Ford (1977b) as:

PEG 4000		Methionine mg/g total N			
Added (mg)	mg/g test protein	Total	Available	Available/ Total	RNV (casein = 100)
0	0	94	43	28	32
1	12·8		52	35	55
5	64		81	54	74
25	320		87	58	84
125	1600		90	60	85

In further tests PEG 4000 was compared with polyvinylpyrollidone (PVP), polyoxyethylene sorbitan oleate (Tween 80) and Lissapol NDB (a non-ionic detergent) all at 100 mg level, DWB. All increased the available methionine two-fold, PVP being marginally best.

Ford and Hewitt (1977) used the same sorghum to supply 80 g crude protein/kg in test diets for rats and chicks without and with additions of PEG 4000. In the rat tests, protein digestibility (D), biological value (BV) and NPU were determined by the balance method of Henry and Toothill (1962) for a single period of 7 days. In the chick tests, D was measured by the "ileal analysis" method of Varnish and Carpenter (1975). The results were:

PEG 4000	Chick test	Rat test		
(g/g crude protein)	D	D	BV	NPU
0	0·44	0·53	0·77	0·42
0·1	0·90	0·92	0·59	0·54
1·0	0·94	0·99	0·58	0·57
LSD (P = 0·05)	0·10	0·07	0·11	0·11

PEG 4000 substantially increased protein digestibility in both rats and chicks. Ford and Hewitt commented that the high BV value in the absence of PEG 4000 was undoubtedly spurious since at low levels of protein, deficient in an essential amino acid, unduly high BV values are obtained (Henry and Kon, 1957).

Halloran and Maunder (1971) compared two brown (bird-resistant) sorghums, a "low-tannin" red sorghum, and maize in an acceptability test with commercial broiler cockerels. All grains were grown at the same time on one Texas farm. Each grain was fed at 57% of the diet formula and two diets were offered each pen; the feeders were switched daily. There were no differences in acceptance among diets with the red sorghum and the two brown sorghums. When the maize diet and the red sorghum diet were compared, chicks chose 57% of the maize diet and 43% of the red sorghum diet. When each grain was fed as the sole grain source, all produced similar weight gains but the brown sorghums showed a poorer feed conversion than the other grains. Compared to maize, a yellow endosperm and a heteroyellow (bronze) sorghum gave slightly higher weight gains and poorer feed conversions; a red sorghum gave slightly lower gains and poorer feed conversions.

Peischel *et al.* (1976b) studied the performance of Meat-Nick strain chicks from 7 to 21 days old on diets containing (A) 100 sorghum; (B) 50 sorghum/50 maize; and (C) 100 maize. Ten (undesignated) varieties of sorghum were used. Weight gains and feed conversions were not significantly different among total treatments. Diet (A) produced the poorest growth; diet (B) produced better growth than (A) or (C). There was a highly significant correlation between gain during the first 14 days of the trial and the tannic acid content (determination method not stated) of the diets.

Swine

Noland *et al.* (1964) compared (A) AKS 614, a bird-resistant sorghum, (protein content 9·9%) with (B) RS 610, non-bird-resistant (protein content 9·7%), (C) a commercial grain sorghum (protein content 6·4%) and (D) maize, in feeding trials with growing-finishing pigs. The rations as fed contained equal levels of protein. The pigs fed (A) made slower gains than pigs fed any of the other three rations.

When field-dried sorghum DeKalb E 56A was compared with field-dried AKS 614, the latter was inferior in % digestibility of dry matter, crude protein, ether extract, in gross energy and in apparent N retention. Differences in the digestibility of the crude fibre and N-free extract were not significant.

In 1975, Noland *et al.* reported than when each of 12 varieties of grain sorghum grown under the same conditions were fed in low protein diets to pigs of 9 kg initial weight, the highest digestion coefficients were obtained for those varieties with a yellow or red seedcoat colour and yellow endosperm; the lowest coefficient for varieties with a brown seedcoat and white endosperm; intermediate values for varieties with a brown seedcoat and a yellow endosperm. Tannin percentages ranged from 0·21 to 0·65.

Campabadal *et al.* (1976) reported briefly feeding trials in which (A) maize; (B) bird-resistant sorghum (variety not stated); (C) non-bird-resistant sorghum (variety not stated); and (D) bird-resistant sorghum with 0·10% methionine were fed to growing pigs. No significant differences in daily gain or feed intake were observed but pigs on diet (B) required more feed/unit of gain. Diet (D) effected an improvement in feed conversion, presumably in relation to diet (B). Dry matter and crude protein digestibility were depressed in pigs fed diets (B) and (D) but N retention was higher for (D).

In a second trial, diets (A) maize; (B) bird-resistant sorghum; (C), (B) plus 0·10% methionine; (D) non-bird-resistant sorghum; and (E), (D) plus 0·10% methionine were compared. No statistically significant differences were found between diets (A) and (C) for daily gain and both diets gave higher gains than diet (B) which required more feed/unit gain than the other diets.

Ruminants

Klosterman *et al.* (1968) compared AKS 614 bird-resistant sorghum with a finely ground No. 2 shelled maize both as silage and as grain in rations for fattening cattle. Steers fed maize silage or maize grain rations gained weight faster and showed a lower (better) F/G than those fed on the sorghum rations whether as silage or grain.

Fox *et al.* (1970) reported feed lot experiments with steers also using AKS 614 bird-resistant grain sorghum in comparison with maize, both being fed either as silage or as grain. The steers fed the bird-resistant sorghum made lower daily gains and showed considerably worse F/G. However, none of the carcass traits evaluated in comparison between the maize and AKS 614 grain rations were significantly different. These findings were similar to those of Davis and Stallcup (1966) and Stallcup and Davis (1962) who reported that the energy value and protein digestibility of AKS 614 was lower than that of non-bird-resistant sorghum and maize when fed to steers.

McGinty and Riggs (1968) compared the coefficients of digestibility for several varieties of ground sorghum grain with steers. The varieties included Kaoliang and Darset 28. The polyphenol content of the two latter varieties was not stated but they are often said to be bird-resistant (Hale, 1970). Two experiments were conducted for the coefficient of digestibility (COD) for dry matter in the first of which the COD for dry matter of the two bird-resistant varieties was significantly lower than that for the other varieties tested; the differences were not significantly greater in the second experiment. In both experiments the COD for protein were significantly lower in the case of the two bird-resistant varieties. The COD for organic matter for the varieties was of the same order as that for the dry matter. *In vitro* data had indicated no differences in the digestibility of the endosperm from the Kaoliang, Darset 28 and the non-bird-resistant type TX 09 Feterita. However, when the pericarp from the Kaoliang and Darset 28 was added to the endosperm of any of these three varieties, digestibility was reduced.

Daniels and Flynn (1972) compared a bird-resistant sorghum, AKS 614 processed by one of four methods with ground maize in calf starter rations. The diets were calculated to be isonitrogenous and isocaloric but as equal quantities of the variously processed sorghum grains were added to the diets, the crude protein level of the rations in fact differed to some extent, as shown below.

Sorghum diets	Protein content in diet (%)
(A) Ground	17·6
(B) Dry rolled	18·9
(C) Cooked in an autoclave for 1 h at 250°F under 18 lb pressure and then rolled when hot	17·2
(D) Expanded-extruded through a Brady farm cooker at 220°F	16·4

The calves fed (D) grew significantly more slowly from birth to 45 days of age but this was partly due to the flake size reducing feed intake. No differences in feed intake or in daily gain occurred among the calves on rations (A), (B) and (C). After 45 days calves on ration (C) grew significantly more rapidly up to 112 days than those on rations (A), (B) and (D) indicating that the starch fraction was more digestible through the cooking. No differences in average daily gain were observed among the calves on rations (A), (B) and (D) and the total daily intake of starter ration and milk were similar for all calves. Daniels and Flynn considered that bird-resistant grain sorghum could be used satisfactorily in calf starter rations but that steam cooking of the sorghum would improve its digestibility.

Loyacano *et al.* (1973, 1975) compared a bird-resistant grain sorghum, Funk's BR 79, in steer finishing diets (A) 100 maize; (B) 50 maize/50 sorghum; (C) 25 maize/75 sorghum; and (D) 100 sorghum. The level of sorghum in the ration had a highly significant effect on average daily gains and a significant effect on feed conversion, both worsening as the level of sorghum in the ration increased in (C) and (D). In a second trial, rations (A) and (D) were retained and ration (E) 50 maize/50 sorghum; and (F) 100 sorghum, both with sufficient cottonseed meal to make their digestibile protein level equal to (A) were included. Again, significant differences were found in average daily gains and in feed conversions as shown below:

Diet	Average daily gain (kg)	F/G
(A)	1·06	3·4
(D)	0·98	5·1
(E)	1·05	4·3
(F)	0·88	5·0

The second trial indicated (Loyacano *et al.*, 1975) that the level of digestible protein in the sorghum rations was not the limiting factor. Although calves on diets (E) and (F) gained weight as rapidly as calves on diet (A) they required respectively 45 and 24% more feed/kg of weight gain. The results indicated that cattle would readily consume bird-resistant sorghum rations but would not perform as well as when fed maize rations.

McCollough (1972b) compared eight hybrid sorghums and three hybrid maizes, all planted at Kansas under similar conditions, dry-rolled and incorporated into all-concentrate rations formulated to meet minimum NRC requirements for both feeding and digestion trials. One of the sorghums, ACCO 1023, of dark red pericarp and white endosperm, was described as bird-resistant. The proximate analysis of the grains showed that this sorghum was the highest in crude protein (14·1%).

In a 126-day feed lot trial with Angus steers, using the grains referred to above, McCollough *et al.* (1972a) found that the steers fed ACCO 1023 gained the least, had the lowest % dressed weight and the poorest feed efficiency.

Maxson (W. E.) *et al.* (1973) compared the energy values for steers, of maize, bird-resistant sorghum, (DeKalb BR 64), and non-bird-resistant sorghum (DeKalb E 57). The grains were added to the rations at equal weights but the percentage protein in the rations differed slightly, being highest for the E 57 (13·7%) and lowest for the BR 64 (12·3%). The tannin contents (Burns, 1963) were: maize 0·51%, E 57 0·54% and BR 64 2·15%. The feed efficiencies were not significantly different, but the trend showed maize to be the most efficient and BR 64 to be the least. This confirmed the finding of Bertrand and Lutrick (1971) who reported poor feed efficiency with bird-resistant sorghum grain.

Brommelsiek *et al.* (1975) reported the effects of bird-resistant sorghum grain (BR 64) fed to steers without and with additions of methionine hydroxyanalogue (MHA) or choline chloride. The ME in the grain was 2·25, 2·36 and 3·06 Mcal/kg dry weight for the diets supplemented with 0, 15 g choline chloride or 30 mg MHA/head/day respectively. Lower levels of choline chloride and MHA supplementation had no effect on ME.

Tannins in Finger Millet

Ramachandra *et al.* (1977) determined the total phenols (Folin-Denis, Swain and Hillis, 1959) and tannin (V-HCl, Burns, 1971) of 19 Indian, 10 African and 3 Indian × African cross-bred finger millet cultivars obtained from the Millet Research Centre, University of Agricultural Sciences, Hebbal, Bangalore, India. The seeds ranged in colour from white to very dark brown and, as will be seen from Table 4.54, white seeds had low values and dark brown seeds high values; differences were highly significant. The *in vitro* protein digestibility (IVPD) (micro-Kjeldahl N in the sample before and after pepsin hydrolysis (AOAC, 1965 × 6·25) of 13 of the cultivars is also shown in Table 4.54.

Low-tannin cultivars showed IVPD values and high-tannin cultivars showed low IVPD values but cultivars with intermediate tannin levels, for example, 14 1A 3 26 and 91 4 7 18, had IVPD values comparable with low tannin cultivars.

When nine amounts of tannic acid graded from 0·0 to 5·0 were added to IE 246 (CE 0·14%), IVPD was reduced correspondingly from 83·0 to 43·3%.

When the seeds were dehulled by soaking for 1 min in conc. H_2SO_4 followed by washing in water, and rubbing off the pericarp by hand, 80% of the total phenols and nearly 90% of the tannins were removed from Hamsa, Purna, IE 929 and IE 927, with corresponding increases in IVPD. The effect of adding tannic acid to dehulled pearl millet flour also lowered the IVPD but not to the same extent as in the whole seed samples. Ramachandra *et al.* suggested that this difference could be due to the loss on dehulling of other polyphenolic compounds present in the seed coat.

Purna, BC 4840, HPB 1 8, 14 1A 3 26 and IE 929 were separated into three fractions (Virupaksha *et al.*, 1975 modification of Landry and Moureaux, 1970), albumin–globulin, prolamine and glutelin, and the micro-Kjeldahl N and CE determined. The results are presented in Table 4.55.

Virupaksha *et al.* (1975) had reported the albumin–globulin and glutelin fractions as having a better distribution of essential amino acids than the prolamine. Ramachandra *et al.* found the glutelin contained the highest tannin content of the three fractions and suggested that this association was likely to affect the digestibility of the glutelin and the nutritive value of pearl millet protein.

Reduction in Polyphenol Content

Many attempts have been made to reduce the antinutritional effects of sorghum polyphenols but because, until recently, the chemistry remained obscure, most of the methods proposed were derived empirically.

Price *et al.* (1977) reported that sorghum tannin appears to be reactive and easily modified to forms which do not respond to chemical tests for tannin. Treatments they have tested include steam and dry heat, ammoniation (both aqueous and anhydrous with differing pressures and exposure times) and aqueous solutions of NaOH and K_2CO_3. Treatments in the absence of water were relatively less effective than comparable treatments with added moisture and most of the treatments decreased the level of tannin, as measured by chemical tests, to 10–50% of the original level.

On ammoniation, the N content of the grain increased only if the sorghum was of the high tannin type. Heat treatments decreased protein digestibility of the grain whereas ammoniation had little effect. Weight gains and F/G were generally improved by the treatment of the high-tannin sorghum, especially by treatment with NH_4OH which produced weight gains equivalent to those of untreated low-tannin sorghum, but the treatments generally decreased the weight gains observed with low-tannin sorghum. Price *et al.* emphasised the need for much additional work, not least to evaluate the effect of such treatments on the nutritional value of the sorghum for human consumption.

The Council for Scientific and Industrial Research (1975) filed a (South African) patent specification covering the use of a non-acidic carbonyl-containing compound, more particularly formaldehyde, to steep sorghum (or other grain) and expose the steeped grain to formaldehyde vapours. The treatment involves reducing the polyphenol content of the grain to a value of less than 0·5%. The unwanted polyphenols are converted to other compounds by, e.g. neutralization.*

Other Antinutrient Factors

Phytic Acid

Phytic acid in wheat, rye and triticale was discussed by Hulse and Laing (1974). Oberleas recently (1973) reviewed the occurence of phytates in food and, citing IUPAC-IUB (1968) stated that the proper chemical designation of phytic acid is myoinositol 1,2,3,4,5,6-hexakis (dihydrogen phosphate). There are no known reagents that specifically identify phytic acid, nor does it display a characteristic absorption spectrum. Hence analytical methods depend upon estimating inositol or phosphate, or upon establishing a stoichiometric relation between phytate and a cation that can be measured easily. Several accepted analytical methods employed appeared to Oberleas to be adequate to determine phytate within the limits of nutritional or toxicological significance.

The importance of phytic acid in nutrition lies in its property of forming insoluble or nearly insoluble compounds with mineral elements, including Ca, Fe, Mg and Zn, the resultant phytates being excreted in the faeces. Diets high in phytic acid and poor in Ca, Fe and Zn, produced mineral deficiency symptoms in experimental animals and in children (Gontzea and Sutzescu, 1968). Phytic acid should not be regarded as a source of available P for man or monogastric animals (Oberleas, 1973).

Phytic acid occurs primarily in the outer seed coats (bran) and germ of plant seeds (Oberleas, 1973). In 1959, Wang *et al.* examined the distribution of phytin P (phytin is the name given to the mixed Ca and Mg salt of phytic acid) and of phytic acid in seven samples of sorghum. Tempered sorghum kernels were fractionated by a process of debranning, grinding, screening, air and gravity separation. The whole grain (ground in a Wiley mill with a 20-mesh screen) and the fractions obtained; (1) germ (embryo); (2) bran (seedcoat layers); (3) grits (endosperm); and (4) mill fines (endosperm mixed with germ and bran fragments), were analysed for total acid-soluble P (AOAC, 1950), phytin P (Pons *et al.*, 1953) and P (McCance and Widdowson, 1935). The phytic acid equivalent was calculated from

* (Authors' note: at the time of writing, no open publications have been traced dealing with the hazards involved in the use of formaldehyde, even at low concentrations, nor on the effect of the proposed treatment on the nutritional value for human consumption.)

phytin P using a factor of 3·55 (established by determining the P and inositol of phytic acid isolated from the grain). The results are given in Table 4.56.

The phytic acid equivalent content of the whole grain ranged from 0·71% (Cody 358) to 1·31% (Western Blackhull). Of the fractions, the germ contained the greatest percentage, then the bran; the endosperm (grits) contained the least. Wang *et al.* (1959) concluded that the phytin content of sorghum was similar to maize.

The quantity of phytic acid in cereal products depends to a large extent on the milling process and the extent to which bran and germ are separated from the endosperm. The P and phytic P contents of sorghum, common millet, pearl millet, foxtail millet, and finger millet reported by various workers are presented together in Table 4.57. The results were obtained by a variety of analytical methods, mainly on unidentified samples.

Courtois and Perles (1954) determined the total P (Fleury and Leclerc, 1943), P soluble in trichloracetic acid (Courtois and Barré, 1953) and phytic P (McCance and Widdowson, 1935) of five samples of *Pennisetum* spp., (1) and (2) sanio, and (3), (4) and (5) souna and seven samples of sorghum (6) teing (or nteing)—probably *S. gambicum* (Snowden), (7) kongossan (probably *S. guineense* Staph.), (8) voyende (probably *S. guineese* var.), (9) mboratel (probably *S. excertum* Snowden), (10) White Fela (*S. cernuum* Host.), (11) Yellow-white Fela (*S. cernuum* Host.) and (12) Red Fela (*S. cernuum* Host.). The results are given in detail in Table 4.58 and summarized in Table 4.57.

These results, in range, are of the same order as those of other workers (see Table 4.57) but considerable variation was shown, e.g. phytic P 368 mg/100 g for teing sorghum and 181 mg/100 g for Yellow-white Fela. Phytin P represented 45% of the total P in *Pennisetum* sanio, sample 2; and 93% in voyende and Red Fela sorghum.

Phytic acid and the phytates may be decomposed by the enzyme phytase to produce inositol and phosphoric acid. Phytase is unequally distributed through the grain in somewhat the same manner as phytic acid.

Giri (1938) stated the phytase activity to be sorghum 0·76, pearl millet 0·17, uncooked finger millet 0·01, whole wheat 0·12 and rice 0·03.

Courtois and Perlès (1954) reported the phytase and glycerophosphatase activities of the sorghums and *Pennisetum* spp. which they examined (see Table 4·58) as being lower than samples of rye and wheat (using the same methods) for which however they did not give values. Courtois and Perlès reported that on germination the phytase activities of the sorghums Yellow-white Fela, kongossan and mboratel varied greatly and were, respectively 11·4, 3·4 and 1·4%. In examining the Senegal sorghum and millets for their phytin P content, they also determined the Ca and Mg contents of the ash of a sample of *Pennisetum* spp. (sanio No. 1) and two samples of sorghum, teing and mboratel. Ca and Mg were determined together by a described method employing titration with ethylenediaminetetracetic acid (Complexon). Mg was determined by precipitating the Ca and titrating the Mg;

Ca was calculated by difference. In summary, Courtois and Perlès concluded that the sorghums and pennisetums of Senegal were unlikely to provide adequate amounts of assimilable Ca in human diets when fed as the basal diet.

Oberleas (1973) pointed out that phytate formation depends on pH as well as on the presence of secondary cations such as Ca. A wide variety of dietary deficiencies result depending on which element became first-limiting under the conditions prevailing. Certainly Ca, Fe, Zn, Mn and Co can be converted to insoluble phytates and thus rendered nutritionally unavailable.

Cyanogenic Glycosides

The cyanogenic glycosides were defined by Conn (1973) as glycosides of α-hydroxynitriles (cyanohydrins). Their occurence, biosynthesis and mode of action were fully discussed at an IDRC sponsored workshop (Nestel and MacIntyre, 1973). They are widely distributed in the higher plants, and the cyanogenic glucoside dhurrin occurs in most sorghum varieties though the quantity depends upon the variety and the environmental conditions. When hydrolysed, dhurrin yields equal parts of hydrocyanic acid (prussic acid) and *p*-hydroxybenzaldehyde. The occurrence of cyanogenic glycosides in the leaves of sorghum and the consequent risk of poisoning from their use as fodder has long been recognized (Doggett, 1970, Wall and Ross, 1970, and the Rockefeller Foundation Bibliographies 1967a, 1973).

The seeds (caryopses) of sorghum appear to be comparatively free from cyanogenic glycosides. Akazawa *et al.* (1960) reported that dhurrin (*p*-hydroxymandelonitrile-β-D-glucoside) was localized in the aerial parts of the sorghum plant and was not found in the seeds. However, 3-day old etiolated seedlings (from which light had been excluded) contained as much as 15 μmol of dhurrin/g of fresh weight. The 5-day old green seedlings contained approximately the same amount but the amount decreased as the plant matured.

Lloyd and Gray (1970) assayed the plant parts, at intervals in their life cycle, of Piper and Greenleaf Sudangrass (*Sorghum bicolor* var. *sudanese* (L.) Moench) and Suki I, (a sorghum (*Sorghum bicolor* (L.) Moench) × Sudangrass hybrid), for "hydrocyanic acid potential" (HCN-p) using the sodium picrate method developed by Anderson 1960 (unpublished mimeograph, University of Wisconsin, Dept of Biochemistry, College of Agriculture and Agricultural Experimental Station). Little or no HCN-p was found in the seed at the time of planting but at harvesting HCN-p was present in all cultivars though at levels below 60 p.p.m. The highest concentrations occurring in the tops, one week after emergence were 2249, 2708 and 4238 p.p.m. for Piper, Greenleaf and Suki I respectively. Concentrations decreased as the plant matured.

Courtois and Perlès (1954) in examining the phytic acid P content of Senegal millet (*Pennisetum* spp.) and sorghum in sprouted and non-sprouted grains, observed that using the sodium picrate method, the pennisetums did not liberate HCN whether sprouted or not, nor did the non-sprouted sorghum. However, the sprouted sorghum liberated HCN after 2 days of germination and Courtois and Perlès (1954) suggested a possibility of toxicity from the preparation of sorghum beer if the preparation area was inadequately ventilated to permit the volatized HCN to escape.

Amino Acid Imbalance

The problems of amino acid imbalance in general have received considerable attention and have been discussed and reviewed by Bender (1965), Carpenter and de Muelenaere (1965), Harper (1969), Harper and Rogers (1965), Sanahuja (1971) and Muramatsu *et al.* (1972).

A joint FAO/WHO *ad hoc* Expert Committee (WHO 1973a) published estimated amino acid requirements for infants, children and adults and recommended an amino acid (chemical) scoring method as a guide to protein nutritional quality. They cautioned, however, that the amino acid composition gives no indication of amino acid availability and verification of nutritional quality by biological tests of protein quality are always necessary. The Committee stated that the results of animal experiments indicated the main effect of amino acid imbalance was to reduce food intake, citing Harper *et al.* (1970). Gopalan (1961) provided evidence to suggest that the high leucine content of sorghum apparently increased the need in man for tryptophan and isoleucine; Bressani *et al.* (1958) reported a similar effect from maize.

Table 4.59 presents mean contents in the protein of isoleucine, leucine and tryptophan of sorghum and the millets with maize and wheat as reported by FAO (1970). It will be observed that maize and sorghum have the highest leucine/isoleucine ratio at 3·4 but that in general, sorghum has a higher tryptophan content than maize. Pickett (1971) reported that the leucine/isoleucine ratios among sorghums in the World Collection ranged from 2·7 to 3·8.

Ganapathy and Chitre (1976) conducted rat growth experiments using male albino weanlings to study the effect of the arginine/lysine ratio in casein, wheat and foxtail millet diets, earlier studies having suggested that the arginine/lysine ratio of 5/1 in foxtail millet might play a part in the utilization of its protein.

Diets were fed at the 10% protein level with fat, salt, and vitamin mix; starch was used to adjust the protein levels. When the amino acid composition of the wheat and foxtail millet diets was equalized with that of the casein diet by the addition of synthetic amino acids, weight gain approximately equalled that from the casein diet. When additions of amino acids were made to the wheat and casein diets to make the levels equivalent to those of the foxtail millet diet, weight gain was reduced.

The diets, with the arginine/lysine quotient are shown in Table 4.60. Ganapathy and Chitre (1976) considered that the results indicated a true metabolic antagonism between arginine and lysine in the protein of foxtail millet.

Kies and Fox (1972) compared the nutritive value (by N balance tests) for the human adult of wheat-based diets with a normal level of leucine and with additions of leucine to raise the level to that found in maize. They also studied the effect on such diets of supplementation with lysine, niacin and tryptophan. Lysine supplementation was shown to be beneficial though the amount added in the tests could not wholly overcome the adverse effect of added leucine. Tryptophan supplementation was not significantly effective and niacin supplementation was effective only at the high (added) leucine level.

Kies and Fox concluded that the high leucine content of maize was an important factor in its lower nutritional performance compared to wheat in human feeding trials.

Nicol and Phillips (1961) reported that adult Nigerian men receiving a sorghum basal diet (leucine/isoleucine ratio 2·3) which provided 4 g N intake were in negative N balance (−0·3 g) but the same subjects when receiving an isonitrogenous rice diet were in N equilibrium. To establish whether this was caused by (1) poor digestibility of the crude fibre in the sorghum or (2) by an excess of leucine in the sorghum, further tests were conducted with Nigerian adult males. The basal rice diet (A) leucine/isoleucine ratio 1·5, was compared with (B); (A) plus 1 g L-leucine added daily in aqueous solution in three equal doses to give a leucine/isoleucine ratio 2·4. The addition of leucine to the rice basal diet did not result in any significant loss of N in terms of N balance. Nicol and Phillips concluded that the poor performance on the sorghum diet was due to poor digestibility of the sorghum protein rather than to an excess of leucine.

Pellagra

Gopalan and Narasinga Rao (1972) reported that pellagra had been known as a human disease for more than 2 centuries but serious investigations did not start until the disease reached epidemic proportions in the southern US about 1900. During the decade 1919–1929, Goldberg and his associates, reviewed by Terris (1964), established pellagra as a disease caused by deficiency of a heat-stable factor which they designated pellagra-preventive or PP factor. Later, this was identified as the vitamin niacin. The signs of pellagra were sometimes referred to as the four successive "Ds": dermatitis, diarrhoea, dementia and death. Some of the symptoms might however arise from other vitamin deficiencies, for example, riboflavin and pyridoxine.

The pathogenesis of the disease has not been fully elucidated and is outside the scope of this review, but some recent papers on pellagra and its relation to metabolism include Ghafoorunissa and Narasinga Rao (1973, 1971), Gopalan and Narasinga Rao (1972, 1971), Krishnaswamy and

Ramanamurthy (1970), Raghuramulu *et al.* (1965a), Ramanamurthy and Srikantia (1970), Srikantia *et al.* (1968a), and Vasantha (1970ab).

Pellagra was for a long time associated with diets based predominantly on maize (Gopalan and Srikantia, 1960). The same authors stated that pellagra was rare in rice-eating populations; they had observed only four cases of classical pellagra among the rice-eating population of Coonoor, India in 10 years. In Hyderabad, however, 500 miles to the north, where sorghum was a basic cereal, pellagra was frequently encountered.

Since maize and sorghum both contain comparatively high levels of leucine in their amino acid composition, Gopalan and Srikantia postulated that the relative excess of leucine might play a part in the pathogenesis of pellagra. They studied the effect of leucine on the urinary excretion of N-methyl nicotinamide (NMN) (Carpenter and Kodicek, 1950) in healthy human subjects and in pellagra sufferers by feeding a daily supplement of 5 g L-leucine over 7 days. Urinary excretion of NMN increased by nearly 50% in all the subjects but returned to normal when administration of leucine ceased. The administration of 2 g daily of DL-isoleucine in addition to the 5 g L-leucine in two healthy subjects did not alter the NMN excretion pattern.

The blood-niacin levels (Sweeney and Hall, 1951) were also determined in two healthy subjects after administration of 15 mg of niacin. Urinary NMN increased by 8·3 mg and 7·5 mg suggesting that the average increase of 1·7 mg in NMN excretion brought about by the administration of 5 g L-leucine reflected approximately an additional 3·5 mg of niacin metabolized daily.

The effect on three pellagrins of substituting sorghum (of unstated origin) for rice in a wheat–rice diet resulted in an immediate increase in urinary NMN which returned to its original level when the sorghum was replaced by rice. The diets were isonitrogenous and isocaloric and the niacin levels of the diets were essentially similar. The daily amount of leucine in the sorghum diet was calculated to be about 5·3 g.

When 5 g L-leucine daily was administered to four patients with pellagra for 7 days, no appreciable effects were observed. When, however, larger amounts, 10 g L-leucine twice daily in one case and three times daily in another, were included in the diets based on sorghum and providing 45 g protein daily, the signs of pellagra were exacerbated and both subjects, who had appeared mentally normal initially, became uncontrollable and violent. Niacin therapy restored the mental condition to normal. The increase in NMN excretion brought about by feeding 20 g L-leucine daily was not significantly greater than the increase when 5 g L-leucine was administered to other pellagrins and to healthy subjects.

Truswell *et al.* (1963) did not agree with Gopalan and Srikantia (1960) that a relative excess of leucine in the diet played a part in the pathogenesis of pellagra. Two normal women maintained in a metabolic ward on constant diets of mixed foods providing 35 g protein, 250 mg tryptophan and 6 mg niacin daily, and with extra thiamine, riboflavin and pyridoxine, did not show increased excretion of NMN when (A) 5 g L-leucine was added to the midday meal each day for 1 week, (B) 10 g/day was given divided equally between breakfast and supper for another week, with adjustment periods preceding and intervening. N balances were slightly more positive in the leucine periods.

Two men accustomed to eating large quantities of maize were also studied in a metabolic ward. On diets of 420 g and 520 g maize meal respectively, with sugar, low-niacin fruits, thiamine, riboflavin, pyridoxine and minerals, additions of 2 g L-leucine with each of three daily meals over 8 days did not increase NMN excretion, and in a further trial, carried out primarily for another purpose, 12 adults in two equal groups consumed their usual diets plus 10 g of L-leucine or glycine added to their lunch for 10 days. There was a 7-day break, then supplementation for a further 10 days. The mean NMN excretions were $6{\cdot}0 \pm 2{\cdot}1$ mg in the leucine group and $5{\cdot}2 \pm 3{\cdot}4$ mg in the glycine group and indicated that the addition of large amounts of leucine to an adequate diet did not appreciably alter NMN excretion.

The effect of adding leucine or methionine to a low-protein diet on NMN excretion in rats was studied using adult female rats (six to a group). In the basal diet the first-limiting amino acid was threonine, followed closely by tryptophan. After feeding the basal diet (A) for 7 days, it was followed in consecutive 7-day periods by the basal diet supplemented in turn by (B) 0·1% DL-leucine, (C) 0·4% DL-leucine and (D) 0·4% DL-methionine. The mean NMN values in μg/day were (A) 364, (B), 330, (C) 278 and (D) 341.

Truswell *et al.* (1963) and Truswell (1963) concluded that a relative excess of leucine in the diet did not play a significant role in the pathogenesis of pellagra.

Belavady and Gopalan (1966) estimated the niacin content of three lots of sorghum, of unstated origin, by the chemical methods of Friedemann and Frazier (1950), and Swaminathan (1942b). These extracts were also examined microbiologically with *Lactobacillus arabinosus* and a further microbiological estimation was carried out by the method of the Association of Vitamin Chemists (1947). The different extraction procedures gave very similar results. The average values, (mg/100 g sorghum) obtained microbiologically, were 2·6, 3·2 and 3·4 and chemically 2·4, 3·3 and 3·6 respectively for the three lots.

Kodicek *et al.* (1959) had suggested that the lime treatment of maize practised in Central America increased the availability of niacin and so reduced the incidence of pellagra, consequently sorghum was heated with calcium oxide for 45 min. After standing overnight, the supernatant was discarded and the sorghum washed three times with water, dried at between 60 and 70°, and ground. Treated and untreated sorghum came from the same batches.

Chromatograms of the acidic methanol extracts of untreated and treated sorghum did not reveal the presence of bound niacin. (These results were at variance with those of Ghosh *et al.*, (1963) who reported that hydrolysis of sorghum with N HCl yielded niacin values, as measured by *Leuconostoc mesenterioides* ATC equivalent to only 3·6 μg/g while the bound niacin released by alkali hydrolysis was 18·6 μg/g sorghum). Lime-treated sorghum, compared to the untreated, showed losses of 2·5–3·0% N (method unstated) and 30–40% niacin.

Female weanling rats on lime-treated sorghum diets showed poorer growth performance than those on the untreated sorghum. Niacin supplementation at 3·0 mg/100 g diet improved the growth performance of the rats on the lime-treated sorghum diets, but had no marked effect on those receiving untreated sorghum. The niacin content provided by the sorghum in those diets was about 1·2–1·4 mg/100 g diet and this was probably adequate since the niacin requirement for rats on maize diets had been shown to be about 1·5 mg/100 g diet (Krehl *et al.*, 1946a Hankes *et al.*, 1948).

Pups aged 8–10 weeks and weighing 1·0–1·5 kg, two to a group, receiving diets in which all the protein was derived from sorghum, whether (A) treated, or (B) untreated, failed to grow and died. Those receiving diets which were isonitrogenous but contained (C) and (D) 3·5% casein, also failed to grow but survived rather longer. Pups whose diet (F) contained 8% casein and treated sorghum, had periods of anorexia after 4 months on the diet; when one died the other was treated with niacin which resulted in an increase in body weight. The pups on (E), the 8% casein diet and untreated sorghum, survived but did not grow as well as the pup on the control diet of 18% casein.

Absorption and retention of absorbed (Kjeldahl) N and urinary excretion of NMN (Carpenter and Kodicek, 1950) were similar for pups on diets (E) and (F), but by 13 weeks the haemoglobin levels in the blood of pups on (E) were higher than those on (F), the treated sorghum, and continued to be so until the end of 21 weeks.

Belavady and Gopalan (1966) concluded that the niacin in sorghum was available to rats and pups and that the prevalence of pellagra in a sorghum-eating community could not be attributed to a niacin deficiency due to the unavailable form of niacin in sorghum.

Denton (1928), cited by Gopalan and Narasinga Rao (1972), had pointed out the similarities between the condition known as black tongue in dogs and human pellagra.

Belavady and Gopalan (1965) reported that nine adult dogs fed *ad lib.* a diet consisting of 65% sorghum of unstated origin, and including all vitamins except niacin, developed signs of black tongue in varying degrees of severity over 30–80 days. In dogs given niacin with the sorghum, the signs regressed. A further group of six dogs on a maize diet also showed signs of the disease.

(The pathological features of these experiments and those of Belavady *et al.*, (1967) are described by Madhavan *et al.*, 1968).

In experiments with pups Belavady *et al.* (1967) induced black tongue in all animals receiving (A), a niacin-free, 21% casein diet, in which the leucine content of 1·8% (microbiological assay) derived from the casein was increased to 3·6%, and in three out of five pups receiving (B), an 18% casein diet of leucine content 2·7% and 300 μg niacin/kg/day.

They commented that it was not the leucine content alone which was responsible, as pups would thrive on diets containing 40% casein (approximately 4% leucine). They suggested that the incidence of pellagra (the human equivalent of black tongue in dogs) among sorghum eaters was due to the relatively high amounts of leucine in a diet otherwise either deficient or only marginally adequate in protein, calories and vitamins.

Gopalan *et al.* (1969) further reported that among nine pups receiving a diet based on Opaque-2 maize, grown at Hyderabad, of leucine content (microbiological assay) 7·75 g/100 g (Kjeldahl) protein, only one developed signs of black tongue; those receiving an isonitrogenous diet based on Deccan-hybrid maize, also grown at Hyderabad, leucine content 12·30 g/100 g protein, all developed signs of black tongue as did four of five pups on a diet based on Opaque-2 maize in which the level of leucine had been raised to that of the Deccan-hybrid. Niacin therapy was successful where applied. Three out of the four pups receiving a diet based on Deccan-hybrid but supplemented with lysine to the level of the Opaque-2 maize also developed signs of black tongue.

Belavady *et al.* (1968) described niacin deficiency induced in adult rhesus monkeys (*Macaca mulatta*) fed diets containing either 81% sorghum, said to contain an adequate level of tryptophan, 0·8 g/100 g protein, but of otherwise unstated composition, or the same level of maize. The deficiency signs included changes in the pigmentation of the skin with loss of hair, loss of body weight, anorexia and diarrhoea. There was a fall in serum albumin, haemoglobin and blood pyridine nucleotides. The pathologic features consisted of chronic atrophic gastritis and degenerative changes in the mucosa of the large and small intestine. The condition could be reversed by the administration of niacin.

Belavady and Udayasekhara Rao (1973) reported that monkeys weighing between 3 and 4 kg and fed a 12·5% protein diet (wheat 5/casein 1) in which the calculated leucine/isoleucine quotient was about 1·7, developed signs of niacin deficiency when 1·5 g of L-leucine was administered daily over a period of 30 weeks. They responded to treatment with niacin.

The signs of deficiency were mild compared to those observed in older and heavier animals receiving a maize or sorghum diet (Belavady *et al.*, 1968).

Krishnaswamy and Gopalan (1971) cited Spolter and Harper (1961) as reporting that the growth retardation in rats induced by the addition of large amounts of leucine to marginal protein diets (Harper *et al.*, 1954, 1955) could be overcome by the addition of isoleucine.

Twenty-six patients with pellagra were maintained on a diet which provided about 2000 kcal/day and 54 g of protein, about 46 g of which was derived from sorghum of unstated composition. The diet also included pigeon pea, bread, sugar, oil, milk and vegetables. The amount of leucine in the diet was calculated to be 4·8 g and isoleucine 2·5 g. Since rice eaters rarely developed pellagra, the leucine/isoleucine ratio in rice, 4/3 was taken as "safe" and the amount of isoleucine supplement was calculated and doubled. Sixteen patients who received 5 g DL-isoleucine/day were completely cured by about 15 days; 10 patients on the basal sorghum diet alone did not show any improvement.

The abnormal EEG pattern of pellagrins, which is reversed by niacinamide (Srikantia *et al.*, 1968b) was also improved by the administration of isoleucine 10–15 days after treatment commenced though it had not become completely normal by that time.

Raghuramulu *et al.* (1965b) reported that when young and adult rats of both sexes were fed 9% casein diets to which L-leucine was added at the 1·5% level, urinary excretion of quinolinic acid (Henderson, 1949) and of NMN (Carpenter and Kodicek, 1950) was increased; the first more markedly in young rats, the second more markedly in adult rats. The addition of 0·2% isoleucine to the diet counteracted the effect in young rats; isoleucine was not administered to the adult rats. N (Kjeldahl) excretion increased in adult rats but not in young rats.

Adult rats (3 male and 3 female) were fed two diets, interchangeably, in which all the protein, at the 9% level, was from (A) 90% sorghum and (B) 71% wheat. The diet intake/rat/day, intake of niacin (Friedemann and Frazier, 1950), and tryptophan (microbiological assay) and weight gain/rat/10 days are shown below:

	Diet	Diet intake/ rat/day	Weight gain/ rat/10 days	Intake	
				Niacin	Tryptophan
		(g)	(g)	(mg)	(mg)
(A)	Sorghum	19·4	11·8	1·2	37·2
(B)	Wheat	19·9	15·5	1·3	62·0

There was a higher level of excretion of NMN and total niacin from the sorghum diet than from the wheat but this was not statistically significant. The weight gain from the sorghum diet would appear to be somewhat lower than from the wheat diet.

Krishnamachari (1974) reported that when 5 g L-leucine was administered to healthy adult males for 6 consecutive days it brought about increases in serum copper levels similar to those observed in adult pellagrins. The simultaneous administration of 5 g DL-isoleucine for 5 days prevented the increase in serum copper. Administration of 5 g DL-valine instead of leucine had no effect on copper levels. Balance studies indicated that excess dietary leucine increased intestinal absorption of copper but the mechanism was still unclear.*

* (Authors' note: the possible influence of leucine/isoleucine ratios on niacin adequacy and pellagra-like symptoms is clearly of serious interest. The reported phenomenon seems to have originated from one institution and does not appear to have been discovered elsewhere. The

Fluorosis and Urolithiasis

Krishnamachari and Krishnaswamy (1973) reported that genu vagum frequently appeared as a complication in endemic fluorosis (fluoride toxicity). This complication appeared to be seen more frequently in subjects whose staple was sorghum than those whose staple was rice (National Institute of Nutrition, 1974). Lakshmaiah and Srikantia (1977) therefore conducted experiments with eight healthy human volunteers in which similar vegetarian diets were fed with the only variable rice or sorghum. In the first experiment a crossover design was adopted and in the second a double crossover design was used. Urinary fluoride was measured by Orion Ion Analyser Fluoride Electrode and found to be lower from sorghum-based diets. Fluoride retention was significantly higher when the diet was based on sorghum than when it was based on rice but the mechanism needs to be elucidated.

Deosthale *et al.* (1977) determined the contents of Mo (Sandell, 1959), Cu (Gubler *et al.*, 1952) and Zn (Kagi and Vallee, 1958) in samples of sorghum, pearl millet and rice from areas of Andhra Pradesh, India, where fluorosis is and is not prevalent. (See Table 4.61.) Wide variations in the contents of all three minerals in all three grains were noted. Both sorghum and pearl millet contained about 60% more Mo in grain from the fluorosis areas than from the nonfluorosis areas. No significant differences in Cu content among the grains were found between areas but Zn content was higher in sorghum, lower in pearl millet and not significantly different in rice, from fluorosis areas compared to nonfluorosis areas.

Gopalan *et al.* (1971) published diet surveys indicating that an average Indian adult consumes 400 g of cereal/day. On that basis, the daily intake of Cu and Zn calculated from the data presented would not meet the WHO (1974) recommended allowance; the Mo intake would be much higher (Schroeder *et al.*, 1970) except from some samples of milled rice. Deosthale *et al.* commented that more work is needed to elucidate the interrelations of the micronutrients in fluorosis.

Kovalsky *et al.* (1961), cited by Deosthale and Gopalan (1974) having shown that uric acid production was affected by very high intakes of Mo, and as, in certain parts of India, urolithiasis (formation of urinary calculi, "stones") is common in "millet"—eating populations (Patwardhan, 1961a cited by Deosthale and Gopalan), the latter authors compared the effect in four adult males of diets containing (A) sorghum of low Mo content (Sandell, 1959) 0·021 mg/100 g, (B) sorghum of high Mo content 0·139 mg/100 g and (C) diet (B) + supplemental ammonium molybdate. The diets supplied daily 11·92 MJ (2·850 Mcal), 50 g protein, and 7 g total

authors have not been able to discover how the incidence of pellagra among people who subsist upon poor sorghum diets compares with the frequency of pellagrins among those who subsist at closely similar caloric, protein and vitamin intakes from diets in which other cereals predominate).

minerals; total S intake was 350 mg/day. Total dietary Mo from diet (A) was 166 µg/day, from diet (B) 540 µg/day and from diet (C) 1540 µg/day; dietary Mo from sources other than sorghum was constant at about 100 µg.

On all three diets, excretion of Ca, P and inorganic S was unchanged. The mean uric acid excretion (Vasantgadkar *et al.*, 1963) in urine was essentially unaltered being, in µg/day, from (A) 503, (B) 457 and (C) 488, suggesting that uric acid metabolism was altered in man at only very high levels of Mo intake. However, the urinary excretion of Cu (spectrophotometric estimation) was increased with increasing dietary Mo from, in µg/day, 24 from diet (A), 42 from diet (B) and 77 from diet (C).

Copper balances were then measured in the four subjects on the low and high Mo diets. Urinary Cu excretion again increased on the high Mo diet, but faecal Cu was not affected.

Deosthale and Gopalan suggested that the increased urinary excretion of Cu caused by high levels of Mo in sorghum required investigation since copper deficiency in animals had been associated with anaemia and osteoporesis, citing Davis (1950).

Selenium

Hadjimarkos (1962) cited McClure (1960) as having reported that diets containing 76–94% of pearl millet and foxtail millet, with or without 18% Cerelose, induced dental caries in rats. Hadjimarkos had already suggested (1961) that selenium (Se) present in a rat diet at a level of 0·34 p.p.m. could have increased susceptibility in rats. He therefore determined the Se contents by neutron activation analysis of Japanese barnyard millet, Early Fortune, white and yellow common millet the exact origins of which were unknown but which were grown in the US. The Se contents, in mg/100 g were: Japanese barnyard millet 0·02; Early Fortune 0·02; white common millet 0·02, and yellow common millet 0·04. This last concentration was 2–3 × higher than a stock caries-producing diet. The higher level in the yellow common millet possibly resulted from a comparatively high Se content in the area where it was grown.

Unknown Toxic Substances

Ayyar and Narayanaswamy (1949) reported an outbreak of illness following the eating of varagu (kodo millet) in both the cooked and the raw form, in the Madras region in 1946. Within 20 min of the food being consumed, the symptoms observed were tremors, giddiness, perspiration and inability to speak or swallow. These symptoms disappeared after 24 h and there were no fatalities. Monkeys refused to take the food but 56 g of kodo millet produced tremors and paralysis in dogs and they died within 24 h.

The fat was extracted with petroleum ether (60–95°C), chloroform or ether, and 1·5 g of fat (corresponding to 50 g of the kodo millet) were fatal to dogs. The fat had the following average figures: melting point 42°C, refractive index (60°C) 1·4650, iodine value 93·6 and saponification value 170·7.

The defatted residue was found to be nontoxic and after treatment with dilute acid and alkali the fat was no longer toxic.

The liquid decanted after shaking the millet with petroleum ether, developed a characteristic red colour when shaken with sulphuric acid but the red colour was also given by the fat after treatment with dilute acid when it was no longer toxic. The colour might therefore have been due to a decomposition product of the toxic factor and not to the factor itself. The fat obtained from nontoxic varieties of kodo millet did not give the sulphuric acid reaction, and was harmless.

The poison did not seem to be an alkaloid, as measured by the Stas Otto test (no reference given), nor a glucoside as it was not extracted with acid, water or 90% alcohol. It seemed to be adsorbed chromatographically on silica column; (further work was said to be in progress but no reference has been traced). Fat (1 g) from the poisonous kodo millet was fatal to dogs and monkeys when injected intramuscularly, and crows were extremely susceptible whether the poison in the fat was taken orally or injected intramuscularly.

Portères (1959d) discussed the toxic properties of *Paspalum scrobiculatum* L. var. *polystachyum* Stapf. a cereal found in West Africa growing and harvested with rice and bearing various local names, translatable as, e.g. "black rice" and as meaning "set on one side; abandoned". It was sometimes eaten, but only by "those who are competent", the effects being trembling and "open eyes". Kodo (*P. scrobiculatum* var. *frumentacea* Stapf.) grown in India, seemed to have somewhat similar effects. Portères asked, but did not answer, the question "Has the toxicity of var. *polystachyum* in West Africa the same origin as that of var. *frumentacea* in India?"

Agarwala *et al.* (1964) stated that kodo millet was distributed throughout India in both the wild and cultivated forms. It would grow on the poorest soil, and, while eaten by a large number of people, was not regarded as a wholesome food. They cited Watt (1908) who referred to the poisonous form as being known in Gujarat as "mina" and the nonpoisonous as "mitha" and who stated that poisonous grains were produced by damp, cloudy weather at harvest time, a damp season and damp soil.

Agarwala *et al.* (1964) fed hay raised from an allegedly toxic variety of kodo seeds to calves without ill effects over 10 weeks of feeding. The straw and the seeds of the hay were free from pathogenic fungi. Rats fed whole kodo seeds without dehusking did as well as, if not better than, those on the control diet. There was no significant difference among treatments when male and female growing albino rats were fed diets in which they received, in %, (A) 70 maize, 25 chickpea flour, 3 wheat bran and 2 minerals, and diets in which the maize, chickpeas and wheat mix was replaced by ground kodo seeds to the extent of (B) 25% (C) 50% and (D) 75%.

In metabolism trials (method of Krishnan, 1950) using similar diets, the rats in all groups maintained positive balances of N, Ca and P, with no statistically significant differences among the diets. Differences in average protein digestibility were also not significant.

Iswarriah (1951) cited by Agarwala *et al.* had reported that the poisonous nature of kodo seeds was due to fungus infestation and Agarwala *et al.* therefore concluded that kodo grains raised under normal conditions might be used like any other grain in the feeding of experimental animals.

Ergot

Ergot results from the infection of grain or grasses by a fungus which prevents the seeds of the host plant from developing and replaces many kernels with hard seed-like fungus bodies known as sclerotia. The size and shape of the sclerotia vary with the host plant. Ergot has been recognized for many years as a source of damage to crops and of toxicity to humans and livestock fed infected grain (Hulse and Laing, 1974).

Edmunds *et al.* (1970) stated that on sorghum, ergot was caused by the pathogen *Sphacelia sorghi* McRae. The first sign of ergot was secretion of a pinkish, sticky "honeydew" and the condition was sometimes called sugary disease. Fully developed sclerotia take the form of hard, grey, horn-shaped structures.

King (1972) stated that the disease was not normally serious but cited Rachie (1970a) as saying that at times it had caused extensive losses in parts of India. Humid weather at the time of flowering favoured infection.

Bhide and Hegde (1957) reported that ergot was recorded on pearl millet for the first time in India in October 1956 in the States of Bombay and Mysore. The sclerotia were hard, horny, with pointed ends, slightly bent in the middle, and measured 3–5 mm. Shinde and Bhide (1958) ascribed the ergot of pearl millet to *Claviceps microcephala* (Wallr.) Tul.

Kumararaj and Bhide (1962) examined 144 varieties, strains and selections of pearl millet maintained at the Agricultural Research Station, Niphad in Maharashtra State, India, as cultures, and found all susceptible to ergot, though certain varieties bore fewer (less than five/ear, versus ten or more) and smaller (2 versus 3 mm) sclerotia. In some susceptible cultivars, the sclerotia were as long as 10 mm.

Sivaprakasam *et al.* (1971) raised four varieties of pearl millet HB 1, HB 3, HB 4 and X 3, under greenhouse conditions and artificially inoculated them with *Claviceps microcephala* (Wallr.) Tul "honeydew" suspended in water (Dwarakanath Reddy *et al.*, 1969). Alkaloid content of the infected ovary was determined at the honeydew stage (10 days after inoculation) and in the sclerotia 20 days after inoculation (Mukerji and De, 1944).

The fungus produced very high amounts of the alkaloid ergotinine at both stages, HB 1 showed the highest content 0·92% in the sclerotia and HB 3 0·48% at the honeydew stage. This high level of alkaloid formation indicated a high risk of mammalian toxicity from infected pearl millet.

Shone *et al.* (1959) reported that ergot-infected pearl millet fed to pregnant sows completely inhibited the normal development of the mammary glands.

In 1960 Shone reported serious losses in piglets caused by agalactia (milklessness) in sows fed pearl millet in Rhodesia. The ergot content of the pearl millet varied from bag to bag and within a bag. Female mice and rats were similarly affected by feeding ergot-infected pearl millet but not cows, ewes or female guinea pigs.

Mantle (1969) reported inhibition of lactation in mice following feeding with ergot sclerotia, *Claviceps fusiformis* (Loveless), and an alkaloid component, derived from pearl millet.

Doggett (1974a) stated that ergot was a disease of major importance since even partial infection of the crop might render the whole harvest from that particular field unfit for human consumption. "True resistance to this disease must evidently be found and used, but delayed pollination, often marked when male steriles flower under poor pollen shedding conditions, is undoubtedly a factor contributing to disease build-up. Prompt and plentiful pollination must be a major consideration in developing millets which can live with this disease unscathed".

Sulaiman *et al.* (1966) reported that aureofungin (5 p.p.m.), Diathane M-20 (1000 p.p.m.), streptomycin sulphate (500 p.p.m.) and oxytetracycline-HCl (500 p.p.m.) inhibited sclerotial germination and development of *Claviceps microcephala* on pearl millet after treatment for 1 h, and captan (2500 p.p.m.) after 2 h. Antibiotics were not effective as soil treatments possibly because they were adsorbed by the soil. As fungicidal sprays, aureofungin at 5 p.p.m. gave 75% control with the least number of sclerotia developing/plant.

An outbreak of poisoning, characterized by nausea, vomiting, giddiness and somnolence which occurred in Rajasthan, India, during 1975, was attributed to the consumption of pearl millet visibly contaminated by ergot (Krishnamachari and Bhat, 1976). A comparison of the pearl millet samples collected from affected and unaffected households indicated that the nontoxic level of alkaloid would be around 28 μg/kg of body weight. Those households which identified the ergot infected pearl millet as being the cause of the illness and which discontinued its use and fed the millet to their camels (used as farm animals in that area) found that some camels suffered from sleepiness and refused the feed.

Bhat and Roy (1976) reported experiments in which ergot infected pearl millet and its alkaloids were fed by various routes to rhesus monkeys. The concentration of the alkaloid required to bring about toxic symptoms in monkeys, almost 10 mg/kg body weight, was much higher than levels, about 5 mg/kg, which severely affected humans. Bhat and Roy therefore concluded that a safe limit for the consumption of ergot infected pearl millet by man could not be derived from experiments on monkeys.

Mycotoxins

Mycotoxins are products of metabolism in various mould fungi which grow upon grains, grain legumes, and oilseeds. The incidence and control of mycotoxins has been reviewed by Scott (1973). Of the mycotoxins, the aflatoxins, formed by *Aspergillus flavus*, are among the most important as a hazard to human health and have been the most intensively studied. The occurrence of *Aspergillus flavus* and/or aflatoxin on sorghum was reported by several investigators.

One grain sorghum processing plant in the United States was sampled for 47 weeks during 1964 and 1965. Aflatoxin analysis gave positive results from only four samples of grain sorghum, with a range of 3–5 parts/10^{12} (p.p.b.) (Watson and Yahl, 1971).

Shotwell *et al.* (1969) by thin-layer chromatography (TLC) analysed 533 grain sorghum samples, of all marketing grades available in the United States, for aflatoxin. The sensitivity limit was 2–5 p.p.b. of the metabolite. Six samples, in the poorest grades, were positive by TLC but none was confirmed when fed to ducklings which are sensitive to 1–2 p.p.b. (In the duckling test, extracts dissolved in propylene glycol were administered to day-old ducklings via stomach tube over a period of 4 days. After 6 days a sample of liver tissue was examined histologically).

Schroeder and Boller (1973) collected samples of field crops, including sorghum, in Texas during 1969, 1970 and 1971, deliberately choosing low-quality lots. Aflatoxins were not found in sorghum in 1969, but in 1970 6% of 114 samples were positive with a range of 3–20 p.p.b., and in 1971, 16% of 25 samples were positive with a range of 4·0 p.p.b. *Aspergillus flavus* was only a minor constituent of the mycoflora. Schroeder and Boller did not indicate how growing and harvesting conditions in Texas might have differed during the 3 years of the survey.

Baseden and Aldrick (1970) examined, for aflatoxins, samples of grain sorghum, grown in the Darwin-Katherine and Kimberley regions of Northern Australia, and which had visibly deteriorated during storage. Aflatoxin B_1 in concentrations of 0·1–1·0 μg/g of substrate was found in bulk and bagged sorghum which had been exposed to rain. Fungal growth and aflatoxins were confined to the exposed layers, about 3 in deep in the bulked grain.

Under laboratory conditions *Aspergillus flavus* strains isolated from sorghum grown in monsoonal northern Australia were proved to be toxigenic by both TLC and the duckling test (purified aflatoxin administered orally in mash). A toxigenic strain of *A. flavus* grew on sorghum grain held at relative humidities of 84·5% and higher, with grain moisture of 14·8% (wet-weight basis) and above. Other fungi present included *Aspergillus niger, Penicillium spp. and Curcularia* spp.

From Queensland, Australia, Connole and Hill (1970) reported that a culture of *Aspergillus flavus* isolated from a sample of mouldy grain sorghum produced aflatoxin B_1 by cultural and clinical methods. It was pathogenic to ducklings, causing bile duct proliferation and death. The sorghum was suspected of having caused abortion in pregnant sows and contained 8 p.p.m. toxin as determined by TLC. Feeding the toxin at that level to two pregnant sows failed to induce abortion. When cultured, the mouldy sorghum also showed the presence of *Aspergillus niger*, *Penicillium* sp. and *Cladosporium* sp.

Lafont and Lafont (1970) examined, chromatographically, contamination with aflatoxin B_1 of samples of mixed animal feed from two factories in France, and the raw materials of which they were composed, taken direct from the storage silos.

Detectable levels (above 0·25 μg/kg) of sample of aflatoxin B_1, were found in three out of 12 lots of sorghum (25%) but none was heavily contaminated (levels above 100 μg/kg) compared with soy cake where 24 of 51 lots (47%) contained detectable levels and seven lots (14%) were heavily contaminated.

Richardson *et al.* (1967) compared normal and mouldy grains in the diets of turkey poults. The grains, including sorghum, of unstated origin, were allowed to mould with naturally occurring fungal flora at 43°C and 28°C following essentially the method of Richardson *et al.* (1962). There was no examination of the grains for the presence or absence of aflatoxin or any other mycotoxin.

In the first trial, the diets were based on 35% soybean meal and contained 58·9% of the test grain, mould induced at 43°C, or without mould. Poults fed the control, nonmouldy sorghum, gained weight faster than those fed the mouldy sorghum but the difference was barely significant.

In a second trial, the diets were based on 20·0% purified soybean protein and 71·5% sorghum, mould induced at 43°C or nonmouldy without and with additions of 0·5% arginine and/or 2·0% glycine, and of 1·0% lysine. All the nonmouldy sorghum diets gave significantly greater weight gains than the mould infected sorghum diets. The addition of arginine, or glycine, or the two together, had little, if any, effect on the growth of poults on either diet. When lysine was the only amino acid added, there was an average increase of 78 g in the group receiving the mouldy sorghum compared to a 59 g increase in the group receiving the nonmouldy sorghum. Inducing mould at 43°C and at 28°C made no significant differences to the performance of the poults. No abnormalities were observed in poults fed the mouldy grains.

Fusarium spp. and *Cladosporium* spp. were the most frequently found, and most toxic, fungi occurring in millet (not otherwise designated but possibly common millet) which had overwintered under snow and which caused many human fatalities in Russia in 1944. The disease was described as alimentary toxic aleukia (Joffe, 1965, cited by Scott, 1973).

Bilay (1960) cited by Scott (1973) stated that a single feeding of millet infected with one strain of *Fusarium sporotrichiodes* Sherb. caused death in several experimental animals. *Fusarium tricinctum* (Cda.) Sacc. had been found to be a major toxin-producing species in mouldy maize in Wisconsin

(Bamburg *et al.*, 1969, Smalley *et al.* 1970 cited by Scott, 1973) and this species was thought to be closely related to the *Fusarium* species believed to be responsible for the alimentary toxic aleukia which occurred in Russia (Scott, 1973).

Martin (unpublished report, 1974) identified a toxin from the *Fusarium* species in sorghum beer in Lesotho. In this connection it is of interest to note that although treatment of maize with lime water, as carried out in the making ot tortillas, masa and pozol in Mexico, reduced the content of aflatoxin, the fermentation process used in pozol did not result in an appreciable loss of aflatoxin (Ulloa and Herrera, 1970 cited by Scott, 1973).

Insect Infestation

Vankat Rao *et al.* (1960) reported feeding male and female weanling albino rats diets of uninfested and insect-infested sorghum. The protein content of the diets was about 8·7% of which sorghum contributed 7·7% and chickpeas 1·0%. The infested sorghum contained 25% of kernels damaged by *Sitophilus oryzae* L; the infested chickpeas contained 25% of kernels damaged by *Bruchus chinensis* L. The origin of the grain samples was not stated.

Diet (A) contained uninfested sorghum and chickpeas; diet (B) was diet (A) plus 53 mg of pure uric acid, the level present in diet (C), consisting of infested sorghum and chickpeas.

The mean weight gain on diet (A) was slightly, but not significantly higher than on diets (B) and (C). The general condition of all the rats was similar at the end of 4 months. There were no significant differences in liver moisture and protein contents or in the haemoglobin and red blood cell count among the three diets. Liver lipid of rats on diet (C) was significantly higher than diets (A) and (B). Histological examination of the livers of rats on diet (C) after 6 months showed different degrees of centrilobular fatty infiltration but those from diets (A) and (B) showed no such change.

Personal communications to the authors from research workers in India indicate that significant protein losses may result in sorghum and other cereals in consequence of insect infestation. Because in many developing countries grains are sold by volume rather than by weight in the market, heavy infestation by boring insects causes serious loss in the quantity of nutritional material purchased and consumed.

Tolerance Levels

It will be observed that no "safe" tolerance levels have been found in the literature for potentially injurious substances which may contaminate sorghum and the millets. What constitutes a "safe" tolerence limit is influenced by the total quantity of the food consumed and the creature for which it is intended. Young infants are more sensitive than healthy adults.

One can only urge, in the case of ergot and the mycotoxins, the development of varieties and the implementation of agronomic, storage, handling and inspection practices which minimize infection. Where tolerance levels are recommended or prescribed, the detailed methods of sampling, inspection and analysis should also be prescribed.

Conclusion

From the foregoing it is evident that the apparent nutritional quality, as determined by proximate chemical analysis, can be seriously impaired by many intrinsic and extrinsic adverse substances, of which for the most part a complete understanding is yet to be gained. The following chapter treats of the manner in which various methods by which sorghum and the millets may be processed and transformed can affect adverse factors, including polyphenols and phytic acid, and essential nutrients such as protein, vitamins and minerals.

CHAPTER 5

Grain Processing

Introduction

It is unusual, in any human society, for cereals to be eaten as uncooked whole seeds. Even for animal feed, cereal grains are often ground, flaked or partly cooked to increase their digestibility. For human food the seeds are customarily milled before being cooked. Dry-milling embraces a wide range of technologies from simple grinding of the whole seed between stones or in a pestle and mortar, to the complex continuous systems of precision rollers, sifters and air classifiers found in modern wheat flour mills.

It is the purpose of most milling processes, even simple traditional household methods, to bring about some separation of the three cereal seed components: germ, endosperm and seed coats (bran). Though there is evidence to suggest the nutritional desirability of retaining at least some of the bran fractions (see Chapter 2), a marked preference for cereal flour composed largely of endosperm is not unusual among people who subsist upon sorghum and the millets.

This chapter presents the effects of various primary and secondary processing conditions upon the composition and/or nutritional quality of the cereals under review. It does not deal comprehensively with alternative processing technologies that have been or might be applied to sorghum and the millets. For the most part the technologies referred to, particularly milling technologies, include: (a) traditional, village and household processes; (b) adapted milling technologies borrowed from wheat, maize and rice milling; and (c) methods specifically designed for sorghum and/or the millets; many of the latter being still in the experimental or developmental stage.

It is difficult if not impossible to compare results among authors since no standardized grading system comparable to that for wheat in North America, exists for sorghum or the millets. Even within a standard grade of wheat, the composition and proportion of the resultant mill fractions are influenced significantly by processing conditions including the spacing and relative velocities of the grinding surfaces, the nature and extent of the purifying system, the condition of the grain before it enters the mill, the number and nature of the fractions extracted, and the care taken to isolate them. Consequently, most of the results reported on sorghum and the millets are empirical, highly variable and only generally indicative of the efficiency of the milling process used.

From a nutritional standpoint, particularly for grains high in polyphenols, the most desirable decortication (or dehulling) process would remove only the outer pericarp and in some instances the testa layer and leave the high-protein germ and aleurone layers attached to the endosperm. Some form of abrasion milling seems best suited for this purpose though much remains to be learned about the optimum conditions for sorghum and the millets of different grain characteristics such as the predominantly corneous vs those with floury endosperms.

Corneous endosperms are harder and produce higher yields of grits (semolina) with lower ash and lipid content. Floury endosperms tend to chip during abrasion and to shatter during milling. Consequently roller-milling, which first breaks open the grain with corrugated rollers then crushes the liberated endosperm between a series of smooth reduction rollers, the seed coats being removed by screening and aspiration, will produce a different set of fractions than a progressive abrasion system that successively removes layers starting with the outer seed coat and progressing to the middle of the endosperm.

Abrasion-milling, as its name implies, employs a rough surface such as carborundum or hard stone to rub off progressively the various layers of the grain. The combination of the rotation of a horizontal abrasive surface and the static pressure within a body of grain, imparts a circulatory motion within the mass of the grain and subjects each individual kernel to abrasion for a period of time dependent upon the design of the abrasion unit and the rate of flow of grain. Abrasion systems rely mainly on carborundum in the form of discs, cones or rolls, and have been reviewed by Reichert (1977) (PhD Thesis). A wire brush surface may be used together with or in place of a carborundum or stone surface. Abrasive resinoid discs are also used.

Attrition-type dehullers usually consist of two metal or stone discs, one or both of which rotate in a horizontal or vertical plane. The attrition process can be modified by introducing a variety of impact or cutting surfaces, such as metal pins or blades, into the surface of one or both of the rotors or the rotor and stator.

Additional methods of dehulling include a variety of wet systems in which the grain is moistened either by damping or soaking. The seed coats are then removed by a large pestle and mortar sometimes after sun-drying. In some instances the seed coats are removed by winnowing, sometimes by washing in water.

Though it is difficult to categorize the processes reported according to precise technological principles, the authors' best count suggests that 11 different abrasion, three attrition, four break and reduction and three pulverization methods have been reviewed. In several reported instances more than one principle is involved. For a more detailed account of grain milling technologies the reader is referred to Lockwood (1960), Ziegler and Greer (1971), Araullo *et al.* (1976) and Watson (1967).

Efficiency of milling can be defined only in relation to an ideal end product. Consequently, one considered efficient as judged by the colour of the principal end product may be unsatisfactory in terms of essential nutrient composition.

The text that follows includes dry- and wet-milling; various forms of cooking; composite flours in which sorghum and the millets are combined with other cereals, legumes or oilseeds; malting and brewing; and processing for animal feed. As stated elsewhere this publication emphasizes human nutrition and only includes references to animal feed that might have a bearing on human nutrition. It is interesting to observe from the literature reviewed that more attention has been given to the quality of processed sorghum for animal feed than for human consumption. It is hoped that the future will see closer cooperation among crop scientists, nutritional biochemists and food technologists in producing sorghum and millet foods best suited to the needs of the many who are nutritionally dependent on them.

Sorghum and the Millets

Traditional and Household Processes

The Applied Nutrition Unit of the London School of Hygiene and Tropical Medicine (1960) examined sorghum and millet food prepared in, mainly, village communities, and analysed by the (UK) Government Laboratory for moisture (loss after drying for 5 h at 100°C), nitrogen (AOAC, 1950), fat (petroleum ether extraction, AOAC, 1950), fibre (Analytical Methods Committee, 1943; (UK) Fertilizer and Feeding Stuffs Regulations, 1932) ash (samples incinerated to constant weight at 600°C), acid-insoluble ash (portion insoluble in 1/1 HCl), calcium (microcalcium oxalate-permanganate titration, AOAC, 1950), iron (orthophenanthroline colour, Pringle, 1946), phosphorus (molybdenum blue colorimetric method, AOAC, 1950), phytic acid phosphorus (Common, 1940), thiamine (chemical assay, Society of Public Analysts Committee Methods, 1951) and niacin and riboflavin (microbiological assay, Society of Public Analysts Committee Methods, 1946). The data are presented in Table 5.1.

Sampling presented obvious problems, not least being the tendency for some deterioration from mould growth. In samples sun-dried before despatch, some reduction in the riboflavin content occurred.

The flour and nyelling (coarse endosperm) from pearl millet in the Gambia was prepared by the traditional method of pounding the grain in a wooden mortar with a small amount of water for about 10 min following which the bran was winnowed off. The washed separated grain was left to stand for several hours (often overnight). At this moisture content of between 20 and 30%, a considerable degree of "malting" occurred before the final pounding into flour. The pounded grain was frequently sieved, or winnowed, to separate the flour from the coarse endosperm (nyelling). The nyelling was completely reduced to a fine flour or used in dishes requiring coarser material. During pounding and winnowing, some evaporation took place so that both flour and nyelling were drier than the original malted grain; typical moisture contents were: grain 30%, flour 27% and nyelling 24%.

Normally the flour and nyelling are used as soon as prepared and not dried before use. The samples analysed were sun-dried before shipment which probably accounts for the unexpectedly low riboflavin values (Table 5.1).

The traditional method of preparing pearl millet flour in Northern Nigeria is similar to the method reported for the Gambia except that, after washing to remove the bran, the grain is left damp for only 10–15 min before being pounded into flour. The malting process does not proceed so far as in the Gambian method and the individual grains remain fairly hard, making the pounding process very heavy work. It was reported that it normally takes one woman about 4 h to reduce 7 kg of grain to flour. In the Gambia, about 1 h is sufficient time for the pounding of malted flour.

In Ethiopia, the traditional bread njera is made from teff alone or from mixtures or sorghum and teff. The grain is ground dry between stones and then winnowed. The flour is mixed with water, a little sour dough from a previous batch is added and the mixture is left to stand for 2–3 days. The flour and water batter is cooked for 2–3 min on a greased hot plate. The samples analysed were shipped by air and analysed about 30 h after being cooked. Mould growth had begun but was not thought to be sufficient to affect composition.

Ethiopian beer "tala" is made from red sorghum. It is much more dilute (1·39% solids) than sorghum beer made in South Africa (5–8% total solids) and the analyses suggest it is a poor source of the B vitamins.

Muller (1970) suggested a rheological classification of traditional Ghanaian and Nigerian cereal foods.

(1) Beverages (low viscosity liquids, water content exceeding 94%)
 sorghum beer (see pp. 453–467)
 pito (Hausa beer drunk warm)
 burukutu (beer liquid decanted)
 kunnu tsaki (beer sediment stirred before drinking)
 aliha (maize used instead of sorghum, sweetened or coloured with caramel)
 iced kenkey (nonalcoholic)

(2) Porridges (moisture content about 90%)
 ogi (see Banigo and Muller, 1972a)
 koko (similar to ogi but fermented as a dough)
 ekuegbemi (prepared from maize)
 tuo (prepared from sorghum or millet)

garin acha: powdered findi (*Digitaria exilis*) (fermented with sugar or honey and eaten raw)
ablemamu akasa (similar to koko but prepared from roasted maize)
kunnu tsamia (boiled ground millet and tamarind water)

(3) Dumplings (moisture content 65–80%)
hura (made from millet)
eko (boiled and gelatinized ogi)
agidi (boiled and gelatinized ogi)
banku (boiled koko dough)
fula (made from millet)
kenkey (made from fermented maize dough)
akporhe (a type of kenkey)
sweet kenkey
nsihu, kokui, abolo } not described

(4) Baked or fried products (moisture content below 50%)
bread
atshomo (fried sweet wheat flour pastry)
togbei (fried sweet wheat flour pastry)

At a recent sorghum workshop (1978) sponsored by IDRC in Nairobi, representatives from Ethiopia, India, Kenya, Nigeria, Sudan, Tanzania and Uganda reported on the current use of sorghum and millets in their respective countries. With few exceptions the basic recipes for the traditional foods of the different countries were very similar (Vogel and Graham, 1979).

Various factors appear to constrain a more widespread acceptance and use of sorghum and millets:

(a) tedious and lengthy traditional methods of household preparation. These include decortication, germination, wet-milling, dry-milling, coarse grinding and roasting;
(b) lack of commercial mills for decortication, grinding and fractionation; and
(c) lack of commercially processed sorghum and millet foods.

The following outline of household processing methods for sorghum and millets was compiled.

(1) Thin porridges

All the countries traditionally prepare a thin porridge from dry-milled flour, wet-processed flour or from cooked paste obtained by pounding soaked grain. These may be fermented as in India, Sudan and Uganda or non-fermented (all but Sudan and Nigeria). In Kenya the porridge is served to children and in Ethiopia porridge is made from wheat or teff.

Common criteria of quality are colour (light colour preferred), mouth feel (not sticky or gluey), and consistency (smooth, pourable consistency preferred). The flavour and aroma range from bland to sour depending on whether or not the porridge is fermented. Undesirable flavours include raw starch and bitterness from dark coloured varieties of sorghum.

(2) Thick porridges

All the countries consume a stiff porridge which in some instances may be moulded or shaped and eaten either hot or cold. In Nigeria and Sudan, it is made from very fine flour.

Quality criteria are the same as for thin porridge. Regarding consistency, the porridge should be stiff enough to mould, smooth and free from cracks.

(3) Beverages

Beer and in some cases, unfermented beverages, are made in these countries. For beer, the reddish brown sorghum varieties are preferred. The grain is germinated, dried and pounded into flour and mixed with water and fermented. In Kenya, sprouted millet is used with sorghum flour in making beer and in Uganda, wood ash is added to the sorghum. Some beers are filtered and other simply sipped through a long straw.

In Nigeria and Sudan, unfermented beverages are made from sorghum; hulu mur (dark red drink) from Sudan is a typical example.

(4) Cooked dehusked grain

In India, Nigeria and Sudan, the sorghum may be dehusked, cooked and served like rice. The broken kernels may also be included.

(5) Unleavened bread

In India and Ethiopia, unfermented breads made from whole sorghum and millet grains are baked in a frying pan.

(6) Fermented bread

In Nigeria, Sudan and Ethiopia, a fermented flour paste is cooked on a griddle. In Nigeria, a masa pan or cuplike frying pan is used. The batters are very similar though the cooking techniques vary. The enjera of Ethiopia is poured on the griddle while the kisra of Sudan is spread very thin.

(8) Deep-fried doughs

In Nigeria, Sudan and India, doughs from sorghum and millets are shaped and cooked in deep fat. The shape and texture of the finished foods are quite different from country to country.

Muller (1970) considered that eventually the manufacture of cereal products in West Africa will be mechanized and this will necessitate equipment such as fermenters, mixers, pumps and conveyors. The choice of equipment will therefore require a knowledge of the structure of each grain and the physical properties of the end products.

Ogi in Nigeria is made by the fermentation of maize, sorghum or millet followed by grinding, wet-sieving and boiling. Smooth in texture and with a sour taste, it is an important traditional weaning food and a breakfast cereal for adults (Banigo and Muller, 1972a).

Akinrele (1970), Oke (1967), Akinrele and Bassir (1967) all reported nutritional losses in the processing of ogi. Banigo and Muller (1972a) reported the results of processing white maize, sorghum and pearl millet all from Ghana, into ogi.

The method of making ogi in the laboratory was as under:

(1) 500 g of the cleaned cereal in 2·5 litres distilled water was fermented at 30°C for 3–4 days.

(2) The fermented grains were separated from the steeping water by a coarse sieve and then ground in a Waring Blendor followed by a Premier colloid mill.

(3) The resultant slurry was washed through a fine wire sieve.

(4) The throughs were allowed to settle, the sediment filtered through cloth and squeezed to remove excess water.

(5) 20 g of ogi was stirred and cooked in 100 ml water to produce ogi porridge of about 90% moisture content.

The total solids contents of the steeping water and combined ogi wash water were determined (Pearson, 1962) as a percentage of the dry weight of the cereal used, and the moisture contents of the original cereal, fermented grain, overtails and ogi (AACC, 1962) expressed as a percentage of the cereal (DWB) are shown in Table 5.2.

The total solids from maize and sorghum represented about 83% of the original grain, from pearl millet only about 61%. Much of this loss occurred in the wash water, some in the steeping water.

The proximate analysis (AACC, 1962) of the original grains, of the laboratory ogi described, and of a maize ogi made by laboratory wet-milling with a pestle and mortar, are shown in Table 5.3.

Ash content fell but there was no significant change in protein, fat and carbohydrate during the Waring Blendor-Colloid mill process. There was a greater loss of nutrients from maize during pestle and mortar milling. Akinrele (1970), Oke (1967) reported nutrient losses in maize ogi of up to 50%. Protein losses were particularly marked.

Six lots of each of the three grains were steeped at 30°C for 6 days. The volume of unabsorbed steeping water, the moisture content and the total weight of the steeped grain were determined for 6 consecutive days. In all three cereals, moisture content increased from about 13% to about 32% with 24 h of steeping. Maize and sorghum remained fairly constant at 40% between the fourth and sixth day. Pearl millet moisture content increased to about 50% on the sixth day. The amino acid composition (Technicon auto-analyser following acid hydrolysis, excluding tryptophan) of the cereals and the laboratory made ogi, expressed as residues/1000, is shown in Table 5.4. Most of the amino acids were unchanged by ogi fermentation but lysine content was almost halved. (Banigo and Muller, 1972a).

Samples of Nigerian made and laboratory prepared ogi were subjected to assessment by taste panels comprising Nigerian students familiar with ogi, using declared control (Peryman, 1958), paired comparison (Cornwell, 1959) and nine-point hedonic scale (Wolfe, 1958). There were no significant differences between samples of commercial ogi and laboratory ogi from the same cereal. These results agreed with those of Banigo and Muller (1972b). Preference tests on ogi boiled in its wash water, a traditional practice in some parts of Nigeria suggested a slight dislike for pearl millet ogi which could however have been due to unfamiliarity with this product by the particular Nigerian tasters.

Banigo (1969) reported that the acidity of maize ogi was correlated with acceptability by taste panel tests, and a pH of about 3·7 seemed the most desirable. Banigo and Muller (1972b) from a study of the carboxylic acids of ogi fermentation analysed by paper, gas–liquid and thin-layer chromatography identified 11 acids, lactic, acetic and butyric being the most predominant.

Acharya *et al.* (1942) examined the effect of parching (or roasting) on the biological value (BV) of the proteins of several cereals and pulses including sorghum, finger millet and maize.

Samples were bought in local markets and parched by a typical method: the grain was

(1) sprinkled with water or saline solution;
(2) mixed with about 4 × its own volume of preheated (235–240°C) sand in a frying pan over an open fire;
(3) parched by rapid mixing in the frying pan with a ladle; the temperature of the grain rose to about 132–136°C in 2–3 min; and
(4) separated from the sand by sieving.

The diets compared included (Table 5.5): (A) N-free diet, (B) sorghum, (C) finger millet and (D) maize fed to provide about 0·9% N, DWB. The results of the rat feeding trials including BV (Chick *et al.*, 1935a,b) are shown in Table 5.6.

Parching raised significantly the BV of the sorghum and finger millet but the increase in the case of maize was not statistically significant.* The increase in BV was not accompanied by an increase in digestibility coefficient, which, in the case of finger millet decreased after parching.

FAO (1970) reported a personal communication from Dreyer (1963) that sorghum cooked and dried at 50°C, fed to rats, 10 to the group, at a 10% protein level in the diet, gave a BV of 73·2, digestibility of 76·3 and a (calculated) NPU of 55·8.

Bandemer and Evans (1963) at Michigan State University, compared the

* (Authors' note: statistical proof was not provided.)

amino acid composition (ion-exchange chromatography, Moore and Stein, 1948, 1951; cystine, modified Schram *et al.*, 1954; tryptophan not determined) of, among other seeds, millet (not identified but possibly common millet). The seeds were spread on an aluminium tray to a depth of 0·25 in and heated in an oven at 250°C for 45 min.

Heating resulted in comparatively small losses of most of the amino acids. The protein content (Kjeldahl N × 6·25) of the ground, air-dried samples and the distribution of N in the hydrolysates are shown below:

	Millet	
	Unheated	Heated
Protein (%)	14·0	14·2
Nitrogen		
(% total N)		
Amino acid N	72·4	69·0
Ammonia N	17·2	15·7

Watson (1971) determined the chemical composition (AOAC, 1965), of four samples of Ghanaian Massa fried millet cake. The mean nutrient content (DWB) of the fried millet cake was: (%) moisture 45·0, calories (bomb calorimeter) 276, protein 6·2, fat 18·2, carbohydrate 72·0, fibre 2·4 and ash 1·3; (mg/100 g) Ca 20, P 209 and Fe 16.

Bressani *et al.* (1977) compared white sorghum and maize in the making of tortillas by the same process. The grains were cooked in 1% (w/v) lime solution until soft. The cooked grain was then washed and ground wet into a dough from which tortillas were made and dehydrated.

The sorghum cooked in about half the time (30 min) taken by the maize. At optimum cooking times, losses of solids were similar for the two grains but overcooking led to higher losses of solids in sorghum than in maize. Sorghum tortillas were acceptable in taste but were greyer in colour and more elastic than maize tortillas.

The protein contents and PER values (in parentheses) of the tortillas were: sorghum 8·5% (0·6) and maize 9·3% (1·0).

Waniska (1976) reported the influence of several sorghum kernel characters on the quality of gruel and chapatis. Specific characters included pericarp colour and thickness, endosperm type, colour and texture, including waxy versus nonwaxy types. Coloured pericarp and coloured testa (related to high CE) produced more darkly-coloured gruel and chapatis.

Mahajan and Pushpamma (personal communication, 1977) reported the effect of baking (biscuits) and frying (muruku) using sorghum, as judged by available lysine and tryptophan content, and PER in rat feeding trials at 7% protein level. Compared to the raw sorghum diet, there was destruction of lysine on baking and greater destruction on frying. The PER values were reduced by half on baking (1·0 raw, 0·4 baked) but four out of the six rats on the fried product died after 6 weeks and the remaining two were in poor condition. Supplementation with lysine improved weight and PER of rats on the biscuit diet but the rats on the fried product continued to lose weight.

Pushpamma and Devi (personal communication, 1977) studied sorghum and legume mixtures (market samples) for children in Andhra Pradesh, India, and examined the effect of roasting and boiling on the nutritive value of such mixtures. The mixtures, with their amino acid content (mg/g N) (auto-analyser) are shown in Table 5.7 compared with the WHO (1973a) recommended allowance. Details of the composition of the mixtures as formulated for feeding trials with weanling female rats, six to the group fed at 10% protein level, are shown in Table 5.8 with the PER (average weight gain/average protein intake) (micro-Kjeldahl N, no factor stated but probably 6·25), percentage digestibility and biological value (derived from N balance studies against a N-free diet). The pulses were cooked in boiling water or roasted for 5 min at 103°C.

The highest PER values were recorded for diet (F) sorghum and urd (black gram) whether raw or heat treated with very similar values for (D) sorghum and cowpeas. The lowest PER was for (E) sorghum and horse gram (*Dolichos biflorus*), raw. Both heat treatments improved the PER of all diets (apart from (D) and (F)). Roasting produced relatively more weight gain in all rats than boiling with the exception of diet (G) sorghum and soybean.

Digestibility was improved by both heat treatments in diets (C), (E), (F) and (G) but roasting somewhat reduced digestibility of (A) sorghum and chickpeas. Both heat treatments improved BV except in the case of (B) sorghum and mung beans, where the BV was reduced, especially by boiling.

Okoh *et al.* (personal communication, 1975) reported the chemical composition of the traditional Nigerian food fura-nono.

Fura is prepared by pounding pearl millet into flour with a pestle and mortar, compressing into balls, and boiling for 20–30 min. The cooked dough balls are broken and mixed with water, and ground spices (pepper, ginger, black pepper and cloves) are added. Small pieces are rolled into small balls in a calabash. The maximum keeping time is 3 days.

Skimmed milk (nono) is beaten with a wooden stirrer (maburgi) in a calabash and sometimes kuka, water soaked baobab seeds, is added, which makes the nono more sour and thicker.

Fura-nono is prepared by mixing equal parts of fura and nono.

Cooked fura and nono were prepared by a traditional method at Hanwa, a village near Zaria, by a local Fulani woman using laboratory ingredients. The proximate analysis, Fe, P, Ca and S (AOAC, 1965), Na and K (flame photometric), Mg (atomic absorption spectroscopy), milk fat (Gerber test, Ling, 1963), thiamine and riboflavin (Association of Vitamin Chemists, 1966) and ascorbic acid (Evered, 1960) are shown in Table 5.9.

Losses in mineral elements in fura did not occur because nearly all the cooking water was returned to the fura during pounding. The cooked fura-nono samples showed an increase in K and Ca compared with raw fura-nono laboratory samples. Increase in Fe content may have come from metal containers. Losses in thiamine (22%) and riboflavin (10%) occurred in cooked fura. There were no particular losses of minerals. Except for P, during the preparation of nono, there were no losses of minerals, thiamine or riboflavin.

Amino acid contents of the fura-nono, (a) raw without kuka, (b) cooked with kuka and (c) cooked without kuka, determined at Columbia (University), are shown in Table 5.10.

All amino acids, except arginine, were higher in the cooked sample compared with the raw, possibly due to microbial synthesis during fermentation.

The essential amino acids (mg/g N) compared with the FAO/WHO provisional pattern are shown in Table 5.10. Cystine, and to a lesser extent, lysine appear as the limiting amino acids in cooked fura-nono mixture.

Soni and Sharma (1974) compared the analytical values for total Fe (Bothwell and Finch, 1962) and ionizable Fe (Shackleton and McCance, 1936) of freshly prepared Indian foods including pearl millet roti with the calculated values of Aykroyd *et al.* (1966). The methods of preparation were those quoted by the Nutrition Research Laboratories (1966), Pasricha and Rebello (1969) and Singh (1969). The tap water used contained 0·4 mg Fe/l. The comparative values are shown below:

	Pearl millet roti (mg/100 g)
Total Fe	
analysed	11·2
calculated	8·3
Ionized Fe	
analysed	6·9
calculated	3·3

De (1936b) using a spectrophotometric method (De, 1936a) reported the carotene content of a sample of finger millet obtained from the local market as 0·06 mg/100 g when powdered and as 0·03 mg/100 g when powdered and boiled for 45 min. Loss of carotene on boiling was also observed in cooked lentils and rice and was attributed to leaching into the boiling water.

Ilany *et al.* (1969) studied alternative methods of preparing products comparable to the Japanese miso, made by fermenting a mixture of cooked soybeans, koji and salt; (koji is made from steamed rice fermented with *Aspergillus oryzae* for 6–12 months). Sorghum was tested as a substitute for the rice.

The taste panel rated the sorghum and wheat miso-type products as "good"; compared with "very good" for the Japanese miso, and for the miso-type products in which rice, maize, barley, oats, potatoes and sweet potatoes were used as the carbohydrate source. The chemical composition of the Japanese miso and the sorghum miso-type is shown in Table 5.11.

Diamant and Laxer (1967) in a short report from the same institute as Ilany *et al.* (1969) referred to miso-type products in which sorghum and the other carbohydrate sources were compared. PER determinations with weanling rats showed that none of the miso samples fed at a 9% protein level could sustain adequate growth in young rats. Thin-layer chromatographic analysis of the maize-miso suggested inadequate methionine, tryptophan and arginine.

Traditional Milling

Carr (1961) described the traditional grinding implements used in Rhodesia, the pestle and mortar, the grinding stone (or quern) and the winnowing basket, and also how sorghum is milled by village people.

Red sorghum was placed in the mortar with about 20% of its weight of water, stamped with the bottom of the pestle for about 20 min, winnowed and restamped. The pounded grain was dried for about 10 min, slightly roasted in a pan then ground by quern and sifted through a fine sieve, the coarser particles being repeatedly winnowed and reground. The "offals" discarded from winnowing contained 36% and the final meal 15% moisture; at equal moisture (10% H_2O) the original grain produced: meal 66%, offals 29% and waste 5%.

Finger millet (rapoko) was ground by pestle without water and winnowed, sifted, winnowed and reground. The ground millet yielded: meal 80%, offals 18% and waste 2%. Carr (1961) stated that finger millet is used mainly for brewing and less commonly as a meal to make porridge (sadza). Pearl millet (munga) was ground as described for sorghum, with 20% of its weight of water added. After stamping, the grain was winnowed, lightly roasted, and ground on a quern without sieving. The extraction rate for pearl millet was about 75%.

The analyses for moisture, Kjeldahl N, total lipid (Kent Jones and Amos, 1947), crude fibre (Fertilizers and Feeding Stuffs Regulations, 1932, (UK)), ash (ignited to constant weight at temperature less than 500°C), Ca, P and Fe (Carr, 1956), thiamine (thiochrome method, Ridyard, 1949) and riboflavin (fluorimetric, Association of Vitamin Chemists Inc., 1951) are shown in Table 5.12.

The traditional grinding of sorghum caused relatively high losses of important nutrients. Losses from grinding pearl millet were comparatively lower. Grinding by stone quern led to an increase in iron content (not revealed in Table 5.12) probably by contamination from the ferric iron present in the stone.

Muller (1970) described two types of mill used in Nigeria and Ghana, the stone mill and the pestle and mortar mill.

The stone mill consists of a stone slab, often fluted, usually placed on levelling stones so that it slopes away from the operator who sits, kneels or squats behind it, and rolls the grain with a round roller. A dish or mat is placed in front of the base stone to receive the ground material.

The pestle and mortar are carved from wood. The mortar is about 60–70 cm high with a diameter of about 30 cm. The pestle weighs about 3 kg, is about 1·2 m long, 6 cm in diameter and has bulbous ends, one more pointed than the other. The grain is pounded by two or three women, each working at about 60 strokes a mintue. There are three types of stroke: (1) up and down; (2) down, across and up and (3) down, "rotary scrape" and up. Strokes (2) and (3) knock the side of the mortar and prevent bridging (uneven build-up around the walls). The pestle and mortar mill is generally used for softer grains.

Muller described sorghum milled by pestle and mortar near Kano in Nigeria. About 2–3 kg of grain with about 250 ml of water were gently pounded with the pointed end of the pestle. The husk (bran) separated from the grain by winnowing on a plate or mat, was discarded, and the grain returned to the mortar where it was pounded vigorously with the flattened ends of the pestle. The material was then sieved. The overtails were ground on a stone mill and the resulting flour combined with the throughs of the sifting operation. The sieving analysis is shown in Table 5.13.

Sorghum

Traditional vs Laboratory Milling

John and Muller (1973) reported the effect of milling sorghum to varying extraction rates by the traditional pestle and mortar method, followed by grinding, and by roller-milling in a Bühler laboratory mill fitted with 3 break and 3 reduction rolls, on the proximate analysis and thiamine content (modified AACC, no date).

The results in Table 5.14 show that the nutrient content decreased with decreasing extraction rate and that the nutrient loss was greater in the Bühler mill than the traditional pestle and mortar. Roller-milling removed more germ and bran than the traditional process.

Favier *et al.* (1972) reported the effect on the protein (Kjeldahl $N \times 6{\cdot}25$), lipid (petroleum ether extract), carbohydrate ("glucides") (by difference), nondigestible carbohydrate (formic acid insoluble matter, Guillemet and Jacquot, 1943), ash (incineration at 550°C for 6–8 h), Ca, K, (flame photometry), P (colorimetric, ammonium phosphovanado molybdate), phytic P (Holt, 1955), thiamine (Deibel *et al.*, 1957), riboflavin (Snell and Strong, 1939) and niacin (Snell and Wright, 1941) of a market sample of Cameroon sorghum (1) traditionally milled by pestle and mortar, (2) ground in a motor driven disc mill after decortication (removal of the seed coats) by pestle and mortar and (3) conversion into kourou (a white flour).

Favier *et al.* (1972) stated that, in the Cameroon, some ethnic groups prepare a wholemeal from sorghum by grinding without decortication. In the towns, and among those of the Islamic faith, the bran is removed and, after grinding, a flour and a semolina relatively white in colour are separated. The semolina may be used as such or reduced to flour. Two pestle and mortar operations observed in different regions gave fairly similar results:

	% of whole grain	
	(1)	(2)
Bran	19·7	15·8
Decorticated grain	76·8	73·9

The chemical composition of the sorghum sample analysed by Favier *et al.* and the analyses of the variously milled products, DWB, appear in Table 5.15.

There was a greater loss of raw material from the traditional pestle and mortar than from the disc mill mainly because grain is thrown out of the mortar during pounding. Heat generated in the flour by the motor caused loss of thiamine and other B vitamins.

To make kourou, decorticated grain was steeped in water from 12 h to 4 days, ground wet several times on a stone mill (sloping stone base and roller), followed by washing and straining, the final residue being a white flour used immediately to make a gruel or porridge or sun dried for future use. The resultant chemical composition appears in Table 5.15.

A comparison of the recovery of the various nutrients from pestle and mortar pounding, disc mill grinding, and the preparation of kourou are shown in Table 5.16. Only 41% of the protein and 13% of the lipid present in the original grain were retained in the preparation of kourou (semolina of kourou—fresh kourou); losses in minerals and vitamins were even greater.

The effect of sun-drying on the chemical composition of sorghum flour and kourou is shown in Table 5.17. The increase in ash content in the sun-dried kourou probably resulted from dust contamination. There was an apparent loss of thiamine, and riboflavin content was lower in the flour but higher in the kourou which Favier *et al.* attributed to synthesis during fermentation while the kourou was wet.

The flour or semolina may be cooked to a dough or a gruel. To make a dough, semolina or flour is mixed with water and boiled and stirred for 10 or more minutes while the rest of the flour is added. The dough formed into a ball about the size of a fist is eaten at the principal meal together with vegetables, meat or fish. The gruel is made by throwing the flour or kourou into an excess of boiling water, and boiling to the desired consistency.

The effect of these cooking procedures on thiamine, riboflavin and niacin is shown in Table 5.18.

Laboratory Grinding Methods

Kurien *et al.* (1960a) separated the husk and endosperm of kernels of sorghum designated as Kanavi 325, and a market sample, using a gravitational method. Cleaned grain was soaked for 48 h in water containing 0·05% potassium metabisulphite, ground in a micro-pulverizer using a 0·27 sieve and mixed with the soak water. The slurry was sieved through a 100-mesh sieve to remove the endosperm. The residue remaining on the sieve was washed until free from starch, as indicated by a weak iodine test, and suspended in water. The specific gravity of the water was adjusted with NaCl to float the germ to the surface. The germ was washed and added to the endosperm, and the fibre dried at 90°C in an air oven. The filtrate containing the endosperm and the germ was centrifuged and the clear supernatant removed. The endosperm with germ was air dried at 45–50°C. The whole grain, the endosperm with germ, and the supernatant were analysed for protein, Ca and P (AOAC, 1950). The results are given in Table 5.19.

Normand *et al.* (1965), using a laboratory tangential abrasion mill developed for rice by Hogan *et al.* (1964), removed successive peripheral layers from commercial samples of TX 601, RS 610, and a mixed sample, designated "elevator run". The protein contents (AOAC, 1960, $N \times 6{\cdot}25$) of the fractions removed are shown in Table 5.20.

As abrasion progressed from the outer to the inner portions of the kernels, the protein content, with the exception of the first fraction, decreased. Fraction I, 6–12% by weight of the original grain sorghum kernel, consisted almost entirely of bran with some embryo. In RS 610, fraction III was highest in protein, probably because a greater proportion was separated in fraction I.

Grain sorghum endosperm, when freed of germ and bran, appeared to contain a fraction with an average protein content of over 18% and representing one-quarter of the weight of the original kernel.

A commercial No. 2 yellow sorghum, similar in composition to that used by Anderson (1969b) was used to prepare a refined endosperm fraction by Anderson *et al.* (1969a). A (1) laboratory Strong-Scott barley pearler and a (2) Squires rice huller were used to remove the bran from the whole kernel before milling for grits or flour. To remove the germ, dehulled grain was impact milled in (3) an Alpine Kolloplex pin mill Model 160Z, (4) a laboratory (NU) brush degerminator developed at the Northern Regional Research Laboratory (Weinecke and Montgomery, 1965) and (5) a NU solid rotor degerminator, in which the brush was replaced by a solid steel rotor with four equally spaced horizontal bars fastened to it. The (6) Alpine Kolloplex was also used to impact whole grain to produce endosperm, germ and hull. After treatment, the ground material was dried and screened into size fractions, the fractions were aspirated to remove the hull, and the germ fractions separated from the +34 and coarser aspirated grit by flotation in a sodium nitrate solution of 1·30 sp gr. The yield and crude fat (DWB) for the six treatments are shown in Table 5.21.

The best low fat endosperm fractions were obtained from the NU solid-rotor degerminator and the NU brush degerminator.

Anderson and Burbridge (1971) described an integrated process for dry milling grain sorghum using the yellow hybrid TE 77 and the white hybrid Funk G 766 of similar composition to that shown in Anderson (1969b), with a yellow endosperm hybrid DeKalb C42Y of proximate analysis

(AACC, 1962) moisture 10·4%, protein 11·3%, crude fat 3·1%, crude fibre 1·9% and ash 1·4%.

The grain was tempered to moisten the hull, passed through the CeCoCo rice huller (previously shown to be effective at removing sorghum bran, Anderson, 1969a) adjusted to remove about 18% of the grain. The "tails" material going through the dehuller and exiting at tail-gate amounted to 80–85% of the material fed. The "throughs" material passing through the enclosing screen was generally 15–20% of the feed grain. The tails were passed through an aspirator to remove the hulls which went to the feed fraction. The grain was degerminated, after tempering to 22·5% moisture, in an Alpine Kolloplex mill. Material from the impact mill was dried to about 15% moisture and sized on 14W, 20W and 34W screens. The "throughs" material was also sized on 14W and 20W mesh screen. The combined +14 and +20 fractions were passed through their respective aspirators to remove hulls. Following aspiration +14 and +20 grit–germ fractions were passed over their respective gravity tables to separate the germ from the grits. The +34 tail fraction after aspiration was combined with the −34 tail fraction and further reduced in a hammer mill to produce a sorghum flour product. This fraction was found to contain about 3% fat. The −20 material from throughs stream was rather high in fat and fibre and was added to the feed. The results of milling TE 77 sorghum by an integrated process are shown in Table 5.22.

The prime products from the flow are the −14 and −20 grits and the flour. Germ and feed fractions are regarded as by-products.

Anderson *et al.* (1977) applied the NRRC integrated milling process (Anderson and Burbridge, 1971) to a sample of the high-lysine hybrid sorghum P 721. The kernel endosperm was soft and floury compared to that of normal sorghum. The fat, fibre and protein (AACC, Methods 30–25, 32–15 and 46–10, 1962) and lysine (auto-analyser) of the whole grain and five fractions are shown in Table 5.23. Compared to normal sorghum, yield of grits was lower, probably because of unevenness in kernel size and the floury character of the endosperm; fat content was double; fibre content was higher; protein content and lysine in protein of the grits and flour was decreased. Lysine content was highest in the germ and feed fractions.

Shepherd *et al.* (1971–72) reported that six East African sorghum varieties, Karachi, Lulu, Mbangala, Mgindu, Msumbiji and Songea, when pin-milled on a laboratory scale and then separated into seven fractions produced good looking grits of 65% yield or greater, Mbangala giving the best product and highest yield at 71%. The proximate analysis of the grains and the "finer looking products", DWB, was:

	Grain (%)	Products (%)
Protein (N × 6·25)	9·2–12·5	8·5–12·1
Fat	2·9– 3·4	1·2– 2·2
Fibre	1·8– 2·1	1·2– 1·9
Ash	1·6– 1·8	0·8– 1·0

The means were not given.

Raghavendra Rao and Desikachar (1964) described the pearling, or polishing, of a locally (Mysore) grown sample of sorghum, tempered to about 15% moisture, in a laboratory McGill rice mill. The degree of polish, expressed in terms of the amount of bran obtained during milling, and the proximate analysis, with Ca, P and thiamine (AOAC, 1950), all DWB, are shown in Table 5.24.

About 10% polish appeared optimum, giving minimum loss in weight of seed, adequate removal of bran and crude fibre, and minimum loss of protein and thiamine. The germ end was retained in most of the grains. The polished sorghum took longer than rice to cook to a soft acceptable consistency, and pressure cooking was needed to obtain a soft product. On grinding, the sorghum flour was attractive in colour and Raghavendra Rao and Desikachar considered that it could be used for roti or bhakri (dry pancake), dosai (fermented pancake) or any deep fried Indian cereal food.

Rooney and Sullins (1969) described a laboratory procedure for milling 100–200 g sorghum samples into grits, flour, germ and bran fractions. A Strong-Scott barley pearler fitted with a 2·4 mm (No. 8 US) screen was modified by substituting a wire brush for the carborundum wheel. The stock from the mill was sifted over a nest of Tyler screens. The overs of the No. 12 screen were termed coarse grits; the overs of the No. 16, 20 and 30 screens were combined and separated by controlled air flotation in a model D South Dakota seed blower, the bran, germ, and germ-rich fractions being removed successively by increasing the air flow. The fraction remaining was labelled fine grits. The germ-rich fraction contained germ and fine grits which were inseparable by air flotation.

Sorghums fractionated included: (1) Martin, a red pigmented variety of hard texture, (2) TX 2536, a yellow endosperm variety and (3) DeKalb G 600, a commercial yellow endosperm variety, all grown under comparable conditions at Lubbock, Texas in 1967. Moisture content was 12–13%. Mean yield and recovery are shown in Table 5.25.

Some small particles of germ and pericarp appeared in the flour, though the bran fractions were comparatively free from contamination by other fractions. The protein (N × 6·25), ether extract, and ash, (AACC, 1962) in composites of the fractions are shown in Table 5.26.

The grits were light in colour, indicating complete bran removal; TX 2536 and G 600 grits were more acceptable in appearance than those from Martin. Of the total protein in the grain, 95·0% and 95·7% were recovered in the milled fractions of Martin and G 600 respectively. Of the lipids, 84·4% of the ether extract in Martin and 82·0% in G 600 were recovered.*

The above laboratory procedure described by Rooney and Sullins (1969) was applied by Maxson *et al.* (1971) to sorghum in two phases.

In the first phase, 11 varieties of sorghum grown at the Texas A&M University Research and Extension Center, Lubbock in 1967, representing a

* (Authors' note: no statement made on TX 2536).

range of kernel colour, hardness and endosperm type were used to study the effect of milling time and the physical properties of the grain on milling performance.

Texture was a subjective rating based on a visual estimation of the proportions of corneous and floury endosperm:

Rating	Proportions of endosperm	
	Corneous	Floury
1	1	0
2	0·75	0·25
3	0·5	0·5
4	0·25	0·75
5	0	1

The description of the varieties used in phase 1 is shown in Table 5.27.

The 1000-seed weight, density and hardness (% over a No. 12 US standard sieve after milling in a Strong-Scott barley pearler) protein (N × 6·25) and ether extract are shown in Table 5.28.

The mean milling data, protein (N × 6·25) and ether extract (AACC, 1962) of the fractions after a milling time of 2 min are shown in Table 5.29. The high yields of coarse and fine grits from the corneous varieties and B 398 (Martin) showed the superiority of these varieties for grit-milling. The low grit yields and the high yields of germ rich grits and fines obtained from the floury endosperm varieties indicated that the fractions were difficult to separate.

In the second phase, seven sorghum varieties of similar kernel size, shape and pigmentation, ranging from "all-corneous" to "all-floury" endosperm, were grown under comparable conditions at Lubbock, Texas in 1968. A description of the seven varieties is shown in Table 5.27.

Table 5.30 presents the kernel size index (Rooney and Sullins, 1970; a high KSI number depicts a low average kernel size), 1000-seed weight, macro test weight, density, hardness, protein (N × 6·25), ether extract, ash (AACC, 1962) and starch (enzymatic, Norris and Rooney, 1970a). The mean milling data and analyses of the fractions are shown in Table 5.31.

In both phases 1 and 2, endosperm texture (relative proportion of corneous to floury) was the property most highly correlated with milling yield. Density and hardness correlated inversely with milling yield. Endosperm texture was negatively correlated with hardness, test weight and density.

The seven varieties used in phase 2 were subjected to break milling (Brabender Quadrumat Sr and Jr mills). The flour yields, protein, ether extract, and starch contents are shown in Table 5.32.

No significant correlation between flour yield, or composition and physical properties of the grain was demonstrated but DeKalb C42Y and SC 283 were among the better varieties both for flour production and grit yields.

Rooney *et al.* (1972) milled six varieties of grain sorghum, five of which: SC 283, B 398, B 3197, TX 09 and NSA 740 were investigated by Maxson *et al.* (1971). The sixth: 70 LH 201, was an experimental hybrid with a white, thin pericarp, and a high proportion of corneous endosperm. Grain moisture content was 10% and the samples were not tempered before milling.

The 150 g samples were milled in a Satake grain testing mill (similar in principle to a Strong-Scott barley pearler), long enough to remove approximately 2% of the initial sample weight, then sifted. The overs of a No. 30 sieve were remilled to provide the next fraction, and the process repeated until more than 45% of the initial sample weight was removed. From most varieties, a total of 24 fractions was obtained.

NSA 740 and TX 09, both floury endosperm types, suffered kernel breakage and milling was discontinued after a little more than 25% of the kernel was removed. The protein (N × 6·25), lipid and ash (AACC, 1962) of blends from the remaining four varieties are presented in Table 5.33.

The highest protein fractions were obtained from SC 283 which had a thin pericarp and an all-corneous endosperm. The highest protein fraction from B 3197 (intermediate endosperm) appeared when about 20% of the kernel was removed: the maxima for B 398, 70 LH 201 and SC 283 occurred on removal of roughly 17·5% of the kernel (Fig. 5.I)

For the most part protein content of the whole grain influenced the protein content of the fractions. About 14% of a fraction containing 24% protein was however obtained from 70 LH 201 of which the initial protein was only 10·8%.

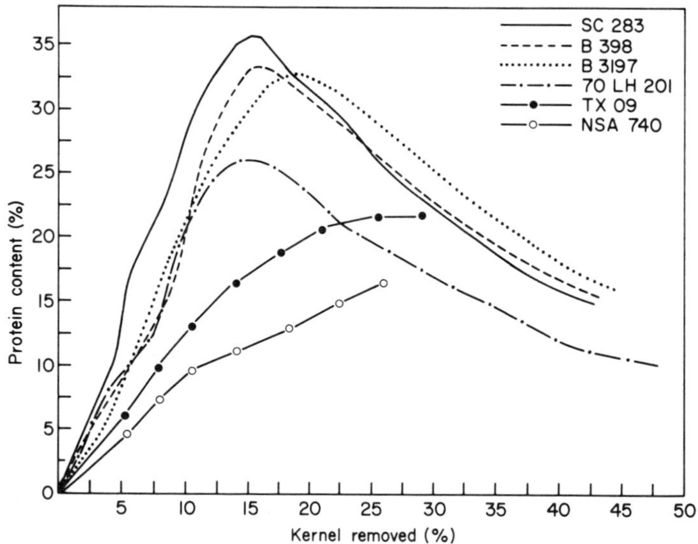

Fig. 5.I Protein contents of the fractions removed by abrasion milling of sorghum in relation to the amount of kernel removed (from Rooney *et al.*, 1972).

The white pericarp samples produced light, off-white flour fractions; the brownish-red to dull pink coloured flour fractions from B 398, of brownish-red pericarp, were considered unsatisfactory.

The amino acid contents (auto-analyser, Spackman *et al.*, 1958) of the fractions from SC 283 and 70 LH 201, shown in Tables 5.34 and 5.35 respectively, were closely similar. Lysine and threonine were highest in the initial fractions, and decreased progressively as fractions were derived from deeper within the kernel.

The sorghum kernel contains a substantial quantity of peripheral endosperm which has a higher protein content than the germ. The high protein fractions consisted predominantly of peripheral endosperm combined with some pericarp and germ. The first fractions collected were high in lysine and low in glutamic acid, probably because they contained a large proportion of germ and pericarp. The amino acid composition of the peripheral endosperm of sorghum was unknown but that of hard wheat is reportedly lower in lysine than the inner endosperm (Kent and Evers, 1969). Shoup *et al.* (1970c) reported that corneous endosperm was lower in lysine and methionine than floury endosperm.

Rooney *et al.* suggested that the integrated milling scheme proposed by Anderson and Burbridge (1971) could be used to obtain fractions by removing the bran in several steps. The removal of 18% of the kernel during debranning might permit an 8–10% yield of a fraction with a high protein content.

Laboratory Roller-milling

Shepherd and Woodhead (1969–70) in East Africa milled two corneous sorghums Konza and Msumbiji in a Bühler laboratory mill to produce bran, granular products and flour. The granular products yield was high, with protein content nearly equal to that of the original grain. Yields and analyses appear in Table 5.36.

Shepherd *et al.* (1970–71) reported the effect of milling by Bühler laboratory mill on a wide selection of grain sorghums. The Bühler mill produced four fractions: break flour and reduction flour for human consumption; and pollard (shorts) and bran for animal feed. The colour of seed coat, endosperm type (floury, corneous or intermediate), average seed weight (100 seeds weighed in triplicate), colour of flour, total flour yield, % of total protein (Kjeldahl N × 6·25) in flour, and yield ×% total protein for a selection of these sorghums are presented in Table 5.37.

There was a relation between endosperm type and kernel weight; the heavier kernels were floury and the lighter ones corneous. Flour colour was in general pink, or off-white with a yellowish tinge. Surprisingly only IS 928, with a greyish purple testa, gave a substantially white flour approaching that from Sagana a sorghum with a floury endosperm, a white pericarp and dark testa that gave the whitest flour produced by roller-milling. The protein content of the flour fractions was much lower than that of the original grain. Total flour yields (break and reduction) ranged from 36·6% from a corneous endosperm to 55·6% from a floury endosperm sorghum. Such yields are uneconomically low (Shepherd, 1974).

Anderson (1969b) reported the proximate analysis (AACC, 1962) of sorghum milled in a Bühler laboratory mill. The results appear in Table 5.38.

The purpose was to produce a flour with good yield, low lipid (1% or less) and ash (below 0·5%) without preliminary dehulling and degermination. Tempering to a moisture content of 19·5% moisture was found to be optimum. A single pass through the Bühler mill resulted in a flour extraction of 51–53% (flour I). To obtain a flour of higher extraction, up to 70%, remilling of the shorts I fraction three times in the Bühler mill was required. About 70% extraction was the limit if the fat content of the flour was to be 1% or less. The proximate analysis of flour I and final flour from the three sorghums is given in Table 5.39. A prepeeling procedure (Freeman and Watson, 1969) was applied to TE 77 and the results are shown in Table 5.40. There was an 8% loss of material from peeling, which Anderson (1969b) suggests was mainly liberated starch, with some bran and solubles.

On a laboratory Bühler mill, Jones and Beckwith (1970) milled three sorghum hybrids: (1) OK 612, a heteroyellow endosperm type, (2) RS 626, an early maturing RS 610 type, resistant to head smut, grown under dryland conditions and (3) TE 77, a common, fall-season hybrid giving a high yield under irrigation. All three were grown under similar conditions at Lubbock, Texas in 1967. The proximate analyses (AOAC, 1965; starch polarimetrically Earle and Milner, 1944) were so similar (Table 5.41) that most are reported as average values of the three sorghums.

Three fractions were obtained from the Bühler mill, (a) bran (containing most of the germ and the hull), (b) shorts (endosperm contaminated by bran) and (c) flour. The proximate analysis (average values of the three fractions) is shown in Table 5.42.

The amino acid composition (mg/g total N) of the whole grain and the mill fractions (auto-analyser Benson and Patterson, 1965; Cavins and Friedman, 1968), are shown in Table 5.43.

The amino acid contents of the flour and whole grain were similar. The bran contained twice as much lysine as the flour. The protein was isolated from the mill fractions by first removing the lipid with *n*-butyl alcohol (Jones and Dimler, 1962) and then successively extracting with water, 1% NaCl, and twice with 60% *t*-butyl alcohol. In some experiments 60% *t*-butyl alcohol at 60°C or 60% ethanol at 60°C replaced the 60% *t*-butyl alcohol at room temperature. The ethanol soluble protein when dialysed before freeze-drying, gave a precipitate that was difficult to redissolve.

The percentage of total nitrogen extracted by the solvents is shown in Table 5.44. The shorts and bran fractions of TE 77 were not extracted.

The amino acid composition (mg/g total N) of the water, NaCl, *t*-BuOH soluble and residue fractions of the flour, and the *t*-BuOH soluble fraction of the shorts and bran are shown in Table 5.45. The *t*-BuOH solubles from flour were relatively low in lysine, arginine, glycine, cystine, histidine and threonine and contained more leucine, glutamic acid, proline and alanine than the water and salt solubles.

The *t*-BuOH extracted more than 90% of the protein from the flour and shorts from the individual hybrids, as shown in Table 5.46.

The alcohol soluble fractions were deep red in colour, TE 77 having the most colour and OK 612 the least. When a *t*-butyl alcohol solution of the protein was passed over a LH-20 Sephadex column two pigments separated, one red and one yellow. The red, but not the yellow, could be removed by adding decolorizing carbon to a *t*-butyl alcohol solution of the protein. The decolorized product had the same amino acid composition and electrophoretic pattern as the original. Electrophoretic patterns (5% polyacrylamide gel) were similar for the three sorghums.

Shoup *et al.* (1969) studied the amino acid composition (auto-analyser, excluding tryptophan, following hydrolysis by procedure of Waggle *et al.*, 1966) and nutritional value of the protein (rat growth study) of Paymaster Kiowa sorghum tempered to 14% moisture content for 16 h then 20% moisture for 15 min immediately before milling with "typical dry-milling equipment". The bran and germ were obtained from corrugated break rolls and separated by aspiration and a gravity table. The grits were reduced by a series of rolls and fractionated by sieves. The yield of milled grits over a 20-mesh wire sieve was 68%. Grits were further milled and the contents of crude fat and crude protein (AOAC, 1960) of the fractions are shown in Table 5.47.

As milling proceeded, softer endosperm components were removed and the protein content of the remaining fraction increased. Percentage protein in the endosperm fraction ranged from 6·5% in a soft floury fraction to 20·6% in a hard corneous product. Colour followed a similar trend: low protein material was white, high protein was tan and speckled.

The amino acid composition (mg/g total N) in the grain, bran and germ, and some of the endosperm fractions, is shown in Table 5.48. All amino acids in the fractions increased as total protein increased but the relative distribution of the amino acid in the protein varied as the protein content of the fraction changed.

In the rat growth studies (female albino rats 21-days old, over 6 weeks) all protein in the diets was derived from the sorghum grain fractions shown below:

	Protein (%)
(A) Bran and germ mixture	10·3
(B) Whole grain	10·1
(C) Flour fraction 3	5·7
(D) Flour fraction 8	9·8
(E) Endosperm fractions 10 and 11	15·1
(F) Endosperm fractions 12, 13 and 14	18·1

The diets were calculated to be isonitrogenous at approximately 9·3% protein on 12% moisture basis, using starch as the diluent. Diet (C), because of the fraction used, was unavoidably lower at 5·6% protein. Rat growth data and PERs are shown in Table 5.49.

At the end of the fourth week significant differences were found among weight gains. Rats on diet (A) gained fastest with those on Diet (B) next. The floury endosperm diets (C) and (D), though producing little gain, were superior to the corneous endosperm diets (E) and (F). Diets (C) and (D) though differing in protein contents gave similar average gains. The PER data suggested that the protein of diet (C) had higher nutritive value than that of diet (D). There was no control diet so PER values are valid only for comparison within these trials. A positive relation existed between weight gain and percentage of lysine supplied by the diet.

At the end of four weeks, lysine was added to diets (B) to (F) to meet adjusted nutritional requirements (National Academy of Sciences—National Research Council, 1963b) as shown in Table 5.50. The weight gains and PERs increased significantly with added lysine. After supplementation the lysine contents and PERs of all diets were closely similar. Among the original, unsupplemented fractions, weight gains and PERs paralleled lysine contents.

Laboratory Peeling and Roller-milling

Shoup *et al.* (1970b) described how an experimental sorghum grain peeler designed at Kansas State University was used with a Miag Multomat experimental mill on two commercial hybrid sorghums, Paymaster Kiowa (67-2), grown in 1966, and Frontier 400C, a darker grain, grown in 1967.

The grain was tempered to between 11·5 and 22·5% moisture before peeling. The yield and proximate analysis (12% H_2O basis) (AACC, 1962) of the whole grain and the products: (1) fines (free bran, germ and endosperm grits), (2) bran and (3) peeled grain are shown in Table 5.51. The peeled grain contained 15·5% moisture which Shoup *et al.* (1970b) stated was close to optimum for germ removal and further milling, the fines 40% and the bran 30%: it was bland in flavour with an off-white colour. Frontier 400C (68-7), the darker of the two grains, had a mottled appearance. Microscopic examination indicated that the aleurone layer remained with the endosperm.

The yield, protein, crude fat, crude fibre and ash content of the products from the Miag Multomat for Paymaster Kiowa peeled and Frontier 400C peeled and nonpeeled are shown in Table 5.52. As revealed here, the higher the proportion of corneous endosperm, the higher the protein content.

Products of the whole grain were more highly pigmented (Agtron F 22 monochromatic reflectance spectrophotometer) than those from either of the peeled grains. Fractions from peeled Frontier 400C were more highly coloured than those from peeled Paymaster Kiowa. Colour intensity increased as the proportion of corneous endosperm increased.

The amino acid composition (mg/g total N) (auto-analyser, Waggle *et al.*, 1966) of the fractions, from Frontier 400C is shown in Table 5.53. As the proportion of floury endosperm decreased, lysine, arginine and glycine decreased; as the proportion of corneous endosperm increased, aspartic acid, serine, glutamic acid, proline, alanine, isoleucine, leucine, tyrosine and phenylalanine increased. Germ protein contained the best nutritional balance of amino acids. These findings agreed with those of Shoup *et al.* (1969).

In continuation of the work described above, Shoup *et al.* (1970c) compared two sorghum hybrids, Paymaster Kiowa and Frontier 400C of equal protein content, 9·5%, (12% moisture basis), when milled by conventional dry-milling equipment. The yield, protein, fat, fibre and ash (AOAC, 1960) of the fractions (1) bran and germ, (2) fines, (3) to (6) floury endosperm and (7) corneous endosperm are shown in Table 5.54. The protein and amino acid (mg/g total N) contents of fractions (3) to (7), and casein are shown in Table 5.55. The results were in general agreement with those of Shoup *et al.* (1969).

Female albino rats, 22-days old, were fed 10 diets over 4 weeks. Two fractions for each hybrid, fraction 3, a low-protein (approximately 5·6%) floury endosperm and fraction 7, a high-protein (approximately 10·2%) corneous endosperm, were fed unsupplemented and supplemented with lysine and methionine, and compared with casein controls at both protein levels. Starch was added to adjust protein content. The proximate analysis of the diets with the levels of amino acid supplementation are shown in Table 5.56. The weight gains, PERs and percentage of the NRC requirements (National Academy of Sciences–National Research Council, 1963b) are shown in Table 5.57.

Within a given hybrid, floury endosperm (5·5% protein) produced higher weight gain than corneous endosperm (9·9% protein) when both were unsupplemented which appeared to reflect differences in the relative proportions of lysine and sulphur amino acids present.

Weight gains and PERs improved markedly with supplementation, the largest gains occurring in the higher protein diets. No significant difference was found in PER between floury and corneous endosperms when both were supplemented to equivalent amino acid contents.

Semi-industrial Milling

Raymond *et al.* (1954) compared a white sorghum (probably mori), a yellow sorghum (probably kaura) both from Nigeria milled in the United Kingdom on a "Maxima" mill equipped with specially fluted rollers. Samples of both white and yellow flour and middlings and fine offal from the white sorghum were analysed for protein (N × 6·25), P (vanadate-molybdate method), phytin P (McCance *et al.*, 1936), thiamine (Accessory Food Committee 1943) and niacin (colorimetrically). The analyses with a laboratory milled Nigerian farafara sorghum flour are shown in Table 5.58.

Narayana Rao *et al.* (1958) in rat growth trials compared sorghum, unpolished and polished in an Engelberg rice huller. The protein, fibre, Ca and P (AOAC, 1950) and thiamine (Swaminathan, 1942a) contents are shown in Table 5.59. The "poor vegetarian" rat diets used were as described by Murthy *et al.* (1950). In the first series two groups of eight rats were fed *ad lib.* and in the second series, two groups were pair-fed. The results are shown below:

	Polished	Unpolished
Ad lib. feeding		
Average daily food intake (g)	8·8	9·2
Average daily weight gain (g)	5·6	8·2
Paired feeding		
Average daily food intake (g)	8·6	8·6
Average daily weight gain (g)	5·5	8·0

The growth rates from unpolished sorghum were significantly higher than for polished sorghum.

After 3 weeks on the growth trials N, Ca and P balances were determined. The results are shown in Table 5.60. N and P retention was significantly higher from the unpolished sorghum but the difference in Ca balance was not significant.

Shepherd *et al.* (1970–71) reported the protein (N × 6·25, 12% moisture basis) and amino acid composition (mg/g total N) (ion-exchange chromatography, Beckman analyser, following acid hydrolysis, excluding cystine, methionine and tryptophan) of sorghums designated Karachi, Lionja, Msumbiji and Konza before and after polishing in a CeCoCo rice polisher. The data are shown in Table 5.61. Konza was the least corneous of the four and showed the greatest reduction in protein content on polishing; the least reduction was shown by Karachi. Isoleucine, leucine, lysine, phenylalanine, threonine and valine were all reduced by polishing.

Shepherd (1974) reported polishing East African white sorghums with corneous endosperms, wherein breakage occurred mainly at the tube and cross cell layer. The outer pericarp tissue broke off in flakes leaving the aleurone attached to the endosperm and the germ intact.

Shepherd (1974) stated that East African brown-coloured bird-resistant sorghums were usually of a floury endosperm-type which possessed insufficient strength to withstand polishing. The friable tube and cross cell layers were found immediately outside the testa. Consequently on polishing, the pigmented testa layer remained with the endosperm.

White East African sorghums contain a corneous endosperm and a thin pericarp, making for good weathering resistance. All floury endosperm sorghums were brown-seeded; white sorghums with a floury endosperm had not been found, possibly Shepherd (1974) suggested, because of their poor resistance to weathering.* Sorghums with tight glumes which did not thresh clean were difficult to polish; the grains emerged in the same state as they entered the machine. Threshing breakage appeared between the head and the glume rather than between the glume and the seed.

* (Authors' note: this statement does not agree with the finding of Maxson *et al.*, 1971).

Kapasi-Kakama (1974) reported that polishing sorghum in a rice polisher reduced protein by about 7% when 75% of the bran was removed. He also reported (1977) the effect of: (a) endosperm texture (corneous, floury or mixed), (b) pericarp thickness (pearly grains had thin pericarps, chalky grains had thin pericarps), (c) pericarp colour (by visual inspection graded 1 (lightest)–4 (darkest)), (d) presence or absence of testa (found by scraping the outer skin of the kernel), (e) mean kernel weight and (f) weathering (by visual inspection), on the yield and attractiveness (visual) of a number of East African sorghums pearled in a CeCoCo model grain polisher.

He reported, in summary,

(1) corneous endosperms gave higher yields (84% or more) than floury (below 74%). The corneous were, visually, the more attractive,
(2) thin pericarp types gave the best results:

Pericarp type	No. in group	% giving yield of 80% or more of pearled product	% with most attractive score
Thin	22	82	59
Intermediate	9	56	44
Thick	14	50	36

(3) grains with a coloured pericarp gave lower yields than those with white or tanned pericarps. The gradings of the 40 sorghums tested were:

Colour class	No. in class	Yield of at least 70% product	% with most attractive score
1	2	100	100
2	21	67	29
3	15	87	27
4	2	0	0

(4) the presence or absence of testa did not affect yield; high and low yields were obtained from samples with testa depending on endosperm texture. All with testas scored lowest for attractiveness,
(5) the thin pericarps were higher in mean kernel weight (30 mg or more) than thick or intermediate pericarps (25 mg or less). Yields were generally higher from grains of highest mean kernel weight,

Mean kernel weight (g)	No. in category	% with yield of pearled product 80% or more and most attractive score
30 or over	16	75
25–30	5	60
less than 25	24	38

(6) weathering reduced product yield and product quality.

Viraktamath *et al.* (1971) removed bran from sorghum with:

(1) a rice huller (a horizontal ribbed rotor in an outer cylinder),
(2) a Gota machine (a horizontal truncated cone rotating in an outer cylindrical frame),
(3) a vertical inverted truncated cone rice polisher.

Machines 1 and 2 did not produce a clean separation. Further trials were conducted with 3, the cone polisher. Passage through two cone polishers removed 7–8% from moistened sorghum. The products and by-products collected as separate fractions were defined as polished sorghum, brokens, bran and aspirated husk or glumes. The proximate analysis of the fractions (AOAC, 1950) is shown in Table 5.62.

Polishing reduced the original fibre content by more than 50%, protein by about 10% and the Ca and P by 20–30%. The germ was largely retained, consequently the fat content was not much reduced.

The polished sorghum was milled in a wheat roller-mill and soji (semolina), maida (fine flour), atta and bran fractions obtained. The percentage recovery, with the protein and fat content, DWB, are shown in Table 5.63.

The pearled sorghum and the broken fractions, ground together and made into bhakri or roti (dry pancakes) were whiter in colour, softer to the palate and more acceptable than those from whole meal sorghum. A blend of 20 pearled sorghum with 80 wheat flour produced acceptable chapatis.

Pilon *et al.* (1977) used a pilot scale hard wheat semolina mill to prepare semolina from two semi-corneous African sorghums, CE 90 from Bambey, Senegal and L 30 from Maradi, Niger. The process resulted in 15% of broken kernels. The semolina fractions obtained represented about 60% of the whole grain.

Industrial Dry-milling

Industrial scale dry-milling of sorghum began in the USA about 1949 (Hahn, 1969, 1970). Little information was available from the six sorghum mills operating in 1969 and most of the published data were from laboratory mills. The commercial objective was a clean separation of endosperm, germ and bran, with the endosperm in the form of clean grits comparatively free from flour (Hahn, 1969). 'Bran' included the outer pericarp, testa, aleurone and nucellar layers.

Hahn (1969) gave the composition of roller-milled sorghum flour as:

	Break flour	70% extraction	90% extraction
Protein (%)	4·5	8·0	8·0
Oil (%)	1·5	2·0	2·8
Fibre (%)	1·0	0·8	1·2
Ash (%)	0·7	0·5	1·0

Break flour represented 10–15% of yield, contained most of the floury endosperm, was low in protein and very "specky". The 90% extraction flour was also "specky" and high in lipid. The 70% extraction had a better appearance. Hahn (1969) stated that sorghum flours were more difficult to grind than whcat but the outer coats could be removed with a barley pearler, the grain being first cleaned and sized. Also, the protein, oil, fibre and ash contents of the pearled grain decreased with increase in bran removed as shown in Table 5.64. After 39% of the kernel weight was removed, the oil and fibre contents of endosperm were roughly as found by Hubbard *et al.* (1950) using hand separation. The protein content was reduced by pearling, supporting the finding of Normand *et al.* (1965) that a higher protein content is found in the outer layers of the grain and suggesting that in grain sorghum both the floury and corneous endosperm are lower in protein than the whole grain when the endosperm is completely free of the "dense peripheral" and aleurone layers.

Hahn (1969) also used specific gravity separators to separate germ and endosperm citing Wang *et al.* (1959) who referred to an unpublished method of Barham by which grain was separated into five fractions:

(1) bran (outermost seed coat layers)
(2) bran fines (remaining seed coat layers)
(3) germ (embryo)
(4) grits (endosperm, and mill fines, i.e. endosperm mixed with germ and bran fragments)

Wang *et al.* (1959) did not give a proximate analysis for the fractions, but Hahn (1969) quoted typical grain sorghum dry-milling results which appear to be related to the method. The values are shown in Table 5.65; the analytical methods and source of the samples were not stated.

The germ was more difficult to separate from waxy sorghum than from nonwaxy using specific gravity separators, and waxy sorghum grits had a slightly higher oil content. Germ isolated by this technique contained about 20% oil, roughly equivalent to 70% purity as judged by hand separation.

Crawford *et al.* (1942) reported the average proximate and mineral composition of commercially roller-milled South African sorghum and its milling products. The analyses are shown in Table 5.66. Although the straight-run product might be expected to have the same composition as the original grain, this was not the case and Crawford *et al.* suggested that part of the bran and germ must have been removed. No explanation was offered for the higher content of Fe in the straight-run meal compared to the whole grain. Household meal, usually after being refined and malted, is a common breakfast food.

Goldberg *et al.* (1946) reported from South Africa the thiamine content (Goldberg *et al.*, 1945) of sorghum obtained from commercial mills (the exact milling method is not stated but probably roller mills) and prepared traditionally by stamping or grinding and winnowing. The mean results from the commercial mills are shown in Table 5.67. The coarse-ground meal and bran were not sold for human food, and the straight-run meal was used principally for the manufacture of sorghum beer.

The mean thiamine content (mg/100 g) (no moisture figure was given) from traditionally milled samples was:

	Thiamine
Original whole grain	0·37
First grinding or stamping and sifting	0·40
Husks from first grinding	0·47
Second grinding or stamping and sifting	0·29
Husks from second sifting	0·38

Millets

Traditional Milling

Kadkol *et al.* (1954b) reported on the composition of seeds of kodo (haraka) millet obtained from the Millet Specialist to the Government of Madras, Coimbatore and grown at Mysore. The grain was moistened and dehusked by hand pounding and winnowing. Both whole and dehusked grain were powdered to pass through a 60-mesh sieve.

The proximate analysis, minerals (AOAC, 1950) and thiamine (Swaminathan, 1942a), in comparison with wheat, are shown in Table 5.68. The dehusked grain was significantly inferior to wheat when assayed in rat growth tests (see p. 203).

Gast and Adrian (1965), and Adrian *et al.* (1967) discussed the effect of traditional and mechanical milling on the nutritive value of pearl millet. This work was further developed by Adrian *et al.* (1975) who reported the effect on the nutritional value of pearl millet, as assayed by rat feeding trials, of three milling methods:

(1) traditional pestle and mortar pounding
(2) mechanical commercial milling by SOTRAMIL (Société de transformation du mil, Zinder, Niger) where millet flour has been produced commercially for some time
(3) mechanical milling by SEPIAL (Société d'études et d'exploitation de procédés pour l'industrie alimentaire, Clichy) stated to be applicable to all cereals in which the endosperm (albumen in French) is not floury, such as durum wheat and pearl millet.

Rat growth trials were conducted over 28 days with male Wistar rats mean weight 45 g. The nonisonitrogenous rations consisted of about 92% of the cereal product with 5% groundnut oil, and 3% mineral and vitamin mix.

Pestle and mortar grinding (Adrian *et al.*, 1967) produced an edible fraction representing 85% of the whole grain, consisting of 39% coarse semolina, 16% fine semolina and 30% flour. The protein (N × 5·83) and lysine content of (A) whole grain, (B) the semolinas and (C) the flour, with rat weight gains, PERs and liver protein content are shown in Table 5.69. The protein and lysine (% protein) were higher in the flour and lower in the semolinas than in the original grain. Consequently the PERs were similar for ration (C), flour and whole grain (A), but lower for the semolinas (B). Liver protein was lowest from ration (B) both as weight/organ and as mg/g liver.

The SOTRAMIL uses decorticating (abrasive) rolls followed by a hammer mill (Ultrafine) to give a flour of 65–75% extraction rate, a mean granularity (particle size) of 40 μm. (Adrian compares this with roller-milled white flour with a mean granularity of 125 μm). The composition and results of rat growth trials for (D) the whole grain and (E) 75% extraction flour are shown in Table 5.69.

The protein content of the flour was lower than that of the grain, but lysine (% protein) only slightly lower after milling. The PER of diet (E), flour, was slightly lower than that of (D), whole grain, but the growth rate from diet (E), is lower by 37% than that of diet (D) and liver protein was also reduced by 18%.

The SEPIAL process makes use of "peeling-decortication". Adrian *et al.* (1975) point out that the terms as used in the SEPIAL patent do not carry quite the same meaning as when they are normally used. In the SEPIAL procedure, conditioned grains are "peeled" in an apparatus with a vertical axis with paddles the action of which is sufficient to detach the moistened pericarp leaving the remaining layers intact. The pericarp is then separated from the grains by brushes or aspiration. The "decorticating" is carried out by a vigorous rubbing in a brushing machine. Because of the preceding "peeling", this brushing is sufficient to remove the remaining outer coats. This procedure gives a "protein-rich fraction" and "decorticated grain". The latter, representing about 84% of the whole grain, may be sold as such, or reduced to semolina and flour in a hammer mill or roller mill. The products of the SEPIAL process, as % of the whole grain were:

	(%)
Whole grain	100
"Peeled" grain	92·5
"Decorticated" grain	84·0
"Protein-rich fraction"	8·5

Under the microscope the "protein-rich fraction" was seen to consist of the aleurone and adjacent layers with a fair proportion of the germ and scutellum. It appeared high in lipid content. The microscopic observations were supported by chemical analysis shown in Table 5.70.

In chemical composition, the peeled grain differed little from the whole grain (except for a reduction in lipid content). "Decorticated" grain contained less fibre, ash, vitamins and lysine than the whole grain (Table 5.70). The results of the rat growth trials for whole grain (F) and the fractions (G) "peeled" grain, (H) "decorticated" grain and (I) "protein-rich fraction" are presented in Table 5.69. The diets were essentially isonitrogeneous (about 8·5%), the cereals in diets (F), (G) and (H) representing 92% of the ration, the "protein-rich fraction" representing 61·5% of the diet in ration (I).

Taking the PER of the whole grain diet (F) as 100, the PER of diet (G) was 95, of (H) 85 and of (I) 139. Liver protein paralleled the PER values.

A digestibility trial was carried out over 7 days with adult male rats of mean weight 100–120 mg. The whole grain (J) and the products (K) "decorticated" grain and (L) "protein-rich fraction" were identical to those employed in the growth trials. Total digestibility of the dry matter and of the protein of the ration was estimated using the formula:

$$\text{Total digestibility} = \frac{(\text{intake—faecal output})}{\text{intake}} \times 100$$

The results are presented in Table 5.71.

The cereal in diets (J) and (K), 65% and 92% respectively, provided roughly equivalent levels of nondigestible matter, 4·8 and 5·0% respectively of cellulose plus lignin (formic acid insoluble) and pentosans. Total digestibility and protein digestibility were comparable. Total digestibility and protein digestibility of diet (L) in which the cereal fraction was 45% of the diet and nondigestible matter 8·1%, were lower than for the whole grain (J), by 8% and 13% respectively.

The differences in digestibility found among diets (J), (K), and (L) were discussed in relation to the nondigestible content of the three diets and summarized in Table 5.72. Adrian *et al.* (1975) referred to the work of Goussault *et al.* (1972) who reported the finding that the presence of bran reduced the total digestibility of pearl millet, and themselves suggested that the nondigestible components of diet (L) affected total digestibility and protein digestibility quite differently than those of diets (J) and (K). The SEPIAL procedure produced an edible fraction representing 85% of the whole grain with a nutritive value nearly equal to that of the whole grain and of better digestibility. There was also a by-product, rich in protein, but high in fat and of lower digestibility. Adrian expressed the opinion that industrial milling procedures can produce pearl millet fractions comparable in protein quality to those from traditional pestle and mortar pounding.

Traditional vs Mechanical Milling

De Wit and Schweigart (1970) described how a rice mill was used to remove the bran by abrasion from cleaned pearl millet followed by reduction in a hammer or pin-disc mill.

In trials at the State Hospital at Runtu, between 80 and 85% was the highest extraction rate found acceptable. The effect of extraction rate on proximate analysis, Ca, Fe, thiamine, niacin and riboflavin (analytical methods not stated) in comparison with a home prepared (pestle and mortar) meal, and home prepared porridge is shown in Table 5.73.* In general, the proportion of protein and other nutrients decreased with lower extraction rates. The home prepared meal and porridge would appear to have been taken from different grain samples than those mechanically milled. Though it is not certain the results suggest the home prepared meal and porridge were of a comparatively high extraction rate.

Thirumala Rao (1969) reported the oil content of foxtail millet grown in Anantapur (Andhra Pradesh, India) (1) traditionally hand-pounded and (2) milled in a rice huller and polisher. The yields of husk, bran and kernel from the two methods are shown below:

	Hand-pounding (%)	Milled (%)
Bran and husk	28	28
Kernel	68	67
Losses	4	5

The protein, oil and ash contents (analytical methods not stated) of fractions of bran passing through a 40-mesh sieve, retained by passing through 60-mesh and retained by 80-mesh, when pooled, are shown in Table 5.74.

Nicol and Phillips (1978) reported that in feeding trials in the metabolic compound of the Federal Nigerian Nutrition Unit laboratories with six young Nigerian men, no significant difference in N balance was found between diets containing (A) hammer-milled whole meal sorghum or (B) flour prepared in the home by traditional pounding and winnowing though differences around mean values were high. The difference in mean BV was small, (A) 0·74 and (B) 0·78, but there was a considerable difference in NPU, (A) 0·58 and (B) 0·69, attributable to the lower true digestibility of (A) 0·79, compared to (B) 0·89. The dietary crude fibre intake (g/day) was (A) 12·5 and (B) 11·6. Nicol and Phillips stated that while the fibre derived from the milled wholemeal sorghum (% total fibre) was only 6% higher than that derived from the home-pounded, winnowed, sorghum, this difference had a considerable effect on the TD and NPU of the two diets.

Laboratory Milling

Kurien *et al.* (1959) separated the husk and the endosperm of four strains of finger millet: Aruna, ES 11, K 11 and H 22 by a wet-processing technique. Washed grain was soaked for 24 h, then crushed in a stone grinder. The ground material was suspended in the soak water and filtered through a 100-mesh sieve. The residue remaining on the sieve was blended twice, with water, for 5 min in a Waring Blendor and the blend passed through a 100-mesh sieve. The residual husk remaining on the sieve was almost free from

* (Authors' note: no details were given of how these last were prepared nor how they were related to the mechanically milled samples.)

starchy endosperm; it was dried in an air oven at 90°C and powdered. The filtrate containing the starchy endosperm was allowed to settle and the clear supernatant decanted. The sediment (endosperm) was dried in a current of air at 50–60°C. The whole grain flour, husk, endosperm and the dried solids from the supernatant were analysed for protein ($N \times 6{\cdot}25$), Ca and P (AOAC, 1950). The mean and range values are shown in Table 5.75. A large proportion of the total N (28%) and Ca (49%) but only a small proportion of the P (14%) was found in the husk. Kurien *et al.* stated that refined finger millet flour, prepared by wet-processing was widely used in feeding weaned infants in South India, though it seems a poor source of protein, Ca and P, a large quantity of the nutrients being discarded in the water.

As gastric juice contains HCl, the effect was studied of pre-soaking the whole grain of ES 11 in 0·05 N HCl solution. This procedure did not release the nutrients present in the husk.

Kurien and Desikachar (1962) found that moistening finger millet with 3–7% water for about 2 h, followed by grinding in a Wiley mill, produced fractions high in husk content.

A technique of wet-processing was evolved by which the grain, original moisture content 11%, was steamed with 5% extra water for various periods. After cooling to room temperature, the grain was milled (Wiley mill) using first a 2 mm screen. The ground product was pressed through a 60-mesh sieve and designated fraction I. The residue was again Wiley milled using a 1 mm screen and after passing through a 60-mesh sieve was designated fraction II. A steaming period of 2 min was found to give maximum flour yield, 35% of fraction I and 34% of fraction II. The residue (+60 from the second milling) was highly coloured and contained most of the husk. Further milling converted the husk to powder. The residue was therefore soaked in water for 2–3 h, homogenized in a Waring Blendor and sieved through a 60-mesh sieve. The process was repeated on the residue retained on the sieve. The combined slurry was centrifuged and the starchy sediment dried. The protein ($N \times 6{\cdot}25$), Ca, P, thiamine and crude fibre contents (analytical methods not stated) of fractions I and II, singly and combined and of the wet-processed residue, are shown in Table 5.76.

Whole finger millet soaked in water for 24 h was wet-processed as described for the treatment of the residue and the effect on flour yield and recovery of protein, Ca, P and thiamine from the three processes, (1) wet-processing of whole grain, (2) dry-milling and (3) dry-milling followed by wet-processing is shown in Table 5.77. Wet-processing gave a higher yield of flour but recovery of nutrients was lower compared to dry-milling because of the loss of water solubles.

The dry-milling process, after heat conditioning of the grain, was used to make malted finger millet flour.

Refined finger millet flour prepared by the method described by Kurien and Desikachar (1962) was compared with whole finger millet flour in feeding trials with 36 boys in an orphanage and reported by Kurien and Doraiswamy (1967). The chemical composition (AOAC, 1950) of the two finger millet flours is given in Table 5.78. The flours, fed as dumplings and served with cooked rice at lunch and dinner, provided about 50% of the cereal content in a diet similar to that eaten by children of low income groups in South India. At the end of 5·5 months, no significant differences between the two groups were observed in height, weight, nutritional status; red blood cell and haemoglobin content of the blood (methods of Subrahmanyan *et al.*, 1961).

After 4 months, the metabolism of N, Ca and P (methods of Murthy *et al.*, 1954) in eight pairs of children on the same diet was studied over a 7-day period. No significant difference in N, Ca and P retention between the two groups was found. There was, however, a significant increase in the apparent digestibility of the proteins of the refined flour diet compared to the whole grain flour. The results appear in Table 5.79.

Kurien and Desikachar (1966) treated a market sample of finger millet with various levels of added moisture before milling. The compared samples included: (1) untreated, (2) 3% water for 10 min, (3) 5% water for 30 min, (4) 10% water for 45 min, (5) 5% water, then steamed for 2 min, (6) steamed for 2 min and (7) dried to 3% moisture, then 5% added water for 15 min. The fractions of the three break rolls, the three reduction rolls, and the residues (shorts and husks) were collected separately from the Bühler laboratory mill.

Treatment 3 gave maximum yield of flour (break and reduction rolls) with minimum colour (as measured in a photovolt reflectance meter using tristimulus green filter). Treatment 2, moisture below 5%, gave a coloured flour with high fibre content. Treatment 4, 10% moisture and steaming and treatments 5 and 6, steaming, reduced flour yield; treatment 7 gave a high yield (74%) of flour, containing a small amount of pulverized husk. The flour yield and chemical composition (AOAC, 1955) of the flour from treatments 1–5, and of whole finger millet grain are shown in Table 5.80. The refined flour from treatment 3 was rich in Ca but the protein content ($N \times 6{\cdot}25$) was only 4% compared to 8·8% in the whole grain. The P content of the flour was considerably lower than that of the original grain, indicating an uneven distribution of Ca and P between the endosperm and the outer layers of the grain.

The chemical composition of all the milling fractions, including shorts and husk, of treatment 3 is shown in Table 5.81. The protein, Ca and P contents were lowest in the first two break fractions and highest from the third reduction roll. Shorts and husk fractions were highest in protein, Ca and P. The shorts and husk fractions were mixed, soaked in three parts of water for 1 h, agitated in a Waring Blendor for 4–5 min at 2500–3000 rev/min and filtered through a 70-mesh sieve. The slurry passing through the sieve was centrifuged in a Westphalia clarifier and the residue air dried. In a parallel experiment the slurry was acidified to pH 4 before centrifuging to precipitate the proteins. The chemical composition of the edible fractions resulting from these treatments is shown in Table 5.82.

The plain (nonacidified) fraction was higher than the acidified in Ca (204 vs 115 mg/100 g) and P (779 vs 199). The protein contents were, respectively, 10·6 and 11·8%. Kurien and Desikachar (1966) commented that wet-processing and drying of the husk and shorts increased the nutritive value of the milled flour but the cost of treatment might well be uneconomical.

Kurien (1967b) reported the proximate analysis, Ca and P (AOAC, 1955) and feeding value for rats of a market sample of finger millet:

(A) whole meal flour from a laboratory Wiley hammer-mill.
(B) refined flour from a Wiley mill following treatment with 5% H_2O for 15 min, steaming for 2 min, combining fractions from (i) a 2 mm screen and (ii) a 1 mm screen that passed through a 60-mesh sieve.
(C) composite flour: refined flour combined with processed residue from the refining process. The residual husk was soaked in water for 2 h, agitated in a blender and then sieved through a 60-mesh vibrating sieve. The slurry was centrifuged and the starch fraction dried at 90°C, powdered and mixed with the refined flour. The analytical values are shown in Table 5.83.

The rat growth trials were conducted with weanling albino rats over 8 weeks using poor vegetarian diets in which the flours (A), (B) or (C) constituted 78·5% of the ration. The average daily food intake, initial body weight and weekly gain are shown below:

	A Whole meal	B Refined flour	C Composite flour
Average daily food intake (g)	10·2	6·6	7·3
Average initial body weight (g)	49·6	49·5	49·4
Average weekly weight gain (g)	12·1	3·7	6·0

The composite flour diet (C) produced significantly higher weight gain than refined flour diet (B) but significantly lower gains than whole meal diet (A).

The availability of Ca from flours (A), (B) and (C) was studied by feeding weanling albino rats a synthetic diet containing 100 mg Ca/kg of diet derived solely from the flours. Calcium-free casein was used to bring the protein content of the diets up to 20%. The findings are presented in Table 5.84. Average intakes of Ca were similar from all three diets but average retention from diet (A) was 47% compared to 68% and 74% from (B) and (C) respectively. Calcium in the refined and composite flours was derived solely from the endosperm and may be more readily available than Ca from the husk.

Rice Cone Polisher

Desikachar (1977) reported that pearl millet could be polished in a vertical cone polisher normally used for rice milling following the techniques described for sorghum (Viraktamath *et al.*, 1971) but with a finer mesh wire screen. The chemical composition (analytical methods not reported) of the products are shown in Table 5.85.

Sorghum and Millets

Traditional vs Mechanical Milling

Reichert and Youngs (1977) reported the proportion of protein (Hewlett Packard Model 185B CHN analyser, % N × 6·25), oil (Troeng, 1955), crude fibre (Stringham *et al.*, 1974), and ash (AOAC, 1960, method 13.006) in samples of red-brown sorghum and pearl millet received from Maiduguri, Nigeria.

The following processes of transformation were compared:

(1) traditional dehulling using a wooden pestle in a metal beaker, 20% by weight of water added before pounding in the mortar, air-drying and winnowing to remove the bran and other fines. The process was repeated twice to remove approximately (a) 10, (b) 25 and (c) 45% kernel.
(2) mechanical dehulling in a Strong-Scott barley pearler.
(3) mechanical dehulling with a Palyi compact milling system (attrition type).
(4) mechanical dehulling with a Hill grain thresher (abrasion type).

The comparative dehulling efficiencies of the last three were discussed by Reichert and Youngs (1976) who stated that (2) and (4) were superior to (3).

Reichert and Youngs (1977) reported the proximate constituents in sorghum after manual dissection. The kernels were soaked in distilled water for 2 h and all pericarp layers and the germ removed. The millet kernel could not be dissected even after prolonged soaking. The values, DWB, are shown in Table 5.86.

The effect of the four types of milling on the protein, oil and ash contents of sorghum and pearl millet in relation to the amount of kernel removed is illustrated in Fig. 5.II. The three mechanical dehullers all reduced the oil and ash contents to a marked degree and the protein to a lesser degree, but among the dehullers there were only small differences in losses of protein, ash and oil when the grain was dehulled to the same extraction level. Sorghum and millet dehulled by traditional methods contained higher oil and ash contents than the mechanically dehulled grains (the germ is firmly embedded in sorghum endosperm and is not readily removed by pounding). Traditionally and mechanically dehulled grains did not differ as markedly in protein content as in oil and ash contents. At 75% extraction the crude

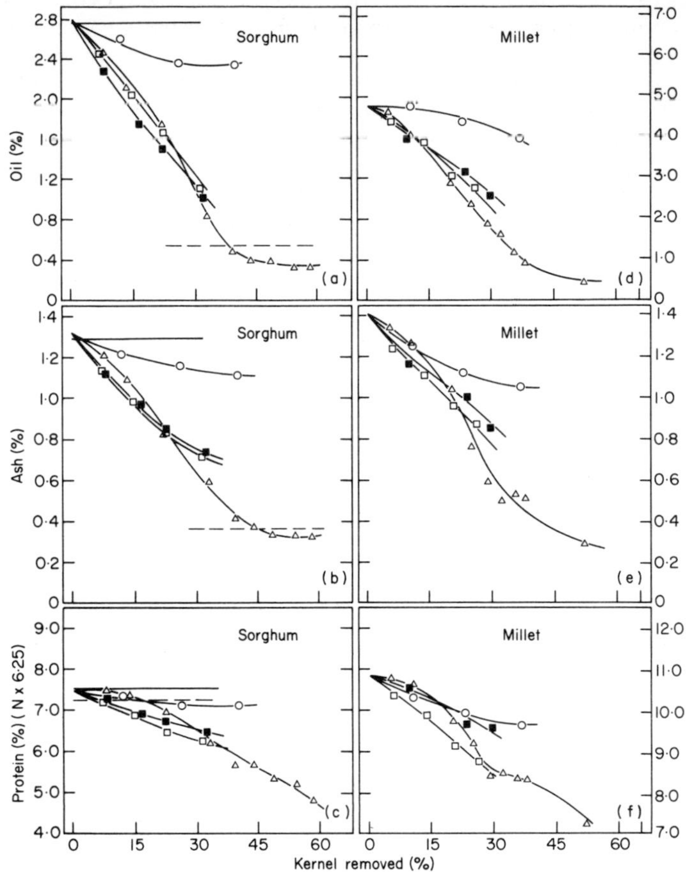

Fig. 5.II Effect of dehulling method on protein, oil and ash contents of sorghum and pearl millet in relation to the amount of kernel removed.
(□) Hill grain thresher, (△) laboratory pearler, (■) Palyi compact mill, (○) traditional mortar and pestle method, horizontal lines indicate composition of sorghum endosperm (- - - -) and sorghum endosperm plus germ (——) (from Reichert and Youngs, 1977).

fibre values were: 0·8, 0·9, 1·2 and 1·2% for sorghum, and 0·6, 0·7, 1·0 and 1·1% for millet in dehulling with the barley pearler, abrasion-type mill, attrition-type mill, and by the traditional method, respectively. The whole sorghum and millet had crude fibre contents of 2·4 and 2·0% respectively. Table 5.87 compares the proximate composition of sorghum and millet dehulled with the Hill grain thresher and by the traditional mortar and pestle method.

Sumner *et al.* (1974b) compared samples of the flour of red sorghum, white sorghum and pearl millet originating from Nigeria. The samples were:

(1) millet, dehulled mechanically (Hill thresher) and ground in a hammer mill at the Prairie Research Laboratory, Canada.
(2) millet, traditionally milled by soaked process. Millet is wetted, pounded in a mortar for about 25 min and the bran winnowed off. Dehulled millet is soaked overnight in sour milk or water with tamarind beans, then washed in water for 3 min, twice, dried on a grass mat for 30 min, ground. (Sumner *et al.*, 1974a).
(3) millet from the same source traditionally milled by the nonsoaked process. Millet is wetted, pounded in a mortar for about 25 min, the bran winnowed off, pounded again lightly for 5 min and winnowed, sun-dried in a single layer on a grass mat for about 15 min, then ground. (Sumner *et al.*, 1974a).
(4) red sorghum mechanically milled in Nigeria (Palyi mill).
(5) red sorghum from the same source traditionally milled. Sorghum is wetted, pounded in a mortar for 20 min and winnowed, washed in water three times for about 5 min, dried on a mat for 30 min and ground. (Sumner *et al.*, 1974a).
(6) white sorghum mechanically milled as (1).
(7) white sorghum mechanically milled in Nigeria (Palyi mill).
(8) white sorghum from the same source as 7 traditionally milled as 6.

The moisture and protein (AOAC, 1965), fat, crude fibre and ash (AACC, 1962), and N-free extract (by difference) of the samples are shown in Table 5.88.

The nitrogen solubility index (American Oil Chemists Society Ba 11-65, 1969, with modifications) was measured of samples of traditionally milled and mechanically (Palyi) milled samples of red and white sorghum flour and millet flour received from Nigeria on a different occasion, and compared with those for wheat flour and a cowpea flour. The latter two showed solubility curves of the normal shape for cereal and legume flours, that is, high solubilities at extreme pHs with pronounced dip as the isoelectric point was approached. The sorghum and millet flours showed nitrogen solubilities which were relatively low at all pH levels and less sharp isoelectric points. The white sorghum may contain at least two major protein types, each with its own isoelectric point.

Mechanical Milling

Reichert and Youngs (1976) compared the Palyi compact mill (attrition-type milling) with two abrasion units, a laboratory scale Strong-Scott barley pearler and a Hill grain thresher, the last modified to provide a pearling action. Efficiency was measured by flour colour (Hitachi Perkin-Elmer spectrophotometer with a diffuse reflectance attachment) vs material removed, kernel cracking (sifting through a Ro-Tap shaker) and throughput.

The Palyi compact mill as described by deMan *et al.* (1973) was modified to include an adjustable cover plate to control the exit of grains from the cylindrical head. In the barley pearler the metal screen surrounding the single carborundum stone was covered with a thin piece of rubber to prevent small seeds passing through. The abrasion mill contained 13 carborundum stones of 12 in diameter spaced equally along a vertical shaft (driver), driven at speeds up to approximately 2000 rev/min. The abrasion discs and shaft were enclosed in a rubber-lined case. The grain was fed by gravity.

In the experiments described by Reichert and Youngs (1976) the mill was operated on a batch basis, 18 kg of grain being loaded into the machine and the discs rotated at 1050 rev/min for 1 min for sorghum and 2 min for millet. The grain was then released and passed through the air separator on the Palyi attrition mill to remove the fines. The dehulled grain was sampled and reintroduced into the abrasion mill; this procedure was repeated four times to produce cumulative extraction levels of approximately 70%. Differences in flour reflectance between whole and dehulled sorghums were determined at wavelengths between 350 and 750 nm.

Differences were greatest close to 450 nm which was chosen as the wavelength of comparison for both sorghum and millets.

Flour colour (reflectance) was plotted against % of kernel removed. The laboratory pearler and the Hill abrasion mill were similar in removing most colour with least loss of material. The attrition mill was, in this respect, least efficient. To put the results into perspective, the three methods were compared with an acceptable sample produced in Nigeria by pestle and mortar. The percentage of sorghum kernel to be removed by each milling method to produce reflectance at 450 nm equivalent to the Nigerian sample is shown below:

	% removed
pearler	21
abrasion	23
attrition	44

The percentage of pearl millet kernel to be removed by each decortication method to achieve equivalent reflectance were closely similar to those quoted above for sorghum.

Grain hardness measured by the pearling test of Maxson *et al.* (1971) showed that at all retention times in the pearler, more material was removed as fines from sorghum than millet indicating that millet was harder than sorghum and less likely to crack under the pressure of carborundum stones or attrition plates. The Palyi attrition mill cracked more of the sorghum kernels at all extraction levels than either of the two other machines. More cracking led to material removal from all parts of the kernel rather than from the peripheral areas only and hence to less efficient dehulling. In removing 25% of the kernel the following approximate percentages of kernels were cracked:

	(%)
pearler	23
Hill grain thresher	36
Palyi compact mill	76

If 25% of the kernel were removed, the throughputs to be expected would be:

	Sorghum kg/h	Millet kg/h
Palyi compact mill (attrition)	90	27
Hill grain thresher (abrasion)	328	162

Reichert and Youngs (1976) concluded that on the basis of relative efficiencies in terms of colour removal, kernel cracking and throughput, the Hill grain thresher (abrasion mill) was to be preferred to the attrition mill. The abrasion mill also had advantages in relative size, maintenance requirements and relative simplicity for a "village scale" milling operation.

Reichert *et al.* (1976) reported that after field trials conducted at Maiduguri, Nigeria using white sorghum and white and yellow maize, the grain thresher (abrasion mill) was found to operate most efficiently when the grain was passed through the mill twice with the throughput and grain level set for optimum production runs. After discharge from the thresher the fines were removed from the seeds by aspiration. At a speed of 890 rev/min the extraction rate was 77%.

Chibber *et al.* (1978) at Purdue described a study done in cooperation with scientists at the Prairie Regional Laboratories in Saskatoon. A high-tannin (BR 64) and low-tannin (RS 626) sorghum grown at Purdue were passed several times through a Hill thresher abrasion mill (Reichert and Youngs, 1976). With each passage about 12% by weight of the grain was removed.

Chibber *et al.* analysed for protein (micro-Kjeldahl N × 6·25), amino acid composition (Spackman *et al.*, 1958) and catechin equivalent (Burns, 1971). The results appear in Tables 5.89 and 5.90. The protein of the various products was fractionated by the method of Landry and Moureaux (1970) and these results appear in Tables 5.91 and 5.92.

During two dehullings: about 24% by weight of each grain was removed; from BR 64 roughly 74% of the tannin was removed, accompanied by a loss of 1·2% in protein content. The tannin content of RS 626 (low-tannin) fell by 58% and the protein fell from 12·8 to 11·9%. After the first dehulling, the lysine (% of protein) increased slightly but fell with subsequent treatments. Chibber *et al.* suggested that the Hill thresher caused both chipping and shearing of the kernel and removed not only the outer bran layers but also some of the endosperm and embryo. To this they attributed the decrease in lysine.*

Chibber *et al.* (1978) commented upon the marked differences between the high- and low-tannin sorghums in the soluble protein fractions extracted. With successive dehulling the sum of fractions I, II and III from the high tannin sample increased from 19% (whole grain) to 49% (after 3 dehullings); in contrast the sum of fractions I, II and III from the low tannin sample remained relatively constant (approximately 50% of soluble N) from the whole grain and through three dehullings. From the tabulated data Chibber *et al.* concluded:

(1) the albumin–globulin fraction is largely associated with the external portion (derived from the embryo);
(2) the glutelins are uniformly distributed;
(3) the kafirins (predominantly protein bodies) are more concentrated in the seed's interior; and,
(4) after three dehullings the residual tannin contents and the solubility patterns were similar.

"These observations suggest that tannins in sorghum associate predominantly with kafirins, particularly cross-linked kafirins (fraction III)...the kafirin–tannin complexes behave as glutelin proteins with respect to their solubility characteristics." "An excellent negative correlation between these two protein constituents (kafirin and glutelin) in BR 64 strongly supports our contention that tannin–kafirin complexes behave as glutelins according to solubility characteristics."†

Perten (1977) reported that soft wheat (from France), pearl millet (from Senegal) and sorghum (from Senegal) gave different results when tempered to 16% moisture content and then milled on a Bühler laboratory roller mill MLU 202 with three break and three reduction rolls. The results are shown in Table 5.93. To increase flour yield, the residue was reground on reduction rolls. The extraction rate, protein (N × 6·25) for sorghum and millet, for wheat (N × 5·7), ash and fat (ICC method) are shown in Table 5.94. The millet and sorghum brans were more easily pulverized than the wheat bran and appeared more prominently in the fine flour than in the coarse.

To remove the outer layers (pericarp) of the sorghum and millet, a separate decortication process, the Eurafric M164 decorticator (Anon, 1969a) was used. This consists of a metal carborundum coated cone with a conical rotor; the degree of decortication is regulated by adjusting the distance between the rotor and the cone. On leaving the cone, the bran and

* (Authors' note: it is believed that both BR 64 and RS 626 contain a soft floury endosperm. It appears that with any abrasion dehuller soft floury endosperms suffer more chipping than the corneous types.)

† (Authors' note: the last statement appears to need clarification. Evidence suggests that the tannins are concentrated in the pericarp and/or testa whereas the kafirins are concentrated near the centre of the endosperm. Consequently it is unlikely that tannins and kafirins are associated in the grain. Presumably "association" occurs during the solvent extraction process.)

decorticated grain are thrown into a cylindrical sieve with brushes; the coarse bran is separated from the decorticated grain by aspiration and the fine bran is passed through the 0·8 mm mesh of the sieve. Not more than 6–8% decortication was achieved in a single pass without crushing the grain. A decortication rate of 20–25%, requiring three successive passes was necessary to produce flour considered acceptable by local consumers. The analyses of the pearl millet and its products are shown in Table 5.95. No similar data was given for sorghum. After decortication, the grain was reduced in an attrition or hammer mill equipped with fine sieves.

DeMan *et al.* (1973) reported on the dehulling of sorghum and millet samples obtained from the USA, Senegal and Niger using a Palyi compact (attrition type) milling system. The system consisted of three dehulling units, each with one dehuller and one air separator. The uncleaned and untempered grains were dehulled in the sequence of three dehullers and the fines from all dehullers combined and received from a single cyclone. The hulls from each dehuller were sent to a separate cyclone. The products obtained were dehulled grains, fines and three fractions of hulls. The proximate analyses (AACC, no date) of the grains, the yields and the products from the process are shown in Table 5.96.

DeMan *et al.* (1973) commented that the results on the US sorghum compared closely with the composition of pearled sorghum reported by Hahn (1969). The Senegal and Niger sorghums gave generally similar results.

A grain of the US millet (probably common millet) was dissected by scalpel. The composition of the endosperm is shown below:

	(%)
Moisture	9·3
Protein (N × 6·25)	13·2
Fat	4·3
Crude fibre	4·6
Ash	1·5

If the above values are compared with those in Table 5.96, it appears that the Palyi process has removed material in addition to the hull from the kernel. The additional material formed part of the fines with a reported ash content of 6·5%.

The results obtained with the Senegal and Niger millets (probably pearl millet) were similar to one another and both behaved differently from the US millet in the milling process. There was more breakage with the African millets and the fibre content was much lower than in the US millet.*

Goussault *et al.* (1972) compared the digestibility of bran fractions from wheat, pearl millet, and sorghum in rat feeding trials. The wheat bran was obtained by conventional roller-milling; the millet and sorghum brans from a SOTRAMIL industrial unit at Zinder, Niger. (a) Formic acid insoluble

* (Authors' note: the high fat and protein contents in the fines from the three millets also reported by Perten (1977) suggest the presence of a significant proportion of germ.)

matter (Guillemet and Jacquot, 1943) corresponding to the sum of cellulose and lignin (Charlet-Lery *et al.*, 1952) and (b) pentosans, using a colorimetric reagent for furfural (Cerning, 1970) were determined. The composition of the brans is shown in Table 5.97. The brans were compared in diets for Wistar rats, of average body weight 140 g, at levels to provide comparable amounts of indigestible matter. The basic balanced diet (A) included 15% casein, 74% sucrose and 5% groundnut oil with salt and vitamin mixtures. In the experimental diets (B) 10% wheat bran, (C) 20% millet bran and (D) 20% sorghum bran were respectively included in place of an equal quantity of sucrose. The protein (Kjeldahl N × 6·25), formic acid insoluble matter, and pentosan composition of the diets together with the nutritional and digestibility data are presented in Table 5.98.

The protein content of the diets was high relative to the rat's nutritional needs, and growth rate was also high. Diet (C), millet, with the highest level of protein, produced the highest gain. The food conversion efficiencies of diets (B) and (C) were somewhat lower than (A), with (D) similar to (A), but the PERs were significantly lower than the control: (B) 87%, (C) 83% and (D) 75% of diet (A).

Faecal output from the three bran diets was markedly higher than from casein, being approximately and respectively (B) wheat 2·6 ×, (C) millet 3·4 ×, and (D) sorghum 3·0 × the output from (A) casein.

The coefficient of digestive utilization:

$$\text{CDU} = \frac{\text{food intake—faecal output}}{\text{food intake}} \times 100$$

was lower on diets (B) 93·9%, (C) 92·7% and (D) 93·0% than on diet (A) 97·5%. The decrease in total digestibility appeared to be governed more by the level of the formic acid insoluble fraction than by the botanical origin of the fraction.

Bran intake increased the faecal N; faeces from diet (B) contained 1·5 ×, (C) 2·2 × and (D) 2·5 × the N present in the faeces from diet (A). There was a corresponding fall in nitrogenous CDU (equivalent to Net Protein Utilization) but the differences among diets were relatively small probably because of their high protein level.

Goussault *et al.* suggested that the nitrogenous CDU was not a function of the indigestible carbohydrate but seemed to be linked, in a way not yet understood, to the pentosans. The digestibility of the pentosans of (B) wheat bran, 60%, was significantly higher than the pentosans of (C) millet, 45·6%, and (D) sorghum, 40·7%. The richer the diet in pentosans or the higher their digestibility, the more satisfactory appeared the nitrogenous CDU, though no specific relation between the two was demonstrable.

Though the intake of pentosans differed among the diets, (B) 625, (C) 468 and (D) 360 mg/rat/day, the faecal loss was relatively constant, (B) 250, (C) 253 and (D) 213 mg/rat/day. Goussault *et al.* suggested there might be a fixed amount of indigestible pentosans in the cereals, and a proportion, varying among cereals, which was broken down during the digestive process. They commented that the digestibility of pentosans presented a certain uniformity regardless of botanical species or experimental animal, for example, in the rat, digestibility of cereal pentosans ranged from 41 to 60%; in sheep, the pentosans of different fodders, including sorghum, ranged between 40 and 80% (Balwani *et al.*, 1969; Lyford *et al.*, 1963). This suggested that digestibility was not in direct relation to the activity of the digestive flora which are infinitely more active in ruminants than in monogastric animals.

The relation "pentosan: formic acid insoluble" was 1·8 for wheat bran, 1·1 for millet bran and 0·8 for sorghum bran, wheat being much richer in pentosans than millet and sorghum. The brans of millet and sorghum appeared more active than wheat bran in increasing nitrogenous faecal waste. The average differences (in mg) in nitrogenous faecal waste between controls and experimental animals expressed on the basis of the ingestion of the same quantity of cellulosic elements from wheat bran, millet bran and sorghum bran were respectively:

	B Wheat bran	C Millet bran	D Sorghum bran
100 mg formic acid insoluble matter	3·1	6·4	7·9
100 mg pentosans	1·3	4·3	6·7

Goussault *et al.* referring also to Samson (1970) and other unpublished data, suggested that the removal of some of the bran from sorghum and pearl millet would be nutritionally advantageous.

Air Classification

Stringfellow and Peplinski (1966) reported the results of air classifying sorghum flours from varieties representing different hardnesses, following laboratory experiments at the Northern Regional Research Laboratories, Peoria, Illinois, in which commercial samples or sorghum, maize and soy flours were fractionated by fine grinding and air classifying (Pfeifer *et al.*, 1960) and further commercial samples of sorghum flours and grits were similarly examined (Peplinski *et al.*, 1963). The origin of the sorghums, all grown during the 1962 crop year, and moisture, protein, fat and ash content, on an "as is" moisture basis, are shown in Table 5.99 (the analytical methods were not stated). The relative hardness of the samples was determined by pearling 200 g of the grain as received for 30 sec in a Strong-Scott barley pearler, followed by screening through US No. 14 and No. 20 screens. Relative value of hardness was indicated by the percentage retained on a US 14 mesh screen. The grains, in decreasing order of hardness, with yields, are shown in Table 5.100. The differences among the grain shown in the pearling test are greater than the differences shown by protein content. Dwarf yellow milo was second highest in protein content but on pearling behaved more like the soft variety Combine 7078.

The Kansas commercial mix was debranned and degermed (Strong-Scott laboratory barley pearler with abrasive wheel) and the resultant grits ground three times at 18 000 rpm in an Alpine pin mill then classified by a Pillsbury laboratory model air classifier into five fractions. Particle size distribution was measured for the coarser fraction in a Micromerograph air sedimentation apparatus and for the finer fractions in a Coulter counter. The yield and chemical composition of the fractions with the mass median diameter (MMD) as a measure of average particle size for each fraction are shown in Table 5.101.

Protein shift, defined as the amount of protein shifted into the high protein fractions plus that shifted out of the lower protein fractions, expressed as % of total protein in the material fractionated, totalled 28. Triumph hard red winter wheat and sorghum Combine 7078, both similarly fractionated into five parts, gave similar patterns except that the hard red winter wheat consistently gave a next-to-fines fraction that was higher in protein than the original flour whereas the comparable fraction II in sorghum was lower in protein than the original flour.

After debranning and degerming samples of the four varieties and the commercial mix, the grits were divided into two lots, one lot was ground and air classified directly, the other was Bühler milled to produce floury and corneous fractions each of which was then ground and air classified. Grits and the floury fractions were each reground by three passes and the corneous fraction by five passes through the Alpine pin mill at 18 000 rpm. Five classified fractions were separated from each flour, the cut points being approximately 15, 18, 22, and 30 μm. The resulting fractions are shown in Table 5.102 which illustrates the wide difference in response to fine grinding and air classification. Corneous endosperm was the hardest fraction and required more grinding. Combine 7078 produced a fraction containing 18·9% protein, highest for all the flours tested, the corneous endosperm fraction lowest in protein (6·8%), the greatest protein shift (28%), and the second lowest coarse residue (12·6 %). Floury endosperm responded best to air classification, protein shift varying between 26 and 41% with low protein fractions down to 3·5%. Comparing flours from grits and floury endosperm, the highest protein fractions were obtained from the grits of the softest sorghums, Dwarf yellow milo and Combine 7078, the lowest protein fractions from the floury endosperm of the same varieties.

Processing for Starch Production

Sorghum is wet milled primarily to produce dry starch or starch hydrolysed to glucose. The basic process has been described by Watson (1970) and Doggett (1970).

Clean sorghum is steeped in water for about 60 h, then crushed to loosen the embryos which float to the surface of the starch suspension. Edible oil is extracted from the embryo and the low fat residue is converted to animal feed. The remaining grist is reground, separated through fine bolting screens, the fibre and coarse fractions going to animal feed. The starch and water-insoluble protein (gluten) is centrifuged, the gluten fraction, containing 60–65% protein, is also used in feed. The starch is recentrifuged, washed and either dried or hydrolysed to glucose.

Rooney (1973) stated that in 1973 no sorghum was wet-milled in the USA although between 1950 and 1970, 6–8 million bushels were milled annually. The facilities used for wet-milling sorghum had been turned over to maize, probably because of technological, marketing and economic difficulties. Prime yields of starch are harder to recover from sorghum than from maize, also the demand for sorghum as a feed grain in the USA increased its price to a level uncompetitive with maize for wet-milling. Sorghum starch needs bleaching to remove polyphenolic pigments not present in maize. The types known as "white milos" originally processed for starch in the USA were in fact kafir types with a light-tan-coloured pericarp. Peeling these sorghums did not improve starch colour.

Freeman and Watson (1971) reported that bleaching the starch extracted by commercial wet milling from the waxy hybrid variety CP 622 with NaOH pH 10 and stirring for 30 min at 53°C (128°F) improved the colour compared to untreated starch but further bleaching with $NaClO_2$ was needed to produce the degree of whiteness demanded. Bleaching added to the processing cost and the viscosity of the bleached sorghum starch tended to fall during storage. Freeman and Watson stated that apart from these considerations, sorghum starch was technologically equivalent to maize starch.

Blessin *et al.* (1971) described a laboratory method of alkali dehulling of sorghum grain to reduce pigment in the pericarp before wet milling. The sorghums RS 626, TE 77 and a commercial US No. 2 yellow were dehulled with 20% NaOH at 160°F (71°C) for 6 min. By hand dissection, the pericarp contents were, respectively, 7·5, 7·8 and 7·8%. The alkali treatment removed the pericarp and left the endosperm and germ intact, the average yield of dehulled sorghum being 92%. Microscopic examination showed that most of the aleurone layer remained and ash, ether extract and protein levels were not significantly affected, though the fibre content of the dehulled grain was about half that of the whole grain.

The yield of starch from dehulled No. 2 yellow grain, processed by a laboratory wet-milling procedure (Anderson 1963) was about 5–8 percentage points lower than that from the whole grain, but protein content in the starch was about the same as from the whole grain (Table 5.103). The colour of the alkali dehulled starch appeared whiter than starch from the whole grain. Alkali dehulling increased the recovery of germ and its oil content. The germ appeared clean. Recovery of oil was 45% from the whole grain and 65% from the dehulled.

Norris and Rooney (1970b) using a laboratory technique, compared the wet-milling properties of Martin (BTX 398), Combine Kafir 60 (BTX 3197), R 7078, RTX 414 and their hybrids RS 608, RS 625, RS 610 and RS 626.

The starch content (modified fungal amylase—acid hydrolysis AOAC,

1965) of the whole grain of the parents and hybrids ranged from 66 to 68%, the differences being not significant (Table 5.104). However, differences in starch recovery were significant (Table 5.105). Hybrids of Martin with R 7078 and RTX 414 gave significantly lower starch recovery and higher protein content in the starch than hybrids of Combine Kafir 60 with R 7078 and RTX 414. Martin and Combine Kafir 60 contained significantly greater numbers of peripheral endosperm cells (which contaminate starch) than R 7078 and RTX 414. Peripheral endosperm cell content was significantly and positively correlated with protein content of the starch and negatively with starch yield and starch recovery. The protein content of the peripheral endosperm cells of the hybrids was not significantly influenced by the parents.

Seeley (1958) cited Texas Agricultural Experiment Station Bulletin 743 (1951) as reporting the proximate analysis, vitamin and mineral composition of sorghum gluten feed and sorghum gluten meal. The values are shown in Table 5.106.

Reiners *et al.* (1973) reported, on an "as is" basis, the dry substance (vacuum oven at 100°C), protein (Kjeldahl N × 6·25), total fat (Alexander *et al.*, 1967 but using carbon tetrachloride–methanol mixture), ash, fibre (American Oil Chemists Society Methods Ba 5-49 and Ba 6-61 respectively), and S (oxidation in a Parr bomb) of an average of monthly samples taken over one year from the CPC International Inc. plant for wet-milling sorghum at Corpus Christi, Texas. The results are shown in Table 5.107, compared with the values from the Atlas of Nutritional Data (National Academy of Sciences, 1971).

Reiners *et al.* also reported the amino acid composition (auto-analyser following acid hydrolysis; tryptophan by microbiological assay, cystine and methionine on performic acid oxidized protein) of gluten meal, gluten feed and germ meal. The average results are shown in Table 5.108. The same results are shown in Table 5.109 but reported on an "as is" basis compared to results on gluten meal from the Atlas of Nutritional Data (national Academy of Sciences, 1971) and on gluten feed and germ meal from Lyman *et al.* (1956), the latter assayed microbiologically. The lysine contents determined by auto-analyser were appreciably lower than those reported by Lyman *et al.* or in the Atlas of Nutritional Data. A sample of gluten feed was therefore assayed microbiologically and found to contain 350 mg/100 g sample compared to 410 mg/100 g by auto-analyser both on an "as is" basis. This seemed to confirm the low values found by auto-analyser. The tryptophan values were also lower than those reported earlier but an attempt to verify the values colorimetrically failed because of the development of interfering colours.

Montgomery and van Assen (1968) described a laboratory method for the wet-milling of sorghum and sorghum malt developed as part of investigations into the brewing of sorghum beer. The isolation of the starch included neutral chemical steeping in sodium diethyldithiocarbamate (during which the water soluble red pigment of short red sorghum appeared to dissolve completely) followed by conditioning in 5% aqueous NaCl solution at 20–25°C for 4 h. The grain was then ground in an electrically driven Fryma mill between parallel stone plates in aqueous 0·3 % NaCl. The remainder of the laboratory wet milling procedure followed that developed by Montgomery *et al.* (1964) for maize, the essential difference between the two methods being the introduction of the conditioning in 5% aqueous NaCl which improved settling of the sorghum starch and improved separation of the hulls and fibre, protein and gluten fractions.

Starches from short red sorghum, 3-day and 6-day malts were obtained in yields of 89 to 92% of the total starch. The pH of the starch was 5·9–6·0 in water at 25°C. Protein content (Earle and Milner, 1944, modified with respect to alcohol washing of the grain by AACC, method 76–20) was 0·20–0·23% of the starch, and ash from 0·11 to 0·20% of the starch.

Freeman and Bocan (1973) examined the suitability of pearl millet as a crop for wet-milling. Two types were compared, Tiflate, a tall variety grown in Texas and harvested in 1968, and Tift 23DB, a dwarf (fertile) maintainer breeding line. Both were processed by a laboratory wet milling procedure (Watson and Hirata, 1955) and the starch properties, protein solubility, composition of the whole grain and milling fractions determined (Freeman *et al.*, 1967). The % protein, starch and crude fat content, DWB, of the two pearl millets compared to commercial maize and sorghum are shown in Table 5.110.

The protein content of Tiflate was particularly high. Freeman and Bocan suggested that the difference between Tiflate and Tift 23DB could be a function of difference in yield but the yield was not stated. They also suggested that the higher oil content of Tift 23DB compared to Tiflate could be due to genetic differences or to a higher proportion of immature kernels in Tiflate. The fatty acid composition (gas chromatography, Beadle *et al.*, 1965) is shown in Table 5.111, compared to averages for maize and sorghum.

Freeman and Bocan suggested that the higher level of linoleic and lower level of oleic acid in Tiflate, compared with Tift 23DB, may result from differences in environmental temperature when the grains were maturing. Vegetable oils synthesized at low temperatures tended to contain more unsaturated fatty acids than oils synthesized at high temperatures (Harris and James, 1969). The protein solubility fractions of the two millets are shown in Table 5.112.

The very small millet kernels were difficult to break by a wet milling technique which was satisfactory for maize and sorghum. Much of the millet germ oil was probably not recovered in the germ fractions. The yield and composition of the wet-milling fractions of the two millets compared to a maize and a sorghum sample are shown in Table 5.113. The millets produced about twice the quantity of steep water solubles as did maize and sorghum. The wet-milling process would need to be modified to obtain the best results from pearl millet. The properties of the starches from the two millets compared to maize and sorghum are shown in Table 5.114.

Freeman and Bocan reported that in general, millet starch was similar to maize and sorghum starches. The major differences were the smaller granule size of pearl millet, and a slightly lower tendency of cooked millet starch to thicken on cooling. Because of its lower cold paste viscosity, pearl millet starch would be less suitable as a thickener than maize. The lower gelatinization temperature of Tiflate starch compared to Tift 23DB might also have resulted from lower environmental temperatures when the grain matured.

Lorenz and Hinze (1976) compared the functional characteristics of the starches of the common millets Abarr, Leonard, Big Red (PI 346 946), Black Russian (PI 346 937), Akron Proso (73 21 1) and (73 1055), with the foxtail millets, Golden German and Chinese, and with HRS wheat (Chris) and rye (Prolific) all grown at Colorado State University Agricultural Experiment Station during the 1975 crop year. The millet starches were isolated by steeping the grain at 6°C for 24 h in distilled water buffered at pH 6·5/0·02 M acetate) and rendered 0·01 M in mercuric chloride. After steeping, the excess water was decanted, the grain washed with distilled water and wet milled in a Waring Blendor for 3 min. The suspension was screened through a bolting cloth, the starch recovered by centrifugation, and air dried. The wheat and rye starches were prepared by the method of Lorenz (1976). The millet starches showed significantly higher water binding capacity values and gelatinization temperatures than the wheat starch. The swelling power values at 90°C of the millet starches were similar to those of wheat (except for Big Red common millet starch which was significantly higher), and lower than those of rye (except for Big Red starch). Solubilities of the millet starches were lower than those of wheat, except for Big Red starch. Amylograph viscosities of millet starches were higher than those of the wheat starch; rye was not reported.

Composite Flours

Introduction

The descriptive name "composite flours" came into the cereal scientist's vocabularly about 10 years ago. Though difficult to define with precision, in most contexts "composite flours" refers to mixtures in which cereal flours, predominantly wheat flour, are combined with other starch and protein sources including those derived from other cereals, roots, food legumes and oilseeds. The subject has been reviewed by Hulse (1974) and FAO (1973).

In many of the less-developed countries where bread consumption is increasing, the climate and soil conditions are not ideal for the production of wheat varieties which by traditional standards would be considered good for bread making. Consequently, many of these countries are spending increasing amounts of foreign currency to import wheat from countries such as Canada, USA and Australia, where bread wheats are produced in surplus. Many of these same less developed, wheat importing countries are better able to produce other cereals such as sorghum, millet, or maize, and/or root starches such as cassava and sweet potatoes, and potential protein supplements from oilseeds and grain legumes. Consequently a growing interest in bread made from composite flours is evident.

Studies of composite flours have for the most part concentrated upon composites of wheat flour used to make European and North American types of bread. The term "composite flours" ought not to be so restricted but should comprehend any mixture of different flours used to produce a traditional or an adopted exotic cereal food. Composites based mainly on wheat flour may be used in Middle Eastern and Asian flat breads; maize composites can be used in tortillas. The many applications of composites of sorghum, and/or millets with other flours provide the main burden of the text that follows.

In 1970 the Tropical Products Institute published the proceedings of a symposium on the use of nonwheat flour in bread and baked goods. In 1973, FAO issued the revised second edition of Documentation Package, Vol. 1, describing their composite flour programme. Where papers in this package contain information on the nutritive value of products, they will be discussed below. The research work reported has included relatively little on the use of sorghum and the millets and has concentrated more on the technological than the nutritional problems involved.

Bakery Products

The possibility of using sorghum in bakery products was however considered as long ago as 1914 when Francis published from the Oklahoma Agricultural and Mechanical College, a booklet of recipes and methods for breakfast foods, muffins, breads, griddle cakes and cookies using sorghum alone, or combined with wheat or rye flour.

Borasio (1937) reported that sorghum flour from Eritrea containing 6·3% protein, 0·7% fibre and 1·7% lipid could be used at a level of 5–10% together with a strong wheat flour in breadmaking.

More recently, Futrell and Abdullahi (1967) formulated pancake, muffin, bread, and cookie recipes using sorghum and wheat flour. Hart *et al.* (1970) reported acceptable pan breads from (US) sorghum containing various gums, including 4000 centipoise methyl cellulose.

De Ruiter (1973) discussed the use of nonwheat flours in bakery products and published formulae for cookies (literally soft cakes) with sorghum flour and biscuits (literally hard cakes) using millet flour. Perten (1977) reported from the Sudan the satisfactory production of French-type bread from composites of 85% wheat flour (87% extraction) and 15% sorghum flour. Bhatia *et al.* (1968) reported from Mysore, India acceptable bread from composites of 80% wheat flour and 20% sorghum flour, and stated that the calculated protein content of the composite was essentially the same as, and the lipid, mineral and fibre content higher than, bread from 100% wheat flour.

In 1916 Langworthy and Holmes reported from the US Office of Home Economics on the human digestibility of feterita, kafir, dwarf milo and kaoliang grain sorghums milled in a laboratory mill. The fraction used passed through a 16-mesh sieve, the amounts of bran retained on the sieve and considered too coarse to use were: kafir 21%, milo 19%, feterita 15% and kaoliang 5%. These differences may have resulted from differences in kernel structure and the manner in which they fractured when milled. The protein contents were ranked inversely with coarse bran removed, the kafir meal being lowest and the kaoliang highest in protein.*

Bread was baked from sorghum, wheat or maize meal, mixed with $NaHCO_3$, molasses, salt, ginger, lard and water. The bread provided about 82% of the dietary protein; apple sauce, potatoes and butter provided the balance of the diet. A second diet consisted of a boiled porridge containing only cereal meal and salt, plus apple sauce, butter and cane sugar syrup. The sorghum provided about 98% of the dietary protein.

Seven subjects, young men who continued their normal routine over the 3-day, nine meal test period, were fed unrestricted quantities of bread or porridge ("mush"). No record of body weights was kept nor was nitrogen equilibrium measured. Table 5.115 presents the average daily results from those who received bread.

At the end the subjects seemed to be in normal condition, with improved peristaltic action of the intestine. There were a few complaints of hunger and nervous headache. The results are self-evident. The protein of first wheat and second maize was more readily digested than the four sorghums, kaoliang being markedly inferior to all the rest. Daily average digestibility values obtained from those fed the porridge are shown in Table 5.116. The subjects reported that their physical condition remained normal. Nevertheless the low level of sorghum protein digestibility, particularly of the kaoliang, is worthy of note.

Subrahmanyan *et al.* (1954c) conducted feeding tests in a boarding home, over 8 weeks, with 42 children, 21 to each group, ranging in age from 5 to 12 years in which the diets were based on (A) rice, (B) vermicelli made from sorghum flour, wheat maida and cassava flour in the proportion 2/1/1. (Earlier experiments using all sorghum flour failed to produce an acceptable product.) Finely ground whole sorghum meal and cassava flour were mixed with boiling water, after which the wheat maida was added. The dough was passed through a vermicelli press, and after drying, lightly roasted. Either rice or composite sorghum vermicelli, constituted about 60% of the diet which also included finger millet and pulses (about 10% of the diet), small amounts of meat, milk, leafy vegetables and condiments.

The average gain in height and weight of the children is shown below:

	A Rice	B Composite sorghum vermicelli
Average height gain (cm)	+1·07	+1·32
Average weight gain (kg)	+0·20	+0·69

* (Authors' note: compare with Perten (1977) who found coarse bran in millet was significantly lower in protein than the whole grain.)

The vermicelli group on diet (B) registered a significantly larger weight gain with no member losing weight. Among the group on rice diet (A), three subjects lost weight. Subrahmanyan *et al.* commented that both diets were similar in calorie, protein and mineral content (calculated from tables). The apparent superiority of the composite flour vermicelli could not be attributed to any one ingredient.

Perten *et al.* (1972) prepared a comprehensive report to the Government of Senegal on the manufacture of French type bread incorporating sorghum and pearl millet. No direct information was given on the nutritive value of the resultant breads and the protein contents of the flours varied with the samples and milling method. Perten *et al.* used the factor N × 5·7 for the wheat flour protein and N × 6·25 for the sorghum and millets. The mean percentage value for the protein using the factor N × 5·7, DWB, for all three flours is shown below.

	Protein (N × 5·7) (%)
Wheat flour	10·6
Sorghum flour	9·1
Pearl millet	9·0

Perten *et al.* reported without comment one exceptionally low millet protein content of 3·7% (N × 6·25). The next lowest was 7·2%, the highest being 13·0%. If the 3·7% protein value is excluded the average millet protein in the table above rises to 9·6%. Thus a composite of 70 wheat with 30 sorghum or millet flour would contain only slightly less crude protein than 100% wheat flour. No comparison of overall nutritional quality was presented.

Pringle *et al.* (1969) described the production of bread from composite flours using mechanical dough development (the Chorleywood Bread Process, Axford *et al.*, 1963). Among the blends was a composite flour of:

	Protein (N × 5·75, 13 % moisture basis) (%)	Parts by weight
Canadian wheat flour	17	64
Defatted soy flour	50	6
Pearl millet flour 70% Sorghum flour 30% milled at Zinder, Niger	8	30

Miller and Mumford (1969) compared the composite flour blends described above in rat feeding trials (method of Miller and Bender 1955). The values for calories derived from protein, NPU_{op}, $ND_pCal\%$, and NPU_{st} for the above blend compared to wheat flour alone follow:

	Canadian wheat flour	Composite flour blend
Calories from protein (%)	21	21
NPU_{op}	32	44
$ND_pCal\%$	6·7	9·2
NPU_{st}	44	64

The blend containing sorghum, millet and soy flour was superior to the wheat flour alone although protein contents were similar.

Bushuk and Hulse (1974) reported the use of low powdered sheeting rolls to develop bread dough and eliminate the need for long bulk fermentation; the bread quality was comparable to that obtained from the Chorleywood Bread Process (Axford *et al.*, 1963). Satisfactory bread (by laboratory standards) was obtained from composite flour blends containing up to 20% of African pearl millet or of US common millet combined with 80% of Canadian wheat flour.

The protein contents were:

	Protein (N × 5·7, 14 % moisture basis) (%)
Canadian wheat flour	13·6
African pearl millet	7·6
US common millet	8·5

The protein contents of the composite flours were:

	(%)
Canadian wheat flour	13·6
Pearl millet blend	12·4
Common millet blend	12·6

Sumner and Nielsen (1976) prepared a traditional Nigerian-type bread using a 80/20 wheat/sorghum composite flour blend. Proof time of the dough was increased from 2 to 2·5 h and the resulting loaf was compared with a 100% wheat flour bread. Although loaf volumes were similar, the specific volume of the composite flour bread was slightly depressed. Major differences were a darker brown external and internal colour, an increased coarseness and firmness and a stronger cereal flavour in the 80/20 wheat/sorghum bread.

Awadalla and Slump (1974) examined the nutritional value of whole and decorticated native Egyptian common millet and whole wheat from the Dutch market. The millet was either (a) cleaned and ground to pass through a 60-mesh sieve or (b) decorticated at 120 rotations/sec followed by grinding as (a). The proximate analysis and minerals (AOAC 1965; crude fibre enzymatically, van de Kamer and van Ginkel, 1952) of the cereals are shown in Table 5.117.

The amino acid composition (mg/100 g sample) (auto-analyser following acid hydrolysis, Spackman *et al.*, 1958; methionine and cystine after performic acid oxidation, Moore, 1963; tryptophan after autoclaving with barium hydroxide, Slump and Schreuder, 1969) is presented in Table 5.118 and, as mg/g total N, with the amino acid score (WHO, 1973a) in Table 5.119. The protein content of the millet is lower than that of the wheat but the "protein quality" (mg essential amino acids/g total N) is higher in millet than in wheat. Lysine was the first-limiting amino acid in all three products.

NPU and true digestibility (TD) were determined in rat feeding trials (Miller and Bender 1955). The diets were (A) whole wheat, (B) 80% whole wheat, 20% millet, (C) 80% whole wheat, 20% decorticated millet, and (D) wheat starch. All the protein in (A), (B) and (C) was derived from the cereals, the adjustment to a uniform 8% protein level being made by starch. NPU, TD and BV (biological value = NPU/TD × 100) are shown in Table 5.120. The protein quality being essentially similar, Awadalla and Slump concluded that the decrease in protein quantity arising from the use of such mixtures is not unacceptable.

Awadalla (1974) reported technological studies using 20% whole or decorticated millet flour with 80% of Dutch wheat flour (treated with $KBrO_3$ at the mill) of 100%, 80% and 70% extraction rate in (Dutch) pan bread and Egyptian flat bread. The baking quality was impaired in both Dutch and Egyptian bread as judged by loaf volume and internal and external characteristics but was improved by the addition of improvers ($KBrO_3$, fat, calcium stearoyl lactylate) the greatest effect being obtained with 80% extraction wheat flour. The results appeared not to be affected by the type of millet used.

Snack Foods

Elliott and McPherson (1971) described the development of a fried sorghum wafer similar to a fried thin potato chip.*

Sumner *et al.* (1974b) evaluated flours from cowpeas, white sorghum and pearl millet originating from Nigeria in making a fried snack food. The samples used were:

(1) red cowpeas; traditionally ground in Nigeria by wetting the cowpeas in a mortar, pounding for about 6 min, drying on a grass mat for about 30 min, winnowing 3–6 min, washing 5–10 times taking about 10 min, drying on a mat for 30 min, and grinding (Sumner *et al.*, 1974a).

(2) white cowpeas; traditionally ground as above for 1.

* (Authors' note: The technology used, microwave drying, and the final product would probably be of more interest to the industrialized than the less developed countries.)

(3) white sorghum, dehulled by abrasion (Hill grain thresher) and ground in a hammer mill.
(4) millet; mechanically milled as (3) above.

The proximate analysis (AACC, 1962; AOAC, 1965) and amino acid composition (auto-analyser; tryptophan by microbiological assay) are shown in Table 5.121.

The basic recipe produced an "excellent fried snack" and was used as a standard:

	(g)
Red cowpea flour	100
Baking powder	5
or Potash	0·8
Salt	1·3
Water (ml)	85

The dry ingredients were thoroughly mixed with water. Using a modified grease gun or a metal cake decorator, the batter was extruded into hot fat at 165°C and deep fried for about 1 min.

In combination with the cereals, red cowpea and white cowpea flours were so similar in behaviour as to be interchangeable. Acceptable products were made from (A) red cowpea flour 33%, millet flour 67%, (B) red cowpea flour 50%, white sorghum flour 50% and (C) red cowpea flour 40%, white sorghum flour 60%. The calculated protein contents, DWB, of the blends were: (A) 14·9%, (B) 16·0% and (C) 14·3%.

Sumner and Nielsen (1976) reported that snacks based on recipes (A) and (B) using white cowpea flour and a small amount of sugar and ground red pepper were made and sold by a Northern Nigeria food seller in the local Maiduguri market.

Pasta Products

Sumner *et al.* (1974a) reported that wheat flour noodles form an important part of the diet in the Maiduguri area of Northern Nigeria. Known as talya they are made from wheat flour and water, formed into a dough, kneaded, rolled and cut into noodles, generally on a small hand-operated machine. They are cooked in a very small amount of water which is completely absorbed by the noodle during cooking. Consequently no soluble protein or other solids are leached out during cooking. The noodles are generally used in soups and stews; in some cases milk may be added (Sumner *et al.*, 1975).

In evolving formulae incorporating cowpea, pearl millet or sorghum flours with wheat flour, Sumner *et al.* (1975) reported that the relative proportions of the cowpea, sorghum or millet flours were not critical but a minimum of 36% wheat flour was needed to produce an acceptable product.

The flours described above (Sumner *et al.*, 1974b) were used in the manufacture of noodles: (1) traditionally milled red cowpea flour, (2) traditionally milled white cowpea flour, (3) white sorghum abrasion milled (Hill thresher) at the Prairie Regional Laboratory, (4) abrasion milled millet flour, and (5) commercially milled Golden Penny wheat flour. The proximate analysis of all samples and their amino acid composition are shown in Table 5.121.

Several of the composite flours were of good cooking quality and improved nutritional value compared to a commercial wheat noodle, a "Golden Penny" (Nigeria commercial) wheat flour noodle made in Nigeria (GPN) and a Golden Penny wheat flour noodle made in the College of Home Economics, Saskatoon, (GPS), which was used as the standard. The calculated protein content and amino acid score (WHO 1973a) of two of these blends, No. 44 containing 52% wheat flour, 32% cowpea flour and 16% sorghum flour, and No. 54 containing 42% wheat, 36% cowpea and 22% millet are shown in Table 5.122.

The flour–water noodles were cut on a hand operated noodle machine to a 2 mm width, air dried for 24 h and packaged in plastic bags at approximately 10% moisture. The cooking methods and tests were adapted from Paulsen (1961); cooking quality from Matsuo and Irvine (1969); water absorption and protein cooking loss from Matsuo and Irvine (1974). The effect of cooking on the composition of the five noodles is presented in Table 5.123.

Noodles based on blends No. 44 (wheat-sorghum-cowpeas) and No. 54 (wheat-millet-cowpeas) but using white cowpeas instead of red, were made in the test kitchen at Maiduguri and subjected to consumer preference tests among 15 local housewives. No significant difference in preference was found among composite flour noodles and wheat flour control noodles but several adverse comments were made on the odour and flavour of blend No. 54 (millet) noodles.

Miche *et al.* (1976) reported the use of four varieties of sorghum in pasta production:

(1) CE 90 from Senegal, semi-corneous endosperm, white pericarp, no coloured testa.
(2) S 29 from Upper Volta, corneous endosperm, white pericarp, no colourted testa.
(3) L 30 from Niger, semi-floury endosperm, red brown spots on pericarp and on testa.
(4) INRA 450, French feed grain, floury endosperm and brown-red pericarp.

Samples (5 kg) tempered to 16·5% H_2O were processed in a laboratory mill with fluted break rolls, a three-sieved plansifter and a purifier working dull to sharp. Between 5 and 10% of the bran and germ fractions was removed.*

* (Authors' note: a detailed mill flow diagram is provided in the original paper.)

The protein (Kjeldahl N × 6·25), lipid (diethyl ether extraction) and ash (incineration at 800°C for 45 min) of S 29 and its milled fractions are shown in Table 5.124.

The pigment content (AACC, method 14·50), phenolic compounds (optical density at 320 nm of an acetone (80%), water (20% extract), absorbance at 400 nm (water extract, Matsuo and Irvine, 1967), peroxidase activity (Honold and Stahmann, 1968), polyphenoloxidase activity (Matsuo and Irvine, 1967), lipoxidase activity (Nicolas *et al*., 1974) and catalase activity (micromoles of oxygen disengaged by the decomposition of a hydrogen peroxide solution) of the four sorghum varieties are shown in Table 5.125.

The distribution of the same components among the whole grain, bran, flour and semolina is shown in Table 5.126.

Varieties (3) L 30 and (4) INRA 450 were unsuitable for pasta production because of very poor colour on milling. Variety (2) S 29 gave a light coloured pasta but was somewhat gritty or sandy to the palate. Variety (1) CE 90 was considered best in eating and cooking characteristics, flours being better for cooking than semolinas.

Colour was measured using a Bausch-Lomb Spectronic 20 spectrophotometer (Alause and Feillet, 1970) to give a brown index, (BI) = 100 log (reflectance 550 nm), and a yellow index, (YI) = 100 log (reflectance 480 nm) − BI. These values are shown in Tables 5.125 and 5.126. A good colour was characterized by low BI and high YI values. In the four sorghum varieties, the BI ranged from 19 to 35 and the YI from 5·3 to 6·7. Sorghum pasta was grey compared to the yellow of wheat pasta. Pasta with good technological and cooking qualities was obtained from a blend of 75% sorghum and 25% durum wheat but the unsatisfactory colour remained an unsolved problem.

In the view of Miche *et al.* the ideal sorghum variety for pasta would contain neither testa nor phenolic compounds; it would contain a thin, translucent pale coloured pericarp and a floury endosperm high in yellow carotenoids.

Desikachar (1977) reported that sorghum and maize flours required more water than wheat flour to make a chapati dough. Mixtures that included 30–40% wheat flour or 15% urd dhal (*Vigna mungo*) were easier to roll into thin sheets.* Mixing with boiling water to partially gelatinize the starch also improved dough cohesion. Mixing 30% sorghum flour into whole wheat meal improved the chapati by reducing "chewiness". Pearl millet imparted a strong flavour considered unacceptable.

Pearled sorghum flour mixed with boiling water was suitable for noodles or vermicelli, but millet flour gave highly coloured, sticky, strongly flavoured, unacceptable products.

* (Authors' note: Conversion of grain legumes to dhal is a unique practice under Indian food habits. Traditional home methods for preparing dhal consist in loosening the husk by water or oil application followed by sun drying. The husk is removed by pounding and winnowing. Yield of dhal is about 80% of the weight of whole grain. Dhal is ground for flour and by-products.)

Anderson *et al.* (1969b) reported that sorghum grits processed by either roll- or extrusion-cooking were similar in water absorption and viscosity characteristics to maize meal processed for use in CSM which contained partially gelatinized maize meal, soy flour, SMP, vitamins, and minerals and was provided by the Food for Peace Program (USA). The composition of sorghum grits, DWB (%), was: moisture 12, protein 10·5, fat 0·9, crude fibre 0·3, ash 0·3.

Crabtree and Dendy (1977) milled a Canadian hard wheat and a white millet (probably foxtail millet), together in a Bühler laboratory mill. The grains were separately conditioned to a moisture content of 16%. The blends ranged in 5% increments from wheat 100% to wheat 75/millet 25. Extraction rates varied from 69·3% to 74·1%. The resultant composite flour up to 10% millet addition gave bread by a bulk fermentation process of satisfactory volume and colour.

As the proportion of millet increased, flour and bread colour darkened; the water absorption, dough strength and extensibility decreased and the peak hot paste viscosity increased (in Brabender Units 100% wheat flour equalled 800 BU, 100% millet flour 2000 BU).

Subbra Rao *et al.* (1953) processed wheat, sorghum and yellow maize to be substituted for and cooked like rice. Among a variety of methods including precooking and decorticating, the only satisfactory method was by Subrahmanyan *et al.* (1950): the milled flour was mixed with cassava flour and hot water before extrusion and drying to a form similar in appearance to rice.

Tripathi and Daté (1975) reported from Kanpur, India that wheat flour could be satisfactorily replaced by 10% maize flour, 5% sorghum flour or 5% pearl millet flour in European type bread.

Malting and Brewing

In several African countries sorghum, and to a lesser extent pearl millet, is fermented into beer. According to Novellie (1968) the beer making process may depend upon two fermentations involving (1) bacteria that produce lactic acid, (2) yeasts that produce alcohol. The malting process that generates the fermentable mono and disaccharides is dependent upon the α- and β- amylases and the maltose splitting enzyme, maltase, that develop in the sorghum during germination. Novellie suggests that the malting process also contributes nutrients that stimulate growth and the rate of fermentation by the microorganisms involved.

Novellie (1977) reviewed sorghum and millet fermentation in a well referenced paper.

The Republic of South Africa produces annually 10^9 litres of commercial sorghum beer. Home brewing, for which there are no reliable statistics, may produce an equivalent volume (Novellie, 1977). Sorghum beer differs from European beer as summarized by Novellie (1977) in Table 5.127.

The modern brewing process has been described by Schwartz (1956) and more recently by Novellie (1968). A mixture of sorghum malt and water is first soured by thermophilic lactobacilli, then boiled with the grain adjunct, sorghum or maize brewers' grits, cooled to 60°C and saccharified with more sorghum malt. The mixture is strained to remove the coarsest particles and fermented with a top fermenting yeast, *Saccharomyces cerevisiae*. The beer is sour (lactic acid), pH 3·3–3·5, opaque because of suspended solids and yeast, still fermenting and lightly alcoholic, 3–3·5% alcohol by weight (Novellie, 1968).

Sorghum beer from Southern Africa is comparable to merissa from the Sudan, bouza from Ethiopia and pombe from East Africa (van der Walt, 1956).

Malting

Siddappa (1954) reported that the best sorghum variety for the preparation of malt extract in India was yellow Periyamanjal grown in the Coimbatore, Bellary and Nandyal districts. After analyzing nearly 500 batches manufactured during 1944–48, the standards suggested for malt extract to be used with shark liver oil were:

Cholam malt-extract is a thick, viscous, amber-coloured liquid with sweet taste and characteristic aroma, miscible with water in all proportions.

Specific gravity 28°C	1·40–1·45
Refractive index 28°C	1·49–1·50
Total solids (by refractometer at 28°C)	80–82
Total sugars as maltose (%)	65–75
Acidity as acetic acid (%)	0·6–0·8
Albuminoids (N × 6·25)	2·8–3·2
Vitamin A (IU/g)	175–225
Thiamine (IU/100 g)	100–125
Diastatic activity	low

The diastatic activity of the malted sorghum was between 4–8° Lintner and of the extract between 1–2°.

Novellie (1977) reported that it is possible to improve the digestibility of high tannin sorghums by treatment of the grain before processing. The method is the subject of a patent specification in the Republic of South Africa 75/4957, Swaziland RP/5/1977, Botswana P261 24/5/77 and is based on the neutralization of unwanted polyphenols in the grain by treating the grain with a predetermined amount of a nonacidic carbonyl containing compound, preferably formaldehyde.

Chandrasekhara and Swaminathan (1953c) prepared finger millet malt from the variety H 22. The grain was steeped for 24 h, germinated for 72 h at 22–28°C, then dried at 45–50°C in a current of air. The vegetative portion was removed and the malted grain powdered to pass through a 60-mesh sieve.

Amylase activity (method of Chandrasekhara and Swaminathan 1953a) was maximum at 60°C. The reducing sugar content (Chandrasekhara and Swaminathan, 1953a) produced by mashing finger millet malt alone and with cooked wheat flour in the ratio of 1/1 by weight is given below:

Reducing sugar as maltose % of mash liquor

	Malt alone	Malt and cooked wheat flour
at the end of 1 h at 55°C	3·6	8·0
at the end of 1 h at 65°C	4·8	9·0

Wheat, sorghum, finger millet and cassava were compared as starchy substrates for the finger millet malt in mashing. The yield of reducing sugars and the quality of malt extract obtained are recorded in Table 5.128. Wheat and sorghum were more suitable than either finger millet alone or cassava.

The crude protein (N × 6·25) of the different flours and the derived malt extracts are shown in Table 5.129. The protein content of the malt extract was related to the protein content of the original flour. The malt extracted from the finger millet malt and wheat flour was considered suitable for the preparation of malted milk powder. According to the British Pharmacopoeia, the protein content of malt extract should be not less than 4·5%.

Chandrasekhara and Swaminathan (1953a) reported that finger millet when malted is a good source of both α- and β-amylases, the amylase content increasing with the period of germination. Malt from the two cultivars examined (R 009 and H 22) was richer in α-amylase than comparable wheat and barley malts, and higher in β-amylase than sorghum malt.

Amylase activity increased rapidly with the progress of germination, the optimum germination being between 72–96 h. Both α- and β-amylase from finger millet were most active at pH 5·0; activity increased to a maximum at 60°C and declined as the temperature increased. The α and β-amylase enzymes were completely inactivated in 15 min at 70°C and pH 6·5.

Chandrasekhara and Swaminathan (1953b) also reported that H 22 finger millet showed no proteolytic activity in the grain but protease increased with germination. The protease could be extracted from the malt by acetate buffer (pH 4·4) or 50% glycerol but not by water. The protease hydrolysed casein and gelatin, activity being optimum at pH 4·4 and 37°C; the critical inactivation temperature was 50°C. The protease could be activated by hydrogen sulphide and cyanide, slightly inhibited by ascorbic acid and completely inhibited by copper.

Chandrasekhara and Swaminathan (1954) reported slight glycero- and pyrophosphatase activity in finger millet before germination and a considerable increase after germination. The enzyme could be extracted from the powdered grain or malt by acetate buffer (pH 5) or 1% NaCl solution. The optimum pH was 5·6 for glycerophosphatase and 5·0 for pyrophosphatase. For both enzymes the optimum temperature for activity was about 45°C and the critical inactivation temperature about 57°C. They were not affected by either Mg or Zn but strongly inhibited by fluoride.

Chandrasekhara and Swaminathan (1956, 1957) prepared malt from the CO 4 variety of pearl millet following the method of the same authors (1953ac) used for finger millet.

The diastatic activity of pearl millet was lower than that of finger millet (Chandrasekhara and Swaminathan, 1953ac). The ungerminated grain showed negligible activity; germination led to a marked increase in α-amylase activity but to only a slight increase in β-amylase activity. Optimum pH for both α- and β-amylase was 4·8 and optimum temperature 60°C. The critical inactivation temperature for α-amylase at pH 4·5 was 45°C.

Narayana and Murthy Reddy (1973) at the University of Agricultural Sciences, Bangalore reported the free amino acid content (extraction Murthy Reddy and Narayana, 1973, auto-analyser) of malt extract of a cultivar Aruna of finger millet. The results appear in Table 5.130.

Hussain *et al.* (1966) at Lahore sought to develop an economical method for preparing infant foods from cereals by increasing their digestibility through germination. The cleaned seeds of wheat, maize and sorghum were (1) steeped in lime water (pH 8) at 28°C for 24 h, (2) washed and steeped in water at 28°C for two days, (3) spread over jute matting; germination continued until the sprouts attained an average length of 2·5 cm, (4) steeped in 15% H_2SO_4 for 24 h, (5) washed in warm water, (6) heated at 78°C for 6 h. The composition of the three resulting malt flours in dry matter (105°C for 24 h), thiamine (thiochrome method), nitrogen (micro-Kjeldahl), soluble nitrogen (extracted with 10% trichloracetic acid), and soluble sugars is shown in Table 5.131.

Malt flour with the highest levels of soluble nitrogen, soluble sugars and thiamine was obtained from wheat, the next highest from maize and the lowest from sorghum.

Perissé *et al.* (1959) described how sorghum beer, known as dam in the Moba tongue, was prepared by a Moba woman in Togo. The grain was steeped in an earthenware jar for about 14 h, twice washed in water, over a period of 24 h, then spread in a layer about 5 cm thick, and covered with moistened leaves. Germination continued for about 48 h during which the temperature rose to 37°C.

The germinated grains were dried in the sun and the malt was ground on a stone mill. The malt flour was mixed with water at ambient temperature and the supernatant liquid strained from the residue and put on one side. The residue was mixed with water and boiled for about 1 h. The boiling liquid was combined with the supernatant liquid, the temperature of the mixture being about 70°C. As the liquid cooled fermentation started spontaneously.

The liquid was decanted and the residue (brewers' grains) filtered twice through a cone of plaited straw. The two liquids, the filtrate and the first decanted supernatant were boiled together for 1·5 h. The liquid was allowed to cool then filtered. This cooled wort was inoculated with yeast from an earlier brew carried on a handful of vegetable fibres which were later removed and dried for subsequent use. It was determined that the fibres carried the yeasts *Saccharomyces rouxii* and *Saccharomyces oviformis* in addition to various moulds and coliforms.

After 8 h of fermentation the beer was considered ready for consumption. It was golden brown in colour, slightly cloudy and contained in suspension some yeasts and small particles of cellulosic material. It is not normally kept for more than 2–3 days. The deposit remaining in the fermentation jar was used to make a sauce of variable composition which may be eaten with cooked sorghum or pearl millet dough (pâté).

The composition of the sorghum grain (one month after harvest), the germinated sorghum (with moulds), the wort and the beer are shown in Table 5.132. (The analyses were made by the Institut Pasteur in Paris, methods not stated). The beer was higher in riboflavin and vitamin B_{12} but lower in all other nutrients analysed than the original grain.

Close and Naves (1958) reported the amino acid composition (ion-exchange chromatography), tryptophan (method of Dickman and Crockett, 1956) of a market sample of the germinated and ungerminated sorghum Kivu, grown for beer making in the Belgian Congo.

During germination total nitrogen fell from 1·8% to 1·7%; lysine from 156 to 138 mg/g total N, threonine from 202 to 193 mg/g total N; leucine rose from 798 to 848 mg/g total N with isoleucine remaining roughly constant. Close and Naves recommended that N × 5·35 is a more accurate conversion factor than N × 6·25.

Because of the importance of sorghum malt to the Bantu beer industry in S. Africa, a number of studies on the amylases were reported (Novellie, 1959; 1960ab; 1962abc; Dyer and Novellie, 1966; Botes *et al.*, 1967ab; Daiber and Novellie, 1968). Sorghum grain before germination contains only traces of α- and β-amylases and optimum amylase production in sorghum occurs at temperatures much higher than those needed for barley. In barley malt β-amylase is the major amylase. In sorghum the ratio of α- to β-amylase may vary from 2/1 to 3/1 (Novellie, 1960ab; Dyer and Novellie, 1966).

The sorghum amylases have been purified and studied in detail (Novellie, 1960ab; Botes *et al.*, 1967ab). The amylases of sorghum malt may be soluble, partially insoluble or totally insoluble depending on the level of polyphenols in the grain.

Daiber (1975) reported that polyphenols in some sorghum hybrids render the amylases insoluble and completely inhibit their activity. Such malts are useless for brewing. High polyphenol malts also inhibit the thermophilic lactobacilli and the lactic acid production connected with the souring process of sorghum beer brewing (Watson, 1975). These inhibiting effects may be prevented by a pre-malting treatment of the grain (see p. 454). Protein digestibility (method not stated) was raised from 60% before treatment to 80% after treatment by this method (Dreyer and Concannon, 1975).

In addition to the amylases, Watson *et al.* (1974), Watson and Novellie (1974) and Adams *et al.* (1975) have studied other enzymes in sorghum including α-glucosidase.

Sorghum maltase differs in its solubility characteristics from barley maltase (Watson and Novellie, 1974). It is present in ungerminated sorghum and does not increase greatly during germination. The sorghum saccharification system, α-amylase plus β-amylase plus maltase is tolerant of acidity and functions well at pH 4 (Novellie, 1977).

Watson and Novellie (1976) discussed the development of amylase during germination and malting of both "normal" (Barnard's Red) and bird-resistant (high-tannin SSK 2) sorghum cultivars. They referred to earlier publications by Novellie (1959, 1960ab, 1962a, 1968) and stated that combined sorghum α- and β-amylase activity increased rapidly and at roughly the same rate during the first 2–4 days of malting; β-amylase constituted 25–30% of total amylase activity.

Watson and Novellie reported that α- and β-amylase were almost completely removed from germinating low tannin sorghum by aqueous extraction. On the other hand, the diastatic activity (combined α-plus β-amylase) of aqueous extracts of bird-resistant, high-tannin sorghum was virtually zero until the fourth day of germination, following which the % extracted increased until the eighth day when the extract contained about 50% of the total diastatic activity.

When extracted by 2% peptone, the combined α-plus β-amylase activity of both low- and high-tannin sorghum increased rapidly during the first 3 days of malting. Watson and Novellie proposed that tannins rendered the amylases insoluble in water. Peptone, they suggested, combined preferentially with the tannins.

They also reported that during the early stages of malting, extraction of maltase from bird resistant grain was increased by the presence of both Triton X 205, Haas, USA (Rohm and Haas, USA) and peptone. They quoted Baijal *et al.* (1972) "Triton is known to increase protein extractability by its action on phenolic compounds". They also quoted Daiber (1975) who stated that there is a considerable reduction in total polyphenol content of bird resistant sorghums during malting; the process of malting may release compounds that combine with tannins.

Watson and Novellie (1976), quoting Daiber (1975), Watson *et al.* (1974) and Watson and Novellie (1974), stated that while maltase of both normal and bird-resistant sorghum may be insoluble in water, it may nevertheless be active in the insoluble state with tannins present. Consequently maltase can function even in high-tannin sorghum beers. The ability of maltase to convert maltose to dextrose depends upon an adequate conversion of starch to maltose by the sorghum amylases both of which may be inhibited by polyphenols.

Goldberg and Thorp (1946) reported from South Africa the thiamine (thiochrome, Goldberg *et al.*, 1945), niacin (microbiologically, Krehl *et al.*, 1944) and riboflavin contents (fluorimetric, Andrews, 1943) of samples of grain and malt of maize, sorghum, finger millet and pearl millet, (genetic origins were not specified). The results are shown in Table 5.133. After malting there appeared to be an increase in riboflavin in all cereals: niacin increased in maize and the millets but declined in sorghum.

Sorghum beer was described by Goldberg and Thorp as a thin, sweetish-sour cereal gruel with an alcohol content of about 3%. The strained residue provided the inoculum from which second beer (sibebe), and third beer (nsipo) were prepared. A watery third beer may be used in a number of fermented cereal foods, the vitamin content of which appears in Table 5.134.

It was suggested that though differences in processing technologies were evident between village household and industrial processes of sorghum beer production, analysable nutrient contents did not differ very greatly. Values for leting, a lactic acid fermentation product usually made of maize meal and sorghum malt are also shown in Table 5.134. Leting may also be made from millets.

Von Holdt and Brand (1960b) identified the saccharides present in the short red sorghum (K 2 Red grade) before and after malting (Table 5.135). There appeared to be an inexplicable loss of roughly 14·8% of the original grain during malting.

Dreyer (1968) reported the biological value (Mitchell 1924, 1943 modified) of the protein of 136 samples of South African foods. Analyses of sorghum malt, sorghum beer and its ingredients are given in Table 5.136 in comparison with maize malt and wheat bread.

The comparatively low digestibility of the sorghum beer and its ingredients was attributed to the fact that sorghum provided about 72% of the total protein. Malting increased the digestibility of sorghum protein by 7%; no comparable improvement was shown on malting maize. The effect of malting on sorghum was attributed to the fact that the corneous protein matrix of the endosperm is more effectively digested by the phytoenzyme liberated during the malting than by the enzymes of the gastrointestinal tract.*

Daiber and Achtig (1970) established methods to measure the grain and malting quality of cultivars available in the Republic of South Africa. Tests on the grain included moisture determination, water uptake capacity (absorption), the ability to germinate (germination capacity) and to germinate rapidly (germination energy), water sensitivity (grains may not germinate in excess moisture), particle size distribution, hardness (Brabender hardness tester), compressibility (Instron tester), specific gravity, grain size distribution, nitrogen determination. The malting method was standardized.

Preliminary results on 120 grain samples of 12 cultivars grown in different environments in the 1967/68 and the 1968/69 seasons indicated

* (Authors' note: it is not stated whether or not the sorghum was "high-tannin". If a high-tannin sorghum was used, this might account for the low protein digestibility.).

that both environment and genetic history influenced grain and malting quality.

Aucamp *et al.* (1961) reported the protein (N × 6·25), fat, fibre and ash (AOAC, 1950), the B vitamins (microbiologically), thiamine (Snell, 1950), niacin (Snell, 1950 and Barton-Wright, 1952) and riboflavin (British Pharmaceutical Codex, 1954) of eight samples of typical sorghums grown in South Africa, including short red, white, bird-proof, Martin and Hegari, obtained from the Potchefstroom Agricultural College, and of eight commercial samples of malt prepared either by the open floor process or by the indoor pneumatic method (Schwartz, 1956). A breakfast food prepared from sorghum and consisting of a mixture of malt and grain flour with some additional bran was also analysed. The values are given in Table 5.137.

The values for niacin and riboflavin are very similar to those for sorghum and malt reported by Goldberg and Thorp (1946) but the thiamine values in the present study were higher than theirs which averaged 0·3 mg/100 g for grain and 0·17 mg/100 g for malt. The changes found in the vitamin content of malt, DWB, from sorghum are shown in Table 5.138. The sprouted grain was either dried in the sun or at temperatures below 50°C. In the preparation of breakfast foods, the malt may be "cured" at much higher (unstated) temperatures. The drop in thiamine with curing is the most marked of the results presented.*

Brewing

Aucamp *et al.* (1961) reported the composition including the B vitamins (microbiologically thiamine, Snell, 1950; niacin, Snell, 1950 and Barton-Wright, 1952 and riboflavin (British Pharmaceutical Codex, 1954) of sorghum beers brewed in municipal South African breweries between 1954 and 1960. During this period, breweries were replacing unmalted sorghum grain with maize meal or grits as adjunct. As sorghum malt of higher diastatic power became available, the ratio of malt to adjunct decreased. The wide variability in composition shown in Table 5.139 reflects the variability in ingredient proportions. Maize grits are preferred to sorghum as adjunct partly because they are cheaper and partly because they give a beer of better body and texture. The preference for maize grits reduced the niacin content of the beer as there is little or no synthesis of this vitamin during brewing. The five beers in the study with the highest niacin contents (> 500 μg/ml) were brewed either wholly from sorghum malt and grain, or with less than 10% of maize products. During brewing there is a small increase in thiamine (16–32%) and a substantial increase (70–80%) in riboflavin (Table 5.139). Novellie (1977) confirmed the lower vitamin content when maize grits replace sorghum (Table 5.140).

* (Authors' note: if, as Aucamp *et al.* suggest, curing is common in baby food preparation, the higher thiamine in the baby food, Table 5.138, may be the result of adding grain flour and bran.).

Representative samples of sorghum grown in South Africa, described as short red, white, and bird-proof sorghum, and a commercial sample of malt from short red sorghum, were analysed for amino acid composition chromatographically except for tryptophan, cysteine + cystine and methionine which were assayed microbiologically (Horn and Schwartz, 1961).

Of the total nitrogen in the grain samples (short red 1·4%, white 1·8%, bird-proof 1·6% and short red malt 1·5%) 78–86% was present as free and combined amino acids. The proportions of the amino acids in the three grain types of sorghum and the malt were very similar but Novellie and O'Donovan (unpublished results) found considerable change in the free and combined acids after malting.

Results obtained by Horn and Schwartz agree with those of Close and Naves (1958) from the Belgian Congo, except that the former quote arginine to range from 237–244 and the latter 312–400 mg/g total N.

Novellie (1977), citing Dreyer and Novellie (1962) stated that protein digestibility, as measured by rat trials, was increased by malting. Novellie (1977) also stated that the polyphenol content of bird-proof sorghum varieties was reduced by malting citing Daiber (1975) and Watson and Novellie (1976).

Doggett (personal communication 1978) describes how certain people of East Africa permit moistened sorghum to germinate in a bed of wood ash before converting it to beer. While no chemical evidence to support the following contention has been found, it seems probable that under such alkaline conditions some proportion of the polyphenols present would be hydrolysed. Whether this is so, together with the effect of the alkaline environment upon other nutrients in sorghum, seems worthy of investigation.

Though the small size of the grains raises technical problems in large scale malting, both pearl millet and finger millet can give diastatic activities equivalent to sorghum. Their diastatic activity expressed in SDU (Sorghum Diastatic Units, Novellie, 1959) is shown in Table 5.141.

Schröder (1972) stated that sorghum beer is low in nitrogen. Total crude protein of beers from 14 South African breweries during 1970–71 ranged from 0·65 to 1·0% of which 0·09–0·17% was soluble protein. In malt, free amino acids comprised from 6 to 7%, and the free plus combined amino acids from 48 to 86% of total nitrogen. The free amino acid content of sorghum beer, 5–17%, is higher than that of the malt, indicating that some protein hydrolysis occurs during brewing (Bantu Beer Unit Annual Report, 1971).

Amino acids in wort can be classified into four groups according to their rate of absorption by yeast (Pierce, 1966): (1) those taken up rapidly; (2) those taken up more slowly; (3) those absorbed after a considerable period and (4) proline only, which is not absorbed by yeast under brewing conditions. Table 5.142 shows the amino acids found in the sorghum malt wort, each labelled according to its rate of disappearance.

Scheigart *et al.* (1972) compared with maize grits, four sorghum varieties used for brewing in South Africa; DC 36, NK 222, NK 150 and Barnard Red using (1) a Miag Universal laboratory rice pearling machine, (2) slight pearling followed by impact milling and a Forsberg separation table and (3) a degerming apparatus based on a model designed by the USDA Northern Regional Laboratory.

No precise data were given but:

(a) percentage yield of a low oil fraction varied among varieties, NK 222 being highest;
(b) all milling methods investigated were satisfactory;
(c) protein content of the final product (11–13%) was acceptable but higher than that of maize grits;
(d) losses of thiamine, niacin and riboflavin were very high;
(e) the by-product was rich in vitamins, oil and protein.

Pilot scale and commercial brewing experiments were successful. The beer was well received by a Bantu taste panel and technologically the pearled sorghum grits behaved very similarly to maize grits. Niacin was higher in the sorghum than the maize beer.

Von Holdt and Brand (1960a) determined the starch content of the South African grown sorghums short red, Framida, Martin, white and bird-proof (perchloric acid extraction, precipitation with iodine and estimation of the carbohydrate with anthrone) which gave an overall recovery of 97·7% on samples of pure sorghum starch. Starch contents of sorghum grains and malts are shown in Table 5.143.

The starch content of maize grits and maize meals ranged from 71·8 to 81·7% and of the spent grains from different breweries from 14·7 to 50·3%. In a trial brew, 40·7% of the starch in the raw materials escaped degradation during mashing, 32·8% appearing in the final beer and 7·9% in the spent grains.

Novellie (1966) reported the effects of temperature, time, pH, diastatic power, malt/grain ratio and the nature of the substrate on the production of sugar by sorghum malts. With malts of diastatic power up to 25 SDU/g, the substrate was in excess; with diastatic powers of 45 SDU/g, or greater, the amylases were in excess. The optimum malt/grain ratio lay between 1/3 and 1/2 at pH 6·3 which reflects the ratio of ungelatinized to gelatinized starch and the affinity of the amylases for these two forms of substrate.

O'Donovan and Novellie (1966) reported that the average fusel oil content (described modification of AOAC, 1960, method for alcoholic beverages) was high (227 p.p.m.) compared to European type beer (110 p.p.m.). The fusel oil in sorghum beer increased with age. In some tribal processes, sugar added to the (unpasteurized) beer prolongs fermentation, produces a stronger beverage, and increases the fusel oil content. In one experiment with sugar added, the alcohol content was 9·5% v/v and fusel oil above 600 p.p.m. Increased yeast in the wort reduced fusel oil from 290 to 190 p.p.m. without affecting alcohol content. Increased yeast with added ammonium dihydrogen phosphate (1%) reduced the fusel oil content to 90 p.p.m. and ethanol content from 3·5 to 2·5% v/v.

Increased yeast (a) with and (b) without added ammonium salts, reduced fusel oil content (a) two-thirds and (b) one-third.

Johannsen (1969) used male and female Wistar albino rats on a stock ration to compare as drinking water, (A) a solution of "fusel oil" containing ethyl alcohol 6%, ethyl acetate 0·004%, isoamyl alcohol 0·12%, β-phenyl ethyl alcohol 0·12%, isobutyl alcohol 0·2% and acetic acid 0·2% (equivalent to double the concentration normally found in sorghum beer) (O'Donovan, unpublished data); (B) a solution of 2% isoamyl alcohol; and (C) tap water.

After 56 weeks, there were no significant differences in body weight, liver weight, alcohol dehydrogenase (ADH), glutamic oxalacetic transaminase (GOT), glutamic pyruvic transaminase (GPT) activities or liver protein. Histological examination of the livers, kidneys, hearts, spleens and lungs did not show any significant abnormalities. Johannsen commented that no significant dietary imbalances or toxic effects were observed from (A) or (B).

Among many African communities, sorghum beer and wine play an important sociological role, being served on festive occasions, at social gatherings and during community work projects. According to Winter (1964), among certain population groups fermented sorghum beverages represent 20% of total cereal consumption. Chevassus-Agnes *et al.* (1976) described in detail the traditional methods by which sorghum beer (amgba) and sorghum wine (affouk) are made in the Cameroon, together with an evaluation of their nutritional quality. Observations were made of amgba beer being made by three different groups of Baya women during 1968, 1970 and 1971.

Brewing of Amgba

The brewing process includes malting, mashing, wort boiling and fermentation. Sorghum ("mouskouari" or "djigari") is steeped in water for 12–72 h when the H_2O content rises to about 40%. The steeped grain is piled in thick heaps, the temperature rises quickly and germination continues for 24–36 h. In warmer regions, the women spread the steeped grain onto a clean surface of dried leaves or flattened ground to a depth of 3–5 cm. The grain is moistened and covered with a layer of leaves. The average duration of germination is four days.

Elsewhere (among the Lamé, Moundang, Toupouri and certain Baya) after the grain has germinated it is placed in compact piles: the temperature rises, saccharification of starch increases and stops when the inactivation temperature of the amylases is reached. The malt is sun dried (to 15–20% H_2O) and ground to a mixture of flour and grits.

The coarse flour is mixed with water and a gelatinous or viscous substance (gombo (*Hibiscus esculenta*) or the sap of certain trees, particularly *Triumfetta* sp.) to precipitate suspended particles. The supernatant is decanted, the residue containing undissolved malt flour is heated to the consistency of a thick paste and the supernatant added back to the main mash. When mashing is complete the wort is filtered to remove spent grains.

In Northern Cameroon, the wort is allowed to stand for a few hours or up to a full day when lactic acid fermentation produces an acid flavour, essential in "kafir beer". The wort is then boiled, some protein coagulates and can be skimmed off. The wort is cooled then allowed to ferment for between 12–48 h. Among the Baya population, "affouk" is added to the fermenting liquid when the temperature of the wort reaches approximately 30°C. The resulting beer is clear, fizzy, semi-sparkling and is consumed shortly after preparation. In many regions, the barm (the yeasty foam on the fermenting beer) is retrieved and sun dried and is later used in brewing or as an ingredient in the diet. The preparation of amgba is outlined in Fig. 5.III.

"Affouk" Sorghum Wine

The preparation of affouk is much simpler than that of amgba. The sorghum wine, most commonly known as kpata, bears various other names: affouk among the Baya, affoukou, poukou or vone among the Mboum, do or do'di among the Dourou, bouèrou among the Namchi and the Voko, tidéré among the Moundang and finally balda-babaran among the Giziga. The total preparation time for sorghum wine is 5–6 days.

After washing, the grain is steeped for 1–3 days and then drained. As with beer, the cooking time varies inversely with the water temperature. The Dourou allow the grain to dry for 1 day. The grain is then ground to a coarse flour and mixed with water to form a paste which is left to stand 1–4 days, or until a sour lactic acid taste develops. The paste is toasted or cooked to a firm consistency, then cooled. Following a process similar to amgba, the sorghum malt is ground or in some communities kept moist for 10–15 days to allow mould formation. (The Namchi, the Dourou and the Mboum insist that mouldy grey-green malt is essential to make a "good wine".) The ground malt is mixed with the cooked paste and a small amount of water. The malt amylases and microorganisms present quickly bring about liquefaction and the mixture is allowed to ferment for approximately 48 h (1 day at least, 5 days at most). A small amount of malt may be added after 12 or 24 h to increase diastatic activity.

When fermentation is complete, water is added and the mixture is filtered. A lesser amount of water will give a more concentrated beverage, called "bolo" among the Baya. The filtered liquid is drunk warm. The filtered sediment may be used as animal feed or sun dried and stored as flour. The preparation of affouk is outlined in Fig. 5.IV.

Table 5.144 provides data on the nutritional composition of amgba and affouk compared to sorghum and malt. Nutritional losses and gains resulting from the preparation of sorghum beer and wine are shown in Table 5.145.

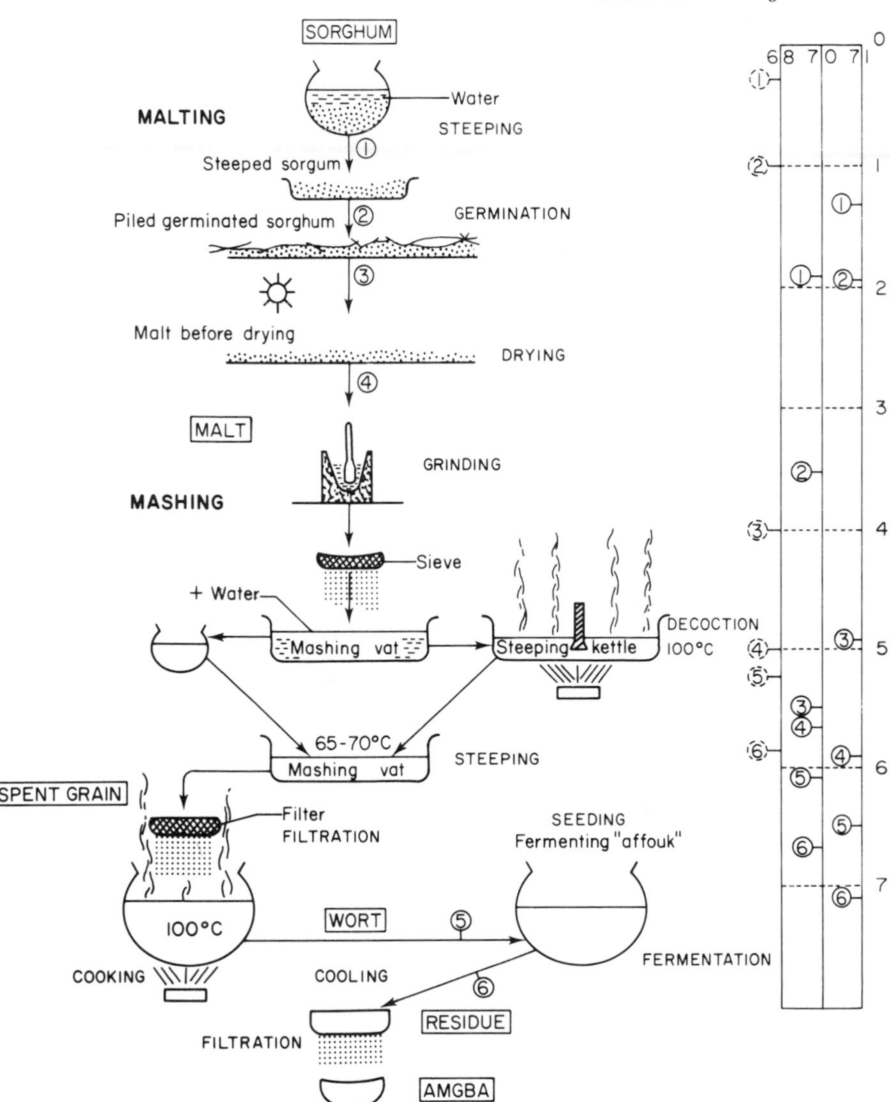

Fig. 5.III Traditional preparation of "Amgba" (from Chevassus-Agnes *et al.*, 1976). Figures at extreme right indicate approximate number of days for each stage of the process.

In the preparation of amgba and affouk, significant losses of solids are incurred during grinding and decantation. Amgba and affouk contain respectively only 20 and 31% of the original sorghum solids. In terms of calories and protein, it is more economical to consume the sorghum flour than sorghum beer or sorghum wine. While the sifted flour contains 75% of the calories and 69% of the protein of the original grain, equivalent figures for the beer are 20% of the calories (29% if the calories from the alcohol are

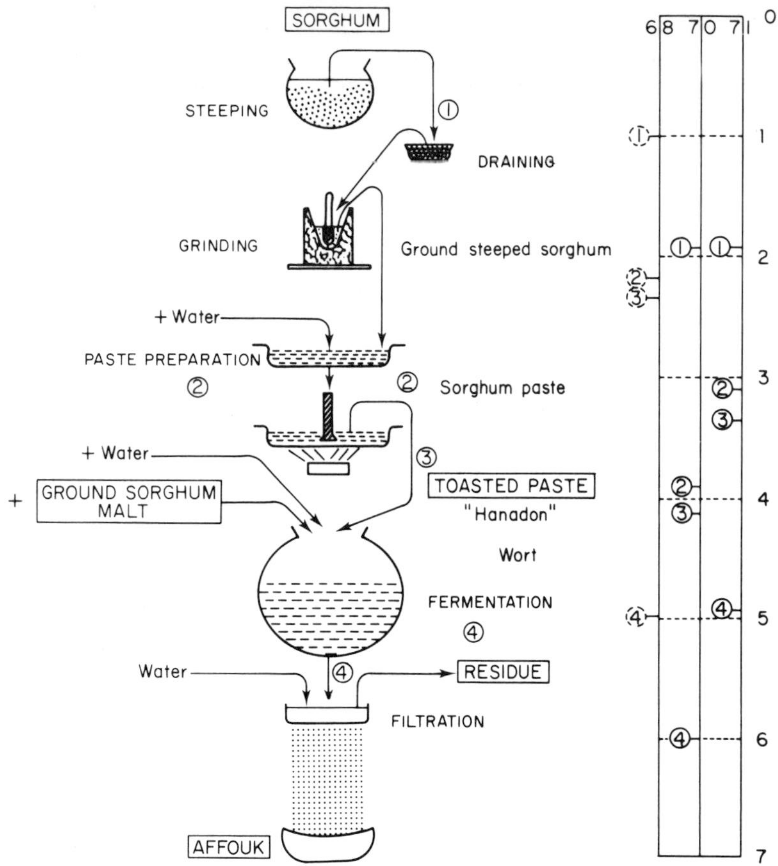

Fig. 5.IV Traditional preparation of "Affouk" (from Chevassus-Agnes *et al.*, 1976). Figures at extreme right indicate approximate number of days for each stage of the process.

included), and 19% of the protein. The preparation of wine also results in nutritional losses, as only 33% of the sorghum calories (55% if the calories from the alcohol are included) and 41% of the sorghum protein are retained.

The amgba contains approximately one-third of the Ca and one-third of the P present in the starting material. Ca and P are present in slightly larger amounts in wine. It is also interesting to note that sorghum wine is comparatively a better source of Ca and P than is sifted sorghum flour.

While fermentation decreases the dietary calorie and protein contribution of sorghum, this same process enhances the vitamin/calorie ratio. The increase in the vitamin content of the beer and wine is attributed to microbial synthesis. Because thiamine producing yeasts are left in the beer, the sorghum beer is found to have a higher concentration of this vitamin. Thiamine is present in lesser amounts in sorghum wine.

The riboflavin content of sorghum wine and beer is greater than that of either the sorghum flour or the grain; the amount synthesized during germination and fermentation more than compensates for losses due to heat and light.

Processing for Animal Feed

It is not the primary purpose of this publication to review the effect of processing on the nutritive value of sorghum and the millets intended for animal feed. The text includes only such recent data on processed animal feeds as may have a bearing on the use of sorghum and millets in human nutrition. Most of the research reported is from the United States. The extensive literature on processing sorghum grain for ruminants has been reviewed by Hale (1970) and more briefly by Riggs (1970) and Totusek and White (1968). Quisenberry and Tanksley Jr (1970) reviewed processing methods related to poultry and swine nutrition.

When fed as whole grain to cattle the dense, hard endosperm and waxy bran of sorghum is relatively unavailable to rumen bacteria and a large proportion appears unchanged in the faeces. Hence, some form of grinding or processing is essential for most animals other than sheep which are able to masticate the grain more thoroughly. (Hale, 1970). Before 1960, little attention was given to sorghum processing other than grinding or crushing (Hale, 1970). More recently, alternative methods have been described.

Totusek and White (1968) regard rate of weight gain and feed efficiency (weight gain/unit weight of feed intake) as the most useful criteria of processing efficiency. A processing method which results in lower feed intake at constant rate of gain reflects an increase in energy value of the grain processed.

Less interest has been shown in processing sorghum for feeding swine than for ruminants. Quisenberry and Tanksley Jr (1970) suggested that for swine the grain should be ground fine to moderately fine since coarsely ground grain is not so well digested; very finely ground grain is dusty and appears less acceptable to swine.

Lawrence (1967) comparing equipment used in the UK reported that sorghum ground in a hammer mill was significantly higher in digestible energy content for pigs that when coarse-rolled (crimped) in a Peerless roller mill. The growth rate, feed conversion efficiency and nitrogen retention were all significantly higher for more finely hammer-milled grain than for the coarser crimped grain. The following are the more common processing methods.

Soaking

Sorghum may be soaked for 12–24 h before grinding or rolling to soften the endosperm and waxy covering. Husted *et al.* (1968) reported a slight improvement in TDN (true digestibility of nitrogen) from water soaked grain than from dry-rolled but Ely and Duitsman (1967) cited by Hale (1970) found no advantage from water-soaking in fattening steers.

Steam-rolling

Sorghum grain may be softened by treatment with live steam for 3–5 min prior to rolling or flaking. (Hale 1970, Totusek and White, 1968).

Steam-flaking

During steam-flaking sorghum receives greater exposure to steam than in steam-rolling. Hale (1970) states that the grain is subjected to steam at low pressure and high moisture, the grain leaving the tempering chamber at approximately 100°C with a moisture content of 18–20%. The grain is then rolled to thin flakes with starch gelatinization not higher than 50%. The thin flakes were best fed intact to cattle; grinding the flakes reduced feed intake. (Hale, 1970).

Pressure-cooked Flaked

Pressure-cooking is carried out with steam at about 50 lb/in^2 for 1–1·5 min. The temperature of the grain is reduced to below 200°F (93°C) and the moisture content to about 20% before flaking. A tougher flake results from this process than from ordinary steam-flaking (Hale, 1970, Totusek and White, 1968).

Expansion–Extrusion

Prasad *et al.* (1975) described comparative tests in which sorghum was processed in a Wenger X-100 cooker-extruder.

High Moisture Processing: Reconstitution

High moisture processing is applicable:

(a) to early harvested grain with moisture content between 20 and 40%.

(b) normally harvested grain to which water is added to raise the moisture content to above 20%.

The grain is subsequently stored under anaerobic conditions for about 21 days. It must be ground or rolled before feeding either before or after storage (Hale, 1970, Totusek and White, 1968).

Walker and Lichtenwalner (1977) reported the effect of reconstitution (adding water to raise the moisture content of whole dry grain to approximately 30% followed by sealing the wet grain in air-tight containers and storing at 30°C for 21 days before grinding) on the protein solubility (Landry and Moureaux, 1970) of (1) a heteroyellow, elevator run, sorghum, (2) dwarf waxy Redlan (SA 413), and (3) nonwaxy Redlan (TX 378). The results are presented in Table 5.146. Reconstitution decreased the total amount of soluble protein and tended to shift a greater percentage of the protein from the alcohol soluble and glutelin fractions to the water soluble fraction.

Mean apparent ruminal digestibility measured by the *in situ* technique of Van Hellen and Ellis (1973) at 24, 48 and 72 h, for the five protein fractions ranged from 40·2% (fraction IV at 24 h) to 84·9% (fraction I at 72 h). Digestibility of the water soluble fraction I was significantly higher than the other fractions. The effect of the waxy gene on apparent ruminal digestibility for the five fractions is shown in Table 5.147. In total the proteins in waxy grain were significantly more digestible than nonwaxy proteins; reconstitution did not significantly affect total ruminal protein digestibility.

Popping

Hale (1970) stated that limited feeding trials with popped sorghum, (similar to popped maize) suggested that because of its low density, steers could not eat enough to gain weight satisfactorily.*

Reeve and Walker (1969) popped grain sorghum (of unstated origin) under laboratory conditions by exposure to temperatures of 246°C in a stream of dry air in a surface combustion pilot toaster. Samples of the raw and popped grains were prepared for histological study by modifications of the method described by Larkin *et al.* (1952). Microscopic examination was made under normal and polarized light.

Localized rupture of the cell wall occurred in the expanded endosperm. The spongy expanded endosperm consisted of intact cells within which the gelatinized starch formed a characteristic "soap bubble" structure, typical of popped sorghum, popped maize and popped dent maize. Endosperm cell walls near the aleurone layer remained mainly intact though some swelling was evident. Near the scutellum, the endosperm starch granules remained ungelatinized. The scutellum did not expand and popping did not disorganize the structure of the embryo. The more completely gelatinized starch network in the endosperm stained blue with dilute iodine, ungelatinized starch varied from blue-black to purple. Except close to the scutellum, the corneous endosperm completely enclosed a narrow central core of floury endosperm.

Riggs *et al.* (1970b) described a process for popping sorghum grain in which six gas infrared burners were suspended above a reciprocating steel table. The grain passed under the burners and was discharged at about

* (Authors' note: though popping is not normally used to manufacture animal feed, the technique is sufficiently close in principle to expansion–extrusion cooking to justify inclusion in this section.)

178°C. The total product was designated "normal run" and consisted of grains completely popped and others partially popped or unpopped. Four fractions were therefore studied:

(1) original grain
(2) "normal run"
(3) completely popped
(4) unpopped and partially popped

The proportion of (3) increased with initial moisture content.

Initial moisture of whole grain (%)	Completely popped grain (%)
11·3	27·4
12·4	33·9
14·7	43·2
16·8	43·5

The bulk density ranged from 779 g/l for original grain to 98 g/l for completely popped.

The popped sorghum was fed to beef cattle, in mixtures consisting of grain 92%, cottonseed meal 7%, minerals and salt 1%. The proximate analysis of the feed mixtures (AOAC, 1960, DWB), feed efficiency, and coefficients of apparent digestibility are shown in Table 5.148 (Riggs *et al.*, 1970b).

The popping treatment produced significantly higher digestibility coefficients for all constituents other than lipid (ether extract), fibre, crude and true protein. Daily gain was not significantly different among treatments; treatments (2), (3) and (4) showed significantly lower feed intakes than treatment (1).

Micronizing

"Micronizing" is a process in which sorghum is subjected to dry heat from gas-fired micro-wave infrared generators, then rolled to a density of about 0·5 kg/l (Schake *et al.*, 1972; Riggs *et al.*, 1970a).

Rooney *et al.* (1977) reported that in a laboratory micronizer, sorghum showed great variation in bulk density, extent of gelatinization and degree of expansion. Waxy sorghums had the greatest expansion followed by heterowaxy and nonwaxy types. Micronized pearled sorghum had light, white flakes with a bland flavour and more gelatinized starch than micronized unpearled grain.

Baird (1976) reported that roasting bird resistant sorghum gave a small improvement in average daily weight gain and F/G ratio over nonheat treated sorghum in diets for feeder pigs. Roasting maize had no improving effect. The diets also included soybean, and roasted and rolled soybean improved performance compared to soybean meal.

Comparison of Processing Methods

Totusek and White (1968) and Hale (1970) presented tables, based on the literature reviewed, in which the processing methods described above were compared in terms of daily weight gain, daily feed intake, and feed/lb of gain for cattle. Because of the wide variation in the composition of the ingredients processed by the methods discussed, the comparisons may be regarded as only indicative of the relative merits of each processing technology.

In summary the compared results indicated, in terms of feed efficiency:

(1) Dry-rolling was about 1% more efficient than fine-grinding.
(2) Fine-grinding was about 5% better than coarse-grinding.
(3) Steam-rolling was about 2% more efficient than dry-rolling.
(4) Pelleted ground sorghum was 9% more efficient than unpelleted ground sorghum, and pelleted rolled sorghum 7% more efficient than unpelleted rolled sorghum. However, pelleting resulted in a slight lowering of carcass grade and a 1% decrease in percentage dressed weight which largely offset the improved rate of gain and feed efficiency.
(5) Steam-flaking improved feed efficiency compared to dry-rolled or ground grain by 3% and 7% respectively. Steam-flaking also increased feed intake.
(6) High moisture harvesting improved feed efficiency by 10% compared to dry sorghum.
(7) Reconstituted grain improved feed efficiency by 9% compared to dry sorghum.
(8) High moisture harvested grain was superior to reconstituted sorghum especially when both products were ground. Rolling was superior to grinding, and the rolled reconstituted grain was utilized almost as efficiently as the rolled high moisture harvested grain.
(9) Reconstituted grain was best stored whole and ground after storage. Sorghum stored in the ground form and then reconstituted was less efficient than dry sorghum.
(10) Pressure cooked sorghum was similar in digestibility (for cattle) to steam processed flaked grain.
(11) Adequate processing appeared to improve total digestibility of sorghum by cattle, but with the possible exception of reconstitution, seemed to have little effect on protein digestibility.

Riggs (1970) reviewed work carried out mainly between 1958 through 1970 at Texas A & M University on the utilization of sorghum by livestock. He stated that lambs responded to all processing treatments in a similar fashion and no advantages were obtained over those from coarse-cracking or dry-rolling.

Hinders and Freeman (1969) compared the % gelatinization (loss of birefringence of starch granules) and the starch availability (gas production from yeast fermentation of sugars produced by amyloglucosidase digestion of starch, adapted from Sandstedt *et al.*, 1962) of the grain sorghums (1) Funk 3761, red waxy, (2) Funk G 766, white nonwaxy with a relatively uniform distribution of corneous starch through the kernel, (3) TX 660, red with relatively softer, floury endosperm, and (4) red "elevator run" grain, subjected to pressure cooking and micronization. Starch was separated from the grain by laboratory wet-milling. The results are shown in Table 5.149.

Elevator run red grain 4 was also steam flaked.

	ml gas produced/h
Elevator run, red	
Ground	2·9
Steam flaked	12·7
Steamed, dry and ground	2·5

Sullins and Rooney (1971) examined by light microscope the changes which occurred when elevator run sorghum was processed as follows:

(1) dry-ground (hammer-milled)
(2) dry-ground (reconstituted) (moisture content 30%, stored in air tight structures for 21 days).
(3) dry-ground, reconstituted, and reground
(4) whole grain reconstituted, then ground.

Mean particle size distribution (modified sieving method of Florence 1968) of the grain from treatment (3) and (4) was smaller than from (1) and (2).

Microscopic analysis indicated that reconstitution affected the kernel structure at the subcellular level notably as a general disruption of the endosperm, especially in the peripheral area. The protein matrix was partially disrupted causing a release of free starch granules and protein bodies, believed to be caused by enzymatic attack during reconstitution, similar to what occurs during malting. The degree of disruption was lower towards the centre of the kernel. Storage under anaerobic conditions for 2, 4, 6, 8 and 10 months produced no additional breakdown of the endosperm of the reconstituted grain. The release of starch and protein, combined with the decrease in particle size, probably accounts for the increased feed eficiency (for cattle) of reconstituted grain observed by Berry (1970). Riggs and McGinty (1970) also reported that in cattle feeding trials reconstituting sorghum (RS 610) followed by grinding increased the digestibility of the protein by 16–22% and dry matter and organic matter by 17–29%, compared to dry ground grain.

Schake *et al.* (1972) reported a commercial feed lot evaluation of four methods of processing sorghum in feeding steers. In trial 1, (A) steam-flaked grain, (B) whole grain, reconstituted, rolled, (C) grain dry-rolled, and reconstituted were compared. Steers gained weight at similar rates and produced carcasses with similar characteristics. In trial 2, (A) steam flaked grain and (D) micronized grain were compared and similar results to trial 1 were observed.

Riggs and McGinty (1970) reported that RS 610 sorghum harvested when the grain reached (A) 32% moisture, when fed to cattle in ground form produced weight gains equal to the same grain harvested at (B) 14–16% moisture. Cattle fed (A) required 11–26% less dry matter from grain and 7–15% less total dry matter/unit of weight gain than cattle fed (B).

Osman *et al.* (1970) compared the *in vitro* starch digestion of barley and sorghum grain (not otherwise identified) using hog pancreatin or lyophilized bovine pancreas. Total starch (modified AOAC, 1960) and % starch digestion (mg maltose/100 mg grain starch incubated, DWB) were reported. The hog pancreatin digestion of three grain samples is shown below:

	Sorghum (%)	Barley (%)
Sample (1)	16·8	22·0
Sample (2)	17·0	21·8
Sample (3)	16·3	21·3

Steaming or pressure-cooking at 1·4 kg/cm^2 for 1 min without flaking decreased *in vitro* starch digestion of both grains compared to the untreated grains. Steam pressures exceeding 2·8 kg/cm^2 improved starch digestion, and flaking the grains after steaming or pressure-cooking markedly increased digestion of both grains, better digestion being obtained from the flatter flakes. Both enzyme sources gave similar results. The findings are summarized in Table 5.150.

Osman *et al.* commented that the increased susceptibility to enzymatic digestion of barley starch compared to sorghum starch may lie in the nature of the protein matrix surrounding sorghum starch granules (Rooney and Clark, 1968) or in the presence of "some inhibitor associated with the nonstarch fractions of sorghum grain".

Frederic *et al.* (1973) followed up the studies of Osman *et al.* (1970) and compared the effect on the enzymatic digestion (lyophilized bovine pancreas) of the starch of barley and "milo" sorghum subjected to various combinations of moisture, heat and pressure.

Enzymatic starch degradation was greatest for treatments involving the application of moisture, heat and pressure. Starch digestibility of pressure cooked grains was increased by each increment in steam pressure from 1·4 to 7·0 kg/cm^2. Flaking after steam processing or pressure-cooking increased starch degradation from 14% in untreated sorghum to 40% in dry-flaked grain; to 50% in steam processed flat-flaked grain, and to 60% in pressure-cooked flat-flaked grain. The optimum cooking pressure for both sorghum and barley, with flat-flaking, appeared to be 4·2 kg/cm^2 which gave 57% enzymatic starch degradation. The starch digestibility of dry, steamed or soaked sorghum was increased by hydraulic pressure, the optimum or "critical" pressure being 140 kg/cm^2. Pressures exceeding 140 kg/cm^2 did not cause further increase. The results appear below:

Grain	Starch digestibility (%)
dry (8% H_2O)	22
steamed (20% H_2O)	33
soaked (36% H_2O)	31

Frederick *et al.* (1973) concluded that there is a critical pressure at which enzymatic digestion is maximum and that this critical pressure appears to be altered by the moisture content of the grain and the temperature of the contact surfaces used to press the whole grain.

Creger *et al.* (1969) reported experiments in which sorghum was soaked in water with and without bacterial amylase and protease for a period of 21 days in the absence of air, dried, ground and fed as the grain source in a standard chick diet. In general it appeared that soaking with or without enzymes did not improve rate of growth; in some instances growth was depressed.

Five varieties of sorghum from the South Plains area of the USA were micronized by subjecting the whole grains to 2·5–3·0 μm infrared radiation for approximately 20 sec, followed by rolling and flaking (Yang and Tsai, 1973). The processed grain was examined by (1) chemical analysis, (2) microscopic examination, (3) *in vitro* starch degradation with amyloglucosidase-yeast, (4) *in vivo* rumen digestibility and (5) rat growth assay. It was concluded that micronizing ruptured the cell wall of grain sorghum and increased the availability of nutrients. The overall nutritional value of grain sorghum appeared improved even though 10–30% of the lysine present appeared to have been destroyed during micronizing.

Pigs fed sorghum (A) dry-ground, (B) steam-flaked, (C) reconstituted or (D) micronized, did well on all treatments and differences among treatments were small, though steam-flaked and reconstituted grains showed significantly higher apparent digestibility of both protein and gross energy than those fed dry-ground or micronized grain (Tanksley and Osbourn, 1969).

Brzozowski *et al.* (1972) compared sorghum (A) dry-ground, (B) pelleted, (C) micronized and (D) steam-flaked, in diets for pigs. The results appear in Table 5.151. There were no significant differences among treatments in digestible energy but (D) steam-flaked, gave a higher protein digestibility than the other treatments. (A) gave significantly lower metabolizable energy than the other three treatments. No significant differences were found in N intake, absorption or retention, though faecal N loss was significantly less for pigs on (C) and (D).

Harbers (1975) by SEM examined structural changes in the starch granules of samples of mill run sorghums (1) steam-flaked, (2) micronized and (3) popped, and the effect of amylolysis of that starch by (a) purified hog pancreatic α-amylase and (b) rumen microflora obtained from fistulated steers. Steam-flaking either left starch granules intact or produced homogeneous, gelatinized clumps of granules. Protein bodies remained intact but adhered to one another. Matrix protein stretched between starch granules. Rolling the steam-flaked grain pressed ruptured starch granules together and crushed them into small irregular pieces.

Popped starch granules expanded from the hylus to the edge and formed large, thin sheets that fused with adjacent granules. Nonexpanded starch granules did not show the same characteristics of starch granule damage produced by steam flaking. Protein bodies remained intact and matrix protein disappeared from among the starch sheets.

Micronizing expanded most of the starch granules, forming sheets. Starch granules near the surface remained intact resembling those obtained from popping. Protein bodies remained generally intact though some seemed to disappear.

Amylolysis, both by purified hog α-amylase and bovine rumen occurred mainly along the edges of ruptured granules from steam flaked or micronized sorghum. In popped sorghum starch, amylolysis began along exposed edges of the starch sheets. In general, the processing methods examined appeared to render sorghum starch more vulnerable to amylolysis than in the case with the intact granules of unprocessed whole grain.

McNeill *et al.* (1975) reported the chemical and physical properties of an undefined sorghum grain grown at one location and processed as follows:

(1) Dry-ground through a hammer mill.

(2) Reconstituted to 30% moisture, stored in air tight plastic bags for 21 days and ground in the same hammer mill as (1).

(3) Steamed at atmospheric pressure for approximately 20 min followed by flaking.

(4) Micronized by dry heat for 2–3 min followed by loose-rolling sufficient only to break the intact structure of the kernel.

The processed grains were fractionated into total carbohydrates ($CaCl_2$ soluble), ethanol soluble carbohydrates and starch (difference between total carbohydrate and ethanol soluble carbohydrates) by the procedure of Dimler (1964). Colorimetric determination of the carbohydrates was as described by Norris (1971). The results, DWB, are shown in Table 5.152. The solubilities of total carbohydrate (CHO) and starch in autoclaved rumen fluid and in 0·1 N HCl are shown in Table 5.153.

The susceptibility of the grain from the four processes to amyloglucosidase (Diazyme L-30) by a described procedure is presented in Table 5.154.

The relative viscosities of the starch–water suspensions of the processed grains before and after removal of available pre-gelatinized starch were determined with the Brabender amylograph using a modification of the procedure of Sandstedt and Abbott (1961). The highest viscosity peak was shown by (2) reconstituted, followed by (1) dry-ground, (3) steam-flaked and (4) micronized in that order. The addition of α-amylase (Rhyozyme-33) to the slurry had no apparent effect on the pasting temperatures, curve rises, peak viscosities or decline of the curves except in the case of steam flaked grain in which the addition of enzyme reduced peak viscosity. Microscopic examination of the starch granules under normal illumination and under plane polarized light showed no significant loss of birefringence for treatments (1) dry-ground and (2) reconstituted, but in both (3) steam-flaked and (4) micronized, the starch granules were greatly distended and nearly completely gelatinized as indicated by loss of birefringence.

The object of the work was to relate chemical and physical properties of processed sorghum to the pattern of *in vivo* (steers) carbohydrate utilization reported by McNeill *et al.* (1971) using similarly processed grain and by Potter *et al.* (1971). McNeill *et al.* (1975) concluded that processing methods which cause a release of starch granules from the sorghum protein matrix and increase their susceptibility to enzyme activity will increase carbohydrate utilization.

Flynn and Stallcup (1973) reported digestion trials with steers fed AKS 663 sorghum (A) dry-rolled, (B) high-moisture ensiled. The results were:

	A	B
	Coefficients of digestibility	
Dry matter (%)	45·3	69·2
Calories (%)	43·5	66·6
Crude protein (%)	9·8	45·3

At the end of the trial animals on diet (B) were in positive N balance; those on diet (A) in negative balance.*

Allee (1976) evaluated the effects of several processing methods on the nutritional value of milo sorghum for growing weaned pigs. The methods included, pelleting, extruding, steam-flaking, micronizing, and near-anaerobic high-moisture storage. The test diet included 68·8% processed or unprocessed milo, 26·7% sorghum meal plus mineral, vitamins and antibiotics. Pelleting was the only process that improved feed efficiency.

Shiau *et al.* (1976) examined two sorghums, C 42Y and 48R in pig feeding experiments. The sorghums were fed (A) raw and (B) micronized. Micronization increased *in vitro* starch availability, improved feed efficiency by roughly 16%, improved average daily gains by about 14%, improved crude protein digestibility, gross energy digestibility and dry matter digestibility. Of the essential amino acids, only lysine was affected, between 15 and 24% being destroyed during micronization.

*(Authors' note: Flynn and Stallcup stated that the tannic acid content of AKS 663 was determined from both treatments but did not state the values. AKS 663 is a bird-resistant sorghum, 2·3% tannin content was reported by Harris (1973) and "above 3·5%" by Harris and Burns 1973. Tannic acid as such is not known to exist in any sorghum—see Chapter 4).

Prasad *et al.* (1975) studied the processing of sorghum grain: (A) dry-rolled, (B) expanded (in a Wenger X-100) dried, cooled and crumbled, (C) pressure-cooked.

The processed grains in a ration also containing hay and soybean meal were examined for digestibility and N balance and *in vitro* for available lysine (Roach *et al.*, 1967) with chromatography, Moore and Stein (1954b) for lysine; pancreatin digestion (Melnic *et al.*, 1946 and Riesen *et al.*, 1947); and starch availability (Diazyme L-30 hydrolysis, Liang *et al.*, 1970).

In a second experiment sorghum from the same source was (D) dry-rolled, (E) steamed and flaked, (F) extruded (under heat and pressure, Hale and Theurer, 1972) and (G) expanded (in a Wenger X-50) dried, cooled and "crumbled". The grains, and a hay ration containing soybean meal were compared in a N balance trial.

The processed grains were examined *in vitro* for available lysine (Kakade and Liener, 1969); pancreatin digestion; starch availability (as in the first experiment, but substituting Diazyme L-100); and protein availability (by a simulated ruminant digestion technique, SRDT, utilizing strained rumen fluid from an animal fed a high-grain ration).

The proximate analyses of feeds (AOAC 1970) from both experiments are shown in Table 5.155. Nutrient digestibilities for rations and sorghum grains appear in Table 5.156 for experiment 1, and Table 5.157 for experiment 2. The effect of processing on N retention and *in vitro* lysine, protein and starch availability in both experiments are shown in Table 5.158.

In the first experiment treatment (B), expanded grains gave significantly higher digestibility of nitrogen-free extract than treatments (A) and (C) but there were no significant differences among treatments in protein digestibility or gross energy. N balance was inversely related to protein digestibility in both the hay ration and grain. *In vitro*, starch availability was increased by treatment (B). Expanding and pressure cooking decreased available lysine.

Crude protein and ether extract digestibilities in the hay ration and in the grain did not differ significantly among treatments. Digestibility of nitrogen free extract in the ration and in the grain was significantly higher from treatments (E) (F) and (G) than from (D), dry-rolled, but the difference between (D) and (G) was not significant. The *in vitro* tests showed a detrimental effect on protein from the moist heat treatments (E) and (G) compared to (D) dry-rolled and (F) extruded. N retention was greater, but not significantly, from (E) and (G) than (A) and (F). Compared to (D) dry-rolling, starch availability was significantly increased by treatments (E) and (F) and still further by (G).

In a finishing trial, with steers (I) flaked, and (J) extruded sorghum gave better feed:gain ratios than treatment (H) dry-rolled, though weight gains did not differ significantly. Results appear in Table 5.159.

Netemeyer *et al.* (1977) reported that in a feeding trial with 30 lactating dairy cows, rations containing (A) finely ground air-dried (90% dry matter),

(B) water-reconstituted (70% dry matter), and (C) acid-reconstituted (70% dry matter, 2% organic acids) sorghum grain, gave similar apparent digestibility of ration components and efficiency of feed utilization. Total milk yield, solids-corrected milk, and feed intake were not affected by grain treatment.

Shadid *et al.* (1976) compared five sorghum hybrids (A) red seed, heterowaxy endosperm, (B) red seed, normal endosperm, (C) heteroyellow seed, normal endosperm, (D) yellow seed, normal endosperm and (E) brown seed (high-tanning, bird-resistant), normal endosperm, (1) ground raw through a hammer mill or (2) micronized followed by hammer-milling, and fed in 16% protein rations, including soybean meal, for 42 days to pigs of 16 kg initial weight. No significant differences in average daily gain or feed intake were found among the hybrids though pigs fed (E) required significantly more feed/unit gain than pigs fed (A), (C) and (D). Micronizing resulted in significantly better F/G ratios than grinding. No statistically significant hybrid × processing interaction for F/G was found but (E) was improved most by micronizing compared to raw-grinding.

Conclusion

This ends the review of the influence of various methods of processing upon composition and nutritive quality. There is a larger literature that covers the technological consequences of processing for human and animal consumption, for alcohol production, for starch and other carbohydrates for edible and technical uses. This mainly technological literature has not been comprehensively reviewed.

A strong trend is evident in the SAT to commercial processing of sorghum and the millets ranging from small-scale village technologies to larger commercial processes. The need is apparent for closer cooperation between plant breeders and technologists to provide grain types that combine desirable agronomic properties with characteristics that facilitate efficient processing. To this end the various related initiatives, particularly of the International Association of Cereal Chemists (ICC), are to be welcomed and encouraged.

CHAPTER 6

Interrelations with Other Nutrient Sources

Introduction

Much has been published on the supplementation and fortification of cereals by protein sources, minerals and vitamins. Supplementation of bread and other baked products with proteins derived from fish, egg and milk, pulses and oilseeds, and synthetic amino acids was reviewed by Hulse (1974) and, with particular reference to triticale, wheat and rye, by Hulse and Laing (1974). Parpia and Swaminathan (1972, published 1975) reviewed the supplementation of cereals and cereal diets, with grain legumes and limiting amino acids.

Venkat Rao *et al.* (1964a) reviewed the mutual supplementation of cereal and legume proteins in typical Indian diets.

In the text that follows, the nutritional interrelations among sorghum, each of the millets, and other nutrient sources are discussed. "Interrelations" better describes what is reported than "supplementation" since many of the diets examined contain several different protein sources rather than one deliberately supplemented by another. Since much of the work reported was conducted along similar lines, it has been possible to prepare summary tables in which the authors of the original papers are listed in alphabetical order of first author. It is hoped this will enable the reader to refer more conveniently from the text to the tables and from the tables to the bibliography.

Though we, the authors, have organized the following text according to the cereal (i.e. sorghum or millet) that appears dominant, many papers describe mixtures of several cereals. Consequently, the reader is advised to make use of the subject index to ensure that all references likely to be of interest are traced.

Jalil and Tahir (1971) surveyed the world's plant protein resources but in somewhat less detail than Autret *et al.* (1969) The latter authors grouped countries under their main source of dietary protein. The groups were:

(1) those in which animal protein was the main source
(2) wheat countries
(3) millet and sorghum countries

(4) maize countries
(5) countries utilizing various cereals (wheat, maize, rice)
(6) rice countries
(7) countries where roots, tubers and plantains were the staple food

The millet and sorghum countries were placed in the Sudanese zone of Africa. Interestingly enough, in spite of a low level of income, the rate of protein calories was high, 10·9–14·5%, due to the facts that in addition to the protein of the millets and sorghum, pulses also formed an important part of the diet, and the nomadic stock breeders in the population also ate important quantities of animal products, mainly dairy produce. The protein score of 59–63 was comparable to that of the wheat countries. In most cases the first limiting amino acid in the diet was isoleucine, but the sulphur-containing amino acids (cystine and methionine) and lysine were close. The leucine contributed by sorghum was very high.

In Table 6.1, the values referring to the amino acid composition were taken by Autret *et al.* (1969) from FAO data (published as FAO, 1970), the protein score was calculated with whole hen's egg as the reference protein, the chemical score of essential amino acids and the protein score of the limiting amino acid were calculated by the method of Block and Mitchell (1946), the rate of protein calories was calculated using the conversion factor, 1 g of protein equals 4 calories, the conversion factor 6·25 was used to convert nitrogen to protein. Autret *et al.* commented that this factor was probably too high and introduced a systematic error which was not negligible in the evaluation of protein content and the protein/calorie ratio.

Operative Net Protein Utilization (NPU_{op}) and the calories derived from the net dietary proteins as % of total calories (ND_pCal %) or more simply, net protein calories, were calculated by the method of Miller and Payne (1961). National food balance sheets of the Indicative World Plan were the basis for the compilation. Hence, for example, wheat consumption was indicated but not that of wheat products. The effect of cooking or processing on the nutritive value of the agricultural products was not taken into account. Protein requirements were calculated according to a Joint FAO/WHO Expert Group (WHO, 1965).

Table 6.1 lists the countries which have millet and sorghum as the main source of protein. Table 6.2 includes other countries, such as India, in which sorghum and the millets form a proportion of the diet.

Autret *et al.* concluded that, considering the data presented, it was possible to rank the diets in the order, first countries with animal proteins; second, countries with wheat, and millets and sorghums; third, maize countries; fourth, rice countries; and fifth, countries with roots and tubers. The diets ranked to a very large extent according to the quality of the protein of the staple food. Autret *et al.* proposed that for wheat, and millets and sorghum, priority in research should be given to finding high lysine varieties rather than to providing high protein, since the diets already appeared to have a satisfactory protein/calorie ratio.

It is important to recognize the inherent weakness of considering the diets of whole countries because of the wide variations among different segments of the population. In India, for example, regarded by Autret *et al.* as a rice-eating country, the sorghum and millet eaters are far more numerous than those of the Sahel, but their importance tends to be obscured by the mass of India's more than 5×10^8 total population.

Early work in the supplementary relation of the proteins of certain seeds, evaluated by rat growth, reproduction and lactation trials was reported from the Johns Hopkins University, Baltimore, by McCollum *et al.* (1919). In 9% protein diets, in which "millet" furnished 3% of the dietary protein with soybean, wheat, rye, maize and rolled oats, or pea, separately providing 6% of the protein, in most cases the proteins of two seeds failed to adequately supplement one another. The 6% pea/3% millet diet gave better growth and reproduction curves than the same protein level from any of the cereal grains alone. Rolled oats 6% and millet 3% also showed a supplementary relation but the improvement was inferior to peas and millet. Wheat and millet, rye and millet, and maize and millet, showed virtually no supplementary effect.

On a 9% protein ($N \times 6{\cdot}25$) diet all derived from "millet" seed, rats grew slowly and did not reproduce; the addition of 2% protein from potato (giving an 11% protein diet) improved growth rates and some young were produced though there was a high mortality rate and those that survived showed early signs of aging (McCollum *et al.*, 1921).

Jayaraj *et al.* (1976) stated that duodenal ulcer is seen frequently in areas in India, particularly in the south, where the staple diet is milled polished rice and sometimes cassava, and infrequently in the north where the diet is unrefined wheat.

They reported that rats prefed on a South Indian diet showed a much higher incidence of ulceration in the stomach after pyloric ligation than those on a North Indian (Punjabi) diet. The composition of the diets, expressed as % by weight was, respectively: (A) North Indian diet (%) unrefined wheat flour 41·2, rice flour 5·7, maize flour 8·6; pulses, chickpeas 1·7, mung beans 1·7, urd 1·7, lentil 0·5; green vegetables (leafy), amaranth 4·0; other vegetables, potato 6·2, "Ladies fingers" 6·2, banana powder 0·6; vegetable fat (dalda) 4·4, sugar 9·5, maize starch 5·0, common salt 0·2, SMP 2·9; (B) South Indian diet, rice flour 78·5, pigeon pea flour 5·0, groundnut oil 5·0, leafy vegetables 2·1, nonleafy vegetables 8·2, common salt 0·3 and SMP 0·9.

Removing vegetables and pulses from the North Indian diet deprived it of its protective effect.

Unmilled rice, unrefined wheat and finger millet, (high buffer content), placed in the stomach after pyloric ligation were protective but milled rice, cassava, sorghum and maize, (low buffer content), were not protective. Refined wheat (low buffer content) also gave a protective effect.

In Africa, in dry areas where maize is the main constituent of the diet, there is a low incidence of duodenal ulcer.

Sorghum

Pulses

Sur *et al.* (1955) reported that when wheat or sorghum (both of unstated composition) were fed as 92·8% of the diet to weanling rats over 8 weeks, sorghum was inferior to wheat in promoting growth. However, when wheat and sorghum were fed as 78·5% of the diet with 5% pigeon peas, 2·1% leafy vegetables, 8·2% nonleafy vegetables, 5% groundnut oil, 0·3% common salt and a small quantity of SMP 0·9%, there was no difference in growth rates. This latter diet, often referred to as a "poor sorghum diet", formed the basis for many experiments conducted in India.

Phansalkar *et al.* (1957) reported the supplementary effect of pulses and leafy vegetables to sorghum in a series of rat growth trials. The pulses, fed as decorticated grams, were: chickpeas, urd (*Phaseolus mungo*), mung beans and pigeon peas. The leafy vegetables, agathi (*Sesbania grandiflora*), amaranth (*Amaranthus gangeticus*), murungu (*Moringa oleifera*), and parpukeerai (*Portulaca olearacea*) were selected as they were readily available for most of the year, were cheap, and on a fresh weight basis, contained 5–8% protein.

From 10% protein diets (9% protein from cereal, 1% from powdered vegetable) fed over 28 days, Phansalkar *et al.* reported that among the leafy vegetables, only amaranth gave a significant supplementary effect.

In the subsequent experimental (10% protein) diets, sorghum was fed (a) as the sole protein source; (b) to provide 7% protein with 3% from a pulse; or (c) at 6% protein with the pulse providing 3% and amaranth 1% protein, this last proportion 6/3/1 being described as a typical Indian diet. Since the protein content of sorghum was less than 10% separated sorghum protein was added in (a) above. All diets contained a vitamin and salt mixture. The composition of the diets and their effect on weight gain and PER are shown in Table 6.20. The improvements to sorghum from diets (F) amaranth, (H) urd and amaranth and (I) mung beans and amaranth were said not to be significant. When the diet (J) containing sorghum, pigeon peas and amaranth, was compared with a skim milk powder (SMP) (10% protein) diet, weight gain and PER were not significantly different.

At the end of 4 weeks feeding on a stock diet and experimental diet, the haemoglobin (Hawk *et al.*, 1947) and plasma proteins (Reinhold, 1953) contents were compared. The results were reported with the diets shown in a simplified form, not fully comparable with those used in the PER determinations. Sorghum alone did not produce as high haemoglobin or plasma protein levels as the stock diet nor did substituting amaranth for part of the sorghum significantly improve the levels. In contrast, the sorghum 7 plus pulse 3 mixture caused a significant rise in haemoglobin and in plasma protein levels. The sorghum 6, pulse 3, amaranth 1 combination was not superior to the sorghum 7, pulse 3 mixture after 4 weeks but after 8 weeks, haemoglobin and plasma protein further increased and the sorghum, pulse, amaranth diet compared favourably with the SMP diet. It was observed that for the same dietary protein level, PER values were significantly higher in summer than in winter though no significant differences in food intake were recorded.

Ramachandran *et al.* (1960) reported the vitamin B_{12} (cyanocobalamin) activity (microbiological assay) of the diet, faeces and liver of male weanling albino rats maintained on vegetable diets similar to those described by Phansalkar *et al.* (1957) over 8 weeks compared with a stock diet of SMP, wheat flour, cabbage and germinated pigeon peas. Since vitamin B_{12} activity was known to be labile to alkali (no reference), interference due to other substances was determined separately (Hoffmann *et al.*, 1949) and the value expressed as alkali stable vitamin B_{12} activity. The "true" vitamin B_{12} activity was expressed as the total value minus the alkali stable factors. The B_{12} activity of the diets and their effect on the B_{12} activity in faeces and livers of rats after 8 weeks of feeding are shown in Table 6.3.

Only the whole egg, SMP and stock diet contained significant amounts of vitamin B_{12} activity.* Except for (D) whole egg, all the diets gave significant increases in faeces vitamin B_{12} activity compared to the weanlings.

Ramachandran *et al.* suggested that the results with diet (B) which contained negligible B_{12} activity in whole liver indicated that the diet probably promoted synthesis of B_{12}-like factors.

A diet consisting of wheat, pigeon peas and amaranth was also fed against a stock diet for 24 weeks and no significant differences were found between the two diets in true B_{12} activity.

Ramachandran *et al.* referred to the work of Wokes *et al.* (1955) who studied 147 British subjects consuming pure vegetarian (vegan) diets over 4–8 years. Minor deficiency symptoms were observed in 28 subjects and severe deficiency symptoms in 11. The remainder showed no deficiency symptoms. Ramachandran *et al.* concluded that their findings indicated that, at least in the rat, subsistence on vegetable proteins for long periods need not necessarily lead to vitamin B_{12} deficiency.

Sirinit *et al.* (1965) described rat growth experiments to determine the nutritional value of locally grown Haitian cereal–legume blends. The legumes included *Phaseolus vulgaris* beans locally known as "pois rouge", "pois noir" and "pois blanc", and *Phaseolus lunatus*, known as "pois beurre". All were heated to destroy trypsin inhibitors before being combined with varying proportions of sorghum, rice, or maize. The procedure of Campbell (1963), 10% protein diet, 10 animals/group, fed for 28 days, was adopted. Details of the combinations in which sorghum was used, and the average weight gain and PER are shown in Table 6.20.

The results suggested that only rice plus 20% blanc beans was comparable to the 10% casein control, but the sorghum and bean blends were comparable to Incaparina.

The nutrient contribution of 70 sorghum plus 30 bean was calculated

* (Authors' note: the values in faeces are probably for true B_{12} activity but it is not clear in the original text).

from INCAP-ICNND food consumption tables (Leung and Flores, 1961) and is shown in Table 6.4 expressed as percentages of National Research Council (1964) recommended dietary allowances. The calculation assumes that 75% of the daily caloric requirement is supplied by the blend. It appears that significant amounts of iron, thiamine, riboflavin and niacin are provided by the blend.

Daniel *et al.* (1968a) compared the supplementary value of chickpeas, pigeon peas, soy flour and SMP in "poor Indian diets" based on sorghum, pearl millet and finger millet, obtained from the local market. Details of the experimental diets and their effect on rat weight gain and PER are presented in Tables 6.20 for sorghum, 6.23 for pearl millet and 6.31 for finger millet. The growth rates of rats fed sorghum and pearl millet diets were significantly improved by the addition of vitamins and minerals.

Daniel *et al.* concluded that incorporation of 15–16% chickpeas or pigeon peas, 6% soybean, or 9% SMP (to provide 2·5% extra protein) would, with vitamins and minerals, produce highly significant increases in growth rates.

Kurien *et al.* (1971a) described rat growth experiments in which pigeon peas were incorporated in poor vegetable diets based on sorghum. The effect on protein content, weight gain, protein intake and PER are shown in Table 6.20. The incorporation of 16·7% pigeon pea flour, plus vitamins and minerals, produced satisfactory growth rates.

Pushpamma *et al.* (personal communication, 1978) reported on the nutritional quality of sorghum and legume based food mixtures for young children. N balance studies were carried out on six healthy children, three boys and three girls between 5–6 years old. The cereal–pulse combinations tested were (A) sorghum plus pigeon peas, (B) sorghum plus urd (*Phaseolus mungo*) and (C) rice plus pigeon peas.

The total daily protein allowance was calculated at the rate of 1·5 g/kg body weight/day. After 3 days of adaptation, faeces and urine were collected. The results of the N balance studies appear in Table 6.5. The children were weighed before and after the study. All children either maintained their weight or increased in weight. The percentage absorption and retention of N was significantly higher with (C) followed by (B) and (A).

Narayanaswamy *et al.* (1975) supplemented sorghum and finger millet diets with black-eyed cowpeas. The finger millet and sorghum were ground to flour and the cowpeas cooked in water, dried at 50–60°C and ground.

In series I, the proportions of finger millet and cowpeas were adjusted to provide protein ratios of (A) 4/0, (B) 3/1, (C) 2/2 and (D) 0/4. In series II, the cereal and the pulse were combined in the proportions (E) 10/0, (F) 9/1, (G) 8/2 and (H) 7/3. The amino acid composition (microbiological assay, Barton-Wright, 1952) of the diets in series I and the amino acid score based on egg protein and also on WHO (1973a) reference pattern for children, are shown in Table 6.6.

The lysine deficiency in finger millet and sorghum, indicated by both amino acid (AA) scores was corrected when the ratio of cereal to cowpea protein was 1/1, at which combination the sulphur amino acids or threonine became limiting. The weight gains, protein intake and PER of female weanling rats are shown in Tables 6.20 and 6.31. The addition of cowpeas to either cereal markedly increased the PER and rate of weight gain. Food intake in relation to weight gain is also shown in Tables 6.20 and 6.31.

Orok and Bowland (1974), with male and female weanling rats, compared Nigerian solvent extracted groundnut meal in conjunction with Nigerian yellow maize or Nigerian sorghum, with North American solvent extracted soybean meal plus Canadian yellow maize. The seven isocaloric diets were fed at three protein levels, 20%, 16% and 12%.

There were no marked differences in protein or amino acid content of the two maizes and the sorghum; the groundnut meal was higher than the sorghum meal in arginine and phenylalanine but lower in lysine, methionine, threonine and isoleucine (as % of sample).

The groundnut meal appeared inferior to soybean meal as a protein supplement when added at isonitrogenous levels to maize and sorghum. Neither an increase in the protein level of the groundnut meal diets from 12 to 20% nor the addition of supplemental lysine and methionine was sufficient to make it equivalent to the soybean diets.

Jansen (1974) reviewed the amino acid fortification or supplementation (using the terms interchangeably) of cereals, including sorghum and the millets. He defined amino acid fortification as "the addition of purified amino acids to various cereals, whether intact grains or in various processed forms, with the purpose of improving the biological value or quality of the cereal protein". In Jansen's view, only lysine, threonine, tryptophan, methionine and cystine are of practical importance. He calculated the AA score for these amino acids against whole egg protein for sorghum supplemented with soy flour (8·1% N), fish meal (12·0% N), groundnut flour (GNF) (9·4% N) and cottonseed flour (8·0% N) at levels from 0 to 32 parts of supplement per 100 parts of sorghum. The basic data on the N content and amino acid composition of the sorghum and the supplements were obtained from FAO (1970) and Orr and Watt (1957). Table 6.7 presents the AA scores.

At all levels of addition of cottonseed flour and GNF, lysine remained the first-limiting amino acid. The sulphur amino acids became limiting at 8, 16 and 32 parts/100 of fishmeal, and at 16 and 32 parts of soy flour.

Swaminathan and Daniel (1973) also reviewed the subject but more briefly than Jansen (1974).

Howe and Gilfillan (1970) cited by Jansen (1974) demonstrated that calcium is the first-limiting nutrient in most cereals fed to weanling rats. The growth rate of rats on cereal diets was not stimulated by the addition of amino acids in the absence of calcium and other nutrients.

Oke (1975, 1977) suggested a method for assessing optimum supplementation of a cereal based diet with grain legumes by plotting the amino acid score of the limiting amino acids of a mixed cereal legume diet against the ratio of the mixture. The intercept indicates the ratio of the mixture that will provide optimal nutritive value. The amino acid composition and amino acid score of the individual foodstuffs were taken from FAO (1970).

Applying the Oke method, optimal combinations appeared as 75 millet (probably pearl millet)/25 cowpeas; and 70 sorghum/30 cowpeas. The AA scores for these mixtures are shown in Table 6.8. Oke suggested that cowpeas and groundnut were the most suitable supplements for millet and sorghum in Nigeria.* Sultana and Pushpamma (personal communication 1977) reported the effect of supplementing sorghum with varying levels of soybean on PER and N balance in rat feeding trials. The diets were isonitrogenous at 10% protein level. Their composition and the results are shown in Table 6.9.

The weight gain from (A), sorghum alone, was significantly lower than from diets (B), (C) and (D), soybean supplemented. Sultana and Pushpamma stated that the biological value (BV) of (D) was higher than that from (C) but no data were presented.

Other Cereals

Mitra *et al.* (1948) investigated the biological value for six adult human male subjects of a rice based diet and diets in which part of the rice was replaced by a mixture of other cereals. Basically, diet (A) consisted of 550–800 g rice (varying with the subject's weight), plus 75–110 g pigeon peas as gruel, 100–120 g potatoes, 140–150 g vegetable marrow and 125–150 g leafy vegetables. No milk or flesh foods were consumed but the test diet included 35–40 g cooking fat and ghee. In diet (B) about 40% of the rice was replaced by wheat flour and in the remaining diets, 25% of the wheat flour was replaced by (C) barley, (D) maize, (E) sorghum, (F) finger millet and (G) pearl millet. The subjects were fed a low N diet for the estimation of endogenous N. The experiments extended over a year during which time the supplies of cereals inevitably changed; though every attempt was made to maintain consistency of quality within each cereal. The mean N intake, total N excretion and N balance, together with the mean relative BV (Chick et al. 1935a) and the digestibility coefficients are presented in Table 6.10.

Kurien *et al.* (1960d) compared mixed diets based on, in parts by weight, (A) rice 4, (B) rice 3/sorghum 1, (C) rice 2/sorghum 2, and (D) sorghum 4, in metabolic trials with 7 boys aged 10–11 years resident in a boarding home in Mysore City. The subjects were fed each diet for 15 days, the first 10 being used for adjustment and the last five for collection of urine and faeces. Protein content of the rice and sorghum, composition of the diets and their effect on N, Ca and P balance are shown in Table 6.11. The diets

* (Authors' note: it is probable that the suitability of groundnut in Nigeria relates more to its availability than to its value as a protein complementary to cereals.)

were well accepted and all subjects remained in positive N, Ca and P balance. As the proportion of sorghum increased, digestibility coefficients and N retention declined, Ca and P intake increased, and P retention increased but Ca retention declined.

Rice and sorghum of a similar composition to the cereals used by Kurien *et al.* (1960d), were used in rat growth trials by Kurien *et al.* (1960e). The composition of the diets and their effect on weight gain are shown in Table 6.20. Substitution of rice by sorghum slightly increased the protein content of the diet. Though growth rate appeared to increase when sorghum replaced rice, the only significant difference among the diets was between (A) and (B).

The red blood cell counts (Neubauer's haemocytometer) in rats fed (B) and (C) were slightly higher than those on (A) but differences in haemoglobin content (Sahli-Hellige haemometer) and serum protein (King and Wooton, 1956) were not significant.

The average liver lipid content (Hawk and Elvehjem, 1953) of rats on all diets was within normal range (3·64–3·95%). Average liver protein (N × 6·25) of diet (D) was significantly lower than diets (A), (B) or (C). The overall nutritive value of the diet was not affected by up to 50% substitution of rice by sorghum.

Saxena *et al.* (1966) investigated the effect of a rice polishing concentrate in poor diets based on sorghum, on the N, Ca and P metabolism of rats. Results appear in Table 6.20.

The concentrate was prepared daily, and 100 g of rice polishings provided 100 cm^3 of concentrate. The Ca (AOAC, 1950), P (Fisk and Subbarow, 1925), phytin P (McCance and Widdowson, 1935), amino acids (Giri *et al.*, 1952), histidine (Bhattacharya *et al.*, 1957), tryptophan (chemically, Guest, 1939, Graham *et al.*, 1947), thiamine (Harris and Wang, 1941), riboflavin (fluorometric, Arnold, 1945), niacin (Association of Vitamin Chemists, 1951), choline (calorimetrically, Schmidt *et al.*, 1952) and inositol (chromatographically, Hubscher and Hawthorne, 1957) contents are shown in Table 6.12.

Details of the diets and their effect on average food intake, and weight gain are shown in Table 6.20. Rats on diet (B), including 12·5% concentrate, showed greater retention of N, Ca and P than those on diet (A). Saxena *et al.* (1966) commented upon the beneficial effects of the B-vitamins and essential amino acids present in the concentrate.

Saxena *et al.* (1968) examined the effect of similar diets on the weight gain of rats and reported that the concentrate increased the protein content of the diet and the rate of weight gain (Table 6.20). Addition of the concentrate also increased haemoglobin (Sahli-Hellige haemometer), red blood cell count (Neubauer's haemocytometer, Hepler, 1950), and liver protein and reduced liver lipid (absolute alcohol and ether, Tyner *et al.*, 1950). Saxena *et al.* commented that, overall, rice polishing concentrate had a beneficial effect on poor sorghum diets.

Adrian and Frangne (1970) studied the effect of adding pearl millet from

Niger or sorghum from North Africa to yellow maize grown in France in growth trials with rats. The diets, rat weight gains, food intake and PER are summarized in Table 6.20 for sorghum and 6.23 for pearl millet. The addition of pearl millet to maize improved weight gain and PER; the addition of sorghum reduced weight gain and PER. Supplementation of the maize–sorghum diet with amino acids indicated that lysine was the first-limiting and tryptophan the second-limiting amino acid.

Doraiswamy *et al.* (1971) reported a feeding trial which extended over 6 months and involved seven groups of girls aged 6-11 years (15 girls to the group) resident in a boarding home at Srivilliputtur, Tamil Nadu. The basal diet (460 g/child/day) was based on wheat and rice (140 g of each), the remaining 180 g consisting of small quantities of pulses, vegetables, condiments and spices, salt, SMP, groundnut oil and jaggery. To this basal diet was added 50 g of (A) maize starch, (B) sorghum, (C) sorghum plus vitamins A and D, thiamine, riboflavin and niacin, ascorbic acid and calcium carbonate, (D) as (C) plus 1·25 g lysine, (E) pigeon pea dhal, (F) as (E) plus the vitamins and minerals as in (C), (G) low cost protein food, Bal-Ahar, a mixture of sorghum 65, low fat groundnut flour 25, chickpea flour 10, plus vitamins and minerals as in (C) and (F).* Diets (D), (F) and (G) resulted in highly significant improvements in height and weight gain, and in the nutritional status (as judged by the score card of the Indian Council of Medical Research, 1948) compared to diets (A) and (B).

Oilseeds and Other Vegetable Proteins

Venkat Rao *et al.* (1958) in rat growth trials reported the supplementary value of low fat groundnut flour (GNF) to poor vegetarian diets based on sorghum and finger millet. The composition of the diets, food intake levels and weight gains are shown in Tables 6.20 for sorghum and 6.31 for finger millet.

The substitution of GNF for part of the cereal, even at the 10% level, increased the protein content of the diet and markedly improved rate of weight gain. Consumer acceptibility tests (no details were given) indicated that GNF could be incorporated up to a level of 20% with sorghum or finger millet in familiar food preparations.

Hariharan *et al.* (1965) in rat feeding trials studied the effect of supplementing poor Indian diets based on sorghum and pearl millet with (a) low fat GNF unfortified or fortified with Ca salts and vitamins and (b) SMP fortified with vitamins. The diets, weight gains, protein intake and PERs are shown in Tables 6.20 for sorghum and 6.23 for pearl millet.

Supplementation of the poor sorghum diet with Ca salts and vitamins, diets (B) and (C), produced a significant increase in weight gain indicative of related deficiencies in the poor sorghum diet (A).

Fortified GNF providing about 2·5% extra protein, diet (E), and unfortified GNF providing 5·0% extra protein, diet (G), did not further

* (Authors' note: for an explanation of the preparation of dhal refer to p. 452.)

improve growth rates significantly. However, fortified GNF providing 5% extra protein, diet (H), improved weight gain equivalent to fortified SMP which contributed 2·5% and 5% extra protein, diets (F) and (I).

The PER of diet (B) was significantly higher than that of diet (A). The PER of diet (F), supplementation with SMP to provide 11% dietary protein was significantly higher than the equivalent protein level derived from fortified GNF, diet (E), but at a dietary protein level of 13·6% there was no significant difference between diet (H), supplementation with fortified GNF and diet (I), supplementation with fortified SMP.

In the case of pearl millet, supplementation with Ca salts and vitamins, diets (B) and (C), significantly improved weight gain compared to poor millet diet (A). Fortified GNF or SMP providing about 2·5% extra protein, and unfortified GNF providing 5% extra protein, also improved weight gain compared to diet (A). There was a further significant increase in weight gain from diets (H) fortified GNF and (I) fortified SMP, providing 5% extra protein in the diet.

The PERs of diets (B) and (C) were significantly higher than diet (A). The PERs from fortified GNF, (E) and (H) were significantly greater than the PER from unfortified GNF, (D) and (G).

When fortified GNF and SMP provided 5% extra protein, there was no significant difference in PER between the two cereals, sorghum and pearl millet.

Peyrot and Adrian (1970) in weanling and adult rat feeding trials, sought to determine the limiting factor in mixtures of sorghum and groundnut cake (GNC) depending on whether the cereal was "poor" or "rich" in protein. Both sorghums were red varieties. The "poor" protein (7·4%) type probably originated in the South of France while the "rich" protein (10·4%) type came from eastern Niger and corresponded to the local variety, Janjare.

GNC products of two protein levels, "rich" 52·8% and "poor" 47·5%, were used. The "rich" GNC was mixed with the "rich" sorghum, diets (A), (B), (C), (G), (H) and (I) and the "poor" GNC with the "poor" sorghum, diets (D), (E), (F), (J), (K) and (L). The results of rat growth trials are presented in Table 6.20.

Judged by weight gains, methionine was first-limiting in the "poor" sorghum ration for adult rats; lysine was first-limiting in the "rich" sorghum for both weanlings and adults.

The rations were isonitrogenous and the levels of sorghum and GNC varied according to protein content; 810 g of "poor" sorghum and 575 g of "rich" sorghum were needed to provide 6% protein in the diet.

The "poor" sorghum diets, therefore contained more fibre from the sorghum and the GNC, and this was reflected in the comparatively poor total digestibility of these diets. Digestibility of diets (A), (B) and (C) was 85%; of (G), (H) and (I) 87% compared to (D), (E) and (F) 93%; (J), (K) and (L) 95%. The addition of lysine and methionine did not appear to influence protein digestibility.

In the "rich" sorghum plus GNC diet, lysine was first-limiting for both young rapidly growing rats and adult rats approaching the maintenance stage. In the "poor" sorghum plus GNC diet, lysine and methionine appeared equally deficient and first-limiting for the weanlings; methionine was first-limiting for adult rats.

Peyrot and Adrian concluded that the limiting amino acid in sorghum diets will vary according to whether the sorghum is high or low in protein, and on whether it is fed to young or adult rats.

Daniel *et al.* (1968b) in rat feeding trials compared:

(A) an Indian poor sorghum diet;
(B) diet (A) plus tricalcium phosphate (TCP) and vitamins;
(C) diet (A) plus 17% screw pressed coconut cake;
(D) diet (C) plus TCP and vitamins; and
(E) diet (A) plus 7·0% SMP, TCP and vitamins.

The coconut and SMP supplements increased the protein content by about 2%.

The composition of the diets and results appear in Table 6.20. Rate of weight gain increased with each supplement from (A) lowest to (E) highest. It was estimated that the coconut would add about 2.0% crude fibre to the diet. Metabolic studies with school children of 11–12 years (Rama Rao *et al.*, 1965) in which 100 g of coconut meal was consumed with a N-free diet produced no evidence of digestive upset. Daniel *et al.* suggested that coconut meal is a useful protein supplement for children.

Ali *et al.* (1964a) in rat trials examined the supplementary value to sorghum of (A) 0, (B) 5, (C) 10 and (D) 15% of a blend of 2 parts cottonseed flour/1 part fish flour prepared by the method of Ali *et al.* (1964b). The protein ($N \times 6{\cdot}25$) contents of the diets were (A) 7·8, (B) 10·2, (C) 12·6 and (D) 15·0%.

Ali *et al.* stated that a 15% addition of the protein concentrate would be required to raise the ND_p Cals percentage to a level adequate to meet the needs of nursing mothers, using the criteria of Platt *et al.* (1961).

Talwalkar and Patel (1970a) by rat growth and N balance studies reported the biological evaluation of two leafy vegetables, ambadi (*Hibiscus cannabinus*) and methi (*Trigonella foenumgraecum*), and their supplementary effect in powdered form on sorghum. Details are presented in Table 6.20. Methi, diet (C), produced higher weight gain and PER than ambadi, diet (B).

In the N balance studies, (E) ambadi, (F) methi, and (D) sorghum were fed separately to provide 5% protein and compared with (G) sorghum providing 3% and ambadi 2% protein, and (H) sorghum providing 3% and methi 2% protein. A stock casein diet providing 18% protein and a N-free diet were also compared. The BV for ambadi, 74, was comparable to cabbage, 72, and amaranth, 71, as reported by Aykroyd *et al.* (1966); methi was higher at 83. The BV of diet (G) sorghum and ambadi was not significantly higher than diet (D) sorghum alone, but the BV of (H), combination of sorghum and methi, was higher than the BV of its constituent proteins compared separately.

The average food consumption of rats on diet (E) ambadi alone, was lower than on (F) methi, possibly due to the bitter taste of ambadi. It was also observed that animals on diets (G) and (H) lost hair; this did not occur with diet (D) sorghum alone. Talwalkar and Patel suggested this may have been caused by an inhibitory factor in the leafy vegetable powders rather than by any nutrient deficiency. Hair growth was restored when the affected animals were fed the stock diet for 3 weeks.

Using essentially similar diets (A), (B), (C), to those described by Talwalkar and Patel (1970a) but including a fourth control diet (D) containing 5% protein from sorghum and 2% protein from casein, the same two authors (1970b) examined the effect of supplementing sorghum with ambadi and methi in regenerating liver and plasma proteins, haemoglobin and erythrocytes (red blood cells, RBC) of rats.

Liver protein level on a stock diet (E) was first lowered by feeding a N-free diet (F). Diet (A), sorghum alone, reduced liver protein still further. Diets (B) and (C) almost restored the normal value, and diet (D) regenerated liver tissue to the level of 83% of normal value.

The globulins from the separated blood serum were precipitated by Na_2SO_4, the albumins determined (Howe, 1921), and the globulins calculated from total protein and albumin contents.

The lower total plasma protein from diet (F) resulted principally from the smaller albumin fraction. Sorghum alone (A) had virtually no restorative effect, but each of the supplemented diets increased total protein, (C) methi more than (B) ambadi. Only casein diet (D) stimulated albumin formation. Talwalkar and Patel stated that this finding supports the generalization that vegetable protein favours globulin synthesis. No reference was cited in support of the generalization.

To study the effect of the diets on regeneration of haemoglobin and RBC, experimental anaemia was induced in the rats by injections of phenylhydrazine.

Sorghum alone (A) restored about 92% of the normal value of haemoglobin and 80% of the normal RBC after 12 days feeding. The other diets attained normal haemoglobin values after 12 days and normal RBC within 16 days. Except for diet (D) casein, the haemoglobin level reached maximum value in 12 days and fell back by the sixteenth day. Ambadi appeared to be superior to methi as a supplement for sorghum and Talwalkar and Patel suggested the findings would justify the use of ambadi in the diets of anaemia patients in South India.

Rajalakshmi and Maliwal (1975) reported experiments in which weanling albino rats, eight to the group, matched for age and sex, were fed over 4 weeks a basal diet consisting of sorghum, groundnut oil, salt mixture and a niacin-free vitamin mixture (A), alone or supplemented/100 g diet, by (B) niacin 1·5 mg, (C) lysine 120 mg, (D) isoleucine 459 mg and (E) fenugreek seeds (methi) (*Trigonella foenumgraecum*) 10 g. The relevant average weight gains in grams (possibly/week but this is not stated) were (A) 8, (B) 13, (C) 17, (D) 10 and (E) 18. Fenugreek gave an improvement over the basal diet similar to that from lysine. The diets were made isonitrogenous with maize starch but the level of protein was not stated.

Proteins of Animal Origin

Sure *et al.* (1957) reported the effect of defatted fish flour made from a mixture of carp, smelts and whitings (Levin and Finn, 1955) substituted at low levels, 1%, 3%, and 5% for sorghum of 10·4% protein content in diets of weanling rats. The diets, weight gains and PER are summarized in Table 6.20. Optimum PER was obtained from diet (C) with some increased growth from diet (D).

Adrian and Jacquot (1961) cited G. Roebben, who, in a report to the FAO in 1960, recommended a mixture of 75 parts millet/20 parts groundnut flour (GNF)/5 parts fish flour as optimum from the standpoint of nutrition, cost, and acceptability.* While defatted GNF was fairly consistent in quality and had a reasonable storage life, the nutritional quality of fish flour varied with the method of production, and in tropical climates deteriorated quickly in storage.

Adrian and Jacquot, in rat feeding experiments, studied whether lysine and methionine supplementation was nutritionally equivalent to fish flour. The products used were being distributed in Dakar by ORANA. The sorghum was probably imported from America; the GNF was obtained by solvent extraction and the fish flour was made by Azote-Union (Morocco). The protein contents of the groundnut flour and fish flour, the composition of the diets and their effect on weight gain and PER are summarized in Table 6.20. The lysine and methionine additions in diets (D), (E), (F) and (G) were equivalent to the loss (determined by analysis) resulting from the omission of fish flour.

Diets (D) and (E) supplemented with methionine gave results practically identical with (B), GNF alone. The addition of lysine (F) and (G), improved growth and PER compared to (B), (D) and (E) with no difference between (F) lysine and (G) lysine plus methionine. The best results were obtained from diet (C) containing fish flour.

By plotting mean weight gain against days of feeding, Adrian and Jacquot deduced that initially the addition of lysine to GNF was equivalent to the addition of fish flour, but after about 35 days, fish flour produced greater weight gains than lysine.

The livers of the rats were all satisfactory in appearance except for those from the unsupplemented sorghum which were yellow brown, indicative of excess fat. The addition of methionine, which had no effect on rat growth, appeared to have had a favourable effect on the condition of the liver.

Adrian and Jacquot concluded that a mixture of 18% GNF with 0·1% L-lysine would be an alternative to 15% GNF plus 3% fish flour in cereal based weaning foods in Africa. They pointed out that fish flour contained vitamin B_{12}, Ca and P, not provided by lysine. Fish flour would be

*(Authors' note: although it is not stated, this was probably related to an African study.)

preferable because of its acceptability by Africans but it suffers the disadvantage of instability.

The advantage of lysine is its consistent quality; its main disadvantage is the difficulty of distributing and incorporating the comparatively small quantities needed, particularly among rural families, added to the very high cost in foreign currency.

Shurpalekar *et al.* (1962a) with weanling rats compared a fish flour fortified with vitamins (prepared by the method of Shurpalekar *et al.*, 1962c) and SMP, fortified with vitamins A and D, as supplements to poor Indian diets based on sorghum. The supplements provided about 2·5% extra protein and additional calcium to the diet.

The composition of the diets and the weight gains are summarized in Table 6.20. Both (B) with fortified fish flour and (C) with fortified SMP produced significantly greater weight gains than the poor sorghum diet (A). Diet (B) was slightly better than (C). Diets (B) and (C) significantly increased haemoglobin, red blood cell counts and liver protein content, compared to (A). Liver lipid content was not different among diets.

Desai *et al.* (1968) compared the PERs of three protein supplements (see Table 6.37):

(C) fish flour (process of Sen *et al.*, 1966);
(D) 1/2/1 blend of fish, GNF and chickpea flour; and
(E) SMP plus thiamine, niacin, vitamins A and D.

The fish flour contained added Ca, thiamine, niacin, riboflavin and vitamins A and D; all three supplements were equal in Ca and vitamin content. The PERs of (C) 3·2 and (E) 3·0 were roughly equivalent and significantly higher than (D) 2·6.

All three foods were subsequently fed as supplements to poor Indian vegetarian diets based on sorghum. The effects on PER are shown in Table 6.20.

The weight gains of the rats on diets (A) basal sorghum diet, and (B) basal sorghum diet with vitamins and minerals, not specifically listed, were not significantly different from one another but were significantly increased in diets (C) with fish flour, (D) with protein blend and (E) with fortified SMP, which provided 2·5% extra protein. There was no significant difference in the PER of the three supplemented diets, nor in the haemoglobin content of rats fed on the different diets. The livers of the rats on the poor sorghum diet (A), showed mild parenchymal damage of the protein deficiency type and mild periportal fatty infiltration. The livers of animals fed (C), (D) and (E) were normal.

Kurien *et al.* (1960f) reported on the metabolism of N, Ca and P in boys, aged between 10 and 11 years in a Muslim orphanage, who were fed a poor vegetarian diet in which sorghum provided 360 g of a total food intake of 580 g/child/day. The diet included 28 g meat and 6·4 g SMP. Kurien *et al.* reported the daily nutrient intake as: 1725 kcal; 43·2 g protein ($N \times 6{\cdot}25$), 41·8 g lipid, 294 g carbohydrate, 441 mg Ca, 1091 mg P (total), 750 mg phytate P, 1·47 mg thiamine, 9·83 mg niacin and 1265 IU vitamin A (mainly carotene).

The reported chemical composition (%) of the sorghum (AOAC, 1960), DWB, was: protein (N × 6·25) 8·8, lipid (ether extract) 2·5, ash 1·9, starch 75·8, total sugars 1·5, crude fibre 3·2, pentosan and other hemicelluloses (by difference) 6·4, (mg/100 g) Ca 37, P (total) 279 and phytate P 221. (Protein, Ca and P (AOAC, 1950); phytate P (McCance and Widdowson, 1935); other nutrients calculated from Aykroyd *et al.*, 1956).

The experimental diet was fed for 15 days, urine and faeces were collected during the last 5 days. Nitrogen, Ca and P in food, urine and faeces were estimated (Murthy *et al.*, 1955). All the subjects maintained a positive N balance, the average daily intake and retention being 6·9 g and 0·9 g respectively. N excretion was high (3·1 g) and the apparent digestibility of the protein 55·4%. Average daily intake and retention of Ca were 0·44 and 0·07 g respectively. The average daily intake and retention of P were 1·0 g and 0·31 g respectively; all the subjects maintained positive Ca and P balance. Although N balance was maintained, in the view of Widdowson and McCance (1954), the intake of N was not adequate.

Daniel *et al.* (1966) reported the effect of supplementing (A) a poor sorghum diet with (B) L-lysine and (C) L-lysine and DL-threonine compared with (D) SMP diet, on the digestibility coefficient and biological value of the protein (Tasker *et al.*, 1962b) and NPU_{op} (Platt *et al.*, 1961). The subjects were eight girls of 11-12 years of age in a boarding home in Mysore City. Lysine and threonine were added to the levels roughly present in egg protein.

The essential amino acid content of the diets (method of Krishnamurthy *et al.*, 1960) are given in Table 6.13 together with mean daily intakes and absorption of essential amino acids (mg/kg body weight) and amino acid requirements (Nakagawa *et al.*, 1962). The diets were fed in five 10-day periods; faeces and urine for N determination (micro-Kjeldahl) were collected during the latter 5 days. The order of feeding was (A) sorghum diet, (B), (C), (D) SMP diet and (E) low protein diet (Parthasarathy *et al.*, 1963).

The sorghum diet provided 1·3 g protein/kg body weight which, except for lysine, met the children's essential amino acid requirements as recommended by Nakagawa *et al.* (1962). Nakagawa *et al.* (1962) as cited by Daniel *et al.* (1966) used pure amino acids which were completely absorbed. From the results in Table 6·13 lysine appears highly limiting.

The effects on mean daily N intake, excretion and balance, and on apparent digestibility, true digestibility, BV and NPU_{op} are shown in Table 6.14. N retention increased significantly among diets from diet (A) 9% of intake and (B) 13% to (C) 22% and (D) 33% of intake. There was no significant difference among diets (A), (B), (C) in true digestibility but diet (D) was significantly higher.

The BVs of each diet increased significantly from (A) to (D). The NPU_{op} of diets (A) and (B) were close; diet (C) was significantly higher than (A) and (B) but lower than diet (D). Net available protein is presented in Table 6·14 and compared to WHO (1965) reference protein requirements. Diets (A) and (B) were below, but diet (C) met the WHO requirement.

Microbial Protein

Protein derived from microorganisms has attracted much scientific attention in recent years. Yeasts, dried or autolysed, have provided the most widespread source of supplementary protein and their supplementary value to bread was reviewed by Hulse and Laing (1974).

When the nucleic acids of yeasts are metabolized in human beings and some monogastric animals, the final product is uric acid which, in excess, may cause gout or the formation of stones in the urinary tract. FAO (1971) set the maximum tolerable intake of nucleic acids at 2 g/day, equivalent to 10–30 g of unicellular protein.

Sur *et al.* (1954a) reported the supplementary effect of food yeast (*Torula utilis*) in poor vegetarian diets, as used by Murthy *et al.* (1950), in rat growth trials.

The cereals used were described as jowar (*Andropogon sorghum*), milo (*Andropogon sorghum* var. *durra*) and ragi (finger millet). The chemical composition of the yeast and the effect of replacing the different cereals in the diets by 2·5% and 5% yeast are shown in Tables 6.20 for sorghum and 6.31 for finger millet.

The yeast added showed greater effects upon the "milo" sorghum, types I and II, and the finger millet than the "jowar" sorghum, type III.

Sur *et al.* (1954b) compared diets in which the same samples of "jowar" sorghum were supplemented with yeast at a constant cereal protein to yeast ratio of 3/1. Rats were fed diets at 5% protein, at which most lost weight, and 8% protein. The results appear in Table 6.20. The PERs were significantly increased by the addition of yeast.

Narayanaswamy *et al.* (1972c) reported the supplementary value of yeast grown on petroleum hydrocarbons at the Indian Institute of Petroleum, Dehra Dun to poor Indian vegetarian diets based on sorghum and similar to those used in many previous rat feeding trials. The lysine, threonine and sulphur containing amino acids in the yeast and the poor sorghum diets (microbiological assay, Barton-Wright, 1952) are shown in Table 6.15. The yeast markedly increased the lysine and threonine contents of the poor sorghum diet.

The composition of the diets and the effect on PER are shown in Table 6.20.

The weight gains on diet (B) sorghum diet plus salt and vitamin mixture and diet (C) poor sorghum diet with petroleum grown yeast, were significantly greater than on diet (A) poor sorghum diet alone. Diet (D) poor sorghum diet plus petroleum grown yeast, and salt and vitamin mixture, produced a further significant increase in weight gain. The PER of all three supplementary diets showed a significant and progressive improvement compared to diet (A).*

* (Authors' note: protein in the diet was not constant.)

Multipurpose Foods

Subrahmanyan *et al.* (1954d) with weanling rats compared the nutritive value of three synthetic grains prepared by the method of Subrahmanyan *et al.* (1954e) from cassava and groundnut flour in the proportions: (a) 90/10, (b) 85/15 and (c) 80/20 as replacements for 25% of the sorghum in poor Indian vegetarian diets. The protein contents of the three synthetic grains are reported in Table 6.20 with the composition of the poor sorghum diets, average weekly food intake and weight gain.

Supplementation with synthetic grains II and III produced a significant improvement in weight gain compared to sorghum alone, diet (A), and synthetic grain I, diet (B).

Joseph *et al.* (1959b) studied the effect of supplementing a poor sorghum diet with Indian MPF (a mixture of protein sources) on the composition of the body and liver of rats, and on weight gain over 8 and 12 weeks (Table 6.20).

Weight gains on diets (B) MPF, (C) MPF and SMP and (D) SMP were significantly higher than diet (A) but diets (B), (C) and (D) were equivalent.

At the end of 12 weeks haemoglobin (Sahli-Hellige haemometer), red blood cell counts (Neubauer's haemocytometer) and liver N were not significantly different among diets.

The liver fat contents (Hawk and Elvehjem, 1953) on diet (A) 4·4% and (B) 4·0% were significantly higher than diets (C) 3·2% and (D) 3·3%. Carcass N contents of rats fed the different diets were not significantly different indicating that increase in body weight was accompanied by a corresponding increase in tissue protein.

Kuppuswamy *et al.* (1957) reported that when added at 12·5% to a poor vegetarian diet, both Indian and American MPFs (latter not identified) caused increase in rat weight gains. (See Table 6.20.)

Tasker *et al.* (1964) supplemented a poor sorghum diet with a protein food (Krishnamurthy *et al.*, 1958) based on a 1/1/2 blend of screw pressed coconut meal, chickpea flour and groundnut flour, plus Ca, vitamins A and D, thiamine and riboflavin. The diets and resulting rat weight gains are summarized in Table 6.20. Supplementation increased dietary protein and significantly increased weight gain.

Panemangalore *et al.* (1965) supplemented a sorghum vegetarian diet with two protein foods, I and II (Guttikar *et al.*, 1965ab). The protein foods, the diets and their effect on rat weight gain and F/G ratio are summarized in Table 6.20.

The supplemented diets (B), (C), (D), (E) and (F) all significantly increased weight gain, serum proteins (King and Wooton, 1956), haemoglobin and red blood cell count (Joseph *et al.*, 1959b). There were no significant differences among the supplemented diets.

The rats fed poor sorghum diet (A) showed significantly higher liver fat and significantly lower carcass protein contents than rats on the remaining diets. The histological examination of the liver (Paul Jayaraj *et al.*, 1960) of rats on diet (A) showed a mild degree of parenchymal damage typical of protein deficiency. Liver cells showed a reduction in cytoplasmic proteins with mild generalized and periportal fatty infiltration. Necrosis of the liver cells was absent. The livers of rats on the other diets were normal.

Desai *et al.* (1970b) replaced 50% of the cereal in poor Indian diets with low cost balanced foods (LCBF). The composition of three of these foods, LCBF III, LCBF IV and LCBF V, the diets in which they were incorporated, and their effect on rat weight gain are shown in Table 6.20. Addition of LCBF III, IV and V markedly increased weight gain.

Narayanaswamy *et al.* (1971) studied the supplementary value of: (a) a low cost protein food (LCPF, wheat flour 70/soy flour 30), (b) chickpea flour and (c) SMP fortified with vitamins A and D, and thiamine in a poor Indian sorghum diet. The composition of the diets, rat weight gains and PER are shown in Table 6.20. Weight gain and PER significantly increased when the vitamin premix and salt mixture were added, diet (B). Diet (C) with 1·5% extra protein from LCPF significantly improved weight gain and PER. Diet (D) with 3·0% more protein than (A) was not superior to (C).

Supplementation with different levels of chickpea flour, diets (E) and (G) significantly increased weight gain but (G) was not better than (F). PER and weight gain were further increased by salt and vitamin additions to (G) and (E) (diets (F) and (H)).

There was no significant difference among diets (C), (D), (H), (I), and (J) in weight gain, nor between diets (D) and (J) in PER. Diet (I), 6% fortified SMP gave the highest PER.

Narayanaswamy *et al.* concluded that the LCPF (70/40 wheat/soy flour) improved the poor sorghum diet.

A different LCPF (wheat flour 70/soy flour 15/groundnut flour 15) was compared with SMP both supplements fortified with vitamins A and D in poor sorghum vegetarian diets by Narayanaswamy *et al.* (1972a). The composition of the diets, rat weight gains and PERs are summarized in Table 6.20. All additions significantly increased weight gain and PER compared with the control diet (A) but there were no significant differences in PER between LCPF diets (C) and (D) nor between SMP diets (E) and (F) though the SMP diets were superior to the LCPF.

Narayanaswamy *et al.* suggested that a daily supplement of 50–60 g LCPF for preschool children would inexpensively provide 10–12 g extra protein and half the daily requirement of essential vitamins and minerals.

For results of similar trials with a poor finger millet diet see Narayansawamy *et al.* 1972b (Table 6.31).

Synthetic Amino Acids

Doraiswamy *et al.* (1968) supplemented a poor sorghum diet with L-lysine and over 6 months measured N retention and growth among 48 girls aged 7-12 years living in a boarding home in Mysore. Mean daily food intake is shown in Table 6.16.

Group (A) received the poor diet which from natural sources contained 250 mg lysine/g total N. Group (B) received an additional 1·25 g L-lysine HC1 daily. After 3 months, N metabolism was studied with six pairs of children from groups (A) and (B). The results after 6 months are shown in Table 6.16, all differences being significant.

The nutritional status (Doraiswamy *et al.*, 1962) of the children was reported as under:

No. of subjects	*Diet A* Sorghum	*Diet B* Sorghum + lysine
Improved	6	12
Stationary	13	12
Deteriorated	5	
Total	24	24

Supplementation of the poor sorghum diet with lysine produced improvement in the height, weight, nutritional status and N balance of the children.

Howe *et al.* (1965) supplemented sorghum with synthetic amino acids. The diets, rat weight gains and adjusted PERs are shown in Table 6.20. With added lysine and threonine, the cereal protein was equivalent to the casein diet. Threonine was the second-limiting amino acid.

Daniel *et al.* (1965a) supplemented market samples of sorghum with L-lysine and DL-threonine to egg protein levels. The diets and rat growth data appear in Table 6.20. The PER of (A) sorghum increased significantly with added lysine (B) and again significantly with (C) added lysine and threonine. Increase in PER was less marked when lysine and threonine were added to (D) poor sorghum diet. The addition of vitamins and minerals to the poor sorghum diets (G), (H) and (I) increased the effect on PER of added lysine and threonine.

Narayanaswamy *et al.* (1970a) supplemented poor sorghum vegetarian diets with L-lysine, DL-threonine and DL-methionine to about the same levels as in human milk protein.

The experimental diets, amino acids including cystine (microbiological assay, Barton-Wright, 1952) and rat growth data are summarized in Table 6.20. Lysine added alone, diet (B), significantly increased PER over (A). Added lysine plus threonine plus methionine, diet (H), increased PER further.

Pushpamma (1968) and Pushpamma *et al.* (1972) reported the essential amino acid content (auto-analyser following acid hydrolysis, Blackburn, 1968); tryptophan (Lombard and DeLange, 1965) of market samples of sorghum, pearl millet and finger millet and compared the grains in rat diets (1) unsupplemented and supplemented with lysine and/or tryptophan and (2) with SMP, fish protein concentrate (FPC), isolated soybean protein, and dried brewers' yeast. The diets, rat weight gains and PERs are presented in Tables 6.20 for sorghum, 6.23 for pearl millet and 6.31 for finger millet. Amino acid contents (mg/g N) with the AA score (WHO, 1973a) calculated by the authors appear in Table 6.17. The weight gains of rats fed finger millet diets were in order of increasing total lysine content, regardless of the source of supplementation. All the supplemented pearl millet diets were superior to the casein control but no supplement was best with all cereals.

FPC was best with finger millet and sorghum, probably because of its relatively high lysine content. SMP was poorest for the sorghum diet. Brewers' yeast was poorer than expected from the analytical data, possibly because of poor digestibility (Tannenbaum *et al.*, 1966).

On the evidence of this paper, finger millet protein appears nutritionally superior to sorghum and pearl millet.

Harden *et al.* (1976) in two series of experiments added various supplements to a composite US elevator sorghum sample. Details of diets, rat weight gains, F/G ratios and NPU (Miller and Bender, 1955) appear in Table 6.20.

The protein-free basal diet contained 83% maize starch plus oil, minerals and vitamins. In the test diets, dietary protein and amino acid supplements were added to the basal diet at the expense of maize starch to provide: (series I) 8·5% protein from casein or sorghum with and without lysine plus threonine, methionine and tryptophan; (series II) 10% protein from casein or sorghum with and without lysine and/or isoleucine.

All supplements significantly increased weight gain over (A) and (C) but none equalled casein. In series I, lysine plus threonine was better than lysine plus methionine indicating that threonine is second-limiting. These results agreed with those of Pond *et al.* (1958), Jensen *et al.* (1965) and Howe *et al.* (1965) but differed from Waggle *et al.* (1967a). In series II, the addition of (D) 0·4% lysine and (F) lysine plus isoleucine was not significantly different, indicating (1) lysine is first-limiting (2) niacin was adequate and no benefit resulted from reducing the leucine/isoleucine ratio.

In a third experiment, the test diets contained 8·0% protein provided by (B) sorghum (9·6% protein), (C) wheat (14·8% protein), (D) maize (10·5% protein) or (E) soybean oil meal (47·6 % protein). The effect on food intake, weight gain and NPU is presented in Table 6·18. There were no significant differences in NPU among diets (B), (C) and (D) but all were significantly lower than from diet (E).

In a fourth experiment, rations similar to those used for swine were formulated to contain 74·4% sorghum, wheat or maize, each supplemented with 22·4% soybean oil meal, 0·5% salt and 1·0% vitamin premix. The results are presented in Table 6·19. Analysis of variance showed there was no significant difference in weight gain or food intake among the experimental diets but the NPU from (D) maize–soy was significantly higher than those of (B) and (C).

In a fifth experiment, the basal protein-free diet was supplemented by the swine-type cereal–soy rations to provide 10% protein. Rats fed the sorghum–soy ration gained less than those fed wheat–soy or maize–soy but one way analysis of variance showed no significant differences in weight gains, food intakes or NPU values.

Pearl Millet

Pulses

Phansalkar *et al.* (1957) reported supplementing pearl millet with essentially the same decorticated pulses and dried vegetables reported under sorghum (Phansalkar *et al.*, 1957, p. 482 and Table 6.20). Of the vegetables alone only amaranth improved the diet. Subsequent diets, all adjusted with maize starch to 10% protein, included (A) pearl millet alone compared with a series in which pearl millet provided 7% and a pulse 3% of the protein. These included millet plus (B) chickpea, (C) urd, (D) mung bean and (E) pigeon pea. In (F) millet provided 9% and amaranth 1% of protein. In (G), (H), (I) and (J), said to correspond to an average poor Indian diet, millet provided 6%, amaranth 1% and pulses 3% of protein. (G), (H), (I) and (J) follow the same pulse sequence as (B), (C), (D) and (E).

The composition of diets, rat weight gains and PERs appear in Table 6.23. All additions to millet increased weight gain and PER following the pattern of the results with sorghum (Table 6.20). All diets were supplemented with minerals and vitamins. Omission of the vitamins from the millet, pigeon pea, amaranth diet had no effect, but PER dropped, not significantly, with omission of the minerals.

The haemoglobin (Hawk *et al.*, 1947) and plasma protein (Reinhold, 1953) contents were compared, after feeding the experimental diets in comparison with a stock diet. The two blood indices were lowest from millet alone and millet plus amaranth. Plasma protein levels increased significantly with the addition of the pulse.

After 8 weeks on the 6/3/1 millet/pulse/amaranth protein formula, haemoglobin and plasma proteins were equivalent to the stock diet and to diet (L) in which SMP provided all the protein.

Phansalkar *et al.* (1958) reported the effect of 10% protein diets provided by (A) SMP, (B) pearl millet 6%, pigeon peas 3%, and amaranth 1% of protein on the body weight, blood plasma and packed cell volume, haemoglobin and plasma protein concentration, and total circulating haemoglobin and plasma protein of adult rats over 8 weeks.

After 8 weeks on a protein-free diet, rats lost body weight; blood and plasma volumes, total circulating haemoglobin and plasma protein all declined. During repletion on diets (A) and (B), blood proteins returned to normal on both diets but at a slower rate of regeneration on diet (B) than on diet (A).

After 3 weeks on diets (A) and (B) total circulating haemoglobin and plasma proteins were restored to normal but body weight was not, suggesting that regeneration of blood proteins takes priority over restoration of body weight.

Ramachandran *et al.* (1960) reported on the vitamin B_{12} activity of a diet based on pearl millet, pigeon peas and amaranth fed to weanling rats. The findings are discussed under Sorghum (p. 483).

Daniel *et al.* (1968a) compared the supplementary value of chickpeas, soy flour and SMP to poor Indian pearl millet diets. The diets, rat weight gains and PERs are presented in Table 6.23 and discussed under Sorghum on p. 484.

Other Cereals

Kurien *et al.* (1961bc) described substitution of rice by pearl millet in experiments similar to those with rice and sorghum (Kurien *et al.* 1960d).

Eight boys, aged 11-12 years living in a boarding home in Mysore City, received diets based on, in parts by weight, (A) rice 4, (B) rice 3/pearl millet 1, (C) rice 2/pearl millet 2, and (D) pearl millet 4. Composition of the cereals and the effects on protein, Ca and P balance are shown in Table 6.21. Results were essentially similar to those for sorghum. All the children maintained positive N balance but the mean apparent digestibility of the protein and the mean N retention decreased as the proportion of pearl millet increased. All subjects remained in positive Ca balance but the mean daily retention decreased as the proportion of pearl millet increased, possibly due to higher phytate P in the millet. Both intake and retention of P increased as pearl millet increased in the diet.

Rice and pearl millet, similar in composition to those used in N balance tests with children (Kurien *et al.*, 1961b), were used by Kurien *et al.* (1962) in rat growth trials. Details are shown in Table 6.23. The protein content of the diet increased with additions of pearl millet and the three diets containing pearl millet (B), (C) and (D) all produced increased weight gains compared to diet (A) but there was no significant difference among diets (B), (C) and (D). There were no significant differences in the haemoglobin and red blood cell counts of the blood, in serum protein, nor in the mean moisture, protein and fat contents of the liver and carcass of rats on any of the diets.

Oilseeds

Rama Rao *et al.* (1953b) reported the supplementary effect of groundnut cake (GNC) and of chickpeas to pearl millet as measured by rat growth trials. The protein contents (AOAC, no date) of the constituents and the effect on PER are shown in Table 6.23. Both chickpeas and GNC improved the PER of the pearl millet.

Hariharan *et al.* (1965) studied the effect of supplementing poor Indian diets based on pearl millet with (a) low fat groundnut flour (GNF), unfortified or fortified with Ca salts and vitamins and (b) SMP fortified with vitamins, as judged by rat growth and PER. The protein content of the two GNF and the SMP is shown under sorghum, (Table 6.20), and the findings discussed on p. 488. The effect of the diets on weight gain, protein intake and PER appear in Table 6.23.

Ali *et al.* (1964c) with rats examined the supplementary value at levels of

(A) 0, (B) 5, (C) 10 and (D) 15% of a blend of cottonseed flour 2 parts/fish flour 1 part (prepared by the method of Ali *et al.*, 1964b) to pearl millet. As the level of the blend increased, the protein content ($N \times 6{\cdot}25$) (%) of the diets increased (A) 11·8, (B) 14·0, (C) 16·2, (D) 18·4. Ali *et al.* stated that a 10% addition of the blend was sufficient to raise the ND_p Cals % to a level sufficient to meet the needs of nursing mothers, using the criteria of Platt *et al.* (1961).

Skim Milk Powder

In 1937 Aykroyd and Krishnan reported adding SMP to the diet of 120 boys in a hostel in South India. The basic diet of about 740 g/"consumption unit" per day was based approximately on pearl millet 225 g, milled parboiled rice 284 g, and pigeon peas 77 g. For 3 months children in group A received the basic diet plus 28 g of SMP reconstituted with 8 parts of water; group B received no milk but sufficient millet to make the diets approximately isocaloric. At the end of 3 months, mean height and weight increases were measured. In a subsequent 3-month trial, group B received the milk supplement and group A did not. Results appear in Table 6.22. Differences in height and weight gains between diets with and without SMP were significant in both trials.

Somewhat similar trials in two other hostels in which the only cereal was rice confirmed the improving effect of SMP addition. In all three hostels an improvement from the addition of milk was observed in the general health of the children manifesting itself as a reduction in minor ailments and in increased vitality.

Multipurpose Food

Hanif *et al.* (1965) reported that, in rat feeding trials, a protein food when added to pearl millet at 5% and 10% levels, significantly improved the ND_p Cals % and NPU_{op} (Platt *et al.*, 1961). A 15% addition gave no further improvement. As ND_p Cals % increased from 6·4 to 9·9% the mixture became adequate to meet the protein needs of toddlers, adolescents and lactating mothers who require respectively (Platt *et al.*, 1961) 7·8, 8·0 and 9·5 ND_p Cal %.

The protein food contained guar (*Cyanopsis* spp.), chickpeas and fish flour but no proportions were stated. The protein content of the diet increased from 11·8% to 16·8% as the additions of protein food increased.

Synthetic Amino Acids

Daniel *et al.* (1965a) reported supplementing pearl millet with L-lysine and DL-threonine. The amino acid supplementation raised the content of lysine and threonine in the pearl millet to that of egg protein. Details of the diets, rat weight gains and PERs are presented in Table 6.23. The addition of lysine, diet (B), doubled the PER, the addition of threonine, diet (C), did not result in a further significant increase. The poor pearl millet diets required the addition of vitamins and minerals, (G), (H) and (I), before the full effect of lysine and threonine supplementation was realized.

Kurien *et al.* (1971b) studied the supplementation of poor pearl millet vegetarian diets with L-lysine HC1, DL-threonine and DL-methionine. The amino acids were added at levels equivalent to human milk proteins (WHO 1965). The experimental diets, proportions of the three amino acids and cystine (microbiological assay, Barton-Wright, 1952), rat weight gains and PERs are shown in Table 6·23.

Supplementation with lysine alone, diet (B), significantly increased PER. Further increases in PER from the addition of methionine or threonine or mixtures with lysine were not significant.

Pushpamma *et al.* (1972) compared market samples of pearl millet with sorghum and finger millet in rat feeding trials. The composition of the diets and their effect on PER are presented in Table 6.23. The amino acid composition of pearl millet and the results of the trials in comparison with the other cereals are discussed under Sorghum, p. 498.

Finger Millet

Pulses

Swaminathan (1938b) reported the nutritive value of rat diets based on finger millet with various combinations of SMP, pigeon peas, chickpeas, mung beans, urd, lentil and soybean. The diets, digestibility coefficients, biological values both by N balance and weight gain methods are shown in Table 6.24. Biological values by both evaluation methods increased with the substitution of 2% SMP for 2% of the pulses.

Patwardhan (1961b) in a literature review reported that the PER of finger millet, fed as the whole grain, was not increased by supplementation with chickpeas or dried powdered amaranth. The PERs were, respectively 2·0 and 2·1.

Daniel *et al.* (1968a) compared the additions of chickpeas, pigeon peas and SMP to poor Indian finger millet diets. The experimental diets, weight gains and PERs are presented in Table 6.31 and discussed under Sorghum on p. 484.

Daniel *et al.* (1970) supplemented poor finger millet diets with cooked, dried and ground pigeon peas. The lysine and threonine contents (microbiological assay, Barton-Wright, 1952), protein contents, weight gain, protein intake, and PER are shown in Table 6.31.

The inclusion of 16·7% of pigeon peas, diet (E), significantly increased growth rate compared with diet (A). No additional significant increase resulted from the addition of vitamins and minerals, diet (F).

Daniel *et al.* (1974) reported on the nutritional complementarity of finger millet and chickpeas. The diets, amino acid contents, rat weight gains and

PER are summarized in Table 6.31. A blend of finger millet 80/chickpeas 20 contained about 9·6% protein, and gave a PER of 2·9. Finger millet and chickpeas separately each adjusted to 6·0% protein with maize starch gave PERs of about 1·2.

Jansen (1974) using FAO (1970) and Orr and Watt (1957) data calculated the amino acid score for lysine, threonine, tryptophan, and methionine plus cystine against whole egg protein for finger millet, supplemented with soy flour (8·1% N), fish meal (12·0% N), groundnut flour (9·4% N) and cottonseed flour (8·0% N) at levels from 0 to 32 parts of supplement/100 parts of finger millet. The data are presented in Table 6.25. At all levels of addition of cottonseed flour and groundnut flour lysine was the limiting amino acid. Methionine plus cystine became limiting at 32 parts of soy flour and at 16 and 32 parts of fish meal per 100 parts of finger millet.

Narayanaswamy *et al.* (1975) supplemented finger millet diets with black-eyed cowpeas. The diets, rat weight gains and PERs are shown in Table 6.31. Lysine, total sulphur amino acids, threonine and tryptophan in the diets, and the findings in comparison with sorghum are discussed under Sorghum, p. 484.

Other Cereals

Joseph *et al.* (1959a) in N, Ca and P metabolism trials with eight girls aged 9–10 years resident in a boarding home in Mysore City employed rice and finger millet of the same protein contents as reported by Kurien *et al.* (1958) (see footnote “l” to Table 6.31). The composition of the diets and their effects on N, Ca and P balance are also shown in Table 6.26. The children remained in positive N balance but N retention was reduced as the proportion of finger millet in the diet was increased.

All the subjects were in positive Ca and P balance. As the proportion of finger millet increased, the mean intake and retention of Ca and the mean intake of P increased. Retention of P did not change significantly. It is suggested that a partial replacement of rice by finger millet could reduce the Ca deficiency of the rice diet.

Mason (1944) cited by Mason *et al.* (1946) reported that though the poor rice diet of South India appeared deficient in fat, the addition of fat reduced rat weight gain. Mason *et al.* (1946) over 9 weeks fed nine male and nine female rats, in pairs, diets based on rice to which varying amounts of butter and/or finger millet were added. The mean percentage growth rates are shown in Table 6.27.

Analysis of variance showed that both the negative effect of butter and the positive effect of finger millet on rat growth were highly significant but there was no interaction between butter and the millet; finger millet did not overcome the growth reducing effect of the fat.

The effect of partial or complete replacement of rice by finger millet in a poor vegetarian diet on the growth, and composition of blood and liver in rats was studied by Kurien *et al.* (1958). The composition of the diets, average daily food intake and average weekly weight gain over 13 weeks are shown in Table 6.31. Growth rates on diets (B) and (C), rice plus finger millet and (D) finger millet alone were greater than on diet (A) rice alone. (B), (C) and (D) were not significantly different.

Serum protein (King, 1951) was not significantly different among diets but diets (B), (C) and (D) all gave significantly higher haemoglobin contents (Sahli-Hellige haemometer, Hepler, 1950) and red blood cell counts (Neubauer's haemocytometer, standard procedures).

There was no difference in average liver protein (N × 6·25%) content among the four diets, but diets (B), (C) and (D) produced significantly higher liver fat (Tyner *et al*., 1950).

Overall, the replacement of even 25% of rice by finger millet in a poor vegetarian diet appeared to improve the nutritive value of the diet.

Oilseeds

Subrahmanyan *et al.* (1954a) compared the supplementary value of (a) cassava flour, (b) sweet potato flour, (c) a mixture of sweet potato flour (80%) and groundnut cake flour (GNC) (20%) and (d) a mixture of cassava flour (80%) and GNC (20%) as 25% replacement of the cereal in poor finger millet vegetarian diets.

The protein contents of the supplements, the diets and average food intake and weekly rat weight gains are shown in Table 6.31. There was no significant difference in weight gain among diets (A), (B) and (C). Diets (D) and(E) containing groundnut cake flour produced significantly higher gains than diet (A).

Hariharan *et al.* (1967) reported experiments with finger millet in poor vegetarian diets similar to those reported in 1965 with sorghum and pearl millet. The diets and the effect of supplementation with Ca salts and vitamins, fortified and unfortified GNF, or fortified SMP on rat weight gain and PER are shown in Table 6.31.

Supplementation of the poor finger millet diet with Ca salts and vitamins did not significantly increase growth rates. Supplementation with GNF (unfortified or fortified) to provide 2·5% or 5% extra protein in the diet, significantly increased weight gain. The PER of the finger millet diet increased slightly with added vitamins and minerals. There was no significant difference in the PER between diets containing 2·5% extra protein from fortified GNF or SMP, but the PER from 5% extra SMP protein was significantly higher than 5% protein from GNF. Fortified and unfortified GNF appeared equally effective in supplementing the poor finger millet diet.

Edible screw-pressed coconut cake from a Mysore mill was compared with SMP in poor Indian finger millet diets (Daniel *et al*., 1968b). The coconut meal and SMP provided about 2·0% extra protein in the diets. The protein contents of the coconut cake, unfortified and fortified with Ca and vitamins, the fortified SMP, the diets and results of rat feeding trials are shown in Table 6.31. The weight gain on diet (B) finger millet plus minerals and vitamins was not significantly higher than (A) finger millet alone. The inclusion of coconut meal, diets (C) unfortified and (D) fortified significantly improved weight gain over (A) and (B) but (C) and (D) were not significantly different. The highest weight gain was from fortified SMP, diet (E).

Daniel *et al.* suggested that the finger millet was adequate in minerals and vitamins and that coconut meal is an effective protein supplement for children.

Subrahmanyan *et al.* (1954b) added alcohol-extracted cottonseed flour to poor finger millet vegetarian diets. Extraction reduced the gossypol content from 2·4% to 0·28%. Protein contents of the supplement and diets, food intake and rat weight gains are shown in Table 6·31. Diet (A), poor finger millet diet, was compared with diet (B), poor finger millet diet plus 10% cottonseed flour.*

The improvement in weight gain from the addition of cottonseed flour was highly significant.

Fish Flour

Shurpalekar *et al.* (1962a) compared a fish flour fortified with vitamins (prepared by the method of Shurpalekar *et al.*, 1962b) and SMP, fortified with vitamins A and D as supplements to poor Indian finger millet diets. The diets, and rat weight gains are summarized in Table 6.31. Diets (B) fish flour and (C) SMP both produced significant increases in weight gain, haemoglobin and red blood cell counts compared to diet (A) but there were no significant differences in liver fat content among diets (A), (B) and (C).

Yeast Protein

Sur *et al.* (1954a) added food yeast (*Torula utilis*) to poor vegetarian diets similar to those reported by Murthy *et al.* (1950). In a further series of tests (Sur *et al.*, 1954b), the same samples of finger millet and food yeast were fed in diets at a 5·0% level of protein. Both series of results appear in Table 6·31. The food yeast at the 5·0% protein level improved the feed/gain ratio of the finger millet diets.

Narayanaswamy *et al.* (1970b) added yeast grown on petroleum hydrocarbons to poor Indian vegetarian finger millet diets. The protein content of the yeast, the diets and rat weight gains and PERs are shown in Table 6.31. The added yeast increased lysine and threonine contents and at the 6% level (2·0% extra protein) with vitamins and mineral mix, diet (D), significantly increased PER.

*(Authors' note: this appears from the text to be a straight addition, not a substitution, but it is not clearly stated.)

Multipurpose Foods

Subrahmanyan *et al.* (1954d) compared the nutritive value of three synthetic grains prepared by the method of Subrahmanyan *et al.* (1954e) from cassava and groundnut flour in the proportions (a) 90/10, (b) 85/15 and (c) 80/20 as replacement for finger millet in poor Indian vegetarian diets. The protein contents of the synthetic grains, and the diets are shown in Table 6.31. Weight gains were not significantly increased by additions of the synthetic grains.

An Indian multipurpose food (MPF) fortified with Ca and vitamins A and D and riboflavin was compared with an American MPF, when added in poor Indian finger millet diets by Kuppuswamy *et al.* (1957). The addition of both the MPF significantly increased weight gains compared to the control (Table 6.31).

Tasker *et al.* (1964) reported the supplementary effect of a protein food (Krishnamurthy *et al.*, 1958) based on a 1/1/2 blend of screw-pressed coconut meal, chickpea and groundnut flour fortified with vitamins A and D, thiamine and riboflavin when added to a poor vegetarian finger millet diet. The diets and rat weight gains are summarized in Table 6.31. Supplementation increased both protein content of the diet and the rate of weight gain.

Guttikar *et al.* (1968) supplemented a poor vegetarian finger millet diet with two protein foods I and II as described by Guttikar *et al.* (1965ab). The diets, rat weight gains and F/G ratios are summarized in Table 6.31. Supplementation significantly increased weight gain, haemoglobin and red blood cell count and improved the F/G ratio. The livers of rats on the basal finger millet diet were significantly lower in protein and higher in fat than those fed the protein enriched diets. Carcass lipid and protein contents were not significantly different among diets.

The livers of rats fed the poor finger millet diet showed mild to moderate degrees of parenchymal damage, a reduction in cellular cytoplasmic protein and varying degrees of cytoplasmic vacuolation, all typical of protein deficiency. Necrosis of the liver cells was not observed. The liver sections of rats fed the supplemented diets were normal.

No significant differences were observed among diets in the mean retention of protein/100 g increase in body weight (method of Venkat Rao *et al.*, 1964b).

Kurien *et al.* (1969) examined the supplementary value of an Indian multipurpose food (MPF) based on a 3/1 blend of groundnut and chickpea fortified with Ca salts, vitamins A and D and riboflavin, to poor finger millet diets fed to rats, under conditions of adequate and inadequate calorie intake. The protein contents of the Indian MPF and the diets, rat weight gains and PERs are shown in Table 6.31.

The rats were fed weighed amounts of the diets (A) 8 g (adequate calorie intake), (C) 6 g (calorie intake restricted to 75%), (E) 4 g (calorie intake restricted to 50%),/rat/day over 4 weeks. The quantity of MPF in (B), (D) and (F) was maintained at 0·8 g/rat/day by replacing an equivalent amount of the basal diet. At each caloric intake supplementation significantly increased weight gain and improved F/G ratios though only at adequate calories was the PER significantly improved. PER declined with reduction in caloric intakes. The results indicate that MPF exerts a significant supplementary value when the calorie intake is restricted to 75% but only a slight supplementary effect on a 50% caloric restriction.

Desai *et al.* (1970b) replaced 50% of the cereal in poor Indian finger millet diets with low cost balanced foods, designated LCBF III, IV and V. The composition of the LCBF, the supplemented diets and rat weight gains are shown in Table 6.31. Addition of all LCBF supplements significantly increased weight gains.

Narayanaswamy *et al.* (1972b) added (1) LCPF (wheat flour 70/soy flour 15/groundnut flour 15) and (2) fortified SMP to a poor finger millet vegetarian diet. The composition of the supplements, the diets, rat weight gains and PERs are shown in Table 6.31. Diets (C) and (E) provided 1·5% and diets (D) and (F) 3·0% extra protein. There was no significant increase in weight gain, and only a slight increase in PER, between diets (A) and (B). Diets (C), (D), (E) and (F) produced highly significant increases in weight gain and PER compared to (A) and (B). PERs were not significantly different between diets (C) and (D), nor between (E) and(F); PERs of (E) and (F) were significantly higher than (C) and (D).

The LCPF compared well with SMP as a protein supplement in poor finger millet diets.

Narayanaswamy *et al.* (1974) studied the supplementary effect of three protein enriched cereal foods (PECF), blends of wheat flour, soybean flour and groundnut flour, on poor vegetarian finger millet diets.

The composition of the PECF, the diets, weight gains and PERs are summarized in Table 6.31. Diets (B), (D), and (F) provided 1·5% and (C), (E) and (G) 3·0% extra protein in the diets. All supplements significantly increased rat weight gains compared to diet (A) and the mean growth rates and PERs from diets (B), (C), (D) and (E) were significantly higher than diets (F) and (G). The results confirmed the recommendations of the Protein Advisory Group (1971) that PECF used to supplement the diets of preschool children should have a minimum protein content of 20%.

Synthetic Amino Acids

Daniel *et al.* (1965b) supplemented (A) a poor finger millet diet with (B) L-lysine, and (C) DL-threonine and together with (D) a skim milk powder diet, compared the digestibility coefficient, biological value of the protein (Tasker *et al.*, 1962) and NPU_{op} (Platt *et al.*, 1961). The test subjects were 8 girls, 11-12 years of age, living in a boarding home in Mysore City. The mean daily food intakes are shown in Table 6.28. The lysine and threonine additions raised the levels to the equivalent of egg protein.

The diets were fed during five 8-day periods, faeces and urine for N determination (micro-Kjeldahl) being collected during the latter 4 days. The order of feeding was: (A) finger millet diet, (B), (C), (D) SMP and (E) low protein diet (Parthasarathy *et al.*, 1963).

Amino acid contents of diets (A) and (D) (Krishnamurthy *et al.*, 1960) mean daily intake and absorption of essential amino acids compared with amino acid requirements (Nakagawa *et al.*, 1962) appear in Table 6.28. The finger millet diet provided 28·7 g protein (N × 6·25)/day derived from: finger millet 23·5 g, pigeon peas 3·4 g and SMP 1·3 g.

At the level fed (1·3 g protein/kg body weight) the finger millet diet was deficient in lysine but according to the requirements proposed by Nakagawa *et al.* (1962) was adequate in all other essential amino acids. The data in Table 6.29 indicate that lysine was significantly first-limiting.

Details of the diets, mean daily N intakes, excretion and balance, apparent digestibility, true digestibility, BV and NPU_{op} are shown in Table 6.29. On diet (A) N retention was 5·8% of intake. Retention increased significantly in (B) over (A), (C) over (B), and (D) 33% over (C) 24·4%. True digestibility coefficient of (C) was 73% and of (D) 88%.

The BV of (D) 85·3 was significantly higher than (B) 81·2 and (B) significantly higher than (A) 67. The NPU_{op} of diets (B) and (C) were significantly higher than diet (A). Diet (D) 74·8 was much higher than (B) and (C).

The net available protein from the different diets compared to WHO (1965) reference protein requirements is given in Table 6.29 from which it appears that diets (A) and (B) were deficient but diet (C) met the requirement. Lysine and threonine were first- and second-limiting amino acids.

Doraiswamy *et al.* (1969) described the effects of supplementing (A) poor finger millet diets with (B) lysine, (C) leaf protein and (D) low fat sesame flour on the growth over 6 months of 80 boys in a boarding home near Mysore. Nitrogen metabolism was also studied with 6 boys fed the diets shown in Table 6.30.

The lysine added, 0·66 g/child/day, diet (B), provided the total 60 mg/kg of body weight required (Nakagawa *et al.*, 1962). The leaf protein was prepared by the methods of Davys and Pirie (1963) and Morrison and Pirie (1961) and provided in diet (C) lysine roughly equivalent to diet (B). The sesame flour, prepared by processing sesame seeds in an oil expeller, in diet (D), provided protein similar to that in diet (C). The protein and lysine contents of the leaf protein and sesame flour and the effect of the diets on height, weight, haemoblobin and red blood cell counts (RBC) (Doraiswamy *et al.*, 1962), N balance and apparent digestibility are shown in Table 6.30. By all criteria diets (B), (C) and (D) were significant for height and weight but not for haemoglobin and RBC values. The height and weight increases from (C) leaf protein were significantly greater than (B) and (D).

The general nutritional status (Indian Council of Medical Research 1948) was significantly better from diets (B), (C) and (D) than diet (A) but differences among (B), (C) and (D) were not significant.

Diet	Number improved	Number stationary	Number deteriorated
(A) finger millet	3	11	6
(B) + lysine	11	9	0
(C) + leaf protein	13	7	0
(D) + sesame flour	8	11	1

Diet (A) was significantly inferior to the other three diets in N retention and apparent digestibility. Differences in N retention among diets (B), (C) and (D) were not significant. In apparent digestibility, diet (B) was significantly higher than (A) but lower than (C) and (D) between which the difference was not significant.

All diets were lower in calories (calculated) and diets (A) and (B) lower in protein than recommended by the Indian Council of Medical Research (1944). The low apparent digestibility of the protein of diet (A) agreed with the findings of Subrahmanyan *et al.* (1955) who reported that only 50% of the protein in finger millet diets was digestible.

The best growth responses were obtained from (C) leaf protein, which provided both additional lysine and protein. Diet (D), sesame flour, supplied additional protein but only half the lysine provided by (B) and (C).

Leela *et al.* (1965) examined the effect of supplementing market samples of finger millet and a poor vegetarian finger millet diet with L-lysine, DL-threonine and SMP. The amino acid additions raised the total to the equivalent of egg protein.

Details of the diets, rat weight gains and PERs are summarized in Table 6.31. PERs were determined both at 6% and 8% levels of dietary protein. The results indicated that finger millet protein is adequate in all essential amino acids except lysine and threonine. The PER of finger millet protein at the 6% protein levels, diet (A) increased significantly on supplementation with lysine, (B) or lysine and threonine (C). The mean liver fat content of rats fed 6% finger millet protein, (A), was significantly higher than that of rats fed 6% milk proteins, (H). Supplementation of the finger millet diet with lysine and threonine, (B) and (C), decreased liver fat content but not significantly.

The PER of the poor finger millet diet fed at the 6% protein level, (D), was significantly higher than that of the finger millet diet (A) at the same level. This may be due to the fact that the lysine content of the proteins of the poor finger millet diet was higher at 256 mg/g total N than that of the cereal itself (212 mg/g total N). Supplementation of the poor finger millet diet with lysine and threonine produced significant increases in PER when fed at both the 6% and 8% protein levels. Addition of vitamins and minerals to the poor finger millet diet did not significantly improve PER.

The PERs of the poor finger millet diet supplemented with lysine and threonine, fed at the 6% and 8% protein levels, (F) and (F1), were nearly equal to those obtained with the poor finger millet diet supplemented with 10% SMP fed at the same protein levels, (G) and (G1).

Desai *et al.* (1970a) supplemented poor vegetarian finger millet diets with L-lysine HC1, DL-threonine and DL-methionine. The lysine, threonine and methionine were added at levels designed to raise their dietary content to about the same level as in human milk protein (WHO, 1965). The experimental diets, their contents of the three amino acids and cystine (microbiological assay, Barton-Wright, 1952), rat weight gains and PERs are summarized in Table 6.31. Supplementation with lysine, diet (B) significantly increased PER over the basic poor diet (A). Addition of (E) lysine plus methionine or (F) lysine plus threonine caused a further significant increase, (H) lysine plus methionine plus threonine, produced the highest PER, significantly higher than (E) and (F).

Pushpamma *et al.* (1972) compared market samples of finger millet, sorghum, pearl millet and maize, in rat feeding trials. The diets, and their effect on PER are presented in Table 6.31. The amino acid composition of finger millet and the results of the trials in comparison with the other cereals are considered under sorghum, p. 488.

Venkataraman *et al.* (1977) reported experiments in which the green algae (*Scenedesmus acutus*) was added at two levels to diets based on finger millet fed to male weanling rats for a period of 4 weeks. The composition of the diets and the effect on weight gain and PER are shown in Table 6.31. The supplementation, which increased the protein content of the diet from 6·0 to 8·6%, significantly increased the PER compared to the finger millet diet alone. Evaluations for the safety of algae as food, as stipulated by the Protein Advisory Group (1972ab) had not, at the time of writing the paper, been completed.

Foxtail Millet

Pulses

Ganapathi *et al.* (1958) in 10% protein diets, compared the biological value of foxtail millet alone and with added horse gram (*Dolichos biflorus*) or cowpea. The diets, weight gains and PERs are shown in Table 6.32. Both pulses improved the nutritive value of foxtail millet, cowpea to a greater extent than horse gram.

Patwardhan (1961b) provided brief details on the effect on rat PER of foxtail millet, fed as whole grain or with the husk removed, and supplemented with chickpeas and the dried powder of amaranth. Removal of the husk reduced the PER from 0·8 (whole grain) to 0·5. Supplementation of the dehusked grain with chickpeas and amaranth raised PER to 2·2.

Other cereals

Mangay *et al.* (1957) reported on mixtures of foxtail millet grown in Tennessee, USA with Jarvis maize. The diets, their effect on total protein, niacin and tryptophan intakes, rat weight gains and PERs are shown in Table 6.32. Animals on diet (Cc) maize addition only, gained less weight than those on the basal vitamin-free casein diet (Ac). Supplementation with niacin (Dc) improved growth and PER, and diet (Ec) 40% foxtail millet addition, also improved growth and PER to satisfactory levels. Increasing the content of millet to 80% (Kc) did not improve growth over (Ec). Though the diets containing foxtail millet were lower in niacin than the comparable maize diets, the addition of niacin to the foxtail millet diets had no significant effect on rat growth.

The highest growth was obtained from (Gc), basal plus 40% maize plus 40% foxtail millet but the PER was lower than that from (Dc) 40% maize plus niacin and (Ec) 40% foxtail millet.

Since supplementation of the maize diets with either niacin or millet improved growth and PER, a further test was conducted to establish whether the apparent supplementary effect of foxtail millet was due primarily to its niacin content. Groups of rats were fed the basal 9% casein diet plus 40% maize (Acm) supplemented with different levels of millet replacing equivalent quantities of sucrose (Table 6.32). Growth rates were compared with groups receiving the basal diet supplemented with additions of niacin equivalent to the millet supplements. The results are shown in Table 6.32. Greatest weight gain resulted from diets (Gcm) and (Icm) in which millet at 20% and 40% of the diet replaced an equivalent amount of sucrose. The addition of niacin alone to maize did not improve growth until the level of 0·24 mg/100 g was reached. Though the growth obtained with millet supplementation was superior to niacin supplementation in every case, the PER ratios were more variable, e.g. diet (Hcm) with niacin gave a weight gain of 25·8 g and a PER of 3·2 compared to (Icm) with a weight gain of 29·8 g and a PER of 2·0.

In subsequent tests, foxtail millet was added to the basal 9% casein and 40% maize diet at levels of 5%, 10% and 20% compared with the basal diet supplemented with tryptophan at 11, 22 and 44 mg/100 g (equivalent to that contributed by the foxtail millet at each level). The results are shown in Table 6.32. At equivalent tryptophan levels the addition of tryptophan produced weight gains equal to foxtail millet. The addition of niacin plus tryptophan gave no further improvement. These findings suggested to Mangay *et al.* that the improving effect of foxtail millet in the 9% casein plus 40% maize diet results from its tryptophan content. As in the previous test, the addition of foxtail millet, while increasing weight gain, tended to reduce PER.

A further trial was conducted in which the diets consisted of 89% maize or foxtail millet and excluded casein. To this niacin at 3 mg/100 g of the diet, tryptophan at 400 mg/100 g of the diet and L-lysine HC1 at 1% of diet were added singly and together (Table 6.32).

The growth of rats on diet (E) in which foxtail millet was the only source of protein was poor, and similar to those on diet (A) maize alone, but the

PER of (E) was lower, 1·1, against 1·5 from diet (A). Addition of lysine to foxtail millet, diet (I) improved weight gain and PER.

When maize was the only source of protein in the diet, the addition of (J) 2%, (K) 10% and (L) 5% foxtail millet to 89% maize and of niacin and/or tryptophan to 80% maize diets had no effect on weight gain or PER. The addition of lysine to a diet of 80% maize, 10% foxtail millet, niacin and tryptophan, (P), improved weight gain and PER, the addition of lysine with niacin but without tryptophan, (O), gave no improvement.

When rats were fed, over 21 days, a diet of 40% maize and 40% foxtail millet alone, and supplemented with lysine, 1% of diet, niacin 3 mg/100 g diet and tryptophan 400 mg/100 g diet, lysine increased weight gain from 14 g to 66 g and PER from 1·1 to 3·2 but neither niacin nor tryptophan, singly or combined with one another or with lysine caused any improvement.

Mangay *et al.* suggested that as the foxtail millet content of a maize diet is increased, the diet becomes more limiting in lysine and less limiting in tryptophan. The use of a millet of high tryptophan content, such as the foxtail millet used in these experiments, 126 mg/g total N should be encouraged in maize-eating regions where pellagra is endemic.

Multipurpose Food

Joseph *et al.* (1957) compared Indian multipurpose food (MPF) and American MPF in rat experiments basically similar to those of Kuppuswamy *et al.* (1957) but using poor vegetarian diets based on foxtail millet. The findings appear in Table 6.32. There was no significant difference between Indian and American MPF as supplements to the poor foxtail millet diet and both produced significantly higher weight gains than the foxtail millet diet alone.

Common and Little Millet

Pulses

Patwardhan (1961b) in a literature review reported the PER of little millet increased from 1·1 (whole grain alone) to 1·8 when supplemented with chickpeas and amaranth.

Other Cereals

Kundaji and Radhakrishna Rao (1954) reported on the additions of common millet (wari) to rice and pigeon pea diets. The diets and their effect on food intake, rat weight gain and PER appear in Table 6.33. In a second trial adult rats received a 5·7% protein diet; details of the diets also appear in Table 6.33. The BV (Chick *et al.*, 1935ab, Swaminathan, 1937abc) and digestibility coefficient of diet (A) (1) were 71·8% and 90% respectively.

Oilseeds

Adrian *et al.* (1956) reported the addition of groundnut flour (GNF) to *Panicum* millet, probably common millet. The diets, rat weight gains and PERs appear in Table 6.33. Diet (A) was equivalent to a typical African diet, the remainder represented different types of supplementation, compared with (F) a casein control diet. Diets (D) GNF, (E) GNF plus fish hydrolysate and (F) were essentially isonitrogenous. Diet (A) alone did not support normal growth.

Supplementation of millet with lysine, diet (B) gave a PER comparable to the casein diet though the weight gain was much lower. GNF in diets (D) and (E) increased weight gain but produced lower PERs than the lysine-supplemented diet (B).

Fish Flour

Sure *et al.* (1957) reported the effect of defatted fish flour made from a mixture of carp, smelt and whitings (Levin and Finn, 1955) substituted at levels of 1%, 3% and 5% for "millet" (not otherwise characterized) of 9·0% protein content in diets of weanling rats. The diets, weight gains and PERs are summarized in Table 6.33. Diet (D), millet plus 5% fish flour, produced greatest weight gain and PER.

Multipurpose Food

Joseph *et al.* (1957) compared Indian multipurpose food (MPF) and American MPF in rat feeding trials using poor vegetarian diets based on little millet. The results are presented in Table 6.33. American MPF gave significantly higher weight gains than Indian MPF; and both MPF gave significantly higher weight gains than little millet alone.

Synthetic Amino Acids

Rat growth experiments were conducted by Howe *et al.* (1965) in which "white millet" was supplemented with amino acids. The diets and their effect on weight gain and adjusted PERs are shown in Table 6.33. Supplementation with lysine, threonine and tryptophan gave growth rates and PERs equivalent to that of casein. The data indicated that lysine was the first- and threonine the second-limiting amino acid in "white millet".

Kodo and Lesser Millets

Pulses

Patwardhan (1961b) in the course of a review reported that the PER of whole grain kodo millet increased from 0·7 to 1·9 when supplemented by chickpeas and dried powdered amaranth.

Rajalakshmi and Majmudar (1966) reported rat feeding trials in which kodo millet was fed in diets (A) alone, and with (B) pigeon peas, (C) chickpeas, (D) moth bean (*Phaseolus aconitifolius*), (E) field bean (*Dolichos lablab*), (F) cowpeas, (*Vigna catiang*) (G) peas (*Pisum sativum*) and (H), a mixture of pulses, excluding the field bean, substituted at a level of 20% by weight for the kodo millet. All grains were bought in the local market and milled at a flour mill.

Mortality was very high among the rats on diet (E) field bean. Rajalakshmi and Majmudar suggested the possible adverse effect of a trypsin inhibitor in the field bean. Results might have been different had the diets been fed cooked and not raw. Field beans were excluded from the mixture fed in diet (H) after the initial 2 weeks.

Details of the diets with the protein content of the diets, as calculated from food tables (Aykroyd *et al.*, 1963), not analysed, the effect on weight gain and PER are summarized in Table 6.34. All the supplemented diets improved the nutritive value of the kodo millet which, fed alone, diet (A), did not sustain growth. Diet (G), kodo millet plus peas, was best and (B), kodo millet with pigeon peas, was poorest.

The (calculated) protein contents of the supplemented diets were about 2·0% higher than that of kodo millet alone.

Other cereals

Kundaji and Radhakrishna Rao (1954) studied the effect of "unfamiliar cereals", including kodo millet and *Amaranthus paniculatus* (rajgira) when added to rice or pigeon pea diets. The diets, rat weight gains, food intakes, protein intakes, and PERs are shown in Table 6.34.

In a second trial adult rats received 5·7% protein diets, details of which are also shown in Table 6·34. The BV (Chick *et al.*, 1935ab, Swaminathan, 1937abc) and digestibility coefficient of the diets were respectively A(1) 91·5 amd 89·6; and B(1) 86·2 and 88·3.

Balanced Diets

In this section reports of foods containing sorghum or one of the millets and intended for weaning infants and/or as supplements to general diet are reviewed.

Chandrasekhara *et al.* (1957) reported that a sorghum malt food (24% protein), made from sorghum malt blended with low fat groundnut flour, or SMP, and fortified with vitamins and minerals, when fed at 11% protein was equivalent in PER to a milk food (24% protein) consisting of a blend of SMP, starch, sugar, hydrogenated fat, fortified with vitamins A and D, Fe and Ca. The sorghum malt food incorporated at the 10% level in a poor rice diet showed higher supplementary value than whole milk powder at the same level and satisfactorily supported reproduction and lactation in albino rats.

Daniel *et al.* (1967) studied low cost balanced foods based on blends of finger millet, low fat groundnut flour (GNF), full fat soy flour (SF), chickpea flour and SMP, with and without fortification with L-lysine HC1 and DL-methionine. Composition of the blends, protein, lysine, methionine, cystine, threonine in protein, rat weight gains and PERs are shown in Table 6.37. The PER of diet (A) increased on supplementation with amino acids, diet (B), to the equivalent of diet (K) SMP. The PER did not change when, in diet (C), SMP was replaced by soy flour but was significantly lower in diet (E) where chickpea replaced SMP.

Fortification of the two blends with amino acids, diet (D) and (F) increased PER to 2·7 and 2·4 respectively. The PER of diet (G) which contained 20% chickpeas did not differ significantly from that of diet (E) which contained 10% chickpeas. Diet (I) which contained sesame seed meal gave a low PER which increased significantly on fortification with amino acids, diet (J).

All the low cost balanced foods promoted a weekly growth of 13·4–18·0 g when providing 10% protein in the rat diet. Daniel *et al.* (1967) suggested that if the foods replaced about half the cereals in a conventional finger millet diet, the protein content of the diet would increase from about 8% to about 14% and would meet the protein requirements of weaned infants and preschool children. These low cost balanced foods are sometimes referred to as Bal-Ahar which in Hindi means "children's food".

Swaminathan *et al.* (1972) in a review of protein enriched cereal foods for preschool children in developing countries lists certain compositions which included sorghum and "millet" flour:

> "Argentirina" (Argentine) contained (%) sorghum flour 20, millet flour 17, wheat flour 10, groundnut flour 40, and white bean flour 10, yeast 2, calcium carbonate 1 and 4500 IU vitamin A/100 g; protein content 27·8%. Kapsiotis (1967) reported favourable acceptance.
>
> "Faffa" (Ethiopia) contained teff, peas, chickpea, lentil and SMP; protein content 15%. Agren *et al.* (1969) reported that the food was acceptable to children.
>
> "Aliment de Sevrage" (weaning food, Senegal) contained millet flour, groundnut flour, SMP, sugar, vitamin A and calcium salts; protein content 20%. The food was readily acceptable (Robinson, 1966).
>
> INCAP (Guatemala) mixture 9, contained (%) maize meal 28, sorghum flour 28, and cottonseed flour 38 with yeast 3, calcium carbonate 1 and 4500 IU vitamin A; protein content 27·5%. Bressani and Elias (1968) reported satisfactory results in feeding children suffering from protein-calorie malnutrition.

Chamberlin and Stickney (1973) offer some general guidelines in formulating foods for young children including a low cost diet for Northern Nigeria (Table 6.35). They specify the nutritional constraints for children weighing 10 kg, 20 kg and adults weighing 60 kg (Table 6.36).

Gopaldas *et al.* (1974) described indigenous (Indian) weaning mixtures which could replace instant sweetened CSM (corn(maize)–soya–milk) powder imported from USA. The criteria for the production of the mixes were:

(1) year round availability of the food commodities in the Ratlam District, Madhya Pradesh,
(2) cost to be within 25 paise/child/day,
(3) mixture to yield at least 300 calories and 12 g protein/child/day,
(4) minimum shelf life of 8 weeks,
(5) palatable and acceptable to mother and child,
(6) tolerance of alternative formulations by the preschool child,
(7) utilization of village level technology for processing,
(8) adaptability of the mixes to the traditional dietary pattern and,
(9) adequate standards for hygiene and quality control in the production.

These criteria were met by several formulations based on combinations of maize, sorghum, wheat, chickpeas, mung beans, groundnut and jaggery. The village "roaster" and "pounder" were employed to dry-roast, mill and mix together the raw cereals and pulses. The jaggery and groundnut were hand pounded and then mixed with the ground flours. After sieving the mixes were stored in plastic lined tins or steel trunks. The mixes were easily incorporated into traditional dishes such as rabdi (a thin gruel), roti, sattu, kheer, halwa and panjeeri.

Typical formulations were (parts, presumably by weight), (A) sorghum 4, chickpeas 2, groundnut 1, jaggery 2 and (B) sorghum 60, chickpeas 17, groundnut 9, jaggery 14.

A "least cost" weaning mix based on sorghum had the following composition: (parts) sorghum 60, chickpeas 17, groundnut 14, jaggery 9, protein 12·25 g/100 g, calories 376/100 g.*

Prasannappa *et al.* (1976) described two blends including sorghum, finger millet and wheat intended to supplement the diets of preschool and school children. The composition of the diets, essential amino acid content, rat weight gains and PERs are shown in Table 6.37. The cereals and chickpea flour were roasted, then powdered and blended with the remaining ingredients. The blends provided about 16 g protein and 388 kcal/100 g. Prasannappa *et al.* calculated that a daily supplement of 55 g would provide about 9 g of protein, 200 kcal and half the daily requirements of Ca, vitamin A, thiamine and riboflavin.

From the amino acid reference pattern (WHO 1973a) methionine plus cystine appeared first-limiting. Roasting reduced available lysine slightly.

The corrected PER values of the unroasted blends were (A) 2·1 and (C) 2·2 and of the roasted (B) 2·0 and (D) 2·3 compared with 2·5 for casein. (A) and (B) contained millet and wheat, (C) and (D) millet, sorghum and wheat.

Roasting of the ingredients helped to reduce insect infestation and improve flavour. The roasted products were stored satisfactorily for 4 months at room temperature. Prepared as dumplings or porridge, the foods were readily accepted by children.

*(Authors' note: Gopaldas *et al.* did not state how the nutritional values were arrived at).

Conclusion

As was emphazied in Chapter 1, the poorer sections of communities in the SAT not only have a lower dietary intake with a lower proportion of protein calories but are subject to wide regional, seasonal and individual variations in food supplies. The interrelations of the basic cereals with other available nutrient sources, and the optimum combinations of such foods, are clearly of great importance in reducing nutritional imbalances.

It is equally evident that the reliable knowledge available concerning caloric and nutrient availability among the main consumers of sorghum and the millets is too sparse and uncertain to permit any definitive judgement on nutritional well-being to be made. Wise men and women will, we believe, err on the side of caution rather than dogmatism.

Tables

Chapter 1

Table 1.1
Dietary energy supply of the population of SAT countries of Africa as % of requirement.[a]

	Dietary energy supply[b]	
SAT countries	Kcal/person/day	% of requirement[c]
Gambia	2490	104
Haute Volta	1710	72
Mali	2060	88
Mauritania	1970	85
Niger	2080	89
Senegal	2370	100
Tchad	2110	89
Ethiopia	2160	93
Tanzania	2260	98
Zambia	2590	112
Mozambique	2050	88
Rwanda	1960	84
Burundi	2040	88
Sudan	2160	94

[a] UN World Food Conference (1974).
[b] 1969 – 1971 average.
[c] Revised standards of average requirements (physiological requirements plus 10% for waste at household level).

Table 1.2
Provisional amino acid scoring pattern.[a]

	Suggested level	
Amino Acid	mg/g protein	mg/g N
Ile	40	250
Leu	70	440
Lys	55	340
Met, Cys	35	220
Phe, Tyr	60	380
Thr	40	250
Trp	10	60
Val	50	310
Total	360	2250

[a] WHO (1973a).

Table 1.3
Safe level of protein intake of 100% sorghum diet relative to milk protein.

Age group	Safe level of protein based on egg and milk protein		Adjusted level for a 100% sorghum diet (50% digestibility)[a]	
	g prot./kg/day	g prot./person/day	g prot./kg/day	g prot./person/day
Infants				
6–11 months	1·53	14	7·65	70
Children				
1–3 years	1·19	16	5·95	80
4–6 years	1·01	20	5·05	100
7–9 years	0·88	25	4·40	125
Male adolescents				
10–12 years	0·81	30	4·05	150
13–15 years	0·72	37	3·60	185
16–19 years	0·60	38	3·00	190
Female adolescents				
10–12 years	0·76	29	3·80	145
13–15 years	0·63	31	3·15	155
16–19 years	0·55	30	2·75	150
Adult man	0·75	37	2·85	185
Adult woman	0·52	29	2·60	145
Pregnant woman		+9		+45
Lactating woman		+17		+85

[a] Recommended level of egg or milk protein $\times \frac{100}{\text{AA score}} \times \frac{100}{\text{protein digestibility}}$

Table 1.4
Summary of tables (range of means).

	Sorghum	Pearl millet	Common millet	Foxtail millet	Little millet	Finger millet	Japanese barnyard millet	Kodo millet
Proximate analyses (%)								
Protein (N × 5·7)	7·1–14·2	7·8–19·3	9·9–17·0	9·1–14·4	6·8–12·6	3·8–10·9	9·6–10·9	6·6–12·1
Lipid	2·4–6·5	1·5–6·8	1·9–4·9	2·5–6·8	5·3–6·8	1·0–4·6	3·0–5·1	3·4–6·6
Carbohydrate	70–90	62–89	61–80	63–72	63–71	74–88	56	73–74
Fibre	1·2–3·5	1·4–7·3	6·3–9·0	6·3–8·1	5·7–7·6	3·0–7·5	13.9	6·2–10·5
Amino acids (mg/g N)								
Lys	71–212	109–297	89–266	115–157	114–123	160–262	106	188–214
Lys amino acid score	21–62	32–87	26–78	33–46	33–36	47–77	31	55–63
Ile/Leu	1·9–5·0	1·6–3·8	2·4–4·3	1·8–3·0	1·6	1·5–2·9	1·4	1·5–3·3
Minerals (mg/100 g)								
Calcium	11–586	30–62	15[a]	34–617	32–598	220–855	25	25–620
Phosphorus	167–751	248–950	182[a]	310–1153	233–1102	131–904	285	247–670
Iron	0·9–20·0	1·1–38·0	5–9[a]	4–13	28–50[a]	6·4–15	—	12
Vitamins (mg/100 g)								
Thiamine	0·24–0·54	0·34–0·41	0·04–0·78[a]	0·43[a]	—	0·19–0·62	0·35	0·15–0·5[a]
Niacin	2·9–6·4	1·5–3·9	0·3–2·3[a]	3·7[a]	—	0·13–2·5	1·84	0·4–1·0[a]
Riboflavin	0·1–0·2	0·2–0·24	0·07–0·38[a]	0·12[a]	—	0·06–0·16	0·03	0·03–0·07[a]

[a] Results from comparatively few samples.

Chapter 2

Table 2.1
Area harvested (10^3 ha), yield (kg/ha) and production (10^3 t) for 1976 of sorghum and millets. F against figure = FAO estimate. U = unofficial figure. Dash indicates no values presented.[a]

	Sorghum			Millets		
	Area Harvested (10^3 ha)	Yield (kg/ha)	Production (10^3 t)	Area Harvested (10^3 ha)	Yield (kg/ha)	Production (10^3 t)
World	43 929	1 179	51 812	72 808	707	51 461
Africa	13 939	704	9 813	16 325	650	10 615
Algeria	2 F	2 000	3 F	—	—	—
Angola	—	—	—	93 F	860	80 F
Benin	84	670	57	25 F	600	15 F
Botswana	100 F	555	56	10 F	500	5 F
Burundi	110 F	1 136	125 F	30 F	900	27 F
Cameroon	—	—	—	430 F	907	390 F
Cent. African Republic	—	—	—	67 F	642	43 F
Chad	—	—	—	900 F	592	533
Egypt	—	—	—	206 U	3 883	800 U
Ethiopia	770	1 120	863	336	980	329
Gambia	—	—	—	33	667	22
Ghana	210 U	386	81 U	215 U	330	71 U
Guinea	11 F	455	5 F	—	—	—
Guinea-Bissau	7 F	714	5 F	13 F	577	8 F
Ivory Coast	42 F	714	30 F	65 F	769	50 F
Kenya	—	—	—	360 F	1 000	360 F
Lesotho	85 F	824	70 F	—	—	—
Libya	—	—	—	1 F	1 000	1 F
Madagascar	3	767	2	—	—	—
Malawi	120 F	875	105 F	—	—	—
Mali	—	—	—	1 244	646	804
Mauritania	—	—	—	180 F	333	60 U
Morocco	49	388	19	4 F	744	3 F
Mozambique	250 F	1 000	250 U	20 F	400	8 F
Namibia	8 F	400	3 F	55 F	355	20 F
Niger	633	486	308	2 532	472	1 195 F
Nigeria	5 940 F	620	3 680 U	5 000 F	640	3 200 F
Rhodesia	70 F	716	50 F	390 F	564	220 F
Rwanda	135 F	1 037	140 F	5 F	600	3 F
Senegal	—	—	—	955	581	555
Sierra Leone	7 F	1 529	11 F	9 F	1 000	9 F
Somalia	250 F	480	120 F	—	—	—
S. Africa	281 U	925	260	22 F	682	15 F
Sudan	2 600 F	692	1 800 F	1 200 F	375	450 F
Swaziland	4 F	714	3 F	—	—	—
Tanzania	600 F	767	460 F	200 F	650	130 F
Togo	—	—	—	199	595	119
Tunisia	14 F	536	8 F	—	—	—
Uganda	359	1 500	538	508	1 279	650 F
Upper Volta	1 138	630	717 F	911	406	370 F
Zaire	—	—	—	70 F	600	42 U
Zambia	58 F	800	46 F	38 F	800	30 F
N. and Cent. America	7 760	2 884	22 382	—	—	—
Costa Rica	20 U	1 800	36 U	—	—	—
Cuba	1 F	1 100	1 F	—	—	—
Dominican Rep.	5 U	3 612	19 U	—	—	—
El Salvador	125 U	1 314	164	—	—	—
Guatemala	57	1 680	96	—	—	—

Table 2.1 *cont.*

	Sorghum			Millets		
	Area Harvested (10^3 ha)	Yield (kg/ha)	Production (10^3 t)	Area Harvested (10^3 ha)	Yield (kg/ha)	Production (10^3 t)
Haiti	223 F	1 009	225 F	—	—	—
Honduras	57 U	821	47	—	—	—
Mexico	1 180 U	2 839	3 350 U	—	—	—
Neth. Antilles	5 F	1 400	7 F	—	—	—
Nicaragua	67	826	55	—	—	—
USA	6 020	3 053	18 382	—	—	—
S. America	2 421	2 696	6 525	241	1 275	307
Argentina	1 834	2 835	5 200 U	231	1 274	294
Brazil	193 U	2 539	490 U	—	—	—
Colombia	174 U	2 464	428 U	10 F	1 300	13 F
Paraguay	6 F	1 107	6 F	—	—	—
Peru	15 U	3 000	45 U	—	—	—
Uruguay	54 U	2 185	118 U	—	—	—
Venezuela	145 U	1 641	238 U	—	—	—
Asia	18 956	591	11 202	53 191	676	35 966
Afghanistan	—	—	—	20 F	1 150	23 F
Bangladesh	1 F	800	1 F	—	786	—
Bhutan	—	—	—	4 F	1 250	5 F
Burma	—	—	—	210 F	324	68 U
China	7 F	3 042	22 F	30 804 F	812	25 007 F
India	16 000 F	544	8 700 F	20 500 F	468	9 600 F
Iran	10 U	1 000	10 U	20 F	1 250	25 F
Iraq	6	600	4	2 F	1 500	3 F
Israel	8 F	2 075	17	—	—	—
Japan	1 F	1 000	1 F	3 F	1 333	4 F
Jordan	—	1 477	—	—	—	—
Korea DPR	65 F	923	60 F	500 F	820	410 F
Korea Rep.	6 F	833	5 F	27 F	963	26 F
Lebanon	1 F	833	1 F	—	—	—
Maldives	—	—	—	—	1 596	—
Nepal	—	—	—	125 F	1 144	143 U
Pakistan	523	621	325 U	696	503	350 F
Saudi Arabia	140 F	1 429	200 F	135 F	1 111	150 F
Sikkim	—	—	—	7 F	286	2 F
Sri Lanka	6 F	1 000	6 F	41 F	488	20 F
Syria	—	—	—	30 F	833	25 F
Thailand	130 F	1 777	231	—	—	—
Turkey	—	—	—	25 F	1 600	40 F
Vietnam Rep.	12 F	1 000	12 F	—	—	—
Yemen Arab Rep.	2 040	788	1 608	—	—	—
Yemen PDR	—	—	—	42 F	1 548	65 F
Europe	151	3 079	465	27	1 384	38
Albania	26 F	1 000	26 F	—	—	—
Austria	—	—	—	1 F	2 308	3 F
Bulgaria	—	—	—	1 F	500	1 F
Czechoslovakia	—	347	—	—	2 514	—
France	78	3 210	251	1	1 222	1
Greece	—	1 000	—	1 F	1 500	1 F
Hungary	2 F	2 500	5 F	5 F	2 200	11
Italy	3	4 182	14	—	—	—
Poland	—	—	—	10 U	800	8 F
Portugal	—	—	—	6 F	1 552	9 F

cont.

Table 2.1 *cont.*

	Sorghum			Millets		
	Area Harvested (10^3 ha)	Yield (kg/ha)	Production (10^3 t)	Area Harvested (10^3 ha)	Yield (kg/ha)	Production (10^3 t)
Spain	36	4 377	158	1 F	1 583	1 F
Yugoslavia	5 F	2 200	11 F	2 F	1 500	3 F
Oceania	1 590	2 224	1 124	25	1 400	35
Australia	504	2 222	1 120 U	25 F	1 400	35 F
Fiji	—	3 702	2 U	—	—	—
New Caledonia	—	1 433	—	—	—	—
Pacific Is.	—	2 500	—	—	—	—
Papua New Guinea	1 F	2 273	3 F	—	—	—
USSR	196	1 531	300 F	2 999	1 501	4 500 F
Developed countries	6 937	2 914	20 214	61	1 176	72
N. America	6 020	3 053	18 382	—	—	—
W. Europe	123	3 534	434	11	1 603	18
Oceania	504	2 222	1 120	25	1 400	35
Other developed	290	957	278	25	760	19
Developing countries	36 683	850	31 173	38 428	558	21 452
Africa	11 058	701	7 753	14 896	628	9 349
Latin America	4 161	2 530	10 525	241	1 275	307
Near East	4 797	755	3 623	1 681	941	1 582
Far East	16 666	556	9 268	21 610	473	10 214
Other developing	2	2 666	4	—	—	—
Centrally planned	308	1 378	425	34 319	872	29 937
Asian CPE	84	1 115	94	31 304	812	25 417
Europe USSR	224	1 476	331	3 015	1 499	4 520

[a] Data from Food and Agriculture Organization Yearbook (1976).

Table 2.2
Consumption and utilization of sorghum and millets by country (mean of 1966–1967 and 1968–1969).[a]

Countries	Consumption	Human nutrition	Animal feed	Fermented beverages and industrial uses	Seed and losses
	(10^3 t)	% by weight of total consumption			
W. Europe					
EEC	1 635	0·6	96·8	0·2	2·4
Denmark	365	—	93·7	1·6	4·7
Spain	682	—	97·9	—	2·1
UK	393	—	99·7	—	1·3
Sweden	268	—	91·0	—	9·0
Total	3 685	1·0	95·7	0·2	3·1

Table 2.2 *cont.*

Countries	Consumption	Human nutrition	Animal feed	Fermented beverages and industrial uses	Seed and losses
	(10^3 t)	% by weight of total consumption			
N. America					
Canada	1 472	—	95·9	—	4·1
USA	15 190	0·6	97·8	1·3	0·4
Total	16 662	0·5	97·6	1·1	0·7
Cent. America					
Haiti	177	90·3	5·1	—	4·6
Mexico	1 450	—	98·2	—	1·8
Salvador	114	43·9	50·0	—	6·1
Total	1 883	13·3	83·9	—	2·8
S. America					
Argentina	1 086	—	94·4	—	3·6
Total	1 115	—	96·2	—	3·8
Africa					
Cameroons	374	86·1	—	1·9	12·0
Ivory Coast	55	90·9	—	—	9·1
Dahomey	70	88·5	—	—	11·5
Ethiopia	2 393	79·9	—	10·0	10·1
Ghana	176	86·9	—	—	13·1
Kenya	328	79·2	7·0	—	13·8
Nigeria	6 501	93·4	—	—	6·6
United Arab Rep.	849	84·0	4·9	—	11·1
Rhodesia	277	89·1	—	4·3	6·6
Sudan	1 441	96·6	1·9	—	1·5
S. Africa	276	20·7	3·3	66·3	9·8
Tanzania	1 119	79·1	2·0	7·0	12·0
Zambia	274	68·2	—	25·5	6·2
Total	20 214	85·2	1·0	5·7	8·0
Asia					
India	17 943	94·1	—	—	5·9
Israel	402	—	98·0	—	2·0
Japan	2 772	4·0	94·0	—	4·0
Pakistan	607	100·0	—	—	—
Turkey	350	—	71·4	—	28·6
Total	23 258	78·1	15·0	—	6·9
Oceania					
Australia	312	—	99·0	—	1·0
Total	313	—	99·0	—	1·0
World Total	67 130	53·3	39·4	2·0	5·3

[a] Data from Arnould and Miche (1971) citing Document GR 70/7, Thirteenth Session of the Cereals Study Group (1970).

Table 2.3
Average yield and calculated nutritive value ha of sorghum and pearl millet in Nigeria.[a]

	Sorghum	Pearl Millet
Yield (kg/ha)	672	560
kcal (10^6)	2·4	2·0
Protein (N × 6·25) (kg)	70·0	61·8
Amino acids (10^6 mg)		
Lys	2·0	2·1
Met + Cys	2·5	2·4
Thr	2·7	2·5
Trp	0·8	1·4
Minerals (10^3 mg)		
Calcium	215·6	140·3
Iron	30·4	16·8
Vitamins (10^3 mg)		
Thiamine	3·5	1·7
Niacin	23·5	11·1
Riboflavin	0·8	0·8

[a] Idusogie (1971).

Table 2.4
Total production and yield rates (1976–77) for sorghum, pearl millet and small millets (excluding finger millet) in the major producing States, and all-India values.[a]
Dash indicates no values presented.

Name of State	Sorghum		Pearl millet		Small millets (foxtail millet, kodo millet)	
	Production (10^3 t)	Yield (kg/ha)	Production (10^3 t)	Yield (kg/ha)	Production (10^3 t)	Yield (kg/ha)
Andhra Pradesh	1 010	493	247	440	269	356
Bihar	—	—	—	—	64	348
Gujarat	581	556	1 263	808	93	767
Haryana	37	201	550	574	—	—
Karnataka	1 180	683	243	579	115	405
Madhya Pradesh	1 304	667	122	625	289	177
Maharashtra	4 710	731	681	385	93	433
Orissa	—	—	—	—	55	271
Punjab	—	—	145	924	—	—
Rajasthan	358	428	1 321	366	25	363
Tamil Nadu	700	908	359	861	308	715
Uttar Pradesh	492	682	770	749	247	589
All-India	10 396	659	5 728	534	1 637	370

[a] Swaminathan (personal communication, 1977).

Table 2.5
Proximate composition, DWB, of whole kernel of sorghums and fractions of kafir, milo and feterita sorghum and maize.[a]
Dash indicates no values presented.

	Proportion of kernel (%)	Protein (N × 6·25) (%)	Ether extract (%)	Carbo-hydrates (%)	Starch (%)	Pentosans (%)	Fibre (%)	Ash (%)
Kafir								
Whole kernel	100	12·7	4·1	79·6[b]	61·9	3·3	1·8	1·8
Bran	6·1	4·8	6·8	70·2	—	18·4	16·2	2·0
Corneous endo.	48·9	14·5	0·7	83·8	68·8	0·7	0·7	0·3
Starchy endo.	35·0	11·7	0·8	86·4	70·4	1·9	0·8	0·3
Germ	10·0	19·3	31·5	32·2	—	6·1	3·8	13·2
Milo								
Whole kernel	100	14·0	3·5	78·7[b]	68·5	3·9	1·9	1·9
Bran	5·5	7·1	4·3	70·2	1·6	21·3	15·4	3·1
Corneous endo.	54·7	15·1	0·1	83·5	72·2	0·7	0·7	0·6
Starchy endo.	28·7	8·9	0·3	89·3	82·5	4·3	0·8	0·7
Germ	11·1	20·8	19·9	40·7	75·8	4·7	9·1	9·5
Feterita								
Whole kernel	100	16·7	3·1	76·2[b]	64·2	3·4	2·2	1·8
Bran	6·6	6·8	5·7	70·9	3·9	15·8	13·6	2·9
Corneous endo.	61·1	19·7	0·3	77·1	60·4	2·1	2·1	0·7
Starchy endo.	25·1	10·6	0·6	85·4	1·5	8·6	2·4	1·0
Germ	7·3	20·8	25·4	33·0	2·2	6·9	8·5	11·3
Maize								
Whole kernel	100	—	—	—	—	—	—	—
Hulls	7·4	4·0	0·9	94·4	—	—	—	0·8
Corneous endo.	55·6	11·8	1·1	86·6	—	—	—	0·4
Starchy endo.	25·5	7·8	0·2	91·7	—	—	—	0·3
Germ	11·5	19·8	34·8	35·5	—	—	—	9·9

[a] Bidwell *et al.* (1922).
[b] Nitrogen-free extract by difference.

Table 2.6
Proximate analysis and B vitamin content of whole grain and fractions of sorghum (DWB) mean and range (in parentheses), or average of composite.[a]
Blank spaces indicate no values presented.

	Whole kernel	Endosperm	Germ	Bran
As % of whole kernel	100	82·3	9·8	7·9
		(81·1–84·6)	(7·8–12·1)	(7·3–9·3)
Protein (N × 6·25%)	12·3	12·3	18·9	6·7
	(11·5–13·2)	(11·2–13·0)	(18·0–19·2)	(5·2–7·6)
As % total protein		80·9	14·9	4·0
Oil (%)	3·6[b]	0·6	28·1	4·9[b]
	(3·2–3·9)	(0·4–0·8)	(26·9–30·6)	(3·7–6·0)
As % total oil		13·2	76·2	10·6
Crude wax (%)	0·3			
	(0·2–0·4)			

cont.

Table 2.6 *cont.*

	Whole kernel	Endosperm	Germ	Bran
Starch (%)	73·8	82·5	13·4[c]	34·6[c]
	(72·3–75·1)	(81·3–83·0)		
As % total starch		94·4[c]	1·8[c]	3·8[c]
Ash (%)	1·6[c]	0·4	10·4[c]	2·0[c]
	(1·6–1·7)	(0·3–0·4)		
As % total ash	.	20 6	68·6	10·8
Niacin (mg/100 g)	4·53	4·37	8·07	4·40
	(2·81–6·8)	(2·69–7·05)	(4·98–10·83)	(2·74–6·26)
As % total niacin		75·6	17·1	7·3
Riboflavin (mg/100 g)	0·13	0·09	0·39	0·40
	(0·08–0·2)	(0·05–0·13)	(0·34–0·43)	(0·28–0·48)
As % total riboflavin		50·5	27·7	21·8
Pantothenic acid (mg/100 g)	1·04	0·87	3·22	1·00
	(0·9–0·15)	(0·68–1·09)	(2·59–5·11)	(0·84–1·20)
As % total pantothenic acid		64·8	28·1	7·1
Biotin (mg/100 g)	0·02	0·011	0·057	0·035
	(0·018–0·022)	(0·010–0·013)	(0·046–0·070)	(0·031–0·042)
As % total biotin		52·7	31·6	15·7
Pyridoxine (mg/100 g)	0·47	0·40	0·72	0·44
	(0·4–0·62)	(0·36–0·47)	(0·65–0·82)	(0·39–0·51)
As % total pyridoxine		75·8	16·3	8·0

[a] Hubbard *et al.* (1950).
[b] Oil includes wax.
[c] Analysis of composite not true average.

Table 2.7
Composition of whole grain sorghums.[a]

	% Whole kernel DWB		
	Endosperm	Germ	Pericarp (bran)
IS 0062	85·3	10·9	4·3
IS 3982	86·0	9·4	4·3
IS 6992	81·7	9·5	8·7
IS 2283	83·6	10·3	6·5
Average	84·2	10·1	6·0

[a] Jambunathan and Mertz (1973a).

Table 2.8
1000-kernel weight, protein and amino acid content in normal and shrunken heads of SC 1030 5 4 sorghum.[a]

	Normal	Shrunken
1000-kernel weight (g)	31·7	19·3
Protein (N × 6·25) (%) (DWB)	9·7	11·7
Amino acids (mg/g total N)		
Ile	215	257
Leu	808	726
Lys	146	219
Met	99	111
Phe	323	323
Tyr	262	257
Thr	210	229
Trp	74	87
Val	291	351
Arg	278	338
His	154	151

Table 2.8 *cont.*

	Normal	Shrunken
Ala	616	520
Asp	462	499
Glu	1462	1156
Gly	221	277
Pro	526	434
Ser	295	292
Ammonia	189	132

[a] Sullins *et al.* (1975).

Table 2.9
Factors recommended for calculating energy values of foods by the Atwater system.[a]

	Protein		Fat		Carbohydrate	
	kcal/g	kJ/g	kcal/g	kJ/g	kcal/g	kJ/g
Maize meal						
Whole ground	2·73	11·42	8·37	35·02	4·03	16·86
Brown rice	3·41	14·27	8·37	35·02	4·12	17·24
Polished rice	3·82	15·98	8·37	35·02	4·16	17·41
Sorghum	2·50	10·46	8·37	35·02	4·03	16·86
Wheat						
Extraction %						
97–100	3·59	15·02	8·37	35·02	3·78	15·82
83–93	3·78	15·82	8·37	35·02	3·95	16·53
70–74	4·05	16·95	8·37	35·02	4·12	17·24
Other cereals, refined	3·87	16·19	8·37	35·02	4·12	17·24

[a] WHO (1973a).

Table 2.10
Contribution of cereals (as % of total food) to the intake of thiamine, riboflavin and niacin in different countries.[a]

	Thiamine		Riboflavin		Niacin	
Country	Total intake[b] (mg/head/day)	% from cereals	Total intake[b] (mg/head/day)	% from cereals	Total intake[b] (mg/head/day)	% from cereals
UK	1·26	30	1·72	8	13·8	30
Israel	1·22	68	1·31	54	11·2	50
USA	2·22	42	2·70	18	25·6	32
Guatemala	1·40	64	0·60	50	11·0	53
Venezuela	0·69–0·80	35	0·92	15	8·9–10·8	20
Brazil	1·00	2	1·0	2	9·7	3
Bolivia	0·76	41	0·65	25	12·9–14·1	28
India	1·06	82	0·60	50	12·4	90
Pakistan	2·14	70	0·90	42	17·7	85
Phillipines	0·78–1·06	60	0·65	35	13·8–17·9	65
Iran	2·30	82	0·76	44	23·0	84
Jordan	0·90–1·71	65	0·70–1·60	40	10·2–16·8	70
Cameroon	0·95–1·12	10	0·58–0·65	5	12·7–15·1	10
Ivory Coast	1·72	1	0·90	?	15·4	60
Kenya	0·92	62	1·29	20	7·4	25
Senegal	2·35	82	1·58	60	21·5	75

[a] WHO (1967).
[b] Authors' note: it is not known why in some instances the intake is given as a single average and sometimes, apparently, as a range.

Table 2.11
Proximate analyses of sorghum, DWB, (mean with ranges in parentheses where available) (from multiple sources). Blank spaces indicate no data presented.

Reference	Location	No. of samples	Prot. N × 6·25) (%) mean	Lipid (%) mean	CBH[a] (%) mean	Fibre (%) mean	Ash (%) mean
Adrian and Sayerse (1957)	India	17	12·9 (11·2–15·6)	3·8 (3·2–4·7)	75·7 (74·7–77·8)		1·8 (1·7–1·9)
Anada Rao and Reddy (1973)		16	10·6 (8·9–11·9)				
Aragon and Bressani (1965)	Cent. America	6	11·6 (10·2–12·9)	3·6 (3·5–3·9)		3·5 (3·0–3·7)	1·5 (1·4–1·7)
Arora and Luthra (1972)	India	19	11·0 (8·3–13·1)				2·6 (1·1–4·1)
Balasubramanian *et al.* (1952ab) Balasubramanian and Ramachandran (1957)		6	8·5 (7·2–9·5)				2·3 (1·7–2·9)
Barham *et al.* (1946)	N. America	14	13·5 (10·7–15·1)	3·8 (3·1–4·5)		2·5 (1·7–4·9)	2·1 (1·5–3·1)
Blondel (1970)	Africa	10	10·1 (8·4–12·5)				
Bono and Vidal (1962)		18	11·5 (9·0–12·9)				
Brawand and Hossner (1976)	USA	46	10·7 (8·8–13·2)				
Bressani and Rios (1962)	Cent. America	25	10·9 (8·8–14·5)	4·0 (3·5–5·2)		3·0 (2·3–3·5)	3·0 (1·6–4·4)
Breuer and Dohm (1972)	USA	27	13·7 (8·8–16·9)				
Burleson *et al.* (1956)		3	8·3 (6·6–10·3)				
Busson *et al.* (1962)	France and Africa	19	14·2 (9·7–18·1)				
Capote *et al.* (1972)	S. America	103	13·5	2·7		2·9	
Chakravorty (1967)	India	42	9·9 (7·5–14·2)				

Deosthale *et al.* (1969)		66	10·5 (5·5–21·5)				
Deosthale *et al.* (1972a)	India	24	9·4 (8·1–10·7)				1·5 (1·3–1·6)
Devadas *et al.* (1966)		5	8·9 (8·1–10·4)				1·5 (1·2–2·0)
Deyoe *et al.* (1970)	USA	2	11·5 (10·8–12·3)	3·4 (3·1–3·6)	81·1 (80·9–81·2)	2·4 (2·2–2·7)	1·6 (1·5–1·8)
Ferreira Oriá and de Lima (1975)	S. America	10	11·7 (10·7–14·2)	3·2 (2·9–3·7)	81·9 (77·7–85·5)	1·6 (1·1–2·2)	1·6 (1·4–1·8)
Fetuga (1977)	Africa	4	10·0 (8·5–11·6)	3·1 (2·8–3·4)	82·8 (80·2–85·0)	2·3 (1·4–3·3)	1·7 (1·5–2·0)
Haji-Hashim and Tipton (1973)	USA	10	11·2 (10·7–12·1)				
Heller and Seiglinger (1944)	N. America	77	11·2 (8·0–16·7)	3·0 (1·6–4·3)	82·2 (76·2–85·4)	1·6 (1·1–3·4)	1·7 (1·3–2·2)
Hillier *et al.* (1959)	USA	6	12·9 (12·3–13·8)	3·3 (2·3–4·4)	80·0 (79·2–80·6)	2·0 (1·9–2·1)	1·8 (1·6–1·9)
Horan and Heider (1946)		101	12·2 (7·7–15·6)	3·6 (2·5–4·5)	71·8 (48·0–81·8)		
Ilori and Conrad (1976)		7	11·9 (9·7–14·1)				
Kandel (1954)	Europe	12	13·1 (10·0–18·8)	3·8 (3·2–4·1)	90·2 (89·1–91·3)	2·0 (1·8–2·2)	1·8 (1·5–2·1)
Karim and Rooney (1972a)	USA	31	14·0 (11·4–17·8)	3·7 (2·7–5·4)			
Khattab *et al.* (1972)	Africa	8	12·0 (10·2–13·5)				2·1 (1·4–2·7)
Lamb *et al.* (1966)	USA	3	11·1 (10·5–11·4)			2·3 (2·0–2·7)	1·4 (1·2–1·9)
Littler (1967)	Australia	16	7·8 (5·6–10·3)				
Lutrick (1970)	USA	10	9·9 (8·3–11·4)				
Mali and Gupta (1974)	India	16	13·6 (11·5–16·9)	3·4 (2·7–4·4)	69·9 (66·2–72·4)	1·2 (1·0–1·4)	1·9 (1·7–2·4)

cont.

Table 2.11 *cont.*

Reference	Location	No. of samples	Prot. (N × 6·25) (%) mean	Lipid (%) mean	CBH[a] (%) mean	Fibre (%) mean	Ash (%) mean
Mehansho and Besrat (personal communication, 1974)	Africa	44	10·5 (7·5–13·1)				
Miller *et al.* (1964)	USA	582	9·2 (6·0–12·8)				
McCollough (1972)		8	11·8 (10·7–14·1)	2·7 (2·3–3·2)	81·6 (78·3–83·5)	1·9 (1·5–2·9)	1·9 (1·6–2·2)
McCollough *et al.* (1972a)		9	10·6 (9·9–11·2)	3·1 (3·0–3·3)	82·5 (82·0–83·4)	2·0 (1·8–2·2)	1·6 (1·6–1·9)
Naik and Abhyanker (1955)	India	16	8·9 (6·3–17·3)	3·4 (2·1–4·2)	84·7 (82·8–88·3)	1·4 (0·9–2·1)	1·7 (1·4–1·9)
NAS (1971)	N. America	13	12·3 (10·6–14·6)	3·6 (2·9–4·6)	79·9 (77·0–82·4)	2·3 (1·8–3·5)	2·0 (1·6–3·4)
Nawar *et al.* (1970)	USA	10	11·2 (9·3–12·8)				
Ozment *et al.* (1963)		4	10·7 (8·1–13·0)	3·3 (3·1–3·6)	81·5 (79·5–84·1)	3·0 (2·9–3·1)	1·7 (1·4–1·8)
Price and Parsons (1975)		1	13·1	3·6	78·0	3·1	2·2
Purdue University (1974)	mixed locations	249	11·0 (7·1–15·1)				
Rai (1964)	Africa	11	10·2 (8·2–13·0)				
Rai (1965a)		24	10·5 (7·3–12·6)				
Reddy and Hussain (1968)	India	7	11·1 (10·0–12·0)				
Roy and Wright (1973)		11	10·3 (8·1–11·9)				
Schaffert *et al.* (1972a)	USA	35	12·7 (10·2–15·0)				
Shepherd *et al.* (1971–72)	Africa	103	13·2 (8·9–16·6)	3·7 (2·1–5·1)		2·2 (0·8–3·5)	1·7 (0·9–2·7)

Singh and Bains (1973)	India	28	11·2 (9·7–12·6)				
Stemler *et al.* (1976)	Mixed locations	68	15·2 (12·1–21·5)				
Sur *et al.* (1955)	India	6	11·3 (11·0–11·7)	3·3 (3·2–3·5)			1·9 (1·8–2·2)
Suryanarayana Rao *et al.* (1968)	Mixed locations	57	10·9 (8·0–13·5)				
Thayer *et al.* (1957)	USA	16	11·5 (9·3–17·0)	3·1 (2·3–4·5)	81·3 (75·2–84·6)	2·2 (1·5–3·6)	1·8 (1·2–2·5)
Viraktamath *et al.* (1972)	India	14	11·1 (9·0–13·2)	2·4 (1·9–3·8)			1·8 (1·2–2·6)
Warsi and Wright (1973)		18	12·6 (10·6–13·7)				
Wehmeyer (1969)	Africa	79	12·2 (7·7–16·1)	3·1 (2·7–4·0)	81·6 (69·8–87·0)	2·7 (2·2–3·6)	1·5 (1·1–2·0)
Wessels (1970a)		24	10·5 (7·8–13·0)				
Wessels (1970b)		5	10·7 (9·0–13·0)				
Yañez *et al.* (1973)	S. America	5	10·3 (9·2–11·0)	6·5 (5·5–7·0)	77·1 (75·8–78·4)	3·0 (2·3–4·2)	2·8 (1·6–3·6)
Yousif and Magboul (1972)	Africa	18	11·9 (8·3–15·3)	3·8 (3·3–4·5)	81·9 (78·8–85·8)	1·8 (1·3–2·8)	1·7 (1·4–2·0)
Representative analytical values and ranges (not means)[b]			11·0 (7·3–15·6)	3·0 (0·5–5·2)	70·0 (53·3–79·4)	3·0 (1·2–6·6)	1·8 (1·1–4·5)
Range of means			(7·8–15·2)	(2·4–6·5)	(69·9–90·2)	(1·2–3·5)	(1·4–3·0)
Range of ranges			(5·5–21·5)	(0·5–7·0)	(48·0–91·3)	(0·8–6·6)	(0·9–4·5)
Range of prot. means at N × 5·7			(7·1–13·9)				
Range of prot. ranges at N × 5·7			(5·0–12·6)				

[a] Carbohydrate.
[b] Data from sources mentioned in text.

Table 2.12
Representative values (/100 g of edible portion) of nutrients in various cereals.[a]

Cereals	Moisture (ml)	Calories	Prot. (g)	Fat (g)	CBH[b] (g)	Fibre (g)	Ca (mg)	Fe (mg)	Thiamine (mg)	Riboflavin (mg)	Niacinamide (mg)
Sorghum spp.	12	355	10·4	3·4	71	2·0	32	4·5	0·50	0·12	3·5
Pearl millet	12	363	11·0	5·0	69	2·0	25	3·0	0·3	0·15	2·0
Finger millet	12	336	6·0	1·5	75	3·0	350	5·0	0·3	0·10	1·4
Common millet and foxtail millet	11	355	10·0	2·5	73	2·0	20	5·0	0·6	0·1	1·0
Kodo millet	12	353	11·5	1·3	74	0·4	35	1·7	0·15	ND[c]	ND[c]
Digitaria exilis	10	330	7·7	1·8	71	6·8	30	3·4	0·3	0·1	3·0
Teff	11	345	8·5	2·2	73	2·2	110	90·0	0·5	0·1	2·0
Job's tears	12	365	14·0	4·0	68	0·7	20	4·0	0·3	0·2	3·0
Wheat	13	344	11·5	2·0	70	2·0	30	3·5	0·4	0·1	5·0
Maize (whole and parboiled)	12	363	10·0	4·5	71	2·0	12	2·5	0·35	0·13	2·0
Rice (parboiled)	12	354	8·0	1·5	77	0·5	10	2·0	0·25	0·05	2·0

[a] Platt (1962).
[b] Carbohydrate.
[c] ND = not determined.

Table 2.13
Amino acid content and score of sorghum (WHO 1973a) (means with ranges in parentheses) (from multiple sources). Blank spaces indicate no data presented.

Reference	Location	Ile: No. of samples	Ile: Prot. (N × 6·25) (%)	Ile: mg/100 g sample	Ile: mg/g total N	Ile: AA score	Leu: No. of samples	Leu: Prot. (N × 6·25) (%)	Leu: mg/100 g sample	Leu: mg/g total N	Leu: AA score	Leu/Ile quotient
Adda (1958)	Africa	25	11·8 (9·7–13·7)				25	11·8 (9·7–13·7)		1300 (625–1544)	340	
Adrian and Sayerse (1957)	India	17			309[a] (266–337)	124	17			1559[a] (1506–1650)	354	5·0
Aragon H. and Bressani (1965)	Cent. America	6			389[a] (372–406)	156	6			759[a] (750–774)	173	1·9
Austin *et al.* (1972)	India	96	12·6 (7·8–21·0)									
Badi *et al.* (1976)	USA	2			234 (231–237)	94	2			819 (819–819)	186	3·5
Balasubramanian *et al.* (1952a,b) / Balasubramanian and Ramachandran (1957)	India	6	8·5 (7·2–9·5)		381[a] (299–437)	152	6	8·5 (7·2–9·5)		808[a] (671–901)	184	2·1
Bressani and Rios (1962)	Cent. America	25	9·5 (7·7–12·9)		295[a] (260–332)	118	25	9·5 (7·7–12·9)		894[a] (805–989)	203	3·0
Breuer and Dohm (1972) (exp. III and IV)	USA	18	11·4 (10·5–13·4)	443 (340–620)			18	11·4 (10·5–13·4)	1507 (1060–2470)			
Busson *et al.* (1962)	Africa	12	15·6 (12·4–18·2)		254 (238–269)	102	12	15·6 (12·4–18·2)		855 (806–938)	194	3·4
Campbell and Pickett (1968)	USA	37	14·3 (11·7–17·1)									
Chitre and Vallury (1956 a,b)	India	1		608			1		1348			2·2
Chitre *et al.* (1956)		3	13·6 (12·8–14·1)	628[a] (566–664)			3	13·6 (12·8–14·1)	1303[a] (1208–1450)			2·1
Deosthale and Mohan (1970)		68	8·9 (4·9–13·8)				68	8·9 (4·9–13·8)	657 (455–841)			
Deosthale *et al.* (1970a)		332										
Deosthale *et al.* (1972a)		24			336[a] (289–357)	134	24			912[a] (744–975)	207	2·7
Deyoe and Shellenberger (1965)	USA	30	10·4 (8·6–12·5)	392[a] (275–537)	237[a] (179–299)	95	30	10·4 (8·6–12·5)	1360[a] (978–2161)	816[a] (637–961)	185	3·4
Deyoe *et al.* (1970)		2			291 (279–303)	116	2			932 (924–939)	212	3·2
Eastoe and Taylor (1974)	Africa	3	9·8 (8·7–10·4)	473 (420–500)	325 (325–325)	130	3	9·8 (8·7–10·4)	1497 (1330–1600)	1027 (1013–1044)	233	3·2
Featherston *et al.* (1975a)	Puerto Rico	2	11·7 (11·5–11·9)	480 (470–490)	257 (257–257)	103	2	11·7 (11·5–11·9)	1615 (1570–1660)	864 (825–902)	196	3·4
Ferriera Oria and de Lima (1975)	S. America	10			117 (96–131)	47	10			486 (408–766)	110	4·1

[a] Microbiological assay: values are in general higher than those obtained by ion-exchange chromatography.

cont.

Table 2.13 *cont.*

Reference	Location	Ile					Leu					Leu/Ile quotient
		No. of samples	Prot. (N × 6·25) (%)	mg/100 g sample	mg/g total N	AA score	No. of samples	Prot. (N × 6·25) (%)	mg/100 g sample	mg/g total N	AA score	
FAO (1970)	Various			397	245 (178–299)	98			1348	832 (637–961)	189	3·4
Gupta and Gupta (1972)	India	8	11·1 (9·4–13·0)				8	11·1 (9·4–13·0)		874[a] (835–946)	199	
Gupta and Gupta (1975)		6	11·9 (10·0–13·9)				6	11·9 (10·0–13·9)	1388[a] (1180–1570)	735[a] (693–860)	167	
Ilori and Conrad (1976)	USA	7	9·5	377 (330–430)			7	9·5	1309 (1200–1410)		298	3·5
Khattab *et al.* (1972)	Africa	8		307 (190–620)	172 (93–353)	69	8		1281 (900–2020)	721 (441–1154)	164	4·2
Mali and Gupta (1974)	India	16	13·6 (11·5–16·9)				16	13·6 (11·5–16·9)	1823[a] (1570–2200)	840[a] (779–899)	191	
Martearena and Pérez (1977)	S. America	31	(8·6–16·5)									
Mehansho and Besrat (1974) (personal communication)	Africa	44		409 (265–529)	243 (214–296)	97	44		1369 (858–1770)	818 (715–936)	186	3·4
Nagarajan (1974)	India	4	9·1 (8·6–9·7)									
NAS (1971)	N. America	5		520 (400–600)			5		1500 (1000–1900)			2·9
Nawar *et al.* (1970)		10	11·6 (10·0–13·1)		268 (230–285)	107	10	11·6 (10·0–13·1)		861 (680–940)	196	3·2
Pond *et al.* (1958)		1	9·1									
Schaffert (1972)		31	12·6		255	102	31	12·6		892	203	3·5
Shepherd *et al.* (1971–72)	Africa	61		262 (187–369)		105	61		875 (675–1225)		199	3·3
Shoup *et al.* (1970a)	USA	7	10·3 (9·2–11·5)	416 (370–500)			7	10·3 (9·2–11·5)	1420 (1200–1690)			3·4
Singh and Axtell (1973a)		5	13·5 (12·7–14·8)									
Singh and Axtell (1973b)	India	5	14·8 (12·4–17·3)									
Singh and Axtell (1973c)	USA and Puerto Rico	9	14·6 (11·8–16·3)		251 (241–264)	100	9	14·6 (11·8–16·3)		873 (816–926)	198	3·5
Singh and Axtell (1973d)		19	14·5 (11·8–17·7)									

Srinivasan *et al.* (1972)	Mixed	522	12·6		255	102	522	12·6		892	203	3·5
Stephenson *et al.* (1971)	USA	24		406			24		1608			
				(290–463)					(1233–1909)			
Waggle *et al.* (1967a)		3	10·3	400	244	98	3	10·3	1390	844	192	3·5
			(8·3–12·1)	(380–430)	(219–287)			(8·3–12·1)	(1110–1640)	(837–850)		
Waggle *et al.* (1967b)		6	11·3	473	260	104	6	11·3	1593	874	199	3·4
			(10·6–12·1)	(440–510)	(256–264)			(10·6–12·1)	(1430–1760)	(832–906)		
Warsi and Wright (1973)	India	2	10·3	437	268	107	2	10.3	1269	768	175	2·9
			(8·9–11·6)	(401–472)	(255–281)			(8·9–11·6)	(819–1514)	(716–819)		
Yañez *et al.* (1973)	S. America	5		448	296	118	5		1324	950	216	3·2
				(410–480)	(269–325)				(1150–1470)	(875–1025)		
Range of means			(8·5–15·6)		(117–389)					(486–1559)		
Range of ranges			(4·9–18·2)		(93–437)					(408–1650)		
Range of protein means at N × 5·7			(7·7–14·2)									
Range of protein ranges at N × 5·7			(4·5–16·6)									

[a] Microbiological assay: values are in general higher than those obtained by ion-exchange chromatography.

cont.

Table 2.13 *cont.*

Reference	Location	Lys				
		No. of samples	Prot. (N × 6·25) (%)	mg/100 g sample	mg/g total N	AA score
Adda (1958)	Africa	25	11·8 (9·7–13·7)		148 (112–181)	44
Adrian and Sayerse (1957)	India	17			146[a] (113–179)	43
Aragon H. and Bressani (1965)	Cent. America	6			152[a] (141–167)	45
Austin *et al.* (1972)	India	96	12.6 (7·8–21·0)	205 (103–434)	103 (45–211)	30
Badi *et al.* (1976)	USA	2			147 (137–156)	43
Balasubramanian *et al.* (1952 a,b)	India	6	8·5 (7·2–9·5)		212[a] (175–244)	62
Balasubramanian and Ramachandran (1957)						
Bressani and Rios (1962)	Cent. America	25	9·5 (7·7–12·9)		184[a] (145–229)	54
Breuer and Dohm (1972) (exp. III and IV)	USA	18	11·4 (10·5–13·4)	254 (210–310)		
Busson *et al.* (1962)	Africa	12	15·6 (12·4–18·2)		144 (119–169)	42
Campbell and Pickett (1968)	USA	37	14·3 (11·7–17·1)		110 (58–163)	32
Chitre and Vallury (1956 a,b)	India	1		402		
Chitre *et al.* (1956)		3	13·6 (12·8–14·1)	433[a] (404–462)		
Deosthale and Mohan (1970)		68			122 (72–266)	36
Deosthale *et al.* (1970a)		332				
Deosthale *et al.* (1972a)		24			155[a] (131–165)	46
Deyoe and Shellenberger (1965)	USA	30	10·4 (8·6–12·5)	203 (152–252)	124 (98–163)	36
Deyoe *et al.* (1970)		2			135 (134–136)	40
Eastoe and Taylor (1974)	Africa	3	9·8 (8·7–10·4)	240 (220–270)	165 (150–175)	49
Featherston *et al.* (1975a)	Puerto Rico	2	11·7 (11·5–11·9)	245 (240–250)	132 (129–134)	39

Ferriera Oriá and de Lima (1975)	S. America	10			76 (50–96)	22
FAO (1970)	Various			204	126 (98–175)	37
Gupta and Gupta (1972)	India	8	11·1 (9·4–13·0)		118[a] (95–132)	35
Gupta and Gupta (1975)		6	11·9 (10·0–13·9)	200[a] (170–230)	105[a] (86–127)	31
Ilori and Conrad (1976)	USA	7		229 (190–280)		
Khattab *et al.* (1972)	Africa	8		126 (90–200)	71 (46–114)	21
Mali and Gupta (1974)	India	16	13·6 (11·5–16·9)	247[a] (190–280)	115[a] (93–146)	34
Martearena and Pérez (1977)					(83–213)	
Mehansho and Besrat (1974) (personal communication)	Africa	44		216 (179–296)	130 (102–160)	38
Nagarajan (1974)	India	4	9·1 (8·6–9·7)		139 (116–162)	41
NAS (1971)	USA	5		220 (200–300)		
Nawar *et al.* (1970)		10	11·6 (10·0–13·1)		146 (125–170)	43
Pond *et al.* (1958)		1	9·1	200		
Schaffert *et al.* (1972a)		35		258 (237–324)	128 (114–146)	38
Shepherd *et al.* (1971–72)	Africa	61			131 (69–175)	39
Shoup *et al.* (1970a)	USA	7	10·3 (8·3–12·1)	229 (210–250)		
Singh and Axtell (1973a)		5	13·5 (12·7–14·8)		131 (116–144)	39
Singh and Axtell (1973b)	India	5	14·8 (12·4–17·3)	343 (300–387)	145 (138–153)	43
Singh and Axtell (1973c)	USA and Puerto Rico	10	14·6 (11·8–16·3)	357 (277–439)	153 (126–171)	45

[a] Microbiological assay: values are in general higher than those obtained by ion-exchange chromatography.

cont.

Table 2.13 *cont.*

Reference	Location	Lys				
		No. of samples	Prot. (N × 6·25) (%)	mg/100 g sample	mg/g total (N	AA score
Singh and Axtell (1973d)	USA and Puerto Rico	19	14·5 (11·8–17·7)	330 (246–439)	141 (109–171)	41
Srinivasan *et al.* (1972)	Mixed	522	12·6		134	39
Stephenson *et al.* (1971)	USA	24		284 (243–317)		
Waggle *et al.* (1967a)		3	10·3 (8·3–12·1)	233 (200–270)	144 (137–150)	42
Waggle *et al.* (1967b)		6	11·3 (10·6–12·1)	243 (240–250)	133 (126–136)	39
Warsi and Wright (1973)	India	2	10.3 (8·9–11·6)	223 (211–234)	137 (126–147)	40
Yañez *et al.* (1973)	S. America	5		244 (210–270)	164 (156–169)	48
Range of means					(71–212)	
Range of ranges					(45–244)	

Reference	Location	Met				Cys				AA score (Met + Cys)
		No. of samples	Prot. (N × 6·25) (%)	mg/100 g sample	mg/g total N	No. of samples	Prot. (N × 6·25) (%)	mg/100 g sample	mg/g total N	
Adrian and Sayers (1957)	India	17			172[a] (156–194)	17			99[a] (87–128)	123
Badi *et al.* (1976)	USA	2			81 (75–87)	2			128 (106–150)	95
Balasubramanian *et al.* (1952a,b); Balasubramanian and Ramachandran (1957)	India	6	8·5 (7·2–9·5)		103[a] (94–112)					
Bressani and Rios (1962)	Cent. America	25	9·5 (7·7–12·9)		101[a] (80–127)					
Breuer and Dohm (1972) (exp. III and IV)	USA	18	11·4 (10·5–13·4)	176 (150–200)		18	11·4 (10·5–13·4)	574[b] (470–700)		
Busson *et al.* (1962)	Africa	12	15·6 (12·4–18·2)		116 (100–144)	12	15·6 (12·4–18·2)		128 (106–144)	111
Chitre and Vallury (1956a,b)	India	1		259		1		155		
Chitre *et al.* (1956)		3	13·6 (12·8–14·1)	331 (278–394)						
Deosthale *et al.* (1972a)		24			138[a] (123–146)					
Deyoe and Shellenberger (1965)	USA	30	10·4 (8·6–12·5)	137 (91–226)	84 (51–123)	30	10·4 (8·6–12·5)	105 (58–153)	62 (31–86)	66
Deyoe *et al.* (1970)		2			88 (81–95)	2			101 (90–112)	86
Eastoe and Taylor (1974)	Africa	3	9·8 (8·7–10·4)	143 (130–160)	98 (81–106)	3	9·8 (8·7–10·4)	170 (140–190)	116 (106–125)	97
Featherston *et al.* (1975a)	Puerto Rico	2	11·7 (11·5–11·9)	120 (80–160)	63 (40–85)	2	11·7 (11·5–11·9)	255 (220–290)	136 (115–156)	90
Ferreira Oriá and de Lima (1975)	S. America	10			39 (15–64)	10			115 (50–181)	70
FAO (1970)	Various			141	87 (19–144)			152	94 (31–189)	82
Gupta and Gupta (1972)	India	8	11·1 (9·4–13·0)		108 (75–218)					

[a] Microbiological assay: values are in general higher than those obtained by ion-exchange chromatography.
[b] Cysteine + cystine.

cont.

Table 2.13 *cont.*

Reference	Location	Met: No. of samples	Met: Prot. (N × 6·25) (%)	Met: mg/100 g sample	Met: mg/g total N	Cys: No. of samples	Cys: Prot. (N × 6·25) (%)	Cys: mg/100 g sample	Cys: mg/g total N	AA score (Met + Cys)
Gupta and Gupta (1975)	India	6	11·9 (10·0–13·9)	205 (150–260)	108 (94–121)					
Ilori and Conrad (1976)	USA	7		71 (50–100)		7		61 (50–80)		
Khattab *et al.* (1972)	Africa					8		91 70–100)	51 (35–67)	
Mali and Gupta (1974)	India	16	13·6 (11·5–16·9)	233 (200–290)	108 (89–124)					
Mehansho and Besrat (1974)	Africa	44		151 (98–214)	90 (64–125)					
NAS (1971)	USA	5		160 (100–200)						
Nawar *et al.* (1970)		10	11·6 (10·0–13·1)		91 (75–110)	10	11·6 (10·0–13·1)		61 (30–85)	69
Schaffert (1972)		31			112	31			94	94
Shoup *et al.* (1970a)	USA	7	10·3 (8·3–12·1)	176 (160–200)		7	10·3 (8·3–12·1)	213 (200–240)		
Singh and Axtell (1973c)	USA and Puerto Rico	9	14·6 (11·8–16·3)		129 (107–156)	9	14·6 (11·8–16·3)		117 (112–125)	112
Srinivasan *et al.* (1972)	Mixed	522	12·6		64	522	12·6		57	55
Stephenson *et al.* (1971)	USA	24		200 (154–242)		24		213 (164–242)		
Waggle *et al.* (1967a)		3	10·3 (8·3–12·1)	170 (140–210)	102 (94–106)	3	10·3 (8·3–12·1)	210 (170–250)	129 (125–131)	105
Waggle *et al.* (1967b)		6	11·3 (10·6–12·1)	133 (120–140)	75 (65–83)	6	11·3 (10·6–12·1)	173 (160–180)	98 (92–109)	79
Warsi and Wright (1973)	India	2	10·3 (8·9–11·6)	238 (212–264)	145 (142–148)	2	10·3 (8·9–11·6)	258 (202–313)	155 (141–169)	136
Yañez *et al.* (1973)	S. America	5		152 (140–170)	104 (100–106)	5		140 (110–170)	90 (81–94)	88
Range of means					(39–172)				(51–155)	
Range of ranges					(15–218)				(30–189)	

Reference	Location	Phe No. of samples	Phe Prot. (N × 6·25) (%)	Phe mg/100 g sample	Phe mg/g total N	Tyr No. of samples	Tyr Prot. (N × 6·25) (%)	Tyr mg/100 g sample	Tyr mg/g total N	AA score (Phe + Tyr)
Adda (1958)	Africa	25	11·8 (9·7–13·7)		312 (256–362)					
Adrian and Sayerse (1957)	India	17			289[a] (247–312)					
Badi *et al.* (1976)	USA	2			303 (300–306)	2			219 (212–225)	137
Balasubramanian *et al.* (1952a,b) / Balasubramanian and Ramachandran (1957)	India	6	8·5 (7·2–9·5)		319[a] (262–369)					
Bressani and Rios (1962)	Cent. America	25			270[a] (201–345)	25			171[a] (137–210)	116
Breuer and Dohm (1972) (exp. III and IV)	USA	18	11·4 (10·5–13·4)	528 (400–680)		18	11·4 (10·5–13·4)	440 (330–570)		
Busson *et al.* (1962)	Africa	12	15·6 (12·4–18·2)		321 (300–350)	12	15·6 (12·4–18·2)		260 (238–275)	153
Chitre and Vallury (1956a,b)	USA India	1		425						
Chitre *et al.* (1956)		3	13·6 (12·8–14·1)	523[a] (357–678)						
Deyoe and Shellenberger (1965)	USA	30	10·4 (8·6–12·5)	490 (366–635)	298 (234–344)	30	10·4 (8·6–12·5)	172 (114–283)	102 (72–154)	105
Deyoe *et al.* (1970)		2			363 (359–367)	2			279 (276–281)	169
Eastoe and Taylor (1974)	Africa	3	9·8 (8·7–10·4)	597 (540–650)	413 (394–425)	3	9·8 (8·7–10·4)	480 (430–510)	331 (325–338)	196
Featherston *et al.* (1975a)	Puerto Rico	2	11·7 (11·5–11·9)	620 (570–670)	332 (300–364)	2	11·7 (11·5–11·9)	375 (340–410)	200 (179–221)	140
Ferreira Oriá and de Lima (1975)	S. America	10			179 (130–221)	10			136 (122–149)	83
FAO (1970)	Various			496	306 (234–369)			271	167 (72–288)	124

[a] Microbiological assay: values are in general higher than those obtained by ion-exchange chromatography.

cont.

Table 2.13 *cont.*

Reference	Location	Phe: No. of samples	Phe: Prot. (N × 6·25) (%)	Phe: mg/100 g sample	Phe: mg/g total N	Tyr: No. of samples	Tyr: Prot. (N × 6·25) (%)	Tyr: mg/100 g sample	Tyr: mg/g total N	AA score (Phe + Tyr)
Ilori and Conrad (1976)	USA	7		481 (440–530)		7		417 (400–430)		
Khattab *et al.* (1972)	Africa	8		357 (290–430)	196 (147–271)	6		149 (90–180)	81 (49–108)	73
Mehansho and Besrat (1974)	Africa	44		509 (314–658)	302 (259–342)					
NAS (1971)	USA	5		580 (400–700)						
Nawar *et al.* (1970)		10	11·6 (10·5–13·4)		339 (290–370)	10	11·6 (10·5–13·4)		256 (210–270)	157
Schaffert (1972)		31			324	31			281	159
Shepherd *et al.* (1971–72)	Africa	61			337 (250–475)	61			244 (156–300)	153
Shoup *et al.* (1970a)	USA	7	10·3 (8·3–12·1)	549 (480–640)		7	10·3 (8·3–12·1)	443 (380–530)		
Singh and Axtell (1973c)	USA and Puerto Rico	9	14·6 (11·8–16·3)		327 (312–349)	9	14·6 (11·8–16·3)		268 (252–288)	157
Srinivasan *et al.* (1972)	Mixed	522	12·6		324	522	12·6		281	159
Stephenson *et al.* (1971)	USA	24		538 (429–683)		24		471 (369–970)		
Waggle *et al.* (1967a)		3	10·3 (8·3–12·1)	530 (430–620)	323 (319–325)	3	10·3 (8·3–12·1)	427 (340–510)	258 (256–262)	153
Waggle *et al.* (1967b)		6	11·3 (10·6–12·1)	593 (540–640)	327 (319–333)	6	11·3 (10·6–12·1)	240 (230–250)	133 (130–136)	121
Warsi and Wright (1973)	India	2	10·3 (8·9–11·6)	385 (344–426)	235 (230–240)	2	10·3 (8·9–11·6)	288 (262–314)	177 (170–183)	108
Yañez *et al.* (1973)	S. America	5		478 (410–530)	342 (287–406)	5		220 (150–280)	161 (100–194)	132
Range of means					(179–413)				(81–331)	
Range of ranges					(130–475)				(49–338)	

Reference	Location	Thr No. of samples	Thr Prot. (N × 6·25) (%)	Thr mg/100 g sample	Thr mg/g total N	Thr AA score	Trp No. of samples	Trp Prot. (N × 6·25) (%)	Trp mg/100 g sample	Trp mg/g total N	Trp AA score
Adda (1958)	Africa	25	11·8 (9·7–13·7)		198 (150–262)	79					
Adrian and Sayerse (1957)	India	17			231[a] (162–269)	92	17			70[a] (59–78)	117
Aragon H. and Bressani (1965)	Cent. America						6			29[a] (26–31)	48
Badi *et al.* (1976)	USA	2			206 (187–225)	82					
Balasubramanian *et al.* (1952a,b) Balasubramanian and Ramachandran (1957)	India	6	8·5 (7·2–9·5)		242[a] (214–284)	97	6	8·5 (7·2–9·5)		77[a] (64·4–84·4)	128
Bressani and Rios (1962)	Africa	25	9·5 (7·7–12·9)		240[a] (200–282)	96	25	9·5 (7·7–12·9)		44[a] (32–58)	73
Breuer and Dohm (1972) (exp. III and IV)	USA	18	11·4 (10·5–13·4)	379 (300–470)							
Busson *et al.* (1962)	Africa	12	15·6 (12·4–18·2)		208 (188–231)	83	12	15·6 (12·4–18·2)		77 (63–144)	128
Chitre and Vallury (1956a,b)	India	1		332			1		47		
Chitre *et al.* (1956)		3	13·6 (12·8–14·1)	385[a] (348–447)			3	13·6 (12·8–14·1)	41 (25–50)		
Deosthale *et al.* (1972a)							24			64[a] (53–71)	107
Deyoe and Shellenberger (1965)	USA	30	10·4 (8·6–12·5)	306 (235–378)	187 (149–232)	75					
Deyoe *et al.* (1970)		2			228 (223–232)	91					
Eastoe and Taylor (1974)	Africa	3	9·8 (8·7–10·4)	403 (360–430)	279 (275–281)	112					
Featherston *et al.* (1975a)	Puerto Rico	2	11·7 (11·5–11·9)	380 (370–390)	205 (197–212)	82					
Ferreira Oriá and de Lima (1975)	S. America	10			124 (101–142)	50	10			77 (59–94)	128
FAO (1970)	Various			306	189 (139–233)	76			123	63[a] (56–144)	105

[a] Microbiological assay: values are in general higher than those obtained by ion-exchange chromatography.

cont.

Table 2.13 *cont.*

Reference	Location	Thr No. of samples	Thr Prot. (N × 6·25) (%)	Thr mg/100 g sample	Thr mg/g total N	Thr AA score	Trp No. of samples	Trp Prot. (N × 6·25) (%)	Trp mg/100 g sample	Trp mg/g total N	Trp AA score
Gupta and Gupta (1972)	India						8	11·1 (9·4–13·0)		67 (63–73)	112
Gupta and Gupta (1975)							6	11·9 (10·0–13·9)	135 (120–150)	71 (66–79)	118
Ilori and Conrad (1976)	USA	7		310 (280–330)							
Khattab *et al.* (1972)	Africa	8		344 (280–460)	196 (148–262)	78					
Mali and Gupta (1974)	India	16	13·6 (11·5–16·9)	338[a] (270–450)	155[a] (136–198)	62	16	13·6 (11·5–16·9)	126 (70–170)	58 (39–69) (7–43)[a]	97
Martearena and Pérez (1977)	S. America	31	(8·6–16·5)								
Mehansho and Besrat (1974) (personal communication)	Africa	44		322 (235–415)	189 (127–227)	76					
NAS (1971)	USA	5		400 (300–500)			5		140 (100–200)		
Nawar *et al.* (1970)		10	11·6 (10·0–13·1)		216 (200–235)	86					
Pond *et al.* (1958)		1	9·1	300							
Schaffert (1972)		31			204	82	31			82	137
Shepherd *et al.* (1971–72)	Africa	61			200 (156–256)	80					
Shoup *et al.* (1970a)	USA	7	10·3 (8·3–12·1)	357 (320–410)							
Singh and Axtell (1973c)	USA and Puerto Rico	9	14·8 (12·4–17·3)		189 (173–207)	76	9	14·8 (12·4–17·3)		61 (57–66)	102
Srinivasan *et al.* (1972)	Mixed	522	12·6		204	82	522	12·6		82	136·7
Stephenson *et al.* (1971)	USA	24		343 (276–424)							
Waggle *et al.* (1967a)		3	10·3 (8·3–12·1)	357 (300–420)	215 (206–219)	86					
Waggle *et al.* (1967b)		6	11·3 (10·6–12·1)	353 (330–380)	199 (198–199)	80					
Warsi and Wright (1973)	India	2	10·3 (8·9–11·6)	232 (196–267)	141 (137–144)	56					
Yañez *et al.* (1973)	S. America	5		384 (360–420)	255 (231–312)	102					
Range of means					(214–279)				(29–82)		
Range of ranges					(101–312)				(7–144)		

Reference	Location	His				Ala			
		No. of samples	Prot. (N × 6·25) (%)	mg/100 g sample	mg/g total N	No. of samples	Prot. (N × 6·25) (%)	mg/100 g sample	mg/g total N
Adda (1958)	Africa	25	11·8 (9·7–13·7)		128 (112–162)				
Adrian and Sayerse (1957)	India	4			133[a] (119–153)				
Badi *et al.* (1976)	USA	2			125 (106–144)	2			
Balasubramanian *et al.* (1952a,b) Balasubramanian and Ramachandran (1957)	India	6	8·5 (7·2–9·5)		98[a] (87–112)				
Bressani and Rios (1962)	Cent. America	25	9·5 (7·7–12·9)		205[a] (148–245)				
Breuer and Dohm (1972) (exp. III and IV)	USA	18	11·4 (10·5–13·4)	289 (260–330)		18	11·4 (10·5–13·4)	1062 (930–1220)	
Busson *et al.* (1962)	Africa	12	15·6 (12·4–18·2)		141 (125–175)	12	15·6 (12·4–18·2)		597 (550–621)
Chitre and Vallury (1956a)	India	1		349					
Chitre *et al.* (1956)		3	13·6 (12·8–14·1)	354 (270–437)					
Deyoe and Shellenberger (1965)	USA	30	10·4 (8·6–12·5)	209 (163–261)	128 (103–146)	30	10·4 (8·6–12·5)	945 (700–1228)	576 (456–667)
Deyoe *et al.* (1970)		2			143 (141–145)	2			673 (661–684)
Eastoe and Taylor (1974)	Africa	3	9·8 (8·7–10·4)	270 (240–290)	185 (181–188)	3	9·8 (8·7–10·4)	1043 (920–1120)	717 (706–731)
Featherston *et al.* (1975a)	Puerto Rico	2	11·7 (11·5–11·9)	270 (270–270)	147 (144–149)	2	11·7 (11·5–11·9)	1135 (1120–1150)	606 (589–622)
Ferreira Oriá and de Lima (1975)	S. America	10			67 (54–79)	10			416 (366–476)
FAO (1970)	Various			217	134 (103–194)			946	584 (456–668)

[a] Microbiological assay: values are in general higher than those obtained by ion-exchange chromatography.

cont.

Table 2.13 *cont.*

Reference	Location	His: No. of samples	His: Prot. (N × 6·25) (%)	His: mg/100 g sample	His: mg/g total N	Ala: No. of samples	Ala: Prot. (N × 6·25) (%)	Ala: mg/100 g sample	Ala: mg/g total N
Ilori and Conrad (1976)	USA	7	9·5	220 (180–250)					
Khattab *et al.* (1972)	Africa	8		112 (70–200)	64 (40–120)				
Mehansho and Besrat (1974)	Africa	44		202 (147–256)	121 (103–164)				
NAS (1971)	USA	5		240 (200–300)					
Nawar *et al.* (1970)		10	11·6 (10·0–13·1)		140 (126–171)				
Schaffert (1972)		31	12·6		126	31	12·6		618
Shepherd *et al.* (1971–72)	Africa	61			131 (87–169)	61			600 (437–850)
Shoup *et al.* (1970a)	USA	7	10·3 (8·3–12·1)	236 (210–260)		7	10·3 (8·3–12·1)	1004 (830–1200)	
Singh and Axtell (1973c)	USA and Puerto Rico	9	14·6 (11·8–16·3)		127 (116–139)	9	14·6 (11·8–16·3)		592 (553–636)
Srinivasan *et al.* (1972)	Mixed	522	12·6		126	522	12·6		618
Stephenson *et al.* (1971)	USA	24		237 (193–286)		24		1130 (951–1279)	
Waggle *et al.* (1967a)		3	10·3 (8·3–12·1)	233 (180–290)	141 (137–150)	3	10·3 (8·3–12·1)	993 (780–1150)	602 (587–625)
Waggle *et al.* (1967b)		6	11·3 (10·6–12·1)	253 (250–260)	140 (134–147)	6	11·3 (10·6–12·1)	1110 (1000–1230)	609 (586–632)
Warsi and Wright (1973)	India	2	10·3 (8·9–11·6)	192 (186–197)	117 (107–127)	2	10·3 (8·9–11·6)	835 (702–968)	507 (491–523)
Yañez *et al.* (1973)	S. America	5		262 (230–300)	172 (162–187)	5		1026 (910–1140)	690 (625–744)
Range of means					(64–205)				(416–717)
Range of ranges					(40–245)				(366–850)

Reference	Location	Val: No. of samples	Val: Prot. (N × 6·25) (%)	Val: mg/100 g sample	Val: mg/g total N	Val: AA score	Arg: No. of samples	Arg: Prot. (N × 6·25) (%)	Arg: mg/100 g sample	Arg: mg/g total N
Adda (1958)	Africa	25	11·8 (9·7–13·7)		378 (312–431)	122	25	11·8 (9·7–13·7)		205 (181–238)
Adrian and Sayerse (1957)	India	17			302[a] (275–325)	97	4			217[a] (200–244)
Badi *et al.* (1976)	USA	2			281 (281–281)	91	2			225 (200–250)
Balasubramanian *et al.* (1952a,b) / Balasubramanian and Ramachandran (1957)	India	6	8·5 (7·2–9·5)		369[a] (326–398)	119	6	8·5 (7·2–9·5)		425[a] (347–564)
Bressani and Rios (1962)	Cent. America	25	9·5 (7·7–12·9)		375[a] (318–436)	121	25	9·5 (7·7–12·9)		295[a] (242–373)
Breuer and Dohm (1972) (exp. III and IV)	USA						18	11·4 (10·5–13·4)	392 (340–490)	
Busson *et al.* (1962)	Africa	12	15·6 (12·4–18·2)		337 (319–356)	109	12	15·6 (12·4–18·2)		260 (231–306)
Chitre and Vallury (1956a,b)	India	1		633			1		847	
Chitre *et al.* (1956)		3	13·6 (12·8–14·1)	611[a] (548–686)			3	13·6 (12·8–14·1)	829 (781–911)	
Deyoe and Shellenberger (1965)	USA	30	10·4 (8·6–12·5)	506 (382–646)	308 (248–365)	99	30	10·4 (8·6–12·5)	278 (218–338)	169 (129–212)
Deyoe *et al.* (1970)		2			348 (294–401)	112	2			247 (242–252)
Eastoe and Taylor (1974)	Africa	3	9·8 (8·7–10·4)	583 (520–620)	400 (394–406)	129	3	9·8 (8·7–10·4)	443 (400–480)	309 (294–319)
Featherston *et al.* (1975a)	Puerto Rico	2	11·7 (11·5–11·9)	610 (590–630)	324 (319–329)	105	2	11·7 (11·5–11·9)	430 (420–440)	
Ferreira Oriá and de Lima (1975)	S. America	10			230 (180–267)	74	10			108 (92–124)
FAO (1970)	Various			507	313 (248–365)	101			311	192 (101–306)

[a] Microbiological assay: values are in general higher than those obtained by ion-exchange chromatography.

cont.

Table 2.13 *cont.*

Reference	Location	Asp				Glu			
		No. of samples	Prot. (N × 6·25) (%)	mg/100 g sample	mg/g total N	No. of samples	Prot. (N × 6·25) (%)	mg/100 g sample	mg/g total N
Adda (1958)	Africa	25	11·8 (9·7–13·7)		263 (225–319)	25	11·8 (9·7–13·7)		102 (881–1225)
Badi *et al.* (1976)	USA	2			422 (412–431)	2			1347 (1300–1394)
Breuer and Dohm (1972) (exp. III and IV)	USA	18	11·4 (10·5–13·4)	856 (650–1000)		18	11·4 (10·5–13·4)	2575 (2110–3160)	
Busson *et al.* (1962)	Africa	12	15·6 (12·4–18·2)		457 (419–506)	12	15·6 (12·4–18·2)		1303 (1231–1444)
Deyoe and Shellenberger (1965)	USA	30	10·4 (8·6–12·5)	645 (461–852)	394 (300–479)	30	10·4 (8·6–12·5)	2170 (1584–2933)	1322 (1062–1553)
Deyo *et al.* (1970)		2			482 (464–499)	2			1531 (1515–1547)
Eastoe and Taylor (1974)	Africa	3	9·8 (8·7–10·4)	757 (660–830)	521 (506–544)	3	9·8 (8·7–10·4)	2333 (2040–2510)	1604 (1575–1638)
Featherston *et al.* (1975a)	Puerto Rico	2	11·7 (11·5–11·9)	865 (860–870)	461 (453–469)	2	11·7 (11·5–11·9)	2455 (2210–2700)	1308. (1202–1414)
Ferreira Oriá and de Lima (1975)	S. America	10			281 (249–314)	10			970 (861–1164)
FAO (1970)	Various			638	406 (292–506)			2141	1322 (1063–1573)
Khattab *et al.* (1972)	Africa	7		470 (321–630)	267 (173–378)	6		1752 (1080–2010)	1051 (833–1327)
Schaffert (1972)		31	12·6		489	31	12·6		1451
Shepherd *et al.* (1971–72)	Africa	61			456 (356–631)	61			1581 (1275–2256)
Shoup *et al.* (1970a)	USA	7	10·3 (8·3–12·1)	731 (640–860)		7	10·3 (8·3–12·1)	2294 (2030–2660)	
Singh and Axtell (1973c)	USA and Puerto Rico	9	14·6 (11·8–16·3)		435 (407–466)	9	14·6 (11·8–16·3)		1485 (1384–1570)
Srinivasan *et al.* (1972)	Mixed	522	12·6		489	522	12·6		1451
Stephenson *et al.* (1971)	USA	24		806 (708–908)		24		3020 (2731–3598)	
Waggle *et al.* (1967a)		3	10·3 (8·3–12·1)	743 (600–880)	448 (444–450)	3	10·3 (8·3–12·1)	2263 (1740–2730)	1366 (1306–1412)
Waggle *et al.* (1967b)		6	11·3 (10·6–12·1)	773 (730–820)	430 (427–432)	6	11·3 (10·6–12·1)	2500 (2250–2780)	1383 (1351–1424)
Warsi and Wright (1973)	India	2	10·3 (8·9–11·6)	568 (462–674)	344 (323–365)	2	10·3 (8·9–11·6)	2483 (2081–2884)	1505 (1454–1556)
Yañez *et al.* (1973)	S. America	5		724 (630–760)	490 (469–519)	5		2264 (1920–2650)	1527 (1387–1656)
Range of mean					(263–490)				(102–1604)
Range of ranges					(173–631)				(833–2256)

Reference	Location	Val					Arg			
		No. of samples	Prot. (N × 6·25) (%)	mg/100 g sample	mg/g total N	AA score	No. of samples	Prot. (N × 6·25) (%)	mg/100 g sample	mg/g total N
Ilori and Conrad (1976)	USA	7	9·5	506 (460–560)			7	9·5	349 (310–410)	
Khattab *et al.* (1972)	Africa	7		424 (320–510)	235 (192–277)	76				
Mehansho and Besrat (1974)	Africa	44		514 (348–713)	308 (231–362)	99	44		339 (205–442)	203 (132–248)
NAS (1971)	USA	5		600 (500–700)			5		380 (300–500)	
Nawar *et al.* (1970)		10	11·6 (10·0–13·1)		332 (280–350)	107				
Schaffert (1972)		31	12·6		334	108	31	12·6		224
Shepherd *et al.* (1971–72)	Africa	61			325 (212–506)	105	61			225 (112–312)
Shoup *et al.* (1970a)	USA	7	10·3 (8·3–12·1)	493 (390–570)			7	10·3 (8·3–12·1)	411 (370–450)	
Singh and Axtell (1973c)	USA and Puerto Rico	9	14·6 (11·8–16·3)		304 (284–329)	98	9	14·6 (11·8–16·3)		227 (207–251)
Srinivasan *et al.* (1972)	Mixed	522	12·6		334	108	522	12·6		224
Stephenson *et al.* (1971)	USA	24		513 (404–602)			24		389 (281–453)	
Waggle *et al.* (1967a)		3	10·3 (8·3–12·1)	477 (390–560)	289 (287–294)	93	3	10·3 (8·3–12·1)	417 (320–510)	252 (244–262)
Waggle *et al.* (1967b)		6	11·3 (10·6–12·1)	590 (560–630)	328 (326–330)	106	6	11·3 (10·6–12·1)	383 (380–390)	215 (203–228)
Warsi and Wright (1973)	India	2	10·3 (8·9–11·6)	475 (408–541)	289 (285–292)	93	2	10·3 (8·9–11·6)	386 (371–401)	238 (216–259)
Yañez *et al.* (1973)	S. America	5		552 (490–610)	372 (319–400)	120	5		346 (290–390)	231 (212–250)
Range of means					(230–400)					(108–425)
Range of ranges					(180–506)					(92–564)

[a] Microbiological assay: values are in general higher than those obtained by ion-exchange chromatography.

cont.

Table 2.13 *cont.*

Reference	Location	Gly: No. of samples	Gly: Prot. (N × 6·25) (%)	Gly: mg/100 g sample	Gly: mg/g total N	Pro: No. of samples	Pro: Prot. (N × 6·25) (%)	Pro: mg/100 g sample	Pro: mg/g total N
Badi *et al.* (1976)	USA	2			200 (187–212)	2			500 (450–550)
Breuer and Dohm (1972) (exp. III and IV)	USA	18	11·4 (10·5–13·4)	369 (320–410)		18	11·4 (10·5–13·4)	792 (610–1030)	
Busson *et al.* (1962)	Africa	12	15·6 (12·4–18·2)		204 (181–231)	12	15·6 (12·4–18·2)		548 (500–606)
Deyoe and Shellenberger (1965)	USA	30	10·4 (8·6–12·5)	308 (241–372)	187 (152–221)	30	10·4 (8·6–12·5)	792 (575–1025)	482 (374–557)
Deyoe *et al.* (1970)		2			207 (207–207)	2			610 (526–676)
Eastoe and Taylor (1974)	Africa	3	9·8 (8·7–10·4)	353 (320–380)	241 (231–250)	3	9·8 (8·7–10·4)	1103 (990–1170)	761 (750–769)
Featherston *et al.* (1975a)	Puerto Rico	2	11·7 (11·5–11·9)	390 (370–410)	208 (193–222)	2	11·7 (11·5–11·9)	1095 (950–1240)	581 (514–648)
Ferreira Oriá and de Lima (1975)	S. America	10			146 (122–165)	10			475 (434–520)
FAO (1970)	Various			301	186 (119–225)			821	507 (372–606)
Khattab *et al.* (1972)	Africa					8		452 (380–580)	256 (186–382)
Schaffert (1972)		31	12·6		192	31	12·6		510
Shepherd *et al.* (1971–72)	Africa	61			194 (150–244)	61			537 (356–712)
Shoup *et al.* (1970a)	USA	7	10·3 (8·3–12·1)	334 (310–370)		7	10·3 (8·3–12·1)	886 (760–1030)	
Singh and Axtell (1973c)	USA and Puerto Rico	9	14·6 (11·8–16·3)		200 (184–221)	9	14·6 (11·8–16·3)		464 (415–497)
Srinivasan *et al.* (1972)	Mixed	522	12·6		191	522	12·6		510
Stephenson *et al.* (1971)	USA	24		367 (325–448)		24		969 (839–1143)	
Waggle *et al.* (1967a)		3	10·3 (8·3–12·1)	327 (270–380)	198 (194–200)	3	10·3 (8·3–12·1)	850 (670–1000)	515 (500–525)
Waggle *et al.* (1967b)		6	11·3 (10·6–12·1)	363 (360–370)	205 (191–221)	6	11·3 (10·6–12·1)	923 (840–1020)	509 (494–524)
Warsi and Wright (1973)	India	2	10·3 (8·9–11·6)	398 (339–457)	242 (237–247)	2	10·3 (8·9–11·6)	940 (734–1146)	566 (513–619)
Yañez *et al.* (1973)	S. America	5		372 (330–430)	249 (212–319)	5		738 (540–840)	510 (375–575)
Range of means					(146–249)				(256–761)
Range of ranges					(119–319)				(186–712)

Reference	Location	Ser				Ammonia			
		No. of samples	Prot. (N × 6·25) (%)	mg/100 g sample	mg/g total N	No. of samples	Prot. (N × 6·25) (%)	mg/100 g sample	mg/g total N
Badi *et al.* (1976)	USA	2			282 (269–294)	2			175 (162–187)
Breuer and Dohm (1972) (exp. III and IV)	USA	18	11·4 (10·5–13·4)	452 (350–540)		18	11·4 (10·5–13·4)	262 (200–330)	
Busson *et al.* (1962)	Africa	12	15·6 (12·4–18·2)		281 (287–294)				
Deyoe and Shellenberger (1965)	USA	30	10·4 (8·6–12·5)	420 (310–531)	256 (202–346)				
Deyoe *et al.* (1970)		2			320 (319–321)				
Eastoe and Taylor (1974)	Africa	3	9·8 (8·7–10·4)	567 (510–600)	388 (381–394)				
Featherston *et al.* (1975a)	Puerto Rico	2	11·7 (11·5–11·9)	545 (540–550)	290 (283–296)				
Ferreira Oriá and de Lima (1975)	S. America	10			166 (146–179)	10			550 (453–661)
FAO (1970)	Various			416	257 (166–346)				
Khattab *et al.* (1972)	Africa	7		341 (130–500)	185 (75–300)				
Schaffert (1972)		31	12·6		282				
Shepherd *et al.* (1971–72)	Africa	61			275 (194–419)	61			169 (75–250)
Shoup *et al.* (1970a)	USA	7	10·3 (9·2–11·5)	500 (440–580)		7	10.3 (9·2–11·5)	320 (280–380)	
Singh and Axtell (1973c)	USA and Puerto Rico	9	14·6 (11·8–16·3)		251 (231–269)				
Singh and Axtell (1973d)									
Srinivasan *et al.* (1972)	Mixed	522	12·6		282				
Stephenson *et al.* (1971)	USA	24		492 (405–588)					
Waggle *et al.* (1967a)		3	10·3 (8·3–12·1)	510 (410–580)	308 (300–319)				
Waggle *et al.* (1967b)		6	11·3 (10·6–12·1)	490 (460–520)	274 (271–276)				
Warsi and Wright (1973)	India	2	10·3 (8·9–11·6)	564 (503–625)	345 (338–352)	2	10·3 (8·9–11·6)	211 (156–265)	126 (109–143)
Yañez *et al.* (1973)	S. America	5		384 (290–480)	276 (181–337)				
Range of means					(166–388)				(126–550)
Range of ranges					(75–419)				(75–661)

[a] Microbiological assay: values are in general higher than those obtained by ion-exchange chromatography.

cont.

Table 2.14
Amino acid composition (mol % total)[a] of different anatomical parts of the kernel of FF 5683 sorghum and the free amino acid pool.[b]

Amino acid	Entire grain kernel	Germ	Pericarp	Endosperm	Amino acid pool[c]
Ile	2·84	3·51	3·92	4·06	1·16
Leu	12·15	7·77	8·97	14·19	3·99
Lys	1·61	3·82	3·00	1·20	3·49
Met	1·45	1·07	1·12	1·25	0·74
Half-Cys	1·09	1·07	1·61	0·83	0·58
Phe	3·81	3·92	3·95	4·35	1·84
Tyr	2·57	1·81	2·14	2·13	2·29
Thr	3·69	4·51	4·86	3·40	2·32
Val	4·25	6·62	6·35	5·26	2·34
Arg	2·62	6·88	4·14	2·21	6·52
His	1·71	2·58	1·69	1·71	0·69
Ala	14·44	10·52	11·29	12·77	10·87
Asp	6·61	7·95	8·33	5·63	21·87
Glu	18·58	14·03	12·43	19·48	11·72
Gly	6·42	11·83	12·63	5·07	5·67
Pro	9·71	5·19	6·44	9·94	8·02
Ser	6·48	6·95	7·16	5·52	15·91

[a] Authors' note: the amino acid composition is expressed as presented by the auto-analyser and cannot be converted to mg/g total N for comparison with other data on the basis of the information provided.
[b] Haikerwal and Mathieson (1971b).
[c] Amino acid pool constitutes 1·28% of the total N.

Table 2.15
Nutrient content of sorghum cultivars grown at Hyderabad, India.[a]

Cultivar	Protein (N × 6·25) (%)	Lysine mg/g total N	Leucine mg/g total N
CSV[b] 1	7·8	145	NR[d]
CSV 2	9·5	125	734
CSV 3	10·9	119	494
CSV 4	10·2	136	652
CSV 5	9·9	161	731
CSV 6	11·5	112	667
CSV 7R	8·7	130	787
Local (Aispuri)	8·3	133	800
Local (PJ16K)	10·4	125	750
CSH[c] 1	9·9	148	557
CSH 2	8·7	NR	681
CSH 4	9·8	147	612
CSH 5	9·8	155	748
CSH 6	10·5	NR[d]	637
IS 511	12·2	142	862
IS 859	13·2	116	925
IS 2031	13·4	138	981
IS 2954	14·3	131	912
IS 3797	14·0	134	906
IS 11758 (high-lysine, Ethiopia)	17·2	196	NR[d]
IS 11167 (high-lysine, Ethiopia)	15·7	208	NR[d]

[a] N. G. P. Rao (personal communication, 1977).
[b] CSV = Coordinated Sorghum Variety.
[c] CSH = Coordinated Sorghum Hybrid.
[d] NR = not reported.

Table 2.16
Average values (ranges in parentheses) of sorghum genotypes for agronomic traits.[a]

Grain yield[b] (14% moisture, kg/are)[c]	69·1	(55·4–91·7)
Prot. (%)	12·7	(10·2–15·0)
Prot. yield[b] (kg/are)[c]	8·7	(6·0–11·3)
Lys content (mg/g total N)	128	(114–146)
(mg/100 g sample)	258	(237–324)
yield[b] (kg/ha)	17·0	(13·6–21·4)
Oil (%)	3·32	(2·62–4·07)
100-seed weight (g)	2·75	(1·75–3·78)
Prot./seed (mg/seed)	3·53	(1·92–5·41)
Height (cm)	200	(99–323)
Days to median flower	79·4	(67·6–93·3)

[a] Schaffert *et al.* (1972a).
[b] Atlas omitted, (low-yielding genotype).
[c] are = 100 m^2.

Table 2.17
Average values of locations and years for sorghum genotypes of the agronomic traits studied.
Blank spaces indicate no data presented.[a]

Trait	Location[b]						Year	
	(1)	(2)	(3)	(4)	(5)	(6)	1970	1971
Grain yield (kg/are)[c]	72·9		72·7	71·7	61·3	66·9	70·4	67·8
Prot. (%)	13·7	13·2	12·4	13·2	11·2	12·6	12·4	13·1
Prot. yield (kg/are)[c]	10·0		9·0	9·3	6·8	8·4	8·6	8·8
Lys content								
(mg/g total N)	121	150	125	117	131	124	126	131
(mg/100 g sample)	266	316	248	247	228	248	245	222
yield (kg/ha)	19·2		17·9	17·4	14·0	16·5	16·8	17·1
Oil (%)	3·33	3·07	3·39	3·44	3·29	3·44	3·43	3·22
100-seed weight (g)	2·83	2·45	2·90	2·84	2·81	2·66	2·80	2·69
Prot./seed (mg/seed)	3·91	3·27	3·65	3·79	3·21	3·39	3·36	3·72
Ht (cm)	180·7		222·3	200·2	203·8	193·1	185·9	214·1
Days to median flower	77·8		76·8	79·0	80·9	82·4	78·7	80·1

[a] Schaffert *et al.* (1972a).
[b] (1) = Texas,
(2) = Nebraska,
(3) = Purdue Agronomy Farm normal fertility,
(4) = Purdue Agronomy Farm high fertility,
(5) = Southern Indiana Purdue Agricultural Centre,
(6) = Pinney Purdue Agriculture Centre.
[c] are = 100 m^2.

Table 2.18
Nitrogen, total, and in different protein fractions in fat and moisture-free samples of sorghum, and content of Lys and Trp.[a]

	VC 10 2	M 35 1	IS 84	IS 3922 405	K5	IS 3691
Total N (%)	1·99	2·00	1·49	1·89	1·71	1·68
Albumin N	0·18	0·105	0·060	0·121	0·075	0·045
as % total N	9·0	5·2	4·0	6·52	4·38	2·68
Globulin N	0·26	0·30	0·24	0·27	0·26	0·26
as % total N	12·9	15·0	16·3	14·3	15·2	15·2
Prolamine N	0·57	0·54	0·47	0·66	0·63	0·72
as % total N	28·5	27·0	31·7	34·9	33·8	43·1
Glutelin N	0·78	0·72	0·54	0·57	0·50	0·44
as % total N	39·5	36·0	35·9	30·0	29·2	26·2
Lys (mg/g total N)	165	146	120	124	92	91
Trp (mg/g total N)	67	72	70	59	46	31
Total N extracted	1·79	1·66	1·31	1·62	1·47	1·47
% Total N extracted	89·9	83·2	87·9	85·7	86·0	87·5

[a] Naik (1968).

Table 2.19
Concentration of Lys and Trp (mg/g total N) in protein fractions of sorghum varieties.[a]

	VC 10 2		M 35 1		IS 84		IS 3922 405		K 5		IS 3691	
	Lys	Trp	Lys	Trp	Lys	Trp	Lys	Trp	Lys	Trp	Lys	Trp
Albumin	309	127	295	191	250	208	381	111	280	136	266	180
Globulin	254	142	202	151	241	161	247	172	199	83	187	78
Prolamine	14	15	13	16	16	15	12	11	11	11	14	7
Glutelin	234	80	239	72	129	77	208	63	198	60	166	46

[a] Naik (1968).

Table 2.20
Distribution of Lys and Trp in different protein fractions (mg/100 g flour) of sorghum varieties.[a]

	VC 10 2	M 35 1	IS 84	IS 3922 405	K 5	IS 3691
Total Lys	330	292	175	237	155	150
Albumin	56	31	15	35	21	12
as % total Lys	17	11	8	14	13	8
Globulin	65	61	58	67	51	48
as % total Lys	19	22	31	28	33	32
Prolamine	8	7	7	8	8	10
as % total Lys	2·6	2·4	4	3	5	6·6
Glutelin	184	173	69	118	64	73
as % total Lys	55	59	40	50	41	49
Total Trp	134	146	104	110	80	52
Albumin	23·1	19·5	12·5	11·3	10·2	8·1
as % total Trp	17	15	12	10	13	16
Globulin	36·4	45·5	31·0	46·6	21·2	20·2
as % total Trp	28	31	31	42	24	40
Prolamine	8·4	8·5	7·0	7·6	7·2	5·0
as % total Trp	6	6	7	7	9	10
Glutelin	63·0	52·0	41·2	42·0	30·0	20·1
as % total Trp	48	35	41	38	37	40

[a] Naik (1968).

Table 2.21
Protein content and amino acid composition of the protein fractions of VC 10 2 and IS 3691 varieties of sorghum.[a]

	VC 10 2				IS 3691			
Protein (%)	12·4				10·5			
Prolamine (%) in prot.	28·5				43·1			
Amino acid content in prot. fractions (mg/g total N)								
	Alb.	Glob.	Prol.	Glut.[b]	Alb.	Glob.	Prol.	Glut.[b]
Ile	280	207	230	276	194	214	252	247
Leu	432	339	1021	559	426	368	1099	589
Lys	304	245	15	230	277	195	17	174
Met	176	146	55	76	182	168	44	69
Cys	212	70	20	84	179	94	26	90
Phe	184	301	321	251	257	310	340	283
Tyr	200	257	276	151	235	301	261	136
Thr	261	270	116	120	308	241	89	114
Trp	127	142	15	80	180	78	7	46
Val	635	401	321	407	554	335	364	492
Arg	536	405	51	339	620	417	54	320
His	201	89	39	221	180	116	27	242
Ala	589	366	515	541	462	377	611	521
Asp	1032	476	275	256	988	527	286	390
Glu	896	988	1761	1136	1026	932	1689	1277
Gly	551	343	102	213	617	381	117	180
Pro	412	321	653	513	384	340	711	476
Ser	452	365	199	326	386	408	215	342
Ammonia	75	26	42	42	82	24	27	52

[a] Ahuja *et al.* (1970).
[b] Albumin, globulin, prolamine, glutelin.

Table 2.22
Percentages of protein fractions, extracted by (1) percolation method and (2) shaking method, of sorghum, pearl millet, finger millet, maize, rice and wheat.[a]

Method	Albumins		Globulins		Prolamines		Glutelins		Residual		Total protein extracted	
	(1) (%)	(2) (%)	(1) (%)	(2) (%)	(1) (%)	(2) (%)	(1) (%)	(2) (%)	(1) (%)	(2) (%)	(1) (%)	(2) (%)
Sorghum												
Indian sample	10·9	11·5	8·4	11·0	8·9	6·2	25·0	12·8	46·8	58·5	53·2	41·5
US commercial sample	14·4	10·0	5·0	9·0	4·3	4·6	31·1	14·6	45·2	61·8	54·8	38·2
Pearl millet	17·1	16·0	11·6	14·2	26·9	26·2	25·8	16·0	18·4	27·6	81·4	72·4
Finger millet	15·8	13·0	8·0	13·3	9·1	6·0	33·0	26·2	34·1	41·5	65·9	58·5
Maize	14·2	12·3	8·1	12·5	24·3	19·3	41·6	35·8	11·8	20·0	88·2	80·0
Rice	2·8	2·3	8·6	9·0	2·6	2·3	60·4	48·7	23·6	37·7	76·4	62·3
Wheat	12·0	10·6	14·2	16·3	35·0	32·8	ND[b]	21·6		18·7		81·3

[a] Pushpamma (1968).
[b] ND = could not be collected.

Table 2.23
Solubility fractions of defatted endosperm protein of five varieties of grain sorghum.[a]

	Prot. content (N × 6·25 DWB) in endosperm (%)	Prot. fraction: Albumin (%)	Globulin (%)	Prolamine (%)	Glutelin (%)	Total prot. extracted (%)	Extraction efficiency (%)
M 35 1 Siruguppa	9·9	0·56	0·73	3·24	3·72	8·25	83·0
as % of total extracted		6·8	8·9	39·3	45·1		
BS 81 3 Annigeri	10·6	0·57	0·77	5·93	3·65	10·92	103·4
as % of total extracted		5·2	7·1	54·3	33·4		
CSH 1 Bijapur	18·1	1·40	1·16	7·88	4·86	15·30	84·4
as % of total extracted		9·1	7·6	51·5	31·8		
160 Cernum	17·1	0·89	1·59	7·06	5·93	15·47	90·7
as % of total extracted		5·8	10·3	45·6	38·3		
361 Dochna	19·0	0·25	0·29	11·18	3·61	15·33	80·7
as % of total extracted		1·6	1·9	72·9	23·5		

[a] Virupaksha and Sastry (1968).

Table 2.24
Amino acid composition, mean and ranges (in parentheses) (mg/g total N) of six genetic and three local varieties of sorghum.[a]

	Genetic varieties	Hybrid varieties
Ile	310 (239–388)	265 (230–309)
Leu	1095 (794–1306)	912 (809–1106)
Lys	133 (89–196)	115 (103–129)
Met	88 (46–114)	85 (51–120)
Half-Cys	58 (33–83)	65 (44–86)
Phe	343 (285–414)	321 (311–333)
Tyr	134 (77–178)	126 (119–134)
Thr	209 (196–222)	209 (196–224)
Val	334 (262–412)	335 (316–354)
Arg	240 (170–253)	173 (159–184)
His	128 (78–157)	117 (63–136)
Ala	751 (642–876)	735 (635–845)
Asp	546 (431–622)	432 (386–491)
Glu	1707 (1444–2006)	1542 (1452–1700)
Gly	188 (142–219)	198 (174–243)
Pro	777 (481–912)	661 (539–820)
Ser	278 (221–307)	281 (263–291)
Prot. (N × 6·25 DWB) (%)	15·0	13·4

[a] Virupaksha and Sastry (1968).

Table 2.25
Amino acid composition (mg/g total N) in protein fractions of CSH 1 Bijapur and 160 Cernum.[a] Blank spaces indicate no data presented.

	Endosperm meal	CSH 1 Bijapur			160 Cernum[b]	
		Globulin	Prolamine	Glutelin	Globulin	Prolamine
Ile	307	216	315	254	175	238
Leu	1036	420	958	781	318	1133
Lys	106	210	9	195	320	23
Met	94	140	83		66	61
Half-Cys	68	124	Trace	76	105	Trace
Phe	400	298	365	306	178	297
Tyr	290	251	323	202	133	228
Thr	238	304		305	250	134
Val	453	401	368	344	253	246
Arg	203	384	41	369	487	91
His	135	91	42	195	80	46
Ala	786	421	873	588	398	792
Asp	391	543	420	567	604	436
Glu	1859	988	1567	1505	668	1769
Gly	204	391	80	333	272	39
Pro	644	333	727	929	301	785
Ser	281	347	208	336	282	191

[a] Virupaksha and Sastry (1968).
[b] No data presented for the glutelin fraction.

Table 2.26
Total protein and fractionation results as % of total protein, of five samples of grain sorghum and Opaque-2 maize.[a]

Fraction as % total protein	Opaque-2 maize	CO 6	CO 88	MO 6	MO 81	MO 88
Albumin (water-soluble)	9·0	4·9	4·5	4·5	2·2	2·7
Globulin (5% NaCl solution)	11·2	5·8	3·5	4·3	3·6	2·5
Prolamine (80% ethanol + 0·2% sodium acetate solution)	9·5	8·5	5·0	7·5	4·3	3·9
Glutelin (0·2% NaOH solution)	56·4	20·8	17·9	14·9	17·4	17·3
Total sol. prot.	86·1	40·0	30·9	31·2	27·5	26·4
Nonsol. residue	10·6	58·6	62·6	52·8	62·3	59·3
Total prot. recovered (%)	96·7	98·6	93·5	84·0	89·8	85·7
Initial prot. of grain (N × 6·25 DWB) (%)	9·8	8·3	8·4	12·8	13·5	14·8

[a] Skoch *et al.* (1970).

Table 2.27
Average amino acid composition (mg/g total N) of sorghum (S) grain and Opaque-2 maize (O-2M) and of their solubility fractions.[a]

Amino acid[b]	Whole kernel		Albumin (water-soluble)		Globulin (NaCl-soluble)		Prolamine (ethanol-acetate-soluble)		Glutelin (NaOH-soluble)		Nonsol. residue	
	S	O-2M	S	O-2M	S	O-2M	S	O-2M	S	O-2M	S	O-2M
Ile	244	207	252	219	267	252	234	245	276	242	264	249
Leu	851	580	427	331	482	461	993	1076	645	660	1002	653
Lys	117	259	279	338	288	374	28	33	171	183	31	224
Met[c]	96	116	47	70	84	401	74	38		94	83	77
Cys[c]	102	156	56	102	132	540		706		45	22	50
Phe	317	259	226	162	265	287	351	409	310	318	357	312
Tyr	249	239	217	173	217	226	284	298	262	292	259	184
Thr	191	248	359	309	313	273	161	176	259	236	172	336
Val	259	298	306	309	338	362	254	226	317	287	199	411
Arg	207	410	331	399	573	566	114	122	331	409	106	312
His	129	220	116	159	132	195	51	74	196	205	100	219
Ala	603	406	565	509	402	426	629	521	474	417	679	437
Asp	399	562	716	612	611	572	429	352	492	437	379	489
Glu	1393	1086	912	924	857	912	1568	1426	1171	1212	1561	940
Gly	176	308	480	484	355	362	112	114	323	341	138	367
Pro	490	527	405	669	345	269	528	604	577	633	519	578
Ser	269	294	327	287	331	319	246	311	299	250	277	294
Ammonia	186	132	155	139	121	110	186	154	152	115	196	126

[a] Skoch *et al.* (1970).
[b] Values corrected to 100% recovery of Kjeldahl protein.
[c] Values estimated by performic acid oxidation; missing values not determined.

Table 2.28
Recovery[a] of each amino acid from the fractionated protein of sorghum and Opaque-2 maize.[b]

Amino acid[a]	Sorghum av. (%)	Opaque-2 maize (%)
Ile	99·4	110·2
Leu	95·0	105·1
Lys	63·0	76·5
Phe	97·6	112·5
Tyr	94·8	103·9
Thr	95·9	96·0
Val	89·7	96·4
Arg	78·1	89·5
His	84·4	81·2
Ala	92·9	101·8
Asp	99.3	78·0
Glu	91·8	99·4
Gly	102·2	103·2
Pro	95·3	104·6
Ser	94·8	87·0
Ammonia	89·8	86·3

[a] Recovery = The ratio (%) of the total g of an amino acid from the five protein fractions to g amino acid in the entire protein of the grain.
[b] Skoch *et al.* (1970).

Table 2.29
N distribution (% of total N) in the solubility fractions of whole kernels of normal (NL) and high-Lys (*hl*) sorghums.[a]

Fraction	Redlan (NL)	Redlan × [b]IS 11758 F$_2$ kernels (NL)	(HL)	IS 11758 (HL)	IS 11167 (HL)
I Saline (Albumin, globulin, free amino acids)	10·0	15·3	22·4	26·0	25·3
II Isopropanol (kafirin)	15·7	26·4	13·7	10·3	15·2
III Isopropanol + 2-mercaptoethanol (alcohol-soluble glutelin)	31·3	26·5	20·2	19·6	19·3
IV Borate buffer + 2-mercaptoethanol	4·5	4·3	4·3	6·5	4·5
V Borate buffer + 2-mercaptoethanol + sodium lauryl sulphate	29·3	22·5	33·5	27·2	29·5
Total N extracted (%)	90·8	95·0	94·1	89·6	93·3
Protein (%)	13·53	13·0	15·6	18·5	16·3
Lys (g/100g protein)	1·56	1·85	3·10	3·27	3·10
(mg/g total N)	97	116	194	204	194

[a] Jambunathan *et al.* (1975).
[b] Av. values of kernels from five different heads (55073–55077).

Table 2.30
Total N and % soluble lysine in the solubility fractions of powdered defatted kernels of low pigment and high-pigment sorghums.[a]

Fraction	IS 0062 (low-pigment) Total N (%)	Soluble lysine (%)	IS 2283 (high-pigment) Total N (%)	Soluble lysine (%)
I 0·5M NaCl	16·8	46·8	3·7	0·8
II 70% isopropanol (v/v)	18·4	0·8	1·9	0·1
III 70% isopropanol with 0·6% 2-mercaptoethanol	21·6	0·5	21·1	1·1
IV Borate buffer (0·5% pH 10) with 0·6% 2-mercaptoethanol	4·9	4·2	13·6	22·7
V Borate buffer with 0·5% sodium lauryl sulfate (w/v) and 0·6% 2-mercaptoethanol	26·0	47·7	47·0	75·3
Total	87·7		87·3	

[a] Jambunathan *et al.* (1972).

Table 2.31
Amino acid composition (mg/g total N) of TE 77 sorghum prolamine and prolamine isolates.[a]

Amino acid	Whole prolamine	95% ethanol fraction	Fast component
Ile	300	237	87
Leu	1200	1012	719
Lys	6	0	0
Met	62	37	625
Half Cys	25	50	394
Phe	400	325	106
Tyr	344	281	487
Thr	162	143	237
Val	312	431	294
Arg	62	75	269
His	56	56	0
Ala	775	662	619
Asp	406	369	181
Glu	1875	1550	1556
Gly	69	69	212
Pro	625	362	506
Ser	256	250	206
Ammonia	219	237	175

[a] Beckwith and Jones (1972).

Table 2.32
Selected amino acid content (mg/g total N) of prolamines and glutelins in grain sorghums.[a]

	Prolamine	Glutelin
Leu	1200	931
Lys	6	131
Half Cys	25	75
His	56	125
Arg	62	225
Glu	1875	1456

[a] Beckwith (1972).

Table 2.33
Chemical composition (DWB) of whole kernel (WK) and protein bodies (PB) of sorghum, pearl millet and finger millet.[a]

	Sorghum		Pearl millet		Finger millet	
	WK	PB	WK	PB	WK	PB
N (%)	1·24	10·95	1·58	11·39	1·11	10·16
P (mg/100 g)	330	370	330	690	220	430
Ca (mg/100 g)	10	20	10	10	60	90
K (mg/100 g)	400	20	460	90	580	60
Mg (mg/100 g)	150	10	140	70	170	60

[a] Adams *et al.* (1976).

Table 2.34
Distribution of protease activity, phytase activity, and protein concentration in the sol. and insol. protein fractions of protein body preparations of sorghum, pearl millet and finger millet.[a]

	Protease (μmol Gly equivalent released/h/ml)	Phytase (μmol P released/h/ml)	Protein (mg/ml)
Sorghum			
Sol.	0·38	0	0·30
Insol.	2·72	0·10	10·90
Pearl millet			
Sol.	2·68	0	3·16
Insol.	5·12	0·12	11·56
Finger millet			
Sol.	1·72	0	0·96
Insol.	3·04	0	12·88

[a] Adams *et al.* (1976).

Table 2.35
Fatty acid composition of sorghum, pearl millet, maize, rice, triticale and wheat.[a]

	Moisture	Total lipid	Fatty acid[b]									
			Saturated					Unsaturated				
			Sum	Myristic	Palmitic	Stearic	Arachidic	Sum	Palmitoleic	Oleic	Linoleic	Linolenic
			g/100 g food edible portion									
Sorghum												
Whole grain	11	3·3	0·48	0·01	0·44	0·03	0	2·74	0·04	1·15	1·46	0·09
Germ oil	0	100	17·2	—	17·2	—	0	77·91	—	26·1	49·9	1·91
Pearl millet												
Whole grain	11·8	4·1	0·86	0	0·68	0·16	0·02	2·67	0·02	0·83	1·69	0·13
Maize												
Whole grain, raw	13·8	4·1	0·47	0	0·40	0·06	0·01	3·07	0·01	0·91	2·12	0·03
Flour	12	2·6	0·30	0	0·25	0·04	0·01	2·00	—	0·64	1·34	0·02
Germ	0	30·8	3·93	0	3·30	0·54	0·09	25·57	0·04	7·58	17·7	0·25
Rice												
Brown, dry form	12	2·3	0·62	0·03	0·54	0·04	0·01	1·36	0·01	0·54	0·78	0·03
White, polished, dry form	12	0·8	0·21	0·01	0·19	—	—	0·47	—	0·19	0·27	0·01
Bran oil	0	100	19·50	0·69	16·7	1·58	0·53	74·56	0·26	39·2	33·5	1·60
Triticale												
Whole grain flour	14	34	0·49	0·02	0·45	0·02	0	1·91	0	0·28	1·48	0·15
White flour	14	1·6	0·23	0·01	0·21	0·01	0	0·87	0	0·13	0·67	0·07
Wheat												
Whole grain (mean of 4 types)	14	2·4	0·34	—	0·32	0·02	—	1·39	0·01	0·24	1·06	0·08
Flour (mean of 4 types)	14	1·4	0·22	—	0·20	0·01	—	0·76	—	0·10	0·64	0·03
Germ oil	0	100	17·54	—	15·7	0·41	1·43	76·88	0·79	15·0	54·4	6·69

[a] Weihrauch *et al.* (1976).
[b] Dashes denote ≤ 0·005 g.

Table 2.36
Conversion factors (fatty acids in 1 g lipid) used to calculate the fatty acid content of lipids of cereal grains.[a]

	Factor
Maize	
Commercial maize	0·86
Oil	0·95
Starch	0·70
Sorghum, Pearl millet	As maize
Wheat	
Whole grain	0·72
Flour	0·67
Starch	0·60
Shorts	0·80
Bran	0·82
Germ	0·93
Germ oil	0·95
Triticale	As wheat
Rice	
Brown rice, or bran	0·92
Milled	0·85
Oil	0·94
Starch	0·70

[a] Weihrauch *et al.* (1976).

Table 2.37
Proximate analysis, total lipids composition, fatty acid composition of total lipids and fatty acid (FA) content of total lipid extract of sorghum, maize, triticale and wheat.[a]

	Sorghum	Maize	Triticale	Wheat
	S. Dakota 106	P3764-3X	N. Dakota 203	Polk
Proximate analysis (%)				
Moisture	9·2	8·3	9·2	9·6
Protein	11·9	8·9	19·6	18·1
Lipid	3·3	5·8	2·4	2·3
Carbohydrate (by difference)	70·8	72·6	63·7	65·8
Crude fibre	2·8	3·1	2·6	2·0
Ash	2·0	1·3	2·5	2·2
Lipid composition (%)				
Neutral lipid	86·2	91·9	66·9	61·9
Glycolipid	3·1	2·1	16·0	21·6
Phospholipid	10·7	6·0	17·1	16·5
FA composition of total lipids (% by weight)				
Myristic (C14/0)	0·20	0·44	0·14	0·19
Palmitic (C16/0)	19·48	13·37	16·57	7·16
Palmitoleic (C 16/1)	0·80	0·15	0·24	0·33
Stearic (C 18/0)	1·70	1·93	0·62	1·44
Oleic (C18/1)	28·13	23·99	13·66	20·38
Linoleic (C18/2)	44·73	59·29	63·78	57·66
Linolenic (C18/3)	4·96	0·83	4·99	2·84
FA content of the total lipid extract				
Fatty acid (%)	76·7	90·8	81·6	75·3
FA (mg/100 g grain)				
Myristic	5·1	23·2	2·7	3·3
Palmitic	493·2	704·3	324·3	297·0
Palmitoleic	20·3	7·9	4·7	5·7
Stearic	43·0	101·7	12·1	24·9
Oleic	712·3	1263·8	267·3	352·8
Linoleic	1132·6	3123·4	1248·2	998·1
Linolenic	125·6	43·7	97·7	49·2
Total	2532	5268	1957	1731

[a] Price and Parsons (1975).

Table 2.38
Characteristics of maize and sorghum lipids in fractions obtained by wet-milling.[a]

	Sorghum				Maize			
	Germ	Starch	Gluten	Fine fibre	Germ	Starch	Gluten	Fine fibre
Amount of fat (% of dry matter)	52·1	0·9	6·9	3·2	56·8	0·6	7·0	1·3
Free fatty acid (% as oleic)	2·0	91·3	21·7	19·9	2·5	71·5	22·2	13·8
Saponification equivalent	294	280	309	320	294	284	324	366
Iodine value (Wijs, 0·5 h)	122	94	98	113	126	103	129	108
Unsaponifiable matter (modified Kerr-Sorber, %)	1·7	2·5	8·0	10·4	2·9	3·2	13·6	22·8

[a] Baldwin and Sniegowski (1951).

Table 2.39
Fatty acid composition of lipids from sorghum and maize fibres.[a]

	Fine fibre	
	Sorghum (%)	Maize (%)
Nonconjugated acids		
Dienoic	38·8	41·4
Trienoic	0·0	0·7
Mono-unsaturated	44·5	41·1
Saturated (primarily C_{16} and C_{18} in sorghum)	14·3	13·1
Conjugated acids		
Dienoic	1·8	3·1
Trienoic	0·5	0·3
Tetraenoic	0·1	0·3

[a] Baldwin and Sniegowski (1951).

Table 2.40
Fatty acid composition of lipids from sorghum and maize germ, starch and gluten.[a]

	Germ		Starch		Gluten	
	Sorghum (%)	Maize (%)	Sorghum (%)	Maize (%)	Sorghum (%)	Maize (%)
Myristic	0·6	0·2	0·4	0·7	2·2	0·6
Palmitic	10·2	9·9	31·6	26·3	14·0	11·4
Stearic	4·1	2·9	5·4	2·8	6·0	3·5
C_{20-22} (saturated)	Trace	0·2	0·1	1·5	Trace	0·2
Hexadecenoic	0·4	0·5	0·6	0·8	0·5	0·3
Oleic	32·2	30·1	22·0	20·7	29·6	28·1
Linoleic	51·5	56·2	39·9	47·2	45·4	53·3
Polyunsaturated (as linolenic)	1·0	0·0	0·0	0·0	2·3	2·6

[a] Baldwin and Sniegowski (1951).

Table 2.41
Protein, oil, oleic acid and linoleic acid content (% in grain) of cultivated and wild sorghums.[a]

	Protein	Oil	Oleic acid	Linoleic acid
Bicolor	15·3	4·3	27·7	48·8
Durra	14·5	3·7	27·4	51·1
Caudatum	14·2	3·3	34·6	40·8
Guinea	13·6	4·2	25·1	52·8
Kafir	12·1	4·1	34·5	44·6
Wild	21·5	3·6	32·9	45·2

[a] Stemler *et al.* (1976).

Table 2.42
Percentage composition of neutral lipids (free and bound) of sorghum.[a]

	JS 2/53		JS 73/53	
Lipid component	Free	Bound	Free	Bound
Partial glycerides	3·2	5·8	3·5	6·3
Sterols	5·7	6·4	6·3	4·9
Pigments	5·8	0·7	4·2	2·3
Free fatty acids	7·2	40·2	8·7	47·4
Triglycerides	63·0	44·5	66·0	31·9
Unidentified	6·9		4·7	
Sterol esters and hydrocarbons	8·2	2·4	7·6	7·2

[a] Paul *et al.* (1972).

Table 2.43
Range of physico-chemical characteristics and fatty acid composition of the germ oil of sorghum varieties grown in the Republic of Argentina.[a]

	Minimum	Maximum
Yield of oil (% of germ)	18·5	30·8
Specific gravity (25/4°C)	0·9128	0·9159
Refractive index (25°C)	1·4710	1·4731
Saponification value	185·8	191·5
Iodine value (Hannus)	116·9	125·5
Unsaponifiable (%)	2·29	3·37
Iodine value of unsaponifiable	76·7	107·6
Tocopherols on unsaponifiable (mg α-tocopherol/100 g oil)	31·2	105·2
Tocopherols on oil, as-is (mg α-tocopherol/100 g oil)	40·3	87·4

[a] Bertoni *et al.* (1963).

Table 2.44
Proximate analysis (range) of the residue after hexane extraction of sorghums grown in various provinces of Argentina.[a]

	Minimum (%)	Maximum (%)
Moisture (100–105°C)	9·1	12·2
Protein (N × 5·7)	17·0	25·1
Carbohydrates (as starch)	23·0	42·3
Pentosans	9·8	16·3
Crude fibre	4·3	7·7
Ash (500–550°C)	1·9	13·5
Not determined	4·5	11·8

[a] Bertoni *et al.* (1963).

Table 2.45
Mean (ranges in parentheses) of wax and oil extracted from 10 sorghum varieties, and characteristics of the mixed fatty acids of the various oils (maize average of three types.[a]

	Sorghum	Maize
Wax		
Extracted (%)	0·3 (0·2–0·5)	0·02 (0·01–0·03)
Melting point (°C)	82 (80–84)	
Oil		
Extracted (%)	2·7 (2·4–3·0)	4·3 (4·0–4·7)
Refractive index (at 25°C)	1·4697 (1·4686–1·4718)	1·4707 (1·4700–1·4714)
Nonsaponifiable (%)	2·6 (2·0–3·2)	2·5 (2·1–2·8)
Mixed fatty acids		
Melting point (°C)	29·02 (26·2–31·2)	25·9
Neutralization equivalent	278·8 (278·5–279·4)	278·8
Iodine value	121·85 (116·5–126·9)	128·6
Thiocyanogen value	81·78 (80·0–83·4)	81·9
Linoleic acid (%)	47·6 (42·5–51·7)	55·3
Oleic acid (%)	39·86 (36·8–44·0)	31·6
Palmitic acid (%)	7·65 (6·0–9·5)	8·3
Stearic acid (%)	4·85 (3·0–6·0)	4·5

[a] Kummerow (1946b).

Table 2.46
Extraction of wax-free ground Western Blackhull Kafir with various solvents.[a]

Solvent	Extract (%)
Acetone	3·66
Alcohol, absolute	5·46
Benzene	3·56
Carbon tetrachloride	3·12
Chloroform	3·62
Dioxane	3·63
Ethyl ether	3·22
Skellysolve F (petroleum ether, b.p. 30–60°)	2·28
Skellysolve B (essentially n-hexane b.p. 60–68°)	2·89
Tetrachloroethane	7·15
Trichlorethylene	4·39

[a] Kummerow (1946b).

Table 2.47
Protein (N × 6·25) (DWB) and amylose content (mean with ranges in parentheses) of 100 lines of sorghum.[a]

Type	No. of samples	Protein (%)	Amylose (% of sample)
With lustre	10	12·6 (11·0–14·8)	18·1 (15·8–23·1)
Persistent subcoat	10	13·3 (10·7–14·3)	17·1 (8·2–21·7)
Completely corneous	5	14·1 (12·3–15·8)	17·7 (13·3–20·6)
Almost corneous	5	12·6 (10·0–14·7)	17·4 (12·5–20·6)
Intermediate	5	13·0 (11·0–15·7)	18·4 (15·5–21·7)
Almost floury	5	13·7 (11·6–16·0)	17·8 (15·8–20·2)
Completely floury	5	13·2 (10·4–16·2)	17·7 (16·2–19·1)
Grain Colour			
White	5	13·6 (11·6–16·0)	19·4 (17·3–21·3)
Straw	5	13·3 (11·5–14·8)	17·4 (16·2–19·5)
Yellow	5	11·8 (10·7–13·0)	18·5 (16·2–20·9)
Light brown	7	12·4 (10·2–16·2)	15·4 (14·4–18·4)
Brown	8	13·1 (11·4–15·9)	17·5 (16·6–19·9)
Reddish-brown	5	12·3 (9·1–15·4)	19·3 (15·8–21·7)
Light red	5	13·6 (12·8–15·1)	21·3 (17·7–23·1)
Red	5	13·2 (11·4–15·0)	21·5 (16·6–31·3)
Grey	5	14·5 (13·2–15·6)	18·3 (15·8–19·9)
Purple	1	14·1	19·9
Waxy endosperm	4	12·3 (11·6–13·5)	11·9 (7·1–19·1)

[a] Williams, personal communication (1978).

Table 2.48
Amino acid composition (mg/g total N) of grain sorghum varieties and hybrids.[a]

Amino acid	USA					Sudan							
	C 42Y	Bulk Hybrid	CP 622	ACCO 1023	Mean	Gezera	Dabar	Gassabi	Mugud	Gedarif	Diraira	Dwarf	Mean
Ile	231	236	229	229	231	224	227	226	229	229	221	222	226
Leu	819	819	770	845	813	873	844	818	796	886	865	714	828
Lys	159	140	159	111	142	104	131	131	147	107	116	188	132
Met	87	77	66	87	79	91	103	107	90	83	96	97	95
Half-Cys	149	108	129	89	119	126	157	139	140	117	131	191	143
Phe	297	306	309	299	303	306	304	311	298	311	294	291	302
Tyr	224	213	244	241	230	252	242	244	229	247	237	230	240
Thr	190	227	212	201	208	181	191	194	196	177	188	209	191
Val	282	282	314	293	293	279	298	287	318	254	337	306	290
Arg	252	199	257	227	234	212	216	241	276	211	207	299	238
His	145	107	165	131	137	131	126	135	147	136	127	145	135
Ala	536	569	534	594	558	569	546	554	525	569	548	529	549
Asp	409	434	416	419	419	391	416	394	417	400	392	445	408
Glu	1300	1392	1338	1432	1365	1437	1366	1375	1336	1446	1422	1227	1373
Gly	189	212	196	169	191	171	186	196	209	167	180	239	193
Pro	549	449	468	462	482	500	493	496	494	500	492	472	493
Ser	269	296	282	272	280	269	272	273	273	272	269	281	273
Ammonia	164	184	161	150	165	133	131	129	132	136	128	163	136
Protein estimated from amino acid recovery (%)	9·2	10·6	8·8	13·3		13·1	9·9	8·9	8·6	12·2	10·4	7·4	

[a] Hoseney *et al.* (1974).

Table 2.49
Kernel characteristics, physical properties, and proximate analysis (DWB) of four sorghum lines.[a]

	NSA 740	Kafir 60	SC 301	TX 615
Endosperm texture	Floury	Intermediate	Corneous	Intermediate
Endo. type	Normal	Normal	Normal	Waxy
1000-kernel weight (g)	22·0	24·0	20·0	26·0
Test weight (lb/bu)	46·2	53·1	53·7	56·4
Hardness (%)	7·2	15·7	33·5	17·1
Density (g/cm³)	1·223	1·370	1·400	1·325
Protein (%)	16·6	11·8	12·9	12·7
Lipid (%)	3·4	2·6	2·8	2·6
Ash (%)	1·9	1·3	1·6	1·8
Starch (%)	66·7	75·2	73·4	73·8
Amylose (% of total starch dry weight)	28·7	29·0	29·0	0·2

[a] Sullins and Rooney (1974).

Table 2.50
Kernel characteristics and proximate analysis (DWB) of waxy and nonwaxy sorghums.[a]

	TX 3197	TX 615	Texioca 54	Texioca 63	TX 378	SA 413
Endosperm type	Nonwaxy Kafir	Waxy Kafir	Waxy Kafir	Waxy Kafir	Nonwaxy Redlan	Waxy Redlan
Test weight (lb/bu)	57·3	57·0	58·1	56·2	58·0	55·3
1000-kernel weight (g)	23·7	24·2	22·5	22·1	26·3	23·8
Hardness value	21	21	11	16	11	13
Protein (N × 6·25) (%)	11·5	11·3	12·1	11·2	12·7	11·6
Ash (%)	2·0	2·2	2·1	2·4	2·0	2·1
Lipid (%)	2·8	3·6	2·9	3·6	3·0	2·3

[a] Sullins and Rooney (1975).

Table 2.51
Chemical, physical and organolepic attributes of sorghums of differing endosperm characteristics.[a]

Suppliers code	Endosperm characteristics	Total starch (%)	Amylose in total starch (%)	Starch granule density (g/cm³)	Mean starch granule diameter (μm)	Organoleptic[b] score
B 607 (Texas)	Waxy	64·20	4·80	1·39	15·00	1·70
Texioca 54 (Texas)	Waxy	69·98	4·76	1·16	12·50	2·80
F_1 (hybrid of above)	Waxy	65·88	0·79	1·30	13·25	1·80
B 3197 (Texas)	Regular	69·74	27·72	1·14	15·00	3·20
TX 2520 (Texas)	Waxy	66·37	5·74	1·39	15·00	2·40
F_1 (hybrid of above)	Intermediate	70·61	27·22	1·50	12·75	2·40
B 398 (Texas)	Regular	67·71	29·15	1·51	15·00	2·50
TX 2536 (Texas)	Yellow	66·52	24·82	1·48	9·00	2·40
F_1 (hybrid of above)	Intermediate	68·73	23·64	1·56	17·50	2·50
	Yellow	69·44	28·08	1·65	13·25	3·00
	Yellow	67·52	27·13	1·61	13·25	2·50
G 600 (hybrid of above)	Yellow	70·60	34·87	1·90	15·00	2·40
B 378 (Texas)	Regular	69·12	25·23	1·63	13·00	2·40
TX 415 (Texas)	Regular	69·84	25·93	1·75	15·00	2·70
RS 671 (hybrid of above)	Regular	70·57	25·97	1·94	11·75	3·10
Pop sorghum	Regular	65·26	30·66	2·01	8·25	2·40
SA 5023 9 1 2 (Texas)	Sugary	57·39	26·80	2·21	15·00	2·30

[a] Miller and Burns (1970).
[b] Organoleptic score: 1 = excellent; 2 = good; 3 = fair; 4 = poor.

Table 2.52
Mean pentosan content (ranges in parentheses) of sorghum grain with different endosperm characteristics.[a]

Endosperm characteristics	Pentosan (DWB, %)	Number of varieties
Yellow	3·76 (3·21–5·57)	7
Waxy	2·77 (2·51–3·03)	2
Sugary	3·84 (3·21–4·48)	2
Corneous	3·51 (3·21–4·01)	6
Floury	3·56 (2·87–4·87)	7
Normal and intermediate	3·40 (2·8–4·1)	7

[a] Karim and Rooney (1972a).

Table 2.53
Distribution of pentosans within the kernel of sorghum B 398.[a]

Parts of the kernel	Pentosan (DWB, %)
Whole kernel	3·72
Pericarp	21·54
Endosperm	0·99
Germ	6·17
Fine fraction	11·16
Recovery	87·73

[a] Karim and Rooney (1972a).

Table 2.54
Ranges and overall mean and range of physical and chemical properties (DWB) of grain from 31 varieties of sorghum of different endosperm characteristics.[a]

Endo. char.	No. of varieties	Kernel size index	1000-kernel weight (g)	Density (g/cm^3)	Test wt (lb/bu)	Hardness	Prot. (N × 6·25) (%)	Lipid (%)	Starch (%)
Yellow	7	50·3–67·7	26·6–59·2	1·32–1·36	57·3–61·1	16·4–29·6	11·4–14·0	2·7–3·7	69·7–75·9
Corneous	6	68·5–74·9	19·0–28·5	1·31–1·40	57·4–63·5	17·9–63·8	13·8–17·8	2·8–4·1	65·1–70·5
Floury	7	63·3–83·0	11·6–28·8	1·22–1·36	51·5–59·1	0·1–30·8	12·6–14·6	3·3–4·2	62·3–67·5
Sugary	2	60·6–69·1	22·4–33·1	1·29–1·30	53·5–55·3	5·0–29·1	14·7–17·3	5·2–5·4	60·9–64·1
Waxy	2	61·4–67·4	27·6–34·6	1·32–1·33	57·1–58·2	4·3–43·5	13·3–15·7	3·4–3·7	68·8–72·6
Normal and intermediate	7	53·8–70·1	24·5–48·7	1·32–1·38	55·8–61·9	1·2–44·1	12·4–15·6	3·2–4·2	68·1–71·4
Overall mean	31	66·2	29·1	1·33	58·2	19·8	14·0	3·7	68·8
Overall range	31	50·3–83·0	11·6–59·2	1·22–1·40	51·5–63·5	0·1–63·8	11·4–17·8	2·7–5·4	60·9–75·9

[a] Karim and Rooney (1972a).

Table 2.55
Ranges and overall mean of dry-milling and wet-milling properties of grain from 31 varieties of sorghum differing in endosperm characteristics (DWB).[a]

Endo. char.	No. of varieties	Coarse grits (%)	Fine grits (%)	Coarse and fine grits (%)	Starch yield (%)	Starch recovery (%)	Starch protein (%)	Peripheral endo. content (%)
Yellow	7	49·0–59·5	15·1–26·0	67·1–76·2	47·1–54·6	66·2–75·1	0·8–1·5	0·6–1·5
Corneous	6	43·6–79·3	4·1–33·5	74·7–83·2	41·5–49·3	59·9–70·0	1·7–2·6	1·8–2·9
Floury	7	0·1–55·8	17·3–58·5	31·0–73·0	46·9–53·6	71·9–83·9	1·4–2·1	0·5–1·4
Sugary	2	37·4–74·0	5·9–29·3	66·7–79·8	21·3–30·8	35·0–48·1	1·8–2·1	0·5–1·4
Waxy	2	35·1–38·5	32·5–39·9	71·0–75·0	43·8–49·7	63·7–68·5	1·1–1·5	1·3–2·4
Normal and intermediate	7	19·0–72·7	6·7–25·2	43·9–79·4	46·0–54·4	66·5–77·6	0·9–3·0	0·8–1·9
Overall mean	31	44·6	23·3	67·9	48·5	70·2	1·6	1·4
Overall range	31	0·1–79·3	4·1–58·5	31·0–83·2	21·3–54·6	35·0–83·9	0·8–3·0	0·3–2·9

[a] Karim and Rooney (1972a).

Table 2.56
Properties of water-sol. and alkali-sol. pentosans obtained from various anatomical portions of sorghum kernel (DWB).[a]

Pentosan type	Part of kernel	Yield (%)	CBH[b] (%)	Prot. (N × 6·25) (%)
Water-sol.	Whole grain	0·90	76·2	10·5
	Endosperm	0·16	69·6	3·2
	Pericarp	0·62	68·7	6·8
Alkali-sol.	Whole grain	0·42	74·0	1·3
	Endosperm	0·09	89·7	3·7
	Pericarp	20·00	76·2	5·2

[a] Karim and Rooney (1972b).
[b] Carbohydrate.

Table 2.57
Sugar composition of sorghum pentosans obtained from various parts of the kernel (% of total separated by paper chromatography).[a]

Type of pentosan	Part of kernel	Galactose	Glucose	Arabinose	Xylose	Arabinose/ xylose	Pentose/ hexose
Water-sol.	Whole grain	9·36	68·73	16·68	5·22	3·19	0·28
	Endo.	4·98	85·14	6·47	3·41	1·89	0·11
	Pericarp	13·67	30·19	37·75	18·38	2·05	1·28
Alkali-sol.	Whole grain	7·60	55·42	22·54	14·42	1·56	0·59
	Endo.	3·85	73·46	14·53	8·16	1·78	0·29
	Pericarp	5·22	37·16	32·37	25·25	1·28	1·36

[a] Karim and Rooney (1972b).

Table 2.58
Content of inorganic elements in sorghum (mg/100 g) (from multiple sources). Blank spaces indicate no data presented.

Reference	Location	No. of samples	Ca	P	Fe
Adrian and Sayerse (1957)[a]	India	17	24·6 (15·7–34·8)	384 (343–429)	
Barham *et al.* (1946)[b]	N. America	14		430 (335–652)	1·0 (0·5–1·7)
Brawand and Hossner (1976)[c]	USA	46	20·0 (10·0–40·0)	280 (240–310)	
Bressani and Rios (1962)	Cent. America	25	20·9 (12·8–54·9)	566 (277–128)	
Capote *et al.* (1972)[d]	S. America	103		280	
Deosthale *et al.* (1972a)[e]	India	15	12·8 (11·2–15·5)	247 (215–261)	5·8 (5·0–7·3)
Devadas *et al.* (1966)	India	5	102·2 (71·0–131·0)	167 (142–189)	4·0 (3·5–5·0)
Ferreira Oriá and de Lima (1975)	S. America	10	11·0 (7·5–17·1)	296 (229–364)	12·1 (7·1–16·9)
Heller and Seiglinger (1944)	N. America	78	21·8 (13·3–55·6)	335 (250–611)	
Kandel (1954)	Europe	12	585·5 (360·3–731·9)	349 (293–405)	
Kramer and Matz (1969)[f]	N. America	1	20·0	400	20·0
Mali and Gupta (1974)	India	16	16·5 (10·2–22·0)	387 (280–500)	7·2 (5·2–9·3)
McCollough (1972)[g]	USA	8	41·2 (29·0–77·0)	349 (301–383)	6·6 (5·6–8·0)
Naik and Abhyanker (1955)[h]	India	16	26·9 (14·0–73·0)	751 (520–1006)	11·0 (4·0–25·0)
National Academy of Sciences (1971)[i]	N. America	8	45·0 (20·0–110·0)	335 (300–380)	1·0 (1·0–1·0)
National Institute of Nutrition,[j] Hyderabad (1974) (personal communication)	India	26	12·7 (7·8–18·6)	348 (307–395)	3·5 (2·7–4·8)
Sur *et al.* (1955)	India	6	88·4 (57·8–144·0)	197 (170–231)	2·3 (1·2–3·0)
Thayer *et al.* (1957)	USA	16	40·8 (29·3–45·8)	285 (208–283)	
Viraktamath *et al.* (1972)	India	14		258 (160–389)	
Wehmeyer (1969)[k]	Africa	79	24·0 (13·5–40·3)	245 (129–355)	5·2 (2·5–11·1)
Yañez *et al.* (1973)	S. America	5	213·6 (90·0–391·0)	566 (527–658)	6·6 (5·1–8·5)
Yousif and Magboul (1972)	Africa	18	25·9 (15·0–48·0)	303 (132–396)	0·9 (0·6–1·6)
Representative analytical values and ranges (not means)[m]			32·0 (10·0–53·0)	350 (100–520)	4·0 (1·0–12·4)
Range of means			11·0–585·5	167–751	0·9–20·0
Range of ranges			7·5–731·9	100–1276	0·5–25·0

The following sources also reported, mg/100 g

[a] K 386 (295–437), Na 88 (56–122).

[b] Cu 1·2 (0·7–2·5), Mn 1·4 (0·8–2·6), Na 22 (10–41).

[c] Mg 140 (110–160), K 420 (390–450).

[d] K 447, Zn 7·6.

[e] Cu 0·5 (0·3–0·9), Mg 141 (105–163), Mn 1·1 (0·9–1·3), Mo 0·005 (0·004–0·007), Zn 0·13 (0·11–0·16).

[f] Al 0·4, Bo 1·0, Cr 0·04, Cu 0·08, K 300, Mg 200, Mn 20, Mo 0·1, Na 0·08, Ni 0·08, Pb 1·0, Si 4, Sn 0·04, Ti 0·04, Zn 8·0.

[g] Mg 152 (137–190), Mn 1·0 (0·8–1·2), Zn 4·7 (2·0–7·7).

[h] Al 6 (2–17), Cu 0·2 (0·2–0·3), K 300 (52–485), Mg 137 (93–212), Mn 1·5 (0·1–1·8), Na 112 (29–162), Si 61 (27–147), Zn 0·3 (0·1–0·8).

[i] Cl 110, Co 0·04, Cu 0·3, K 380, Mg 170, Mn 0·8, Na 70, S 180.

[j] Cu 0·4 (0·27–0·77), Mo 0·06 (0·05–0·08), Phytin P240 (213–278), Zn 1·86 (1·40–2·60).

[k] Cu 0·5 (0·2–0·9), K 395 (278–557), Mg 126 (87–193), Zn 2·9 (1·5–6·5).

[l] Al 3·0, Bo (0·13–0·33), Cl (510–580), Co (0·001–0·12), Cr 0·05, Cu 1·1 (0·2–1·9), K 360 (217–480), Mg 190 (20–250), Mn 2·0 (0·0–5·0), Mo (0·06–0·4), Na 5·0 (1·9–18·0), Ni 0·17, Pb 0·11, S (90–117), Sn 0·05, Ti 0·10, Zn 2·9 (2·5–3·7).

[m] Data from various sources mentioned in the text.

Table 2.59
Vitamin content of sorghum (mg/100 g) (from multiple sources). Blank spaces indicate no data presented.

Reference	Location	No. of samples	Vitamin B								
			Thiamine	Niacin	Riboflavin	Biotin	Pantothenic acid	Pyridoxine	Folic acid	Choline	Carotene
Adrian and Sayerse (1957)[a]	India	17	0·54 (0·29–0·67)	6·40 (4·18–8·65)	0·15 (0·13–0·20)						
Bressani and Rios (1962)	Cent. America	25	0·33 (0·13–0·62)	3·37 (2·33–4·30)	0·22 (0·09–0·34)						
Chitre *et al.* (1955)	India	7	0·33 (0·25–0·47)	2·91 (2·24–3·80)	0·13 (0·10–0·18)						
Deosthale *et al.* (1969)		66									0·04 (0·01–0·07)
Feher (1971)	Europe	14									0·08 (0·0–0·63)
Gross and Heller (1943)	USA	36									0·23 (0·12–0·51)
Kapoor and Naik (1970)	India	21									0·07 (0·03–0·14)
Knox *et al.* (1944)[a]	USA	67		3·9 (2·0–8·4)	0·16 (0·1–0·2)		1·3 (1·0–1·6)				
Lamb *et al.* (1966)		3	0·24 (0·22–0·27)	2·9 (2·2–4·2)	0·14 (0·12–0·16)						
National Academy of Sciences (1971)[c]	N. America	2		4·34 (3·13–5·60)	0·16 (0·09–0·22)	0·04 (0·03–0·07)	1·21 (0·55–1·57)	0·40 (0·22–0·57)	2·20 (2·20–2·20)		0·30 (0·30–0·30)

cont.

Table 2.59 *cont.*

Reference	Location	No. of samples	Vitamin B: Thiamine	Niacin	Riboflavin	Biotin	Pantothenic acid	Pyridoxine	Folic acid	Choline	Carotene
Passmore and Sundararajan (1941)	India	6	0·35 (0·27–0·49)								
Sur *et al.* (1955)		6	0·30 (0·28–0·33)								
Suryanarayana Rao *et al.* (1968)[d]	Various	57									0·02 (0·00–0·10)
Tanner *et al.* (1947)	USA	42		5·58 (2·77–9·19)	0·13 (0·07–0·20)	0·03 (0·01–0·05)	0·81 (0·46–1·48)	0·58 (0·21–0·86)			
Wehmeyer (1969)	Africa	79	0·48 (0·35–0·65)	4·21 (2·58–6·87)	0·20 (0·09–0·29)						
Yousif and Magboul (1972)		6	0·34 (0·23–0·43)	4·67 (3·30–6·10)	0·09 (0·06–0·12)						
Representative analytical values and ranges (not means)[e]			0·50 (0·13–0·88)	4·8 (1·9–10·2)	0·15 (0·04–0·57)		1·25 (0·33–2·42)	0·6 (0·2–1·03)	(0·02[b]–0·2[b])	0·76 (0·42–0·95)	0·13 (0·02–0·4)
Range of means			0·24–0·54	2·9–6·40	0·09–0·22	0·03–0·04	0·81–1·25	0·40–0·6	2·20	0·76	0·02–0·30
Range of ranges			0·13–0·88	1·9–10·2	0·04–0·57	0·01–0·07	0·33–2·42	0·2–1·03	0·02–2·20	0·42–0·95	0·00–0·63

[a] Microbiological assay.
[b] Described as folates.
[c] This source also reported vitamin A, IU/g, mean 3·05 range 1·10–5·00.
[d] This source also reported Provitamin A activity IU/100 g, mean 34·1, range 0·0–163·0.
[e] Data from various sources mentioned in the text.

Table 2.60
Niacin, bound niacin and niacinamide contents (mg/100 g) in samples of sorghum, pearl millet, maize and chickpea.[a]

	Niacin	Bound Niacin	Niacinamide
Sorghum	0·28	1·86	0·08
Pearl millet	0·08[b]	1·81	0·11[b]
Maize	0·17[b]	3·45	0·21
Chickpea	0·62	nil	1·21

[a] Ghosh *et al.* (1963).
[b] Amount too small to be determined accurately.

Table 2.61
Content of niacin in sorghum, pearl millet and little millet extracted by various solvents.[a]

	(A) Water in the cold		(B) Hot 0·1N HCl		(C) 0·1N boiling HCl + 1N NaOH		(D) Autoclaving 2N HCl	
	(mg/100 g)	(% of total)	(mg/100 g)	(% of total)	(mg/100 g)	(% of total)	(mg/100 g)	(% of total)
Sorghum								
RS 610 (France)	0·8	28	1·77	61	2·35	82	2·88	100
AKS 614 (France)	0·56	20	1·64	57	1·98	69	2·86	100
White (Niger)	1·68	28	3·52	60	4·86	83	5·9	100
White (Morocco)	1·2	24	2·68	58	3·8	82	4·6	100
Pearl millet								
(Niger)	0·95	31	2·1	69	2·78	91	3·05	100
(Cameroon)	0·75	40	1·42	76	1·74	93	1·88	100
Little millet	1·4	13	4·97	46	9·03	83	10·88	100

[a] Adrian *et al.* (1970).

Table 2.62
Moisture, protein content, digestibility coefficient (DC), biological value (BV) and (calculated) net available protein of cereals.[a]

Cereal	Moisture (%)	Crude prot. (%)	BV (%)	DC (%)	Available or net prot. (%)
Sorghum	11·2	10·3	83	91	7·76
Pearl millet	12·1	10·5	83	89	7·74
Finger millet	12·6	7·1	89	80	5·07
Raw polished rice	12·5	6·9	80	97	5·34
Whole wheat	12·3	12·6	66	93	7·75

[a] Swaminathan (1937a).

Table 2.63
Moisture, protein content and PER of cereals.[a]

Cereals	Moisture (%)	Crude prot. (%)	PER (4 weeks) (%)	PER (8 weeks) (%)
Sorghum	12·5	9·8	0·9	0·8
Pearl millet	12·6	10·0	1·3	1·1
Finger millet	12·6	7·4	1·0	0·7
Raw milled rice	12·9	6·3	2·2	1·7
Whole wheat	12·2	12·3	1·6	1·3

[a] Swaminathan (1937b).

Table 2.64
Total (Kjeldahl) N, protein N, and nonprotein N of sorghum and three millets, and loss of food N in digestion.[a]

Cereal	Total N (%)	Prot. (N × 6·25) (%)	Prot. N (%)	Nonprot. N (%)	Nonprot. N (as % of total N)	Loss of food N in digestion (%)
Sorghum	1·36	8·5	1·29	0·07	5·1	9
Pearl millet	2·20	13·7	2·05	0·15	6·8	11
Finger millet	1·13	7·1	1·06	0·07	6·2	20
Foxtail millet	1·64	10·2	1·53	0·11	6·7	9

[a] Swaminathan (1938a).

Table 2.65
Calcium (Ca) and phosphorus (P) intake, Ca/P ratio and % Ca and P retention of albino rats on finger millet diets.[a]

Total in diet		Ca/P ratio	Mean retention	
Ca (%)	P (%)		Ca (%)	P (%)
0·240	0·175	1·31	67·9	57·6
0·140	0·113	1·24	87·2	79·1
0·072	0·064	1·13	84·1	70·4

[a] Giri (1940).

Table 2.66
Protein efficiency ratios (PER) and biological values (BV) of sorghum and millets.[a]

	Level of protein in diet (%)	PER (range)	BV (range %)
Sorghum	5–6	0·2–0·89	83·0–83·1
	9–11	1·4–1·8	—
Pearl millet	5–6	1·2	83·0
	9–11	1·4–1·8	—
Finger millet	5–6	0·7–1·0	89·0–90·5
	9–11	1·99	—
Foxtail millet	5–6	—	77·0
	9–11	0·5–0·8	—
Common millet	9–11	1·2	56·0
Little millet	9–11	0·9–1·1	—

[a] Kuppuswamy *et al.* (1958) as cited by Narayana Rao *et al.* (1961).

Table 2.67
***In vitro* digestibility of sorghum, wheat and rice grown in South Gujarat.**[a]

	Sorghum	Wheat	Rice
Total prot. (%)	10·0	16·2	8·81
Amino N (mg/100 g flour)			
0·2% pepsin			
1h	156·5	171·5	126·0
2h	168·0	185·6	136·5
3h	182·0	203·0	157·5
0·5% pepsin			
1h	325·5	350·0	262·5
2h	385·0	409·6	273·0
3h	406·1	455·0	308·0
1·0% pepsin			
1h	217·0	245·0	192·5
2h	241·5	273·0	199·5
3h	255·5	296·5	227·5
0·5% pepsin 3h + 0·25 g trypsin 3h Total 6h	432·5	483·5	385·0
0·5% pepsin 3h + 0·25 g trypsin 3h + 0·1 g erepsin 2h Total 8h	855·0	1225·0	700·0
Prot. digested (%)	8·55	7·56	7·95

[a] Vasi and Desai (1976).

Table 2·68
True digestibility (TD) for pigs and rats of total N and the single amino acid, for casein, maize and sorghum.[a]

	Casein	Maize	Sorghum
TD for total N (pigs)	99·4	90·2	85·3
TD for total N (rats)	101·1	87·6	91·5
TD for			
Ile	99·1	88·5	83·3
Leu	99·3	92·8	87·0
Lys	99·7	89·3	71·5
Met	100·3	93·6	84·6
Cys	100·1	91·4	78·6
Phe	98·8	92·2	85·4
Tyr	98·7	93·0	86·5
Thr	98·8	89·8	82·7
Val	99·1	89·9	84·8
Arg	99·2	95·2	87·2
His	99·7	93·5	85·2
Ala	97·9	93·0	87·2
Asp	98·9	91·1	83·0
Glu	99·6	92·2	88·1
Gly	97·7	89·6	77·7
Ser	100·9	96·9	90·4

[a] Eggum (1970).

Table 2.69
Effect of nutritional supplements to cereal grains on 35-day weight gain in rats.[a]

	35-day weight gain (g)				
	Rice	Wheat	Maize	Sorghum	Millet
Protein (%)	6·9	15·0	10·1	9·7	11·6
Diets					
(A) Basal + 4% maize starch	5	52	11	23	22
(B) Basal + 4% salt mixture[b] replacing maize starch	20	92	39	39	26
(C) Basal + vitamin mix[c]	44	67	31	29	20
(D) Lysine HCl					
(i) 0·05% rice	5				
(ii) 10·1% wheat and maize		36	18		
(iii) 0·2% sorghum				42	
(iv) 0·3% millet					30
(E) Basal + salts and vitamins	42	97	33	37	24
(F) Basal + salts and Lys (as D)	24	131	57	96	100
(G) Basal + vitamins and Lys (as D)	46	73	49	56	70
(H) Basal + salts + vitamins + Lys (as D)	68	130	61	128	132

[a] Howe and Gilfillan (1970).
[b] Hegsted *et al.* (1941).
[c] Howe and Bosshardt (1960).

Table 2.70
Amino acid composition in sorghum composites of protein 8·4% and 11·8% (moisture content of 8·4% and 8·2% respectively).[a]

	Low-prot.		High-prot.	
Amino acids	(mg/100 g sample)	(mg/g total N)	(mg/100 g sample)	(mg/g total N)
Ile	310	251	460	247
Leu	1040	845	1700	908
Lys	190	155	210	113
Met	90	76	140	73
Half-Cys	111	90	180	97
Phe	390	315	600	320
Tyr	200	160	330	177
Thr	250	204	350	185
Val	390	313	570	304
Arg	270	221	350	188
His	180	142	250	131
Ala	710	571	1150	616
Asp	510	416	750	402
Glu	1630	1318	2570	1372
Gly	250	203	320	169
Pro	640	521	960	512
Ser	340	274	490	259

[a] Waggle *et al.* (1966).

Table 2.71
Composition of diets and average weekly weight gain of rats, over 4 weeks fed high- and low-protein sorghum diets.[a]

Diets	(%)	Prot. value of diet (%)	Average weekly weight gain (g)
(A) Low-prot. sorghum	96		
Gly	0·073	7·6	17·6
Glu	0·073		
(B) High-prot. sorghum	64·2		
Starch	31·65		
Na_3PO_4	0·15	7·6	4·4
Gly	0·073		
Glu	0·073		
(C) High-prot. sorghum	64·2		
Starch	31·65		
Na_3PO_4	0·15		
Gly	0·039	7·6	11·0
Lys	0·048		
His	0·010		
Arg	0·030		
(D) High-prot. sorghum	96		
Gly	0·073	11·5	19·4
Glu	0·073		
(E) Low-prot. sorghum	87·5		
Casein	8·7	14·0	117·8
(F) High-prot. sorghum	58·5		
Starch	28·85	14·0	99·4
Casein	8·7		
Na_3PO_4	0·15		

[a] Waggle *et al.* (1966).

Table 2.72
Body and organ weights of rats fed (A) maize, (B) rice, (C) wheat, (D) cassava and (E) sorghum diets compared to (F) stock diet.[a]

	(A) Maize meal	(B) Rice flour	(C) Wheat flour (white)	(D) Cassava flour	(E) Sorghum Kafir 44 14	(F) Stock diet
3-day forced feeding experiment						
No. of rats	20	19	16	15	9	22
Body weight gain (g)	+7·1	+6·2	+6·9	−0·8	+2·3	—
Liver (g)	3·43[c]	3·47[c]	3·33	2·98	3·27	2·98
Pancreas (mg)	330[c]	379	415	381	398	381
Submaxillary gland (mg)	64[c]	68[b]	69[b]	47[c]	75	78
Kidney (mg)	331[c]	344[b]	332	346	345	372
Gastrocnemius muscle (mg)	350[c]	392	376	346[c]	385	401
Spleen (mg)	231[c]	308	247[c]	185[c]	269[c]	373
Testis (mg)	435	457	467	410	492	403
7-day *ad lib.* experiment						
No. of rats	5	5	5	5	5	3
Body weight gain (g)	+4·6	+5·9	+2·4	−6·5	+2·1	+26·3
Liver (g)	3·11[c]	3·07[c]	2·94[c]	2·52[c]	2·82[c]	4·02
Pancreas (mg)	330[b]	379	415	381[b]	398	381
Submaxillary gland (mg)	64[c]	74	57[c]	52[c]	57[c]	99
Kidney (mg)	298[c]	298[c]	282[c]	274[c]	304[c]	455
Gastrocnemius muscle (mg)	270[c]	335[b]	315[c]	247[c]	295[c]	400
Spleen (mg)	229[b]	240[b]	248[b]	176[c]	243[b]	710
Testis (mg)	436	489	519	376	476	551

[a] Sidransky (1960).
[b] Significance between P 0·01 and 0·05.
[c] Significance < P 0·01.

Table 2.73
Total liver lipid, glycogen and protein of rats fed (A) maize, (B) rice, (C) wheat, (D) cassava and (E) sorghum diets compared to (F) stock diet.[a] [b]

	(A) Maize meal	(B) Rice flour	(C) Wheat flour (white)	(D) Cassava flour	(E) Sorghum Kafir 44 14	(F) Stock diet
3-day forced-feeding experiment						
No. of rats	19	18	16	15	8	18
Total liver lipid mg/liver	259[d]	210[d]	200[d]	208[d]	195[d]	130
Liver glycogen mg/liver	188[d]	164[d]	157[d]	103[d]	124[d]	3
Liver protein mg/liver	451[d]	498[c]	500[c]	414[d]	566	572
7-day *ad lib.* experiment						
No. of rats	5	5	5	5	5	3
Total liver lipid mg/liver	113[d]	164	153	123[d]	171	172
Liver glycogen mg/liver	317	337	323	247	250	141
Liver protein mg/liver	372[d]	384[d]	363[d]	308[d]	379[d]	768

[a] Sidransky (1960).
[b] Authors' note: the original paper shows no indication of significance for liver glycogen.
[c] Significance between P 0·01 and 0·05.
[d] Significance P < 0·01.

Table 2.74
Summary of reproduction data for rats fed three sorghums and wheat.[a]

	Wheat		Shallu		Martin		Kafir	
Generation of female bred	P-1	F-1	P-1	F-1	P-1	F-1	P-1	F-1
Generation of offspring	F-1	F-2	F-1	F-2	F-1	F-2	F-1	F-2
No. of females	10	20	10	20	10	20	13	20
Total No. of litters from two breedings	19	32	18	20	18	23	19	8
No. of litters with survival	14	23	9	9	17	17	12	4
Total young born	199	310	175	173	169	208	179	65
Total young dead	70	128	109	108	42	107	99	41
Young dead in complete litter loss	56	88	89	88	15	52	70	31
Total young surviving at 3 weeks	129	182	66	65	127	101	80	24
Av. young born/litter	10·5	9·7	9·7	8·7	9·4	9·1	9·4	8·1
Av. size of surviving litters	10·2	9·7	9·6	9·5	9·1	9·2	9·1	8·5

[a] Lamb *et al.* (1966).

Table 2.75
Food intake, weight gain and body composition of rats on 10 sorghum grain diets and casein.[a]

	IS numbers										Casein	
	0083	0182	0948	0957	1220	2031	2232	3472	3552	8231	Prot. 10%	Prot. 20%
Digestibility (%)	60·2	60·3	50·1	74·6	67·2	74·3	77·3	48·9	76·8	52·9	87·1	88·1
Food intake g/day	8·5	8·1	8·4	8·5	9·6	10·1	8·6	7·2	10·0	8·3	11·8	12·8
N intake mg/day	0·13	0·13	0·13	0·14	0·14	0·16	0·14	0·11	0·16	0·13	0·19	0·41
Daily gain (g)	0·8	0·8	0·5	0·8	1·1	1·9	0·9	0·3	1·9	0·7	3·5	5·7
Body N stored (g)	0·42	0·32	0·22	0·45	0·53	0·95	0·43	0·11	0·95	0·32	2·01	3·50
Dry liver (fat free) (g)	0·5	0·5	0·6	0·7	0·8	0·8	0·6	0·5	0·8	0·5	1·2	1·9
Total N (mg)	72	68	66	95	81	102	80	59	104	68	161	247
Lipid (%)	26·7	25·0	30·5	22·4	25·0	19·7	22·4	40·5	22·1	33·9	14·0	11·3

[a] Nawar *et al.* (1970).

Table 2.76
Relative nutritive value (RNV) of protein and net available protein in 10 sorghum lines (ranges in parentheses).[a]

Sorghum (IS number)	Prot. content (%)	RNV (%)	Net available prot. (%)
0271	11·7	30	3·5
0718	10·4	21	2·2
0948	9·3	32	3·0
0957	12·4	34	4·2
1220	9·8	31	3·0
3796	10·4	39	4·1
4328	12·3	32	3·9
88	11·8	30	3·5
8120	12·8	26	3·3
8382	11·1	31	3·4

[a] Nawar *et al.* (1970).

Table 2.77
Diets and results in rat feeding trials Experiment 2 sorghum Martin B 398.[a]

Diets (%)	(A)	(B)	(C)	(D)
Martin B 398	88·55	79·45	77·30	—
Maize starch	—	—	—	34·15
Sucrose	—	—	—	34·15
Casein	—	—	11·25	20·25
AA[b] mixture	—	7·60	—	—
Constant nonprot. ingredients	11·45	11·45	11·45	11·45
$NaHCO_3$	—	1·50	—	—
Rat feeding trials				
Wt gain (g)	0·58	3·00	4·75	4·41
Feed intake (g)	8·3	10·0	12·2	10·4
Feed efficiency g gain/g feed	0·07	0·30	0·40	0·42
True prot. digestibility (%)	86·1	89·4	92·1	100·0
Nonprot. organic matter digestibility (%)	91·6	91·9	91·5	99·0

[a] Breuer and Dohm (1972).
[b] Formulated to provide in mg/100 g diet, DL-Met 300, L-Lys 1000, L-Cys 200, L-Trp 200, L-Thr 500, L-Val 500, L-Ile 500, L-Glu acid 4000, L-Asp acid 400.

Table 2.78
Composition of sorghum diets in rat feeding trials.[a]

Ingredients (%)	Sorghum prot. content (%)	Diets (A)	(B)	(C)	(D)	(E)	(F)	(G)
IS 0819	11·7	81·16	—	—	—	—	—	—
IS 2190	14·1	—	67·42	—	—	—	—	—
IS 2197	11·7	—	—	81·27	—	—	—	—
IS 2478	12·2	—	—	—	77·74	—	—	—
IS 2481	12·5	—	—	—	—	75·82	—	—
IS 2486	11·6	—	—	—	—	—	81·97	—
RS 610	9·7	—	—	—	—	—	—	92·88
Cerelose		15·94	29·58	15·73	19·26	21·18	15·03	4·12
Constant (nonprot.) ingredients		3·00	3·00	3·00	3·00	3·00	3·00	3·00
Analysed values prot. (%)		9·5	9·5	9·5	9·5	9·5	9·5	9·5
Gross energy (kcal/g)		3·74	3·69	3·78	3·87	3·71	3·64	3·89
Experiment I								
Daily wt gain (g)		1·1	0·6	0·7	1·1	0·6	0·6	0·5
Feed efficiency gain/feed		0·11	0·06	0·08	0·10	0·06	0·06	0·06
Experiment II								
Dry matter digestibility (%)		83·2	85·2	85·2	84·8	85·1	84·5	85·3
Crude prot. digestibility (%)		71·3	72·6	73·7	71·5	73·7	74·5	70·7
N intake retained (%)		10·3	9·1	9·3	10·9	9·0	8·1	7·9

[a] Ilori and Conrad (1976).

Table 2.79
Mean metabolizable energy (ME) and feed efficiency of chick diets in which the cereals replaced 25% of glucose monohydrate in the reference diet.[a]

	Reference diet	Maize	Sorghum	Millet
Wt gain[b] (g)	301	303	309	307
Feed efficiency[b] F/G	2·1	2·1	2·1	2·2
ME of test diets (MJ/kg DM)	16·6	16·4	16·2	16·2
Energy value of substituted cereals (MJ/kg DM)				
Gross energy (GE)	—	18·6	18·3	18·6
ME	—	14·9	14·0	14·1
ME_n	—	14·1	13·3	13·3
ME % of GE	—	79·8	76·3	76·1

[a] Fetuga (1977).
[b] Observations on 2-week old chicks over 14 days.

Table 2.80
Average adjusted body moisture content/group (g) (as criterion of N retention) of chicks fed diets of DC 500 F and Excell 505 sorghum, (a) unsupplemented and (b) supplemented with amino acids.[a]

	DC 500 F	Excell 505	Mean	Difference
(N range in diet 1·45–1·86%)				
No supp.	156·7	174·6	165·7	17·9[c]
L-Lys HCl 0·3%	160·2	178·8	169·5	18·6[c]
L-Arg HCl 0·4%	162·0	169·3	165·7	7·3
DL-Met 0·2%	167·1	176·1	171·6	9·0[b]
L-Trp 0·1%	163·9	176·4	170·2	12·5[c]
DL-Thr 0·2%	157·0	177·1	167·1	20·1[c]
Gly 0·4%	158·8	175·3	167·1	16·5[c]

[a] Wessels (1970a).
[b] Significant at $P < 0·05$.
[c] Significant at $P < 0·01$.

Table 2.81
Performance of White Leghorn and Rhode Island Red chicks on various cereal-based diets.[a]

Cereal	Time taken to gain slaughter wt (days)	Total wt gain (kg)	Total feed consumed (kg)	F/G quotient
White Leghorn				
Maize	105	1·29	6·11	4·73
Barley	105	1·33	6·43	4·83
Sorghum	105	1·35	6·25	4·63
Pearl millet	98	1·31	5·22	3·98
Rhode Island Red				
Maize	98	1·35	5·36	3·98
Barley	98	1·42	6·15	4·33
Sorghum	98	1·36	5·47	4·02
Pearl millet	98	1·48	5·52	3·72

[a] Singh and Barsaul (1977).

Table 2.82
Composition of basal diets, (A) maize, (B) sorghum, used in chick feeding trials.[a]

	Trial 1		Trial 2		Trial 3	
	(A)	(B)	(A)	(B)	(A)	(B)
Grain	66·0	58·0	69·0	66·0	65·0	59·0
Glucose	—	8·0	—	3·0	—	6·0
Soybean meal	26·0	26·0	23·0	23·0	27·0	27·0
Vitamins and minerals	8·0	8·0	8·0	8·0	8·0	8·0
Prot. content						
Total (assayed)	18·1	18·1	16·7	17·0	17·4	17·8
Cereal-derived (calculated)	5·6	5·6	6·3	6·3	5·3	5·3

[a] Bornstein and Lipstein (1971).

Table 2.83
Proximate analysis of sorghum grains used in chick feeding trials.[a]

	Moisture (%)	Prot. (%)	Fat (%)	NFE (%)	Crude fibre (%)	Ash (%)	cal/lb metabolizable energy
Milo (average value)	3·0	10·5	2·5	2·0	70·5	11·5	1480
Y 8	3·2	8·4	2·7	1·7	73·0	11·0	1548
Y 10	2·8	10·4	2·6	1·6	71·3	11·2	1535
Redlan	2·8	11·6	2·6	1·3	71·1	10·5	1549
DeKalb E 56 A	2·9	7·2	2·7	1·2	74·4	11·6	1521

[a] Ozment *et al.* (1963).

Table 2.84
Availability, mean and range, of amino acids in 24 varieties of sorghum as evaluated by the chick.[a]

	Mean	Range
Ile	83·5	50·9 (Ark 61002)–95·2 (TE 68)
Leu	86·4	48·3 (Ark 61002)–95·7 (TE 68)
Lys	81·9	58·3 (RS 617)–92·5 (TE 68)
Met	88·6	55·1 (Ark 61002)–98·8 (RS 624)
Cys	91·1	67·5 (RS 617)–98·3 (RS 640)
Phe	84·1	43·7 (Ark 61002)–95·4 (TE 68)
Tyr	85·8	48·6 (Ark 61002)–95·0 (TE 68)
Thr	77·9	46·8 (Ark 61002)–91·8 (TE 68)
Val	82·4	52·4 (Ark 61002)–94·6 (TE 68 DeKalb F 63)
Arg	85·2	59·5 (Ark 61002)–95·5 (TE 68)
His	77·3	43·4 (Ark 61002)–90·2 (TE 68)
Ala	85·4	53·9 (Ark 61002)–94·2 (TE 68)
Asp	84·8	56·3 (RS 617)–94·5 (TE 68)
Glu	87·9	53·4 (Ark 61002)–96·1 (TE 68)
Gly	75·8	40·8 (Ark 61002)–90·7 (AK 300 1R)
Pro	78·3	19·4 (Ark 61002)–93·3 (RS 640)
Ser	81·6	42·6 (Ark 61002)–93·4 (TE 68)

[a] Stephenson *et al.* (1971).

Table 2.85
Proximate analysis of maize, sorghum and millet feeds (DWB) with digestibility coefficients (DC), gross energy (GE), digestible energy (DE) and metabolizable energy (ME) when fed to growing pigs.[a]

	Maize	Sorghum	Millet
Dry matter (%)	91·8	89·7	92·4
Crude prot. (%)	9·8	11·2	9·3
Ether extract (%)	3·8	3·4	4·1
Crude fibre (%)	2·2	2·9	2·4
NFE (%)	82·6	80·1	82·8
Ash (%)	1·6	2·4	1·4
Ca (mg/100 g)	40	70	400
P (mg/100 g)	340	380	230
Experiment I DCs			
Dry matter	93·5	90·7	89·8
Crude prot.	91·8	89·4	89·8
Ether extract	65·4	62·9	63·4
Crude fibre	64·6	61·3	60·7
NFE	94·8	93·5	91·8
Total digestible nutrients	94	92	92
Experiment II			
GE	4346	4284	4146
DE	4042	3812	3943
ME	3825	3711	3752
ME_n	3731	3627	3666

[a] Oyenuga and Fetuga (1975).

Table 2.86
Effect on body weight gain, feed intake, F/G quotient, digestibility, N retention, digestibility of energy (DE) and metabolizable energy (ME) of growing-finishing diets for swine containing maize (M) and sorghum (S).[a]

	Commercial control	Diets (A) (%)	(B) (%)	(C) (%)	(D) (%)
		M 100 S 0	M 75 S 25	M 50 S 50	M 0 S 100
Wt gain (kg)	63	70	68	70	69
Feed intake (kg)	226	213	215	228	227
F/G	3·6	3·0	3·2	3·2	3·3
Digestibility (%)					
DM	78·6	87·3	86·3	87·3	87·2
Protein	81·5	86·1	86·3	86·8	85·2
Fat	73·1	77·7	59·6	66·6	61·0
Fibre	39·2	41·3	47·5	58·0	54·8
NFE	86·0	92·4	92·3	92·1	93·1
N retention	37·0	44·2	39·6	42·1	36·4
DE kcal	2908	3477	3417	3451	3441
(as %)	—	(100)	(98·3)	(99·3)	(99·0)
ME kcal	2250	2917	2847	2842	2829
(as %)	—	(100)	(97·6)	(97·4)	(97·0)

[a] Han and Ha (1977).

Table 2.87
Effect of feeding Opaque-2 maize and sorghum by three methods and in two forms to finishing swine.[a]

	Opaque-2 maize		Sorghum	
	Meal	Pellet	Meal	Pellet
Experiment 1				
Ad lib. feeding				
Av. daily gain (kg)	0·71	0·68	0·63	0·68
Av. daily feed (kg)	2·28	2·11	2·55	2·64
Gain/feed	0·311	0·322	0·247	0·296
Experiment 2				
Group-fed				
Av. daily gain (kg)	0·79	0·85	0·75	0·85
Av. daily feed (kg)	3·04	2·92	3·01	2·95
Gain/feed	0·260	0·291	0·249	0·288
Individually-fed				
Av. daily gain (kg)	0·77	0·75	0·74	0·79
Av. daily feed (kg)	2·82	2·51	2·92	2·91
Gain/feed	0·274	0·301	0·255	0·271

[a] Jensen *et al.* (1969).

Table 2.88
Chemical analysis (%) of six sorghum varieties (as-is basis) and summary of results of pig feeding trials.[a]

	DeKalb F 62 A	Kafir 44 14	Darset	RS 610	Redlan	Amak R 12
Moisture	9·5	9·0	9·0	8·6	8·9	8·6
Prot.	11·9	11·3	11·2	11·6	12·6	11·9
Fat	2·9	3·3	4·0	3·3	2·1	2·4
NFE	72·1	72·9	72·5	73·1	73·0	73·4
Fibre	1·9	1·8	1·7	1·7	1·9	1·9
Ash	1·7	1·6	1·5	1·7	1·5	1·7
Feeding trials						
Av. daily gain (lb)	1·66	1·66	1·64	1·72	1·71	1·70
Av. feed/lb gain						
Sorghum	3·192	3·145	3·364	3·146	3·210	3·149
Prot. supplement	0·597	0·536	0·652	0·607	0·522	0·554
Total	3·79	3·68	4·02	3·75	3·73	3·70
Av. % prot. in ration as consumed	15·73	14·97	15·17	15·51	15·85	15·54

[a] Hillier *et al.* (1959).

Table 2.89
Proximate analyses and gross energy content of diets containing sorghums differing in endosperm texture, floury (F), intermediate (I) and corneous (C), and starch type, normal (N) and waxy (W) (10% moisture basis) with results of feeding trials on pig performance.[a]

	(A) NSA 740 (F–N)	(B) TX 3197 (I–N)	(C) SC 301 6 (C–N)	(D) BTX 615 (I–W)
Prot. (N × 6·25) (%)	21·1	16·7	18·4	16·5
Ether extract (%)	3·6	3·2	3·3	2·8
Crude fibre (%)	3·7	2·9	2·9	3·2
Ash (%)	5·1	4·4	4·5	4·4
Gross energy (kcal/g)	3·95	3·94	3·91	3·92
Digestible energy (kcal/g)	3·55	3·70	3·60	3·68
Metabolizable energy (unadjusted kcal/g)	3·46	3·62	3·51	3·60
(N-corrected kcal/g)	3·36	3·53	3·40	3·51
N retention (g/day)	26·64	26·81	32·20	26·58
Gain (kg/day)	0·83	0·96	0·93	1·07
Feed/gain	2·87	2·63	2·69	2·35

[a] Cohen and Tanksley, Jr (1973a).

Table 2.90
Description and 48 h Nylon bag dry matter digestibility (NBDMD) of eight varieties of sorghum grown in Puerto Rico in 1966.[a]

Varieties	Kernel description	% floury endosperm	% NBDMD	
			Ground	Whole
IS 1309	Dented	90	67·6	32·7
IS 3515	Tortoise-shape	75	66·5	20·7
IS 12577	Round	95	62·1	3·8
IS 3579	Large round	80	61·0	32·4
IS 6269	Yellow endosperm	40	56·5	17·6
RS 626	Round	30	54·4	20·9
SA 5875 6 1	Round, waxy endosperm	30	53·4	17·6
B TX 406	Round	40	52·3	16·2

[a] Miller *et al.* (1972).

Table 2.91
Description and 72-hour NBDMD of 22 varieties of sorghum grown in Puerto Rico in 1967.[a]

Varieties	Seed colour	% floury endosperm	% NBDMD	
			Ground	Whole
IS 8893	White	70	80·6	9·1
IS 3515	White	75	77·3	10·1
IS 9086	White	95	76·3	8·8
IS 9021	Purple	50	72·7	2·6
IS 8549	Purple	90	71·7	6·0
IS 8874	White	85	71·1	6·8
IS 8899	White	80	70·1	6·9
IS 8918	Purple	60	69·7	3·6
B TX 3197 × IS 1309	Purple	50	69·3	2·8
IS 1309	Purple	70	69·2	19·5
B TX 3197 × IS 3579	Brown	50	68·9	5·3
IS 9034	Purple	95	68·9	4·6
IS 3579	White	65	68·5	11·8
IS 8188	Purple	95	67·7	6·8
IS 8580	Purple	80	66·7	1·3
B TX 3197 × IS 3515	Brown	50	66·3	1·3
IS 8880	Purple	90	65·9	1·6
IS 8852	Purple	50	64·3	2·2
TX 414	Red	50	62·9	8·7
B TX 3197	White	50	62·3	8·5
IS 7173	White	0	60·0	2·5
B TX 406	Red	50	56·2	10·9

[a] Miller *et al.* (1972).

Table 2.92

Yield, pericarp colour, endosperm description, proximate composition and feed efficiency for steers of nine grain sorghums.[c]

	Funks G 766W	ACCO R 109	Northrup King 275	Northrup King 280	RS 671	Corpustar CP 622	Funks G 522	Northrup King X 4087	Funks 3135
Pericarp colour	White	Red-bronze	Red	Red	Red	White	Red-bronze	White	Red
Endo. descr.	Yellow	Yellow	White 0·25 waxy[b]	White 0·25 waxy[b]	White	White All waxy[b]	Yellow	Yellow	White 0·67 waxy[b]
Yield, bu/ac	115	133	115	114	127	90	120	118	99
Proximate analysis, DWB (%)									
Crude prot.	10·3	10·3	10·8	9·9	11·2	11·2	10·7	10·3	10·9
Ether extract	3·3	3·0	3·1	3·1	3·0	3·0	3·2	3·0	3·2
Crude fibre	2·2	2·1	2·1	2·0	1·9	1·9	1·8	2·2	2·1
NFE	82·5	82·0	82·3	83·4	82·2	82·3	82·8	82·7	81·9
Ash	1·6	1·6	1·6	1·6	1·6	1·6	1·6	1·7	1·9
Gross energy (kcal/g)	4·459	4·513	4·349	4·484	4·446	4·425	4·482	4·424	4·508
Feed efficiency	7·24	7·34	7·95	7·90	8·34	7·27	7·29	8·03	8·08

[a] McCollough *et al.* (1972a).

[b] Waxy refers to amylopectin type of starch (α 1–4, α 1–6 linkage).

Table 2.93

Pericarp colour, endosperm description, proximate composition, mineral matter, gross energy (GE) and feed efficiency (FE) of eight grain sorghum hybrids and three maize hybrids.[a]

	Sorghum								Maize		
	Funks G 766W	ACCO R 109	DeKalb E 57	Northrup King 222	RS 671	ACCO[b] 1023	Asgrow Jumbo C	Funks 3135	Funks 2438	Hulting X9770	Funks 24554
Pericarp colour	White	Red-bronze	Red	Red-bronze	Red	Dark red	Red	Red	Yellow	Yellow	Yellow corn
Endo. descr.	Yellow	Yellow	Yellow	Yellow	White	White	White	White 0·67 waxy[c]	High-oil	Yellow dent	High-Lys Opaque-2 (floury endo.)
Proximate composition (DWB) (%)											
Crude protein	10·7	11·0	10·7	12·4	11·8	14·1	11·8	12·0	12·2	10·7	10·9
Ether extract	3·15	2·85	2·32	2·34	2·45	3·09	2·84	2·85	5·65	3·66	5·22
NFE	82·4	82·7	83·5	82·0	81·8	78·3	82·3	80·2	78·1	81·6	79·5
Crude fibre	1·76	1·64	1·70	1·51	1·96	2·40	1·49	2·91	2·20	2·21	2·41
Ash	1·94	1·75	1·80	1·70	1·97	2·16	1·56	2·12	1·80	1·79	1·96
Mineral matter (mg/100 g)											
Ca	29	36	43	29	35	37	44	77	26	31	26
P	383	345	380	327	301	376	308	373	382	367	358
Fe	5·59	7·02	6·07	5·86	6·62	6·62	7·40	7·98	6·44	6·87	5·84
Mg	136·6	144·2	172·3	144·1	138·2	141·3	153·9	189·8	145·6	138·8	133·5
Mn	0·82	0·76	0·87	1·17	1·12	1·11	1·10	1·21	1·05	1·02	0·86
Zn	2·69	2·20	2·23	2·01	7·03	7·67	7·04	6·79	7·73	3·42	7·45
GE (kcal/g)	4·635	4·635	4·668	4·608	4·517	4·734	4·516	4·406	4·742	4·439	4·484
FE	8·07	8·54[d]	7·68[d]	8·32[d]	8·84[d]	10·21[d]	9·21[d]	6·95[e]	8·96[d]	7·75[d]	6·75[e]

[a] McCollough (1972).
[b] Bird-resistant.
[c] Waxy refers to amylopectin type of starch; 0·67 of the contribution of waxy gene from waxy parents.
[d] Means with different superscripts differ significantly $P < 0{\cdot}05$.
[e] Feed efficiencies not significantly different from one another but fed under different conditions from the remaining samples.

Table 2.94
Yield, protein and moisture for laboratory milled sorghum and pearl millet samples.[a]

Fractions	Sorghum			Pearl millet		
	Yield (%)	Moisture (%)	Prot. (N × 6·25) (%)	Yield (%)	Moisture (%)	Prot. (N × 6·25) (%)
Whole grain		10·0	9·6		10·6	12·3
Bran	46·9	11·5	11·5	42·0	11·2	16·4
Low grade flour	14·8	12·2	9·3	16·6	12·2	11·3
+9 xx flour	33·1	12·5	10·0	31·1	12·6	9·2
−9 xx flour	6·3	12·1	5·9	10·3	12·4	7·2
Total extraction (Whole grain minus bran)	54·2			58·0		

[a] Badi *et al.* (1976).

Table 2.95
Amino acid composition of laboratory milled fractions of pearl millet.[ab]

	Fractions			
	Bran (mg/g total N)	Low-grade flour (mg/g total N)	+9 xx flour (mg/g total N)	−9 xx flour (mg/g total N)
Ile	250	256	262	262
Leu	569	656	656	650
Lys	256	162	150	169
Met	131	144	156	175
Half-Cys	181	206	219	175
Phe	306	331	337	331
Tyr	212	212	219	225
Thr	256	244	244	244
Val	337	331	319	344
Arg	437	306	262	281
His	162	144	137	144
Ala	487	500	506	487
Asp	544	475	475	494
Glu	1112	1281	1306	1281
Gly	256	187	175	181
Pro	350	412	412	406
Ser	306	294	294	287
Ammonia	106	112	119	125

[a] Based on recovery Kjeldahl protein; bran 94·97: low grade flour 101·43; +9 xx flour 100·28; and −9 xx flour 96·65.
[b] Badi *et al.* (1976).

Table 2.96
Chemical constants and mean (ranges in parentheses) of 25 lines of pearl millet flour lipid.[a]

Group	Total lipid (% of dry matter)	Acid value	Saponification value	Iodine value
African	5·27	5·95	277·7	85·5
	(4·95–5·70)	(4·75–6·63)	(200·9–317·8)	(74·1–97·9)
American	5·74	6·03	274·6	89·6
	(4·85–6·70)	(2·33–11·95)	(200·9–331·1)	(73·7–101·8)
Indian inbreds	5·59	5·59	265·6	85·0
	(5·35–5·92)	(2·20–10·91)	(190·1–311·5)	(58·4–101·2)
Indian varieties	5·31	5·47	242·9	73·7
	(4·22–6·77)	(2·33–7·68)	(206·1–288·2)	(58·9–88·2)
Hybrids	6·02	3·00	256·6	97·4
	(4·65–7·40)	(2·02–5·82)	(216·4–302·8)	(68·4–107·9)
Pooled av.	5·62	5·13	263·5	86·6
	(4·22–7·40)	(2·02–11·95)	(190·1–331·1)	(58·4–107·9)

[a] Sharma and Goswami (1969).

Table 2.97
Fatty acid composition of oil extracted from 65 pearl millet lines of differing origin grown at Tifton, Georgia, single values or mean (ranges in parentheses).[a]

	Fatty Acids (%) total lipid						
Origin	Palmitic	Palmitoleic	Stearic	Oleic	Linoleic	Linolenic	Arachidic
Nigeria (31 lines)	20·2	0·58	4·14	26·1	44·8	3·5	0·71
	(17·7–23·2)	(0·32–0·82)	(1·96–5·22)	(21·3–30·6)	(40·3–50·0)	(2·3–4·6)	(0·48–0·91)
Sudan (one line)	23·3	0·57	4·64	27·1	40·4	3·31	0·73
Ethiopia (one line)	23·4	0·83	2·85	24·3	44·1	4·17	0·32
S. Africa (five lines)	21·4	0·51	2·66	24·6	45·6	4·73	0·57
	(16·7–25·0)	(0·37–0·62)	(1·83–4·36)	(23·2–30·3)	(43·3–48·9)	(4·01–5·80)	(0·32–0·77)
India (26 lines)	20·5	0·50	3·92	25·5	45·2	3·70	0·58
	(18·8–23·0)	(0·36–0·79)	(2·00–7·95)	(20·2–29·5)	(41·2–50·3)	(2·41–4·89)	(0·30–0·95)
USA (one line)	19·3	0·37	2·56	21·2	51·7	4·41	0·52

[a] Jellum and Powell (1971).

Table 2.98
Proportions and 1000-seed weight of sieve fractions of three races of pearl millet.[a]

	Fractions				
	< 1·0 mm	1·0–1·5 mm	1·5–2·0 mm	2·0–2·5 mm	> 2·5 mm
1000-weight (g)					
Zongo	1·55	2·99	5·56	9·92	14·46
P3 Kolo	1·60	3·00	6·04	9·36	14·20
Tamangagi		3·17	6·18	9·68	15·52
Mean	1·58	3·05	5·93	9·65	14·73
Distribution of fractions (% of total sample)					
Zongo	2·7	12·3	51·8	22·8	10·4
P3 Kolo	2·3	11·8	48·3	35·1	2·5
Tamangagi		5·8	29·85	51·15	5·1
Mean	2·5	10·0	43·3	36·3	6·0

[a] Samson and Adrian (1971).

Table 2.99
Planimeter values of longitudinal cuts of large and small pearl millet grain, and wheat.[a]

	Millet				Wheat	
	Small grain (3·0 mg) Surface[b] area	(%)	Large grain (14·5 mg) Surface[b] area	(%)	Surface[b] area	(%)
Endo.	13·2	53	27·6	58	57·2	86
Germ	6·3	25	8·2	17	5·75	8·5
Scut.	5·35	22	11·9	25	3·8	5·5
Total	24·85	100	47·7	100	66·75	100

[a] Samson and Adrian (1971).
[b] cm^2—magnification × 26.

Table 2.100
Means of proximate composition, B vitamins, and amino acids in sieve fractions of three races of pearl millet with % of fraction (in parentheses).[a]

	Fraction			
	(1) 1·0–1·5 mm	(2) 1·5–2·0 mm	(3) 2·0–2·5 mm	(4) >2·5 mm
Mean 1000—seed weight (g)	3·05	5·93	9·65	14·73
Moisture (%)	7·5	7·3 (97)	7·1 (95)	7·7 (102)
Prot. (N × 5·83) (%)	7·9	8·4 (107)	8·3 (105)	9·2 (117)
Formic acid insol. (%)	4·2	3·0 (71)	2·95 (70)	3·0 (71)
Pentosans (%)	4·21	3·54 (84)	3·19 (76)	3·00 (71)
Formic acid insol + pentosans	8·41	6·54 (78)	6·14 (73)	6·00 (71)
Ash (%)	1·67	1·40 (84)	1·29 (77)	1·35 (81)
B vitamins (mg/100 g)				
Thiamine	0·34	0·38 (112)	0·37 (109)	0·44 (129)
Niacin	1·59	1·50 (94)	1·62 (102)	1·78 (112)
Riboflavin	0·27	0·25 (92)	0·24 (89)	0·23 (85)
Pantothenic acid	1·04	0·88 (82)	0·82 (79)	0·80 (77)
Amino acids (mg/g total N)				
Ile	247	234 (95)	269 (109)	295 (119)
Leu	470	455 (97)	539 (115)	591 (126)
Lys	189	162 (86)	170 (89)	136 (72)
Met	121	117 (96)	131 (108)	129 (107)
Thr	207	189 (91)	216 (104)	224 (108)
Trp	69	68 (98)	76 (111)	90 (132)
Val	312	293 (94)	342 (110)	368 (118)

[a] Samson and Adrian (1971).

Table 2.101
Chemical composition of four pearl millets.[a]

	HB 3	HB 4	HB 1	Local variety
Grain yield (kg/ha)	2539	3296	2101	1514
100-seed weight (g)	7·4	9·1	6·7	6·2
Grain colour	Slate grey with green tinge	Greenish yellow	Slate grey	Slate grey
Prot. (%)	9·3[b]	15·7	9·0–12·3[c]	12·2[c]
Ether extract (%)	5·5	5·0	4·8	4·0
Crude fibre (%)	2·0	1·6	2·0	
Ash (%)	1·25	1·06	2·02	2·02

[a] Murty (1969).
[b] Limited moisture areas.
[c] Assured moisture areas with 80 kg N/ha.

Table 2.102
Mean chemical composition, ranges (in parentheses), and 1000-seed weight on 14 improved pearl millet hybrids.[a]

	Mean
Moisture (%)	10·1 (9·0–11·0)
Prot. (%)	14·5 (11·3–19·6)
Fat (%)	3·6 (3·0–4·6)
Starch (%)	65·5 (59·3–69·5)
Crude fibre (%)	2·1 (1·3–3·0)
Ash (%)	1·9 (1·5–2·6)
Total sugar (%)	2·2 (2·0–2·7)
Ca (mg/100 g)[b]	50 (35–62)
P (mg/100 g)[b]	297 (245–348)
Phytin P (mg/100 g)[b]	90 (35–221)
Fe (mg/100 g)[b]	3·4 (2·1–5·2)
Yellow pigment as carotene (mg/100 g)	0·54 (0·42–0·65)
1000-seed weight (g)	7·7 (4·6–11·9)
Prot. in 1000 seeds/1000-seed weight	1·12 (0·67–1·77)

[a] Uprety and Austin (1972).
[b] DWB.

Table 2.103
Distribution of protein, Ca and P between the husk and endosperm of pearl millet.[a]

	Components			
	Whole grain (%)	Husk (%)	Endosperm (%)	Dried solids from supernatant liquid (%)
Component as % of whole grain	(100)	11·8	83·6	3·8
Prot. (N × 6·25) (%)	12·9	12·4	2·4	24·4
As % prot. in whole grain		11·5	15·8	72·0
Ca (mg/100 g)	55	168	17	534
As % Ca in whole grain		36·1	25·4	36·9
P (mg/100 g)	358	442	240	3404
As % P in whole grain		14·6	56·1	28·7

[a] Kurien *et al.* (1961a).

Table 2.104
Composition of UP, isolated variety of pearl millet.[a]

Specific gravity 30/30°C	0·9163
Refractive index 40°C	1·4656
Acid value	14·3
Saponification value	201·3
Iodine value	112·3
Hehner value	93·4
Acetyl value	8·4
Unsaponifiable matter (%)	2·08

[a] Agarawal and Sinha (1964).

Table 2.105
Sugar content (DWB) (%) of two varieties of pearl millet 100 days after sowing.[a]

	A1/3 (long-eared)	S 530 (bristle-eared)
Free reducing sugars	3·43	2·23
Total water-soluble sugars	5·43	4·56
Nonreducing sugars	2·10	2·33
Starch	82·75	86·05
Free glucose	2·47	1·44
Free fructose	0·96	0·79
Bound glucose	0·68	0·68
Bound fructose	1·42	1·65

[a] Bhatia *et al.* (1972).

Table 2.106
Protein, carbohydrate and sugars content, means, ranges in parentheses, of seven high-yielding varieties of pearl millet.[a]

	Mean
Protein (%)	11·7 (8·4–14·5)
Total carbohydrate (%)	68·2 (65·4–71·2)
Starch (%)	60·2 (56·3–63·7)
Amylose (%)	20·5 (18·3–24·6)
Total sugars (mg maltose/10 g)	282 (244–331)
Reducing sugars (mg maltose/10 g)	48 (39–57)
Nonreducing sugars (mg sucrose/10 g)	203 (181–233)
Diastatic activity (mg maltose/10 g)	433 (338–551)

[a] Singh and Popli (1973).

Table 2.107
Mean protein and amino acid content (mg/g total N) in 17 pearl millet lines of high protein content (ranges in parentheses).[a]

	Mean
Prot. (N × 6·25) (%)	21·2 (14·4–27·1)
Lys	130 (114–154)
Thr	184 (139–208)
Met	119 (84–148)
Trp	78 (61–86)

[a] Deosthale *et al.* (1971).

Table 2.108
Distribution of protein in the protein fractions of 12 varieties of pearl millet, mean and ranges (in parentheses).[a]

	Mean	
Total prot. (g/100 g of defatted material)	14·5 (9·2–21·0)	
	Prot. in different fractions	
	g/100 g defatted material	As % of total prot.
Albumin	2·0 (1·1–4·4)	15·1 (6·1–26·5)
Globulin	1·2 (0·5–2·2)	8·7 (3·5–14·7)
Prolamine	4·4 (2·3–6·1)	30·2 (21·3–38·0)
Glutelin	4·3 (2·3–6·3)	30·3 (23·8–37·7)

[a] Sawhney and Naik (1969).

Table 2.109
Distribution of amino acids (mg/g total N) in the protein fractions of HB 1 variety of pearl millet.[a]

	Albumin	Globulin	Prolamine	Glutelin
Total prot. (g/100 g grain)	1·8	2·3	2·6	3·0
Ile	257	155	207	239
Leu	382	351	754	636
Lys	245	236	104	134
Met	121	132	64	86
Cys	182	164	89	77
Phe	239	259	239	309
Tyr	205	269	289	277
Thr	132	165	216	182
Trp	95	34	177	43
Val	352	459	202	257
Arg	272	301	190	307
His	136	139	134	102
Ala	507	376	588	651
Asp	779	714	467	397
Glu	1022	1169	1390	1355
Gly	459	257	77	171
Pro	551	466	639	589
Ser	301	351	382	334
Ammonia	57	51	71	76
% recovery of total prot.	100·76	96·8	100·5	99·6

[a] Sawhney and Naik (1969).

Table 2.110
Mean distribution and concentration of S in different protein fractions of 12 varieties of pearl millet (ranges in parentheses).[a]

	Mean
Total S (mg/100 g defatted material)	172 (125–203)

	S in different prot. fractions		
	mg S/100 g defatted material	As % total S	g S/100 g prot. fraction
Albumin	35·8 (18·9–45·4)	20·9 (10·5–24·3)	1·9 (0·9–3·4)
Globulin	37·2 (26·1–50·0)	21·8 (14·8–28·2)	3·5 (1·34–6·5)
Prolamine	46·8 (30·4–71·6)	27·1 (18·4–40·7)	1·1 (0·6–2·05)
Glutelin	66·6 (22·7–92·5)	37·9 (8·1–49·0)	1·61 (0·7–3·0)

[a] Sawhney and Naik (1969).

Table 2.111
Chemical composition of pearl millet fed in boys' diets.[a]

	(%)
Moisture	12·2
Protein (N × 6·25)	11·3
Ether extractives	3·8
Crude fibre	3·2
Starch	60·7
Total sugars	1·2
Pentosans and other hemicelluloses (by difference)	5·8
Ash	1·8
Minerals (mg/100 g)	
Ca	46
P	314
Phytate P	224
Calories (keal/100 g)	331

[a] Kurien *et al.* (1961b).

Table 2.112
Proximate composition, minerals and thiamine content of pearl millet and wheat fed in rat diets.[a]

	Pearl millet	Wheat
Moisture (%)	11·5	10·6
Prot. (N × 6·25) (%)	13·1	12·2
Ether extractives (%)	5·2	1·5
Fibre (%)	1·2	1·3
CBH[b] (by difference) (%)	66·5	72·8
Mineral matter (%)	2·5	1·6
Ca (mg/100 g)	60	48
P (mg/100 g)	310	340
Fe (mg/100 g)	8	6
Thiamine (μg/100 g)	350	510

[a] Rama Rao *et al.* (1953a).
[b] Carbohydrate.

Table 2.113
Average rat weight changes, haemoglobin (Hb) levels, and plasma protein levels, from diets containing CO 3 pearl millet, CO 3 finger millet and CO 3 foxtail millet.[a]

	Wt changes (g)	Hb levels (g/100 cm³ blood)	Plasma prot. (g/100 cm³ blood)
Casein	+85·3	14·87	5·85
Pearl millet			
Raw	+27·3	11·72	4·90
Autoclaved	−5·0	9·81	2·98
Finger millet			
Raw	+14·8	10·51	3·87
Autoclaved	−3·6	8·87	2·48
Foxtail millet			
Raw	+2·5	8·47	2·98
Autoclaved	+7·4	10·92	3·44

[a] Chitre and Vallury (1956a).

Table 2.114
Essential amino acids of CO 3 pearl millet, CO 3 finger millet and CO 3 foxtail millet in rat diets.[a]

	Pearl millet (mg/100 g diet)	Finger millet (mg/100 g diet)	Foxtail millet (mg/100 g diet)
Ile	643	229	497
Leu	1059	357	1210
Lys	427	124	226
Met	296	283	242
Cys	343	234	206
Phe	449	197	454
Thr	—	—	292
Trp	120	42	37
Val	696	259	510
Arg	884	573	1000
His	389	210	422

[a]Chitre and Vallury (1956a).

Table 2.115
Average glycogen, fat and protein (g/100 g fresh liver) from diets containing CO 3 pearl millet, CO 3 finger millet and CO 3 foxtail millet.[a]

	Glycogen (% liver)	Fat (% liver)	Prot. (% liver)
Casein	2·6	3·5	22·9
Pearl millet			
Raw	5·4	5·4	17·6
Autoclaved	6·8	7·6	14·4
Finger millet			
Raw	8·3	6·8	12·4
Autoclaved	8·6	8·4	12·1
Foxtail millet			
Raw	5·1	5·1	13·9
Autoclaved	4·5	4·8	17·0

[a]Chitre and Vallury (1956b).

Table 2.116
Effect of supplementation with L-lysine HCl on weight gain and PER of rats fed teff and pearl millet.[a]

Diet	Lys (mg/g total N)	Wt gain (g)	PER
(A) Teff	194	50·3	1·95
(B) Teff + 0·25% L-Lys HCl	312	104·3	2·78
(C) Teff + 0·42%, L-Lys HCl	394	138·5	3·27
(D) Casein control		91·2	3·47
(E) Pearl millet	181	36·2	1·83
(F) Pearl millet + 0·25% L-Lys HCl	300	103·9	2·94
(G) Pearl millet + 0·50% L-Lys HCl	419	118·0	3·28
(H) Casein control		91·7	3·55

[a]Jansen *et al.* (1962).

Table 2.117
Chemical composition of two varieties and three hybrids of pearl millet, and effect of diets containing them on rat weight gain and PER.[a]

	(A) A1/3	(B) T 55	(C) 23A × BIL 3B	(D) 23A × BIL 1	(E) 101A × BIL 3B	
Moisture (%)	8·9	8·1	7·8	8·5	8·6	
Crude prot. (%)	7·5	10·5	7·5	9·7	8·4	
Crude fat (%)	6·2	6·1	5·6	5·9	5·7	
Cellulose (%)	1·8	1·9	2·0	1·9	1·9	
NFE (%)	82·5	80·3	82·5	80·9	82·1	
Lys (mg/g total N)	220	210	180	210	240	
Met (mg/g total N)	100	100	110	130	130	
Trp (mg/g total N)	120	120	130	110	120	
Diets (6·4% prot. level)						Casein
Prot. intake (g/4 weeks)	8·7	8·9	8·3	8·4	8·4	8·7
Wt gain (g/4 weeks)	8·9	8·7	7·8	9·4	10·1	16·4
PER	1·01	0·98	0·94	1·11	1·21	1·88

[a] Goswami *et al.* (1969c).

Table 2.118
Proximate analyses of pearl millet (DWB) (mean with ranges in parentheses where available).[a] Blank spaces indicate no data presented.

Reference	Location	No. of samples	Prot. (N × 6·25) (%)	Lipid (%)	CBH[b] (%)	Fibre (%)	Ash (%)
Adrian and Sayerse (1957)	Africa	5	11·6 (9·9–13·5)	5·6 (5·2–6·1)	75·8 (75·0–77·0)		1·7 (1·6–1·8)
Austin *et al.* (1971)	India	40	16·1 (13·1–18·5)				
Balasubramanian *et al.* (1952ab); Balasubramanian and Ramachandran (1957)	India	2	12·6 (12·4–12·8)				3·6 (3·2–3·8)
Blondel (1970)	Africa	8	8·6 (6·7–13·6)				
Burton *et al.* (1972)	USA	180	16·0 (8·8–20·9)				
Deosthale *et al.* (1971)	India	5	9·4 (7·2–10·6)				1·7 (1·5–2·0)
Fetuga (1977)	Africa	4	10·7 (8·9–12·2)	4·3 (2·9–5·3)	80·0 (76·8–83·8)	2·4 (1·8–3·1)	2·9 (2·1–3·7)
Freeman (personal communication) to Burton *et al.* (1972)	USA	1	17·4				
Gandry and Bideau (1974)	Africa	8	11·1 (8·7–13·3)	5·1 (4·7–5·3)		1·5 (1·5–1·5)	1·9 (1·7–2·0)
Goswami *et al.* (1969c)	India	5	9·2 (8·1–11·4)	6·4 (6·1–6·8)	89·1 (87·4–90·6)		
Goswami *et al.* (1969ab) (1970ab)	Various	163	11·8 (9·4–15·2)	5·2 (2·8–8·0)			2·0 (1·6–2·5)
Khanna (1966)	India		11·4 (10·2–12·3)	6·0 (5·8–6·0)	79·2 (78·3–80·4)	1·4 (0·4–1·8)	1·9 (1·8–2·0)
Kinra *et al.* (1970)		6	10·2 (9·2–11·0)				
Kumar *et al.* (1973)		3	10·7 (9·2–13·0)				
Magboul (1974—personal communication)	Africa	5	11·7 (10·4–12·0)	6·2 (5·9–6·8)	78·8 (76·8–82·3)	2·0 (1·2–4·1)	2·1 (1·5–2·9)

Table 2.118 *cont.*

Reference	Location	No. of samples	Prot. (N × 6·25) (%)	Lipid (%)	CBH[b] (%)	Fibre (%)	Ash (%)
Murty (1969)	India	4	11·7 (9·0–15·7)	4·9 (4·1–5·5)		2·0 (1·6–2·0)	1·6 (1·1–2·0)
Narayanamurti and Aiyar (1930)			10·1	6·8	78·1	2·5	2·5
Popli and Singh (1972)	India	7	11·7 (8·4–14·5)	5·9 (5·2–6·3)	68·2 (65·4–71·2)	1·4 (1·0–1·7)	2·1 (1·8–2·5)
Rama Rao and Swaminathan (1953)		6	15·1 (14·0–16·2)	6·1 (5·7–6·5)			2·3 (1·7–2·5)
Samson and Adrian (1971)	Africa	3	9·1 (8·9–9·6)	1·5 (1·4–1·5)			
Sawhney and Naik (1969)	India	12	13·5 (7·9–20·1)	5·5 (4·6–6·4)	67·5 (59·8–74·5)	1·4 (1·0–2·2)	2·5 (2·0–3·2)
Sawhney and Naik (1969)		4	13·1 (12·6–13·4)				
Sharma and Goswami (1969)	Various	25		5·6 (4·2–7·4)			
Shepherd *et al.* (1971–72)	Africa	94	12·7 (9·3–14·5)	5·9 (4·8–9·2)	78·2	1·6 (1·0–3·8)	1·6 (1·1–2·4)
Singh and Popli (1974)	India	7	11·7 (8·4–14·5)				
Solpico and Yambao (1966)	Philip-pines	2	13·0 (12·8–13·2)	4·2 (4·1–4·3)	70·9 (69·3–72·5)	7·3 (5·7–7·1)	3·6 (3·4–3·7)
Tomer (1970)	India	8	9·2 (8·7–9·5)				
Uprety and Austin (1972)		14	16·1 (12·6–21·8)	4·0 (3·3–5·1)	72·9 (66·0–77·3)	2·3 (1·4–3·3)	2·1 (1·7–2·9)
Representative analytical values and ranges (not means) data from sources mentioned in text.			11·0 (8·8–20·9)	5·0 (2·3–6·25)	69·0 (66·0–83·0)	2·0 (0·7–3·9)	1·8 (0·8–3·4)
Range of means			8·6–17·4	1·5–6·8	61·5–89·1	1·4–7·3	1·6–3·6
Range of ranges			6·7–21·8	1·4–9·2	59·8–90·6	0·4–7·1	0·8–3·8
Range of prot. means at N × 5·7			7·8–15·9				
Range of prot. ranges at N × 5·7			6·1–19·9				

[a] Data derived from multiple sources.
[b] Carbohydrate.

Table 2.119
Amino acid content and score of pearl millet.[a] (Mean with ranges in parentheses) (from multiple sources). Blank spaces indicate no data presented.

Reference	Location	Ile					Leu					Leu/Ile quotient
		No. of samples	Prot. (N × 6·25) (%)	mg/100 g sample	mg/g total N	AA score	No. of samples	Prot. (N × 6·25) (%)	mg/100 g sample	mg/g total N	AA score	
Adda (1958)	Africa	16	11·9 (7·7–15·1)				16	11·9 (7·7–15·1)		927 (819–1075)		2·5
Adrian and Sayerse (1957)		5			284[b] (187–369)	114	5			1071[b] (1025–1150)	243	3·8
Badi *et al.* (1976)	USA	1	13·8		244	98	1	13·8		612	139	2·5
Balasubramanian *et al.* (1952a,b); Balasubramanian and Ramachandran (1957)	India	2	12·6 (12·4–12·8)	772[b] (757–786)	368[b] (367–369)	147	2	12·6 (12·4–12·8)	1247[b] (1246–1247)	595[b] (583–607)	135	1·6
Burton *et al.* (1972)	Mixed	24		670 (410–960)			24			1570 (920–2200)		2·3
Busson *et al.* (1962)	Africa	17	9·8 (8·6–10·9)		262 (250–275)	105	17	9·8 (8·6–10·9)		581 (569–594)	132	2·2
		1	11·2		275	110	1	11·2		619	141	2·3
Chitre and Vallury (1956a)	India	1		643			1		1059			
Chitre *et al.* (1956)		1	13·1	534[b]			1	13·1	1230[b]			
Deosthale *et al.* (1971)	Africa	22	15·2 (12·1–17·2)				22	15·2 (12·1–17·2)		613[b] (577–647)	139	
Deosthale *et al.* (1972b)	India	30	9·5 (8·4–11·1)	494[b] (450–560)	255[b] (244–269)	102	30	9·5 (8·4–11·1)	1084[b] (950–1300)	555[b] (506–606)	126	2·2
FAO (1970)	Various			397	256 (198–306)	102			927	598 (456–719)	136	2·3
Ganry and Bideau (1974)	Africa	8			284 (269–300)	114	8			664 (631–687)	151	2·3
Jansen *et al.* (1962)		1	10·5		193	77				456	104	1·3

Pokhriyal *et al.* (1977)	India	8	13·1	242	97			541	123	2·2
			(11·0–14·7)	(193–285)				(426–601)		
Popli and Singh (1972)		7	11·7	319	128	7	11·7	537	122	1·7
			(8·4–14·5)	(261–407)			(8·4–14·5)	(499–579)		
Samson and Adrian (1971)	Africa	3	9·1	255	102	3	9·1	500	114	2·0
			(8·9–9·6)	(250–260)			(8·9–9·6)	(493–509)		
Sawhney and Naik (1969)	India	3	13·8	267	107	3	13·8	650	148	2·4
			(9·6–20·1)	(197–376)			(9·6–20·1)	(508–792)		
Shepherd *et al.* (1971–72)	Africa	6	11·8	306	122	6	11·8	744	169	2·4
			(9·9–13·4)	(237–362)			(9·9–13·4)	(562–875)		
Taira (1962a)				350[b]	140			635[b]		1·8
Range of means			8·9–21·2	224–368				500–1570		
Range of ranges			6·4–27·1	187–407				426–2200		
Range of protein means at N × 5·7			8·1–19·3							
Range of protein ranges at N × 5·7			7·0–24·7							

[a] WHO (1973a).
[b] Microbiological assay: values are in general higher than those obtained by ion-exchange chromatography.

cont.

Table 2.119 *cont.*

Reference	Location	Met: No. of samples	Met: Prot. (N × 6·25) (%)	Met: mg/100 g sample	Met: mg/g total N	Cys: No. of samples	Cys: Prot. (N × 6·25) (%)	Cys: mg/100 g sample	Cys: mg/g total N	AA score (Met + Cys)
Adrian and Sayerse (1957)		5			152[b] (137–169)	5			78[b] (71–84)	105
Badi *et al.* (1976)	USA	1	13·8		119	1	13·8		156	125
Balasubramanian *et al.* (1952a,b) } Balasubramanian and Ramachandran (1957) }	India	2	12·6 (12·4–12·8)	237[b] (224–249)	113[b] (106–119)					
Burton *et al.* (1972)	Mixed			210 (80–370)				1720 (1020–2460)		
Busson *et al.* (1962)	Africa	17	9·8 (8·6–10·9)		150 (144–162)	17	9·8 (8·6–10·9)		156 (144–169)	139
		1	11·2		150	1	11·2		150	136
Chitre and Vallury (1956a)	India	1		296		1		343		
Chitre and Vallury (1956b)		1		348		1		404		
Chitre *et al.* (1956)		1	13·1	325						
Deosthale *et al.* (1971)	Mixed	17	21·2 (14·4–27·1)		119[b] (84–148)					
Deosthale *et al.* (1972b)	India	30	9·5 (8·4–11·1)	306[b] (270–330)	163[b] (144–175)					
FAO (1970)	Various			239	154 (84–246)			229	148 (69–169)	137
Ganry and Bideau (1974)	Africa	8			93 (50–112)					
Goswami *et al.* (1969c)	India	5			114 (100–130)					
Jansen *et al.* (1962)					199				84	129
Pokhriyal *et al.* (1977)					118 (84–162)				129 (111–163)	112
Popli and Singh (1972)		7	11·7 (8·4–14·5)		127 (111–151)					
Samson and Adrian (1971)	Africa	3	9·1 (8·9–9·6)		125 (121–130)					
Sawhney and Naik (1969)	India	3	13·8 (9·6–20·1)		102 (94–107)	3	13·8 (9·6–20·1)		173 (145–208)	125
Taira (1962a)					112[b]				44[b]	71
Range of means					93–163				44–173	
Range of ranges					50–246				69–208	

Reference	Location	Thr					Trp				
		No. of samples	Prot. (N × 6·25) (%)	mg/100 g sample	mg/g total N	AA score	No. of samples	Prot. (N × 6·25) (%)	mg/100 g sample	mg/g total N	AA score
Adda (1958)	Africa	16	11·9 (7·7–15·1)		272 (244–300)	109	16	11·9 (7·7–15·1)		99 (81–119)	165
Adrian (1969)			11·1 (9·1–13·5)					11·1 (9·1–13·5)	192[b] (134–210)	110[b] (91–131)	183
Adrian and Sayerse (1957)		5			249[b] (150–344)	100	5			130[b] (100–153)	217
Badi *et al.* (1976)	USA	1	13·8		256	102					
Balasubramanian *et al.* (1952a,b)	India	2	12·6 (12·4–12·8)	496[b] (472–519)	237[b] (221–253)	95	2	12·6 (12·4–12·8)	251[b] (245–257)	120[b] (119–120)	200
Balasubramanian and Ramachandran (1957)	India										
Burton *et al.* (1972)	Mixed			510 (340–680)							
Busson *et al.* (1962)	Africa	17	9·8 (8·6–10·9)		256 (250–262)	102	17	9·8 (8·6–10·9)		119 (112–131)	198
		1	11·2		250	100	1	11·2		125	208
Chitre and Vallury (1956a)	India						1		120		
Chitre *et al.* (1956)		1	13·1	379[b]			1	13·1	140[b]		
Deosthale *et al.* (1971)	Africa	22	15·2 (12·1–17·2)		168[b] (131–205)	67					
Deosthale *et al.* (1971)	Mixed	17	21·2 (14·4–27·1)		184[b] (130–208)	74	17	21·2 (14·4–27·1)		78[b] (61–86)	130
Deosthale *et al.* (1972b)	India	30	9·5 (8·4–11·1)	432[b] (400–470)	227[b] (219–237)	91	30	9·5 (8·4–11·1)	134[b] (116–159)	70[b] (63–75)	117
FAO (1970)	Various			374	241 (156–263)	96			189	122 (99–125)	203
Ganry and Bideau (1974)	Africa	8			267 (231–300)	107					

[b] Microbiological assay: values are in general higher than those obtained by ion-exchange chromatography.

cont.

Table 2.119 *cont.*

Reference	Location	Thr					Trp				
		No. of samples	Prot. (N × 6·25) (%)	mg/100 g sample	mg/g total N	AA score	No. of samples	Prot. (N × 6·25) (%)	mg/100 g sample	mg/g total N	AA score
Goswami *et al.* (1969c)	India						5		120 (110–130)		
Jansen *et al.* (1962)					156	62				101	168
Pokhriyal *et al.* (1977)					141 (124–152)	56					
Popli and Singh (1972)		7	11·7 (8·4–14·5)		220 (198–261)	88	7	11·7 (8·4–14·5)		77 (57–110)	128
Samson and Adrian (1971)	Africa	3	9·1 (8·9–9·6)		206 (200–213)	82	3	9·1 (8·9–9·6)		72 (69–76)	120
Sawhney and Naik (1969)	India	3	13·8 (9·6–20·1)		219 (189–270)	88	15			69 (45–111)	115
		4		152 (138–159)	72 (68–74)	29					
Shepherd *et al.* (1971–72)	Africa	5	11·8 (9·9–13·4)		250 (219–306)	100					
Taira (1962a)					300[b]	120				87[b]	145
Range of means					72–300					69–130	
Range of ranges					68–344					45–153	

Reference	Location	Lys				
		No. of samples	Prot. (N × 6·25) (%)	mg/100 g sample	mg/g total N	AA score
Adda (1958)	Africa	16	11·9		164	48
			(7·7–15·1)		(144–188)	
Adrian and Sayerse (1957)		5			165[b]	49
					(152–184)	
Badi *et al.* (1976)	USA	1	13·8		225	66
Balasubramanian *et al.* (1952a,b) } Balasubramanian and Ramachandran (1957) }	India	2	12·6	497[b]	238[b]	70
			(12·4–12·8)	(490–503)	(231–244)	
Burton *et al.* (1972)	Mixed			410		
				(310–540)		
Busson *et al.* (1962)	Africa	17	9·8		237	70
			(8·6–10·9)		(225–244)	
		1	11·2		244	72
Chitre and Vallury (1956a)	India	1		427		
Chitre *et al.* (1956)		1	13·1	492[b]		
Deosthale *et al.* (1971)	Africa	22	15·2		163[b]	48
			(12·1–17·2)		(141–187)	
		15	10·1		153[b]	45
			(6·4–13·2)		(122–199)	
	India	40	12·3		177[b]	52
			(7·2–14·9)		(136–310)	
		14	12·7		166[b]	49
			(11·5–14·9)		(146–194)	
	Mixed	17	21·2		131[b]	39
			(14·4–27·1)		(114–154)	
Deosthale *et al.* (1972b)	India	30	9·5	318[b]	169[b]	50
			(8·4–11·1)	(290–350)	(156–181)	
FAO (1970)	Various			332	214	63
					(100–244)	
Ganry and Bideau (1974)	Africa	8		329	204	60
				(310–350)	(175–244)	
Goswami *et al.* (1969c)	India	5			212	62

[b] Microbiological assay: values are in general higher than those obtained by ion-exchange chromatography.

cont.

Table 2.119 *cont.*

Reference	Location	No. of samples	Prot. (N × 6·25) (%)	Lys mg/100 g sample	Lys mg/g total N	AA score
					(180–240)	
Jansen *et al.* (1962)					181	53
Nagarajan (1974)	India	4	8·9 (7·2–10·6)		297 (181–475)	87
Pokhriyal *et al.* (1977)					192 (166–216)	56
Popli and Singh (1972)		7	11·7 (8·4–14·5)		165 (131–226)	49
Samson and Adrian (1971)	Africa	3	9·1 (8·9–9·6)		170 (165–174)	50
Sawhney and Naik (1969)	India	15			167 (89–239)	49
		4		421 (409–432)	203 (192–211)	60
Shepherd *et al.* (1971–72)	Africa	6	11·8 (9·9–13·4)		206 (162–244)	60
Swaminathan *et al.* (1969)	India	12	15·4 (10·2–23·0)		109 (56–189)	32
Swaminathan *et al.* (1971)		9	13·0 (11·3–15·3)		192 (135–227)	56
Taira (1962a)					181[b]	53
Range of means					109–297	
Range of ranges					56–475	

Reference	Location	Phe				Tyr				AA score (Phe + Tyr)
		No. of samples	Prot. (N × 6·25) (%)	mg/100 g sample	mg/g total N	No. of samples	Prot. (N × 6·25) (%)	mg/100 g sample	mg/g total N	
Adda (1958)	Africa	16	11·9		288					
			(7·7–15·1)		(269–306)					
Adrian and Sayerse (1957)		5			280^b					
					(216–316)					
Badi *et al.* (1976)	USA	1	13·8		312	1	13·8		219	140
Balasubramanian *et al.* (1952a,b)	India	2	12·6	564^b	269					
Balasubramanian and Ramachandran (1957)			(12·4–12·8)	(528–600)	(256–281)					
Burton *et al.* (1972)	Mixed			760				480		
				(440–1120)				(300–690)		
Busson *et al.* (1962)	Africa	17	9·8		300	17	9·8		212	135
			(8·6–10·9)		(288–319)		(8·6–10·9)		(200–219)	
		1	11·2		319	1	11·2		219	142
Chitre and Vallury (1956a)	India	1		449						
Chitre *et al.* (1956)		1	13·1	416^b						
FAO (1970)	Various			467	301			315	203	133
					(216–331)				(88–231)	
Ganry and Bideau (1974)	Africa	8			333	8			238	150
					(325–362)				(231–244)	
Jansen *et al.* (1962)					216				88	80
Pokhriyal *et al.* (1977)					266				187	119
					(215–322)				(151–226)	
Popli and Singh (1972)		7	11·7		335					
			(8·4–14·5)		(294–415)					
Sawhney and Naik (1969)	India	3	13·8		343	3	13·8		302	170
			(9·6–20·1)		(317–371)		(9·6–20·1)		(208–386)	
Shepherd *et al.* (1971–72)	Africa	6	11·8		319	6	11·8		212	140
			(9·9–13·4)		(269–381)		(9·9–13·4)		(194–237)	
Taira (1962a)					231^b				106^b	89
Range of means					231–335				106–302	
Range of ranges					216–415				88–386	

^{b} Microbiological assay: values are in general higher than those obtained by ion-exchange chromatography.

cont.

Table 2.119 *cont.*

Reference	Location	Val: No. of samples	Val: Prot. (N × 6·25) (%)	Val: mg/100 g sample	Val: mg/g total N	Val: AA score	Arg: No. of samples	Arg: Prot. (N × 6·25) (%)	Arg: mg/100 g sample	Arg: mg/g total N
Adda (1958)	Africa	16	11·9		299	96	16	11·9		200
			(7·7–15·1)		(269–338)			(7·7–15·1)		(175–231)
Adrian and Sayerse (1957)		5			321[b]	104	5			267[b]
					(250–387)					(219–328)
Badi *et al.* (1976)	USA	1	13·8		325	105	1	13·8		375
Balasubramanian *et al.* (1952a,b)	India	2	12·6	823[b]	393[b]	127	2	12·6	1060[b]	507[b]
Balasubramanian and Ramachandran (1957)			(12·4–12·8)	(819–827)	(387–399)			(12·4–12·8)	(1050–1070)	(491–522)
Burton *et al.* (1972)	Mixed			200						
				(80–240)						
Busson *et al.* (1962)	Africa	17	9·8		356	115	17	9·8		350
			(8·6–10·9)		(344–362)			(8·6–10·9)		(338–362)
		1	11·2		356		1	11·2		325
Chitre and Vallury (1956a)	India	1		696			1		884	
Chitre *et al.* (1956)		1	13·1	630[b]			1	13·1	911	
FAO (1970)	Various			535	345	111			512	330
					(269–375)					(200–362)
Ganry and Bideau (1974)	Africa	8			338	109	8			331
					(256–362)					(306–375)
Jansen *et al.* (1962)					281	91				217
Pokhriyal *et al.* (1977)					287	93				297
					(239–364)					(240–386)
Popli and Singh (1972)		7	11·7		404	130				
			(8·4–14·5)		(365–447)					
Samson and Adrian (1971)	Africa	3	9·1		321	104				
			(8·9–9·6)		(315–325)					
Sawhney and Naik (1969)	India	3	13·8		383	124	3	13·8		239
			(9·6–20·1)		(377–388)			(9·6–20·1)		(196–294)
Shepherd *et al.* (1971–72)	Africa	6	11·8		437	141	6	11·8		287
			(9·9–13·4)		(337–519)			(9·9–13·4)		(231–400)
Taira (1962a)					400[b]	129				281[b]
Range of means					299–437					200–507
Range of ranges					250–519					175–522

Reference	Location	His: No. of samples	His: Prot. (N × 6·25) (%)	His: mg/100 g sample	His: mg/g total N	Ala: No. of samples	Ala: Prot. (N × 6·25) (%)	Ala: mg/100 g sample	Ala: mg/g total N
Adda (1958)	Africa	16	11·9 (7·7–15·1)		111 (94–131)				
Adrian and Sayerse (1957)		5			129[b] (114–148)				
Badi *et al.* (1976)	USA	1	13·8		162	1	13·8		487
Balasubramanian *et al.* (1952a,b)	India	2	12·6 (12·4–12·8)	228[b] (208–247)	110[b] (100–119)				
Balasubramanian and Ramachandran (1957)	India								
Burton *et al.* (1972)	Mixed			370 (230–480)				1140 (120–1690)	
Busson *et al.* (1962)	Africa	17	9·8 (8·6–10·9)		162 (156–169)	17	9·8 (8·6–10·9)		469 (456–488)
		1	11·2		156	1	11·2		481
Chitre and Vallury (1956a	India	1		389					
Chitre *et al.* (1956)		1	13·1	458					
FAO (1970)	Various			237	153 (85–170)			769	496 (456–631)
Ganry and Bideau (1974)	Africa	8			166 (156–181)	8			483 (462–494)
Jansen *et al.* (1962)					130				449
Pokhriyal *et al.* (1977)					144 (115–196)				449 (409–492)
Sawhney and Naik (1969)	India	3	13·8 (9·6–20·1)		140 (94–179)	3	13·8 (9·6–20·1)		491 (456–551)
Shepherd *et al.* (1971–72)	Africa	6	11·8 (9·9–13·4)		131 (87–162)	6	11·8 (9·9–13·4)		656 (562–756)
Taira (1962a)					119[b]				481[b]
Range of means					110–116				469–656
Range of ranges					85–196				456–756

[b] Microbiological assay: values are in general higher than those obtained by ion-exchange chromatography.

cont.

Table 2.119 *cont.*

Reference	Location	Asp				Glu			
		No. of samples	Prot. (N × 6·25) (%)	mg/100 g sample	mg/g total N	No. of samples	Prot. (N × 6·25) (%)	mg/100 g sample	mg/g total N
Adda (1958)	Africa	16	11·9 (7·7–15·1)		304 (263–344)	16	11·9 (7·7–15·1)		771 (713–825)
Badi *et al.* (1976)	USA	1	13·8		512	1	13·8		1187
Burton *et al.* (1972)	Mixed			1080 (670–1470)				2640 (380–4000)	
Busson *et al.* (1962)	Africa	17	9·8 (8·6–10·9)		512 (500–525)	17	9·8 (8·6–10·9)		1125 (1088–1194)
		1	11·2		494	1	11·2		1212
FAO (1970)	Various			777	501 (388–619)			1801	1162 (1087–1369)
Ganry and Bideau (1974)	Africa	8			524 (512–544)	8			1278 (1219–1325)
Pokhriyal *et al.* (1977)					305 (264–342)				851 (726–924)
Sawhney and Naik (1969)	India	3	13·8 (9·6–20·1)		530 (342–708)	3	13·8 (9·6–20·1)		1383 (1100–1585)
Shepherd *et al.* (1971–72)	Africa	6	11·8 (9·9–13·4)		644 (531–775)	6	11·8 (9·9–13·4)		1587 (1294–1837)
Taira (1962a)					444[b]				1287[b]
Range of means					304–644				771–1587
Range of ranges					263–775				713–1837

Reference	Location	Gly				Pro			
		No. of samples	Prot. (N × 6·25) (%)	mg/100 g sample	mg/g total N	No. of samples	Prot. (N × 6·25) (%)	mg/100 g sample	mg/g total N
Badi *et al.* (1976)	USA	1	13·8		231	1	13·8		369
Burton *et al.* (1962)	Africa			440 (330–570)				900 (490–1200)	
Busson *et al.* (1962)	Africa	17	9·8 (8·6–10·9)		250 (238–262)	17	9·8 (8·6–10·9)		381 (356–412)
		1	11·2		237	1	11·2		406
FAO (1970)	Various			364	235 (144–262)			595	384 (281–444)
Ganry and Bideau (1974)	Africa	8			228 (206–262)	8			448 (431–481)
Pokhriyal *et al* (1977)					173 (134–197)				304 (244–352)
Sawhney and Naik (1969)	India	3	13·8 (9·6–20·1)		342 (289–401)	3	13·8 (9·6–20·1)		607 (532–735)
Shepherd *et al.* (1971–72)	Africa	6	11·8 (9·9–13·4)		206 (175–237)	6	11·8 (9·9–13·4)		450 (369–525)
Taira (1962a)					219[b]				887[b]
Range of means					173–342				369–887
Range of ranges					134–401				281–735

[b] Microbiological assay: values are in general higher than those obtained by ion-exchange chromatography.

cont.

Table 2.119 *cont.*

Reference	Location	Ser				Ammonia			
		No. of samples	Prot. (N × 6·25) (%)	mg/100 g sample	mg/g total N	No. of samples	Prot. (N × 6·25) (%)	mg/100 g sample	mg/g total N
Badi *et al.* (1976)	USA	1	13·8		306	1	13·8		156
Burton *et al.* (1972)	Mixed			610 (390–860)					
Busson *et al.* (1962)	Africa	17	9·8 (8·6–10·9)		287 (281–300)				
		1	11·2		287				
FAO (1970)	Various			471	304 (338–413)				
Ganry and Bideau (1974)	Africa	8			323 (312–344)				
Pokhriyal *et al.* (1977)					232 (190–274)				
Sawhney and Naik (1969)	India	3	13·8 (9·6–20·1)		320 (254–366)	3	13·8 (9·6–20·1)		78 (61–92)
Shepherd *et al.* (1971–72)	Africa	6	11·8 (9·9–13·4)		350 (306–481)	6	11·8 (9·9–13·4)		131 (112–150)
Taira (1962a)					262				
Range of means					232–350				(78–156)
Range of ranges					190–481				61–156

[a] WHO (1973a).
[b] Microbiological assay: values are in general higher than those obtained by ion-exchange chromatography.

Table 2.120
Content of inorganic elements of pearl miller (mg/100 g).[a]
Blank spaces indicate no data presented.

Reference	Location	No. of samples	Ca	P	Fe
Adrian and Sayerse (1957)[b]	Africa	5	30 (17–47)	384 (344–415)	
Deosthale *et al.* (1971)[c]	India	5	53 (44–63)	360 (320–410)	7 (5–9)
Desai and Zender (1972)[d]		12	34 (28–40)	298 (260–332)	38 (25–46)
Ganry and Bideau (1974)	Africa	8	30 (25–39)	248 (245–259)	2 (1–2)
Goswami *et al* (1969a,b) (1970a,b)	Various	163	35 (20–53)	812 (650–960)	6 (2–12)
Khanna (1966)	India	1	34 (31–39)	950 (938–949)	
Magboul, personal communication (1974)	Africa	5	36 (19–68)	388 (129–782)	1·1 (1–2)
Popli and Singh (1972)	India	7	62 (48–94)	310 (283–340)	
Rama Rao and Swaminathan (1953)		6	41 (33–56)	385 (303–438)	11 (9–12)
Sawhney and Naik (1969)[e]		12	54 (14–102)	321 (279–373)	8 (4–12)
Shukla and Bhatia (1971)[f]		12			13 (11–15)
Uprety and Austin (1972)[g]		14	50 (35–62)	297 (245–348)	3 (2–5)
Representative analytical values and ranges (not means) data from sources mentioned in text[h]			(13–50)	(290–670)	(3–11)
Range of means			30–62	248–950	1·1–38
Range of ranges			13–102	129–960	1–46

[a] Data derived from multiple sources.
The following sources also reported, mg/100 g:
[b] K 387 (344–409), Na 91 (49–126),
[c] Cu 0·5 (0·4–0·6), Mg 142 (125–165), Mn 0·8 (0·7–0·9), Mo 0·019 (0·014–0·024), Zn 0·2 (0·1–0·2),
[d] K 692 (579–806), Mg 7 (4–9), Mn 20 (16–24),
[e] S 156 (117–183) (12 samples), 180 (152–202) (4 samples),
[f] Mn 3·4 (2·7–4·1) and
[g] Phytin P 90 (35–221).
[h] Pinta and Busson (1963) also reported single values (arc spectrography): Al 1·66, Ba 0·04, Be $<$ 0·05, Bi $<$ 0·05, B 0·2, Cl 94, Cr $\leqslant$ 0·03, Co $\leqslant$ 0·05, Ga $<$ 0·01, Ge 0·01, Pb 0·02, Li $\leqslant$ 0·01, Ni 0·11, Rb 0·32, Ag $<$ 0·005, Sr 0·02, Sn $<$ 0·004, Ti 0·02 and V $<$ 0·01.

Table 2.121
Vitamin content of pearl millet (mg/100 g) (from multiple sources).
Blank spaces indicate no data presented.

			Vit. B				Vit. A
Reference	Location	No. of samples	Thiamine	Niacin	Riboflavin	Pantothenic acid	Carotene
Adrian and Sayerse (1957)	Africa	5	0·38[a] (0·31–0·42)	3·02[a] (1·70–5·62)	0·20[a] (0·16–0·24)	1·31 (1·10–1·60)	
Chitre *et al.* (1955)	India	2	0·34 (0·28–0·39)	3·88 (3·32–4·43)	0·19 (0·19–0·19)		
Curtis *et al.* (1966)	Africa	6					0·12 (0·09–0·20)
Kapoor and Naik (1970)	India	38					0·03 (0·01–0·08)
Rama Rao and Swaminathan (1953)		6	0·41 (0·35–0·50)				
Samson and Adrian (1971)	Africa	3	0·37 (0·36–0·37)	1·54 (1·46–1·6)	0·24 (0·22–0·26)	0·87 (0·83–0·9)	
Uprety and Austin (1972)	India	14					0·54 (0·42–0·65)
Representative analytical[b] values and ranges (not means) data from sources mentioned in text.			(0·29–0·5)	(1·0–5·15)	(0·07–0·26)		
Range of means			0·34–0·41	1·54–3·88	0·19–0·24	0·87–1·31	0·03–0·54
Range of ranges			0·28–0·50	1·0–5·62	0·07–0·26	0·10–1·60	0·01–0·65

[a] Microbiological assay.
[b] Single representative values were reported: choline 38·2 mg/100 g; folic acid total 0·063, free 0·018 mg/100 g.

Table 2.122
Mean distribution[a] of proteins (ranges in parentheses) in different fractions of six millets.[b]

	Finger millet	Foxtail millet	Common millet	Little millet	Kodo millet	Japanese barnyard millet
Albumin	12·7 (10·2–15·7)	4·4 (3·2–5·6)	9·0 (8·3–9·7)	7·7 (5·3–11·3)	22·0 (16·0–28·5)	2·8 (2·0–3·2)
Globulin	17·9 (14·3–20·4)	18·0 (12·2–23·5)	10·9 (10·5–11·3)	25·3 (19·3–32·7)	15·3 (14·4–18·9)	3·0 (2·6–3·4)
Prolamine	21·5 (15·0–24·8)	38·5 (37·8–39·7)	31·0 (25·1–36·9)	31·3 (19·2–40·4)	25·3 (17·5–35·6)	5·7 (2·6–8·1)
Glutelin	12·2 (7·4–17·0)	22·7 (21·0–23·7)	8·0 (7·7–8·3)	6·8 (5·6–7·8)	5·1 (4·6–7·9)	7·5 (7·1–8·1)
Recovery of prot. (%)	62·3 (54·7–67·3)	82·6 (74·2–89·5)	63·9 (63·6–64·2)	71·8 (67·7–74·5)	68·9 (61·0–75·4)	20·6 (20·2–21·2)

[a] Expressed as % of total protein.
[b] Indira and Naik (1971).

Table 2.123
Distribution of Lys and Trp in different protein fractions of five species of millet (mg/g total N).[a]

	Albumin		Globulin		Prolamine		Glutelin	
	Lys	Trp	Lys	Trp	Lys	Trp	Lys	Trp
Finger Millet								
IE 903	119	208	65	65	54	101	477	212
IE 831	110	106	61	56	29	73	567	261
Foxtail Millet								
ISe 711	291	136	133	154	20	37	348	105
ISe 3	133	46	86	31	30	34	309	116
Common Millet								
IPm 640	138	132	486	132	18	17	651	220
IPm 1639	276	92	566	117	11	22	764	227
Little Millet								
IPe 420	364	59	139	14	21	22	653	184
IPe 108	147	34	33	13	42	54	420	226
Kodo Millet								
IPs 141	199	56	416	42	32	32	355	454
IPs 158	277	45	417	71	38	21	332	434

[a] Indira and Naik (1971).

Table 2.124
Amino acid composition of the solubility fractions of the protein of finger millet (mg/g total N).[a]

	Albumin	Globulin	Prolamine	Glutelin
Ile	325	237	337	275
Leu	512	356	719	525
Lys	519	337	10	200
Met	137	87	125	169
Cys	51	137	62	125

cont.

Table 2.124 *cont.*

	Albumin	Globulin	Prolamine	Glutelin
Phe	325	337	394	281
Tyr	244	137	319	237
Thr	331	219	250	244
Trp	94	87	156	125
Val	481	419	325	400
Arg	337	900	75	550
His	194	281	125	194
Ala	469	319	387	431
Asp	650	487	256	525
Glu	700	1344	2237	1356
Gly	337	400	54	444
Pro	350	325	537	462
Ser	381	437	319	275

[a] Taira (1965).

Table 2.125
Protein content and colour of finger millet varieties.[a]

		Prot. (N × 6·25) (DWB) (%)	
Variety	Seed colour	Whole seed	Endo.
ECW 854	White	11·0	6·3
EC 4840	Brown	9·1	5·2
HPW 27 4	White	8·6	4·4
ECW 955	White	8·5	5·5
Hamsa	White	8·2	5·3
HPB 7 6	Brown	8·2	4·4
HX 799	White	8·1	4·8
HPB 20 5	Brown	7·8	4·5
HPW 83 4	White	7·5	4·5
HPB 1 8	Brown	7·4	3·7
Purna	Qrown	7·3	4·4
HPB 23 6	Brown	6·8	3·5

[a] Virupaksha *et al.* (1975).

Table 2.126
Solubility fractions (% of N in sample) of whole seed (WS) and endosperm flours of six varieties of finger millet.[a]

	Fractions					
Variety	I	II	III	IV	V	Total N extracted
ECW 854 (white)						
WS	12·9	14·6	36·0	2·7	12·4	78·6
Endo.	14·7	10·1	31·3	2·0	13·5	71·6
Hamsa (white)						
WS	10·3	7·0	36·2	2·5	21·1	77·1
Endo.	8·4	4·5	30·9	2·3	13·4	59·5
HPB 83 4 (brown)						
WS	12·9	10·3	35·9	3·0	21·8	83·9
Endo.	12·4	8·2	32·0	2·8	20·9	76·3
EC 4840 (brown)						
WS	13·0	10·6	24·6	3·3	28·2	79·7
Endo.	9·8	7·2	25·3	2·4	23·9	68·6
Purna (brown)						
WS	11·7	8·7	32·6	2·5	19·2	74·7
Endo.	9·9	5·1	24·0	2·8	24·6	66·4
HPB 20 5 (brown)						
WS	11·8	10·3	28·8	2·7	24·6	78·2
Endo.	12·0	4·6	27·9	2·7	24·4	71·6

[a] Virupaksha *et al.* (1975).

Table 2.127

Amino acid composition (mg/g total N) of the whole seed (WS) protein, endosperm protein and three solubility fractions of Hamsa (white) and Purna (brown) varieties of finger millet.[a]

	Hamsa					Purna				
			Fractions					Fractions		
Amino acid	WS prot.	Endo. prot.	(A) (Albumin-globulin)	(B) (Prolamine)	(C) (Glutelin)	WS prot.	Endo. prot.	(A) (Albumin-globulin)	(B) (Prolamine)	(C) (Glutelin)
Ile	282	324	168	318	269	331	326	205	343	284
Leu	883	899	328	767	499	946	860	392	859	594
Lys	238	201	378	29	451	217	247	385	26	425
Met	126	151	Traces	144	84	140	171	Traces	138	81
Half-Cys	74	52	184	84[b]	153	105	42	13[b]	131[b]	77[b]
Phe	369	372	129	409	268	437	390	162	554	299
Tyr	188	190	141	283	186	244	221	154	313	244
Thr	312	301	270	309	288	361	319	271	333	273
Val	492	552	407	470	299	521	553	306	468	407
Arg	322	245	586	108	540	267	288	539	94	475
His	169	79	150	117	197	147	107	145	127	299
Ala	532	524	409	378	370	549	540	493	434	403
Asp	451	535	480	281	471	548	579	537	295	494
Glu	2024	1957	746	1996	1143	801	1821	739	2047	1312
Gly	299	315	328	111	274	329	336	392	112	280
Pro	385	338	203	549	350	501	308	272	568	392
Ser	431	384	316	357	367	491	399	364	444	334
Protein %	8·2					7·3				

[a] Virupaksha *et al.* (1975).
[b] As cysteic acid.

Table 2.128
PER, blood and liver composition of rats fed isonitrogenous diets of B 11 finger millet.[a]

	(A) Casein	(B) Finger millet	(C) Finger millet (defatted)
Av. initial weight (g)	58·6	58·8	58·6
Av. weight gain (g)	86·0	29·8	31·0
Av. daily food intake (g)	9·47	6·59	5·94
PER	1·90	0·95	0·95
Blood composition			
Hb (%)	11·4	10·8	10·5
Serum prot. (g/100 ml)	6·30	5·71	5·33
Total cholesterol (mg/100 ml)	95	65	65
Phospholipids (mg/100 ml)	380	435	380
Liver composition			
Weight (g)	5·21	2·83	2·94
Moisture (%)	66·8	65·0	64·2
Protein (%)	26·0	24·7	20·1
Total lipids (%)	10·7	12·9	13·0
Total cholesterol[b]	0·338	0·378	0·370

[a] Pore and Magar (1976a).
[b] In the original the unit of measurement is given as "g %". This could possibly have been intended to read "mg%", i.e. "mg/100 g".

Table 2.129

Proximate analyses of finger millet, DWB (mean with ranges[a] in parentheses) (from multiple sources). Blank spaces indicate no data presented.

Reference	Location	No. of samples	Prot. (N × 6·25) (%)	Lipid (%)	CBH[b] (%)	Fibre (%)	Ash (%)
Balakrishna Rao *et al.* (1973)	India	15	7·4 (4·7–9·9)			4·2 (3·1–5·7)	2·8 (2·0–4·1)
Indira and Naik (1971)		5	8·0 (6·2–9·5)	1·9 (1·4–2·5)			
Joseph *et al.* (1958ab)			7·7	1·5	84·4	3·6	2·9
Kadkol and Swaminathan (1954)		8	10·0 (8·2–12·2)	1·6 (1·4–1·9)	82·4 (79·3–84·4)	3·6 (3·2–3·9)	2·4 (1·8–3·2)
Kamalanathan *et al.* (1971)		2	9·1 (7·3–10·8)				
Kempanna and Kavallappa (1968)		19	10·9 (8·7–12·7)	1·5 (1·3–1·7)			2·9 (2·5–3·4)
Krishnamurthy (1968)		25	6·0 (4·5–7·6)				
Mahudeswaran *et al.* (1966)		10	9·7 (7·9–11·6)				
Mallanna and Rajashekara (1969)		1	9·7			4·6	2·6
Niyogi *et al.* (1934)		1	8·4	1·7	83·8	3·4	2·7
Pore and Magar (1976a)		1	8·2	1·9	87·5		2·4
Pore and Magar (1977)		36	9·6 (5·8–12·8)	2·0 (1·3–2·7)	85·7 (81·3–89·4)		2·7 (2·1–3·7)
Shepherd *et al.* (1971–72)	Africa	30	8·6 (7·5–9·5)	1·0 (0·7–1·7)	84·7	3·4 (2·6–4·2)	2·3 (1·7–2·8)

cont.

Table 2.129 *cont.*

Reference	Location	No. of samples	Prot. (N × 6·25) (%)	Lipid (%)	CBH[b] (%)	Fibre (%)	Ash %
Solpico and Yambao (1966)	Philippines	2	10·5 (9·1–11·9)	4·6 (4·5–4·6)	73·5 (73·0–73·9)	7·5 (6·8–8·2)	3·9 (3·6–4·2)
Sree Ramulu and Mariakulandai (1964)	India	30	9·2 (7·7–11·4)				
Stabursvik and Heide (1974)	Africa	20	9·5 (6·8–13·0)				
Subrahmanyan *et al.* (1955)	India		7·7	1·5	84·5	4·0	2·5
Venkataramana and Krishna Rao (1961)		22	9·9 (0·9–2·1)				
Representative analytical values and ranges (not means)[c]			6·0 (5·8–8·5)	1·5 (0·8–5·75)	75·0 (70·8–83·0)	3·0 (2·0–6·8)	3·0 (1·8–4·2)
Range of means			6·0–10·9	1·0–4·6	73·5–87·5	3·0–7·5	2·3–3·9
Range of ranges			0·9–13·0	0·7–5·75	70·8–89·4	2·0–8·2	1·7–4·2
Range of protein means at N × 5·7			5·5–9·9				
Range of protein ranges at N × 5·7			0·9–11·9				

[a] Where available.
[b] Carbohydrate.
[c] Data from sources mentioned in text.

Table 2.130

Amino acid content and score of finger millet[a] (mean with ranges in parentheses) (from multiple sources). Blank spaces indicate no data presented.

Reference	Location	Ile No. of samples	Ile Prot. (N × 6·25) (%)	Ile mg/100 g sample	Ile mg/g total N	Ile AA score	Leu No. of samples	Leu Prot. (N × 6·25) (%)	Leu mg/100 g sample	Leu mg/g total N	Leu AA score	Leu/Ile quotient
Balasubramanian *et al.* (1952a,b)	India	3	4·2	273[b]	402[b]	161	3	4·2	404[b]	594[b]	135	1·5
Balasubramanian and Ramachandran (1957)			(3·8–4·5)	(254–306)	(386–419)			(3·8–4·5)	(338–478)	(534–654)		
Chitre and Vallury (1956a)		1		229			1		357			
Deosthale *et al.* (1970b)		20	8·3									
			(5·6–11·6)									
FAO (1970)	Various	1		324	275		1		701	594		
		4			531[b]	212[b]	4			783[b]	178[b]	2·2
					(513–549)					(706–858)		
Indira and Naik (1971)	India	1	7·7		285	114	1	7·7		576	131	2·0
Nagarajan (1974)		5	8·6									
			(7·4–9·5)									
Shepherd *et al.* (1971–72)	Africa	6			300	120	6			662	150	2·2
					(262–350)					(637–675)		
Stabursvik and Heide (1974)		33	9·1		280	112	33	9·1		624	142	2·2
			(4·0–15·3)		(235–319)			(4·0–15·3)		(499–712)		
Taira (1963b)	Japan		12·0		262[b]	105[b]		12·0		612[b]	139[b]	2·3
			(9·0–16·0)					(9·0–16·0)				
Virupaksha *et al.* (1975)	India	12	8·2		297	119	12	8·2		848	193	2·9
			(6·8–11·0)		(239–403)			(6·8–11·0)		(678–1011)		
Range of means			4·2–12·0		262–531					576–848		
Range of ranges			3·8–16·0		235–549					499–1011		
Range of protein means at N × 5·7			3·8–10·9									
Range of protein ranges at N × 5·7			3·5–14·6									

[a] WHO (1973a).
[b] Microbiological assay: values are in general higher than those obtained by ion-exchange chromatography.

cont.

Table 2.130 *cont.*

Reference	Location	No. of samples	Lys: Prot. (N × 6·25) (%)	Lys: mg/100 g sample	Lys: mg/g total N	Lys: AA score
Balasubramanian *et al.* (1952) / Balasubramanian and Ramachandran (1957)	India	3	4·2	144[b]	212[b]	62
			(3·8–4·5)	(138–154)	(206–219)	
Chitre and Vallury (1956a)		1		124		
Deosthale *et al.* (1970b)		20	8·3		169[b]	50
			(5·6–11·6)		(131–238)	
FAO (1970)	Various	1		213	181	
		4			256[b]	75[b]
					(186–290)	
Indira and Naik (1971)	India	5			162	48
					(133–188)	
Mallanna and Rajashekara (1969)		1			250[b]	73
Nagarajan (1974)		5	8·6		160	47
			(7·4–9·5)		(131–187)	
Shepherd *et al.* (1971–72)	Africa	6			187	55
					(169–206)	
Stabursvik and Heide (1974)		33	9·1		166	48
			(4·0–15·3)		(112–246)	
Taira (1963b)	Japan		12·0		262[b]	77[b]
			(9·0–16·0)			
Virupaksha *et al.* (1975)	India	12	8·2		246	72
			(6·8–11·0)		(181–345)	
Range of means					160–262	
Range of ranges					112–345	

Reference	Location	Met: No. of samples	Met: Prot. (N × 6·25) (%)	Met: mg/100 g sample	Met: mg/g total N	Cys: No. of samples	Cys: Prot. (N × 6·25) (%)	Cys: mg/100 g sample	Cys: mg/g total N	AA score (MET + Cys)
Balasubramanian *et al.* (1952)	India	3	4·2	128^b	189^b					
Balasubramanian and Ramachandran (1957)			(3·8–4·5)	(119–145)	(162–219)					
Chitre and Vallury (1956a)		1		283		1		234		
Chitre and Vallury (1956b)		1		325		1		260		
FAO (1970)	Various	1		229	194	1		192	163	
		4			226	4			119^b	157^b
					(153–287)					
Indira and Naik (1971)	India	1	7·7		83	1	7·7		89	78
Mallanna and Rajashekara (1969)		1			181^b					
Stabursvik and Heide (1974)		33	9·1		183	33	9·1		118	137
			(4·0–15·3)		(88–227)		(4·0–15·3)		(66–144)	
Taira (1963b)	Japan		12·0		169^b		12·0		44^b	97^b
			(9·0–16·0)				(9·0–16·0)			
Virupaksha *et al.* (1975)	India	2	8·2		182	12	8·2		107	131
			(6·8–11·0)		(113–266)		(6·8–11·0)		(56–183)	
Range of means					83–226				44–163	
Range of ranges					83–287				56–183	

Reference	Location	Phe: No. of samples	Phe: Prot. (N × 6·26) (%)	Phe: mg/100 g sample	Phe: mg/g total N	Tyr: No. of samples	Tyr: Prot. (N × 6·25) (%)	Tyr: mg/100 g sample	Tyr: mg/g total N	AA score (Phe + Tyr)
Balasubramanian *et al.* (1952)	India	3	4·2	189^b	273^b					
Balasubramanian and Ramachandran (1957)			(3·8–4·5)	(118–229)	(187–325)					
Chitre and Vallury (1956a)		1		197						
FAO (1970)	Various	1		383	325					
		4			303			266	225	
					(119–434)					
Indira and Naik (1971)	India	1	7·7		359	1	7·7		352	187
Shepherd *et al.* (1971–72)	Africa	6			344	6			231	151
					(312–362)				(200–250)	
Stabursvik and Heide (1974)		33	9·1		330	33	9·1		222	145
			(4·0–15·3)		(278–361)		(4·0–15·3)		(172–257)	
Taira (1963b)	Japan		12·0		350^b		12·0		125^b	125^b
			(9·0–16·0)				(9·0–16·0)			
Virupaksha *et al.* (1975)	India	12	8·2		399	12	8·2		231	166
			(6·8–11·0)		(321–527)		(6·8–11·0)		(188–342)	
Range of means					273–399				125–352	
Range of ranges					119–527				172–352	

^{b} Microbiological assay: values are in general higher than those obtained by ion-exchange chromatography.

cont.

Table 2.130 *cont.*

Reference	Location	Thr: No. of samples	Thr: Prot. (N × 6·25) (%)	Thr: mg/100 g sample	Thr: mg/g total N	Thr: AA score	Trp: No. of samples	Trp: Prot. (N × 6·25) (%)	Trp: mg/100 g sample	Trp: mg/g total N	Trp: AA score
Balasubramanian *et al.* (1952); Balasubramanian and Ramachandran (1957)	India	3	4·2 (3·8–4·5)	149[b] (120–196)	218[b] (189–271)	87	3	4·2 (3·8–4·5)	66[b] (62–69)	97[b] (91–103)	162
Chitre and Vallury (1956a)							1		42		
FAO (1970)	Various	1		310	263				107	91	
		4			218[b] (189–171)	87[b]	4			105[b]	175[b]
Indira and Naik (1971)	India	1	7·7		234	94	5	7·7		82 (72–90)	136
Shepherd *et al.* (1971–72)	Africa	6			275 (262–287)	110					
Stabursvik and Heide (1974)		33	9·1 (4·0–15·3)		260 (202–291)	104					
Taira (1963b)	Japan		12·0 (9·0–16·0)		294[b]	118[b]		12·0 (9·0–16·0)		62[b] (34–137)	103[b]
Virupaksha *et al.* (1975)	India	12	8·2 (6·8–11·0)		316 (245–361)	126					
Range of means					218–316					62–105	
Range of ranges					189–361					34–137	

Reference	Location	Val: No. of samples	Val: Prot. (N × 6·25) (%)	Val: mg/100 g sample	Val: mg/g total N	Val: AA score
Balasubramanian *et al.* (1952); Balasubramanian and Ramachandran (1957)	India	3	4·2 (3·8–4·5)	285[b] (229–339)	418[b] (361–464)	135
Chitre and Vallury (1956a)		1		259		
FAO (1970)	Various	1		487	413	
		4			552[b] (478–609)	178[b]
Indira and Naik (1971)	India	1	7·7		364	117
Shepherd *et al.* (1971–72)	Africa	6			487 (450–525)	157
Stabursvik and Heide (1974)		33	9·1 (4·0–15·3)		402 (348–447)	130
Taira (1963b)	Japan		12·0 (9·0–16·0)		369[b]	119[b]
Virupaksha *et al.* (1975)	India	12	8·2 (6·8–11·0)		509 (425–650)	164
Range of means					364–552	
Range of ranges					348–650	

Reference	Location	Arg No. of samples	Arg Prot. (N × 6·25) (%)	Arg mg/100 g sample	Arg mg/g total N	His No. of samples	His Prot. (N × 6·25) (%)	His mg/100 g sample	His mg/g total N
Balasubramanian *et al.* (1952) / Balasubramanian and Ramachandran (1957)	India	3	4·2 (3·8–4·5)	225[b] (141–311)	326[b] (222–425)	3	4·2 (3·8–4·5)	61[b] (58–64)	92[b] (87–94)
Chitre and Vallury (1956a)		1		573		1		210	
FAO (1970)	Various	1		331	281	1		163	138
		4			327[b] (223–425)	4			119 (115–121)
Indira and Naik (1971)	India	1	7·7		514	1	7·7		162
Shepherd *et al.* (1971–72)	Africa	6			300 (231–356)	6			162 (150–169)
Stabursvik and Heide (1974)		33	9·1 (4·0–15·3)		266 (210–371)	33	9·1 (4·0–15·3)		148 (134–163)
Taira (1963b)	Japan		12·0 (9·0–16·0)		237[b]		12·0 (9·0–16·0)		144
Virupaksha *et al.* (1975)	India	12	8·2 (6·8–11·0)		327 (240–431)	12	8·2 (6·8–11·0)		174 (103–249)
Range of means					237–514				92–174
Range of ranges					210–514				87–249

Reference	Location	Ala No. of samples	Ala Prot. (N × 6·25) (%)	Ala mg/100 g sample	Ala mg/g total N	Asp No. of samples	Asp Prot. (N × 6·25) (%)	Asp mg/100 g sample	Asp mg/g total N
FAO (1970)	Various	1		458	388	1		479	406
Indira and Naik (1971)	India	1	7·7		405	1	7·7		614
Shepherd *et al.* (1971–72)	Africa	6			506 (437–537)	6			406 (375–437)
Stabursvik and Heide (1974)		33	9·1 (4·0–15·3)		370 (318–398)	33	9·1 (4·0–15·3)		455 (361–1059)
Taira (1963b)	Japan		12·0 (9·0–16·0)		506[b]		12·0 (9·0–16·0)		437[b]
Virupaksha *et al.* (1975)	India	12	8·2 (6·8–11·0)		495 (392–562)	12	8·2 (6·8–11·0)		484 (392–626)
Range of means					370–506				406–614
Range of ranges					318–562				361–1059

[b] Microbiological assay: values are in general higher than those obtained by ion-exchange chromatography.

cont.

Table 2.130 *cont.*

Reference	Location	Glu: No. of samples	Glu: Prot. (N × 6·26) (%)	Glu: mg/100 g sample	Glu: mg/g total N	Gly: No. of samples	Gly: Prot. (N × 6·25) (%)	Gly: mg/100 g sample	Gly: mg/g total N
FAO (1970)	Various	1		1497	1269	1		295	250
Indira and Naik (1971)	India	1	7·7		1385	1	7·7		370
Shepherd *et al.* (1971–72)	Africa	6			1787 (1694–1819)	6			225 (200–244)
Stabursvik and Heide (1974)		33	9·1 (4·0–15·3)		1393 (1012–1716)	33	9·1 (4·0–15·3)		243 (179–329)
Taira (1963b)	Japan		12·0 (9·0–16·0)		1175[b]		12·0 (9·0–16·0)		356[b]
Virupaksha *et al.* (1975)	India	12	8·2 (6·8–11·0)		1914 (1446–2364)	12	8·2 (6·8–11·0)		293 (233–359)
Range of means					1269–1914				225–370
Range of ranges					1012–2364				179–370

Reference	Location	Pro: No. of samples	Pro: Prot. (N × 6·25) (%)	Pro: mg/100 g sample	Pro: mg/g total N	Ser: No. of samples	Ser: Prot. (N × 6·25) (%)	Ser: mg/100 g sample	Ser: mg/g total N	Ammonia: No. of samples	Ammonia: Prot. (N × 6·25) (%)	Ammonia: mg/100 g sample	Ammonia: mg/g total N
FAO (1970)	Various	1		517	438	1		376	319				
Indira and Naik (1971)	India	1	7·7		426	1	7·7		421	1	7·7		103
Shepherd *et al.* (1971–72)	Africa	6			475 (437–500)	6			344 (300–400)	6			137 (119–150)
Starbursvik and Heide (1974)						33	9·1 (4·0–15·3)		358 (296–539)				
Taira (1963b)	Japan		12·0 (9·0–16·0)		631[b]		12·0 (9·0–16·0)		362[b]				
Virupaksha *et al.* (1975)	India	12	8·2 (6·8–11·0)		381 (262–605)	12	8·2 (6·8–11·0)		437 (345–544)				
Range of means					381–631				319–437				103–137
Range of ranges					262–605				296–544				103–150

[b] Microbiological assay: values are in general higher than those obtained by ion-exchange chromatography.

Table 2.131
Content of inorganic elements in finger millet (mg/100 g) (from multiple sources). Blank spaces indicate no data presented.

Reference	Location	No. of samples	Ca	P	Fe
Balakrishna Rao *et al.* (1973)	India	15	450 (203–690)	340 (227–470)	9 (3–20)
Deosthale *et. al.* (1970b)		13	407 (253–661)	261 (204–330)	6·4 (1·3–17·6)
Indira and Naik (1971)		5	320 (272–352)	262 (242–284)	13 (11–18)
Joseph *et al.* (1958ab)			389	239	
Kadkol annd Swaminathan (1954)		8	395 (301–495)	275 (228–370)	6·4 (5·9–6·9)
Kamalanathan *et al.* (1971)[b]		2	381 (344–417)		15 (12–17)
Kempanna and Kavallappa (1968)		19	660[a] (455–930)	713[a] (640–890)	
Krishnamurthy (1968)		25	364 (289–445)		
Kumaraswamy and Venkataramanan (1974)		10	220 (164–369)		
Mallanna and Rajashekara (1969)		1	344	251	7·8
Pore and Magar (1976a)		1	284	131	9·2
Pore and Magar (1977)[c]		36	338 (259–520)	257 (125–404)	7·7 (3·4–19·0)
Sree Ramulu and Mariakulandai (1964)		30	855 (613–1116)	807 (639–1002)	
Subrahmanyan *et al.* (1955)			502	325	
Venkataramana and Krishna Rao (1961)		22		904[a] (690–1200)	
Representative analytical values and[d] ranges (not means).[e]			257–528	60–311	(4·5–9·2)
Range of means			220–855	131–904	6·4–15
Range of ranges			164–1116	60–1200	1·3–20

[a] Ca expressed as CaO, P as P_2O_5.
The following sources also reported, mg/100 g.
[b] Phytin P 227 (208–246).
[c] Cu 1·0 (0·5–1·5), Mg 173 (105–251), Mn 1·7 (0·6–3·6), Mo 0·01 (0·009–0·02).
[d] Pinta and Busson (1963) reported mostly from single samples, K, Mg, Na (Flame spectrophotometry), Zn (polarography) and otherwise (arc spectrography) mg/100 g sample:
Al 0·4, Ba 2·2, Be < 0·05, Bi < 0·05, Bo 0·05, Cl 70·8, Co 0·01, Cr 0·02, Cu 0·5, Ga < 0·01, K 260–330, Ge < 0·10, Li 0·2, Mg 140, Mn 1·9, Mo 0·2, Na 20–67, Ni 0·02, Pb 0·6, Ru 0·2, S 122, Sn 0·006, Sr 3·3, Ti 0·03, V 0·04 and Zn 1·5. They commented that the level of Ba and Sr found in a single sample appeared to be relatively high.
[e] Data from sources mentioned in text.

Table 2.132
Vitamin content of finger millet (mg/100 g) (from multiple sources).

Reference	Location	No. of samples	Vit. B Thiamine	Vit. B Niacin	Vit. B Riboflavin
Chitre *et al.* (1955)	India	1	0·19	2·5	0·1
Deosthale *et al.* (1970b)		10	0·23 (0·11–0·61)	0·60[a] (0·27–0·87)	0·06[a] (0·02–0·07)
Kadkol and Swaminathan (1954)		8	0·62 (0·57–0·67)		
Kamalanathan *et al.* (1971)		2	0·42 (0·41–0·42)		
Mahudeswaran and Ayyamperumal (1970)		20	0·52 (0·46–0·57)	0·13 (0·10–0·15)	0·16 (0·11–0·22)
Passmore and Sundararajan (1941)		10	0·42 (0·27–0·65)		
Representative analytical values[b] and ranges (not means).[c]			(0·1–0·47)	(1·14–1·5)	(0·05–0·19)
Range of means			0·19–0·62	0·13–2·5	0·06–0·16
Range of ranges			0·1–0·67	0·10–2·5	0·02–0·22

[a] Microbiological assay.
[b] Single representative values for choline 16·9 mg/100 g; and vitamin A carotene 0·037–0·090 mg/100 g.
[c] Data from sources mentioned in text.

Table 2.133
Analyses and Osborne fractions of common millet and foxtail millet.[a]

	Common Millet		Foxtail Millet	
	% in grain	% of total prot.	% in grain	% of total prot.
Moisture (%)	9·14		9·01	
Total prot. (N × 6·25) (%)	16·9	100	10·3	100
Ash (%)	1·31		1·79	
Prot. fractions (% of grain)				
Water-sol.	0·86	5·07	0·62	6·02
Salt (10% NaCl-sol.)	0·84	4·95	0·61	5·92
Boiling (80% alcohol-sol.)	9·01	53·18	4·34	42·13
Alkali (0·5% NaOH-sol.)	4·81	28·39	3·59	34·85
Nonprotein				
Water-sol. N (× 6·25)	0·9	5·37	0·7	7·08
Insol. N in residue (× 6·25)	0·35	2·06	0·29	2·81
Total	16·8	99·02	10·2	98·81

[a] Lo (1941).

Table 2.134
Amino acid composition (mg/g total N) of the protein in the solubility fractions of Japanese barnyard millet (JBM) and foxtail millet (FM).[a]

	Albumin		Globulin		Prolamine		Glutelin	
	JBM	FM	JBM	FM	JBM	FM	JBM	FM
Ile	350	344	244	262	431	406	444	419
Leu	525	519	362	519	856	1250	631	725
Lys	456	462	375	375	26	20	381	437
Met	162	169	112	119	156	150	231	237
Cys	81	94	231	144	37	55	41	44
Phe	325	312	237	312	406	394	394	350
Tyr	275	275	225	259	275	212	275	275
Thr	344	362	219	244	262	281	344	344
Trp	156	162	94	87	69	144	106	144
Val	450	462	462	462	412	337	469	412
Arg	406	450	900	1012	125	119	469	381
His	200	200	231	219	106	119	181	156
Ala	425	406	319	369	794	794	556	506
Asp	631	675	450	556	394	462	637	762
Glu	731	725	1200	1150	1706	1587	906	896
Gly	419	387	375	362	100	100	269	275
Pro	437	412	294	362	781	919	487	450
Ser	362	306	387	406	362	269	312	350

[a] Taira (1962b).

Table 2.135

Nitrogen balance data, biological value (BV) and coefficient of digestibility (DC) of the protein of casein, foxtail millet and foxtail millet plus lysine.[a] Blank spaces indicate no data presented.

Diets	N						BV (% of diet)	DC (% of diet)
	Intake (mg/day/rat)	Output (mg/day/rat)		Retention (mg/day/rat)	Retention (% of N intake)	Faecal (% of N intake)		
		Urinary	Faecal					
(A) Casein								
Ad lib. feeding	63·1	27·5	9·3	26·3	41·7	14·7	100	83
Isocaloric feeding	45·6	18·7	7·1	19·8	43·5	15·6		
(B) Foxtail millet								
Ad lib. feeding	41·5	18·1	15·4	8·0	19·3	37·2	54	68
Isocaloric feeding	41·2	18·1	14·6	8·5	20·2	35·4		
(C) Foxtail millet plus 0·47% Lys HCl								
Ad lib. feeding	66·3	31·5	14·0	20·8	31·4	21·4	80	80
Isocaloric feeding	46·8	20·0	11·7	15·1	32·5	25·0		

[a] Ganapathy *et al.* (1957).

Table 2.136
Proximate analyses of foxtail millet (DWB) (mean with ranges[a] in parentheses) (from multiple sources). Blank spaces indicate no data presented.

Reference	Location	No. of samples	Prot. (N × 6·25) (%)	Lipid (%)	CBH (%)	Fibre (%)	Ash (%)
Indira and Naik (1971)	India	3	11·3 (10·0–12·3)	4·4 (4·3–4·5)			
Kametaka (1952)	Japan	1	11·4	2·9			1·4
Lo (1941)	USA	4	11·3 (11·3–11·3)				2·0 (1·9–2·0)
Ramanathan *et al.* (1975)	India	3	15·8 (12·1–18·7)	6·8 (4·9–9·3)	65·4 (62·9–67·1)	6·3 (6·2–6·5)	5·7 (4·9–7·1)
Solpico and Yambao (1966)	Philippines	3	11·1 (8·5–13·0)	4·2 (3·4–4·9)	72·4 (71·1–73·6)	8·1 (7·0–8·8)	4·4 (4·1–4·6)
Representative analytical values (ranges only)[c]			10·0 (6·0–14·0)	2·5 (1·2–5·2)	63·0 (60·0–75·0)	7·0 (2·0–8·6)	3·0 (1·5–3·6)
Range of means			10·0–15·8	2·5–6·8	63·0–72·4	6·3–8·1	1·4–5·7
Range of ranges			6·0–18·7	1·2–9·3	60·0–75·0	2·0–8·8	1·4–7·1
Range of Protein means at N × 5·7			9·1–14·4				
Range of Protein ranges at N × 5·7			5·5–17·0				

[a] Where available.
[b] Carbohydrate.
[c] Data from various sources referred to in the text.

Table 2.137

Amino acid content and score of foxtail millet[a] (mean with ranges in parentheses) (from multiple sources). Blank spaces indicate no data presented.

Reference	Location	Ile: No. of samples	Ile: Prot. (N × 6·25) (%)	Ile: mg/100 g sample	Ile: mg/g total N	Ile: AA score	Leu: No. of samples	Leu: Prot. (N × 6·25) (%)	Leu: mg/100 g sample	Leu: mg/g total N	Leu: AA score	Leu/Ile quotient
Chitre *et al.* (1956)	India	2	10·7 (10·4–10·9)	611[b] (603–619)			2	10·7 (10·4–10·9)	1506[b] (1358–1653)			2·5
Chitre and Ganapathy (1956)				720	387	155			1336	719	163	1·8
Chitre and Vallury (1956a)		1		497			1		1210			
FAO (1970)	Various	1		803	475[b]	190[b]	1		1764	1044	237[b]	2·2
Taira (1963a)	Japan	4			337[b] (250–369)	135[b]	4			743[b] (718–775)	169[b]	2·2
Taira (1968)		10	13·0 (10·6–15·2)		257[b] (244–275)	102	10	13·0 (10·6–15·2)		781[b] (712–881)	177	3·0
Range of means			10·7–13·0		257–475					719–1044		
Range of ranges			10·4–15·2		244–369					712–881		
Range of protein means at N × 5·7			9·7–11·8									
Range of protein ranges at N × 5·7			9·5–13·9									

Reference	Location	Lys: No. of samples	Lys: Prot. (N × 6·25) (%)	Lys: mg/100 g sample	Lys: mg/g total N	Lys: AA score
Chitre *et al.* (1956)	India	2	10·7 (10·4–10·9)	279[b] (266–292)		
Chitre and Ganapathy (1956)				249	134	39
Chitre and Vallury (1956a)		1		226		
FAO (1970)	Various			233[b]	138[b]	41
Indira and Naik (1971)	India	3			157 (143–169)	46
Taira (1963a)	Japan	4			137[b] (94–187)	40
Taira (1968)		10	13·0 (10·6–15·2)		115[b] (112–150)	33
Range of means					115–157	
Range of ranges					94–187	

[a] WHO (1973a).
[b] Microbiological assay: values are in general higher than those obtained by ion-exchange chromotography.

cont.

Table 2.137 *cont.*

Reference	Location	Met: No. of samples	Met: Prot. (N × 6·25) (%)	Met: mg/100 g sample	Met: mg/g total N	Cys: No. of samples	Cys: Prot. (N × 6·25) (%)	Cys: mg/100 g sample	Cys: mg/g total N	AA score (Met + Cys)
Chitre *et al.* (1956)	India	2	10·7 (10·4–10·9)	371 (348–394)						
Chitre and Ganapathy (1956)				267	144					
Chitre and Vallury (1956a)		1		242		1		206		
Chitre and Vallury (1956b)		1		302		1		258		
FAO (1970)	Various			296[b]	175[b]					
Taira (1963a)	Japan	4			171[b] (162–181)	4			86[b] (62–112)	117
Taira (1968)		10	13·0 (10·6–15·2)		157[b] (150–169)	10	13·0 (10·6–15·2)		92[b] (88–100)	113
Range of means					144–175				86–92	
Range of ranges					150–181				62–112	

Reference	Location	Phe: No. of samples	Phe: Prot. (N × 6·25) (%)	Phe: mg/100 g sample	Phe: mg/g total N	Tyr: No. of samples	Tyr: Prot. (N × 6·25) (%)	Tyr: mg/100 g sample	Tyr: mg/g total N	AA score (Phe + Tyr)
Chitre *et al.* (1956)	India	2	10·7 (10·4–10·9)	565[b] (548–582)						
Chitre and Ganapathy (1956)				501	269					
Chitre and Vallury (1956a)		1		454						
FAO (1970)	Various			708	419[b]					
Taira (1963a)	Japan	4			305[b] (262–344)	4			137[b] (125–150)	147
Taira (1968)		10	13·0 (10·6–15·2)		310[b] (288–338)	10	13·0 (10·6–15·2)		189[b] (175–213)	131
Range of means					269–419				137–189	
Range of ranges					262–344				125–213	

Reference	Location	Thr: No. of samples	Thr: Prot. (N × 6·25) (%)	Thr: mg/100 g sample	Thr: mg/g total N	Thr: AA score	Trp: No. of samples	Trp: Prot. (N × 6·25) (%)	Trp: mg/100 g sample	Trp: mg/g total N	Trp: AA score
Chitre *et al.* (1956)	India	2	10·7 (10·4–10·9)	375 (361–388)			2	10·7 (10·4–10·9)	86 (78–93)		
Chitre and Ganapathy (1956)				322	172	69			41	22	36
Chitre and Vallury (1956a)		1		292			1		37		
FAO (1970)	Various			328[b]	194[b]	78			103	61[b]	102
Indira and Naik (1971)	India						3			62 (51–70)	103
Taira (1963a)	Japan	4			275[b] (262–287)	110	4			103[b] (100–106)	172
Taira (1968)		10	13·0 (10·6–15·2)		261 (250–269)	104	4	13·0 (10·6–15·2)		110[b] (94–119)	183
Range of means					172–275					22–110	
Range of ranges					250–287					51–119	

Reference	Location	Val: No. of samples	Val: Prot. (N × 6·25) (%)	Val: mg/100 g sample	Val: mg/g total N	Val: AA score
Chitre *et al.* (1956)	India	2	10·7 (10·4–10·9)	637[b] (618–656)		
Chitre and Ganapathy (1956)				563	303	97
Chitre and Vallury (1956a)		1		510		
FAO (1970)	Various			728[b]	431[b]	139
Taira (1963a)	Japan	4			237[b] (306–319)	76
Taira (1968)		10	13·0 (10·6–15·2)		313[b] (300–331)	101
Range of means					237–431	
Range of ranges					300–331	

[b] Microbiological assay: values are in general higher than those obtained by ion-exchange chromotography.

cont.

Table 2.137 *cont.*

Reference	Location	Arg				His			
		No. of samples	Prot. (N × 6·25) (%)	mg/100 g sample	mg/g total N	No. of samples	Prot. (N × 6·25) (%)	mg/100 g sample	mg/g total N
Chitre *et al.* (1956)	India	2	10·7 (10·4–10·9)	767 (736–797)		2	10·7 (10·4–10·9)	422 (370–473)	
Chitre and Ganapathy (1956)				1104	594			466	251
Chitre and Vallury (1956a)		1		1000		1		422	
FAO (1970)	Various			380[b]	225[b]			221[b]	131[b]
Taira (1963a)	Japan	4			186[b] (150–206)	4			133[b] (119–169)
Taira (1968)		10	13·0 (10·6–15·2)		171[b] (156–200)	10	13·0 (10·6–15·2)		114[b] (106–125)
Range of means					171–594				114–251
Range of ranges					150–206				106–169

Reference	Location	Ala				Asp			
		No. of samples	Prot. (N × 6·25) (%)	mg/100 g sample	mg/g total N	No. of samples	Prot. (N × 6·25) (%)	mg/100 g sample	mg/g total N
Taira (1963a)	Japan	4			540[b] (525–556)	4			458[b] (444–469)
Taira (1968)		10	13·0 (10·6–15·2)		572[b] (550–619)	10	13·0 (10·6–15·2)		402[b] (381–431)
Range of means					540–572				402–458
Range of ranges					525–619				381–469

Reference	Location	Glu				Gly			
		No. of samples	Prot. (N × 6·25) (%)	mg/100 g sample	mg/g total N	No. of samples	Prot. (N × 6·25) (%)	mg/100 g sample	mg/g total N
Taira (1963a)	Japan	4			1242[b] (1225–1250)	4			174[b] (162–187)
Taira (1968)		10	13·0 (10·6–15·2)		1102[b] (950–1244)	10	13·0 (10·6–15·2)		192[b] (156–244)
Range of means					1102–1242				174–192
Range of ranges					950–1250				156–244

Reference	Location	Pro				Ser			
		No. of samples	Prot. (N × 6·25) (%)	mg/100 g sample	mg/g total N	No. of samples	Prot. (N × 6·25) (%)	mg/100 g sample	mg/g total N
Taira (1963a)	Japan	4			670[b] (644–712)	4			359[b] (337–381)
Taira (1968)		10	13·0 (10·6–15·2)		650[b] (556–775)	10	13·0 (10·6–15·2)		367[b] (344–388)
Range of means					650–670				359–367
Range of ranges					644–775				337–388

[b] Microbiological assay: values are in general higher than those obtained by ion-exchange chromotography.

Table 2.138
Content of inorganic elements in foxtail millet (mg/100 g) (ranges in parentheses) (from multiple sources). Blank spaces indicate no data presented.

Reference	Location	No. of samples	Ca	P	Fe
Indira and Naik (1971)	India	3	34 (23–45)	310 (291–321)	13 (12–14)
Ramanathan *et al.* (1975)		3	617[a] (600–630)	1153[a] (1050–1290)	
Representative analytical[b] values.[c]			(23–98)	(200–315)	(4·0–6·4)

[a] Ca reported as CaO; P as P_2O_5.
[b] Pinta and Busson (1963) reported K (flame spectrophotometry) in a single sample 430 mg/100 g.
[c] Data from various sources referred to in last text ranges only.

Table 2.139
Amino acid composition of the dehulled grain (various fractions) of common millet protein. Blank spaces indicate no data presented.[a]

Amino acids (mg/g total N)	Dehulled grain	H_2O solubles	NaCl solubles	t-BuOH solubles	Residue
Ile	256	194	169	281	219
Leu	762	375	325	912	581
Lys	94	406	275	6	125
Met	137	106	44	150	44
Cys	31	150	256	75	12
Phe	344	206	206	406	256
Tyr	250	194	187	269	319
Thr	187	287	200	169	250
Trp	50				
Val	337	319	312	312	337
Arg	200	569	831	81	300
His	131	162	162	100	256
Ala	681	431	419	800	537
Asp	387	287	462	331	456
Glu	1331	1325	1269	1562	1137
Gly	131	394	337	62	269
Pro	456	306	287	412	381
Ser	394	281	369	456	306
Ammonia	181	94	69	187	181

[a] Jones *et al.* (1970).

Table 2.140
Amino acid composition of the prolamine of maize (zein) and common millet (panicin).[a]

	Zein (maize) (mg/g total N)	Panicin (common millet) (mg/g total N)
Ile	28	27
Leu	142	73
Lys	0	1
Met	6	24
Half-Cys	4	9
Phe	38	32
Tyr	28	20
Thr	21	17
Val	33	32
Arg	30	28
His	17	39
Ala	100	103
Asp	34	21
Glu	153	137
Gly	13	16
Pro	81	63
Ser	48	46
Ammonia	189	255
Total	965	943

[a] Waldschmidt-Leitz and Metzner (1962).

Table 2.141
Amino acid score[a] and composition[b] of triticale, soy-TVP, and common millet fed in adult male and female rat diets.

	(A) Triticale		(B) Soy-TVP		(C) Common millet	
	AA score	(mg/g total N)	AA score	(mg/g total N)	AA score	(mg/g total N)
Ile	88	220	122	306	108	269
Leu	91	401	115	506	182	801
Lys	55	186	116	393	28	95
Met }	66	80	57	74	58	128
Half-Cys[c] }		(65)		(51)		(0)
Phe }		266		343		339
(Phe + Tyr)	101		143		119	
Tyr }		119		201		114
Thr	75	187	108	269	82	205
Trp[d]	(120)	(72)	(143)	(86)	(63)	(38)
Val	85	263	105	325	100	310
Arg		291		469		177
His		141		167		132
Ala		222		271		665
Asp		378		745		377
Glu		1956		1634		1392
Gly		249		263		138
Pro		622		364		413
Ser		279		452		418
N content (mg N/g sample)		25·6		77·5		18·7

[a] (WHO 1973a).
[b] Kies *et al.* (1975).
[c] Cystine values probably low owing to destruction by method used.
[d] Estimated from handbook values.

Table 2.142
Comparative protein value of (A) triticale, (B) soy-TVP and (C) common millet for human adults.[a]

Diet	Protein source intake		Total N intake	N excretion		N balance
				Urine	Faeces	
	(g source/day)	(g N/day)	(g N/day)	(g N/day)	(g N/day)	(g N/day)
(A) Triticale	156·07	4·0	4·8	4·24	1·04	−0·58
(B) Soy-TVP	52·98	4·0	4·8	4·18	1·04	−0·42
(C) Common millet	213·68	4·0	4·8	3·99	1·80	−0·99

[a] Kies *et al.* (1975).

Table 2.143
Proximate analyses of common millet, DWB (mean with ranges[a] in parentheses) (from multiple sources). Blank spaces indicate no data presented.

Reference	Location	No. of samples	Prot. (N × 6·25) (%)	Lipid (%)	CBH[b] (%)	Fibre (%)	Ash (%)
Goodearl (1943)	USA	1	16·2	4·7	68·3	7·9	3·1
Indira and Naik (1971)	India	2	11·9 (11·6–12·2)	2·3 (1·9–2·7)			
Jones *et al.* (1970)	USA	1	12·5	4·9	80·1	0·7	
Lo (1941)		4	18·6 (18·6–18·7)			1·4 (1·4–1·5)	
Solpico and Yambao (1966)	Philippines	1	13·4	3·6	72·9	6·3	3·7
Vatagin and Oksenenko (1971)	Russia	6	10·9 (10·3–12·0)	3·9 (3·7–4·1)	60·6 (59·9–61·1)	8·0 (7·9–8·2)	2·8 (2·7–2·9)
Representative analytical value[cd] (range only).			11·0 (6·4–12·8)	3·5 (2·9–4·9)	63·0 (57·0–71·0)	9·0 (4·6–10·7)	3·7 (1·4–5·0)
Range of means			10·9–18·6	2·3–4·9	60·6–80·1	0·7–9·0	1·4–3·7
Range of ranges			6·4–18·7	1·9–4·9	57·0–80·1	0·7–10·7	1·4–5·0
Range of prot. means at N × 5·7			9·9–17·0				
Range of prot. ranges at N × 5·7			5·8–17·1				

[a] Where available.
[b] Carbohydrate.
[c] Data from various sources referred to in the text.
[d] Common and little millet combined, not separable.

Table 2.144

Amino acid content and score of common millet[a] (mean with ranges in parentheses) (from multiple sources). Blank spaces indicate no data presented.

Reference	Location	Ile					Leu					
		No. of samples	Prot. (N × 6·25) (%)	mg/100 g sample	mg/g total N	AA score	No. of samples	Prot. (N × 6·25) (%)	mg/100 g sample	mg/g total N	AA score	Leu/Ile quotient
FAO (1970)	Various	2		562[b]	405[b] (390–419)	162	2		1059[b]	762[b] (681–843)	173	1·9
Janicki *et al.* (1973)	Poland	2	12·8 (12·4–13·2)		210 (208–212)	84	2	12·8 (12·4–13·2)		665 (651–679)	151	3·4
Jones *et al.* (1970)	USA	1			256	102	1			762	173	3·0
Kovalev *et al.* (1974)	Russia	5			193 (163–231)	77	5			827 (742–962)	188	4·3
NAS (1971)	N America			450		180			1080		245	2·4
Rakhimbaev (1967)	Russia						4			963 (903–1022)		
Taira (1963a)	Japan	2	11·4 (10·9–11·9)		256[b] (250–262)	102	2	11·4 (10·9–11·9)		753[b] (725–781)	171	2·9
Range of means			11·4–12·8		193–405					665–963		
Range of ranges			10·9–13·2		163–419					651–1022		
Range of protein means at N × 5·7			10·4–11·7									
Range of protein ranges at N × 5·7			9·9–12·0									

[a] WHO (1973a).
[b] Microbiological assay: values are in general higher than those obtained by ion-exchange chromotography.

cont.

Table 2.144 *cont.*

Reference	Location	Lys: No. of samples	Lys: Prot. (N × 6·25) (%)	Lys: mg/100 g sample	Lys: mg/g total N	Lys: AA score
FAO (1970)	Various	2		262[b]	189[b] (113–265)	56
Indira and Naik (1971)	India	2			266 (259–272)	78
Janicki *et al.* (1973)	Poland	2	12·8 (12·4–13·2)		89 (81–98)	26
Jones *et al.* (1970)	USA	1			94	28
Kovalev *et al.* (1974)	Russia	5			111 (95–132)	33
NAS (1971)	USA			230		68
Rakhimbaev (1967)	Russia	4			135 (101–179)	40
Taira (1963a)	Japan	2	11·4 (10·9–11·9)		137[b] (112–162)	40
Range of means					89–266	
Range of ranges					81–272	

Reference	Location	Met: No. of samples	Met: Prot. (N × 6·25) (%)	Met: mg/100 g sample	Met: mg/g total N	Cys: No. of samples	Cys: Prot. (N × 6·25) (%)	Cys: mg/100 g sample	Cys: mg/g total N	AA score (Met + Cys)
FAO (1970)	Various	2		222[b]	160[b] (144–175)					
Janicki *et al.* (1973)	Poland	2	12·8 (12·4–13·2)		84 (78–91)	2	12·8 (12·4–13·2)		46[c] (46–46)	59
Jones *et al.* (1970)	USA	1			137	1			31	76
Kovalev *et al.* (1974)	Russia	5			86 (73–114)	5			122[c] (106–153)	95
NAS (1971)	USA			280				250		24
Rakhimbaev (1967)	Russia	4			120 (79–149)	4			94[c] (66–131)	97
Taira (1963a)	Japan					2	11·4 (10·9–11·9)		172[c] (144–200)	128
Range of means					84–160				31–172	
Range of ranges					73–175				31–200	

Reference	Location	Phe: No. of samples	Phe: Prot. (N × 6·25) (%)	Phe: mg/100 g sample	Phe: mg/g total N	Tyr: No. of samples	Tyr: Prot. (N × 6·25) (%)	Tyr: mg/100 g sample	Tyr: mg/g total N	AA score (Phe + Tyr)
FAO (1970)	Various	2		427[b]	307[b] (300–313)					
Janicki *et al.* (1973)	Poland	2	12·8 (12·4–13·2)		279 (273–285)	2	12·8 (12·4–13·2)		112 (100–125)	103
Jones *et al.* (1970)	USA	1			344	1			250	156
Kovalev *et al.* (1974)	Russia	5			277 (248–306)	5			179 (158–199)	120
NAS (1971)	USA			540						
Rakhimbaev (1967)	Russia	4			349 (308–408)	4			202 (146–235)	145
Taira (1963a)	Japan	2	11·4 (10·9–11·9)		303[b] (281–325)	2	11·4 (10·9–11·9)		159[b] (156–162)	122
Range of means					277–349				112–250	
Range of ranges					248–408				100–250	

Reference	Location	Thr: No. of samples	Thr: Prot. (N × 6·25) (%)	Thr: mg/100 g sample	Thr: mg/g total N	Thr: AA score	Trp: No. of samples	Trp: Prot. (N × 6·25) (%)	Trp: mg/100 g sample	Trp: mg/g total N	Trp: AA score
FAO (1970)	Various	2		204	147[b] (81–213)	59			68[b]	49[b] (35–62)	82
Indira and Naik (1971)	India						2			71 (69–74)	118
Janicki *et al.* (1973)	Poland	2	12·8 (12·4–13·2)		154 (148–161)	62					
Jones *et al.* (1970)	USA	1			187	75	1			50	83
Kovalev *et al.* (1974)	Russia	5			146 (108–150)	58	5			105 (92–116)	175
NAS (1971)	USA			360		144			170		283
Rakhimbaev (1967)	Russia	4			279 (199–332)	112	4			96 (77–113)	160
Taira (1963a)	Japan	2	11·4 (10·9–11·9)		247[b] (244–250)	99	2	11·4 (10·9–11·9)		87[b] (81–94)	79
Range of means					146–279					49–105	
Range of ranges					81–332					35–116	

[b] Microbiological assay: values are in general higher than those obtained by ion-exchange chromotography.
[c] Cysteine + cystine.

cont.

Table 2.144 *cont.*

Reference	Location	Val: No. of samples	Val: Prot. (N × 6·25) (%)	Val: mg/100 g sample	Val: mg/g total N	Val: AA score
FAO (1970)	Various	2		566[b]	407[b] (351–432)	131
Janicki *et al.* (1973)	Poland	2	12·8 (12·4–13·2)		252 (247–258)	101
Jones *et al.* (1970)	USA	1			337	109
Kovalev *et al.* (1974)	Russia	5			296 (261–356)	95
NAS (1971)	USA			540		174
Rakhimbaev (1967)	Russia	4			378 (352–406)	122
Taira (1963a)	Japan	2	11·4 (10·9–11·9)		321[b] (306–337)	104
Range of means					252–407	
Range of ranges					247–432	

Reference	Location	Arg: No. of samples	Arg: Prot. (N × 6·25) (%)	Arg: mg/100 g sample	Arg: mg/g total N	His: No. of samples	His: Prot. (N × 6·25) (%)	His: mg/100 g sample	His: mg/g total N
FAO (1970)	Various	2		409	294[b]	2		165[b]	119[b]
Janicki *et al.* (1973)	Poland	2	12·8 (12·4–13·2)		168 (163–174)	2	12·8 (12·4–13·2)		110 (110–110)
Jones *et al.* (1970)	USA	1			569	1			131
Kovalev *et al.* (1974)	Russia	5			182 (143–226)	5			172 (148–197)
NAS (1971)	USA			330				180	
Rakhimbaev (1967)	Russia	4			296 (236–364)	4			181 (145–231)
Taira (1963a)	Japan	2	11·4 (10·9–11·9)		196[b] (187–206)	2	11·4 (10·9–11·9)		132[b] (125–137)
Range of means					168–569				110–181
Range of ranges					143–569				110–231

Reference	Location	Ala No. of samples	Ala Prot. (N × 6·25) (%)	Ala mg/100 g sample	Ala mg/g total N	Asp No. of samples	Asp Prot. (N × 6·25) (%)	Asp mg/100 g sample	Asp mg/g total N
Janicki *et al.* (1973)	Poland	2	12·8 (12·4–13·2)		241 (235–248)	2	12·8 (12·4–13·2)		339 (333–345)
Jones *et al.* (1970)	USA	1			681	1			387
Kovalev *et al.* (1974)	Russia	5			652 (560–716)	5			298 (227–355)
Rakhimbaev (1967)	Russia	4			627 (591–657)	4			229 (197–274)
Taira (1963a)	Japan	2	11·4 (10·9–11·9)		763[b] (757–769)	2	11·4 (10·9–11·9)		394[b] (394–394)
Range of means					241–763				229–394
Range of ranges					235–769				197–394

Reference	Location	Glu No. of samples	Glu Prot. (N × 6·25) (%)	Glu mg/100 g sample	Glu mg/g total N	Gly No. of samples	Gly Prot. (N × 6·25) (%)	Gly mg/100 g sample	Gly mg/g total N
Janicki *et al.* (1973)	Poland	2	12·8 (12·4–13·2)		1160 (1131–1189)	2	12·8 (12·4–13·2)		124 (121–128)
Jones *et al.* (1970)	USA	1			1331	1			131
Kovalev *et al.* (1974)	Russia	5			930 (862–975)	5			107 (75–154)
NAS (1971)	USA							300	
Rakhimbaev (1967)	Russia	4			1181 (1149–1232)	4			152 (79–197)
Taira (1963a)	Japan	2	11·4 (10·9–11·9)		1393[b] (1256–1531)	2	11·4 (10·9–11·9)		159[b] (156–162)
Range of means					930–1393				107–159
Range of ranges					862–1531				75–197

[b] Microbiological assay: values are in general higher than those obtained by ion-exchange chromotography.

cont.

Table 2.144 *cont.*

Reference	Location	Pro				Ser				Ammonia			
		No. of samples	Prot. (N × 6·25) (%)	mg/100 g sample	mg/g total N	No. of samples	Prot. (N × 6·25) (%)	mg/100 g sample	mg/g total N	No. of samples	Prot. (N × 6·25) (%)	mg/100 g sample	mg/g total N
Janicki *et al.* (1973)	Poland	2	12·8 (12·4–13·2)		388 (373–403)	2	12·8 (12·4–13·2)		301 (300–301)	2	12·8 (12·4–13·2)		114 (112–117)
Jones *et al.* (1970)	USA	1			456	1			394	1			181
Kovalev *et al.* (1974)	Russia	5			511 (427–558)	5			369 (315–436)				
Rakhimbaev. (1967)	Russia	4			328 (242–431)	4			394 (330–450)				
Taira (1963a)	Japan	2	11·4 (10·9–11·9)		647[b] (600–694)	2	11·4 (10·9–11·9)		434[b] (418–450)				
Range of means					328–647				301–434				
Range of ranges					242–694				300–450				

[b] Microbiological assay: values are in general higher than those obtained by ion-exchange chromotography.

Table 2.145

Proximate analyses and inorganic elements of little millet (DWB) (mean with ranges[a] in parentheses) (from multiple sources). Blank spaces indicate no data presented.

Reference	Location	No. of samples	Protein (N × 6·25) (%)	Lipid (%)	CBH[b] (%)	Fibre (%)	Ash (%)
Indira and Naik (1971)[c]	India	4	7·5 (6·9–7·8)	5·3 (4·7–6·1)			
Ramanathan *et al.* (1975)[d]		5	13·8 (10·8–16·4)	6·8 (4·1–9·0)	66·3 (62·6–71·0)	7·0 (5·7–7·6)	5·9 (4·6–10·1)
Range of means			7·5–13·8	5·3–6·8	66·3	7·0	5·9
Range of ranges			6·9–16·4	4·1–9·0	62·6–71·0	3·7–7·6	4·6–10·1
Range of protein means at N × 5·7			6·8–12·6				
Range of protein ranges at N × 5·7			6·3–15·0				

[a] Where available.
[b] Carbohydrate.
[c] Also reported, mg/100 g sample, Ca 32 (24–39), P 233 (229–241), Fe 28 (24–31), S 150 (142–167).
[d] Also reported, mg/100 g sample, Ca as CaO 598 (250–920), P as P_2O_5 1102 (560–1720), Mg as MgO 233. (150–430), K as K_2O 2212 (870–4220).
Other representative analytical values from various sources mentioned in the text, ranges only, mg/100 g sample, Ca 15–30, P 320–364, Fe 40–71.

Table 2.146

Amino acid content and score[a] of little millet (mean with ranges in parentheses) (from two sources). Blank spaces indicate no data presented.

	Source A[b]		Source B[c]	
	AA score	mg/g total N	AA score	mg/g total N
Ile[d]			166	416[e]
Leu[d]			154	679[e]
Lys	56	123 (94–138)	33	114[e]
Met				142[e]
Cys				
Phe				297[e]
Tyr				
Thr			84	212[e]
Trp	65	39 (32–42)	58	35[e]
Val			122	379[e]
Arg				291[e]
His				117[e]
Protein (N × 6·25) (%)				12·5
(N × 5·7) (%)				11·4

[a] WHO (1973a).
[b] Ramachandran and Phansalkar (1956) (1 sample).
[c] Indira and Naik (1971) (4 samples).
[d] Leu/Ile 1·6.
[e] Microbiological assay: values are in general higher than those obtained by ion-exchange chromatography.

Table 2.147

Proximate analyses and inorganic elements of Japanese barnyard millet (DWB) (means with ranges in parentheses) (from multiple sources). Blank spaces indicate no data presented.

Reference	Location	No. of samples	Prot. (N × 6·25) (%)	Lipid (%)	CBH[b] (%)	Fibre (%)	Ash (%)
Indira and Naik (1971)[c]	India	3	12·0 (11·2–12·7)	3·0 (2·5–3·6)			
Representative analytical[de] values			10·5	5·1	55·7	13·9	4·6
Range of protein means at N × 5·7			9·6–10·9				
Range of protein ranges at N × 5·7			10·2–11·6				

[a] Where available.
[b] Carbohydrate.
[c] This source reported, mg/100 g sample, Ca 25 (21–29), P 285 (282–288), Fe 104 (26–153) and S 140 (135–146).
[d] Also reported, mg/100 g sample, ranges only, Ca 14–19, P 121–440 and Fe 2·9–5·0. Pinta and Busson, (1963) reported, mg/100 g sample, K (flame spectrophotometry) 330–370 and from a single sample (arc spectrography), Cu 1·7, Mn 6·6.
Data from various sources mentioned in the text.

Table 2.148
Amino acid content and score[a] of Japanese barnyard millet (mean with ranges in parentheses) (from multiple sources). Blank spaces indicate no data presented.

	Source A[b]			Source B[c]		Source C[d]	
	AA score	mg/100 g sample	mg/g N	AA score	mg/g N	AA score	mg/g N
Ile	115	484[e]	288 256[f]			114	284[e,f] (281–287)
Leu	165	1218	725 513[f]			164	721[f] (712–731)
Lys	31	178 231[f]	106	31	106 (86–126)	31	106[f] (100–112)
Met	131	190	113			129	112[f] (100–125)
Cys		294	175				171[f] (106–237)
Phe	135	608	362 256[f]			137	369[f] (344–395)
Tyr		252	150				150[f] (144–156)
Thr	92	388	231 169[f]			91	228[f] (225–231)
Trp	105	106	63 63[f]	52	31 (26–33)	104	62[f] (62–62)
Val	125	652	388 300[f]			124	384[f] (381–387)
Arg		388	231 225[f]				227[f] (218–237)
His		200	119 94[f]				118[f] (112–125)
Ala		1060	631				581[f] (569–594)
Asp		640	381				381[f] (375–387)
Glu		2520	1500				1496[f] (1431–1562)
Gly		252	150				146[f] (137–156)
Pro		1060	631				631[f] (587–675)
Ser		598	356				353[f] (337–369)
Protein (N × 6·25) (%)							9·8 (9·7–10·0)
(N × 5·7) (%)							8·9 (8·8–9·1)

[a] WHO (1973a).
[b] FAO (1970) (1 sample).
[c] Indira and Naik (1971) (3 samples) India.
[d] Taira (1963a) (2 samples) Japan.
[e] Leu/Ile 2·5.
[f] Microbiological assay: values are in general higher than those obtained by ion-exchange chromatography.

Table 2.149
Proximate analyses and inorganic elements of kodo millet (DWB) (mean with ranges in parentheses)[a] (from multiple sources). Blank spaces indicate no data presented.

Reference	Location	No. of samples	Protein (N × 6·25) (%)	Lipid (%)	CBH[b] (%)	Fibre (%)	Ash (%)
Indira and Naik (1971)[c]	India	4	12·1 (11·1–13·1)	4·3 (3·6–4·9)			
Majmudar and Kuhnte (1955)		2	6·6 (6·2–6·9)	3·4 (3·2–3·6)	74·2 (72·8–75·6)	10·5 (10·0–11·0)	4·1 (4·1–4·1)
Ramanathan *et al.* (1975)[d]		1	10·9	6·6	72·5	6·2	3·7
Representative analytical values[e,f]			11·5	1·5	74·0 (68·6–74·6)	0·4	2·9
Range of means			6·6–12·1				
Range of ranges			6·2–13·1				
Range of means at N × 5·7			6·0–11·0				
Range of ranges at N × 5·7			5·6–12·0				

[a] Where available.
[b] Carbohydrate.
[c] This source also reported, mg/100 g sample, Ca 25 (15–45), P 247 (220–271), Fe 12 (10–15) and S 150 (120–172).
[d] This source also reported, mg/100 g sample, Ca as CaO 620, P as P_2O_5 670, K as K_2O 3460 and Mg as MgO 430.
[e] This source also reported, mg/100 g sample, ranges Ca 25–39, P 118–242 and Fe 1·7–12·9.
[f] Data from various sources in the text.

Table 2.150
Amino acid content and score[a] of kodo millet (mean and ranges in parentheses) (from multiple sources). Blank spaces indicate no data presented.

	Source A[b]			Source B[c]		
	AA score	mg/100 g sample	mg/g N	AA score		mg/g N
Ile[d]	75	171	188	79		197
	192[e]		479[e] (477–481)			
Leu[d]	95	381	419	149		655
	166[e]		731[e] (669–792)			
Lys	55	171	188	62		212 (193–236) 4 samples
	63[e]		214[e] (206–221)			
Met }		85	94	82		115
			160[e] (119–200)			
Cys }						65[f]
Phe }		341	375			352
	154		427[e] (285–569)		161	
Tyr }		194	213			259
Thr	78	176	194		67	167
	78[e]		196[e] (154–238)			
Trp	63[e]	35[e]	38[e] (30–46)		108	65 (58–71)
Val	77	216	238	94		292
	143[e]		442[e] (430–450)			
Arg		205	225			247
			300[e]			
His		86	94			132
			125[e]			
Ala		341	375			334
Asp		347	381			402
Glu		694	763			2119
Gly		171	188			290
Pro		301	331			577

cont.

Table 2.150 *cont.*

	Source A[b]			Source B[c]	
	AA score	mg/100 g sample	mg/g N	AA score	mg/g N
Ser		228	250		261
Ammonia					87
Protein (N × 6·25) (%)				11·2	
(N × 5·7) (%)				10·2	

[a] (WHO 1973a).
[b] FAO (1970) (1 sample chemical, 2 samples microbiological).
[c] Indira and Naik (1971) (1 sample).
[d] Leu/Ile. FAO (1970) 2·2; Indira and Naik (1971) 3·3.
[e] Microbiological assay: values are in general higher than those obtained by ion-exchange chromatography.
[f] Cysteine + cystine.

Table 2.151
Mean chemical composition (DWB) (ranges in parentheses) of whole and dehusked *Digitaria exilis* grain (8 samples).[a]

	Whole grain %	Dehusked grain %
Crude prot.	8·7 (5·1–10·4)	8·3 (7·3–9·6)
Lipids	3·5 (2·1–5·2)	2·0 (0·6–3·8)
CBH[b]	73·6 (62·7–80·0	85·1 (82·2–87·3)
Fibre	8·5 (4·6–11·3)	1·8 (0·4–2·9)
Ash	3·8 (1·8–6·0)	1·8 (0·5–4·4)

[a] Portères (1955) citing several sources.
[b] Carbohydrate.

Table 2.152
Amino acid content[a] (one sample each) chromatographic analysis, of *Digitaria exilis* and *Digitaria iburua* and *AA scores*.[b]

	Digitaria exilis			*Digitaria iburua*		
	AA score	(mg/g total N)	(mg/100 g total grain)	AA score	(mg/g total N)	(mg/100 g total grain)
Ile[c]	100	250	315	107	269	508
Leu[c]	139	613	772	168	738	1395
Lys	48	163	205	35	119	225
Met }	239	350	441	148	188	355
Cys }		175	221		138	261
Phe }	143	319	402	174	425	803
Tyr }		225	284		238	450
Thr	100	250	315	82	206	389
Trp	147	88	111	190	114	215
Val	117	363	457	105	325	614
Arg		238	300		206	389
His		131	165		138	261
Ala		563	709		619	1170
Asp		406	512	381	720	
Glu		1263	1591		1519	2871

cont.

Table 2.162 *cont.*

	Digitaria exilis			*Digitaria iburua*		
	AA score	(mg/g total N)	(mg/100 g total grain)	AA score	(mg/g total N)	(mg/100 g total grain)
Gly		200	252		175	331
Pro		444	559		456	862
Ser		319	402		356	673

[a] FAO (1970).
[b] Calculated according to WHO (1973a).
[c] Leu/Ile. *Digitaria exilis* 2·4; *Digitaria iburua* 2·7.

Table 2.153
Amino acid content of teff[a] (mean and ranges in parentheses) mean of 'combined' teff[b] and range of six samples and AA score.[c]

	Jansen (1962)		FAO (1970)		
	AA score	mg/g total N	AA score	(mg/100 g total grain)	(mg/g total N)
Ile[d]	101	254 (231–262)	96	378	241 (223–258)
Leu[d]	110	483 (406–512)	105	724	461 (408–527)
Lys	57	194 (156–219)	51	273	174 (127–217)
Met } Cys }	150	330 (300–375)	136[e]		
Met				246	157 (141–181)
Cys				226	144[e] (110–185)
Phe } Tyr }	116	441 (375–531)	114		
Phe				474	302 (279–335)
Tyr				207	132 (109–218)
Thr	84	209 (187–219)	85	334	213 (195–264)
Trp	135	81 (94–112)	155	146	93[e] (75–107)
Val	106	328 (262–337)	109	491	313 (259–350)
Arg		221 (156–250)		337	215 (160–277)
His		134 (94–137)		188	120 (99–152)
Ala				474	302 (251–403)
Asp				600	382 (352–440)
Glu				2471	1574 (1472–1695)
Gly				363	231 (195–277)
Pro				487	310 (283–349)
Ser				413	263 (244–295)

[a] Chromatographic analysis and microbiological assay (FAO, 1970).
[b] Jansen (1962).
[c] WHO (1973a).
[d] Leu/Ile 1·1.
[e] Microbiological assay: values are in general higher than those obtained by ion-exchange chromatography.

Table 2.154

Mineral composition of teff seeds compared with spring wheat, winter wheat, winter barley and grain sorghum, direct reading emission (mg/100 g).[a]

Chemical element	Purple teff[b]	Purple teff[c]	White teff[b]	White teff[c]	Spring wheat grand mean	Winter wheat grand mean	Winter Barley grand mean	Grain sorghum grand mean
Ca	140	207	160	187	< 100	< 100	< 100	< 100
P	415	453	480	440	510	400	480	520
Fe	12·70	23·63	10·60	12·47	7·85	4·00	3·50	6·65
Al	2·25	12·40	< 1	1·43	< 1	< 1	< 1	< 1
Ba	1·5	2·30	2·17	2·53	0·75	0·60	0·70	< 0·1
B	1·60	1·33	1·37	1·30	1·20	1·15	1·10	1·65
Co	0·05	0·06	0·05	0·08	0·06	0·06	0·03	0·03
Cu	4·67	5·86	2·30	4·90	2·00	1·10	1·40	2·35
Mg	155	190	183	190	150	120	130	180
Mn	1·55	2·50	1·67	4·43	5·30	3·60	1·20	2·90
Mo	0·06	0·09	0·07	0·09	0·06	0·06	0·04	0·05
K	570	223	330	< 100	370	330	440	440
SiO	trace	157	trace	90	trace	trace	trace	trace
Na	25·20	20·07	22·70	19·73	19·50	16·85	39·20	14·15
Sr	< 0·1	< 0·1	< 0·1	< 0·1	< 0·1	< 0·1	< 0·1	< 0·1
Zn	5·60	7·57	6·33	7·23	6·00	3·95	4·50	4·40

[a] Mengesha (1966).
[b] Pure strain.
[c] 12 mixed strains.

Table 2.155
Nitrogen, DWB and amino acid composition of two varieties of teff[a] and amino acid score.[b]

	Variety A				Variety B	
	Whole grain		Flour		Whole grain	
		AA score		AA score		AA score
N (%)	2·05		2·02		1·82	
Amino acids (mg/g total N)						
Ile[c]	229	92	236	94	262	105
Leu[c]	445	101	450	104	527	120
Lys	127	37	134	39	217	64
Met }	147	119	147	120	174	129
Cys }	114		117		110	
Phe }	280	102	290	106	336	126
Tyr }	109		115		143	
Thr	210	84	229	92	264	106
Trp	105	175	100	167	75	125
Val	311	100	374	121	350	113
Arg	186		204		277	
His	103		115		152	
Ala	311		329		404	
Asp	352		354		441	
Glu	1597		1555		1695	
Gly	196		220		277	
Pro	283		297		349	
Ser	250		244		296	

[a] Kihlberg and Ericson (1963).
[b] WHO (1973a).
[c] Leu/Ile. Variety A 1·9, Variety B 2·0.

Table 2.156
Effect of amino acid supplementation of teff on rat weight gain, N efficiency ratio (NER) and fat content of liver.[a]

Diet	Av. wt gain (g/day)	NER (g wt gain/ g N intake)	Fat content of liver (DWB %)
Series I			
(A) Teff flour A (TF-A)	1·34	11·17	21·9
(B) TF-A + 0·3% L-Lys HCl	4·35	20·04	18·1
(C) TF-A + 0·3% DL-Thr	1·29	9·92	19·2
(D) TF-A + 0·5% L-Lys HCl + 0·3% DL-Thr	5·33	24·08	11·2
Series II			
(E) Whole teff A	1·20	8·21	30·7
(F) TF-A	1·32	9·18	22·4
(G) TF-A + 0·5% L-Lys HCl + 0·3% DL-Thr	5·00	21·52	10·9
(H) Casein	3·67	19·45	16·0
Series III			
(J) Whole teff B	1·64	10·98	25·4
(K) TF-B + 0·5% L-Lys HCl + 0·3% DL-Thr	4·75	22·96	15·0
(L) TF-B + 0·5% L-Lys HCl + 0·2% L-Ile	3·49	18·81	10·3
(M) Diet (K) + 0·2% L-Ile	4·56	22·96	12·8
(N) Diet (K) + 0·15% DL-Trp	4·95	23·52	10·7

[a] Kihlberg and Ericson (1963).

Table 2.157
Distribution of protein, Ca and P between the husk and endosperm of *Amaranthus paniculatus* seeds.[a]

	Whole grain	Husk	Endo.	Dried solids from the supernatant
% of whole grain	100	20·4	67·4	12·1
Prot. (%)	14·8	24·6	7·7	31·0
% of prot. in whole grain	100	33·9	35·1	25·4
Ca (mg/100 g)	221	34·7	116	566
% of Ca in whole grain	100	32·0	35·3	31·0
P (mg/100 g)	646	651	230	2691
% of P in whole grain	100	20·6	24·0	50·4

[a] Kurien (1967a).

Table 2.158
Amino acid content[a] (ranges in parentheses) of Job's tears and AA score.[b] Blank spaces indicate no data presented.

	AA score	Column chromatographic method		Microbiological method[c]	
		(mg/100 g total grain)	(mg/g total N)	(mg/100 g total grain)	(mg/g total N)
Ile[d]	110	608	275 (244–306)		363
Leu[d]	195	1901	860 (850–869)		1200
Lys	38	285	129 (119–138)		141
Met }	104	318	144 (125–163)		166
Cys }		188	85 (63–106)		113
Phe }	130	636	288 (269–306)		294
Tyr }		457	207 (144–269)		
Thr	74	409	185 (181–188)		213
Trp	75	99	45 (26–63)	57	26
Val	124	851	385 (356–413)		363
Arg		568	257 (225–288)		247
His		305	138 (131–144)		122
Ala		1319	597 (569–625)		
Asp		877	397 (381–413)		
Glu		3275	1482 (1444–1519)		
Gly		409	185 (169–200)		
Pro		1589	719 (531–906)		
Ser		630	285 (275–294)		

[a] FAO (1970).
[b] WHO (1973a).
[c] Microbiological method; values are in general higher than those obtained by ion-exchange chromatography.
[d] Leu/Ile 1·8.

Chapter 3

Table 3.1
Mean oil and protein percentages and 100-seed weight of 14 varieties of sorghum.[a]

	Prot.[b] (%)	Oil[b] (%)	100-seed weight[b] (g)
Male parents			
Norghum	12·0	3·38	2·46
TX 7078	12·6	3·17	2·85
TX 04	12·7	2·92	2·65
TX 07	12·1	3·30	2·13
TX 74	11·5	3·35	2·55
Redbine 60	13·3	2·93	2·80
Plainsman	11·1	3·01	2·39
Caprock	11·4	3·15	2·66
Mean of male parents	12·1	3·15	2·56
Female parents			
Reliance	14·4	3·31	2·39
Martin	13·3	3·45	3·01
Combine Kafir 60	11·8	3·05	3·02
Westland	11·3	3·10	2·50
Wheatland	11·9	3·05	3·52
Redlan	11·3	2·77	3·14
Mean of female parents	12·3	3·12	2·93
Mean of all parents	12·2	3·14	2·72

[a] Reich and Atkins (1971).
[b] LSD_{05} (among individual parent means, % oil = 0·45; % protein = 2·1; and 100-seed weight = 0·58.

Table 3.2
Ranges of grain yield, protein in seed and yield, and Lys in protein and yield for parents and hybrids from four sorghum cultivars (female parents) and 14 sorghum restorer lines.[a]

	Ranges of means			
	Parents		Hybrids	
Grain yield[b] (kg/ha)	4601 (IS 2942)	10 363 (Female parent possibly Wheatland)	5854 (Combine Kafir 60 × Inbred sel. 2)	17 288 (Wheatland × IS 5437)
Prot. in seed (N × 6·25) (%)	9·9 (IS 5437)	15·7 (IS 2942)	9·1 (Combine Kafir 60 × IS 115)	14·3 (Martin × Inbred sel. 2)
Lys (% of prot.)	1·60 (Female parent possibly Redlan)	2·19 (IS 3977)	1·69 (Martin × IS 855)	2·27 (Combine Kafir 60 × IS 115)
Prot. yield (kg/ha)	663 (IS 3977)	1114 (IS 8295)	702 (Combine Kafir 60 × Inbred sel. 2)	1876 (Wheatland × IS 5437)
Lys yield (kg/ha)	12·7 (IS 855)	21·5 (IS 8295)	13·3 (Combine Kafir 60 × Inbred sel. 2)	37·3 (Wheatland × IS 5437)

[a] Abifarin and Pickett (1970).
[b] The grain yield of RS 610 grown as a check was 7645 kg/ha.

Table 3.3
Means of grain sorghum F_1 hybrids with one parent in common and the mean of the common parent for yield, protein and lysine.[a]

Parent	Grain yield		Prot. (N × 6·25)		Lys	
	Hybrid mean (kg/ha)	Parent mean (kg/ha)	Hybrid mean (%)	Parent mean (%)	Hybrid mean (mg/g total N)	Parent mean (mg/g total N)
Female						
Martin	6474	4546	13·6	15·0	96	86
CK 60	6985	5081	12·4	11·3	99	104
Wheatland	7483	5353	12·6	13·2	94	88
Redlan	7695	5665	12·7	12·2	98	107
Male						
IS 0271	6391	3592	12·7	13·4	94	102
IS 0508	7701	5736	10·7	11·3	106	94
IS 0718	7520	5290	12·2	11·8	93	97
IS 1220	8497	8143	13·3	11·2	105	124
IS 2319	7551	6734	13·1	14·0	98	123
IS 3935	7451	4410	12·3	13·2	94	103
IS 6898	6722	5189	12·7	13·8	86	102
IS 8005	6750	3088	13·9	13·3	99	85
IS 8120	7020	4326	14·3	15·5	87	79
IS 8168	8329	5697	12·1	12·5	91	104
IS 8255	5921	5394	13·6	12·0	92	101
IS 8301	6036	3832	12·7	14·9	102	92

[a] Collins and Pickett (1972a).

Table 3.4
Mean content of protein, Lys, Thr and Leu, hardness, 100-seed weight (ranges in parentheses) and specific gravity (SG) of samples of grain sorghum.[a]

	Mean
Prot.[b] (%)	12·2 (7·6–22·2)
Lys[b] (mg/100 g grain)	252 (100–500)
(mg/g total N)	131 (72–184)
Thr[c] (mg/100 g grain)	242 (200–730)
(mg/g total N)	179 (100–278)
Leu[d] (mg/100 g grain)	1578 (780–2400)
(mg/g total N)	788 (426–985)
Hardness[e] (kg)	5·65 (0·04–11·38)
100-seed weight[f] (g)	2·42 (0·14–4·8)
SG[g]	60 NF 20 F

[a] Tripathi *et al.* (1971).
[b] 123 samples.
[c] 57 samples.
[d] 64 samples.
[e] 81 samples.
[f] 82 samples.
[g] F = floating; NF = non-floating.

Table 3.5
Taxonomic group and characteristics of 24 lines of sorghum.[a]

Code No.	IS No.	Group	Flowering time (days)	Plant height (cm)	100-seed weight (g)	Endosperm		Grain size[d]	Prot. (N × 6·25) (DWB) (%)	Leu (mg/g total N)	Lys (mg/g total N)
						Texture[b]	Colour[c]				
1	3922	Caffrorum	65·0	65·0	4·60	4·0	3·0	5·5	12·3	893	114
2	3126	Nervosum-broomcorn	74·5	142·5	2·35	1·0	1·0	1·0	17·9	971	141
3	8353	Caffrorum-durra	59·0	62·5	4·15	4·0	1·0	6·5	17·4	831	150
4	7005	Nigricans-bicolour	65·5	234·0	1·95	4·5	1·0	2·0	13·4	851	166
5	8850	Caudatum-kaura	104·0	216·4	2·70	1·0	1·0	4·0	9·0	856	156
6	10202	Durra	60·0	222·5	5·05	4·0	1·5	6·5	12·8	922	187
7	8892	Nigricans	93·5	194·4	2·75	1·0	1·0	4·0	9·1	751	155
8	3766	Nervosum-kaoliang	62·5	171·9	2·55	2·0	1·0	3·5	13·3	927	138
9	7461	Caudatum-kaura	70·5	225·5	3·55	4·0	3·0	7·0	12·6	887	128
10	4400	Durra	75·0	183·8	3·85	3·0	1·0	5·0	10·6	877	174
11	675	Durra-bicolor	78·0	170·0	3·75	4·0	1·0	6·0	8·0	781	204
12	3817	Guineense	78·5	284·0	3·45	2·0	1·0	2·5	12·6	885	154
13	3935	Caudatum-kaura	54·5	95·7	4·30	4·0	2·0	5·5	11·3	878	151
14	3911	Roxburghii	93·0	200·8	2·95	1·0	1·0	2·5	14·3	946	190
15	8191	Caffrorum-birdproof	69·5	157·0	2·65	4·0	1·0	6·0	16·8	866	161
16	2549	Zera-zera	68·0	152·0	2·75	3·0	1·0	3·0	9·4	806	196
17	2939	Caffrorum	57·0	49·7	4·15	4·0	2·0	5·5	11·2	784	152
18	3775	Nervosum-kaoliang	57·0	223·8	2·65	1·5	1·0	4·0	12·3	930	182
19	5600	Durra	74·0	173·1	3·50	3·0	1·0	3·0	9·8	823	179
20	968	Caudatum-kaura	69·0	144·3	5·05	4·0	3·0	9·0	11·9	883	119
21	3688	Caffrorum	52·0	91·1	3·70	4·0	3·0	6·0	12·5	787	130
22	6820	Conspicuum	89·0	250·0	3·40	1·0	1·0	3·5	12·1	919	138
23	932	Durra	58·5	176·5	3·75	4·0	1·0	4·0	11·6	1125	218
24	2039	Caffrorum-bicolor	53·0	168·0	2·55	1·0	1·0	3·0	13·0	882	134

[a] Govil and Murty (1973b).
[b] Score: 1—starchy to 5—highly glutinous;
[c] Score: 1—white to 6—purple;
[d] Score: 1—small to 10—bold.

Table 3.6
Sorghum varieties used in diallel and line × tester analysis.[a]

Parents[b]	Origin	Classificatory status	Protein (N × 6·25) (DWB) (%)	Lys (mg/100 grain)	Lys (mg/g total N)
(1) IS 3922	USA	Caffrorum feterita	12·0	241	126
(2) IS 3924	USA	Caffrorum feterita	14·8	277	117
(3) IS 9837	Sudan	Roxburghii shallu	16·4	280	107
(4) IS 10525	USA	Caffrorum	12·5	228	110
(5) IS 10526	USA	Caffrorum	13·8	221	100
(6) IS 10670	Nigeria	Caudatum-kafir	12·5	247	122
(7) IS 10890	USA	Caffrorum	14·2	276	121
(8) IS 3703	Ethiopia	Milo-kaura	15·0	257	107
(9) IS 8191	India (Assam)	Caffrorum-birdproof	17·3	368	133
(10) IS 9985	Nigeria	Caudatum-kaura	16·3	234	89
(11) IS 10202	Egypt	Durra	16·1	284	110
(12) IS 10521	USA	Milo	16·0	284	111
	Low (L)		< 13·0	< 260	< 112
	Medium (M)		13·0–16·0	260–300	112–125
	High (H)		> 16·0	> 300	> 125

[a] Rana and Murty (1975).
[b] (1) to (7) diallel parents; IS 9837 and IS 10526—female parents; and rest of the 10—male parents in line × tester mating.

Table 3.7
Average performance of hybrids and heterosis over superior parents (SP) for protein and Lys in sorghum.[a]

Cross	Prot. (N × 6·25 DWB) (%)		Lys (mg/100 g grain)		Lys (% prot.)	
	Hybrid mean	Heterosis (% SP)	Hybrid mean	Heterosis (% SP)	Hybrid mean	Heterosis (% SP)
(1) IS 3924 × IS 3922	13·35	− 10·10	248·5	− 10·45	1·86	− 7·69
(2) IS 3924 × IS 10525	14·82	− 02·00	315·0	13·35	2·12	13·67
(3) IS 3924 × IS 10526	14·36	− 3·30	295·0	6·31	2·04	9·38
(4) IS 3924 × IS 10670	12·01	− 19·19	171·0	− 38·38	1·43	− 27·04
(5) IS 3924 × IS 10890	15·27	− 3·91	259·5	− 6·49	1·73	− 10·59
(6) IS 9837 × IS 3703	16·61	1·16	259·5	− 7·49	1·56	− 9·30
(7) IS 9837 × IS 3922	13·06	− 20·46	276·5	− 1·43	2·11	4·71
(8) IS 9837 × IS 3924	14·38	− 12·42	271·0	− 3·39	1·88	0·80
(9) IS 9837 × IS 8191	13·76	− 20·32	286·0	− 22·39	2·07	− 2·82
(10) IS 9837 × IS 9985	15·54	− 5·36	279·5	− 0·36	1·78	4·40
(11) IS 9837 × IS 10202	13·98	− 14·86	237·5	− 16·37	1·75	− 0·85
(12) IS 9837 × IS 10521	13·12	− 20·09	270·5	− 4·75	2·07	16·95
(13) IS 9837 × IS 10525	13·62	− 17·05	260·5	− 7·37	1·91	8·21
(14) IS 9837 × IS 10526	13·74	− 16·32	271·0	− 3·39	1·97	15·54
(15) IS 9837 × IS 10670	14·46	− 11·94	251·0	− 10·51	1·72	− 12·24
(16) IS 9837 × IS 10890	15·01	− 8·58	286·5	2·14	1·89	− 2·32
(17) IS 10525 × IS 3922	13·30	6·06	288·0	19·25	2·15	6·70
(18) IS 10525 × IS 10526	13·00	− 5·52	317·0	38·70	2·46	39·37
(19) IS 10525 × IS 10670	14·96	19·30	341·5	38·26	2·27	15·82
(20) IS 10526 × IS 3703	13·62	− 8·96	261·5	1·55	1·91	11·05
(21) IS 10526 × IS 3922	15·46	− 12·35	277·5	14·91	1·79	− 11·17
(22) IS 10526 × IS 8191	14·27	− 17·37	296·0	19·67	2·07	− 2·82
(23) IS 10526 × IS 9985	15·27	− 6·20	284·0	21·36	1·85	15·26
(24) IS 10526 × IS 10202	13·99	− 12·94	252·5	− 11·09	1·81	2·55
(25) IS 10526 × IS 10521	14·10	− 11·76	264·0	− 7·04	1·87	5·65
(26) IS 10526 × IS 10670	17·09	24·00	324·0	31·17	1·89	− 3·57
(27) IS 10670 × IS 3922	14·32	14·19	273·0	10·53	1·90	− 5·21
(28) IS 10890 × IS 3922	13·12	− 7·80	319·5	15·76	2·44	21·09
(29) IS 10890 × IS 10525	16·73	17·57	176·5	− 36·05	1·04	− 46·25
(30) IS 10890 × IS 10526	14·00	− 1·62	182·0	− 34·06	1·29	− 33·33
(31) IS 10890 × IS 10670	14·87	4·49	313·0	13·41	2·11	7·65

[a] Rana and Murty (1975).

Table 3.8
Means (ranges in parentheses) for protein, amino acids, β-carotene and yield attributes in 55 sorghum genotypes.[a]

	Parents	Hybrids
Yield (g/plant)	37·8 (6·9–88·4)	58·3 (1·3–120·4)
100-seed weight (g)	2·8 (1·1–3·5)	3·2 (1·8–4·1)
Grain hardness (kg)	8·0 (3·7–10·9)	9·0 (3·7–11·2)
Prot. (N × 6·25) (%)	14·1 (12·2–18·3)	14·5 (11·6–19·2)
AAs (mg/g total N)		
Ile	241 (222–255)	247 (220–357)
Leu	878 (831–931)	942 (864–1436)
Lys	116 (94–133)	104 (84–143)
Met	68 (39–92)	66 (21–102)
Cys	64 (56–84)	61 (38–205)
Phe	324 (300–362)	335 (306–491)
Tyr	274 (264–287)	277 (246–421)
Thr	188 (171–204)	185 (149–267)
Trp	42 (42–57)[b]	57 (45–61)
Val	316 (257–349)	318 (258–419)
Arg	214 (181–236)	208 (177–272)
His	130 (101–155)	124 (87–167)
Ala	596 (541–670)	608 (458–909)
Asp	441 (397–479)	442 (325–644)
Glu	1467 (1347–1600)	1554 (1086–2322)
Gly	186 (150–206)	173 (142–230)
Pro	514 (459–561)	561 (484–849)
Ser	247 (126–309)	237 (106–359)
Vitamin (mg/100 g)		
β-Carotene	(0·013–0·060)	(0·008–0·064)

[a] Nanda and Rao (1975a).
[b] Values converted from original text which shows mean as equal to lower range value.

Table 3.9
Chemical composition and seed characteristics of whole grain samples of two high-Lys lines compared to normal sorghum.[a]

Composition	High-Lys lines IS 11167	IS 11758	Normal[b] sorghum
Prot. composition			
Prot. (%)	15·70	17·20	12·70
Lys (g/100 g protein)	3·33	3·13	2·05
Lys (% of sample)	0·52	0·54	0·26
Prot./seed (mg)	4·38	4·21	3·53
Lys/seed (mg)	0·15	0·13	0·07
Chemical composition			
Catechin equivalent value	0·34	0·37	0·38
Oil (%)	5·81	6·61	3·32
Seed characteristics			
Germ (%)	14·60	16·30	10·10
100-seed weight (g)	2·78	2·45	2·75
CBH[c] composition (% of sample)			
Reducing sugars	0·38	0·32	0·34
Sucrose	3·08	2·61	1·03
Total sugars	3·46	2·93	1·34
WSP	0·91	1·01	1·11
Starch	58·90	57·80	60·80
Amylose (% of starch)	25·00	26·20	25·00
Total CBH (% of sample)	63·27	61·74	63·25

[a] Singh and Axtell (1973a).
[b] Protein composition, oil % and 100-seed weight average of 31 genotypes over six locations and 2 years (Schaffert, 1972); catechin equivalents average of 300 low-tannin lines in the World Collection (Axtell *et al.*, 1974); % germ average of four lines (Jambunathan and Mertz, 1973a); and carbohydrate composition, values of IS 8313.
[c] Carbohydrate.

Table 3.10
Amino acid composition and protein in the defatted endosperm tissue from segregating F_2[a] corneous (C) and floury (F) kernels.[b]

Amino acid	PP3R*ms*$_3$ × IS 11167 C	F	PP3R*ms*$_3$ × IS 11758 C	F
	(mg/g N)			
Ile	244	260	263	249
Leu	982	896	996	789
Lys	75	136	81	162
Met	100	95	132	121
Cys	129	95	114	110
Phe	341	341	345	327
Tyr	267	270	284	265
Thr	160	188	172	194
Trp	57	76	58	109
Val	272	311	294	299
Arg	159	229	173	277
His	124	119	121	127
Ala	634	603	656	542
Asp	347	468	384	452
Glu	1667	1499	1749	1301
Gly	136	186	148	212
Pro	495	432	522	399
Ser	234	256	240	249
Prot. (N × 6·25) (DWB) (%)	9·7	10·7	9·7	12·1
Leu/Ile	4·03	3·44	3·93	3·16

[a] From crosses between normal (PP3R*ms*$_3$) and high-lysine sorghum lines.
[b] Singh and Axtell (1973a).

Table 3.11

Amino acid content in defatted whole kernel of average sorghum and high-Lys sorghum lines IS 11167 and IS 11758.[a]

Amino acid	Average sorghum[b] (mg/g N)	IS 11167 Laf. (mg/g N)	IS 11167 PR (mg/g N)	IS 11758 Laf. (mg/g N)	IS 11758 PR (mg/g N)
Ile	255	254	270	255	259
Leu	892	747	869	839	810
Lys	134	213	211	187	204
Met	112	108		128	
Cys	94	114		121	
Phe	324	322	329	345	334
Tyr	281	249	257	274	261
Thr	204	211	229	204	191
Trp	82[c]	74		76	
Val	334	328	319	319	328
Arg	224	316	327	292	307
His	126	147	151	133	134
Ala	618	527	568	584	566
Asp	489	439	492	426	495
Glu	1451	1244	1496	1397	1406
Gly	192	254	259	237	231
Pro	510	426	434	432	406
Ser	282	263	302	267	260
Prot. (N × 6·25) (DWB) (%)	12·6	14·3	16·5	16·6	17·8
Leu/Ile	3·50	2·95	3·22	3·29	3·12

[a] Grown at Lafayette (Laf.) and Puerto Rico (PR) (Singh and Axtell, 1973c).
[b] Weighted average for 522 lines (Srinivasan *et al.*, 1972).
[c] Weighted average for 3 lines.

Table 3.12

Amino acid content in defatted endosperm of average sorghum and high-Lys. sorghum lines IS 11167 and IS 11758.[a]

Amino acid	Average sorghum[b] (mg/g N)	IS 11167 Laf. (mg/g N)	IS 11167 PR (mg/g N)	IS 11758 Laf. (mg/g N)	IS 11758 PR (mg/g N)
Ile	287	261	261	269	271
Leu	1044	771	912	899	866
Lys	125	188	154	153	152
Met	84				
Cys	109				
Phe	375	326	352	359	349
Tyr	331	262	274	281	279
Thr	225	218	192	203	185
Val	356	334	306	319	306
Arg	225	294	277	239	239
His	159	137	120	124	116
Ala	684	527	597	604	586
Asp	512	472	463	437	405
Glu	1684	1245	1493	1457	1532
Gly	184	259	206	217	189
Pro	566	474	432	482	446
Ser	309	272	277	271	248
Prot. (N × 6·25) (DWB) (%)	9·3	12·1	14·1	13·9	16·5
Leu/Ile	4·60	2·95	3·49	3·34	3·27

[a] Grown at Lafayette (Laf.) and Puerto Rico (PR) (Singh and Axtell, 1973c).
[b] Average of two lines (Jambunathan and Mertz 1972).

Table 3.13
Protein fractionation[a] of normal and high-Lys sorghum.[b]

	Normal sorghum	High-Lys sorghum
Lys content		
% of protein	1·8	3·3
mg/g total N	112	206
Protein distribution (%)		
Fraction I		
Albumins and globulins (NaCl 0·5 M, 4°C, Water)	15·3	22·4
Fraction II		
Prolamine (Isopropanol 70%, v/v, 20°C)	26·4	13·7
Fraction III		
Prolamine (Isopropanol 70%, v/v, + 2-mercaptoethanol 0·6% v/v, 20°C)	26·5	20·2
Fraction IV		
Glutelin-like (Borate buffer pH 10 + 2-mercaptoethanol 0·6%, v/v, 20°C)	4·3	4·3
Fraction V		
Glutelin and residue (Borate buffer pH 10 + 2-mercaptoethanol 0·6%, v/v + sodium dodecyl sulphate 0·5%, w/v)	22·5	33·5

[a] Landry and Moureaux (1970).
[b] Mertz (1974).

Table 3.14
Percent protein and lysine of whole kernels for some selected F_3 segregates between normal and high-lysine sorghum crosses.[a]

Pedigree	Grain type[b]	Protein	Lys in sample (mg/100 g)	Lys in prot. (mg/g total N)
(SC120 14 × IS 16210)-1-G-1	M	6·9	225	206
-2	M	6·6	225	212
-3	M	6·5	205	194
-4	M	7·5	221	181
-5	OP	7·0	217	194
-6	OD	9·6	395	256
(SC120 14 × IS 16210)-2	M	7·8	245	194
(SC423 14 × IS 16210)-1	M	7·6	213	175
-3	M	6·4	195	187
-4	M	6·5	195	187
SC170 6 17 × IS 16210	M	9·9	218	137
P74A S424 × IS 16210	M	13·5	278	125
P74A S454 × IS 16210	M	12·0	226	119
P74A S461 × IS 16210	M	6·8	226	206
(P74A S463 × IS 16210)-2	OP	7·3	226	194
-3	OP	7·1	228	200
-5	OP	7·2	233	200
-6	OP	8·8	260	181
High-Lys[c]		15·1	481	200
Normal sorghum[d]		12·1	241	125

[a] Singh (1976).
[b] M = modified, OP = opaque plump, and OD = opaque dented.
[c] Average of four high-Lys lines.
[d] Average of three normal lines.

Table 3.15
Percent protein and lysine of whole kernels for some selected F_5 segregates between normal and high-Lys sorghum crosses.[a]

Pedigree	Grain type[b]	Protein (%)	Lys in	
			sample (mg/100 g)	prot. (mg/g total N)
(SC423 14 × IS 16210)-1-1	N	10·4	230	142
-2	M	9·8	240	152
(P74A S424 × IS 16210)-1-1	M	9·6	237	154
-2	N+M+OP	10·6	217	128
-3	N	10·5	236	140
(CK60B × IS 16199)-3	N+M+OP	12·4	250	126
(SC120 14 × IS 16210)-4	N	12·5	238	119
-13	N	12·6	268	133
-31	N	13·1	258	123
(SC120 6 17 × IS 16210)-36	OP	14·8	230	97
-39	OP	15·5	252	101
(P74A S461 × IS 16210)-53	OP	16·4	300	114
(CK60B × IS 16199)-60	OP	12·8	244	119
-62	OP	20·0	306	96
(SC120 14 × IS 16210)-64	OD	19·4	581	187
-69	OD	17·3	582	210

[a] CIMMYT (1976).
[b] N = normal, M = modified, OP = opaque plump and OD = opaque dented.

Table 3.16
Protein and lysine content of defatted embryo and endosperm and the proportion of embryo and endosperm.[a]

Kernel class	Part of kernel	Prot. %	Lys (mg/g total N)	Ppn. (% of seed)
P 721 opaque	Embryo	27·7	336	11·4
Normal sib	Embryo	23·8	335	10·4
P 721 opaque	Endo.	12·5	129	88·6[b]
Normal sib	Endo.	11·7	78	89·6[b]

[a] Axtell (1976).
[b] Pericarp not removed from endosperm.

Table 3.17
Amino acid and protein content of defatted kernels of P 721 opaque and normal sib sorghum lines.[a]

AAs	Whole kernels			Endosperm		
	P 721 opaque[b]	Normal sib[b]	Av.[c] sorghum	Av.[d] sorghum	P 721 opaque[e]	Normal sib[e]
	(mg/g total N)					
Ile	256	263	255	287	265	262
Leu	717	854	892	1044	805	887
Lys	202	149	134	125	135	75
Met	127	123	112	84	125	101
Cys	121	123	94	109	115	112
Phe	320	383	324	375	369	345
Tyr	276	294	281	331	300	289

cont.

Table 3.17 *cont.*

AAs	Whole kernels				Endosperm	
	P 721 opaque[b]	Normal sib[b]	Av.[c] sorghum	Av.[d] sorghum	P 721 opaque[e]	Normal sib[e]
	(mg/g total N)					
Thr	211	229	204	225	185	171
Trp			82[h]			
Val	345	340	334	356	340	315
Arg	331	254	224	225	215	166
His	468	146	126	159	135	117
Ala	460	538	618	684	470	534
Asp	491	457	489	512	455	358
Glu	1209	1376	1451	1684	1210	1351
Gly	225	213	192	184	165	144
Pro	472	543	510	565	585	614
Ser	258	287	282	309	210	235
% prot.	14·3	12·2	12·6	9·3	12·5	11·7
Leu/Ile	2·8	3·2	3·5	4·6	3·0	3·4

[a] Axtell (1976).
[b] The amino acid values for whole kernels and endosperm are the av. of three determinations from separate hydrolysates.
[c] Weighted av. of 522 lines (Srinivasan *et al.*, 1972).
[d] Av. of two lines (Jambunathan and Mertz, 1972).
[e] From single determination only.
[f] Av. of two determinations from separate hydrolysates.
[g] Weighted av. of three lines.
[h] Weighted av. of nine lines.

Table 3.18
Protein content, U/P ratio,[a] lysine content[b] and 100-seed weight from bulked rows of a number of selections from crosses[c] and subsequent intercrosses with high U/P ratios.[d]

Row number 77 BS	1977 BS (bulked rows)			
	Prot. (%)	U/P ratio	Lys mg/g N	100-seed weight (g)
Q 50662	15·1	4·2	164	1·60
Q 50665	16·4	4·2	164	1·70
Q 50687	15·3	3·9	152	1·60
Q 50710	12·4	3·7	144	2·30
Q 50718	10·3	3·7	144	2·55
Q 50823	12·5	3·7	144	2·15
Q 50902	9·7	3·7	144	2·15
Q 50890	10·2	4·1	160	1·65
Q 50992	9·9	3·6	140	2·13
Q 52809	10·9	3·6	140	1·78
Checks: P 721	13·9	4·3	168	1·89
CSV 5	12·7	3·3	128	2·24

[a] Udy intensity reading divided by % prot.
[b] Calculated by the authors using the equation: Lys % = (U/P × 6·44) − 0·083.
[c] Made at Purdue University (includes high-lysine Ethiopian mutant).
[d] Riley (personal communication, 1978).

Table 3.19

Protein quality estimations, grouped according to U/P ratio,[a] and phenotypic correlations[b] of different generations of crosses or intercrosses involving *hl* derived lines and P 721.[c]

Generation progeny group	U/P ratio group	Means: No. of heads	% prot.	U/P ratio	Lys[d] mg/g N	100 seeds (g)	Phenotypic Correlations: U/P ratio vs seed wt	% prot. vs seed wt	% prot. vs U/P ratio
I									
F_3 *hl* derived and P 721 crosses	(a) Above 4·0	20	13·6	4·3	168	2·05	NS	S	S
	(b) 3·5–4·0	64	11·2	3·7	144	2·26	NS	NS	S
	(c) Below 3·5	108	10·7	3·1	120	2·54	S	NS	NS
	(d) Total	192	11·2	3·4	132	2·39	S	NS	S
II									
F_3 *hl* derived intercrosses	(a) Above 4·0	0							
	(b) 3·5–4·0	2	9·8	3·5	136	2·30			
	(c) Below 3·5	33	11·7	3·0	116	2·40	NS	NS	NS
	(d) Total	35	11·6	3·0	116	2·30	NS	NS	NS
III									
F_5 crosses with P 721	(a) Above 4·0	18	11·7	4·3	168	2·30	NS	NS	NS
	(b) 3·5–4·0	36	10·7	3·7	144	2·30	NS	NS	NS
	(c) Below 3·5	66	10·9	2·9	111	2·80	S	NS	S
	(d) Total	120	10·9	3·4	132	2·60	S	NS	NS
IV									
F_1 *hl* derived and P 721 second intercross	(a) Above 4·0	5	12·3	4·5	176	2·13	NS	NS	NS
	(b) 3·5–4·0	39	9·4	3·6	140	2·35	NS	NS	NS
	(c) Below 3·5	69	10·1	3·1	120	2·57	NS	NS	S
	(d) Total	112	9·9	3·4	132	2·48	S	NS	NS
All selections	Total	482	10·9	3·4	132	2·47	S	NS	S
		(Rows)							
Checks P 721		10	10·8	3·5	136	2·18			
CSH 1		10	8·9	3·1	120	3·20			

[a] Udy intensity reading, divided by % prot.
[b] S = significant; NS = not significant.
[c] All groups grown during 1977 rainy season, June–November. Riley (personal communication, 1978).
[d] Calculated by the authors using the equation: Lys % = (U/P × 6·44) − 0·083.

Table 3.20
Mean comparison of high-lysine and normal genotype of sorghum for % protein (N × 6·25) and lysine in the defatted endosperm and germ at 21 days after flowering (DAF), 31 DAF and 61 DAF.[a]

	High-Lys				Normal			
	IS 11758	IS 11167	YM3	P 7210	P 721N	BG5	BG6	BG10
21 DAF								
Endo.								
% prot.	13·9	13·8	12·8	9·7	11·2	12·4	12·4	12·0
Lys (mg/100 g sample)	393	357	397	262	235	356	354	344
(mg/g total N)	177	162	194	169	131	179	179	179
Germ								
% prot.	15·7	18·2	18·9	16·5	16·2	19·0	18·5	17·7
Lys (mg/100 g sample)	530	574	592	592	570	729	743	729
(mg/g total N)	211	197	196	224	220	239	251	257
31 DAF								
Endo.								
% prot.	13·1	13·9	14·3	10·2	11·0	11·6	11·3	10·7
Lys (mg/100 g sample)	401	469	433	270	152	304	235	331
(mg/g total N)	191	211	189	166	88	163	129	193
Germ								
% prot.	14·5	14·9	16·6	15·9	17·5	18·9	18·1	16·6
Lys (mg/100 g sample)	565	572	532	664	702	805	737	692
(mg/g total N)	243	224	200	261	251	266	255	260
61 DAF								
Endo.								
% prot.	14·6	15·7	15·7	10·6	12·6	12·2	11·8	11·3
Lys (mg/100 g sample)	298	345	373	236	194	172	166	171
(mg/g total N)	127	138	149	139	96	88	88	94
Germ								
% prot.	20·0	19·6	22·8	21·0	23·5	22·0	21·0	20·4
Lys (mg/100 g sample)	1014	941	1171	976	1067	1092	1103	1079
(mg/g total N)	317	301	321	290	284	311	328	330

[a] Ejeta and Axtell (1977).

Table 3.21

Mean comparison of high-Lys and normal genotypes for % starch, amylose, total sugars, reducing sugars, sucrose and water soluble polysaccharides (WSP) in the endosperm at 31 days after flowering.[a]

Types	Genotypes	Starch % of sample	Amylose % of starch	Total sugars mg/g of sample	Reducing sugars mg/g of sample	Sucrose mg/g of sample	WSP mg/g of sample
High-Lys	IS 11758	46·5	16·5	184·3	80·3	104·0	8·6
	IS 11167	49·3	19·2	169·0	65·3	103·7	7·6
	YM 3	43·0	16·9	164·7	74·9	93·8	17·9
P 721 opaque		69·4	24·9	27·0	23·6	3·4	5·3
Normal	P 721 normal	72·4	24·1	19·4	14·7	4·7	4·9
	BG 5	67·7	22·2	40·2	26·1	14·1	8·4
	BG 6	71·7	23·2	16·7	12·1	4·6	4·7
	BG 10	65·6	22·6	45·7	39·4	6·4	3·6

[a] Ejeta and Axtell (1977).

Table 3.22

Biological quality of isonitrogenous high-Lys and normal sorghum lines in rat diets.[a]

Source	Composition of grain		Composition of feed					
				Lys				
	Prot. (N × 6·25) (%)	Lys (mg/g total N)	Prot. %	(g/100 g prot.)	(mg/100 g sample)	Wt gain (g)	F/G	PER
High-Lys lines								
IS 11167	16·6	210	10·1	2·81	284	34·5	5·6	1·78
IS 11758	18·0	211	10·0	3·15	315	48·8	4·9	2·06
Normal lines								
IS 2319	12·7	144	10·9	2·25	245	25·3	7·5	1·24
IS 2520	13·3	116	11·2	1·76	197	10·3	13·3	0·61
IS 1269	14·8	131	10·7	2·01	215	14·0	13·3	0·74
Mean of normal lines	13·6	130	10·9	2·01	219	16·5	11·3	0·86
Casein	91·0	500	13·3	7·36	979	85·8	3·5	2·20

[a] Singh and Axtell (1973a).

Table 3.23
Biological quality of 94% whole grain rations of high-lysine and normal sorghum, Opaque-2 maize and normal maize in rat diets.[a]

Source	Composition of grain: Prot. (N × 6·25) (%)	Composition of grain: Lys (mg/g total N)	Composition of feed: Prot. %	Composition of feed: Lys (g/100 g prot.)	Composition of feed: Lys (mg/100 g sample)	Wt gain (g)	F/G
High-Lys line							
IS 11758	18·4	211	18·4	3·17	583	94·2	3·0
Normal sorghum							
IS 2319	12·7	144	12·6	2·23	281	28·5	6·8
IS 1484	14·0	121	12·6	2·01	253	19·2	8·5
Mean for normal sorghum	13·3	132	12·6	2·12	267	23·9	7·6
Opaque-2 maize	12·5	250	12·1	3·89	471	91·5	3·4
Normal maize	9·4	17'	8·6	2·90	249	30·2	7·4
Casein	91·0	500	17·2	6·42	1104	181·2	2·0

[a] Singh and Axtell (1973a).

Table 3.24
Biological quality of high-Lys, sugary and normal sorghum lines in rat diets (21 days).

Genotype	Composition of grain				Composition of feed						
		Lys				Lys					
	Prot. (N × 6·25) (%)	mg/g total N	mg/100 g sample	Oil (%)	Prot. (%)	% of prot.	mg/100 g sample	Wt gain (g)	Feed consumed (g)	F/G	PER
High-Lys lines											
IS 11758	16·0	187	480	6·97	13·1	2·59	339	37·2	175·7	4·7	1·61
Sugary lines											
IS 4526	16·2	169	439	5·00	11·5	2·60	299	26·7	168·5	6·5	1·37
IS 4668	14·9	136	325	3·38	12·5	1·97	246	16·3	140·3	8·9	1·04
IS 5376	16·3	142	370	4·64	12·1	2·28	276	16·7	136·5	8·5	1·00
IS 5614	14·8	166	392	4·90	12·7	2·51	319	25·8	170·3	6·7	1·18
IS 5623	15·1	171	412	5·06	12·1	2·51	304	—	—	—	—
Normal lines											
IS 2319	14·2	146	332	4·70	12·3	2·17	267	20·0	148·0	7·6	1·10
IS 5568	14·4	117	269	3·75	12·1	1·71	207	13·3	137·5	10·8	0·80
IS 2520	13·3	116	246	3·50	11·8	1·75	206	7·3	113·5	17·1	0·54
IS 0057	14·3	114	323	3·05	11·6	2·21	256	13·7	139·0	11·1	0·85
IS 8313	17·7	109	310	4·64	11·6	1·80	209	6·8	125·3	21·3	0·47
IS 6901	16·0	126	323	2·94	10·4	1·95	203	12·7	144·5	11·7	0·84
Mean of normal lines	15·0	126	301	3·76	11·6	1·93	225	12·3	134·6	13·3	0·77
Casein	91·0	500	7280	—	15·3	7·05	1079	90·8	227·7	2·5	2·59
Grand mean	21·1	171	883	4·38	12·2	2·55	324	23·9	151·9	9·8	1·12

[a] Singh and Axtell (1973d).

Table 3.25
Biological quality of high-Lys, sugary and normal sorghum lines in rat diets (28 days).

Genotype	Composition of grain				Composition of feed			Wt gain (g)	Feed consumed (g)	F/G	PER
		Lys				Lys					
	Prot. (N × 6·25) (%)	mg/g total N	mg/100 g sample	Oil (%)	Prot. %	% of prot.	mg/100 g sample				
High-Lys lines											
IS 11167	16·6	210	558	5·81	10·1	2·81	284	34·5	193·0	5·6	1·78
IS 11758	18·0	211	608	6·25	10·0	3·15	315	48·8	234·8	4·9	2·06
Sugary lines											
IS 4526	14·6	166	388	4·72	10·7	2·73	292	31·8	200·2	6·5	1·45
IS 4668	14·3	154	353	3·68	11·2	2·02	226	16·7	159·2	9·8	0·92
IS 5376	13·6	148	322	4·45	10·7	2·35	251	26·3	193·5	8·0	1·25
IS 5614	14·0	154	346	4·00	10·8	2·01	217	25·3	179·2	7·4	1·30
IS 5623	11·8	147	277	3·47	10·7	1·98	211	27·0	194·3	7·3	1·32
Normal lines											
IS 2319	12·7	144	292	4·70	10·9	2·25	245	25·3	187·8	7·5	1·24
IS 2520	13·3	116	246	3·50	11·2	1·76	197	10·3	150·0	13·3	0·61
IS 1269	14·8	131	310	4·21	10·7	2·01	215	14·0	179·8	13·5	0·74
Mean of normal lines	13·6	130	283	4·14	10·9	2·01	219	16·5	172·5	11·3	0·86
Casein	91·0	500	7280	—	13·3	7·36	979	85·8	291·2	3·5	2·20
Grand mean	21·3	189	402	4·48	10·9	2·76	312	31·5	196·6	9·8	1·35

[a] Singh and Axtell (1973d).

Table 3.26
Protein, Lys, methionine and cystine content of normal and high-Lys sorghum used in chick studies.[a]

AAs	Experiment 1: RS 671 (normal) (mg/100 g sample)	Experiment 1: RS 671 (normal) (mg/g total N)	Experiment 1: IS 11758 (high-Lys) (mg/100 g sample)	Experiment 1: IS 11758 (high-Lys) (mg/g total N)	Experiments 2 and 3: RS 610 (normal) (mg/100 g sample)	Experiments 2 and 3: RS 610 (normal) (mg/g total N)	Experiments 2 and 3: IS 11758 (high-Lys) (mg/100 g sample)	Experiments 2 and 3: IS 11758 (high-Lys) (mg/g total N)
Lys	250	134	480	220	240	129	450	229
Met	160	85	180	82	80	40	170	87
Cys	290	156	280	130	220	115	250	130
Prot. (%)	11·5		13·6		11·9		12·2	

[a] Featherston *et al.* (1975a).

Table 3.27
Composition of diets and effect on chick weight gain and feed conversion of sorghum RS 671 (normal) and IS 11758 (high-Lys) fed at different protein levels, Experiment I.[a]

Ingredients (%)	Diets (A)	(B)	(C)	(D)	(E)	(F)	(G)	(H)
RS 671	92·05		91·85		86·80	70·65		66·75
IS 11758		92·05		91·85			70·65	
Soybean meal					5·05	21·20	21·20	25·10
DL-Met			0·20	0·20	0·20	0·20	0·20	0·20
Soybean oil	3·00	3·00	3·00	3·00	3·00	3·00	3·00	3·00
Constant (nonprot.) ingredients	4·95	4·95	4·95	4·95	4·95	4·95	4·95	4·95
Total prot. (%)	10·6	12·5	10·8	12·7	12·7	18·6	20·0	20·0
Chick trials								
Wt gain (g)	13·9	37·6	14·3	41·2	58·1	245·9	283·0	284·4
Feed/gain	6·7	3·6	6·0	3·4	2·9	1·6	1·5	1·5

[a] Featherston *et al.* (1975a).

Table 3.28
Composition of diets and effect on chick weight gain and feed conversion of sorghum RS 610 (normal) and IS 11758 (high-Lys) fed at 15% protein level, Experiment 2.[a]

Ingredients (%)	Diets (A)	(B)	(C)	(D)	(E)	(F)
RS 610	77·12		79·17	79·17		
IS 11758		75·47			77·44	77·44
Safflower meal (42·9% prot.)	13·50	13·50				
Soybean meal (49% prot.)			11·30	11·30	11·30	11·30
DL-Met				0·15		0·15
Glucose monohydrate		1·65	0·15		1·88	1·73
Soybean oil	5·00	5·00	5·00	5·00	5·00	5·00
Constant (nonprot.) ingredients	4·38	4·38	4·38	4·38	4·38	4·38
Chick trials						
Wt gain (g)	24·8	60·6	100·1	105·0	186·2	199·8
Feed/gain	3·5	2·8	2·2	2·1	2·0	1·9

[a] Featherston *et al.* (1975a).

Table 3.29
Composition of diets and effect on chick weight gain and conversion of RS 610 without and with Lys supplementation compared to IS 11758, Experiment 3.[a]

Ingredients %	Diets (A)	(B)	(C)
RS 610	91·36	91·36	
IS 11758			89·40
L-Lys HCl		0·21	
Diammonium citrate	0·26		0·26
Glucose monohydrate	0·04	0·09	2·00
Soybean oil	3·00	3·00	3·00
Constant (nonprot.) ingredients	5·34	5·34	5·34
Total prot. (%)	10·90	10·90	10·90
Dietary Lys (%)	0·22	0·39	0·39
Chick trials			
Wt gain (g)	14·5	48·0	50·8
Feed/gain	6·0	3·0	3·2

[a] Featherston *et al.* (1975a).

Table 3.30
Phenotypic characteristics, origin, protein and Lys of five sugary endosperm lines.[a]

IS No.	Kernel and endo. phenotype	Origin	Prot. (N × 6·25) (DWB) (%)	Lys (mg/g total N)	Lys (mg/100 g sample)
4526	Plump, major part of endo. floury with thin outer layer of glassy endo.	Maharashtra, India	15·3	138	338
4668	Moderately shrivelled, endo, glassy and translucent	Maharashtra, India	13·4	153	328
5376	Shrivelled, endo. glassy and translucent	Madras, India	17·3	140	387
5614	Shrivelled, endo. glassy and translucent		15·5	145	360
5623	Shrivelled, endo. glassy and translucent	Mysore, India	12·4	151	300

[a] Singh and Axtell (1973b).

Table 3.31
Carbohydrate analysis of whole kernels of high-Lys and sugary sorghum lines (DWB %).[a]

Entry	Reducing sugars	Sucrose	Total sugars	WSP[b]	Starch	Amylose (% of starch)	Total CBH[c]
High-Lys lines							
IS 11167	0·38	3·08	3·46	0·91	58·9	25·0	63·27
IS 11758	0·32	2·61	2·93	1·01	57·8	26·2	61·74
Sugary lines							
IS 4526	0·26	1·70	1·96	1·27	58·5	40·6	61·73
IS 4668	2·52	0·68	3·20	29·40	34·0	30·4	66·60
IS 5376	2·78	1·06	3·84	34·20	25·4	31·7	63·44
IS 5614	0·57	2·10	2·67	39·60	22·2	37·4	64·47
IS 5623	2·69	1·70	4·39	35·70	26·3	33·3	66·39
Normal check							
IS 8313	0·34	1·03	1·34	1·11	60·8	25·0	63·25

[a] Singh and Axtell (1973b).
[b] Water-soluble polysaccharides.
[c] Carbohydrates.

Table 3.32
Carbohydrate analysis (DWB) of the endosperm of plump and mutant kernels of segregating heads.[a]

	Genotype			
	PP3Rms_3 × IS 4668		PP3Rms_3 × IS 5376	
	Plump kernel	Mutant kernel	Plump kernel	Mutant kernel
Reducing sugars (%)	0·23	0·64	0·29	0·44
Sucrose (%)	0·08	0·69	0·10	0·56
Total sugars (%)	0·31	1·33	0·39	1·00
WSP (%)	0·79	22·60	0·99	21·30
Starch (%)	69·1	49·1	69·1	48·6
Amylose (% of starch)	26·2	26·9	27·5	25·6
Total carbohydrates (%)	70·2	73·0	70·4	74·2

[a] Singh and Axtell (1973c).

Table 3.33
Proximate composition, minerals, vitamins (as-is basis) and amino acid content in the protein of sorghum, peasant and experimental cultivation, in Senegal.[a]

	Peasant cultivation Bassi	Experimental cultivation		
		S. guineensis	*S. durra*	*S. caffra*
Moisture (%)	9·1	9·4	9·2	8·9
Prot. (N × 5·83) (%)	9·5	13·2	12·3	9·8
Fat (%)	3·7	4·3	2·9	3·0
Ash (%)	1·71	1·73	1·70	1·53
Total reducing substances (%)	68·5	68·0	67·8	70·9
Inorganic elements (mg/100 g)				
Ca	23·2	20·4	31·6	14·3
P	312·0	389·0	367·0	326·0
Ratio, Ca/P	0·074	0·050	0·084	0·44
K	397·0	267·0	383·0	355·0
Na	51·3	71·0	110·5	87·0
Ratio, K/Na	7·74	3·75	4·45	4·10
Vitamins (mg/100 g)				
Thiamine	0·26	0·61	0·60	—
Niacin	3·8	4·85	7·85	6·75
Riboflavin	0·12	0·14	0·18	0·12
Pantothenic acid	1·05	1·25	1·10	1·30
AAs (mg/g total N)				
Ile	312	337	266	322
Leu	1506	1506	1650	1575
Lys	113	151	142	179
Met	175	162	156	194
Cys	128	87	91	91
Phe	247	312	300	297
Thr	162	237	269	256
Trp	72	78	72	59
Val	300	306	325	275
Arg	200	219	206	244
His	132	128	119	153

[a] Adrian and Sayerse (1957).

Table 3.34
Agronomic traits of high-lysine sorghum lines grown at Lafayette (Laf.) and Puerto Rico (PR).[a]

Trait	IS 11167		IS 11758		Normal sorghum[b]
	Laf.	PR	Laf.	PR	
Protein (N × 6·25) (%)	14·8	16·6	16·0	18·4	12·7
Lysine					
(mg/g N)	207	210	187	204	128
(mg/100 g sample)	490	558	480	600	258
Oil (%)	—	5·81	6·97	6·25	3·32
100-seed weight (g)	2·55	3·01	2·50	2·40	2·75
Protein/seed (mg)	3·77	5·00	4·00	4·42	3·53
Lysine/seed (mg)	0·125	0·168	0·120	0·144	0·071
% germ in seed	14·0	15·3	15·4	17·2	10·10[c]
Height (cm)	395·0	250·0	385·0	245·0	200·0
B and R reaction[d]	—	B	—	B	—

[a] Singh and Axtell, 1973c.
[b] Average of 31 genotypes over six locations and two years (Schaffert, 1972).
[c] Average of four lines (Jambunathan and Mertz, 1973a).
[d] B, Non-restorer; R, Restorer.

Table 3.35
Agronomic traits of sugary sorghum lines grown at Lafayette (Laf.) and in Puerto Rico (PR).[a]

	IS 4526		IS 4668		IS 5376		IS 5614		IS 5623		
Trait	Laf.	PR	Laf.	PR	Laf.	PR	Laf.	PR	Laf.	PR	Av. sorghum[b]
Prot. (N × 6·25) (%)	16·2	15·0	14·9	14·3	16·3	13·6	14·8	14·0	15·1	11·8	12·7
Lys											
(mg/g N)	169	166	136	126	142	148	166	154	171	147	128
(mg/100 g sample)	439	398	325	287	370	322	392	346	412	277	258
Oil (%)	5·00	4·72	3·38	3·68	4·64	4·45	4·90	4·00	5·06	3·47	3·32
100-seed wt (g)	1·57	2·36	2·02	2·58	1·61	2·34	1·45	2·49	2·04	2·64	2·75
Prot./seed (mg)	2·54	3·54	3·01	3·69	2·62	3·18	2·15	3·49	3·08	3·12	3·53
Lys/seed (mg)	0·069	0·094	0·066	0·074	0·060	0·075	0·057	0·086	0·080	0·073	0·071
% germ in seed	8·8	12·5	4·7	7·8	5·7	8·9	5·0	9·2	9·4	9·8	10·10[c]
Ht (cm)	270	—	275	—	255	—	250	—	300	—	200
Days to 50% flowering	103	—	96	—	98	—	99	—	98	—	—
B and R reaction[d]	—	B	—	R	—	R	—	B	—	B	—

[a] Singh and Axtell (1973c).
[b] Av. of 31 genotypes over six locations and 2 years (Schaffert, 1972).
[c] Av. of 4 lines (Jambunathan and Mertz, 1973a).
[d] B = Non-restorer; R = Restorer.

Table 3.36
Mean average grain protein (Udy) for 10 brown-seeded grain sorghum hybrids grown at four locations[a] and mean protein yield.[b]

Hybrid	Grain prot. (%)	Mean prot. yield (kg/ha)
ACCO R 1093	12·1	554
RS 700	11·7	395
Niagra Shoo Bird	11·3	506
Frontier 409	11·1	480
AKS 614	10·7	476
Excel Bird Go A	11·1	493
Funk BR 79	11·1	515
DeKalb BR 64	11·0	445
NK Savanna 3	11·0	464
KB Golden Grain BR	10·7	510

[a] Louisiana (1972–73).
[b] Haji-Hashim and Tipton (1973).

Table 3.37
Mean (ranges in parentheses) of amino acids of 11 botanical types of sorghum originating in Africa.[a]

	Mean
N (DWB) (%)	2·5 (2·0–3·0)
AAs (mg/g N)	
Ile	254 (237–269)
Leu	855 (806–937)
Lys	144 (119–169)
Met	116 (100–144)
Cys	128 (106–144)
Phe	321 (300–350)
Tyr	260 (237–275)
Thr	208 (187–231)
Trp	77 (62–144)
Val	337 (319–356)
Arg	260 (237–306)
His	141 (125–175)
Ala	597 (550–631)
Asp	457 (419–506)
Glu	1303 (1231–1444)
Gly	204 (181–231)
Pro	548 (500–606)
Ser	281 (287–294)

[a] Busson *et al.* (1962).

Table 3.38
Protein concentration and protein yield (mean with ranges in parentheses) from analysis of seed returned by cooperators at 8–12 locations.[a]

Identification No.	No. of locations reporting	Prot. (%)		Prot. yield (kg/ha)		Grain yield[b] rank
		Mean	Rank	Mean	Rank	
954114	11	10·9 (10·8–13·5)	11	354 (59–778)	17	22
954255	11	11·5 (10·3–12·5)	7	397 (51–797)	12	12
954063	11	10·4 (8·7–12·6)	18	480 (107–956)	4	5
932127	12	10·2 (8·8–12·3)	26	425 (128–834)	7	8
RS 610[c]	12	10·3 (8·5–13·2)	21	479 (106–771)	5	4
BR 64[c]	12	9·9 (8·1–11·4)	24	566 (177–1498)	2	2
IS 7822	12	10·2 (9·5–11·3)	22	404 (112–967)	10	9
954062	12	10·9 (8·4–11·7)	11	454 (99–966)	6	6
932296	11	10·4 (8·8–11·7)	18	357 (62–743)	16	11
IS 8361	10	11·6 (9·1–15·0)	6	370 (82–634)	14	14
932075	10	10·7 (9·2–11·8)	14	364 (91–857)	15	18
956031	11	10·5 (9·5–11·6)	15	410 (70–991)	9	13
954130	11	11·7 (9·0–13·8)	4	404 (66–867)	10	15
954206	10	10·4 (9·1–12·2)	18	415 (124–716)	8	7
954164	11	11·8 (8·3–13·6)	2	378 (29–723)	13	10
NK 300[c]	11	10·8 (8·8–13·2)	13	710 (193–1776)	1	1
IS 2319	10	12·7 (11·0–15·1)	1	297 (95–874)	22	23
954104	10	11·0 (8·5–12·6)	10	345 (124–718)	19	19
932062	9	11·2 (9·4–13·6)	9	288 (71–957)	23	17
IS 9198	8	10·5 (7·1–12·1)	15	223 (19–768)	24	24
IS 7579	8	10·5 (8·5–12·4)	15	346 (167–738)	18	16
IS 9569	9	11·8 (8·7–14·8)	2	322 (98–1051)	21	21
932027	8	11·7 (8·3–13·8)	4	329 (118–923)	20	20
IS 4225	9	11·3 (9·5–12·9)	8	507 (31–1904)	3	3
Average		11·0 (7·1–15·1)		401 (19–1904)		

[a] Purdue University, (1974).
[b] Correlation of prot. yield with grain yield: $r = 0{\cdot}933$.
[c] RS 610, BR 64, NK 300 are commercial US hybrids, all other entries are lines.

Table 3.39
Protein, yield and N accumulation of 11 varieties of sorghum grown in the Sudan in 1960.[a]

	Prot. (N × 6·25) (%)	Yield (kg/ha)	Yield[b] Rank	N accumulation (kg/ha)
Gassabi	13·0	1117	10	23·6
Mugud	12·6	—	—	—
Abu Dereira	11·5	1436	9	26·4
Mugbash	11·5	1559	8	28·8
Dinderawi	10·1	2079	3	33·8
Feterita Matuk	9·2	2076	4	32·9
Wad Aker	9·6	1805	7	27·4
Combine Kafir	9·2	2098	2	31·2
Um Benein	8·9	1948	6	27·9
Plainsman	8·7	2233	1	31·2
Al Fadni	8·2	1986	5	26·2

[a] Rai (1964).
[b] Converted from kg/feddan; 1 feddan = 0·42 ha.

Table 3.40
Mean effects on grain yield and grain protein of Um Benein sorghum, continuously mono-cropped, of different levels of N fertilizer application 1959–1961.[a]

Year	N application (kg/ha) (%)	Crude prot. (N × 6·25) (%)	Yield (kg/ha)	N accumulation (kg/ha)
1959	0	12·6	2193	44·3
	23·8	12·6	2412	48·8
	47·6	12·1	2519	48·8
	95·2	11·9	2750	52·6
Mean		12·3	2469	48·6
1960	0	10·2	3143	51·4
	23·8	10·9	3464	60·5
	47·6	11·5	3238	59·8
	95·2	12·1	3329	64·3
Mean		11·2	3293	59·0
1961	0	7·3	1231	14·5
	23·8	8·1	1952	25·5
	47·6	8·7	2031	27·9
	95·2	9·9	2057	32·9
Mean		8·5	1819	25·2

[a] Rai (1965a).

Table 3.41
Mean effect on grain yield and grain protein of Wad Aker sorghum, rotational-cropped, of different levels of N fertilizer application.[a]

Year	N application (kg/ha)	Crude prot. (N × 6·25) (%)	Yield (kg/ha)	N accumulation (kg/ha)
1959	0	10·0	2286	36
	23·8	9·4	2883	42
	47·6	9·6	3252	50
	95·2	9·7	3502	54
Mean		9·7	2981	46
1960	0	8·2	1809	24
	42·9	9·1	2059	30
	85·7	10·6	2278	39
	171·4	12·0	2376	45
Mean		10·0	2140	35
1961	0	11·4	2585	43
	42·9	11·5	3059	53
	85·7	11·8	3243	57
	171·4	11·9	3531	63
Mean		11·7	3114	54

[a] Rai (1965a).

Table 3.42
Effect of date of sowing, and fertilizer on grain yield, N and P in the grain of Wad Aker sorghum at maturity.[a]

Treatments				Grain			
				N		P	
Date of sowing	P_2O_5 (kg/ha)	N (kg/ha)	Yield (kg/ha)	(%)	(kg/ha)	(%)	(kg/ha)
7.7.61	0	0	2936	1·58	46·2	0·22	6·43
		42·9	3695	1·53	56·4	0·20	7·14
		85·7	3948	1·59	62·9	0·20	7·86
		171·4	4433	1·72	76·0	0·21	9·05
	85·7	0	3909	1·28	50·0	0·24	9·52
		42·9	4762	1·51	71·9	0·21	9·76
		85·7	5086	1·53	77·6	0·22	11·19
		171·4	5829	1·58	92·1	0·22	12·6
13.8.61	0	0	2048	2·22	45·5	0·25	5·24
		42·9	2019	2·13	42·9	0·27	5·48
		85·7	2090	2·23	46·7	0·27	5·71
		171·4	2060	2·17	44·7	0·24	5·00
	85·7	0	1448	2·24	32·4	0·31	4·52
		42·9	1905	2·21	42·1	0·29	5·48
		85·7	1845	2·22	41·0	0·29	5·24
		171·4	1800	2·15	38·8	0·27	4·76

[a] Rai (1965b).

Table 3.43
Effect on grain yield, grain protein and protein/ha of sorghum under five levels of N fertilizer at sowing.[a]

	Location					
	Kotiary			Nioro		
N application (kg/ha)	Grain yield (kg/ha)	Prot. (N × 6·25) (DWB) (%)	Prot. yield (kg/ha)	Grain yield (kg/ha)	Prot. (N × 6·25) (DWB) (%)	Prot. yield (kg/ha)
0	1804	8·4	152	866	10·3	89
As % of control	100	100	100	100	100	100
50	3013	8·4	253	3127	10·2	319
As % of control	167	100	167	361	100	357
100	3501	9·3	326	2747	11·1	305
As % of control	194	110	215	317	108	341
150	3865	9·3	359	2598	12·5	325
As % of control	214	111	237	300	122	364
200	3804	9·5	361	2538	12·3	312
As % of control	211	113	239	293	120	350

[a] Two locations in Senegal (1968) (from Blondel, 1970).

Table 3.44
Locational and varietal differences in protein (%) and amino acids (mg/g N) content of sorghum.[a]

	Prot.	Ile	Leu	Lys	Met	Trp
Varieties (mean of 3 locations)						
Swarna	9·0	314	894	160	141	58
CSH 1	8·1	356	938	160	146	71
CSH 2	8·8	349	931	159	131	66
Ganeri 2	9·4	289	744	131	123	53
CO 9	8·8	357	938	164	145	68
Shenoli 4 2	10·4	348	963	155	144	69
G 3	10·7	338	975	144	131	64
PJ 8K	9·5	336	913	165	143	60
Locations (mean of 8 varieties)						
Coimbatore	10·5	344	894	144	124	60
Karad	6·4	378	1044	196	174	81
Parbhani	11·2	286	794	124	113	51

[a] Deosthale *et al.* (1972a).

Table 3.45
Locational and varietal differences in total ash and eight inorganic elements of five varieties of sorghum grown at three locations.[a]

	Total ash (%)	Ca	P	Fe	Cu (mg/100 g)	Mg	Mn	Mo	Zn
Varieties (mean of three locations)									
Swarna	1·5	13·0	261	5·8	0·31	157	1·11	0·004	0·12
CSH 1	1·3	15·5	247	5·3	0·52	147	1·09	0·007	0·11
CSH 2	1·4	12·1	254	5·0	0·37	105	1·08	0·006	0·14
Ganeri 2	1·5	11·2	215	7·3	0·43	331	0·95	0·004	0·16
CO 9	1·6	12·4	256	5·7	0·93	163	1·29	0·004	0·12
Location (mean of five varieties)									
Coimbatore	1·4	9·5	304	3·7	0·41	110	1·43	0·008	0·15
Karad	1·2	18·2	183	6·9	0·83	124	0·70	0·002	0·12
Parbhani	1·8	13·8	253	6·9	0·30	188	1·18	0·005	0·10

[a] Deosthale *et al.* (1972a).

Table 3.46
Effect of N fertilizer on protein and amino acid content (mg/g N) of two varieties of sorghum.[a]

	Grain yield (kg/ha)	Prot. (%)	Leu	Lys	Met	Trp
N fertilizer (kg/ha)						
N_0	750	10·6	789	199	91	79
N_{50}	1735	10·3	769	172	94	69
N_{100}	2325	10·0	741	181	89	66
N_{150}	2535	11·5	747	172	87	71
N_{200}	2680	11·6	791	149	94	62
Variety						
CSH 1	2344	9·6	811	176	96	69
Swarna	1666	12·0	724	173	86	69

[a] Deosthale *et al.* (1972a).

Table 3.47
Effect of P_2O_5 and K_2O levels on grain yield (? kg/ha) and mineral composition (mg/100 g sample) of CSH 1 sorghum.[a]

Treatment	Grain yield (kg/ha)	Ash (%)	Ca	P	Fe	Cu	Mg	Mn	Mo	Zn
P_2O_5 (kg/ha)										
0	2960	1·0	9·6	118	2·7	0·32	74	0·48	0·015	0·10
40	3253	1·1	6·2	133	3·1	0·32	84	0·52	0·018	0·08
80	3363	1·1	8·0	163	2·9	0·33	126	0·57	0·018	0·12
120	3467	1·1	11·3	154	3·0	0·30	87	0·58	0·019	0·13
K_2O (kg/ha)										
0	2853	1·1	8·7	142	2·9	0·36	97	0·52	0·17	0·10
30	3415	1·1	9·1	142	2·8	0·32	99	0·53	0·017	0·10
60	3515	1·1	8·6	142	3·0	0·28	83	0·56	0·018	0·12

[a] Deosthale *et al.* (1972a).

Table 3.48
Density, protein content, grain yield and protein yield of CSH 1 sorghum under varying conditions of N fertilization.[a]

	Grain density (g/cm^3)	Prot. ($N \times 6{\cdot}25$)[b] (%)	% increase over control	Grain yield[b] (kg/ha)	% increase over control	Prot. yield (kg/ha)[c]
1968						
Control N_0 (0 kg N/ha)	1·3	9·3	—	3449	—	360 (?321)
N_{40} (40 kg N/ha); M_1[d]	1·4	10·7	15	4244	23	455
N_{40}; M_2[d]	1·4	10·5	13	4158	21	439
N_{40}; M_3[d]	1·4	11·3	21	4185	21	474
N_{80} (80 kg N/ha); M_1	1·4	11·6	25	4444	29	519
N_{80}; M_2	1·4	11·1	19	4273	24	474
N_{80}; M_3	1·4	11·8	27	4003	16	472
N_{80}; M_2; 0·2% $ZnSO_4$ spray	1·4	11·1	19	3955	15	438
N_{80}; M_2; 2·0% $FeSO_4$ spray	1·4	11·3	21	3906	13	452 (?442)
1969						
Treatments						
N_0 (0 kg N/ha)	1·3	9·1	—	2575	—	336 (?236)
N_{60} (60 kg N/ha)	1·3	10·6	16	3896	51	416
N_{120} (120 kg						

cont.

Table 3.48 *cont.*

	Grain density (g/cm^3)	Prot. (N × 6·25)[b] (%)	% increase over control	Grain yield[b] (kg/ha)	% increase over control	Prot. yield (kg/ha)[c]
N/ha)	1·3	10·9	20	3863	20	423
N_{180} (180 kg N/ha)	1·4	11·4	25	3699	44	421
Methods						
M_1	1·3	11·0	21	3992	55	438
M_2	1·3	10·6	16	3695	44	393
M_3	1·4	11·4	25	3528	37	401
M_4	1·3	11·2	23	3756	46	422
M_5	1·3	10·7	18	4127	60	441

[a] Warsi and Wright (1973).
[b] 14% moisture basis.
[c] The prot. yield (kg/ha) is presumably prot. content × grain yield ÷ 100 but not all the calculations agree.
[d] M_1—all at planting.
M_2—half at planting, half top-dressed at knee-high stage through soil.
M_3—as M_2 but through foliage.
M_4—one-third at planting, one-third at knee-high stage and one-third at flower initiation.
M_5—six split applications at 12-day intervals.

Table 3.49
Protein solubility fractions[a] of CSH 1 sorghum as influenced by rates and methods of N application.[b]

Treatment	Alb. (%N)	Glob. (%N)	Prol. (%N)	Glut. (%N)	Total (%N)	% total N
Control N_0	0·134	0·212	0·346	0·440	1·43	
% of total N	9·4	14·8	24·2	30·8	—	79·2
N_{60} M_1	0·136	0·218	0·489	0·512	1·69	
% of total N	9·6	14·8	24·2	30·8	—	79·4
N_{60} M_3	0·126	0·235	0·563	0·525	1·76	
% of total N	7·2	13·2	32·6	30·1	—	83·1
N_{120} M_1	0·130	0·230	0·569	0·561	1·76	
% of total N	7·4	13·1	32·3	31·8	—	84·6
N_{120} M_3	0·126	0·226	0·603	0·540	1·78	
% of total N	6·9	12·7	33·8	30·3	—	83·7
N_{180} M_1	0·140	0·245	0·640	0·578	1·85	
% of total N	7·6	13·2	34·5	31·2	—	86·5
N_{180} M_3	0·148	0·220	0·686	0·528	1·85	
% of total N	7·9	12·0	36·9	28·3	—	85·1

[a] Albumin, globulin, prolamine, glutelin.
[b] Warsi and Wright (1973).

Table 3.50
Amino acid composition in the grain and protein of CSH 1 sorghum at two levels of fertility.[a]

	N_0 (prot. 8·9%)		N_{180} (prot. 11·6%)	
	In grain (mg/100 g)	In prot. (mg/g total N)	In grain (mg/100 g)	In prot. (mg/g total N)
Ile	401	281	472	255
Leu	1024	716	1514	819
Lys	211	147	234	126
Met	212	148	264	142
Cys	202	141	313	169
Phe	344	240	426	230
Tyr	262	183	314	170
Thr	196	137	267	144
Val	408	285	541	292
Arg	371	259	401	216
His	186	127	197	107
Ala	702	491	968	523
Asp	462	323	674	365
Glu	2081	1454	2884	1556
Gly	339	237	457	247
Pro	734	513	1146	619
Ser	503	352	625	338
Ammonia	156	109	265	143

[a] Warsi and Wright (1973).

Table 3.51
Grain yield and grain protein in CSH 1 and Swarna sorghum grown under differing levels of N fertilizer and plant population, mean of two years.[a]

	Grain yield (kg/ha)	Prot. (%)
Variety		
CSH 1	4479	10·8
Swarna	3260	11·6
N (kg/ha)		
N_0	2313	9·6
N_{60}	3757	11·4
N_{120}	4867	11·7
N_{180}	4557	12·7
Plant population/ha		
P272,000	4122	10·1
P136,000	3850	11·5
P91,000	3656	12·0

[a] Singh and Bains (1973).

Table 3.52
Grain yield, 1000-seed weight and protein content of four samples of sorghum under differing levels of N fertilizer.[a]

	Grain yield (kg/ha)	1000-seed wt (g)	Prot. content (N × 6·25) (%)
Swarna			
N_0	1625	26·8	9·8
N_{50}	1980	28·9	9·7
N_{100}	2915	29·1	10·8
N_{150}	3020	29·1	11·5
CSH 1			
N_0	1795	26·1	8·9
N_{50}	2430	27·9	10·0
N_{100}	3210	28·8	11·2
N_{150}	3300	28·8	11·7
PJ7R			
N_0	1252	26·8	9·3
N_{50}	2530	28·6	10·1
N_{100}	2015	28·7	11·4
N_{150}	2103	29·3	11·8
M35 1			
N_0	945	27·1	9·5
N_{50}	2324	28·9	10·4
N_{100}	1620	29·3	11·6
N_{150}	1625	29·4	11·9

[a] Anada Rao and Reddy (1973).

Table 3.53
Grain yield, N content in grain of CSH 1 sorghum under different fertilizer treatments.[a]

Treatments	Av. 1967–68 grain yield[b] (kg/ha)	Grain N 1968 (%)
N (kg/ha)		
N_0	2860	1·4
N_{60}	4220	1·7
N_{120}	4590	1·8
P (kg/ha)		
P_0	3160	1·6
P_{26}	4620	1·7
N_0P_0	2270	1·3
N_0P_{26}	3450	1·6
$N_{60}P_0$	3530	1·6
$N_{60}P_{26}$	4910	1·7
$N_{120}P_0$	3670	1·9
$N_{120}P_{26}$	5500	1·8
N × P	NS	NS[b]

[a] Roy and Wright (1973).
[b] NS = not significant. 14% moisture basis.

Table 3.54
Effect of N fertilizer application and plant density of grain yield, 1000-seed weight and protein content of CSH-1 sorghum.[a]

N level (kg/ha)	1000-seed wt (g)	Grain: Yield/hill (g)	Grain: Yield/ha (kg)	Prot. (%)
N_{112}	30·79	50·21	5135	10·0
N_{140}	32·22	54·33	5269	11·0
N_{168}	34·19	59·00	5613	11·4
N_{196}	35·40	62·71	5992	12·0
Plant density				
S_1 (1/hill)	34·61	47·31	4993	11·8
S_2 (2/hill)	33·48	64·25	5804	11·0
S_3 (3/hill)	31·36	58·12	5712	10·5

[a] Reddy and Hussain (1968).

Table 3.55
Protein in the grain, and leucine, lysine, methionine and tryptophan in the protein of two varieties of sorghum as affected by N fertilizer applications (kg/ha).[a]

	CSH 1				Swarna			
	N_0	N_{75}	N_{150}	N_{225}	N_0	N_{75}	N_{150}	N_{225}
Prot. (N × 6·25) (%)								
Crude	10·0	11·4	12·5	10·9	9·8	10·9	12·0	11·4
% increase		14	25	9		11	20	16
True	8·9	10·4	11·1	9·6	8·8	9·6	10·6	9·9
% increase		17	25	8		9	20	13
AAs in prot. (mg/g N)								
Leu	871	841	835	919	837	859	887	67
Lys	95	111	109	127	109	129	127	132
Met	76	92	103	93	75	218	111	94
Trp	66	62	72	72	66	62	67	62

[a] Gupta and Gupta (1972).

Table 3.56
Protein fractions of two varieties of sorghum (DWB and fat-free) as affected by N fertilizer application (kg/ha).[a]

	CSH 1				Swarna			
	N_0	N_{75}	N_{150}	N_{225}	N_0	N_{75}	N_{150}	N_{225}
Total prot. in grain (N × 6·25) (%)	9·4	11·0	12·9	10·8	10·0	10·8	12·0	11·3
Total fractionated prot. (%) (A)	7·4	8·0	10·0	9·0	7·5	7·9	9·3	9·8
Extraction efficiency (%)	79	72	77	83	74	73	77	86
Prot. fractions[b]								
Alb. in grain (%)	1·0	1·0	1·1	1·0	1·1	1·1	1·1	1·1
As % of total prot.[c]	11·1	11·5	8·8	9·7	10·7	10·5	9·1	10·0
[d]As % of (A)	13·6	12·0	11·4	11·6	14·5	14·4	11·7	11·6
Glob. in grain (%)	0·9	0·9	0·8	1·1	1·1	1·1	1·1	1·1
As % of total prot.[c]	8·6	7·9	7·1	10·5	10·7	10·7	9·1	10·0
[d]As % of (A)	11·7	10·9	8·2	12·6	14·5	13·8	11·7	11·6

cont.

Table 3.56 *cont.*

	CSH 1				Swarna			
	N_0	N_{75}	N_{150}	N_{225}	N_0	N_{75}	N_{150}	N_{225}
Prol. in grain (%)	2·4	2·8	4·0	3·3	2·4	2·4	3·8	4·0
As % of total prot.[c]	26·9	25·0	31·1	30·2	23·2	23·6	31·7	34·9
[d]As % of (A)	32·9	34·5	40·2	36·4	31·4	30·9	40·6	40·4
Glut. in grain (%)	3·1	3·4	4·0	3·5	3·0	3·2	3·3	3·6
As % of total prot.[c]	41·8	30·9	31·1	32·7	29·4	29·9	27·8	31·5
[d]As % of (A)	41·8	42·6	40·2	39·3	39·6	40·9	35·9	36·4

[a] Gupta and Gupta (1972).
[b] Albumin, globulin, prolamine, glutelin.
[c] Authors' note: Gupta and Gupta's calculations appear to be % of total prot. in grain and not of total fractionated prot. (A) as stated in their table. Gupta and Gupta's values corrected to nearest one place of decimals appear in rows labelled "c".
[d] The authors' calculated values of each fraction as % of total fractionated prot. appear in rows labelled "d".

Table 3.57
Grain yield and protein content, Leu, Lys, Met and Trp in the grain and in the protein of two varieties of sorghum as affected by urea foliar spray.[ab]

	CSH 1			Swarna		
	N_0	N_2	N_4	N_0	N_2	N_4
Grain yield (g/pot)	14·8	25·0	23·0	16·7	26·7	25·2
Prot. (N × 6·25) (%)						
Crude	10·0	12·5	11·3	10·2	13·3	13·9
True	9·0	11·4	11·3	9·1	12·1	12·7
AAs in grain (mg/100 g)						
Leu	1380	1440	1280	1180	1480	1570
Lys	200	230	180	170	230	190
Met	150	210	180	190	260	240
Trp	130	140	120	120	150	150
AAs in prot. (mg/g N)						
Leu	860	719	704	727	693	706
Lys	127	73	97	106	106	86
Met	94	104	101	117	121	108
Trp	79	71	66	73	72	67

[a] N_0—water.
N_2—2% urea.
N_4—4% urea.
[b] Gupta and Gupta (1975).

Table 3.58
Effect of N fertilizer levels on grain protein of two sorghum cultivars with two cropping treatments.[a]

	Cropping treatment							
	Sorghum following sorghum				Sorghum following 'panicum'			
	Alpha		TX 610		Alpha		TX 610	
	Prot. in grain (%)	Prot. yield (kg/ha)	Prot. in grain (%)	Prot. yield (kg/ha)	Prot. in grain (%)	Prot. yield (kg/ha)	Prot. in grain (%)	Prot. yield (kg/ha)
Elemental N applied (kg/ha)								
0	6·6	138	5·6	134	9·4	372	7·8	390
28	6·6	195	5·9	202	9·3	389	8·1	461
56	7·1	268	6·0	255	9·3	421	8·4	477
84	7·8	343	7·1	336	10·3	496	8·9	547

[a] Littler (1967).

Table 3.59
Means[a] and ranges (in parentheses) of protein content (N × 6·25%) in six sorghum varieties grown under four N treatments.[b]

N_2 treatment kg/ha	ES7	T671	P846	P579	F64	C44B	Mean
0	9·0	11·2	11·1	12·5	12·0	12·6	11·4
	(6·9–10·9)	(8·1–14·2)	(9·3–12·7)	(8·0–15·1)	(10·4–13·6)	(11·0–14·2)	
68	10·7	9·2	11·0	9·5	11·6	9·6	10·3
	(10·2–11·5)	(7·3–10·9)	(8·0–14·8)	(7·2–13·2)	(9·0–16·1)	(8·0–11·2)	
136	10·5	11·6	10·9	11·0	11·3	12·0	11·2
	(9·0–11·5)	(8·6–15·0)	(7·0–13·6)	(7·4–14·7)	(8·5–15·1)	(8·8–15·1)	
273	10·4	12·7	11·5	11·7	11·9	11·7	11·6
	(9·6–11·3)	(10·8–16·1)	(9·6–12·6)	(9·5–13·0)	(11·0–13·5)	(10·3–13·2)	
Mean	10·1	11·2	11·1	11·2	11·7	11·5	11·1

[a] Means of four values except for C44B with only two.
[b] Kondos (personal communication, 1978).

Table 3.60
Mean values[a] with ranges[b] (in parentheses) (mg/g total N) of 10 amino acids within and between six sorghum varieties.[c]

	E57	T671	P846	P579	F64	C44B	Mean	Max.[d] range	Max.[e] diff. % of mean
Ile	215	194	224	215	243	211	217	106	49·0
	(162–237)	(162–225)	(212–237)	(212–237)	(225–269)	(194–237)			
Leu	737	714	753	750	695	757	734	206	28·1
	(687–812)	(625–812)	(750–812)	(687–812)	(625–750)	(750–812)			
Lys	159	147	152	150	143	154	151	75	49·6
	(131–181)	(119–169)	(150–156)	(112–187)	(125–162)	(119–175)			
Met	92	96	92	92	104	102	96	50	52·0
	(75–106)	(75–106)	(81–112)	(75–112)	(87–119)	(87–125)			
Phe	287	246	259	232	287	286	266	112	42·2
	(231–319)	(219–262)	(237–275)	(212–312)	(250–325)	(269–300)			
Thr	197	221	204	196	213	212	204	156	76·4
	(169–225)	(175–281)	(125–225)	(125–225)	(187–250)	(181–194)			
Trp	62	66	80	71	69	76	71	31	44·2
	(56–75)	(62–75)	(75–87)	(56–81)	(56–75)	(69–81)			
Val	234	231	236	254	262	228	241	119	51·0
	(219–244)	(212–256)	(219–250)	(187–306)	(231–287)	(212–250)			
Arg	266	259	244	258	248	261	256	75	29·3
	(244–300)	(237–256)	(237–300)	(231–300)	(244–250)	(244–287)			
His	158	138	146	143	143	150	146	62	42·7
	(137–175)	(125–156)	(125–175)	(112–169)	(137–150)	(125–175)			

[a] Mean of four analyses; two only for C44B.
[b] Range in the level within the variety.
[c] Kondos (personal communication, 1978).
[d] Maximum differences among amino acid values among varieties.
[e] Maximum differences expressed as percent of overall mean.

Table 3.61
Effect of fertilizer treatments on proximate composition and amino acid content in protein of Hegari sorghum[a] and weight gain and PER of weanling rats fed sorghum diets.[b]

	(1) NPK[c]	(2) NPK + minor elements	(3) Organic fertilizer	(4) Organic fertilizer + minor elements	(5) Minor elements	(6) Control
Proximate composition (10% moisture basis)						
N 1961 (%)	1·86	1·78	1·68	1·67	1·54	1·47
1962 (%)	1·79	1·77	1·60	1·68	1·62	1·65
Ether extract 1961 (%)	3·22	3·18	3·50	3·31	3·22	3·14
Crude fibre 1961 (%)	2·96	3·25	3·20	3·30	3·20	2·74
Ash 1961 (%)	1·39	1·35	1·51	1·46	1·25	1·33
AAs (mg/g N)						
Ile	372	373	387	393	404	406
Leu	758	753	774	755	762	750
Lys	144	141	151	153	167	153
Trp	28	28	26	30	31	29
Rat feeding trials (over 35 days)						
Wt gain (g) 1961	12·5	13·5	10·9	16·9	23·5	18·9
PER 1961	0·65	0·65	0·59	0·80	0·93	0·79
Wt gain (g) 1962	18·7	19·4	15·7	20·9	25·4	23·2
PER 1962	0·75	0·74	0·62	0·83	0·92	0·87

[a] Grown 1961–62 (Guatemala).
[b] Aragón H. and Bressani (1965).
[c] NPK—N, 19%; superphosphate, 13%; and K, 7%.

Table 3.62
Average amino acid content in the grain and in the protein of sorghum grain grown under varying levels of N fertilization.[a]

	N fertilization level (kg/ha)			N fertilization level (kg/ha)		
	0	89·7	134·6	0	89·7	134·6
AAs	(in mg/100 g sample)			(in mg/g N)		
Ile	370	510	540	252	265	264
Leu	1200	1730	1850	812	897	911
Lys	220	250	250	157	130	126
Met	110	140	150	77	73	74
Half-Cys	140	190	200	98	97	99
Phe	460	640	680	316	331	334
Tyr	190	260	270	130	134	135
Thr	300	370	390	206	194	195
Val	480	630	670	326	328	329
Arg	340	400	420	234	207	206
His	210	270	280	144	139	137
Ala	860	1190	1280	581	616	629
Asp	610	830	880	424	429	435
Glu	1910	2700	2890	1285	1400	1427
Gly	320	370	400	266	194	197
Pro	720	1000	1050	489	516	521
Serine	400	520	560	274	271	277
Prot. (N × 6·25 DWB) (%)	9·1	12·0	12·6	9·1	12·0	12·6

[a] Waggle *et al.* (1967b).

Table 3.63
Average amino acid content in the grain and in the protein of sorghum grain grown at three locations in Kansas.[a]

	Location			Location		
	Obelin	Newton	Centralia	Obelin	Newton	Centralia
AAs	(in mg/100 grain)			(in mg/g N)		
Ile	470	510	440	262	264	256
Leu	1590	1760	1430	882	906	832
Lys	240	240	250	136	127	134
Met	140	120	140	76	65	82
Half-Cys	160	180	180	92	94	109
Phe	600	640	540	332	331	319
Tyr	240	250	230	133	130	136
Thr	350	380	330	199	197	199
Val	580	630	560	326	327	330
Arg	380	390	380	214	203	227
His	250	260	250	141	134	147
Ala	1100	1230	1000	609	632	586
Asp	770	820	730	430	427	432
Glu	2470	2780	2250	1373	1424	1351
Gly	360	360	370	204	191	221
Pro	910	1020	840	509	524	494
Serine	490	520	460	276	271	274
Prot. DWB (N × 6·25) (%)	11·2	12·1	10·6	11·2	12·1	10·6

[a] Waggle *et al.* (1967b).

Table 3.64
Weight of 100 seeds, grain yield/panicle, protein, lysine, protein × lysine of 18 inbred lines of sorghum, and RS610 (Test 1 only) under varying levels of fertility.[a]

Genotype	100-seed wt (g)		Grain wt/panicle (g)		Prot. (N × 6·25) (%)		Lys (mg/g total N)		Lys × prot.	
	Test 1	Test 2	Test 1	Test 2	Test 1	Test 2	Test 1	Test 2	Test 1	Test 2
2190	3·8	3·5	30·4	24·5	16·3	17·1	107	84	1744	1443
6269	0·7	0·7	6·1	9·6	16·3	16·0	157	107	2567	1710
1897	3·4	3·2	51·1	10·7	15·5	16·8	102	75	1581	1260
1177	2·1	1·3	25·6	25·4	15·3	15·3	104	61	1591	933
2833	2·2	2·5	23·2	24·4	15·3	16·0	131	89	2004	1424
0819	2·9	2·7	69·9	31·0	14·7	15·3	131	107	1926	1637
5228	1·0	0·7	11·8	5·7	14·4	15·6	146	163	2102	2543
0692	1·9	2·3	23·7	31·5	14·3	13·3	132	94	1888	1250
5075	1·2	1·3	18·4	19·8	14·2	14·2	132	88	1874	1250
1054	2·7	2·7	53·9	41·2	14·1	14·7	124	116	1748	1705
5217	1·8	1·8	37·9	25·1	14·0	14·8	123	69	1722	1021
5383	2·9	2·4	59·8	45·9	14·0	13·7	111	103	1442	1411
2016	2·3	1·9	46·0	36·3	13·6	14·4	124	97	1686	1397
1102	2·4	2·0	57·8	41·4	12·9	14·3	107	92	1380	1316
0305	1·5	1·0	32·7	21·1	12·6	13·4	135	132	1663	1769
1499	0·9	1·3	11·5	20·2	12·3	12·1	134	58	1648	702
RS 610	2·6	—	75·1	—	12·1	—	122	—	1476	—
5206	2·7	2·7	51·1	40·6	12·0	12·9	109	78	1308	1006
1121	2·2	2·3	49·3	49·6	11·7	12·2	127	96	1486	1171
Mean	2·2	2·0	38·7	28·0	14·0	14·6	124	95	1736	1387
LSD 0·05	1·10	0·90	58·48	22·84	3·39	3·28	93	65	315	213
0·01	1·45	1·21	77·29	30·76	4·49	4·41	122	88	548	388

[a]Campbell and Pickett (1968).

Table 3.65
Yield and protein content of DeKalb BR 64 sorghum under varying levels of ammonium nitrate.[a]

	1969		1970	
Ammonium nitrate (kg/ha)	Grain yield (kg/ha)	Grain prot. (N × 6·25) (%)	Grain yield (kg/ha)	Grain prot. (%)
0	4974	9·1	4304	8·3
112	5595	9·1	5165	8·7
224	5500	9·9	5691	10·1
336	5404	10·9	5691	10·3
448	5069	11·3	5309	11·4

[a] Lutrick (1970).

Table 3.66
Average yield and uptake of phosphorus of sorghum CO 1 and CO 18.[a]

Treatment No.	Amount of phosphorus P_2O_5 kg/h supplied as		Grain yield		P_2O_5 uptake in grain	
	Super-phosphate	Compost	CO 1 (kg/ha)	CO 18 (kg/ha)	CO 1 (kg/ha)	CO 18 (kg/ha)
(1)	0	0	656	911	2·3	7·7
(2)	16·8		771	1366	3·9	13·9
(3)	33·6		851	2083	5·0	18·5
(4)	50·4		969	2277	7·2	20·5
(5)	67·2		885	2000	6·3	18·2
(6)		16·8	791	1549	4·7	16·0
(7)		33·6	782	1366	5·5	12·3
(8)		50·4	809	1605	3·7	18·3
(9)		67·2	775	1366	3·6	11·9
(10)	16·8	16·8	894	1570	4·1	17·3
(11)	16·8	33·6	1012	2305	6·1	22·1
(12)	16·8	50·4	894	1776	3·4	13·7
(13)	33·6	16·8	1019	2316	8·2	22·2
(14)	33·6	33·6	907	1749	4·6	16·6
(15)	50·4	16·8	918	1879	4·3	14·1

[a] Kamalam (1964).

Table 3.67
Mean moisture, protein and β-carotene (ranges in parentheses) of two varieties of sorghum grown at 33 locations.[a]

	IS 511 (46)	IS 3797 (370)
Moisture (%)	9·96 (7·5–12·9)	9·62 (7·1–11·0)
Prot. (N × 6·25) (%)	9·41 (6·4–19·4)	9·45 (5·0–13·0)
β-carotene (mg/100 g grain)	0·04 (0·01–0·07)	0·03 (0·01–0·06)
As % of total yellow pigment	12·7 (6·0–20·4)	11·5 (3·0–16·6)

[a] Deosthale *et al.* (1969).

Table 3.68
Protein (% in grain) of 5 pearl millet parents in diallel crosses.[a]

	Mean values				
	Location I K1 4	PT 819/4	PT 852/2	PT 870	PT 888
K1 4	12·4	16·1	13·2	14·2	13·5
PT 819/4		14·9	11·5	14·2	15·3
PT 852/2			9·6	11·1	17·8
PT 870				14·3	11·3
PT 888					14·9
	Location 2				
	K1 4	PT 819/4	PT 852/2	PT 870	PT 888
K1 4	14·6	15·0	12·8	14·4	12·7
PT 819/4		12·4	16·1	14·8	15·0
PT 852·2			13·2	10·9	11·8
PT 870				12·6	15·8
PT 888					16·3

[a] Mahadevappa (1967).

Table 3.69
Proximate composition, inorganic elements, vitamins ("as-is" basis), and amino acid content in the protein of pearl millet, peasant and experimental cultivation in Senegal.[a]

	Peasant cultivation		Experimental cultivation		
	Souna	Sanio	*P. pycnostachyum*	*P. nigritarum*	*P. pycnostachyum* × *nigritarum*
Moisture %	9·4	8·9	9·4	8·3	9·0
Prot. (N × 5·83) (%)	9·0	9·9	10·3	12·4	11·3
Fat (%)	5·2	4·9	4·7	5·6	5·0
Total reducing substances (%)	68·2	68·3	69·3	69·0	70·1
Ash (%)	1·5	1·6	1·6	1·7	1·6
Inorganic elements (mg/100 g)					
Ca	15·1	17·45	25·0	42·7	38·0
P	343·0	350·0	311·5	359·0	378·0
Ratio, Ca/P	0·05	0·04	0·08	0·12	0·10
K	312·0	339·0	371·0	374·0	365·0
Na	76·9	44·6	114·5	110·0	68·5
Ratio, K/Na	4·1	7·6	3·3	3·4	5·3
Vitamins (mg/100 g)					
Thiamine	0·28	0·36	0·37	0·34	0·38
Riboflavin	0·15	0·15	0·19	0·18	0·22
Niacin	2·10	1·55	3·05	5·15	1·90
Pantothenic acid	1·20	1·00	1·45	1·25	1·05

Table 3.69 *cont.*

	Peasant cultivation		Experimental cultivation		
	Souna	Sanio	*P. pycnostachyum*	*P. nigritarum*	*P. pycnostachyum* × *nigritarum*
AAs (mg/g N)					
Ile	369	250	303	187	312
Leu	1044	1050	1087	1025	1150
Lys	152	184	161	165	165
Met	150	144	137	169	162
Cys	71	84	78	78	—
Phe	216	269	297	316	300
Thr	184	150	344	262	303
Trp	119	100	131	147	153
Val	250	294	325	387	350
Arg	234	219	237	319	328
His	119	129	114	148	133

[a] Adrian and Sayerse (1957).

Table 3.70
Proximate composition of pearl millet under two levels of urea fertilization and two methods of application.[a]

	Foliar application (kg/ha)			Soil application (kg/ha)		
	0	11·2	22·4	0	22·4	44·8
1000-seed wt (g)	7·30	7·64	7·89	7·72	7·70	7·99
Moisture (%)	10·4	10·3	10·4	10·5	10·3	10·3
Prot. (%)	9·2	10·4	10·9	9·4	10·3	11·0
Fat (%)	5·2	5·4	5·4	5·3	5·4	5·4
Crude fibre (%)	1·5	1·6	0·4	1·4	1·5	1·4
Ash (%)	1·7	1·8	1·7	1·6	1·7	1·6
CBH[b]	72·1	70·5	70·2	71·8	70·9	70·2
Ca (mg/100 g)	31	35	31	28	31	28
P_2O_5 (mg/100 g)	850	850	850	840	850	850

[a] Khanna (1966).
[b] Carbohydrate.

Table 3.71
Effect on grain yield, grain protein and protein/ha of local pearl millet[a] under five levels of N fertilizer at sowing.[b]

N application (kg/ha)	Yield (kg/ha)	Prot. (N × 6·25) (DWB) (%)	Prot. yield (kg/ha)
0	1100	6·7	74
50	1250	7·5	94
100	1350	8·3	112
150	1700	8·2	139
200	1500	8·1	121

[a] Grown at Kotiary, Senegal (1967).
[b] Blondel (1970).

Table 3.72
Effect on grain yield, grain protein and protein/ha of pearl millet, PC 28,[a] under two levels of N fertilizer.[b]

N application (kg/ha)	Yield (kg/ha)	Prot. (N × 6·25) (DWB) (%)	Prot. yield (kg/ha)
0	781	7·3	56
50 at sowing	1333	8·9	119
50 at sowing + 50 at half bloom	2178	13·6	296

[a] Bambey, Senegal (1968).
[b] Blondel (1970).

Table 3.73
Effect on yield, and proximate analysis[a] of pearl millet Souna III, of four levels of N fertilizer, with and without millet straw compost.[b]

	Yield t/ha	Prot. (N × 6·25) (%)	Ash (%)	Lipid (%)	Cellulose (%)	Ca (mg/100 g)	P (mg/100 g)	Fe (mg/100 g)
N_0								
With compost	1·8	8·6	1·6	4·5	1·4	23	226	1·6
Without compost	1·5	8·0	1·7	4·9	1·4	27	227	1·4
N_{60}								
With compost	2·6	10·4	1·7	4·8	1·4	25	226	1·6
Without compost	2·3	9·8	1·7	4·9	1·4	26	227	1·5
N_{90}								
With compost	2·9	10·6	1·7	4·8	1·4	24	226	1·4
Without compost	2·6	10·2	1·8	4·3	1·4	24	226	1·4
N_{150}								
With compost	2·9	12·2	1·8	4·6	1·4	35	238	1·8
Without compost	3·1	12·0	1·8	4·6	1·4	36	226	1·1

[a] At approx. 8% H_2O basis.
[b] Ganry and Bideau (1974).

Table 3.74
Effect on amino acid composition in protein of pearl millet Souna III and Lys in grain, of four levels of fertilizer, without and with millet straw compost.[a]

	N_0		N_{60}		N_{90}		N_{150}	
AAs	With compost[b]	Without compost[b]	With compost[b]	Without compost[b]	With compost[b]	Without compost[b]	With compost[b]	Without compost[b]
Ile	281	269	269	300	287	294	294	281
Leu	637	631	675	675	662	662	687	681
Lys	225	244	212	206	194	194	175	181
Met	112	50	50	106	87	125	100	112
Phe	325	325	337	337	325	331	325	362
Tyr	244	244	237	244	237	237	231	231
Thr	275	300	269	231	275	262	262	262
Val	350	344	362	362	350	350	331	256
Arg	350	375	331	325	331	325	306	306
His	175	181	169	169	162	162	156	156
Ala	462	481	481	487	487	481	494	487
Asp	519	544	525	525	531	512	519	519
Glu	1225	1219	1287	1269	1300	1275	1325	1325
Gly	244	262	225	219	231	225	206	212
Pro	450	444	444	431	444	456	481	437
Ser	319	344	319	319	331	312	325	312
Lys (mg/100 g grain)	310	310	350	320	330	320	340	350

[a] Ganry and Bideau (1974).
[b] mg/g N. Converted from the % pure protein values by the present authors using the factor 62·5.

Table 3.75
Effect on rat growth, protein intake and PER of four levels of N fertilizer applied to pearl millet, with and without millet straw compost.[a]

Fertilizer	Wt gain/ animal/day (g)	Prot. intake/ animal/day (g)	PER[b]
N_0			
With compost	0·60	0·39	1·55
Without compost	0·80	0·45	1·77
N_{60}			
With compost	1·50	0·90	1·58
Without compost	1·55	0·83	1·63
N_{90}			
With compost	1·06	0·72	1·47
Without compost	0·70	0·58	1·20
N_{150}			
With compost	0·83	0·69	1·15
Without compost	0·73	0·60	1·16

[a] Ganry and Bideau (1974).
[b] Authors' note: since the protein intake was not constant, i.e. diets were not isonitrogenous, these are not true PER values.

Table 3.76
Effects of fertilizer on grain yield and crude protein of pearl millet grain.[a]

Treatment (kg/ha)	Monsoon 1956 Yield (kg/ha)	Monsoon 1956 Prot. (N × 6·25) (%)	Monsoon 1956 Prot. (kg/ha)	Summer 1957 Yield (kg/ha)	Summer 1957 Prot. (N × 6·25) (%)	Summer 1957 Prot. (kg/ha)	Monsoon 1957 Yield (kg/ha)	Monsoon 1957 Prot. (N × 6·25) (%)	Monsoon 1957 Prot. (kg/ha)
No manure	414	13·2	54	361	13·2	48	576	13·2	76
Ammonium chloride									
22·4	504	14·3	72	768	13·7	105	914	14·1	129
44·8	557	14·4	80	987	13·9	137	1153	14·3	164
Ammonium sulphate									
22·4	461	13·6	63	762	13·7	104	793	13·6	107
44·8	573	13·9	73	1009	13·8	140	1043	13·8	144
Ammonium sulphate nitrate									
22·4	578	13·8	79	806	13·8	111	860	13·6	118
44·8	818	14·0	114	1161	13·9	161	1105	13·9	153
Ammonium nitrate									
22·4	610	14·0	85	771	13·7	105	939	13·8	129
44·8	737	14·9	110	1108	13·8	152	1132	14·3	162
Urea									
22·4	594	13·6	81	828	13·5	112	799	13·6	109
44·8	785	13·7	108	1014	13·7	139	965	13·7	132
Groundnut cake									
44·8	256	13·5	50	368	13·5	94	793	13·4	106
Farmyard manure									
44·8	389	13·7	53	389	13·5	60[b]	714	13·7	98

[a] Shah and Mehta (1959b).
[b] Authors' note: the figure of 60 kg prot./ha quoted by Shah and Mehta for farmyard manure should read 52 if yield of grain and prot. % are correct.

Table 3.77
Effect of different levels of ammonium sulphate application on protein, total sulphur, Trp and Lys content of HB 1 pearl millet.[a]

	N (kg/ha)			
	Control N_0	N_{44}	N_{88}	N_{176}
Prot. in grain (%)	12·6	13·0	13·4	13·3
S (mg/100 g grain)	152	165	202	200
Trp				
(mg/100 g grain)	138	153	159	157
(mg/g N)	68	73	74	73
Lys				
(mg/100 g grain)	412	429	432	409
(mg/g N)	211	206	202	192

[a] Sawhney and Naik (1969).

Table 3.78
Effect of different fertilizer treatments on grain yield and grain protein in pearl millet.[a]

Treatments	Grain yield (kg/ha)	Grain prot. (%)
Foliar application (kg N/ha)		
0 (N_0)	668	9·2
11·2 (N_{11})	821	10·3
22·4 (N_{22})	1028	10·9
Soil application (kg N/ha)		
0 (N_0)	698	9·2
22·4 (N_{22})	819	10·3
44·8 (N_{45})	1025	11·0

[a] Kinra *et al.* (1970).

Table 3.79
Effect on test weight and protein content in the grain of pearl millet of two methods of sowing, differing fertilizer treatments and two seed reates.[a]

Treatment	Test weight (g)	Prot. in grain (%)
Sowing		
Line	6·7	8·9
Broadcast	7·9	9·4
Fertilizer		
(1) 55 kg N/ha as $(NH_4)_2SO_4$ at sowing	7·4	9·2
(2) 55 kg N/ha as municipal compost at sowing	7·1	8·7
(3) 27·5 kg N/ha as $(NH_4)_2SO_4$ and 27·5 kg N/ha as compost both at sowing	7·6	9·5
(4) 41·25 kg N/ha as compost at sowing and 13·75 kg N/ha as $(NH_4)_2SO_4$ at top dressing	7·2	9·3
Seeding rates (kg seeds/ha)		
2·5	7·5	9·2
5·0	7·1	9·1

[a] Tomer (1970).

Table 3.80
Effect of N fertilizer levels on grain yield, protein and amino acid content of five varieties of pearl millet.[a]

	Grain yield (kg/ha)	Prot. (N × 6·25) (DWB) (%)	Ile		Leu		Lys		Met		Thr		Trp	
			(mg/100 g grain)	(mg/g N)	(mg/100 g grain)	(mg/g N)	(mg/100 g grain)	(mg/g N)	(mg/100 g grain)	(mg/g N)	(mg/100 g grain)	(mg/g N)	(mg/100 g grain)	(mg/g N)
N level (kg/ha)														
0	1690	9·4	370	244	810	531	290	194	260	175	360	244	89	63
40	1780	10·5	420	256	910	544	300	181	290	175	380	231	113	67
80	2050	11·8	470	250	1040	550	320	169	290	156	410	219	126	69
120	2480	13·0	530	250	1210	569	320	150	330	156	470	225	146	75
160	2150	13·6	570	262	1180	544	350	162	340	156	480	219	159	73
200	2160	13·8	600	269	1350	581	340	156	350	156	490	225	169	76
Variety														
HB 1	2000	11·8	480	250	1120	587	290	156	270	144	430	225	144	75
HB 3	2090	13·4	560	262	1300	606	350	162	310	150	470	219	159	74
HB 4	2560	11·4	450	244	950	525	310	175	310	175	400	225	116	63
KL 559	1750	11·7	470	250	1000	506	310	169	310	169	420	231	122	64
D 356	1830	11·8	510	269	1050	550	330	181	330	175	440	237	127	72

[a] Deosthale *et al.* (1972b).

Table 3.81
Protein content in three varieties of pearl millet as affected by two levels of N fertilizer and two methods of weed control.[a]

Treatment	Prot. (N × 6·25) (%) 1963	1964
Varieties		
RSK	13·0	10·4
Chadi	10·1	9·2
Local	11·8	9·9
N dosage (kg/ha)		
0	10·3	8·7
20	11·3	10·0
40	13·1	10·8
Methods of weed control		
None	11·0	8·9
Hand weeding	12·0	10·4
2, 4-D	11·8	10·3

[a] Kumar *et al.* (1973).

Table 3.82
Effect of N fertilizer levels on Fe and Mn concentration in the grain of two pearl millet varieties.[a]

	Fe (mg/100 g)		Mn (mg/100 g)	
Treatment	Local	Hybrid	Local	Hybrid
N (kg/ha)				
0	11·4	11·6	2·67	2·97
30	13·7	14·4	3·08	3·35
60	13·6	13·4	3·07	3·53
90	14·9	13·0	3·57	3·57
120	14·7	13·8	4·10	3·93
150	13·6	12·8	3·22	3·53
Mean	13·6	13·2	3·28	3·48

[a] Shukla and Bhatia (1971).

Table 3.83
Effect of four fertilizers on N content (%) and inorganic elements of two samples of pearl millets.[a]

	HB 1						Pusa Moti					
	Control 0	$(NH_4)_2SO_4$	NH_4Cl	$(NH_4)_3SO_4NO_3$	Urea	Mean	Control 0	$(NH_4)_2SO_4$	NH_4Cl	$(NH_4)_3SO_4NO_3$	Urea	Mean
N (%)	2·44	2·49	2·45	2·47	2·52	2·47	2·40	2·44	2·43	2·43	2·48	2·44
Inorganic elements (mg/100 g)												
Ca	30	36	33	40	37	35	28	35	31	38	35	33
P	270	333	298	300	310	302	260	323	288	293	305	294
Fe	28	46	34	40	45	39	25	44	31	36	42	36
Mg	4	7	6	9	8	7	4	6	5	8	6	6
Mn	18	24	20	21	21	21	16	20	18	18	19	18
K	666	790	717	752	806	746	579	682	637	635	657	638

[a] Desai and Zende (1972).

Table 3.84

Composition in N, P_2O_5 and K_2O_5 DWB of pearl millet grain under different fertilizer treatments, with and without basal dressing.[a]

Treatment (all 50·4 kg/ha)	With basal dressing				Without basal dressing			
	Grain yield (kg/ha)	N (%)	P_2O_5 (mg/100 g)	K_2O (mg/100 g)	Grain yield (kg/ha)	N (%)	P_2O_5 (mg/100 g)	K_2O (mg/100 g)
(1) No manure	685	1·14	760	720	590	1·64	850	500
(2) N	638	2·10	1120	720	563	1·60	920	750
(3) $N+P_2O_5$	1507	2·05	770	770	1351	1·90	960	920
(4) $N+P_2O_5+K_2O$	1649	2·06	910	530	1570	1·59	200	500
(5) $N+K_2O$	801	1·99	900	790	773	1·70	860	890
(6) $K_2O+P_2O_5$	1373	1·40	1200	770	1292	2·10	970	890
(7) K_2O	923	1·88	960	590	829	1·20	850	690
(8) P_2O_5	1066	1·80	900	450	986	1·20	740	660
(9) N (as farmyard manure)	868	1·50	850	670	801	1·35	820	480
(10) N (as green manure)	737	1·30	1100	480	685	1·20	750	720
(11) N (as compost)	726	1·30	800	530	629	0·94	690	640

[a] Venkataramana and Krishna Rao (1961).

Table 3.85
Effect on N, P, K, and Ca contents of CO 1 and CO 7 Finger Millet under differing levels of P fertilization.[a]

Treatment	CO 1					CO 7				
	Moisture (%)	N (%)	P (mg/100 g)	K (mg/100 g)	Ca (mg/100 g)	Moisture (%)	N (%)	P (mg/100 g)	K (mg/100 g)	Ca (mg/100 g)
(1) No manure	10·9	1·13	570	1270	850	12·6	1·21	680	1250	750
Farmyard manure to supply P_2O_5 kg/ha										
(2) 11·2	11·6	1·22	680	1570	770	11·8	1·45	680	1290	570
(3) 22·4	10·3	1·10	780	1590	570	10·5	1·23	750	1340	870
(4) 33·6	11·0	1·40	750	1670	580	11·3	1·29	720	1400	820
(5) 44·8	11·0	1·25	650	1730	810	10·6	1·29	690	1360	770
Superphosphate to supply P_2O_5 kg/ha										
(6) 11·2	11·5	1·34	600	1460	830	11·4	1·31	680	1350	680
(7) 22·4	8·8	1·33	710	1810	850	8·7	1·16	730	1320	750
(8) 33·6	11·4	1·25	730	1810	750	10·7	1·29	770	1310	780
(9) 44·8	11·3	1·35	740	1500	720	11·2	1·55	890	1510	830
Farmyard manure (FYM) and superphosphate (S) to supply P_2O_5 kg/ha										
(10) 11·2 (FYM) + 11·2 (S)	11·4	1·30	690	1950	760	11·4	1·62	790	1550	940
(11) 11·2 (FYM) + 22·4 (S)	10·5	1·22	730	1840	620	9·7	1·45	880	1630	790
(12) 11·2 (FYM) + 33·6 (S)	8·7	1·15	750	1370	560	10·3	1·40	890	1430	790
(13) 22·4 (FYM) + 11·2 (S)	11·8	1·52	710	1740	740	11·5	1·37	700	1740	930
(14) 22·4 (FYM) + 22·4 (S)	9·9	1·49	650	1590	610	11·3	1·32	700	1470	990
(15) 33·6 (FYM) + 11·2 (S)	9·6	1·16	640	1600	680	9·1	1·21	700	1520	950

[a] Sree Ramulu and Mariakulandai (1964).

Table 3.86
The effect of five fertilizer levels on the protein and calcium contents of four varieties of finger millet.[a]

	Fertilizer levels				
	N_0	N_{44}	N_{90}	N_{134}	$N_{90}+P_{44}$
Prot. (N × 6·25) (%)					
H 22	4·7	5·2	6·7	7·6	6·2
Anna Purna	5·0	6·2	6·9	7·2	5·4
Caveri	5·4	5·8	6·1	6·1	5·6
Purna	5·6	6·5	7·0	7·0	7·0
Mean	5·2	5·9	6·7	7·0	6·0
Ca (mg/100 g)					
H 22	445	391	380	353	433
Anna Purna	373	304	296	311	309
Caveri	366	356	366	368	378
Purna	289	306	344	334	399
Mean	368	339	346	342	380

[a] Krishnamurthy (1968).

Table 3.87
Effect of N application and row spacings on yield (kg/ha) and protein (N × 6·25 DWB) of finger millet.[a]

	N_0		N_{50}		N_{100}		N_{150}		Mean	
Row spacings (cm)	Prot. (%)	Yield (kg/ha)	Prot. (%)	Yield (kg/ha)	Prot. (%)	Yield (kg/ha)	Prot. (%)	Yield (kg/ha)	Prot. (%)	Yield (kg/ha)
S15	6·8	2461	7·7	2476	8·6	2744	9·9	3330	8·3	2753
S30	7·8	3839	8·1	3870	9·4	3411	10·2	3856	8·9	3744
S45	8·1	3791	9·3	4043	10·6	3701	12·0	4132	10·0	3917
S60	8·8	3394	10·2	3606	12·1	3411	13·0	3983	11·1	3599
Mean	7·9	3371	8·8	3499	10·2	3317	11·3	3825	9·6	3503

[a] Stabursvik and Heide (1974).

Table 3.88
Amino acid composition of finger millet, mean of two individual samples from different plots, and differing crude protein content.[a]

Crude prot. (N × 6·25) (%)	6·0	7·5	9·0	11·6	13·5
AAs (mg/g N)					
Ile	262	278	301	296	297
Leu	573	622	684	691	691
Lys	188	166	151	127	112
Met	221	220	211	174	149
Cys	129	128	127	116	101
Phe	327	349	361	346	329
Tyr	207	224	226	239	232
Thr	279	282	291	271	248
Trp[b]	94	94	94	94	94
Val	387	398	422	438	432
Arg	299	274	261	224	210
His	141	145	147	142	139
Ala	380	382	392	382	361
Asp	429	414	397	379	361
Glu	1194	1306	1462	1594	1612
Gly	271	246	226	197	179
Pro	411	419	428	427	420
Ser	339	367	369	362	346

[a] Stabursvik and Heide (1974).
[b] Approximate literature value. Not determined.

Table 3.89
Pot Experiment I. Crude protein content and amino acid composition of finger millet under different levels of N and S fertilizer.[a]

N (p.p.m.)	10			50			250				
S (p.p.m.)	5	25	125	5	25	125	5	(12·5)	25	125	Local market samples
S/N ratio	1/2	5/2	25/2	1/10	1/2	5/2	1/50	(1/20)	1/10	1/2	
Crude prot. (N × 6·25) (DWB) (%)	5·16	5·16	4·80	4·32	4·10	4·00	7·32	(7·08)	7·22	7·04	7·4
AAs (mg/g N)											
Ile	236	244	235	252	249	237	246	(279)	272	281	279
Leu	519	521	509	538	536	499	544	(589)	625	601	641
Lys	226	223	224	242	233	246	167	(177)	187	191	189
Met	221 } 361	219 } 356	214 } 357	223 } 367	227 } 367	227 } 367	88 } 154	(129) } 227	224 } 362	222 } 362	189 } 315
Cys	140 }	137 }	143 }	144 }	140 }	140 }	66 }	(98) }	138 }	140 }	126 }
Phe	311	314	297	312	327	294	278	(314)	334	339	344
Tyr	183	188	172	187	188	188	194	(221)	238	231	204
Thr	276	289	275	279	285	274	202	(236)	260	274	275
Trp	94	94	94	94	94	94	94	(94)	94	94	94
Val	356	358	354	366	364	348	355	(390)	402	395	395
Arg	346	362	371	351	344	339	226	(359)	282	290	274
His	149	146	144	163	161	156	155	(150)	151	146	152
Ala	369	384	377	396	398	377	318	(352)	369	394	379
Asp	464	501	488	482	482	498	1059	(631)	452	437	396
Glu	1059	1047	1034	1066	1097	1012	1344	(1369)	1359	1325	1316
Gly	307	321	315	324	314	329	225	(251)	266	277	255
Pro	406	406	406	406	406	406	406	(406)	406	406	394
Ser	336	336	329	338	336	336	296	(324)	326	321	343

[a] Stabursvik and Heide (1974).
[b] Approximate literature values. Not determined.

Table 3.90

Pot Experiment II. Crude protein content and amino acid composition (mg/g total N) of finger millet seed protein under different levels of N and S fertilizer.[a]

N (p.p.m.)	250				500				1000			
S (p.p.m.)	12·5	25	50	125	25	50	100	250	50	100	200	500
S/N ratio	1/20	1/10	1/5	1/2	1/20	1/10	1/5	1/2	1/20	1/10	1/5	1/2
Crude prot. (N × 6·25) (DWB) (%)	9·06	9·38	9·50	10·31	10·44	11·44	12·00	10·81	14·81	15·31	14·81	14·50
AAs (mg/g N)												
Ile	278	302	301	309	282	273	301	300	319	309	307	304
Leu	601	681	669	700	634	662	681	675	712	700	697	681
Lys	146	147	138	140	136	136	134	137	115	125	124	129
Met	115 } 199	209 } 333	216 } 345	199 } 323	122 } 208	152 } 259	166 } 277	183 } 301	137 } 236	147 } 247	154 } 259	161 } 266
Cys	84	124	129	124	86	107	111	118	99	100	105	105
Phe	294	338	352	361	324	340	344	343	342	342	349	335
Tyr	216	235	235	256	239	244	244	252	257	239	231	255
Thr	221	266	269	270	223	256	245	257	241	241	239	249
Trp[b]	94	94	94	94	94	94	94	94	94	94	94	94
Val	396	427	423	432	406	416	418	431	436	447	439	431
Arg	231	226	217	244	224	222	224	212	216	228	239	249
His	134	145	144	152	138	147	151	157	144	146	156	149
Ala	327	379	380	382	334	360	358	377	367	364	359	374
Asp	564	392	383	377	473	366	374	371	410	396	386	390
Glu	1453	1497	1478	1531	1556	1562	1581	1550	1619	1716	1656	1609
Gly	207	226	230	224	201	206	204	211	190	196	196	201
Pro[b]	406	406	406	406	406	406	406	406	406	406	406	406
Ser	332	362	364	368	331	357	359	369	366	349	535	539

[a] Stabursvik and Heide (1974).
[b] Approximate literature values. Not determined.

Table 3.91
Pot Experiments I and II. Effects of N and S fertilizers on N and S content of finger millet.[a]

N fertilizer level (p.p.m.)		Fertilizer S/N ratio	Seed N (%)	Seed S (%)	Seed S/N ratio	AA/S as % of total S
10		1/2	0·83	0·15	1/6	50
50		1/10	0·69	0·10	1/7	60
		1/2	0·66	0·11	1/6	52
250	I	1/50[b]	1·17	0·06	1/20	86
		1/20[c]	1·11	0·07	1/16	87
		1/10	1·15	0·12	1/10	81
		1/2	1·13	0·15	1/8	64
250	II	1/20[c]	1·45	0·08	1/18	85
		1/10	1·50	0·13	1/12	90
		1/5	1·52	0·14	1/11	88
		1/2	1·65	0·19	1/9	68
500		1/20[d]	1·67	0·13	1/13	63
		1/10	1·83	0·16	1/11	70
		1/5	1·92	0·18	1/11	69
		1/2	1·73	0·18	1/10	68
1000		1/20	2·37	0·18	1/13	74
		1/10	2·45	0·20	1/12	71
		1/5	2·37	0·19	1/12	76
		1/2	2·32	0·21	1/11	69

[a] Stabursvik and Heide (1974).
[b] Very severe S deficiency.
[c] Pronounced S deficiency.
[d] Less pronounced S deficiency.

Table 3.92
Pot Experiments I and II. Effects of N and S fertilizers on methionine and cystine (mg/pot).[a]

		S/N ratios				
N (p.p.m.)		1/50	1/20	1/10	1/5	1/2
10		—	—	—	—	2·9
50		—	—	16·8	—	18·1
250	I	9·0[b]	43·1[c]	97·7	—	101·7
250	II	—	49·8[c]	102·4	113·1	89·9
500		—	97·7[d]	134·4	151·3	159·5
1000		—	129·8	148·8	150·2	122·5[e]

[a] Stabursvik and Heide (1974).
[b] Very severe S deficiency.
[c] Pronounced S deficiency.
[d] Less pronounced S deficiency.
[e] Strongly negative effect of excessive S and N.

Table 3.93
Basic analytical data of soils from (1) Thiruthuraipoondi and (2) Thiruchendur (coastal sand), (3) Cuddalore and (4) Coimbatore (inland sand).[a]

	Area			
	(1)	(2)	(3)	(4)
Mechanical composition				
Coarse sand	2·6	39·0	34·2	54·2
Fine sand	88·4	59·2	51·8	39·8
Silt + clay + loss on solution	9·0	1·8	14·0	6·0
Total nutrients (%)				
N	0·03	0·02	0·02	0·02
P_2O_5	0·012	0·124	0·056	0·036
K_2O	0·11	0·06	0·09	0·07
Available nutrients (kg/ha)				
N	148	106	131	134
P_2O_5	2·0	2·5	4·0	4·0
K_2O	160	310	107	87
Other properties				
Organic C (%)	0·015	0·015	0·015	0·057
pH	7·3	6·0	9·0	8·2
EC (mmhos/cm)	0·3	4·5	0·2	0·2
CEC (me/100 g)	4·9	1·7	3·1	2·5

[a] Kanaka Doss *et al.* (1975).

Table 3.94
Grain yield and uptake of nutrients by grain of CO 7 finger millet grown in pot culture soils from (1) Thiruthuraipoondi, (2) Thuruchendur, (3) Cuddalore and (4) Coimbatore under various fertilizer treatments.[a]

Treatments	Yield (g/pot)	Uptake in grain (mg/pot)		
		N	P_2O_5	K_2O
Soil type (1)				
(1) Control	5·78	213	24	10
(2) $(NH_4)_2SO_4$ + superphos.	6·79	247	31	14
(3) $(NH_4)_2SO_4$ + dicalc. phos.	8·21	276	39	17
(4) Urea + superphos.	13·77	347	66	30
(5) Urea + dicalc. phos.	12·41	625	60	25
(6) Sod. nitrate + superphos.	9·14	307	41	18
(7) Sod. nitrate + dicalc. phos.	8·64	387	48	16
(8) Groundnut cake + superphos.	6·70	243	33	15
(9) Groundnut cake + dicalc. phos.	6·38	357	29	14
Soil type (2)				
(1) Control	2·28	70	10	4
(2) $(NH_4)_2SO_4$ + superphos.	4·29	156	19	9
(3) $(NH_4)_2SO_4$ + dicalc. phos.	3·03	102	12	6
(4) Urea + superphos.	10·75	331	52	22
(5) Urea + dicalc. phos.	2·94	123	11	6
(6) Sod. nitrate + superphos.	6·37	268	24	12
(7) Sod. nitrate + dicalc. phos.	5·39	196	19	10

cont.

Table 3.94 *cont.*

Treatments	Yield (g/pot)	Uptake in grain (mg/pot)		
		N	P_2O_5	K_2O
(8) Groundnut cake + superphos.	4·12	150	12	6
(9) Groundnut cake + dicalc. phos.	4·44	633	15	6
Soil type (3)				
(1) Control	4·24	589	13	—
(2) $(NH_4)_2SO_4$ + superphos.	10·46	408	28	24
(3) $(NH_4)_2SO_4$ + dicalc. phos	10·19	357	32	23
(4) Urea + superphos.	9·79	96	34	23
(5) Urea + dicalc. phos.	4·70	197	9	11
(6) Sod. nitrate + superphos.	9·36	394	31	22
(7) Sod. nitrate + dicalc. phos.	6·62	278	22	15
(8) Groundnut cake + superphos.	8·48	285	31	23
(9) Groundnut cake + dicalc. phos.	8·16	205	30	22
Soil type (4)				
(1) Control	4·19	235	16	8
(2) $(NH_4)_2SO_4$ + superphos.	5·31	268	20	12
(3) $(NH_4)_2SO_4$ + dicalc. phos.	5·37	211	20	13
(4) Urea + superphos.	6·00	286	19	13
(5) Urea + dicalc. phos.	9·91	561	39	20
(6) Sod. nitrate + superphos.	5·88	165	21	10
(7) Sod. nitrate + dicalc. phos.	3·44	116	12	7
(8) Groundnut cake + superphos.	5·76	258	18	10
(9) Groundnut cake + dicalc. phos.	5·60	188	11	11

[a] Kanaka Doss *et al.* (1975).

Table 3.95
Proximate analysis, mean values over 3 years, from N, P, K fertilizers and manure applied to Gor'kovstoe 43 common millet.[a]

Fertilizer	Prot. (%)	Fat (%)	CBH[b] (%)	Fibre (%)	Ash (%)	Yield (kg/ha)
Control	10·3	3·8	61·1	8·2	2·7	1250
N_{40}(40 kg N/ha)	10·6	3·7	60·9	7·9	2·8	1230
P_{60}(60 kg P_2O_5/ha)	10·8	3·9	60·4	7·9	2·9	1540
K_{60}(60 kg K_2O/ha)	10·5	3·7	60·7	8·2	2·8	1360
N_{40} P_{60} K_{60}	10·9	4·1	60·6	7·9	2·9	1520
Manure (20 t/ha)	12·0	3·9	59·9	7·9	2·9	1710

[a] Vatagin and Oksenko (1971).
[b] Carbohydrate.

Chapter 4

Table 4.1
Analysis of "tannin" content of sorghum seeds by three methods.

Designation of cultivar	Method 1[a]	Method 2[a]	Method 3[a]
BR 44[b]	0·76, 0·70	0·85, 0·80	0·80, 0·73
BR 54[b]	1·61, 1·49, 1·64	1·14, 1·24, 1·29	1·68, 1·62, 1·73
BR 64[b]	2·66, 2·48	2·12, 2·20	2·09, 2·15
NK 300[b]	0·92, 0·88	0·70, 0·75	0·90, 0·94
NK 300[c]	0·11, 0·15	—	0·02, 0·03
IS 2319[d]	0·14, 0·12	—	0·07, 0·06
IS 3648[b]	2·95, 2·82, 2·69	2·38, 2·20, 2·41	2·80, 2·64, 2·73
IS 3924[c]	0·09	—	0·04
IS 4225[b]	0·66, 0·59, 0·69	0·55, 0·50, 0·58	0·85, 0·74, 0·76
IS 9115[b]	0·53, 0·58, 0·59	0·54, 0·49, 0·46	0·71, 0·67, 0·61
IS 9119[e]	0·14	—	0·06
IS 11167[e]	0·20	—	0·22
IS 15991[b]	0·31	—	0·16
954114[c]	0·04	—	0·05
P 954206[b]	0·04	—	0·02
RS 671[c]	0·02, 0·08	0·06	0·06

[a] Standard was the polymeric procyanidin isolated from sorghum NK 300 supplied by Purdue University, Lafayette. Figures given are % age of dry wt of the whole sorghum seed.
[b] Coloured peri. with coloured testa.
[c] White peri. with white or no testa.
[d] White peri. with coloured testa.
[e] Coloured peri. with white or no testa.

Table 4.2
Analysis of "tannin" content of sorghum seeds by four methods.

Designation of cultivar	Method 1[a]	Method 2[a]	Method 3[a]	Method 4
IS 3648	2·95, 2·82, 2·69	2·38, 2·20, 2·41	2·80, 2·64, 2·73	1·38, 1·2
IS 4225	0·66, 0·59, 0·69	0·55, 0·50, 0·58	0·85, 0·74, 0·76	0·69
IS 9115	0·53, 0·58, 0·59	0·54, 0·49, 0·46	0·71, 0·67, 0·61	0·40
BR 44	0·76, 0·70	0·85, 0·80	0·80, 0·73, 0·76	—
BR 54	1·61, 1·49, 1·64	1·14. 1·24, 1·29	1·68, 1·62, 1·73	0·57
RS 671	0·02, 0·08	0·06	0·06	—

[a] Standard was the polymeric procyanidin isolated from sorghum NK 300 supplied by Purdue University.
Method 1—vanillin—hydrochloric acid method
Method 2—ultra-violet absorption method
Method 3—cyanidin colouration method
Method 4—toluene-α-thiol method (see p. 315).

Table 4.3
Analysis of "tannin" content of sorghum seeds by three methods using different standards.

Designation of cultivar	Method 1		Method 2		Method 3	
	Catechin[a]	Polymer[a]	Catechin[a]	Polymer[a]	B-2[a]	Polymer[a]
IS 3648	11·00	2·95	2·97	2·38	5·8	2·8
IS 4225	2·5	0·66	0·70	0·55	1·6	0·85
IS 9115	1·64	0·53	0·68	0·54	1·33	0·71
BR 44	2·87	0·76	1·2	0·85	1·5	0·8
BR 54	6·0	1·61	1·42	1·14	3·14	1·68

[a] Standards in analytical procedure.

Table 4.4
Mean tannin content (mg/100 mg) and 95 % confidence interval (in parentheses) of four varieties of sorghum as determined by each of six methods of analysis.[a]

Method of Analysis	Data expressed as	GA 615	B 398 (Martin)	B 3197 (Kafir 60)	TX 2536
(1a) FAS 5 g urea	Tannic acid equivalents	0·16 (0·12–0·19)	0·07 (0·04–0·09)	0·04 (0·03–0·05)	0·04 (0·03–0·05)
(1b) FAS 50 g urea	Tannic acid equivalents	0·76 (0·71–0·81)	0·22 (0·18–0·25)	0·23 (0·20–0·27)	0·20 (0·17–0·24)
(2) Vanillin HCl	Catechin equivalents	3·10 (2·99–3·22)	0·55 (0·52–0·59)	0·29 (0·27–0·31)	0·21 (0·19–0·24)
(6) Methanolic HCl	Pigment concentrate equivalents	7·87 (6·74–9·15)	2·54 (2·23–2·89)	1·66 (1·50–1·84)	1·20 (1·05–1·36)
(7) Bate-Smith & Rasper	Pigment concentrate equivalents	19·78 (18·73–20·90)	13·15 (12·62–13·74)	5·56 (5·12–6·03)	5·03 (4·63–5·46)
(9) Folin-Denis 50 g urea	Tannic acid equivalents	8·64 (7·83–9·52)	8·02 (7·21–8·91)	6·56 (5·92–7·25)	6·91 (6·12–7·79)
(10) Folin-Denis	Tannic acid equivalents	1·61 (1·43–1·81)	0·53 (0·48–0·59)	0·37 (0·34–0·40)	0·36 (0·32–0·40)

[a] Maxson and Rooney (1972a).

Table 4.5
Summary of tannin values obtained from three sets of grain sorghum samples analysed by the FAS and MV-HC1 methods.[a]

	FAS method (tannic acid equivalents mg/100 mg) (DWB) Mean	MV-HC1 method (CE) (mg/100 mg) (DWB) Mean
Set I (GA 615)	0·89 (0·77–1·10)	3·41 (2·71–4·32)
SD	0·06	0·19
Set II (with and without pigmented testa)	0·64 (0·27–1·05)	1·76 (0·14–2·86)
SD	0·05	0·09
Set III (colour variation)	0·51 (0·28–1·15)	1·13 (0·08–3·25)
SD	0·04	0·10

[a] Maxson and Rooney (1972a).

Table 4.6
Tannin content and description of grain used to study the effect of pericarp colour, plant colour and pigmented testa on tannin content.[a]

Grain sorghum phenotype	Group[b]	Description: Peri. colour	Description: Testa[c]	Tannin content (mg/100 mg): FAS tannic acid equiv.	Tannin content (mg/100 mg): MV-HC1 catechin equiv.
(BT × 406 × IS 12610[2])	I	White	A	0·34	0·16
F6	I	Red	A	0·34	1·05
	I	Lemon-yellow	A	0·28	0·33
(BT × 406 × IS 12663[3])	II	White	A	0·30	0·22
F6	II	Red	A	0·35	0·74
	II	Lemon-yellow	A	0·36	0·27
(BT × 406 × IS 2579[4]) F4	III	White (red plant)	P	0·75	2·82
	III	White (red plant)	A	0·30	0·34
	III	White (tan plant)	P	0·77	2·53
	III	White (tan plant)	A	0·34	0·11
(BT × 406 × IS 1133[5])	IV	Red	pt	0·35	0·67
F3	IV	Red	A	0·36	0·41
	IV	White	pt	0·36	0·73
	IV	White	A	0·31	0·18
Shallu × DDF	V	Brown	P	1·12	2·83
Shallu × TX 09	V	Brown	P	0·88	2·32
(BT × 406 × IS 12682[5])	VI	White	A	0·32	0·08
F3	VI	White	A	0·31	0·25
(BT × 406 × IS 12617[5]) F3	VII	Dark brown	P	0·96	3·23
663	VII	Dark brown	P	1·15	3·25
SD				0·04	0·10

[a] Maxson *et al.* (1972).
[b] Group I through Group IV were selections from Sorghum Conversion Programme and were near-isogenic lines.
[c] Testa: P = present; A = absent; pt = thin testa that did not completely surround the endo.

Table 4.7
Visual and colorimetric estimation of tannin content of sorghum grain.[a]

Grain	Estimation methods: Visual (variation I): Colour	Rank	Tannin content	Standard spectrophotometric methods: Corrected vanillin CE	Prussian Blue (water extract) CE
BR 54	Deep blue	1	High	3·2	0·58
NK 300	Blue	2	High	2·2	0·56
IS 8164	Turquoise	3	Mod. high	1·4	0·36
IS 15612	Dk green	4	Mod. high	1·0	0·33
IS 15991	Green	5	Intermed.	0·4	0·04
RS 610	Lime green	6	Low	0·0	0·04
IS 954063	Lime green	6	Low	0·00	0·00

[a] Price and Butler (1977).

Table 4.8
Tannin content (%) and grain colour of varieties and line hybrids grown in Czechoslovakia.[a]

Variety	Tannin content (% of dry matter)	Grain[b] colour
NK 135	0·09	Red-yellow
CK 60	0·10	—
Solary A	0·19	White-red
Solary F	0·25	Red-yellow
Solary C	0·26	Yellow-brown
Early Hegari	0·40	Off-white
Hybar MV 342	0·46	Red-yellow
Solary BR	0·46	Yellow-red-brown
Solary L	0·47	White-yellow
Hybar MV 560	0·55	Yellow-brown
NK 300	0·57	Yellow-brown
ZH 12	0·57	Red-yellow-brown
Solary D	0·59	Red-brown
ZH 6	0·59	—
Solary B	0·61	Yellow-brown
NK 120	0·63	—
Sudanska trava	0·63	Yellow-brown
ZH 3	0·66	Yellow-brown
ZH 26	0·66	Yellow-brown
ZH 32	0·66	Brown-red-yellow
Solary T	0·68	Brown-red-yellow
Solary NT	0·69	—

[a] Nasinec *et al.* (1966).
[b] Authors' note: it is our belief that in the Czech language the position of the word indicates colour gradation, e.g. yellow-brown is more yellow, red-brown is more red.

Table 4.9
1000-seed weight, proportion of embryo and pericarp, and of endosperm, with % tannin content, in five Czech grown varieties of sorghum.[a]

Variety	1000-seed wt (g)	Proportions in whole grain (%)		Tannin in dry matter (%)
Early Hegari	20·6	Whole grain	100·00	0·39
		Embryo and peri.	14·25	1·54
		Endo.	85·75	8·08
Solary A	21·2	Whole grain	100·00	0·18
		Embryo and peri.	9·15	0·24
		Endo.	90·85	0·03
Solary ZH 6	25·8	Whole grain	100·00	0·60
		Embryo and peri.	15·83	2·70
		Endo.	84·17	0·10
Solary T	21·6	Whole grain	100·00	0·69
		Embryo and peri.	12·93	0·92
		Endo.	87·07	0·16
NK 135	19·6	Whole grain	100·00	0·09
		Embryo and peri.	11·71	0·17
		Endo.	88·29	0·01

[a] Nasinec *et al.* (1966).

Table 4.10
Tannic acid content of 14 varieties of S. African sorghums.[a]

Variety	Tannic acid (% of sample)
NK 222	0·24
X 19109	0·25
SSK 6	0·25
Excell 505	0·26
SSK 10	0·27
K 5043 (1)	0·32
TE 77	0·34
Pawney	0·35
TE 93x	0·36
DC 39	0·37
NK 300	0·37
TE 44C	0·38
K 5043 (2)	0·42
DC 500 F	0·46
SSK 8	0·94

[a] DuPreez and Wessels (1970).

Table 4.11
Tannin content of 19 fodder varieties of sorghum grown at Hissar, India (DWB).[a]

Variety	Tannin mg/g
JS 73/53	0·23
IS 1503	0·27
IS 5470	0·32
IS 4906	0·36
IS 5459	0·36
JS 263	0·41
T 48	1·27
T 122 C	1·32
IS 3214	1·36
C 25	1·73
T 26	1·86
IS 3289	2·23
IS 5585	2·27
C 429 1	2·36
C 406 8	2·59
C 40	2·82
JS 20	2·82
IS 1087	2·91
C 433	3·73

[a] Arora and Luthra (1972).

Table 4.12
Tannin content (catechin equivalent, CE) of 44 Ethiopian sorghum varieties.[a]

Variety	Tannin CE/100 g (dry sample)
Gambella 1096	0·17
WS 1767	0·17
Alemaya 70	0·17
Alwegere 812	0·17
WS 1521	0·20
Gambella 1107	0·20
Woldya 157	0·20
Damota 93	0·25
Hirna 547	0·25
Jij Wegere 935	0·30
Jij Dasola 781	0·30
Asbeteferi 332	0·30
Dodessa 36	0·30
Debesso 356	0·30
Alemaya 296	0·30
Adi Ugri 149	0·35
WS 1520	0·35
Dodessa 1057	0·35
Netch Muyra	0·35
WS 1584	0·35
Awash 1050	0·35
Gato 1001	0·40
CK 60	0·40
Bedele 1106	0·40
Babile 1033	0·40
Key Muyra	0·40
WS 1158	0·45
Alemaya 477	0·45
WS 1297	0·45
Hirna 576	0·45
WS 1821	0·45
Hirna 305	0·50
Erwegere 794	0·50
WS 1509	0·60
ASAO 122	0·60
Hirna 335	0·60
Martin	0·70
WS 1641	2·20
WS 1763	2·40
Gato 994	2·65
Netch Shure	2·90
Abay 145	4·00
Alemaya 108	5·00
Leoti 696	5·20

[a] Mehansho and Besrat (unpublished report, 1974).

Table 4.13
Weight gain of rats fed low-pigment (L-P) and high-pigment (H-P) sorghum diets on an equal weight, and isonitrogenous basis, over 28 days.[a]

	Equal wt basis				Isonitrogenous basis	
	(A) IS 0062 (L-P)	(B) IS 3982 (L-P)	(C) IS 6992 (H-P)	(D) IS 2283 (H-P)	(E) IS 8165 (H-P)	(F) IS 2319 (L-P)
Total av. gain/rat (g)	34·8	28·7	7·3	−0·6	11·6	19·6
Total av. feed consumed/rat (g)	225·7	197·3	163	156	181	171·6
Total av. prot. consumed/rat (g)	24·0	23·6	14·3	11·4	15·6	15·1
Tannin (catechin equivalents) (%)	0·52	0·47	4·62	6·88	2·69	0·51
Protein Efficiency Ratio					0·71[b]	1·28[b]

[a] Jambunathan and Mertz (1973a).
[b] Significantly different from each other at the 1% level.

Table 4.14
Nitrogen distribution[a] (as % of soluble N) in whole kernels (WK) and endosperms (E) of three low-pigment (L-P) and three high-pigment (H-P) sorghums.[b]

Fraction	IS 0062 (L-P)		IS 3982 (L-P)		IS 2319 (L-P)	IS 6992 (H-P)		IS 2283 (H-P)		IS 8165 (H-P)
	WK	E	WK	E	WK	WK	E	WK	E	WK
I Saline (albumins and globulins)	16·8	8·0	15·4	6·6	16·1	5·6	2·8	4·1	2·9	8·5
II Isopropyl alcohol (kafirins)	18·4	19·9	10·6	13·6	14·8	5·9	9·2	2·5	13·3	6·2
III Isopropyl alcohol + 2-mercaptoethanol	18·9	35·1	18·2	29·7	14·9	11·5	27·0	13·8	28·1	7·6
IV Borate buffer + 2-mercaptoethanol (new class)	6·1	6·4	6·4	9·2	4·4	15·0	10·6	17·3	11·2	8·1
V Borate buffer + 2-mercaptoethanol + sodium dodecyl sulphate (glutelin)	29·9	26·5	34·2	24·0	40·5	54·8	45·0	49·3	35·7	57·3
Total N extracted	90·1	95·9	84·8	83·1	90·7	92·8	94·6	87·0	91·2	87·7

[a] Landry and Moureaux (1970).
[b] Jambunathan and Mertz (1973a).

Table 4.15
Nitrogen distribution (% of soluble N) in whole kernel (WK) and alkali dehulled kernels of two high-pigment sorghums.[a]

Fraction	IS 6992		IS 2283	
	WK	Dehulled	WK	Dehulled
I Saline	6·0	9·3	4·1	5·8
II Isopropyl alcohol	3·8	9·8	2·5	5·2
III Isopropyl alcohol + 2-mercaptoethanol	6·9	20·9	13·8	18·0
IV Borate buffer + 2-mercaptoethanol	12·8	7·0	17·3	7·5
V Borate buffer + 2-mercaptoethanol + sodium dodecyl sulphate	59·9	43·9	49·3	50·4
Total N extracted	89·4	90·9	87·0	86·9

[a] Jambunathan and Mertz (1973a).
[b] Authors' note: the differences between the data for the two different samples of the whole kernel IS 6992 in the above and in Table 4.14 is marked.

Table 4.16
Protein and N distribution in fractions of sorghum endosperm.[a]

	P 721 N	P 721 O	IS 11167	IS 4225
Prot. (N × 6·25) (g/100 g endo.)	12·0	10·6	10·5	9·4
Fractions (% soluble N)				
I (albumins and globulins)	9·0	28·6	23·1	6·2
II (kafirin)	25·1	9·9	10·7	10·2
III (cross-lined kafirin)	25·1	15·3	19·0	18·7
IV (glutelin-like)	6·8	4·1	4·8	9·4
V (glutelin)	34·0	42·1	42·4	55·5
Total N extracted (%)	98·6	97·9	91·4	80·6

[a] Guiragossian *et al.* (1978).

Table 4.17
Amino acid composition in mg/g total N of defatted sorghum kernels, endosperms and embryos.[a]

	Cultivar											
	Whole kernel				Endo.				Embryo			
AAs	P 721 N	P 721 O	IS 11167	IS 4225	P 721 N	P 721 O	IS 11167	IS 4225	P 721 N	P 721 O	IS 11167	IS 4225
Ile	237	244	244	237	287	256	306	269	206	206	237	244
Leu	887	762	694	775	1056	862	944	944	406	437	506	587
Lys	125	184	200	150	100	162	169	125	400	394	450	425
Met	106	100	87	94	119	119	106	75[b]	112	106	131	94
Cys	100	94	75	69[b]	100	100	106	62[b]	94	119	106	87
Thr	194	206	212	194	156	206	237	206	319	281	250	306
Phe	344	306	312	300	406	350	394	344	269	275	319	325
Tyr	287	262	250	256	331	269	312	294	212	212	244	262
His	137	144	137	131	144	144	150	137	194	194	225	219
Ala	575	525	494	519	656	581	637	631	394	369	462	506
Val	319	319	325	294	350	350	381	312	431	462	450	456
Arg	231	281	294	225	225	269	262	62[b]	681	756	800	637
Asp	400	469	500	431	356	462	569	256	606	494	531	444
Glu	1200	1256	1081	1256	1350	1344	1406	1550	1069	1100	1106	1119
Gly	181	219	237	181	225	231	237	162	406	444	450	481
Pro	531	475	481	462	706	512	575	519	369	319	481	425
Ser	269	262	281	262	200	262	312	281	231	306	319	369
Ammonia	231	200	169	300	412	237	294	231	131	156	162	112
Total recovery (%)	101·5	100·7	97·2	98·2	114·9	107·5	121·4	103·4	104·5	106·1	115·7	113·6

[a] Guiragossian *et al.* (1978).
[b] Low values not in agreement with analyses of similar low- and high-tannin sorghums by Jambunathan and Mertz (1973a). Their values are near to those reported for P 721 N above.

Table 4.18
Amino acid composition (mg/g total N) of Landry and Moureaux fractions of defatted sorghum endosperm.[a]

Fraction	Cultivar	Ile	Leu	Lys	Met	(Cys$_2$)	Thr	Phe	Tyr	His	Ala	Val	Arg	Asp	Glu	Gly	Pro	Ser	NH$_3$	Total Recovery (%)
I	P 721 N	162	312	281	112	119	206	200	162	131	319	287	531	525	719	306	225	250	312	82·6
	P 721 O	175	312	294	75	100	206	206	175	137	331	294	537	550	725	312	231	250	250	82·6
	IS 11167	156	256	237	44	62	125	169	144	106	262	225	331	425	550	219	294	162	344	65·8
	Corrected	206	319	300	56	75	156	212	181	131	325	281	412	531	687	275	362	200	425	82·2
	IS 4225	100	137	200	50	37	100	75	87	156	244	144	219	612	712	144	400	162	337	62·7
	Corrected	131	181	262	69	50	131	100	119	206	319	187	287	806	937	187	525	212	444	62·7
II	P 721 N	256	1062	6	37	25	119	369	262	56	625	250	69	387	1550	75	531	212	275	98·7
	P 721 O	244	100	19	31	31	137	337	275	50	612	250	87	400	1469	87	531	225	187	95·6
	IS 11167	212	900	19	25	25	144	256	256	31	625	256	62	412	1494	94	475	262	275	93·2
	IS 4225	262	950	19	31	25	137	375	281	19	625	275	75	437	1550	94	581	250	200	100·3
III	P 721 N	225	994	12	81	81	144	344	287	62	600	244	106	325	1456	94	619	250	250	98·8
	P 721 O	225	994	6	75	44	106	312	250	44	544	237	81	281	1400	81	531	162	219	89·5
	IS 11167	256	1056	6	50	31	125	381	262	56	619	244	69	337	1506	69	600	244	187	97·5
	IS 4225	244	1044	6	31	6	87	356	231	106	581	244	81	281	1425	69	381	150	275	88·5
IV	P 721 N	162	481	81	69	62	244	144	175	362	250	331	250	200	1012	331	875	231	244	87·9
	P 721 O	162	450	106	87	25	194	150	150	312	250	312	237	200	962	325	719	206	156	80·1
	IS 11167	137	337	87	62	19	181	131	137	194	206	237	187	250	712	250	512	275	94	64·2
	Corrected	187	462	119	87	25	244	181	187	262	281	325	256	344	975	344	700	375	131	87·8

	IS 4225	162	287	37	69	25	144	119	144	69	206	212	144	219	694	231	525	194	125	57·7
	Corrected	244	437	56	106	37	219	181	219	106	312	325	219	331	1056	350	800	294	187	87·7
V	P 721 N	300	806	137	94	25	194	331	300	137	650	381	219	406	1606	200	581	244	312	110·8
	P 721 O	194	794	175	150	25	225	331	325	131	650	425	256	425	1606	219	500	287	362	113·3
	IS 11167	225	725	125	87	19	119	287	187	94	506	294	169	350	1112	175	531	162	275	86·8
	Corrected	287	919	162	112	25	150	362	237	119	644	375	212	444	1412	225	675	181	350	110·3
	IS 4225	231	800	81	87	12	169	306	250	94	531	275	162	369	1219	125	400	237	369	91·5
	Corrected	281	969	100	106	12	206	369	300	112	644	331	194	444	1475	150	481	287	444	110·5

[a] Guiragossian *et al.* (1978).

[b] Total recovery of P 721 N amino acids (last column) divided by total recovery of IS 11167 amino acids (last column) × IS 11167 amino acid level.

Table 4.19
Protein and estimated tannic acid in sorghum samples.[a]

Variety/hybrid	Prot. (N × 6·25) (air-dried basis) (%)	Estimated tannic acid (1–5)	Seed coat colour
IS 229	17·2	Trace	Lt pink
IS 3906	15·6	3	Lt straw, red spots
IS 503	14·6	3	Dp brown and straw
IS 2469	14·4	3	Dp brown
IS 474	14·2	4	Dp straw
MS × HC 39,142 Coddy Line II	13·8	4	Pinkish-brown
IS 3843	13·7	4	Brownish-straw
IS DH 61 R	13·3	3	Dp brown and straw
IS 3485	12·6	3	Brownish-straw
IS 3691 3[b]	12·5	1	Yellowish-light brown
IS 1107	12·4	3	Dp brown
IS 2931 8 2	12·0	1	Yellowish-dull white
IS 2908	11·8	3	Reddish-brown
MS × HC 39,142 Coddy Line III	11·8	6	Brownish-grey
IS 954	11·7	Trace	Yellowish-red
IS 3677	11·6	3	Pinkish-deep brown
IS 3441	11·5	5	Pinkish-brown
MS × Nythin	11·5	4	Lt pink
MS × HC 39,142 Coddy Line I	11·4	4	Straw
IS 3539[c]	11·4	3	Brownish-straw
IS 84 1	11·3	3	Straw, red spots
IS 3691 3[b]	11·2	1	Dp straw, red spots
IS 4299	11·1	4	Straw
IS 3691 8	11·1	3	Dp straw
IS 2299	11·0	4	Brown
MS × PJ 16 K	11·0	2	White
IS 2392	10·7	1	Lt pink
IS 3523 1	10·5	4	Straw
IS 944 Line II	10·3	Trace	Pinkish-straw
MS × D 340	10·2	3	White
MS × 2918	10·2	3	White
MS × Yel. End. Hegari	10·2	2	White
IS 429	10·0	4	Straw
MS × Mayo	10·0	3	White
MS × 2267	9·9	3	White
IS 944 Line I	9·9	Trace	Yellowish-straw
IS 2933 2	9·9	4	Straw
MS × Sel 124 3	9·7	8	Brownish-straw
MS Yel. End (IP)	9·4	2	White
IS 520	9·3	5	Dp straw
MS × 2300	9·2	3	White
IS 3539[c]	9·1	1	White
Local check BP 53	9·1	1	White
IS 2942 2	8·9	4	Straw
MS × BP 53	8·8	3	White
MS × Tarano	8·5	2	White
MS × Zera Zera	8·3	2	White
MS × Aispuri	8·2	3	White
Bazaar Samples			
I	9·3	2	Lt pinkish-brown
II	9·5	2	Pinkish-brown
III	9·8	3	Bt pinkish-brown
IV	10·5	2	Dp pinkish-brown
BP 53	8·2	2	White

[a] Chakravorty (1969).
[bc] Identical variety numbers.

Table 4.20
Colour classification and mean CE (ranges in parentheses) of sorghum samples.[a]

	Catechin equiv.							
	International sorghum numbers				Purdue base and commercial hybrids			
Peri. colour	No. of samples	Below 1·0	No. of samples	Above 1·0	No. of samples	Below 1·0	No. of samples	Above 1·0
White	5	0·35 (0·22–0·56)						
Straw	66	0·46 (0·11–0·88)	10	2·76 (1·12–4·89)	8	0·48 (0·20–0·82)	3	3·15 (1·14–6·97)
Yellow	16	0·27 (0·07–0·68)	3	2·53 (1·47–3·38)	6	0·44 (0·09–0·86)		
Orange	4	0·64 (0·42–0·81)	4	3·88 (3·23–4·31)			1	1·23
Lt brown	21	0·65 (0·38–0·96)	38	3·06 (1·01–8·82)	11	0·64 (0·38–0·92)	10	4·71 (1·19–7·31)
Brown			1	4·40				
Red brown			4	4·74 (2·56–7·06)				

[a] Oswalt and Srinivasan (1973).

Table 4.21
Tannic acid (TA) and total astringents content[a] of sorghum grain at five stages of maturity.[b]

Maturity stage	Bird-resistant Lindsey BR 75		Non-bird resistant DeKalb E 57		Non-bird resistant Asgrow Ranger B	
	TA (%)	Total astring. (%)	TA (%)	Total astring. (%)	TA (%)	Total astring. (%)
Milk stage	0·68	0·94	0·03	0·20	0·07	0·24
Dough stage	0·50	0·72	0·09	0·25	0·07	0·24
22% moisture	0·44	0·66	0·09	0·24	0·09	0·26
17% moisture	0·57	0·74	0·05	0·18	0·07	0·24
16% moisture	0·53	0·66	0·06	0·19	0·08	0·23

[a] Unstated method, Louisiana State University.
[b] Tipton *et al.* (1970).

Table 4.22
Influence of sorghum variety on content of inhibitor.[a]

	Extract (mg)	Dextrinization time (min)
	Control (no inhibitor)	21·5
Schrock	100	172
Leoti Red	100	120
Leoti Red (Hays, Kansas)	100	43·5
Westland	500	20·5
Atlas	500	20·5
Blackhull Kafir	500	20·5
Early Sumac	200	135

[a] Miller and Kneen (1947).

Table 4.23
Distribution of trypsin inhibiting activity (TIU) in extracts of sorghum.[a]

	Volume (ml)	TIU/ml	Spec. act. (TIU/mg prot.)
Aqueous extract	32·0	4·6	0·315
pH 4·0 supernat.	27·5	5·1	0·447
pH 4·0 supernat. dialysed	10·2	1·9	0·705
pH 4·0 supernat. heat treated	9·2	5·5	0·441

[a] Filho (1974).

Table 4.24
Effect of peptone on the solubility in water of sorghum malt amylases and on the solubility NaCl on sorghum grain α-glucosidase.[a]

	Enzyme activities					
	Malt amylases[b]		Grain α-glucosidase (nmol/min/mg)		% sol.	
Cultivar	Water	Peptone 2%	NaCl, 1M	NaCl, 1M + peptone 5%	Amylases	α-glucosidase
DC 109	43·5	45·3	0·97	1·06	96	92
DC 59	28·7	31·1	0·95	1·10	92	86
NK 300	32·9	45·6	0·72	0·87	72	83
Barnard Red	56·8	59·1	0·99	1·08	96	92
SSK 2	12·8	62·6	0·36	0·82	20	44
SSK 52	2·2	46·9	0·08	0·75	5	11
C 39	30·8	35·8	0·90	1·05	86	86
Viva 101	1·3	64·9	0·05	0·74	2	7

[a] Watson *et al.* (1974).
[b] Sorghum diastatic units.

Table 4.25
Comparison of polyphenol determinations using different analytical procedures.[a]

Method	Standard substance	% standard substance		
		Non-bird resistant grain	Bird-resistant grain	Bird-resistant malt
Urea—FAS[b]	Tannic acid	0·56	0·60	0·52
Modified V-HCl[b]	Catechin equiv.	0·50	1·13	1·35
	Tannic acid	0·025	0·056	0·068
DMF-FAC[c]	Tannic acid			
Variation I		0·11	0·70	1·30
Variation II		0·11	0·74	1·79
Modified DMF-FAC	Tannic acid			
Sample: DMF = 1/10		0·19	1·12	1·60
Sample: DMF = 1/20		0·22	1·07	1·56

[a] Daiber (1975).
[b] Maxson and Rooney (1972a).
[c] Jerumanis (1972).

Table 4.26
Inhibition of some enzymic preparations by finely milled sorghum grain.[ab]

Enzyme, % conc. activity determined	Inhibition: % enzyme activity					
	Barnard Red	DC 37	NK 300	SSK 2	SSK 52	Maize endo.
α-amylase (0·35%)						
Reducing substance[c]	−15·9[e]	10·4	10·3	83·3	85·9	−10·8
Dextrinization[d]	−8·7	16·1	21·3	85·3	88·1	0·7
Barley diastase (1·6%)						
Reducing substance	0	11·4	12·5	70·6	72·9	−1·4
Dextrinization	6·2	21·6	12·5	71·6	87·2	3·5
Sorghum malt extract (5%)						
Reducing substance	−0·8	29·7	46·8	92·2	98·7	−0·2
Dextrinization			25·3			
Sorghum malt extract (7·5%)						
Reducing substance	4·7	15·5	6·8	46·9	59·2	0·2
Dextrinization	−0·8	10·2	13·2	57·3	72·0	1·0

[a] 20 mg grain/2 ml enzyme solution.
[b] Daiber (1975).
[c] Ferri-cyanide determination (Novellie, 1960a).
[d] α-Amylase assay of Briggs (1967).
[e] Negative values: more enzyme activity was found than in the control.

Table 4.27
Enzyme inhibition and polyphenol content within grain sorghum classes (47 cultivars from one farm).[a]

Sorghum class	No. of cultivars	Enzyme[b] inhibition (%)	Total[c] polyphenols (%)	Sol.[d] malt enzymes (%)	Sol.[e] grain maltase (%)
KR	3	−2·4±2.7[f]	0·25±0·03	100·6±3·2	95·7±2·8
KM	28	−1·6±4·6[f]	0·18±0·06	90·3±6·1	91·8±6·3
KW	2	1·1±0·4	0·08±0·03	93·4±6·6	91·6±6·9
KF	14	68·0±22·6	1·40±0·36	25·4±32·7	50·6±30·1

[a] Daiber (1975).
[b] α-Amylase inhibition determined by residual dextrinization activity (Briggs, 1967).
[c] Method of Jerumanis (1972).
[d] Diastatic malt enzymes soluble in water or 2% aqueous peptone (Novellie, 1960a).
[e] Method of Watson *et al.* (1974).
[f] Negative values: more enzyme activity was found than in the controls.

Table 4.28
Bird-resistant cultivars (KF) class: relation between enzyme inhibition and polyphenol content.[a]

Cultivar	Enzyme[b] inhib. (%)	Total[c] polyph. (%)	Difference polyph. calc. found	Sol.[d] malt diastatic power (%)	Sol.[e] grain maltase (%)
SSK 52	98·7	1·72	+0·18	1·3	14·3
Brawis	96·3	1·70	+0·16	0·6	10·0
SSK 2	92·2	1·76	+0·03	7·8	54·0
Viva	86·4	2·06	−0·37[f]	1·4	4·6
DC 52	78·2	1·41	+0·14	19·5	44·0
SSK 2	77·6	1·64	−0·10	17·8	53·0
SSK 22	74·0	1·59	−0·11	89·3	48·6
DC 99	70·3	1·15	+0·26	51·1	78·9
DC 50 A	58·4	1·21	0	1·6	80·0
FS 79	57·2	1·41	−0·22	87·4	12·0
DC 50	55·8	1·16	+0·01	57·1	82·9
NK 300	46·8	1·06	−0·05	62·6	87·6
C 41	36·0	0·92	−0·09	33·6	55·6
DC 75	24·7	0·87	−0·23	64·7	83·3
Mean	68·0	1·40		35·4	50·6
SD	22·6	0·36		32·7	30·1

[a] Daiber (1975).
[b] α-Amylase inhibition determined by residual dextrinization activity (Briggs, 1967).
[c] Method of Jerumanis (1972).
[d] Diastatic malt enzymes soluble in water or 2% aqueous peptone (Novellie, 1960a).
[e] Method of Watson *et al.* (1974).
[f] Negative values: more enzyme activity was found than in the controls.

Table 4.29
Tannin content and IVDM of 49 sorghum cultivars.[a]

Cultivar	Tannin content (as tannic acid) (%)	Prot. digestibility (%)
C 45	1·54	12·9
C 19	1·26	16·2
C 17	1·21	17·6
C 29	1·1	20·4
C 16	0·92	24·3
C 18	1·10	26·2
C 20	1·3	27·0
C 33	1·06	32·2
C 21	1·15	34·0
C 48	0·93	34·0
C 34	1·06	35·6
C 11	0·97	36·7
C 23	1·25	38·5
C 5	1·02	43·8
C 6	1·11	45·3
C 44	1·0	50·9
C 31	0·93	60·3
C 9	0·95	61·9
C 2	0·65	67·7
C 1	0·28	73·1
C 38	0·34	73·2
C 41	0·28	72·9
C 30	0·33	73·9
C 24	0·34	74·1
C 37	0·36	74·2
C 7	0·33	74·3
C 32	0·33	74·8
C 35	0·34	74·9
C 46	0·40	75·2
C 40	0·37	75·4
C 36	0·36	75·5
C 15	0·29	76·5
C 22	0·46	76·4
C 47	NR[b]	76·6
C 13	0·37	76·9
C 39	0·29	77·4
C 4	0·32	77·5
C 3	0·35	77·8
C 10	0·28	78·9
C 25	0·36	79·3
C 28	0·39	79·4
C 43	0·35	79·9
C 14	0·36	80·4
C 8	0·42	80·8
C 12	0·33	81·2
C 26	0·39	81·8
C 49	0·34	82·0
C 27	0·3	82·1

[a] Van Wyk *et al.* (1973).
[b] Not reported.

Table 4.30
Yield, protein content (N × 6·25) and IVDMD percentages (DWB) of the grain of selected commercial sorghum hybrids.[a]

Variety	Yield		Prot.		IVDMD	
	% of plant	(kg/ha)	(%)	(kg/ha)	(%)	(kg/ha)
DeKalb						
A 25	43	7350	13·3	970	50	3680
C 42a	44	8580	9·7	830	70	6000
BR 64	40	6350	10·0	630	50	3150
Pioneer						
866	42	9910	9·8	970	71	7070
846	41	8160	9·7	800	67	5500
Taylor Evans						
Exp 11105	40	9960	8·3	820	67	6740
RS 610	41	7360	9·7	720	67	4940
Averages	41	8230	10·1	820	63	5300
CV(%)	4·9	12·1	5·8	11·1	5·9	13·8
LSD (0·05)	NSD	NSD	1·4	NSD	5·6	1530

[a] Oswalt (1973a).

Table 4.31
Grain yield, seed colour, bird damage, grain tannin content, grain IVDMD and protein content (method not stated) of sorghum varieties.[ab]

Hybrid	kg/ha[c]	Seed colour	Bird damage Athens 1970 (V-HCL) (%)	Grain tannin cont. Blairsville 1970 (%)	Grain IVDMD Blairsville 1969 (%)	Seed prot. cont. Blairsville 1969 (%)
Funk BR 79	6000	Brown	0	5·7	49	10·7
DeKalb BR 64	5900	Brown	3	5·3	55	9·1
GA 615	5700	Brown	0	5·2	55	11·9
Asgrow Double TX	5600	Red	68	0·2	66	9·8
Excel Bird-Go	5600	Brown	0	6·0	50	9·0
Pioneer 828	5400	Red	68	0·2	69	10·8
AKS 663	5300	Brown	0	2·3	57	8·7
RS 700	5200	Tan	7	3·3	59	11·3
ACCO R 1090	5100	Bronze	73	0·2	70	10·6
Dorman BR 100	5100	Brown	0	4·5	57	10·8
DeKalb F 64	5000	Bronze	87	0·2	69	11·1
ACCO R 109	5000	Bronze	70	0·2	69	10·4
AKS 614	5000	Brown	3	3·0	57	10·5
DeKalb E 57	4900	Bronze	50	0·2	69	10·3
ACCO R 1093	4900	Brown	0	5·2	52	11·9
Funk G 522	4700	Bronze	78	0·2	72	10·0
Asgrow Bravia R	4600	Brown	0	6·9	58	9·4
ACCO R 1029	4600	Bronze	70	0·2	70	11·7
McNair 546	4500	Brown	0	3·0	54	10·3

[a] Grown at different locations in Georgia (1969–1971).
[b] Harris (1973).
[c] Authors' note: converted from bu/ac assuming 55 lb/bushel.

Table 4.32
Relationship of tannin content and IVDMD.[a]

Sample	Prot. (N × 6·25) (%)	Tannin CE (%)	IVDMD (%)
2752	15·7	5·9	49·60
1456	9·6	4·4	57·60
3606	9·4	5·0	58·96
529	11·8	4·8	58·96
310	13·7	5·3	59·36
3367-2	15·7	4·8	60·80
MP Chari	13·8	3·4	61·20
186	11·6	3·0	63·20
2677	10·3	3·2	66·00
7155	15·1	6·2	68·16
JS 20	14·0	4·3	69·20
JS 263	10·9	0·2	88·32
M 35 1	10·7	0·3	87·04
CSH 1	10·9	0·1	92·80
148	8·3	0·2	94·16
Swarna	13·1	0·3	94·48
303	14·4	0·1	95·60

[a] Arora and Luthra (1974).

Table 4.33
Description and composition of sorghum grain with contrasting kernel characteristics, tannin in the diets[a] and performance of weanling rats fed casein control and sorghum diets.[b]

Pedigree or line number	B_1B_2S Genotype	Peri. colour	Testa[c]	Prot. (DWB) (%)	Lys. (% prot.)	Tannin in diet (mg/100 mg)		Performance of rats			
						FAS	MV-HCl	Weight gain	Feed intake	PER	DMD
RTX09 × Shallu	*BBS*	Brown	p	12·3	2·06	0·68	2·09	24·5	198·6	1·23	85·0
Shallu × BTX378	*BBS*	Brown	p	12·5	1·90	0·63	1·83	20·4	180·0	1·13	85·2
Shallu × BTX398	*BBS*	Brown	p	11·9	2·00	0·63	1·80	24·0	179·2	1·35	88·4
Shallu × BTX 3197	*BBS*	Brown	p	10·9	2·03	0·63	1·97	25·0	204·0	1·26	86·5
BTX3197 × DDF	*BBS*	Brown	p	11·0	2·04	0·65	2·03	25·0	190·7	1·34	91·1
Double Dwarf Feterita	*BBs*	White	p	13·7	1·77	0·48	1·73	22·9	189·7	1·21	89·7
Shallu	*BbS*	White	a	12·2	2·05	0·21	0·11	26·6	191·6	1·37	91·6
BTX3197	*bBS*	White	a	11·7	1·85	0·21	0·17	25·2	188·7	1·32	91·9
RTX09	*bBs*	White	a	13·0	1·98	0·27	0·21	24·8	192·8	1·28	92·9
BTX378	*bBS*	Red	a	12·6	1·71	0·28	0·75	20·9	174·8	1·19	91·9
BTX398	*bBS*	Red	a	13·0	1·93	0·20	0·52	22·7	177·8	1·28	91·8
ATX3197 × RTX414	*bBS*	Red	a	10·8	2·00	0·26	0·62	24·1	194·8	1·27	91·7
69 Chil 109 14E	—	White	p	11·6	2·11	0·41	3·29	30·3	207·8	1·46	90·2
69Lu7878 SC423	—	White	p	11·5	2·26	0·59	2·08	32·2	218·5	1·48	89·7
	—	White	a	11·3	2·04	0·25	0·18	25·6	193·3	1·45	91·5
69 Chil 112 12E	—	Red	p	12·3	2·12	0·26	0·49	29·9	202·7	1·47	92·1
	—	Red	a	13·5	1·88	0·24	0·27	27·9	199·9	1·39	92·8
	—	White	p	11·7	2·44	0·28	0·56	39·1	228·4	1·71	90·5
	—	White	a	11·8	2·20	0·23	0·14	31·8	208·4	1·52	92·1
10 % Casein Control	—	—	—	—	—	—	—	61·4	244·1	2·50	94·5
5 % Casein Control	—	—	—	—	—	—	—	16·1	164·8	1·90	95·0
Correlation with Lys						0·05	0·07	0·87 **	0·84 **	0·83**	—0·09
Corelation with FAS tannin							0·82**	0·21	−0·01	−0·32	−0·86**
Correlation with MV-HCl tannin								0·07	0·10	−0·18	−0·67**

[a] 9% protein from sorghum, 1% protein from casein.
[b] Maxson (E.D.) *et al.* (1973).
[c] a = pigmented testa absent.
p = pigmented testa present.

Table 4.34
Mean values of performance data for rats fed samples of sorghum grain differing in pericarp colour and presence or absence of a pigmented testa.[a]

Peri. colour	Testa[b]	No. samples fed	Lys (% prot.)	Tannin in diet (mg/100 mg) FAS	Tannin in diet (mg/100 mg) MV-HCl	Wt gain	Feed intake	PER	IVDMD
White	a	5	2·02	0·23	0·16	26·8	195·0	1·39	92·0
White	p	4	2·15	0·44	1·92	31·2	211·1	1·46	90·4
Red	a	4	1·88	0·25	0·54	23·9	186·8	1·28	92·0
Red	p	1	2·12	0·26	0·49	29·9	202·7	1·47	92·1
Brown	p	5	2·01	0·64	1·94	23·8	190·5	1·26	82·0

[a] Maxson (E.D.) *et al.* (1973).
[b] p = pigmented testa present, a = pigmented testa absent.

Table 4.35
Growth response and general body condition of rats on low- and high-tannin (CE) sorghum isonitrogenous diets.[a]

Samples	Prot. in sample (%)	Lys as % prot. in sample	CE of sample (%)	Av. wt gain (g)	Feed consumed (g)	PER	As % casein	General body condition
(A) 05274	10·8	1·91	0·3	6·6	137·1	0·54	18·6	Average
(B) 925120	9·9	2·08	2·08	5·7	142·1	0·45	15·5	Poor
(C) Rx 025440	9·6	2·05	3·08	−0·5	121·2	—	—	Poor
(D) (A)+Lys	—	—	—	31·4	167·7	2·09	72·1	Normal
(E) (B)+Lys	—	—	—	14·4	148·1	1·08	37·2	Average
(F) (C)+Lys	—	—	—	17·2	143·9	1·37	47·2	Fair
Casein	—	—	—	62·4	240·9	2·90	—	—

[a] With or without additions of L-Lys HCl to total 1% of diet.
[b] Jambunathan and Mertz (1973b).

Table 4.36
Protein, Lys and CE of bulked sorghum samples used in rat diets.[a]

	Groups		
	I	II	III
Coloured testa	Absent	Present	Present
No. of samples	50	63	24
Av. V-HC1 CE	0·32	0·57	2·53
Av. MV-HC1 CE	0·63	3·13	5·01
Prot. (% of sample)	10·8	10·6	10·5
Lys (% of protein)	1·48	1·49	1·64
Lys (% of sample)	0·16	0·16	0·17

[a] Cummings and Axtell (1973a).

Table 4.37
Vanillin-HC1 CE, protein and Lys analysis of diets, F/G and PER from rat feeding trials.[a]

	Groups		
	I	II	III
Vanillin HCl (CE)	0·24	0·52	3·20
Prot. (% of sample)	11·7	11·4	11·3
Lys (% of prot.)	1·84	1·83	1·74
F/G	14·59	16·27	34·92
PER	0·58	0·63	0·26

[a] Cummings and Axtell (1973a).

Table 4.38
Effect of dehulling and Lys supplementation of IS 8260 on weanling rat growth over 14 days.[a]

Diets	Feed consumption (g)	Weight gain (g)	F/G	PER
(H) Whole grain	88·8	2·3	44·97	0·32
(G) Whole grain, lys added	77·5	3·1	36·47	0·37
(A) Dehulled	85·6	12·2	7·45	1·17
(B) Dehulled, lys added	120·4	40·7	2·99	2·70

[a] Cummings and Axtell (1973b).

Table 4.39
Mean values of feed consumption, weight gain, F/G and PER for rats on diets varying in tannin (CE) over 28 days.[a]

Diet	CE level	Feed consumption (g)	Wt gain (g)	F/G	PER
(B)	0·14	319·3	99·1	3·31	2·45
(C)	0·57	226·4	50·4	4·59	1·76
(D)	1·20	194·6	30·9	6·71	1·24
(E)	1·49	175·1	22·6	9·00	0·99
(F)	2·02	177·2	18·9	11·51	0·88
(G)	3·66	164·5	9·0	31·99	0·45

[a] Cummings and Axtell (1973b).

Table 4.40
Rat weight gain as related to CE values, and IVDMD of 110 grain sorghum samples.[a]

	Protein (%)	Wt gain (mean) (g)	(% check)	IVDMD (mean) (%)
Low CE group Mean CE 0·58, range 0·00–0·99 66 samples	10·7	5·72	63	77
High CE group Mean CE 4·13, range 1·00–8·10 44 samples	9·7	2·15	17	59
All samples Mean CE 1·90, range 0·00–8·10 110 samples	10·3	4·29	44	69

[a] Oswalt (1973a).

Table 4.41
Comparisons of means of grain sorghum genotypes grouped by crude protein levels, and high and low CE concentrations, with rat gain, IVDMD and lysine.[a]

No. of samples	CE	Prot. (%)	Rat wt gain (g)	(% check)	IVDMD (%)	Lys[b] % sample	% prot.
		Diets below 8·9% crude prot.					
12	0·40	8·37	4·10	47	70	0·20	3·00
17	4·00	8·24	0·13	1	55	0·23	2·82
		Diets from 9·0 to 9·9% crude prot.					
15	0·26	9·63	4·87	46	74	0·24	2·59
13	4·96	9·48	2·39	17	52	0·22	2·64
		Diets from 10·0 to 10·9% crude prot.					
11	0·51	10·47	5·19	71	73	0·25	2·75
8	4·09	10·34	5·20	39	68	0·24	2·96
		Diets from 11·0 to 11·9% crude prot.					
13	0·34	11·26	6·86	66	84	0·28	2·45
		Diets from 12·0 to 14·0% crude prot.					
15	0·42	13·19	6·94	80	80	0·30	2·50
6	2·21	12·58	3·33	30	71	0·26	2·75

[a] Oswalt (1973a).
[b] Rat requirement of Lys is 1% of sample (Mertz 1969).

Table 4.42
Digestibility studies with rats fed sorghums RS 671, BR 64 and extracted BR 64 diets.[a]

	RS 671	BR 64	BR 64 extracted
Feed consumed (g)	50·66	58·11	57·33
Faeces wt (g)	6·17	12·00	6·93
Faecal N (%)	2·96	3·50	2·67
Total faecal N (mg)	182·60	420·80	184·90
Dietary N absorbed (%)	79·80	49·00	77·20
Dietary calories absorbed (%)	88·20	78·70	88·60

[a] Featherston and Rogler (1975).

Table 4.43

N and tannin contents of six selected South African grown sorghum cultivars, extracted with 60% (v/v) ethanol and unextracted, and effect on weanling rat growth rate, digestibility of the protein and liver lipid content.[a][b]

Cultivar	N content of cultivar	(%) of exp. diet (grain sorghum N only)	1% HC1 (v/v) methanol-extractable tannin content (tannic acid equivalents)		Rel. growth rate (g/100 g body mass/week)	Digestibility of prot. (%)	Liver lipid content (% of dried liver)
			of cultivar	of ration			
Unextracted							
I 1	1·541	0·835	1·32(4)	0·72(4)	14·08(4)	75·69(3)	10·3(2)
CI 43	1·536	0·852	1·39(3)	0·77(3)	14·83(3)	66·96(4)	11·1(3)
CI 34	1·581	0·870	0·31(6)	0·17(6)	16·09(1)	87·73(1)	10·8(4)
CI 23	1·701	0·868	0·87(5)	0·45(5)	15·75(2)	85·92(2)	9·9(1)
C 26	1·177	0·847	1·94(1)	1·37(1)	12·90(5)	51·89(5)	13·7(6)
C 45	1·309	0·859	1·60(2)	0·85(2)	12·17(6)	45·21(6)	13·5(5)
Extracted with 60% (v/v) ethanol							
I 1	1·579	0·842	0·80	0·42	14·91(3)	75·98(3)	12·7(5)
CI 43	1·543	0·848	0·76	0·43	13·91(4)	70·94(4)	11·4(2)
CI 34	1·560	0·878	0·22	0·15	15·89(1)	91·92(1)	11·7(4)
CI 23	1·716	0·842	0·68	0·34	15·54(2)	80·56(2)	11·5(3)
C 26	1·194	0·870	0·98	0·72	13·70(5)	67·66(6)	13·3(6)
C 45	1·318	0·849	0·74	0·48	13·18(6)	67·90(5)	10·7(1)

[a] Dreyer and van Niekerk (1973).
[b] Figures in parentheses indicate ranking.

Table 4.44
Protein digestibility (%), *in vivo* (rat) and *in vitro*, and tannin content (%) (as tannic acid) of 13 sorghum cultivars.[a]

Cultivar	Digestibility (%)		Tannin content[b] (as tannic acid) (%)
	In vivo	*In vitro*	
C 45	69·1	12·9	1·545
C 18	80·1	26·2	1·105
C 21	77·6	34·0	1·150
C 23	73·9	38·5	1·255
C 31	86·6	60·3	0·930
C 1	95·2	73·1	0·285
C 37	98·9	74·2	0·360
C 46	93·4	75·2	0·405
C 22	97·6	76·4	0·465
C 4	95·9	77·5	0·325
C 25	99·1	79·3	0·360
C 8	97·8	80·8	0·420
C 27	94·7	82·1	0·300

[a] Dreyer (1974).
[b] Data extracted from Dreyer and van Niekerk (1973).

Table 4.45
Digestible energy (DE) and metabolizable energy (ME) content (DWB) for rats of six varieties of sorghum varying in tannin content.[a]

Variety	Seed colour	Tannin (%)	AA absorption[b] (%)	DE[c] (kcal/g)	ME[d] (kcal/g)
ARK 61002	Brown	0·45	50	3·49	3·30
AKS 614	Brown	1·08	83	3·73	3·56
RS 617	Brown	0·55	60	3·87	3·72
RS 608	Yellow	0·84	91	3·92	3·74
TE 66	Yellow	0·88	89	4·19	4·01
ARK 62002	Yellow	0·76	87	4·13	3·95

[a] May and Nelson (1973).
[b] Chick availability (Stephenson *et al.*, 1971).
[c] *r* of DE to tannin 0·38.
[d] *r* of DE to tannin 0·39, corrected for N balance.

Table 4.46
Chemical composition (DWB) of sorghum samples and diets, and effect of proanthocyanidins on the nutritive value.[a]

Sorghum sample	INRA 450	NK 222	INRA 450	INRA 450	NK 120	INRA 450	AKS 614
Year	1967	1970	1971	1970	1970	1971	1970
Source[b]	CAL	ENSA	ENSA	ENSA	CAL	CAL	ENSA
Prot. (%) (N × 6·25)	14·1	8·0	13·2	11·4	11·1	10·2	11·3
Diet (g)							
Sorghum	880	940	957	950	950	910	900
Fish flour	40	46	19	36	36	75	80
L-Ile	3·0	0	0	2·5	5·0	0	0
L-Lys	6·6	6·0	6·0	6·0	6·0	6·0	6·0
Met	0·8	2·0	2·0	2·0	2·0	2·0	2·0
L-Thr	2·0	1·0	1·0	1·0	1·0	1·0	1·0
L-Trp	0	0·5	0·5	0·5	0·5	0·5	0·5
Min. mix	30	35	40	45	45	35	35
Vit. mix	20	20	20	20	20	20	20
Vitd. maize oil	20	20	20	20	20	20	20

cont.

Table 4.46 *cont.*

Sorghum sample	INRA 450	NK 222	INRA 450	INRA 450	NK 120	INRA 450	AKS 614
Year	1967	1970	1971	1970	1970	1971	1970
Source[b]	CAL	ENSA	ENSA	ENSA	CAL	CAL	ENSA
Prot. (%) (N × 6·25)	14·1	8·0	13·2	11·4	11·1	10·2	11·3
Prot. (%) (N × 6·25)	15·4	12·0	14·1	13·7	13·6	15·0	16·5
Digestible prot. (%)	13·5	9·7	8·7	8·8	7·4	8·7	8·2
LA[c] (content in procyanidins)	0·01	0·05	0·31	0·35	0·375	0·45	0·60
V	—	—	0·14	0·145	0·17	0·25	0·48
V/LA	—	—	0·45	0·41	0·45	0·55	0·80
Total ingested energy[d]	66·7	68·8	65·7	64·5	65·6	71·0	75·3
Ingested metabolizable energy[d]	58·0	60·8	54·0	52·4	52·4	56·5	58·1
N consumed[d]	383	308	340	317	321	389	448
N absorbed[d]	333	250	210	204	174	226	222
N retained[d]	174	167	147	116	108	139	130
CDU E[e]	89·8	90·1	83·5	83·1	81·1	81·5	79·0
CDU N[e]	87·7	81·3	61·9	64·2	54·1	58·1	49·5

[a] Martin-Tanguy *et al.* (1976).
[b] CAL: Coopérative Agricole Lauragaise, Castelnaudary, (Aude).
ENSA: Station d'Amélioration des Plantes, Ecole nationale supérieure agronomique de Montpellier (Hérault).
[c] Quantities LA and V are expressed in optical densities and measured for 8 g of fresh grains.
[d] mg/rat/day.
[e] CDU E = coefficient of digestive utilization (energy).
CDU N = coefficient of digestive utilization (nitrogen).

Table 4.47
Analysis of grain sorghums and metabolizable energy for chicks.[a]

Variety	Tannin (%)	Fibre (%)	Prot. (%)	ME kcal/lb
Combine Kafir 60	0·2	1·57	11·6	1591
Martin Milo	0·4	1·54	13·2	1415
Redbine 60	0·4	1·75	15·7	1445
GA 609	1·3	1·99	14·5	1392
Combine Sagrain	2·0	2·09	13·5	1478
NK 230 (discontinued)	2·0	1·83	15·9	1456

[a] Chang and Fuller (1964).

Table 4.48
Effect of grain sorghums on growth, feed consumption and F/G.[a]

Diet	Tannin in diet (%)	8-week gain[b] (g)	F/G
Basal	—	1693	1·88
Combine Kafir 60	0·1	1646	1·95
Martin Milo	0·2	1654	1·91
Redbine 60	0·2	1629	1·93
GA 609	0·65	1610	2·02
Combine Sagrain	1·0	1585	2·05
NK 230	1·0	1530	2·10

[a] Chang and Fuller (1964).
[b] Contributed by the various grain sorghum varieties.

Table 4.49
Sorghum varieties, in ascending order of tannin content, showing seed colour, contents of tannin, protein, fat and fibre and effect on weight gain and F/G of chicks.[a]

Variety or hybrid	Seed colour	Tannin (%)	Prot. (%)	Fat (%)	Fibre (%)	8-week body weight (lb)	Feed consumption (lb)	F/G
G 600	Yellow	0·25	7·8	2·0	2·4	3·78	7·79	2·12
C 44B	Red	0·27	8·5	2·4	1·9	3·65	7·85	2·21
Frontier 401	Lt. red	0·28	8·9	3·2	2·5	3·71	7·83	2·17
Frontier 400C	Red	0·32	10·3	3·0	2·4	3·71	7·72	2·14
E 57	Bronze	0·33	8·1	2·6	1·9	3·81	8·00	2·16
NK 255	Bronze	0·33	8·0	2·8	2·2	3·78	7·94	2·16
NK 222(1)	Lt. yellow	0·35	9·2	3·0	1·9	3·68	7·76	2·17
NK 222(2)	Lt. yellow	0·37	12·0	2·2	2·5	3·65	7·70	2·17
RS 610	Red	0·37	10·0	2·9	3·0	3·68	8·24	2·30
PAG 515	Red	0·37	10·7	3·1	2·1	3·78	7·82	2·13
Ranger A	Red	0·39	9·4	3·0	1·9	3·67	7·49	2·10
Pioneer 820	Dk. red	0·39	8·5	2·6	2·0	3·70	7·55	2·10
AMAK R12	Red	0·40	8·4	2·7	2·5	3·67	7·67	2·15
NK 275	Lt. red	0·41	8·6	2·8	2·2	3·65	7·74	2·18
Rico	Red	0·46	9·3	2·5	2·1	3·62	7·64	2·17
GA 615	Red brown	1·65	8·6	3·1	2·9	3·42	7·46	2·25

[a] Fuller *et al.* (1966).

Table 4.50
Protein and tannic acid (TA) contents of chick diets.[a]

Diet	Prot. in diet (%)	TA equivalent in grain (%)	TA equivalent in diet (%)	Apparent digestibility (%)
(A) RS 610	9·0	0·37	0·33	73·0
(B) NK 300	8·1	0·66	0·59	41·5
(C) BR 804	8·8	1·22	1·10	25·6
(D) BR 64	7·0	1·57	1·41	22·2
(E) (A)+TA 0·26%	9·0		0·59	74·1
(F) (A)+TA 0·75%	9·0		1·10	67·1
(G) (A)+TA 1·08%	9·0		1·41	62·6
(H) Protein free	—		—	—
(I) (H)+TA 1·41%	—		1·41	—

[a] Rostagno *et al.* (1973b).

Table 4.51
Protein and tannic acid (TA) content of chick diets, chick weight gains and feed conversion.[a]

	Prot. in grain (air-dried) (%)	Prot. in diet (%)	TA equivalent (%)	Wt gain (g)	F/G
(A) RS 610	10·5	15·7	0·27	202	1·9
(B) BR 64	8·9	14·5	1·13	113	2·81
(C) NK 300	9·7	15·1	0·48	146	2·46
(D) Maize	9·4	15·0	—	282	1·9
(E) RS 610	—	14·5	0·23	187	2·01
(F) NK 300+TA		14·5	0·43	122	2·61
(G) (A)+TA 0·65%		15·7	1·13	149	2·20
(H) (C)+TA 0·65%		15·1	1·13	69	3·50
(I) (D)+TA 1·13%		15·0	1·13	170	2·14

[a] Rostagno *et al.* (1973a).

Table 4.52
Protein and tannic acid equivalents[a] of sorghum varieties.[b]

Variety	Prot. (air-dried basis) (%)	Tannic acid equiv. (%)
RS 671 (1)	9·4	0·37
RS 671 (2)	10·8	0·56
RS 671—extracted (1)	9·8	0·27
RS 671—extracted (2)	10·8	0·34
IS 8260 (1)	7·6	2·66
IS 8260 (2)	8·7	2·27
IS 8260—extracted (1)	7·6	0·25
IS 8260—extracted (2)	6·7	0·34
RS 610	9·8	0·33
IS 6992	9·8	1·60
IS 6992—extracted	8·3	0·42
BR 64	8·7	2·26
BR 64—extracted	9·0	0·49

[a] Burns (1963).
[b] Armstrong *et al.* (1974a).

Table 4.53
Colour and tannin equivalents (DWB) of sorghum with effect of tannin and dry matter digestion on amino acid availability and energy utilization by chicks.[a]

Variety	Colour		Tannin equiv. (DWB)	Dry matter digestion (%)	AA availability (avg. of 17)	Gross energy (DWB) kcal/g	ME (DWB) kcal/g	ME/GE %
	Seed coat[b]	Endo.[c]						
ARK 61002	B	W	0·77	70	60	4·47	3·18	71
Funk BR 76	B	W	0·56	74	66	4·35	3·26	75
Savannah	B	W	0·54	73	67	4·34	3·19	74
AKS 618	B	W	0·52	78	71	4·41	3·40	77
RS 617	B	W	0·49	80	74	4·46	3·50	78
AKS 663	B	Y	0·43	81	76	4·38	3·58	82
GA 615	B	W	0·59	87	83	4·40	3·84	87
Niagara Shoo Bird	B	W	0·54	91	86	4·40	3·92	89
DeKalb E 57	W	Y	0·15	87	88	4·35	3·78	87
RS 608	R	W	0·16	84	88	4·45	3·79	85
TE 66	R	W	0·16	88	91	4·40	3·85	87
Niagara ORO 7	R	Y	0·16	88	94	4·40	3·82	87

[a] Nelson *et al.* (1975).
[b] B—brown, W—white, R—red.
[c] W—white, Y—yellow.

Table 4.54
Content of total phenols and tannins, catechin equivalent (CE), DWB, in 32 pearl millet cultivars (differing seed colour[a]) with *in vitro* protein digestibility (IVPD) of 13 of these cultivars.[b]

Cultivar	Seed colour	Total phenols (%)	Tannins CE (%)	IVPD (%)
Indian				
Hamsa	W	0·08	0·06	85·1
HPW 27 4	W	0·09	0·04	
HPW 83 4	W	0·07	0·04	
ECW 854	W	0·06	0·06	
ECW 955	W	0·09	0·05	
HX 799	W	0·09	0·04	
Purna	B	0·74	0·85	78·0
HPB 20 5	B	0·53	0·57	86·5
HPB 1 8	B	0·71	1·03	81·3
HPB 7 6	B	0·62	0·84	
HPB 23 6	B	0·57	0·76	
EC 4840	B	0·63	0·92	86·3
IE 246	DB	0·37	0·14	83·0
IE 328	DB	0·42	0·26	
IE 736	DB	0·37	0·12	
IE 497	DB	0·55	0·36	
IE 860	DB	0·96	1·05	84·2
IE 121	B	0·69	0·57	88·1
IE 395	B	0·64	0·52	
African				
IE 927	B	2·44	3·47	55·4
IE 929	B	2·38	3·42	57·4
IE 974	B	0·92	0·85	
IE 976	B	0·66	0·76	
IE 978	B	0·54	0·50	
IE 979	B	1·00	1·14	85·5
IE 1029	B	0·85	0·93	
IE 1038	B	0·58	0·46	
IE 1039	B	0·86	1·04	
IE 1065	B	0·61	0·52	
Indian × African				
14-1A-3-26	B	1·08	1·46	83·8
91-4-7-18	B	1·48	2·02	84·3
HX 929-43-7-3	W	0·10	0·03	

[a] W—white, B—brown, DB—dark brown.
[b] Ramachandra *et al.* (1977).

Table 4.55
Estimation of N and tannins (CE) in three Landry and Moureaux fractions of undefatted whole seed flour of five pearl millet cultivars.[a]

	Purna		EC 4840		HPB 1 8		14 1A 3 26		IE 929	
	N (%)	CE (%)	N (%)	CE (%)	N (%)	CE (%)	N (%)	CE (%)	N (%)	CE (%)
Albumin-globulin (0·5 M NaCl, distilled water)	25·1	Tr	21·8	Tr	21·7	Tr	18·8	Tr	8·7	Tr
Prolamine (70% isopropyl alcohol + 0·6% 2-mercaptoethanol)	38·8	0·02	36·8	0·02	37·4	0·03	34·1	0·03	41·9	0·04
Glutelin (borate buffer pH 10·0, containing 0·5 M NaCl, 0·6% 2-mercaptoethanol, and 0·5% sodium dodecyl sulphate)	15·2	0·21	14·9	0·17	19·7	0·36	14·7	0·36	15·2	0·74

[a] Ramachandra *et al.* (1977).

Table 4.56
Distribution of phytin P and phytic acid in the sorghum grains and their fractions.[a]

Sample	Fraction	Total acid soluble P (mg/100 g)	Phytin P (mg/100 g)	Phytin P in total acid soluble P (%)	Phytic acid equiv. (mg/100 g)
Westland 287-1	Whole grain	350	250	72	890
	Germ	1230	1090	88	3870
	Bran	450	350	78	1240
	Grit	50	30	58	110
	Mill fines	170	130	74	460
Westland 290-2	Whole grain	330	270	81	960
	Germ	2190	1910	87	6780
	Bran	590	490	83	1730
	Grit	50	20	48	70
	Mill fines	130	90	64	320
Blackhull 259-2	Whole grain	350	250	72	890
	Germ	2810	1770	96	6280
	Bran	390	340	87	1210
	Grit	100	80	85	290
	Mill fines	190	170	89	600
Western Blackhull 285-1	Whole grain	410	370	91	1310
	Germ	1980	1810	91	6420
	Bran	250	200	80	710
	Grit	80	40	53	140
	Mill fines	140	100	71	360
Midland	Whole grain	430	350	78	1170
	Germ	1920	1690	88	5990
	Bran	460	390	86	1380
	Grit	90	40	50	140
	Mill fines	210	170	83	600
Pink Kafir 421	Whole grain	410	330	80	1170
	Germ	1270	1030	81	3650
	Bran	250	190	76	670
	Grit	260	180	70	640
	Mill fines	400	310	78	1100
Cody 358	Whole grain	260	200	78	710
	Germ	640	540	84	1920
	Bran	380	310	81	1100
	Grit	110	70	64	250
	Mill fines	530	470	89	1570

[a] Wang *et al.* (1959).

Table 4.57
P and phytin P, and phytase activity in sorghum and millets.

Cereal	Total P (mg/100 g)	Acid-sol. P (mg/100 g)	Phytin P (mg/100 g)	Phytin P (% of total P)	Phytase activity	Reference
Sorghum	460		130	30		Oke (1965ab)
	232–365		206–280	77–88		Gontzea and Sutzecu (1968)
	179		138	76		Achutha Murthy *et al.* (1965)
	240–611	192–382	181–317		<1	Courtois and Perles (1954)
		260–430	200–370	72–91 (acid-soluble P)		Wang *et al.* (1959)
	295–347		225–255	74–76		Becker (1950)
	374		228	60	0·08	Giri (1938)
	233		206	88		Sundararajan (1938)
Pennisetum spp.	264–367	144–288	129–284	—	<1	Courtois and Perles (1954)
Pearl millet	230		114	50	0·17	Giri (1938)
	328		246	75		Sundararajan (1938)
	344		150	44		Oke (1965b)
Finger millet grain	245		172	70	0·01	Giri (1938)
cooked	244		150	61		Giri (1938)
grain	273		246	90		Sundararajan (1938)
Foxtail millet	230–350		153–220	50–66		Gontzea and Sutzecu (1968)
	300		200	69		Achutha Murthy *et al.* (1965)
	320		213	67		Sundararajan (1938)
Common millet	145		83	57		Gontzea and Sutzecu (1968)
Wheat	320–360		170–280	47–86		Gontzea and Sutzecu (1968)
Wheat bran	1210–1609		1170–1439	89–97		Gontzea and Sutzecu (1968)

Table 4.58
Total P, acid-soluble P and phytin P of samples of *Sorghum* spp. and *Pennisetum* spp.[a]

Cereal	Total P (mg/100 g)	Acid-soluble P (mg/100 g)	Phytin P (mg/100 g)	Phytin P (% of total P)	Glycero-phosphatase activity (%)	Phytase activity (%)
Sorghum spp.						
Teing	611	382	368	60	22·5	<1
Kongossan	314	300	273	87	33·6	<1
Voyendé	340	334	316	93	32·4	<1
Mboratel	337	288	260	77	39·4	<1
White Fela	375	255	242	65	49·6	<1
Yellow-white Fela	240	192	181	75	39·0	<1
Red Fela	339	330	317	93	34·4	<1
Pennisetum spp.						
Sanio 1	264	188	177	67	28·2	<1
2	284	144	129	45	27·6	<1
Souna 3	318	241	236	74	23·6	<1
4	367	288	284	77	28·8	<1
5	327	284	281	86	23·6	<1

[a] Courtois and Perlès (1954).

Table 4.59
Mean contents of Ile and Leu (column chromatography) and Trp (column chromatography or microbiological assay (M)) of various cereals.[a]

	Ile (mg/g total N)	Leu (mg/g total N)	Trp (mg/g total N)	Leu/Ile
Sorghum	245	832	63(M)	3·40
Pearl millet	256	598	106(M)	2·33
Finger millet	275	594	105(M)	2·16
Foxtail millet	475	1044	61(M)	2·20
Common millet	405	762	49(M)	1·88
Kodo millet	188	419	38(M)	2·22
Digitaria exilis	250	613	88	2·45
Digitaria iburua	269	738	114	2·75
Coix lachryma-jobi (Job's tears)	275	860	45	3·13
Teff	241	461	93(M)	1·91
Maize	230	783	38(M)	3·40
Rice (brown)	238	514	78(M)	2·16
(polished)	262	514	84(M)	1·96
Wheat (whole grain)	204	417	68(M)	2·04

[a] FAO (1970).

Table 4.60
Effect of Arg/Lys ratio in 10% protein diets based on casein, wheat and foxtail millet on growth of weanling rats.[a]

	AA content (% in diet)		A/L quotient	Wt gain (g)	
	Arg	Lys		First 4 weeks	Second 4 weeks
(A) Casein	0·41	0·69	0·6	30·5	29·5
(B) Casein + 2·03% L-Arg HCl	3·45	0·69	5·0	19·5	10·5
(C) (B) + 2·0% L-Lys HCl	3·45	2·69	1·3	—	29·0[c]
(D) Wheat	0·30	0·33	0·9	20·0	22·0
(E) Wheat + 1·35 L-Arg HCl	1·65	0·33	5·0	10·5	0·5
(F) (E) + 1·0% L-Lys HCl	1·65	1·33	1·2	—	19·5[c]
(G) Wheat + AA (mix 1[b])	0·41	0·69	0·6	27·5	—
(H) Wheat + AA (mix 2[b])	1·00	0·33	3·0	11·0	9·0
(I) (H) + 0·36 L-Lys HCl	1·00	0·69	1·4	—	18·0[c]
(J) Foxtail millet	1·00	0·21	4·8	2·5	Died
(K) Foxtail millet + AA (mix 3[b])	1·00	0·69	1·4	26·5	—
(L) Foxtail millet + AA (mix 4[b])	1·00	0·33	3·0	2·5	2·0
(M) (L) + 0·38% L-Lys HCl	1·00	0·71	1·4	—	27·5[c]

[a] Ganapathy and Chitre (1976)
[b] Amino acid Mix 1 to bring wheat diet to level of casein diet, (% of diet); L-Arg 0·11, DL-Ile 0·19, Leu 0·59, L-Lys HCl 0·36, DL-Met 0·06, DL-Phe 0·17, DL-Thr 0·05, DL-Trp 0·07, DL-Val 0·36.
Mix 2 to bring wheat diet to approximate level of foxtail millet diet, (% of diet); L-Arg 0·7, DL-Ile 0·16, Leu 0·53, DL-Phe 0·08, DL-Val 0·15.
Mix 3 to bring foxtail millet diet to level of casein diet, (% of diet); DL-Ile 0·03, Leu 0·06, L-Lys 0·48, DL-Met 0·12, DL-Phe 0·09, DL-Thr 0·11, DL-Trp 0·14, DL-Val 0·21.
Mix 4 to bring finger millet diet to level of wheat diet, (% of diet); DL-His HCl 0·45, L-Lys HCl 0·12, DL-Met 0·06, DL-Thr 0·06, DL-Trp 0·07.
[c] Wt gain in 4 weeks after Lys supplementation.

Table 4.61
Mean Mo, Cu and Zn contents of sorghum, pearl millet and rice from non-fluorosis (NF) and fluorosis (F) areas of Andhra Pradesh.[a]

Cereal	Area	No. of samples	Mo mg/100 g	Cu mg/100 g	Zn mg/100 g
Sorghum	NF	47	0·07	0·40	1·1
	F	52	0·12	0·48	1·9
Pearl millet	NF	15	0·08	0·66	3·9
	F	55	0·12	0·57	2·2
Rice (milled)	NF	32	0·04	0·33	1·3
	F	10	0·05	0·27	1·4
Rice (brown)	NF	8	0·09	0·37	1·5
	F	39	0·09	0·46	1·6

[a] Deosthale *et al.* (1977).

Table 5.1
Proximate analysis, minerals and vitamins (DWB) in samples of locally produced flours and foods. Blank spaces indicate no values presented.[a]

	Dry matter	N	Fat	Fibre	Ash	Acid-insoluble ash	Ca	P	Phytin P	Fe	Thiamine	Niacin	Riboflavin
	(%)						(mg/100 g)						
Sorghum													
Kenya													
Meal	88·8	1·5		2·4	2·3						0·34	3·96	0·14
Northern Nigeria													
Whole grain, Kaura	90·8	1·6	2·6	3·7	1·3		20	168	87	4·5	0·40	3·08	0·13
Whole grain, Jardawa	92·1	1·6	3·5	3·0	1·6		11	312	200	4·9	0·37	4·48	0·12
Flour from Jarwada prepared in N. Nigeria	88·5	1·6	1·2	0·7	0·8		12	148	70	3·4	0·20	1·21	0·06
North Ghana													
Machine milled flour	87·3	1·8	3·4	2·3	2·3		9		250	19·5	0·30	3·55	0·13
Stone ground flour	89·2	1·7	4·1	2·0	2·6		9		314	20·1	0·35	4·48	0·12
Pounded flour	88·9	1·6	3·1	1·8	1·5		13		311	11·2	0·28	3·26	0·90
Machine milled dry flour	89·5			1·6							0·37	3·46	0·11
Machine milled after soaking	56·0			2·5							0·09	5·71	0·36
Stone ground dry	89·8			1·6							0·39	3·45	0·11
Stone ground after soaking	59·9			2·0							0·08	9·85	0·50
Pearl millet													
The Gambia													
Whole grain	86·8	1·7	3·6	1·6	1·4		18	154		2·8	0·17	1·50	0·20
Flour from above prepared by pounding (dry matter content of freshly prepared flour 73·5%)	91·4	1·5	2·2	1·5	1·6		26			1·3	0·07	0·81	0·10
Nyelling from whole grain above (dry matter content of freshly prepared hyelling 75·6%)	94·2	1·5	1·9	2·1	1·8		18			2·2	0·08	0·10	0·08
Northern Nigeria													
Whole grain	92·9	1·9	4·3	2·1	1·7		13	257	131	4·1	0·37	1·51	0·09
Flour from above prepared by pounding	89·8	1·6	3·4	0·9	1·1		16	170	78	6·0	0·27	0·89	0·06

cont.

Table 5.1 *cont.*

	Dry matter	N	Fat	Fibre	Ash	Acid-insoluble ash	Ca	P	Phytin P	Fe	Thiamine	Niacin	Riboflavin
	(%)						(mg/100 g)						
North Ghana													
Machine milled flour	90·7	1·8	4·8	2·3	1·9	0·8	14	347	242	18·8	0·39	2·12	0·21
Stone ground flour	92·1	1·8	7·2	3·0	2·4	0·2	16	403	237	19·6	0·25	1·60	0·22
Pounded flour	92·0	1·9	6·6	2·5	2·0	0·2	25	404	308	18·5	0·40	1·52	0·20
Finger millet													
Uganda													
Sifted meal	87·3	1·0	0·9	2·7	4·1		423	255		90·5	0·17	0·92	0·08
Kenya													
Flour from malted grain	90·9	1·1		4·1	3·6						0·23	1·47	0·12
Teff													
Ethiopia													
Whole grain, white, market sample, uncleaned	88·8	1·6	2·5	2·5	3·6	0·7	123			101·3	10·53	2·36	0·12
Njera made from black teff	36·7	1·9	4·9	3·3	4·6		134			122·6	0·41	3·27	0·35
Njera made from white teff	32·5	2·5	5·5	2·5	4·3		206			98·5	0·68	5·29	0·74
Njera made from sorghum/teff, 2/1	38·6	11·8	4·4	2·1	3·0		98			12·9	0·52	3·63	0·26
Digitaria exilis													
The Gambia													
Whole grain	89·8	1·4	2·0	7·6	3·1		34	149		3·8	0·31	3·25	0·11
'Clean grain' pounded and winnowed, prepared from above	93·7	1·5	1·2	1·2	1·6		38	127		1·3	0·33	2·77	0·13
Beverages													
Ethiopian beer 'tala' made from red sorghum: alcohol 2·1%, total solids 1·39%		0·04									0·01	0·20	0·01

[a] Applied Nutrition Unit (1960).

Table 5.2
Material and protein balance, as % of original grain (DWB) of laboratory prepared ogi.[a]

	Material balance			Prot. balance		
	Maize	Sorghum	Millet	Maize	Sorghum	Millet
Grain	100	100	100	12·0	11·3	10·3
Steeping water	1·5	1·6	4·8	0·5	0·3	2·4
Fermented grain	97·4	94·9	78·8	11·5	10·8	7·8
Overtails	0·9	0·5	1·6	0·08	0·04	0·16
Ogi wash water	5·5	5·3	11·4	1·2	0·9	1·7
Ogi (uncooked)	82·3	83·6	61·1	9·5	9·4	5·6

[a] Banigo and Muller (1972a).

Table 5.3
Proximate composition of maize, sorghum and pearl millet, and their ogi samples (DWB).[a]

	Maize		Sorghum		Millet		Maize	
	Cereal	Ogi[b]	Cereal	Ogi[b]	Cereal	Ogi[b]	Cereal	Ogi[b]
Moisture (%)	10·8	41·0	11·8	41·0	12·4	41·3	10·7	47·4
Crude prot. (%)	12·0	11·5	11·3	11·3	10·3	9·1	7·2	3·6
Fat (%)	4·4	3·9	3·9	2·7	4·8	7·5	5·0	1·6
Crude fibre (%)	1·5	1·1	1·5	1·7	1·5	1·5	1·7	0·2
CBH[d] (%) (by difference)	81	83	82	84	82	81	85	93
Ash (%)	1·5	0·4	1·7	0·4	1·8	0·6	1·4	0·2
Calories	410	414	407	405	411	430	413	410

[a] Banigo and Muller (1972a).
[b] Wet-milled with Waring Blendor-Colloid mill combination.
[c] Wet-milled with a pestle and mortar.
[d] Carbohydrate.

Table 5.4
Amino acids (as residues/10^3) of cereals and their ogi samples. Blank spaces indicate no values presented.[a]

	Maize		Sorghum		Millet	
	Cereal	Ogi	Cereal	Ogi	Cereal	Ogi
Ile	32	42	40	43	46	58
Leu	134	158	146	142	109	124
Lys	23	13	25	13	29	18
Met	20	10		11	}	23
Cys	w/v[b]	w/v[b]	11	10	19 }	
Phe	41	w/n[c]	w/n[c]	35	38	w/n[c]
Tyr	34	35	32	31	23	21
Thr + Ser	102	105	92	78	103	99
Val	69	51	58	62	69	70
Arg	34	30	30	20	38	32
His	24	42	28	20	21	25
Ala	113	120	125	131	111	133
Asp	58	65	69	77	91	78
Glu	168	185	192	175	169	190
Gly	59	59	56	52	62	52
Pro	87	83	97	100	71	79

[a] Banigo and Muller (1972a).
[b] Cys appeared to run with Val on the chromatogram.
[c] Phe appeared to run with NH_3 on the chromotogram.

Table 5.5
Composition of rat feeding diets of unparched and parched sorghum, finger millet and maize.[a]

	(A) N free	(B) Sorghum		(C) Finger millet		(D) Maize	
	diet	unparched	parched	unparched (g)	parched	unparched	parched
Sorghum		54·2	51·7				
Finger millet				72·0	70·0		
Maize						55·4	40·3
Starch	75·0	20·8	23·3	3·0	5·0	19·6	34·7
Cane sugar	10·0	10·0	10·0	10·0	10·0	10·0	10·0
Oils and salt mixture	15·0	15·0	15·0	15·0	15·0	15·0	15·0
Nitrogen in diet	0·4[b]	0·9	0·9	0·9	0·9	0·9	0·9
(% DWB)	0·5[c]						

[a] Acharya *et al.* (1942).
[b] When fed against diets B and D.
[c] When fed against diet C.

Table 5.6
Biological value (BV) and digestibility coefficient (DE) of unparched and parched sorghum, finger millet and maize.[a]

Diet	N intake	Urinary N (mg)	Faecal N	Food N absorbed	Body N saved	BV %	DE %
(A) N-free	3·4	33·4	16·5				
(B) Sorghum							
Unparched	77·6	43·8	23·3	67·8	56·4	83·1	91·4
Parched	76·9	39·4	22·1	68·1	62·1	90·9	92·7
(A) N-free	6·7	29·6	15·2				
(C) Finger millet							
Unparched	86·0	36·8	20·9	73·4	66·2	89·9	92·6
Parched	82·4	33·7	23·9	67·0	62·9	93·9	85·5
(A) N-free	2·7	24·6	14·4				
(D) Maize							
Unparched	54·8	42·2	24·8	41·7	25·1	60·1	80·3
Parched	69·0	44·6	27·0	53·8	34·2	63·9	80·8

[a] Acharya *et al.* (1942).

Table 5.7
Composition of sorghum and pulse mixtures formulated to provide protein in the ratio 2 sorghum/1 pulse and their AA composition, (mg/g total N) against recommended[a] allowance.[b]

Mixtures	Ile	Leu	Lys	Met + Cys	Phe + Tyr	Thr	Val
Sorghum							
+ chickpeas	262	731	300	162	537	231	325
+ mung beans	250	719	306	169	506	206	325
+ pigeon peas	225	687	281	150	587	212	300
+ cowpeas	344	512	294	337	506	225	231
+ horse gram *Dolichos biflorus*)	269	731	312	162	662	244	219
+ urd	275	762	300	175	562	231	350
+ soybean	294	750	306	181	662	262	356
Recommended allowance[a]	250	437	344	219	381	250	312

[a] WHO (1973a).
[b] Pushpamma and Devi (personal communication, 1977).

Table 5.8
Effect of roasting or boiling on nutritive values of sorghum and pulse mixtures at 10% protein level in rat feeding trials.[a]

Diet[b]	(%)	Protein Efficiency Ratio			% digestibility			Biological value		
		Raw	Roasted	Boiled	Raw	Roasted	Boiled	Raw	Roasted	Boiled
(A) Sorghum	64·0	1·6	2·4	2·4	87·2	84·2	97·5	83·7	88·7	95·0
Chickpeas	16·5									
Maize starch	2·0									
(B) Sorghum	65·0	1·5	2·4	2·0	79·0	91·3	80·9	94·0	90·9	86·0
Mung beans	15·0									
Maize starch	2·5									
(C) Sorghum	65·0	1·4	2·3	1·9	67·0	89·8	95·5	89·7	92·0	89·4
Pigeon peas	16·5									
Maize starch	1·0									
(D) Sorghum	64·0	2·3	2·3	2·3	81·3	86·4	90·2	76·0	83·4	88·9
Cowpeas	15·0									
Maize starch	3·5									
(E) Sorghum	65·0	0·6	2·3	2·3	61·0	82·5	87·1	77·0	90·0	91·4
Horse gram (*Dolichos biflorus*)	16·5									
Maize starch	1·0									
(F) Sorghum	66·0	2·4	2·5	2·4	79·0	90·6	93·1	78·5	93·0	89·6
Urd	15·0									
Maize starch	1·5									
(G) Sorghum	64·0	1·5	1·8	2·5	80·3	90·8	86·7	71·3	88·2	85·9
Soybean	8·5									
Maize starch	10·0									

[a] Pushpamma and Devi (personal communication, 1977).
[b] All diets contained oil 10·0%, tricalcium phosphate 1·5%, amaranth 5·0% and salt 1·0%.

Table 5.9
Proximate composition, minerals and vitamins, of fura-nono (DWB). Blank spaces indicate no values presented.[a]

	Nono withou kuka		Nono with kuka		Fura		Fura-nono without kuka		Fura-nono with kuka	
	Raw	Cooked	Raw	Cooked	Raw	Cooked	Raw	Cooked	Raw	Cooked
kcal/100 g		18·5	1	15·5		401·6		404·6		408·6
Dry matter (%)	8·2	4·4	6·7	3·8	41·7	47·8	29·6	28·7	21·6	21·2
Crude prot. (%)	26·8	50·0	22·4	28·9	9·0	9·0	10·7	12·3	9·2	9·2
Fat (%)	42·7	1·3	37·2	5·3	1·9	1·3	5·7	3·4	4·7	1·5
CBH[b] (%) (by difference)	25·6	47·8	34·4	55·3	88·5	88·6	81·3	81·2	83·7	86·6
Crude fibre (%)					1·2	1·5	1·0	1·3	1·2	1·2
Ash (%)	4·9	0·9	6·0	10·5	1·0	0·9	1·3	1·8	1·2	1·5
Minerals (mg/100 g)										
Ca	64	66	55	48	53	50	95	172	62	90
P	105	45	40	35	125	125	175	150	175	150
Fe	11	11	0	9	11	11	21	11	18	11
Mg	7	7	7	6	48	61	54	66	55	65
K	79	80	80	73	200	200	275	355	285	355
Na	28	25	18	16	48	54	68	84	64	66
S	26	17	24	19	167	186	167	187	170	169
Vitamins (mg/100 g)										
Thiamine	0·01	0·02	0·01	0·02	0·41	0·32	0·42	0·32	0·41	0·33
Riboflavin	0·10	0·09	0·08	0·08	0·14	0·13	0·15	0·14	0·15	0·14

[a] Okoh *et al.* (personal communication, 1975).
[b] Carbohydrate.

Table 5.10
Amino acid composition (mg/g total N) of fura-nono.[a]

	Cooked fura-nono with kuka	Raw fura-nono without kuka	Cooked fura-nono without kuka	FAO/WHO prov. pattern (1973a)
		(mg/g total N)		
Ile	278	320	303	250
Leu	685	722	672	440
Lys	232	180	272	340
Met	174	155	187	{220
Cys	58	28	60	
Phe	336	352	332	{380
Tyr	232	201	235	
Thr	277	257	268	250
Val	341	390	366	310
Arg	194	209	77	
His	84	90	142	
Ala	485	529	444	
Asp	523	508	520	
Glu	1280	1322	1325	
Gly	187	167	175	
Pro	511	480	537	
Ser	362	320	350	
OH-proline	6	<6	<6	

[a] Okoh *et al.* (personal communication, 1975).

Table 5.11
Chemical composition of Japanese miso and sorghum miso-type products.[a]

	Japanese miso	Sorghum miso-type
Moisture (%)	52·6	58·9
Kjeldahl N		
Total (%)	2·0	2·2
Soluble (%)	1·3	1·2
Fat (ether extract) (%)	6·8	2·5
Reducing sugars (as glucose)	13·9	8·6
Chloride (as NaCl) (%)	12·2	9·5
pH	5·1	5·4

[a] Ilany *et al.* (1969).

Table 5.12
Proximate composition and Ca, P, Fe, thiamine and riboflavin content of meals traditionally ground in Rhodesia (DWB).[a]

	Sorghum		Pearl millet		Finger millet	
	Meal	Offals	Meal	Offals	Meal	Whole grain
Moisture (%)	15·4	36·0	16·0	24·6	13·0	11·9
N (%)	1·9	1·9	1·2	2·1	1·2	1·2
Total lipid (%)	3·1	6·6	4·2	11·6	2·3	2·2
CBH[b] (%) (by difference)	82·3	76·6	84·1	62·7	83·5	84·7
Crude fibre (%)	0·7	2·9	0·7	8·1	3·3	3·1
Ash (%)	3·0	2·8	3·9	5·3	3·8	3·2
Minerals (mg/100 g)						
Ca	16	41	20	50	300	300
P	176	373	202	631	220	245
Fe	18	13	46	34	35	11
Vitamins (mg/100 g)						
Thiamine	0·31	0·67	0·22	0·50	0·34	0·36
Riboflavin	0·06	0·18	0·26	0·51	0·16	0·19

[a] Carr (1961).
[b] Carbohydrate.

Table 5.13
Sieving analysis (%) of traditionally ground sorghum in Nigeria.[a]

Mesh size	5	18	36	40	60	80	100	<100
Original sorghum	87·8	12·3						
Sorghum husk after winnowing	29·6	41·7	17·6	1·0	7·5	1·0	1·0	1·0
Shelled sorghum	53·6	46·4						
Stock after pestle and mortar winnowing	0·6	65·7	8·6	0	8·0	8·6	5·5	3·1
Flour from above stock				2·0	52·9	4·9	17·7	23·3
Overtails from above stock	60·9	16·3	11·5	1·0	4·8	1·0	2·0	2·9
Final stone milled stock	3·2	18·9	0·3	8·2	2·2	2·5	14·8	49·9

[a] Muller (1970).

Table 5.14
Proximate analysis and thiamine content of roller-milled and traditionally milled sorghum flour of various extractions.[a]

	Extraction	Prot.	Fat	Fibre	Ash	Thiamine
			(%)			(mg/100 g)
Roller-milled	100	8·2	2·6	1·3	1·3	0·36
	90	6·0	1·3	0·8	0·7	0·20
	70	3·5	0·3	0·3	0·5	0·09
	50	2·0	0·3	0·3	0·5	0·05
	30	1·0	0·2	0·1	0·2	0·03
Pestle-and-mortar-milled	100	8·2	2·6	1·3	1·3	0·36
	90	7·3	2·2	1·1	1·1	0·32
	70	5·5	1·7	0·9	0·9	0·26
	50	3·4	1·2	0·6	0·5	0·18
	30	1·6	0·6	0·3	0·3	0·09

[a] John and Muller (1973).

Table 5.15

Nutrient content (DWB) of traditionally milled and disc-milled sorghum, and of kourou. Blank spaces indicate no values presented.[a]

	Traditional milling						Disc-milling		Kourou		
	Whole sorghum	Bran	Decorticated sorghum	Semolina	Fresh flour	Sun-dried flour	Flour	Semolina	Fresh kourou	Dried kourou	Kourou semolina
Moisture (%)	13·2	43·8	29·0	23·0	30·0	13·2	19·4	19·6	48·8	13·0	46·5
Calories	383	253	411	409	413	413	412	412	414	412	409
Prot. (N × 6·25) (%)	9·6	10·4	9·5	9·5	9·6	9·4	8·5	12·8	7·9	7·4	11·3
Lipid (%)	3·0	5·7	1·4	0·1	1·8	1·7	1·3	2·1	1·1	0·6	0·1
Total CBH[b] (%)	85·6	79·3	88·0	90·1	87·4	88·0	89·3	83·7	90·7	91·8	88·4
Formic acid insol. matter (%)	2·5	8·9	1·0	1·5	0·9	0·9	0·7	1·8	0·4	0·5	1·2
Ash (%)	1·8	4·6	1·0	0·3	1·2	0·9	0·9	1·4	0·3	0·2	0·1
Ca (mg/100 g)	12·8	15·9	13·5	6·6	11·8	10·4	11·6	16·9	5·6	3·0	3·4
P (mg/100 g)	335	644	192	42	227	208	159	262	56	62	35
Phytic P (mg/100 g)	194	417	116	32		135			36	36	28
Phytic P/total P	0·58	0·65	0·60	0·70		0·65			0·64	0·62	0·80
Ca/P	0·04	0·02	0·07	0·15	0·05	0·05	0·07	0·06	0·10	0·05	0·10
Na (mg/100 g)	5·7	8·6	4·2	4·6	5·9	4·7	4·8	4·2	4·6	4·8	5·5
K (mg/100 g)	361	682	165	78	199	184	190	262	16	18	21
Thiamine (mg/100 g)	0·54	1·05	0·30	0·04	0·37	0·32	0·21	0·04	0·07	0·07	0·01
Niacin (mg/100 g)	4·6	9·9	1·9	0·5	2·2	2·2	1·2	2·9	1·6	1·5	0·8
Riboflavin (mg/100 g)	0·10	0·20	0·05	0·03	0·05	0·04	0·04	0·07	0·05	0·07	0·01

[a] Favier *et al.* (1972).
[b] Carbohydrate.

Table 5.16

Effect of traditional milling, disc-milling and preparation of kourou on nutritive value of sorghum (DWB). Blank spaces indicate no values presented.[a]

	Traditional milling (% of nutrient by weight recovered from each fraction)						Disc-milling (% of nutrient by weight recovered from each fraction)			Preparation of kourou (% of nutrient by weight recovered from each fraction)				
	Whole sorghum	Bran	Decorticated sorghum	Semolina	Fresh flour	Sun-dried flour	Decorticated sorghum	Sieved flour	Semolina	Whole sorghum	Decorticated sorghum	Steeped sorghum	Semolina of kourou	Fresh kourou
Dry matter	100	18	76	13	57	55	100	79	19	100	77	62	14	33
Calories	100	12	80	13	61	59	100	79	19	100	82	66	15	36
Prot. (N × 6·25)	100	20	75	12	57	54	100	71	26	100	76	60	16	25
Lipid	100	35	36	0·4	35	31	100	75	29	100	34	30	1	12
Total CBH[b]	100	17	77	13	58	56	100	80	18	100	79	63	14	36
Cellulose	100	65	29	8	20	20	100	58	36	100	29		7	5
Ash	100	47	43	2	37	27	100	68	27	100	48	29	1	2
Ca	100	23	80	7	53	45	100	67	24	100	87	27	3	6
P	100	35	43	2	53	34	100	66	26	100	46	26	1	6
Phytic P	100	39	45	2		38				100	46	19	2	
Na	100	27	55	11	59	45	100	90	20	100	62	54	13	26
K	100	35	36	3	32	29	100	91	31	100	35		1	1
Thiamine	100	35	42	1	38	32	100	57	3	100	41	29	0·4	5
Niacin	100	39	32	1	28	26	100	48	29	100	38	26	2	10
Riboflavin	100	37	35	3	41	23	100	64	29	100	37	35	2	21

[a] Favier *et al.* (1972).
[b] Carbohydrate.

Table 5.17
Effect of sun-drying on the nutrient content of sorghum flour and kourou (DWB). Blank spaces indicate no values presented.[a]

	Fresh flour	Sun-dried flour	% variation	Fresh kourou	Sun-dried kourou	% variation
Calories	413	413	0	414	412	0
Prot. (%)	9·6	9·4	−2	7·3	7·4	+1
Lipid (%)	1·8	1·7	−6	1·1	0·6	−45
Total CBH[b] (%)	87·4	88·0	+0·7	91·5	91·8	0
Non-digestible CBH[b] (%)	0·9	0·9	0	0·4	0·5	+25
Ash (%)	1·2	0·9	−25	0·12	0·16	+33
Ca (mg/100 g)	11·8	10·4	−12	2·8	3·0	+7
P (total) (mg/100 g)	227	208	−8	61	62	+2
Na (mg/100 g)		4·6		4·4	4·8	+9
K (mg/100 g)		184		16	18	+12
Thiamine (mg/100 g)	0·37	0·32	−12	0·07	0·07	−8
Niacin (mg/100 g)	2·2	2·2	0	1·4	1·5	+7
Riboflavin (mg/100 g)	0·05	0·04	−23	0·06	0·07	+21

[a] Favier *et al.* (1972).
[b] Carbohydrate.

Table 5.18
Effect of cooking on the vitamin content of traditionally cooked sorghum flour, dough and gruel (DWB).[a]

	Flour	Dough	Gruel
Thiamine			
mg/100 g	0·23	0·18	0·20
% difference	0	−22	−12
Niacin			
mg/100 g	2·13	2·05	2·13
% difference	0	−4	0
Riboflavin			
mg/100 g	0·05	0·05	0·05
% difference	0	+2	−10

[a] Favier *et al.* (1972).

Table 5.19
Protein, Ca and P content of whole grain and fractions of sorghum (mean with ranges in parentheses (DWB)).[a]

	Whole grain	Fibrous seed coat	Endo. and germ	Dried solids from supernatant
As % of whole grain		10·3 (9·6–10·8)	84·9 (84·2–85·6)	3·7 (3·2–4·2)
Prot. (N × 6·25) (%)	8·8 (7·7–9·6)	15·0 (12·0–17·1)	5·8 (5·3–6·1)	62·5 (59·3–65·4)
As % of prot. in whole grain		17·6 (16·9–18·1)	56·1 (54·2–57·7)	26·1 (26·1–28·5)
Ca (mg/100 g)	41·3 (37–49)	82·0 (70–90)	18·3 (16–22)	466·7 (405–501)
As % of Ca in whole grain		20·2 (18·7–21·6)	37·3 (36·2–38·3)	41·3 (38·4–42·9)
P (mg/100 g)	233 (187–267)	292·7 (228–398)	90·3 (71–115)	3487 (2391–4495)
As % of P in whole grain		12·7 (11·1–14·3)	32·8 (29·1–36·8)	53·6 (48·2–58·7)

[a] Kurien *et al.* (1960a).

Table 5.20
Weight and protein content (N × 6·25) of successively removed fractions of three grain sorghum hybrids (DWB).[a]

	Kernel wt removed (%)	Total cumulative wt removed (%)	Prot. % of fraction	% total prot. removed in each fraction (N × 6·25)	Cumulative % prot. removed
Elevator run					
Fraction I	9·7	9·7	10·1	8·5	8·5
II	7·5	17·2	20·1	13·2	21·8
III	7·8	25·0	19·0	12·9	34·7
IV	9·0	34·1	15·9	12·6	47·3
Prot. content: original kernel, 11·4%; residual kernel, 9·7%					
Texas 601					
Fraction I	6·7	6·7	9·8	5·4	5·4
II	7·5	14·2	27·1	16·9	22·3
III	5·8	20·0	26·9	12·9	35·2
IV	6·6	26·6	22·2	12·2	47·4
V	5·7	32·3	18·0	8·5	55·9
Prot. content: original kernel, 12·1%; residual kernel, 8·1%					
RS 610					
Fraction I	11·5	11·5	12·9	12·9	12·9
II	8·9	20·4	23·1	17·9	30·8
III	8·9	29·3	24·1	18·7	49·5
IV	8·1	37·5	17·4	12·3	61·8
Prot. content: original kernel, 11·5%; residual kernel, 7·7%					

[a] Normand *et al.* (1965).

Table 5.21

Yield and crude fat (DWB) obtained by dry-milling grain sorghum under varied conditions in different machines.[a]

Sorghum fraction	Impact only (900 rev/min)		Impact only (1500 rev/min)		Squires and impact (900 rev/min)		Pearling and impact (900 rev/min)		NU brush machine		NU solid-rotor machine	
	Yield	Crude fat	Yield	Crude fat	Yield	Crude fat	Yield	Crude fat	Yield	Crude fat	Yield	Crude fat
	(%)											
+8	5	2·5	3	1·2	<1	3·5	<1	2·4	3	1·4	<1	1·4
+14	52	0·8	49	0·6	35	0·8	34	0·8	56	0·4	50	0·5
+20	16	0·5	20	0·5	38	0·6	39	0·5	14	0·4	25	0·4
+34	1	1·1	1	0·7	4	1·0	3	0·6	1	0·6	3	0·6
+50	2	3·5	1	3·7	2	2·7	2	2·3	1	10·4	1	4·2
−50	3	2·0	4	2·1	2	2·4	2	1·9	9	8·3	4	2·9
Hull	17	11·2	15	9·5	15	10·3	14	8·7	10	8·6	9	6·6
Germ	4	14·1	7	17·1	4	20·4	6	23·7	6	23·8	8	26·6

[a] Anderson *et al.* (1969a).

Table 5.22
Analyses of milled TE 77 sorghum fractions obtained by an integrated process. Blank spaces indicate no values presented.[a]

Fraction	Yield (%)	Moisture (%)	Crude fat (%)	Crude fibre (%)
+14 grits	50	5·9	0·8	0·2
+20 grits	23	5·9	0·5	0·2
Flour	4	7·3	3	0·9
Germ	7	5·2	15	
Feed	16	8·1	9	7·8

[a] Anderson and Burbridge (1971).

Table 5.23
Yield and chemical composition (DWB) of dry-milled fractions of P 721 hybrid sorghum.[a]

	Yield (%)	Lipid (%)	Fibre (%)	Prot. (%)	Lys (mg/g total N)
Whole grain		4·2	2·5	13·7	195
+14 grits	23·1	1·5	1·2	12·7	116
+20 grits	31·3	1·0	0·9	12·5	116
Flour	12·1	2·3	1·0	11·9	182
Germ	16·2	10·5	4·0	18·2	266
Feed	17·3	5·5	6·0	13·7	219

[a] Anderson *et al.* (1977).

Table 5.24
Effect of degree of polish on nutrient constituents of sorghum (DWB).[a]

	Degree of polish (% of bran removed)					
	0	4·5	7·8	12·0	15·0	18·4
Moisture in grain immediately after milling (%)	11·3	12·4	13·0	13·0	13·1	13·0
Protein (N × 6·25) (%)	12·0	12·0	11·5	11·4	11·4	11·2
Ether extractives (%)	2·6	2·2	1·9	1·3	1·1	1·0
Crude fibre (%)	3·1	2·0	1·7	1·1	0·9	0·6
Ash (%)	2·0	1·5	1·4	1·3	1·1	1·0
Ca (mg/100 g)	40	35	31	26	22	19
P (mg/100 g)	254	219	218	160	135	134
Thiamine (mg/100 g)	0·42	0·41	0·39	0·36	0·34	0·31

[a] Raghavendra Rao and Desikachar (1964).

Table 5.25
Mean milling yield and recovery of milled fractions obtained from three sorghum varieties.[a]

Fractions	Martin 5 (%)	DeKalb G 600 13 (%)	TX 2536 9 (%)
	No. of replicates		
Coarse grits	65·8	51·0	57·4
Fine grits	8·4	19·0	13·2
Germ rich grits	2·1	4·7	5·0
Germ	1·2	1·5	1·1
Bran	4·6	3·3	4·5
Flour	16·4	18·1	16·4
Recovery	97·4	97·6	97·6

[a] Rooney and Sullins (1969).

Table 5.26
Proximate composition of the milling fractions of three varieties of sorghum. Blank spaces indicate no values presented.[a]

Fraction	Ether extract			Prot. (N × 6·25) (%)			Ash		
	Martin	TX 2536	G 600	Martin	TX 2536	G 600	Martin	TX 2536	G 600
Whole grain	4·1	3·4	3·0	15·1	12·1	11·5	1·5	1·5	1·6
Coarse grits	0·4	0·5	0·4	14·1	11·5	10·6	0·2	0·3	0·2
Fine grits	0·4	0·4	0·4	14·7	10·6	10·0	0·3	0·4	0·3
Germ rich grits	4·1		4·5	17·2		12·2	0·9		2·2
Germ	16·2		15·8	19·5		15·7	5·4		8·2
Bran	7·5		9·2	13·0		15·7	3·1		4·8
Flour	14·3	7·2	7·5	15·7	12·9	13·0	5·1	4·2	3·9

[a] Rooney and Sullins (1969).

Table 5.27
Sorghum varieties used in laboratory milling, two phases.[a]

Variety	Endo. Texture rating	Endo. Description[b]	Peri. Colour[c]	Peri. Thickness
Phase 1				
NSA 740	5	AF	W	Thick
SC 84 6	5	AF	B	Thick
TX 09	4	R	W	Thick
7078	3	R	R	Thick
TX 2538	3	Y	W	Thick
B 398 (Martin)	2	R	R	Intermed.
DeKalb hybrid G 600	2	Y	W	Thin
TX 2536	2	Y	W	Thin
SC 303 6	1	C	W	Thin
SC 283 6	1	C	W	Thin
SC 283 6	1	C	R	Thin

cont.

Table 5.27 *cont.*

Variety	Endo. Texture rating	Endo. Description[b]	Peri. Colour[c]	Peri. Thickness
Phase 2				
NSA 740	5	AF	W	Thick
TX 09	4	F	W	Thick
DeKalb hybrid C42Y	3	I	W	Intermed.
B 3197	3	I	W	Intermed.
SC 110	3	I	W	Thin
SC 170	2	C	W	Thin
SC 283	1	AC	W	Thin

[a] Maxson *et al.* (1971).
[b] R—regular, C—corneous, F—floury, Y—yellow, AF—all floury, AC—all corneous, I—intermediate.
[c] W—white, B—brown, R—red.

Table 5.28
Macro-test weight, 1000-seed weight, density, hardness, protein and ether extract of 11 varieties of sorghum.[a]

Variety	Macro-test wt (lb/bu)	1000-seed wt (g)	Density (g/cm^3)	Hardness	Prot. (N × 6·25) (%) (DWB)	Ether extract (%) (DWB)
NSA 740	52·6	25·1	1·2	1·0	14·3	3·6
SC 84 6	54·1	19·6	1·3	0·0	13·1	3·3
TX 09	56·5	31·0	1·3	1·3	13·8	3·2
7078	59·2	27·5	1·3	13·1	12·3	3·8
TX 2538	57·3	59·2	1·3	22·2	13·8	3·7
B 398 (Martin)	60·4	27·5	1·4	36·7	15·1	4·1
DeKalb hybrid G 600	60·5	32·9	1·4	26·6	11·5	3·0
TX 2536	60·6	28·9	1·3	30·6	12·2	3·4
SC 303 6	59·7	19·0	1·4	37·4	14·0	3·1
SC 283 6 white	62·1	25·0	1·4	60·9	13·8	3·6
SC 283 6 red	63·5	24·4	1·4	65·5	13·9	3·4

[a] Maxson *et al.* (1971).

Table 5.29
Mean milling composition, "as-is" basis protein, and petroleum ether extract content (DWB) calculated from protein values of grits.[a]

Variety	Endo. texture rating	Coarse grits (%)	Germ rich grits (%)	Coarse & fine grits (%)	Fines (%)	Coarse & fine grits: Prot. (N × 6·25) (%)	Coarse & fine grits: Ether extract (%)
NSA 740	5	1·1	26·2	39·6	29·8	15·6	1·3
SC 84 6	5	0·1	38·4	30·2	30·0	13·5	1·4
TX 09	4	8·2	39·9	32·9	24·5	13·6	0·6
7078	3	23·7	19·6	53·3	22·9	12·0	0·7
TX 2538	3	28·2	17·6	48·8	29·7	14·7	0·4
B 398 (Martin)	2	53·2	3·2	72·3	19·1	14·5	0·4
DeKalb hybrid G 600	2	38·4	10·2	65·3	23·0	10·3	0·3
TX 2536	2	30·7	5·1	70·2	21·5	11·1	0·6
SC 303 6	1	45·7	5·4	75·0	16·1	13·9	0·2
SC 283 6 white	1	71·5	0·7	82·1	14·8	12·1	0·2
SC 283 6 red	1	70·7	1·1	80·3	16·4	12·4	0·3

[a] Coarse and fine grits of 11 sorghum varieties, laboratory milled for 2 min (Maxson *et al.*, 1971).

Table 5.30

Kernel size index (KSI), 1000-seed weight, macro-test weight, density, hardness, protein, ether extract, starch and ash (DWB) of seven varieties of sorghum.[a]

Variety	Endo. texture rating	KSI	1000-seed wt (%)	Macro-test wt (lb/bu)	Density (g/cm^3)	Hardness	Prot. (N × 6·25)	Ether extract	Starch	Ash
							(%)			
NSA 740	5	62·7	24·9	48·1	1·2	0·9	17·5	3·6	66·8	1·9
TX 09	4	59·8	32·9	51·8	1·3	2·0	16·5	3·2	70·4	1·8
DeKalb hybrid C42Y	3	62·7	32·4	59·2	1·3	11·8	12·2	3·3	73·7	1·3
B 3197	3	65·8	28·6	57·7	1·3	12·7	15·5	3·1	72·1	1·5
SC 110	3	66·0	26·5	57·6	1·3	14·0	14·8	3·4	71·9	1·5
SC 170	2	65·6	27·2	57·3	1·3	23·5	16·5	3·4	69·0	1·6
SC 283	1	71·4	23·0	61·9	1·4	29·6	15·2	2·9	72·8	1·3

[a] Maxson *et al.* (1971).

Table 5.31

Mean milling composition, protein, ether extract, ash content in total grits, of seven sorghum varieties.[a]

Variety	Endo. texture rating	Milling fraction yield ("as-is" basis)					Total grits[b] (DWB)		
		Coarse grits	Germ rich grits	Coarse & fine grits	Total grits	Fines	Prot. (N × 6·25)	Ether extract	Ash
		(%)							
NSA 740	5	3·6	59·9	3·6	56·5	26·0	21·2	0·8	3·7
TX 09	4	20·9	36·3	36·1	69·0	21·6	18·5	0·6	1·4
DeKalb hybrid C42Y	3	47·9	14·4	62·7	75·3	18·3	12·1	0·6	0·6
B 3197	3	51·8	13·5	62·4	72·7	18·6	15·5	0·8	0·9
SC 110	3	50·2	18·0	63·7	79·7	14·0	15·3	0·7	0·6
SC 170	2	65·5	9·0	73·9	80·8	13·5	15·8	1·0	0·5
SC 283	1	74·6	3·3	81·5	84·1	10·7	14·6	0·7	0·4

[a] Maxson *et al.*, 1971).
[b] Total grits includes coarse grits, fine grits and grits from $NaNO_3$ flotation.

Table 5.32
Flour yield, "as-is" basis, protein, ether extract and starch (DWB) of seven sorghum varieties.[a]

Variety	Yield	Prot. (N × 6·25)	Ether extract	Starch
	(%)			
NSA 740	59·4	15·4	1·2	80·3
TX 09	63·8	11·2	1·2	86·5
C42Y	68·1	10·8	1·1	86·3
B 3197	61·0	12·6	1·5	84·4
SC 110	61·1	12·6	0·9	86·6
SC 170	62·5	13·4	1·2	81·9
SC 283	67·9	12·9	1·3	84·1

[a] Maxson *et al.* (1971).

Table 5.33
Composition in protein, lipid and ash (DWB) of blends obtained by compositing selected fractions from successive abrasion milling on four varieties of sorghum grain. Blank spaces indicate no values presented.[a]

Description of grain	Variety	Cumulative weight of kernel removed (%)	Quantity (% based on original grain wt)	Blends obtained by compositing selected fractions		
				Protein (N × 6·25) (%)	Lipid (%)	Ash (%)
All-corneous endo.	SC 283	Whole grain	100	15·2	2·9	1·3
thin, white peri.		7·9–14·2	6·2	29·8	13·6	4·5
		14·2–33·1	18·9	27·4	6·8	3·1
		Residue	58·1	8·6	0·4	0·4
Intermed. endo.	B 398	Whole grain	100	14·7	4·1	1·5
brownish-red,		9·8–15·5	5·7	27·4	14·0	3·8
medium-thick peri.		15·5–23·9	8·4	30·6	9·8	2·8
		23·9–33·4	9·5	24·1	6·1	2·0
		Residue	56·5	8·2	0·7	0·3
Intermed. endo.	B 3197	Whole grain	100	15·5	3·1	1·5
thick, white peri.		8·0–17·2	9·2	28·9	11·2	3·8
		17·2–25·3	8·1	31·6	8·0	3·0
		25·3–36·8	11·5	24·6	4·9	2·1
		Residue	53·3	8·9	0·8	0·4
Intermed. to	70 LH 201	Whole grain	100	10·8		
corneous endo.:	(exp. hybrid)	8·4–13·9	5·5	24·4	13·6	4·0
white, thin peri.		13·9–22·4	8·5	24·4	9·7	2·9
		Residue	52·2	6·6	0·6	0·3

[a] Rooney *et al.* (1972).

Table 5.34
Amino acid distribution in the protein (mg/g total N) of the fraction of SC 283 grain removed by successive abrasion milling.[a]

	Whole grain	Total quantity of original grain removed (%)						
		3·3	7·9	11·1	15·9	18·6	31·1	41·9
Fraction No.	0	1	4	7	10	12	19	24
Prot. (N × 6·25) (%) (DWB)	15·2	8·1	22·1	31·9	35·3	32·1	20·6	15·4
Ile	214	215	221	229	224	232	242	244
Leu	803	492	678	763	806	1032	906	934
Lys	99	207	184	149	138	126	91	73
Met	107	77	93	102	106	101	112	114
Phe	280	223	282	282	292	308	315	325
Tyr	239	200	229	237	246	259	267	268
Thr	173	246	206	186	173	179	185	191
Val	288	307	302	299	294	304	318	309
Arg	222	346	367	327	295	282	221	182
His	111	77	121	125	126	127	118	114
Ala	552	431	517	546	564	602	618	638
Asp	383	492	432	422	397	413	403	402
Glu	1286	669	1091	1246	1267	1347	1431	1447
Gly	156	384	249	194	164	164	154	150
Pro	466	337	412	446	454	499	518	544
Ser	235	307	274	258	246	251	272	252

[a] Rooney *et al.* (1972).

Table 5.35
Amino acid distribution in the protein (mg/g total N) of the fraction of 70 LH 201 grain removed by successive abrasion milling.[a]

	Whole grain	Total quantity of original grain removed (%)									
		3·9	8·4	12·6	13·9	15·9	20·3	22·4	27·5	30·7	42·6
Fraction No.	0	1	3	6	7	8	11	12	15	17	24
Prot. (N × 6·25) (%) (DWB)	10·8	7·6	16·2	24·7	25·6	25·6	22·1	21·2	18·4	16·2	11·5
Ile	266	214	229	258	246	242	258	256	257	258	267
Leu	931	504	653	870	846	849	937	927	939	951	978
Lys	126	261	232	158	139	130	116	116	97	101	89
Met	96	86	97	108	97	101	103	100	97	97	94
Phe	350	564	296	339	319	326	366	359	351	363	360
Tyr	279	213	241	272	259	261	317	284	282	285	285
Thr	218	259	236	221	206	199	196	211	202	192	193
Val	332	334	340	356	316	307	311	336	317	332	315
Arg	247	397	377	305	264	249	235	231	213	211	193
His	138	131	151	142	129	126	123	129	117	122	82
Ala	625	427	504	611	584	586	619	632	624	631	621
Asp	433	521	491	471	428	416	427	434	413	412	404
Glu	1466	768	1083	1416	1360	1339	1445	1466	1466	1481	1502
Gly	206	318	304	240	204	191	206	181	172	169	169
Pro	544	359	427	523	446	503	524	562	529	554	569
Ser	295	302	290	301	267	263	279	285	276	281	284

[a] Rooney *et al.* (1972).

Table 5.36
Yield and protein (N × 6·25) analyses of two corneous sorghums.[a]

	Konza	Msumbiji
Bran		
Yield (%)	18·0	15·6
Prot. (%)	9·9	13·2
% of total prot.	16·2	19·3
Granulars		
Yield (%)	75·3	82·0
Prot. (%)	11·6	10·2
% of total prot.	79·5	79·2
Flour		
Yield (%)	6·7	2·5
Prot. (%)	7·1	6·5
% of total prot.	4·3	1·5
Prot. in umilled grain (calculated to weighted average moisture content of mill fractions) (%)	10·9	10·1
(DWB) (%)	12·3	11·7

[a] Shepherd and Woodhead (1969–70).

Table 5.37
Physical characteristics, flour yield and protein content in roller-milled flour (Bühler laboratory mill) of 19 samples of sorghum.[a]

Designation	Seed coat colour	Endo. type[b]	100-seed wt (mg)	Flour colour	Total flour yield (break and reduction) (%)	% of total prot. in flour (N × 6·25)	% total prot. in flour × yield of total flour
3D × 59	White, dark brown patches	I	22·1	Lt. pink	48·1	34·6	1664
IS 3404	Tan, purple spots	I	19·1	Pink	47·5	31·2	1482
E 6952	White, reddish-brown spots	C	32·7	Lt. pink	40·7	27·6	1223
IS 9369	Reddish-brown	F	31·7	Lt. pink	50·7	33·6	1704
E 6955	White, red spots	C	32·1	Pink	37·1	27·1	1005
IS 2771	White, brown patches	C	13·6	Yel.-white	50·1	35·1	1759
E 6051	White, purple spots	F	25·5	Pink	55·6	39·6	2202
E 1236	Cream coloured	C	19·3	Yel.-white	40·1	27·7	1111
IS 2341	White, brown patches	F	27·3	Pink	54·9	40·7	2234
E 6900	Yellowish-brown, purple spots	C	22·5	Lt. pink	43·5	28·9	1257
E 2334	White, brown spots	F	30·9	Pink	52·3	39·2	2050
IS 9671	Grey, red patches	C	23·7	Pink	45·7	31·6	1444
E 253	White, red patches	I	22·6	Lt. pink	39·8	25·9	1031
IS 3565	Mixed white and brown	I	20·3	Pink	52·9	43·3	2291
IS 928	Greyish-purple	I	22·5	White	43·4	28·8	1250
E 6435	Yellow, brown patches	C	32·3	Lt. pink	36·6	20·2	739
E 6992	White, brown patches	F	23·7	Pink	50·8	41·9	2129
E 6954	White, purple spots	I	27·2	Pink	51·4	35·8	1840
IS 9886	White, brown patches	F	34·1	Pink	52·4	34·4	1803

[a] Shepherd *et al.* (1970–71).
[b] C—corneous, F—floury, I—intermediate.

Table 5.38
Proximate analysis of three sorghum hybrids.[a]

	Commercial No. 2 yellow hybrid mix	Yellow hybrid TE 77	White hybrid Funk G 766
Moisture (%)	12·3	11·8	9·3
Prot. (%)	8·6	9·1	9·1
Crude fat (%)	3·2	2·9	3·1
Crude fibre (%)	1·4	1·7	1·7
Ash (%)	1·3	1·1	1·2

[a] Anderson (1969b).

Table 5.39
Extraction rate and proximate analysis of three sorghums after laboratory roller-milling. Blank spaces indicate no values presented.[a]

	Commercial No. 2 yellow		Yellow TE 77		White Funk G 766	
	Flour I	Final flour	Flour I	Final flour	Flour I	Final flour
	(%)					
Extraction	51·0	68·4	43·4	62·7	48·2	66·0
Moisture	15·0	13·1	14·0	12·7	13·6	12·5
Prot.		8·5	7·0	9·4	5·9	8·1
Crude fat	0·9	0·9	0·9	0·9	1·1	1·2
Crude fibre			0·1	0·3	0·2	0·5
Ash	0·5	0·5	0·4	0·4	0·5	0·5

[a] Anderson (1969b).

Table 5.40
Raw material and products from roller-milling peeled grain sorghum hybrid TE 77.[a]

Fractions	Extraction % of original grain	Crude fat (%)	Crude fibre (%)	Ash (%)
Original grain		2·9	1·7	1·1
Peeled grain	88	2·9	0·9	1·0
Bran from peeling	4	3·8	36·7	2·0
Loss[b]	8			
Flour I	47	0·6	0·1	0·3
Shorts II	11	0·9	0·6	0·3
Shorts flour	22	1·4	0·5	0·5
Germ and bran from milling	8	19·7	10·6	6·6

[a] Anderson (1969b).
[b] Including starch released by peeling, some bran and solubles.

Table 5.41
Composition (DWB) of whole grain sorghums. Blank spaces indicate no values presented.[a]

	OK 612	RS 626	TE 77
Prot. (N × 6·25) (%)	11·6	11·5	11·7
Starch (%)	75·9	76·3	75·9
Fat (hexane-soluble material) (%)	3·3	3·1	3·4
Fibre (%)	1·9	1·8	
Ash (%)	1·3	1·2	

[a] Jones and Beckwith (1970).

Table 5.42
Proximate analysis of milled fractions[a] of sorghum.[b]

Fraction	Prot.	Fat	Fibre	Ash	Weight
			(%)		
Bran	14·6	13·3	7·5	4·2	
Shorts	17·0	1·0	1·4	0·4	
Flour	9·8	1·3	0·4	0·6	
Distribution of constituents in total grain					
Bran	22	72	74	62	18
Shorts	20	4	11	5	14
Flour	58	29	15	34	69

[a] Average of three hybrids.
[b] Jones and Beckwith (1970).

Table 5.43
Amino acid content (mg/g total N) of sorghum mill fractions.[a]

	Grain	Flour	Shorts	Bran
Ile	244	281	300	237
Leu	906	1019	1050	569
Lys	112	106	50	237
Met	81	75	69	87
Half-Cys	44	25	44	75
Phe	331	337	350	262
Tyr	287	312	306	231
Thr	219	219	219	262
Val	306	325	306	306
Arg	200	175	131	387
His	131	131	106	156
Ala	619	656	687	450
Asp	437	462	437	544
Glu	1556	1531	1694	1025
Gly	200	200	156	331
Pro	562	556	644	319
Ser	287	306	300	306
Ammonia	206	237	194	150

[a] Jones and Beckwith (1970).

Table 5.44
Solvent extraction of sorghum mill fractions. Blank spaces indicate no values presented.[a]

%of total N extracted by:	Bran[b]	Shorts[b]	Flour
H_2O	12	0·5	3
1% NaCl	10	0·8	3
t-BuOH (60% room temp.)	14	44	37
EtOH (60% room temp.)			3
t-BuOH (60% at 60°C)			35
EtOH (60% at 60°C)			40

[a] Jones and Beckwith (1970).
[b] The shorts and bran fractions of TE 77 were not extracted.

Table 5.45
Amino acid composition (mg/g total N) of sorghum extracts.[a]

	Flour				Shorts	Bran
	H_2O soluble	NaCl soluble	*t*-BuOH soluble	Residue	*t*-BuOH soluble	*t*-BuOH soluble
Ile	269	219	300	269	262	250
Leu	469	400	1200	931	1287	1012
Lys	394	294	6	131	0	69
Met	125	87	62	87	87	37
Half-Cys	87	181	25	75	0	0
Phe	244	256	400	344	400	337
Tyr	231	206	344	319	362	306
Thr	375	287	162	219	150	187
Val	375	331	312	319	319	306
Arg	431	675	6	225	87	119
His	118	156	56	125	56	75
Ala	462	375	775	569	644	600
Asp	650	494	406	350	387	406
Glu	887	856	1875	1456	1987	1587
Gly	437	400	69	150	50	144
Pro	375	281	625	681	687	525
Ser	300	356	256	275	231	256
Ammonia	87	94	219	200	212	269

[a] Jones and Beckwith (1970).

Table 5.46
Protein analysis of sorghum solids in extract from different milling fractions.[a]

	% protein (N × 6·25)		
Sol. in	Flour	Shorts	Bran
Water			
OK 612	12	5	52
RS 626	25	2	61
TE 77[b]	9		
NaCl			
OK 612	23	5	54
RS 626	20	3	61
TE 77[b]	13		
t-BuOH			
OK 612	95	91	43
RS 626	90	91	44
TE 77[b]	90		

[a] Jones and Beckwith (1970).
[b] The shorts and bran fractions of TE 77 were not extracted.

Table 5.47
Yield, protein and fat content (DWB) of fractions obtained from dry-milling sorghum grain.[a]

Fraction	Descr.	Fraction of whole grain (%)	Prot. in fraction (N × 6·25) (%)	Fat in fraction (%)
1	Bran and germ	19·3	12·8	10·4
2	Fines	12·0	8·0	1·9
3	Flour	12·3	6·5	0·5
4	Flour	3·0	7·2	1·1
5	Endo.	4·7	8·9	0·4
6	Endo.	14·8	12·1	1·0
7	Flour	7·2	8·6	1·2
8	Flour	6·8	11·1	1·3
9	Endo.	8·9	14·2	0·6
10	Endo.	2·1	15·7	1·1
11	Endo.	3·7	17·3	0·8
12	Endo.	2·2	19·7	1·1
13	Endo.	1·0	18·8	1·5
14	Endo.	1·9	20·6	1·9

[a] Shoup *et al.* (1969).

Table 5.48
Amino acid composition (mg/g total N) (DWB) and distribution in protein of sorghum grain and milled products.[a]

AAs	Sorghum grain	Bran and germ	Endo. fractions			
		1	3	8	10, 11	12, 13, 14
Protein (N × 6·25) (%)[b]	10·1	10·3	5·7	9·8	15·1	18·1
Ile	254	206	256	287	281	276
Leu	832	452	852	990	1017	1005
Lys	131	277	114	81	60	58
Met	109	109	114	109	98	96
Cys	127	126	135	119	107	1081
Phe	338	239	328	362	366	364
Tyr	273	168	252	274	262	261
Thr	208	221	235	199	190	182
Trp	84	131	197	117	71	83
Val	310	308	285	330	321	330
Arg	243	435	196	181	149	156
His	140	147	136	135	126	124
Ala	605	412	635	684	701	716
Asp	437	502	421	404	401	397
Glu	1395	832	1387	1582	1622	1571
Gly	188	319	192	157	138	144
Pro	524		561	560	574	566
Ser	298	278	291	285	276	279
Ammonia	222	132	239	228	226	241

[a] Shoup *et al.* (1969).
[b] 12% moisture basis.

Table 5.49
Rat growth data.[a]

Diet	Prot. (N × 6·25)[b]		Without Lys supp. (0–4 weeks)		% of NRC requirement, adjusted[c] (0–4 weeks)			With Lys supp. (5–6 weeks)	
	Fraction (%)	Diet (%)	Av. gain (g)	PER	Lys (%)	Met (%)	Val (%)	Av. gain (g)	PER
(A) Bran and germ	10·3	9·3	20·4	2·1	100	89·3	145·4		
(B) Sorghum grain	10·1	9·3	5·2	1·0	44·0	88·1	135·0	24·4	2·6
(C) Endo. fraction 3	5·7	5·6	1·8	0·6	34·6	88·2	107·0	11·8	2·6
(D) Endo. fraction 8	9·8	9·3	1·8	0·4	25·5	89·3	145·4	24·0	2·8
(E) Endo. fractions 10, 11	15·1	9·3	0·6	0·1	20·2	78·6	145·4	21·4	2·6
(F) Endo. fractions 12, 13, 14	18·1	9·3	0·6	0·2	21·0	78·6	148·4	23·4	2·9

[a] Shoup *et al.* (1969).
[b] 12% moisture basis.
[c] National Research Council requirements based on 20% dietary protein; adjusted requirements based on 9·3/20 of NRC requirement for diets (A), (B), (D), (E), (F) and 5·6/20 of NRC requirement for diet (C).

Table 5.50
Supplemental levels of Lys, calculations on "as-is" moisture basis.[a]

Diet	Amount of sorghum fraction in diet	Lys in milled fraction	Lys in diet (0–4 weeks)	Lys supp.	Lys in diet (5–6 weeks)	Adjusted NRC requirement[b]
			(%)			
(A)	92·2	0·47	0·43		0·43	0·43
(B)	88·1	0·21	0·19	0·24	0·43	0·43
(C)	94·5	0·10	0·10	0·16	0·26	0·26
(D)	94·5	0·12	0·12	0·31	0·43	0·43
(E)	62·0	0·14	0·09	0·34	0·43	0·43
(F)	52·0	0·17	0·09	0·34	0·43	0·43

[a] Shoup *et al.* (1969).
[b] National Research Council requirements based on 20% dietary protein; adjusted requirements based on 9·3/20 of NRC requirement for diets (A), (B), (D), (E), (F) and 5·6/20 of NRC requirement for diet (C).

Table 5.51
Yield and composition of sorghum fractions[a] from the brush peeler.[b]

Fractions	Fraction No.	Yield (% of whole grain)	Prot.	Fat	Fibre	Ash
			(% of fraction)			
Paymaster Kiowa (67 2)						
Whole grain		100	10·1	2·3	2·0	1·4
Fines	1	11·7	9·5	6·4	5·9	3·1
Bran (aspirated)	2	4·7	8·9	6·4	7·9	3·2
Peeled grain	3	83·7	10·1	2·7	1·0	1·2
Frontier 400C (68 7)						
Whole grain		100	10·8	2·8	1·9	1·4
Fines	1	11·9	9·1	3·6	5·1	1·8
Bran (aspirated)	2	4·8	7·2	5·6	9·4	2·4
Peeled grain	3	83·3	11·7	3·5	1·0	1·4

[a] 12% moisture basis.
[b] Shoup *et al.* (1970b).

Table 5.52

Yield, protein (N × 6·25), fat, and ash content[a] of sorghum fractions from a Miag Multomat experimental flour mill.[b]

Fractions	Fraction No.	Paymaster Kiowa (67 2) Peeled				Frontier 400C (68 7) Peeled				Frontier 400C (68 7) Nonpeeled			
		Yield (% of whole grain)	Prot.	Fat	Ash	Yield (% of whole grain)	Prot.	Fat	Ash	Yield (% of whole grain)	Prot.	Fat	Ash
			(% of fraction)				(% of fraction)				(% of fraction)		
Peeled grain	3	83·7				83·3				100			
1 Break	4	2·5	3·9	0·9	0·4	3·0	4·6	1·4	0·4	4·2	4·1	1·7	0·4
2 Break	5	1·3	4·0	1·0	0·3	1·5	4·6	1·0	0·3	1·9	4·7	1·5	0·4
Grader	6	1·5	3·8	1·0	0·3	1·0	4·6	0·8	0·3	1·2	4·6	1·5	0·5
3 Break	7	2·3	4·9	1·0	0·4	2·3	5·5	1·1	0·4	2·9	5·4	1·7	0·5
1 Middlings	8	8·0	5·9	0·6	0·3	5·9	6·6	0·7	0·2	6·6	6·8	0·9	0·3
2 Middlings	9	2·8	5·7	0·6	0·3	6·5	6·5	0·9	0·3	5·7	6·4	1·0	0·3
3 Middlings	10	4·3	6·5	0·3	0·3	8·5	7·3	0·8	0·3	9·0	7·4	1·0	0·4
1 M Redust	11	2·0	7·1	0·5	0·3	1·7	8·0	1·1	0·3	1·7	7·9	1·0	0·4
4 Middlings	12	3·0	7·4	0·5	0·3	9·1	8·0	0·7	0·3	9·5	8·1	0·9	0·3
5 Middlings	13	19·1	8·8	0·2	0·3	16·1	12·5	1·4	0·6	18·4	12·1	1·5	0·7
Red Dog	14	19·1	12·8	0·7	0·3	6·2	20·0	1·3	0·7	7·6	18·6	2·4	1·1
Reduction shorts	15	7·6	12·5	1·5	0·7	8·8	16·8	0·6	0·4	9·9	15·7	1·6	0·6
Bran and germ	16	10·3	13·4	13·4	5·5	12·7	14·8	14·8	6·2	21·4	13·5	13·5	5·2

[a] 12% moisture basis.
[b] Shoup *et al.* (1970b).

Table 5.53
Amino acid composition (mg/g total N) (DWB) of milled sorghum grain products.[a]

Amino acid	Whole grain	Fines bran	Bran	Peeled grain	1BK	2BK	Grader	3BK	1M	2M	3M	1M Red	4M	5M	Red Dog	Redn shorts	Germ and bran
Prot. (%) (N × 6·25)	12·3	10·4	8·2	12·5	5·2	5·2	5·2	6·2	7·5	7·4	8·3	9·0	9·1	14·2	22·7	19·1	16·8
Ile	252	242	207	252	204	223	241	239	257	251	243	251	247	263	271	281	206
Leu	842	796	567	909	634	704	742	812	940	910	923	915	943	964	1086	1126	582
Lys	115	163	261	105	162	152	164	231	104	99	94	90	75	86	67	59	272
Met	102	91	92	95	95	109	110	109	102	104	98	99	106	94	91	82	121
Cys	108	116	126	107	133	128	147	143	123	119	119	120	113	109	96	111	119
Phe	310	316	254	328	240	274	299	302	342	332	321	321	331	326	377	391	265
Tyr	247	237	179	247	187	210	239	238	265	259	246	241	255	253	286	300	200
Thr	199	224	231	194	206	201	219	211	203	206	191	188	192	184	184	198	199
Trp	ND[b]																
Val	278	329	314	309	302	292	326	311	316	174	299	306	345	312	321	331	316
Arg	206	255	302	198	226	234	259	212	199	179	166	179	154	189	160	154	414
His	123	132	137	155	136	130	143	134	138	124	126	125	117	146	127	126	169
Ala	584	563	439	606	449	466	497	538	627	606	604	612	622	660	720	728	449
Asp	409	457	463	396	361	386	406	440	398	396	382	375	425	316	414	419	441
Glu	1418	1247	909	1394	1008	1123	1166	1245	1479	1393	1436	1426	1437	1485	1679	1727	979
Gly	176	242	286	166	214	195	218	196	177	173	159	156	133	143	142	135	242
Pro	519	481	359	520	434	482	494	489	550	537	569	533	546	562	568	587	351
Ser	274	299	297	279	274	250	279	284	291	289	275	269	275	272	299	302	267
Ammonia	185	166	205	177	267	143	154	217	184	194	176	186	189	214	219	219	139
Recovery (g AA N recovered/100 g Kjeldahl N)	90·5	91·9	88·5	91·2	88·2	82·1	87·2	93·9	94·5	90·6	89·9	90·3	90·5	95·2	100·4	101·7	87·5

[a] Shoup *et al.* (1970b).
[b] Not determined.

Table 5.54
Yield and proximate composition of fractions obtained from dry-milling sorghum Paymaster Kiowa (PK) and Frontier 400C (F), DWB. Blank spaces indicate no values presented.[a]

Fr. descr.	Yield (% of whole grain)		Prot. (% of fraction)		Fat (% of fraction)		Fibre (% of fraction)		Ash (% of fraction)	
	PK	F	PK	F	PK	F	PK	F	PK	F
1 Bran and germ	27·3	25·3								
2 Fines	18·9	19·6								
3 Floury endo.	12·5	12·2	7·4	6·8	0·9	1·0	0·9	0·8	0·5	0·5
4 Floury endo.	5·1	5·1	9·5	0·6	1·0	0·9	1·1	1·4	0·5	0·5
5 Floury endo.	9·9	10·3	9·9	8·8	1·0	1·1	0·7	1·1	0·5	0·5
6 Floury endo.	8·4	8·2	14·0	12·6	0·9	1·1	0·7	1·1	0·5	0·5
7 Corneous endo.	14·3	16·6	16·0	15·5	1·1	1·2	1·1	1·3	0·6	0·5

[a] Shoup *et al.* (1970c).

Table 5.55
Amino acid distribution in protein (mg/g total N) of Paymaster Kiowa (PK) and Frontier 400C (F) sorghum milled fractions and casein. Blank spaces indicate no values presented.[a]

	Fraction 3		Fraction 4		Fraction 5		Fraction 6		Fraction 7		Casein
	PK	F	PK	F	PK	F	PK	F	PK	F	
Prot. (N × 6·25) (%) (DWB)	7·4	6·8	9·5	10·6	9·9	8·8	14·0	12·6	16·0	15·5	94·3
Ile	249	239	249	257	269	261	264	266	269	278	362
Leu	852	809	971	1033	1001	1029	1018	1020	1025	1052	655
Lys	107	112	83	84	81	96	62	72	66	72	561
Met	101	107							86	97	137
Cys	111	133							92	111	33
Phe	321	317	331	341	362	337	364	371	376	382	352
Tyr	255	251	259	276	277	278	286	277	281	283	396
Thr	193	197	196	204	197	211	187	201	189	209	295
Val	267	304	274	292	301			304	316	337	437
Arg	178	188	163	169	144	192	134	159	142	164	256
His	124	131	127	135	122	105	113	136	121	128	201
Ala	584	557	651	676	666	676	701	721	684	887	207
Asp	383	384	371	389	396	395	402	373	399	461	502
Glu	1358	1292	1479	1553	1558	1569	1577	1605	1599	1691	1697
Gly	163	174	156	160	147	172	134	150	136	160	126
Pro	504	511	558	575	564	580	592	589	554	679	789
Ser	278	269	284	307	291	299	288	294	289	307	397
Ammonia	184	172	192	211	172	202	184	197	197	181	116
N recovery	83·3	85·3	92·3	97·5	92·5	95·9	91·7	97·0	94·4	99·8	103·4

[a] Shoup *et al.* (1970c).

Table 5.56
Designation and proximate analysis of sorghum experimental rat diets.[a]

Diet	Hybrid	Milling fraction	AA supp.	Moisture	Prot. (N × 6·25)	Fat	Ash
			(%)				
(A)	Paymaster Kiowa	3		11·2	5·6	4·2	3·5
(B)	Paymaster Kiowa	3	+0·33 Lys +0·12 Met	11·3	5·8	4·3	3·4
(C)	Paymaster Kiowa	7		10·6	9·8	4·3	3·4
(D)	Paymaster Kiowa	7	+0·64 Lys +0·25 Met	10·7	10·4	4·3	3·4
(E)	Hybrid Frontier 400C	3		10·7	5·4	4·3	3·4
(F)	Frontier 400C	3	+0·33 Lys +0·12 Met	10·8	5·8	4·4	3·4
(G)	Frontier 400C	7		10·5	10·0	4·2	3·4
(H)	Frontier 400C	7	+0·64 Lys +0·25 Met	10·8	10·7	4·0	3·5
(I)	Casein			10·0	5·8	4·0	3·2
(J)	Casein			9·9	10·0	4·0	3·2

[a] Shoup *et al.* (1970c).

Table 5.57
Rat weight gains, PER, and % Lys and Met requirements applied by Paymaster Kiowa (PK) and Frontier 400C (F) sorghum diets.[a]

Diet	Prot. in diet (%)	Hybrid	Milled fraction	% requirement met[b] Lys	Met	Gain (g)	PER[c] (g)
(A)	5·6	PK	3 (floury)	23	71	5·5	0·5
(B)	5·8	PK	3 (floury)	98	114	35·5	2·6
(C)	9·8	PK	7 (corneous)	14	55	2·5	0·2
(D)	10·4	PK	7 (corneous)	99	105	75·7	2·3
(E)	5·4	F	3 (floury)	23	76	8·8	0·8
(F)	5·8	F	3 (floury)	100	118	36·7	2·5
(G)	10·0	F	7 (corneous)	15	65	4·5	0·3
(H)	10·7	F	7 (corneous)	100	110	91·7	2·5
(I)	5·8	PK	Cassein (control)	111	49	37·0	2·5
(J)	10·0	F	Cassein (control)	112	51	62·2	2·3

[a] Shoup *et al.* (1970c).
[b] National Research Council requirements are based on 12% dietary protein. Percentages shown are adjusted 5·6/12 of the requirements for diets (A), (B), (E), (F) and (I) and 10/12 of requirements for diets (C), (D), (G), (H) and (J).
[c] Grammes of gain/g of prot. consumed, corrected as follows:
—for the low prot., floury endo. diets (A), (B), (E) and (F) used the factor 2·5 (assumed PER casein) divided by 2·5 (PER of the controlled casein diet (I)).
—for the high prot., corneous endo. diets (C), (D), (G) and (H) used the factor 2·5 (assumed PER for casein) divided by 2·3 (PER of the controlled casein diet (J)).
Note: PER presumes isonitrogenous diets. Therefore diets (A), (B), (E), (F) and (I) are comparable; (C), (D), (G), (H) and (J) are comparable.

Table 5.58
Chemical composition of milled Nigerian sorghum. Blank spaces indicate no values presented.[a]

	Commercially milled			Laboratory milled at 25% moisture	
	White (mori) flour	Yellow (kaura) flour	Middlings	White (mori) flour	Farafara flour
Moisture (loss at 100°C) (%)	12·7	12·8	12·4	25 (Assumed)	25 (Assumed)
Prot. (N × 6·25) (%)	7·7	6·9	10·9	7·9	7·9
Crude fibre	trace	trace	5·5		
P (total) (mg/100 g)	218	189	350		
Phytin P (mg/100 g)	34·6	27·7	30·4		
Vitamins (mg/100 g)					
Thiamine	0·23	0·20	0·54	0·26	0·24
Niacin	3·15	3·30	7·74	2·0	2·3
Riboflavin	0·08	0·10	0·15	0·05	0·05

[a] Raymond *et al.* (1954).

Table 5.59
Chemical composition of polished and unpolished sorghum.[a]

	Sorghum sample	
	Polished	Unpolished
Prot. (%)	7·8	9·5
Crude fibre (%)	0·4	2·1
Ca (mg/100 g)	21	39
P (mg/100 g)	124	276
Thiamine (mg/100 g)	0·21	0·35

[a] Narayana Rao *et al.* (1958).

Table 5.60
N, Ca and P metabolism (daily averages in mg) of rats fed on diets based on polished and unpolished sorghum.[a]

	Polished	Unpolished
N		
Intake	155·2	180
Urinary excretion	80·4	80·8
Faecal excretion	39·2	53·7
Balance	35·7	45·6
Ca		
Intake	6·53	9·26
Urinary excretion	0·69	0·86
Faecal excretion	2·48	4·85
Balance	3·36	3·55
P		
Intake	16·87	31·57
Urinary excretion	2·61	9·18
Faecal excretion	7·06	12·50
Balance	7·20	9·89

[a] Narayana Rao *et al.* (1958).

Table 5.61
Protein and amino acid composition (mg/g total N) of four samples of East African sorghums.[a]

	Konza		Msumbiji		Lionja		Karachi	
	Whole	Polished	Whole	Polished	Whole	Polished	Whole	Polished
Prot. (N × 6·25) (%)[b]	10·5	9·4	10·4	9·9	9·8	9·3	9·9	9·7
Ile	269	250	281	319	275	237	275	262
Leu	1037	906	1031	1044	962	987	1219	894
Lys	112	106	150	131	150	125	144	112
Phe	350	350	381	394	350	306	356	331
Tyr	287	281	281	262	250	231	206	250
Thr	200	194	231	225	212	175	206	194
Val	281	581	362	356	344	306	325	312
Arg	206	212	269	269	262	219	250	244
His	125	112	212	137	156	125	137	119
Ala	681	631	700	731	644	575	650	619
Asp	444	412	475	456	469	412	450	412
Glu	1719	1587	1575	1894	1425	1294	1687	1644
Gly	187	194	225	212	212	181	200	181
Pro	556	494	631	925	612	550	500	512
Ser	262	287	350	325	300	250	300	275
Ammonia	162	181	194	175	175	169	162	156

[a] Shepherd *et al.* (1970–71).
[b] 12% moisture basis.

Table 5.62
Chemical composition (DWB) of products of sorghum milled in a commercial Dandekar type rice mill. Blank spaces indicate no values presented.[a]

	Protein (N × 6·25)	Fat (ether extract)	Fibre	Ash	Ca (mg/100 g)	P (mg/100 g)
	(%)					
Unpolished sorghum	11·3	4·2	1·8	1·7	49	303
Polished sorghum						
Cone[b] I	11·0	3·8	1·0	1·5	45	270
II	10·7	3·2	0·8	1·3	39	276
III	10·5	2·2	0·7	1·1	39	210
Big brokens	10·5	1·8	0·8	1·2		231
Small brokens	10·4	4·7	0·8	1·2		190
Bran						
Cone I	11·6	8·4	11·9	5·7		517
II	13·0	9·8	6·4	5·2		931
III	13·7	2·8	4·8	4·8		718
Husk	13·0	8·0	8·6	7·2		1383

[a] Viraktamath *et al.* (1971).
[b] The degree of polish from the three cones was: I 3·7%, II 7·0%, III 9·5%.

Table 5.63
Composition of sorghum fractions from roller mill (DWB). Blank spaces indicate no values presented.[a]

	Recovery	Prot. (N × 6·25)	Fat
		(%)	
Fine soji (semolina)	33·5	10·9	2·1
Maida (fine flour)	33·7	6·9	2·5
Atta (whole grain flour)	26·5	10·3	8·8
Bran	4·1	11·9	1·4
Sweepings	0·8		
Losses	1·4		

[a] Viraktamath *et al.* (1971).

Table 5.64
Composition of pearled grain sorghum kernels.[a]

	Kernel wt removed (%)				
	0	6·4	11·2	16·1	39·0
Prot. (%)	9·6	9·4	9·5	9·1	6·9
Oil (%)	3·4	3·0	2·5	2·0	0·6
Fibre (%)	2·2	1·3	1·0	0·8	0·7
Ash (%)	1·5	1·2	1·1	0·9	0·4

[a] Hahn (1969).

Table 5.65
Yield and proximate analysis of dry-milled nonwaxy and waxy sorghum.[a]

	Yield	Prot.	Oil	Fibre	Ash
			(%)		
Nonwaxy sorghum					
Whole grain	100	9·6	3·4	2·2	1·5
Grit	67	9·6	0·6	0·8	0·5
Bran	12	8·9	5·5	8·6	2·4
Germ	11	15·1	20·0	2·6	8·2
Fines	10	7·1	2·4	1·3	1·0
Waxy sorghum					
Whole grain	100	9·8	3·2	2·2	1·6
Grit	65	9·8	1·0	0·8	0·6
Bran	15	7·1	4·8	11·3	2·1
Germ	14	15·3	18·7	3·0	8·5
Fines	6	6·5	2·3	1·5	1·0

[a] Hahn (1969).

Table 5.66
Proximate composition and mineral content of South African sorghum and its milling products. Blank space indicates no value presented.[a]

	Whole grain	Straight-run meal	Household meal
Moisture (%)	12·0	11·9	11·9
Prot. (N × 6·25) (%)	10·4	9·6	9·5
Fat (%)	3·4	2·7	2·7
CBH[b] (by difference) (%)	70·9	72·2	73·3
Fibre (%)	2·0	2·3	1·5
Ash (%)	1·3	1·3	1·1
Minerals (mg/100 g)			
Ca	32	34	38
P	232	238	225
Fe	7	25	9
Mg	122		115

[a] Crawford *et al.* (1942).
[b] Carbohydrate.

Table 5.67
The thiamine content DWB of milled sorghum products.[a]

	Moisture content (%)	Thiamine content (mg/100 g)
Coarse ground meal	11·4	0·33
Straight-run meal	9·8	0·31
Household meal	11·1	0·27
Refined breakfast meal	12·3	0·21
Bran	10·9	0·43

[a] Goldberg *et al.* (1946).

Table 5.68
Chemical composition of kodo millet, whole and dehusked, in comparison with wheat (DWB).[a]

	Kodo millet		Wheat
	Whole grain	Dehusked	
Prot. (N × 6·25) (%)	12·0	13·1	14·7
Fat (ether extract) (%)	4·7	1·5	1·6
Fibre (%)	11·3	0·4	1·5
CBH[b] (by difference) (%)	67·0	83·8	80·6
Mineral matter (%)	5·0	1·1	1·7
Ca (mg/100 g)	56	40	52
P (mg/100 g)	321	137	359
Fe (mg/100 g)	7	2	6
Thiamine (mg/100 g)	0·45	0·17	0·55

[a] Kadkol *et al.* (1954b).
[b] Carbohydrate.

Table 5.69

Protein values of the fractions of pearl millet obtained by traditional mortar and pestle pounding, by the SOTRAMIL milling process and by the SEPIAL milling process. Blank spaces indicate no values presented.[a]

	Traditional mortar and pestle pounding			SOTRAMIL milling process		SEPIAL milling process			
	(A) Whole grain	(B) Semolina	(C) Flour	(D) Whole grain	(E) Flour	(F) Whole grain	(G) "Peeled" grain	(H) "Decorticated" grain	(I) "Protein-rich fraction"
% of whole grain	100	55	30	100	75	100	92·5	84	8·5
Biochemical composition									
Cellulose and lignin (%)	4·8	2·1	3·8	5·1	1·1	4·1	2·7	2·6	6·3
Pentosans (%)						3·2	2·8	2·2	11·9
Prot. (N × 5·83) (%)	11·0	9·5	11·2	10·8	8·4	9·0	9·2	9·4	14·0
Lys (% of total prot.)	2·59	1·64	2·71	2·84	2·62	3·27	2·87	2·86	4·75
Rations									
Cereal product (%)	92	92	92	92	92	92	92	92	61·5
Protein content (%)						8·3	8·4	8·7	8·4
Prot. efficiency of rations									
Ration consumed (g/rat/day)	8·7	7·2	8·0	9·4	7·9	8·0	7·8	7·8	9·7
Growth rate (g/rat/day)	1·5	0·6	1·4	1·7	1·1	0·9	0·8	0·8	1·5
PER	1·7	1·0	1·7	1·8	1·7	1·3	1·2	1·1	1·8
	(100)	(56)	(102)	(100)	(96)	(100)	(95)	(88)	(139)
Live wt of liver (%)	5·2	4·3	4·2	4·3	4·4	4·6	4·3	4·3	4·4
Liver prot. (mg)									
—/organ	98·1	60·2	83·2	101·5	85·5	72·0	67·0	66·6	95·1
—/g	18·7	14·0	18·8	23·7	19·3	25·3	26·3	26·1	27·8
	(100)	(75)	(106)	(100)	(82)	(100)	(104)	(103)	(110)

[a] Adrian *et al.* (1975).

Table 5.70
Average chemical composition of the pearl millet fractions obtained by SEPIAL milling process. Blank spaces indicate no values presented.[a]

	(F) Whole grain	(G) "Peeled" grain	(H) "Decorticated" grain	(I) "Protein-rich fraction"
% of whole grain	100	92·5	84·0	8·5
Biochemical composition				
Ash (%)	1·7	1·6	1·3	4·1
Cellulose and lignin (%)	3·4	3·0	2·2	4·7
Prot. (N × 5·83) (%)	9·1	8·7	8·4	12·4
Lys (% of total prot.)	2·70	2·80	2·20	4·40
Met (% of total prot.)	2·05	2·25	2·20	1·60
Lipid (%)	6·6			14·2
Niacin (mg/100 g)	1·65	1·65	1·10	6·60

[a] Adrian *et al.* (1975).

Table 5.71
Digestibility (dig.) of pearl millet fractions obtained by the SEPIAL milling process.[a]

	(J) Whole grain	(K) "Decorticated" grain	(L) "Prot.-rich fraction"
Rations			
Cereal product (%)	65	92	45
Cellulose and lignin (%)	2·7	2·5	2·8
Pentosans (%)	2·1	2·5	5·3
Prot. (%)	6·9	9·3	7·2
Digestibility			
Ration consumed (g/rat/day)	12·9	15·6	15·2
Faeces (g/rat/day)	0·8	0·9	2·2
Total dig. (%)	93·7	94·3	85·8
	(100)	(101)	(92)
Prot. dig. (%)	84·7	88·6	74·0
	(100)	(105)	(87)

[a] Adrian *et al.* (1975).

Table 5.72
Digestibility (dig.) of pearl millet rations in relation to the nondigestible content of the ration.[a]

	Control ration	Experimental rations[b]		
		Whole grain	"Decorticated" grain	"Prot.-rich fraction"
Composition of rations				
Cellulose and lignin (%)	0	2·7	2·5	2·8
Pentosans (%)	0	2·1	2·5	5·3
Total dig.				
In %	96·1	93·8	94·3	85·8
Diff. between control and exp. ration	0	2·3	1·8	10·3
Prot. dig.				
In %	94	84·7	88·6	74·0
Diff. between control and exp. ration	0	9·3	5·4	20·0
Decr. in dig. expressed as % of:				
Cellulose and lignin:				
Total dig.		0·8	0·9	3·6
Prot. dig.		3·5	3·7	7·0
Pentosans:				
Total dig.		1·1	0·7	1·9
Prot. dig.		4·4	2·1	3·7

[a] Adrian *et al.* (1975).
[b] Presented in Table 5.71.

Table 5.73
Effect of milling on nutrients in samples of pearl millet (DWB).[a]

	Extraction rate (%)									
	Sample 1		Sample 2						Home prepared meal	Home prepared porridge
	100	76·5	100	95·7	90·1	85·2	80·0	75·1		
Prot. (%)	16·8	13·7	13·3	13·3	13·2	12·2	11·6	11·0	14·8	14·9
Fat (%)	5·7	4·4	4·4	4·3	3·9	3·7	3·2	2·2	3·2	2·7
Fibre (%)	4·5	2·0	1·3	1·1	0·7	0·6	0·5	0·4	1·6	2·0
Ca (mg/100 g)	38·7	10·4	18·0	20·0	8·4	8·0	10·0	8·4	22·2	30·9[b]
Fe (mg/100 g)	6·2	2·8	10·7	5·1	3·9	3·9	3·5	3·5	4·9	14·4
Thiamine (mg/100 g)	0·66	0·45	0·58	0·40	0·40	0·35	0·32	0·28	0·29	0·46
Niacin (mg/100 g)	2·47	0·86	1·16	1·44	1·01	0·82	0·77	0·65	1·63	1·42
Riboflavin (mg/100 g)	0·26	0·12	0·13	0·13	0·10	0·09	0·07	0·06	0·21	0·07

[a] De Wit and Schweigart (1970).
[b] Cooked in a clay pot.

Table 5.74
Protein, oil and ash contents of foxtail millet fractions of bran of mesh size −40 +80.[a]

	Hand-pounded	Milled
Oil (%)	16·2	16·7
Prot. (N × 6·25) (%)	12·6	14·4
Total ash (%)	11·5	11·5
Insol. ash (%)	4·2	4·1

[a] Thirumala Rao (1969).

Table 5.75
Protein, calcium and phosphorus contents of the whole grain, husk and endosperm of four strains of finger millet (mean with ranges in parentheses (DWB)).[a]

	Whole grain	Husk	Endo. (after decanting supernatant)	Dried solids from supernatant
% of whole grain	100	13·4 (12·9–14·4)	81·3 (80–82)	5·7 (5·4–6·0)
Prot. (N × 6·25) (%)	7·0 (6·7–7·7)	14·8 (14·0–16·7)	3·2 (2·4–3·9)	45·5 (37·6–51·4)
As % prot. in whole grain		28·2 (26·7–29·8)	36·7 (29·2–46·6)	36·9 (29·8–44·2)
Ca (mg/100 g)	345 (337–382)	1254 (1188–1373)	57·8 (52·5–70·4)	2234 (2050–2440)
As % Ca in whole grain		48·8 (46·4–50·2)	13·6 (12·7–15·1)	36·4 (34·4–38·7)
P (mg/100 g)	237 (186–279)	246 (177–300)	84·2 (74·5–93·1)	2437 (1500–3210)
As % P in whole grain		13·9 (12·0–15·8)	29·8 (23·3–40·9)	57·6 (43·5–63·8)

[a] Kurien *et al.* (1959).

Table 5.76
Composition of fractions from dry-milling and wet-processing of finger millet. Blank spaces indicate no values presented.[a]

	Whole grain	Fraction I	Fraction II	Combined fractions	Fraction from wet-processing of residue
Prot. (N × 6·25) (%)	7·2	3·6	9·1	6·1	8·4
Crude fibre (%)	2·6	0·2	0·6	0·4	0·3
Ca (mg/100 g)	359	215	267	230	133
P (mg/100 g)	211	36	237	127	338
Thiamine (mg/100 g)	0·55			0·72	

[a] Kurien and Desikachar (1962).

Table 5.77
Percentage extraction of nutrients by wet-processing and dry-milling of finger millet. Blank space indicates no value presented.[a]

	Wet-processing	Dry-milling	Dry-milling followed by wet-processing of residue
Total flour yield	81	70	82·5
Prot.	37	54	74·2

Table 5.77 *cont.*

	Wet-processing	Dry-milling	Dry-milling followed by wet-processing of residue
Ca	14	45	49·5
P	30	42	62·2
Thiamine	13	91	

[a] Kurien and Desikachar (1962).

Table 5.78
Chemical composition of raw and refined finger millet flours (DWB).[a]

	Whole flour	Refined flour
Dry matter (%)	88·2	87·3
Prot. (N × 6·25) (%)	8·0	6·8
Fat (ether extract) (%)	1·7	1·1
CBH[b] (by difference) (%)	83·5	89·9
Crude fibre (%)	3·6	1·0
Ash (%)	3·2	1·2
Ca (mg/100 g)	388	260
P (mg/100 g)	261	160

[a] Kurien and Doraiswamy (1967).
[b] Carbohydrate.

Table 5.79
Apparent digestibility, N, Ca and P balance of boys fed whole and refined finger millet flour diets.[a]

	Whole finger millet flour diet	Refined finger millet flour diet
N balance (g)	1·3	1·6
Apparent dig. of prot. (%)	56	64
Ca balance (g)	224	175
P balance (g)	222	223

[a] Kurien and Doraiswamy (1967).

Table 5.80
Flour yield and chemical composition of whole finger millet grain and laboratory (Bühler) milled flours from various conditioning treatments (DWB).[a]

	Whole grain	Treatment 1	2	3	4	5
Flour yield (%)		64	72	67	62	53
Crude prot. (N × 6·25) (%)	8·8	4·5	3·9	3·9	3·9	3·8
Ether extract (%)	1·8	0·4	0·3	0·3	0·3	0·3
CBH[b] (by difference) (%)	82·6	92·2	93·8	94·0	94·2	94·3
Crude fibre (%)	3·6	1·3	0·7	0·6	0·4	0·4
Ash (%)	3·2	1·5	1·3	1·2	1·2	1·2
Ca (mg/100 g)	468	286	242	245	238	244
P (mg/100 g)	273	90	47	45	48	55
Thiamine (mg/100 g)	0·63	0·29	0·29	0·31	0·31	0·35
Reflectance reading	47	65	75	78	80	80

[a] Kurien and Desikachar (1966).
[b] Carbohydrate.

Table 5.81
Chemical composition of individual fractions obtained by (Bühler) laboratory milling finger millet, treatment 3 (DWB). Blank spaces indicate no values presented.[a]

	Break rolls			Reduction rolls			Residue	
	1	2	3	1	2	3	Shorts	Husk
Crude prot. (N × 6·25) (%)	2·5	3·4	5·5	4·1	5·3	8·1	13·8	15·9
Ether extract (%)	0·2	0·2	0·3	0·3	0·3	0·3	1·9	1·1
CBH[b] (by difference) (%)	95·9	95·1	92·4	93·7	92·3	89·0	73·7	66·7
Crude fibre (%)	0·5	0·6	0·8	0·7	0·8	0·9	5·7	9·5
Ash (%)	0·9	0·7	1·0	1·2	1·3	1·7	4·9	6·8
Ca (mg/100 g)	168	159	212	252	308	342	762	709
P (mg/100 g)	36	41	69	47	68	125	482	935
Thiamine (mg/100 g)	0·23	0·33	0·58	0·30	0·37	0·49	0·50	0·45
Reflectance reading	79	79	75	78	79	63		

[a] Kurien and Desikachar (1966).
[b] Carbohydrate.

Table 5.82
Chemical composition of flour recovered by wet-processing of shorts and husk of finger millet.[a]

	Processing	
	Plain	Acidified
Yield		
As % of starting material	22·0	22·5
As % of whole grain	7·5	7·6
Crude prot. (N × 6·25) (%)	10·6	11·8
Ether extract (%)	4·0	4·4
CBH[b] (by difference) (%)	82·5	81·0
Crude fibre (%)	1·4	1·5
Ash (%)	1·5	1·3
Ca (mg/100 g)	204	115
P (mg/100 g)	779	199
Reflectance reading	61	63

[a] Kurien and Desikachar (1966).
[b] Carbohydrate.

Table 5.83
Proximate analysis, Ca and P of whole, refined and composite finger millet flours (DWB).[a]

	(A) Whole meal flour	(B) Refined flour	(C) Composite flour
Dry matter (%)	88·6	87·2	87·7
Prot. (N × 6·25) (%)	8·7	3·9	6·1
Fat (ether extract) (%)	1·7	1·1	1·1
Crude fibre (%)	3·5	0·6	0·8
Ash (%)	3·1	1·0	1·1
CBH[b] (by difference) (%)	83·0	93·4	90·9
Ca (mg/100 g)	466	258	239
P (mg/100 g)	272	68	103

[a] Kurien (1967b).
[b] Carbohydrate.

Table 5.84
Availability of Ca from whole, refined and composite finger millet flours.[a]

	(A) Whole meal flour	(B) Refined flour	(C) Composite flour
Average daily food intake (g)	10·1	9·3	10·0
Ca/P ratio in diet	1:5·8	1:2·8	1:4·6
Average daily Ca intake (mg)	11·14	10·23	10·97
Ca excreted			
Faeces (mg)	5·37	2·76	2·35
Urinary (mg)	0·50	0·53	0·51
Total (mg)	5·87	3·29	2·86
Ca retained (mg)	5·27	6·94	8·11
% Ca retained	47·3	67·8	73·9

[a] Kurien (1967b).

Table 5.85
Chemical composition of polished pearl millet grains.[a]

	Whole grain	Polished (8% polish) grain	Bran	Husk
Moisture (%)	11·6	13·2	11·2	14·9
Prot. (%)	12·5	12·1	12·2	5·5
Lipid (%)	5·6	4·8	5·3	2·6
Crude fibre (%)	3·0	2·0	12·0	25·0
Total ash (%)	2·8	1·2	6·7	6·3
P (mg/100 g)	343	320	1450	324

[a] Desikachar (1977).

Table 5.86
Proximate constituents (DWB) in hand dissected Nigerian sorghum kernel.[a]

Component	% by weight	Prot. (N × 6·25)	Oil	Crude fibre	Ash
			(%)		
Whole sorghum	100·0	7·5	2·8	2·4	1·3
Peri.	4·9	6·3	3·2	22·9	1·9
As % in whole kernel		4·0	5·7	53·4	7·1
Germ	6·9	16·5	30·4	5·0	12·7
As % in whole kernel		14·5	76·9	16·4	68·2
Endo.	88·2	7·2	0·5	0·7	0·4
As % in whole kernel		81·5	17·5	30·2	24·7

[a] Reichert and Youngs (1977).

Table 5.87
Summary of proximate composition of sorghum and millet dehulled with the Hill grain thresher and by the traditional mortar and pestle method.[a]

	Sorghum			Millet		
	Whole sorghum	Trad. dehulled	Hill grain thresher dehulled	Whole millet	Trad. dehulled	Hill grain thresher dehulled
Extraction level (%)	100	71	77	100	75	74
Prot. (N × 6·25) (%)	7·5	7·1	6·4	10·8	9·8	8·7
Ash (%)	1·3	1·1	0·8	1·4	1·1	0·9
Oil (%)	2·8	2·3	1·6	4·8	4·4	2·6
Crude fibre (%)	2·4	1·1	1·0	2·0	1·1	0·7

[a] Reichert and Youngs (personal communication, 1978).

Table 5.88
Proximate analysis of pearl millet and sorghum samples traditionally and mechanically milled.[a]

	Moisture	Prot.	Fat	N-free extract	Crude fibre	Ash
				(%)		
Millet						
(1) Mech. milled (PRL)	8·3	9·1	6·0	67·3	1·2	2·4
(2) Trad. milled soaked (Nigeria)	11·5	6·9	4·1	73·1	1·4	0·5
(3) Trad. milled nonsoaked (Nigeria)	9·2	6·4	3·1	73·6	2·0	0·9
Red sorghum						
(4) Mech. milled (Nigeria)	8·5	6·6	2·4	74·0	1·6	1·4
(5) Trad. milled (Nigeria)	13·4	7·4	1·8	74·7	0·9	1·2
White sorghum						
(6) Mech. milled (PRL)	9·9	7·7	2·7	73·3	0·9	1·4
(7) Mech. milled (Nigeria)	8·4	6·0	2·5	74·6	1·4	1·5
(8) Trad. milled (Nigeria)	10·9	9·9	2·5	70·8	1·1	1·7

[a] Sumner *et al.* (1974b).

Table 5.89
Content and recovery of proteins and tannins in dehulled sorghum.[a]

Sample	% dehulled	Prot. content[b] (%)	Prot. recovery[c] (%)	Tannin content[d]	Tannin remaining[c] (%)
BR 64 normal	0	9·4	100·0	4·5	100·0
	12·3	8·6	80·2	4·0	77·0
	24·2	8·2	66·3	1·6	26·4
	37·0	8·2	55·0	0·2	2·4
RS 626 high-tannin	0	12·8	100·0	0·5	100·0
	11·3	12·6	87·0	0·3	57·5
	23·4	11·9	71·4	0·3	42·5
	36·0	11·3	56·4	0·2	22·5

[a] Chibber *et al.* (1978).
[b] Expressed as g protein/100 g dehulled sample.
[c] Based on 100 g whole sample.
[d] Expressed as catechin equivalents/g sample.

Table 5.90
Amino acid composition of whole and dehulled sorghum grain (mg/g total N).[a]

	Sample							
	BR 64 % dehulled				RS 626 % dehulled			
Amino acid	0	12·3	24·2	37·0	0	11·3	23·4	36·0
Ile	225	219	206	206	212	231	219	244
Leu	750	706	706	737	769	819	812	931
Lys	137	150	112	81	87	106	75	69
Met	137	144	137	131	9	112	87	119
Cys	125	81	100	44	81	69	12	44
Phe	306	275	262	262	294	312	287	331
Tyr	244	231	237	212	244	256	225	269
Thr	194	194	156	150	169	169	144	169
Val	287	281	250	237	269	287	262	294
Arg	250	262	206	156	206	250	156	169
His	137	156	137	131	125	131	112	131
Ala	512	481	469	469	544	550	537	587
Asp	381	375	331	331	394	394	337	369
Glu	1200	1144	1137	1150	1275	1287	1275	1412
Gly	212	206	175	144	162	162	131	137
Pro	500	437	462	481	481	506	475	500
Ser	269	250	225	212	244	250	219	256
Ammonia	237	225	387	537	162	337	231	200

[a] Chibber *et al.* (1978).

Table 5.91
Nitrogen[a] distribution in dehulled sorghums.[b]

	Sample							
	BR 64 % dehulled				RS 626 % dehulled			
Prot. fraction	0	12·3	24·2	37·0	0	11·3	23·4	36·0
I Albumins and globulins	4·3	3·9	5·0	5·4	13·4	13·6	11·3	5·6
II Kafirin (prolamine)	3·1	2·6	7·1	9·9	8·8	13·1	15·4	17·2
III Cross-linked kafirin (kafirin-like)	11·9	17·4	24·9	33·6	37·2	34·6	30·9	36·3
IV Glutelin-like	17·7	20·4	17·4	12·7	6·6	6·4	6·8	6·2
V Glutelin	63·0	55·7	45·6	38·4	34·0	32·3	35·6	34·7
Total N extracted (%)	90·4	89·4	92·7	88·8	84·8	90·1	88·9	88·6

[a] Chibber *et al.* (1978).
[b] % of soluble N.

Table 5.92
Content and recovery of individual protein fractions in dehulled sorghum.[a]

	Sample							
	BR 64				RS 626			
Prot. fr.	% dehulled	% of Soluble prot.[b]	% of sample[c]	% recovery[d]	% dehulled	% of soluble prot.[b]	% of sample[c]	% recovery[d]
Albumins and globulins (I)	0	4·3	0·4	100·0	0	13·4	1·7	100·0
	12·3	3·9	0·3	72·8	11·3	13·6	1·7	88·2
	24·2	5·0	0·4	77·1	23·4	11·3	1·3	60·0
	37·0	5·4	0·4	69·2	36·0	5·6	0·6	23·5
Kafirins (II and III)	0	15·0	1·4	100·0	0	46·0	5·9	100·0
	12·3	20·0	1·7	107·0	11·3	47·7	6·0	90·3
	24·2	32·0	2·6	141·3	23·4	46·3	5·5	71·8
	37·0	43·5	3·6	160·0	36·0	53·5	6·1	66·0
Glutelins (IV and V)	0	80·7	7·6	100·0	0	40·6	5·2	100·0
	12·3	76·1	6·5	75·6	11·3	38·7	4·9	83·0
	24·2	63·0	5·2	51·7	23·4	42·4	5·1	74·5
	37·0	51·1	4·2	34·9	36·0	40·9	4·6	56·8

[a] Chibber *et al.* (1978).
[b] From Table 5.90
[c] Based on 100 g dehulled sample.
[d] Based on 100 g whole sample.

Table 5.93
Milling performance of wheat, pearl millet and sorghum.[a]

	Wheat	Pearl millet	Sorghum
1000-seed wt (g)	42·2	5·4	30·0
Flour (%)	69	57	47
Bran (%)	28	17	15
Residue (%)	3	26	38

[a] Perten (1977).

Table 5.94
Chemical composition (DWB) of reground wheat, pearl millet and sorghum products.[a]

	Extraction	Prot. (N × 5·7)	Fat	Ash
	(%)			
Wheat				
Flour	69	10·1	2·2	0·5
I Residue	3	15·1	6·2	3·1
Residue flour	(54)	14·2	6·0	2·0
II Residue	(46)	14·7	6·2	4·0
Millet				
Flour	57	7·1	5·2	1·4
I Residue	26	12·4	8·3	2·6
Residue flour	(72)	12·7	10·1	3·0
II Residue	(28)	10·9	4·5	1·5
Sorghum				
Flour	47	7·2	3·3	1·3
I Residue	38	12·4	5·4	2·1
Residue flour	(37)	11·0	5·8	2·8
II Residue	(63)	13·1	4·5	1·9

[a] Perten (1977).

Table 5.95
Chemical composition (DWB) of decorticated products of pearl millet.[a]

	%	Prot. (N × 5·7)	Fat	Fibre	Ash
		(%)			
Millet	100	8·9	6·9	3·0	2·1
Coarse bran					
Passage I	0·5	8·2	4·4	12·5	2·6
II	0·5	8·2	5·8	8·6	2·7
III	0·5	7·9	5·9	7'3	2·2
IV	0·5	7·7	5·5	6·8	2·1
Fine bran					
Passage I	5·5	12·7	11·4	6·2	10·6
II	6·5	13·1	13·0	5·4	4·6
III	5·5	13·3	14·5	5·6	4·1
IV	6·5	12·6	13·8	3·5	3·7
Decorticated grain					
Passage I	94	8·6	6·8	2·6	1·7
II	87	8·4	6·2	2·0	1·5
III	81	7·9	5·7	2·0	1·3
IV	74	7·6	4·8	1·7	1·1
Aspirated bran	1	10·9	12·2	9·4	3·9
Cleaned grain	73	7·6	4·8	1·6	1·0

[a] Perten (1977).

Table 5.96
Yield and proximate composition (DWB) of products from US, Senegal and Niger sorghums and millets.[a]

	Weight (kg)	Prot.	Fat	Fibre	Ash
			(%)		
US sorghum	45	12·0	4·1	2·7	1·9
Dehulled	38	12·4	3·3	1·0	1·6
By-product 1	0·5	5·5	6·0	18·8	6·2
2	1·4	10·2	8·5	9·4	4·2
3	0·5	12·0	9·8	8·8	4·3
Fines	4	11·5	8·7	8·1	3·7
Senegal sorghum	45	12·0	2·7	2·0	1·7
Dehulled	35	12·5	1·8	0·9	1·3
By-product 1	0·5	5·5	5·7	13·1	2·8
2	2·5	10·9	6·1	5·4	3·7
3	1	13·4	7·2	4·7	4·4
Fines	5	12·0	6·7	5·9	3·6
Niger sorghum	45	12·8	3·7	2·1	2·0
Dehulled	36	13·1	2·8	1·1	1·4
By-product 1	1	6·4	5·4	14·3	3·7
2	1·4	11·3	8·3	7·0	3·9
3	0·5	13·6	8·6	8·1	3·7
Fines	5	12·1	7·8	6·5	3·7
US millet	45	13·6	4·5	10·8	3·4
Dehulled	35	14·1	3·5	3·9	1·1
By-product 1	0·5	6·2	2·8	41·9	16·0
2	7	8·9	2·2	41·6	9·3
3	0·5	14·6	16·9	20·8	10·1
Fines	1·4	18·7	18·7	12·7	7·1
Senegal millet	45	12·7	5·3	1·8	2·0
Dehulled	34	10·7	3·0	0·9	1·1
By-product 1	1	21·0	14·0	9·1	5·8
2	2·2	18·4	12·1	5·8	4·2
3	1	17·5	9·9	4·7	3·5
Fines	6	18·9	12·7	3·8	4·2
Niger millet	45	12·0	5·9	1·7	2·1
Dehulled	35	9·8	3·0	1·0	1·9
By-product 1	1	16·0	17·1	8·6	5·7
2	1	14·4	14·6	6·3	4·7
3	0·5	13·8	10·6	8·2	4·0
Fines	8	15·2	14·2	2·7	4·5

[a] deMan *et al.* (1973).

Table 5.97
Composition (DWB) of wheat, pearl millet and sorghum brans from industrial milling.[a]

	Bran		
	Wheat	Pearl millet	Sorghum
Moisture (%)	11·7	6·0	12·8
Pentosans (%)	31·2	10·4	9·3
Formic acid insol. (%)	17·8	9·4	10·7
Ash (%)	8·7	3·6	1·5

[a] Goussault *et al.* (1972).

Table 5.98
Protein (N × 6·25), formic acid insoluble and pentosan contents of wheat, pearl millet and sorghum brans, nutritional effectiveness, and digestibility for rats.[a]

	Diets			
	(A) Casein	(B) Wheat bran	(C) Millet bran	(D) Sorghum bran
Prot. (N × 6·25) (%)	12·2	14·4	15·7	14·6
Formic acid insol. (%)	1·0	2·6	2·8	2·8
Pentosans (%)	0	2·8	2·0	1·6
Dry ingestion (g/rat/day)	16·1	16·7	18·0	16·6
(% of control A)		(103)	(111)	(103)
N intake (mg/rat/day)	318	387	450	421
(% of control A)		(122)	(144)	(134)
Pentosans intake (mg/rat/day)	0	625	468	360
Wt gain (g/rat/day)	4·6	5·0	5·6	4·7
PER	2·4	2·1	2·0	1·8
(% of control A)		(87)	(83)	(75)
Food conversion eff.	3·5	3·3	3·2	3·6
(% of control A)		(95)	(92)	(102)
Coeff. of dig. utilization (CDU)	97·5	93·9	92·7	93·0

[a] Goussault *et al.* (1972).

Table 5.99
Chemical composition of five sorghum varieties.[a]

Sample	Moisture	Prot.	Fat	Ash
	(%)			
Kansas commercial mix	11·1	9·0	2·9	1·4
Oklahoma				
Martin	13·5	7·8	3·0	1·3
Plainsman	14·0	7·6	2·8	1·3
Dwarf Yellow Milo	13·8	8·6	2·7	1·4
Combine 7078	15·0	7·8	2·7	1·5

[a] Stringfellow and Peplinski (1966).

Table 5.100
Results of pearling and screening tests to determine the relative hardness of five sorghum varieties.[a]

	Yields		
	On 14	−14 +20	Through 20
	(%)		
Martin (hardest)	70	12	18
Commercial mix	66	12	22
Plainsman	58	12	30
Dwarf Yellow Milo	49	17	34
Combine 7078	43	15	42

[a] Stringfellow and Peplinski (1966).

Table 5.101
Fractionation of ground grits from Kansas commercial mix. Blank spaces indicate no values presented.[a]

Flour or fraction	Prot. (%)	Fat (%)	Ash (%)	MMD[b] (μ)	Yield (%)
As milled grits	7·5	0·7	0·5		
Reground					
Fraction I	11·8	2·2	2·1	12	8·0
II	5·9	1·0	0·8	15	13·2
III	4·6	0·5	0·4	18	25·4
IV	6·6	0·5	0·3	21	17·7
V	9·3	0·3	0·3	54	35·7

[a] Stringfellow and Peplinski (1966).
[b] Mass median diameter.

Table 5.102
Fractionation response of grits and flours from different varieties of sorghum.[a]

Variety	Protein[b] Flour or grits	High-prot. fine fraction	Low-prot. fraction	Coarse residue	Yield coarse residue	Total prot. shifted
	(%)					
Grits (ground 3 × 18 000 rev/min)						
Martin	8·0	10·9	4·9	9·5	40·6	26
Commercial mix	7·5	11·8	4·6	9·3	35·7	28
Plainsman	6·5	11·1	4·2	8·4	32·6	29
Dwarf Yellow Milo	8·0	18·6	5·1	10·2	27·3	35
Combine 7078	8·0	17·2	5·0	9·1	21·0	31
Floury endo. (ground 3 × 18 000 rev/min)						
Martin	7·0	12·5	5·0	8·6	22·0	26
Commercial mix	6·3	11·4	4·0	7·5	22·0	29
Plainsman	6·0	13·5	3·6	7·3	18·1	35
Dwarf Yellow Milo	5·7	15·4	3·5	6·7	14·2	40
Combine 7078	6·2	16·6	3·6	5·9	12·4	41
Corneous endo. (ground 5 × 18 000 rev/min)						
Martin	10·7	14·6	8·8	14·0	11·9	17
Commercial mix	11·1	13·5	8·4	13·8	15·9	19
Plainsman	9·2	13·2	6·9	11·5	14·3	20
Dwarf Yellow Milo	11·9	16·5	9·6	16·2	15·6	20
Combine 7078	9·6	18·9	6·8	11·9	12·6	28

[a] Stringfellow and Peplinski (1966).
[b] 14% moisture basis.

Table 5.103
Comparison of starch and germ fractions recovered from whole and alkali dehulled No. 2 yellow sorghum grain (DWB).[a]

Fraction	Whole grain	Dehulled 2 min	Dehulled 4 min	Dehulled 6 min
	(%)			
Starch				
Recovery[b]	81	73	74	76
Yield	63	57	58	60
Protein	0·4	0·4	0·4	0·5
Germ				
Oil in germ	33·8	50·3	51·9	50·5
Oil recovery[c]	45	65	65	58

[a] Blessin *et al.* (1971).
[b] Based on starch presnt in whole grain.
[c] Based on oil present in whole grain.

Table 5.104
Composition (DWB) of the grain from parents and hybrids of sorghum.[a]

	Starch	Prot.	Ether extract	Ash
	(%)			
Parents				
BTX 398[b]	65·8	13·8	3·3	1·6
BTX 3197[c]	68·4	13·2	2·7	1·6
R 7078	66·8	12·4	2·7	1·5
RTX 414	67·7	13·0	2·6	1·7
Mean	67·2	13·1	2·8	1·6
Hybrids				
RS 608	66·3	13·4	2·9	1·5
RS 625	67·2	14·0	2·3	1·5
RS 610	67·8	11·3	2·8	1·5
RS 626	66·6	11·8	3·1	1·6
Mean	67·0	12·6	2·8	1·5

[a] Norris and Rooney (1970b).
[b] Martin.
[c] Combine Kafir 60.

Table 5.105
Wet-milling attributes of parents and hybrids.[a]

	Starch yield	Starch recovery	PEC[b] content	Prot. content of starch	Prot. content of PEC
	(%)				
Parents					
BTX 398[c]	44·6	67·7	1·9	1·9	26·0
BTX 3197[d]	48·4	70·8	1·6	1·0	21·8
R 7078	48·2	72·2	1·2	0·9	25·8
RTX 414	53·4	78·9	0·8	0·9	30·4

Table 5.105 *cont.*

	Starch yield	Starch recovery	PEC[b] content	Prot. content of starch	Prot. content of PEC
			(%)		
Hybrids					
RS 608	49·5	74·7	1·4	1·2	25·6
RS 625	49·1	73·1	1·3	1·1	24·5
RS 610	52·9	78·0	1·2	0·8	24·7
RS 626	53·8	80·8	1·4	1·0	25·5

[a] Norris and Rooney (1970b).
[b] PEC = peripheral endo. cells.
[c] Martin.
[d] Combine Kafir 60.

Table 5.106
Chemical, mineral and vitamin composition of sorghum gluten feed and sorghum gluten meal. Blank space indicates no value presented.[a]

	Gluten feed	Gluten meal
Crude prot. (N × 6·25) (%)	25·0	41·7
Fat (%)	3·4	4·1
N-free extract (%)	48·4	40·3
Fibre (%)	6·3	2·8
Ash (%)	7·7	0·7
Minerals (mg/100 g)		
Ca	90	20
P	590	170
Vitamins (mg/100 g)		
Thiamine	0·58	
Niacin	10·2	4·97
Riboflavin	1·2	0·42
Pantothenic acid	2·2	0·53

[a] Seeley (1958).

Table 5.107
Average proximate composition of 12 samples of gluten meal and gluten feed and 5[a] **samples of germ meal. Blank spaces indicate no values presented.**[b]

	Gluten meal		Gluten feed		Germ meal (oil cake)
	Reiners *et al.*	NAS	Reiners *et al.*	NAS	Reiners *et al.*
Dry matter (%)	88·7	88·7	89·2	89·2	98·1
Prot. (%)	47·6	50·0	24·6	27·5	17·2
Total fat (%)	7·0	4·9[c]	4·8[d]	10·1[e]	
Ash (%)	3·7	1·4	8·3	8·0	1·7
Fibre (%)	4·9	3·3	8·5	7·2	13·2
S (%)	0·62		0·59		0·30

[a] Analysed by Reiners *et al.* with National Academy of Sciences (1971) values (NAS) where available (DWB).
[b] Reiners *et al.* (1973).
[c] Ether extract.
[d] Four samples.
[e] Three samples.

Table 5.108
Amino acid composition (mg/total N) of gluten meal, gluten feed and germ meal.[a]

	Gluten meal	Gluten feed	Germ meal
	42·2% prot.	21·9% prot.	16·9% prot.
Ile	267	210	211
Leu	961	581	479
Lys	80	131	152
Met	113	119	172
Cys	89	126	136
Phe	365	241	248
Tyr	274	159	157
Thr	167	236	208
Trp	37	24	35
Val	324	333	321
Arg	165	217	300
His	117	156	176
Ala	647	564	343
Asp	384	314	342
Glu	1559	983	857
Gly	157	297	311
Pro	511	426	313
Ser	254	243	236

[a] Reiners *et al.* (1973).

Table 5.109
Amino acid composition (mg/100 g sample), "as-is" basis, of gluten meal, gluten feed and germ meal.[a] Blank spaces indicate no values presented.[b]

	Gluten meal 11·3% moisture 42·2% prot.		Gluten feed 10·8% moisture 21·9% prot.		Germ meal 1·9% moisture 16·9% prot.	
	Reiners *et. al*	NAS	Reiners *et. al.*	Lyman *et al.*	Reiners *et al.*	Lyman *et al.*
Ile	1800	2260	700	1000	600	700
Leu	6400	7280	2000	2500	1300	1400
Lys	500	790	500	700	400	700
Met	800	590	400	400	500	300
Cys	600		400		400	
Phe	2400	2560	800	1000	700	700
Tyr	1800		600		400	
Thr	1100	1370	800	800	600	600
Trp	250	390	80	200	100	180
Val	2200	2460	1100	1300	900	1100
Arg	1100	1180	800	900	800	1100
His	800	790	500	600	500	600
Ala	4300		1900		900	
Asp	2600		1100		900	
Glu	10400		3400		2400	
Gly	1100		1000		900	
Pro	3500		1500		900	
Ser	1700		800		700	

[a] Reiners *et al.* (1973), National Academy of Sciences (1971) (NAS) and Lyman *et al.* (1956).
[b] Reiners *et al.* (1973).

Table 5.110
Protein, starch and crude fat content (DWB) of two pearl millet samples compared to commercial maize and sorghum. Blank spaces indicate no values presented.[a]

	Prot.	Starch	Crude fat
		(%)	
Av. commercial maize processed by wet-milling	9·5	72·0	4·5
Av. commercial sorghum processed by wet-milling	10·7	73·8	3·7
Tiflate millet	17·4	58·8	4·9
Tift 23DB millet (dwarf)	14·2	64·6	6·9
Pearl millet[b]			
Av.	13·6		5·4
Range	(8·1–20·9)		(2·8–8·0)

[a] Freeman and Bocan (1973).
[b] Published results but no source stated; prot. for 370 entries, fat 167 entries.

Table 5.111
Fatty acid composition of oil from Tiflate and Tift 23DB pearl millet kernels in comparison with maize and grain sorghum oils. Blank spaces indicate no values presented.[a]

	Palmitic	Stearic	Oleic	Linoleic	Linolenic	Arachidic
			(%)			
Pearl millet						
Tiflate	17·1	4·0	19·6	55·0	3·8	0·4
Tift 23DB	18·6	5·4	28·2	45·2	2·1	0·5
Range of 65 selfed lines[b]	16·7–25·0	1·8–8·0	20·2–30·6	40·3–51·7	2·3–5·0	0·3–1·0
Maize[c]						
Average for 81 Georgia inbreds	15·6	2·4	37·5	43·4	1·2	
Average for 72 Illinois inbreds	12·4	1·7	29·3	55·4	1·0	
Sorghum[d]						
Average	12·3	0·8	34·3	49·9	2·7	
Range	10·1–14·2	0·2–1·3	27·8–41·8	42·0–56·1	1·1–4·7	

[a] Freeman and Bocan (1973).
[b] Jellum and Powell (1971).
[c] Thornton *et al.* (1969).
[d] Freeman and Heatherwick (unpublished).

Table 5.112
Protein solubility fractions (%) of Tiflate and Tift 23DB pearl millet grains.[a]

	Tiflate	Tift 23DB
Albumins (water-sol.)	16·9	13·2
Globulins (salt-sol.)	6·7	9·4
Prolamines (alcohol-sol.)	45·2	40·7
Glutelin (alkali-sol.)	10·5	11·9
Insol.	16·8	16·7
Total extracted	96·1	91·7

[a] Freeman and Bocan (1973).

Table 5.113
Tield[a] and composition[b] from Tiflate and Tift 23DB millet in comparison with maize and sorghum. Blank spaces indicate no values presented.[c]

	Pearl millet		Maize[d]	Sorghum[e]
	Tiflate	Tift 23DB	Funk G 4444	(Red Milo)
Germ				
Yield	8·9	6·1	9·3	6·2
Starch	11·8	8·9	13·1	18·6
Prot.	11·0	9·7	13·7	11·8
Fat	35·1	56·0	44·1	38·8
Fibre (+5028 nylon)				
Yield	7·0	7·6	9·6	7·4
Starch	16·5	10·5	20·3	30·6
Prot.	12·8	10·8	10·1	17·6
Fat	3·2	8·7	2·0	2·4
Tailings (−5028 nylon, +53 nitex)				
Yield	2·1	1·0	1·1	0·8
Starch	35·6	32·8	46·2	25·3
Prot.	42·0	26·3	19·0	39·2
Fat	0·8	3·0	0·7	
Gluten				
Yield	10·7	13·5	10·2	10·6
Yield at 70% protein	7·7	8·3	7·2	7·7
Starch	40·0	48·1	38·5	42·8
Prot.	43·6	32·0	46·1	46·7
Fat	7·5	10·5	4·8	5·1
Sqeegee				
Yield	2·6	5·6	2·1	1·2
Starch	76·1	74·8	90·7	81·6
Prot.	14·8	20·7	5·7	14·0
Fat	0·1	1·6	0·4	0·6
Starch				
Yield	46·7	51·9	58·1	63·2
Yield with gluten at 70% prot.	52·3	62·7	63·2	67·3
Prot.	0·7	0·7	0·3	0·4
Fat	0·2	0·1	0·1	
Solubles				
Yield	18·7	14·8	7·8	6·6
Prot.	45·6	46·6	45·8	43·7

[a] % of dry substance in whole grain represented by individual fractions.
[b] % of dry substance in individual fractions.
[c] Freeman and Bocan (1973).
[d] Freeman and Watson (1969).
[e] Freeman and Heatherwick (unpublished).

Table 5.114
Properties of Tiflate and Tift 23DB pearl millet starches in comparison with commercial maize and sorghum starches.[a]

	Pearl millet starch		Maize starch	Sorghum (Red Milo) starch
	Tiflate	Tift 23DB		
Intrinsic viscosity	1·74	1·50	1·78	1·78
Iodine affinity (% DWB)	4·84	5·05	5·11	4·94
Swelling power				
65°C	7	2	7	4
75°C	10	9	12	11
85°C	12	12·5	14	13
Solubles (%) (DWB)				
65°C	1·4	0·4	1·4	0·9
75°C	2·3	4·1	4·7	4·1
85°C	3·2	6·1	6·3	5·5
Gelatinization temp. (°C)				
2%	59	67	58	61
50%	64	72	66	67
98%	70	75	72	72
Brabender viscosity (35 g)				
Peak	660	600	640	610
95°C start	660	580	610	590
95°C 60 min	400	600	700	505
50°C start	660	760	960	750
50°C 60 min	600	690	940	730

[a] Freeman and Bocan (1973).

Table 5.115
Average daily digestibility values from subjects fed grain sorghum breads in a simple mixed diet.[a]

Bread	Prot. from bread (g)	CBH[b] from bread (g)	Digestibility of prot. (%)	Digestibility of CBH (%)
Kafir	35	230	51	96
Feterita	31	185	51	97
Milo	36	251	40	96
Kaoliang	41	288	20	96
Maize	38	314	60	96
Wheat	53	269	77	95

[a] Langworthy and Holmes (1916).
[b] Carbohydrate.

Table 5.116
Average daily digestibility values from subjects fed grain sorghum porridges ("mushes") in a simple mixed diet.[a]

	Prot. from porridge (g)	CBH[b] from porridge (g)	Digestibility of prot. (%)	Digestibility of CBH (%)
Kafir	42	227	48	96
Feterita	42	259	48	99
Milo	39	224	34	98
Kaoliang	28	168	4	96

[a] Langworthy and Holmes (1916).
[b] Carbohydrate.

Table 5.117
Chemical composition (DWB) of whole and decorticated Egyptian native millet as compared with wheat. Blank spaces indicate no values presented.[a]

	Egyptian millet		Wheat
	Whole	Decorticated	
Moisture (%)	10·7	11·8	10·6
Prot. (N × 6·25) (%)	10·0	10·9	12·5
Fat (%)	5·1	2·9	2·1
CBH[b] (by difference) (%)	76·4	84·5	67·3
Fibre (%)	5·7	0·7	13·1
Ash (%)	2·8	1·0	5·0
Minerals (mg/100 g)			
Ca	40	10	70
Fe	2·8	1·3	7·1
Cu	0·8	0·1	0·5
Mg	190	90	160
Mn	0·9	0·4	3·6
Ni	0·15	0·0	
K	390	200	470
Na	20	10	50
Zn	2·5	0·8	2·9

[a] Awadalla and Slump (1974).
[b] Carbohydrate.

Table 5.118
Amino acid content (mg/100 g sample) of Egyptian native millet as compared with that of wheat and wheat and millet mixtures.[a]

	Whole millet	Decorticated millet	Whole wheat	80% wheat and 20% whole millet	80% wheat and 20% decorticated millet
				(calculated)	
Ile	410	470	440	434	446
Leu	1150	1430	780	854	910
Lys	260	220	330	316	308
Met	180	170	200	196	134
Cys	170	180	260	242	244
Phe	460	570	520	508	530
Tyr	380	450	360	634	378
Thr	330	350	360	364	358
Trp	100	120	130	124	128
Val	560	550	540	544	542
Arg	420	400	550	534	520
His	210	230	250	242	246
Ala	830	990	440	518	550
Asp	720	700	610	632	628
Glu	1920	2270	3610	3270	3343
Gly	340	290	460	436	426
Pro	690	920	1150	1058	1150
Ser	460	480	580	550	580
N	1430	1540	1840	1518	1840

[a] Awadalla and Slump (1974).

Table 5.119
Amino acid content in protein (mg/g total N) of Egyptian native millet as compared to whole wheat. Blank spaces indicate no values presented.[a]

	Whole millet	AA score[b]	Whole wheat	AA score[b]	Decorticated millet	AA score[b]
Ile	287	115	238	95	305	122
Leu	804	183	425	97	929	211
Lys	182	54	181	53	143	42
Met	129	113	106	114	110	103
Cys	119		144		117	
Phe	322	155	281	125	370	174
Tyr	266		194		292	
Thr	231	92	227	91	194	78
Trp	70	117	69	115	78	130
Val	392	126	294	95	357	115
Arg	294		300		260	
His	147		138		149	
Ala	580		238		643	
Asp	503		331		455	
Glu	1336		1719		1474	
Gly	238		250		188	
Pro	483		625		597	
Ser	322		313		312	

[a] Awadalla and Slump (1974).
[b] WHO (1973a).

Table 5.120
Net Protein Utilization (NPU), true digestibility (TD), biological value (BV) of wheat and mixtures of wheat and millet.[a]

Diet	NPU	TD	BV
(A) Whole wheat	53	85	62
(B) Mixture of 80% whole wheat and 20% whole millet	50	85	59
(C) Mixture of 80% whole wheat and 20% decorticated millet	51	88	58

[a] Awadalla and Slump (1974).

Table 5.121
Proximate analysis (DWB) and amino acid composition of cowpeas, sorghum and millet flour used in preparation of snack foods.[a]

	(1) Red cowpeas trad. milled	(2) White cowpeas trad. milled	(3) White sorghum mech. milled	(4) Millet mech. milled	(5) Golden Penny wheat flour
Moisture (%)	9·6	11·0	9·9	8·3	13·6
Prot. (%)[b]	24·2	27·3	7·7	10·3	12·4
Fat (%)	3·7	2·3	3·1	7·0	2·0
N-free extract (%)	66·6	65·5	86·5	78·5	84·2
Crude fibre (%)	2·1	1·6	1·1	1·4	0·7
Ash (%)	3·4	3·3	1·6	2·8	0·7
AAs (mg/g total N)[c]					
Ile	310	280	270	255	216
Leu	670	600	1040	709	531
Lys	580	530	240(70)	281(82)	186(55)
Met + Cys	130(59)	70(32)	170(78)	268	269
Phe + Tyr	790	720	790	770	700
Thr	340	300	280	228(91)	190(76)
Trp	84	81	113	144	72
Val	380	340	360	321	259
AA score (WHO 1973a)	59	32	70	82	55

[a] Sumner *et al.* (1974b, revised, 1977).
[b] Wheat protein (N × 5·7), other flour (N × 6·25).
[c] Data in parentheses are AA scores of those amino acids present in amounts lower than WHO (1973a).

Table 5.122
Composition (DWB) and protein quality of composite flour for noodles.[a]

Blend No.	Golden Penny wheat flour	Cowpea flour	Sorghum flour	Millet flour	Protein content[b]	AA score
	(%)					
Wheat						
Commercial	100				11·7	
GPN	100				12·2	
GPS	100				12·4	55
Sorghum						
44	52	32	16		15·5	89
Millet						
54	42	36		22	16·3	89

[a] Sumner *et al.* (1975, revised, 1977).
[b] Wheat prot. (N × 5·7), other flour (N × 6·25).

Table 5.123
Effect of cooking on noodle composition (DWB).[a]

Blend No.	Raw noodle prot.	Cooking loss	Prot. cooking loss	Cooked noodle prot.
	(%)			
Wheat				
Commercial	11·7	5·2	7·9	11·9
GPN	12·2	7·0	10·7	12·4
GPS	12·4	16·1	9·3	13·0
Sorghum				
44	15·5	7·8	8·3	16·2
Millet				
54	16·3	9·5	7·1	17·2

[a] Sumner *et al.* (1975, revised, 1977).
[b] Wheat prot. (N × 5·7), other flour (N × 6·25).

Table 5.124
Extraction and composition (DWB) of sorghum S 29 milling products. Blank spaces indicate no values presented.[a]

	Extraction	Moisture	Prot. (N × 6·25)	Lipid (ether extract)	Ash content
	(%)				
Whole grain	100	11·9	13·9	4·7	1·8
Scoured grain	90·5	11·8	14·1	4·1	1·5
Bran	10·9	16·0	16·2		3·5
Flour	5·6	15·7	9·6		1·2
Total semolina	74·0	16·1	14·0	2·4	0·9
Purified semolina					
SSSE Type A (higher quality)	54·8	16·1	13·8		0·8
Purified semolina					
SSSE Type B (lower quality)	19·0	15·6	16·1		1·4
Filter flour	0·2	13·2	15·0		2·0

[a] Miche *et al.* (1976).

Table 5.125
Colour and effect of oxidizing systems on pigments of four sorghum varieties.[a]

	S 29	L 30	INRA 450	CE 90
Milling extraction	100	100	100	100
Pigments:				
Absorbance at 400 nm	0·25	0·42	0·65	0·17
"Phenolic" compounds:				
absorbance at 320 nm	0·46	0·88	1·26	1·28
Water extract:				
absorbance at 400 nm	0·51	0·66	1·17	0·61
Peroxidase activity:				
$^{\Delta}$absorbance/min/g (DWB)	13·6	20	19·7	10·1
Polyphenol oxidase activity:				
$^{\Delta}$absorbance/min/g (DWB)	1·2	2·0	2·4	1·4
Catalase activity:				
μ mol O_2/min/g (DWB)	25	107	93	49
Lipoxidase activity:				
μ mol O_2/min/g (DWB)	0	0	0	0
Brown index[b] (dried disc)	20·5	23	34·7	19·3
Yellow index (dried disc)	6	5·3	6·7	5·8
Dominant wavelength	479	482	486	478
Purity	20	16	21	22
Luminance	20·4	22·4	33·1	19·4

[a] Miche *et al.* (1976).
[b] Authors' note: it is interesting that the Brown Index appears to parallel (a) the total of pigments and phenolics and (b) water extract. Also, "phenolic" compounds were lower in INRA 450 (brown-red peri.) than CE 90 (white peri. no coloured testa).

Table 5.126
Histological distribution of browness, oxidases and pigments in sorghum CE 90.[a]

	Whole grain	Bran	Flour	Semolina
Milling extraction	100	10·9	5·6	54·8
Pigments:				
absorbance at 400 nm	0·17	0·28	0·10	0·11
"Phenolic" compounds:				
absorbance at 320 nm	1·3	1·5	0·89	0·84
Water extract:				
absorbance at 400 nm	0·61	1·40	0·20	0·26
Peroxidase activity				
$^{\Delta}$absorbance/min/g (DWB)	10	130	4·8	3·2
Polyphenoloxidase activity				
$^{\Delta}$absorbance/min/g (DWB)	1·4	4·7	1·1	0·5
Catalase activity:				
μ mol O_2/min/g (DWB)	49	226	43	27
Lipoxidase activity:				
μ mol O_2/min/g (DWB)	0	0	0	0
Brown index[b] (dried disc)	19·3	34·3	11·5	12·2
Yellow index (dried disc)	5·8	10·5	3·8	4·5

[a] Miche *et al.* (1976).
[b] Authors' note: the Brown Index appears to parallel (a) pigments and phenolics and (b) water extract. As CE 90 is a white peri. non-coloured testa sorghum, pigments may not be phenolic in origin.

Table 5.127
Comparison of sorghum beer with European beer.[a]

Beer	Raw material used			Fermentation process	
	Malt	Adjunct	Hops	Lactic	Alcoholic
Sorghum	Sorghum	Maize or sorghum	None	Yes	Yes
European	Barley	Cereals or syrups	Yes	No	Yes

[a] Novellie (1977).

Table 5.128
Yield of reducing sugars and quality of malt extract.[a]

	Reducing sugars as maltose (g)	Malt extract quality		
		Colour	Taste	Aroma
Finger millet malt alone	700	Lt. brown	Sweet	Strong malt flavour
Wheat and malt	1475	Lt. brown	Sweet	Moderate
Sorghum and malt	1132	Lt. brown	Sweet	Moderate
Finger millet and malt	1029	Dk. brown	Sweet	Moderate
Cassava and malt	1090	Dk. brown	Sweet with slight bitter taste	Mild

[a] Chandrasekhara and Swaminathan (1953c).

Table 5.129
Reducing sugars and crude protein (N × 6·25) of malt extracts.[a]

Flour mashed with finger millet malt	Crude prot. of flour	Reducing sugars as maltose	Prot. in malt extract
		(%)	
Finger millet	7·2	78·3	2·5
Wheat	14·2	73·5	3·9
Sorghum	9·2	72·0	2·2
Cassava	1·8	74·0	1·7

[a] Chandrasekhara and Swaminathan (1953c).

Table 5.130
Free amino acid content in sample of Aruna finger millet malt extract.[a]

	(mg/100 g sample)
Ile	58
Leu	118
Lys	177
Met	336
Cys + Val	103
Phe	59
Tyr	64

Table 5.130 *cont.*

	(mg/100 g sample)
Thr	132
Arg	41
His	60
Ala	34
Asp	260
Glu	302
Gly	214
Pro	49
Ser	92

[a] Narayana and Murthy Reddy (1973).

Table 5.131
Dry matter, thiamine, N and soluble sugars (DWB) in seed flour and malt flour of wheat, maize and sorghum.[a]

	Wheat		Maize		Sorghum	
	Seed flour	Malt flour	Seed flour	Malt flour	Seed flour	Malt flour
Dry matter (%)	91·0	91·5	84·3	90·9	85·5	86·2
N (mg/100 g)						
Total	2072	2106	1490	1501	2064	2060
Soluble	198	365	96	285	82	197
Soluble sugars (mg/100 g)	210	5450	189	3245	146	2552
Thiamine (mg/100 g)	0·56	0·46	0·50	0·36	0·43	0·29

[a] Hussain *et al.* (1966).

Table 5.132
Composition of sorghum, sorghum malt, wort and beer traditionally brewed in Togo. Blank spaces indicate no values presented.[a]

	Values/100 g		Values/100 cc	
	Sorghum grain	Sorghum malt	Wort	Beer
Moisture (g)	15·3	15·0		
Calories	320	320	40	35
Prot. (g)	8·0	8·1	0·3	0·3
Fat (g)	2·3	1·6		
CBH[b] (g)	72·1	73·4	10·3	3·3
Ash (g)	2·3	1·9	0·2	0·2
Ca (mg)	33	28	1·6	1·4
Thiamine (mg)	0·56	0·44	0·03	0·03
Niacin (mg)	4·9	7·1	0·6	0·5
Ribofl. (mg)	0·08	0·25	0·02	0·04
Vit. B_{12} (γ) μg	traces	traces	0·01	0·03
Pantoth. acid (mg)	0·53	0·96	0·07	0·09
Alcohol (g)				3·03

[a] Perisse *et al.* (1959).
[b] Carbohydrate.

Table 5.133
Effect of germination on thiamine, niacin and riboflavin content of cereals (DWB).[a]

	Thiamine	Niacin	Riboflavin
		(mg/100 g)	
Maize			
Grain	0·35	2·0	0·15
Malt	0·23	2·9	0·20
Sorghum			
Grain	0·33	3·5	0·13
Malt	0·17	3·4	0·24
Finger millet			
Grain	0·19	0·95	0·11
Malt	0·16	1·37	0·20
Pearl millet			
Grain	0·34	2·36	0·15
Malt	0·21	3·02	0·32

[a] Goldberg and Thorp (1946).

Table 5.134
Vitamin contents of sorghum beer and some related products. Blank spaces indicate no data presented.[a]

		Thiamine		Niacin		Riboflavin	
	Solids (%)	Wet μg/ml	Dry mg/100 g	Wet μg/ml	Dry mg/100 g	Wet μg/ml	Dry mg/100 g
Meal-malt mixture	89·7		0·24		3·25		0·19
Sorghum beer (municipal)	8·8	0·57	0·65	4·8	5·46	0·51	0·58
Sorghum beer (mine breweries)	5·8	0·25	0·44	2·7	4·71	0·35	0·59
Sorghum beer (home brewed)	8·6	0·36	0·41	4·5	5·28	0·56	0·65
Sorghum beer residue	33·8	0·18	0·05	12·7	3·83	0·65	0·19
Sibebe (second beer)	5·0	0·22	0·44	2·8	5·62	0·46	0·92
Nsipo (third beer)	3·4	0·23	0·68	1·65	4·85	0·38	1·12
Mkuku	21·7	0·75	0·35	6·1	2·82	0·50	0·23
Leting	9·5	0·30	0·32	3·64	3·83	0·34	0·36

[a] Goldberg and Thorp (1946).

Table 5.135
Saccharide content of K 2 Red grade of sorghum before and after malting (DWB).[a]

	Grain	Malt
	(g/100 g malt)	
Fructose	0·03	1·57
Glucose	0·04	5·48
Sucrose	0·89	5·58
Maltose and maltose oligosaccharides	nil	3·33

[a] Von Holdt and Brand (1960b).

Table 5.136
Digestibility (rats) of the protein of maize meal and malt, sorghum meal and malt, sorghum beer ingredients, and sorghum beer.[a]

	N content of test diet[b] (%)	Digestibility of prot. (%)
Maize meal (sifted, granulated, pressure-cooked for 20 min at 15 lb/in^2, roller-dried, milled	1·2	96·3
Maize malt (pressure-cooked for 20 min at 15 lb/in^2, roller-dried, milled)	1·1	93·6
Sorghum (commercial sample, cooked in manner similar to that employed in brewing of sorghum beer, roller-dried, milled)	1·3	73·5
Sorghum malt (commercial sample, cooked in manner similar to that employed in brewing of sorghum beer, roller-dried, milled)	1·4	80·2
Sorghum beer ingredients (sorghum and malt and maize meal mixed in proportions for brewing (1/1·6/1·0) cooked, roller-dried, milled)	1·6	84·0
Sorghum beer (strained, evaporated under vacuum, roller-dried, milled	1·6	81·3
Sorghum beer 'waste' (dried in circulation oven at 50°C, milled)	1·6	78·5
Wheaten bread (80% extraction rate, laboratory-baked, dried in air circulation oven at 40°C, ground in hammer mill)	1·3	99·8
Wheaten bread (90% extraction rate, commercial sample, freeze-dried, ground in hammer mill)	1·6	96·7

[a] Dreyer (1968).
[b] Air-dried basis.

Table 5.137
Chemical composition of sorghum and sorghum malt mean and ranges in parentheses) and of sorghum breakfast food.[a]

	Grain (8 samples)	Brewer's malt (8 samples)	Breakfast food (1 sample)
Prot. (N × 6·25) (%)	10·7 (8·5–11·9)	11·0 (10·0–12·2)	9·5
Fat (%)	3·6 (1·6–4·6)	2·8 (1·4–3·6)	3·6
Fibre (%)	4·7 (3·4–7·3)	4·2 (2·5–5·0)	3·5
Ash (%)	1·7 (1·4–2·2)	1·6 (1·3–2·0)	3·0
CaO (as % of ash)	3·1 (2·1–4·1)	2·6 (0·9–4·6)	2·9
P_2O_5 (as % of ash)	40·7 (31·6–47·7)	44·6 (36·1–51·0)	24·1
Thiamine (mg/100 g)	0·58 (0·46–0·68)	0·5 (0·32–0·85)	0·7
Niacin (mg/100 g)	3·2 (2·3–4·0)	4·1 (3·5–4·7)	3·5
Riboflavin (mg/100 g)	0·2 (0·12–0·31)	0·26 (0·2–0·34)	0·17

[a] Aucamp *et al.* (1961).

Table 5.138
Vitamin contents (mg/100 g) of sorghum malt.[a]

	Thiamine	Niacin	Riboflavin
Series I			
Grain	0·55	4·0	0·12
Malt, dried only	0·52	3·7	0·20
Malt, cured	0·25	3·8	0·22
Series II			
Grain	0·62	4·0	0·16
Malt, dried only	0·85	3·7	0·22
Malt, cured	0·35	4·0	0·24

[a] Aucamp *et al.* (1961).

Table 5.139
Mean composition (ranges in parentheses) of 21 samples of sorghum beer from municipal South African breweries.[a]

	Mean	No. of analyses
pH	3·4 (3·2–3·7)	10
Alcohol (% w/v)	3·0 (1·8–3·9)	17
Solids (% w/v)		
Total	5·4 (3·0–8·0)	17
Insol.	3·7 (2·3–6·1)	6
N (% w/v)		
Total	0·093 (0·059–0·137)	16
Sol.	0·014 (0·010–0·017)	9
Thiamine (μg/100 ml)	93 (20–230)	21
Riboflavin (μg/100 ml)	56 (27–170)	21
Niacin (μg/100 ml)	315 (130–660)	21
Ascorbic acid (mg/100 ml)	0·04 (0·01–0·15)	7

[a] Aucamp *et al.* (1961).

Table 5.140
Vitamin content (μg/100 g) of sorghum beer (mean and ranges in parentheses).[a]

	Year brewed			
	1975 Maize as adjunct	1975 Sorghum as adjunct	1972	1963[b]
Thiamine	27 (23–35)	47 (39–60)	35 (28–43)	32
Niacin	222 (180–270)	400 (340–450)	270 (180–360)	189

Table 5.140 *cont.*

	Year brewed			
	1975 Maize adjunct	1975 Sorghum as adjunct	1972	1963[b]
Riboflavin	27 (22–34)	35 (28–43)	27 (19–36)	36

[a] Novellie (1977).
[b] No range quoted for 1963.

Table 5.141
Diastatic power of pearl millet and finger millet on malting at 30°C. Blank spaces indicate no values presented.[a]

	Pearl millet	Finger millet
Days of germination	(SDU/g DWB)	
1	22·0	7·2
2		26·4
3	53·4	41·4
4	65·4	51·2
6	65·1	

[a] Novellie (1977).

Table 5.142
Amino acids, mg/100 ml, found after 2 h mashing of malt (5% in 10% starch solution) at pH 4 and 60°C.[a]

		Amount found in wort of malt			
Amino acid	Group[b]	M67/152	M69/75	M69/96	M70/34
Ile	2	0·91	1·71	0·94	0·88
Leu	2	3·44	5·53	3·71	3·45
Lys	1	0·96	1·28	1·18	0·96
Met	2	0·52	0·72	0·78	0·48
Phe	3	3·38	3·13	2·51	2·05
Tyr	3	2·89	3·85	2·93	2·69
Thr	1	1·82	2·12	1·62	1·31
Trp	3	0·89	1·06	0·83	0·74
Val	2	1·94	3·22	1·81	1·67
Arg	1	2·67	1·94	2·25	2·23
His	2	2·07	2·04	2·30	2·11
Ala	3	2·89	5·09	2·97	3·05
Asp	1	1·41	2·07	1·54	1·72
Asparagine	1	9·63	10·11	8·43	11·38
Glu	1	3·20	3·93	3·19	4·57
Glutamine	1	19·16	30·72	31·12	16·05
Gly	3	0·58	1·00	0·66	0·70
Pro	4	13·40	7·40	11·64	16·09
Ser	1	2·49	2·69	2·45	2·11
Ammonia	3	0·37	0·28	0·30	0·38
AAs (mg/g)		75·57	94·77	84·40	75·87
Crude prot. (%)		10·0	10·2	10·6	12·4
Diastatic power (KDU)		35·1	44·3	47·5	39·4

[a] Schröder (1972).
[b] Groups ranged from 1 absorbed rapidly by yeast to 4 not absorbed by yeast under brewing conditions.

Table 5.143
Starch contents of sorghum grains and malts (DWB). Blank spaces indicate no values presented.[a]

	Starch content	
	Grain (%)	Malt (Range %)
Short red	69·0	45·9–61·2[b]
Framida	61·1	
Martin	63·2	51·0[c]
White	64·5	
Birdproof	68·6	52·4–55·0[d]

[a] Von Holdt and Brand (1960a).
[b] Five samples.
[c] One sample.
[d] Three samples.

Table 5.144
Mean composition (ranges for thiamine in parentheses) of amgba and affouk (DWB).[a]

	Sorghum	Malt	Amgba (beer)	Affouk (wine)	Residue
Moisture (%)	9·6	24·7	90·7	86·3	69·2
Calories	381	380	394[b]	408[b]	241
Prot. (N × 6·25) (%)	9·4	9·8	8·7	12·6	17·7
Lys (% total prot.)	3·3	3·7	7·2	4·6	2·4
Lipid (%)	2·8	2·2	0·3	3·3	3·2
CBH[c] (%) (by difference)	85·6	86·2	86·1	80·4	77·5
Cellulose (%)	2·3	3·7	0·3	0·9	11·1
Ash (%)	2·14	1·75	4·06	3·7	1·6
Minerals (mg/100 g)					
Ca	11·0	9·3	20·7	30·1	8·6
P	319	327	630	684	260
Phytic P	166	85	112	320	85
Vitamins (mg/100 g)					
Thiamine	0·41 (0·17–0·54)	0·43 (0·17–0·56)	3·44 (1·69–5·24)	0·89 (0·31–1·10)	0·10 (0·07–0·12)
Riboflavin	0·10	0·23	0·76	0·44	0·18
Niacin	4·3	5·3	8·0	5·9	6·9
Alcohol			2·7° GL 2·13 g	5° GL 3·9 g	

[a] Chevassus-Agnes *et al.* (1976).
[b] Includes calories provided by the alcohol.
[c] Carbohydrate.

Table 5.145
Comparison of nutritional composition of sifted flour, amgba and affouk as % of whole gain. Blank spaces indicate no values presented.[a]

	Sorghum	Sifted flour[b]	Amgba (beer)	Affoul (wine)
Solids	100	70	20	31
Calories (alcohol not included)	100	75	20	33
Total calories (alcohol included)	100	75	29	55
Prot.	100	69	19	41
Lys	100		41	56
Lipid	100	35	0	28
Total CBH[c]	100	72	19	29
Cellulose	100	28	0	12
Ash	100	39	36	42
Ca	100	59	35	81
Total P	100	41	38	65
Phytic P	100		13	47
Thiamine	100	39	160	65
Riboflavin	100	34	145	141
Niacin	100	30	35	41

[a] Chevassus-Agnes *et al.* (1976).
[b] Obtained by traditional milling methods, Favier *et al.* (1972).
[c] Carbohydrate.

Table 5.146
Protein distribution (Landry and Moureaux fractions) in three cultivars of dry and reconstituted sorghum grain.[a]

	Dry grain			Reconstituted grain		
Crude prot. (%) (DWB)	Hetero yellow 11·0	Redlan waxy 11·6	Redlan nonwaxy 12·7	Hetero yellow 11·0	Redlan waxy 11·6	Redlan nonwaxy 12·7
	% of total crude prot.					
Fraction						
I 0·5 M NaCl H_2O (albumins and globulins)	18·5	16·3	16·5	19·6	19·3	14·4
II 70% isopropyl alcohol (v/v) (kafirins)	5·3	6·2	6·2	6·9	4·0	5·0
III 70% isopropyl alcohol (v/v)+ 0·6% 2-mercaptoethanol (v/v)	19·4	22·3	18·1	21·8	25·0	22·8
IV Borate buffer with NaCl (pH 10, 0·5 M)+ 0·6% 2-mercaptoethanol (v/v)	4·6	4·5	4·8	3·0	3·1	2·8
V Borate buffer with NaCl (pH 10, 0·5 M)+ 0·6% 2-mercaptoethanol (v/v) 0·5% sodium lauryl sulphate (w/v) (glutelin)	29·3	34·6	32·9	25·1	30·9	28·9
Total	77·0	84·0	78·4	76·4	82·2	74·0

[a] Walker and Lichtenwalner (1977).

Table 5.147
Effect of the waxy gene of sorghum on ruminal digestibility of dry matter and five (Landry and Moureaux) protein factions.[a]

Variety	Dry matter	Fraction I	II	III	IV	V	Total
Redlan waxy	61·3	79·6	45·9	55·0	49·4	65·4	53·1
Redlan nonwaxy	75·8	74·4	43·5	43·9	57·6	61·8	44·8

[a] Walker and Lichtenwalner (1977).

Table 5.148
Proximate analyses (DWB), feed efficiency and coefficients of apparent digestibility of popped sorghum fed to beef cattle.[a]

	(1) Original grain	(2) Normal run	(3) Completely popped	(4) Partially popped or unpopped
Proximate analyses				
Prot. (%)	14·4	14·6	14·4	13·7
Ether extract (%)	3·1	3·9	3·9	3·9
N-free extract (%)	77·8	77·0	76·6	76·9
Fibre (%)	2·0	1·7	1·9	2·0
Ash (%)	2·7	2·8	3·3	3·6
Feed efficiency				
Initial weight (kg)	343·2	341·4	342·7	344·1
Daily gain (kg)	1·4	1·2	1·2	1·2
Daily feed intake (kg)	9·6	6·8	6·9	7·9
Feed/gain ratio	6·9	5·5	5·9	6·4
Coefficients of apparent digestibility				
Dry matter	57·3	74·6	79·3	76·0
Organic matter	57·5	75·5	80·3	77·0
N-free extract	61·3	82·6	88·8	83·5
Nonprot. organic matter	59·8	80·4	85·9	81·3
Ether extract	70·1	71·1	66·4	68·0
Fibre	31·0	33·7	31·2	26·8
Crude prot.	39·5	38·9	37·6	41·0
True prot.	68·8	67·0	65·5	70·9

[a] Riggs *et al.* (1970b).

Table 5.149
Percent gelatinization and starch availability of flour grain sorghums. Blank spaces indicate no values presented.[a]

	(1) Funk 3761	(2) Funk 766	(3) TX 660	(4) Elevator run
% gelatinization				
Pressure cooked	40	40		30
Micronized	25	25	12	
Starch availability (ml gas produced/h)				
Pressure cooked	19·0	11·4		9·7
Micronized	9·8	9·7	4·7	
Ground	4·8	4·1	3·8[b]	

[a] Hinders and Freeman (1969).
[b] Authors' note: it is uncertain from the text whether the figure of 3·8 for gas produced from ground relates to type 3 or type 4.

Table 5.150
Effect of flaking on starch digestion of pressure cooked[a] barley and sorghum. Blank spaces indicate no values presented.[b]

Pressure (kg/cm^2)	Unflaked	Poor flaked	Intermed. flaked	Flat flaked
		(%)		
Sorghum (untreated 16·7%)				
1·4	14·0	14·2	25·8	28·5
2·8	14·7		26·6	29·6
4·2	22·6	27·7	42·4	40·9
5·6	29·1	38·0		47·8
Barley (untreated 21·7%)				
1·4	17·3		29·0	33·8
2·8	18·6	30·4	43·6	47·0
4·2	34·3	38·4	41·1	48·3

[a] Cooked for 1 min after reaching given pressure.
[b] Osman *et al.* (1970).

Table 5.151
Effect of four processing methods on digestibility of sorghum grain in swine.[a]

Treatments of sorghum grain	Apparent coeff. of digestibility		Metabolizable energy
	Gross energy	Prot.	
(A) Dry grinding	82·2	74·3	3·5
(B) Pelleting	83·2	75·5	3·6
(C) Micronizing	84·7	76·5	3·6
(D) Steam flaking	85·3	81·6	3·6

[a] Brzozowski *et al.* (1972).

Table 5.152
Effect of processing method on carbohydrate (CBH) content of sorghum grain.[a]

Treatment	Total CBH (%)	Ethanol-sol. CBH (%)	Starch (%)
Dry-ground	71·7	2·9	68·8
Reconstituted	66·4	1·6	64·8
Steam-flaked	71·2	4·4	66·8
Micronized	69·1	5·5	63·6

[a] McNeill *et al.* (1975).

Table 5.153
Effect of processing method on total carbohydrate (CBH) and starch solubility of sorghum grain.[a]

	Solubilities (%)			
	Total CBH		Starch	
Treatment	In rumen fluid	In 0·1N HCl	In rumen fluid	In 0·1N HCl
Dry ground	3·6	31·0	0·4	0·1
Reconstituted	2·3	24·8	0·5	0·3
Steam flaked	4·9	36·4	0·2	0·1
Micronized	6·1	29·6	1·2	0·3

[a] McNeill *et al.* (1975).

Table 5.154
Susceptibility of processed sorghum grain to amyloglucosidase.[a]

Treatment	mg glucose released/g dry matter
Dry ground	118·6
Reconstituted	139·3
Steam flaked	615·5
Micronized	232·7

[a] McNeill *et al.* (1975).

Table 5.155
Proximate analysis of processed sorghum feeds (DWB).[a]

	Crude prot.		Ether extract		N-free extract		Crude fibre		Ash	
	Exp. 1	Exp. 2	Exp. 1	Exp. 2	Exp. 1	Exp. 2	Exp. 1	Exp. 2	Exp. 1	Exp. 2
	(%)									
Prairie hay	6·0	5·3	1·8	1·7	50·4	51·1	33·7	34·7	8·1	7·2
Soybean meal	51·6	52·1	0·7	1·0	34·1	33·7	6·9	6·6	6·8	6·6
Sorghum grain										
Rolled	10·9	11·5	2·8	2·3	82·7	81·7	1·9	2·3	1·7	2·2
Expanded	10·7	12·2	1·6	1·4	84·2	82·0	1·8	2·3	1·7	2·1
Pressure cooked	10·0		2·1		84·0		2·0		1·8	
Flaked		11·0		1·4		84·4		1·7		1·5
Extruded		11·2		1·7		82·3		2·6		2·2

[a] Prasad *et al.* (1975).

Table 5.156
Apparent nutrient digestibilities of rations and orghum grain, Experiment 1.[a]

	Crude protein	Ether extract	N-free extract	Crude fibre	Gross energy
Ration					
Dry-rolled	51·8	52·6	70·8	59·6	62·4
Expanded	48·3	28·7	77·9	40·1	63·6
Pressure cooked	44·8	39·4	70·6	50·8	60·0
Hay	65·6	13·1	64·1	63·7	58·6
Grain					
Dry-rolled	37·9	72·3	74·2	[b]	65·6
Expanded	30·8	42·2	84·6	[b]	67·8
Pressure cooked	22·4	56·8	73·8	[b]	61·1

[a] Prasad *et al.* (1975).
[b] Because of large changes in digestibility of crude fibre in hay when grain was added to the ration and because of small amounts of crude fibre in grain, determining crude fibre digestibility of grains by difference would be meaningless.

Table 5.157
Apparent nutrient digestibilities of rations and sorghum grain, Experiment 2.[a]

	Crude fibre	Ether extract	N-free extract	Crude fibre
Ration				
Dry-rolled	49·3	50·9	81·5	49·9
Flaked	47·5	32·0	88·5	45·3
Extruded	51·5	44·4	89·9	53·1
Expanded	49·6	42·2	85·6	40·1
Hay	58·7	7·4	71·6	66·5
Grain				
Dry-rolled	33·2	67·1	84·1	[b]
Flaked	27·0	37·7	94·4	[b]
Extruded	35·7	59·8	96·6	[b]
Expanded	31·0	57·3	89·8	[b]

[a] Prasad *et al.* (1975).
[b] Because of large changes in digestibility of crude fibre in hay when grain was added to the ration, and because of small amounts of crude fibre in grain, determining crude fibre digestibility of grains by differences would be meaningless.

Table 5.158
Effect of processing sorghum grain on N retention, Lys, protein and starch availability, Experiments 1 and 2.[a]

Treatment	N balance		Available Lys		Protein hydrolysis			Starch availability[a]	
	Exp. 1 (g/day)	Exp. 2 (g/day)	Exp. 1 (μmol/g dry grain)	Exp. 2	Exp. 1	Exp. 2		Exp. 1	Exp. 2
						pancreatin digestion	SRDT[b]		
					(%)	(%)	(%)		
Dry rolled	15·9	11·6	15·8	14·6	56·0	58·0	44·0	62	44
Expanded	19·7	18·5	14·8	5·0	50·7	48·7	35·1	519	434
Pressure cooked	20·5		8·4		48·1			191	
Flaked		17·5		5·9		51·8	37·8		186
Extruded		12·9		8·0		56·1	49·9		169

[a] Prasad *et al.* (1975).
[b] Reducing activity increase, mg glucose equivalent/g grain.
[c] Simulated ruminant digestion technique.

Table 5.159
Effect of grain processing on weight gain and feed intake of finishing steers.[a]

	(H) Dry-rolled	(I) Flaked	(J) Extruded
Av. daily gain (kg)	1·3	1·2	1·3
Av. daily feed[b] consumption (kg)	11·6	10·3	10·3
Feed/gain ration	9·2	8·4	8·0

[a] Prasad *et al.* (1975).
[b] Air-dried basis.

Table 6.1
Protein value of the diet of countries in which sorghum and the millets are the major source of protein in the diet.[a]

	Niger	Senegal	Sudan	The Gambia	Ethiopia	Tanzania	Upper Volta	Chad	Mali	Nigeria
Calories	2160	2280	2030	2300	2040	2100	2020	2180	2120	2180
Total protein (g)	78·2	64·5	69·3	60·5	68·8	58·6	69·9	75·8	64·3	59·4
Prot./cal. ratio	14·5	11·3	13·7	10·5	13·5	11·2	13·9	13·9	12·1	10·9
NPU_{op}	52	56	53	57	53	54	51	51	52	54
ND_p Cal (%)	7·6	6·4	7·2	6·0	7·1	6·0	7·0	7·1	6·2	5·9
AA score										
Ile	63	63	65	62	62	60	60	61	62	59
Leu	110	106	116	109	99	116	114	119	121	105
Lys	66	71	79	65	72	68	61	64	60	62
Met + Cys	68	64	62	64	65	60	60	60	63	61
Phe	83	80	85	80	85	82	84	84	82	81
Tyr	77	76	81	73	72	75	74	75	74	73
Thr	74	69	75	68	73	71	68	69	69	69
Trp	86	75	78	77	76	69	74	73	76	77
Val	73	71	73	72	70	69	69	70	71	68
Protein sources (%)[b]										
Millets and sorghum	65	31	41	39	45[c]	38	56	56	55	48
Animal products	18	28	33	20	22	16	8	16	17	9
Pulses	15		14		18	18	27	24	12	17
Rice		13		25						
Maize	8				8	18	6			
Roots and tubers										17

[a] Autret *et al.* (1969).
[b] It will be noted that the protein source percentages do not total 100.
[c] Including teff.

Table 6.2
Protein value of the diet of countries in which sorghum and the millets provide part of the protein but are not the predominant source of protein in the diet.[a]

	India	Ghana	Cameroon	Togo	Cent. African Rep.
Calories	1980	2160	2130	2210	2170
Total protein (g)	52·2	48·6	54·5	49·1	44·8
Prot./cal. ratio	10·5	9·0	10·2	8·9	8·2
NPU_{op}	53	56	65	56	55
ND_p Cal (%)	5·7	5·1	5·6	4·9	4·5
AA score					
Ile	61	59	59	58	56
Leu	93	91	104	102	85
Lys	70	75	66	65	73
Met + Cys	58	62	61	59	59
Phe	85	78	81	80	74
Tyr	73	77	79	76	71
Thr	69	71	70	68	67
Trp	72	72	72	67	71
Val	69	68	68	66	62
Protein sources (%)					
Rice	22				
Animal products	10	22	18	12	21
Wheat	15				
Pulses	24	13		22	27
Maize		11	16	16	
Roots and tubers		32	17	19	27
Millets and sorghum	16	13	24	22	12

[a] Autret *et al.* (1969).
[b] It will be noed that the protein source percentages do not total 100.

Table 6.3
Vitamin B_{12} (cyanocobalamin) activity in faeces and livers of rats on different diets after 8 weeks of feeding. Blank spaces indicate no values presented.[a]

Diet	Total B_{12} act in diet (μg/100 g)	True B_{12} act in diet (μg/100 g)	B_{12} activity in faeces (μg/rat/day)	Body wt (g)	Liver wt (g)	True B_{12} act. in whole liver (μg)
Weanlings (at start of exp.)			1·61	45	1·9	0·40
(A) Sorghum (6% of prot.) + pigeon peas (3% of prot.) + amaranth (1% of prot.)	1·71	0·03	3·37			
(B) Pearl millet (6% of prot.) + pigeon peas (3% of prot.) + amaranth (1% of prot.)	1·28		3·40	201	8·0	0·88
(C) SMP	0·68	0·59	2·23	213	7·5	1·11
(D) Whole egg	1·15	1·15	1·83	233	9·4	2·62
(E) Stock diet	1·41	0·48	4·28	186	7·3	0·92

[a] Ramachandran *et al.* (1960).

Table 6.4
Nutrient contribution of sorghum—bean blend[a] assuming 75% of the daily calorie requirement is supplied by the blend.[b c]

	70 sorghum + 30 bean				
	Infants	Children		Women	Men
	1 year	1–3 years	3–6 years	18–35 years	18–35 years
Wt to provide 75% of total calories (g)	203	331	440	616	880
	(%)	(%)	(%)	(%)	(%)
Prot.	143	146	155	150	177
Ca	10	14	20	28	40
Fe	126	206	219	205	438
Vit. A	1	1	2	1	2
Vit. D	0	0	0	0	0
Ascorbic acid	7	7	8	9	11
Thiamine	255	196	330	345	329
Riboflavin	47	58	61	65	72
Niacin	99	107	116	199	135

[a] Nutrient content based on INCAP-ICNND data.
[b] Figures represent % of NRC (1964) dietary allowances.
[c] Sirinit *et al.* (1965).

Table 6.5
Mean nitrogen balance data for preschool children fed on cereal pulse diets.[a]

Diets	N			
	Intake	Total excretion	Absorption	Retention
	(g)			
(A) Sorghum + pigeon peas	8·2	2·5	7·2	5·6
(B) Sorghum + urd	7·5	2·1	6·9	5·4
(C) Rice + pigeon peas	8·9	1·7	8·4	7·1

[a] Pushpamma *et al.* (personal communication, 1978).

Table 6.6
Amino acid composition (mg/g total N) and AA scores of different combinations of sorghum and cowpeas, and finger millet and cowpeas.[a]

Diets	Lys	Total S AA	Thr	Trp	AA score Egg prot.	AA score WHO (1973a)
Egg prot.	406	344	312	106		
WHO (1973a)	337	219	250	62		
Sorghum and cowpea prot.						
(A) 4/0	194	187	206	62	48	59
(B) 3/1	250	169	200	62	49	74
(C) 2/2	306	150	200	56	44	69
(D) not shown						
Finger millet and cowpea prot.						
(A) 4/0	200	319	225	81	50	59
(B) 3/1	412	112	187	44	33	51
(C) 2/2	250	269	212	69	60	76
(D) 0/4	306	219	206	62	44	82

[a] Narayanaswamy *et al.* (1975).

Table 6.7
Amino acid scores of four amino acids in sorghum supplemented with soy flour, cottonseed flour, fish meal and groundnut flour.[a]

AA	AA score[b] Parts of supplement/100 parts sorghum by weight				
	0	4	8	16	32
			Plus soy flour		
Lys	38	50	59	70	82
Thr	78	82	85	89	93
Met + Cys	66	66	66	66	66
Trp	107	111	113	116	120
			Plus cottonseed flour		
Lys	38	46	52	60	69
Thr	78	81	83	86	90
Met + Cys	66	67	69	71	73
Trp	107	113	118	124	131
			Plus fish meal		
Lys	38	62	77	95	111
Thr	78	84	88	93	97
Met + Cys	66	71	74	77	81
Trp	107	101	98	93	89
			Plus groundnut flour		
Lys	38	45	50	56	64
Thr	78	78	78	79	79
Met + Cys	66	66	66	66	66
Trp	107	108	108	109	109

[a] Jansen (1974).
[b] Values listed are AA scores for individual AAs as compared to whole egg. AA scores calculated as follows:

$$\text{AA score} = \frac{\text{mg AA/g essential AA N} \times 100}{\text{mg AA in whole egg/g essential AA N in whole egg}}$$

Table 6.8
Amino acid scores for millet and sorghum based diets supplemented with cowpeas.[a]

AAs	Millet	75 millet + 25 cowpeas	Sorghum	70 sorghum + 30 cowpeas	Cowpeas
Ile	81	82	82	83	84
Leu	135	129	109	172	111
Lys	61	80	38	67	136
Met + Cys	103	91	66	62	54
Phe	104	109	112	116	125
Tyr	97	94	84	85	87
Thr	93	94	78	84	97
Trp	163	147	107	106	101
Val	100	98	96	95	92

[a] Oke (1975, 1977).

Table 6.9
Effect of soybean supplementation of sorghum based diets (10% protein) in rat feeding trials.[ab]

Diet[c] (%)		Total food intake (g)	Total prot. intake (g)	Total wt. gain (g)	F/G	PER
(A) Sorghum	91·0	165·7	16·6	18·1	9·1	1·3
(B) Sorghum Soybean	72·7 5·3	178·2	17·8	25·2	7·0	1·7
(C) Sorghum Soybean	54·5 10·3	206·0	20·6	31·4	6·6	1·8
(D) Sorghum Soybean	36·4 15·6	206·0	20·6	40·9	5·0	2·3

[a] The number and sex of the rats and the duration of the trial were not stated.
[b] Sultana and Pushpamma (personal communication, 1977).
[c] All diets contained (%) tricalcium phosphate 1·0, salt 1·0, amaranth 4·0, corn starch (A) 3·0, (B), (C) and (D) 8·0.

Table 6.10
N balance, BV and digestibility coefficients (DC) of rice based diets for six adult human male subjects.[a]

Diet	N intake from diet (g)	Total N excretion (g)	N balance (g)	BV (%)	DC[b] (%)
(A) Rice	12·2	8·1	+4·2	67	81
(B) Rice, wheat	14·1	10·4	+3·7	55	79
(C) Rice, wheat, barley	13·8	9·8	+4·0	60	79
(D) Rice, wheat, maize	13·9	10·1	+3·8	57	79
(E) Rice, wheat, sorghum	13·0	9·7	+3·3	54	82
(F) Rice, wheat, finger millet	13·9	9·5	+4·4	60	83
(G) Rice, wheat, pearl millet	13·9	9·8	+4·0	57	82

[a] Mitra *et al.* (1948).
[b] $DC = \frac{\text{N intake} - \text{faecal N}}{\text{N intake}} \times 100.$

Table 6.11
Results of metabolic trials with seven boys on mixed diets based on rice and sorghum.[a]

Diet[b]	Content in diet				Metabolic results[c]					
					N		Ca		P	
	Prot. (N × 6·25) (%)	Ca (mg/100 g)	P (mg/100 g)	Phytate P (mg/100 g)	MDR (g)	DC (%)	MDI (mg)	MDR (mg)	MDI (mg)	MDR (mg)
(A) Rice 4	6·9	61	128	63	1·8	74·7	355	123	744	169
(B) Rice 3/sorghum 1	6·9	66	144	80	1·5	69·3	381	109	835	204
(C) Rice 2/sorghum 2	7·1	71	160	96	1·3	63·7	410	97	928	233
(D) Sorghum 4	7·4	76	188	129	0·9	55·4	441	74	1091	309

[a] Kurien *et al.* (1960d).
[b] Diets contained approximately 62% cereal (rice, sorghum) with small amounts of pulses, vegetables, meat and milk. Prot. contents (N × 6·25) (DWB, %) rice 7·4, sorghum 8·8.
[c] DC—digestibility coefficient, MDR—mean daily retention, MDI—mean daily intake.

Table 6.12
Essential amino acids, some minerals, and B vitamin contents of rice polishing concentrate (DWB). Blank spaces indicate no values presented.[a]

	Amount present (mg/100 g)	AA score (WHO 1973a)		Amount present (mg/100 g)
AAs			Minerals	
Ile	273	109	Ca	48
Leu	339	77	P	871
Lys	216	63	Phytin P	638
Met	147		B vitamins	
Cys			Thiamine	2·26
Phe	322		Niacin	301
Tyr			Riboflavin	0·17
Thr	119	47	Choline	122
Trp			Inositol	398
Val	202	65		
Arg	427			
His	146			

[a] Saxena *et al.* (1966).

Table 6.13

Amino acid (AA) content,[a] mean daily intake and absorption of essential AAs by children from diets A and D as compared with AA requirements.[b]

AAs	Diet (A) Sorghum			Diet (D) SMP			
	AA content (mg/g total N)	Intake (mg/kg)	Absorption[c] (mg/kg)	AA content (mg/g total N)	Intake (mg/kg)	Absorption (mg/kg)	AA requirement[d] (mg/kg)
Ile	375	79	58	394	79	69	30
Leu	800	168	124	606	122	107	45
Lys[e]	244	51	38	469	94	82	60
Met	125	26	19	150	30	26	27
Cys	100	21	15	56	11	10	
Phe	306	64	48	306	62	45	27
Thr[e]	244	51	38	281	57	49	35
Trp	69	14	11	87	18	14	9
Val	362	76	56	431	87	76	33
Arg	412	86	64	244	49	43	
His	106	22	16	162	33	29	

[a] Krishnamurthy *et al.* (1960).
[b] Daniel *et al.* (1966).
[c] Intake × digestibility coefficient/100.
[d] Data of Nakagawa *et al.* (1962).
[e] Lys content of diets B and C 412 mg/g total N; thr content of diet C 306 mg/g total N.

Table 6.14

Effect of supplementation of poor sorghum diets with L-Lys HCl and DL-Thr on biological value of the protein in girls' diets.[a]

Diets[c]	N							Prot. (N × 6·25)		
	Intake (g)	Total excretion (g)	Balance (g)	AD[b] (%)	TD (%)	BV (%)	NPU_{op} (%)	Intake (g/kg)	NAP[d] (g/kg)	WHO[e] (1965) (g/kg)
(A) Sorghum	4·79	4·4	0·4	56·0	73·6	66·9	49·3	1·31	0·64	0·72
(B) Sorghum + L-Lys HCl	4·87	4·2	0·6	54·8	72·6	71·4	52·0	1·33	0·69	0·72
(C) Sorghum + L-Lys HCl + DL-Thr	4·96	3·9	1·1	59·5	76·5	78·3	59·8	1·35	0·81	0·72
(D) SMP	4·62	3·1	1·5	69·5	87·8	84·0	73·7	1·26	0·93	0·72

[a] Daniel *et al.* (1966).

[b] AD—apparent digestibility, TD—trude digestibility.

[c] Mean daily intake of foodstuffs (g): (A) sorghum 300, sugar 150, groundnut oil (fortified with vit. A and D) 24, pigeon peas 15, condiments 8, SMP 5; providing 30·0 prot. from sorghum 24·7, pigeon peas 3·4, SMP 1·8. (B) as (A) plus L-Lys HCl 0·98; (C) as (A) plus L-Lys HCl 0·98, DL-Thr 0·86; (D) sugar 19, groundnut oil (fortified with vit. A and D) 43, maize starch 290 and SMP 70. All the diets contained (g) common salt 12, tamarind pulp 8, vegetables 80, onion 20, salt mixture 5 and vitaminized starch 5.

[d] Net Available Prot. = (prot. intake + NPU)/100.

[e] WHO (1965) reference prot. requirement.

Table 6.15
Lys, Thr and S containing amino acids (mg/g total N) in yeast and poor sorghum diets.[a]

AAs	Yeast	Diet (A) (Poor sorghum diets)	Diet (C) (Diet A plus 6% yeast)
Lys	462	206	244
Thr	331	194	219
Met	87	112	106
Cys	31	69	62

[a] Narayanaswamy *et al.* (1972c).

Table 6.16
Effect of supplementing a poor sorghum diet with L-Lys HCl on N retention and growth of girls.[a]

Diet[b]	N						
	Intake	Total excretion	Balance	Ht inc.	Wt inc.	Hb inc.	RBC[c] inc.
	(g/child/day)			(cm)	(kg)	(g/100 ml blood)	(10^6/mm^3)
(A) Sorghum (providing 41 g prot. (N × 6·25) daily; Lys content 250 mg/g total N)	6·5	5·6	0·9	1·8	0·8	0·51	0·08
(B) Sorghum + Lys (Lys content 406 mg/g total N)	6·7	4·9	1·8	2·9	1·7	0·89	0·20

[a] Doraiswamy *et al.* (1968).
[b] Sorghum diet, (A) mean intake (g/child/day), sorghum 300, pigeon peas 40, vegetables 80, plantain (ripe) 30, SMP 7, jaggery 22, groundnut oil 12, condiments and spices 4, coconut 7, tamarind fruit pulp 4, common salt 20 and onion 3. (B) as (A) plus L-Lys HCl 1·25 g in sugar syrup.
[c] Red blood cells.

Table 6.17
Amino acid (AA) content and protein quality evaluation[a] of market samples of sorghum, pearl millet and finger millet.[b]

AAs	Sorghum		Pearl millet		Finger millet	
	mg/g total N	AA score	mg/g total N	AA score	mg/g total N	AA score
Ile	256	102	256	102	294	118
Leu	844	192	594	135	656	149
Lys	156	45	187	55	250	73
S-containing AAs	225	102	294	134	381	173
Phe+Tyr	612	161	500	131	650	171
Thr	219	88	231	92	306	122
Trp	62	103	106	177	112	187
Val	312	101	325	105	369	119
Arg	262		294		362	
His	137		137		181	

[a] Prot. quality evaluations:
—against egg (AA composition from Orr and Watt, 1957) sorghum 38, pearl millet 47, finger millet 63;
—against FAO pattern (WHO, 1965) sorghum 38, pearl millet 61, finger millet 68;
—essential AA index (EAA) (Oser, 1959) sorghum 69, pearl millet 77, finger millet 89;
—biological value (BV = 1·09 (EAA)—11·73) sorghum 64, pearl millet 72, finger millet 85.
[b] Pushpamma *et al.* (1972) AA scores WHO (1973a) calculated by the authors.

Table 6.18
Food intake, weight gain and NPU of rats fed a protein-free diet supplemented with 8·0% protein from sorghum, wheat, maize or soybean oil meal over 10 days.[a]

Diets	Food intake (g)	Wt gain (g)	F/G ratio	NPU (%)
(A) Prot.-free	40·4	−11·7		
(B) Sorghum	71·3	4·2	17·0	28·6
(C) Wheat	79·7	7·9	10·1	38·3
(D) Maize	71·1	3·8	18·7	28·8
(E) Soybean oil meal	91·5	23·9	3·8	74·2

[a] Harden *et al.* (1976).

Table 6.19
Food intake, weight gain and NPU of weanling rats fed a protein-free basal diet supplemented with swine-type cereal-soy rations containing sorghum, wheat or maize over 10 days.[a]

Diets	Prot. level (%)	Food intake (g)	Wt gain (g)	F/G ratio	NPU (%)
(A) Protein-free	18·2	41·9	−10·3		
(B) Sorghum-soy swine-type ration	18·2	126·9	63·4	2·0	47·2
(C) Wheat-soy swine-type ration	18·0	124·4	65·3	1·9	44·7
(D) Maize-soy swine-type ration	20·0	123·8	66·2	1·9	51·6

[a] Harden *et al.* (1976).

Table 6.20

Effect of protein (N × 6·25) supplementation of sorghum based diets (DWB) on growth of weanling rats. Blank spaces indicate no values presented.

Reference		Sorghum diet (%)	Nonprot. supp. (%)	Prot. supp. (%)		Prot. content of diet (%)	Av. feed intake/wk (g)	Av. wt gain/wk (g)	Av. F/G	Av. prot. intake/wk (g)	Av. PER
Adrian and Frangne (1970)[b]	(A)			Maize	82·0	8·4	56	4·9	11·4[a]	4·7[a]	1·1[a]
	(B)	16·0		Maize	60·0	8·4	57	3·4	16·8[a]	4·2[a]	0·8[a]
	(C)	16·0		Maize	60·0	8·4	49	2·4	20·4[a]	4·0[a]	0·6[a]
				L-Trp	0·15						
	(D)	16·0		Maize	60·0	8·4	64	9·2	7·0[a]	5·1[a]	1·8[a]
				L-Lys HCl	0·2						
	(E)	16·0		Maize	60·0	8·4	69	12·2	5·6[a]	5·3[a]	2·3[a]
				L-Trp	0·15						
				L-Lys HCl	0·2						
Adrian and Jacquot (1961)[c]	(A)	93·0				8·3	47	1·3	36·1	3·9	0·3
	(B)	80·0		Groundnut flour		14·4	83	13·5	6·1	11·9	1·1
				(GNF)	13·8						
	(C)	80·0		GNF	10·0	14·4	92	21·3	4·3	13·3	1·6
				Fish flour	2·2						
	(D)	80·0		GNF	13·8	14·4	86	14·0	6·1	12·5	1·2
				L-Met	0·05						
	(E)	80·0		GNF	13·8	14·4	76	13·3	5·7	11·0	1·2
				DL-Met	0·05						
	(F)	80·0		GNF	13·8	14·4	97	20·0	4·8	14·0	1·4
				L-Lys	0·11						
	(G)	80·0		GNF	13·8	14·4	94	19·0	4·9	13·5	1·4
				L-Met	0·05						
				L-Lys	0·11						
Daniel *et al.* (1965a)	(A)					8·2		2·5		2·6	0·9
	(B)			L-Lys HCl	0·26	8·2		8·6		3·5	2·4
	(C)			L-Lys HCl	0·26	8·4		13·7		4·7	2·9
				DL-Thr	0·1						
	(D)[d]	78·5				9·8		6·6		4·0	1·6
	(E)[d]	78·5		L-Lys HCl	0·22	10·0		8·1		4·4	1·8
	(F)[d]	78·5		L-Lys HCl	0·22	10·1		9·4		5·3	1·8
				DL-Thr	0·09						
	(G)[d]	75·5	3·0[e]			9·4		9·5		4·8	2·0
	(H)[d]	75·5	3·0[e]	L-Lys HCl	0·22	9·7		15·2		5·6	2·7
	(I)[d]	75·5	3·0[e]	L-Lys HCl	0·22	9·8		16·8		6·0	2·8
				DL-Thr	0·09						

Daniel *et al.* (1968a)	(A)[d]	78·5				10·3		8·4		4·4	1·9
	(B)[d]	75·5	3·0[e]			10·2		11·8		4·9	2·4
	(C)[d]	63·5		Chickpeas	15·0	12·9		15·8		7·0	2·2
	(D)[d]	60·5	3·0	Chickpeas	15·0	12·8		18·4		7·7	2·4
	(E)[d]	63·5		Pigeon peas	15·0	12·9		15·7		7·8	2·0
	(F)[d]	60·5	3·0	Pigeon peas	15·0	12·8		18·1		8·1	2·2
	(G)[d]	73·5		Soy flour	5·0	12·9		16·3		7·2	2·3
	(H)[d]	70·5	3·0	Soy flour	5·0	12·9		18·8		7·7	2·4
	(I)[d]	68·5		Soy flour	10·0	15·5		18·4		8·4	2·0
	(J)[d]	65·5	3·0	Soy flour	10·0	15·5		21·5		9·5	2·2
	(K)[d]	71·5		SMP	7·0	12·8		19·5		7·9	2·5
	(L)[d]	68·5	3·0	SMP	7·0	12·8		19·8		8·0	2·5
	(M)[d]	64·5		SMP	14·0	15·5		23·5		10·2	2·3
	(N)[d]	61·5	3·0	SMP	14·0	15·5		24·0		10·3	2·3
Daniel *et al.* (1968b)	(A)[d]	78·5				10·1	64	8·8	7·3[a]	6·5[a]	
	(B)[d]	78·0	0·5[f]			10·0	69	11·2	6·2[a]	6·9[a]	
	(C)[d]	61·5		Unfortified coconut meal[f]	17·0	11·9	76	14·9	5·1[a]	9·0[a]	
	(D)[d]	61·5		Fortified coconut meal[f]	17·0	11·8	80	17·4	4·6[a]	9·4[a]	
	(E)[d]	71·5		Fortified SMP[f]	7·0	11·9	85	20·1	4·2[a]	10·1[a]	
Desai *et al.* (1968)	(A)[d]	78·5				10·2		11·0		5·5	2·0
	(B)[d]	78·5	Salt and vitamin mix			10·1		12·5		5·6	2·2
	(C)[d]	75·1		Fish flour[g]	3·4	12·5		15·9		7·6	2·1
	(D)[d]	73·4		Protein food[g]	5·1	12·7		16·2		7·8	2·1
	(E)[d]	71·3		Fortified SMP[g]	7·2	12·6		15·6		8·0	1·9
Desai *et al.* (1970b)	(A)[d]	78·5				12·4	50	7·8	6·4[a]	6·2[a]	
	(B)[d]	39·25		LCBF IV[h]	39·25	17·8	71	14·6	4·9[a]	12·6[a]	
	(A)[d]	Rice flour 78·5				8·4	52	8·5	6·1[a]	4·4[a]	
	(B)[d]	Rice flour 39·25		LCBF III[h]	39·25	16·0	73	17·4	4·2[a]	11·7[a]	
	(C)[d]	LCBF III[h] 78·5				21·7	76	20·4	3·7[a]	16·5[a]	
	(D)[d]	Rice flour 39·25		LCBF IV[h]	39·25	16·8	60	14·1	4·2[a]	4·1[a]	
	(E)[d]	LCBF IV[h] 78·5				23·1	71	17·3	4·1[a]	16·4[a]	

cont.

Table 6.20 *cont.*

Reference		Sorghum diet (%)	Nonprot. supp. (%)	Prot. supp. (%)		Prot. content of diet (%)	Av. feed intake/wk (g)	Av. wt gain/wk (g)	Av. F/G	Av. prot. intake/wk (g)	Av. PER
	(F)[d]	Rice flour 78·5				8·2	52	7·1	7·3[a]	4·3[a]	
	(G)[d]	Rice flour 39·25		LCBF V[h]	39·25	16·3	70	14·3	4·9[a]	11·4[a]	
											Av. NPU (%)
Harden *et al.* (1976)	*Series I*										
	(A)	Prot. free					35	−7·2	−4·9[a]		
	(B)	Casein				8·5	69	21·3	3·2[a]		94·4
	(C)					8·5	50	4·0	12·5[a]		29·8
	(D)			L-Lys HCl	0·4	8·5	55	18·3	3·0[a]		47·6
				L-Thr	0·15						
	(E)			L-Lys HCl	0·4	8·5	57	10·7	5·3[a]		30·2
				L-Met	0·3						
	(F)			L-Lys HCl	0·4	8·5	68	18·1	3·7[a]		36·9
				L-Thr	0·15						
				L-Met	0·3						
	(G)			L-Lys HCl	0·4	8·5	66	19·5	3·4[a]		37·7
				L-Thr	0·15						
				L-Met	0·3						
				L-Trp	0·1						
	(H)			L-Lys HCl	0·4	8·5	67	20·2	3·3[a]		45·2
				L-Thr	0·15						
				L-Met	0·3						
				L-Trp	0·1						
				L-Ile	1·0						
		Series II									
	(A)	Prot. free					35	−8·4	−4·2[a]		
	(B)	Casein				10·0	81	29·6	2·7[a]		
	(C)					10·0	52	2·1	24·8[a]		
	(D)			L-Lys HCl	0·4	10·0	58	12·2	4·7[a]		
	(E)			L-Ile	1·0	10·0	50	1·7	29·4[a]		
	(F)			L-Lys HCl	0·4	10·0	59	11·1	5·3[a]		
				L-Ile	1·0						

											Average PER
Hariharan *et al.* (1965)	(A)[d]	78·5				9·0		9·1		5·6	1·6
	(B)[d]	78·5	0·5[i]			8·9		10·8		5·6	1·9
	(C)[d]	78·5	1·0[i]			8·9		10·8		5·7	1·9
	(D)[d]	73·8		Groundnut flour (GNF)[i]	4·7	11·6		11·7		7·2	1·6
	(E)[d]	73·5		Fortified GNF[i]	5·0	11·6		11·5		7·3	1·6
	(F)[d]	71·3		Fortified SMP[i]	7·2	11·0		13·8		7·0	2·0
	(G)[d]	69·1		GNF	9·4	13·5		11·6		8·3	1·4
	(H)[d]	68·5		Fortified GNF	10·0	13·6		15·4		9·1	1·7
	(I)[d]	64·1		Fortified SMP[i]	14·4	13·7		14·4		8·7	1·6
											(Adjusted)
Howe *et al.* (1965)	(A)					9·0		4·5			0·7
	(B)			L-Lys HCl	0·1	9·0		11·0			1·4
	(C)			L-Lys HCl	0·2	9·0		16·2			1·8
	(D)			L-Lys HCl	0·2	9·0		16·2			1·7
				DL-Thr	0·2						
	(E)	Casein[j]				9·0		21·5			2·5
	(F)			L-Lys HCl	0·4	9·0		24·2			2·3
	(G)			L-Lys HCl	0·4	9·0		30·5			2·4
				DL-Thr	0·3						
	(H)	Casein[j]				9·0		22·2			2·5
Joseph *et al.* (1959b)	(A)[d]	78·5				9·9	74	6·8	10·9[a]		
	(B)[d]	66·0		Indian MPF[k]	12·5	14·3	73	11·5	6·3[a]		
	(C)[d]	66·0		Indian MPF with 20% added SMP	12·5	14·1	76	12·1	6·3[a]		
	(D)[d]	66·0		SMP	12·5	13·4	78	11·8	6·6[a]		
Kuppuswamy *et al.* (1957)	(A)[d]	78·5					57	8·4	6·8[a]		
	(B)[d]	66·0		Indian MPF[k]	12·5		71	14·0	5·1[a]		
	(C)[d]	66·0		American MPF[k]	12·5		71	14·4	5·0[a]		
Kurien *et al.* (1960e)[l]	(A)[d]			Rice	78·5	8·4	54	5·7	9·5[a]	4·5[a]	
	(B)[d]	19·25		Rice	59·25	8·6	53	6·6	8·0[a]	4·5[a]	
	(C)[d]	39·25		Rice	39·25	8·9	55	6·3	8·7[a]	4·9[a]	
	(D)[d]	78·5				9·4	56	6·5	8·6[a]	5·3[a]	
Kurien *et al.* (1971a)	(A)[m]	83·5				10·8		4·4		3·7	1·2
	(B)[m]	80·5	3·0[e]			10·8		7·0		4·9	1·4
	(C)[m]	75·0		Pigeon peas	8·5	12·0		8·0		5·7	1·4
	(D)[m]	72·0	3·0	Pigeon peas	8·5	11·9		11·3		6·0	1·9
	(E)[m]	66·8		Pigeon peas	16·7	13·1		9·6		6·0	1·6

cont.

Table 6.20 *cont.*

Reference		Sorghum diet (%)	Nonprot. supp. (%)	Prot. supp. (%)		Prot. content of diet (%)	Av. feed intake/wk (g)	Av. wt gain/wk (g)	Av. F/G	Av. prot. intake/wk (g)	Av. PER
	(F)[m]	63·8	3·0	Pigeon peas	16·7	13·1		13·8		6·8	2·0
	(G)[m]	58·5		Pigeon peas	25·0	14·2		10·2		6·5	1·6
	(H)[m]	55·5	3·0	Pigeon peas	25·0	14·2		15·4		7·5	2·0
Narayanaswamy *et al.* (1970a)	(A)[dn]	75·5	3·0[e]			10·1		11·2		5·5	2·0
	(B)[dn]	75·5	3·0	L-Lys HCl	0·3	10·2		17·9		6·0	3·0
	(C)[dn]	75·5	3·0	DL-Met	0·09	10·1		11·7		5·3	2·2
	(D)[dn]	75·5	3·0	DL-Thr	0·1	10·2		11·3		5·8	1·9
	(E)[dn]	75·5	3·0	L-Lys HCl DL-Met	0·3 0·09	10·2		18·6		6·1	3·0
	(F)[dn]	75·5	3·0	L-Lys HCl DL-Thr	0·3 0·1	10·2		19·1		6·1	3·1
	(G)[dn]	75·5	3·0	DL-Met DL-Thr	0·09 0·1	10·2		10·9		5·9	1·8
	(H)[dn]	75·5	3·0	L-Lys HCl DL-Met DL-Thr	0·3 0·09 0·1	10·2		20·0		6·2	3·2
Narayanaswamy *et al.* (1971)	(A)[d]	78·5				10·4		6·2		3·5	1·8
	(B)[d]	75·5	3·0[e]			10·4		9·7		3·9	2·5
	(C)[d]	65·5	3·0	LCPF[o]	10·0	11·9		16·4		6·1	2·7
	(D)[d]	55·5	3·0	LCPF	20·0	13·4		17·6		6·2	2·8
	(E)[d]	68·5		Chickpeas[o]	10·0	11·4		10·7		5·0	2·1
	(F)[d]	65·5	3·0	Chickpeas	10·0	11·2		13·5		5·2	2·6
	(G)[d]	58·5		Chickpeas	20·0	13·2		13·3		5·9	2·2
	(H)[d]	55·5	3·0	Chickpeas	20·0	13·2		16·5		6·6	2·5
	(I)[d]	69·5	3·0	Fortified SMP[o]	6·0	11·9		16·6		5·3	3·1
	(J)[d]	63·5	3·0	Fortified SMP	12·0	13·4		17·8		6·0	3·0
Narayanaswamy *et al.* (1972a)	(A)[d]	78·5				10·8		9·7		5·8	1·7
	(B)[d]	75·5	3·0[e]			10·7		14·6		6·0	2·4
	(C)[d]	65·5	3·0	LCPF[p]	10·0	12·1		16·2		6·1	2·6
	(D)[d]	55·5	3·0	LCPF	20·0	13·4		18·3		6·8	2·7
	(E)[d]	69·5	3·0	Fortified SMP[p]	6·0	12·1		17·9		5·7	3·1
	(F)[d]	63·5	3·0	Fortified SMP	12·0	13·5		19·1		6·2	3·1
Narayanaswamy *et al.* (1972c)[q]	(A)[d]	78·5				10·3		7·5		5·1	1·5
	(B)[d]	75·5	3·0[e]			10·2		11·3		5·4	2·1
	(C)[d]	72·5		Petroleum grown yeast	6·0	12·2		10·8		6·2	1·8
	(D)[d]	69·5	3·0	Petroleum grown yeast	6·0	12·2		17·0		7·5	2·3

Narayanaswamy *et al.* (1975)		*Series I*									
	(A)	87·0	3·0[e]			10·0		3·6	9·2	3·3	1·1
	(B)	65·2	3·0	Cowpeas	11·4	10·0		9·7	4·6	5·4	1·8
	(C)	43·5	3·0	Cowpeas	22·8	10·0		12·4	5·6	5·7	2·2
	(D)		3·0	Cowpeas	45·0	10·0		13·1	4·6	6·0	2·2
		Series II									
	(E)[d]	90·0	3·0			10·2		3·2	12·0	4·0	0·8
	(F)[d]	81·0	3·0	Cowpeas	9·0	11·4		11·0	4·4	5·5	2·0
	(G)[d]	72·0	3·0	Cowpeas	18·0	12·7		13·6	4·4	6·5	2·1
	(H)[d]	63·0	3·0	Cowpeas	27·0	14·0		14·8	3·8	7·3	2·0
Panemangalore *et al.* (1965)	(A)[d]					8·7	51	6·0	8·5	4·4[a]	
	(B)[d]			Prot. food I[r]		10·9	68	12·1	5·6	7·4[a]	
	(C)[d]			Prot. food I		11·0	65	11·9	5·5	7·1[a]	
				L-Lys							
				DL-Met							
	(D)[d]			Prot. food II[r]		10·9	66	11·1	6·0	7·2[a]	
	(E)[d]			Prot. food II		11·0	62	11·1	5·6	6·8[a]	
				L-Lys							
				DL-Met							
	(F)[d]			SMP		10·9	65	12·6	5·1	7·1[a]	
Peyrot and Adrian (1970)[s]		(% *prot.*)		(% *prot.*)							
	(A)	50 (poor)		Groundnut cake		12·0		16·1		12·5	1·3
				(GNC)	50						
	(B)	50 (poor)		GNC	50	12·0		18·5		13·4	1·4
				L-Lys HCl of diet	0·2%						
	(C)	50 (poor)		GNC	50	12·0		18·3		12·7	1·4
				L-Met of diet	0·1%						
	(D)	50 (rich)		GNC	50	12·0		21·5		9·5	2·3
	(E)	50 (rich)		GNC	50	12·0		34·6		12·6	2·7
				L-Lys HCl of diet	0·2%						
	(F)	50 (rich)		GNC	50	12·0		23·0		10·6	2·2
				L-Met of diet	0·1%						
	(G)	50 (poor)		GNC	50	9·0		10·3		13·6	0·7
	(H)	50 (poor)		GNC	50	9·0		9·5		14·1	0·7
				L-Lys HCl of diet	0·2%						
	(I)	50 (poor)		GNC	50	9·0		14·4		13·9	1·0
				L-Met of diet	0·1%						
	(J)	50 (rich)		GNC	50	9·0		21·5		11·2	1·9
	(K)	50 (rich)		GNC	50	9·0		30·7		12·6	2·4
				L-Lys HCl of diet	0·2%						
	(L)	50 (rich)		GNC	50	9·0		21·6		11·6	1·9
				L-Met of diet	0·1%						

cont.

Table 6.20 *cont.*

Reference		Sorghum diet (%)	Nonprot. supp. (%)	Prot. supp. (%)		Prot. content of diet (%)	Av. feed intake/wk (g)	Av. wt gain/wk (g)	Av. F/G	Av. prot. intake/wk (g)	Av. PER
Phansalkar *et al.* (1957)		(*% prot.*)		(*% prot.*)							
	(A)	9·8				9·8		14·0			1·6
	(B)	7·0		Chickpeas	3·0	10·2		16·2			1·9
	(C)	7·0		Urd	3·0	10·2		18·2			2·0
	(D)	7·0		Mung beans	3·0	10·5		16·7			1·8
	(E)	7·0		Pigeon peas	3·0	10·2		16·5			1·8
	(F)	9·0		Amaranth	1·0	9·8		14·5			1·7
	(G)	6·0		Chickpeas	3·0						
				Amaranth	1·0	10·2		17·7			1·9
	(H)	6·0		Urd	3·0						
				Amaranth	1·0	10·3		15·7			1·8
	(I)	6·0		Mung beans	3·0						
				Amaranth	1·0	10·5		16·2			1·7
	(J)	6·0		Pigeon peas	3·0						
				Amaranth	1·0	10·3		17·2			1·8
Phansalkar *et al.* (1957)		(*% prot.*)		(*% prot.*)							
	(A)	6·0		Pigeon peas	3·0	10·3		17·2			2·4
				Amaranth	1·0						
	(B1)			SMP	10·5	10·5		21·0			2·6
Pushpamma (1968, Pushpamma *et al.* (1972)[f]	(A)	96·5				9·9	77	10	7·5	7·5	1·3
	(B)	96·5		L-Lys HCl to total 0·75 in diet		9·9	100	22	4·6	10·3	2·1
	(C)	96·5		L-Trp	0·1	9·9	74	10	7·5	7·3	1·3
	(D)	96·5		L-Lys HCl to total 0·75 in diet		9·9	95	25	3·9	9·9	2·5
				L-Trp	0·1						
	(E)	Casein				10·0	85	22	3·9	8·5	2·5
				(*% prot.*)							
	(F)	86·0		SMP	3·0	11·8	93	21	4·5	11·0	1·9
	(G)	86·0		Isolated soybean prot.	3·0	11·8	98	19	5·2	11·5	1·6
	(H)	86·0		Fish prot. concentrate	3·0	11·8	107	28	3·8	12·6	2·2
	(I)	86·0		Dried brewers' yeast	3·0	11·8	94	18	5·2	11·1	1·6
	(J)	Casein				12·5	75	21	3·6	9·4	2·2

Saxena *et al.* (1966)	(A)[d]	78·5				50	5·6	8·9[a]		
	(B)[d]	66·0	Rice polishing concentrate[u]	12·5		52	8·6	6·0[a]		
Saxena *et al.* (1968)					(“*as-is*” *basis*)					
	(A)[d]	78·5			7·7	55	5·6	9·8[a]		
	(B)[d]	66·0	Rice polishing concentrate[u]	12·5	8·7	61	8·6	7·1[a]		
Shurpalekar *et al.* (1962a)	(A)[d]	78·5			10·6	52	7·2	7·2[a]	5·5[a]	
	(B)[d]	75·5	Fortified fish flour	3·0	13·0	64	14·0	4·6[a]	8·3[a]	
	(C)[d]	71·5	Fortified SMP	7·0	13·1	63	13·6	4·6[a]	8·2[a]	
Sirinit *et al.* (1965)										*Ajusted*
	(A)	80·0	Red bean	20·0	10·0		9·6			1·4
	(B)	80·0	White bean	20·0	9·9		14·7			1·7
	(C)	70·0	White bean	30·0	10·3		15·5			1·8
	(D)	70·0	Red bean	30·0	10·1		18·2			2·0
	(E)	80·0	Black bean	20·0	9·8		18·5			1·8
	(F)	70·0	Black bean	30·0	10·0		21·0			1·9
	(G)	Casein			10·0		33·0			2·5
	(H)	Incaparina			10·2		36·2			1·9
Subrahmanyan *et al.* (1954d)[v]	(A)[d]	78·5			11·4	64	4·9	13·1[a]	7·3[a]	
	(B)[d]	58·9	Synthetic grain I	19·6	10·5	62	5·1	12·1[a]	6·5[a]	
	(C)[d]	58·9	Synthetic grain II	19·6	11·2	64	6·3	10·1[a]	7·2[a]	
	(D)[d]	58·9	Synthetic grain III	19·6	11·6	64	7·0	9·1[a]	7·4[a]	
Sur *et al.* (1954a)	(A)[d]	Type I[w] 78·5				38	1·8	21·1[a]		
	(A)[d]	Type II[w] 78·5				52	4·8	10·8[a]		
	(A)[d]	Type III[w] 78·5				53	7·3	7·3[a]		
	(B)[d]	Type I 76·0	Food yeast[w]	2·5		38	2·4	15·8[a]		
	(B)[d]	Type III 76·0	Food yeast	2·5		54	7·7	7·0[a]		
	(C)[d]	Type I 73·5	Food yeast	5·0		59	5·9	10·0[a]		
	(C)[d]	Type II 73·5	Food yeast	5·0		54	5·6	9·6[a]		
	(C)[d]	Type III 73·5	Food yeast	5·0		50	8·8	5·7[a]		

cont.

Table 6.20 *cont.*

Reference		Sorghum diet (%)	Nonprot. supp. (%)	Prot. supp. (%)		Prot. content of diet (%)	Av. feed intake/wk (g)	Av. wt gain/wk (g)	Av. F/G	Av. prot. intake/wk (g)	Av. PER
Sur *et al.* (1954b)	(A)[d]	Type III[w]				5·0					0·2
	(B)[d]	Food yeast[w]				5·0					1·2
	(C)[d]	Type III		Food yeast[w] (protein ratio sorghum 3/yeast 1)		5·0					0·9
	(D)[d]	Type III				8·0					1·2
	(E)[d]	Food yeast				8·0					1·8
	(F)[d]	Type III		Food yeast (protein ratio sorghum 3/yeast 1)		8·0					1·8
Sure *et al.* (1957)	(A)	77·0				8·0	48	1·9	25·3[a]	3·9	0·5
	(B)	76·0		Fish flour[x]	1·0	8·7	60	5·6	10·7[a]	5·2	1·1
	(C)	74·0		Fish flour	3·0	10·0	78	11·3	6·9[a]	7·9	1·4
	(D)	72·0		Fish flour	5·0	11·4	85	14·2	6·0[a]	9·7	1·5
Talwalkar and Patel (1970a)		(% *prot.*)		(% *prot.*)							
	(A)	7·0				7·0		2·4		2·6	0·9
	(B)	5·0		Ambadi (*Hibiscus cannabinus*)	2·0	7·0		2·7		2·8	1·0
	(C)	5·0		Methi (*Trigonella foenumgraecum*)	2·0	7·0		5·5		3·5	1·5
Tasker *et al.* (1964)	(A)[d]	78·5				9·6	50	7·6	6·6	4·8[a]	
	(B)[d]	66·0		Protein food[y]	12·5	14·4	58	14·6	4·0	8·3[a]	
Venkat Rao *et al.* (1958)[z]	(A)[d]	78·5				11·0	57	6·5	8·8[a]	6·3[a]	
	(B)[d]	68·5		Low-fat GNF	10·0	16·1	57	8·7	6·5[a]	9·2[a]	
	(C)	58·5		Low-fat GNF	20·0	21·2	58	9·6	6·0[a]	12·3[a]	

[a] Calculated by the authors from av. wt gain, food intake and prot. content in the diet.
[b] Prot. contents (N × 6·25) (%) sorghum 9·4, maize 10·5.
[c] Prot. contents (N × 6·25) (%) sorghum 9·0, groundnut flour (ORANA) (DWB) 55·9, fish flour (ORANA) (DWB) 79·2.
[d] All diets contained (%) pigeon peas 5·0, groundnut oil 5·0, SMP 0·9, common salt 0·3, green leafy vegetables (*Amaranthus gangeticus*) 2·1, other vegetables (brinjal and potato) 8·2.
[e] Salt mix (Hubbel *et al.*, 1937) 2·0%, vit. premix (Chapman *et al.*, 1959) 1·0%.
[f] Diet (B) Ca salts and vitamins providing same quantities as 10 g fortified coconut meal. Prot. contents (N × 6·25) (DWB, %) unfortified coconut meal 22·5, fortified coconut meal 22·0, fortified SMP 36·4.
[g] Prot. contents (N × 6·25) (%) fortified Bombay Duck fish flour 75·8, fortified prot. food, blend of 1/2/1 fish, groundnut and chickpea flours 52·3, fortified SMP 36·0.
[h] Low cost balanced food (LCBF) III; finger millet flour 65/cottonseed flour 25/chickpea flour 10; prot. content (N × 6·25) (DWB, %) 20·9. LCBF IV; sorghum flour 65/cottonseed flour 25/chickpea flour 10, prot. 22·6. LCBF V; finger millet 65/groundnut flour 25/chickpea flour 10; prot. 21·4.

[i] Diets (B) and (C) Ca salt and vit. premix to provide same quantities as in 5·0 g and 10·0 g respectively fortified groundnut flour (GNF). Prot. contents (N × 6·25) (DWB, %) unfortified GNF 56·8, fortified GNF 53·8 and fortified SMP 36·0.

[j] Casein prot. (N × 6·25) 10 %.

[k] Indian multipurpose food (MPF); low-fat groundnut flour (GNF) 75/roasted chickpea flour 25; fortified with Ca, vit. A and D and riboflavin. Composition of American MPF not stated.

[l] Prot. contents (N × 6·25) (DWB, %) sorghum 8·8, rice 7·4.

[m] All diets as[d] but excluding pigeon peas. Lys and Thr contents in diets (microbiological assay, Barton-Wright, 1952) (mg/g total N) respectively, (A) and (B) 175, 187; (C) and (D) 225, 200; (E) and (F) 275, 212; (G) and (H) 325, 225.

[n] AA composition (microbiological assay, Barton-Wright, 1952) of diets (mg/g total N). Lys (A) 200 (B) 394 (C) 200 (D) 200 (E) 394 (F) 394 (G) 200 (H) 394, egg prot. 400, human milk prot. 394. Thr (A) 225 (B) 225 (C) 225 (D) 287 (E) 225 (F) 287 (G) 287 (H) 287, egg prot. 319, human milk prot. 287. Met (A) 100 (B) 100 (C) 156 (D) 100 (E) 156 (F) 100 (G) 156 (H) 156, egg prot. 194, human milk prot. 137. Cys all diets 112, egg prot. 150, human milk prot. 131.

[o] Low cost prot. food (LCPF) wheat flour 70/soy flour 30, prot. content (N × 6·25) (DWB, %) 24·4; chickpea flour prot. 24·2; SMP prot. 34·9.

[p] Low cost prot. food (LCPF) wheat flour 70/soy flour 15/groundnut flour 15, prot. content (N × 6·25) (DWB, %) 24·5; fortified SMP prot. 35·1.

[q] Petroleum grown yeast prot. content (N × 6·25) (DWB, %) 50·0. AA composition of yeast and diets (microbiological assay, Barton-Wright, 1952) (mg/g total N) Lys yeast 462, (A) 206 (C) 244. Thr yeast 331, (A) 194 (C) 219. Met yeast 112, (A) 106 (C) 200, Cys yeast 31, (A) 69 (C) 62.

[r] Prot. food I; groundnut 40/chickpea 40/sesame flours 20. II groundnut 40/soy 30/sesame flours 30 fortified with Ca salt, vit. A and D, thiamine and riboflavin.

[s] Diets (A), (B), (C), (G), (H), (I) contained sorghum of 7·4 % prot. (N × 6·25), groundnut cake (GNC) 47·5 % protein; diets (D), (E), (F), (J), (K), (L) contained sorghum of 10·4 % prot., GNC 52·8 % prot. Lys and Met contents (no methods stated) (mg/g total N) respectively: sorghum 7·4 % prot. 160, 112; 10·4 % prot. 131, 96; GNC 47·5 % prot. 219, 50; 52·8 % prot. 219, 50.

[t] Prot. contents (N × 6·25) (%) sorghum 10·3, maize 9·7, SMP 35·6, fish prot. concentrate (FPC) 83·0, isolated soybean prot. 88·0, dried brewers' yeast 40·0.

[u] Composition of rice polishing concentrate in text p. 487.

[v] Prot. contents (N × 6·25) (DWB, %) sorghum 11·2, synthetic grains I 7·1, II 9·8, III 12·4.

[w] Prot. contents (N × 6·25) (%) sorghum type I 8·2, type II 9·3, type III 9·4, food yeast (*Torula utilis*) 60·4.

[x] Defatted fish flour made from a mixture of carp, smelts and whitings (Levin and Finn, 1955).

[y] Prot. food (Krishnamurthy *et al.*, (1958) screw pressed coconut meal 1/chickpea flour 1/groundnut flour 2, fortified with Ca vit. A and D, thiamine and riboflavin.

[z] Prot. contents (N × 6·25) (DWB, %) sorghum 10·9, low-fat groundnut flour (GNF) 54·3.

Table 6.21
Results of metabolic trials with eight boys on mixed diets of rice and pearl millet.[a]

Diet[b] (parts by wt)	Prot. (N × 6·25) mean daily			AD[c]	Calcium mean daily			Phosphorus mean daily		
	Intake (g)	Excr. (g)	Balance (g)	(%)	Intake (mg)	Excr. (mg)	Balance (mg)	Intake (mg)	Excr. (mg)	Balance (mg)
(A) Rice 4	6·9	4·9	+2·0	75·3	352	233	+119	726	564	+162
(B) Rise 3/pearl millet 1	7·3	5·4	+1·9	73·1	378	261	+117	867	643	+224
(C) Rice 2/pearl millet 2	7·8	6·3	+1·5	64·4	418	305	+113	1029	732	+297
(D) Pearl millet 4	8·7	7·6	+1·1	52·9	479	287	+92	1346	990	+356

[a] Kurien *et al.* (1961b).
[b] Prot. contents (N × 6·25), (DWB) (%) rice 8·8, pearl millet 12·8.
[c] Apparent digestibility.

Table 6.22
Effect of skimmed milk powder in supplementing typical Indian diets on height and weight gains of young boys.[a]

	Wt inc. (kg)	Ht inc. (cm)
First trial		
Group A (receiving milk)	2·16	1·55
Group B (no milk)	0·97	0·89
Second trial		
Group B (receiving milk)	1·39	1·75
Group A (no milk)	0·50	1·09

[a] Aykroyd and Krishnan (1937).

Table 6·23

Effect of protein (N × 6·25) supplementation of diets (DWB) based on pearl millet on growth of weanling rat. Blank spaces indicate no values presented.

Reference		Pearl millet diet (%)	Nonprot. supp. (%)	Prot. supp.	Prot. supp. (%)	Prot. content of diet (%)	Av. feed intake/wk (g)	Av. wt gain/wk (g)	Av. F/G	Av. prot. intake/wk (g)	Av. PER
Adrian and Frangne (1970)[b]	(A)			Maize	82·0	8·4	56	4·9	11·4[a]	4·7[a]	1·1[a]
	(B)	16·0		Maize	60·0	8·4	53	5·7	9·2[a]	4·4[a]	1·3[a]
Daniel *et al.* (1965a)	(A)					10·2		6·2		3·6	1·7
	(B)			L-Lys HCl	0·35	10·7		19·1		5·6	3·4
	(C)			L-Lys HCl	0·35	10·8		21·0		5·9	3·5
				DL-Thr	0·14						
	(D)[c]	78·5				12·2		8·2		5·0	1·7
	(E)[c]	78·5		L-Lys HCl	0·27	12·4		11·0		5·8	1·9
	(F)[c]	78·5		L-Lys HCl	0·27	12·4		11·6		5·8	2·0
				DL-Thr	0·1						
	(G)[c]	75·5	3·0[d]			12·0		13·0		6·0	2·2
	(H)[c]	75·5	3·0[d]	L-Lys HCl	0·27	12·1		22·9		7·2	3·2
	(I)[c]	75·5	3·0[d]	L-Lys HCl	0·27	12·2		24·5		7·6	3·2
				DL-Thr	0·1						
Daniel *et al.* (1968a)	(A)[c]	78·5				12·1		10·8		5·3	2·0
	(B)[c]	75·5	3·0[d]			12·0		14·7		5·5	2·6
	(C)[c]	63·5		Chickpeas	15·0	14·1		14·5		5·6	2·6
	(D)[c]	60·5	3·0	Chickpeas	15·0	14·0		21·7		7·6	2·8
	(E)[c]	63·5		Pigeon peas	15·0	14·1		14·1		6·2	2·2
	(F)[c]	60·5	3·0	Pigeon peas	15·0	14·0		20·0		7·1	2·8
	(G)[c]	73·5		Soy flour	5·0	14·1		15·6		6·3	2·5
	(H)[c]	70·5	3·0	Soy flour	5·0	14·0		21·7		7·2	3·0
	(I)[c]	68·5		Soy flour	10·0	16·2		17·2		7·2	2·4
	(J)[c]	65·5	3·0	Soy flour	10·0	16·1		25·2		8·6	2·9
	(K)[c]	71·5		SMP	7·0	14·2		21·7		7·5	2·9
	(L)[c]	68·5	3·0	SMP	7·0	14·1		24·7		8·0	3·1
	(M)[c]	64·5		SMP	14·0	16·4		26·1		9·7	2·7
	(N)[c]	61·5	3·0	SMP	14·0	16·3		28·4		9·4	3·0
Hariharan *et al.* (1965)	(A)[c]	78·5				11·0		12·3		7·4	1·7
	(B)[c]	78·5	0·5[e]			11·0		14·9		7·6	2·0
	(C)[c]	78·5	1·0[e]			11·4		15·6		7·5	2·1
	(D)[c]	73·8		Groundnut flour		13·3		14·6		8·5	1·7
				(GNF)[e]	4·7						

cont.

Table 6.23 *cont.*

Reference		Pearl millet diet (%)	Nonprot. supp. (%)	Prot. supp. (%)		Prot. content of diet (%)	Av. feed intake/wk (g)	Av. wt gain/wk (g)	Av. F/G	Av. prot. intake/wk (g)	Av. PER
	(E)[c]	73·5		Fortified GNF[e]	5·0	13·1		17·1		8·9	1·9
	(F)[c]	71·3		Fortified SMP[e]	7·2	13·1		16·5		8·8	1·9
	(G)[c]	69·1		GNF	9·4	15·9		16·3		10·2	1·6
	(H)[c]	68·5		Fortified GNF	10·0	15·8		17·8		10·5	1·7
	(I)[c]	64·1		Fortified SMP	14·4	15·6		19·6		10·7	1·8
Kurien *et al.* (1962)[f]	(A)[c]			Rice	78·5	9·8	57	5·8	9·8[a]	5·6[a]	
	(B)[c]	19·25		Rice	59·25	10·9	56	7·4	7·6[a]	6·1[a]	
	(C)[c]	39·25		Rice	39·25	11·7	55	7·3	7·5[a]	6·4[a]	
	(D)[c]	78·5				13·5	57	7·3	7·8[a]	7·7[a]	
Kurien *et al.* (1971b)	(A)[g]	75·5	3·0[d]			12·3		16·3		6·8	2·4
	(B)[g]	75·5	3·0	L-Lys HCl	0·3	12·3		25·0		7·4	3·4
	(C)[g]	75·5	3·0	DL-Met	0·1	12·3		16·3		6·1	2·7
	(D)[g]	75·5	3·0	DL-Thr	0·17	12·3		17·3		6·1	2·8
	(E)[g]	75·5	3·0	L-Lys HCl	0·3						
				DL-Met	0·1	12·3		28·0		8·0	3·5
	(F)[g]	75·5	3·0	L-Lys HCl	0·3						
				DL-Thr	0·17	12·3		23·8		6·9	3·4
	(G)[g]	75·5	3·0	DL-Met	0·1						
				DL-Thr	0·17	12·3		19·1		7·1	2·7
	(H)[g]	75·5	3·0	L-Lys HCl	0·3						
				DL-Met	0·1	12·3		28·5		8·1	3·5
				DL-Thr	0·17						
Phansalkar *et al.* (1957)		(*% prot.*)		(*% prot.*)							
	(A)	9·4				9·4		12·2			1·6
	(B)	7·0		Chickpeas	3·0	9·6		18·0			2·2
	(C)	7·0		Urd	3·0	9·5		17·7			2·1
	(D)	7·0		Mung beans	3·0	9·9		18·7			2·1
	(E)	7·0		Pigeon peas	3·0	9·4		16·5			2·0
	(F)	9·0		Amaranth	1·0	9·7		14·5			1·7
	(G)	6·0		Chickpeas	3·0						
				Amaranth	1·0	9·7		19·5			2·2
	(H)	6·0		Urd	3·0						
				Amaranth	1·0	9·4		16·5			2·0
	(I)	6·0		Mung beans	3·0						
				Amaranth	1·0	9·4		19·5			2·3

	(J)	6·0	Pigeon peas Amaranth	3·0 1·0	9·4		18·5			2·2
Phansalkar *et al.* (1957)		(% *prot.*)	(% *prot.*)							
	A(1)	6·0	Pigeon peas Amaranth	3·0 1·0	10·8		18·2			2·3
	B(1)		SMP	10·5	10·5		21·0			2·6
Pushpamma (1968), Pushpamma *et al.* (1972)[h]	(A)	96·5			11·0	85	17	5·0	9·3	1·8
	(B)	96·5	L-Lys HCl to total in diet	0·75	11·0	101	28	3·6	11·5	2·4
	(C)	96·5	L-Trp	0·1	11·0	81	16	5·0	8·9	1·8
	(D)	96·5	L-Lys HCl to total in diet L-Trp	0·75 0·1	11·0	100·4	27	3·7	11·5	2·3
	(E)	Casein			10·0	85·0	22	3·9	8·5	2·5
			(% *prot.*)							
	(F)	86·0	SMP	3·0	12·5	94	26	3·6	11·8	2·2
	(G)	86·0	Isolated soybean prot.	3·0	12·5	101	29	3·5	12·7	2·3
	(H)	86·0	Fish prot. concentrate	3·0	12·5	97	24	4·0	11·9	2·0
	(I)	86·0	Dried brewers' yeast	3·0	12·5	91	22	4·1	11·4	2·0
	(J)	Casein			12·5	75	21	3·6	9·4	2·2
Rama Rao *et al.* (1953b)[i]		(% *prot.*)	(% *prot.*)							
	(A)	10·0			10·0					1·4
	(B)	Groundnut cake 10·0			10·0					1·4
	(C)	5·0	Groundnut cake	5·0	10·0					1·6
	(D)	10·0			10·0					1·5
	(E)	Chickpeas 10·0			10·0					1·5
	(F)	5·0	Chickpeas	5·0	10·0					1·8

[a] Calculated by the authors from av. wt gain, food intake and prot. content in the diet.
[b] Prot. contents (N × 6·25) (%) pearl millet 10·9, maize 10·5.
[c] All diets contained (%) pigeon peas 5·0, groundnut oil 5·0, SMP 0·9, common salt 0·3, green leafy vegetables (*Amaranthus gangeticus*) 2·1, other vegetables (brinjal and potato) 8·2.
[d] Salt mix 2·0% (Hubbel *et al.*, 1973), vit. premix 1·0% (Chapman *et al.*, 1959).

cont.

Table 6.23 *cont.*

[e] Diets (B) and (C) Ca salt and vit. premix to provide same quantities as in 5 g and 10 g respectively fortified groundnut flour (GNF). Prot. contents ($N \times 6{\cdot}25$) (DWB, %) unfortified GNF 56·8, fortified GNF 53·8, fortified SMP 36·0.

[f] Prot. contents ($N \times 6{\cdot}25$) (DWB, %) pearl millet 9·9, rice 8·8.

[g] All diets as[c] but chickpeas replacing pigeon peas, groundnut oil 4·0%, groundnut oil with vit. A and D 1·0%. AA composition (microbiological assay, Barton-Wright, 1952) of diets (mg/g total N). Lys (A) 231 (B) 394 (C) 231 (D) 231 (E) 394 (G) 231 (H) 394, egg protein 400, human milk prot. 394. Thr (A) 244 (B) 244 (C) 244 (D) 287 (E) 244 (F) 287 (G) 287 (H) 287, egg prot. 319, human milk prot. 287. Met + Cys (A) 218 (B) 218 (C) 268 (D) 218 (E) 268 (F) 218 (G) 268 (H) 268, egg protein 344, human milk prot. 268.

[h] Prot. contents ($N \times 6{\cdot}25$) (%) finger millet 5·6, SMP 35·6, fish prot. concentrate (FPC) 83·0, isolated soybean prot. 88·0, dried brewers' yeast 40·0.

[i] Prot. contents ($N \times 6{\cdot}25$) (%) pearl millet 14·7, groundnut cake 52·7, chickpeas 24·8. Quantities of pearl millet, groundnut cake and chickpeas in diet not recorded.

Table 6·24
Effect of protein (N × 6·25) supplementation of finger millet diets on biological value for rats. Blank spaces indicate no values presented.[a]

Finger millet diet providing prot. (%)	Prot. supp. providing protein (%)		Prot. content of diet (%)	DC (%)	N balance method: BV (%)	Rat growth method: BV (4 weeks) (%)
(A) 4·0	SMP	4·0	8·0	93	94	2·0
(B) 4·0	Pigeon peas	4·0	8·0	83	84	1·0
(C) 4·0	Chick peas	4·0	8·0	84	76	1·4
(D) 4·0	Mung bean	4·0	8·0	86	84	1·5
(E) 4·0	Urd	4·0	8·0	81	76	1·5
(F) 4·0	Soybean	4·0	8·0	82	71	
(G) 4·0	Lentil	4·0	8·0	91	71	
(H) 4·0	SMP	2·0	8·0	92	94	1·5
	Pigeon peas	2·0				
(I) 4·0	SMP	2·0	8·0	87	88	1·9
	Chick peas	2·0				
(J) 4·0	SMP	2·0	8·0	91	87	1·9
	Mung bean	2·0				
(K) 4·0	SMP	2·0	8·0	82	89	1·9
	Urd	2·0				
(L) 4·0	SMP	2·0	8·0	83	87	
	Soybean	2·0				
(M) 4·0	SMP	2·0	8·0	90	84	
	Lentil	2·0				

[a] Swaminathan (1938b).

Table 6.25
Amino acid (AA) scores of 4 amino acids in finger millet supplemented with soy flour, cottonseed flour, fish meal and groundnut flour.[a]

AAs	AA score[b] Parts of supp./100 parts finger millet by weight: 0	4	8	16	32
			Plus soy flour		
Lys	39	51	59	70	82
Thr	78	82	85	89	93
Met + Cys	94	89	85	81	76
Trp	164	157	153	147	141
			Plus cottonseed flour		
Lys	39	47	52	60	69
Thr	78	81	83	86	90
Met + Cys	94	91	90	87	85
Trp	164	161	160	157	155
			Plus fish meal		
Lys	39	63	78	95	111
Thr	78	84	88	93	97
Met + Cys	94	92	91	90	89
Trp	164	144	133	119	105
			Plus groundnut flour		
Lys	39	46	51	57	64
Thr	78	78	79	79	79
Met + Cys	94	89	86	81	77
Trp	164	155	149	141	132

[a] Jansen (1974).
[b] Values listed are AA scores for individual AAs as compared to whole egg. AA scores calculated as follows:

$$\text{AA score} = \frac{\text{mg AA/g essential AA N} \times 100}{\text{mg AA in whole egg/g essential AA N in whole egg}}$$

Table 6.26
Results of metabolic trials with eight girls on mixed diets[a] based on rice and finger millet.[b]

Diet	Content in diet				Metabolic results[c]						
					N			Ca		P	
	Prot. (N × 6·25) (%)	Ca (mg/100 g)	P (mg/100 g)	Phytate P (mg/100 g)	MDI (g)	MDR (g)	ADC (%)	MDI (mg)	MDR (mg)	MDI (mg)	MDR (mg)
(A) Rice 4	27·6	259	464	264	441	1·48	70·7	258	51	464	117
(B) Rice 3/finger millet 1	27·9	471	577	383	446	1·18	66·5	471	106	577	166
(C) Rice 2/finger millet 2	28·0	693	680	502	448	0·89	63·5	693	175	680	125
(D) Finger millet 4	28·2	1151	887	740	451	0·52	53·2	1151	226	887	135

[a] All diets contained approximately 37% pigeon peas, 1% SMP and various leafy and nonleafy vegetables and condiments. The cereals (rice, finger millet) constituted approximately 62% of the diet.
[b] Joseph *et al.* (1959a).
[c] MDR—mean daily retention; MDI—mean daily intake; ADC—apparent digestibility coefficient.

Table 6.27
Percentage growth rates of rats fed a poor rice diet supplemented with butter and finger millet.[a]

Butter in diet (g)	Proportion of finger millet in diet			
	Nil	One-third	One-sixth	Mean
Nil	5·0	9·3	12·5	8·9
0·4	3·1	7·1	8·1	6·1
0·8	0·9	5·6	4·9	3·8
Mean	3·0	7·3	8·5	6·3

[a] Mason *et al.* (1946).

Table 6.28

Amino acid (AA) content,[a] mean daily intake and absorption of essential AAs by children from diets A and D as compared with the AA requirement.[b]

AAs	Diet (A) Finger millet			Diet (D) SMP			
	AA content (mg/g total N)	Intake (mg/kg)	Absorption[c] (mg/kg)	AA content (mg/g total N)	Intake (mg/kg)	Absorption[c] (mg/kg)	AA requirement[d] (mg/kg)
Ile	387	81·2	55·2	394	78·7	69·3	30·0
Leu	581	121·8	82·8	606	121·3	106·7	45·0
Lys[e]	250	52·4	35·6	469	93·8	82·5	60·0
Met	187	39·3	26·7	150	30·0	26·4	
Cys	137	28·8	19·6	56	11·3	9·9	27·0
Phe	262	55·0	37·4	306	61·3	45·1	27·0
Thr[e]	219	45·9	31·2	281	56·2	49·5	35·0
Trp	81	17·0	11·6	81	16·3	14·3	9·0
Val	394	82·5	56·1	431	86·2	75·9	33·0
Arg	325	68·1	46·3	244	48·8	42·9	
His	112	23·0	15·6	162	32·5	28·6	

[a] Krishnamurthy *et al.* (1960).
[b] Daniel *et al.* (1965b).
[c] Intake × digestibility coefficient/100.
[d] Data of Nakagawa *et al.* (1962).
[e] Lys content of diets (B) and (C) 412 mg/g total N; Thr content of diet (C) 306 mg/g total N.
[f] Mean daily intake of foodstuffs (g). (A) finger millet 360, sugar 100, groundnut oil (fortified with vit. A and D) 24, pigeon peas 15, condiments 8, SMP 5, salt mixture 5; (B) as (A) plus L-Lys HCl 0·9; (C) as (A) plus L-Lys HCl 0·9, DL-Thr 0·58; (D) sugar 19, groundnut oil (fortified with vit. A and D) 43, maize starch 290, SMP 70 and salt mixture 2. All diets contained vitaminized starch 5, common salt 12, tamarind pulp 8, vegetables 80 and onions 20.

Table 6.29

Effect of supplementation of poor finger millet diets with L-Lys HCl and DL-Thr on the biological value (BV) of the protein in girls' diets.[a]

Diets	N							Prot. ($N \times 6{\cdot}25$)		
	Intake	Total excretion	Balance	AD[b] (%)	TD[b] (%)	BV (%)	NPU_{op} (%)	Intake (g/kg)	NAP[c] (g/kg)	WHO (1965) (g/kg)
	(g/child/day)									
(A) Finger millet	4·7	4·5	0·3	49·7	67·8	67·0	45·5	1·3	0·60	0·72
(B) Finger millet + L-Lys HCl	4·8	4·1	0·6	51·5	69·4	75·9	52·7	1·3	0·70	0·72
(C) Finger millet + L-Lys HCl + DL-Thr	4·9	3·8	1·0	55·3	73·0	81·2	59·3	1·3	0·80	0·72
(D) SMP	4·5	3·0	1·5	68·4	87·7	85·3	74·8	1·2	0·94	0·72

[a] Daniel *et al.* (1965b).
[b] AD—apparent digestibility; TD—true digestibility.
[c] Net Available Prot. = (prot. intake × NPU)/100.

Table 6.30

Effect of supplementing a poor finger millet diet with L-Lys HCl, leaf protein and low-fat sesame flour on N retention and growth of boys.[a]

Diet[b]	N Intake	N Total excretion	N Balance	AD[c] (%)	Ht inc. (cm)	Wt inc. (kg)	Hb inc. (g/100 ml blood)	RBC[d] ($10^6/mm^3$)
	(g/child/day)							
(A) Finger millet providing prot. 39 g, Lys 1·79 g	6·1	5·3	0·8	55·1	2·2	0·5	0·29	0·06
(B) Finger millet + L-Lys HCl 0·66 g; providing prot. 39 g, Lys 2·29 g	6·3	4·8	1·5	60·2	4·2	1·0	0·64	0·22
(C) Finger millet + leaf prot. 15 g; providing prot. 49 g, Lys 2·34 g	7·7	5·9	1·9	66·0	4·8	1·3	0·87	0·23
(D) Finger millet + low fat sesame flour 24 g; providing prot. 49 g, Lys 2·05 g	7·8	6·2	1·6	64·4	3·5	0·9	0·73	0·19

[a] Doraiswamy *et al.* (1969).

[b] All diets contained, mean intake (g/child/day): finger millet 350, pulses (horse gram, *Dolichos biflorus*) 33, vegetables 106, SMP 10, jaggery (palm sugar) 25, salt 18, condiments and spices 3, tamarind fruit pulp 4 and groundnut oil 15. Additionally diets contained (A) jaggery 24; (B) jaggery 12, sugar 12; and (C) jaggery 10. (The jaggery was added to make the diets isocaloric.)

Prot. (N × 6·5), (DWB) (%) and Lys (microbiological assay, Barton-Wright, 1952) (mg/g total N) contents, leaf prot. 69·3, 344; and sesame flour 45·0, 162.

[c] AD—apparent digestibility.

[d] RBC—red blood cell.

Table 6.31
Effect of protein (N × 6·25) supplementation of diets (DWB) based on finger millet on growth of weanling rats. Blank spaces indicate no values presented.

Reference		Finger millet diet (%)	Nonprot. supp. (%)	Prot. supp.	(%)	Prot. content of diet (%)	Av. feed intake/wk (g)	Av. wt gain/wk (g)	Av. F/G	Av. prot. intake/wk (g)	Av. PER
Daniel *et al.* (1968a)	(A)[a]	78·5				8·1		9·3		4·6	2·0
	(B)[a]	75·5	3·0[b]			8·0		10·4		4·7	2·2
	(C)[a]	63·5		Chickpeas	15·0	10·7		18·4		7·4	2·5
	(D)[a]	60·5	3·0	Chickpeas	15·0	10·7		20·1		7·4	2·7
	(E)[a]	63·5		Pigeon peas	15·0	10·7		17·6		7·3	2·4
	(F)[a]	60·5	3·0	Pigeon peas	15·0	10·7		18·8		7·4	2·5
	(G)[a]	73·5		Soy flour	5·0	10·6		18·7		7·7	2·4
	(H)[a]	70·5	3·0	Soy flour	5·0	10·6		20·2		7·1	2·8
	(I)[a]	68·5		Soy flour	10·0	13·7		20·2		8·0	2·5
	(J)[a]	65·5	3·0	Soy flour	10·0	13·6		22·7		8·2	2·8
	(K)[a]	71·5		SMP	7·0	10·8		21·7		7·1	3·1
	(L)[a]	68·5	3·0	SMP	7·0	10·7		21·0		6·9	3·0
	(M)[a]	64·5		SMP	14·0	13·6		21·9		8·8	2·5
	(N)[a]	61·5	3·0	SMP	14·0	13·5		24·2		9·0	2·7
Daniel *et al.* (1968b)	(A)[a]	78·5				8·0	64	8·3	7·7[d]	5·1[d]	
	(B)[a]	78·0	0·5[c]			7·9	65	8·8	7·4[d]	5·1[d]	
	(C)[a]	63·5		Unfortified coconut meal[c]	15·0	10·1	69	10·9	6·3[d]	7·0[d]	
	(D)[a]	63·5		Fortified coconut meal[c]	15·0	10·0	71	11·4	6·2[d]	7·1[d]	
	(E)[a]	71·5		Fortified SMP[c]	7·0	10·2	79	17·6	4·5[d]	8·0[d]	
Daniel *et al.* (1970)	(A)[e]	83·5				7·1		5·5		3·2	1·7
	(B)[e]	80·5	3·0[b]			7·1		5·9		3·2	1·8
	(C)[e]	75·0		Pigeon peas	8·5	8·6		12·8		5·1	2·5
	(D)[e]	72·0	3·0	Pigeon peas	8·5	8·6		14·8		4·9	3·0
	(E)[e]	66·8		Pigeon peas	16·7	10·1		16·3		6·2	2·6
	(F)[e]	63·8	3·0	Pigeon peas	16·7	10·0		17·1		5·8	2·9
	(G)[e]	58·5		Pigeon peas	25·0	11·5		17·3		7·1	2·4
	(H)[e]	55·5	3·0	Pigeon peas	25·0	11·5		18·6		6·9	2·7
Daniel *et al.* (1974)	(A)[f]	88·2				6·0		3·7		2·8	1·3
	(B)[f]	66·2		Chickpeas	6·1	6·0		5·5		3·0	1·8
	(C)[f]	44·1		Chickpeas	6·2	6·0		7·9		3·0	2·6

Desai *et al.* (1970a)	(D)[f]			Chickpeas	24·4	6·0		2·9		2·4	1·2
	(E)[f]	90·0				6·1		3·7		2·7	1·4
	(F)[f]	81·0		Chickpeas	9·0	7·9		8·6		3·6	2·4
	(G)[f]	72·0		Chickpeas	18·0	9·6		13·7		4·7	2·9
	(H)[f]	63·0		Chickpeas (% *prot.*)	27·0	11·3		15·9		5·6	2·8
	(A)[g]	75·6	3·0[b]			8·5		9·3		4·6	2·0
	(B)[g]	75·6	3·0	L-Lys HCl	2·7	8·5		15·5		4·8	3·2
	(C)[g]	75·6	3·0	DL-Met	0·2	8·5		9·5		4·5	2·1
	(D)[g]	75·6	3·0	DL-Thr	0·7	8·5		10·5		4·5	2·3
	(E)[g]	75·6	3·0	L-Lys HCl DL-Met	2·7 0·2	8·5		19·8		5·6	3·5
	(F)[g]	75·6	3·0	L-Lys HCl DL-Thr	2·7 0·7	8·6		19·9		5·7	3·5
	(G)[g]	75·6	3·0	DL-Met DL-Thr	0·2 0·7	8·6		11·7		4·7	2·5
	(H)[g]	75·6	3·0	L-Lys HCl DL-Met DL-Thr	2·7 0·2 0·7	8·6		24·8		6·5	3·8
Desai *et al.* (1970b)	(A)[a]	78·5				7·6	55	10·1	5·4[d]	4·2[d]	
	(B)[a]	39·25		LCBF III[h]	39·25	15·3	74	17·6	4·2[d]	11·3[d]	
	(C)[a]	LCBF V[h] 78·5				22·4	74	17·8	4·1[d]	16·6[d]	
	(D)[a]	78·5				7·5	53	8·5	6·2[d]	4·0[d]	
	(E)[a]	39·25		LCBF V[h]	39·25	15·8	64	14·5	4·4[d]	10·1[d]	
Guttikar *et al.* (1968)	(A)[a]					7·9	47	10·9	4·3[d]	3·7[d]	
	(B)[a]			Protein food I		10·7	48	12·3	3·9[d]	5·1[d]	
	(C)[a]			Protein food I L-Lys HCl DL-Met		10·5	49	14·4	3·4[d]	5·1[d]	
	(D)[a]			Protein food II		10·6	49	13·3	3·7[d]	5·2[d]	
	(E)[a]			Protein food II L-Lys HCl DL-Met		10·8	49	13·6	3·6[d]	5·3[d]	
	(F)[a]			SMP		10·0	49	14·8	3·3[d]	4·9[d]	
Hariharan *et al.* (1967)	(A)[a]	78·5				8·3		9·5		5·7	1·7
	(B)[a]	78·5	0·5[j]			8·2		10·3		5·5	1·9
	(C)[a]	78·5	1·0[j]			8·3		11·0		5·5	2·0
	(D)[a]	73·8		Groundnut flour (GNF)[j]	4·7	10·5		13·0		7·1	1·8

cont.

Table 6.31 *cont.*

Reference		Finger millet diet (%)	Nonprot. supp. (%)	Prot. supp. (%)		Prot. content of diet (%)	Av. feed intake/wk (g)	Av. wt gain/wk (g)	Av. F/G	Av. prot. intake/wk (g)	Av. PER
	(E)[a]	73·5		Fortified GNP[j]	5·0	10·4		13·7		7·0	2·0
	(F)[a]	71·3		Fortified SMP[j]	7·2	10·3		14·8		7·0	2·1
	(G)[a]	69·1		GNF	9·4	12·4		14·5		8·4	1·7
	(H)[a]	68·5		Fortified GNF	10·0	12·6		14·5		8·5	1·7
	(I)[a]	64·1		Fortified SMP	14·4	12·3		15·2		7·7	2·0
Kuppuswamy *et al.* (1957)	(A)[a]	78·5					57	8·4	6·8[d]		
	(B)[a]	66·0		Indian MPF[k]	12·5		71	13·3	5·3[d]		
	(C)[a]	66·0		American MPF[k]	12·5		71	14·4	4·9[d]		
Kurien *et al.* (1958)	(A)[a]			Rice[l]	78·5	9·0	56	4·1	13·6[d]	5·0[d]	
	(B)[a]	19·25		Rice[l]	59·25	8·8	71	8·0	8·9[d]	6·2[d]	
	(C)[a]	39·25		Rice[l]	39·25	8·6	72	8·7	8·3[d]	6·2[d]	
	(D)[a]	78·5				8·2	74	8·7	8·5[d]	6·1[d]	
Kurien *et al.* (1969)		(kcal/rat/day)		(kcal/rat/day)							
	(A)[a]	32·8					56	8·7	6·4[d]	4·5	
	(B)[a]	29·5		Indian MPF[m]	3·3		56	13·4	4·2[d]	6·4	
	(C)[a]	24·6					42	5·4	7·8[d]	3·4	
	(D)[a]	21·3		Indian MPF	3·3		42	9·2	4·6[d]	5·3	
	(E)[a]	16·4					28	2·6	10·8[d]	2·2	
	(F)[a]	13·1		Indian MPF	3·3		28	5·3	5·3[d]	4·2	
Leela *et al.* (1965)	(A)	97·0	3·0[b]			6·2		2·3		2·1	1·1
	(B)	97·0	3·0	L-Lys HCl to total of prot.	6·6	6·4		6·1		2·8	2·2
	(C)	97·0	3·0	L-Lys HCl to total of prot. DL-Thr to total of prot.	6·6 4·9	6·6		10·1		3·3	3·1
	(D)[a]	62·5	3·0			6·2		3·9		2·7	1·4
	(E)[a]	62·5	3·0	L-Lys HCl to total of protein	6·6	6·4		7·3		3·0	2·4
	(F)[a]	62·5	3·0	L-Lys HCl to total of prot. L-Thr to total of prot.	6·6 4·9	6·5		11·0		3·5	3·1
	(G)[a]	52·0	3·0	SMP	10·0	6·5		11·7		3·8	3·1
	(H)	SMP	3·0			6·3		12·2		3·9	3·1
Leela *et al.* (1966)	(A1)[a]	78·5				7·9		8·5		4·1	2·1
	(B1)[a]	78·5		L-Lys HCl to total of protein	6·6%	8·2		13·2		5·3	2·5

	(C1)[a]	78·5		L-Lys HCl to total of protein DL-Thr to total of protein	6·6 4·9	8·4	15·8		5·4	2·9
	(D1)[a]	75·5	3·0			7·7	8·3		4·0	2·1
	(E1)[a]	75·5	3·0	L-Lys HCl to total of protein	6·6	8·0	12·9		5·0	2·6
	(F1)[a]	75·5	3·0	L-Lys HCl to total of protein DL-Thr to total of protein	6·6 4·9	8·2	15·5		5·4	2·9
	(G1)[a]	65·5	3·0	SMP	10·0	10·9	21·1		7·2	2·9
	(H1)[a]	39·5	3·0			8·2	15·8		5·4	2·9
	(I1)	SMP	3·0			8·0	15·7		5·1	3·1
Narayanaswamy *et al.* (1970b)	(A)[a]	78·5				8·4	10·9		4·7	2·3
	(B)[a]	75·5	3·0[b]			8·3	11·6		4·8	2·4
	(C)[a]	72·5		Petroleum grown yeast[n]	6·0	10·4	17·1		6·5	2·6
	(D)[a]	69·5	3·0	Petroleum grown yeast	6·0	10·4	20·9		6·7	3·1
Narayanaswamy *et al.* (1972b)	(A)[a]	78·5				8·1	8·0		4·2	1·9
	(B)[a]	75·5	3·0[b]			8·0	8·5		4·0	2·1
	(C)[a]	65·5	3·0	LCPF[o]	10·0	9·7	13·1		5·1	2·6
	(D)[a]	55·5	3·0	LCPF	20·0	11·2	15·1		5·8	2·6
	(E)[a]	69·5	3·0	Fortified SMP[o]	6·0	9·6	14·8		5·3	2·8
	(F)[a]	63·5	3·0	Fortified SMP	12·0	11·2	17·1		5·8	3·0
Narayanaswamy *et al.* (1974)	(A)[a]	75·5	3·0[b]			8·3	7·2		3·6	2·0
	(B)[a]	65·5		PECF I[p]	10·0	9·8	14·1		5·1	2·8
	(C)[a]	55·5	3·0	PECF I	20·0	11·4	17·7		6·2	2·9
	(D)[a]	65·5	3·0	PECF II[p]	10·0	9·5	12·9		4·9	2·6
	(E)[a]	55·5	3·0	PECF II	20·0	10·7	14·8		5·6	2·6
	(F)[a]	65·5	3·0	PECF III[p]	10·0	9·1	10·8		4·5	2·4
	(G)[a]	55·5	3·0	PECF III	20·0	10·0	11·6		4·8	2·4
Narayanaswamy *et al.* (1975)		Series I								
	(A)	85·7	3·0[b]			6·0	3·6	12·8	2·8	1·3
	(B)	64·3	3·0	Cowpeas	7·0	6·0	7·4	8·1	3·5	2·1
	(C)	42·8	3·0	Cowpeas	14·0	6·0	9·7	7·7	3·4	2·8
	(D)		3·0	Cowpeas	28·0	6·0	6·1	5·9	3·0	2·0
		Series II								
	(E)	90·0	3·0[b]			6·2	3·7	12·0	2·8	1·3
	(F)	81·0	3·0	Cowpeas	9·0	7·7	12·9	3·4	5·0	2·6

cont.

Table 6.31 *cont.*

Reference		Finger millet diet (%)	Nonprot. supp. (%)	Prot. supp. (%)		Prot. content of diet (%)	Av. feed intake/wk (g)	Av. wt gain/wk (g)	Av. F/G	Av. prot. intake/wk (g)	Av. PER
	(G)	72·0	3·0	Cowpeas	18·0	9·2		16·2	5·0	6·0	2·7
	(H)	63·0	3·0	Cowpeas	27·0	10·8		17·1	4·0	6·3	2·7
Pushpamma (1968), Pushpamma *et al.* (1972)[q]	(A)	96·5				5·4	73	7·0	10·7	3·9	1·7
	(B)	96·5		L-Lys HCl to total in diet	0·75	5·4	86	15·0	5·7	5·1	3·0
	(C)	96·5		L-Trp	0·1	5·4	70	7·0	10·0	3·8	1·9
	(D)	96·5		L-Lys HCl to total in diet L-Trp	0·75 0·1	5·4	77	13·0	6·1	4·6	2·7
	(E)	Casein				5·0	67	9·0	7·5	3·4	2·7
				(*% prot.*)							
	(F)	86·0		SMP	3·0	7·8	88	21·0	4·2	6·8	3·0
	(G)	86·0		Isolated soybean prot.	3·0	7·8	97	18·0	5·4	7·5	2·4
	(H)	86·0		Fish prot. concentrate	3·0	7·8	104	27·0	5·5	8·0	3·8
	(I)	86·0		Dried brewers' yeast	3·0	7·8	89	19·0	4·6	6·9	2·8
	(J)	Casein				8·0	59	12·0	5·1	4·7	2·5
Shurpalekar *et al.* (1962a)	(A)[a]	78·5				8·8	53	7·8	6·8[d]	4·7[d]	
	(B)[a]	75·5		Fortified fish flour	3·0	11·3	64	13·6	4·7[d]	7·0[d]	
	(C)[a]	71·5		Fortified SMP	7·0	11·3	62	12·7	4·9[d]	7·0[d]	
Subrahmanyan *et al.* (1954a)[r]	(A)[a]	78·5				8·9	67	8·4	8·0[d]	6·0[d]	
	(B)[a]	58·9		Cassava flour	19·6	7·8	62	7·6	8·1[d]	4·8[d]	
	(C)[a]	58·9		Sweet potato flour	19·6	7·8	64	7·6	8·4[d]	5·0[d]	
	(D)[a]	58·9	Cassava flour 15·7	Groundnut cake flour	3·9	10·9	73	10·8	6·7[d]	8·0[d]	
	(E)[a]	58·9	Sweet potato flour 15·7	Groundnut cake flour	3·9	10·1	71	10·9	6·5[d]	7·2[d]	
Subrahmanyan *et al.* (1954b)	(A)[a]						62	7·7	8·0[d]		
	(B)[a]			Cottonseed flour[r]	10·0		69	13·7	5·0[d]		

Subrahmanyan *et al.* (1954d)[s]	(A)[a]	78·5			8·1	70	8·8	7·9[d]	5·7[d]	
	(B)[a]	58·9	Synthetic grain I	19·6	8·0	70	9·8	7·1[d]	5·6[d]	
	(C)[a]	58·9	Synthetic grain II	19·6	8·4	69	9·8	7·0[d]	5·8[d]	
	(D)[a]	58·9	Synthetic grain III	19·6	9·1	68	10·2	6·7[d]	6·2[d]	
Sur *et al.* (1954a)[t]	(A)[a]	78·5				77	10·0	7·7[d]		
	(B)[a]	76·0	Food yeast	2·5		78	14·2	5·5[d]		
	(C)[a]	73·5	Food yeast	5·0		72	17·0	4·2[d]		
Sur *et al.* (1954b)[t]	(A)				5·0					0·9
	(B)		Food yeast (prot. ratio finger millet/3 yeast 1)		5·0					1·8
Tasker *et al.* (1964)	(A)				7·6	49	5·6	8·7[d]	3·7[d]	
	(B)		Prot. food[u]		12·5	59	14·7	4·0[d]	7·4[d]	
Venkat Rao *et al.* (1958)	(A)[a]	78·5			9·7	66	6·6	10·0[d]	6·4[d]	
	(B)[a]	68·5	Low-fat GNF	10·0	15·0	66	10·5	6·3[d]	9·9[d]	
	(C)[a]	58·5	Low-fat GNF	20·0	20·3	66	12·1	5·4[d]	13·4[d]	
Venkataraman *et al.* (1977)[w]	(A)	85·0			6·0		2·7		2·0	1·4
	(B)	82·0	Green algae (prot. ratio finger millet 3/algae 1)	4·5	8·6		10·0		4·8	2·1
	(C)	62·0	Green algae (prot. ratio finger millet 1/algae 1)	10·0	8·8		14·1		5·8	2·4

[a] All diets contained (%) pigeon peas 5·0, groundnut oil 5·0, SMP 0·9, common salt 0·3, green leafy vegetables (*Amaranthus gangeticus*) 2·1, other vegetables (brinjal and potato) 8·2.

[b] Salt mix 2·0% (Hubbel *et al.*, 1937), vitamin premix 1·0% (Chapman *et al.*, 1959).

[c] Diet (B) Ca salts and vitamins providing same quantities as 15 g fortified coconut meal. Prot. contents (N × 6·25) (DWB, %) unfortified coconut meal 22·5, fortified coconut meal 22·0, fortified SMP 36·4.

[d] Calculated by the authors from av. weight gain, food intake and prot. content in diet.

[e] All diets as[a] but excluding pigeon peas. AA composition (microbiological assay, Barton-Wright, 1952) of diets (mg/g total N). Lys (A) and (B) 212, (C) and (D) 225, (E) and (F) 237, (G) and (H) 262. Thr (A) and (B) 244, (C) and (D) 250, (E) and (F) 256, (G) and (H) 269.

[f] All diets contained (%) salt mix 2·0, vit. premix 1·0, groundnut oil with vit. A and D 1·0, groundnut oil 6·0, maize starch to 100. AA composition (microbiological assay, Barton-Wright, 1952) of the diets (mg/g total N). Lys (A) 187 (B) 250 (C) 319 (D) 450 (F) 256 (G) 325 (H) 344, egg prot. 400, human milk prot. 394. Thr (A) to (H) 256, egg prot. 319, human milk prot. 287. Met (A) 187 (B) 162 (C) 144 (D) 106 (F) 162 (G) 150 (H) 137, egg prot. 194, human milk prot. 137. Cys (A) 137 (B) 112 (C) 94 (D) 50 (F) 112 (G) and (H) 87, egg prot. 150, human milk prot. 131. Trp (A) 75 (B) 62 (C) 56 (D) 37 (F) and (G) 62 (H) 56, egg prot. 106, human milk prot. 106. Diet (E) not shown in original possibly similar to (A).

[g] All diets as[a] but groundnut 4·0%, groundnut oil with vit. A and D 1·0%. AA composition (microbiological assay, Barton-Wright, 1952) in diets (mg/g total N). Lys (A) 225 (B) 394 (C) 225 (D) 225 (E) 394 (F) 394 (G) 225 (H) 394, egg prot. 400, human milk prot. 394 Thr (A) 244 (B) 244 (C) 244 (D) 287 (E) 244 (F) 287 (G) 287 (H) 287, egg prot. 319, human milk prot. 287. Met + Cys (A) 256 (B) 256 (C) 269 (D) 256 (E) 269 (F) 256 (G) 269 (H) 269, egg prot. 344, human milk prot. 268.

Table 6.31 *cont.*

[h] Low cost balanced food (LCBF) III finger millet flour 65/cottonseed flour 25/chickpea flour 10, prot. content (N × 6·25) (DWB, %) 20·9. LCBF V finger millet 65/groundnut flour 25/chickpea flour 10, prot. content 21·4.

[i] Prot. food I groundnut 40/chickpea 40/sesame flours 20; II groundnut 40/soy 30/sesame flours 30, fortified with Ca salt, vit. A and D, thiamine and riboflavin.

[j] Diets (B) and (C) Ca salts and vit. premix to provide same quantities as in 5 g and 10 g respectively fortified groundnut flour (GNF). Prot. contents (N × 6·25) (DWB, %) unfortified GNF 56·8, fortified GNF 53·8, fortified SMP 36·0.

[k] Indian multipurpose food (MPF) 75 parts low-fat groundnut flour (GNF) 75/roasted chickpea flour 25/fortified with Ca, vit. A and D and riboflavin. Composition of American MPF not stated.

[l] Prot. contents (N × 6·25) (DWB, %) finger millet 7·7, rice 8·2.

[m] Indian multipurpose food (MPF) groundnut 3/chickpea 1, fortified with Ca salts, vit. A and D and riboflavin, prot. content (N × 6·25) (DWB, %) 45·0.

[n] Petroleum grown yeast prot. content (N × 6·25) (DWB, %) 46·5. AA composition of yeast and diets (auto-analyser) (mg/g total N). Lys yeast 456, (A) 225 (C) 269. Thr yeast 350 (A) 219 (C) 250. Met yeast 106, (A) 150 (C) 144. Cys yeast 62, (A) 125 (C) 112.

[o] Low cost prot. food (LCPF) wheat flour 70/soy flour 15/groundnut flour 15, prot. content (N × 6·25), (DWB, %) 24·5, fortified SMP 35·1.

[p] Composition of prot. enriched cereal foods (PECF) I (%) wheat flour 70/soy bean flour 15/groundnut flour 15, prot. content (N × 6·25) 22·6. PECF II (%) wheat flour 80/soy bean flour 10/groundnut flour 10, prot. 19·2. PECF III (%) wheat flour 90/soy bean flour 5/groundnut flour 5, prot. 15·6.

[q] Prot. contents (N × 6·25) (%) finger millet 5·6, SMP 35·6, fish prot. concentrate (FPC) 83·0, isolated soy bean prot. 88·0, dried brewers' yeast 40·0.

[r] Prot. contents (N × 6·25) (DWB, %) tapioca flour 1·9, sweet potato flour 1·7, groundnut cake flour (without red cuticle) 59·0, finger millet 7·9. Cottonseed flour, alcohol-extracted (DWB) prot. 57·3, gossypol (Pons and Guthrie, 1949) 0·28.

[s] Prot. contents (N × 6·25) (DWB, %) finger millet 9·0, synthetic grains I 7·1, II 9·8, III 12·4.

[t] Prot. contents (N × 6·25) (DWB, %) finger millet 9·2, food yeast (*Torula utilis*) 60·4.

[u] Prot. food (Krishnamurthy *et al.*, 1958) screw pressed coconut meal 1/chickpea flour 1/groundnut flour 2 fortified with Ca, vit. A and D and riboflavin.

[v] Prot. contents (N × 6·25) (DWB, %) finger millet 9·3, low-fat groundnut flour (GNF) 54·3.

[w] Prot. contents (N × 6·25) (%) finger millet 7·1, green algae (*Scenedesmus acutus*) 45·0.

Table 6.32

Effect of protein supplementation of diets (DWB) based on foxtail millet, on growth of weanling rats. Blank spaces indicate no values presented.

Reference	Diet	Foxtail millet diet (%)	Nonprot. supp. (mg/100 g diet)	Prot. supp.	Prot. supp. (%)	Prot. content of diet (%)	Niacin intake (mg/100 g diet)	Trp intake intake (mg/100 g diet)	Av. wt gain/wk (g)	Av. Prot. intake/wk (g)	Av. PER
		(% *prot.*)			(% *prot.*)						
Ganapathi *et al.* (1958)	(A)	10·0				10·0					0·6
	(B)	5·0		Horse gram (*Dolichos biflorus*)	5·0	10·0					1·6
	(C)	5·0		Cowpeas	5·0	10·0					2·0
Joseph *et al.* (1957)	(A)[a]	Foxtail millet 78·5							4·8		
	(B)[a]	Foxtail millet 66·0		Indian MPF[a]	12·5				13·5		
	(C)[a]	Foxtail millet 66·0		American MPF[a]	12·5				14·5		
Mangay *et al.* (1957)	(Ac)	basal[b]					0	106	13·0	6·4	2·0
	(Bc)[b]		Niacin 3·0				3·0	106	16·3	7·4	2·1
	(Cc)[b]			Maize	40·0		1·10	133	9·6	6·8	1·4
	(Dc)[b]		Niacin 3·0	Maize	40·0		4·10	133	25·9	10·7	2·4
	(Ec)[b]			Foxtail millet	40·0		0·49	192	26·2	11·7	2·2
	(Fc)[b]		Niacin 3·0	Foxtail millet	40·0		3·49	192	23·3	11·2	2·1
	(Gc)[b]			Foxtail millet Maize	40·0 40·0		1·59	219	28·1	15·9	1·8
	(Hc)[b]		Niacin 3·0	Foxtail millet Maize	40·0 40·0		4·59	219	25·2	15·5	1·6
	(Ic)[b]			Foxtail millet Maize	40·0 20·0		1·04	206	27·4	14·1	1·9
	(Jc)[b]		Niacin 3·0	Foxtail millet Maize	40·0 20·0		4·04	206	25·2	13·6	1·8
	(Kc)[b]			Foxtail millet	80·0		0·98	278	26·6	16·9	1·6
	(Lc)[b]		Niacin 3·0	Foxtail millet	80·0		3·98	278	26·9	16·8	1·6
	(Acm)	basal[c]					1·10	133	15·9	8·1	2·0
	(Bcm)[c]		Niacin 0·06				1·16	133	15·9	6·7	2·4

cont.

Table 6.32 *cont.*

Reference	Foxtail millet diet (%)	Nonprot. supp. (mg/100 g diet)	Prot. supp. (%)	Prot. content of diet (%)	Niacin intake (mg/100 g diet)	Trp intake intake (mg/100 g diet)	Av. wt gain/wk (g)	Av. Prot. intake/wk (g)	Av. PER
(Ccm)[c]			Foxtail millet 5·0		1·16	144	22·6	9·2	2·4
(Dcm)[c]		Niacin 0·12			2·22	133	15·9	7·5	2·1
(Ecm)[c]			Foxtail millet 10·0		1·22	155	26·1	11·2	2·3
(Fcm)[c]		Niacin 0·24			1·34	133	23·7	8·6	2·7
(Gcm)[c]			Foxtail millet 20·0		1·34	177	29·3	13·5	2·2
(Hcm)[c]		Niacin 0·48			1·58	133	25·8	8·0	3·2
(Icm)[c]			Foxtail millet 40·0		1·58	219	29·8	15·3	2·0
(Jcm)	basal[c]				1·10	133	20·2	8·4	2·4
(Kcm)[c]			Foxtail millet 5·0		1·16	144	24·9	9·9	2·5
(Lcm)[c]			Trp 11 mg		1·10	144	25·9	9·2	2·8
(Mcm)[c]		Niacin 0·06	Trp 11 mg		1·16	144	25·1	10·0	2·5
(Ncm)[c]			Foxtail millet 10·0		1·22	155	27·0	10·5	2·6
(Ocm)[c]			Trp 22 mg		1·10	155	24·9	9·4	2·6
(Pcm)[c]		Niacin 0·12	Trp 22 mg		1·22	155	26·7	10·1	2·6
(Qcm)[c]			Foxtail millet 20·0		1·34	177	27·2	12·0	2·3
(Rcm)[c]			Trp 44 mg		1·10	177	26·0	9·8	2·7
(Scm)[c]		Niacin 0·24	Trp 44 mg		1·34	177	26·7	10·4	2·6
(A)[d]	Maize 89·0				2·45	60	5·7	3·7	1·5
(B)[d]	Maize 89·0	Niacin 3·0			5·45	60	7·7	4·5	1·7
(C)[d]	Maize 89·0		Trp 400 mg		2·45	460	5·7	4·1	1·4
(D)[d]	Maize 89·0	Niacin 3·0	Trp 400 mg		5·45	460	6·3	4·0	1·6
(E)[d]	Foxtail millet 89·0				1·09	191	6·0	5·6	1·1
(F)[d]	Foxtail millet 89·0	Niacin 3·0			4·09	191	6·7	6·4	1·0
(G)[d]	Foxtail millet 89·0		Trp 400 mg		1·09	591	4·7	5·8	0·8
(H)[d]	Foxtail millet 89·0	Niacin 3·0	Trp 400 mg		4·09	591	4·3	5·4	0·8

(I)[d]	Foxtail millet 89·0		L-Lys HCl	1000 mg	1·09	191	27·6	8·7	3·2
(J)[d]	Maize 89·0		Foxtail millet	2·0	2·47	64	7·0	4·5	1·5
(K)[d]	Maize 80·0		Foxtail millet	10·0	2·32	76	6·3	4·4	1·4
(L)[d]	Maize 80·0	Niacin 3·0	Foxtail millet	5·0	5·26	65	6·3	4·0	1·6
(M)[d]	Maize 80·0		Foxtail millet Trp	10·0 400 mg	2·32	476	5·0	3·6	1·4
(N)[d]	Maize 80·0	Niacin 3·0	Foxtail millet Trp	10·0 400 mg	5·32	476	4·3	4·0	1·1
(O)[d]	Maize 80·0	Niacin 3·0	Foxtail millet L-Lys HCl	10·0 1000 mg	2·32	76	5·2	3·5	1·5
(P)[d]	Maize 80·0	Niacin 3·0	Foxtail millet L-Lys HCl Trp	10·0 1000 mg 400 mg	5·32	476	17·1	5·7	3·0

[a] All diets contained (%) pigeon peas 5·0, groundnut oil 5·0, SMP 0·9, common salt 0·3, green leafy vegetables (*Amaranthus gangeticus*) 2·1, other vegetables (brinjal and potato) 8·2. Indian multipurpose food (MPF) low-fat groundnut flour 75/roasted chickpea flour 25, fortified with Ca, vit. A and D and riboflavin. Composition of American MPF not stated.

[b] Prot. contents (N × 6·25) (DWB, %) foxtail millet 14·1, maize 12·3. Basal diet included (%) vitamin-free casein 9·0, cottonseed oil 5·0, salt 4·0, L-Cys 0·2, sucrose 71·8, vitamin B mix 0·11. Additions of finger millet and maize replaced equivalent amounts of sucrose.

[c] Basal diet included (%) vitamin-free casein 9·0, maize 40·0.

[d] Basal diet excluded casein.

Table 6.33
Effect of protein supplementztion of diets (DWB) based on common, little, or unspecified "millet" on growth of weanling rats. Blank spaces indicate no values presented.

Reference		Cereal diet (%)	Nonprot. supp. (%)	Prot. supp. (%)		Prot. content of diet (%)	Av. feed intake/wk (g)	Av. wt gain/wk (g)	Av. F/G	Av. prot. intake/wk (g)	Av. PER
Adrian *et al.* (1956)[a]		*Panicum* millet									
	(A)	90·0				9·9		1·7			0·5
	(B)	75·0		DL-Lys HCl	0·6	9·0		12·1			1·7
	(C)	75·0		Groundnut flour (GNF)	10·0	13·6		11·7			1·1
	(D)	75·0		GNF	18·0	18·0		16·5			1·2
	(E)	75·0		GNF	15·4	17·8		19·2			1·3
				Fish hydrolysate	3·0						
	(F)	Casein			18·0	16·8		23·2			1·8
Howe *et al.* (1965)		White millet									*(adjusted)*
	(A)					10·0		1·7			0·3
	(B)			L-Lys HCl	0·3	10·0		18·7			1·8
	(C)			L-Lys HCl	0·5	10·0		22·7			2·1
	(D)			L-Lys HCl	0·7	10·0		19·2			1·9
	(E)			L-Lys HCl	0·7	10·0		19·0			1·8
				L-His HCl	0·2						
	(F)			L-Lys HCl	0·7	10·0		30·2			2·4
				DL-Thr	0·4						
	(G)	Casein				10·0		24·5			2·5
Joseph *et al.* (1957)		Little millet									
	(A)[b]	78·5						3·9			
	(B)[b]	66·0		Indian MPF[b]	12·5			12·9			
	(C)[b]	66·0		American MPF[b]	12·5			14·6			
Kundaji and Radhakrishna Rao (1954)		Common millet (% prot.)		(% prot.)							
	(A)	3·25		Pigeon peas	4·75	8·0	34	2·3	14·8[d]	2·7	0·8
	(B)	Rice 3·25		Pigeon peas	4·75	8·0	34	3·8	8·9[d]	2·7	1·4
	(A1)	2·85		Rice	2·85	5·7					

Sure *et al.* (1957)[c]		"Millet"								
	(A)	78·3			7·0	56	3·2	17·5[d]	3·9	0·8
	(B)	77·3	Fish flour	1·0	7·7	65	5·5	11·8[d]	5·0	1·1
	(C)	75·3	Fish flour	3·0	9·1	78	8·2	9·5[d]	7·1	1·1
	(D)	73·3	Fish flour	5·0	10·4	94	17·4	5·4[d]	9·8	1·8

[a] Prot. contents (N × 6·25) (%) *Panicum* millet 11·0, groundnut flour (GNF) 53·7, fish (enzymatic hydrolysate) 50·6, casein 90·6.

[b] All diets contained (%) pigeon peas 5·0, groundnut oil 5·0, SMP 0·9, common salt 0·3, green leafy vegetables (*Amaranthus gangeticus*) 2·1, other vegetables (brinjal and potato) 8·2. Indian multipurpose food (MPF), low-fat groundnut flour (GNF) 75/roasted chickpea flour 25, fortified with Ca, vit. A and D and riboflavin. Composition of American MPF not stated.

[c] Prot. content of "millet" 9·0%.

[d] Calculated by the authors from av. weight gain divided by food intake.

Table 6.34

Effect of protein supplementation (DWB) of diets based on kodo millet and *Amaranthus paniculatus* on weanling rat growth. Blank spaces indicate no values presented.

Reference		Diet (%)		Prot. supp. (%)		Prot. content of diet (%)	Av. food intake/wk (g)	Av. wt gain/wk (g)	Av. F/G	Av. prot. intake/wk (g)	Av. PER
Kundaji and Radhakrishna Rao (1954)		(*% prot.*)		(*% prot.*)							
	(A)	Kodo millet	3·25	Pigeon peas	4·75	8·0	43	5·4	8·0[a]	3·4	1·6
	(B)	*Amaranthus paniculatus*	3·25	Pigeon peas	4·75	8·0	34	3·5	9·7[a]	2·7	1·2
	(A1)	Kodo millet	2·85	Rice	2·85	5·7					
	(B1)	*Amaranthus paniculatus*	3·25	Rice	2·85	5·7					
Rajalakshmi and Majmudar (1966)				(calculated from tables)							
	(A)	Kodo millet	100·0			8·3		−0·5		3·0	−0·2
	(B)		80·0	Pigeon peas	20·0	11·1		4·0		4·7	0·8
	(C)		80·0	Chickpeas	20·0	10·8		6·2		5·0	1·2
	(D)		80·0	Moth bean *Phaseolus aconitifolius*	20·0	11·3		7·5		5·5	1·4
	(E)		80·0	Field bean *Dolichos lablab*	20·0		Experiment abandoned after 2 weeks owing to high mortality				
	(F)		80·0	Cowpeas *Vigna catiang*	20·0	11·6		6·2		5·2	1·2
	(G)		80·0	Peas *Pisum sativum*	20·0	10·5		8·2		5·0	1·6
	(H)		80·0			11·0		6·0		5·2	1·1
				Chickpeas	5·0						
				Moth bean	5·0						
				Cowpeas	5·0						
				Peas	5·0						

[a] Calculated by the authors from av. wt gain divided by food intake.

Table 6.35
Least cost diets calculated for children weighing 10 kg (C10), 20 kg (C20) and for a 60 kg adult (A60) based on millet and sorghum (g/day).[a]

Foods	A60	C20	C10
Millet[b]	362	201	148
Oil	169	110	63
Cowpeas	4	35	25
Shadow prices[c]			
Calories	0·84	0·80	0·75
Total prot.	0·07	0	0
Lys	0·09	0·17	0·18
S AAs	0	0·05	0·05
Trp	0	0	0
Bulk	0	0	0

[a] Chamberlin and Stickney (1973).
[b] Millet and sorghum were nearly equivalent in the A60 least cost diets. The diet cost will increase by less than 1 % if millet were completely replaced by an appropriate amount of sorghum.
[c] Shadow price: the amount (% change) by which the cost of the least cost diet would change if the value of the constraint were changed by 1 %.

Table 6.36
Nutritional constraints for children weighing 10 kg (C10) and 20 kg (C20) and for an adult weighing 60 kg (A60).[a]

Nutrients	C10	C20	A60
Calories (kcal/day)[b]	1150	1775	2750
Total prot. (g/day)[c]	14	20	36
Essential AAs (mg/day)[d]			
Ile	760	1030	930
Leu	1010	1375	1240
Lys	735	1000	900
Met + Cys	630	855	770
Phe + Tyr	1150	1560	1405
Thr	580	790	715
Trp	180	250	225
Val	830	1130	1015

[a] Chamberlin and Stickney (1973).
[b] Calories constraints based on FAO (1957b). C10 corresponded to 1-year old child, C20 to 5–6-year old children, A60 to average of "reference male" and "reference female".
[c] Protein constraints based on WHO (1965). The required prot./unit body weight was assumed to be 1·4, 1·0, and 0·6 g/kg for C10, C20 and A60 respectively.
[d] Based on the assumption that E/T is 42 % for C10, 40 % for C20, and 20 % for A60. (E/T represents the required proportion (% by wt) of essential AAs in the total prot.)

Table 6.37

Effect of balanced diets (DWB) on growth of weanling rats. Blank spaces indicate no values presented.

Reference		Diet	(%)	Prot. content of balanced food (%)	AA supp.	(mg/g total N)	Prot. content of diet (%)	Av. wt gain/wk (g)	Av. prot. intake/wk (g)	Av. PER (adjusted)
Daniel *et al.* (1967)[a]	(A)	Finger millet	70·0	18·2			10·0	18·0	6·9	2·4
		Groundnut flour (GNF)	20·0							
		SMP	10·0							
	(B)	Finger millet	70·0	18·3	L-Lys HCl	81	10·0	24·6	7·4	3·0
		GNF	20·0		DL-Met	31				
		SMP	10·0							
	(C)	Finger millet	70·0	19·8			10·0	17·3	6·5	2·4
		GNF	20·0							
		Full fat soy flour (SF)	10·0							
	(D)	Finger millet	70·0	20·1	L-Lys HCl	87	10·0	19·3	6·4	2·7
		GNF	20·0		DL-Met	31				
		SF	10·0							
	(E)	Finger millet	70·0	17·0			10·0	14·5	6·5	2·0
		GNF	20·0							
		Chickpea	10·0							
	(F)	Finger millet	70·0	17·2	L-Lys HCl	100	10·0	16·0	6·1	2·4
		GNF	20·0		DL-Met	50				
		Chickpea	10·0							
	(G)	Finger millet	60·0	18·5			10·0	14·4	6·4	2·0
		GNF	20·0							
		Chickpea	20·0							
	(H)	Finger millet	60·0	18·7	L-Lys HCl	87	10·0	16·2	6·2	2·4
		GNF	20·0		DL-Met	31				
		Chickpea	20·0							
	(I)	Finger millet	65·0	18·9			10·0	13·4	6·0	2·0
		GNF	18·0							
		Sesame	7·0							
		Chickpea	10·0							
	(J)	Finger miilet	65·0	19·1	L-Lys HCl	112	10·0	17·7	5·9	2·7

		GNF	18·0		DL-Met	19				
		Sesame	7·0							
		Chickpea	10·0							
	(K)	SMP					10·0	22·3	6·5	3·0
Desai *et al.* (1968)[b]	(A)	Fortified fish flour					10·0	21·8	6·7	3·2
	(B)	Fortified protein food					10·0	17·9	6·8	2·6
	(C)	Fortified SMP					10·0	18·8	6·3	3·0
Prasannappa *et al.* (1976)[c]	(A)	*Raw*		17·2			10·0	9·9	3·4	2·1
		Finger millet	25·0							
		Wheat	15·0							
	(B)	*Roasted*								
		Finger millet	25·0	15·9			10·0	8·4	3·2	2·0
		Wheat	15·0							
	(C)	*Raw*								
		Finger millet	18·0	16·1			10·0	10·6	3·6	2·2
		Sorghum	12·0							
		Wheat	10·0							
	(D)	*Roasted*								
		Finger millet	18·0	16·0			10·0	10·3	3·3	2·3
		Sorghum	12·0							
		Wheat	10·0							
	(E)	Casein					10·0	11·7	3·6	2·5

[a] AA composition (microbiological assay, Barton-Wright, 1952) of diets (mg/g total N). Lys (A) 1262 (B) (D) (F) (H) (J) 344 (C) (G) 256 (E) 244 (I) 231. Met (A) 131 (B) 156 (C) (G) 119 (D) 150 (E) 112 (F) 162 (H) 150 (I) 144 (J) 162. Cys (A) (B) (E) (F) (I) (J) 119 (C) (D) (G) (H) 131. Thr (A) (B) 219 (C) (D) (G) (H) 206 (E) (F) 194 (I) (J) 200.

[b] Prot. contents (N × 6·25) (DWB, %) fortified Bombay Duck fish flour 75·8, fortified prot. food, blend of 1/2/1 fish, groundnut and chickpea flours, 52·3, fortified SMP 36·0.

[c] All diets contained (%) chickpeas 15, groundnut flour 15, jaggery (crude sugar) 23, hydrogenated fat 5, Ca salts 1, vit. mix 1. Prot. contents (N × 6·25) (DWB, %) blend A (raw) 17·2, blend D (roasted) 16·7, blend C (raw) 17·4, blend A (roasted) 16·8. Essential AA composition (microbiological assay, Barton-Wright, 1952) and available Lys (Carpenter, 1960) (mg/g total N). Ile (A) (B) (C) 325, (D) 312. Leu (A) 475 (B) 487 (C) 462 (D) 462. Lys (A) 262 (B) 256 (C) 269 (D) 250. Available Lys (A) 164 (B) 156 (C) 169 (D) 150. Met + Lys (A) 156 (B) 150 (C) 156 (D) 144. Phe (A) 350 (B) 369 (C) (C) 319 (D) 325. Thr (A) (C) 187 (B) (D) 181. Trp (A)–(D) 75. Val (A) 312 (B) 319 (C) (D) 306. Arg (A) 444 (B) (C) (D) 437. His (A) (B) 131 (C) (D) 125.

Appendix

Numerous vernacular names for sorghum and the millets appear in the literature reviewed and there are also many variants in the botanical equivalents cited. Listed in Part I, in alphabetical order, are vernacular names which were found in surveying the literature for the present text with the botanical equivalent preferred at the present time. In Part II, the same information appears but arranged in alphabetical order of the botanical name. Readers interested in the origin, history and usage of local names in India and Africa are referred to the papers by Bono (1973), Gupta and Dutta (1967), Krishnaswamy (1962) and Portères (1955, 1958 abc, 1959 abc). Lists of vernacular names and botanical equivalents appear in Adrian and Jacquot (1964), National Research Council (1961) (Appendix), Rachie (1965, 1974) and Rachie and Peters (1977).

PART I

Colloquial name	Botanical term
Abôra	*Sorghum bicolor* (L.) Moench
Acha	*Digitaria exilis*
Adlay	*Coix lachryma-jobi*[d]
African millet	*Eleusine coracana*
Arikelu	*Paspalum scrobiculatum*
Babala	*Pennisetum typhoides*
Bajra	*Pennisetum typhoides*
Bajri	*Pennisetum typhoides*
Banti	*Echinochloa stagnata*[a]
Barnyard millet	*Echinochloa crus-galli*
Bechna	*Sorghum bicolor* (L.) Moench
Bechna	*Pennisetum typhoides*
Bessna	*Eleusine coracana*
Brown top millet	*Brachiaria ramosa* Stapf
Bulrush millet	*Pennisetum typhoides*
Bultuc	*Pennisetum typhoides*
Cambu	*Pennisetum typhoides*
Chena	*Panicum miliaceum*
Cheno	*Panicum repens*[b]
Cholam	*Sorghum bicolor* (L.) Moench
Cholam irungu	*Sorghum bicolor* (L.) Moench
Common millet	*Panicum miliaceum*
Coracan	*Eleusine coracana*
Cumbu	*Pennisetum typhoides*

Colloquial name	Botanical term
Dagussa	*Eleusine coracana*
Dauro	*Pennisetum maiwa*
Dawa	*Panicum* spp.
Dura	*Sorghum bicolor* (L.) Moench
Durra	*Sorghum bicolor* (L.) Moench
Enélé	*Pennisetum typhoides*
Fela	*Sorghum bicolor* (L.) Moench
Findi	*Digitaria exilis*
Finger millet	*Eleusine coracana*
Fonio	*Digitaria exilis* *Digitaria iburua*
Foxtail millet	*Setaria italica*
German millet	*Setaria italica*
Gero	*Pennisetum typhoides*
Gindhi	*Panicum miliare*
Great millet	*Sorghum bicolor* (L.) Moench
Guinea corn	*Sorghum vulgare* *Sorghum bicolor* (L.) Moench
Haraka	*Paspalum scrobiculatum*
Harik	*Paspalum scrobiculatum*
Heen meneri	*Panicum miliare*
Hog millet	*Panicum miliaceum*
Iburu	*Digitaria iburua*
Indian millet (Sri Lanka)	*Panicum miliaceum*
Indian millet (Rhodesia)	*Eleusine coracana*
Indian milo	*Pennisetum typhoides*
Italian millet	*Setaria italica*
Japanese barnyard millet	*Echinochloa crus-galli*
Jawar	*Sorghum bicolor* (L.) Moench
Job's tears	*Coix lachryma-jobi*
Jola	*Sorghum vulgare* *Sorghum bicolor* (L.) Moench
Jowar	*Sorghum bicolor* (L.) Moench
Juar	*Sorghum bicolor* (L.) Moench
Junera	*Sorghum bicolor* (L.) Moench
Jungle rice	*Echinochloa colona*
Kaffir-corn	*Sorghum bicolor* (L.) Moench

Colloquial name	**Botanical term**
Kangi	*Setaria italica*
Kangni	*Setaria italica*
Kaoliang	*Sorghum bicolor* (L.) Moench
Kaon rice	*Setaria italica*
Koda millet	*Paspalum scrobiculatum*
Kodaka	*Paspalum scrobiculatum*
Kodo millet	*Paspalum scrobiculatum*
Kodra	*Paspalum scrobiculatum*
Kodu	*Paspalum scrobiculatum*
Kongoni	*Setaria italica*
Kongossan	*Sorghum bicolor* (L.) Moench
Korda	*Paspalum scrobiculatum*
Korra	*Setaria italica*
Kurakkan	*Eleusine coracana*
Little millet	*Panicum miliare* *Panicum sumatrense*
Maha meneri	*Panicum miliaceum*
Maicillo	*Sorghum bicolor* (L.) Moench
Maiwa	*Pennisetum maiwa*
Maiz de Guinea	*Sorghum bicolor* (L.) Moench
Mapfunde	*Sorghum bicolor* (L.) Moench
Marooa	*Eleusine coracana*
Mawele	*Pennisetum typhoides*
Mboratel	*Sorghum bicolor* (L.) Moench
Meneri	*Panicum miliaceum*
Mhunga	*Pennisetum typhoides*
Mijo común	*Panicum miliaceum*
Mijo perla	*Pennisetum typhoides*
Mijo ragi (Spanish)	*Eleusine coracana*
Mil à chandelles	*Pennisetum typhoides*
Mil commun	*Panicum miliaceum*
Millet des oiseaux	*Setaria italica*
Millet indigène	*Paspalum scrobiculatum*
Mohango	*Pennisetum* spp.
Mohren hirse	*Sorghum bicolor* (L.) Moench *Andropogon sorghum*
Monya	*Paspalum scrobiculatum*
Mtama	*Sorghum bicolor* (L.) Moench
Munga	*Pennisetum typhoides*
Nachni	*Eleusine coracana*
Nagli	*Eleusine coracana*

Colloquial name	Botanical term
Navane	*Setaria italica*
Njera	*Eleusine coracana*
Nteing	*Sorghum bicolor* (L.) Moench
Nyauti	*Pennisetum typhoides*
Pahi	*Paspalum scrobiculatum*
Pani varagu	*Panicum miliaceum*
Pearl millet	*Pennisetum typhoides*
Petit mil	*Pennisetum* spp.
Pothu meneri	*Brachiaria ramosa*
Proso millet	*Panicum miliaceum*
Ragi	*Eleusine coracana*
Ragi marua	*Eleusine coracana*
Rajgira	*Amaranthus paniculatus*[c]
Rajika	*Eleusine coracana*
Rajkeera	*Amaranthus paniculatus*
Rala	*Setaria italica*
Rapoko	*Eleusine coracana*
Rispen hirse	*Panicum miliaceum*
Rukweza	*Eleusine coracana*
Rupoko	*Eleusine coracana*
Sajjalu	*Pennisetum typhoides*
Sama	*Echinochloa crus-galli*
Samai	*Panicum miliare*
Samalu	*Panicum miliare*
Sanai	*Panicum miliare*
Sanio	*Pennisetum pychnostachyum*
Sanwa	*Echinochloa crus-galli*
Save	*Panicum miliare*
Sawan	*Panicum miliare*
Shama millet	*Echinochloa colona*
Shavan	*Panicum miliare*
Shevan	*Panicum miliare*
Sorgho (French)	*Sorghum bicolor* (L.) Moench
Sorghum	*Sorghum* spp.
Sorgo (Spanish)	*Sorghum* spp.
Souna	*Pennisetum* spp.
Tabsout	*Sorghum bicolor* (L.) Moench
Tafsoût	*Sorghum bicolor* (L.) Moench
Tafsoût el beïda	*Sorghum bicolor* (L.) Moench
Tafsoût el hamra	*Sorghum bicolor* (L.) Moench
Tanahal	*Setaria italica*
Teff	*Eragrostis tef*
Teff d'abyssinie	*Eragrostis tef*

Colloquial name	Botanical term
Teing	*Sorghum bicolor* (L.) Moench
Tenai	*Setaria italica*
Thenai	*Setaria italica*
Thenar	*Setaria italica*
Tocusso	*Eleusine coracana*
Varagu	*Paspalum scrobiculatum*
Vari	*Panicum miliaceum*
Varo	*Panicum miliaceum*
Voyendé	*Sorghum bicolor* (L.) Moench
Wari	*Panicum miliaceum*
Zviyo	*Eleusine coracana*

PART II

Botanical term	Colloquial name
Amaranthus paniculatus[c]	Rajgira
	Rajkeera
Brachiara ramosa	Brown top millet
	Pothu meneri
Coix lachryma-jobi[d]	Adlay
	Job's tears
Digitaria exilis	Acha
	Findi
	Fonio
Digitaria iburua	Fonio
	Iburu
Echinochloa colona	Jungle rice
	Shama millet
Echinochloa crus-galli	Barnyard millet
	Japanese barnyard millet
	Sama
	Sanwa
Echinochloa stagnana[a]	Banti
Eleusine coracana	African millet
	Bessna
	Coracan
	Dagussà
	Finger millet
	Indian millet
	Kurakkan
	Marooa
	Mijo ragi (Spanish)
	Nachni
	Nagli

Botanical term	Colloquial name
Eleusine coracana	Njera
	Ragi
	Ragi marua
	Rajika
	Rapoko
	Rukweza
	Rupoko
	Tocusso
	Zviyo
Eragrostis tef	Teff
	Teff d'abyssinie
Panicum spp.	Dawa
Panicum miliaceum	Chena (Proso millet)
	Common millet
	Hog millet
	Indian millet (Sri Lanka)
	Maha meneri
	Meneri
	Mijo comun
	Mil commun
	Pani varagu
	Proso millet
	Rispen hirse
	Vari
	Varo
	Wari
Panicum miliare	Gindhi
	Heen meneri
	Little millet
	Samai
	Samalu
	Sanai
	Save
	Sawan
	Shavan
	Shevan
Panicum repens[b]	Cheno
Panicum sumatrense	Little millet
Paspalum scrobiculatum	Arikelu
	Haraka
	Harik
	Koda millet
	Kodaka
	Kodo millet
	Kodra
	Kodu
	Korda
	Millet indigène
	Monya
	Pahi
	Varagu

Botanical term	**Colloquial name**
Pennisetum spp.	Mohango
	Petit mil
	Souna
Pennisetum maiwa	Dauro
	Maiwa
Pennisetum pychnostachyum	Sanio
Pennisetum typhoides	Babala
	Bajra
	Bajra
	Bechna
	Bulruch millet
	Bultuc
	Cambu
	Cumbu
	Enélé
	Gero
	Indian milo
	Mawele
	Mhunga
	Mijo perla
	Mil à chandelles
	Munga
	Nyauti
	Pearl millet
	Sajjalu
Setaria italica	Foxtail millet
	German millet
	Italian millet
	Kangi
	Kangni
	Kaon rice
	Kongoni
	Korra
	Millet des oiseaux
	Navane
	Rala
	Tanahal
	Tenai
	Thenai
	Thenar
Sorghum spp.	Sorghum
	Sorgo (Spanish)
Sorghum bicolor (L.) Moench	Abôra
	Bechna
	Cholam
	Cholam irungu
	Dura
	Durra
	Fela

Botanical term	Colloquial name
Sorghum bicolor (L.) Moench	Great millet
	Guinea corn
	Jawar
	Jola
	Jowar
	Juar
	Junera
	Kaffir-corn
	Kaoliang
	Kongossan
	Maicillo
	Maiz de Guinea
	Mapfunde
	Mboratel
	Milo
	Mohren hirse
	Mtama
	Nteing
	Sorgho
	Tabsoùt
	Tafsoût
	Tafsoût el beida
	Tafsoût el hamra
	Teing
	Voyendé

[a] *Echinochloa stagnana* (swamp grass) is gathered, rather than cultivated, for food.
[b] *Panicum repens* is not regarded as a millet and is possibly a misprint for *Panicum miliaceum.*
[c] *Amaranthus paniculatus* is not regarded as a millet and is not even of the family Gramineae but may be used as a grain.
[d] *Coix lachryma-jobi* is not regarded as a millet but the grain has been suggested as suitable for human consumption.

Bibliography

AACC. (1962). "Cereal Laboratory Methods". 7th ed. American Association of Cereal Chemists, St. Paul, Minnesota.

Abifarin, A. O. and Pickett, R. C. (1970). Combining ability and heterosis for yield, lysine and certain plant characters in 18 diverse inbreds and 56 hybrids in *Sorghum bicolor* (L.) Moench. *African Soils/Sols africans* **15**, 399–416.

Accessory Food Committee. (1943). *Biochem. J.* **37**, 436.

Acharya, B. N., Niyogi, S. P. and Patwardhan, V. N. (1942). Balanced diets. III. Effect of parching on the biological value of the proteins of some cereals and pulses. *Indian J. med. Res.* **30**, 73–80.

Achuta Murthy, P. N., Meena Rao, J., Kadkol, S. B., Krishnamurthy, K. S. Rao, M. B., Satyanarayana, M. N. and Thakare, R. D. (1965). Nutritive value of some Indian foodstuffs. *Indian J. med Res.* **53**, 259–268.

Adams, C. A. and Novellie, L. (1975). Acid hydrolases and autolytic properties of protein bodies and spherosomes isolated from ungerminated seeds of *Sorghum bicolor* (L.) Moench. *Pl. Physiol.* **55**, 7–11.

Adams, C. A., Watson, T. G. and Novellie, L. (1975). Lytic bodies from cereals hydrolysing maltose and starch *Phytochemistry* **14**, 953–956.

Adams, C. A., Novellie, L. and Liebenberg, N. v. d. W. (1976). Biochemical properties and ultrastructure of protein bodies isolated from selected cereals. *Cereal Chem.* **53,** 1–12.

Adda, J. (1958). Content of amino acids of varieties of *Pennisetum* and sorghum from Senegal (in French). *Qualitas Pl. Mater. Veg.* **3|4**, 138–141.

Adkins, G. K. and Greenwood, C. T. (1966). The isolation of cereal starches in the laboratory. *Staerke* **18**, 213–218.

Adrian, J. (1959). The microbiological assay of the B group Vitamins (in French). *Cah. tech. Cent. natn. Coord. Etud. Rech. Nutr. Aliment.*

Adrian, J. (1969). Tryptophan and Vitamin PP content of the vegetable food products of intertropical Africa (in French). *Annls Nutr. Aliment.* **23**, 233–252.

Adrian, J. and Frangne, R. (1970). Canary grass. II. Role of canary grass (*Phalaris canariensis* L.) and sorghum (*Sorghum sp.*) in maize diets (in French). *Annls. Nutr. Aliment.* **24**, 1–9.

Adrian, J. and Jacquot, R. (1961). Comparative studies in supplementing a sorghum and peanut basic ration with fish meal and pure amino acids (in French). *Annls. Nutr. Aliment.* **15**, 227–237.

Adrian, J. and Jacquot, R. (1964). "Sorghum and the millets in human and animal nutrition" (in French). Monographies Alimentaires 5. Vigot Freres, Paris.

Adrian, J. and Sayerse, C. (1957). Composition of Senegal millets and sorghums. *Br. J. Nutr.* **11**, 99–105.

Adrian, J., Périssé, J. and Jacquot, R. (1956). Interest and significance of groundnut flour as a protective food for the black African (in French). *Annls. Nutr. Aliment.* **10**, 3–10.

Adrian, J., Frangne, R., Boulenger, P., Davin, A., Gallant, D., Abel, H., Guilbot, A. and Gast, M. (1967). Evaluation and nutritional interest of different milling processes for millet (in French). *Cah. Nutr. Diet.* **2**, 67–77. (also in *Agron. trop.* Nogent **22**, 687–698).

Adrian, J., Murias de Queroz, M. J. and Frangne, R. (1970). Nicotinic acid (Vitamin PP) in cereals and legumes (in French). *Annls. Nutr. Aliment.* **24,** 155–166.

Adrian, J., Goussault, B., Arnal-Peyrot, F. and Samson, M. F. (1975). Techniques of milling millet and sorghum, and protein value of semolina and flours (in French). *Bull. Soc. Scient. Hyg. Aliment.* **63**, 250–264. (Same authors and title as *Agron. trop.* Nogent. 1975, **30**, 43–51.)

Agarwal, P. N. and Sinha, N. S. (1964). Chemical composition of the oil from the seeds of *Pennisetum typhoideum* (Bajra). *Indian J. Agron.* **9**, 288–291.

Agarwala, O. N., Negi, S. S. and Mahadevan, V. (1964). Studies on the toxicity and nutritive value of fungus-free *Paspalum scrobiculatum* grains. *Indian vet. J.* **41**, 43–47.

Agren, G., Hofvander, Y., Selinus, R. and Vahlquist, B. (1969). Faffa: a supplementary cereal-based weaning food in Ethiopia. *In* "Protein enriched cereal foods for world needs". (M. Milner, Ed.). 278–287. American Association of Cereal Chemists, St. Paul, Minnesota 55104.

Agricultural Research Council. (1966). The nutrient requirements of farm livestock No. 3. Pigs. Summaries of estimated requirements. H. M. Stationery Office, London.

Ahmad, Z. and Murty, B. R. (1972). Inheritance of protein and some essential amino acids in a 22 × 22 partial diallel of *Pennisetum typhoides. Indian J. Genet. Pl. Breed.* **32**, 400–407.

Ahmad, Z., Murty, B. R. and Harrinarayana, G. (1972). Protein content in successive generations of partial diallel crosses in dwarf pearl millet. *Indian J. Genet. Pl. Breed.* **32**, 325–330.

Ahuja, V. P., Singh, J. and Naik, M. S. (1970). Amino acid balance of proteins of maize and sorghum. *Indian J. Genet. Pl. Breed.* **30**, 727–731.

Akazawa, T., Miljanich, P. and Conn, E. E. (1960). Studies on cyanogenic glycoside of *Sorghum vulgare. Pl Physiol.* **35**, 535–538.

Akeson, W. R. and Stahmann, M. A. (1964). A pepsin pancreatin digest index of protein quality evaluation. *J. Nutr.* **83**, 257–261.

Akinrele, I. A. (1970). Fermentation studies on maize during the preparation of a traditional African starch-cake food. *J. Sci. Fd Agric.* **21**, 619–625.

Akınrele, I. A. and Bassir, O. (1967). The nutritive value of "Ogi", a Nigerian infant food. *J. trop. Med. Hyg.* **70**, 279–280.

Akotar, S. N. and Deshmukh, V. A. (1974). Effect of BHC on ammonification, nitrification and nitrogen uptake in Jowar. *Punjabrao Krishi Vidyapeeth Res. J.* **3**, 68–71.

Alause, J. and Feillet, P. (1970). Simple method for forecasting the colour of alimentary paste (in Italian). *Techn. Molit.* **21**, 511–517.

Alcantara, P. F., Rigor, E. M., Miller, J. C. and Arganosa, V. G. (1970). The feeding value of grain sorghum for pigs. *Philipp. Agric.* **53**, 588–603.

Alexander, D. E., Silvela, S. L., Collins, F. I. and Rodgers, R. C. (1967). Analysis of oil content of maize by wide-line NMR. *J. Am. Oil Chem. Soc.* **44**, 555–558.

Ali, S. M., Razzaq, A. and Jamil, M. (1964a). Supplementary value of a blend of cottonseed flour and fish flour on certain indigenous cereals. *Pakist. J. scient. Res.* **16**, 135–137.

Ali, S. M., Razzaq, A. and Jamil, M. (1964b). Improvement of the protein value of cottonseed flour by supplementation with fish flour and skimmed milk powder. *Pakist J. Sci.* **16**, 113–115.

Allee, G. L. (1976). Effect of processing methods on nutritional value of milo for weaned pigs. *J. Anim. Sci.* **41**, 248.

Allee, G. L. and Paulsen, G. (1975). Nutritional value of millet as a swine feed. *J. Anim. Sci.* **43**, 306–307.

All-Union Scientific Research Institute of Oils. (1962). Calorimetric method of determining the total Vitamin E content of oils (in Russian). Instruction No. 634. VN11 Zh, Leningrad.

All-Union Vitamin Scientific Research Institute. (1954). "Methods of vitamin assay" (in Russian). V. A. Devyatnin, Ed. Pishchepromizdat, Moscow.

Almgard, G. (1963). High content of iron in teff, *Eragrostis abyssinica* Link., and some other crop species from Ethiopia—a result of contamination. *Lantbr. Högsk. Annlr.* **29**, 215–220.

Ambegaokar, S. D., Seshadri, S., Shah, H. C., Shinde, V. P., Adhikari, H. R. Patel, S. M. and Rao, M. V. R. (1964). Studies in nutritive value of Indian foodstuffs. I. Proximate principles, minerals and vitamins. *J. Nutr. Diet.* **1**, 269–275.

Ambegaokar, S. D., Raj, H., Shinde, V. P. and Rao, M. V. R. (1965). Studies in nutritive value of Indian foodstuffs. II. Amino acid composition of certain leafy vegetables and millets. *J. Nutr. Diet.* **2**, 14–16.

American Oil Chemists Society. (1946). Official and tentative methods.

American Oil Chemists Society. (1949). Report of the Spectroscopy Committee. *J. Am. Oil Chem. Soc.* **26**, 399–404.

American Oil Chemists Society. (1960). Official methods.

American Oil Chemists Society. (1969). Official and tentative methods.

Analytical Methods Committee. (1943). Determination of the crude fibre in national flour. *Analyst* **68**, 276–278.

Ananda Rao, B. and Reddy, P. R. (1973). Dry matter accumulation at important physiological stages, grain yield and protein quality under different levels of nitrogen in sorghum. *Indian J. agric. Sci.* **43**, 138–142.

Anderson, E and Martin, J. H. (1949). World production and consumption of millet and sorghum. *Econ. Bot.* **3**, 265–288.

Anderson, R. A. (1963). Wet-milling properties of grain: Bench-scale study. *Cereal Sci. Today* **8**, 190.

Anderson, R. A. (1969a). Research on improved methods of milling sorghum grain. *In* "Proceedings of Sixth Biennial International Grain Sorghum Research and Utilization Conference". Amarillo, Texas, USA.

Anderson, R. A. (1969b). Producing quality sorghum flour on wheat milling equipment. *N West Miller* **276**, 10–12, 14–15.

Anderson, R. A. and Burbridge, L. H. (1971). Integrated process for dry milling grain sorghum. *N West Miller* **278**, 24–28.

Anderson, R. A., Montgomery, R. R. and Burbridge, L. H. (1969a). Low-fat endosperm fractions from grain sorghum. *Cereal Sci. Today* **14**, 366–368.

Anderson, R. A., Conway, H. F., Pfeifer, V. F. and Griffin, E. L. (1969b). Roll and extrusion-cooking of grain sorghum grits. *Cereal Sci. Today* **14**, 372–375, 381.

Anderson, R. A., Conway, H. F. and Burbridge, L. H. (1977). Yield and chemical composition of fractions from the dry milling of a high-lysine grain sorghum. *Cereal Chem.* **54**, 855–856.

Andrews, D. J. and Majmudar, J. V. (1975). "The pearl millet breeding program". ICRISAT, Begumpet, Hyderabad 500016, India, mimeographed.

Andrews, D. J., Majmudar, J. V. and Doggett, H. (1975). Pearl millet programme. *In*

"Sorghum and millet improvement programme", pp. 1–24. ICRISAT, Begumpet, Hyderabad 500016, India, mimeographed.

Andrews, J. S. (1943). Report of 1942-43. Methods of Analysis. Subcommittee on riboflavin assay. *Cereal Chem.* **20**, 613–625.

Angelov, I. (1965). Effect of sorghum in fattening pigs for meat (in Bulgarian). *Zhivot.-Nauki* **2**, 909–914.

Anon. (1966). Alteration in amino acids of sorghum hybrids grown in different locations. *Nutr. Rev.* **24**, 223–224.

Anon. (1969a). Use and maintenance of the huller-cleaner M 164 for millet and sorghum (in French). *Ombre Baobab* **16**, 2, 8.

Anon. (1969b). Teff in Ethiopia (in German). *Afrika Heute* **9**, 135–137.

Anthony, W. B., Smith, L. A. and Teem, D. H. (1975). Relationship of nutrients in sorghum grain to nutrients in the plant. *J. Anim. Sci.* **40**, 197.

AOAC. (1940). "Official and tentative methods of analysis." 5th ed. Association of Official Agricultural Chemists, Washington, DC.

AOAC. (1945). "Official methods of analysis." 6th ed. Association of Official Agricultural Chemists, Washington, DC.

AOAC. (1950). "Official methods of analysis." 7th ed. Association of Official Agricultural Chemists, Washington, DC.

AOAC. (1955). "Official methods of analysis." 8th ed. Association of Official Agricultural Chemists, Washington, DC.

AOAC. (1960). "Official methods of analysis." 9th ed. Association of Official Agricultural Chemists, Washington, DC.

AOAC. (1965). "Official methods of analysis." 10th ed. Association of Official Agricultural Chemists, Washington, DC.

AOAC. (1970). "Official methods of analysis." 11th ed. Association of Official Analytical Chemists, Washington, DC.

AOAC. (1975). "Official and tentative methods of Association of Analytical Chemists." 12th ed. The Association, Washington, DC.

Aoyama, Y. and Ashida, K. (1972). Effect of excess and deficiency of individual essential amino acids in diets on the liver lipid content of growing rats. *J. Nutr.* **102**, 1025–1032.

Aoyama, Y., Nakanishi, M. and Ashida, K. (1969). Effect of dietary amino acid composition on the accumulation of lipids in the liver of growing rats. *J. Nutr.* **97**, 348–352.

Aoyama, Y., Nakanishi, M. and Ashida, K. (1973). Effect of methionine on liver lipid content and lipid metabolism of rats fed a protein-free diet. *J. Nutr.* **103**, 54–60.

Applied Nutrition Unit. (1960). Miscellaneous information about tropical foods. A Memorandum of the Applied Nutrition Unit, Department of Human Nutrition, London School of Hygiene and Tropical Medicine, London, England.

Aragon, H. R. and Bressani, R. (1965). Effect of fertilization with minor elements on the protein value of corn and sorghum (in Spanish). *Archos venez. Nutr.* **15**, 63–86.

Araullo, E. V., De Padua, D. B. and Graham, M. (Eds). (1976). "Rice postharvest technology." IDRC, Box 8500, Ottawa, Canada K1G 3H9.

Armstrong, W. D., Featherston, W. R. and Rogler, J. C. (1973). Influence of methionine and other dietary additions on the performance of chicks fed bird resistant sorghum grain diets. *Poult. Sci.* **52**, 1592–1599.

Armstrong, W. D., Rogler, J. C. and Featherston, W. R. (1974a). Effect of tannin extraction on performance of chicks fed bird resistant sorghum grain diets. *Poult. Sci.* **53**, 714–720.

Armstrong, W. D., Rogler, J. C. and Featherston, W. R. (1974b). In vitro studies of the protein digestibility of sorghum grain. *Poult. Sci.* **53**, 2224–2227.

Armstrong, W. D., Featherston, W. R. and Rogler, J. C. (1974c). Effects of bird resistant sorghum grain and various commercial tannins on chick performance. *Poult. Sci.* **53**, 2137–2142.

Arnal-Peyrot, F. and Adrian, J. (1974). Metabolism of cereal pentosans in the rat (in French). *Int. J. Vitam. Nutr. Res.* **44,** 543–552.

Arnold, A. (1945). Report of the 1944-45 Committee on Riboflavin Assay. *Cereal Chem.* **22**, 455–461.

Arnon, I. (1972). "Crop production in dry regions. Vol II. Systematic treatment of the principal crops." Leonard Hill, London.

Arnould, J-P. and Miche, J. C. (1971). Review of the economy and utilization of the millets and sorghums in the world (in French). *Agron. trop.* Nogent. **26**, 865–887.

Arora, S. K. and Luthra, Y. P. (1972). Variability of starch and sugar contents in grains of sorghum forages and its correlation with tannin and mineral matter content. *Staerke* **24**, 51–53.

Arora, S. K. and Luthra, Y. P. (1974). The in vitro digestibility of promising Indian varieties of sorghum and in relation with tannin content. *Indian J. Nutr. Diet.* **11**, 233–236.

Association of Vitamin Chemists. (1947). "Methods of Vitamin assay." Interscience Publishers Inc., New York.

Association of Vitamin Chemists. (1951). "Methods of analysis." 2nd ed. Interscience Publishers, New York.

Association of Vitamin Chemists. (1966). "Methods of Vitamin assay." 3rd ed. Interscience Publishers, New York.

Atkin, L., Schultz, A. S., Williams, W. L. and Frey, C. N. (1943). Yeast microbiological methods for determination of vitamins. Pyridoxine. *Ind Engng Chem.* (Analytical Edition). **634**, 6.

Atkinson, R. L., Krueger, K. K., Bradley, J. W., Couch, J. R. and Krueger, W. F. (1974). Relation of protein level and grain source to turkey reproduction. *Nutr. Rep. Int.* **9**, 259–266.

Atkinson, R. L., Bradley, J. W. and Krueger, W. F. (1975). Wheat, milo and corn as ingredients in feeds for young turkeys. *Nutr. Rep. Int.* **11**, 345–349.

Aucamp, M. C., Grieff, J. T., Novellie, L., Papendick, B., Schwartz, H. M. and Steer, A. G. (1961). Kaffircorn malting and brewing studies. VIII. Nutritive value of some kaffircorn products. *J. Sci. Fd Agric.* **12**, 440–456.

Austin, A., Hanslas, V. K., Singh, H. D. and Ram, A. (1971). Protein content and chapati-making properties of some improved bajra (pearl millet) hybrids and varieties. *Bull. Grain Technol.* **9**, 247–251.

Austin, A., Singh, H. D., Hanslas, V. K., and Rao, N. G. P. (1972). Variations in protein and lysine content in *Sorghum vulgare. Acta agron. hung.* **21**, 81–88.

Autret, M., Cresta, M., Périssé, J. and Sizaret, F. (1969). Protein value of different types of diet in the world—their appropriate supplementation. *FAO Nutr. Newsl.* **6**, 1–29.

Awadalla, M. Z. (1974). Native Egyptian millet as supplement of wheat flour in bread. II. Technological studies. *Nutr. Rep. Int.* **9**, 69–78.

Awadalla, M. Z., and Slump, P. (1974). Native Egyptian millet as supplement of wheat flour in bread. I. Nutritional studies *Nutr. Rep. Int.* **9,** 59–68.

Axford, D. W. E., Chamberlain, N., Collins, T. H. and Elton, G. A. H. (1963). The Chorleywood process. *Cereal Sci. Today* **8**, 265.

Axtell, J. D. (1976). Annual report on inheritance and improvement of protein quality and content in *Sorghum bicolor* (L.) Moench. April 1, 1975-March 31, 1976. Report No. 12. Purdue University, Lafayette, Indiana, USA.

Axtell, J. D., and Oswalt, D. L. (1972). Research Progress Report on inheritance and improvement of protein quality and content in *Sorghum bicolor* (L.) Moench. January 1, 1972-December 31, 1972. Report No. 9. Contract CSD/1175. Submitted to Agency for International Development, Department of State, Washington, DC.

Axtell, J. D., Oswalt, D. L., Mertz, E. T., Pickett, R. C., Jambunathan, R. and Srinivasan, G. (1974). Components of nutritional quality in grain sorghum. *In* "Proceedings of the CIMMYT-Purdue Symposium on Protein Quality in Maize, El Batan, Mexico", pp. 374–386. December 4-8, 1972. Dowden, Hutchinson and Ross, Inc., Stroudsburg, Pa., USA.

Aykroyd, W. R. and Krishnan, B. G. (1937). The effect of skimmed milk, soya bean and other foods in supplementing typical Indian diets. *Indian J. med. Res.* **24**, 1093–1106.

Aykroyd, W. R., Patwardhan, V. N. and Ranganathan, S. (1956). "The nutritive value of Indian foods and the planning of satisfactory diets". 5th ed. Manager of Publications, Delhi, India.

Aykroyd, W. R., Gopalan, C. and Balasubramanian, S. C. (1963). "The nutritive value of Indian foods and the planning of satisfactory diets." Special Report Series No. 42. Nutrition Research Laboratories, Hyderabad. Indian Council of Medical Research, New Dehli.

Aykroyd, W. R., Gopalan, C. and Balasubramanian, S.C. (1966). "The nutritive value of Indian foods and the planning of satisfactory diets". Special Report Series No. 42. Revised 6th ed. Indian Council of Medical Research, New Dehli.

Ayyangar, G. N. R. and Krishnaswami, N. (1941). Studies on the histology and colouration of the pericarp of the sorghum grain. *Proc. Indian Acad. Sci.* Sect. B 14, 114–116.

Ayyangar, G. N. R., Ayyar, S., Panduranga Rao, V. and Kunhikoran Nambiar, A. (1936). Inheritance of characters in sorghum—the great millet. IX. Dimpled grains. *Indian J. agric. Sci.* **6**, 938–945.

Ayyar, K. V. S., and Narayanaswamy, K. (1949). Varagu poisoning. *Nature*, Lond. **163**, 912–913.

Bacon, J. S. D. and Edelman, J. (1951). The carbohydrates of the Jerusalem artichoke and other compositae. *Biochem. J.* **48**, 114–126.

Badi, S. M., Hoseney, R. C. and Casady, A. J. (1976). Pearl millet. 1. Characterization by SEM, amino acid analysis, lipid composition, and prolamine solubility. *Cereal Chem.* **53**, 478–487.

Baijal, M., Singh, S., Shukla, R. N. and Sanwal, G. G. (1972). Enzymes of the banana plant: optimum conditions for extraction. *Phytochemistry* **11**, 926–936.

Baird, D. M. (1976). Heat treated soybeans, corn and BR grain sorghum for feeder pigs. *J. Anim. Sci.* **42**, 256.

Balakrishna Rao, K., Mithyantha, M. S., Devi, L. S. and Perur, N. G. (1973). Nutrient content of some new ragi varieties. *J. agric. Sci. Camb.* **7**, 562–565.

Balasubramanian, S. C. and Ramachandran, M. (1957). Amino acid composition of Indian foodstuffs. Part III. Threonine and arginine content of some cereals. *Indian J. med. Res.* **45**, 623–629.

Balasubramanian, S. C., Ramachandran, M., Viswanatha, T. and De, S. S. (1952a). Amino acid composition of Indian foodstuffs. Part I. Tryptophane, leucine, isoleucine and valine content of some cereals. *Indian J. med. Res.* **40**, 73–87.

Balasubramanian, S. C., Ramachandran, M., Viswanatha, T. and De, S. S. (1952b). Amino acid composition of Indian foodstuffs. Part II. Lysine, methionine, phenylalanine and histidine content of some cereals. *Indian J. med. Res.* **40**, 219–234.

Baldwin, A. R. and Sniegowski, M. S. (1951). Fatty acid compositions of lipids from corn and grain sorghum kernels. *J. Am. Oil Chem. Soc.* **28**, 24–27.

Balwani, T. L., Johnson, R. R. and Dehority, B. A. (1969). Comparison of cellulose and pentosan digestibilities in roughage feeds. *J. Dairy Sci.* **52**, 1290–1293.

Bamburg, J. R., Strong, F. M. and Smalley, E. B. (1969). Toxins from moldy cereals. *J. agric. Fd Chem.* **17**, 443–450.

Bandemer, S. L. and Evans, R. J. (1963). The amino acid composition of some seeds. *J. agric. Fd Chem.* **11**, 134–137.

Banigo, E. O. I. (1969). An investigation into the fermentation and enrichment of Ogi. Ph.D. Thesis, University of Leeds, England.

Banigo, E. O. I. and Muller, H. G. (1972a). Manufacture of Ogi (a Nigerian fermented cereal porridge). Comparative evaluation of corn, sorghum and millet. *Can. Inst. Fd Sci. Technol.* J. **5**, 217–221.

Banigo, E. O. I. and Muller, H. G. (1972b). Carboxylic acid patterns in Ogi fermentation. *J. Sci. Fd Agric.* **23**, 101–111.

Bantu Beer Unit. (1971). Annual Report. *Chem.* **175**, Pretoria, South Africa.

Baptist, N. G. and Perera, B. P. M. (1956). Essential amino acids of some tropical cereal millets. *Br. J. Nutr.* **10**, 334–337.

Barbosa, H. M. (1971). Genes and gene combinations associated with protein, lysine, and carbohydrate content in the endosperm of maize. (*Zea mays*, L.) Ph. D. Thesis. Purdue University, Lafayette, Indiana, USA.

Barham, H. N., Wagoner, J. A., Campbell, C. L. and Harclerode, E. H. (1946). The chemical composition of some sorghum grains and the properties of their starches. Technical Bulletin 61. Agricultural Experiment Station, Kansas College of Agriculture and Applied Science, Manhattan, Kansas.

Barnes, R. F., Muller, L. D., Bauman, L. F. and Colenbrander, V. F. (1971). In vitro dry matter disappearance of brown midrib mutants of maize (*Zea mays* L.) *J. Anim. Sci.* **33**, 881–884.

Barton, J. C. (1948). Photometric analysis of phosphate rock. *Analyt. Chem.* **20**, 1068–1073.

Barton-Wright, E. C. (1946). Microbiological assay of amino acids. Tryptophan, leucine, isoleucine, valine, cystine, methionine, lysine, phenylalanine, histidine, arginine, and threonine. *Analyst* **71**, 267.

Barton-Wright, E. C. (1952), "Microbiological assay of the vitamin B complex and amino acids." Sir Isaac Pitman, London.

Baseden, S. C. and Aldrick, S. (1970). Toxigenic *Aspergillus flavus* in northern Australia. *J. Aust. Inst. agric. Sci.* **36**, 237–240.

Basson, W. B. (1967). Determination of total sulphur in biological materials. *Cereal Chem.* **44**, 92–94.

Bates, F.L., French, D. and Rundle, R. E. (1943). Amylose and amylopectin content of starches determined by their iodine complex formation. *J. Am. Chem. Soc.* **65**, 142–148.

Bate-Smith, E. C. (1954a). Astringency in foods. *Food* **23**, 124–127.

Bate-Smith, E. C. (1954b). Leuco-anthocyanins 1. Detection and identification of anthocyanidins formed from leuco-anthocyanins in plant tissue. *Biochem. J.* **58**, 122-126.

Bate-Smith, E. C. (1969). Luteoforol (3′, 4, 4′, 5, 7-pentahydroxyflavan) in *Sorghum vulgare* L. *Phytochemistry* **8**, 1803–1810.

Bate-Smith, E. C. (1973a). Haemanalysis of tannins: the concept of relative astringency. *Phytochemistry* **12**, 907–912.

Bate-Smith, E. C. (1973b). Tannins of herbaceous leguminosae. *Phytochemistry* **12**, 1809–1812.

Bate-Smith, E. C. and Lerner, N. H. (1954). Leuco-anthocyanins. 2. Systematic distribution of leuco-anthocyanins in leaves. *Biochem. J.* **58**, 126–132.

Bate-Smith, E. C. and Rasper, V. (1969). Tannins of grain sorghum: luteoforol (leucoluteolinidin), 3′, 4, 4′, 5, 7-pentahydroxyflavan. *J. Fd Sci.* **34**, 203–209.

Bate-Smith, E. C. and Swain, T. (1953). Identification of leucoanthocyanins as tannins in foods. *Chemy. Ind.* (April 18): 377–378.

Baudet, G. J., Mosse, J., Landry, J. A. and Moureaux, T. G. (1966). Studies on the extraction and composition of rice proteins. *Cereal Chem.* **43**, 145–148.

Beadle, J. B., Just, D. E., Morgan, R. E. and Reiners, R. A. (1965). Composition of corn oil. *J. Am. Oil Chem. Soc.* **42**, 90–95.

Bealing, F. J. and Bacon, J. S. D. (1953). The action of mould enzymes on sucrose. *Biochem. J.* **53**, 277–285.

Beames, R. M., Daniels, L. J. and Sewell, J. O. (1973). The value of protein content of sorghum grain in pig diets. *Aust. J. exp. Agric. Anim. Husb.* **13**, 146–152.

Becker, M. (1950). Phytic acid in animal nutrition (in German). *Landw. Forsch.* **2**, 64–74.

Beckwith, A. C. (1972). Grain sorghum glutelin: isolation and characterization. *J. agric. Fd Chem.* **20**, 761–764.

Beckwith, A. C. and Jones, R. W. (1972). Physical chemical characterization of grain sorghum prolamine fractions and components. *J. agric. Fd Chem.* **20**, 259–261.

Beil, G. M. and Atkins, R. E. (1967). Estimates of general and specific combining ability in F_1 hybrids for grain yield and its components in grain sorghum, *Sorghum vulgare* Pers. *Crop Sci.* **7**, 225–228.

Belavady, B. (1977). Nutritive value of sorghum ICRISAT. International sorghum workshop. National Institute of Nutrition, Jamai Osmania, Hyderabad, India.

Belavady, B. and Gopalan, C. (1965). Production of black tongue in dogs by feeding diets containing jowar (*Sorghum vulgare*). *Lancet* (7424), 1220–1221.

Belavady, B. and Gopalan, C. (1966). Availability of nicotinic acid in jowar (*Sorghum vulgare*). *Indian J. Biochem.* **3**, 44–47.

Belavady, B. and Udayasekhara Rao, P. (1973). Production of nicotinic acid deficiency in monkeys fed leucine supplemented diets. *Int. J. Vitam. Nutr. Res.* **43**, 454–460.

Belavady, B., Madhavan, T. A. and Gopalan, C. (1967). Production of nicotinic acid deficiency (black tongue) in pups fed on diets supplemented with leucine. *Gastroenterology* **53**, 749–753.

Belavady, B., Madhavan, T. V. and Gopalan, C. (1968). Experimental production of niacin deficiency in adult monkeys by feeding jowar diets. *Lab. Invest.* **18**, 94–99.

Belova, S. M. and Denisenko, Ya I. (1965a). Vitamin composition of millet oil. *Prikladnaya Biokhimiya i Mikrobiologiya* **1**, 387–390. (*Appl. Biochem. Microbiol.* **1**, 287–289).

Belova, S. M. and Denisenko, Ya I. (1965b). Determination of linoleic and linolenic acids in millet oil by a spectrophotometric method. *Prikladnaya Biokhimiya i Mikrobiologiya* **1**, 474–476. (*Appl. Biochem. Microbiol.* **1**, 363–365).

Belova, S. M. and Denisenko, Ya I. (1965c). The chemical nature of miliacin (prosol). *Prikladnaya Biokhimiya i Mikrobiologiya* **1**, 664–668. *Appl. Biochem. Microbiol.* **1**, 518–521).

Belova, Z. A., Nechaev, A. P. and Severinenko, S. M. (1970). Content of fatty acids in the lipids of millet (in Russian). *Izv. vyssh. ucheb. Zaved. pishch. Tekhnol.* 32–33.

Bender, A. E. (1965). The balancing of amino acid mixtures and proteins. *Proc. Nutr. Soc.* **24**, 190–196.

Bender, A. E. and Miller, D. S. (1953). Constancy of the N/H_2O ratio of the rat and its use in the determination of net protein value. *Biochem. J.* **53**, vii.

Benson, J. V. and Patterson, J. A. (1965). The accelerated automatic chromatographic analysis of amino acids on a spherical resin. *Analyt. Chem.* **37**, 1108–1110.

Berry, L. D. (1970). Particle size of reconstituted sorghum grain as related to its digestibility and to cattle performance. M. S. Thesis. Texas A&M University, College Station, Texas, USA.

Bertoni, M. H. and Cattáneo, P. C. (1960). *An. Asoc. quim. argent.* **48**, 169.

Bertoni, M. H., De Sutton, G. S. K., Beretta, A. M., Burguete, J. A. and Cattáneo, P. (1963). Argentine sorghum germ oil; chemical composition (in Spanish). *An Asoc. quim. argent.* **51**, 29–42.

Bertrand, J. E. and Lutrick, M. C. (1971). Feeding value of NBR (non-bird-resistant) and BR (bird-resistant) sorghum grain in the ration of beef steers. *Proc. Soil Crop Sci. Soc. Fla.* **31**, 24.

Berulava, I. T. (1950). Nutritive value of *Setaria italica* (in Russian). *Gig. Sanit.* 42–43. (*Chem. Abst.* 1950, **44**, 8009).

Best, C. H. and Ridout, J. H. (1940). The lipotropic action of methionine. *J. Physiol.* **97**, 489–494.

Beza, J. (1967). Aminokwasy w zywieniu zwierzat. PWRiL, Warszawa.

Bhat, R. V. and Roy, D. N. (1976). Toxicity study of ergoty bajra pearl millet in rhesus monkeys. *Indian J. med. Res.* **64**, 1629–1633.

Bhatia, B. S., Chakrabarty, T. K., Mathur, V. K., Siddish, C. H. and Raghavan, P. K. V. (1968). Use of maize and milo flours in the preparation of bread. *Indian Fd Pckr.* **22**, 33–35.

Bhatia, I. S., Singh, R. and Dua, (Mrs) S. (1972). Changes in carbohydrates during growth and development of bajra (*Pennisetum typhoides*), jowar (*Sorghum vulgare*), and Kangui (*Setaria italica*). *J. Sci. Fd Agric.* **23**, 429–440.

Bhattacharya, K. R., Dutta, J. and Ray, D. K. (1957). A new paper chromatographic method for the estimation of histidine. *Ann. Biochem. exp. Med.* **17**, 1–4.

Bhide, V. P. and Hegde, R. K. (1957). Ergot on bajri (*Pennisetum typhoides* (*Burm.*)) Stapf and Hubbard, in Bombay State. *Curr. Sci.* **26**, 116.

Bidwell, G. L. (1918). A physical and chemical study of the Kafir kernel. United States Department of Agriculture Bulletin No. 634.

Bidwell, G. L., Bopst, L. E. and Bowling, J. D. (1922). A physical and chemical study of milo and feterita kernels. United States Department of Agriculture Bulletin No. 1129.

Bilay, V. I. (Ed.) (1960). "Mycotoxicoses of man and agricultural animals." US Joint Publication Research Service, Washington, DC.

Bishop C. and Taylor, T. G. (1963). Studies on the vitamin content of bird seeds. *Vet. Rec.* **75**, 688–691. (*Nutr. Abstr. Rev.* 1964, **34**, 392).

Bishop, L. R. (1972). Analysis committee of the European Brewery Convention. The measurement of total polyphenols in worts and beers. *J. Inst. Brew.* **78**, 36–38.

Blackburn, S. (1968). "Amino acid determination." Marcel Dekker, Inc., New York.

Blessin, C. W. (1971). Processing sorghum grain to improve nutritional quality and industrial use. *In* Seventh Biennial Grain Sorghum Research and Utilization Conference, p. 70.

Blessin, C. W., Vanetten, C. H. and Wiebe, R. (1958). Carotenoid content of the grain from yellow endosperm-type sorghums. *Cereal Chem.* **35**, 359–365.

Blessin, C. W., Dimler, R. J. and Webster, O. J. (1962). Carotenoids of corn and sorghum. II. Carotenoid loss in yellow-endosperm sorghum grain during weathering. *Cereal Chem.* **39**, 389–392.

Blessin, C. W., Vanetten, C. H. and Dimler, R. J. (1963). An examination of anthocyanogens in grain sorghums. *Cereal Chem.* **40**, 241–250.

Blessin, C. W., Anderson, R. A., Deatherage, W. J. and Inglett, G. E. (1971). Effect of alkali dehulling on composition and wet-milling characteristics of sorghum grain. *Cereal Chem.* **48**, 528–532.

Block, R. J. and Bolling, D. (1945). "Amino acid composition of protein and foods". Charles C. Thomas, Illinois, USA.

Block, R. J. and Mitchell, H. H. (1946). The correlation of the amino acid composition of proteins with their nutritive value. *Nutr. Abst. Rev.* **16**, 249–278.

Blondel, D. (1970). "New results on increasing the protein content of the grain of millet (*Pennisetum typhoides*) and sorghum (*Sorghum vulgare*) by nitrogen fertilization in Senegal" (in French). Unpublished report, Centre National de Recherches Agronomiques, Bambey, Senegal.

Boas Fixsen, M. A. and Roscoe, M. H. (1937–1938). Tables of the vitamin content of human and animal foods. *Nutr. Abst. Rev.* **7**, 823–867.

Boas Fixsen, M. A. and Roscoe, M. H. (1940). Tables of the vitamin content of human and animal foods *Nutr. Abst. Rev.* **9**, 795–861.

Bodwell, C. E. (1977). Application of animal data to human protein nutrition: a review. *Cereal Chem.* **54**, 958–983.

Bokorov, T. and Srećković, A. (1964). Effect of feeding with sorghum on weight gain and quality of pigs for bacon and pork (in Serbo-Croat with English summary). *Arhivza Poljoprivredne Nauke* **17**, 134–143.

Bono, M. (1973). Contribution to the morpho-systematics of the annual *Pennisetum* grown for their grains in Francophone West Africa (in French). *Agron. trop.* Nogent. **28**, 229–355.

Bono, M. and Vidal, P. (1962). Protein content of several lines of sorghum cultivated at the CRA, Bambey: variations within the same botanical group and their relation to the vitreousness of the grain (in French). *Agron. trop.* Nogent **17**, 67–74.

Booth, A. N., Masri, M. S., Robbins, D. J., Emerson, O. H., Jones, F. T. and DeEds, F. (1959). The metabolic fate of gallic acid and related compounds. *J. biol. Chem.* **234**, 3014–3016.

Booth, A. N., Robbins, D. J. and DeEds, F. (1961). Effect of dietary gallic acid and progallol on choline requirement of rats. *J. Nutr.* **75**, 104–106.

Booth, V. H. (1957). "Carotene: its determination in biological materials". Heffer & Sons Ltd., Cambridge, England.

Borasio, L. (1937). Sorghum as food and as substitute in baking (in Italian). *Giornale di Risicolture* **27**, 7–12.

Borchers, R. (1962). A note on the digestibility of the starch of high-amylose corn by rats. *Cereal Chem.* **39**, 145–146.

Borchers, R. and Ackerson, C. W. (1947). Trypsin inhibitor. IV. Occurrence in seeds of the leguminosae and other seeds. *Archs Biochem.* **13**, 291–293.

Bornstein, S. and Bartov, I. (1967). Comparisons of sorghum grain (milo) and maize as the principal cereal grain source in poultry rations. I. Their relative feeding value for broilers. *Br. Poult. Sci.* **8**, 213–221.

Bornstein, S. and Lipstein, B. (1971). Comparisons of sorghum grain (milo) and maize as the principal cereal grain source in poultry rations. 4. Relative content of available sulphur amino acids in milo and maize. *Br. Poult. Sci.* **12**, 1–13.

Bornstein, S. and Lipstein, B. (1972). Comparisons of sorghum grain (milo) and maize as the principal cereal grain source in poultry rations. 5. The effect of methionine and linoleic acid supplementations on all-vegetable milo layer diets. *Br. Poult. Sci.* **13**, 91–103.

Botes, D. P., Joubert, F. J. and Novellie, L. (1967a). Kaffircorn malting and brewing studies. XVII. Purification and properties of sorghum malt α-amylase. *J. Sci. Fd Agric.* **18**, 409–415.

Botes, D. P., Joubert, F. J. and Novellie, L. (1967b). Kaffircorn malting and brewing studies. XVIII. Purification and properties of sorghum malt β-amylase. *J. Sci. Fd Agric.* **18**, 415–419.

Bothwell, T. H. and Finch, C. A. (1962). "Iron metabolism". J. & A. Churchill, London.

Bouchet, P. (1963). Millets and sorghum in the Republic of Mali (in French). *Agron. trop.* Nogent **18**, 85–107.

Bragg. D. B., Ivy, C. A. and Stephenson, E. L. (1969). Methods for determining amino acid availability of feeds. *Poult. Sci.* **48**, 2135–2137.

Brawand, H. and Hossner, L. R. (1976). Nutrient content of sorghum leaves and grain as influenced by long-term crop rotation and fertilizer treatment. *Agron. J.* **68**, 277–280.

Bredon, R. M. (1961). Chemical composition of some foods and feeding stuffs in Uganda. Scientific Committee on Human Nutrition Protein Supplies Conference, Uganda Government Department, Veterinary Science and Animal Industries. (cyclostyled).

Brenner, N. W., Vigilante, C. and Owades, J. L. (1956). A study of hop bitters (isohumulones) in beer. *Am. Soc. brew. Chem. Proc.* 48–61.

Bressani, R. and Elias, L. G. (1968). Processed vegetable protein mixtures for human consumption in developing countries. *In* "Advances in Food Research". Vol. 16. pp. 1–103. Academic Press, New York.

Bressani, R. and Rios, B. J. (1962). The chemical and essential amino acid composition of twenty-five selections of grain sorghum. *Cereal Chem.* **39**, 50–58.

Bressani, R., Scrimshaw, N. S., Behar, M. and Vitteri, F. (1958). Supplementation of cereal proteins with amino acids. II. Effect of amino acid supplementation of corn-masa at intermediate levels of protein intake on the nitrogen retention of young children. *J. Nutr.* **66**, 501–513.

Bressani, R., Elias, L. G., Allwood Paredes, A. E. and Huezo, M. T. (1977). Processing of sorghum by lime-cooking for the preparation of tortillas. *In* Proceedings, Symposium. Sorghum and millets as human food, pp. 57–58. International Association of Cereal Chemistry Symposium, Vienna, 1976. Tropical Products Institute, London.

Brethour, J. R. and Duitsman, W. W. (1965). Utilization of fererita-type and waxy endosperm sorghum grains in all-concentrate rations fed to yearling steers. *Kans, agric. Exp. Sta. tech. Bull.* **34**, 482.

Breuer, L. H. Jr. and Dohm, C. K. (1972). Comparative nutritive value of several sorghum grain varieties and hybrids. *J. agric. Fd. Chem.* **20**, 83–86.

Briggs, D. E. (1967). Modified assay for α-amylase in germinating barley. *J. Inst. Brew.* **73**, 361–370.

Brinsmead, R. B., Moore, R. F., Delaney, N. E. and Gunton, J. L. (1970). Performance of grain sorghum strains under irrigation on the Darling Downs and in near South-Western Queensland. *Qd. J. agric. Anim. Sci.* **27**, 199–202.

"British Pharmaceutical Codex." (1954). The Pharmaceutical Press, London.

Brommelsiek, W. A., Shirley, R. L., Easley, J. F. and Bertrand, J. E. (1975). Effect of MHA on ME in sorghum grain diets fed steers. *J. Anim. Sci.* **40**, 197.

Brzozowski, G. R., Tanksley, T. D. Jr and Jungmeyer, C. K. (1972). Effect of four processing methods on digestibility of sorghum grain in swine. *J. Anim. Sci.* **35**, 212.

Burkitt, D. P., Walker, A. R. P. and Painter, N. S. (1974). Dietary fiber and disease. *J. Am. med Ass.* **229**, 1068–1074.

Burleson, C. A., Cowley, W. R. and Otey, G. (1956). Effect of nitrogen fertilization on yield and protein content of grain sorghum in the lower Rio Grande Valley of Texas. *Agron. J.* **48**, 524–525.

Burns, R. E. (1963). "Methods of tannin analysis for forage crop evaluation". University of Georgia Technical Bulletin N. S. 32.

Burns, R. E. (1971). Method for estimation of tannin in grain sorghum. *Agron. J.* **63**, 511–512.

Burton, G. W. and Fortson, J. C. (1966). Inheritance and utilization of five dwarfs in pearl millet (*Pennisetum typhoides*) breeding. *Crop Sci.* **6**, 69–72.

Burton, G. W., Knox, F. E. and Beardsley, D. W. (1964). Effect of age on the chemical composition, palability and digestibility of grass leaves. *Agron. J.* **56**, 160–161.

Burton, G. W., Wallace, A. T. and Rachie, K. O. (1972). Chemical composition and nutritive value of pearl millet (*Pennisetum typhoides* (*Burm.*)) Stapf and E. C. Hubbard, grain. *Crop Sci.* **12**, 187–188.

Burton, H. W. and Milne, F. N. J. (1961). White French millet for chickens. *Qd agric. J.* **87**, 236–238.

Busareva, N. N., Denisenko, Ya. I. and Nechaev, A. P. (1973). Composition of fatty acids in free, bound and firmly bound lipids of millet (in Russian). *Izv. vyssh. ucheb. Zaved. Pishch. Tekhnol.*, 132–134. (*Fd Sci. Technol. Abst.* (1974) **6**, 2M252).

Bushuk, W. and Hulse, J. H. (1974). Dough development by sheeting and its application to bread production from composite flours. *Cereal Sci. Today* **19**, 424–427.

Busson, F., Lunven, P., Lanza, M., Aquaron, R., Gayte-Sorbier, A. and Bono, M. (1962). Contribution to the chemical study of the millets and sorghums: influence of varietal and ecological factors on the amino acid composition of *Pennisetum* millet and of sorghum (in French). *Agron. trop.* Nogent **17**, 752–764.

Butters, B. and Chenery, E. M. (1959). A rapid method for determination of total sulphur in soils and plants. *Analyst* **84**, 239–245.

Cagampang, G. B., Cruz, L. J., Espiritu, S. G., Santiago, R. G. and Juliano, B. O. (1966). Studies on the extraction and composition of rice proteins. *Cereal Chem.* **43**, 145–155.

Cahn, R. S. (1964). An introduction to the sequence rule. A system for the specification of absolute configuration. *J. Chem. Educ.* **41**, 116–125.

Calder, A. (1955). Value of munga (millet) for pig feeding. *Rhodesia agric. J.* **52**, 161–170.

Calder, A. (1960). The value of rupoko (millet) for pig feeding. *Rhodesia agric. J.* **57**, 116–119.

Campabadal, C., Hammell, H. D., Combs, G. E. and Hammell, D. L. (1976). Bird resistant sorghum grain for growing pigs. *J. Anim. Sci.* **43**, 250.

Campbell, A. R. and Pickett, R. C. (1968). Effect of nitrogen fertilization on protein quality and certain other characteristics of nineteen strains of *Sorghum bicolor* (L.) Moench. *Crop Sci.* **8**, 545–547.

Campbell, J. A. (1963). *In* "Evaluation of protein quality". p. 31. Pub. No. 1100. National Academy of Sciences, National Research Council, Washington, DC.

Capote, F. A., Boscan, L., Taborda, F. and Inciarte, F. (1972). Variations in nutritive composition of grain sorghum hybrids and varieties cultivated in Venezuela (in Spanish). *Acta Cient. venez.* **23**, 67.

Carotene panel of subcommittee on vitamin estimation. (1955). Rapid and accurate method for estimation of β-carotene. *In* "Methods of Plant Analysis" Vol. 3. K. Peach and M. Tracey (eds.). Springer-Verlag, Berlin.

Carpenter, K. J. (1960). The estimation of available lysine in animal protein foods. *Biochem. J.* **77**, 604–610.

Carpenter, K. J. and De Muelenaere, H. J. H. (1965). A comparative study of performance on high-protein diets of unbalanced amino acid composition. *Proc. Nutr. Soc.* **24**, 202–209.

Carpenter, K. J. and Kodicek, E. (1950). The fluorimetric estimation of N′-methylnicotinamide and its differentiation from coenzyme 1. *Biochem. J.* **46**, 421–426.

Carr, W. R. (1956). The preparation and analysis of some African foodstuffs. *Cent. Afr. J. Med.* **2**, 334–339.

Carr, W. R. (1961). Observations on the nutritive value of traditionally ground cereals in Southern Rhodesia. *Br. J. Nutr.* **15**, 339–343.

Cartano, A. V. and Juliano, B. O. (1970). Hemicelluloses of milled rice. *J. agric. Fd Chem.* **18**, 40–42.

Casey, P. and Lorenz, K. (1977). Millet—functional and nutritional properties. *Bakers' Dig.* **51**, 45–57.

Castaing, J. and Leuillet, M. (1976). Partial or total replacement of maize by milo corn in pigs subjected to restricted and iso-proteic feeding. *Annls Zootech.* **25**, 425–426.

Castaing, J. and Moal, J. (1973). Substitution of maize by sorghum grain (milo-corn) in growing-finishing pig diets (in French). *Annls Zootech.* **22**, 359.

Cavins, J. F. and Friedman, M. (1968). Automatic integration and computation of amino acid analysis. *Cereal Chem.* **45**, 172–176.

Cerning, J. (1970). Thesis, Faculty of Lille, 23rd October, 1970. No. 80.

Cerning, J. and Guilbot, A. (1973). A specific method for the determination of pentosans in cereals and cereal products. *Cereal Chem.* **50**, 176–184.

Chakravorty, S. C. (1967). Chemical study of sorghum as a staple food in India. II. Protein content in relation to size of jowar grains (*Sorghum vulgare*). *Indian Agriculturist* **11**, 113–116.

Chakravorty, S. C. (1969). Chemical study of sorghum as a staple food in India. Part I. Protein content in relation to colour of jowar grains. *Labdev. J. Sci. Technol.* **7B**, 127–129.

Chamberlain, G. T. (1955). The major and trace element composition of some East African feedingstuffs. *E. Afr. agric. J.* **21**, 103–107.

Chamberlin, J. G. and Stickney, R. E. (1973). Improvement of children's diets in developing countries: an analytical approach to evaluation of alternative strategies. *Nutr. Rep. Int.* **7**, 71–84.

Chandrasekhara, M. R. and Swaminathan, M. (1953a). Enzymes of ragi (*Eleusine coracana*) and ragi malt. I. Amylases. *J. Scient. ind. Res.* **12B**, 51–56.

Chandrasekhara, M. R. and Swaminathan, M. (1953b). Enzymes of ragi (*Eleusine coracana*) and ragi malt. II. Proteases. *J. Scient. ind. Res.* **12B**, 481–484.

Chandrasekhara, M. R. and Swaminathan, M. (1953c). Factors affecting the yield and quality of malt extract from ragi (*Eleusine coracana*) malt. *J. Scient. ind. Res.* **12B**, 610–613.

Chandrasekhara, M. R. and Swaminathan, M. (1954). Enzymes of ragi and ragi malt. III. Pyro and glycerophosphatases. *J. Scient. ind. Res.* **13B**, 492–496.

Chandrasekhara, M. R. and Swaminathan, M. (1956). The amylases of pearl millet (*Pennisetum typhoideum*) malt. *Bull. Central Food Tech. Res. Inst.*, Mysore 5, 189.

Chandrasekhara, M. R. and Swaminathan, M. (1957). The enzymes of pearl millet (*Pennisetum typhoideum*) malt. Part 1. Amylases. *J. Scient. ind. Res.* **16C**, 35–38.

Chandrasekhara, M. R., Swaminathan, M., Sankaran, A. N. and Subrahmanyan, V. (1957). Nutritive value of balanced malt foods. *Indian J. Physiol. all. Sci.* **11**, 27–41. (*Chem. Abstr.* 1958, **52**, 4875).

Chang, S. I. and Fuller, H. L. (1964). Effect of tannin content of grain sorghums on their feeding value for growing chicks. *Poult. Sci.* **43**, 30–36.

Channon, H. J., Manifold, M. C. and Platt, A. P. (1938). The action of cystine and methionine on liver fat deposition. *Biochem. J.* **32**, 969–975.

Chapman, D. G., Castillo, R. and Campbell, J. A. (1959). Evaluation of protein in foods. I. A method for the determination of protein efficiency ratios. *Can. J. Biochem. Physiol.* **37**, 679–686.

Charlet-Lery, G., Francois, A. and Leroy, A. M. (1952). Analysis of foods intended for animals and interpretation of the results (in French). *Annls Zootech.* **1**, 45–61.

Chatterjee, I. B. (1970). Biosynthesis of L-ascorbate in animals. *Meth. Enzym.* **18**, 28–34.

Chaudhuri, D. K. and Kodicek, E. (1960). The availability of bound nicotinic acid to the rat. 4. The effect of treating wheat, rice and barley brans and a purified preparation of bound nicotinic acid with sodium hydroxide. *Br. J. Nutr.* **14**, 35–42.

Chaugale, D. S. Jr, Argikar, G. P., Gopalkrishna, N., Thobi, V. V. and Katwe, G. A. (1955a). Bibliography of jowar (*Sorghum vulgare* Pers.). *Poona agric. Coll. Mag.* **46**, 215–243.

Chaugale, D. S. Jr, Gopalkrishna, N. and Katwe, G. A. (1955b). Bibliography of nagli (*Eleusine coracana*) (Linn.). Gaertn. *Poona agric. Coll. Mag.* **46**, 245–247.

Chevassus-Agnes, S., Favier, J. C. and Joseph, A. (1976). Traditional technology and nutritive value of sorghum beer from Cameroon (in French). *Cah. Nutr. Diet.* **11**, 89-104.

Chibber, B. A. K., Mertz, E. T. and Axtell, J. D. (1978). Effects of dehulling on tannin content, protein distribution, and quality of high and low tannin sorghum. *J. agric. Fd Chem.* **26**, 679–683.

Chick, H., Hutchinson, J. C. D. and Jackson, H. M. (1935a). The biological value of proteins. VI. Further investigation of the balance sheet method. *Biochem. J.* **29**, 1702–1711.

Chick, H., Boas-Fixsen, M. A., Hutchinson, J. C. D. and Jackson, H. M. (1935b). The biological value of proteins. VII. The influence of variation in the level of protein in the diet and of heating the protein on its biological value. *Biochem. J.* **29**, 1712–1719.

Chilton, J. M. (1953). Simultaneous colorimetric determination of copper, cobalt and nickel as diethyldithio carbamate. *Analyt. Chem.* **25**, 1274–1275.

Chitre, R. G. and Ganapathy, S. (1956). Nutritive value of proteins of Italian millet (*Setaria italica*). *J. Scient. ind. Res.* **15C**, 95–99.

Chitre, R. G. and Vallury, S. M. (1956a). Studies on the protein value of cereals and pulses. Part I. Effect of feeding on growth, blood haemoglobin and plasma protein in young rats. *Indian J. med. Res.* **44**, 555–563.

Chitre, R. G. and Vallury, S. M. (1956b). Studies on the protein value of cereals and pulses. Part II. The effect of feeding on total protein, fat, and glycogen content on livers in young rats. *Indian J. med. Res.* **44**, 565–571.

Chitre, R. G., Desai, D. B. and Raut, V. S. (1955). The nutritive value of pure bred strains of cereals and pulses. Part I. Thiamine, riboflavin and nicotinic acid contents of 107 pure bred strains of cereals and pulses. *Indian J. med. Res.* **43**, 575–583.

Chitre, R. G., Desai, D. B., Ganapathy, S., Kumana, J. S. and Vallury, S. M. (1956). Nutritive value of pure bred strains of cereals and pulses. Part III. The essential amino acid content of some pure bred strains of cereals and pulses. *Indian J. med. Res.* **44**, 573–576.

Christensen, K. D. (1974). "Improving plant proteins by nuclear techniques." International Atomic Energy Agency, Vienna.

Christensen, P. J. (1977). Association of dye-binding capacity with protein content in a collection of sorghum varieties from North Cameroon. *In* Axtell, J. D.: Annual Report on inheritance and improvement of protein quality and content in *Sorghum bicolor* (L.) Moench. Report No. 13. pp. 79–80. Purdue University, Lafayette, Indiana. Agency for International Development, Department of State, Washington, DC.

Chughtai, M. I. D. and Waheed Khan, A. (1960). Nutritive value of foodstuffs and planning of satisfactory diets in Pakistan. I. Composition of raw foodstuffs. Division of Biochemistry, Institute of Chemistry, Punjab University, Lahore, Pakistan.

CIMMYT. (1976). Highland cold tolerant sorghum project. IDRC-CIMMYT. Annual progress Report December 1976. IDRC, Box 8500, Ottawa, Canada K1G 3H9.

CIMMYT. (1977). CIMMYT Review 1977. El Batan, Mexico.

Clegg, K. M. (1956). The application of the anthrone reagent to the estimation of starch in cereals. *J. Sci. Fd Agric.* **7**, 40–44.

Clifford, P. A. and Winkler, W. O. (1954). Report of the determination of sodium in foods. Gravimetric and flame photometric methods. *J. Ass. off. agric. Chem.* **37**, 586–600.

Close, J. and Naves, G. (1958). The amino acid composition of sorghum in the Belgian Congo: introduction to the study of the fermented beverages prepared from this cereal (in French). *Annls Nutr. Aliment.* **12**, 41–50.

Coates, J. H. and Simmonds, D. H. (1961). Proteins of wheat and flour, extraction, fractionation and chromatography of the buffer-soluble proteins of flour. *Cereal Chem.* **38**, 256–272.

Cohen, R. S. and Tanksley, T. D. Jr (1973a). Energy and protein digestibility of sorghum grains with different endosperm textures and starch types by growing swine. *J. Anim. Sci.* **37**, 931–935.

Cohen, R. S. and Tanksley, T. D. Jr (1973b). Limiting amino acids in sorghum grain for growing pigs. *J. Anim. Sci.* **37**, 276–277.

Cohen, R. S. and Tanksley, T. D. Jr (1975). Limiting amino acids in sorghum for finishing swine. *J. Anim. Sci.* **41**, 309.

Cohen, R. S. and Tanksley, T. D. Jr (1976). Limiting amino-acids in sorghum for growing and finishing swine. *J. Anim. Sci.* **43**, 1028–1034.

Cohen, R. S. and Tanksley, T. D. Jr (1977). Threonine requirement of growing and finishing swine fed sorghum soybean meal diets. *J. Amin. Sci.* **45**, 1079–1083.

Cole, E. W. (1967). Isolation and chromatographic fractionation of hemicelluloses from wheat flour. *Cereal Chem.* **44**, 411–416.

Collins, F. C. and Pickett, R. C. (1972a). Combining ability for yield, protein, and lysine in an incomplete diallel of *Sorghum bicolor* (L.) Moench. *Crop Sci.* **12**, 5–6.

Collins, F. C. and Pickett, R. C. (1972b). Combining ability for grain yield, percent protein, and g lysine/100 g protein in a nine-parent diallel of *Sorghum bicolor* (L.) Moench. *Crop Sci.* **12**, 423–425.

Common, R. H. (1940). The phytic acid content of some poultry feeding stuffs. *Analyst* **65**, 79–83.

Comstock, R. E. and Robinson, H. F. (1948). The components of genetic variance in populations of biparental progenies and their use in estimating the average degree of dominance. *Biometrics* **4**, 254–266.

Conn, E. E. (1973). Cyanogenic glycosides: their occurrence, biosynthesis, and function. *In* "Chronic cassava toxicity". (B. Nestel and R. MacIntyre Eds) pp. 55–63. IDRC-OIOe. International Development Research Centre, Box 8500, Ottawa, Canada.

Connole, M. D. and Hill, M. W. M. (1970). Aspergillus flavus contaminated sorghum grain as a possible cause of aflatoxicosis in pigs. *Aust. vet J.* **46**, 503–505.

Connor, J. K., Hurwood, I. S., Burton, H. W. and Fuelling, D. E. (1969). Some nutritional aspects of feeding sorghum grain of high tannin content to growing chickens. *Aust. J. exp Agric. Anim. Husb.* **9**, 497–501.

Connor, J. K., Neill, A. R. and Barram, K. M. (1976). The metabolizable energy content for the chicken of maize and sorghum grain hybrids grown at several geographical regions. *Aust. J. exp. Agric. Anim. Husb.* **16**, 699–703.

Cooley, M. L. and Koehn, R. C. (1950). Chromatographic estimation of carotene in feeds and feed ingredients. *Analyt. Chem.* **22**, 322–326.

Coons, C. M. (1968). Selected references on cereal grains in protein nutrition. Human and experimental animal studies of major and minor cereals 1910–1966. ARS 61–5. Washington, DC. United States Department of Agriculture.

Copelin, J. L., Tanksley, T. D. Jr and Knabe, D. A. (1974). Irrigated corn vs sorghums with different endosperm types and protein content for growing finishing swine. *J. Anim. Sci.* **39**, 180.

Copelin, J. L., Sasse, C. E. and Tribble, L. F. (1976a). Availability of lysine in sorghum. *J. Anim. Sci.* **42**, 245.

Copelin, J. L., Tribble, L. F. and Gaskins, C. T. (1976b). The availability of threonine in sorghum *J. Anim. Sci.* **43**, 251.

Copelin, J. L., Sasse, C. E., Gaskins, C. T. and Tribble, L. F. (1976c). The availability of tryptophan in sorghum. *J. Anim. Sci.* **43**, 251.

Copelin, J. L., Gaskins, C. T. and Tribble, L. F. (1978). Availability of tryptophan, lysine and threonine in sorghum for swine. *J. Anim. Sci.* **46**, 133–142.

Cornwell, P. B. (1959). Effects of gamma-radiation on the taste and manufacturing properties of soft wheat. *J. Sci. Fd Agric.* **10**, 409–412.

Coulson, C. B. and Sim, A. K. (1964). Proteins of various species of wheat and closely related genera and their relationship to genetical characteristics. *Nature, Lond.* **202**, 1305–1308.

Coulson, C. B. and Sim, A. K. (1965). Wheat proteins. 1. Fractionation and varietal variations of endosperm proteins of *T. vulgare. J. Sci. Fd Agric.* **16**, 458–464.

Council for Scientific and Industrial Research. (1975). Treatment of cereal grain. (Republic of South Africa). Specification of patent application 75/4957 dated 31.7.75.

Courtois, J. and Barré, R. (1953). Research on combinations of protein and phosphorus derivatives. III. Action of some reactive protein precipitants on phytic acid-protein combinations (in French). *Annls. pharm. fr.* **11**, 653–663.

Courtois, J. E. and Perlès, R. (1954). Study of the plant phosphorus content in different millet grains of Senegal. Possible consequences on the absorption of calcium in nutriment (in French). *Bull. Inst. fr. Afr. noire* **16**, 379–397.

Crabtree, J. and Dendy, D. A. V. (1977). Comilling of wheat and millet grain. *In* Proceedings, Symposium Sorghum and Millets for human food. Tropical Products Institute, London.

Cramer, V. A. et al. (1950). Teor mineral de alguns alimentos brasileiros. *Rev. Nutr. Rio de Janeiro* **1**, 83–93.

Crawford, D. C., Hamersma, P. J. and Marloth, B. W. (1942). The chemical composition of some South African cereals and their milling products. Department of Agriculture and Forestry, Science Bulletin No. 20, South Africa.

Creger, C. R., De Silva, P. C. and Couch, J. R. (1969). Effect of water and enzyme treatments on the amino acid content and nutrient values of milo. *Fed. Proc.* **28**, 811 (abstr.).

Cromwell, G. L., Rogler, J. C., Featherston, W. R. and Pickett, R. A. (1967). Nutritional value of opaque-2 corn for the chick. *Poult. Sci.* **46**, 705-712.

Crook, W. J. and Casady, A. J. (1974). Heritability and interrelationships of grain protein content with other agronomic traits of sorghum. *Crop Sci.* **14**, 622–624.

Crumpacker, D. W. and Allard, R. W. (1962). A diallel cross analysis of heading date in wheat. *Hilgardia*, **32**, 275–318.

Cummings, D. P. (1973). Effect of tannin content and testa of *Sorghum bicolor* (L.) Moench grain on nutritional quality. M.S. Thesis. Purdue University, Lafayette, Indiana, USA.

Cummings, D. P. and Axtell, J. D. (1973a). Effect of tannin content of *Sorghum bicolor* (L.) Moench. *In* "Inheritance and Improvement of Protein Quality and Content in Sorghum". Research Progress Report No. 10, pp. 85–111. Department of Agronomy, Agricultural Experiment Station, Purdue University, Lafayette, Indiana, Agency for International Development, Department of State, Washington, D.C.

Cummings, D. P. and Axtell, J. D. (1973b). Relationships of pigmented tests to nutritional quality of sorghum grains. *In* "Inheritance and Improvement of Protein Quality and Content in Sorghum". Research Progress Report No. 10, pp. 112–122. Department of Agronomy, Agricultural Experiment Station, Purdue University, Lafayette, Indiana, Agency for International Development, Department of State, Washington, DC.

Cummins, D. G. (1971). Relationships between tannin content and forage digestibility in sorghum. *Agron. J.* **63**, 500–502.

Curtis, D. L., Burton, G. W. and Webster, O. J. (1966). Carotenoids in pearl millet. *Crop Sci.* **6**, 300–301.

Cuthbertson, D. P. (1973). Human requirements for minerals and trace elements. *Indian J. Nutr. Diet.* **10**, 31–49.

Czarnocki, J., Sibbald, I. R. and Evans, E. V. (1961). The determination of chromic oxide in samples of feed and excreta by acid digestion and spectrophotometry. *Can. J. Anim. Sci.* **41**, 167–179.

Daiber, K. H. (1975). Enzyme inhibition by polyphenols of sorghum grain and malt. *J. Sci. Fd Agric.* **26**, 1399–1411.

Daiber, K. H. and Achtig, W. (1970). "Specific gravity of kaffircorn malt as a measure of modification". Special Report Chem. 152. Bantu Beer Unit, National Chemical Research Laboratory, Council for Scientific and Industrial Research, Pretoria, South Africa.

Daiber, K. H. and Novellie, L. (1968). Kaffircorn malting and brewing studies. XIX. Gibberellic acid and amylase formation in kaffircorn. *J. Sci. Fd Agric.* **19**, 87–90.

Dakshinamurti, K. (1955). Choline content of some South Indian foodstuffs. *Curr. Sci. (India)* **24,** 194–195.

Dalton, J. L. and Mitchell, H. L. (1959). Grain wax components. Fractionation of sorghum grain wax. *J. agric. Fd Chem.* **7**, 570–573.

Damodaran, M. (1931). The dicarboxylic acid-nitrogen of proteins: with a note on the alcohol-soluble protein from ragi (*Eleusine coracana*). *Biochem. J.* **25**, 2123–2130.

Daniel, V. A., Leela, R., Urs, T. S. S., Venkat Rao, S., Hariharan, K., Rajalakshmi, D., Swaminathan, M. and Parpia, H. A. B. (1965a). Amino acid supplementation of proteins. II. The effect of supplementing kaffircorn and pearl millet and diets based on them with L-lysine and DL-threonine on the nutritive value of their proteins. *J. Nutr. Diet.* **2**, 134–137.

Daniel, V. A., Leela, R., Doraiswamy, T. R., Rajalakshmi, D., Venkat Rao, S., Swaminathan, M. and Parpia, H. A. B. (1965b). The effect of supplementing a poor Indian ragi diet with L-lysine and DL-threonine on the digestibility coefficient, biological value and net utilization of the proteins and on nitrogen retention in children. *J. Nutr. Diet.* **2**, 138–143.

Daniel, V. A., Leela, R., Doraiswamy, T. R., Rajalakshmi, D., Venkat Rao, S., Swaminathan, M. and Parpia, H. A. B. (1966). The effect of supplementing a poor kaffircorn (*Sorghum vulgare*) diet with L-lysine and DL-threonine on the digestibility coefficient, biological value and net utilization of proteins and retention of nitrogen in children. *J. Nutr. Diet.* **3**, 10–14.

Daniel, V. A., Urs, T. S. S. R., Desai, B. L. M., Rao, S. V., Rajalakshmi, D., Swaminathan, M. and Parpia, H. A. B. (1967). Studies of low cost balanced foods suitable for feeding weaned infants in developing countries. 1. The protein efficiency ratio of low cost balanced foods based on ragi, or maize, ground nut, Bengal gram, soya and sesame flours and fortified with limited [*sic*] amino acid. *J. Nutr. Diet.* **4**, 183–188.

Daniel, V. A., Desai, B. L. M., Urs, T. S. S. R., Rao, S. V., Swaminathan, M. and Parpia, H. A. B. (1968a). The supplementary value of Bengal gram, red gram, soya bean, as compared with skim milk powder to poor Indian diets based on ragi, kaffircorn and pearl millet. *J. Nutr. Diet.* **5**, 283–291.

Daniel, V. A., Urs., T. S. S. R., Desai, B. L. M., Rao, S. V. and Swaminathan, M. (1968b). Supplementary value of edible coconut meal to poor Indian diets based on rice, ragi, wheat and sorghum, *J. Nutr. Diet.* **5**, 104–109.

Daniel, V. A., Narayanaswamy, D., Desai, B. L. M., Kurien, S., Swaminathan, M. and Parpia, H. A. B. (1970). Supplementary value of varying levels of red gram

(*Cajanus cajan*) to poor diets based on rice and ragi. *Indian J. Nutr. Diet.* **7**, 358–362.

Daniel, V. A., Kurien, S., Narayana, D. and Swaminathan, M. (1974). Supplementary relations between proteins of Bengal gram, rice and ragi (*Eleusine coracana*). *Indian J. Nutr. Diet.* **11**, 137–143.

Daniels, L. B. and Flynn, C. (1972). Processing bird-resistant grain sorghum for calf starter rations. *Arkans. Fm. Res.* **21**, 11.

Das, M. L. and Guha, B. C. (1960). Isolation and chemical characterization of bound niacin (niacinogen) in cereal grains. *J. biol. Chem.* **235**, 2971–2976.

Davis, A. B. and Harbers, L. H. (1974). Hydrolysis of sorghum grain starch by rumen microorganisms and purified porcine α-amylase as observed by scanning electron microscopy. *J. Anim. Sci.* **38**, 900–907.

Davis, B. J. (1964). Disc electrophoresis. II. Method and application to human serum proteins. *Ann. N.Y. Acad. Sci.* **121**, 404–427.

Davis, G. K. (1950). *In* "Symposium on copper metabolism". (W. D. McElroy and B. Glass, Eds) Johns Hopkins Press, Baltimore, USA.

Davis, G. V. and Stallcup, O. T. (1966). Nutritive value of grain sorghum AKS-614 when fed to steers with and without amino acid supplementation. *J. Dairy Sci.* **49**, 448. (Abstr.).

Davys, M. N. G. and Pirie, N. W. (1963). Batch production of protein from leaves. *J. agric. Engng Res.* **8**, 70–73.

Day, A. D. and Tucker, T. C. (1977). Effects of treated municipal waste-water on growth, fiber, protein, and amino-acid content of sorghum grain. *J. envir. Qual.* **6**, 325–327.

Dayton, S. (1975). Diet and coronary heart disease. *In* "Proceedings of Western Hemisphere Nutrition Congress IV". pp. 64–69. Publishing Sciences Inc.

Dayton, S., Chapman, J. M. and Pearce, M. L. (1970). Cholesterol, atherosclerosis, ischemic heart disease and stroke. *Ann. int. Med.* **72**, 97–109.

De, N. K. (1936a). The carotene content of some Indian vegetable foodstuffs with a preliminary note on its variation due to storage. Parts I-II. *Indian J. med. Res.* **23**, 937–948.

De, N. K. (1936b). Factors affecting the carotene content of certain vegetable foodstuffs. *Indian J. med. Res.* **24**, 201–212.

Deatherage, W. L., MacMasters, M. M. and Rist, C. E. (1955). A partial survey of amylose content in starch from domestic and foreign varieties of corn, wheat and sorghum and from some other starch-bearing plants. *Trans. Am. Ass. Cereal Chem.* **13**, 31–42.

Dechev, I. (1971a). Development, yield and quality of sorghum grain as affected by fertilizing (in Russian). *Rastenievod Nauki*, **8**, 75–85.

Dechev, I. (1971b). Effect of F_1, F_2, and F_4 generations on yield and quality of sorghum (in Bulgarian). *Nauchni Trud. viss. Selskostop. Inst. Vasil Kolarov* **20**, 67–71.

Dechev, I. (1973). The effect of nitrogen dressing on the quality of grain and nutrient uptake by sorghum (in Bulgarian with English summary). *Pochvoznanie i Agrokhimiya* **8**, 71–79.

Dehlavi, A. (1974). "Correlation studies, seasonal production of tannin and inheritance of testa color and tannin content in grain sorghum". Ph. D. Thesis, University of Georgia. *Diss. Abstr.*, 1975, 3708-B.

Deibel, R. H., Evans, J. B. and Niven, C. F. (1957). Microbiological assay for thiamin using *Lacto bacillus viridescens*. *J. Bact.* **74**, 818–821.

deMan, J. M., Banigo, E. O. I., Rasper, V., Gade, H. and Slinger, S. J. (1973).

Dehulling of sorghum and miliet with the Palyi compact milling system. *Can. Inst. Fd Sci. Technol. J.* **6**, 188–193.

Denton, J. (1928). A study of the tissue changes in experimental blacktongue of dogs compared with similar changes in pellagra. *Am. J. Path.* **4**, 341.

Deosthale, Y. G. and Gopalan, C. (1974). The effect of molybdenum levels in sorghum (*Sorghum vulgare*) on uric acid and copper excretion in man. *Br. J. Nutr.* **31**, 351–355.

Deosthale, Y. G. and Mohan, V. S. (1970). Locational differences in protein, lysine, and leucine content of sorghum varieties. *Indian J. agric. Sci.* **40**, 935–941.

Deosthale, Y. G., Mohan, V. S., Radhakrishnan, M. R. and Bhat, S. (1969). Locational variations in β-carotene content of two yellow endosperm varieties of sorghum. *J. Nutr. Diet.* **6**, 244–228.

Deosthale, Y. G., Mohan, V. S. and Rao, K. V. (1970a). Varietal differences in protein, lysine, and leucine content of grain sorghum. *J. agric. Fd Chem.* **18**, 644–646.

Deosthale, Y. G., Nagarajan, V. and Pant, K. C. (1970b). Nutrient composition of some varieties of ragi (*Eleusine coracana*). *Indian J. Nutr. Diet.* **7**, 80–84.

Deosthale, Y. G., Visweswar Rao, K., Nagarajan, V. and Pant, K. C. (1971). Varietal differences in protein and amino acids of grain bajra (*Pennisetum typhoides*). *Indian J. Nutr. Diet.* **8**, 301–308.

Deosthale, Y. G., Nagarajan, V. and Visweswar Rao, K. (1972a). Some factors influencing the nutrient composition of sorghum grain. *Indian J. agric. Sci.* **42**, 100–108.

Deosthale, Y. G., Visweswar Rao, K. and Pant, K. C. (1972b). Influence of the levels of N fertilizer on the yield, protein and amino acids of pearl millet (*Pennisetum typhoides*) (Burm. f.). Stapf & C. E. Hubb. *Indian J. agric. Sci.* **42**, 872–876.

Deosthale, Y. G., Krishnamachari, K. A. V. R. and Belavady, B. (1977). Copper, molybdenum and zinc in rice, sorghum and pearl millet grains from fluorosis and nonfluorosis areas of Andhra Pradesh. *Indian J. agric. Sci.* **47**, 333–335.

De Paula Santos, O. (1950–51). Nutritive value of adlay (*Coix lachryma-jobi*, L.) (in Portuguese). *Annls. Fac. Med. Univ. Sao Paulo* **25**, 323–342.

De Ruiter, D. (1973). Use of non-wheat flours in bakery products (in German). *Getreide, Mehl Brot* **27**, 170–176.

Desai, B. B. and Zende, G. K. (1972). Nutrient composition of hybrid and local bajra crops grown in medium black soil for assessing their nutritional requirements. *J. Indian Soc. Soil Sci.* **20**, 175–182.

Desai, B. L. M., Hariharan, K., Jayaraj, A. P., Venkat Rao, S. and Swaminathan, M. (1968). Studies on the protein efficiency ratio of fish flour (from Bombay duck) and a protein food based on fish, groundnut and Bengal gram flours and their supplementary value to low protein diets. *J. Nutr. Diet.* **5**, 45–51.

Desai, B. L. M., Narayanaswamy, D., Daniel, V. A., Swaminathan, M. and Parpia, H. A. B. (1970a). The improvement of protein value of ragi (*Eleusine coracana*) diet by supplementation with limiting amino acids. *Nutr. Rep. Int.* **2**, 185–191.

Desai, B. L. M., Daniel, V. A., Venkat Rao, S. Swaminathan, M. and Parpia, H. A. B. (1970b). Studies on low cost balanced foods suitable for feeding weaned infants in developing countries. II. Supplementary value of low cost balanced foods based on cereals, cottonseed or peanut flour and Bengal gram flour to poor Indian diets. *Indian J. Nutr. Diet.* **7**, 21–26.

Deshaprabhu, S. B. (1966). "The wealth of India: raw materials". Vol. 7. Publication and Information Directorate, CSIR, New Delhi, India.

Desikachar, H. S. R. (1977) Processing of sorghum and millets for versatile food uses in India. Proceedings, symposium. Sorghum and millets for human food. International Association for Cereal Chemistry, Vienna, 11 May 1976. Tropical Products Institute, London.

Desikachar, H. S. R. and De, S. S. (1947). The cystine and methionine contents of common Indian foodstuffs. *Curr. Sci.* **16**, 284.

Devadas, R. P., Jalaja, R. and Radharukmani, A. (1966). Comparison of nutritive value, cooking qualities and acceptability of four selected strains of hybrid cholam (*Sorghum vulgare*) with a local strain. *J. Nutr. Diet.* **3**, 47–49.

De Wit, J. P. and Schweigart, F. (1970). The potential role of pearl millet as a food in South Africa. *S. Afr. med. J.* **44**, 364–366.

Deyoe, C. W. and Shellenberger, J. A. (1965). Amino acids and protein in sorghum grain. *J. agric. Chem.* **13**, 446–450.

Deyoe, C. W., Sanford, P. E., Murphy, L. M. and Waggle, D. H. (1968). Protein quality studies on grain sorghum. *J. Am. Oil Chem. Soc.* **45**, 1125A.

Deyoe, C. W., Shoup, F. K., Miller, G. D., Bathurst, J., Liang, D., Sanford, P. E. and Murphy, L. S. (1970). Amino acid composition and energy value of immature sorghum grain. *Cereal Chem.* **47**, 363–368.

Diamant, E. J. and Laxer, S. (1967). The preparation of Japanese-type Miso products by fermentation of defatted soybean meal and various cereals. *Israel J. Chem.* **5**, 138.

Dickman, S. R. and Crockett, A. L. (1956). Reactions of xanthydrol. IV. Determination of tryptophan in blood plasma and in proteins. *J. biol. Chem.* **220**, 957–965.

Diggs, B. G., Becker, D. E., Terrill, S. W. and Jensen, A. H. (1965). Energy value of various feeds for the young pig. *J. Anim. Sci.* **24**, 555–558.

Dimler, R. J. (1964). Determination of optical rotation. *In* "Methods in carbohydrate chemistry". (R. L. Whistler, R. J. Smith, J. N. BeMiller and M. L. Wolfrom Eds) Vol. iv. Starch, pp. 133–139. Academic Press Inc., New York.

Doggett, H. (1970). "Sorghum." Longmans, Green & Co. Ltd., London, SBN 582 46647 4.

Doggett, H. (1974a). United Nations Development Program, Global research for the improvement of sorghum and millet. Report for the period August 1973–December 1974. ICRISAT, Begumpet, Hyderabad 500016, India.

Doggett, H. (1974b). Progress in breeding for quality protein in other cereals. *In* "Proceedings of the CIMMYT-Purdue Symposium on Protein Quality in Maize", El Batan, Mexico. 4-8 December 1972. pp. 371–373. Dowden, Hutchinson and Ross, Inc., Stroudsburg, PA., USA.

Doggett, H. (1976). Sorghum (*Sorghum bicolor*) (Gramineae, Andropogoneae). *In* "Evolution of crop plants". (N. W. Simmonds, Ed.) pp. 112–117. Longman, London, New York.

Doraiswamy, T. R., Parthasarathy, H. N., Tasker, P. K., Sankaran, A. N., Rajagopalan, R., Swaminathan, M., Sreenivasan, A. and Subramanyan, V. (1962). Effect of supplementing groundnut flour fortified with vitamins and minerals on the growth and nutritional status of children subsisting on a poor Indian diet based on rice. *Fd Sci.* **11**, 186–189.

Doraiswamy, T. R., Urs, T. S. S. R., Rao, S. V., Swaminathan, M. and Parpia, H. A. B. (1968). Effect of supplementation of poor kaffircorn diet (*Sorghum vulgare*) with L-lysine on nitrogen retention and growth of school children. *J. Nutr. Diet.* **5**, 191–196.

Doraiswamy, T. R., Singh, N. and Daniel, V. A. (1969). Effects of supplementing ragi (*Eleusine coracana*) diets with lysine or leaf protein on the growth and nitrogen metabolism of children. *Br. J. Nutr.* **23**, 737–743.

Doraiswamy, T. R., Daniel, V. A., Rajalakshmi, D., Swaminathan, M. and Parpia, H. A. B. (1971). Effect of supplementing a poor diet based on rice and wheat consumed by school children with vitamins, minerals, lysine and protein-rich foods, on their growth and nutritional status. *Nutr. Rep. Int.* **3**, 67–78.

Doty, W. H., Munson, A. W., Wood, D. E. and Schneider, E. L. (1970). Comparative analysis of variance of the Kjeldahl nitrogen and a neutron activation nitrogen technique. *J. Ass. off. analyt. Chem.* **53**, 801–803.

Drewes, S. E. and Roux, D. G. (1964). Condensed tannins. 18. Stereochemistry of flavin 3, 4-diol tannin precursors (+) molli-(sacacidin) (−) leuco-fisetinidin and (+) leuco-robinetinidin. *Biochem. J.* **90**, 343–350.

Drewes, S. E., Roux, D. G., Eggers, S. H. and Feeney, J. (1967a). Three diastereoisomeric 4, 6- linked bilencofisetinidins from the heartwood of *Acacia mearnsii*. *J. Chem. Soc.* (C), 1217–1227.

Drewes, S. E., Roux, D. G., Saayman, H. M., Eggers, S. H. and Feeney, J. (1967b). Some stereo-chemically identical biflavanols from the bark tannins of *Acacia mearnsii*. *J. Chem. Soc.* (C), 1302–1308.

Dreyer, J. J. (1962). The effect of certain anti-diarrhoeic substances on protein digestibility and metabolic faecal nitrogen in the rat. *Proc. Nutr. Soc. S. Afr.* **3**, 87–90.

Dreyer, J. J. (1968). Biological assessment of protein quality: digestibility of the proteins in certain foodstuffs. *S. Afr. Med. J.* **42**, 1304–1313.

Dreyer, J. J. (1973). Methods of protein evaluation with rats. *In* "Proteins in human nutrition". (J. W. G. Porter and B. A. Rolls Eds) pp. 245–261. Academic Press, London.

Dreyer, J. J. (1974). Report on a comparison of the results of in vitro and in vivo estimations of sorghum protein digestibility. *Nat. Fd Res. Inst.*, CSIR, Pretoria, South Africa. Unpublished report.

Dreyer, J. J. and Concannon, T. R. J. (1975). Digestibility of the protein in certain untreated and chemically treated grain sorghum cultivars. C/Voed 80 National Food Research Institute, Council for Scientific and Industrial Research, Pretoria, South Africa.

Dreyer, J. J. and Novellie, L. (1962). An investigation of the nutritional value of the proteins in kaffir beer and its basic ingredients. Contract Report C.V.6. National Nutritional Research Institute, South African Council for Science and Industrial Research, Pretoria.

Dreyer, J. J. and Van Niekerk, P. J. (1973). Growth-suppressing and related effects of unextracted and ethanol-extracted grains of certain sorghum cultivars. National Food Research Institute, CSIR, Pretoria, South Africa. Unpublished report.

Driedger, A. and Riggs, J. K. (1973). Hybrid hy location effect on digestibility of grain. *J. Anim. Sci.* **35**, 263.

Dubois, M., Gilles, K. A., Hamilton, J. K., Rebers, P. A. and Smith, F. (1956). Colorimetric method for determination of sugars and related substances. *Analyt. Chem.* **28**, 350–356.

Du Preez, J. J. and Wessels, J. P. H. (1970). Kaffircorn and tannin in chick feeding (in Afrikaans). *S. Afr. Soc. Anim. Prod. Proc.* **9**, 109–110.

Du Vigneaud, V., Cohen, M., Chandler, J. P., Schenck, J. R. and Simmonds, S.

(1941). The utilization of the methyl group of methionine in the biological synthesis of choline and creatine. *J. biol. Chem.* **140**, 625–641.

Dwarakanath Reddy, K., Govindaswamy, C. V. and Vidbyasekaran, D. P. (1969). Studies on ergot disease of cumbu (*Pennisetum typhoides*). *Madras agric. J.* **56**, 367–377.

Dyer, T. A. and Novellie, L. (1966). Kaffir-corn malting and brewing studies. XVI. The distribution and activity of α- and β-amylases in germinating kaffircorn. *J. Sci. Fd Agric.* **17**, 449–456.

Earle, F. A. and Milner, R. T. (1944). Improvements in the determination of starch in corn and wheat. *Cereal Chem.* **21**, 567–575.

Eastoe, J. E. and Taylor, R. H. (1974). Composition of protein of sorghum grain grown in Botswana. *J. Sci. Fd Agric.* **25**, 563–569.

Eastwood, M. A., Eastwood, J. and Ward, M. (1976). Epidemiology of bowel disease. *In* "Fiber in Human Nutrition". (G. A. Spiller and R. J. Amen Eds) Chapter 9. Plenum Press, New York.

Eckert, T. E. and Allee, G. L. (1974). Limiting amino acids in milo for growing pig. *J. Anim. Sci.* **39**, 694–698.

Edmunds, L. K., Futrell, M. C. and Frederiksen, R. A. (1970). Sorghum diseases. *In* "Sorghum Production and Utilization". (J. S. Wall and W. M. Ross Eds) pp. 200–234. AVI Publishing Co Inc., Westport, Connecticut, USA.

Edwards, W. M. and Curtis, J. J. (1943). "Grain sorghums, their products and uses". ACE-193, NM-229. USDA Northern Regional Research Laboratory, Peoria, Ill., USA.

Eggum, B. O. (1970). Current methods of nutritional protein evaluation. *In* "Proceedings of a joint IAEA/FAO Symposium on plant protein resources: their improvement through the application of nuclear techniques". pp. 289–302. International Atomic Energy Agency, Vienna.

Eggum, B. O. and Christensen, K. D. (1973). The influence of tannin on protein utilization in feedstuffs with special reference to barley. Proceedings of the Second Research Co-ordination Meeting of the FAO/IAEA/GSF Seed Protein Improvement Programme held at Ibadan, Nigeria. December, 1973.

Eggum, B. O. and Mercer, N. (1964). En biologisk proteinvurdering ved hjaelp af rotter (in Danish). *Ugeskr. Landn.* **109**, 799–801.

Ejeta, G. and Axtell, J. D. (1977). Evaluation of high lysine and normal sorghum varieties for protein quality and carbohydrate composition at three stages of grain development. *In* Axtell, J. D. Annual Report on Inheritance and Improvement of Protein Quality and Content in *Sorghum bicolor* (*L.*) Moench. Report No. 13. pp. 24–55. Purdue University, Lafayette, Indiana, USA. Agency for International Development, Department of State, Washington, DC.

Ekman, P., Emanuelson, H. and Fransson, A. (1949). Investigations concerning the digestibility of protein in poultry. *Ann. R. agric. Coll. Swed.* **16**, 749–777.

Elder, A. H., Lubisich, T. M. and Mecham, D. K. (1953). Studies on the relation of the pentosans extracted by mild acid treatment to milling properties of Pacific Northwest wheat varieties. *Cereal Chem.* **30**, 103–114.

El-Harith, E-H. A., Dickerson, J. W. T. and Walker, R. (1976). On the nutritive value of various starches for the albino rat. *J. Sci. Fd Agric.* **27**, 521–526.

Elliott, J. S. and McPherson, C. M. (1971). Nutrient values of and consumer preference for grain sorghum wafers. *J. Am. diet. Ass.* **58**, 225–229.

Elvehjem, C. A. (1930). A note on the determination of iron in milk and other biological materials. *J. Biol. Chem.* **86**, 463–467.

Ely, D. G. and Duitsman, W. W. (1967). Beef cattle investigations, 1966–67. Bulletin 506. Kansas Agricultural Experiment Station, Kansas, USA.

Erickson, D. O., Haugse, C. N., Dinusson, W. E., Bolin, D. W. and Buchanan, M. L. (1963). Proso: lysine deficient for pigs. *N. Dak. Fm Res.* **22**, 7–8.

Ethiopia Nutrition Survey. (1959). Interdepartmental Committee on Nutrition for National Defense Report. Washington, DC.

Evered, D. F. (1960). The determination of ascorbic acid in highly coloured solutions of N-bromo succinimide. *Analyst* **85**, 515–517.

Eyjolfsson, R. (1970). Recent advances in the chemistry of cyanogenic glycosides. *Fortschr. Chem. org. Nat. Stoffe* **28**, 74–108.

Fabriani, G. (1939). Composition of the ash of the principal African grains (*Triticum durum—Zea mays—Hordeum disticum—Sorghum aethiopicum—Eragostis abyssinica—Pennisetum typhoideum—Eleusine coracana*) (in Italian). *Quad. Nutr. Bologna* **6**, 72–81.

FAO. (1949). "Food composition tables for international use". Food and Agriculture Organization Nutrition Studies No. 3. FAO, Rome.

FAO. (1957). FAO Nutritional Studies No. 16. Protein requirements. FAO, Rome.

FAO. (1963). "Amino acid content of foods (provisional)". FAO, Rome.

FAO. (1966). "Agricultural Development in Nigeria 1965–1980". FAO, Rome.

FAO. (1970). "Amino acid content of foods and biological data on proteins". FAO Nutrition Studies No. 24. FAO, Rome.

FAO. (1971). "Meeting report on single cell protein". Document 3. 14/15 Moscow. FAO, Rome.

FAO. (1973). "Composite flour programme". Documentation Package, Vol. 1. 2nd ed. revised. Food and Agricultural Industries Service, Agricultural Services Division, FAO, Rome.

FAO. (1976). Production yearbook, Vol. 30, FAO, Rome.

FAO. (1977). "Dietary fats and oils in human nutrition". FAO Food and Nutrition paper 3. FAO, Rome.

FAO/WHO. (1975). Energy and protein requirements: recommendations by a joint FAO/WHO informal gathering of experts. *Fd Nutr.* **1**, 11–19. FAO, Rome.

Farrar, G. E. Jr (1935). The determination of iron in biological materials. *J. biol. Chem.* **110**, 685–694.

Fasold, H. and Gundlach, G. (1963). Characterization of peptides and proteins with enzymes. *In* "Methods of Enzymatic Analysis". (H. U. Bergmeyer Ed.) pp. 350–362. Academic Press, New York.

Favier, J. C., Chevassus-Agnes, S., Joseph, A. and Gallon, G. (1972). Traditional technology of sorghum in Cameroun—influence of pestling on nutritional value (in French). *Annls Nutr. Aliment.* **26**, 221–250.

Featherston, W. R. and Rogler, J. C. (1975). Influence of tannins on the utilization of sorghum grain by rats and chicks. *Nutr. Rep. Int.* **11**, 491–497.

Featherston, W. R., Rogler, J. C., Axtell, J. D. and Oswalt, D. L. (1975a). Nutritional value of high lysine sorghum grain for the chick. *Poult. Sci.* **54**, 1220–1225.

Featherston, W. R., Rogler, J. C., Axtell, J. D., Mohan, D. P. and Oswalt, D. L. (1975b). Nutritional value of high lysine sorghum grains for chick. *Poult. Sci.* **54**, 1760–1761.

Feher, K. (1971). Experience in production with some grain sorghums bred in Szeged, and quality examination results. *In* "Proceedings of the fifth meeting of the maize and sorghum section of EUCARPIA". (I. Kovacs, Ed.) pp. 258–264. Akademiai Kiado, Budapest.

Fernandez, R., Lucas, E. and McGinnis, J. (1974). Comparative nutritional value of different cereal grains as protein sources in a modified chick bioassay. *Poult. Sci.* **53**, 39–46.

Ferrar, B. and De Paolis, P. (1960). Application of sorghum in feeding domestic animals. Research on the content of amino acids of the grain of hybrid sorghum (in Italian). *Acta méd. vet., Madr.* **6**, 337–343.

Ferraris, R. (1973). Pearl millet (*Pennisetum typhoides*). Review Series No. 1/1973. Commonwealth Bureau of Pastures and Field Crops, Hurley, Maidenhead, Berks SL6 5LR, U.K.

Ferreira Oriá, H. and De Lima, M. M. G. O. R. (1975). Preliminary bromatological study of sorghum grain raised in Brazil (in Portuguese). *Rev. farm. Bioquim. Universidade de Sao Paulo.* **13**, 323–336.

Ferro, P. V. A. S. and Ham, A. N. (1957). Colorimetric determination of calcium by chloranilic acid. II. A semimicro method with reduced precipitation time. *Am. J. clin. Path.* **28**, 689–692.

Fertiliser and Feeding Stuffs Regulations. (1932). S. R. & O. 1932. No. 658. H. M. Stationery Office, London.

Fetuga, B. L. (1977). The metabolizable energy value of some cereals and cereal byproducts for chickens. *W. Afr. J. biol. appld. Chem.* **20**, 3–15.

Fetuga, B. L. and Oluyemi, J. A. (1976). The metabolizable energy of some tropical tuber meals for chicks. *Poult. Sci.* (in the press).

Fiedler, R., Proksch, A. and Koepf, A. (1973). The determination of total nitrogen in plant materials with an automatic nitrogen analyser. *Analyt. Chim. Acta* **63**, 435–443.

Filho, J. X. (1974). Trypsin inhibitors in sorghum grain. *J. Fd Sci.* **39**, 422–423.

Fiorentini, G. (1942). Contribution to the study of chemical and technological characteristics of the grains of durra from Italian East Africa (in Italian). *Agric. Coloniale* **36**, 134–139.

Fiske, C. H. and Subbarow, Y. (1925). The colorimetric determination of phosphorus. *J. biol. Chem.* **66**, 375–400.

Fleury, P. and Leclerc, M. (1943). La methode nitro-vanado-molybdique de Misson pour le dosage colorimetrique du phosphore. Son interet en biochimie. *Bull. Soc. Chim. Biol.* **25**, 201–205.

Florence, J. D. Jr (1968). Studies of physical characteristics and animal response to reconstituted sorghum grain. M. S. Thesis. Texas A&M University, College Station, Texas, USA.

Flynn, M. F. and Stallcup, O. T. (1973). Digestion trials on high moisture rolled milo and dry rolled milo. *J. Dairy Sci.* **56**, 306.

Folch, J., Lees, M. and Sloane-Stanley, G. H. (1957). A simple method for the isolation and purification of total lipids from animal tissues. *J. biol. Chem.* **226**, 497–509.

Fonseca, J. B. (1970). Evaluation of the protein quality of selected varieties of corn and sorghum for poultry. Ph.D. Thesis, Purdue University, Lafayette, Indiana, USA.

Food and Nutrition Board. 1968. "Recommended dietary allowances." 7th ed. National Research Council Pubn. No. 1694. National Academy of Sciences, Washington, DC.

Ford, J. E. (1962). A microbiological method for assessing the nutritional value of proteins. 2. The measurement of 'available' methionine, leucine, isoleucine, arginine, histidine, tryptophan and valine. *Br. J. Nutr.* **16**, 409–425.

Ford, J. E. (1977a). Availability of methionine and lysine in sorghum grain in relation to tannin content. *Proc. Nutr. Soc.* **36**, A124.

Ford, J. E. (1977b). Influence of polyethylene-glycol and related compounds on nutritional availability of methionine in a high-tannin sorghum and in field beans. *Proc. Nutr. Soc.* **36**, A125.

Ford, J. E. and Hewitt, D. (1977). Influence of polyethylene-glycol on digestibility of protein in high-tannin sorghum in rats and chicks. *Proc. Nutr. Soc.* **36**, A126.

Forsyth, W. G. C. (1952). Cacao polyphenolic substances. I. Fractionation of the fresh bean. II. Changes during fermentation. *Biochem. J.* **51**, 511–520.

Forsyth, W. G. C. and Roberts, J. B. (1960). Cacao polyphenolic substances 5. The structure of cacao 'leucocyanidin 1'*: Biochem. J.* **74**, 374–378.

Foster, H. L. (1972). Comparison of the yield response of groundnuts to single and triple superphosphate in Uganda. *E. Afr. agric. For. J.* **38**, 103–113.

Fournier, P. and Digaud, A. (1948). Chemical composition of sorghum grain (*Sorghum vulgare* Pers (in French). *Bull. Soc. Scient. Hyg. aliment.* **36**, 33–36.

Fox, D. G., Klosterman, E. W., Newland, H. W. and Johnson, R. R. (1970). Net energy of corn and bird resistant grain sorghum rations for steers when fed as grain or silage. *J. Anim. Sci.* **30**, 303–308.

Francis, C. K. (1914). Foods from the grain sorghums. Circular No. 27. Oklahoma Agricultural Experiment Station. Stillwater, Oklahoma, USA.

Francis, C. K. and Smith, O. C. (1916). The starches of the grain sorghums. Bulletin No. 110. Oklahoma Agricultural and Mechanical College, Agricultural Experimental Station, Stillwater, Oklahoma, USA.

Fraser, J. R., Brandon, B. M. and Holmes, D. C. (1956). The proximate analysis of wheat flour carbohydrates. I. Methods and scheme of analysis. *J. Sci. Fd Agric.* **7**, 577–589.

Frederick, H. M., Theurer, B. and Hale, W. H. (1973). Effect of mosture, pressure and temperature on enzymatic starch degradation of barley and sorghum grain. *J. Dairy Sci.* **56**, 595–601.

Freeman, J. E. and Bocan, B. J. (1973). Pearl millet: a potential crop for wet milling. *Cereal Sci. Today.* **18**, 69–73.

Freeman, J. E. and Watson, S. A. (1969). Peeling sorghum grain for wet milling. *Cereal Sci. Today* **14**, 10–15.

Freeman, J. E. and Watson, S. A. (1971). Influence of sorghum endosperm pigments on starch quality. *Cereal Sci. Today* **16**, 378, 380–381.

Freeman, J. E., Yahl, K. R. and Watson, S. A. (1967). A chemical and structural analysis of the caryopsis of an autotetraploid corn. *Crop Sci.* **7**, 655–658.

Freeman, J. E., Kramer, N. W. and Watson, S. A. (1968). Gelatinization of starches from corn (*Zea mays* L.) and sorghum (*Sorghum bicolor* L. Moench): effects of genetic and environmental factors. *Crop Sci.* **8**, 409–413.

Freudenberg, K. and Weinges, K. (1960). Classification and nomenclature of the flavonoides (in German). *Tetrahedron* **8**, 336–349.

Friedmann, T. E. and Frazier, E. I. (1950). The determination of nicotinic acid. *Archs. Biochem.* **26**, 361–374.

Fuller, H. L., Potter, D. K. and Brown, A. R. (1966). The feeding value of grain sorghums in relation to their tannin content. Bulletin NS-176. Georgia Agriculture Experiment Station, Georgia, USA.

Furr, R. D. and Sherrod, L. B. (1968). Variation in protein and mineral content of grain sorghum. *Proc. west. Sect. Am. Soc. Anim. Sci.* **19**, 157–159.

Futrell, M. F. and Abdullahi, R. (1967). Recipes for breads and cookies using sorghum flour. *Sorghum Newsl.* **10**, 93–96.

Fyfe, J. L. and Gilbert, N. (1963). Partial diallel crosses. *Biometrics* **19**, 278–286.

Gallus, H. P. C. and Jennings, A. C. (1971a). Compounds of low molecular weight in washed wheat gluten. *Aust. J. biol. Sci.* **24**, 825–828.

Gallus, H. P. C. and Jennings, A. C. (1971b). Phenolic compounds in wheat flour and dough. *Aust. J. biol. Sci.* **24**, 747–753.

Ganapathy, S. N. and Chitre, R. G. (1976). Factors affecting utilization of millet protein by rats. *Indian J. Nutr. Diet.* **13**, 67–71.

Ganapathy, N. S., Chitre, R. G. and Gokhale, S. K. (1957). The effect of the protein of Italian millet (*Setaria italica*) on nitrogen retention in albino rats. *Indian J. med. Res.* **45**, 395–399.

Ganapathi, S., Narayana Rao, M., Swaminathan, M. and Subrahmanyan, V. (1958). Supplementary relations of the proteins of horse gram and cow pea to those of Italien millet (*Setaria italica*). *Fd Sci.* **7**, 7–8.

Ganapati, S. and Chitre, R. G. (1961). The effect of Italian millet (*Setaria italica*) and field bean (*Dolichos lablab*) proteins on the formation of haemoglobin and plasma proteins in albino rats. *J. postgrad. Med. Bombay* **7**, 164–166. (*Nutr. Abst. Rev.* 1962, **32**, 857.)

Ganry, F. and Bideau, J. (1974). Effect of nitrogenous fertilization on the yields and nutritional value of a millet, Souna III (in French, English summary). *Agron. trop.* Nogent. **29**, 1006–1015.

Gartner, R. J. W. and Twist, J. O. (1968). Mineral content of a variety of sorghum grain. *Aust. J. exp. Agric. Anim. Husb.* **8**, 210–211.

Gartner, R. J. W., Laws, L. and O'Rourke, P. K. (1975). Millet compared with wheat and sorghum in high grain diets for cattle. *Aust. J. exp. Agric. Anim. Husb.* **15**, 446–450.

Gast, M. and Adrian, J. (1965). "Millet and sorghum in Ahaggar; ethnological and nutritional study" (in French). Centre de Recherches Anthropologiques, Préhistoriques et Ethnographiques, Memoires, IV. Arts et Métiers Graphiques, 18, Rue Séguier, Paris.

Gayte-Sorbier, A., Saurat, A., Busson, F. and Lunven, P. (1960). Contribution to the chemical study of *Eleusine coracana* Gaertner (in French). *J. Agric. trop. Bot. appl.* **7**, 383–386.

Gerencser, V., Duduk, V. and Vincze, L. (1966). Nutritive value of sorghum and maize in growers' rations for chickens (in Hungarian). *Allattenyesztes* **15**, 91–100.

Ghafoorunissa, and Narasinga Rao, B. S. (1971). Effect of leucine load on plasma tryptophan and other amino acids in man. *Indian J. med. Res.* **59**, 1861–1868.

Ghafoorunissa, and Narasinga Rao, B. S. (1973). Effect of leucine on enzymes of the tryptophan-niacin metabolic pathway in rat liver and kidney. *Biochem. J.* **134**, 425–430.

Ghosh, H. P., Sarkar, P. K. and Guha, B. C. (1963). Distribution of the bound form of nicotinic acid in natural materials. *J. Nutr.* **79**, 451–453.

Gilbert, G. A. and Spragg, S. P. (1964). Iodimetric determination of amylose, iodine sorption "blue value". *In* "Methods of Carbohydrate Chemistry". (R. L. Whistler, R. J. Smith, J. N. BeMiller and M. L. Wolfrom Eds) Vol. IV. p. 335. Academic Press, New York.

Giri, K. V. (1938). The availability of phosphorus from Indian foodstuffs. *Indian J. med. Res.* **25**, 869–877.

Giri, K. V. (1940). The availability of calcium and phosphorus in cereals. *Indian J. med. Res.* **28**, 101–111.

Giri, K. V., Radhakrishnan, A. N. and Vaidyanathan, C. S. 1952. Some factors influencing the quantitative determination of amino acids separated by circular paper chromatography. *Analyt. Chem.* **24**, 1677–1678.

Glick, Z. and Joslyn, M. A. (1970a). Food intake depression and other metabolic effects of tannic acid in the rat. *J. Nutr.* **100**, 509–515.

Glick, Z. and Joslyn, M. A. (1970b). Effect of tannic acid and related compounds on the absorption and utilization of protein in rats. *J. Nutr.* **100**, 516–520.

Glover, M. E. (1977). Location and effect of polyphenolic compounds in grain sorghum. Presentation in partial fulfillment of the requirements of the University Undergraduate Fellows Program 1976–1977. Texas A&M University, Texas, USA.

Goldberg, L. and Thorp, J. M. (1945). A survey of vitamins in African foodstuffs. III. The thiamin content of maize, kaffircorn, and other cereals. *S. Afr. J. med. Sci.* **10**, 1–8.

Goldberg, L. and Thorp, J. M. (1946). A survey of vitamins in African foodstuffs. VI. Thiamine, riboflavin, and nicotinic acid in sprouted and fermented cereal foods. *S. Afr. J. med. Sci.* **11**, 177–185.

Goldberg, L., Thorp, J. M. and Sussman, S. (1945). A survey of vitamins in African foodstuffs. IV. The thiamin content of beans and other legumes. *S. Afr. J. med. Sci.* **10**, 87–94.

Goldberg, L., Thorp, J. M. and Sussman, S. (1946). A survey of vitamins in African foodstuffs. V. The thiamine content of processed cereals and legumes. *S. Afr. J. med. Sci.*, **11**, 121–134.

Goldsmith, G. A., Gibbens, J., Unglaub, W. G. and Miller, O. N. (1956). Studies of niacin requirements in man. III. Comparative effects of diets containing lime treated and untreated corn in the production of experimental pellagra. *Am. J. clin. Nutr.* **4**, 151–160.

Goldstein, J. and Scott, M. L. (1956). Electrophoretic study of exudative diathesis in chicks. *J. Nutr.* **60**, 349–359.

Goldstein, J. L. and Swain, T. (1963). Changes in tannins in ripening fruits. *Phytochemistry* **2**, 371–383.

Goldstein, J. L. and Swain, T. (1965). The inhibition of enzymes by tannins. *Phytochemistry* **4**, 185–192.

Golenkov, V. F. (1960). Amino acid composition of rye proteins (in Russian). Trudy VNII Zerna No. 38, 201–211.

Gontijo, V. de P. M., Pereira, J. A. A., Costa, P. M. A. and De Mello, H. V. (1976). Replacement of maize by sorghum and its supplementation with lysine and methionine in rations for swine (in Portuguese). *Rev. Soc. bras. de Zootec.* **5**, 83–104.

Gontzea, I. and Sutzescu, P. (1968). "Natural anti-nutritive substances in foodstuffs and forages". K. Karger, Basel, Switzerland.

Good, C. A., Kramer, H. and Somogyi, M. (1933). The determination of glycogen. *J. biol. Chem.* **100**, 485–491.

Goodearl, G. P. (1943). Comparison of proso millet and yellow corn for feeding laying hens. Bulletin No. 329. North Dakota Agricultural Experiment Station, North Dakota, USA.

Goodwin, J. T. (1959). Wet milling. *In* "The Chemistry and Technology of Cereals as Food and Feed". (S. A. Matz, ed.) AVI Publishing Co. Inc., Westport, Connecticut, USA.

Gopalan, C. (1961). A report on some recent studies on protein malnutrition in India. *In* "Meeting protein needs of infants and children". Publication No. 843. p. 211. National Academy of Sciences—National Research Council, Washington, DC.

Gopalan, C. and Narasinga Rao, B. S. (1972). Experimental niacin deficiency. *Meth. Archievm. exp. Path.* **6**, 49–80.

Gopalan, C. and Srikantia, S. G. (1960). Leucine and pellagra. *Lancet*, April 30, 954–957.

Gopalan, C., Belavady, B. and Krishnamurthi, D. (1969). The role of leucine in the pathogenesis of canine black-tongue and pellagra. *Lancet*, November 14, 956–957.

Gopalan, C., Balasubramanian, S. C., Ramasastri, B. V. and Rao, K. V. (1971). Diet Atlas of India. pp. 26–33. National Institute of Nutrition, Indian Council of Medical Research, Hyderabad, India.

Gopaldas, T., Varadarajan, I., Grewal. T. and Shingwekar, A. G. (1974). Development of indigenous multi mixes for the preschool child. *Indian Pediat.* **11**, 501–509.

Gopalkrishna, N. Katwe, G. A. and Chaugale, D. S. Jr (1955a). Bibliography of bajri (*Pennisetum typhoides* (Burm.) S. & H. *Poona agric. Coll. Mag.* **46**, 244–245.

Gopalkrishna, N., Katwe, G. A. and Chaugale, D. S. Jr (1955b). Bibliography [of] other millets [other than *Pennisetum typhoides, Eleusine coracana, Setaria italica*]. *Poona agric. Coll. Mag.* **46**, 250–254.

Gorbet, D. W. and Weibel, D. E. (1972). Inheritance and genetic relationships of six endosperm types in sorghum. *Crop Sci.* **12**, 378–382.

Goswami, A. K., Sehgal, K. L. and Sharma, K. P. (1969a). Chemical composition of Bajra grains. 1. African entries. *J. Nutr. Diet.* **6**, 287–290.

Goswami, A. K., Sharma, K. P. and Gupta, B. K. (1969b). Chemical composition of Bajra grains. 2. American entries. *J. Nutr. Diet.* **6**, 291–294.

Goswami, A. K., Sharma, K. P. and Sehgal, K. L. (1969c). Nutritive value of proteins of pearl millet of high yielding varieties and hybrids. *Br. J. Nutr.* **23**, 913–916.

Goswami, A. K., Sehgal, K. L. and Gupta, B. K. (1970a). Chemical composition of Bajra grains. 3. Indian inbreds. *J. Nutr. Diet.* **7**, 5–9.

Goswami, A. K., Sharma, K. P. and Sehgal, K. L. (1970b). Chemical composition of Bajra grains. 4. Indian varieties. *Indian J. Nutr. Diet.* **7**, 67–70.

Goussault, B., Samson, M-F. and Adrian, J. (1972). Influence of wheat, millet and sorghum brans on the digestibility of the diet (in French). *Inds. aliment. agric.* **89**, 1597–1602.

Government of India. (1973). Integrated agricultural development in drought prone areas. Report by the Task Force on Integrated Rural Development. Planning Commission.

Govil, J. N. and Murty, B. R. (1973a). Combining ability for yield and quality characteristics in grain sorghum. *Indian J. Genet. Pl. Breed.* **33**, 239–251.

Govil, J. N. and Murty, B. R. (1973b). Genetic divergence and nature of heterosis in grain sorghum. *Indian J. Genet. Pl. Breed.* **33**, 252–260.

Graham, C. E., Smith, E. P., Hier, S. W. and Klein, D. (1947). An improved method for the determination of tryptophane with *p*-Dimethylamino-benzaldehyde. *J. biol. Chem.* **168**, 711–716.

Greenaway, W. T. and Johnson, R. M. (1974). Five-minute biuret method for protein content of wheat. *Bakers' Dig.* **48**, 38–39, 72.

Griffing, B. (1956). Concept of general and specific combining ability in relation to diallel crossing system. *Aust. J. biol. Sci.* **9**, 227–232.

Gross, W. and Heller, V. G. (1943). The carotene content of the grain sorghums. *Proc. Okla. Acad. Sci.* 97–98.

Gubler, C. J., Lahey, M. E., Ashenbrucker, H., Cartwright, H. and Wintrobe, M. M. (1952). Studies on copper metabolism. I. A method for the determination of copper in whole blood, red blood cells and plasma. *J. biol Chem.* **196**, 209–220.

Guenthner, E. and Carlson, C. W. (1970). A comparison of triticale, corn, wheat, and milo laying diets *Poult. Sci.* **49**, 1390. (Abstr.).

Guest, G. H. (1939). A note on the colorimetric determination of proline. *Can. J. Res.* **17B**, 132–144.

Guillemet, R. and Jacquot, R. (1943). Chimie Physiologique—Essai de détermination de l'indigestible glucidique. *C. r. hebd. Séanc. Acad. Sci.*, Paris.

Guillemet, R. and Jacquot, R. (1944). Nutritive value of different parts of the wheat grain (in French). *Bull. Soc. Chim. Biol.* **26**, 324–334.

Guiragossian. V., Chibber, B. A. K., Van Scoyoc, S., Jambunathan, R., Mertz, E. T. and Axtell, J. D. (1978). Characteristics of proteins from normal, high-lysine, and high-tannin sorghums. *J. agric. Fd Chem.* **26**, 219–223.

Gupta, A. K. and Gupta, Y. P. (1972). Effect of nitrogen application on the protein quality of *Sorghum vulgare* Pers. *Indian J. agric. Res.* **6**, 191–195.

Gupta, A. K. and Gupta, Y. P. (1974). Distribution of amino acids and protein fractions in the developing grain of *Sorghum vulgare* Pers. *Indian J. agric. Res.* **8**, 162–164.

Gupta, A. K. and Gupta, Y. P. (1975). Distribution of nitrogen in different plant parts of sorghum and amino acid composition of its grain as affected by foliar application of urea. *Indian J. agric. Res.* **9**, 31–36.

Gupta, R. K. and Dutta, B. K. (1967). Vernacular names of the useful plants of north-west Indian arid regions. *J. Agric. trop. Bot. appl.* **14**, 402–453.

Guttikar, M .N., Panemangalore, M., Narayana Rao, M. and Swaminathan, M. (1965a). Studies on processed protein foods based on blends of groundnut, bengalgram, soyabean and sesame flours and fortified with minerals and vitamins. 1. Preparation, chemical composition and shelf life. *J. Nutr. Diet.* **2**, 21–23.

Guttikar, M. N., Panemangalore, M., Narayana Rao, M., Rajalakshmi, D. and Swaminathan, M. (1965b). Studies on processed protein foods based on blends of groundnut, bengalgram, soyabean and sesame flours and fortified with minerals and vitamins. II. Amino acid composition and nutritive value of the proteins. *J. Nutr. Diet.* **2**, 24–27.

Guttikar, M. N., Panemangalore, M., Jayaraj, A. P., Rao, M. N. and Swaminathan, M. (1968). Studies on processed protein foods based on blends of groundnut, bengal gram, soya bean and sesame flours and fortified with minerals and vitamins. IV. Supplementary value to diets based on ragi (*Eleusine coracana*) or maize and tapioca. *J. Nutr. Diet.* **5**, 110–120.

Hackler, L. R. (1977). Methods of measuring protein quality: a review of bioassay procedures. *Cereal Chem.* **54**, 984–995.

Hadjimarkos, D. M. (1961). Effect of selenium on dental caries in the rat. *Archs. oral Biol.* **3**, 143–145.

Hadjimarkos, D. M. (1962). Selenium content of millet and dental caries in the rat. *Nature, Lond.* **193**, 178.

Haenszel, W., Berg, J. W., Segi, M., Kurihara, M. and Locke, M. and F. B. (1973). Large-bowel cancer in Hawaiian Japanese. *J. nat. Cancer Inst.* **51**, 1765–1779.

Hagerman, A. E. and Butler, L. G. (1978). Protein precipitation method for the quantitative determination of tannins. *J. Agric. Fd Chem.* **20**, 809–812.

Hahn, R. R. (1969). Dry milling of grain sorghum. *Cereal Sci. Today.* **14**, 234–237.

Hahn, R. R. (1970). Dry milling and products of grain sorghum. *In* "Sorghum Production and Utilization". (J. S. Wall and W. M. Ross Eds) pp. 573–601. AVI Publishing Co. Inc., Westport, Connecticut, USA.

Haikerwal, M. and Mathieson, A. R. (1971a). Extraction and fractionation of proteins of sorghum kernels. *J. Sci. Fd Agric.* **22**, 142–145.

Haikerwal, M. and Mathieson, A. R. (1971b). Protein content and amino acid composition of sorghum grain. *Cereal Chem.* **48**, 690–699.

Haji-Hashim, A. H. and Tipton, K. W. (1973). Evaluation of brown-seeded grain sorghum hybrids for crude protein. Varieties. *In* Rep. Proj. Dep. Agron. pp. 121–127. La. State Univ. Agric. Mech. Coll. Agric. Exp. Stn., 100 L936, USA.

Hale, W. H. (1970). Sorghum grain in ruminant nutrition. *In* "Sorghum Production and Utilization". (J. S. Wall and W. M. Ross, Eds) pp. 507–533. The AVI Publishing Company Inc., Westport, Connecticut, USA.

Hale, W. H. and Theurer, C. B. (1972). Feed preparation and processing. *In* "Digestive physiology and nutrition of ruminants". Vol. 3. (D. C. Church, Ed.) Practical Nutrition. Oregon State University, Corvallis, USA.

Halloran, H. R. and Maunder, A. B. (1971). Nutritive evaluations with "bird resistant" and yellow endosperm sorghums. *Poult Sci.* **50**, 1582.

Hamlyn, F. G. and Gasser, J. K. R. (1970). Some causes of error in determination of total nitrogen in plant material. *Chemy Ind.* 1142–1143.

Han, I. K. and Ha, J. K. (1977). Studies on the nutritive values of sorghum grain in the rations of growing finishing swine. *Korean J. Anim. Sci.* **19**, 180–188.

Hanif, M., Awais, M. and Ali, S. M. (1965). Supplementary value of a protein food (PFI) based on guar, gram and fish flours to certain indigenous cereals. *Pakis. J. Scient. Res.* **17**, 132–134.

Hankes, L. U., Henderson, L. M., Brickson, W. L. and Elvehjem, C. A. (1948). Effect of amino acids on the growth of rats on niacin-tryptophan-deficient rations. *J. biol. Chem.* **174**, 873–881.

Harbers, L. H. (1974). Starch granule hydrolysis patterns in the rumen. *J. Anim. Sci.* **39**, 998.

Harbers, L. H. (1975). Starch granule structural changes and amylolytic patterns in processed sorghum grain. *J. Anim. Sci.* **41**, 1496–1501.

Harbers, L. H. and Davis, A. B. (1974a). Sorghum grain starch hydrolysis in pigs and rats. *J. Anim. Sci.* **39**, 976.

Harbers. L. H. and Davis, A. B. (1974b). Digestion of sorghum grain endosperm in the rat and pig observed by scanning electron microscopy. *J. Anim. Sci.* **39**, 1009–1105.

Harborne, J. B. (ed.). (1964). "Biochemistry of phenolic compounds". Academic Press, London and New York.

Harborne, J. B. (1967). "Comparative biochemistry of the flavonoids". Academic Press, New York.

Harborne, J. B. (1973). "Phytochemical methods". Chapman and Hall, London.

Harden, M. L., Stanaland, R., Briley, M. and Yang, S. P. (1976). The nutritional quality of proteins in sorghum. *J. Fd Sci.* **41**, 1082–1085.

Hariharan, K., Urs, T. S. S. R., Desai, B. L. M., Rao, S. V., Rajalakshmi, D., Swaminathan, M. and Parpia, H. A. B. (1965). Effect of supplementing poor Indian diets based on kaffir corn, pearl millet and maize with vitamins and minerals and fortified and unfortified groundnut flour on the nutritive value of the diets as judged by the growth of rats and on the protein effeciency ratio. *J. Nutr. Diet.* **2**, 196–201.

Hariharan, K., Desai, B. L. M, Venkat Rao, S. Rajalakshmi, D., Swaminathan, M. and Parpia, H. A. B. (1967). Effect of supplementing poor Indian diets based on wheat, rice and ragi with vitamins, minerals and groundnut flour on the nutritive value of the diets as judged by the growth of albino rats. *J. Nutr. Diet.* **4**, 56–64.

Harinarayana, G. and Murty, B. R. (1970). Inheritance of protein content in *Pennisetum typhoides. Indian J. Genet. Pl. Breed.* **30**, 280–286.

Harlan, J. R. (1972). A new classification of cultivated sorghum. *In* "Sorghum in Seventies". (N. G. P. Rao and L. R. House, Eds) pp. 512–516. Oxford & IBH Publishing Co., New Delhi, Bombay, Calcutta.

Harlan, J. R. and De Wet, J. M. J. (1972). A simplified classification of cultivated sorghum. *Crop Sci.* **12**, 172–176.

Harper, A. E. (1969). Amino acid toxicities and imbalances. *In* "Mammalian protein metabolism". Vol. 3, (H. N. Munro, Ed.) pp. 391–422. Academic Press, New York.

Harper, A. E. and Rogers, Q. R. (1965). Amino acid imbalance. *Proc. Nutr. Soc.* **24**, 173–190.

Harper, A. E., Benton, D. A., Winje, M. E. and Elvehjem, C. A. (1954). Leucine-isoleucine antagonism in the rat. *Archs. Biochem. Biophys.* **51**, 523–524.

Harper, A. E., Benton, D. A. and Elvehjem, C. A. (1955). L-leucine, an isoleucine antagonist in the rat. *Archs. Biochem. Biophys.* **57**, 1–12.

Harper, A. E., Benevenga, N. J. and Wohlheuter, R. M. (1970). Effects of ingestion of disproportionate amounts of amino acids. *Physiol. Rev.* **50**, 428–558.

Harris, G. and Ricketts, R. W. (1958). Flavonoid compounds of hops and of malt concerned in the formation of beer hazes. *Chemy Ind.* (June 7), 686–687.

Harris, H. B. (1973). RS 700, a bird resistant grain sorghum hybrid with improved digestibility. Research Report 150. College of Agriculture Experiment Stations, University of Georgia, USA.

Harris, H. B. and Burns, R. E. (1970). Influence of tannin content on preharvest seed germination in sorghum *Agron. J.* **62**, 835–836.

Harris, H. B. and Burns, R. E. (1973). Relationship between tannin content of sorghum grain and preharvest seed molding. *Agron. J.* **65**, 957–959.

Harris, H. B., Cummins, D. G. and Burns, R. E. (1970). Tannin content and digestibility of sorghum grain as influenced by bagging. *Agron. J.* **62**, 633–635.

Harris, L. J. and Wang, Y. L. (1941). Vitamin methods. 1. An improved procedure for estimating Vitamin B_1 in foodstuffs and biological materials by the thiochrome test including comparisons with biological assays. *Biochem. J.* **35**, 1050–1067.

Harris, P. and James, A. T. (1969). The effect of low temperatures on fatty acid biosynthesis in plants. *Biochem. J.* **112**, 325–330.

Harris, R. W., Greene, D. E., Waldroup, P. W. and Stephenson, E. L. (1966). Comparison of corn, wheat and milo as principal cereal grains in turkey diets. *Poult. Sci.* **45**, 1091.

Harris, W. D. and Popat, P. (1954). Determination of the phosphorus content of lipids. Method for the determination of inorganic phosphorus. *J. Am. Oil Chem. Soc.* **31**, 124–127.

Hart, M. R., Graham, R. P., Gee, M. and Morgan, A. I. Jr (1970). Bread from sorghum and barley flours. *J. Fd Sci.* **35**, 661–665.

Haslam, E. (1966). "Chemistry of vegetable tannins". Academic Press, London and New York.

Haslam, E. (1974). Polyphenol-protein interactions. *Biochem. J.* **139**, 285–288.

Hassan, O. E. M. (1974). Utilization of tropical feedingstuffs in the nutrition of modern commercial laying stock. *Trop. Agric.* **51**, 569–573.

Hassid, W. Z. and Neufeld, E. F. (1964). Whole starches and modified starches. 9. Quantitative determination of starch in plant tissues. *In* "Methods in Carbohydrate Chemistry". (R. L. Whistler, R. J. Smith, J. N. BeMiller and M. L. Wolfrom, Eds) Vol. IV. pp. 33–36. Academic Press, New York and London.

Hawk, E. A. and Elvehjem, C. A. (1953). The effects of vitamins B_{12} and B_{12f} on growth, kidney haemorrhage and liver fat in rats fed purified diets. *J. Nutr.* **49**, 495–504.

Hawk, P. B., Oser, B. L. and Summerson, W. H. (1947). "Practical physiological chemistry". 12th ed. The Blakiston Co.

Hawk, P. B., Oser, B. L. and Summerson, W. H. (1965). "Practical physiological chemistry". McGraw Hill Book Co., London.

Hayman, B. L. (1954). The theory and analysis of diallel crosses. *Genetics* **39**, 789–809.

Health and Welfare Canada. (1976). Dietary Standard for Canada.

Hegsted, D. M. and Worcester, J. (1967). Assessment of protein quality with young rats. Proceedings 7th International Congress of Nutrition 4, 318.

Hegsted, D. M., Neff, R. and Worcester, J. (1968). Determination of the relative nutritive value of proteins. Factors affecting precision and validity. *J. agric. Fd Chem.* **16**, 190–195.

Hellegouarch, Monjour, Giorgi and Toury, (1967). Enquiry on the food consumption of two villages in Senegal at three periods of the year (in French). Organisme de recherches sur l'alimentation et la nutrition africaines (ORANA), 39 avenue Pasteur, BP 2089, Dakar, Senegal.

Heller, V. G. and Green, R. (1926). The chemical and nutritive properties of the grain sorghums. *J. metab. Res.* **7–8**, 205–215.

Heller, V. G. and Seiglinger, J. B. (1944). Chemical composition of Oklahoma grain sorghums. Bulletin, B-274 Oklahoma Agricultural Experiment Station, Oklahoma, USA.

Helper, D. E. (1950). "Manual of Clinical Laboratory Methods". 4th ed. C. C. Thomas, Springfield, Ill., USA.

Henderson, L. M. (1949). Quinolinic acid metabolism. II. Replacement of nicotinic acid for the growth of the rat and neurospora. *J. biol. Chem.* **181**, 677–685.

Henderson, L. M. and Snell, E. E. (1948). A uniform medium for determination of amino acids with various microorganisms. *J. biol. Chem.* **172**, 15–29.

Hennessey, D. J. and Cerecedo, L. R. (1939). The determination of free and phosphorylated thiamin by a modified thiochrome assay. *J. Am. chem. Soc.* **61**, 179–183.

Henry, K. M. and Kon, S. K. (1957). Effect of level of protein intake and age of rat on the biological value of proteins. *Br. J. Nutr.* **11**, 305–313.

Henry, K. M. and Toothill, J. (1962). A comparison of the body-water and nitrogen balance-sheet methods for determining the nutritive value of proteins. *Br. J. Nutr.* **16**, 125–133.

Hepler, O. E. (1950). "Manual of clinical laboratory methods". 4th ed. C. C. Thomas, Springfield, Ill., USA.

Herrett, R. J. (1956). I. Quantitative determination of tannin in grain sorghum. II. Paper chromatography of sorghum tannin. M.Sc. Thesis. Oklahoma State University, USA.

Herstad, O., Sannan, F. and Hoie, J. (1966). The relative feed value of cereals for chicks (in Norwegian). *Meld. Norg. Landbr. Hogsk.* **45**, 1–19.

Hilbert, G. E. and MacMasters, M. M. (1946). Pea starch, a starch of high amylose content. *J. biol. Chem.* **162**, 229–238.

Hill, F. W., Anderson, D. L., Renner, R. and Carew, L. B. Jr (1960). Studies of the metabolizable energy of grain and grain products. *Poult. Sci.* **39**, 573–579.

Hillier, J. C., Martin, J. J. and Waller, G. R. (1959). "The relative value of six varieties of milo for growing and finishing swine". Miscellaneous Publication MP-55. Oklahoma Agricultural Experiment Station, USA.

Hinders, R. G. and Freeman, J. E. (1969). Effects of processing technic on gelatinization and starch availability of grain sorghum types. *Proc. west. Sect. Am. Soc. Anim. Sci.* **20**, 253–255.

Hinton, J. J. C. (1959). The distribution of ash in the wheat kernel. *Cereal Chem.* **36**, 19–31.

Hinton, J. J. C., Peers, F. G. and Shaw, B. (1953). The B- vitamins in wheat: the unique aleurone layer. *Nature, Lond.* **172**, 993–995.

Hodge, J. E. and Hofreiter, B. T. (1962). Determination of reducing sugars and carbohydrates. *In* "Methods in Carbohydrate Chemistry". (R. L. Whistler, M. L. Wolfrom, J. N. BeMiller and F. Shafizadeh, Eds) Vol. I. pp. 380–394. Academic Press, Inc., New York.

Hodson, A. Z. and Norris, L. C. (1939). A fluorometric method for determining the riboflavin content of foodstuffs. *J. biol. Chem.* **131**, 621–630.

Hodson, H. H., Snyder, R. and Kroening, G. H. (1973). Corn vs milo for pig with varying levels of protein and lysine. *J. Anim. Sci.* **37**, 283.

Hoffman, C. E., Stokstad, E. L. R., Hutchings, B. L., Dornbush, A. C. and Jukes, T. H. (1949). The microbiological assay of Vitamin B_{12} with *lactobacillus leichmannii. J. biol. Chem.* **181**, 635–644.

Hogan, J. T., Normand, F. L. and Deobald, H. J. (1964). Method for removal of successive surface layers from brown and milled rice. *Rice J.* **67** (April), 27–34.

Holt, R. (1955). Studies on dried peas. 1. The determination of phytate phosphorus. *J. Sci. Fd Agric.* **6**, 136–142.

Honold, G. R. and Stahmann, M. A. (1968). The oxidase reduction enzymes of wheat. IV. Qualitative and quantitative investigation of the oxidases. *Cereal Chem.* **45**, 99–106.

Hoover, A. A. and Jayasuriya, G. C. N. (1951). Microbiological assay of vitamins. 2. Riboflavin. *Ceylon J. med. Sci.* **8**, 183–189. (*Nutr. Abst. Rev.* 1954, **24**, 324.).

Hopkins, Smith and East. (1903). *In* Illinois Agricultural Experiment Station Bulletin 87, p. 83. Illinois, USA.

Horan, F. E. and Heider, M. F. (1946). A study of sorghum and sorghum starches. *Cereal Chem.* **23**, 492–503.

Horel, J. 1956. Chemicko-technologicke rozbory zemadelskych plodin. SZ N. Prague.

Horigome, T. and Kandatsu, M. (1968). Biological value of proteins allowed to react with phenolic compounds in presence of *O*-diphenol oxidase. *Agric. biol. Chem.* **32**, 1093–1102.

Horn, M. J., Jones, D. B. and Blum, H. E. (1946). Colorimetric determination of methionine in proteins and foods. *J. biol. Chem.* **166**, 313–320.

Horn, P. J. and Schwartz, H. M. (1961). Kaffircorn malting and brewing studies. IX. Amino acid composition of kaffircorn grain and malt. *J. Sci. Fd Agric.* **12**, 457–459.

Hornoiu, M., Stoicea, M., Timariu, S., Petcu, D., Calotoiu, E. and Stavri, J. (1965).

The use of hybrid sorghum in animal and poultry feeding (in Romanian). *Lucr. Stiint. Inst. Cerc. Zooteh.* **22**, 33–46.

Hornoiu, M., Sirbu, M., Saghin, F., Feredean, I., Rosca, N. and Tascenco, V. (1967). Nutritive value of hybrid sorghum grain for pigs and sheep (in Romanian). *Lucr. Stiint. Inst. Cerc. Zooteh.* **25**, 527–533.

Hoseney, R. C., Davis, A. B. and Harbers, L. H. (1974). Pericarp and endosperm structure of sorghum grain shown by scanning electron microscopy. *Cereal Chem.* **51**, 552–558.

Howe, E. E. and Gilfillan, E. W. (1970). Limiting nutrients in cereal grains for the growth of the laboratory rat. *Indian J. Nutr. Diet.* **7**, 17–20.

Howe, E. E., Jansen, G. R. and Gilfillan, E. W. (1965). Amino acid supplementation of cereal grains as related to the world food supply. *Am. J. clin. Nutr.* **16**, 315–320.

Howe, P. E. (1921). The determination of proteins in blood: a micro method. *J. biol. Chem.* **49**, 109–113.

Hubbard, J. E., Hall, H. H. and Earle, F. R. (1950). Composition of the component parts of the sorghum kernel. *Cereal Chem.* **27**, 415–420.

Hubbel, R. B., Mendel, L. B. and Wakeman, A. J. (1937). A new salt mixture for use in experimental diets. *J. Nutr.* **14**, 273–285.

Hubscher, G. and Hawthorne, J. N. (1957). The isolation of inositol monophosphate from liver. *Biochem. J.* **67**, 523–527.

Hulme, A. C. and Narain, R. (1931). CXVII. The ferricyanide method for the determination of reducing sugars. A modification of the Hagedorn-Jensen-Hanes technique. *Biochem. J.* **25**, 1051–1061.

Hulse, J. H. (1974). The protein enrichment of bread and baked products. Composite flours. *In* "New protein foods". (A. A. Altschul, Ed.) Vol. 1A. Technology, pp. 205–229. Academic Press, New York.

Hulse, J. H. (1980). Polyphenols in cereals and legumes. Proceedings of a Symposium, International Development Research Centre, Box 8500, Ottawa, Canada, K1G 3H9 1DRC-145e.

Hulse, J. H. and Laing, E. M. (1974). "Nutritive value of triticale protein". IDRC-021e. International Development Research Centre, Box 8500, Ottawa, Canada, K1G 3H9.

Hulse, J. H., Rachie, K. O. and Billingsley, L. W. (1977). "Nutritional standards and methods of evaluation for food legume breeders". Prepared by the International Working Group on Nutritional Standards and Methods of Evaluation for Food Legume Breeders. IDRC-TS7e. International Development Research Centre, Box 8500, Ottawa, Canada K1G 3H9.

Hurrell, R. F. and Carpenter, K. J. (1975). The use of three dye–binding procedures for the assessment of heat damage to food proteins. *Br. J. Nutr.* **33**, 101–115.

Hussain, S., Khan, A. H., Yasin, M. and Shah, F. H. (1966). Chemical changes during malting of cereals. *Pakist. J. Scient. ind. Res.* **9**, 137–139.

Husted, W. T., Mehen, S., Hale, W. H., Little, M. and Theurer, B. (1968). Digestibility of milo processed by different methods. *J. Anim. Sci.* **27**, 531–534.

ICRISAT. (1976). "Sorghum Improvement". Departmental Annual Report 1975–76. International Crops Research Institute for the Semi-Arid Tropics, 1-11-256, Begumpet, Hyderabad 500016 (AP) India.

IDRC. (1977). "Agriculture, Food and Nutrition Sciences Division: the First Five Years". International Development Research Centre, Box 8500, Ottawa, Canada, K1G 3H9.

Idusogie, E. O. (1971). The nutritive value per acre of selected food crops in Nigeria. *J. Afr. Sci. Ass.* **16**, 17–24.

Ilany, J., Diamant, J., Laxer, Sh. and Pinsky, A. (1969). Japanese miso-type products prepared by using defatted soybean flakes and various carbohydrate containing foods. *Fd Technol. Champaign* **23**, 156–158.

Ilori, J. O. and Conrad, J. H. (1976). Nutritive value of protein in selected sorghum lines as measured by rat performance. *Nutr. Rep. Int.* **13**, 307–314.

Indian Council of Agricultural Research. (1970). "New vistas in crop yields". Agricultural year-book. Indian Council of Agricultural Research, New Delhi.

Indian Council of Medical Research. (1944). Nutrition Advisory Committee—Recommendations. Indian Council of Medical Research, New Delhi.

Indian Council of Medical Research. (1948). Nutrition Advisory Committee. Recommendations on the nutritional assessment schedules. Indian Council of Medical Research, New Delhi.

Indira, R. and Naik, M. S. (1971). Nutrient composition and protein quality of some minor millets. *Indian J. agric. Sci.* **41**, 795–797.

Isaac, R. A. and Kerber, J. D. (1971). Atomic absorption and flame photometry. Techniques and uses in soils, plants and water analysis. *In* "Instrumental methods for analysis of soils and plant tissue". (L. M. Walsh, Ed.). pp. 17–37.

Iswarriah, V. (1951). Final report on the scheme for research into indigenous drugs of India used in veterinary practice with special reference to their toxicology. Indian Council of Agricultural Research, Madras.

IUPAC-IUB. (1968). The nomenclature of cyclitols. *Eur. J. Biochem.* **5**, 1–12.

Ivancenko, D., Dodok, L. and Sadlon, J. (1972). Isolation of protein fractions of sorghum flour (in Slovak). Zbornik Prac Chemickotechnologickej Fakulty SVST. 289–297. (*Fd Sci. Technol. Abstr.* 1974, **6**, M4297).

Jacquot, R. Raulin, J., Adrian, J. and Rerat, A. (1955). Composition and feeding value of Job's tears or Adlay (*Coix lachryma-jobi*) (in Spanish). *Archos. venez. Nutr.* **6**, 3–21. (*Nutr. Abst. Rev.* 1956, **26**, 2885).

Jadhav, P. S., Jain, T. C. and Prasannalakshmi, S. (1975). "Sorghum—Millets—Peas. A bibliography of Indian literature 1969–1973". ICRISAT, 1-11-256, Begumpet, Hyderabad-500016, (AP) India.

Jakushevsky, E. S. (1969). Varietal composition of sorghum and its use for breeding (in Russian). *Trudy prikl. Bot. Genet. Selek*, **41**, 148–178.

Jalil, M. E. and Tahir, W. M. (1971). Review of the world's plant protein resources. Information Bulletin, Near East Cereal Improvement Production Project, FAO 8, 2–3, pp. 27–35. FAO, Rome.

Jambunathan, R. (1977). Biochemistry and nutrition. UNDP/CIMMYT/ICRISAT Policy Advisory Committee, Hyderabad. ICRISAT, 1-11-256 Begumpet, Hyderabad 500016 (AP) India.

Jambunathan, R. and Mertz, E. T. (1972). Amino acid composition of whole kernel and endosperm protein fractions of sorghum. *In* "Research Progress Report on Inheritance and Improvement of Protein Quality and Content in *Sorghum bicolor* (L.) Moench." (J. D. Axtell and D. L. Oswalr, Eds) pp. 43–56. Purdue University, Lafayette, Indiana; agency for International Development, Washington, DC.

Jambunathan, R. and Mertz, E. T. (1973a). Relationship between tannin levels, rat growth, and distribution of proteins in sorghum. *J. agric. Fd Chem*, **21**, 692–696.

Jambunathan, R. and Mertz, E. T. (1973b). Biochemical and nutritional studies in sorghum. *In* "Inheritance and Improvement of Protein Quality and Content in Sorghum". Research Progress Report No. 10, pp. 126–138. Department of

Agronomy, Agricultural Experiment Station, Purdue University, Lafayette, Indiana: Agency for International Development, Department of State, Washington, DC.

Jambunathan, R., Misra, P. S. and Mertz, E. T. (1972). Nutritive value and protein solubility characteristics of normal and pigmented sorghums. *Fedr. Proc.* **31**, 695.

Jambunathan, R., Mertz, E. T. and Axtell, J. D. (1975). Fractionation of soluble proteins of high lysine and normal sorghum grain. *Cereal Chem.* **52**, 119–121.

James, E. M., Norris, F. M. and Wokes, F. (1947). The chemical estimation of nicotinic acid in cereals and other foods. *Analyst* **72**, 327–336.

Janicki, J., Sobkowska, E., Warchalewski, J., Nowakowska, K., Chelkowski, J. and Stasinska, B. (1973). Amino acid composition of cereal and oil-seed. *Nahrung* **17**, 359–365.

Jansen, G. R. (1972). Seeds as a source of protein for humans. *In* "Symposium: Seed Proteins". (G. E. Inglett, Ed.), pp. 19–38, AVI Publishing Co. Inc., Westport, Connecticut, USA.

Jansen, G. R. (1974). The amino acid fortification of cereals. *In* "New Protein Foods". (A. A. Altschul, Ed.) Vol. 1A. Technology. pp. 39–120. Academic Press Inc., New York, USA.

Jansen, G. R., Dimaio, L. R. and Hause, N. L. (1962). Cereal proteins. Amino acid composition and lysine supplementation of teff. *J. agric. Fd Chem.* **10**, 62–64.

Jaros, N. P. (1965). Quantitative and qualitative composition of proteins and starch in grain of millets belonging to different ecological and geographical groups (in Russian). *Trudy prikl. Bot. Genet. Selek.* **37**, 50–58.

Jayaraj, A. P., Tovey, F. I. and Clark, C. G. (1976). The possibility of dietary protective factors in duodenal ulcer II. An investigation into the effect of prefeeding with different diets and of instillation of foodstuffs into the stomach on the incidence of ulcers in pylorus-ligated rats. *Postgrad. med. J.* **52**, 640–644.

Jellum, M. D. (1970). Plant introductions of maize as a source of oil with unusual fatty acid composition. *J. agric. Fd Chem.* **18**, 365–370.

Jellum, M. D. and Powell, J. B. (1971). Fatty acid composition of oil from pearl millet seed. *Agron. J.* **63**, 29–33.

Jensen, A. H., Becker, D. E. and Harmon, B. G. (1965). Nutritional adequacy of milo for the finishing pig. *J. Anim. Sci.* **24**, 398–402.

Jensen, A. H., Baker, D. H., Becker, D. E. and Harmon, B. G. (1969). Comparison of opaque-2 corn, milo and wheat in diets for finishing swine. *J. Anim. Sci.* **29**, 16–19.

Jensen, L. S., Chang, C. H. and Wyatt, R. D. (1976). Influence of carbohydrate source on liver fat accumulation in hens. *Poult. Sci.* **55**, 700–709.

Jerumanis, J. (1972). Variation of polyphenols during melting and mashing [barley] (in German). *Brauwissenschaft* **25**, 313–322.

Jimenez, J. R. (1966). Protein fractionation studies of high lysine corn. Proceedings, High-lysine Corn Conference, p. 74.

Jinks, J. L. (1954). The analysis of continuous variation in a diallel cross of *N. rustica* varieties. *Genetics* **39**, 767–788.

Jinks, J. L. and Hayman, B. I. (1953). The analysis of diallel crosses. *Maize Genet. Newsl.* **27**, 48.

Joffe, A. Z. (1965). Toxin production by cereal fungi causing taxic alimentary aleukia in man. *In* "Mycotoxins in Foodstuffs". (G. N. Wogan, Ed.) pp. 77–85. MIT Press, Cambridge, Mass. USA.

Johannsen, E. (1969). Kaffircorn malting and brewing studies. XXI. The effect of the fusel oils of Bantu beer on rat liver. *S. Afr. med. J.* **43**, 326–328.

Johansen, D. A. (1940). "Plant microtechnique". McGraw Hill Books Inc., New York.

Johari, R. P., Mehta, S. L., Gupta, R. K. and Naik, M. S. (1976). Incorporation of ^{15}N labelled urea and ammonium into proteins and amino acids of sorghum. *Phytochemistry* **15**, 1841–1843.

Johari, R. P., Mehta, S. L. and Naik, M. S. (1977). Changes in protein fractions and leucine-[^{14}C] incorporation during sorghum grain development. *Phytochemistry* **16**, 311–314.

John, S. W. and Muller, H. G. (1973). Thiamine in sorghum. *J. Sci. Fd Agric.* **24**, 490–491.

Johnson, B. C. (1945). The microbiological determination of nicotinic acid, nicotinamide and nicotinuric acid. *J. biol. Chem.* **159**, 227–230.

Johnson, C. H. and Arkley, T. H. (1956). Determination of molybdenum in plant tissue. *Analyt. Chem.* **26**, 572–574.

Johnson, M. C. and Ulrich, A. (1959). Analytical methods for use in plant analysis. *In* Bulletin 766. pp. 66–68. California Agricultural Experiment Station, California, USA.

Johnson, R. M. and Raymond, W. D. (1964). The chemical composition of some tropical food plants. I. Finger and bulrush millets. *Trop. Sci.* **6**, 6–11.

Jones, R. W. and Beckwith, A. C. (1970). Proximate composition and proteins of three grain sorghum hybrids and their dry-mill fractions. *J. agric. Fd Chem.* **18**, 33–36.

Jones, R. W. and Dimler, R. J. (1962). Electrophoretic composition of glutens from air-classified flours. *Cereal Chem.* **39**, 336–340.

Jones, R. W., Taylor, N. W. and Senti, F. R. (1959). Electrophoresis and fractionation of wheat gluten. *Archs. Biochem. Biophys.* **84**, 363–376.

Jones, R. W., Beckwith, A. C., Khoo, U. and Inglett, G. E. (1970). Protein composition of proso millet. *J. agric. Fd Chem.* **18**, 37–39.

Jones, W. T., Lyttleton, J. W. and Clarke, R. T. J. (1970). Bloat in cattle. XXXIII. The soluble proteins of legume forages in New Zealand and their relationship to bloat. *N. Z. J. agric. Res.* **13**, 149–156.

Jordan, R. M., Burkitt, W. H. and Wilson, J. W. (1952). "Sorghum as a feed for lambs". South Dakota Agricultural Experiment Station Bulletin No. 417. (*Nutr. Abst. Rev.* 1954, **24**, 454).

Joseph, K., Narayana Rao, M., Swaminathan, M. and Subrahmanyan, V. (1957). Supplementary value of Indian multipurpose food to poor vegetarian diets based on Italian millet (*Setaria italica*) and little millet (*Panicum miliare*). *Fd. Sci.* **6**, 205–206.

Joseph, K., Kurien, P. P. and Swaminathan, M. (1958a). The metabolism of nitrogen, calcium and phosphorus in children on a poor vegetarian diet based on ragi (*Eleusine coracana*). *Ann. Biochem. exp. Med.* **18**, 195–200.

Joseph, K., Kurien, P. P., Swaminathan, M. and Subrahmanyan, V. (1958b). Metabolism of nitrogen, calcium, and phosphorus in children on a poor vegetarian diet based on ragi (*Eleusine coracana*). *Fd Sci.* **7**, 324–325.

Joseph, K., Kurien, P. P., Swaminathan, M. and Subrahmanyan, V. (1959a). The metabolism of nitrogen, calcium and phosphorus in under-nourished children. 5. The effect of partial or complete replacement of rice in poor vegetarian diets by ragi (*Eleusine coracana*) on the metabolism of nitrogen, calcium and phosphorus. *Br. J. Nutr.* **13**, 213–218.

Joseph, K., Rao, M. N., Swaminathan, M., Sankaran, A. N. and Subrahmanyan, V. (1959b). The effect of supplementing poor jowar diet with Indian multipurpose food on the growth and composition of body and liver of albino rats. *Ann. Biochem. exp. Med.* **19**, 87–94.

Joslyn, M. A. and Glick, Z. (1969). Comparative effects of gallotannic acid and related phenolics on the growth of rats. *J. Nutr.* **98**, 119–126.

Kadkol, S. B. and Swaminathan, M. (1954). Chemical composition of different varieties of ragi (*Eleusine coracana*). *Bull. cent. Fd technol. Res. Inst. Mysore*, **4**, 12–13.

Kadkol, S. B. and Swaminathan, M. (1955). The nutritive value of Italian millet (*Setaria italica*). *Sci. Cult.* **20**, 340–341.

Kadkol, S. B., Sreenivasamurty, V. and Swaminathan, M. (1954a). Nutritive value of the seeds of *Panicum miliare* (little millet). *Bull. cent. Fd. technol. Res. Inst. Mysore* **3**, 247–248.

Kadkol, S. B., Srinivasamurthy, V. and Swaminathan, M. (1954b). Nutritive value of the seeds of *Paspalum scrobiculatum*. *J. scient. ind. Res.* **13B**, 744–745.

Kagi, J. H. R. and Vallee, B. L. (1958). Determination of zinc by direct extraction of urine with diphenyl thio carbazone. *Analyt. Chem.* **30**, 1951–1954.

Kakade, M. L. and Liener, I. E. (1969). Determination of available lysine in proteins. *Analyt. Biochem.* **27**, 273–280.

Kalmykov, K. V., Sergeeva, G. N. and Matsievskaya, T. M. (1954). Digestibility of foxtail millet seed by fowl (in Russian). *Ptitsevodstvo* **9**, 10–13.

Kamalam, N. (1964). Influence of phosphates on the growth, yield and composition of cholam crop. *Madras agric. J.* **51**, 197–206.

Kamalanathan, G., Girija, K., A. and Devadas, R. P. (1971). Possibilities of white ragi *Eleusine coracana* CO-9 in human dietary. *Indian J. Nutr. Diet.* **8**, 315–318.

Kambal, A. E. and Webster, O. J. (1965). Estimates of general and specific combining ability in grain sorghum. *Crop Sci.* **5**, 521–523.

Kambal, A. E. and Webster, O. J. (1966). Manifestations of hybrid vigor in grain sorghum and the relations among the components of yield, weight per bushel, and height. *Crop Sci.* **6**, 513–515.

Kametaka, M. (1952). On the chemical property of Italian millet, especially of the prolamin (in Japanese). *J. agric. Chem. Soc. Japan* **25**, 512–516.

Kanaka Doss, A., Raj, D. and Loganathan, S. (1975). Progressive changes in available nutrient contents in the sandy soils and their effect on yield and uptake of nutrients in ragi (*Eleusine coracana* Gaertn.). *Madras agric. J.* **62**, 138–144.

Kandel, E. R. (1954). Comparative chemical analysis of the Hungarian varieties of sorghum (in Hungarian with English summary). *Annls. Inst. biol. Tihany* **23**, 155–160.

Kanwar, J. S. and Chopra, S. L. (1959). "Practical Agricultural Chemistry". S. Chand & Co., New Delhi.

Kapasi-Kakama, J. (1974). Milling trials on sorghum grain. Paper presented at the 5th Eastern African Cereals Research Conference, March 1974.

Kapasi-Kakama, J. (1977). Some characteristics which influence the yield and quality of pearled sorghum grain. Proceedings, Symposium, Sorghum and millets as human food. International Association for Cereal Chemistry, Vienna, 11 May 1976. Tropical Products Institute, London.

Kapoor, H. C. and Naik, M. S. (1970). Effects of soil and spray applications of urea and storage on the β-carotene content of yellow endosperm sorghum and pearl millet grains. *Indian J. agric. Sci.* **40**, 942–947.

Kapsiotis, G. D. (1967). A list of protein food mixtures. *PAG Bull.* **No. 7**, 71.

Karim, A. and Rooney, L. W. (1972a). Pentosans in sorghum grain. *J. Fd Sci.* **37**, 365–368.

Karim, A. and Rooney, L. W. (1972b). Characterization of pentosans in sorghum grain. *J. Fd. Sci.* **37**, 369–371.

Karper, R. E. (1933). Inheritance of waxy endosperm in sorghum. *J. Hered.* **24**, 257–262.

Karper, R. E. and Quinby, J. R. (1963). Sugary endosperm in sorghum. *J. Hered.* **54**, 121–126.

Katwe, G. A., Gopalkrishna, N. and Chaugale, D. S. Jr (1955). Bibliography of rala (*Setaria italica* Beauv.). *Poona agric. Coll. Mag.* **46**, 248–249.

Keller, E. B., Rachele, J. R. and Du Vigneaud, V. (1948). A study of transmethylation with methionine containing deuterium and C^{14} in the methyl group. *J. biol. Chem.* **177**, 733–738.

Kelley, J. J. and Koeing, V. L. (1963). Electrophoretic analysis of flour proteins from various varieties of wheat. *J. Sci. Fd Agric.* **14**, 29–38.

Kemmerer, A. R. and Heywang, B. W. (1965). A comparison of various varieties of sorghum as substitute for corn in practical chick diets. *Poult. Sci.* **44**, 260–264.

Kempanna, C. and Kavallappa, B. N. (1968). Quantitative assessment for nutritive quality of *Eleusine coracana* (finger millet or ragi). *Mysore J. agric. Sci.* **2**, 324–329.

Kempthorne, O. (1957). "An introduction to genetic statistics". John Wiley & Sons Inc., New York.

Kempthorne, O. and Curnow, R. N. (1961). The partial diallel cross. *Biometrics* **17**, 229–250.

Kenney, M. A. and Puar, M. (1973). Brain lipid and cholesterol in young rats fed millet. *Fed. Proc.* **32**, 901.

Kent, N. L. and Evers, A. D. (1969). Variation in protein composition within the endosperm of hard wheat. *Cereal Chem.* **46**, 293–300.

Kent Jones, D. W. and Amos, A. J. (1947). "Modern Cereal Chemistry". 4th ed. The Northern Publishing Co. Ltd., Liverpool.

Kerr, R. W. (1950). "Chemistry and industry of starch". 2nd ed. Academic Press, New York.

Keys, A. (1970). Coronary heart disease in seven countries. American Heart Association Monograph. 29.

Khanna, M. L. (1966). Composition of bajra (*Pennisetum typhoides*) grain as influenced by fertilization. *Indian J. Agron.* **11**, 247–249.

Khattab, A. H., Karam-Alla, K. A. and Nour, A. A. M. (1972). Amino acid composition of some sorghum grain varieties. *Sudan J. Fd Sci. Technol.* **4** (January), 27–29.

Kies, C. and Fox, H. M. (1970). Protein value of wheat and triticale grain for humans, studied at two levels of protein intake. *Cereal Chem.* **47**, 671–678.

Kies, C. and Fox, H. M. (1971). Comparison of the protein nutritional value of TVP, methionine-enriched TVP, and beef at two levels of intake for human adults. *J. Fd. Sci.* **36**, 841–845.

Kies, C. and Fox, H. M. (1972). Interrelationships of leucine with lysine, tryptophan, and niacin as they influence protein value of cereal grains for humans. *Cereal Chem.* **49**, 223–231.

Kies, C., Fox, H. M. and Nelson, L. (1975). Triticale soy textured vegetable protein and millet based diets as protein suppliers for human adults. *J. Fd. Sc.* **40**, 90–93.

Kihlberg, R. and Ericson, L. E. (1963). Amino acid composition and supplementation of teff. *Nutritio Diet.* **6**, 151–155.

King, E. J. (1932). The colorimetric determination of phosphorus. *Biochem. J.* **26**, 292–297.

King, E. J. (1951). "Microanalysis in medical biochemistry". 2nd ed. J. A. Churchill Ltd., London.

King, E. J. and Wooton, I. D. P. (1956). "Microanalysis in medical biochemistry". Grune and Stratton, Inc., New York. J. & A. Churchill, London.

King, S. B. (1972). Sorghum diseases and their control. *In* "Sorghum in Seventies". (N. G. P. Rao and L. R. House, Eds) pp. 411–434. Oxford and IBH Publishing Co., New Delhi.

Kinra, K. L., Nijhawan, H. L. and Rao, S. B. P. (1970). Foliar versus soil application of nitrogen on pearl millet. *Ann. Arid Zone* **9**, 256–260.

Kleiber, R. G. (1961). "The fire of life". pp. 253–265. John Wiley & Sons, New York, N.Y.

Klett, R. H. (1973). Results of comparison feeding. *In* "Eighth Biennial Grain Sorghum Research and Utilization Conference". pp. 8–14. Grain Sorghum Producers Association, Amarillo, Texas, USA.

Klimenko, V. G. and Zubaidov, U. Z. (1971). Chromatographic and electrophoretic studies of sorghum seed glutelins. *Dokl. Akad. Nauk Tadzhik SSR* **14**, 68–71.

Kline, G. M. and Acree, S. F. (1930). A study of the method for titrating aldose sugars with standard iodine and alkali. *Bur. Stand. J. Res.* **5**, 1063–1084.

Klosterman, E. W., Fox, D. G. and Newland, W. H. (1968). Grain sorghum evaluated for fattening cattle. Ohio Report on Research and Development in *Biol. Agric. Home Econ.* **53**, 67–69.

Kneen, E. and Sandstedt, R. M. (1946). Distribution and general properties of an amylase inhibitor in cereals. *Archs. Biochem.* **9**, 235–249.

Knox, G., Heller, V. G. and Sieglinger, J. B. (1944). Riboflavin, niacin and pantothenic acid contents of grain sorghums. *Fd Res.* **9**, 89–91.

Kodicek, E. (1940). Estimation of nicotinic acid in animal tissues, blood and certain foodstuffs. *Biochem. J.* **34**, (1) Methods 712–723(2) Applications 724–735.

Kodicek, E. and Pepper, C. R. (1948). The microbiological estimation of nicotinic acid and comparison with chemical method. *J. gen. Microbiol.* **2**, 306–314.

Kodicek, E., Braude, R., Kon, S. K. and Mitchell, K. G. (1959). The availability to pigs of nicotinic acid in tortilla baked from maize treated with lime water. *Br. J. Nutr.* **13**, 363–384.

Koening, R. F. and Maranville, J. (1970). Variation of protein content within a line of sorghum. *In* "The physiology of yield and management of sorghum in relation to genetic improvement". Annual Report No. 4. pp. 138–144. Agency for International Development, Department of State, Washington, DC.

Konczacki, Z. A. (1972). Infant malnutrition in sub-Saharan Africa: a problem in socio-economic development. *Can. J. Afr. Studies* **6**, 444.

Konstantinov, S. I. (1975). Use of *Setaria italica* type millet in breeding for higher protein content (in Russian). *Sel. Semenovod (Kiev)* **29**, 29–35.

Konstantinov, S. I., Linnik, V. M. and Nikulina, N. D. (1975). Use of chemical mutagenesis for the production of millet forms with an increased content of protein. *Soviet Genet.* **11**, 40–42.

Konstantinov, S. I., Linnik, V. M. and Nikulina, N. D. (1977). Application of Experimental mutagenesis to millet breeding. *Tsitologiya Genet.* **11**, 231–236.

Kovalev, N. I Makarenko, L. I. and Orlova, S. A. (1974). The effect of processing on the in vitro digestibility of the proteins of millet grits and their amino acid composition (in Russian). *Nahrung* **18**, 517–522.

Kovalsky, V. V., Yarovaya, G. A. and Shmavonyan, D. M. (1961). The changes in purine metabolism of humans and animals under the conditions of molybdenum biogeochemical provinces (in Russian). *Zh. obshch. Biol.* **22**, 179–191.

Kramer, N. W. and Matz, S. A. (1969). Sorghum. *In* "Cereal Science". (S. A. Matz, Ed.) pp. 150–171. AVI Publishing Co. Inc., Westport, Connecticut, USA.

Krehl, W. A. and Strong, F. M. (1944). Studies on the distribution, properties and isolation of a naturally occurring precursor of nicotinic acid. *J. biol. Chem.* **156**, 1–12.

Krehl, W. A., Elvehjem, C. A. and Strong, F. M. (1944). The biological activity of a precursor of nicotinic acid in cereal products. *J. biol. Chem.* **156**, 13–19.

Krehl, W. A., Sarma, P. S., Teply, L. J. and Elvehjem, C. A. (1946a). Factors affecting the dietary niacin and tryptophane requirement of the growing rat. *J. Nutr.* **31**, 85–106.

Krehl, W. A., De la Huega, J., Elvehjem, C. A. and Hart, E. B. (1946b). The distribution of niacinamide and niacin in natural materials. *J. biol. Chem.* **166**, 53–57.

Krishnamachari, K. A. V. R. (1974). Some aspects of copper metabolism in pellagra. *Am. J. clin. Nutr.* **27**, 108–111.

Krishnamachari, K. A. V. R. and Bhat, R. V. (1976). Poisoning by ergoty bajra pearl millet in man. *Indian J. med. Res.* **64**, 1624–1628.

Krishnamachari, K. A. V. R. and Krishnaswamy, K. (1973). Genu vagum and osteoporosis in an area of endemic fluorosis. *Lancet* (ii), 877–879.

Krishnamurthy, K. (1968). Nutritive content of ragi varieties in relation to fertilizer levels. *J. Nutr. Diet.* **5**, 10–12.

Krishnamurthy, K., Tasker, P. K., Indira, K., Rajagopalan, R., Swaminathan, M. and Subrahmanyan, V. (1958). The nutritive value of the proteins of cocoanut meal and a low-cost protein food containing cocoanut and groundnut meals and bengal gram (*Cicer arietinum*). *Ann. Biochem. exp. Med.* **18**, 175–178.

Krishnamurthy, K., Tasker, P. K., Ramakrishnan, T. N., Rajagopalan, R. and Swaminathan, M. (1960). Studies on the nutritive value of sesame seeds. Part I. The amino acid composition of the proteins and the chemical composition of white and black varieties of sesame seed and meal. *Ann. Biochem. exp. Med.* **20**, 73–76.

Krishnan, T. S. (1950). Thesis approved for the Ph.D. Degree of the University of Madras.

Krishnaswamy, N. (1962). Bajra (*Pennisetum typhoides* S. & H.). Cereal Crop Series No. 2. Indian Council of Agricultural Research.

Krishnaswamy, K. (1971). Plasma "cortisol" levels in pellagrins. *Clinica Chim. Acta* **34**, 415–418.

Krishnaswamy, K. (1972). Mental function in pellagra. *Proc. Nutr. Soc. India* **12**, 34–42.

Krishnaswamy, K. and Gopalan, C. (1971). Effect of isoleucine on skin and electroencephalogram in pellagra. *Lancet* (7735), 1167–1169.

Krishnaswamy, K. and Ramanamurthy, P. S. V. (1970). Mental changes and platelet serotonin in pellagrins. *Clinica Chim. Acta* **27**, 301–304.

Kuhn, R. and Winterstein, A. (1932). *Ber. dtsch. chem. Ges.* **65**, 1742.

Kuiken, K. A. and Lyman, C. M. (1948). Availability of amino acids in some foods. *J. Nutr.* **36**, 359–368.

Kulkarni, L. and Sohonie, K. (1956). Non-protein nitrogen in vegetables. *Indian J. med. Res.* **44**, 511–518.

Kumanov, S. (1958). The digestibility of whole, ground and ground sieved grains for poultry (in German). *Archives Tierernahrung* **8**, 147–158. (*Nutr. Abst. Rev.* 1958, 6279).

Kumar, V., Raheya, P. C. and Chaudhary, M. S. (1973). Agronomic studies on bajra (*Pennisetum typhoides*) under dry land agriculture. 2. Nitrogen content of plants as affected by varieties, doses of nitrogen and methods of weed control. *Ann. Arid Zone* **13**, 155–162.

Kumararaj, S. and Bhide, V. P. (1962). Varietal resistance in bajri (*Pennisetum typhoides*) (Burm.) S. & H. to ergot (*Claviceps microcephale* (Wallr.) (Tul.) *Curr. Sci.* **31**, 76.

Kumaraswamy, K. and Venkataramanan, C. R. (1974). Pot culture studies on the response of CO-10 finger millet to phosphorus fertilization. *Indian J. agric. Sci.* **44**, 50–54.

Kummerow, F. A. (1946a). The composition of sorghum grain oil, *Andropogon sorghum* var. *vulgaris*. *Oil Soap* **23**, 167–170.

Kummerow, F. A. (1946b). The composition of the oil extracted from 14 different varieties of *Andropogon sorghum* var. *vulgaris*. *Oil Soap* **23**, 273–275.

Kundaji, T. R. and Radhakrishna Rao, M. V. (1954). Supplementary nutritive value of some subsidiary cereals. *Curr. Sci.* **23**, 93–94.

Kuppuswamy, S., Joseph, K., Rao, M. N., Rao, G. R., Sankaran, A. N., Swaminathan, M. and Subrahmanyan, V. (1957). Supplementary value of Indian multipurpose food to poor vegetarian diets based on different cereals and millets. *Fd Sci.* **6**, 84–86.

Kuppuswamy, S., Srinivasan, M. and Subrahmanyan, V. (1958). "Proteins in foods". Special Report Series No. 33. Indian Council of Medical Research, New Delhi.

Kurien, P. P. (1967a). Distribution of protein, calcium and phosphorus between the husk and endosperm of rajgira seeds (*Amaranthus paniculatus*). *J. Nutr. Diet.* **4**, 152–154.

Kurien, P. P. (1967b). Nutritive value of refined ragi (*Eleusine coracana*) flour. 1. The effect of feeding poor diets based on whole, refined and composite ragi flours on the growth and availability of calcium in albino rats. *J. Nutr. Diet.* **4**, 96–101.

Kurien, P. P. and Desikachar, H. S. R. (1962). Studies on refining of millet flours. I. Ragi (*Eleusine coracana*). *Fd Sci.* **11**, 136–137.

Kurien, P. P. and Desikachar, H. S. R. (1966). Preparation of a refined white flour from ragi (*Eleusine coracana*) using a laboratory mill. *J. Fd Sci. Technol.* **3**, 56–58.

Kurien, P. P. and Doraiswamy, T. R. (1967). Nutritive value of refined ragi (*Eleusine coracana*) flour. II. Effect of replacing cereal in a poor diet with whole or refined ragi flour on the nutritional status and metabolism of nitrogen, calcium and phosphorus in children (boys). *J. Nutr. Diet.* **4**, 102–109.

Kurien, P. P., Sivaramakrishnan, R., Swaminathan, M., Indiramma, K. and Subrahmanyan, V. (1958). The effect of partial or complete replacement of rice in poor vegetarian diet by ragi (*Eleusine coracana*) on growth and composition of blood and liver of rats. *Ann. Biochem. exp. Med.* **18**, 187–194.

Kurien, P. P., Joseph, K., Swaminathan, M. and Subrahmanyan, V. (1959). The distribution of nitrogen, calcium and phosphorus between the husk and endosperm of ragi (*Eleusine coracana*). *Fd Sci.* **8**, 353–355.

Kurien, P. P., Swaminathan, M. and Subrahmanyan, V. (1960a). The distribution of protein, calcium and phosphorus between the fibrous coat and endosperm of jowar (*Sorghum vulgare*). *Fd Sci.* **9**, 334–335.

Kurien, P. P., Narayana Rao, M., Swaminathan, M. and Subrahmanyan, V. (1960b). Nutritive value of ragi (*Eleusine coracana*) and ragi diets. *Fd Sci.* **9**, 49–54.

Kurien, P. P., Narayana Rao, M., Swaminathan, M. and Subrahmanyan, V. (1960c). Chemical composition and nutritive value of jowar (Kaffircorn-Sorghum vulgare) and jowar diets. *Fd Sci.* **9**, 205–210.

Kurien, P. P., Narayana Rao, M., Swaminathan, M. and Subrahmanyan, V. (1960d). The metabolism of nitrogen, calcium and phosphorus in undernourished children. 6. The effect of partial or complete replacement of rice in poor vegetarian diets by kaffircorn (*Sorghum vulgare*). *Br. J. Nutr.* **14**, 339–346.

Kurien, P. P., Narayana Rao, M., Swaminathan, M., Sankaran, A. N. and Subrahmanyan, V. (1960e). The effect of partial or complete replacement of rice in poor vegetable diet by jowar (*Sorghum vulgare*) on the growth and composition of blood and liver of albino rats. *Ann. Biochem. exp. Med.* **20**, 293–298.

Kurien, P. P., Narayana Rao, M., Swaminathan, M., Sankaran, A. N. and Subrahmanyan, V. (1960f). The metabolism of nitrogen, calcium and phosphorus in children on a jowar (*Sorghum vulgare*) diet. *Ann. Biochem. exp. Med.* **20**, 47–52.

Kurien, P. P., Swaminathan, M. and Subrahmanyan, V. (1961a). The chemical composition and nutritive value of Bajra (*Pennisetum typhoideum*) and Bajra diets. *Fd Sci.* **10**, 3–6.

Kurien, P. P., Swaminathan, M. and Subrahmanyan, V. (1961b). The metabolism of nitrogen, calcium and phosphorus in children on a poor Indian diet based on bajra (*Pennisetum typhoideum*). *Ann. Biochem. exp. Med.* **21**, 41–46.

Kurien, P. P., Swaminathan, M. and Subrahmanyan, V. (1961c). The metabolism of nitrogen, calcium and phosphorus in undernourished children. 7. The effect of partial or complete replacement of rice in poor vegetarian diets by pearl millet (*Pennisetum typhoideum*). *Br. J. Nutr.* **15**, 345–347.

Kurien, P. P., Narayana Rao, M., Swaminathan, M., Sankaran, A. N. and Subrahmanyan, V. (1962). The effect of partial or complete replacement of rice in poor rice diet by bajra (*Pennisetum typhoideum*) on the growth and composition of blood, body and liver of albino rats. *Ann. Biochem. exp. Med.* **22**, 245–248.

Kurien, S., Daniel. V. A., Venkat Rao, S., Swaminathan, M. and Parpia, H. A. B. (1969). Effect of calorie restriction on the supplementary value of protein foods to poor vegetarian diets based on rice and ragi (*Eleusine coracana*). *J. Nutr. Diet.* **6**, 111–114.

Kurien, S., Narayanaswamy, D., Daniel, V. A., Swaminathan, M. and Parpia, H. A. B. (1971a). Supplementary value of pigeonpea (*Cajanus cajan*) and chickpea to poor diets based on kaffircorn and wheat. *Nutr. Rep. Int.* **4**, 229–236.

Kurien, S., Narayanaswamy, D., Daniel, V. A., Swaminathan, M., Rajalakshmi, D. and Parpia, H. A. B. (1971b). Improvement of protein value of poor pearl millet (*Pennisetum typhoideum*) diet by supplementation with limiting amino acids. *Nutr. Rep. Int.* **3**, 357–362.

Lafont, P. and Lafont, J. (1970). Contamination of cereal products and animal feeds by aflatoxin (in French). *Fd Cosmetics Toxic.* **8**, 403–408.

Lakshmaiah, N. and Srikantia, S. G. (1977). Fluoride retention in humans on sorghum and rice based diets. *Indian J. med. Res.* **5**, 543–548.

Lakshmiah, N. and Ramasastri, B. V. (1969). Folic acid content of some Indian foods of plant origin. *Nutr. Diet.* **6**, 200–203.

Lal, B. M., Ahluwalia, M., Rao, C. H. and Sikka, K. C. (1972). Top cross study of grain yield and nutritional quality characters in pearl millet. *Indian J. Genet. Pl. Breed.* **32**, 353–359.

Lal, S. B. (1950). Microbiological assay of amino acids in gram and ragi. *Indian J. med. Res.* **38**, 131–137.

Lamb, M. W., Michie, J. M. and Rivers, J. M. (1966). A comparison of the nutritive value of three sorghum grains with that of wheat. *Cereal Chem.* **43**, 447–456.

Landry, J. and Moureaux, T. (1970). Heterogeneity of the glutelins of maize grain: selective extraction and composition in amino acids of the three fractions isolated (in French). *Bull. Soc. Chim. biol.* Paris **52**, 1021–1037.

Langworthy, C. F. and Holmes, A. D. (1916). Studies on the digestibility of the grain sorghums. Bulletin No. 470. United States Department of Agriculture.

Larkin, R. A., MacMasters, M. M., Cull, I. M., Wolf, M. J. and Rist, C. E. (1952). Preparing serial sections of mature corn and wheat kernels. *Stain Technol.* **27**, 107–112.

Larson, J. C. and Maranville, J. W. (1975). The effect of lodging on yield, protein, and test weight of two grain sorghum hybrids. *In* "Research in the Physiology of yield and management of sorghum in relation to genetic improvement". Annual Report No. 8. pp. 82–89. University of Nebraska, USA. Agency for International Development, Department of State, Washington, DC.

Larson, J. C. and Maranville, J. W. (1976). The effects of lodging on yield, protein, and test weight of grain sorghum. *In* "Research in development of improved high-yielding sorghum cultivars". Annual Report No. 9. pp. 55–60. University of Nebraska, USA. Agency for International Development, Department of State, Washington, DC.

Lawrence, T. L. J. (1967). High level cereal diets for the growing/finishing pig. II. The effect of cereal preparation on the performance of pigs fed diets containing high levels of maize, sorghum and barley. *J. agric. Sci.* **69**, 271–281.

Lawrence, T. L. J. (1968). High level cereal diets for the growing/finishing pig. III. A comparison with a control diet of diets containing high levels of maize, flaked maize, sorghum, wheat and barley. *J. agric. Sci. Camb.* **70**, 287–297.

Lee, H. K. (1970). The chemical score improvement method for mixed grains (in Japanese with English summary). *J. Jap. Soc. Fd Nutr.* **23**, 140–145. (*Fd Sci. Technol. Abstr.* 1971, **3**, M1204).

Leela, R., Daniel, V. A., Rao, S. V., Hariharan, K., Rajalakshmi, D., Swaminathan, M. and Parpia, H. A. B. (1965). Amino acid supplementation of proteins. I. The effect of supplementing ragi (*Eleusine coracana*) and ragi diets with lysine. threonine, and skim milk powder on the nutritive value of their proteins. *J. Nutr. Diet.* **2**, 78–82.

Lema De Rocha, G. (1950). Analyses of adlay (in Portugeuse). *Colheitas Mercados* **6**, 12–13.

Leonard, W. H. and Martin, J. H. (1963). "Cereal Crops". (Sorghum pp. 679–739. Millets pp. 740–769.) The MacMillan Co., New York.

Leung, W. W. and Flores, M. (1961). "INCAP/CNND food composition table for use in Latin America". Interdepartmental Committee on Nutrition for National Defense, National Institutes of Health, Bethseda, Maryland, USA.

Levin, E. and Finn, R. K. (1955). A process for dehydrating and defatting tissues at low temperature. *Chem. Engng. Prog.* **51**, 223–226.

Li, T-W. (1930). Biological value of the proteins of barley, rice, kaoling and millet. *Chin. J. Physiol.* **4**, 49–58. (*Chem. Abstr.* 1930, **24**, 2768).

Liang, G. H. L., Heyne, E. G., Chung, J. H. and Koh, Y. O. (1968). the analysis of heritable variation for three agronomic traits in a six-variety diallel of grain sorghum, *Sorghum vulgare* Pers. *Can. J. Genet. Cytol* **10**, 460–469.

Liang, G. H. L., Overley, C. B. and Casady, A. J. (1969a). Interrelations among agromonic characters in grain sorghum (*Sorghum bicolor* Moench.) *Crop Sci.* **9**, 299–302.

Liang, G. H. L., Walter, T. L., Nickell, C. D. and Koh, Y. O. (1969b). Heritability estimates and interrelationships among agronomic traits in grain sorghum, *Sorghum bicolor* (L.) Moench. *Can. J. Genet. Cytol.* **11**, 199–208.

Liang, Y. T., Morrill, J. L., Anstaett, F. R., Dayton, A. D. and Pfost, H. B. (1970). Effect of pressure, moisture and cooking time on susceptibility of corn or sorghum grain starch to enzymatic attack. *J. Dairy Sci.* **53**. 336–341.

Lieverman, M. C., Craft, C. and Wilcox, M. S. (1959). Effect of chilling on the chlorogenic acid and ascorbic acid content of Puerto Rico sweet potatoes. *Proc. Am. Soc. Hort. Sci.* **74**, 642–648.

Leiner, I. E. (Ed.) (1969). "Toxic constituents of plant foodstuffs". Academic Press, New York.

Lin, G. M. and Pomeranz, Y. (1968). Characterization of water-soluble wheat flour pentosans. *J. Fd Sci.* **33**, 599–606.

Ling, E. R. (1963). "A textbook of dairy chemistry". Vol. 1. 3rd ed. Chapman and Hall Ltd., London.

Linkswiler, H., Geschwender, D., Ellison, J. and Fox, H. M. (1958). Availability to man of amino acids from foods. 1. General methods. *J. Nutr.* **65**, 441–454.

Littler, J. W. (1967). Nitrogen raises grain sorghum yield and protein. *Qd agric. J.* **93**, 193–196.

Lo, T-Y. (1941). Proteins of hog and German millets. *J. Chin. chem. Soc. Peiping.* **8**,170–176.

Lockwood, J. F. (1960). "Flour Milling". 4th ed. Henry Simon Ltd., Stockport, Cheshire, England.

Lombard, J. H. and DeLange, D. J. (1965). The chemical determination of tryptophan in foods and mixed diets. *Anal. Biochem.* **10**, 260–265.

Loomis, R. S. and Williams, W. A. (1963). Maximum crop productivity: an estimate. *Crop Sci.* **3**, 67–72.

Loomis, W. D. (1969). Removal of phenolic compounds during the isolation of plant enzymes. *In* "Methods in Enzymology". (J. M. Lowenstein, Ed.) Vol. 13. pp. 555–563.

Loomis, W. D. and Battaile, J. (1966). Plant phenolic compounds and the isolation of plant enzymes. *Phytochemistry* **5**, 423–438.

Lorenz, K. (1976). Physico-chemical properties of lipid-free cereal starches. *J. Fd Sci.* **41**, 1357–1359.

Lorenz, K. (1977). Proso and foxtail millets—scanning electron microscopy and starch characteristics. *Lebensm.-Wiss. u.-Technol.* **10**, 324–327.

Lorenz, K. and Hinze, G. (1976). Functional characteristics of starches from proso and foxtail millets. *J. agric. Fd Chem.* **24**, 911–914.

Loska, S. J. Jr and Shellenberger, J. A. (1949). Determination of the pentosans of wheat and flour and their relation to mineral matter. *Cereal Chem.* **26**, 129–139.

Lowry, O. H. and Lopez, J. A. (1946). The determination of inorganic phosphate in the presence of labile phosphate esters. *J. biol. Chem.* **162**, 421–428.

Lowry, O. H., Rosebrough, N. J., Farr, A. L. and Randal, R. J. (1951). Protein measurement with the Folin phenol reagent. *J. biol. Chem.* **193**, 265–275.

Loyacano, A. F., Pontif, J. E., Nipper, W. A. and Hembry, F. G. (1973). Bird resistant grain sorghum in steer finishing diets. *J. Anim. Sci.* **36**, 222.

Loyacano, A. F., Nipper, W. A., Pontif, J. E. and Hembry, P. G. (1975). Comparisons of corn and a bird-resistant grain sorghum in beef finishing rations. Bulletin No. 686. Agricultural Experiment Station, Baton Rouge, Louisiana State University, USA.

Loyd, R. C. and Gray, E. (1970). Amount and distribution of hydrocyanic acid potential during the life cycle of plants of three sorghum cultivars. *Agron. J.* **62**, 394–397.

Luce, W. G., Peo, E. R. Jr, and Hudman, D. B. (1967). Availability of niacin in corn and milo for swine. *J. Anim. Sci.* **26**, 76–84.

Luce, W. G., Omtvedt, I. T. and Robbins, B. S. (1972). Comparison of wheat and grain sorghum for growing finishing swine. *J. Anim. Sci.* **35**, 947–952.

Lutrick, M. C. (1970). Preliminary report on the response of grain sorghum to applied nitrogen. *Proc. Soil Crop Sci. Soc. Fla.* **30**, 46–50.

Lyford, S. J. Jr, Smart, W. W. G. Jr and Matrone, G. (1963). Digestibility of the alpha-Cellulose and pentosan components of the cellulosic micelle of fescue and alfalfa. *J. Nutr.* **79**, 105–108.

Lyman, C. M., Kuiken, K. A. and Hale, F. (1956). Essential amino acid content of farm feeds. *J. agric. Fd Chem.* **4**, 1008–1013.

Mabry, T. J., Markham, K. R. and Thomas, M. B. (1970). "The systematic identification of flavonoids". Springer-Verlag, New York.

McCance, R. A. and Widdowson, E. M. (1935). Phytin in human nutrition. *Biochem. J.* **29**, 2694–2699.

McCance, R. A., Widdowson, E. M. and Shackleton, L. (1936). *In* Special Report Series No. 213. p. 17. Medical Research Council, H. M. Stationery Office, London.

McCarthy, T. E. and Paille, M. M. Sr. (1959). A rapid determination of methionine in crude protein. *Biochem. Biophys. Res. Commun.* **1**, 29.

McClure, F. J. (1960). Millet (cattail and foxtail) as the cariogenic component of experimental rat diets. *J. Dent. Res.* **39**, 1172–1176.

McClymont, G. L. and Duncan, D. C. (1952). Studies on nutrition of poultry, III. Toxicity of grain sorghum for chickens. *Aust. vet. J.* **28**, 229–233.

McCollough, R. L. (1972). Nutritive value of eight hybrid sorghum grains and three hybrid corns compared in all-concentrate rations. Part 1. Hybrid sorghum and corn characteristics and methods used to nutritionally evaluate them. Bulletin 557, 15–20. Kansas Agricultural Experiment Station, Kansas, USA.

McCollough, R. L. and Brent, B. E. (1972). Nutritive value of eight hybrid sorghum grains and three hybrid corns compared in all-concentrate rations. Part 3. Digestibility of eight hybrid sorghum grains and three hybrid corns. Bulletin 557, 27–31. Kansas Agricultural Experiment Station, Kansas, USA.

McCollough, R. L. and Schalles, R. R. (1972). Two-year summary. Four hybrid sorghum grains fed in all-concentrate rations to steers. Bulletin 557, 32–36. Kansas Agricultural Experiment Station, Kansas, USA.

McCollough, R. L., Riley, J. G., Drake, C. L. and Roth, G. M. (1972a). Feedlot performance of nine hybrid sorghum grains fed to steers winter, 1971–1972. Bulletin 557, 37–42. Kansas Agricultural Experiment Station, Kansas, USA.

McCollough, R. L., Drake, C. L. and Roth, G. M. (1972b). Nutritive value of eight hybrid sorghum grains and three hybrid corns compared in all-concentrate rations. Part 2. Feedlot performance of eight hybrid sorghum grains and three hybrid corns. Bulletin 557, 21–26. Kansas Agricultural Experiment Station, Kansas, USA.

McCollum, E. V., Simmonds, N. and Parsons, H. T. (1919). Supplementary relationships between the proteins of certain seeds. *J. biol. Chem.* **37**, 155–178.

McCollum, E. V., Simmonds, N. and Parsons, H. T. (1921). Supplementary protein value in foods. III. The supplementary dietary relations between the proteins of the cereal grains and the potato. *J. biol. Chem.* **47**, 175–206.

McGinty, D. D. (1969). Variation in digestibility of sorghum varieties. *In* "Proceedings of the Sixth Biennial International Grain Sorghum Research and Utilization Conference". pp. 20–23. Amarillo, Texas, USA.

McGinty, D. D. and Riggs, J. D. (1968). Variation in digestibility of sorghum grain varieties. *J. Anim. Sci.* **27**, 1170. (Abstr.).

MacKenzie, D. H., Basinski, J. J. and Parbery, D. B. (1970). The effect of varieties, nitrogen and stubble treatments on successive cycles of grain and forage sorghums in the Ord River Valley. *Aust. J. exp. Agric, Anim. Husb.* **10**, 111–117.

McKenzie, H. A. and Wallace, H. S. (1954). The Kjeldahl determination of nitrogen—a critical study of digestion conditions—temperature: catalyst and oxidising agent. *Aust. J. Chem.* **7**, 55–70.

McLaughlan, J. M. and Campbell, J. A. (1974). Methodology for evaluation of plant proteins for human use. *In* "Nutritive value of triticale protein". (J. H. Hulse and E. M. Laing). IDRC-021e. pp. 27–39. International Development Research Centre, Box 8500, Ottawa, Canada K1G 3H9.

MacMasters, M. M. (1964). Microscopic techniques for determining starch granule properties. *In* "Methods in Carbohydrate Chemistry". (R. L. Whistler, R. J. Smith, J. N. BeMiller and M. L. Wolfrom, Eds) pp. 233–240. Academic Press, New York.

MacMasters, M. M. and Hilbert, G. E. (1944). Glutinous corn and sorghum starches. *Ind. Engng Chem.* **36**, 958–965.

McMillian, W. W. Wiseman, B. R., Burns, R. E., Harris, H. B. and Greene, G. L. (1972). Bird resistance in diverse germplasm of sorghum. *Agron. J.* **64**, 821–822.

McNeill, J. W., Potter, G. D. and Riggs, J. K. (1971). Ruminal and postruminal carbohydrate utilization in steers fed processed sorghum grain. *J. Anim. Sci.* **33**, 1371–1374.

McNeill, J. W., Potter, G. D., Riggs, J. K. and Rooney, L. W. (1975). Chemical and physical properties of processed sorghum grain carbohydrates. *J. Anim. Sci.* **40**, 335–341.

McPherson, H. T. (1942). Modified procedures for the colorimetric estimation of arginine and histidine. *Biochem. J.* **36**, 59–63.

Madhavan, T. V., Belavady, B. and Gopalan, C. (1968). Pathology of canine black tongue. *J. Path. Bact.* **95**, 259–263.

Maes, E. (1962). Progressive extraction of proteins. *Nature, Lond.* **193,** 880.

Mahadevappa, M. (1967). Investigations on the inheritance of protein content in pearl millet (*Pennisetum typhoides* Stapf. and Hubb.). *Curr. Sci.* **36**, 186–188.

Mahudeswaran, K. and Ayyamperumal, A. (1970). A note on the vitamin content of ragi. Madras *agric. J.* **57**. 289–290.

Mahudeswaran, K., Natarajan, M. and Ramachandran, A. (1966). White ragi—a source for more protein. *Madras agric. J.* **53**, 179–180.

Majmudar, J. V. and Khunte, P. D. (1955). Improved strains of Kodra (*Paspalum scrobiculatum* Linn.) for Gujarat. *Poona agric. Coll. Mag.* **46**, 183–185.

Majumder, A. K., Nandi, B. K., Subramanian, N. and Chatterjee, I. B. (1975).

Growth and ascorbic acid metabolism in rats and guinea pigs fed cereal diets. *J. Nutr.* **105**, 233–239.

Mali, P. C. and Gupta, Y. P. (1974). Chemical composition and protein quality of improved Indian varieties of *Sorghum vulgare* Pers. *Indian J. Nutr. Diet.* **11**, 289–295.

Mallanna, K. N. and Rajashekara, B. G. (1969). Hamsa—a protein-rich, high-yielding white ragi. *Mysore J. agric. Sci.* **3**, 1–6.

Malm, N. R. (1968). Exotic germplasm use in grain sorghum improvement. *Crop Sci.* **8**, 295–298.

Malm, N. R. and Rachie, K. O. (1971). The Setaria millets. A review of the World literature. Experiment Station, College of Agriculture, University of Nebraska, USA.

Mangay, A. S., Pearson, W. N. and Darby, W. J. (1957). Millet (*Setaria italica*): its amino acid and niacin content and supplementary nutritive value for corn (maize). *J. Nutr.* **62**, 377–393.

Mann, F. G. and Saunder, B. C. (1960). "Practical Organic Chemistry". 4th ed. Longmans, Green & Co. Ltd., London.

Mann, H. H. (1950). World cereals today: the millets. *Wld Crops* **2**, 97–101.

Mann, H. S., Singh, P. and Malhotra, S. P. (1976). Pearl millet in India and in arid zones. *Ann. Arid Zone* **15**, 53–62.

Mantle, P. G. (1969). Inhibition of lactation in mice following feeding with ergot sclerotia (*Claviceps fusiformia* (Loveless)) from the bulrush millet (*Pennisetum typhoides* Stapf and Hubb.) and an alkaloid component. *Proc. R. Soc. Series B. Biol. Sci.* **170**, 423. (*Chem. Abstr.* **72**, 98640t).

Maranville, J. W. (1970). Improvement of protein content in grain sorghum. *In* 7th Biennial Grain Sorghum Production, Research, and Utilization Conference. pp. 87–90.

Markuze, Z. (1937). Biological value of the proteins of certain cereals. *Biochem. J.* **31**, 1973–1977.

Martearena, O. F. and Perez, E. H. (1977). Protein, tryptophan and lysine contents in experimental sorghum varieties (*Sorghum vulgare*). *Acta Cient. venez.* **28**, 150.

Martin, J. H. (1970). History and classification of sorghum *Sorghum bicolor* (Linn.) Moench. *In* "Sorghum production and utilization". (J. S. Wall and W. M. Ross, Eds) pp. 1–27. AVI Publishing Co. Inc., Westport, Connecticut, USA.

Martin-Tanguy, J., Vermorel, M., Lenoble, M. and Martin, C. (1976). Importance of sorghum seed tannins in protein conversion in growing rats (in French). *Ann. Biol. anim. Biochim. Biophys.* **16**, 879–890.

Marusev, A. I. and Il'in, V. A. (1965). The best millet cultivars in the South East (in Russian). *Selekts. Semenov.* (**5**), 7–9. (*Pl. Breed. Abst.* 1966, No. 3949).

Mason, E. D. (1944). *In* Report of the Scientific Advisory Board. p. 61. Indian Research Fund Association, New Delhi.

Mason, E. D., Devadas, R. and Frimodt-Moller, J. (1946). The effects on growth in rats of butter and ragi (*Eleusine coracana*), separately and combined, as supplements to the poor rice diet of South India. *Indian J. med. Res.* **34**, 45–48.

Mather, K. (1949). "Biometrical Genetics". Methuen & Co., London.

Matsuo, R. R. and Irvine, G. N. (1967). Macaroni browness. *Cereal Chem.* **44**, 78–85.

Matsuo, R. R. and Irvine, G. N. (1969). Spaghetti tenderness testing apparatus. *Cereal Chem.* **46**, 1–6.

Matsuo, R. R. and Irvine, G. N. (1974). Relationship between the GRL spaghetti tenderness tester and sensory testing of cooked spaghetti. *J. Inst. Con. Sci. Tech. Aliment.* 7, 155–156.

Matz, S. A. (1969). Millet, wild rice, adlay, and rice grass. *In* "Cereal Science". (S. A. Matz, Ed.) pp. 224–236. AVI Publishing Co. Inc., Westport, Connecticut, USA.

Maxson, E. D. and Rooney, L. W. (1972a). Evaluation of methods for tannin analysis in sorghum grain. *Cereal Chem.* **49**, 719–729.

Maxson, E. D. and Rooney, L. W. (1972b). Two methods of tannin analysis for *Sorghum bicolor* (L.) Moench grain. *Crop Sci.* **12**, 253–254.

Maxson, E. D., Fryar, W. B., Rooney, L. W. and Krishnaprasad, M. N. (1971). Milling properties of sorghum grain with different proportions of corneous to floury endosperm. *Cereal Chem.* **48**, 478–490.

Maxson, E. D., Clark, L. E., Rooney, L. W. and Johnson, J. W. (1972). Factors affecting tannin content of sorghum grain as determined by two methods of tannin analysis. *Crop Sci.* **12**, 233–235.

Maxson, E. D., Rooney, L. W., Lewis, R. W. Clark, L. E. and Johnson, J. W. (1973). The relationship between tannin content, enzyme inhibition, rat performance, and characteristics of sorghum grain. *Nutr. Rep. Int.* **8**, 145–152.

Maxson, W. E. and Shirley, R. L. (1973). Milo sorghum diets with added sulfate fed to rats. *Q. Jl. Fla Acad. Sci.* **36**, 159–163.

Maxson, W. E., Shirley, R. L., Bertrand, J. E. and Palmer, A. Z. (1973). Energy values of corn, bird-resistant and non bird-resistant sorghum grain in rations fed to steers. *J. Anim. Sci.* **37**, 1451–1457.

May, M. A. and Nelson, T. S. (1972). Correlation of nitrogen to the heat of combustion of rat urine. *J. Anim. Sci.* **35**, 38–40.

May, M. A. and Nelson, T. S. (1973). Digestible and metabolizable energy content of varieties of milo for rats. *J. Anim. Sci.* **36**, 874–876.

Maynard, L. A. and Loosli, J. K. (1962). "Animal Nutrition". 5th ed. McGraw Hill, New York, N.Y.

Meadows, D. G., Tanksley, T. D. Jr and Hesby, J. H. (1974). Dryland corn vs three sorghums with different endosperm types for growing finishing swine. *J. Anim Sci.* **39**, 185–186.

Mehansho, H. and Besrat, A. (1974). Improvement of protein quality and quality in Ethiopian sorghum grain. (Unpublished report).

Mehra, K. L. (1963). Differentiation of the cultivated and wild *Eleusine* species. *Phyton* **21**, 189–198.

Mejbaum-Katzenellenbogen, W.. and Kudrewiczhubica, Z. (1966). Application of urea, ferric ammonium sulfate and casein for determination of tanning substances in plants. *Acta Biochim. Pol.* **13**, 57.

Melnic, D., Oser, B. L. and Weiss, S. (1946). Rate of enzymic digestion of proteins as a factor in nutrition. *Science* **103**, 326–329.

Mengesha, M. H. (1966). Chemical composition of teff (*Eragrostis tef*) compared with that of wheat, barley, and grain sorghum. *Econ. Bot.* **20**, 268–273.

Mengesha, M. H., Pickett, R. C. and Davis, R. L. (1965). Genetic variability and inter-relationship of characters in teff, *Eragrostis tef* (Zucc.) Trotter. *Crop Sci.* **5**, 155–157.

Mertz, E. T. (1969). Amino acid and protein requirements of fish. *In* "Fish in Research". (O. W. Neuhaus and J. E. Halver, Eds) pp. 233–244. Academic Press Inc., New York.

Mertz, E. T. (1974). Breeding for improved nutritional value. *Cereal Sci. Today* **19**, 385.

Mertz, E. T. and Bressani, R. (1957). Studies on corn proteins. I. A new method of extraction. *Cereal Chem.* **34**, 63–69.

Mertz, E. T., Bates, L. S. and Nelson, O. E. (1964). Mutant gene that changes protein and increases lysine content of maize endosperm. *Science* **145**, 279–280.

Michael, G. (1963). Influence of manuring on the protein quality and protein fractions of food plants (in German). *Qualitas Pl. Mater. veg.* **10**, 248–265.

Miche, J. C., Alary, R., Jeanjean, M. F. and Abecassis, J. (1977). Potential use of sorghum grains in pasta processing. Proceedings, Symposium, "Sorghum and millets as human food". International Association for Cereal Chemistry, Vienna, 11 May 1976. Tropical Products Institute, London.

Miller, B. S. and Kneen, E. (1947). The amylase inhibitor of Leoti sorghum. *Archs. Biochem. Biophys.* **15**, 251–264.

Miller, D. F. (1958). Composition of cereal grains and forages. Publication No. 585. National Academy of Sciences, National Research Council, Washington, DC.

Miller, D. S. and Bender, A. E. (1955). The determination of the net utilization of proteins by a shortened method. *Br. J. Nutr.* **9**, 382–388.

Miller, D. S. and Mumford, P. M. (1969). Communication to the Editor. (Mechanically developed doughs from composite flours.) *In* "Composite flour programme". Documentation Package Vol. 1. 2nd ed. revised. pp. 130–131. Food and Agricultural Industries Service, Agricultural Services Division, FAO, Rome.

Miller, D. S. and Payne, P. R. (1961). Problems in the prediction of protein values of diets. The influence of protein concentration. *Br. J. Nutr.* **15**, 11–19.

Miller, E. L. (1967). Determination of the tryptophan content of feedingstuffs with particular reference to cereals. *J. Sci. Fd Agric.* **18**, 381–386.

Miller, F. R., Lowrey, R. S., Monson, W. G., Burton, G. W. and Cruzado, H. J. (1972). Estimates of dry matter digestibility differences in grain of some *Sorghum bicolor* (L.) Moench varieties. *Crop Sci.* **12**, 563–566.

Miller, J. D., Deyoe, C. W., Walter, T. I. and Smith, F. W. (1964). Variations in protein levels in Kansas sorghum grain. *Agron. J.* **56**, 302–304.

Miller, O. H. and Burns, E. E. (1970). Starch characteristics of selected grain sorghums as related to human foods. *J. Fd Sci.* **35**, 666–668.

Ministry of Agriculture, Fisheries and Food. (1976). "Manual of Nutrition". 8th edition. H. M. Stationery Office, London.

Misra, K. and Seshadri, T. R. (1967). Chemical components of *Sorghum durra* glumes. *Indian J. Chem.* **5**, 409–412.

Misra, P. S., Mertz, E. T. and Glover, D. V. (1975). Studies on corn proteins. VI. Endosperm protein changes in single and double endosperm mutants of maize. *Cereal Chem.* **52**, 161–166.

Misra, R., Misra, U. K. and Venkitasubramanian, T. A. (1972/73). Brain lipid of rats fed millet (*Sorghum vulgare*) proteins. *Biochem. exp. Biol.* **10**, 315–322.

Misra, R., Misra, U. K. and Venkitasubramanian, T. A. (1973a). Plasma lipids of rats fed millet (*Sorghum vulgare*) at various protein levels. *Agric. biol. Chem.* **37**, 55–65.

Misra, R., Misra, U. K. and Venkitasubramanian, T. A. (1973b). Effect of feeding millet (*Sorghum vulgarie*) protein on growth, nucleic acids and proteins of liver and plasma of rats. *Agric. biol. Chem.* **37**, 711–717.

Misra, R., Misra, U. K. and Venkitasubramanian, T. A. (1974a). Lipogenesis from acetate-1-^{14}C in liver of rats fed millet (*Sorghum vulgarie*) at different protein levels. *Nutr. Rep. Int.* **9**, 441–452.

Misra, R., Misra, U.K. and Venkitasubramanian, T. A. (1974b). Lipogenesis in adipose tissue of rats fed millet (*Sorghum vulgarie*) at different protein levels. *Nutr. Rep. Int.* **10**, 55–59.

Misra, R., Misra, U. K. and Venkitasubramanian, T. A. (1974c). Incorporation of palmitate-1-^{14}C into liver lipids of rats fed millet (*Sorghum vulgarie*) at different protein levels. *Nutr. Metab.* **17**, 9–19.

Misra, R., Misra, U. K. and Venkitasubramanian, T. A. (1974d). Effect of feeding defatted millet (*Sorghum vulgarie*) flour at different protein levels on phospholipids of rat liver. *Agric. Biol. Chem.* **38**, 2059–2063.

Misra, R., Misra, U. K. and Venkitsubramanaian, T. A. (1974e). Hepatic subcellular fractions lipids in rats fed millet (*Sorghum vulgarie*) at different protein levels. *Agric. biol. Chem.* **38**, 2113–2123.

Misra, R., Misra, U. K. and Venkitasubramanian, T. A. (1974f). Effect of feeding millet (*Sorghum vulgarie*) protein at various protein levels on adrenal lipids of rats. *Agric. biol. Chem.* **38**, 1657–1660.

Misra, R., Misra, U. K. and Venkitasubramanian, T. A. (1974g). Effect of feeding millet (*Sorghum vulgarie*) protein on heart lipids of rats. *Biochem exp. Biol.* **11**, 53–58.

Misra, R., Misra, U. K. and Venkitasubramanian, T. A. (1974/1975a). Aorta lipids of rats fed millet (*Sorghum vulgarie*) at different protein levels. *Biochem. exp. Biol.* **11**, 183–190.

Misra, R., Misra, U. K. and Venkitasubramanian, T. A. (1974/75b). Incorporation of [U-^{14}C] glucose into liver lipids of rats fed millet (*Sorghum vulgarie*) at different protein concentrations. *Biochem. exp. Biol.* **11**, 343–350.

Mitchell, H. H. (1923/24). A method of determining the biological value of protein. *J. biol. Chem.* **58**, 873–903.

Mitchell, H. H. and Carman, G. C. (1926). The biological value of the nitrogen of mixtures of patent white flour and animal foods. *J. biol. Chem.* **68**, 183–215.

Mitchell, J. H., Kraybill, H. R. and Scheile, F. P. (1943). Quantitative spectral analysis of fats. *Ind. Engng. Chem. Anal. Ed.* **15**, 1–3.

Mitchell, W. D. and Eastwood, M. A. (1976). Dietary fiber and colon function. *In* "Fiber in Human Nutrition". (G. A. Spiller and R. J. Amen, Eds) Chapter 8. Plenum Press, New York.

Mitjavila, S. and Derache, R. (1968). Effect of tannin on the steatogenic effect of ethanol on the liver of the rat (in French). *Fd Cosmetics Toxic.* **6**, 39–44.

Mitjavila, S., Carrera, G. and Derache, R. (1971). Effect of tannic acid on growth, body composition and biological utilization of food in the rat (in French). *Annls Nutr. Aliment.* **25**, 297–310.

Mitra, K., Verma, S. K. and Ahmed, S. (1948). Investigations on biological value of cereal mixtures in a rice eater's diet by human feeding trials. *Indian J. med. Res.* **36**, 261–269.

Mohan, D. P. (1975). "Chemically induced high-lysine mutants in Sorghum bicolor (L.) Moench." Ph.D. Thesis, Purdue University, Lafayette, Indiana, USA.

Mohan, D. P. and Axtell, J. D. (1974). Chemically induced high lysine mutants in *Sorghum bicolor* (L.) Moench. *Agron. Abstr.* **66**.

Mohan, V. S. and Deosthale, Y. G. (1969). Varietal difference in protein quality of cereals and millets. *Proc. Nutr. Soc. India* **7**, 23–30.

Monson, W. G., Lowrey, R. S. and Forbes, I. Jr (1969). In vivo nylon bag vs two-stage in vitro digestion. Comparison of two techniques for estimating dry matter digestibility of forages. *Agron. J.* **61**, 587–589.

Montgomery, E. M. and Van Assen, E. (1968). Kaffircorn malting and brewing studies. XX. Isolation of starch by laboratory wet-milling of Kaffircorn and malted Kaffircorn after a neutral chemical steep. *Staerke* **20**, 60–63.

Montgomery, E. M., Sexson, K. R. and Dimler, R. J. (1964). Isolation of starch from corn and amylo-maize after a neutral chemical steep. *Staerke* 314–320.

Moore, S. (1963). On the determination of cystine as cysteic acid. *J. biol. Chem.* **238**, 235–237.

Moore, S. and Stein, W. H. (1948). Photometric ninhydrin method for use in the chromatography of amino acids. *J. biol. Chem.* **176**, 367–388.

Moore, S. and Stein, W. H. (1951). Chromatography of amino acids on sulfonated polystyrene resins. *J. biol. Chem.* **192**, 663–681.

Moore, S. and Stein, W. H. (1954a). Procedures for the chromatographic determination of amino acids on 4 per cent cross-linked sulfonated polystyrene resins. *J. biol. Chem.* **211**, 892–906.

Moore, S. and Stein, W. H. (1954b). A modified ninhydrin reagent for the photometric determination of amino acids and related compounds. *J. biol. Chem.* **211**, 907–913.

Moore, S., Spackman, D. H. and Stein, W. M. (1958). Chromatography of amino acids on suphonated polystyrene resins. An improved system. *Analyt. Chem.* **30**, 1185–1190.

Moreland, T. A., Topps, J. H. and Michie, W. (1975). Level of dietary protein and toxicity of tannic acid in young turkeys. *Proc. Nutr. Soc.* **34**, 2A.

Morris, J. G. (1970). the survival feeding of pregnant and lactating beef cows on all-sorghum grain rations: the effects of two levels of grain and early weaning of the calves. *J. agric. Sci. Camb.* **75**, 479–484.

Morrison, J. E. and Pirie, N. W. (1961). The large-scale production of protein from leaf extracts. *J. Sci. Fd Agric.* **12**, 1–5.

Morton, J. F. (1970). Tentative correlations of plant usage and esophageal cancer zones. *Econ. Bot.* **24**, 217–226.

Mosquera, F., Santoro, R., Saralegui, W. and Azzarini, A. (1966). The effect of different percentages of milo grain on chick growth during the first twelve weeks of age (in Spanish). *Bln Fac. Agron. Univ. Repub. Montev.* **86**, 3–18.

Mosse, J. (1966). Alcohol soluble proteins of cereal grains. *Fed. Proc.* **25**, 1663–1669.

Mukerji, B. and De, N. K. (1944). A method for the assay of individual ergot sclerotium. *Curr. Sci.* **13**, 128.

Mukherjee, R. and Parthasarathy, D. (1948a). The digestible nutrients of certain cereal grains as determined by experiments on Indian fowls. *Indian J. vet. Sci.* **18**, 41–45. (*Chem. Abstr.* 1949, **43**, 6292).

Mukherjee, R. and Parthasarathy, D. (1948b). Studies of the biological values of the proteins of certain poultry feeds. *Indian J. vet. Sci.* **18**, 51–56. (*Chem. Abstr.* 1949, **43**, 6330g).

Mukuru, Z., Nyquist, W. E. and Axtell, J. D. (1973). Estimation of genetic components, heritability and genetic advance of protein, lysine and oil content in grain sorghum. *In* "Inheritance and Improvement of Protein Quality and Content in Sorghum"—Research progress Report No. 10. pp. 82–84. Department of Agronomy, Agricultural Experiment Station, Purdue University, Lafayette, Indiana; Agency for International Development, Department of State, Washington, DC.

Muller, H. G. (1970). Traditional cereal processing in Nigeria and Ghana. *Ghana J. Agric.* **3**, 187–195.

Munsell, H. E., Williams, L. O., Guild, L. P., Troescher, C. B., Nightingale, G. and Harris, R. S. (1949). Composition of food plants of Central America. I. Honduras. *Fd Res.* **14**, 144–164.

Munsell, H. E., Williams, L. O., Guild, L. P., Kelley, L. T., Harris, R. S., Troescher, C. B., Nightingale, G. and McNally, A. M. (1950a). Composition of food plants of Central America. IV. El Salvador. *Fd Res.* **15**, 263–296.

Munsell, H. E., Williams, L. O., Guild, L. P., Kelley, L. T., Harris, R. S., Troescher, C. B., Nightingale, G. and McNally, A. M. (1950b). Composition of food plants of Central America. VI. Costa Rica. *Fd Res.* **15**, 379–404.

Munsell, H. E., Williams, L. O., Kelley, L. P., Harris, R. S., Troescher, C. B., Nightingale, G. and McNally, A. M. (1950c). Composition of food plants of Central America. VII. Honduras. *Fd Res.* **15**, 421–438.

Munsell, H. E., Williams, L. O., Guild, L. P., Troescher, C. B., Nightingale, G. and Harris, R. S. (1950d). Composition of food plants of Central America. II. Guatemala. *Fd Res.* **15**, 16–33.

Muramatsu, K., Takeuchi, H., Funaki, Y. and Chisuwa, A. (1972). Influence of protein source on growth-depressing effect of excess amino acids in young rats. *Agric. biol. Chem.* **36**, 1269–1276.

Murthy, H. B. N., Swaminathan, M. and Subrahmanyan, V. (1950). Supplementary value of groundnut cake to tapioca and sweet potato. *J. Scient. ind. Res.* **9B**, 173–176.

Murthy, H. B. N., Swaminathan, M. and Subrahmanyan, V. (1954). Effects of partial replacement of rice in a rice diet by tapioca flour on the metabolism of nitrogen, calcium and phosphorus in adult human beings. *Br. J. Nutr.* **8**, 11–16.

Murthy, H. B. N., Reddy, S. K., Swaminathan, M. and Subrahmanyan, V. (1955). The metabolism of nitrogen, calcium and phosphorus in under-nourished children. 1. Adaptation to low intake of calories, protein, calcium and phosphorus. *Br. J. Nutr.* **9**, 203–209.

Murthy Reddy, K. B. S. and Narayana, R. (1973). Ragi malt, a growth factor, for *Vigna sinensis* Endl. callus tissues. *Pl. Cell Physiol. Tokyo.* **14**, 803–814.

Murty, B. R. (1969). New hybrids of bajra. *Indian Fmg* **19**, 13–15.

Murty, B. R., Arunachalam, V. and Saxena, M. B. L. (1967a). Classification and catalogue of a world collection of sorghum. *Indian J. Genet. Pl. Breed.* **27** (Special Number), 1–312.

Murty, B. R., Upadhyay, M. K. and Manchanda, P. L. (1967b). Classification and cataloguing of a world collection of genetic stocks of Pennisetum. *Indian J. Genet. Pl. Breed.* **27** (Special Number).

Nagarajan, V. (1974). Improved agricultural practices for better nutrition. *Swasth Hind* **18**, 120–122.

Nagpal. M. L. and Bhatia, I. S. (1971). Tryptophan content of some Indian foods and feeds. *Indian J. Nutr. Diet.* **8**, 183–185.

Nagy, D., Weidlein, W. and Hixon, R. M. (1941). Factors affecting the solubility of corn proteins. *Cereal Chem.* **18**, 514–523.

Naik, M. S. (1968). Lysine and tryptophan in protein fractions of sorghum. *Indian J. Genet. Pl. Breed.* **28**, 142–146.

Naik, M. S. and Abhyankar, V. S. (1955). Nutritive value of improved strain of jowar (*Sorghum vulgare* Pers.) *Poona agric. Coll. Mag.* **46**, 130–137.

Nair, T. V. R., Sinha, S. K. and Abrol, Y. P. (1974). Protease and nitrate reductase activity in relation to protein content in two sorghum hybrids CSH-2 and CSH-3. *Indian J. Genet. Pl. Breed.* **34**A, 1062–1066.

Nakagawa, I., Takahasi, T., Suzuki, T. and Kobayashi, K. (1962). Amino acid requirements of children: minimal needs of threonine, valine and phenylalanine based on nitrogen balance method. *J. Nutr.* **77**, 61–68.

Nanda, G. S. and Phul, P. S. (1974). Genetic analysis of yield factors and protein content in pearl millet. *Genet. agr.* **28**, 150–161.

Nanda, G. S. and Rao, N. G. P. (1975a). Genetic analysis of some exotic × Indian crosses in sorghum. 9. Nutritional quality and its association with yield. *Indian J. Genet. Pl. Breed.* **35**, 131–135.

Nanda, G. S. and Rao, N. G. P. (1975b). Gene action for content of amino acids in grain sorghum. *Indian J. Genet. Pl. Breed.* **35**, 395–398.

Nandi, B. K., Subramanian, N., Majumder, A. K. and Chatterjee, I. B. (1974). Effect of ascorbic acid on detoxification of histamine under stress conditions. *Biochem. Pharmacol.* **23**, 643–647.

Naphade, D. S. and Ghawghawe, B. G. (1971). The effects of gamma irradiation on an increase in the relative content of protein in sorghum varieties. *Madras agric. J.* **58**, 429–431.

Narayana, N. and Norris, R. V. (1928). Studies in the protein of Indian foodstuffs. Part 1. The proteins of ragi (*Eleusine coracana*). Eleusin the alcohol-soluble protein. *J. Indian Inst. Sci.* **11** A, 91–95.

Narayana, R. and Murthy Reddy, K. B. S. (1973). Free amino acid content of Aruna ragi malt extract. *Curr. Sci.* **42**, 757–758.

Narayanamurti, D. and Aiyar, C. V. R. (1930). Vegetable proteins. II. Typhoidin—the alcohol-soluble protein of *Pennisetum typhoideum*. *J. Indian Chem. Soc.* **7**, 945–952.

Narayana Rao, M., Sur, G., Swaminathan, M. and Subrahamanyan, V. (1958). Effect of milling on the nutritive value of Jowar (*Sorghum vulgare*). *Fd Sci. Mysore* **7**, 254–255.

Narayana Rao, M., Kurien, P. P., Swaminathan, M. and Subrahmanyan, V. (1961). Nutritive value of certain cereals and cereal diets consumed in India. *Fd Sci. Mysore* **10**, 163–175.

Narayanaswamy, D., Desai, B. L. M., Daniel, V. A. Kurien, S., Swaminathan, M. and Parpia, H. A. B. (1970a). Improvement of protein value of poor kaffircorn (*Sorghum vulgare*) diet by supplementation with limiting amino acids. *Nutr. Rep. Int.* **1**, 297–303.

Narayanaswamy, D., Kurien, S., Desai, B. L. M., Daniel, V. A., Venkat Rao, S., Swaminathan, M. and Parpia, H. A. B. (1970b). Supplementary value of yeast grown on petroleum hydrocarbons to poor diets based on rice and ragi (*Eleusine coracana*). *Nutr. Rep. Int.* **1**, 305–312.

Narayanswamy, D., Kurien, S., Daniel, V. A., Swaminathan, M. and Parpia, H. A. B. (1971). Supplementary value of a low cost protein food based on a blend of wheat and soybean flours to poor Indian diets based on wheat and kaffir corn. *Indian J. Nutr. Diet.* **8**, 309–314.

Narayanaswamy, D., Kurien, S., Daniel, V. A., Swaminathan, M. and Parpia, H. A. B. (1972a). Improvement of poor wheat and kaffir corn (*Sorghum vulgare*) diets by supplementation with a low cost protein food (Bal-ahar) based on a blend of wheat, peanut and soybean flours. *Nutr. Rep. Int.* **6**, 157–164.

Narayanaswamy, D., Kurien, S., Daniel, V. A., Swaminathan, M. and Parpia, H. A. B. (1972b). Effect of incorporation of a low cost protein food (Bal-ahar) in poor rice and ragi diets on their overall nutritive value. *Indian J. Nutr. Diet.* **9**, 73–77.

Narayanaswamy, D., Kurien, S., Danial, V. A., Venkat Rao, S., Swaminathan, M.

and Parpia, H. A. B. (1972c). Supplementary value of yeast grown on petroleum hydrocarbons to poor diets based on kaffir corn and wheat. *Pl. Fd Hum. Nutr.* **2**, 167–170.

Narayanaswamy, D., Daniel, V. A., Kurien, S., Rajalakshmi, D. and Swaminathan, M. (1974). Supplementary value of protein enriched cereal foods containing varying amounts of proteins to poor rice and ragi diets. *Indian J. Nutr. Diet.* **11**, 72–76.

Narayanaswamy, D., Daniel, V. A., Kurien, S. and Swaminathan, M. (1975). Supplementary relations between the proteins of cowpea (*Vigna sinensis*) ragi and jowar. *Baroda J. Nutr.* **2**, 31–36.

Nasinec, J., Juza, J. and Cerny, J. (1966). A study of the tannin content in the grain of breeding material of grain sorghum (*Sorghum vulgare* Pers.) (in Czech.) *Genetika Slechteni* **2**, 65–74.

National Academy of Sciences. (1971). Atlas of nutritional data on United States and Canadian feeds. National Academy of Sciences, Washington, DC.

National Academy of Sciences—National Research Council. (1963a). Evaluation of protein quality. "Report of the International Conference Committee on Protein Malnutrition." Publication 1100. National Research Council, Washington, DC.

National Academy of Sciences—National Research Council. (1963b). Committee on animal nutrition. "Nutrient requirements of laboratory animals." Publication 990. National Research Council, Washington, DC.

National Academy of Sciences—National Research Council. (1966). "Nutrient requirements of domestic animals." Biological interrelationships and glossary of energy. Publication 1411. National Research Council, Washington, DC.

National Institute of Nutrition. (Indian). (1974). Annual Report.

National Research Council. (1961). "Progress in meeting protein needs of infants and pre-school children." Publication 843. National Research Council, Washington, DC.

National Research Council. (1964). Recommended dietary allowances. 6th ed. Food and Nutrition Board, National Academy of Sciences, National Research Council, Washington, DC.

National Research Council. (1966). Nutrient requirements for domestic animals. Nutrient requirements for poultry. National Research Council, Washington, DC.

National Research Council. (1968). "Nutrient requirements of domestic animals." No. 2. Nutrient requirements of swine. Publication 1599. National Research Council, Washington, DC.

National Research Council. (1970). "Nutrient requirements of domestic animals." No. 4. Nutrient requirements of beef cattle. National Research Council, Washington, DC.

Nawar, I. A., Clark, H. E., Pickett, R. C. and Hegsted, D. M. (1970. Protein quality of selected lines of *Sorghum vulgare* for the growing rat. *Nutr. Rep. Int.* **1**, 75–81.

Nechaev, A. P., Busareva, N. N., Denisenko, Ya. I. and Kuznetsov, D. I. (1973). Study of changes in fatty acid composition of lipids of millet during storage (in Russian). *Prikl. Biokhim. Mikrobiol.* **9**, 733–736.

Nelson, L. R. and Cummins, D. G. (1975). Effects of tannin content and temperature on storage of propionic acid treated grain sorghum. *Agron. J.* **67**, 71–73.

Nelson, O. E. (1969a). Genetic modifications of protein quality in plants. *Adv. Agron.* **21**, 171–194.

Nelson, O. E. (1969b). The modification by mutation of protein quality in maize. *In*

"Proceedings of a panel meeting on new approaches to breeding for plant protein improvement". pp. 41–55. International Atomic Energy Agency, Vienna.

Nelson, O. E., Mertz, E. T. and Bates, L. S. (1965). Second mutant gene affecting the amino acid pattern of maize endosperm proteins. *Science* **150**, 1469–1470.

Nelson, T. S., Stephenson, E. L., Burgos, A., Floyd, J. and York, J. O. (1975). Effect of tannin content and dry-matter digestion on energy utilization and average amino acid availability of hybrid sorghum grains. *Poult. Sci.* **54**, 1620–1623.

Nesheim, M. C. and Carpenter, K. J. (1967). The digestion of heat-damaged protein. *Br. J. Nutr.* **21**, 399–411.

Nestel, B. and MacIntyre, R. (1973). "Chronic cassava toxicity". Proceedings of an interdisciplinary workshop 29–30 January 1973. IDRC-010e. International Development Research Centre, Box 8500, Ottawa, Canada, K1G 3H9.

Netemeyer, D. T., Bush, L. J. and Adams, G. D. (1977). Feeding value of reconstituted and finely ground sorghum grain for dairy cows. *J. Dairy Sci.* **60**, 748–751.

Nicol, B. M. and Phillips, P. G. (1961). Reference groundnut flour (GNF) and reference dried skimmed milk (DSM) as supplements to the diets of Nigerian men and children. *In* "Progress in meeting protein needs of infants and pre-school children." Publication 843. pp. 157–168. National Academy of Sciences, National Research Council, Washington, DC.

Nicol, B. M. and Phillips, P. G. (1978). Utilization of proteins and amino acids in diets based on cassava (*Manihot utilissima*), rice or sorghum (*Sorghum sativa*) by young Nigerian men of low income. *Br. J. Nutr.* **39**, 271–287.

Nicolas, T., Beaux, Y. and Drapron, R. (1974). Methods for the measurement of lipoxygenase activity—Presentation of a new method (in French). *Annls Technol. agric.* **23**, 287–308.

Niehaus, M. H. and Pickett, R. C. (1966). Heterosis and combining ability in a diallel cross in *Sorghum vulgare* Pers. *Crop Sci.* **6**, 33–36.

Nierenstein, M. (1944). A micromethod for the estimation of tannin. *Analyst* **69**, 91.

Nip, W. K. and Burns, E. E. (1969). Pigment characterization in grain sorghum. I. Red varieties. *Cereal Chem.* **46**, 490–495.

Nip, W. K. and Burns, E. E. (1971). Pigment characterization in grain sorghum. II. White varieties. *Cereal Chem.* **48**, 74–80.

Nishimuta, J. F., Sherrod, L. B. and Furr, R. D. (1969). Digestibility of regular, waxy and white sorghum grain rations by sheep. *Proc. west. Sect. Am. Soc. Anim. Sci.* **20**, 259–263.

Niyogi, S. P., Narayana, N. and Desai, B. G. (1934). The nutritive value of Indian vegetable foodstuffs. V. The nutritive value of ragi (*Eleusine coracana*). *Indian J. med. Res.* **22**, 373–382.

Noland, P. R. and Scott, K. W. (1963). Substituting various grains and rice milling by-products for corn in rations for growing-finishing swine. Bulletin No. 668, 3–16. Arkansas Agricultural Experiment Station, Arkansas, USA.

Noland, P. R., Scott, K. W., Baugus, C. A. and McNeal, X. (1964). Effect of various conditioning treatments and supplements on the feeding value of grain sorghums for swine. Bulletin No. 683, 3–222. Arkansas Agricultural Experiment Station, Arkansas, USA.

Noland, P. R., Sharp, R. N., Campbell, D. R., Johnson, Z. B. and York, J. O. (1975). Utilization of milo varieties by swine. *J. Anim. Sci.* **40**, 180.

Normand, F. L., Hogan, J. T. and Deobald, H. J. (1965). Protein content of successive peripheral layers milled from wheat, barley, grain sorghum, and glutinous rice by tangential abrasion. *Cereal Chem.* **42**, 359–367.

Norris, J. R. (1971). "Chemical, physical and histological characteristics of sorghum grain as related to wet milling properties". Ph.D. dissertation. Texas A&M University, College Station, Texas, USA.

Norris, J. R. and Rooney, L. W. (1970a). Enzymatic determination of starch in sorghum grain. *Cereal Sci. Today* **15**, Abstr.

Norris, J. R. and Rooney, L. W. (1970b). Wet milling properties of four sorghum parents and their hybrids. *Cereal Chem.* **47**, 64–69.

Novellie, L. (1959). Kaffircorn malting and brewing studies. III. Determination of amylases in kaffircorn malts. *J. Sci. Fd Agric.* **10**, 441–449.

Novellie, L. (1960a). Kaffircorn malting and brewing studies. IV. The extraction and nature of the insoluable amylases of kaffircorn malts. *J. Sci. Fd Agric.* **11**, 408–421.

Novellie, L. (1960b). Kaffircorn malting and brewing studies. V. Occurrence of β-amylase in kaffircorn malts. *J. Sci. Fd Agric.* **11**, 457–463.

Novellie, L. (1962a). Kaffircorn malting and brewing studies. XI. Effect of malting conditions on the diastatic power of Kaffircorn malt. *J. Sci. Fd Agric.* **13**, 115–120.

Novellie, L. (1962b). Kaffircorn malting and brewing studies. XII. Effect of malting conditions on malting losses and total amylase activity. *J. Sci. Fd Agric.* **13**, 121–123.

Novellie, L. (1962c). Kaffircorn malting and brewing studies. XIII. Variation of diastatic power with variety, season, maturity and age of grain. *J. Sci. Fd Agric.* **13**, 124–126.

Novellie, L. (1966). Kaffircorn malting and brewing studies. XIV. Mashing with Kaffircorn malt: factors affecting sugar production. *J. Sci. Fd Agric.* **17**, 354–361.

Novellie, L. (1968). Kaffir beer brewing—ancient and modern industry. *Wallerstein* Labs. Commun. **31**, 17–32.

Novellie, L. (1977). Beverages from sorghum and millets. *In* Proceedings, Symposium, Sorghum and millets for human food. pp. 73–77. International Association for Cereal Chemistry, Symposium Vienna, 1976. Tropical Products Institute, London.

Nutrition Research Laboratories. (1966). Menus for low cost balanced diets and school lunch programmes. Nutrition Research Laboratories, Hyderabad, India.

Obara, T. and Kihara, H. (1973). Glucosyl-glycerides of Italian millet (*Setaria italica* Beauvois) (in Japanese with English summary). *J. agric. Chem. Soc. Japan* **47**, 231–236. (*Fd Sci. Technol. Abstr.* 1973, **5**, 12M1466).

Oberleas, D. (1973). Phytates. *In* "Toxicants occurring naturally in foods". Committee on Food Protection. *Fd. Nutr. Bd Nat. Res. Coun.* pp. 363–371. National Academy of Sciences, Washington, D. C.

O'Donovan, M. B. and Novellie, L. (1966). Kaffircorn malting and brewing studies. XV. The fusel oils of kaffir beer. *J. Sci. Fd Agric.* **17**, 362–365.

Okano, K. and Ohara, I. (1935). Coloring matters of kaoliang. II. The presence of phlobaphene and durasontalin (in Japanese). *J. agric. Chem. Soc Japan* **11**, 667–672. (*Chem. Abstr.* 1936, **30**, 2602).

Okano, K., Abe, T. and Ohara, I. (1934). Coloring matters of kaoliang. I. Apegenin (in Japanese). *J. agric. Chem. Soc. Japan* **10**, 889–893. (*Chem. Abstr.* 1935, **29**, 830).

Oke, O. L. (1965a). Chemical studies on some Nigerian cereals. *Cereal Chem.* **42**, 299–302.

Oke, O. L. (1965b). Phytic acid—phosphorus content of Nigerian foodstuffs. *Indian J. med. Res.* **53**, 417–420.

Oke, O. L. (1967). Chemical studies on the Nigerian foodstuff "Ogi". *Fd Technol. Champaign* **21**, 202–204.

Oke, O. L. (1975). A method for assessing optimum supplementation of a cereal based diet with grain legumes. *Nutr. Rep. Int.* **11**, 313–322.

Oke, O. L. (1977). The potential of millet and sorghum as food in Nigeria. Proceedings, Symposium, Sorghum and millets as human food. International Association for Cereal Chemistry, Vienna, 11 May 1976. Tropical Products Institute, London.

Orok, E. J. and Bowland, J. P. (1974). Comparison of Nigerian yellow corn, guinea corn (sorghum) and peanut meal (groundnut cake) with Canadian corn and soybean meal. *Can. J. Anim. Sci.* **54**, 217–228.

Orr, D. E., Tribble, L. F. and Lennon, A. M. (1976). Added lysine in sorghum based young pig diets. *J. Anim. Sci.* **43**, 257.

Orr, M. L. and Watt, B. K. (1957). Amino acid content of foods. Home Economics Research Report No. 4. United States Department of Agriculture, Washington, DC.

Osborne, T. B. (1924). "The vegetable proteins". Longmans Green, New York.

Osborne, T. B. and Mendel, L. B. (1914). Nutritional properties of proteins of maize kernel. *J. biol. Chem.* **18**, 1–16.

Osborne, T. B., Mendel, L. B. and Ferry, E. L. (1919). A method of expressing numerically the growth-promoting value of proteins. *J. biol. Chem.* **37**, 223–229.

Oser, B. L. (1959). "Protein and amino acid nutrition". A. A. Albanese (Ed.). Academic Press Inc., New York.

Osman, H. F., Theurer, B., Hale, W. H. and Mehen, S. M. (1970). Influence of grain processing on in vitro enzymatic starch digestion of barley and sorghum grain. *J. Nutr.* **100**, 1133–1139.

Oswalt, D. L. (1973a). Nutritional quality of *Sorghum bicolor* (L.) Moench grain as estimated by polyphenols, crude protein, amino acid composition and rat performance. *In* "Inheritance and Improvement of Protein Quality and Content in Sorghum". pp. 144–186. Research Progress Report No. 10, Department of Agronomy, Agricultural Experiment Station, Purdue University, Lafayette, Indiana, Agency for International Development, Department of State, Washington, DC.

Oswalt, D. L. (1973b). Nutritional quality of *Sorghum bicolor* (L.) Moench as estimated by polyphenols, crude protein, amino acid composition and rat performance. PhD. Thesis. Purdue University, West Lafayette, Indiana.

Oswalt, D. L. and Srinivasan, G. (1973). Agronomically acceptable line selections from the world collection of grain sorghum for breeding and development programs. *In* "Inheritance and Improvement of Protein Quality and Content in Sorghum". pp. 187–198. Research Progress Report, No. 10, Department of Agronomy, Agricultural Experiment Station, Purdue University, Lafayette, Indiana, Agency for International Development, Department of State, Washington, DC.

Oswalt, D. L., Shaffert, R. E., Pickett, R. C. and Axtell, J. D. (1972). Relationships of chemical compositions to biological values of *Sorghum bicolor* (Linn.) Moench grain. *Agron. Abstr.* **59**, 70.

Oyenuga, V. A. (1967). "Agriculture in Nigeria". FAO, Rome.

Oyenuga, V. A. and Fetuga, B. L. (1975). The apparent digestibility of nutrients and energy values of some oilseed meals and three commonly used cereal grains fed to pigs. *E. Afr. agric. For. J.* **40**, 388–393.

Ozment, D. D., Dunkelgod, K. E., Tonkinson, L. V., Gleaves, E. W., Thayer, R. H. and Davies, F. F. (1963). Comparing milo and corn in broiler diets on an equivalent nutrition intake basis. *Poult. Sci.* **42**, 472–481.

Pai, M. L., Mehta, B. V. and Shah, H. C. (1958). Thiamin content of bajri (*Pennisetum typhoideum*) grown under different manurial treatments. M. S. Baroda University 7, 41–48. (*Chem. Abstr.* 1959, **53**, 22657i).

Palaniappan, S. P. and Vijayakumar, M. R. (1976). Note on the effect of time of harvest on the nutritional quality of grain in two sorghum cultivars. *Indian J. agric. Res.* **10**, 136–138.

Panemangalore, M., Guttikar, M. N., Rao, M. N., Rajalakshmi, D. and Swaminathan, M. (1965). Studies on processed protein foods based on blends of groundnut, bengal gram, soyabean and sesame flours and fortified with minerals and vitamins. III. Supplementary value to a poor Indian kaffir corn diet. *J. Nutr. Diet.* **2**, 28–33.

Pant, K. C. (1975). High nicotinic acid content in two Ethiopian sorghum lines. *J. agric. Fd Chem.* **23**, 608–609.

Parpia, H. A. B. and Swaminathan, M. (1975). Supplementation of cereals and cereal diets with grain legumes and limiting amino acids. Proceedings, Ninth International Congress of Nutrition Vol. 4, 139–148.

Parr Instrument Company. (1960). Oxygen bomb calorimetry and consumption methods. Technical Manual No. 130. pp. 33–36. Illinois, USA.

Parsons, J. G. and Price, P. B. (1974). Search for barley (*Hordeum vulgare* L.) with higher lipid content. *Lipids* **9**, 804–808.

Parthasarathy, H. N., Doraiswamy, T. R., Panemangalore, M., Narayana Rao, M., Chandrasekhar, B. S., Swaminathan, M., Sreenivasan, A. and Subrahmanyan, V. (1963). The effect of fortification of processed soya flour with DL-methionine hydroxy analogue or DL-methionine on the digestibility, biological value, and net protein utilization of the proteins as studied in children. *Can. J. Biochem.* **42**, 377–384.

Partridge, S. M. (1948). Filter-paper partition chromatography of sugars. 1. General description and application to the qualitative analysis of sugars in apple juice, egg white and foetal blood of sheep. *Biochem. J.* **42**, 238–248.

Pasricha, S. and Rebello, L. M. (1969). Some common Indian recipes and their nutritive value. Nutrition Research Laboratories, Hyderabad, India.

Passmore, R. and Sundararajan, A. R. (1941). The vitamin B_1 content of the millets *Eleusine coracana* and *Sorghum vulgare*, whole wheat grown under different manurial conditions, and rice stored underground. *Indian J. med. Res.* **29**, 89–94.

Patel, M. L. and Patel, G. B. (1928). Studies of the jowars of Gujarat. 1. The jowars of the Surat district. Memoirs of the Department of Agriculture, India. Botany Series Vol. 16 (1). Consulted from *J. Hered.* **54**, 121.

Patwardhan, V. N. (1961a). "Nutrition in India". 3rd ed. The Indian Journal of Medical Sciences, Bombay, India.

Patwardhan, V. N. (1961b). Nutritive value of cereal and pulse proteins. *In* "Progress in meeting protein needs of infants and preschool children"—Publication No. 843. pp. 201–210. National Academy of Sciences—National Research Council, Washington, DC.

Paul, Y., Sharma, B. N. and Bhatia, I. S. (1972). Note on lipids in sorghum. *Indian J. agric. Sci.* **42**, 435–436.

Paulis, J. W., James, C. and Wall, J. S. (1969). Comparison of glutelin proteins in normal and high-lysine corn endosperms. *J. agric. Fd Chem.* **17**, 1301–1305.

Paul Jayaraj, A., Joseph, K., Narayana Rao, M. Bhagwan, R. K., Swaminathan, M. and Subrahmanyan, V. (1960). Effect of varying levels of protein and calcium in the diet on the structure and composition of liver of rats. Part 1. The effect of supplementation of diets based on rice and rice-tapioca with calcium on the growth and structure and composition of liver of albino rats. *Ann. Biochem. exp. Med.* **20**, 341–348.

Paulsen, T. M. (1961). A study of macaroni products containing soy flour. *Fd Technol. Champaign* **15**, 118–121.

Pearson, D. (1962). "Chemical analysis of foods". 5th ed. J. & A. Churchill, London.

Peischel, H. A., Lee, D. D., Hall, G. A. B., Costa, P. T. C, Stiles, D. A. and Sanford, P. E. (1976a). Effect of varying levels of added tannic acid on chick growth. *Poult. Sci.* **55**, 2078–2079.

Peischel, H. A., Lee, D. D., Hall, G. A. B., Costa, P. T. C., Stiles, D. A., and Sandford, P. E. (1976b). Nutritive evaluation of sorghum grain varieties fed chicks. *Poult. Sci.* **55**, 2079.

Peischel, H. A., Lee, D. D., Costa,. P. T. C., Hall, G. A. B., Stiles, D. A. and Sanford, P. E. (1976c). Effect of sorghum grain, corn, and fish meal on the performance of laying hens. *Poult. Sci.* **55**, 2078.

Peplinski, A. J., Stringfellow, A. C. and Burbridge, L. H. (1963). Fractionating commercial flours and grits from grain sorghum. *Am Miller Processor* **91**, 10–12, 14.

Périssé, J., Adrian, J., Rerat, A. and Le Berre, S. (1959). Nutrient balance in the conversion of sorghum into beer. Preparation, composition and consumption of a Togo beer (in French). *Annls. Nutr. Aliment.* **13**, 1–15.

Perten, H. (1977). UNDP/FAO sorghum processing project in Sudan. Proceedings, Symposium, Sorghum and millets as human food. International Association for Cereal Chemistry, Vienna, 11 May 1976. Tropical Products Institute, London, England.

Perten, H., Guinet, R. and Abert, P. (1972). Manufacture of French type bread incorporating millet and sorghum flours (in French). Report prepared for the Government of Senegal by the Food and Agriculture Organization of the United Nations. Rome 1972.

Peryman, D. (1958). Sensory difference test. *In* "Flavour Research and Food Acceptance". Reinhold Publishing Corporation, New York.

Peyrot, F. and Adrian, J. (1970). Nature of the limiting factor (amino acid) of rations based on sorghum and groundnut (in French). *Agron. trop.* Nogent. **25**, 44–51.

Pfeifer, V. F., Stringfellow, A. C. and Griffin, E. L. Jr (1960). Fractionating corn, sorghum and soy flours by fine grinding and air classification. *Am. Miller Processor* **88**, 11–13, 24.

Phansalkar, S. V., Ramachandran, M. and Patwardhan, V. N. (1957). Nutritive value of vegetable proteins. I. Protein efficiency ratios of cereals and pulses and the supplementary effect of the addition of a leafy vegetable. *Indian J. med. Res.* **45**, 611–621.

Phansalkar, S. V., Ramachandran, M. and Patwardhan, V. N. (1958). Nutritive value of vegetable proteins. II. The effect of vegetable protein diets on the regeneration of haemoglobin and plasma proteins in protein depleted rats. *Indian J. med. Res.* **46**, 333–344.

Phul, P. S. and Athwal, D. S. (1969). Inheritance of grain size and grain hardness in pearl millet. *Indian J. Genet Pl. Breed.* **29**, 184–191.

Phul, P. S. and Gill, B. S. (1970). Diallel analysis of grain size and protein content in pearl millet. *J. Res. Punjab agric. Univ. Ludhiana* **7**, 285–289.

Phul, P. S., Rana, N. D. and Goswami, A. K. (1969). The effect of heterosis on protein content in pearl millet. *Curr. Sci.* **38**, 247–248.

Pickett, R. C. (1971). Sorghum breeding for improved protein content, amino acid composition, yield, and digestibility. *In* "Improving the nutrient quality of cereals". Report of workshop on breeding and fortification. Annapolis, Maryland. December 1970. pp. A–77–A–80. National Technical Information Service, M. S. Department of Commerce, Springfield Va. 22151.

Pierce, J. S. (1966). Amino acids in malting and brewing. *Technical Quarterly Master Brewers Association of America* **3**, 231–236.

Pierce, W. C., Willis, Conway, Sawyer, Turner, Halnisch and Lauth, (1962). "Quantitative Analysis". 4th ed. John Wiley & Sons Inc. New York.

Piez, K. A. and Morris, L. (1960). A modified procedure for the automatic analysis of amino acids. *Analyt. Biochem.* **1**, 187–201.

Pilon, R., Sitti, A. and Adrian, J. (1977). Nutritional evaluation of two African sorghums ground using a [durum wheat] semolina flow chart (in French). *Techqe. Inds Cereal.* (161). 3–9.

Pinta, M. and Busson, F. (1963). Chemical composition of African sorghums and millets (minerals and trace elements) (in French). *Annls. Nutr. Aliment.* **17**, 103–126.

Pion, R. and Fauconneau, G. (1969). Composition and nutritive value of wheat grain (in French). *Bull. Technqe. Inform.* No. 244 (special), 905–913.

Piper, C. S. (1947). "Soil and Plant Analysis". Adelaide, Australia.

Piper, C. S. (1950). "Soil and Plant Analysis". Adelaide, Australia.

Piper, C. S. (1960). "Soil and Plant Analysis". Interscience Publishers, New York.

Piper, C. S. (1966). "Soil and Plant Analysis". (Reprint for) Asia Hans Publishers, Bombay.

Platt, B. S. (1962). Tables of representative values of foods commonly used in tropical countries. Medical Research Council Special Report Series No. 302 (revised edition of SRS 253). H. M. Stationery Office, London.

Platt, B. S. and Miller, D. S. (1959). The net dietary-protein value (ND_pV) of mixtures of foods, its definition, determination and application. *Proc. Nutr. Soc.* **18**, VII–VIII (abstract of communications).

Platt, B. S., Miller, D. S. and Payne, P. R. (1961). *In* "Recent advances in human nutrition with special reference to clinical medicine". (J. F. Brock, Ed.) J. & A. Churchill Ltd., London.

Pluenneke, R. H., Wilson, R. P. and Merwine, N. C. (1973/1974). Amino acid composition and tannin interrelationships in *Sorghum bicolor*. *J. Miss. Acad. Sci.* **1**, 186.

Pokhriyal, T. G. Chatterjee, S. R. and Abrol, Y. P. (1977). Protein content and amino acid composition of pearl millet. *J. Fd Sci. Technol. Mysore*, **14**, 231–233.

Polidori, F., Galvano, G. and Lanza, A. (1967). Grain sorghum in rations for broilers to replace maize (in Italian). *Riv. Zootec. Agric. Vet.* **5**, 351–360. (*Nutr. Abst. Rev.* 1968, **38**, 6124).

Pomeranz, Y. (1976). "Advances in cereal science and technology." Vol. 1. American Association of Cereal Chemists, 3340 Pilot Knob Road, St. Paul, Minnesota 55121.

Pomeranz, Y. and Moore, R. B. (1975). Reliability of several methods for protein determination in wheat. *Bakers' Dig.* (Feb.): 44–48, 58.

Pond, W. G., Hillier, J. C. and Benton, D. A. (1958). The amino acid adequacy of milo (grain sorghum) for the growth of rats. *J. Nutr.* **65**, 493–502.

Pons, W. A. Jr and Guthrie, J. D. (1949). Determination of free gossypol in cottonseed materials. *J. Am. Oil Chem. Soc.* **26**, 671–676.

Pons, W. A. Jr, Stansbury, M. F. and Hoffpauir, C. L. (1953). An analytical system for determining phosphorus compounds in plant materials. *J. Ass. off. agric. Chem.* **36**, 492–504.

Popli, S. and Singh, R. (1972). Nutrient composition and amino acid pattern of some high yielding varieties of bajra (*Pennisetum typhoides*). *J. Res. Haryana agric. Univ.* **2**, 213–217.

Pore, M. S. and Magar. N. G. (1976a). Effect of ragi feeding on serum cholesterol level. *Indian J. med. Res.* **64**, 909–914.

Pore, M. S. and Magar, N. G. (1976b). Effect of feeding ragi on serum and liver lipids. *Indian J. Biochem. Biophys.* **13**, 188–190.

Pore, M. S. and Magar, N. G. (1977). Nutritive value of hybrid varieties of finger millet. *Indian J. agric. Sci.* **47**, 226–228.

Porter, K. S. and Axtell, J. D. (1977). Modification of the opaque endosperm phenotype of the high lysine sorghum line P-721 (*Sorghum bicolor* (L.) Moench), using the chemical mutagen diethyl sulfate. *In* "Annual Report on Inheritance and Improvement of Protein Quality and Content in *Sorghum bicolor* (L.) Moench". Report No. 13. (J. D. Axtell, Ed.) pp. 56–58. Purdue University, Lafayette, Indiana, Agency for International Development, Department of State, Washington, DC.

Portères, R. (1955). The minor cereals of the genus *Digitaria* in Africa and Europe (in French). *J. Agric. trop. Bot. appl.* **2**, 349–386, 477–510, 620–675.

Portères, R. (1958a). Nomenclature of cereals in Africa. V. Teff of Abyssinia (in French). *J. Agric. trop. Bot. appl.* **5**, 454–463.

Portères, R. (1958b). Nomenclature of cereals in Africa. VI. The millet Eleusine of India and East Africa (*Eleusine coracana* Gaertn). (in French). *J. Agric. trop. Bot. appl.* **5**, 463–486.

Portères, R. (1958c). Nomenclature of cereals in Africa. VII. The sorghums (in French). *J. Agric. trop. Bot. appl.* **5**, 732–761, 1959, **6**, 68–84.

Portères, R. (1959a). Nomenclature of cereals in Africa. X. Candle millet or pennisetum millets (*Pennisetum typhoideum*) (in French). *J. Agric. trop. Bot. appl.* **6**, 290–302.

Portères, R. (1959b). Nomenclature of cereals in Africa. XI. The millet acha (*Digitaria exilis* Stapf.) (in French). *J. Agric. trop. Bot. appl.* **6**, 302–305.

Portères, R. (1959c). Nomenclature of cereals in Africa. XII. The millets *Digitaria* div. sp. (in French). *J. Agric. trop. Bot. appl.* **6**, 305–309.

Portères, R. (1959d). A toxic cereal gathered in West Africa (*Paspalum scrobiculatum* L. var. *polystachyum* Stapf.) (in French). *J. Agric. trop Bot. appl.* **6**, 680–684.

Potter, G. D., McNeill, J. W. and Riggs, J. K. (1971). Utilization of processed sorghum grain proteins by steers. *J. Anim. Sci.* **32**, 540–543.

Potter, L. H. and Matterson, L. D. (1960). The metabolizable energy of feed ingredients for chickens. Progress Report No. 39. Connecticut Agricultural Experiment Station, Connecticut, USA.

Prasad, D. A., Morrill, J. L., Melton, S. L., Dayton, A. D. and Arnett, D. W. (1975). Evaluation of processed sorghum grain and wheat by cattle and by in vitro techniques. *J. Anim. Sci.* **41**, 578–587.

Prasannappa, G., Chandrasekhara, H. N. and Rani Padma, R. (1976). Supplementary foods for preschool children. *Nutr. Rep. Int.* **13**, 71–77.

Preece, I. A. and Mackenzie, K. G. (1962). Non-starchy polysaccharides of cereal grains. 2. Distribution of water-soluble gum-like materials in cereals. *J. Inst. Brew.* **58**, 457–464.

Price, M. L. and Butler, L. G. (1977). Rapid visual estimation and spectrophotometric determination of tannin content of sorghum grain. *J. agric. Fd Chem.* **25**, 1268–1273.

Price, M. L., Butler, L. G., Featherston, W. R. and Rogler, J. G. (1977). Detoxification of high tannin sorghum grain. (In the press).

Price, P. B. and Parsons, J. G. (1974). Lipids of six cultivated barley (*Hordeum vulgare* L.) varieties. *Lipids* **9**, 560–566.

Price, P. B. and Parsons, J. G. (1975). Lipids of 7 cereal grains. *J. Am. Oil Chem. Soc.* **52**, 490–493.

Prine, G. M., Lutrick, M. C and Lipscomb, R. W. (1967). Old crop has new outlook. *Sunshine St. Agric.* **12**, 12–13.

Pringle, W., Williams, A. and Hulse, J. H. (1969). Mechanically developed doughs from composite flours. *In* "Composite flour programme". Documentation Package Vol. 1. 2nd ed. revised. pp. 125–131. Food and Agricultural Services Division, FAO, Rome.

Pringle, W. J. S. (1946). The determination of iron in cereals. *Analyst* **71**, 490–492.

Pro, M. J. (1952). Report on spectrophotometric determination of tannin in wines and whiskies. *J. Ass. off. Agric. Chem.* **35**, 255–257.

Protein Advisory Group. (1971). PAG Guideline No. 8. "Weaning Foods". Protein Advisory Group, United Nations, New York.

Protein Advisory Group. (1972a). PAG Statement of single cell protein. No. 4. WHO, Geneva.

Protein Advisory Group. (1972b). PAG Guideline for preclinical testing of novel sources of protein. PAG Guideline No. 6, WHO, Geneva.

Protein-Calorie Advisory Group. (1975). PAG Guideline (No. 16) on protein methods for cereal breeders as related to human nutritional requirements. *PAG bull.* **5**, 22–48.

Prugar, J. and Vesela, J. (1969). Thiamine content of cereals and leguminous plants cultivated in the CSSR (in German). *Getreide Mehl.* **19**, 94–95.

Pruthi, T. D. and Bhatia, I. S. (1970). Lipids in cereals. 1. *Penniesetum typhoideum. J. Sci. Fd Agric.* **21**, 419–422.

Pundarikakshudu, R. and Seshadri, V. (1969). A comparison of nutritive value of two promising sorghum hybrids and CSH-1. *J. Nutr. Diet.* **6**, 98–99.

Purdue University. (1974). Grain sorghum international protein yield and quality trials. Report No. 1. Station Bulletin No. 52. Department of Agronomy, Agricultural Experiment Station, Purdue University, West Lafayette, Indiana.

Purdue University. (1977). "Chemical and Biological Methods for Grain and Forage Sorghums". (V. Y. Guiragossian, S. W. van Scoyoc and J. D. Axtell Eds). Department of Agronomy, Purdue University, West Lafayette, Indiana 47907. In cooperation with AID.

Purseglove, J. W. (1972). "Tropical Crops". Monocotyledons 1. Longman Group Ltd., London.

Purser, K. W. and Tanksley, T. D. Jr (1976). 3rd and 4th limiting amino acids in sorghum for growing swine. *J. Anim. Sci.* **43**, 257–258.

Pushpamma, S. (1968). Protein quality and nutritive value of three Indian millets. Dissertation Abstracts International B 29 (6): 1931-B. Dissertation, Ph.D. Kansas State University, Kansas, USA.

Pushpamma, S., Parrish, D. B. and Deyoe, C. W. (1972). Improving protein quality of millet, sorghum, and maize diets by supplementation. *Nutr. Rep. Int.* **5**, 93–100.

Qudrat-I-Khuda, M., Mukerjee, B. D., Hossain, M. A. and Khan, N. A. (1960). Cereals and cereal products: properties of certain starch varieties and their sources in East Pakistan. *Pakis. J. scient. agric. Res.* **13**, 159–162.

Quesnel, V. C. (1968). Fractionation and properties of the polymeric leucocyanidin of the seeds of *Theobroma cacao. Phytochemistry* **7**, 1583–1592.

Quisenberry, J. H. and Tanksley, T. D. Jr (1970). *In* "Grain sorghum in poultry and swine nutrition". (J. S. Wall and W. M. Ross, Eds). Sorghum production and utilization. pp. 534–572. AVI Publishing Co. Inc., Westport, Connecticut, USA.

Quisenberry, J. H., Harms, R. H., Malik, D. D., Deaton, J. W., Bradley, J. W. and Murthy, P. V. L. N. (1970). Utilization of sorghum grain in poultry diets. PR-2947. *In* "Grain Sorghum Research in Texas 1970". Consolidated PR-2938-2949. pp. 96–100. Texas A&M University, Texas Agricultural Experiment Station, College Station, Texas, USA.

Qureshi, M. S. (1967). To assess the value of maize and bajra alone and in combination in the starting and growing ration for chicken. *Agriculture Pakist.* **18**, 519–529.

Rabson, R. (1976). Considerations on the use of protein mutants in cross-breeding maize, barley, sorghum. *In* "Induced mutations in cross-breeding". Proceedings of an Advisory Group organized by the Joint FAO/IAEA Division of Atomic Energy in Food and Agriculture, Vienna 13–17 October 1975. International Atomic Energy Agency, Vienna.

Rachie, K. O. (1965). The systematic collection of sorghums, millets and maize in India. Report of the Rockefeller Foundation and the Indian Council of Agricultural Research.

Rachie, K. O. (1966). Utilizing genetic diversity in sorghum and millet improvement. *Indian J. Genet. Pl. Breed.* **26** A, 61–72.

Rachie, K. O. (1970a). Sorghum in Asia. *In* "Sorghum Production and Utilization". (J. S. Wall and W. M. Ross, Eds) pp. 328–381. The AVI Publishing Co. Inc., Westport, Connecticut, USA.

Rachie, K. O. (1970b). Developing germplasm resources in crop plants. Unpublished report, Makerere.

Rachie, K. O. (1974). "The millets and minor cereals". A bibliography of the world literature on millets pre-1930 and 1964–69, and of all literature on other minor cereals. The Scarecrow Press, Inc. Metuchen, New Jersey, USA.

Rachie, K. O. (1975). "The millets: importance, utilization and outlook". International Crops Research Institute for the Semi-arid Tropics, 1-11-256, Begumpet, Hyderabad-500016 (AP) India.

Rachie, K. O. and Peters, L. V. (1977). "The Eleusines". ICRISAT, Begumpet, Hyderabad 500016, India.

Radley, J. A. (1968). "Starch and its derivatives". 4th ed. 2 vol. Chapman and Hall, London.

Raghavendra Rao, S. N. and Desikachar, H. S. R. (1964). Pearling as a method of refining jowar and wheat and its effect on their chemical composition. *J. Fd Sci. Technol.* **1**, 40–42.

Raghuramulu, N., Srikantia, S. G., Rao, B. S. N. and Gopalan, C. (1965a). Nicotinamide nucleotides in the erythrocytes of patients suffering from pellagra. *Biochem. J.* **96**, 837–839.

Raghuramulu, N., Narasinga Rao, B. S. and Gopalan, C. (1965b). Amino acid imbalance and tryptophan-niacin metabolism. I. Effect of excess leucine on the urinary excretion of tryptophan-niacin metabolites in rats. *J. Nutr.* **86**, 100–106.

Rahman, Q. N., Akhtar, N. and Matin Chowdhury, A. (1974). Proximate composition of foodstuffs in Bangladesh. Part 1. Cereals and pulses. *Bangladesh J. sci. ind. Res.* **9**, 129–133.

Rai, K. D. (1964). Study of rain-grown sorghum and maize in the central rainlands of the Sudan. I. Effect of date of sowing, varieties and spacing on crude protein content and nitrogen accumulation. *Indian J. Agron.* **9**, 175–183.

Rai, K. D. (1965a). Study of rain-grown sorghum and maize in the central rainlands of the Sudan. II. Effect of fertilizers on crude protein content and nitrogen accumulation. *Indian J. Agron.* **10**, 139–144.

Rai, K. D. (1965b). Study of rain-grown sorghum and maize in the central rainlands of the Sudan. III. Effect of date of sowing and fertilizers on growth and N and P accumulation. *Indian J. Agron.* **10**, 235–243.

Rajalakshmi, R. and Majmudar, N. (1966). Effect of different legume supplements to kodri (*Paspalum scrobiculatum*) on weight gains and body composition of albino rats. *J. Nutr. Diet.* **3**, 67–71.

Rajalakshmi, R. and Maliwal, B. P. (1975). Effects of different supplements on the nutritive value of maize and jowar to albino rats. *Baroda J. Nutr.* **2**, 21–30.

Rakhimbaev, I. R. (1967). The amino acid composition of the proteins of Kazakhstan millet. *Prikl. Biokhim.· Mikrobiol.* **3**, 17–20. (*Appl. Biochem. Microbiol.* **3**, 10–12).

Ramachandra, G., Virupaksha, T. K. and Shadaksharaswamy, M. (1977). Relationship between tannin levels and in vitro protein digestibility in finger millet. (*Eleusine coracana* Gaertn.). *J. agric. Fd Chem.* **25**, 1101–1104.

Ramachandran, M. and Phansalkar, S. V. (1956). Essential amino acid composition of certain vegetable foodstuffs. *Indian J. med. Res.* **44**, 501–509.

Ramachandran, M. Phansalkar, S. V. and Patwardhan, V. N. (1960). Nutritive value of vegetable protein. III. Biosynthesis of vitamin B_{12} and its utilization in rats on vegetable protein diets. *Indian J. med. Res.* **48**, 243–249.

Ramanamurthy, P. S. V. and Srikantia, S. G. (1970). Effects of leucine on brain serotonin. *J. Neurochem.* **17**, 27–32.

Ramanathan, K. M., Subbiah, S., Francis, H. J. and Krishnamoorthy, K. K. (1975). A note on the nutritive value of certain minor millets. *Madras Agric. J.* **62**, 225–226.

Rama Rao, G. and Swaminathan, M. (1953). Chemical composition of different varieties of Bajra. *Bull. cent. Fd technol. Res. Inst. Mysore* **3**, 68.

Rama Rao, G., Murthy, H. B. N. and Swaminathan, M. (1953a). Nutritive value of pearl millet (*Pennisetum typhoideum*). *Indian J. Physiol. all. Sci.* **7**, 236–239.

Rama Rao, G., Murthy, H. B. N. and Swaminathan, M. (1953b). Supplementary relations of bengalgram and groundnut proteins to bajra (*Pennisetum typhoideum*) proteins. *Bull. cent. Fd technol. Res. Inst. Mysore* **3**, 44.

Rama Rao, G., Doraiswamy, T. R., Indira, K., Mahadeviah, B. and Chandrasekhara, M. R. (1965). Effect of fibre on the utilization of protein in coconut cake: metabolism studies on children. *Indian J. exp. Biol.* **3**, 163–165.

Ramasastri, B. V. and Mohan, V. S. (1969). Nutritive value of foods. *Indian J. med. Res.* **57**, (8 supplement), 1–15.

Ramasastri, B. V. and Srinivasa Rao, P. (1969). *Proc. Nutr. Soc. India* No. 7, 13.

Ramiah, P. V. and Satyanarayana, P. (1936). The quality of crops. I. Nutritive values of different varieties of ragi grains (*Eleusine coracana*). *Proc. Ass. Econ. Biol.* **4**, 13–31. (*Chem. Abstr.* **32**, 6299).

Ramiah, P. V. and Satyanarayana, P. (1937). Biological values of ragi proteins. *Proc. Soc. Biol. Chem.* **1**, 7–8. (*Chem. Abstr.* **31**, 7943).

Rana, B. S. and Murty, B. R. (1975). Heterosis and components of genetic variation for protein and lysine content in some grain sorghums. *Theoretical Appl. Genet.* **45**, 225–230.

Ranganathan, S. (1935). Influence of cereals on calcium, magnesium and phosphorus assimilation. *Indian J. med. Res.* **23**, 229–236.

Ranganathan, S., Sundararajan, A. R. and Swaminathan, M. (1937). Survey of the nutritive value of Indian foodstuffs. Part 1. The chemical composition of 200 common foods. *Indian J. med. Res.* **24**, 689–706.

Rao, C. R. (1952). "Advanced statistical methods in biometrical research". John Wiley & Sons, New York.

Rao, H. K. H., Pundarikakshudu, R. and Meenakshisundaram, P. C. (1972). Qualitative aspects of sorghum hybrids. *Indian J. agric. Sci.* **42**, 1004–1007.

Rao, K. H. and Subramanian, N. (1970). Essential amino acid composition of commonly used Indian pulses by paper chromatography. *J. Fd Sci. Technol.* **7**, 31–34.

Rao, N. G. P. and House, L. R. (1972). "Sorghum in the Seventies". Mohan Primlani, Oxford and IBH Publishing Co., 66 Janpath, New Delhi-1.

Raymond, W. D., Squires, J. A. and Ward, J. B. (1954). the milling of sorghum in Nigeria. *Colon. Pl. Anim. Prod.* **4**, 152–156.

Rayudu, G. V. N., Kadirvel, R., Vohra, P. and Kratzer, F. H. (1970a). Toxicity of tannic acid and its metabolities for chickens. *Poult. Sci.* **49**, 957–960.

Rayudu, G. V. N., Kadirvel, R., Vohra, P. and Kratzer, F. H. (1970b). Effect of various agents in alleviating the toxicity of tannic acid for chickens. *Poult. Sci.* **49**, 1323–1326.

Reddy, P. R. and Husain, M. M. (1968). Influence of nitrogen level and plant density on yield and yeild components in hybrid sorghum. CSH 1. *Indian J. agric. Sci.* **38**, 408–415.

Reeve, R. M. and Walker, H. G. Jr (1969). The microscopic structure of popped cereals. *Cereal Chem.* **46**, 227–241.

Reich, V. H. and Atkins, R. E. (1971). Variation and interrelationships of protein and oil content, and seed weight, in grain sorghum. *Iowa St. J. Sci.* **46**, 13–22.

Reichert, R. D. (1977). Dehulling cereal grains and grain legumes for developing countries. PhD. thesis University of Saskatchewan, Saskatoon, Canada.

Reichert, R. D. and Youngs, C. G. (1976). Dehulling cereal grains and grain legumes for developing countries. I. Quantitative comparison between attrition and abrasive type mills. *Cereal Chem.* **53**, 829–839.

Reichert, R. D. and Youngs, C. G. (1977). Dehulling cereal grains and grain legumes for developing countries. II. Chemical composition of mechanically and traditionally dehulled sorghum and millet. *Cereal Chem.* **54**, 174–178.

Reichert, R. D., Lorer, E. K. and Youngs, C. G. (1976). Processing and utilization of cereal grains and legumes. IDRC Project File 3-P-74-0168. Final Report Phase I. International Development Research Centre, Box 8500, Ottawa, Canada, K1G 3H9.

Reiners, R. A., Hummel, J. B., Pressick, J. C. and Morgan, R. E. (1973). Composition of feed products from wet-milling of grain sorghum. *Cereal Sci. Today* **18**, 378–379, 383.

Reinhold, J. G. (1953). "Standard methods of clinical chemistry". Vol 1. Academic Press, New York.

Reutlinger, S. and Selowsky, M. (1976). "Malnutrition and poverty-magnitude and policy options". World Bank Staff Occasional Papers No. 23. The Johns Hopkins University Press, Baltimore and London.

Ribereau-Gayon, P. (1972). "Plant phenolics". Oliver and Boyd, London.

Richardson, L. R., Wilkes, S., Godwin, J. and Pierce, K. R. (1962). Effect of moldy diet and moldy soybean meal on the growth of chicks and poults. *J. Nutr.* **78**, 301–306.

Richardson, L. R., Hayes, S. and Rigdon, R. H. (1967). The nutritive value of moldy grains and protein concentrates for the growth of poults. *Poult. Sci.* **46**, 168–176.

Ridyard, H. N. (1949). The determination of aneurine (Vitamin B_1) in uncooked wheat products. *Analyst* **74**, 18–24.

Riemenschneider, R. W., Swift, C. E. and Sando, C. E. (1941). The thiocyanogen values of the methyl esters of oleic, linoleic and linolenic acids. *Oil Soap* **18**, 203–206.

Riesen, W. H., Clandinin, D. R. Elvehjem, C. A. and Cravens, W. W. (1947). Liberation of essential amino acids from raw, properly heated and over-heated soybean oil meal. *J. biol. chem.* **167**, 143–153.

Riewe, M. E. and Breuer, L. H. (1967). Utilization of several parent lines of sorghum grain and their hybrid derivatives by sheep. *Agron. Abstr. Am. Soc. Agron.* **59**.

Riggs, J. K. (1970). Utilization of sorghum grain by livestock. PR 2946. *In* "Consolidated PR 2938-2949". pp. 82–95. Texas A&M University. Texas Agricultural Experiment Station, College Station, Texas, USA.

Riggs, J. K. and McGinty, D. D. (1970). Early harvested and reconstituted sorghum grain for cattle. *J. Anim. Sci.* **31**, 991–995.

Riggs, J. K., Sorensen, J. W. Jr and Hobgood, P. (1970a). Dry heat processing of sorghum grain for beef cattle. Bulletin 1096. Texas Agricultural Experiment Station, Texas, USA.

Riggs, J. K., Sorensen, J. W. Jr, Adame, J. L. and Schake, L. M. (1970b). Popped sorghum grain for finishing beef cattle. *J. Anim. Sci.* **30**, 634–638.

Roach, A. G., Sanderson, P. and Williams, D. R. (1967). Comparison of methods for the determination of available lysine value in animal and vegetable protein sources. *J. Sci. Fd Agric.* **18**, 274–278.

Robinson, A. (1966). Weaning food products—Senegal, West Africa. *PAG. Bull.* No. 6, 9.

Robutti, J. L., Hoseney, R. G. and Wassom, C. E. (1974). Modified opaque-2 corn endosperm. II. Structure viewed with a scanning electron microscope. *Cereal Chem.* **51**, 173–180.

Rockefeller Foundation. (1967a). "Sorghum. A bibliography of the world literature covering the years 1930–1963". Biological Sciences Communication Project of the George Washington University, Scarecrow Press, Inc. Metuchen, New Jersey, USA.

Rockefeller Foundation. (1967b). "The millets. A bibliography of the world literature covering the years 1930–1963". Biological Sciences Communication Project of the George Washington University. Scarecrow Press, Inc. Metuchen, New Jersey, USA.

Rockefeller Foundation. (1973). "Sorghum. A bibliography of the world literature 1964–1969". Indian Agricultural Program of the Rockefeller Foundation. Scarecrow Press, Inc. Metuchen, New Jersey, USA.

Rojas, B. A. and Sprague, G. F. (1952). A comparison of variance components in corn yields trials. III. General and specific combining ability and their interaction with locations and years. *Agron. J.* **44**, 426–466.

Rojas, E. and Maranville, J. W. (1975). The relation of potassium to protein and protein quality of the sorghum grain. *In* "Research in the physiology of yield and management of sorghum in relation to genetic improvement". Annual report No. 8. pp. 90–97. Agency for International Development, Department of State, Washington, DC.

Ronalds, J. A. (1974). Determination of the protein content of wheat and barley by direct alkaline distillation. *J. Sci. Fd Agric.* **25**, 179–185.

Rooney, L. W. (1968). The lipids of sorghum grain. 1. Characterization of free and bound lipids from selected grain sorghum varieties. *J. Am. Oil Chem. Soc.* **45**, 1125A.

Rooney, L. W. (1973). A review of the physical properties, composition and structure of sorghum grain as related to utilization. *Indus. Uses Cereals*, 316–342.

Rooney, L. W. and Clark, L. E. (1968). The chemistry and processing of sorghum grain. *Cereal Sci. Today* **13**, 259–261, 264–265, 285–286.

Rooney, L. W. and Sullins, R. D. (1969). A laboratory method for milling small samples of sorghum grain. *Cereal Chem.* **46**, 486–490.

Rooney, L. W. and Sullins, R. D. (1970). Chemical, physical and morphological properties of diploid and tetraploid *Sorghum bicolor* (L.) Moench kernels. *Crop Sci.* **10**, 97–99.

Rooney, L. W. and Sullins, R. D. (1973). Varietal differences in Sorghum—do they exist? *In* "Eighth Biennial Grain Sorghum Research and Utilization Conference". pp. 26–32. Grain Sorghum Producers Association, Amarillo, Texas, USA.

Rooney, L. W. and Sullins, R. D. (1974). Differences in feedlot performance of ruminants fed sorghum grains with different endosperm types; a review. Beef cattle. PR-3242: 106–109. Texas Agricultural Experiment Station, College Station, Texas, USA.

Rooney, L. W., Johnson, J. W. and Rosenow, D. T. (1970). Sorghum quality improvement: types for food. *Cereal Sci. Today* **15**, 240–243.

Rooney, L. W., Fryar, W. B., and Cater, C. M. (1972). Protein and amino acid contents of successive layers removed by abrasive milling of sorghum grain. *Cereal Chem.* **49**, 399–406.

Rooney, L. W., Rusnak, B. A. and Sullins, R. D. (1977). Effect of micronizing on cereal grains and their potential food use. *Cereal Fd Wld* **22**, 457.

Rosenheim, O. (1920). Observations on anthocyanins. 1. The anthocyanins of the young leaves of the grape vine. *Biochem. J.* **14**, 178–188.

Ross, W. M. and Eastin, J. D. (1972). Grain sorghum in the USA. *Fd Crop Abst.* **25**, 169–174.

Ross, W. M. and Webster, O. J. (1970). Culture and use of grain sorghum. Agriculture Handbook No. 385. Agricultural Research Service, United States Department of Agriculture, Washington, DC.

Rostagno, H. S., Rogler, J. C. and Featherston, W. R. (1973a). Studies on the nutritional value of sorghum grains with varying tannin contents for chicks. 2. Amino acid digestibility studies. *Poult. Sci.* **52**, 772–778.

Rostagno, H. S., Featherston, W. R. and Rogler, J. C. (1973b). Studies on the nutritional value of sorghum grains with varying tannin contents for chicks. 1. Growth studies. *Poult. Sci.* **52**, 765–772.

Roy, R. N. and Wright, B. C. (1973). Sorghum growth and nutrient uptake in relation to soil fertility. I. Dry matter accumulation patterns, yield, and N content of grain. *Agron, J.* **65**, 709–711.

Ryan, J. G., Sheldrake, R. and Yadav, S. P. (1975). Human nutritional needs and crop breeding objectives in the semi-arid tropics; a further note. Occasional Paper 8, Economics Department, ICRISAT, Begumpet, Hyderabad 500016, India.

Sahasrabuddhe. (1925). Bulletin of the Department of Agriculture, Bombay, No. 124.

Salomatina, L. G. (1969). Determination of carotenes and tocopherols in millet oil (in Russian). *Izv. vyssh. ucheb. Zaved. pishch. Teknol.* (3), 168–170. (*Fd Sci. Technol. Abstr.* 1970, **2**, 01N12).

Salomatina, L. D. and Olifson, L. E. (1969). Chemical composition and physical properties of millet oil. *Maslozhirovaya Promyshlennost* (4), 9–11.

Salunkhe, D. K., Kadam, S. S. and Chavan, J. K. (1977). Nutritional quality of proteins in grain-sorghum. *Qualitas Pl. Pl. Fd Hum. Nutr.* **27**, 187–205.

Samford, R. A., Riggs, J. K., Rooney, L. W., Potter, G. D., and Coon, J. G. (1970). Ruminal digestibility of sorghum endosperm types. *Proc. west. Sect. Am. Soc. Anim. Sci.* **21**, 123.

Samson, M. F. (1970). Thesis 3rd cycle. Paris.

Samson, M-F., and Adrian, J. (1971). Biochemical composition of the grains of *Pennisetum typhoideum* in relation to the grain size (in French). *Agron. trop.* Nogent. **26**, 1090–1099.

Sanahuja, J. C. (1971). Implications of amino acid imbalance studies in rats. *In* "Amino acid fortification of protein foods". (N. S. Scrimshaw and A. M. Altschul, Eds) pp. 179–182. MIT Press, Cambridge, Massachusetts, USA.

Sandell, E. B. (1959). "Colorimetric determination of traces of metals". 3rd ed. Interscience Publishers Inc., New York.

Sandstedt, R. M. and Abbott, R. C. (1961). A small water-jacketed bowl for the amylograph. *Cereal Sci. Today* **6**, 312.

Sandstedt, R. M., Strahan, D., Ueda, S. and Abbott, R. C. (1962). The digestibility of high amylose corn starches compared to that of other starches. The apparent effect of the ae gene on susceptibility to amylose action. *Cereal Chem.* **39**, 123–131.

Sanford, P. E. (1972). Comparison of feeding broiler-strain chicks yellow endosperm sorghum grain. Sorghum grain and corn as sources of energy. *Poult. Sci.* **51**, 1856.

Sanford, P. E. and Deyoe, C. W. (1974). Performance of laying hens fed new crop field sprouted and old crop nonsprouted sorghum grain. *Poult. Sci.* **53**, 1975. (Abstr.).

Sanford, P. E., Deyoe, C. W. and Shoup, F. K. (1968). Growth performance of sorghum grain and corn diets of equal amino acid composition. *Poult. Sci.* **47**, 1714. (Abstr.).

Sanford, P. E., Camacho, F., Knake, R. P., Deyoe, C. W. and Casady, A. J. (1973). Performance of meat-strain chicks fed pearl millet as a source of energy and protein. *Poult. Sci.* **52**, 2081.

Santoro, R., Azzarini, A., Mosquera, F., Saralegui, W. and Estevez, R. (1966). Effects of different proportions of feterita grain on broiler growth (in Spanish). *Bln. Fac. Agron. Univ. Repub. Montev.* **85**, 3–22.

Sass, J. E. (1945). Schedules for sectioning maize kernels in paraffin. *Stain Technol.* **20**, 93–98.

Sastri, B. W. (1952). The wealth of India: raw materials. Vol. III. Council for Scientific Industrial Research, Delhi, India.

Sastry, L. V. S. and Virupaksha, T. V. (1967). Disc electrophoresis of sorghum seed proteins in polyacrylamide gels. *Analyt. Biochem.* **19**, 505–513.

Sauberlich, H. E., Chang, W-Y. and Salmon, W. D. (1953). The amino acid and protein content of corn as related to variety and nitrogen fertilization. *J. Nutr.* **51**, 241–250.

Sawhney, S. K. and Naik, M. S. (1969). Amino acid composition of protein fractions of pearl millet and the effect of nitrogen fetrilization on its proteins. *Indian J. Genet. Pl. Breed.* **29**, 395–406.

Saxena, K. L., Chakrabarti, C. H. and Nath, M. C. (1966). Effect of rice polishing concentrate on the nitrogen, calcium and phosphorus metabolism of albino rats kept on rice and jowar diets. *J. Nutr. Diet.* **3**, 121–125.

Saxena, K. L., Chakrabarti, C. H. and Nath, M. C. (1968). Effect of rice polishing concentrate on growth and liver and blood composition of albino rats kept on rice and jowar diets. *J. Nutr. Diet.* **5**, 125–129.

Scaut, A. (1962). Coix lacryma-jobi: Chemical composition, digestibility, and its energetic value in swine (in French). *Bull. agric. Congo belge.* **52**, 265–270. (*Chem. Abstr.* 1963, **58**, 13058).

Schaffert, R. E. (1972). Protein quantity, quality and availability in *Sorghum bicolor* (L.) Moench grain. Ph.D. Thesis. Purdue University, Lafayette, Indiana, USA.

Schaffert, R. E., Pickett, R. C., Oswalt, D. L. and Axtell, J. D. (1972a). Genotype-environment interactions for yield, protein, lysine, oil and seed weight in *Sorghum bicolor* (L.) Moench. *In* "Research Progress Report on Inheritance and Improvement of Protein Quality and Content in *Sorghum bicolor* (L.) Moench." (J. D. Axtell and D. L. Oswalt, Eds) pp. 21–42. Purdue University, Lafayette, Indiana, Agency for International Development, Department of State, Washington, DC.

Schaffert, R. E., Oswalt, D. L. and Pickett, R. C. (1972b). Genotype by environment interaction effects for grain yield, protein, lysine, oil and seed weight of *Sorghum bicolor* (L.) Moench. *Agron. Abstr.* **19**.

Schaffert, R. E., Lechtenberg, V. L., Oswalt, D. L., Axtell, J. D., Pickett, R. C. and Rhykerd, C. L. (1974a). Effect of tannin on in vitro dry matter and protein disappearance in sorghum grain. *Crop Sci.* **14**, 640–643.

Schaffert, R. E., Oswalt, D. L. and Axtell, J. D. (1974b). Effect of supplemental protein on the nutritive value of high and low tannin *Sorghum bicolor* (L.) Moench grain for the growing rat. *J. Anim. Sci.* **39**, 500–505.

Schaiberger, G. E. and Ferrari, A. (1960). Automatic enzymatic analysis for L-lysine via decarboxylation. *Ann. N. Y. Acad. Sci.* **87**, 890–893.

Schake, L. M., Riggs, J. K. and Butler, O. D. (1972). Commercial feedlot evaluation of four methods of sorghum grain processing. *J. Anim. Sci.* **34**, 926–930.

Schanderl, S. H. (1970). "Methods in food analysis." (M. A. Joslyn, Ed.) Ch. 22. Academic Press, New York.

Scheigart, F., Vlietstra, H. and Van Twisk, P. (1972). Bantu beer—a return to Kaffir corn? *Fd Inds S. Afr.* (March), 27, 29, 33.

Schertz, K. F. and Clark, L. E. (1967). Controlling dehiscence with plastic bags for band crosses in sorghum. *Crop Sci.* **7**, 540–542.

Schertz, K. F. and Stephens, J. C. (1966). Compilation of gene symbols, recommended revisions and summary of linkages for inherited characters of *Sorghum vulgare* Pers. Technical Monograph 3, Texas A&M University, Texas Agricultural Experiment Station, Texas, USA.

Schlesinger, J. S. (1942). Correlations between crude fibre and ash of wheat shorts. *Cereal Chem.* **19**, 838–839.

Schmidt, G., Hecht, L., Fallot, P., Greenbaum, L. and Thannhauser, S. J. (1952). The amounts of glycerylphosphorylcholine in mammalian tissues. *J. biol. Chem.* **197**, 601–609.

Schmidt, O.Th. (1956). Gallotannins and ellagitannins (in German). *Fortschr. Chem. org. Nat. Stoffe* **13**, 70–136.

Schmidt-Hebbel, H. (1966). Chemistry and technology of foodstuffs (in Spanish). Editorial Salesiana, Santiago, Chile.

Schmidt-Hebbel, H. (1969). Tables of the chemical composition of Chilean foodstuffs. Faculty of Chemistry and Pharmacy, University of Chile.

Schoch, T. J. and Maywald, E. C. (1956). Microscopic examination of modified starches. *Analyt. Chem.* **28**, 382–387.

Schram, E., Moore, S. and Bigwood, E. J. (1954). Chromatographic determination of cysteine or cysteic acid. *Biochem. J.* **57**, 133–137.

Schröder, H. H. E. (1972). Modifications of nitrogenous compounds. Extract from Bantu Beer Unit Annual Report, Chem 217, Pretoria, South Africa.

Schröder, H. H. E. (1973). Nitrogenous compounds. Extract from Bantu Beer Unit of the CSIR, Annual Report, BB33, PO Box 395, Pretoria, South Africa.

Schroeder, H. A., Balasa, J. J. and Tipton, I. H. (1970). Essential trace metals in man: Molybdenum. *J. chron. Dis.* **23**, 481–499.

Schroeder, H. W. and Boller, R. A. (1973). Aflatoxin production of species and strains of the *Aspergillus flavus* group isolated from field crops. *Appl. Microbiol.* **25**, 885–889.

Schwartz, H. M. (1956). Kaffircorn malting and brewing studies. I. The Kaffir beer brewing industry in South Africa. *J. Sci. Fd Agric.* **7**, 101–105.

Scott, P. M. (1973). Mycotoxins in stored grain, feeds, and other cereal products. *In* "Grain Storage, Part of a system". (R. N. Sinha and W. E. Muir, Eds) pp. 343–365. AVI Publishing Co. Inc., Westport, Connecticut, USA.

Scrimshaw, N. S. (1976). Shattuck lecture. Strengths and weaknesses of the Committee approach. An analysis of past and present recommended daily allowances for protein in health and disease. *New England J. Med.* **294**, 136–142, 198–203.

Scrimshaw, N. S. (1977). Through a glass darkly: discerning the practical implications of human dietary protein-energy interrelationships. *Nutr. Rev.* **35**, 321–337.

Seckinger, H. L. and Wolf, M. J. (1973). Sorghum protein ultrastructure as it relates to composition. *Cereal Chem.* **50**, 455–465.

Seeley, R. D. (1958). Milling feeds. *In* "Processed plant protein foodstuffs". (A. M. Altschul, Ed.) pp. 761–787. Chapter 28. Academic Press Inc., New York.

Seidler, S., Kotowski, J. and Swierczynska, Z. (1964). Digestibility and N balance in poultry given maize and sorghum (in Polish). *Zesz. nauk. Wyzsz. Szk. roln. Szczec.* (15), 58–61. (*Nutr. Abstr. Rev.* 1965, **35**, 4957).

Seit-Ablaeva, S. K., Nechaev, A. P., Denisenko, Ya. I. and Yanotovskii, M. P. (1973). Tocopherols of lipids from millet and husked millet and their change during storage (in Russian). *Prikl. Biokhim. Mikrobiol.* (Moscow) **9**, 737–739.

Seljametov, R. A. and Massino, I. V. (1969). Vitamin composition of irrigated sorghum (in Russian). *Vest. Sel'-khoz. Nauki. Mosk.* (12), 47–49. (*Nutr. Abstr. Rev.* 1970, **40**, 4825).

Sen, D. P., Satyanarayana Rao, T. S. and Lahiry, N. L. (1966). Defatting and deodorization of fish protein concentrate from Harpoden nehereus. *J. Fd Sci.* **31**, 344–350.

Sepel, N. A. and Sepel, O. F. (1969). The effect of hybridization and external conditions on grain yield and quality in sorghum (in Russian). *Vest. Sel'-khoz. Nauki.* (*Mosk.*) (7), 50–55. (*Pl. Breed. Abstr.* 1970, **40**, 112).

Shackleton, L. and McCance, R. A. (1936). The ionisable iron in foods. *Biochem. J.* **30**, 582–591.

Shadid, J. D, Sasse, C. E., Krieg, D. R. and Tribble, L. F. (1976). Value of micronizing sorghum hybrids for swine. *J. Anim. Sci.* **42**, 257.

Shah, H. C. and Mehta, B. V. (1959a). Magnesium-phosphorus-crude fat relationship in the seed of pearl millet (*Pennisetum typhoideum* Rich.). *Soil Science* **87**, 320–324.

Shah, H. C. and Mehta, B. V. (1959b). Comparative studies on the effect of ammonium chloride and other fertilizers on the yield and crude protein content of pearl millet. *Indian J. Agron.* **4**, 105–113.

Shannon, J. C. (1968). A procedure for the extraction and fractionation of carbohydrates from immature *Zea mays* kernels. Agricultural Experiment Station. Research Bulletin No. 842. Purdue University, Lafayette, Indiana, USA.

Sharda, D. P., Pradhan, K. and Sagar, V. (1972). The relative value of feeds for swine: Bajra (*Pennisetum typhoideum*) in the diet of growing pigs. *J. Res. Haryana Agric. Univ.* **2**, 222–228.

Sharma, K. P. and Goswami, A. K. (1969). Chemical constants of lipid content of high yielding varieties and hybrids of bajra (*Pennisetum typhoideum*) flour. *J. Nutr. Diet.* **6**, 316–318.

Shay, H. and Gruenstein, M. (1946). A simple and safe method for the gastric instillation of fluids in the rat. *J. Lab clin Med.* **31**, 1384–1386.

Shechter, Y. and De Wet, J. M. J. (1975). Comparative electrophoresis and isozyme analysis of seed proteins from cultivated races of sorghum. *Am. J. Bot.* **62**, 254–261.

Shepherd, A. D. (1974). How grain structure influences sorghum quality. Presented to Fifth E. Afr. Cereals Res. Conf. Malawi.

Shepherd, A. D. and Woodhead, A. H. (1969–70). Sorghum processing. *In* "Annual Report, 1969–70, East African Industrial Research Organization". pp. 28–39.

Shepherd, A. D., Woodhead, A. H. and Okorio, J. F. (1970–71). Sorghum processing. *In* "Annual Report, 1970–71, East African Industrial Research Organization". pp. 36–58.

Shepherd, A. D., Woodhead, A. H. and Kapasi-Kakama, J. (1971–72). Cereal processing. *In* "Annual Report, 1971–72, East African Industrial Research Organization". pp. 23–52.

Sherman, H. C. and Campbell, H. L. (1924). Growth and reproduction on simplified food supply. IV. Improvement in nutrition resulting from an increased proportion of milk in the diet. *J. biol. Chem.* **60**, 5–15.

Sherrod, L. B. and Albin, R. C. (1973). Nutritive value of different sorghum grain types. *J. Anim. Sci.* **36**, 1208.

Sherrod, L. B., Albin, R. C. and Furr, R. D. (1969). Net energy of regular and waxy sorghum grains for finishing steers. *J. Anim. Sci.* **29**, 997.

Shiau, S. Y., Yang, S. P., Tribble, L. F., Lennon, A. M. and Williams, I. L. (1976). Effect of micronizing of sorghum for swine. *J. Anim. Sci.* **43**, 258–259.

Shinde, P. A. and Bhide, V. P. (1958). Ergot of bajri (*Pennisetum typhoides*) in Bombay State. *Curr. Sci.* **27**, 499–500.

Shone, D. K. (1960). Ergot of munga as a cause of agalactia of sows. *Rhodesia agric. J.* **57**, 120–121.

Shone, D. K., Philip, J. R. and Christie, G. J. (1959). Agalactia of sows caused by feeding the ergot of the bulrush millet, *Pennisetum typhoides*. *Vet. Rec.* **71**, 129–132. (*Nutr. Abstr. Rev.* 1959, **29**, 6856).

Shotwell, O. L., Hesseltine, C. W., Burmeister, H. R., Kwolek, W. J., Shannon, G. M. and Hall, H. H. (1969). Survey of cereal grains and soybeans for the presence of aflatoxin: 1. Wheat, grain sorghum, and oats. *Cereal Chem.* **46**, 446–453.

Shoup, F. K. (1970). Factors affecting protein utilization of sorghum grain in feeds and foods. *Diss. Abstr.* B **31**, 3470–3471.

Shoup, F. K., Deyoe, C. W., Campbell, J. and Parrish, D. B. (1969). Amino acid composition and nutritional value of milled sorghum grain products. *Cereal Chem.* **46**, 164–171.

Shoup, F. K., Deyoe, C. W., Sanford, P. E. and Murphy, L. S. (1970a). Nutritive value of six commercial sorghum grain hybrids. *Poult. Sci.* **49**, 168–172.

Shoup, F. K., Deyoe, C. W., Farrell, E. P., Hammond, D. L. and MIller, G. D. (1970b). Sorghum grain dry milling. *Fd Technol. Champaign* **24**, 1028–1032.

Shoup, F. K., Deyoe, C. W., Skoch, L., Shamsuddin, M., Bathurst, J., Miller, G. D., Murphy, L. S. and Parrish, D. B. (1970c). Amino acid composition and nutritional value of milled fractions of sorghum grain. *Cereal Chem.* **47**, 266–273.

Shukla, U. C. and Bhatia, K. N. (1971). Effect of nitrogen levels on the Fe and Mn concentration and uptake in hybrid and local varieties of pearl millet (*Pennisetum typhoides* (Burm. f.) Stapf & C. E. Hubb.). *Indian J. agric. Sci.* **41**, 790–794.

Shuman, A. C. and Plunkett, R. A. (1964). Determination of amylose content of corn starch. *In* "Methods of Carbohydrate Chemistry". Vol. 4, 174–178. Academic Press, New York.

Shurpalekar, S. R., Joseph, A. A., Moorjani, M. N., Lahiry, N. L., Indiramma, K., Swaminathan, M., Sreenivasan, A. and Subrahmanyan, V. (1962a). Supplementary value of fish flour fortified with vitamins to poor Indian diets based on different cereals and millets. *Fd Sci. Mysore* **11**, 49–51.

Shurpalekar, S. R., Joseph, A. A., Lahiry, N. L., Moorjani, M. N., Sankaran, A. N., Swaminathan, M., Sreenivasan, A. and Subrahmanyan, V. (1962b). Supplementary value of fish flour and a protein food containing low fat groundnut flour, Bengalgram flour and fish flour to poor rice diet. *Fd Sci. Mysore* **11**, 45–48.

Sibbald, I. R. and Slinger, S. J. (1963). A biological assay for metabolizable energy together with findings which demonstrate some of the problems associated with the evaluation of fats. *Poult. Sci.* **42**, 313–325.

Sibbald, I. R., Summers, J. D. and Slinger, S. J. (1960). Factors affecting the metabolizable energy content of poultry feeds. *Poult. Sci.* **39**, 544–556.

Sibbald, I. R., Czarnocki, J., Slinger, S. J. and Ashton, G. C. (1963). The prediction of the metabolizable energy content of poultry feedstuffs from a knowledge of their chemical composition. *Poult. Sci.* **42**, 486–492.

Siddappa, G. S. (1954). Standards for cholam malt extract. *J. Scient. ind. Res.* **13** A, 33–34.

Sidransky, H. (1960. Chemical pathology of nutritional deficiency induced by certain plant proteins. *J. Nutr.* **71**, 387–395.

Singh, A. and Bains, S. S. (1973). Yield, grain quality and nutrient uptake of CSH 1 and Swarna sorghum at different levels of N and plant population. *Indian J. Agric. Sci* **43**, 408–413.

Singh, B. (1969). "Indian cookery". Mills and Boon Ltd., London.

Singh, B. B., Hadley, H. H. and Collins, F. I. (1968). Distribution of fatty acids in germinating soybean seed. *Crop Sci.* **8**, 171–173.

Singh, R. and Axtell, J. D. (1973a). High lysine mutant gene (*hl*) that improves protein quality and biological value of grain sorghum. *Crop Sci.* **13**, 535–539.

Singh, R. and Axtell, J. D. (1973b). Survey of world sorghum collection for opaque and sugary lines. *In* "Inheritance and improvement of protein quality and content in sorghum". Research Progress Report No. 10, pp. 1–22. Department of Agronomy, Agricultural Experiment Station, Purdue University, Lafayette, Indiana. Agency for International Development, Department of State, Washington, DC.

Singh, R. and Axtell, J. D. (1973c). Inheritance of high lysine and sugary lines and their amino acid composition. *In* "Inheritance and improvement of protein quality and content in sorghum". Research Progress Report No. 10, pp. 58–81. Department of Agronomy, Agricultural Experiment Station, Purdue University, Lafayette, Indiana. Agency for International Development, Department of State, Washington, DC.

Singh, R. and Axtell, J. D. (1973d). Biological value of high lysine (hl) and sugary (su) mutants. *In* "Inheritance and improvement of protein quality and content in sorghum". Research Progress Report No. 10, pp. 23–57. Department of Agronomy, Agricultural Experiment Station, Purdue University, Lafayette, Indiana. Agency for International Development, Department of State, Washington, DC.

Singh, R. and Popli, S. (1973). Amylose content and amylolytic studies on high yielding varieties of Bajra (*Pennisetum typhoides*). *J. Fd Sci. Technol.* (India) **10**, 31–33.

Singh, R. and Popli, S. (1974). Studies on the protein content and electrophoretic analysis of various protein fractions of high yielding varieties of Bajra (*Pennisetum typhoides*) pearl millet. *J. Fd Sci. Technol.* **11**, 216–220.

Singh, S. D. and Barsaul, C. S. (1977). A note of the effeciency and economics of feeding different cereal grains on growth production in White Leghorn and Rhode Island Red birds. *Indian J. Anim. Sci.* **47**, 159–161.

Singh, S. P. (1976). Modified vitreous endosperm recombinants from crosses of normal and high lysine sorghum. *Crop Sci.* **16**, 296–297.

Singhania, D. L., Rao, N. G. P. and House, L. R. (1970). A note on the inheritance of beta carotene content in sorghum. *Curr. Sci.* **39**, 544–545.

Singleton, V. L. and Kratzer, F. H. (1969). Toxicity and related physiological activity of phenolic substances of plant origin. *J. agric. Fd Chem.* **17**, 497–512.

Singleton, V. L. and Kratzer, F. H. (1973). Plant Phenolics. *In* "Toxicants occurring naturally in foods". Committee on Food Protection. Food and Nutrition Board. pp. 309–345. National Research Council, National Academy of Sciences, Washington, DC.

Sirinit, K., Soliman, A. M., Van Loo, A. T. and King, K. W. (1965). Nutritional value of Haitian cereal-legume blends. *J. Nutr.* **86**, 415–423.

Sivaprakasam, K., Chinnadurai, G. and Krishnamurthy, C. S. (1971). Alkaloid production by *Claviceps microcephala* on some varieties of pearl millet. *Madras agric. J.* **58**, 431–432.

Skeggs, H. R. and Wright, L. D. (1944). The use of *Lactobacillus arabinosus* in the microbiological determination of pantothenic acid. *J. biol. Chem.* **156**, 21–26.

Skoch, L. V., Deyoe, C. W., Shoup, F. K., Bathurst, J. and Liang, D. (1970). Protein fractionation of sorghum grain. *Cereal Chem.* **47**, 472–481.

Slump, P. and Schreuder, H. A. W. (1969). Determination of tryptophan in foods. *Anal. Biochem.* **27**, 182–186.

Smalley, E. B., Marasas, W. F. O., Strong, F. M., Bamburg, J. R., Nichols, R. E. and Kosuri, N. R. (1970). Mycotoxicoses associated with moldy corn. *In* "1st US-Japan Conference on Toxic Microorganisms, 1968". (M. Herzberg, Ed.) pp. 163–173. UJNR Joint Panels on Toxic Micro-Organisms and US Department of Interior, Washington, DC.

Smirnova-Ikonnikova, M. I., Veselova, E. P. and Petrova, T. M. (1965). Rapid method for the quantitative determination of tryptophan in corn and bean crops. *Trudy prikl. Bot. Genet. Selek.* **37**, 169–171.

Smith, A. J. (1967). Sorghum (kaffir corn) as a replacement for maize in rations for growing and laying pullets. *Rhodesia agric. J.* **64**, 67–68.

Smith, A. M. and Agiza, A. H. (1951). The determination of amino acids colorimetrically by the ninhydrin reaction. *Analyst, Lond.* **76**, 623–627.

Smith, D. and Grotelueschen, R. D. (1966). Carbohydrates in grasses. I. Sugar and fructosan composition of the stem bases of several Northern-adapted grasses at seed maturity. *Crop Sci.* **6**, 263–266.

Smith, E. F., Richardson, D., Drake, C. L. and Brent, B. E. (1968a). High protein sorghum grain with no added protein in all concentrate cattle finishing rations; urea and soybean oil meal in all concentrate rations. (Project 253–6, 1967) Bulletin (518), 24–28. Kansas Agricultural Experiment Station, Kansas, USA.

Smith, E. F., Richardson, D., Drake, C. L. and Brent, B. E. (1968b). Sorghum grain as the only protein source in all-concentrate heifer finishing rations; two levels of urea in an all-concentrate ration. Bulletin 518, 29–32. Kansas Agricultural Experiment Station, Kansas, USA.

Smith, K. and Allee, G. L. (1973). Effect of endosperm type on nutritional value of sorghum grain for swine. *J. Anim. Sci.* **37**, 291.

Snehalatha, N. and Reddy, P. R. (1978) Studies on the bioavailability of iron from selected cereals and pulses in iron-deficient rats. *Nutr. Rep. Int.* **17**, 43–48.

Snell, E. E. (1950). Microbiological methods in vitamin research. *In* "Vitamin methods". (P. Gyorgyi, ed.), Vol. 1. pp. 327–505. Academic Press Inc., New York.

Snell, E. E. and Strong, F. M. (1939). A microbiological assay for riboflavin. *Ind. Engng Chem.* (*Analyt. Ed.*) **11**, 346–350.

Snell, E. E. and Wright, L. D. (1941). A microbiological method for the determination of nicotinic acid. *J. biol. Chem.* **139**, 675–686.

Snell, F. D. and Snell, C. T. (1937). "Colorimetric methods of analysis including some turbidimetric and nephelometric methods". Vol. II. D. Van Nostrand Co. Inc.

Snell, F. D. and Snell, C. T. (1953). "Colorimetric methods of analysis". Vol. III. Van Nostrand Co. Inc., New York.

Snowden, J. D. (1936). "The cultivated races of sorghum". Adlard, London.

Society of Public Analysts. (1946). Analytical Methods Committee Report on the microbiological assay of riboflavine and nicotinic acid. *Analyst* **71**, 397–406.

Society of Public Analysts. (1951). Analytical Methods Committee chemical assay of aneurine in foodstuffs. *Analyst* **76**, 127–133.

Solpico, F. O. and Yambao, A. N. (1966). Performance test of millet at the economic garden, Las Baños, Laguna, *Philipp. J. Pl. Ind.* **31**, 219–229.

Somogyi, M. (1952). Notes on sugar determination. *J. biol. Chem.* **195**, 19–23.

Soni, B. L. and Sharma, D. C. (1974). Total and ionizable iron in common Indian cooked foods. *Am. J. clin. Nutr.* **27**, 455–457.

Sorre, M. (1942). Annls Géogr. **51**, 81.

South African Government. (1968).Consolidated regulations relating to the grading of Kaffircorn.

Sowbhagya, C. M. and Bhattacharya, K. R. (1971). A simplified colorimetric method for determination of amylose content in rice. *Staerke* **23**, 53–56.

Spackman, D. H., Stein, W. H. and Moore, S. (1958). Automatic recording apparatus for use in the chromatography of amino acids. *Analyt. Chem.* **30**, 1190–1206.

Sperry, W. M. and Webb, M. (1950). A revision of the Schoenheimer-Sperry method for cholesterol determination. *J. biol. Chem.* **187**, 97–106.

Spies, J. R. (1950). Determination of tryptophan with *p*-dimethyl amino benzaldehyde. *Analyt. Chem.* **22**, 1447–1449.

Spies, J. R. and Chambers, D. C. (1949). Chemical determination of tryptophan in proteins. *Analyt. Chem.* **21**, 1249–1266.

Spolter, P. D. and Harper, A. E. (1961). Leucine-isoleucine antagonism in the rat. *Am. J. Physiol.* **200**, 513–518.

Sprague, G. F. and Tatum, L. A. (1942). General versus specific combining ability in single crosses of corn. *Agron. J.* **34**, 923–933.

Sree Ramulu, K. (1975). Mutation breeding in sorghum (a review). *Z. Pfl. Züeht.* **74**, 1–17.

Sree Ramulu, U. S. and Mariakulandai, A. (1964). The composition of the ragi (*Eleusine coracana*) grain and straw as affected by the application of farmyard manure and superphosphate fertilizer. *Madras agric. J.* **51**, 379–385.

Srikantia, S. G., Rao, B. S. N., Raghuramulu, N. and Gopalan, C. (1968a). Pattern of nicotinamide nucleotides in the erythrocytes of Pellagrins. *Am. J. clin. Nutr.* **21**, 1306–1309.

Srikantia, S. G., Reddy, M. V. and Krishnaswamy, K. (1968b). Electroencephalographic patterns in pellagra. *Electroenceph. clin. Neurophysiol.* **25**, 386–388.

Srinivasa Rao, P. (1971). Studies on the nature of carbohydrate moiety in high yielding varieties of rice. *J. Nutr.* **101**, 879–884.

Srinivasa Rao, P. and Ramasastri, B. V. (1969a). The nutritive value of some indica, japonica and hybrid varieties of rice. *J. Nutr. Diet.* **6**, 204–208.

Srinivasa Rao, P., and Ramasastri, B. V. (1969b). Riboflavin and nicotinic acid content of some foods of plant origin. *J. Nutr. Diet.* **6**, 218–223.

Srinivasan, G., Axtell, J. D. and Jambunathan, R. (1972). Amino acid composition of sorghum grain and relationships between amino acids and protein. *In* "Research Progress Report on inheritance and improvement of protein quality and content in *Sorghum bicolor* (L.) Moench". (J. D. Axtell and D. L. Oswalt, Eds) pp. 71–74. No. 9. Purdue University, West Lafayette, Indiana. Agency for International Development, Department of State, Washington, DC.

Srivastava, K. N. and Mehta, S. L. (1976). Rapid estimation of lysine by high voltage electrophoresis. *Curr. Sci.* **45**, 283–284.

Stabursvik, A. and Heide, O. M. (1974). protein content and amino acid spectrum of finger millet (*Eleusine coracana* (L.) Gaertn.) as influenced by nitrogen and sulphur fertilizers. *Pl. Soil* **41**, 549–571.

Stafford, H. A. (1965). Flavonoids and related phenolic compounds produced in the first internode of *Sorghum vulgare* Pers. in darkness and in light. *Pl. Physiol.* **40**, 130–138.

Stafijcuk, A. A. and Teljatnikov, N. Ya. (1968). Composition, digestibility and feeding value of grain sorghum (in Russian). *Dokl. vses. Akad. Sel'-khoz. Nauk.* (2), 8–10.

Stallcup, O. T. and Davis, G. C. (1962). Nutritive value of AKS 614 grain sorghum. *Arkans. Fm Res.* **11**, 5.

Stanbury, J. B. and Childs, J. A. (1974). Health, nutrition and population. Annex 2. A framework for evaluating long-term strategies for the development of the Sahel-Sudan region. Centre for Policy Alternatives, Massachusetts Institute of Technology, USA.

Steele, B. F., Sauberlich, H. E., Reynolds, M. S. and Bauman, C. A. (1949). Media for *Leuconostoc mesenteroides* P–60 and *Leuconostoc citrovorum* 8081. *J. biol. Chem.* **177**, 533–544.

Steinke, F. H. (1977). Protein Efficiency Ratio pitfalls and causes of variability: a review. *Cereal Chem.* **54**, 949–957.

Stemler, A. B. L., Collins, F. I., De Wet, J. M. J. and Harlan, J. R. (1976). Variation in levels of lipid components and protein in ecogeographic races of *Sorghum bicolor*. *Biochem. System. Ecol.* **4**, 43–45.

Stephens, D. (1970). Soil fertility. *In* "Agriculture in Uganda". (J. D. Jameson, Ed.) pp. 72–89. Oxford University Press, Oxford.

Stephenson, E. L., York, J. O., Bragg, D. B. and Ivy, C. A. (1971). The amino acid content and availability of different strains of grain sorghum to the chick. *Poult. Sci.* **50**, 581–584.

Stringfellow, A. C. and Peplinski, A. J. (1966). Air classification of sorghum flours from varieties representing different hardnesses. *Cereal Sci. Today* **11**, 438–440, 455.

Stringham, G. R., McGregor, D. I. and Pawlowski, S. H. (1974). Chemical and morphological characteristics associated with seedcoat color in rapeseed. Proceedings 4th International Rapeseed Conference, Giessen, Germany.

Strumeyer, D. H. and Malin, M. J. (1969). Identification of the amylase inhibitor from seeds of Leoti sorghum. *Biochim. biophys. Acta* **184**, 643–645.

Strumeyer, D. H. and Malin, M. J. (1975). Condensed tannins in grain sorghum: isolation, fractionation, and characterization. *J. agric. Fd Chem.* **23**, 909–914.

Subba Rao, G. N., Bains, G. S., Bhatia, D. S. and Subrahmanyan, V. (1953). Processing of millets and cereals (other than rice) into rice substitute. *Trans. Am. Assoc. Cereal Chem.* **11**, 167–171.

Subrahmanyan, V., Rama Rao, G. and Swaminathan, M. (1950). Investigations on the preparation, properties and nutritive value of rice substitutes from tubers and millets. *J. sci. ind. Res.* **9**, 259–261.

Subrahmanyan, V., Murthy, H. B. N. and Swaminathan, M. (1954a). Effects of partial replacement of rice, wheat or ragi (*Eleusine coracana*) by tuber flours on the nutritive value of poor vegetarian diets. *Br. J. Nutr.* **8**, 1–10.

Subrahmanyan, V., Krishnamurthy, K., Swaminathan, M., Bhatia, D. S. And Raghunatha Rao, Y. K. (1954b). Supplementary value of cottonseed flour to wheat and ragi diets. *Bull. cen. Fd technol. Res Inst. Mysore* **3**, 225–226.

Subrahmanyan, V., Swaminathan, M., Bhatia, D. S, Bains, G. S., Sur, G., Bhagwan, R. K., Doraiswamy, T. R., Anandaswamy, B. and Sankaran, A. N. (1954c). Effect of replacing rice in the diet by composite jowar (*Sorghum vulgare*) vermicelli on the health and nutritional status of children. *Bull. cent. Fd technol. Res. Inst. Mysore* **3**, 245–247.

Subrahmanyan, V., Kuppuswamy, S., Rao, G. R., Swaminathan, M. and Bhatia, D. S. (1954d). Investigations on grain substitutes. III. The nutritive value of round grains from blends of tapioca and groundnut flour. *Bull. cent. Fd technol. Res. Inst. Mysore* **3**, 187–189.

Subrahmanyan, V., Bhatia, D. S., Bains, G. S., Swaminathan, M. and Rao, Y. K. R. (1954e). Investigations on grain substitutes. 1. Production of round grain from blends of tapioca and groundnut. *Bull. cent. Fd technol. Res. Inst. Mysore*, 180–183.

Subrahmanyan, V., Narayana Rao, M., Rama Rao, G. and Swaminathan, M. (1955). The metabolism of nitrogen, calcium and phosphorus in human adults on a poor vegetarian diet containing ragi (*Eleusine coracana*). *Br. J. Nutr.* **9**, 350–357.

Subrahmanyan, V., Doariswamy, T. R., Bhagwan, R. K., Rajagopalan, R., Kurien, P. P., Sankaran, A. N., Bhatia, D. S. and Swaminathan, M. (1961). The effect of replacing wheat in a poor Indian diet by a blend of whole wheat flour, tapioca flour and low-fat groundnut flour on the growth and nutritional status of children. *Ann. Biochem. exp. Med.* **21**, 7–12.

Subramanian, N. and Srinivasan, M. (1951). Nutritive value of the seeds of *Amaranthus paniculatus* Linn. *Curr. Sci.* **20**, 294–295.

Sufian, S. and Pittwell, L. R. (1968). Iron content of Teff (*Eragrostis abyssinica*). *J. Sci. Fd Agric.* **19**, 439.

Sulaiman, M. Lukade, G. M. and Dawkhar, G. S. (1966). Effect of some fungicides and antibiotics on sclerotial development and germination of ergot on *Pennisetum typhoideum*. *Hindustan Antibiot. Bull.* **9**, 94–96. (*Chem. Abstr.* 1967, **66**, 64618c).

Sullins, R. D. and Rooney, L. W. (1971). Physical changes in the kernel during reconstitution of sorghum grain. *Cereal Chem.* **48**, 567–575.

Sullins, R. D. and Rooney, L. W. (1974). Microscopic evaluation of the digestibility of sorghum lines that differ in endosperm characteristics. *Cereal Chem.* **51**, 134–142.

Sullins, R. D. and Rooney, L. W. (1975). Light and scanning electron microscopic studies of waxy and non-waxy endosperm sorghum varieties. *Cereal Chem.* **52** (3 part 1), 361–366.

Sullins, R. D. and Rooney, L. W. (1977a). Pericarp and endosperm structure of pearl millet (*Pennisetum typhoides*). Paper presented to Symposium "The processing of sorghum and millets for human food". International Association for Cereal Chemistry, Vienna, 11 May 1976. Tropical Products Institute, London.

Sullins, R. D. and Rooney, L. W. (1977b). Structure and physical properties of high lysine sorghum mutants. *Cereal Fd Wld* **22**, 472.

Sullins, R. D., Rooney, L. W. and Rosenow, D. T. (1975). Endosperm structure of high lysine sorghum. *Crop Sci.* **15**, 599–600.

Sumner, A. K. and Nielsen, M. A. (1976). Food legume utilization. Progress Report No. 4. July 1, 1975 to June 30, 1976. IDRC Project File No. 3-P-73-0032. International Development Research Centre, Box 8500, Ottawa, Canada K1G 3H9.

Sumner, A. K., Nielsen, M. A. and Johnston, E. F. (1974a). Food legume utilization—Progress Report No. 1. November 1 1973–April 30 1974. IDRC File No. 3-P-73-0032. International Development Research Centre, Box 8500, Ottawa, Canada K1G 3H9.

Sumner, A. K., Nielsen, M. A. and Johnston, E. F. (1974b). Food legume utilization. Progress Report No. 2. May 1 1974–October 31 1974. IDRC File No. 3-P-73-0032. International Development Research Centre, Box 8500, Ottawa, Canada K1G 3H9.

Sumner, A. K., Nielsen, M. A. and Johnson, E. F. (1975). Food legume utilization. Progress Report No. 3. November 1 1974–June 30 1975. IDRC File No. 3-P-73-0032. International Development Research Centre, Box 8500, Ottawa, Canada K1G 3H9.

Sundararajan, A. R. (1938). Phytin-phosphorus content of Indian foodstuffs. *Indian J. med. Res.* **25**, 685–691.

Sur, G., Reddy, S. K., Swaminathan, M. and Subrahmanyan, V. (1954a). Supplementary value of food yeast (*Torula utilis*) to poor vegetable diets based on cereals. *Bull cent. Fd technol. Res. Inst. Mysore* **3**, 111–112.

Sur, G., Reddy, S., Swaminathan, M. and Subrahmanyan, D. V. (1954b). Supplementary value of the proteins of food yeast to cereal proteins. *Bull cent. Fd technol. Res. Inst. Mysore* **4**, 35–36.

Sur, G., Swaminathan, M. and Subrahmanyan, V. (1955). Studies on the nutritive value of Jowar (Kaffir corn—*Sorghum vulgare*). *Bull. cent. Fd. technol. Res. Inst. Mysore* **4**, 133–134.

Sure, B., Easterlang, L., Dowell, J. and Crudup, M. (1957). The addition of small amounts of defatted fish flour to whole yellow corn, whole wheat, whole and milled rye, grain sorghums and millet. I. Influence on growth and protein efficiency. II. Nutritive value of the minerals in fish flour. *J. Nutr.* **63**, 409–416.

Suryanarayana Rao, K., Rukmini, C. and Mohan, V. S. (1968). β-carotene content of some yellow endosperm varieties of sorghum. *Indian J. agric. Sci.* **38**, 368–372.

Swain, T. and Hillis, W. E. (1959). The phenolic constituents of *Prunus domestica* 1. The quantitative analysis of phenolic constituents. *J. Sci. Fd Agric.* **10**, 63–68.

Swaminathan, M. (1937a). The relative value of the proteins of certain foodstuffs in nutrition. *Indian J. med. Res.* **24**, 767–786.

Swaminathan, M. (1937b). The relative values of the proteins of certain foodstuffs in nutrition. II. The comparative biological values of the proteins of certain cereals, pulses and skim milk powder measured by the growth of young rats. *Indian J. med. Res.* **25**, 57–79.

Swaminathan, M. (1937c). The relative value of the proteins of certain foodstuffs in nutrition. III. The biological value of the proteins of various pulses, oil-seeds, nuts, and skimmed milk, studied by the balance sheet method. *Indian J. med. Res.* **25**, 381–398.

Swaminathan, M. (1938a). The relative amounts of the protein and non-protein nitrogenous constituents occurring in foodstuffs and their significance in the determination of the digestibility co-efficient of proteins. *Indian J. med. Res.* **25**, 847–855.

Swaminathan, M. (1938b). The relative value of the proteins of certain foodstuffs in nutrition. V. Supplementary values of the proteins of *Eleusine coracana* (ragi) and of certain pulses and skimmed milk powder studied by the nitrogen balance and the growth method. *Indian J. med. Res.* **26**, 107–112.

Swaminathan, M. (1942a). An improved method for the estimation of Vitamin B_1 in foods by the thiochrome reaction. *Indian J. med. Res.* **30**, 263–272.

Swaminathan, M. (1942b). A simple procedure for estimating nicotinic acid in biological materials using the cyanogen bromide-aniline reagent. *Indian J. med. Res.* **30**, 397–401.

Swaminathan, M. (1942c). A fluorimetric method for the estimation of riboflavin in foodstuffs. *Indian J. med. Res.* **30**, 23–35.

Swaminathan, M. and Daniel, V. A. (1973). Amino acid imbalance, toxicity and antagonism. *Indian J. Nutr. Diet.* **10**, 148–157.

Swaminathan, M., Daniel, V. A. and Parpia, H. A. B. (1972). Protein-enriched cereal foods for overcoming malnutrition among preschool children in India and other developing contries. *Indian J. Nutr. Diet.* **9**, 22–48.

Swaminathan, M. S., Austin, A., Kaul, A. K. and Naik, M. S. (1969). Genetic and agronomic enrichment of the quantity and quality of proteins in cereals and pulses. *In* "Proceedings of a panel meeting on new approaches to breeding for plant protein improvement". pp. 71–86. International Atomic Energy Agency, Vienna.

Swaminathan, M. S., Naik, M. S., Kaul, A. K. and Austin, A. (1971). Choice of strategy for the genetic upgrading of protein properties in cereals, millets, and pulses. *Indian J. agric. Sci.* **41**, 393–406.

Swanson, A. F. (1928). Seed-coat structure and inheritance of seed color in sorghums. *J. agric. Res.* **37**, 577–588.

Sweeney, J. H. and Hall, W. L. (1951). Chemical differentiation between nicotinic acid and nicotinamide. *Analyt. Chem.* **23**, 983–986.

Synge, R. L. M. (1975). Interactions of polyphenols with proteins in plants and plant products. *Qual. Pl.–Pl. Fds Hum. Nutr.* **24**, 337.

Tagle, M. A. and Donoso, G. (1965). Net protein utilisation determined in short- and long-term experiments with rats. *J. Nutr.* **87**, 173–178.

Taira, H. (1962a). Studies on amino acid contents in plant seed. Amino acid contained in the seed of Gramineae. *Bot. Mag. Tokyo* **75**, 242–243.

Taira, H. (1962b). Studies on amino acid contents in plant seeds. II. Amino acid pattern of seed protein fractions of Gramineae. *Bot. Mag. Tokyo* **75**, 273–277.

Taira, H. (1963a). Studies on amino acid contents in food crops. II. Amino acids in maize, foxtail millet, Japanese barnyard millet, proso millet, sorghum and buckwheat (Japanese domestic) (in Japanese with English summary). Report of the Food Research Institute, Tokyo **17**, 299–303.

Taira, H. (1963b). Studies on amino acid contents in plant seeds. III. Amino acid contained in the seeds of Gramineae (Part 2). *Bot. Mag. Tokyo* **76**, 340–341.

Taira, H. (1965). Studies on amino acid contents in food crops. 9. Amino acid contained in protein fractions of oat (*Avena sativa*) and ragi (*Eleusine coracana*) (in Japanese with English summary). *J. Jap. Soc. Fd Nutr.* **18**, 194–196.

Taira, H. (1966). Studies on amino acid contents in plant seeds IV. Amino acid contained in the seed of Gramineae (Part 3). *Bot. Mag. Tokyo* **79**, 36–48.

Taira, H. (1968). Amino acid composition of different varieties of foxtail millet (*Setaria italica*). *J. agric. Fd Chem.* **16**, 1025–1027.

Takahashi, S., Furuya, S., Jitsukawa, Y. and Morimoto, H. (1968). Studies on nutritive value of feedstuffs for pigs. I. Grains and potatoes (in Japanese). *Bull. Nat. Inst. Anim. Ind. Chiba* (17), 1–7.

Talwalkar, R. T. and Patel, S. M. (1970a). Biological evaluation of proteins of ambadi (*Hibiscus cannabinus*) and methi (*Trigonella foenum-graecum*) and their supplementary effect on jowar (*Sorghum vulgare*). *Indian J. Nutr. Diet.* **7**, 13–16.

Talwalkar, R. T. and Patel, S. M. (1970b). Supplementary effect of ambadi (*Hibiscus cannabinus*) and methi (*trigonella foenum-graecum*) on jowar (*Sorghum vulgare*) in regenerating tissue proteins. *Indian J. Nutr. Diet.* **7**, 74–79.

Tamura, G., Tsunoda, T., Kirimura, J. and Miyazawa, S. (1952). The microbiological determination of amino acids by lactic acid bacteria. 1. Amino acid requirements of lactic acid bacteria and standard curves for amino acid assay. *Nippon Nogei Kagakai Kaishi* **26**, 464–470.

Tanksley, T. D. Jr (1973). Research with sorghum for swine. *In* Eighth Biennial Grain Sorghum Research and Utilization Conference. pp. 20–25. Grain Sorghum Producers Association, Amarillo, Texas, USA.

Tanksley, T. D. Jr and Osbourn, L. K. (1969). Evaluation of methods of processing sorghum grain for growing-fattening swine. *J. Anim. Sci.* **21**, 147.

Tannenbaum, S. R., Mateles, R. I. and Capco, G. R. (1966). "World protein resources." (R. F. Gould, Ed.) Advances in Chemistry Series 57. American Chemical Society, Washington, DC.

Tanner, F. W. Jr, Pfeiffer, S. E. and Curtis, J. J. (1947). B-complex vitamins in grain sorghums. *Cereal Chem.* **24**, 268–274.

Tanner, F. W. Jr, Swanson, A. F. and Curtis, J. J. (1949). Breeding for niacin content in a sorghum cross, Westland x Cody. *Cereal Chem.* **26**, 333–338.

Tashiro, M. and Maki, Z. (1977). An evaluation of the nutritive value of proso millet protein (in Japanese with English summary). Kyoto-Furitsu Daigaku Gakujutsu Hokoku: Rigaku, Seikatsu Kagaku 28, 23–30. (Scientific reports of the Kyoto Prefectural University natural science, living science and welfare science).

Tasker, P. K., Doraiswamy, T. R., Narayana Rao, M., Swaminathan, M. Sreenivasan, A. and Subrahmanyan, V. (1962). The metabolism of nitrogen, calcium and phosphorus in undernourished children. 8. The metabolism of nitrogen, calcium and phosphorus, and the digestibility coefficient and biological value of the proteins and the net protein utilization on poor Indian diets based on rice, maize or a mixture of rice and maize. *Br. J. Nutr.* **16**, 361–368.

Tasker, P. K., Rao, M. N. and Swaminathan, M. (1964). Supplementary value of a processed protein food based on a blend of coconut meal, groundnut flour and Bengal gram flour to poor Indian diets based on different cereals and millets. *J. Nutr. Diet.* **1**, 95–97.

Taurog, A., Entenman, C. and Chaikoff, I. L. (1944). The choline-containing and non choline-containing phospholipids of plasma. *J. biol. Chem.* **156**, 385–391.

Tausky, H. H. and Shorr, E. (1953). A micromethod for the determination of inorganic phosphorus. *J. biol. Chem.* **202**, 675–685.

Terpstra, K. (1961). Digestibility of millet for poultry (in Dutch). *Landbouwk. Tijdschr.'s-Grav.* **73**, 247–250. (*Nutr. Abstr. Rev.* 1961, **3**, 6674).

Terris, M. (1964). Goldberger on pellagra. Louisiana State University Press, Baton Rouge, Louisiana, USA.

Thayer, R. H., Sieglinger, J. B. and Heller, V. G. (1957). Oklahoma grain sorghums for growing chicks. Bulletin No. B-487. Division of Agriculture, Oklahoma A&M College, USA.

Thirumala Rao, S. D. (1969). Oil of Italian millet (*Setaria italica*) bran. *Telhan Patrika* **1**, 27–29.

Thomas, R. L., Sheard, R. W. and Moyer, J. R. (1967). Comparison of conventional and automated procedures for nitrogen, phosporus and potassium analysis of plant material using a single digestion. *Agron. J.* **59**, 240–243.

Thompson, J. E. and Steward, F. C. (1951). Investigations on nitrogen compounds and nitrogen metabolism in plants. II. Variables in two directional paper chromatography of nitrogen compounds: a quantitative procedure. *Pl. Physiol.* **3**, 421–440.

Thompson, J. F., Zacharius, R. P. and Steward, F. C. (1951). Investigations on nitrogen compounds and nitrogen metabolism in plants. I. The reaction of nitrogen compounds with ninhydrin on paper: a quantitative procedure. *Pl. Physiol.* **2**, 375–397.

Thompson, R. S., Jacques, D., Haslam, E. and Tanner, R. J. N. (1972). Plant proanthocyanidins. Part 1. Introduction: the isolation, structure, and distribution in nature of plant procyanidins. *J. chem. Soc.* (Perkin 1), 1387–1399.

Thornton, J. H., Goodrich, R. D. and Meiske, J. C. (1969). Corn maturity. I. Composition of corn grain of various maturities and test weights. *J. Anim. Sci.* **29**, 977–986.

Tilley, J. M. A. and Terry, R. A. (1963). A two-stage technique for the in vitro digestion of forage crops. *J. Br. Grassld Soc.* **18**, 104–111.

Tipton, K. W., Floyd, E. H., Marshall, J. G. and McDevitt, J. B. (1970). Resistance of certain grain sorghum hybrids to bird damage in *Louisiana. Agron. J.* **62**, 211–213.

Tipton, K. W., Mabbayad, B. B., Marshall, J. G., Rabb, J. L., Robinson, D. L. and Sloane, L. W. (1974). Tannin investigations of brown-seeded grain sorghum hybrids. Partial summary of Ph.D. dissertation by B. B. Mabbayad. Louisiana State University, Baton Rouge, Louisiana, USA.

Tkachuk, R. (1977). Calculation of the nitrogen-to-protein conversion factor. *In* "Nutritional Standards and methods of evaluation for food legume breeders" (J. H. Hulse, K. O. Rachie and L. W. Billingsley, Eds) Prepared by the International Working Group on Nutritional Standards and Methods of Evaluation for Food Legume Breeders. IDRC–TS7e. pp. 78–82. International Development Research Centre, Box 8500, Ottawa, Canada K1G 3H9.

Tkachuk, R. and Irvine, G. N. (1969). Amino acid compositions of cereals and oilseed meals. *Cereal Chem.* **46**, 206–218.

Tomer, P. S. (1970). Testweights and protein content in bajra in relation to methods of sowing, manurial treatments and seed rates. *Ann. Arid Zone* **9**, 159–162.

Tonroy, B., Plumlee, M. P., Conrad, J. H. and Cline, T. R. (1973). Apparent digestibility of the phosphorus in sorghum grain and soybean meal for growing swine. *J. Anim. Sci.* **36**, 669–673.

Toth, S. J., Prince, A. L., Wallace, A. and Mikkelsen, D. S. (1948). Rapid quantitative determination of eight mineral elements in plant tissue by a systematic procedure involving use of a flame photometer. *Soil Sci.* **66**, 459–466.

Totusek, R. and White, D. (1968). Methods of processing milo for cattle. *Proc. Texas Nutr. Conf.* **23**, 50–68.

Trei, J., Hale, W. H. and Theurer, B. (1970). Effect of grain processing on in vitro gas production. *J. Anim. Sci.* **30**, 825–831.

Tribble, L. F., Lennon, A. M., Gaskins, C. T. Jr., Clapp, K. L. and Ramsey, C. B. (1972). Various grains in swine rations at same lysine level. *J. Anim. Sci.* **35**, 226.

Tripathi, B. D. and Daté, W. B. (1975). Partial substitution of wheat flour by other flours for bread preparation. I. Use of cereal flours. *Indian Fd Packr* **29**, 62–65.

Tripathi, B. K., Gupta, Y. P. and House, L. R. (1971). Selection for high protein and amino acids in grain sorghum. *Indian J. Genet. Pl. Breed.* **31**, 275–282.

Troeng, S. (1955). Oil determination of oil seed. Gravimetric routine method. *J. Am. Oil Chem. Soc.* **32**, 124–126.

Tropical Products Institute. (1970). Proceedings of a Symposium on the use of non-wheat flour in bread and baked goods manufacture. Publication G62. Tropical Products Institute, 52/62 Grays Inn Road, London WC1X 8LU.

Troyer, J. R. (1964). Leucoanthocyanin formation in buckwheat seedling hypcotyls. *Phytochemistry* **3**, 535–539.

Truswell, A. S. (1963). The role of maize in the pathogenesis of pellagra. Effect of leucine on N-methylnicotinamide excretion in human adults. *S. Afr. med. J.* **37**, 253–256.

Truswell, A. S. (1976). A comparative look at recommended intakes. *Proc. Nutr. Soc.* **35**, (May), 1–14.

Truswell, A. S., Goldsmith, G. A. and Pearson, W. N. (1963). Leucine and pellagra. *Lancet* (i), 778–779.

Tschiderer, K. (1966). Fattening pigs on Austrian sorghum millet (milocorn) (in German). *Bodenkultur* **17**, 187–192.

Tsukunaga, K., Nishino, T. and Tanaka, M. (1932). Tannin in Manchurian sorghum (Kaoliang) seeds and a simple method for the determination of small quantities of tannin (in Japanese with English summary). *J. Sappora Soc. Agric. For.* **24**, 23–49. (*Chem. Abstr.* 1932, **26**, 5449).

Tucker, H. F. and Eckstein, H. C. (1937). The effect of supplementary methionine and cystine on the production of fatty livers by diet. *J. biol. Chem.* **121**, 479–484.

Turek, F., Lettner, F. and Steinacker, G. (1966). Sorghum millet for fattening poultry (in German). *Bodenkultur* **17**, 368–372.

Twist, J. O., Morris, J. G. and Gartner, R. J. W. (1965). Cobalt content of sorghum grain and cereal forages. *Aust. J. Sci.* **28**, 125–126.

Tyner, E. P., Lewis, H. B. and Eckstein, M. C. (1950). Niacin and the ability of cystine to augment deposition of liver fat. *J. biol. Chem.* **187**, 651–654.

Udy, D. C. (1956). Estimation of protein in wheat and flour by ion binding. *Cereal Chem.* **33**, 190–197.

Udy, D. C. (1971). Improved dye method for estimating protein. *J. Am. Oil Chem. Soc.* **48**, 29A–33A.

Ulloa, M. and Herrera, T. (1970). Persistence of aflatoxins during "pozol" fermentation (in Spanish). *Revta lat.-am. Microbiol.* **12**, 19–25.

UN World Food Conference. (1974). Assessment of the World Food Situation, Present and Future. 5–16 November 1974, Rome.

United States Department of Agriculture. (1970). Official grain standards, revised. SRA–C & MS–177.

Uprety, D. C. and Austin, A. (1972). Varietal differences in the nutrient composition of improved bajra (pearl millet) hybrids. *Bull. Grain Technol.* **10**, 249–255.

US Salinity Laboratory. (1954). Diagnosis and improvement of saline and alkaline soils. US Department of Agriculture Handbook 60.

Van de Kamer, J. H. and Van Ginkel, L. (1952). Rapid determination of crude fibers in cereals. *Cereal Chem.* **29**, 239–251.

Van der Walt, J. P. (1956). Kaffircorn malting and brewing studies. II. Studies on the microbiology of Kaffir beer. *J. Sci. Fd Agric.* **7**, 105–113.

Van Hellen, R. W. and Ellis, W. C. (1973). Membranes for human *in situ* digestion techniques. *J. Anim. Sci.* **37**, 358.

Vanschoubroek, F. X., Van Spaendonk, R. L. and Nauwynck, W. (1964). A comparison of the feeding value of maize and sorghum for fattening pigs. *Anim. Prod.* **6**, 357–362.

Van Wyk, C. P., Robbins, D. J. and Dreyer, J. J. (1973). Report on the in vitro determination of the digestibility of the protein in certain locally-grown grain sorghum cultivars. National Food Research Institute, CSIR. Pretoria, South Africa. Unpublished report.

Varnish, S. A. and Carpenter, K. J. (1975). Mechanisms of heat damage in proteins. 6. The digestibility of individual amino acids in heated and propionylated proteins. *Br. J. Nutr.* **34**, 339–349.

Vasal, S. K. (1975). Use of genetic modifiers to obtain normal type kernels with the opaque 2 gene. *In* "High quality protein maize". pp. 197–216. Dowden, Hutchinson and Ross, Inc., Stroudsburg Pa. USA.

Vasantgadkar, P. S., Venkatachalam, P. S. and Tulpule, P. G. (1963). Partition of urinary nitrogen in children with kwashiorkor treated with animal and vegetable proteins. *Am. J. clin. Nutr.* **12**, 150–156.

Vasantha, L. (1970a). Histidine urocanic acid and histidine α-Deaminase in the stratum corneum in pellagrins. *Indian J. med. Res.* **58**, 1079–1084.

Vasantha, L. (1970b). Collagen content and dermal amino acid pattern in pellagra. *Clinica Chim. Acta* **27**, 543–547.

Vasi, I. G. and Desai, G. M. (1976). In vitro digestibility of proteins present in some cereals and pulses grown in South Gujarat nutritional study. *Sci. Cult.* **42**, 126–128.

Vasudeva Rao, M. J. and Goud, J. V. (1976). Genic analysis of percent protein in grains of five sorghum inbreds. *Cereal Research Communication* **4**, 441–448.

Vatagin, A. V. and Oksenenko, N. I. (1971). The effect of fertilizers on the yield and quality of millet (in Russian). *Khim. Sel'skom Khoz.* **9**, 21–22.

Vavich, M. G., Kemmerer, A. R., Nimbkar, B. and Stith, L. S. (1959). Nutritive value of low and high protein sorghum grains for growing chickens. *Poult. Sci.* **38**, 36–40.

Vavilov, N. I. (1951). "The origin, variation, immunity and breeding of cultivated plants". pp. 37–38. Translated from the Russian by K. Starr Chester. The Ronald Press Co., New York.

Venkataraman, L. V., Becker, W. E., Khanum, P. M. and Murthy, I. A. S. (1977). Supplementary value of the proteins of alga *Scenedesmus acutus* to rice, ragi, wheat and peanut proteins. *Nutr. Rep. Int.* **15**, 145–155.

Venkataramana, R. S. and Krishna Rao, D. V. (1961). Chemical composition and nutrient uptake of ragi. *J. Indian Soc. Soil Sci.* **9**, 245–252.

Venkat Rao, S., Pantulu, A. J., Swaminathan, M. and Subrahmanyan, V. (1958). Supplementary value of low-fat groundnut flour to poor vegetarian diets based on jowar (*Sorghum vulgare*) and ragi (*Eleusine coracana*). *Ann. Biochem. exp. Med.* **18**, 33–38.

Venkat Rao, S., Jayaraj, A. P., Bhagavan, R. K., Sankaran, A. N., Swaminathan, M. and Sriramachari, S. (1960). The effect of feeding diets containing insect-infested jowar on the growth and composition of blood and liver of albino rats. *Ann. Biochem. exp. Med.* **20**, 135–142.

Venkat Rao, S., Swaminathan, M. and Parpia, H. A. B. (1964a). Mutual supplementation of dietary proteins for meeting protein needs and overcoming protein shortages in developing countries. *J. Nutr. Diet.* **1**, 128–138.

Venkat Rao, S., Daniel, V. A., Panemangalore, M., Tasker, P. K., Paul Jayaraj, A., Acharya, U. S. V., Parthasarathy, L. and Narayana Rao, M. (1964b). *J. Nutr. Diet.* **1**, 8.

Vermorel, M. (1969). Energy utilization by the growing rat of opaque-2 maize diet adjusted for amino acids (in French). *C. R. Acad. Sci.* **268**, 834–837.

Vermorel, M. (1970). Utilization of energy and nitrogen of the sorghum hybrid INRA 450 in a diet balanced in essential amino acids for the growing rat (in French). *Ann. Biol. Anim. Bioch. Biophys.* **10**, 327–330.

Vermorel, M. and Keller, J. (1967). Energy utilization by the growing rat of the principal cereals in isonitrogenous diets adjusted for amino acids (in French). *Ann. Zootech.* **16**, 223–234.

Viraktamath, C. S., Raghavendra, G. and Desikachar, H. S. R. (1971). Use of rice milling machinery for commercial pearling of grain sorghum (jowar) and culinary uses for pearled sorghum products. *J. Fd Sci. Technol.* **8**, 11–13.

Viraktamath, C. S., Raghavendra, G. and Desikachar, H. S. R. (1972). Varietal differences in chemical composition, physical properties and culinary qualities of some recently developed sorghum strains. *J. Fd Sci. Technol.* **9**, 73–76.

Virupaksha, T. K. and Sastry, L. V. S. (1968). Studies on the protein content and amino acid composition of some varieties of grain sorghum. *J. Agric. Fd Chem.* **16**, 199–203.

Virupaksha, T. K., Ramachandra, G. and Nagaraju, D. (1975). Seed proteins of finger millet and their amino acid composition. *J. Sci. Fd Agric.* **26**, 1237–1246.

Visco, S. (1930). The chemical and biological analysis of African cereals. II. The nutritive value of the grain of *Pennisetum typhoideum* (in Italian). *Boll. Soc. ital. Biol. Sper.* **5**, 185–192. (*Chem. Abstr.* 1930, **24**, 4810).

Vithayathil, P. J. and Murthy, G. S. (1972). New reaction of 0-benzoquinone at the thioether group of methionine. *Nature, New Biology* **236**, 101.

Vogel, S. and Graham, M. (1979). "Sorghum and Millet: Food Production and Use". Report of workshop held in Nairobi, Kenya, 4–7 July 1978. IDRC–123e. International Development Research Centre, Box 8500, Ottawa, Canada K1G 3H9.

Vohra, P., Kratzer, F. H. and Joslyn, M. A. (1966). The growth depressing and toxic effects of tannins to chicks. *Poult. Sci.* **45**, 135–142.

Von Holdt, M. M. and Brand, J. C. (1960a). Kaffircorn malting and brewing studies. VI. Starch content of kaffir beer brewing materials. *J. Sci. Fd Agric.* **11**, 463–467.

Von Holdt, M. M. and Brand, J. C. (1960b). Kaffircorn malting and brewing studies. VII. Changes in the carbohydrates of Kaffircorn during malting. *J. Sci. Fd Agric.* **11**, 467–471.

Von Schaaffhausen, R. (1952). Adlay or Job's tears—a cereal of potentially greater importance. *Econ. Bot.* **6**, 216–227.

Wada, S. and Ariyama, H. (1966). Dietary factors and cholesterol metabolism. I. Effect of the level and type of cereal protein on cholesterol metabolism (in Japanese with English summary). *J. Jap. Soc. Fd Nutr.* **19**, 27–31.

Waggle, D. H. and Deyoe, C. W. (1966). Relationship between protein level and amino acid composition of sorghum grain. *Feedstuffs* **38**, 18–19.

Waggle, D. H., Parrish, D. B. and Deyoe, C. W. (1966). Nutritive value of protein in high and low protein content sorghum grain as measured by rat performance and amino acid assays. *J. Nutr.* **88**, 370–374.

Waggle, D. H., Deyoe, C. W. and Sanford, P. E. (1967a). Relationship of protein level of sorghum grain to its nutritive value as measured by chick performance and amino acid composition. *Poult. Sci.* **46**, 655–664.

Waggle, D. H., Deyoe, C. W. and Smith, F. W. (1967b). Effect of nitrogen fertilization on the amino acid composition and distribution in sorghum grain. *Crop Sci.* **7**, 367–368.

Waldroup, P. W., Greene, D. E., Harris, R. H., Maxey, J. F. and Stephenson, E. L. (1967). Comparison of corn, wheat, and milo in turkey diets. *Poult. Sci.* **46**, 1581–1585.

Waldschmidt-Leitz, E. and Metzner, P. (1962). On the prolamine of wheat, rye, maize and millet (X. Communication on seed protein) (in German with [short] English summary). *Hoppe-Seyler's Z. physiol. Chem.* **329**, 52–61.

Walker, R. D. and Lichtenwalner, R. E. (1977). Effect of reconstitution on protein solubility and digestibility of waxy sorghum. *J. Anim. Sci.* **44**, 843–849.

Wall, J. S. and Blessin, C. W. (1969). Composition and structure of sorghum grains. *Cereal Sci. Today* **14**, 264–270, 276.

Wall, J. S. and Ross, W. K. (1970). Sorghum production and utilization. AVI Publishing Co., Inc., Westport, Connecticut, USA.

Wall, M. and Kelley, E. G. (1943). Determination of pure carotene in plant material. *Ind. Engng. Chem. Anal. Ed.* **15**, 18–20.

Wang, C., Mitchell, H. C. and Barham, H. N. (1959). The phytin content of sorghum grain. *Trans. Kans. Acad. Sci.* **62**, 208–211.

Waniska, R. B. (1976). Methods to assess quality of boiled sorghum gruel and chapaties from sorghum with different kernel characteristics. M.Sc. Thesis, Texas A&M Universtiy, USA.

Warner, J. N. (1952). A method of estimating heritability. *Agron. J.* **44**, 427–430.

Warsi, A. S. and Wright, B. C. (1973). Effects of rates and methods of nitrogen application on the quality of sorghum grain. *Indian J. agric. Sci.* **43**, 722–726.

Watson, J. D. (1971). Investigations on the nutritive value of Ghanaian foodstuffs. *Ghana J. agric. Sci.* **4**, 95–111.

Watson, S. A. (1964). Determination of starch geletinization temperature. *In* "Methods in carbohydrate chemistry". Vol. IV. Starch. (R. L. Whistler R. J. Smith, J. N. BeMiller and M. L. Wolfrom Eds) pp. 240–242. Academic Press, New York and London.

Watson, S. A. (1967). Manufacture of corn and milo starches. *In* "Starch Chemistry and Technology II. Industrial Aspects". (R. L. Whistler and E. F. Paschall, Eds) pp. 1–51. Academic Press, New York.

Watson, S. A. (1970). Wet-milling process and products. *In* "Sorghum production and utilization". (J. S. Wall and W. M. Ross, Eds) pp. 602–626. AVI Publishing Co. Inc., Westport, Connecticut, USA.

Watson, S. A. and Hirata, Y. (1955). The wet milling properties of grain sorghum. *Agron. J.* **47**, 11–15.

Watson, S. A. and Yahl, K. R. (1971). Survey of aflatoxins in commercial supplies of corn and grain sorghum used for wet milling. *Cereal Sci. Today* **16**, 153–155, 163.

Watson, T. G. (1975). Inhibition of microbial fermentations by sorghum grain and malt. *J. appl. Bact.* **38**, 133–142.

Watson, T. G. and Novellie, L. (1974). Extraction of *Sorghum vulgare* and *Hordeum vulgare* α-glucosidase. *Phytochemistry* **13**, 1037–1041.

Watson, T. G. and Novellie, L. (1976). The development of amylase and maltase during the malting of *Sorghum vulgare*. *Agrochemophysics* **8**, 61–64.

Watson, T. G., Daiber, K. H. and Novellie, L. (1974). Extraction of a salt-insoluble α-glucosidase from *Sorghum vulgare* grain. *Phytochemistry* **13**, 901–904.

Watt, G. (1908). Commercial products of India. (being an abridgement of "The Dictionary of the Economic Products of India"), London.

Wehmeyer, A. S. (1969). Composition of kaffircorn (including hybrids). Council for Scientific and Industrial Research, Report C Chem 220, Pretoria, South Africa.

Weihrauch, J. L., Kinsella, J. E. and Watt, B. K. (1976). Comprehensive evaluation of fatty acids in foods. VI. Cereal products nutritional evaluation. *J. Am. diet. Ass.* **68**, 335–340.

Weinecke, L. A. and Montgomery, R. R. (1965). Experimental unit now suitable for scale-up to mill production size. *Am. Miller Processor* **93**, 8–9, 33.

Weinges, K., Kaltenhauser, W., Marx, H-D., Nader, E., Nader, F., Perner, J. and Seiler, D. (1968). Proanthocyanidins. X. Procyanidins of fruit (in German). *Justus Liebigs Annln. Chem.* **711**, 184–204.

Wessels, J. P. H. (1970a). Variation in amino acids in Kaffircorn cultivars available to chickens. *Agroanimalia* **2**, 77–84.

Wessels, J. P. H. (1970b). Variation in amino acids available to chickens in grain sorghum cultivars commonly grown in South Africa. *Agroanimalia* **2**, 199–204.

Wessels, J. P. H. and Bundock, M. J. (1968). Applicability to chickens of the carcass analysis method for determination of net protein utilization. III. The effect of shorter test period, sex and breed upon the body water-body nitrogen relationship in the chicken. *S. Afr. J. agric. Sci.* **11**, 531–535.

WHO. (1962). FAO/WHO Expert Group on calcium requirements. FAO Nutrition Meetings Report Series No. 30. WHO Technical Report Series No. 230, WHO, Geneva.

WHO. (1965). Protein requirements. Joint FAO/WHO Expert Group—WHO Technical Report Series No. 301. FAO Nutrition Meetings Report Series No. 37, WHO, Geneva.

WHO. (1967). Joint FAO/WHO Expert Group on requirements of vitamin A, thiamine, riboflavine, and niacin. FAO Nutrition Meetings Report Series No. 41. WHO Technical Report Series No. 362. WHO, Geneva.

WHO. (1970). Joint FAO/WHO Expert Group on requirements of Ascorbic acid, Vitamin D, Vitamin B_{12}, Folate and Iron. FAO Nutrition Meetings Report Series No. 47, WHO Technical Report Series No. 452. WHO, Geneva.

WHO. (1972). Human development and public health; report of a WHO scientific group. WHO Technical Report Series No. 485. WHO, Geneva.

WHO. (1973a). Energy and protein requirements. Report of a Joint FAO/WHO *Ad Hoc* Expert Committee. WHO Technical Report Series No. 522. FAO Nutrition Meetings Report Series No. 52. WHO, Geneva.

WHO. (1973b). Trace elements in human nutrition. Report of a WHO Expert Committee. WHO Technical Report Series No. 532. WHO, Geneva.

WHO. (1974). Handbook on human nutritional requirements. WHO, Geneva.

Widdowson, E. M. and McCance, R. A. (1954). Studies of the nutritional value of bread and the effect of variation in the extraction rate of flour on the growth of undernourished children. Medical Research Council Special Report No. 287. H. M. Stationery Office, London.

Williams, P. C. (1974). Errors in protein testing and their consequences. *Cereal Sci. Today* **19**, 280–282, 286.

Williams, P. C. (1977). A short summary of methods of testing grains, seeds and related products for protein content. *In* "Nutritional Standards and Methods of Evaluation for Food Legume Breeders". (J. H. Hulse, K. O. Rachie, and L. W. Billingsley, Eds) Prepared by the International Working Group on Nutritional Standards and Methods of Evaluation for Food Legume Breeders. IDRC–TS7e. pp. 73–78. International Development Research Centre, Box 8500, Ottawa, Canada K1G 3H9.

Williams, P. C., Kuzina, F. D. and Hlynka, I. (1970). A rapid colorimetric procedure for estimating the amylose content of starches and flours. *Cereal Chem.* **47**, 411–420.

Williams, R. S. and Hughes, R. E. (1972). Dietary protein, growth and retention of ascorbic acid in guinea pigs. *Br. J. Nutr.* **28**, 167–172.

Winter, G. (1964). The standard of living of the populations of Adamaoua (in French). Direction de la Statistique and ORSTOM, Yaounde.

Winton, A. L. and Winton, K. B. (1945). "The analysis of foods". Chapman and Hall Ltd., London.

Wokes, F., Badenoch, J. and Sinclair, H. M. (1955). Human dietary deficiency of Vitamin B_{12}. *Voeding* **16**, 590–602.

Wolf, M. J. (1963). Wheat starch. *In* "Methods in Carbohydrate Chemistry". (R. L. Whistler, R. J. Smith, J. N. BeMiller and M. L. Wolfrom, Eds) Vol. IV, pp. 6–9. Academic Press, Inc., New York.

Wolf, M. J., MacMasters, M. M., Cannon, J. A., Rosenall, E. C. and Rist, C. E. (1953). Preparation and properties of some hemi-cellulose from corn hulls. *Cereal Chem.* **30**, 451–470.

Wolfe, H. D. (1958). Consumer product testing. *In* "Flavour Research and Food Acceptance". Reinhold Publishing Corporation, New York.

Wolfe, M. and Fowden, L. (1957). Composition of the protein of whole maize seeds. *Cereal Chem.* **34**, 286–295.

Worker, G. F. Jr, and Ruckman, J. (1968). Variations in protein level in grain sorghum grown in the South-West desert. *Agron. J.* **60**, 485–488.

Worzella, W. W., Khalidy, R., Badawi, Y. and Daghir, S. (1965). Inheritance of β-carotene in grain sorghum hybrids. *Crop Sci.* **5**, 591–592.

Wu, Y. V. and Dimler, R. J. (1963). Hydrogen ion equilibria of wheat glutenin and gliadin. *Archs. Biochem. Biophys.* **103**, 310–318.

Wu, Y. V., Cluskey, and Jones, R. W. (1971). Sorghum prolamins: their optical rotatory dispersion, circular dichroism, and infrared spectra. *J. agric. Fd Chem.* **19**, 1139–1143.

Yañez, E., Ballester, D. and Gonzalez, M. (1973). Chemical and amino acid composition and biological quality of five sorghum (*Sorghum vulgare*) varieties (in Spanish). *Agricultura tec.* **33**, 77–81.

Yang, S. P. and Tsai, L. C. (1973). Effect of heat treatment on nutritional value of grain sorghum. *Fedr. Proc. Fedr.. Am. Socs. exp. Biol.* **32**, 941.

Yasumatsu, K., Nakayama, T. O. M. and Chichester, C. O. (1965). Flavonoids of sorghum. *J. Fd Sci.* **30**, 663–667.

Yemm, E. W. and Willis, A. J. (1954). The estimation of carbohydrates in plant extracts by anthrone. *Biochem. J.* **57**, 508–514.

York, J. O. (1976). Inheritance of pericarp and subcoat colors in sorghum. Proceedings 1976 Arkansas Nutrition Conference, Arkansas Feed Manufacturers' Association and Department of Animal Sciences, University of Arkansas, USA.

Young, V. R., Rand, W. M. and Scrimshaw, N. S. (1977). Measuring protein quality in humans: a review and proposed method. *Cereal Chem.* **54**, 929–948.

Yousif, Y. B. and Magboul, B. E. I. (1972). Nutritive value of Sudan foodstuffs. 1. *Sorghum vulgare* (Dura). *Sudan J. Fd Sci. Technol.* **4**, 39–45.

Ziegler, E. and Greer, E. N. (1971). Principles of milling. *In* "Wheat Chemistry and Technology". (Y. Pomeranz, Ed.) Monograph Series Vol. III revised. pp. 115–119. American Association of Cereal Chemists, St. Paul, Minnesota, USA.

Zubaidov, U. (1968a). Seed Albumins of some sorghum species (in Russian). *Dokl. Akad. Nauk Tadzhik SSR* **11**, 57–59. (*Chem. Abstr.* 1968, **69**, No. 65101c).

Zubaidov, U. (1968b). Seed globulins of some sorghum varieties (in Russian). *Dokl. Akad. Nauk Tadzhik. SSR* **11**, 70–72. (*Chem. Abstr.* 1968, **69**, No. 8867h).

Zubaidov, U. (1968c). On protein and non-protein N content in sorghum seeds (in Russian). *Izv. Akad. Nauk. Tadshikskio SSR Otdelenie Biologicheskikh Nauk* **2**, 31–36. (*Biol. Abstr.* 1970, **51**, No. 115688).

Subject Index

Note: Illustrations and diagrams are indicated in **bold**

G

H

I

J

M

Q

R

S

T